THIRD EDITION

ELECTRONIC DEVICES

THOMAS L. FLOYD

MERRILL, AN IMPRINT OF
MACMILLAN PUBLISHING COMPANY
NEW YORK

MAXWELL MACMILLAN CANADA
TORONTO

MAXWELL MACMILLAN INTERNATIONAL
NEW YORK ■ OXFORD ■ SINGAPORE ■ SYDNEY

ONCE AGAIN, TO SHEILA, WITH LOVE.

Cover art: Tech-Graphics
Editor: David Garza
Developmental Editor: Carol Hinklin Robison
Production Editor: Sharon Rudd
Art Coordinator: Mark D. Garrett
Text Designer: Anne Flanagan
Cover Designer: Russ Maselli
Production Buyer: Pamela D. Bennett
Insert art: Tech-Graphics

This book was set in Times Roman by Clarinda Company and was printed and bound by Arcata Graphics/Fairfield. The cover was printed by Lehigh Press, Inc.

Printed in the United States of America

Macmillan Publishing Company
866 Third Avenue
New York, NY 10022

Macmillan Publishing Company is part of the
Maxwell Communication Group of Companies.

Maxwell Macmillan Canada, Inc.
1200 Eglington Avenue East, Suite 200
Don Mills, Ontario M3C 3N1

Library of Congress Cataloging-in-Publication Data
Floyd, Thomas L.
Electronic devices / Thomas L. Floyd.—3rd ed.
p. cm.
Includes index.
ISBN 0-675-22170-6
1. Electronic apparatus and appliances. 2. Solid state electronics. I. Title.
TK7870.F52 1992
621.381—dc20 91-21429 CIP

Printing: 1 2 3 4 5 6 7 8 9 Year: 2 3 4 5

MERRILL'S INTERNATIONAL SERIES IN ENGINEERING TECHNOLOGY

Adamson	*Applied Pascal for Technology,* 0-675-20771-1
	The Complete Dictionary of Electronics, 0-02-300820-2
	Microcomputer Repair, 0-02-300825-3
	Structured BASIC Applied to Technology, 0-675-20772-X
	Structured C for Technology, 0-675-20993-5
	Structured C for Technology (w/disks), 0-675-21289-8
Antonakos	*The 68000 Microprocessor: Hardware and Software Principles and Applications,* 0-675-21043-7
Asser/Stigliano/ Bahrenburg	*Microcomputer Servicing: Practical Systems and Troubleshooting,* 0-675-20907-2
	Microcomputer Theory and Servicing, 0-675-20659-6
	Lab Manual to accompany Microcomputer Theory and Servicing, 0-675-21109-3
Aston	*Principles of Biomedical Instrumentation and Measurement,* 0-675-20943-9
Bateson	*Introduction to Control System Technology, Third Edition,* 0-675-21010-0
Beach/Justice	*DC/AC Circuit Essentials,* 0-675-20193-4
Berlin	*Experiments in Electronic Devices to accompany Floyd's Electronic Devices and Electronic Devices: Electron Flow Version, Third Edition,* 0-02-308422-7
	The Illustrated Electronics Dictionary, 0-675-20451-8
Berlin/Getz	*Experiments in Instrumentation and Measurement,* 0-675-20450-X
	Fundamentals of Operational Amplifiers and Linear Integrated Circuits, 0-675-21002-X
	Principles of Electronic Instrumentation and Measurement, 0-675-20449-6
Berube	*Electronic Devices and Circuits Using MICRO-CAP II,* 0-02-309160-6
Bogart	*Electronic Devices and Circuits, Second Edition,* 0-675-21150-6
Bogart/Brown	*Experiments in Electronic Devices and Circuits, Second Edition,* 0-675-21151-4
Boylestad	*DC/AC: The Basics,* 0-675-20918-8
	Introductory Circuit Analysis, Sixth Edition, 0-675-21181-6
Boylestad/Kousourou	*Experiments in Circuit Analysis, Sixth Edition,* 0-675-21182-4
	Experiments in DC/AC Basics, 0-675-21131-X
Brey	*Microprocessors and Peripherals: Hardware, Software, Interfacing, and Applications, Second Edition,* 0-675-20884-X
	The Intel Microprocessors—8086/8088, 80186, 80286, 80386, and 80486—Architecture, Programming, and Interfacing, Second Edition, 0-675-21309-6
Broberg	*Lab Manual to accompany Electronic Communication Techniques, Second Edition,* 0-675-21257-X
Buchla	*Digital Experiments: Emphasizing Systems and Design, Second Edition,* 0-675-21180-8
	Experiments in Electric Circuits Fundamentals, Second Edition, 0-675-21409-2
	Experiments in Electronics Fundamentals: Circuits, Devices and Applications, Second Edition, 0-675-21407-6
Buchla/McLachlan	*Applied Electronic Instrumentation and Measurement,* 0-675-21162-X

Ciccarelli	*Circuit Modeling: Exercises and Software, Second Edition,* 0-675-21152-2
Cooper	*Introduction to VersaCAD,* 0-675-21164-6
Cox	*Digital Experiments: Emphasizing Troubleshooting, Second Edition,* 0-675-21196-4
Croft	*Getting a Job: Resume Writing, Job Application Letters, and Interview Strategies,* 0-675-20917-X
Davis	*Technical Mathematics,* 0-675-20338-4
	Technical Mathematics with Calculus, 0-675-20965-X
	Study Guide to Accompany Technical Mathematics, 0-675-20966-8
	Study Guide to Accompany Technical Mathematics with Calculus, 0-675-20964-1
Delker	*Experiments in 8085 Microprocessor Programming and Interfacing,* 0-675-20663-4
Floyd	*Digital Fundamentals, Fourth Edition,* 0-675-21217-0
	Electric Circuits Fundamentals, Second Edition, 0-675-21408-4
	Electronic Devices, Third Edition, 0-675-22170-6
	Electronic Devices, Electron Flow Version, 0-02-338540-5
	Electronics Fundamentals: Circuits, Devices, and Applications, Second Edition, 0-675-21310-X
	Fundamentals of Linear Circuits, 0-02-338481-6
	Principles of Electric Circuits, Electron Flow Version, Second Edition, 0-675-21292-8
	Principles of Electric Circuits, Third Edition, 0-675-21062-3
Fuller	*Robotics: Introduction, Programming, and Projects,* 0-675-21078-X
Gaonkar	*Microprocessor Architecture, Programming, and Applications with the 8085/8080A, Second Edition,* 0-675-20675-8
	The Z80 Microprocessor: Architecture, Interfacing, Programming, and Design, 0-675-20540-9
Gillies	*Instrumentation and Measurements for Electronic Technicians,* 0-675-20432-1
Goetsch	*Industrial Supervision: In the Age of High Technology,* 0-675-22137-4
Goetsch/Rickman	*Computer-Aided Drafting with AutoCAD,* 0-675-20915-3
Goody	*Programming and Interfacing the 8086/8088 Microprocessor,* 0-675-21312-6
Hubert	*Electric Machines: Theory, Operation, Applications, Adjustment, and Control,* 0-675-21136-0
Humphries	*Motors and Controls,* 0-675-20235-3
Hutchins	*Introduction to Quality: Management, Assurance and Control,* 0-675-20896-3
Keown	*PSpice and Circuit Analysis,* 0-675-22135-8
Keyser	*Materials Science in Engineering, Fourth Edition,* 0-675-20401-1
Kirkpatrick	*The AutoCAD Book: Drawing, Modeling and Applications, Second Edition,* 0-675-22288-5
	Industrial Blueprint Reading and Sketching, 0-675-20617-0
Kraut	*Fluid Mechanics for Technicians,* 0-675-21330-4
Kulathinal	*Transform Analysis and Electronic Networks with Applications,* 0-675-20765-7
Lamit/Lloyd	*Drafting for Electronics,* 0-675-20200-0
Lamit/Wahler/Higgins	*Workbook in Drafting for Electronics,* 0-675-20417-8
Lamit/Paige	*Computer-Aided Design and Drafting,* 0-675-20475-5

Laviana	*Basic Computer Numerical Control Programming, Second Edition,* 0-675-21298-7
MacKenzie	*The 8051 Microcontroller,* 0-02-373650-X
Maruggi	*Technical Graphics: Electronics Worktext, Second Edition,* 0-675-21378-9
	The Technology of Drafting, 0-675-20762-2
	Workbook for the Technology of Drafting, 0-675-21234-0
McCalla	*Digital Logic and Computer Design,* 0-675-21170-0
McIntyre	*Study Guide to accompany Electronic Devices, Third Edition and Electronic Devices: Electron Flow Version,* 0-02-379296-5
	Study Guide to accompany Electronics Fundamentals, Second Edition, 0-675-21406-8
Miller	*The 68000 Microprocessor Family: Architecture, Programming, and Applications, Second Edition,* 0-02-381560-4
Monaco	*Essential Mathematics for Electronics Technicians,* 0-675-21172-7
	Introduction to Microwave Technology, 0-675-21030-5
	Laboratory Activities in Microwave Technology, 0-675-21031-3
	Preparing for the FCC General Radiotelephone Operator's License Examination, 0-675-21313-4
	Student Resource Manual to accompany Essential Mathematics for Electronics Technicians, 0-675-21173-5
Monssen	*PSpice with Circuit Analysis,* 0-675-21376-2
Mott	*Applied Fluid Mechanics, Third Edition,* 0-675-21026-7
	Machine Elements in Mechanical Design, Second Edition, 0-675-22289-3
Nashelsky/Boylestad	*BASIC Applied to Circuit Analysis,* 0-675-20161-6
Panares	*A Handbook of English for Technical Students,* 0-675-20650-2
Pfeiffer	*Proposal Writing: The Art of Friendly Persuasion,* 0-675-20988-9
	Technical Writing: A Practical Approach, 0-675-21221-9
Pond	*Introduction to Engineering Technology,* 0-675-21003-8
Quinn	*The 6800 Microprocessor,* 0-675-20515-8
Reis	*Digital Electronics Through Project Analysis,* 0-675-21141-7
	Electronic Project Design and Fabrication, Second Edition, 0-02-399230-1
	Laboratory Manual for Digital Electronics Through Project Analysis, 0-675-21254-5
Rolle	*Thermodynamics and Heat Power, Third Edition,* 0-675-21016-X
Rosenblatt/Friedman	*Direct and Alternating Current Machinery, Second Edition,* 0-675-20160-8
Roze	*Technical Communication: The Practical Craft,* 0-675-20641-3
Schoenbeck	*Electronic Communications: Modulation and Transmission, Second Edition,* 0-675-21311-8
Schwartz	*Survey of Electronics, Third Edition,* 0-675-20162-4
Sell	*Basic Technical Drawing,* 0-675-21001-1
Smith	*Statistical Process Control and Quality Improvement,* 0-675-21160-3
Sorak	*Linear Integrated Circuits: Laboratory Experiments,* 0-675-20661-8
Spiegel/Limbrunner	*Applied Statics and Strength of Materials,* 0-675-21123-9
Stanley, B.H.	*Experiments in Electric Circuits, Third Edition,* 0-675-21088-7
Stanley, W.D.	*Operational Amplifiers with Linear Integrated Circuits, Second Edition,* 0-675-20660-X

Subbarao	*16/32-Bit Microprocessors: 68000/68010/68020 Software, Hardware, and Design Applications,* 0-675-21119-0
Tocci	*Electronic Devices: Conventional Flow Version, Third Edition,* 0-675-20063-6
	Fundamentals of Pulse and Digital Circuits, Third Edition, 0-675-20033-4
	Introduction to Electric Circuit Analysis, Second Edition, 0-675-20002-4
Tocci/Oliver	*Fundamentals of Electronic Devices, Fourth Edition,* 0-675-21259-6
Webb	*Programmable Logic Controllers: Principles and Applications, Second Edition,* 0-02-424970-X
Webb/Greshock	*Industrial Control Electronics,* 0-675-20897-1
Weisman	*Basic Technical Writing, Sixth Edition,* 0-675-21256-1
Wolansky/Akers	*Modern Hydraulics: The Basics at Work,* 0-675-20987-0
Wolf	*Statics and Strength of Materials: A Parallel Approach,* 0-675-20622-7

Preface

This third edition of *Electronic Devices* provides thorough, comprehensive, and practical coverage of electronic devices, circuits, and applications. The extensive troubleshooting coverage and innovative system application sections serve as very important and necessary links between theory and the real world.

This book is divided into two basic parts. Chapters 1 through 11 cover discrete devices and circuits, while Chapters 12 through 17 deal mostly with linear integrated circuits, with emphasis on the operational amplifier.

A Brief Overview

Basic semiconductor theory and the concept of the pn junction are introduced in Chapter 1. Various types of diodes and their applications ae covered in Chapters 2 and 3. Bipolar junction transistors (BJTs) and small-signal BJT amplifiers are covered in Chapters 4 through 6. Field-effect transistors (FETs) and small-signal FET amplifiers are studied in Chapters 7 and 8. Power amplifiers are presented in Chapter 9, and factors that determine the frequency response of amplifiers are covered in Chapter 10. In Chapter 11, thyristors and other special semiconductor devices are introduced.

The coverage of linear integrated circuits begins with an introduction to operational amplifiers (op-amps) in Chapter 12. Op-amp operation and configurations are studied in Chapters 13 and 14. Oscillators (both discrete and IC) are covered in Chapter 15, followed by active filters in Chapter 16. Finally, Chapter 17 provides a coverage of voltage regulators.

Organizational Changes

Generally, this third edition retains the basic organization that was so well received in the second edition. Three important modifications, however, are worth noting. First, the chapter on small-signal bipolar amplifiers now immediately follows the chapter on BJT biasing. The chapters on field-effect transistors and small-signal FET amplifiers are also now in consecutive order. Second, the chapter on thyristors and special devices has been moved into the first part of the book in order to consolidate coverage of discrete devices. Also, for the purpose of consolidation of discrete device coverage, the second edition chapter on optoelectronic devices has been merged into Chapter 3 (Special Diodes) and into Chapter 11 (Thyristors and Special Devices). All in all, these changes result in more streamlined and logical organization.

NEW AND IMPROVED FEATURES

Several new features and improvements to existing features have been incorporated into this third edition.

- □ An innovative system application section in each chapter (except Chapter 1)
- □ A functional full-color insert keyed to selected system applications
- □ System-related chapter openers
- □ Improved and increased troubleshooting coverage
- □ Improved organization
- □ More effective use of second color
- □ Standard component values used throughout
- □ An introductory message at the beginning of each section that sets the tone for that section
- □ A practice exercise for each example
- □ More end-of-chapter problems
- □ Margin logos indicating troubleshooting problems and color insert references
- □ Multiple-choice self-tests
- □ Performance-based chapter objectives
- □ More data sheets available in Appendix A
- □ Increased use and reference to data sheets

An improved and expanded ancillary package for this edition includes the following:

- □ Expanded transparency and transparency master package
- □ Computerized and hard copy test bank
- □ System applications worksheet masters
- □ Study Guide
- □ Lab Manual
- □ Instructor's Resource Manual

In addition to these specific features, many areas have been partially or completely rewritten to improve clarity or to expand coverage. All improvements have been carefully incorporated without altering the basic style and format which has been so well received in the previous editions.

ILLUSTRATION OF FEATURES WITHIN EACH CHAPTER

CHAPTER OPENER As shown in Figure P–1, each chapter begins with a two-page opener. The left page contains a listing of the sections within the chapter, the chapter objectives, and a brief introduction. The right page presents a preview of the system application that will be the focus of the last section of the chapter and provides several specialized objectives oriented to this feature.

List of performance-based objectives.

System block diagram with highlighted circuit board.

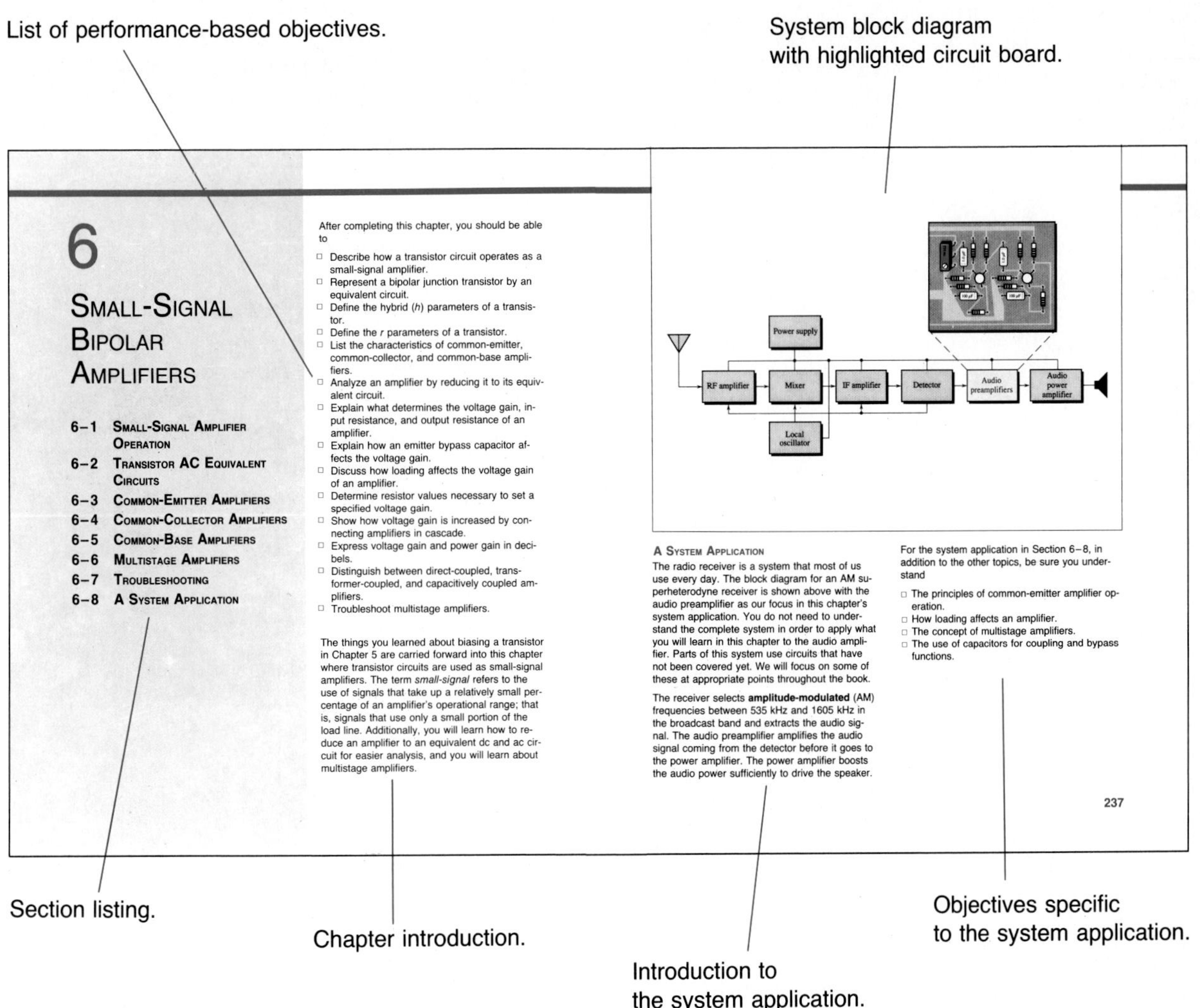

6
SMALL-SIGNAL BIPOLAR AMPLIFIERS
6–1 SMALL-SIGNAL AMPLIFIER OPERATION
6–2 TRANSISTOR AC EQUIVALENT CIRCUITS
6–3 COMMON-EMITTER AMPLIFIERS
6–4 COMMON-COLLECTOR AMPLIFIERS
6–5 COMMON-BASE AMPLIFIERS
6–6 MULTISTAGE AMPLIFIERS
6–7 TROUBLESHOOTING
6–8 A SYSTEM APPLICATION
After completing this chapter, you should be able to
Describe how a transistor circuit operates as a small-signal amplifier.
Represent a bipolar junction transistor by an equivalent circuit.
Define the hybrid (h) parameters of a transistor.
Define the r parameters of a transistor.
List the characteristics of common-emitter, common-collector, and common-base amplifiers.
Analyze an amplifier by reducing it to its equivalent circuit.
Explain what determines the voltage gain, input resistance, and output resistance of an amplifier.
Explain how an emitter bypass capacitor affects the voltage gain.
Discuss how loading affects the voltage gain of an amplifier.
Determine resistor values necessary to set a specified voltage gain.
Show how voltage gain is increased by connecting amplifiers in cascade.
Express voltage gain and power gain in decibels.
Distinguish between direct-coupled, transformer-coupled, and capacitively coupled amplifiers.
Troubleshoot multistage amplifiers.
The things you learned about biasing a transistor in Chapter 5 are carried forward into this chapter where transistor circuits are used as small-signal amplifiers. The term small-signal refers to the use of signals that take up a relatively small percentage of an amplifier's operational range; that is, signals that use only a small portion of the load line. Additionally, you will learn how to reduce an amplifier to an equivalent dc and ac circuit for easier analysis, and you will learn about multistage amplifiers.
Power supply
RF amplifier
Mixer
IF amplifier
Detector
Audio preamplifiers
Audio power amplifier
Local oscillator
A SYSTEM APPLICATION
The radio receiver is a system that most of us use every day. The block diagram for an AM superheterodyne receiver is shown above with the audio preamplifier as our focus in this chapter's system application. You do not need to understand the complete system in order to apply what you will learn in this chapter to the audio amplifier. Parts of this system use circuits that have not been covered yet. We will focus on some of these at appropriate points throughout the book.
The receiver selects amplitude-modulated (AM) frequencies between 535 kHz and 1605 kHz in the broadcast band and extracts the audio signal. The audio preamplifier amplifies the audio signal coming from the detector before it goes to the power amplifier. The power amplifier boosts the audio power sufficiently to drive the speaker.
For the system application in Section 6–8, in addition to the other topics, be sure you understand
The principles of common-emitter amplifier operation.
How loading affects an amplifier.
The concept of multistage amplifiers.
The use of capacitors for coupling and bypass functions.
237

Section listing.

Chapter introduction.

Introduction to the system application.

Objectives specific to the system application.

FIGURE P–1
Chapter opener

SECTION OPENER AND SECTION REVIEW Each section within a chapter begins with a brief introduction that highlights the material to be covered or provides a general overview. Each section ends with a set of review questions that focus on the key concepts presented in the section. Answers to these review questions are given at the end of the chapter. Figure P–2 illustrates these two features.

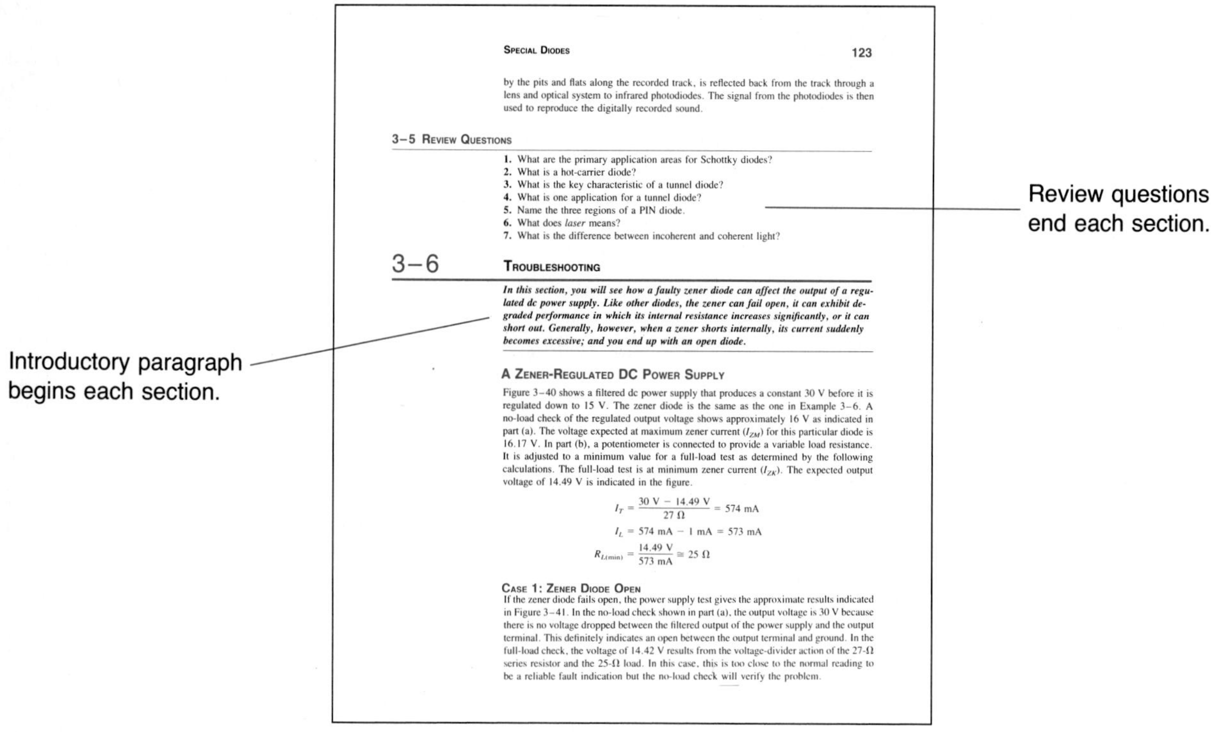

Review questions end each section.
Introductory paragraph begins each section.

SPECIAL DIODES 123

by the pits and flats along the recorded track, is reflected back from the track through a lens and optical system to infrared photodiodes. The signal from the photodiodes is then used to reproduce the digitally recorded sound.

3–5 REVIEW QUESTIONS

1. What are the primary application areas for Schottky diodes?
2. What is a hot-carrier diode?
3. What is the key characteristic of a tunnel diode?
4. What is one application for a tunnel diode?
5. Name the three regions of a PIN diode.
6. What does *laser* means?
7. What is the difference between incoherent and coherent light?

3–6 TROUBLESHOOTING

In this section, you will see how a faulty zener diode can affect the output of a regulated dc power supply. Like other diodes, the zener can fail open, it can exhibit degraded performance in which its internal resistance increases significantly, or it can short out. Generally, however, when a zener shorts internally, its current suddenly becomes excessive; and you end up with an open diode.

A ZENER-REGULATED DC POWER SUPPLY

Figure 3–40 shows a filtered dc power supply that produces a constant 30 V before it is regulated down to 15 V. The zener diode is the same as the one in Example 3–6. A no-load check of the regulated output voltage shows approximately 16 V as indicated in part (a). The voltage expected at maximum zener current (I_{ZM}) for this particular diode is 16.17 V. In part (b), a potentiometer is connected to provide a variable load resistance. It is adjusted to a minimum value for a full-load test as determined by the following calculations. The full-load test is at minimum zener current (I_{ZK}). The expected output voltage of 14.49 V is indicated in the figure.

$$I_T = \frac{30\text{ V} - 14.49\text{ V}}{27\ \Omega} = 574\text{ mA}$$

$$I_L = 574\text{ mA} - 1\text{ mA} = 573\text{ mA}$$

$$R_{L(min)} = \frac{14.49\text{ V}}{573\text{ mA}} \cong 25\ \Omega$$

CASE 1: ZENER DIODE OPEN

If the zener diode fails open, the power supply test gives the approximate results indicated in Figure 3–41. In the no-load check shown in part (a), the output voltage is 30 V because there is no voltage dropped between the filtered output of the power supply and the output terminal. This definitely indicates an open between the output terminal and ground. In the full-load check, the voltage of 14.42 V results from the voltage-divider action of the 27-Ω series resistor and the 25-Ω load. In this case, this is too close to the normal reading to be a reliable fault indication but the no-load check will verify the problem.

FIGURE P–2
Section opener and section review

EXAMPLES AND PRACTICE EXERCISES Frequent examples help to illustrate and clarify basic concepts. At the end of each example is a practice exercise, which is intended to help reinforce or expand on the example in some way. The nature of the practice exercises varies. Some require the student to repeat the procedure demonstrated in the example but with a different set of values or conditions. Others focus on a more limited part of the example or ask questions that encourage further thought beyond the procedure contained in the example. Answers to all practice exercises are given at the end of the chapter. A typical example and practice exercise is shown in Figure P–3.

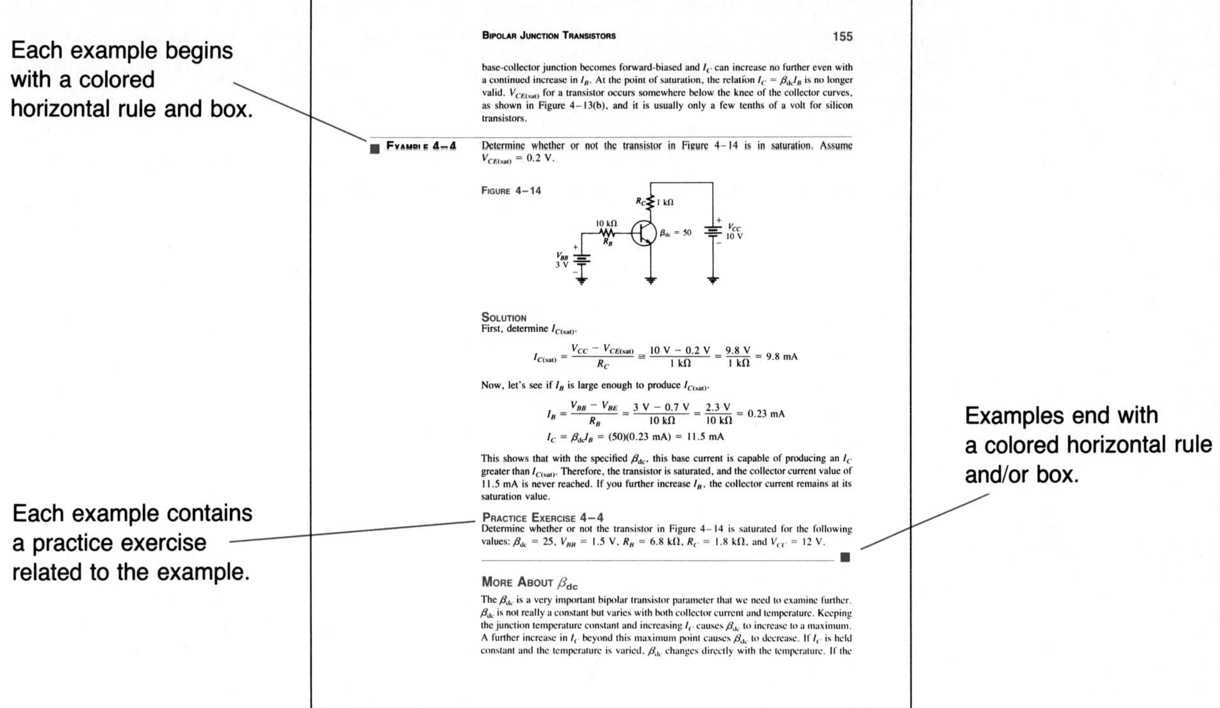
Each example begins with a colored horizontal rule and box.
Each example contains a practice exercise related to the example.
Examples end with a colored horizontal rule and/or box.
R_C 1 kΩ
10 kΩ
R_B
$\beta_{dc} = 50$
V_{CC} 10 V
V_{BB} 3 V

BIPOLAR JUNCTION TRANSISTORS 155

base-collector junction becomes forward-biased and I_C can increase no further even with a continued increase in I_B. At the point of saturation, the relation $I_C = \beta_{dc} I_B$ is no longer valid. $V_{CE(sat)}$ for a transistor occurs somewhere below the knee of the collector curves, as shown in Figure 4–13(b), and it is usually only a few tenths of a volt for silicon transistors.

EXAMPLE 4–4 Determine whether or not the transistor in Figure 4–14 is in saturation. Assume $V_{CE(sat)} = 0.2$ V.

FIGURE 4–14

SOLUTION
First, determine $I_{C(sat)}$.

$$I_{C(sat)} = \frac{V_{CC} - V_{CE(sat)}}{R_C} \cong \frac{10\text{ V} - 0.2\text{ V}}{1\text{ k}\Omega} = \frac{9.8\text{ V}}{1\text{ k}\Omega} = 9.8\text{ mA}$$

Now, let's see if I_B is large enough to produce $I_{C(sat)}$.

$$I_B = \frac{V_{BB} - V_{BE}}{R_B} = \frac{3\text{ V} - 0.7\text{ V}}{10\text{ k}\Omega} = \frac{2.3\text{ V}}{10\text{ k}\Omega} = 0.23\text{ mA}$$

$$I_C = \beta_{dc} I_B = (50)(0.23\text{ mA}) = 11.5\text{ mA}$$

This shows that with the specified β_{dc}, this base current is capable of producing an I_C greater than $I_{C(sat)}$. Therefore, the transistor is saturated, and the collector current value of 11.5 mA is never reached. If you further increase I_B, the collector current remains at its saturation value.

PRACTICE EXERCISE 4–4
Determine whether or not the transistor in Figure 4–14 is saturated for the following values: $\beta_{dc} = 25$, $V_{BB} = 1.5$ V, $R_B = 6.8$ kΩ, $R_C = 1.8$ kΩ, and $V_{CC} = 12$ V.

MORE ABOUT β_{dc}

The β_{dc} is a very important bipolar transistor parameter that we need to examine further. β_{dc} is not really a constant but varies with both collector current and temperature. Keeping the junction temperature constant and increasing I_C causes β_{dc} to increase to a maximum. A further increase in I_C beyond this maximum point causes β_{dc} to decrease. If I_C is held constant and the temperature is varied, β_{dc} changes directly with the temperature. If the

FIGURE P–3
An example and practice exercise

COMPUTER PROGRAMS A few selected examples contain BASIC computer programs that show how the given procedure or analysis can be implemented with a computer. While there are no SPICE programs in this text, *PSpice and Circuit Analysis* by John Keown (available from Merrill) is recommended as a supplement for those who wish to introduce SPICE in their course.

SYSTEM APPLICATION The last section of each chapter (except Chapter 1) is a system application in which a certain circuit board in a "real-world" system is the focus of several on-the-job type activities. Certain activities require the student to troubleshoot the circuit board for specified faults, including interpretation of instrument readings in the color insert. Generally, the type of circuit board relates directly to some or all of the material covered in the chapter. In some cases, however, the student is required to "stretch" a bit and, as a result, to learn something new.

Results and answers for the activities in the system application sections are provided only in the Instructor's Resource Manual, where a set of worksheet masters for appropriate activities is also provided. These can be photocopied for student hand-outs.

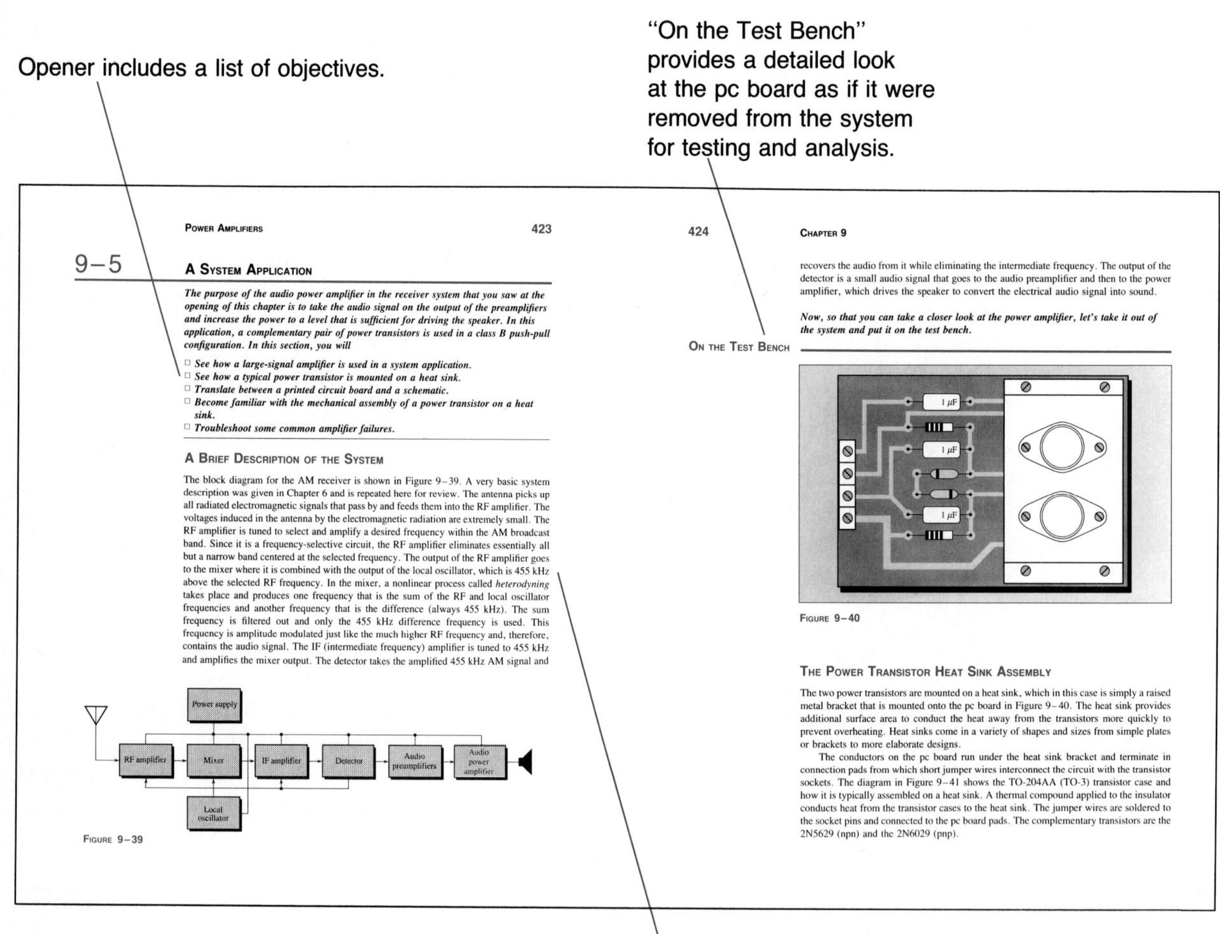

POWER AMPLIFIERS 423

9–5 **A SYSTEM APPLICATION**

The purpose of the audio power amplifier in the receiver system that you saw at the opening of this chapter is to take the audio signal on the output of the preamplifiers and increase the power to a level that is sufficient for driving the speaker. In this application, a complementary pair of power transistors is used in a class B push-pull configuration. In this section, you will

- ***See how a large-signal amplifier is used in a system application.***
- ***See how a typical power transistor is mounted on a heat sink.***
- ***Translate between a printed circuit board and a schematic.***
- ***Become familiar with the mechanical assembly of a power transistor on a heat sink.***
- ***Troubleshoot some common amplifier failures.***

A BRIEF DESCRIPTION OF THE SYSTEM

The block diagram for the AM receiver is shown in Figure 9–39. A very basic system description was given in Chapter 6 and is repeated here for review. The antenna picks up all radiated electromagnetic signals that pass by and feeds them into the RF amplifier. The voltages induced in the antenna by the electromagnetic radiation are extremely small. The RF amplifier is tuned to select and amplify a desired frequency within the AM broadcast band. Since it is a frequency-selective circuit, the RF amplifier eliminates essentially all but a narrow band centered at the selected frequency. The output of the RF amplifier goes to the mixer where it is combined with the output of the local oscillator, which is 455 kHz above the selected RF frequency. In the mixer, a nonlinear process called *heterodyning* takes place and produces one frequency that is the sum of the RF and local oscillator frequencies and another frequency that is the difference (always 455 kHz). The sum frequency is filtered out and only the 455 kHz difference frequency is used. This frequency is amplitude modulated just like the much higher RF frequency and, therefore, contains the audio signal. The IF (intermediate frequency) amplifier is tuned to 455 kHz and amplifies the mixer output. The detector takes the amplified 455 kHz AM signal and

FIGURE 9–39

424 CHAPTER 9

recovers the audio from it while eliminating the intermediate frequency. The output of the detector is a small audio signal that goes to the audio preamplifier and then to the power amplifier, which drives the speaker to convert the electrical audio signal into sound.

Now, so that you can take a closer look at the power amplifier, let's take it out of the system and put it on the test bench.

ON THE TEST BENCH

FIGURE 9–40

THE POWER TRANSISTOR HEAT SINK ASSEMBLY

The two power transistors are mounted on a heat sink, which in this case is simply a raised metal bracket that is mounted onto the pc board in Figure 9–40. The heat sink provides additional surface area to conduct the heat away from the transistors more quickly to prevent overheating. Heat sinks come in a variety of shapes and sizes from simple plates or brackets to more elaborate designs.

The conductors on the pc board run under the heat sink bracket and terminate in connection pads from which short jumper wires interconnect the circuit with the transistor sockets. The diagram in Figure 9–41 shows the TO-204AA (TO-3) transistor case and how it is typically assembled on a heat sink. A thermal compound applied to the insulator conducts heat from the transistor cases to the heat sink. The jumper wires are soldered to the socket pins and connected to the pc board pads. The complementary transistors are the 2N5629 (npn) and the 2N6029 (pnp).

An overall introduction to the system application is provided before a particular pc board circuit is focused on.

FIGURE P–4
A system application section

The overall objectives of the system application are

- ☐ To provide a transition between theoretical concepts and real-world circuitry.
- ☐ To help provide a "physical" sense of the devices and circuits studied in the chapter.
- ☐ To increase student skills with on-the-job type activities.
- ☐ To help answer the question, "Why do I need to know this?"

A typical system application section is shown in Figure P–4.

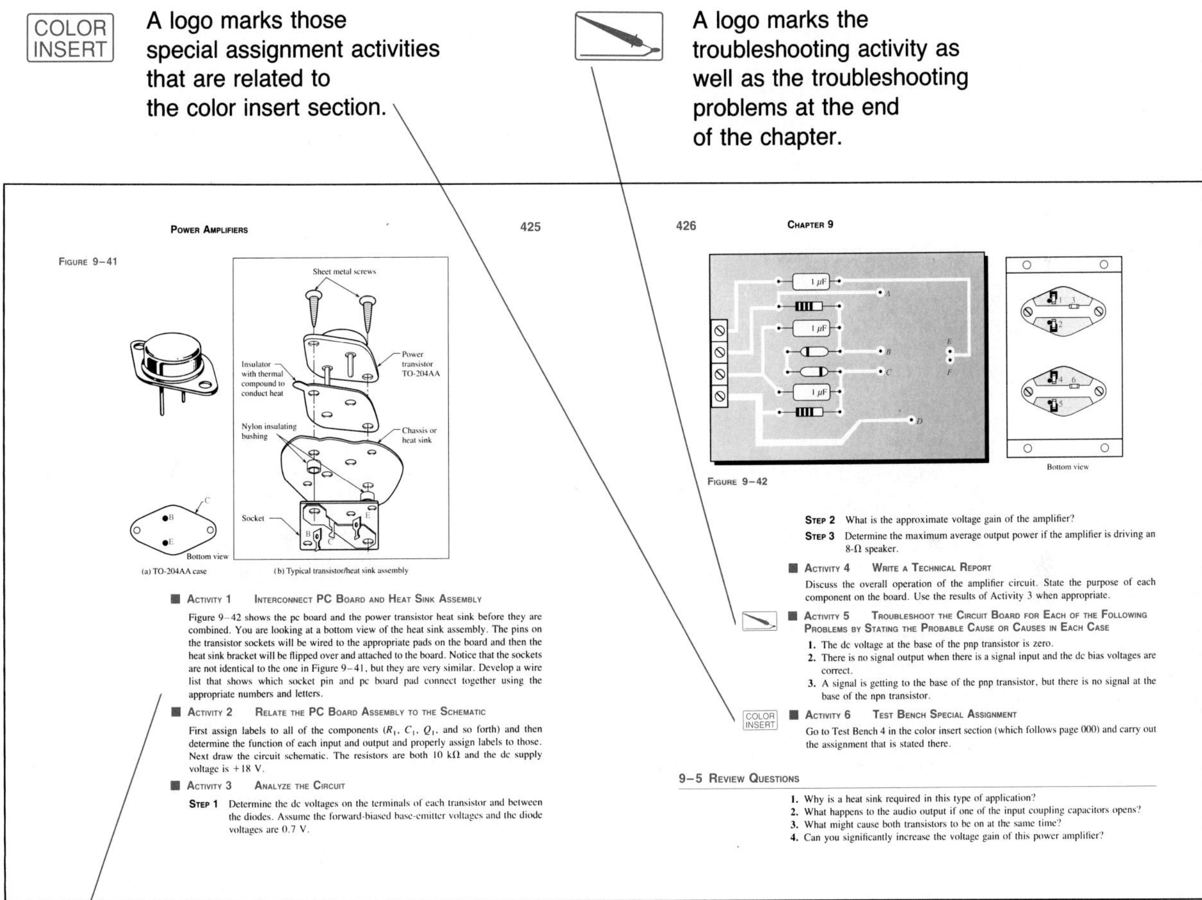

POWER AMPLIFIERS 425

FIGURE 9–41

(a) TO-204AA case (b) Typical transistor/heat sink assembly

■ ACTIVITY 1 INTERCONNECT PC BOARD AND HEAT SINK ASSEMBLY

Figure 9–42 shows the pc board and the power transistor heat sink before they are combined. You are looking at a bottom view of the heat sink assembly. The pins on the transistor sockets will be wired to the appropriate pads on the board and then the heat sink bracket will be flipped over and attached to the board. Notice that the sockets are not identical to the one in Figure 9–41, but they are very similar. Develop a wire list that shows which socket pin and pc board pad connect together using the appropriate numbers and letters.

■ ACTIVITY 2 RELATE THE PC BOARD ASSEMBLY TO THE SCHEMATIC

First assign labels to all of the components (R_1, C_1, Q_1, and so forth) and then determine the function of each input and output and properly assign labels to those. Next draw the circuit schematic. The resistors are both 10 kΩ and the dc supply voltage is +18 V.

■ ACTIVITY 3 ANALYZE THE CIRCUIT

STEP 1 Determine the dc voltages on the terminals of each transistor and between the diodes. Assume the forward-biased base-emitter voltages and the diode voltages are 0.7 V.

426 CHAPTER 9

FIGURE 9–42

STEP 2 What is the approximate voltage gain of the amplifier?

STEP 3 Determine the maximum average output power if the amplifier is driving an 8-Ω speaker.

■ ACTIVITY 4 WRITE A TECHNICAL REPORT

Discuss the overall operation of the amplifier circuit. State the purpose of each component on the board. Use the results of Activity 3 when appropriate.

■ ACTIVITY 5 TROUBLESHOOT THE CIRCUIT BOARD FOR EACH OF THE FOLLOWING PROBLEMS BY STATING THE PROBABLE CAUSE OR CAUSES IN EACH CASE

1. The dc voltage at the base of the pnp transistor is zero.
2. There is no signal output when there is a signal input and the dc bias voltages are correct.
3. A signal is getting to the base of the pnp transistor, but there is no signal at the base of the npn transistor.

COLOR INSERT ■ ACTIVITY 6 TEST BENCH SPECIAL ASSIGNMENT

Go to Test Bench 4 in the color insert section (which follows page 000) and carry out the assignment that is stated there.

9–5 REVIEW QUESTIONS

1. Why is a heat sink required in this type of application?
2. What happens to the audio output if one of the input coupling capacitors opens?
3. What might cause both transistors to be on at the same time?
4. Can you significantly increase the voltage gain of this power amplifier?

A series of activities involves the student in working with pc boards and schematics, circuit analysis, report writing, troubleshooting, and test setups.

FIGURE P–4
Continued

FULL-COLOR INSERT Seven selected system applications are related to the full-color insert using a special assignment activity marked by a color insert logo. The color insert consists of circuit board test set-ups that either require the student to troubleshoot the board based on instrument readings or to determine instrument settings for testing the board for proper operation.

CHAPTER END MATTER At the end of each chapter is a summary, glossary, formula list, multiple-choice self-test, and sectionalized problem set, as well as answers to section review questions and to practice exercises. Terms that appear boldface in the text are defined in the glossary.

SUGGESTIONS FOR USE

Electronic Devices, Third Edition, can be used in several ways to accommodate a variety of scheduling and program needs. A few suggestions are as follows.

OPTION 1 In a two-term electronics course, most, if not all, of this book can be utilized by covering discrete devices and circuits in Chapters 1 through 10 in the first term. If time and/or program requirements permit, Chapter 11 on thyristors may also be covered. In many programs, much of the material in Chapter 11 is covered in a later industrial electronics course. Linear integrated circuits in Chapters 12 through 17 can be covered during the second term.

OPTION 2 With selective presentation of topics, this text can be used for a one-term devices and circuits course by omitting certain topics and maintaining a rigorous schedule. An example of this approach for a course covering only discrete devices (basically Chapters 1 through 11) is as follows: Cover Chapter 1 lightly, cover Chapter 4 selectively, cover Chapter 10 lightly, and omit or selectively cover Chapter 11. The remaining chapters should be covered thoroughly, although selective omissions can be made throughout. For a one-term course that needs some linear IC coverage, additional reductions in Chapters 1 through 11 may be necessary.

SYSTEM APPLICATIONS The system applications are an extremely versatile tool for providing both motivation and real-world experiences in the classroom. The variety of systems is intended to give the student an appreciation for the wide range of applications for electronic devices.

Although these system applications can be treated as optional, it is highly recommended that they be included in your course. System applications can be used as

- □ An integral part of the chapter for the purpose of relating devices to a realistic system and for establishing a useful purpose for the device(s). All or selected activities can be assigned and discussed in class or turned in for a grade.
- □ A separate out-of-class assignment to be turned in for extra credit.
- □ An in-class activity to promote and stimulate discussion and interaction among students and between students and the instructor.
- □ A case in point to help answer the question on the mind of most students: ''Why do I need to know this?''

A NOTE TO THE STUDENT

The material in this preface is intended to help both you and your instructor make the most effective use of this textbook as a teaching and learning tool. Although you should certainly read everything in this preface, this part is especially for you, the student.

I am sure that you realize that knowledge and skills are not obtained easily or without effort. Much hard work is required to properly prepare yourself for any career, and electronics is no exception. You should use this book as more than just a reference. You must really dig in by reading, thinking, and doing. Don't expect every concept or procedure to become immediately clear. Some topics may take several readings, working many problems, and much help from your instructor before you really understand them.

Work through each example step-by-step and then do the associated practice exercise. Answer the review questions at the end of each section. If you don't understand an example or if you can't answer a question, go back into the section until you can. Check your answers at the end of the chapter. The multiple-choice self-tests at the end of each chapter are a good way to check your overall comprehension and retention of the subjects covered. You should do the self-test before you start the problems. Check your answers at the end of the book.

The problem sets at the end of each chapter (except Chapter 1) provide exercises with varying degrees of difficulty. In any technical field, it is very important that you work lots of problems. Working through a problem gives you a level of insight and understanding that reading or classroom lectures alone do not provide. Never think that you fully understand a concept or procedure by simply watching or listening to someone else. In the final analysis, you must do it yourself and you must do it to the best of your ability.

A LOOK BACK

Now, before you begin your study of electronic devices and circuits, let's briefly look back at the beginnings of electronics and some of the important developments that have led to the electronics technology that we have today. It is always good to have a sense of the history of your career field. The names of many of the early pioneers in electricity and electromagnetics still live on in terms of familiar units and quantities. Names such as Ohm, Ampere, Volta, Farad, Henry, Coulomb, Oersted, and Hertz are some of the better known examples. More widely known names such as Franklin and Edison are also very significant in the history of electricity and electronics because of their tremendous contributions.

THE BEGINNING OF ELECTRONICS

The early experiments in electronics involved electric currents in glass vacuum tubes. One of the first to conduct such experiments was a German named Heinrich Geissler (1814–1879). Geissler removed most of the air from a glass tube and found that the tube glowed when there was an electric current through it. Around 1878, British scientist Sir William Crookes (1832–1919) experimented with tubes similar to those of Geissler. In his experiments, Crookes found that the current in the vacuum tubes seemed to consist of particles.

Thomas Edison (1847–1931), experimenting with the carbon-filament light bulb he had invented, made another important finding. He inserted a small metal plate in the bulb.

When the plate was positively charged, there was a current from the filament to the plate. This device was the first thermionic diode. Edison patented it but never used it.

The electron was discovered in the 1890s. The French physicist Jean Baptiste Perrin (1870–1942) demonstrated that the current in a vacuum tube consists of the movement of negatively charged particles in a given direction. Some of the properties of these particles were measured by Sir Joseph Thomson (1856–1940), a British physicist, in experiments he performed between 1895 and 1897. These negatively charged particles later became known as electrons. The charge on the electron was accurately measured by an American physicist, Robert A. Millikan (1868–1953), in 1909. As a result of these discoveries, electrons could be controlled, and the electronic age was ushered in.

Putting the Electron to Work A vacuum tube that allowed electrical current in only one direction was constructed in 1904 by British scientist John A. Fleming. The tube was used to detect electromagnetic waves. Called the Fleming valve, it was the forerunner of the more recent vacuum diode tubes. Major progress in electronics, however, awaited the development of a device that could boost, or amplify, a weak electromagnetic wave or radio signal. This device was the audion, patented in 1907 by Lee deForest, an American. It was a triode vacuum tube capable of amplifying small electrical ac signals.

Two other Americans, Harold Arnold and Irving Langmuir, made great improvements in the triode vacuum tube between 1912 and 1914. About the same time, deForest and Edwin Armstrong, an electrical engineer, used the triode tube in an oscillator circuit. In 1914, the triode was incorporated in the telephone system and made the transcontinental telephone network possible. The tetrode tube was invented in 1916 by Walter Schottky, a German. The tetrode, along with the pentode (invented in 1926 by Dutch engineer Tellegen), greatly improved the triode. The first television picture tube, called the kinescope, was developed in the 1920s by Vladimir Sworykin, an American researcher.

During World War II, several types of microwave tubes were developed that made possible modern microwave radar and other communications systems. In 1939, the magnetron was invented in Britain by Henry Boot and John Randall. In the same year, the klystron microwave tube was developed by two Americans, Russell Varian and his brother Sigurd Varian. The traveling-wave tube (TWT) was invented in 1943 by Rudolf Komphner, an Austrian-American.

Solid-State Electronics The crystal detectors used in early radios were the forerunners of modern solid-state devices. However, the era of solid-state electronics began with the invention of the transistor in 1947 at Bell Labs. The inventors were Walter Brattain, John Bardeen, and William Shockley. Figure P–5 shows these three men.

In the early 1960s, the integrated circuit (IC) was developed. It incorporated many transistors and other components on a single small *chip* of semiconductor material. Integrated circuit technology has been continuously developed and improved, allowing increasingly more complex circuits to be built on smaller chips.

Around 1965, the first *integrated general-purpose operational amplifier* was introduced. This low-cost, highly versatile device incorporated nine transistors and twelve resistors in a small package. It proved to have many advantages over comparable discrete component circuits in terms of reliability and performance. Since this introduction, the IC operational amplifier has become a basic building block for a wide variety of linear systems.

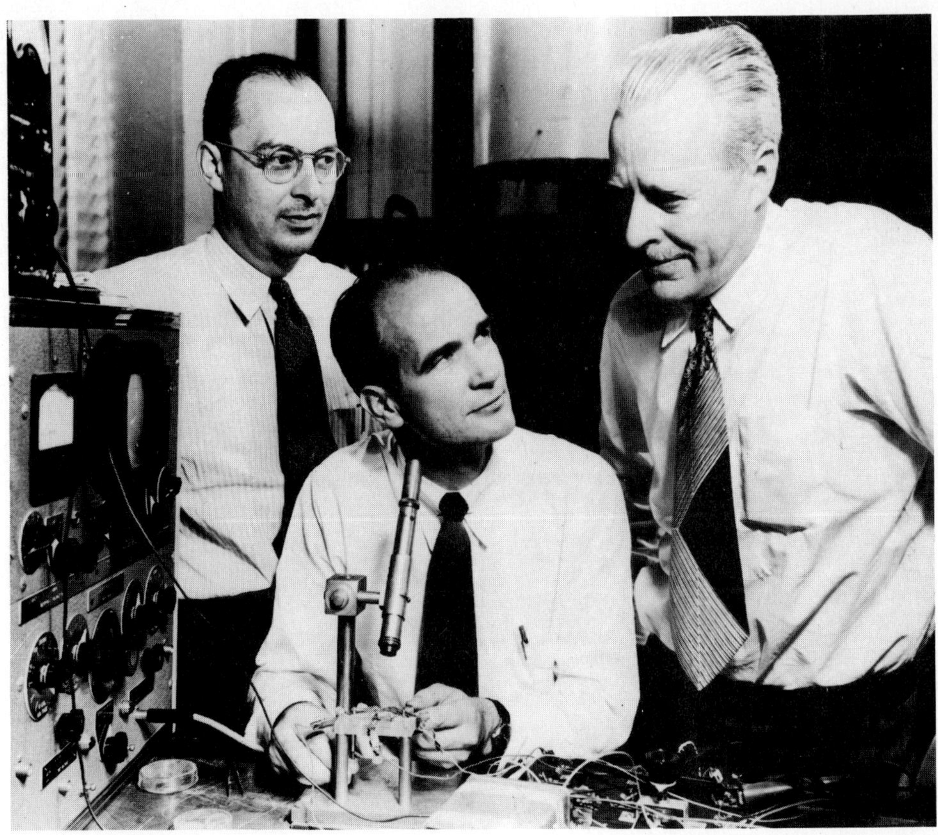

FIGURE P–5
Nobel Prize winners Drs. John Bardeen, William Shockley, and Walter Brattain, shown left to right, with apparatus used in their first investigations that led to the invention of the transistor. The trio received the 1956 Nobel Physics award for their invention of the transistor, which was announced by Bell Laboratories in 1948. (Courtesy of Bell Laboratories)

ACKNOWLEDGMENTS

As with the previous editions, this third edition of *Electronic Devices* was made possible by the efforts and talents of many people. I want to express my appreciation to Carol Robison, Dave Garza, Steve Helba, Sharon Rudd, Mark Garrett, Anne Flanagan, and Russ Maselli of Merrill, an imprint of Macmillan Publishing Company. Also, thanks to Lois Porter for a great job in editing the manuscript.

I am grateful to the many contributors to this third edition. Without your input, an effective revision would be next to impossible. The following are the users of the second edition who provided many excellent suggestions and the instructors who reviewed the manuscript in detail and provided invaluable feedback: Don Arney, Indiana Vocational Technical College; Dave Bechtal, DeVry–Columbus; Howard Berlin, Delaware Technical and Community College; Frank Brattain, Indiana Vocational Technical College;

William Burns, Greenville Technical College; Darrell Cole, Southern Arkansas University–Tech; Mark Coleman, Lamar University–Orange; Jim Davis, Muskingum Technical College; Michael Dotson, Southern Illinois University; George Fredericks, Northeast State Technical Community College; Byron Hall, ITT–Dayton; Roger Hack, Indiana University–Purdue University; Chuck Kenny, Delaware Technical and Community College; Ed Kerly, Delaware Technical and Community College–Terry; Robert Koval, Salem Community College; Don LaFavour, Columbus State Community College; James Lovvorn, RETS Electronic Institute; Wally McIntyre, DeVry–Chicago; Mike Merchant; Rick Miller, Ferris State University; Maurice Nadeau, Central Minnesota Vocational Technical Institute; Mark Porulsky, Milwaukee Area Technical College; Thomas Ratliff, Central Carolina Community College; Robert Reaves, Durham Technical Community College; Mohammed Taher, DeVry–Lombard; and James Walker, Greenville Technical Institute. We depend on you, so keep up the good work.

Also thanks to Kevin Gray of DeVry–Columbus and Gary Snyder of Yuba College for their assistance in checking the accuracy of the manuscript.

Tom Floyd

Contents

4
BIPOLAR JUNCTION TRANSISTORS 142

5
BIPOLAR TRANSISTOR BIASING 190

6
SMALL-SIGNAL BIPOLAR AMPLIFIERS 236

14 BASIC OP-AMP CIRCUITS 650

15 OSCILLATORS 710

16 ACTIVE FILTERS 762

17 VOLTAGE REGULATORS 804

This special 16-page full-color insert (which follows page 422) provides realistic printed circuit boards and test instruments for special Test Bench assignments related to certain System Applications.

1

Semiconductor Materials and PN Junctions

After completing this chapter, you should be able to

- □ Describe an atom.
- □ Define the electron and describe how it fits into the atomic structure.
- □ Discuss the process of ionization.
- □ Identify silicon and germanium by their atomic structure.
- □ Explain how atoms bond together to form crystals.
- □ Show how the energy levels within an atom's structure relate to electrical current in a material.
- □ Define p-type and n-type semiconductors.
- □ Explain how p-type and n-type semiconductors are created by the addition of impurity atoms.
- □ Describe how a pn junction is formed and discuss its characteristics.
- □ Describe a semiconductor diode and explain how it works.
- □ Explain what is meant by the terms *forward bias* and *reverse bias.*

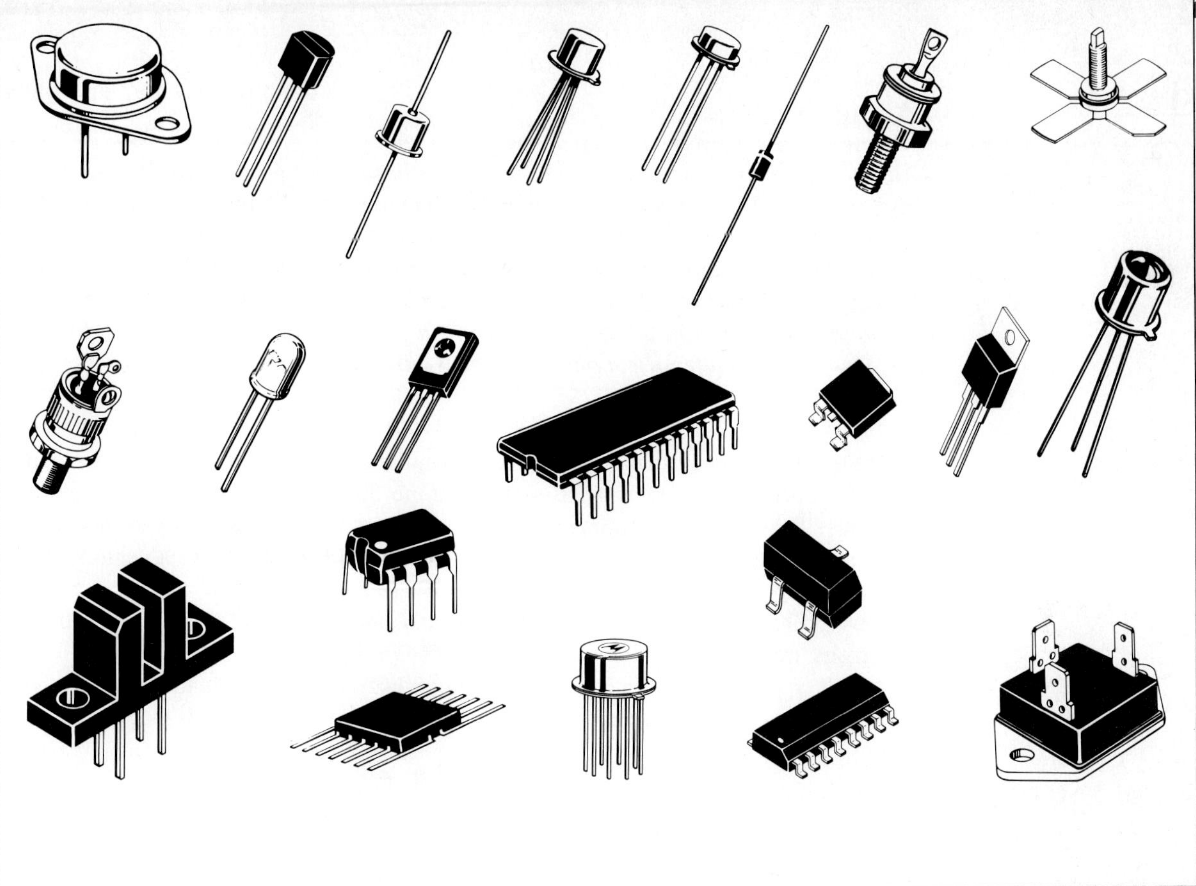

All diodes, transistors, and integrated circuits are made of semiconductor material. These electronic devices are shown above in a variety of common packaging configurations. In electronic systems, individual devices such as these are interconnected to form circuits that are designed to function in a specific way by properly utilizing the characteristics of each device. In the following chapters, you will learn how the various devices are used in certain system applications.

To understand how electronic devices work, you need a basic knowledge of atomic theory and the structure of semiconductor materials. Also the pn junction, which is formed when two types of semiconductor materials are joined, is fundamental to the operation of many types of semiconductor devices.

1–1 ATOMS

All matter is made up of atoms; and all atoms are made up of electrons, protons, and neutrons. In order to better understand how semiconductors work, you need to know something about the atom. In this section, you will learn about the structure of the atom, electron orbits and shells, valence electrons, ions, and the two major semiconductor materials—silicon and germanium. This material is important because the configuration of certain electrons in an atom is the key factor in determining how a given material conducts electrical current.

An **atom** is the smallest particle of an element that retains the characteristics of that element. Each known element has atoms that are different from the atoms of all other elements. This gives each element a unique atomic structure. According to the classical Bohr model, atoms have a planetary type of structure that consists of a central nucleus surrounded by orbiting electrons, as illustrated in Figure 1–1. The **nucleus** consists of positively charged particles called **protons** and uncharged particles called **neutrons.** **Electrons** are the basic particles of negative charge.

Each type of atom has a certain number of electrons and protons that distinguishes it from the atoms of all other elements. For example, the simplest atom is that of hydrogen.

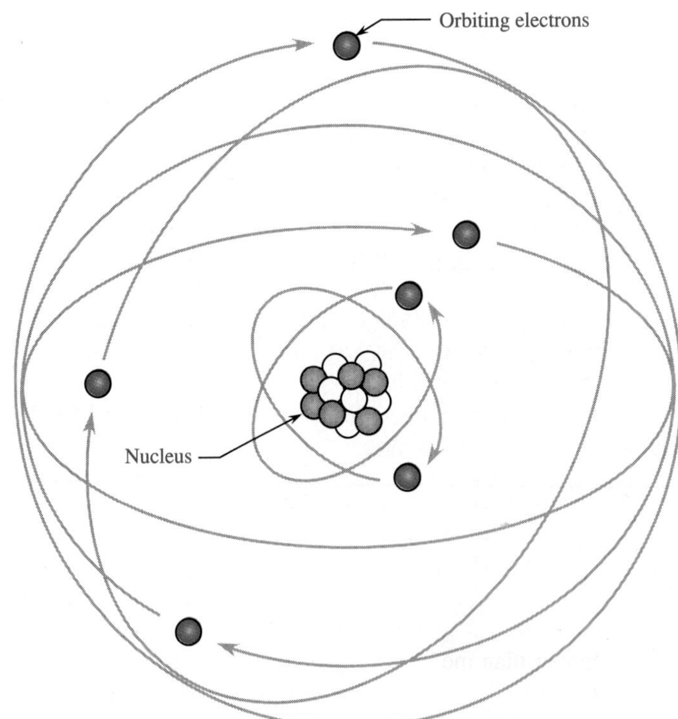

FIGURE 1–1
The Bohr model of an atom.

It has one proton and one electron, as shown in Figure 1–2(a). The helium atom, shown in Figure 1–2(b), has two protons and two neutrons in the nucleus orbited by two electrons.

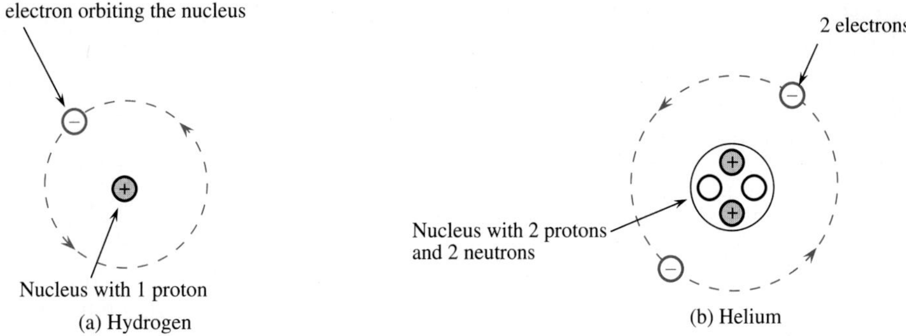

FIGURE 1–2
Hydrogen and helium atoms.

ATOMIC NUMBER AND WEIGHT

All elements are arranged in the periodic table of the elements in order according to their **atomic number,** which equals the number of electrons in an electrically balanced (neutral) atom. The elements can also be arranged by their **atomic weight,** which is approximately the number of protons and neutrons in the nucleus. For example, hydrogen has an atomic number of one and an atomic weight of one. The atomic number of helium is two, and its atomic weight is four. In their normal, or neutral, state, all atoms of a given element have the same number of electrons as protons; the positive charges cancel the negative charges, and the atom has a net charge of zero.

ELECTRON SHELLS AND ORBITS

Electrons orbit the nucleus of an atom at certain distances from the nucleus. Electrons near the nucleus have less energy than those in more distant orbits. It is known that only discrete (separate and distinct) values of electron energies exist within atomic structures. Therefore, electrons must orbit only at discrete distances from the nucleus.

Each discrete distance **(orbit)** from the nucleus corresponds to a certain energy level. In an atom, the orbits are grouped into energy bands known as **shells.** A given atom has a fixed number of shells. Each shell has a fixed maximum number of electrons at permissible energy levels (orbits). The differences in energy levels within a shell are much smaller than the difference in energy between shells. The shells are designated *K, L, M, N,* and so on, with *K* being closest to the nucleus. This concept is illustrated in Figure 1–3, which shows the *K* and *L* shells.

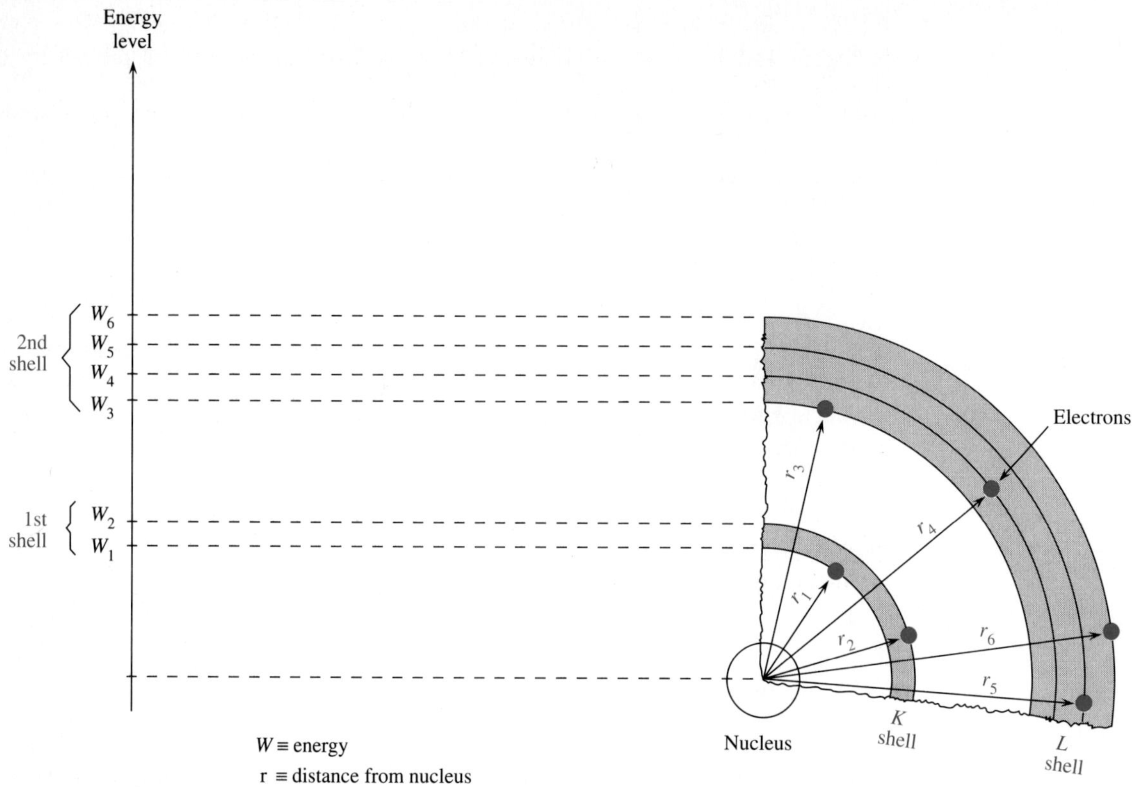

FIGURE 1–3
Energy levels increase as distance from nucleus increases.

VALENCE ELECTRONS

Electrons in orbits farther from the nucleus are less tightly bound to the atom than those closer to the nucleus. This is because the force of attraction between the positively charged nucleus and the negatively charged electron decreases with increasing distance. Electrons with the highest energy levels exist in the outermost shell of an atom and are relatively loosely bound to the atom. These valence electrons contribute to chemical reactions and bonding within the structure of a material. The **valence** of an atom is the number of electrons in its outermost shell.

IONIZATION

When an atom absorbs energy from a heat source or from light, for example, the energy levels of the electrons are raised. When an electron gains energy, it moves to an orbit farther from the nucleus. Since the valence electrons possess more energy and are more loosely bound to the atom than inner electrons, they can jump to higher orbits more easily when external energy is absorbed.

If a valence electron acquires a sufficient amount of energy, it can be completely removed from the outer shell and the atom's influence. The departure of a valence electron leaves a previously neutral atom with an excess of positive charge (more protons than electrons). The process of losing a valence electron is known as **ionization** and the resulting positively charged atom is called a *positive ion.* For example, the chemical symbol for hydrogen is H. When it loses its valence electron and becomes a positive ion, it is designed H^+. The escaped valence electron is called a **free electron.** When a free electron falls into the outer shell of a neutral hydrogen atom, the atom becomes negatively charged (more electrons than protons) and is called a *negative ion,* designated. H^-.

SILICON AND GERMANIUM ATOMS

Two types of semiconductor materials used to manufacture diodes, transistors, and other devices are **silicon** and **germanium.** Both the silicon and the germanium atoms have four valence electrons. They differ in that silicon has 14 protons in its nucleus, and germanium has 32. Figure 1–4 shows the atomic structure for both materials. Silicon is by far the most widely used of the two materials.

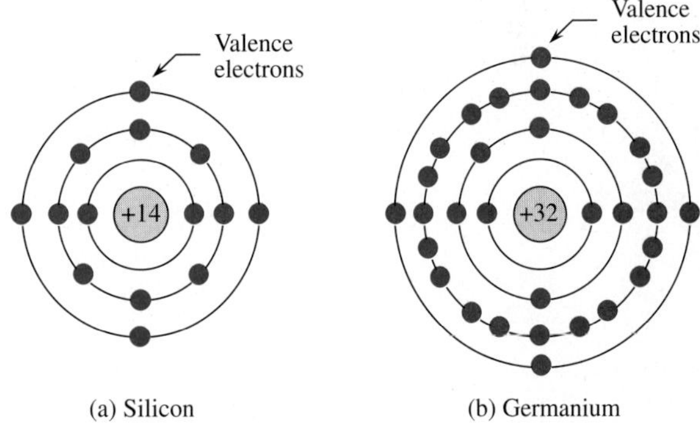

FIGURE 1–4
Silicon and germanium atoms.

1–1 REVIEW QUESTIONS

1. Describe an atom.
2. What is an electron?
3. What is a valence electron?
4. What is a free electron?
5. How are ions formed?
6. Name two semiconductor materials.

1–2 COVALENT BONDS

When silicon atoms combine to form a solid material, they arrange themselves in a fixed pattern called a* crystal. *The atoms within the crystal structure are held together by covalent bonds, which are created by the interaction of the valence electrons of the atoms. A solid chunk of silicon is a crystalline material.

Figure 1–5 shows how each silicon atom positions itself with four adjacent atoms. A silicon atom with its four valence electrons shares an electron with each of its four

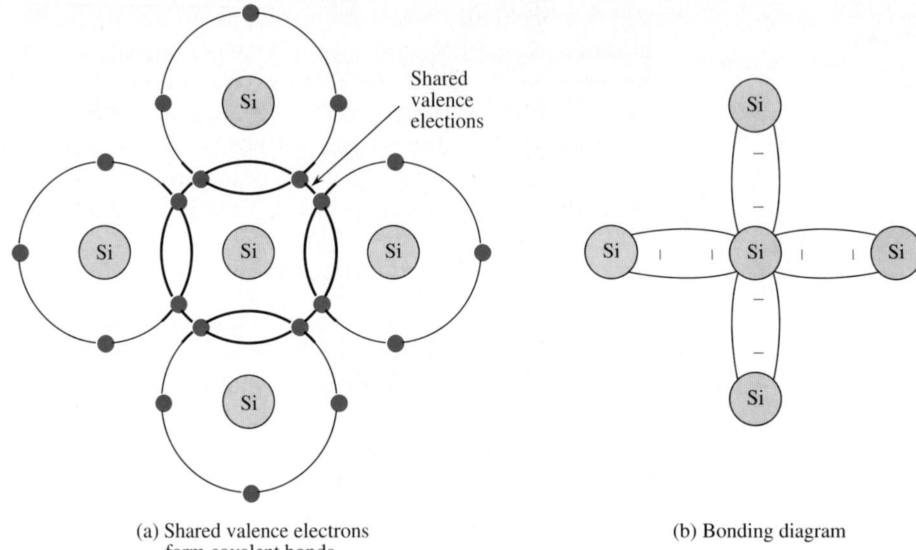

(a) Shared valence electrons form covalent bonds.

(b) Bonding diagram

FIGURE 1–5
Covalent bonds in silicon.

neighbors. This effectively creates eight valence electrons for each atom and produces a state of chemical stability. Also, this sharing of valence electrons produces the **covalent** bonds that hold the atoms together; each shared electron is attracted equally by two adjacent atoms which share it. Covalent bonding of a pure **(intrinsic)** silicon crystal is shown in Figure 1–6. Bonding for germanium is similar because it also has four valence electrons.

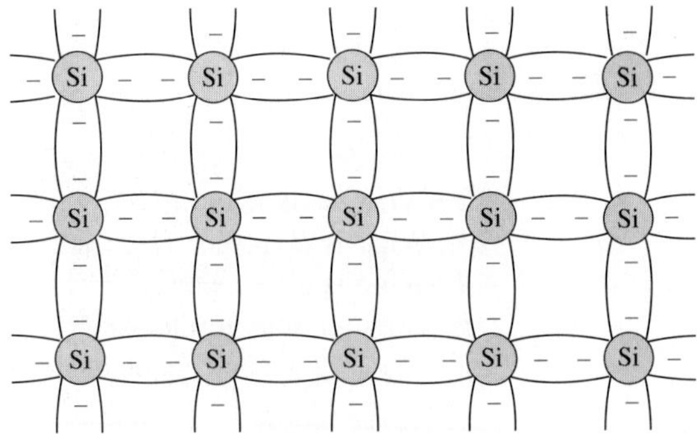

FIGURE 1–6
Covalent bonds in a pure (intrinsic) silicon crystal.

1–2 REVIEW QUESTIONS

1. What is a crystal?
2. How are covalent bonds formed?
3. What is meant by the term *intrinsic?*
4. How many valence electrons are there in each atom within a silicon crystal?

1–3 CONDUCTION IN SEMICONDUCTOR CRYSTALS

How a material conducts electrical current is very important in understanding how electronic devices operate. You can't really understand the operation of a device such as a diode or transistor without knowing something about the basic current phenomenon. In this section you will see how conduction occurs and why some materials are better conductors than others.

As you have learned, the electrons of an atom can exist only within prescribed energy bands. Each shell around the nucleus corresponds to a certain energy band and is separated from adjacent shells by energy gaps, in which no electrons can exist. This is shown in Figure 1–7 for an unexcited (no external heat energy) silicon crystal. This condition occurs *only* at absolute 0° temperature.

FIGURE 1–7
Energy band diagram for unexcited silicon crystal. There are no electrons in the conduction band.

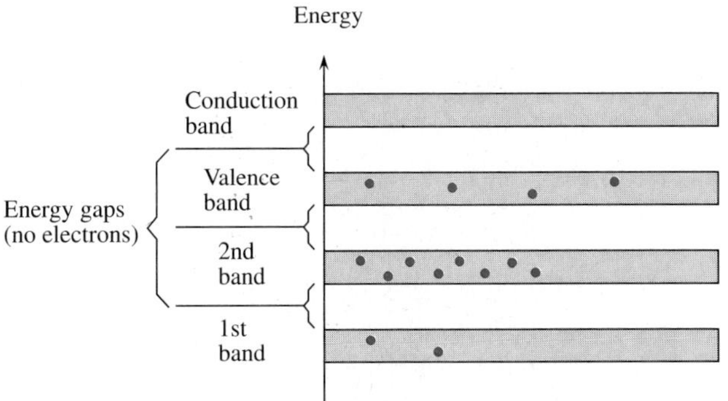

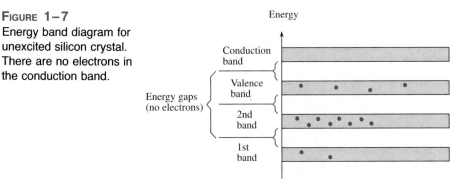

CONDUCTION ELECTRONS AND HOLES

A pure silicon crystal at room temperature derives heat (thermal) energy from the surrounding air, causing some valence electrons to gain sufficient energy to jump the gap from the valence band into the conduction band, becoming free electrons, also called **conduction electrons.** This is illustrated in the energy diagram of Figure 1–8(a) and in the bonding diagram of Figure 1–8(b).

When an electron jumps to the conduction band, a vacancy is left in the valence band. This vacancy is called a **hole.** For every electron raised to the conduction band by thermal or light energy, there is one hole left in the valence band, creating what is called an **electron-hole pair. Recombination** occurs when a conduction-band electron loses energy and falls back into a hole in the valence band.

To summarize, a piece of pure silicon at room temperature has, at any instant, a number of conduction-band (free) electrons that are unattached to any atom and are essentially drifting randomly throughout the material. There is also an equal number of holes in the valence band created when these electrons jump into the conduction band. This is illustrated in Figure 1–9.

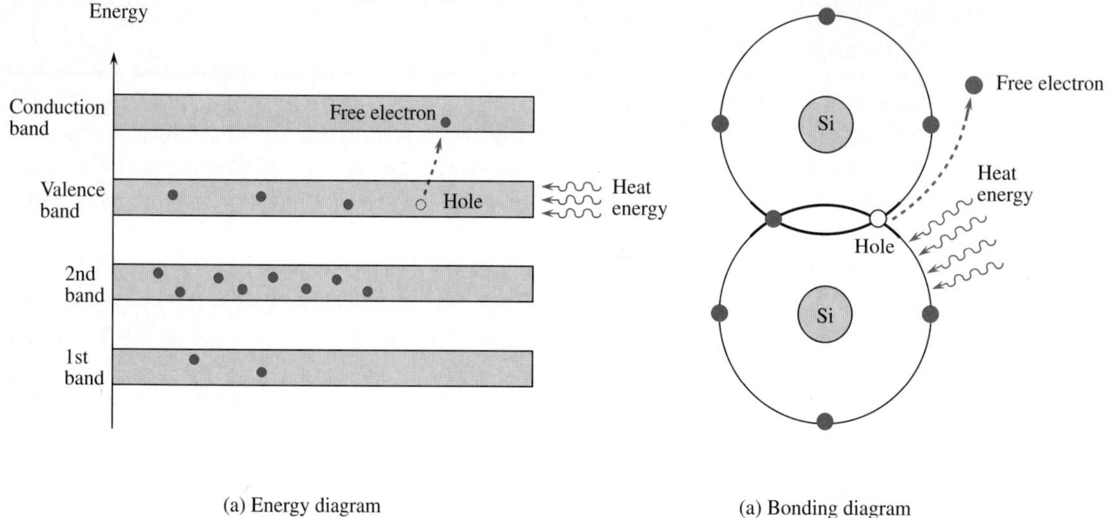

FIGURE 1–8
Creation of an electron-hole pair in an excited silicon atom. An electron in the conduction band is a free electron.

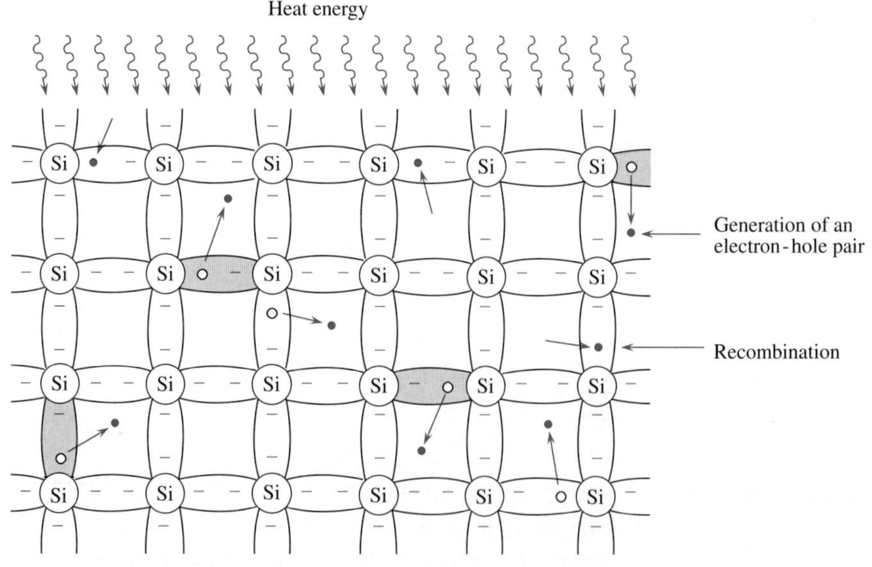

FIGURE 1–9
Electron-hole pairs in a silicon crystal. Free electrons are being generated continuously while some recombine with holes.

Germanium Versus Silicon

The situation in a germanium crystal is similar to that in silicon except that, because of its atomic structure, pure germanium has more free electrons than silicon and therefore a higher conductivity. Silicon, however, is the favored semiconductor material and is far more widely used than germanium. One reason for this is that silicon can be used at a much higher temperature than germanium.

Electron and Hole Current

When a voltage is applied across a piece of intrinsic silicon, as shown in Figure 1–10, the thermally generated free electrons in the conduction band are easily attracted toward the positive end. This movement of free electrons is one type of **current** in a semiconductor material, called *electron current*.

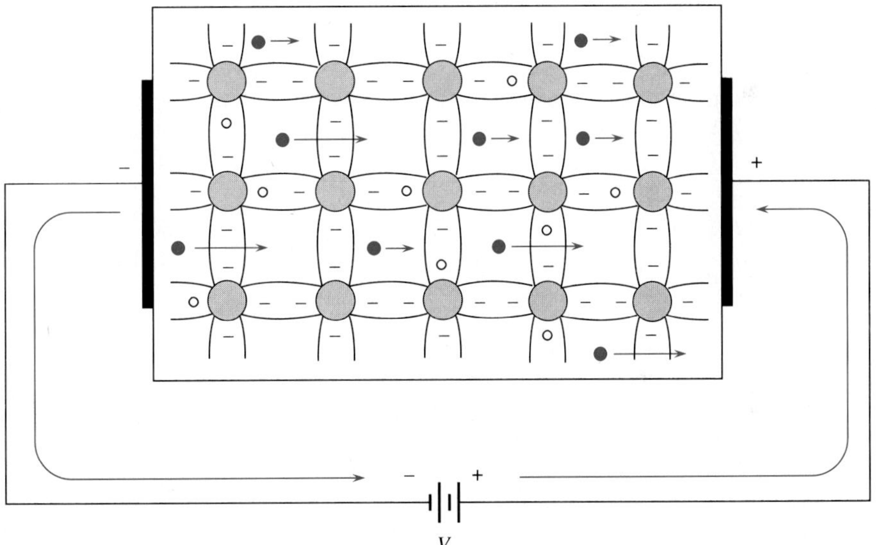

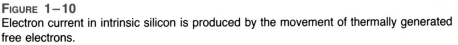

Figure 1–10
Electron current in intrinsic silicon is produced by the movement of thermally generated free electrons.

Another type of current occurs at the valence level, where the holes created by the free electrons exist. Electrons remaining in the valence band are still attached to their atoms and are not free to move randomly in the crystal structure. However, a valence electron can move into a nearby hole, with little change in its energy level, thus leaving another hole where it came from. Effectively the hole has moved from one place to another in the crystal structure, as illustrated in Figure 1–11. This is called *hole current*.

FIGURE 1–11
Hole current in intrinsic silicon.

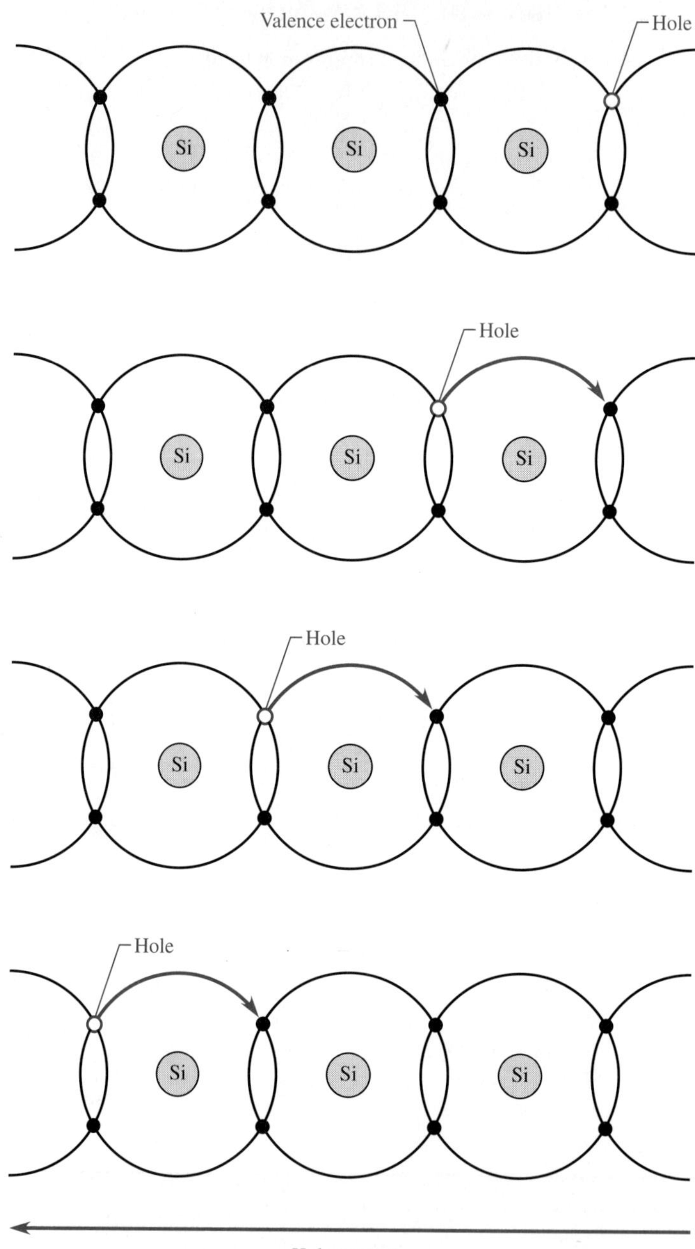

SEMICONDUCTORS, CONDUCTORS, AND INSULATORS

In a pure (intrinsic) **semiconductor,** there is a relatively small number of free electrons; so neither silicon nor germanium is very useful in its intrinsic state. They are neither **insulators** nor good **conductors** because current in a material depends directly on the number of free electrons.

A comparison of the energy bands in Figure 1–12 for the three types of materials shows the essential differences among them regarding conduction. The energy gap for an insulator is so wide that very few electrons acquire enough energy to jump into the conduction band. The valence band and the conduction band in a conductor (like copper) overlap so that there are always many conduction electrons. Even without the application of external energy, many valence electrons in a conductor already have enough energy to jump into the conduction band. The semiconductor, as the figure shows, has an energy gap that is much narrower than that in an insulator.

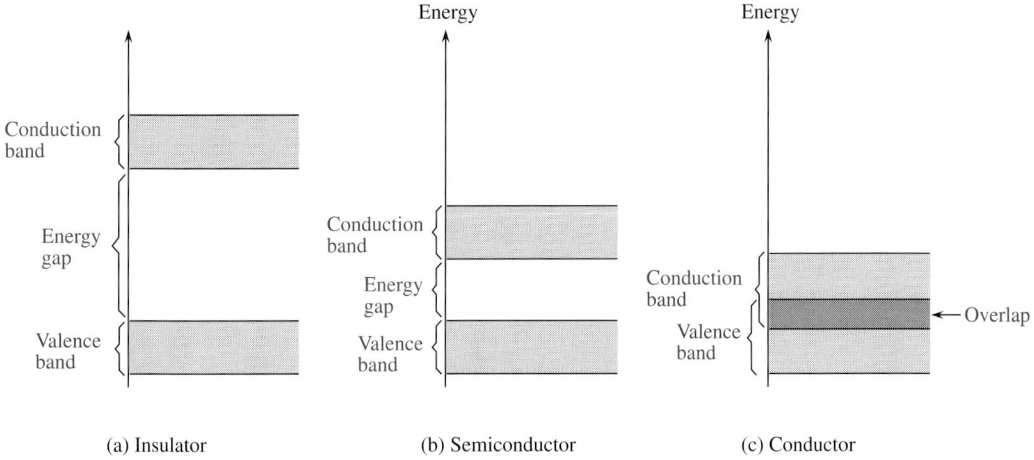

FIGURE 1–12
Energy diagrams for three categories of materials.

1–3 REVIEW QUESTIONS

1. Are free electrons in the valence band or in the conduction band?
2. Which electrons are responsible for current in a material?
3. What is a hole?
4. At what energy level does hole current occur?
5. Why does a semiconductor have fewer free electrons than a conductor?

1–4 N-TYPE AND P-TYPE SEMICONDUCTORS

Semiconductor materials do not conduct current well and are of very little value in their intrinsic state. This is because of the limited number of free electrons in the conduction band and holes in the valence band. Pure silicon (or germanium) must be modified by increasing the free electrons and holes to increase its conductivity and make it useful in electronic devices. This is done by adding impurities to the intrinsic material as you will learn in this section. The two types of semiconductor materials that you will learn about are the key building blocks for all types of electronic devices.

DOPING

The conductivity of silicon and germanium can be drastically increased by the controlled addition of impurities to the pure semiconductor material. This process, called **doping,** increases the number of current carriers (electrons or holes), thus increasing the conductivity and decreasing the resistivity. The two categories of impurities are *n-type* and *p-type*.

N-TYPE SEMICONDUCTOR

To increase the number of conduction-band electrons in pure silicon, **pentavalent** impurity atoms are added. These are atoms with five valence electrons such as arsenic, phosphorus, and antimony.

As illustrated in Figure 1–13, each pentavalent atom (antimony, in this case) forms covalent bonds with four adjacent silicon atoms. Four of the antimony atom's valence electrons are used to form the covalent bonds, leaving one extra electron. This extra electron becomes a conduction electron because it is not attached to any atom. The number of conduction electrons can be controlled by the number of impurity atoms added to the silicon. A conduction electron created by this doping process does not leave a hole in the valence band.

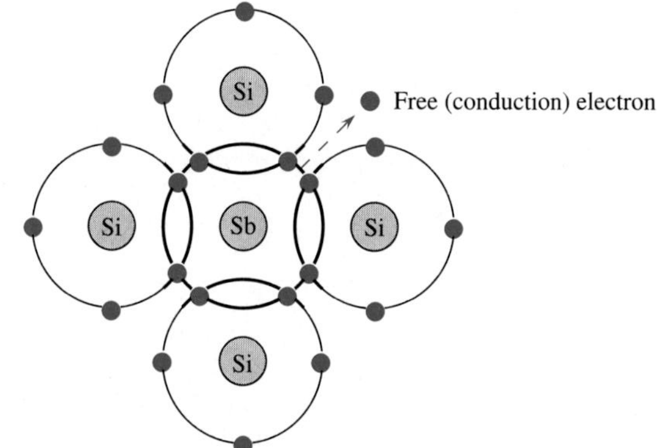

FIGURE 1–13
Pentavalent impurity atom in a silicon crystal. An antimony (Sb) impurity atom is shown in the center. The extra electron becomes a free electron.

Since most of the current carriers are electrons, silicon (or germanium) doped in this way is an n-type semiconductor material where the *n* stands for the negative charge on an electron. The electrons are called the **majority carriers** in n-type material. Although the majority of current carriers in n-type material are electrons, there are a few holes that are created when electron-hole pairs are thermally generated. These holes are *not* produced by the addition of the pentavalent impurity atoms. Holes in an n-type material are called **minority carriers.**

P-TYPE SEMICONDUCTOR

To increase the number of holes in pure silicon, **trivalent** impurity atoms are added. These are atoms with three valence electrons such as aluminum, boron, and gallium. As illustrated in Figure 1–14, each trivalent atom (boron, in this case) forms covalent bonds

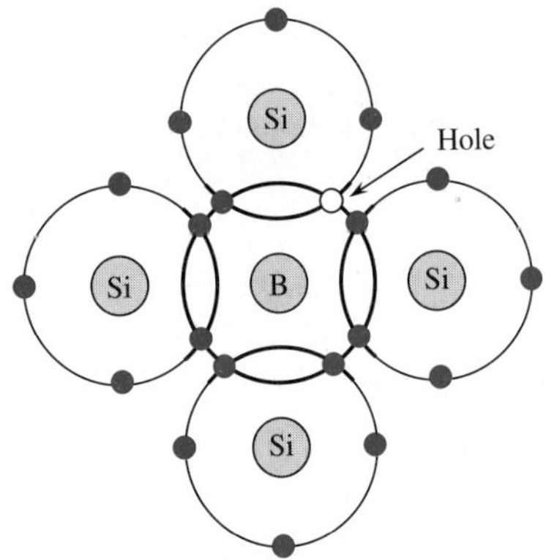

FIGURE 1–14
Trivalent impurity atom in a silicon crystal. A boron (B) impurity atom is shown in the center.

with four adjacent silicon atoms. All three of the boron atom's valence electrons are used in the covalent bonds; and, since four electrons are required, a hole is formed with each trivalent atom. The number of holes can be controlled by the number of trivalent impurity atoms added to the silicon. A hole created by this doping process is *not* accompanied by a conduction (free) electron.

Since most of the current carriers are holes, silicon (or germanium) doped in this way is called a p-type semiconductor material because holes can be thought of as positive charges. The holes are the majority carriers in p-type material. Although the great majority of current carriers in p-type material are holes, there are a few free electrons that are created when electron-hole pairs are thermally generated. These free electrons are *not* produced by the addition of the trivalent impurity atoms. Electrons in p-type material are the minority carriers.

1–4 REVIEW QUESTIONS

1. Define doping.
2. What is the difference between a pentavalent atom and a trivalent atom?
3. How is an n-type semiconductor formed?
4. How is a p-type semiconductor formed?
5. What is the majority carrier in an n-type semiconductor?
6. What is the majority carrier in a p-type semiconductor?
7. By what process are the majority carriers produced?
8. By what process are the minority carriers produced?

1–5 PN JUNCTIONS

If we take a block of silicon and dope half of it with a pentavalent impurity and the other half with a trivalent impurity, a boundary called the* pn junction *is formed between the resulting n-type and the p-type portions. Amazingly, this pn junction is the thing that allows diodes and transistors to work. This section and the next one will give you some insight into the pn junction and provide a springboard into the next chapter where you will learn about an important electronic device called the diode.

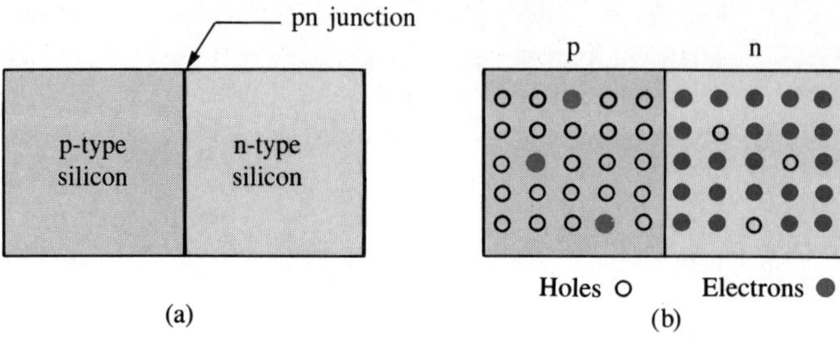

FIGURE 1–15
Basic pn structure at the instant of junction formation. Both majority and minority carriers are shown.

Figure 1–15(a) shows a pn junction formed between the two regions when a piece of silicon is doped so that half is n-type and the other half is p-type. This basic structure forms a semiconductor **diode.** The n region has many free electrons (majority carriers) and only a few thermally generated holes (minority carriers). The p region has many holes (majority carriers) and only a few thermally generated free electrons (minority carriers). This is shown in Figure 1–15(b). The pn junction is fundamental to the operation of diodes, transistors, and other solid-state devices.

THE DEPLETION LAYER

With no external voltage, the conduction electrons in the n region are aimlessly drifting in all directions. At the instant of junction formation, some of the conduction electrons near the junction diffuse across into the p region and recombine with holes near the junction. For each electron that crosses the junction and recombines with a hole, a pentavalent atom is left with a net positive charge in the n region near the junction, making it a positive ion. Also, when the electron recombines with a hole in the p region, a trivalent atom acquires a net negative charge, making it a negative ion.

As a result of this recombination process, a large number of positive and negative ions build up near the pn junction. As this buildup occurs, the conduction electrons in the n region must overcome both the attraction of the positive ions and the repulsion of the negative ions in order to migrate into the p region. Thus, as the ion layers build up, the area on both sides of the junction becomes essentially depleted of any conduction electrons or holes and is known as the **depletion layer.** This condition is illustrated in Figure 1–16. When an equilibrium condition is reached, the depletion layer has widened to a point where no further electrons can cross the pn junction.

FIGURE 1–16
PN junction equilibrium condition. The few electrons in the p region (solid dots) and the few holes (open circles) in the n region are the minority carriers created by thermally produced electron-hole pairs.

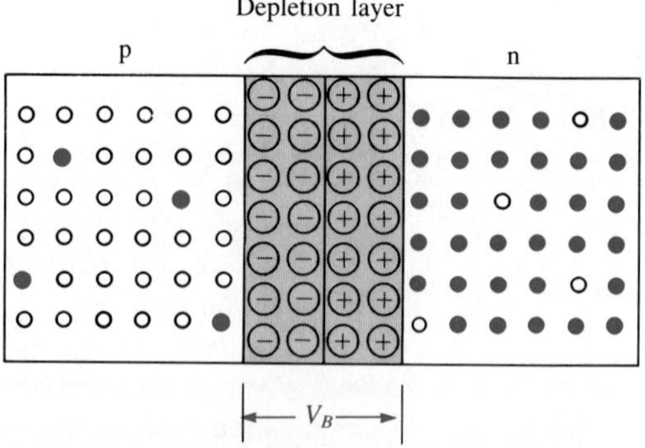

The existence of the positive and negative ions on opposite sides of the junction creates a **barrier potential** (V_B) across the depletion layer, as indicated in Figure 1–16. At 25°C, the barrier potential is approximately 0.7 V for silicon and 0.3 V for germanium. As the junction temperature increases, the barrier potential decreases, and vice versa.

The barrier potential is significant because it determines the amount of external voltage that must be applied across the pn junction to produce current. You will see this more clearly when we discuss biasing of the pn junction.

ENERGY DIAGRAM OF THE PN JUNCTION

Now, we will look at the operation of the pn junction in terms of its energy level. First consider the pn junction at the instant of its formation. The energy bands of the trivalent impurity atoms in the p-type material are at a slightly higher level than those of the pentavalent impurity atom in the n-type material, as shown in Figure 1–17. This is because the core attraction for the valence electrons (+3) in the trivalent atom is less than the core attraction for the valence electrons (+5) in the pentavalent atom. So, the trivalent valence electrons are in a slightly higher orbit and, thus, at a higher energy level.

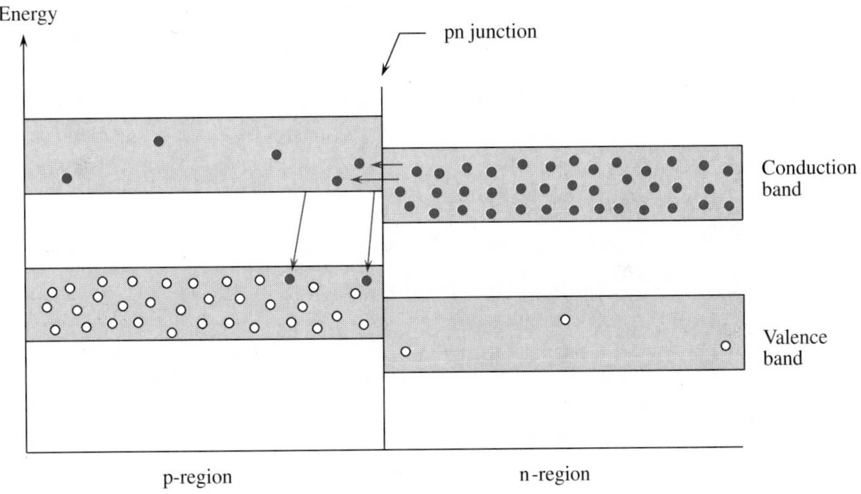

FIGURE 1–17
PN junction energy diagram as diffusion across the junction begins.

Notice in Figure 1–17 that there is some overlap of the conduction bands in the p and n regions and also some overlap of the valence bands in the p and n regions. This permits the electrons of higher energy near the top of the n-region conduction band to begin diffusing across the junction into the lower levels of the p-region conduction band. As soon as an electron diffuses across the junction, it recombines with a hole in the valence band. As diffusion continues, the depletion layer begins to form. Also, the energy bands in the n region "shift" down as the electrons with higher energy are lost to diffusion. When the top of the n-region conduction band reaches the same level as the bottom of the p-region conduction band, diffusion ceases and the equilibrium condition is reached. This

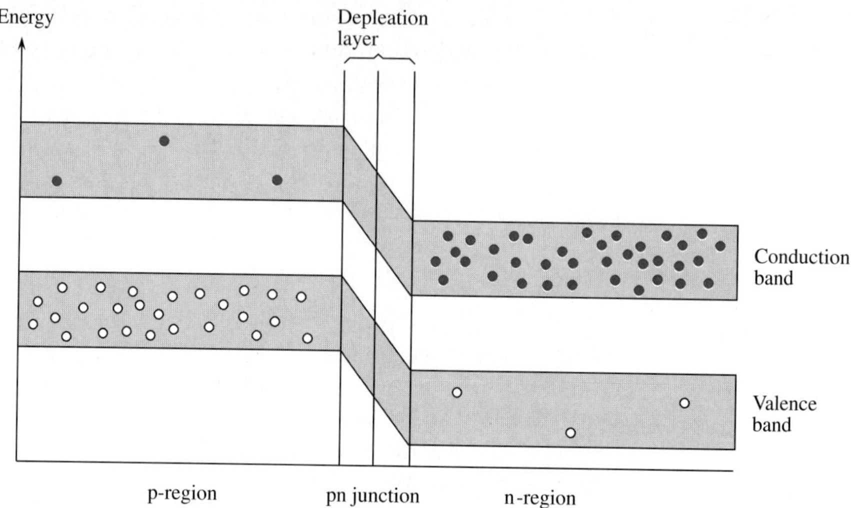

FIGURE 1–18
Energy diagram at equilibrium. The n-region bands are shifted down from their original levels because the higher-energy electrons are lost to diffusion.

is shown in terms of energy levels in Figure 1–18. There is an energy gradient across the depletion layer rather than an abrupt change in energy level.

1–5 REVIEW QUESTIONS

1. What is a pn junction?
2. What makes up the depletion layer?
3. Are there any majority carriers in the depletion layer?
4. What is the value of the barrier potential in silicon?

1–6 BIASING THE PN JUNCTION

As you have learned, there is no current across a pn junction at equilibrium. The primary usefulness of the pn junction diode is its ability to allow current in only one direction and to prevent current in the other direction as determined by the bias. There are two bias conditions for a pn junction: forward and reverse. Either of these conditions is created by connecting an external dc voltage in the proper direction across the diode.

FORWARD BIAS

The term **bias** in electronics refers to a fixed dc voltage that sets the operating conditions for a semiconductor device. **Forward bias** is the condition that permits current across a pn junction. Figure 1–19 shows a dc voltage connected in a direction to forward-bias the diode. Notice that the negative **terminal** of the battery is connected to the n region (called the **cathode**), and the positive terminal is connected to the p region (called the **anode**).

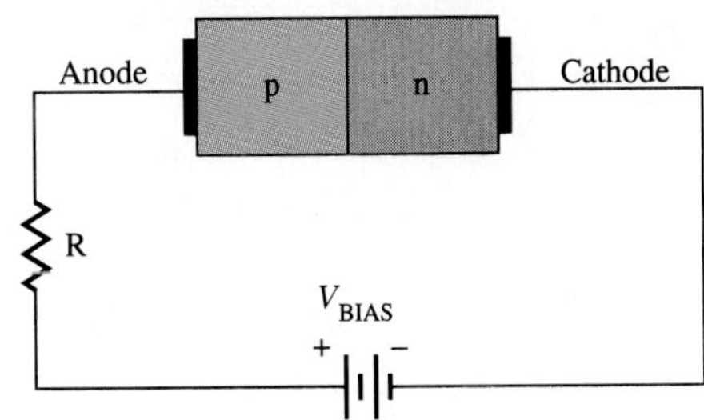

FIGURE 1–19
Forward-bias connection. The purpose of the resistor is to limit the current to prevent damage to the diode.

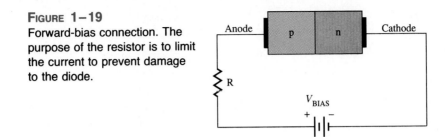

This is how forward bias works: The negative terminal of the battery pushes the conduction electrons in the n region toward the junction, while the positive terminal pushes the holes in the p region also toward the junction. (Recall that like charges repel each other.) When it overcomes the barrier potential, the external bias voltage source provides the n-region electrons with enough energy to penetrate the depletion layer and cross the junction, where they combine with the p-region holes. As electrons leave the n region, more flow in from the negative terminal of the battery. So, current through the n region is the movement of conduction electrons (majority carriers) toward the junction.

When the conduction electrons enter the p region and combine with holes, they become valence electrons. They then move as valence electrons from hole to hole toward the positive anode connection. The movement of these valence electrons essentially creates a movement of holes in the opposite direction as you learned earlier. So, current in the p region is the movement of holes (majority carriers) toward the junction. Figure 1–20 illustrates electron flow in a forward-biased diode.

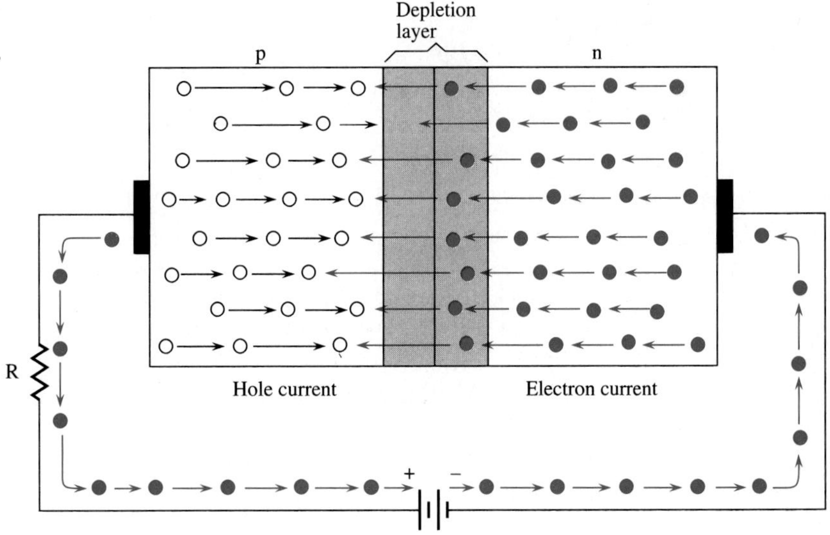

FIGURE 1–20
Electron flow in a pn junction diode.

EFFECT OF BARRIER POTENTIAL ON FORWARD BIAS The barrier potential of the depletion layer can be envisioned as acting as a small battery that opposes bias, as illustrated in Figure 1–21(a). Keep in mind that the barrier potential is not a voltage source and cannot be

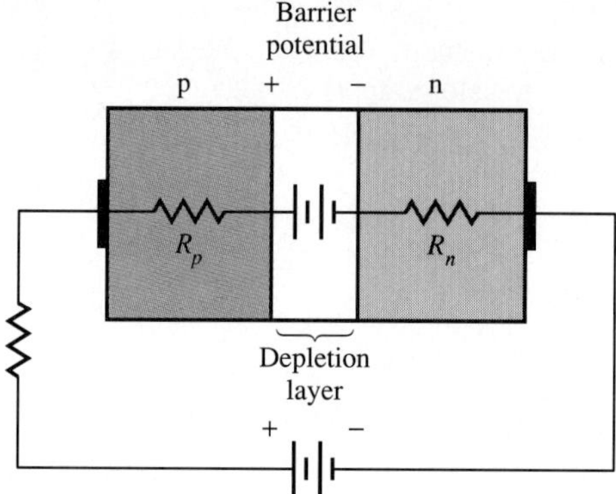

(a) Barrier potential and bulk resistance equivalent

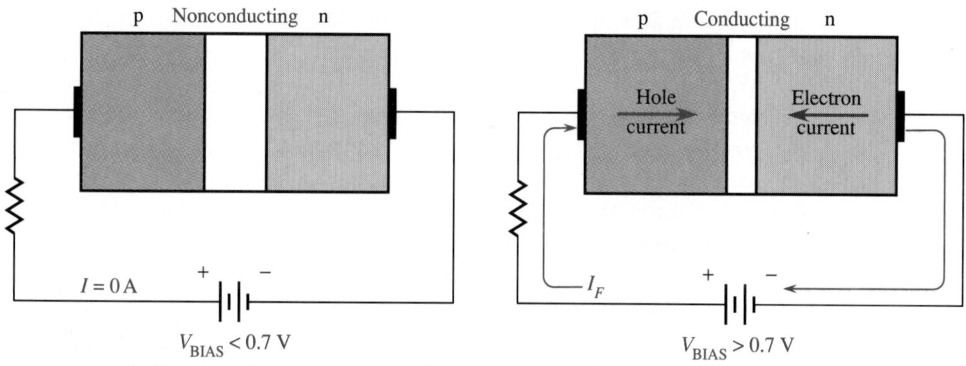

(b) The bias voltage must overcome the barrier potential for the diode to conduct current.

FIGURE 1–21
Diode barrier potential and bulk resistance.

measured with a voltmeter; it only has the effect of a battery when forward bias is applied because the external bias voltage must overcome the barrier potential before the diode conducts, as illustrated in Figure 1–21(b). Conduction occurs at approximately 0.7 V for silicon and 0.3 V for germanium. Once the diode is conducting in the forward direction, the voltage drop across it remains at approximately the barrier potential and changes very little with changes in forward current (I_F) except for bulk resistance effects. The bulk resistances are usually only a few ohms and result in only a small voltage drop when the diode conducts. Often this drop can be neglected. R_p and R_n represent the bulk resistances of the p and n regions.

ENERGY DIAGRAM FOR FORWARD BIAS Forward bias raises the energy levels of the conduction electrons in the n region, allowing them to move into the p region and combine with holes in the valence band. This condition is shown in Figure 1–22.

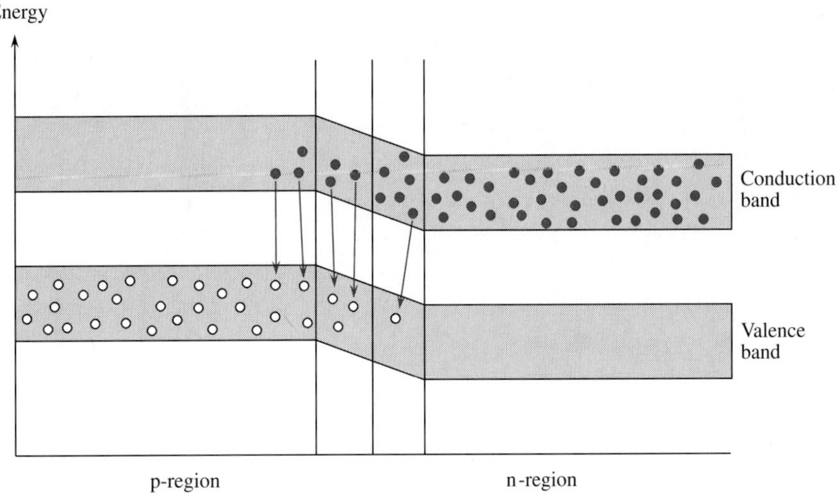

FIGURE 1–22
Energy diagram for forward bias showing recombination in the depletion layer and in the p region as conduction electrons move across junction.

REVERSE BIAS

Reverse bias is the condition that prevents current across the pn junction. Figure 1–23 shows a dc voltage source connected to reverse-bias the diode. Notice that the negative terminal of the battery is connected to the p region, and the positive terminal is connected to the n region. The negative terminal of the battery attracts holes in the p region away from the pn junction, while the positive terminal also attracts electrons away from the junction. As electrons and holes move away from the junction, the depletion layer widens; more positive ions are created in the n region, and more negative ions are created in the p region, as shown in Figure 1–24.

FIGURE 1–23
Reverse-bias connection.

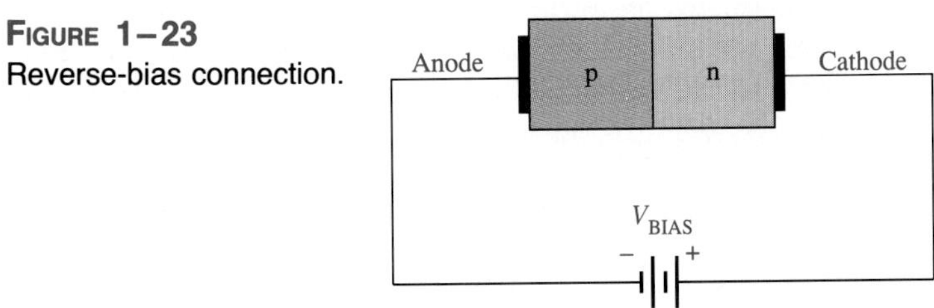

The depletion layer widens until the potential difference across it equals the external bias voltage. At this point, the holes and electrons stop moving away from the junction and majority current ceases, as indicated in Figure 1–24(b). The initial movement of majority carriers away from the junction is called *transient current* and lasts only for a very short time upon application of reverse bias.

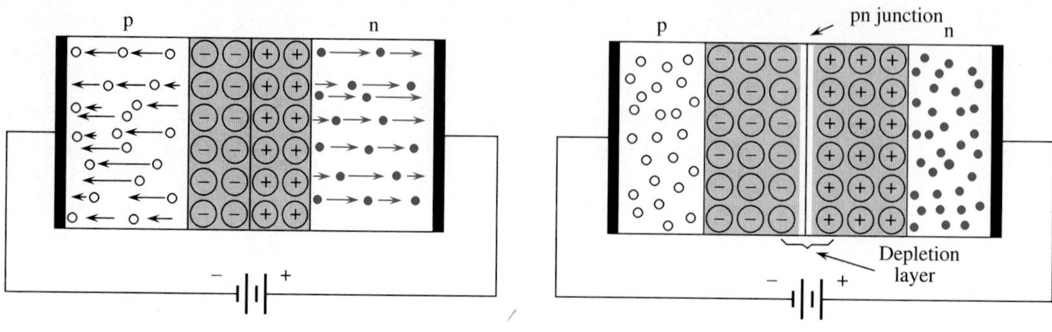

(a) Transient current as depletion layer widens

(b) Current ceases when barrier potential equals bias voltage

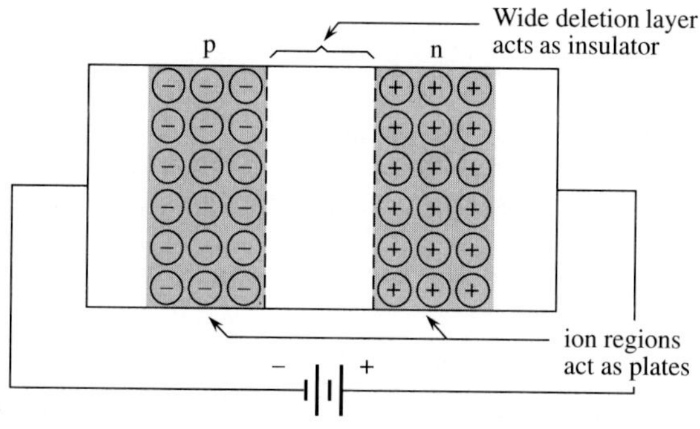

(c) Depletion layer widens as reverse bias increases

FIGURE 1–24
Reverse bias.

When the diode is reverse-biased, the depletion layer effectively acts as an insulator between the layers of oppositely charged ions. This effectively forms a capacitor, as illustrated in Figure 1–24(c). Since the depletion layer widens with increased reverse-biased voltage, the capacitance decreases and vice versa. This internal capacitance is called the *depletion-layer capacitance* and has some very practical applications that you will learn about later.

MINORITY CURRENT As you have learned, majority current very quickly becomes zero when reverse bias is applied. There is, however, a very small current produced by minority carriers during reverse bias. Germanium, as a rule, has a greater reverse current than silicon. This current in silicon is typically in the μA or nA range. A relatively small number of thermally produced electron-hole pairs exist in the depletion layer. Under the influence of the external voltage, some electrons manage to diffuse across the pn junction before recombination. This process establishes a small minority carrier current throughout the material. The reverse current is dependent primarily on the junction temperature and not on the amount of reverse-biased voltage. A temperature increase causes an increase in reverse current.

REVERSE BREAKDOWN If the external reverse-bias voltage is increased to a large enough value, avalanche breakdown occurs. Here is what happens: Assume that one minority conduction-band electron acquires enough energy from the external source to accelerate it toward the positive end of the diode. During its travel, it collides with an atom and imparts enough energy to knock a valence electron into the conduction band. There are now two conduction-band electrons. Each will collide with an atom, knocking two more valence electrons into the conduction band. There are now four conduction-band electrons which, in turn, knock four more into the conduction band. This rapid multiplication of conduction-band electrons, known as an *avalanche effect,* results in a rapid buildup of reverse current. Most diodes are normally not operated in reverse breakdown and can be damaged by the resulting excessive power if they are. However, a particular type of diode (to be studied later), known as a zener diode, is designed for reverse-breakdown operation.

1–6 REVIEW QUESTIONS

1. What is the difference between reverse bias and forward bias in terms of current in the diode?
2. Which bias condition produces majority carrier current?
3. What minimum value does the bias voltage have to be in order to forward-bias a silicon diode?
4. Explain how to forward-bias a diode.
5. Explain how to reverse-bias a diode.
6. What happens to the depletion layer when the reverse-bias voltage is increased?
7. What happens when the reverse-bias voltage becomes excessively high?

SUMMARY

- □ According to the classical Bohr model, the atom is viewed as having a planetary-type structure with electrons orbiting at various distances around the central nucleus.
- □ The nucleus of an atom consists of protons and neutrons. The protons have a positive charge and the neutrons are uncharged. The number of protons and neutrons is the atomic weight of the atom.
- □ Electrons have a negative charge and orbit around the nucleus at distances that depend on their energy level. An atom has discrete bands of energy called *shells* in which the electrons orbit. Atomic structure allows a certain maximum number of electrons in each shell. These shells are designated *K, L, M,* and so on. The number of electrons in a neutral atom is determined by the number of protons. In their natural state, all atoms are neutral because they have an equal number of protons and electrons.
- □ The outermost shell or band of an atom is called the *valence band,* and electrons that orbit in this band are called *valence electrons*. These electrons have the highest energy of all those in the atom. If a valence electron acquires enough energy from an outside source such as heat, it can jump out of the valence band and break away from its atom.

- Both silicon and germanium atoms have four valence electrons. Silicon is the most widely used semiconductor material.
- Atoms bond together to form a solid material called a *crystal*. The bonds that hold a crystal together are called *covalent bonds*. Within the crystal structure, the valence electrons that manage to escape from their parent atom are called *conduction electrons* or *free electrons*. They have more energy than the electrons in the valence band and are free to drift throughout the material. When an electron breaks away to become free, it leaves a hole in the valence band creating what is called an *electron-hole pair*. These electron-hole pairs are thermally produced because the electron acquired enough energy from external heat to break away.
- A free electron will eventually lose energy and fall back into a hole. This is called *recombination*. But, electron-hole pairs are continuously being thermally generated so there are always free electrons in the material.
- When a voltage is applied across the semiconductor, the thermally produced free electrons move in a net direction and form the current. This is one type of current in an intrinsic (pure) semiconductor.
- Another type of current is the hole current. This occurs as valence electrons move from hole to hole creating, in effect, a movement of holes in the opposite direction.
- Materials that are conductors have a large number of free electrons and conduct current very well. Insulating materials have very few free electrons and do not conduct current at all under normal circumstances. Semiconductive materials fall in between conductors and insulators in their ability to conduct current.
- An n-type material is created by adding impurity atoms that have five valence electrons. These impurities are pentavalent atoms and the process is called *doping*. A p-type semiconductor is created by adding impurity atoms with only three valence electrons. These impurities are trivalent atoms.
- The majority carriers in an n-type semiconductor are free electrons acquired by the doping process, and the minority carriers are holes produced by thermally generated electron-hole pairs. The majority carriers in a p-type semiconductor are holes acquired by the doping process, and the minority carriers are free electrons produced by thermally generated electron-hole pairs.
- A pn junction is formed when part of a material is n-type and part of it is p-type. A depletion layer forms near the junction that is devoid of any majority carriers. The depletion layer is formed by ionization.
- Current flows across the pn junction only when it is forward-biased. Current does not flow when there is no bias or when there is reverse bias. There is a very small current in reverse bias due to the thermally generated minority carriers, but this can usually be neglected. A semiconductor diode is formed by a single pn junction.

GLOSSARY

Anode The more positive terminal of a diode and certain other electronic devices.

Atom The smallest particle of an element that possesses the unique characteristics of that element.

Atomic number The number of electrons in a neutral atom.

Atomic weight The number of protons and neutrons in the nucleus of an atom.

Barrier potential The effective voltage across a pn junction.

Bias The application of a dc voltage to a diode to make it either conduct or block current.

Cathode The more negative terminal of a diode and certain other electronic devices.

Conduction electron A free electron.

Conductor A material that conducts electrical current very well.

Covalent Related to the bonding of two or more atoms by the interaction of their valence electrons.

Crystal The pattern or arrangement of atoms forming a solid material.

Current The rate of flow of electrons.

Depletion layer The area near a pn junction that has no majority carriers.

Diode A two-terminal electronic device that permits current in only one direction.

Doping The process of imparting impurities to an intrinsic semiconductor material in order to control its conduction characteristics.

Electron The basic particle of negative electrical charge.

Electron-hole pair The conduction electron and the hole created when the electron leaves the valence band.

Foward bias The condition in which a pn junction conducts current.

Free electron An electron that has acquired enough energy to break away from the valence band of the parent atom. Also called a *conduction electron*.

Germanium A semiconductor material.

Hole The absence of an electron in the valence band of an atom.

Insulator A material that does not conduct current.

Intrinsic The pure or natural state of a material.

Ionization The removal or addition of an electron from or to a neutral atom so that the resulting atom (called an ion) has a net positive or negative charge.

Majority carrier The most numerous charge carrier in a semiconductor material (either free electrons or holes).

Minority carrier The least numerous charge carrier in a semiconductor material (either free electrons or holes).

Neutron An uncharged particle found in the nucleus of an atom.

Nucleus The central part of an atom containing protons and neutrons.

Orbit The path an electron takes as it circles around the nucleus of an atom.

Pentavalent atom An atom with five valence electrons.

PN junction The boundary between two different types of semiconductor materials.

Proton The basic particle of positive charge.

Recombination The process of a free (conduction band) electron falling into a hole in the valence band of an atom.

Reverse bias The condition in which a pn junction blocks current.

Semiconductor A material that lies between conductors and insulators in its conductive properties.

Shell An energy band in which electrons orbit the nucleus of an atom.

Silicon A semiconductor material.

Terminal The external contact point on an electronic circuit or device.

Trivalent atom An atom with three valence electrons.

Valence Related to the outer shell of an atom.

SELF-TEST

1. Every known element has
(a) the same type of atoms **(b)** the same number of atoms
(c) a unique type of atom **(d)** several different types of atoms

2. An atom consists of
(a) one nucleus and only one electron
(b) one nucleus and one or more electrons
(c) protons, electrons, and neutrons
(d) b and c

3. The nucleus of an atom is made up of
(a) protons and neutrons **(b)** electrons
(c) electrons and protons **(d)** electrons and neutrons

4. The atomic number of silicon is
(a) 8 **(b)** 2 **(c)** 4 **(d)** 14

5. The atomic number of germanium is
(a) 8 **(b)** 2 **(c)** 4 **(d)** 32

6. The valence shell in a silicon atom has the letter designation of
(a) *A* **(b)** *K* **(c)** *L* **(d)** *M*

7. Valence electrons are
(a) in the closest orbit to the nucleus
(b) in the most distant orbit from the nucleus
(c) in various orbits around the nucleus
(d) not associated with a particular atom

8. A positive ion is formed when
(a) a valence electron breaks away from the atom
(b) there are more holes than electrons in the outer orbit
(c) two atoms bond together
(d) an atom gains an extra valence electron

9. The most widely used semiconductor material in electronic devices is

(a) germanium **(b)** carbon

(c) copper **(d)** silicon

10. The energy band in which free electrons exist is the

(a) first band **(b)** second band

(c) conduction band **(d)** valence band

11. Electron-hole pairs are produced by

(a) recombination **(b)** thermal energy

(c) ionization **(d)** doping

12. Recombination is when

(a) an electron falls into a hole

(b) a positive and a negative ion bond together

(c) a valence electron becomes a conduction electron

(d) a crystal is formed

13. In a semiconductor crystal, the atoms are held together by

(a) the interaction of valence electrons

(b) forces of attraction

(c) covalent bonds

(d) all of the above

14. Each atom in a silicon crystal has

(a) four valence electrons

(b) four conduction electrons

(c) eight valence electrons, four of its own and four shared

(d) none, because all are shared with other atoms

15. The current in a semiconductor is produced by

(a) electrons only **(b)** holes only

(c) negative ions **(d)** both electrons and holes

16. In an intrinsic semiconductor

(a) there are no free electrons

(b) the free electrons are thermally produced

(c) there are only holes

(d) there are as many electrons as there are holes

(e) b and d

17. The difference between an insulator and a semiconductor is

(a) a wider energy gap between the valence band and the conduction band

(b) the number of free electrons

(c) the atomic structure

(d) all of the above

18. The process of adding an impurity to an intrinsic semiconductor is called
(a) doping **(b)** recombination
(c) atomic modification **(d)** ionization

19. A trivalent impurity is added to silicon to create
(a) germanium **(b)** a p-type semiconductor
(c) an n-type semiconductor **(d)** a depletion layer

20. The purpose of a pentavalent impurity is to
(a) reduce the conductivity of silicon
(b) increase the number of holes
(c) increase the number of free electrons
(d) create minority carriers

21. The majority carriers in an n-type semiconductor are
(a) holes **(b)** valence electrons
(c) conduction electrons **(d)** protons

22. Holes in an n-type semiconductor are
(a) minority carriers that are thermally produced
(b) minority carriers that are produced by doping
(c) majority carriers that are thermally produced
(d) majority carriers that are produced by doping

23. A pn junction is formed by
(a) the recombination of electrons and holes
(b) ionization
(c) the boundary of a p-type and an n-type material
(d) the collision of a proton and a neutron

24. The depletion layer is created by
(a) ionization **(b)** diffusion
(c) recombination **(d)** all of the above

25. The depletion layer consists of
(a) nothing but minority carriers **(b)** ions
(c) no majority carriers **(d)** b and c

26. The term *bias* means
(a) the ratio of majority carriers to minority carriers
(b) the amount of current across the pn junction
(c) a dc voltage that is applied to control the operation of a device
(d) none of the above

27. To forward-bias a pn junction diode,

(a) an external voltage is applied that is positive at the anode and negative at the cathode

(b) an external voltage is applied that is negative at the anode and positive at the cathode

(c) an external voltage is applied that is positive at the p-region and negative at the n-region

(d) a and c

28. When a pn junction is forward-biased,

(a) the only current is hole current

(b) the only current is electron current

(c) the only current is produced by majority carriers

(d) the current is produced by both holes and electrons

29. Although current is blocked in reverse bias,

(a) there is some current due to majority carriers

(b) there is a very small current due to minority carriers

(c) there is an avalanche current

30. For a silicon diode, the value of the forward bias typically

(a) must be greater than 0.3 V

(b) must be greater than 0.7 V

(c) depends on the width of the depletion layer

(d) depends on the concentration of majority carriers

ANSWERS TO REVIEW QUESTIONS

SECTION 1–1

1. An atom is the smallest particle that retains the characteristics of its element.
2. An electron is the basic particle of negative electrical charge.
3. A valence electron is an electron in the outermost shell of an atom.
4. A free electron is one that has broken out of the valence band and entered the conduction band.
5. When a neutral atom loses an electron, the atom becomes a positive ion. When a neutral atom gains an electron, the atom becomes a negative ion.
6. Silicon and germanium are semiconductor materials.

SECTION 1–2

1. A crystal is a solid material formed by atoms bonding together.
2. Covalent bonds are formed by the sharing of valence electrons with neighboring atoms.

3. An intrinsic material is one that is in a pure state.
4. There are eight shared valence electrons in each atom of a silicon crystal.

SECTION 1–3

1. Free electrons are in the conduction band.
2. Free (conduction) electrons are responsible for current in a material.
3. A hole is the absence of an electron in the valence band.
4. Hole current occurs at the valence level.
5. A semiconductor has fewer free electrons than does a conductor because the energy gap between the valence band and the conduction band in a semiconductor is greater than in a conductor (actually the bands overlap in a conductor).

SECTION 1–4

1. Doping is the process of adding impurity atoms to a semiconductor in order to modify its conductive properties.
2. A pentavalent atom has five valence electrons and a trivalent atom has three valence electrons.
3. An n-type material is formed by the addition of pentavalent impurity atoms to the intrinsic semiconductor material.
4. A p-type material is formed by the addition of trivalent impurity atoms to the intrinsic semiconductor material.
5. The majority carrier in an n-type semiconductor is the free electron.
6. The majority carrier in a p-type semiconductor is the hole.
7. Majority carriers are produced by doping.
8. Minority carriers are thermally produced when electron-hole pairs are generated.

SECTION 1–5

1. A pn junction is the boundary between n-type and p-type semiconductor materials.
2. The depletion layer consists of positive and negative ions.
3. There are no majority carriers in the depletion layer.
4. The barrier potential in silicon is approximately 0.7 V.

SECTION 1–6

1. Current flows in forward bias and is blocked in reverse bias.
2. Forward bias produces majority carrier current.
3. The bias voltage must be at least 0.7 V to forward-bias a silicon diode.
4. To forward-bias a diode, the anode or p-region must be positive with respect to the cathode or n-region.
5. To reverse-bias a diode, the anode or p-region must be negative with respect to the cathode or n-region.
6. The depletion layer widens in reverse bias.
7. When the reverse-bias voltage exceeds a breakdown value, the diode will go into avalanche breakdown and destroy itself.

2

Diodes and Applications

After completing this chapter, you should be able to

- □ Explain what the diode characteristic curve indicates.
- □ Define the terms *anode* and *cathode.*
- □ Identify the terminals of a diode.
- □ Explain how a half-wave rectifier works.
- □ Determine the average value of a half-wave voltage.
- □ Discuss full-wave rectification.
- □ Determine the average value of a full-wave voltage.
- □ Explain how a center-tapped full-wave rectifier works.
- □ Explain how a full-wave bridge rectifier works.
- □ Determine the peak inverse voltage (PIV) across diodes in a rectifier.
- □ Describe how a capacitor filter smooths out a rectified voltage.
- □ Define *ripple voltage.*
- □ Explain how the value of the filter capacitor determines the amount of ripple voltage.
- □ Describe how an inductor-capacitor filter improves the dc output voltage.
- □ Explain how diode limiters work.
- □ Explain how diode clamping circuits work.
- □ Describe the operation of voltage doublers, triplers, and quadruplers.
- □ Interpret a diode data sheet.
- □ Troubleshoot diode circuits.

In the last chapter, you learned that a semiconductor diode is a device with a single pn junction. The importance of the diode in electronic circuits cannot be overemphasized. Its ability to conduct current in one direction while blocking current in the other direction is essential to the operation of many types of circuits. One circuit in particular is the ac rectifier, which is covered in this chapter. Other important applications are circuits such as diode limiters, diode clampers, and diode voltage multipliers.

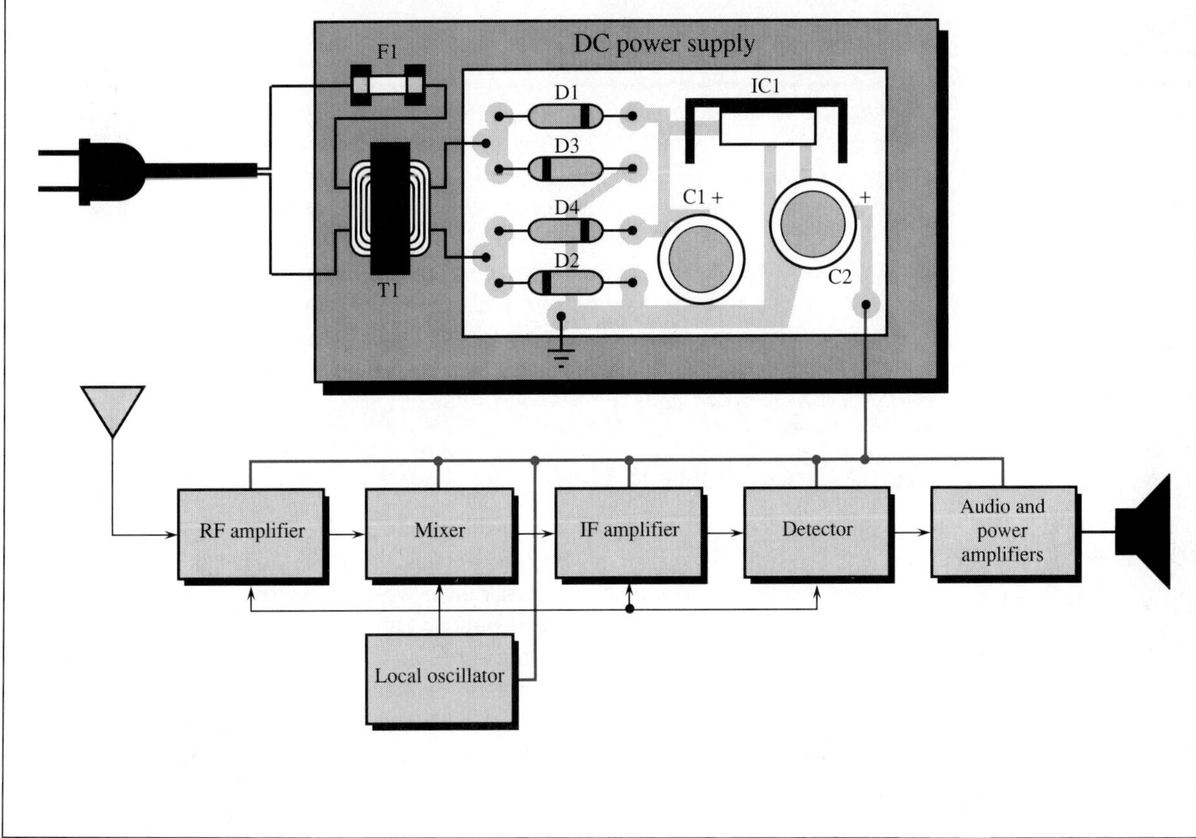

A System Application

The **power supply** is an important part of any electronic system because it supplies the dc voltage and current necessary for all the other circuits in the system to operate. You will learn how a typical power supply in an electronic system works. The diagram on this page shows a dc power supply as part of a radio receiver system. As you can see, the output of the power supply goes to all parts of the system and provides the necessary bias voltage and operating power for the diodes, transistors, and other devices in the amplifiers and other circuitry to function properly.

For the system application in Section 2–9, in addition to the other topics, be sure you understand

- ☐ How a diode works.
- ☐ The parameters and ratings of diodes.
- ☐ How to read a diode data sheet.
- ☐ What rectification of ac voltage means.
- ☐ How rectifiers work.
- ☐ What a filter is.
- ☐ Why filters are important in rectifiers.

2–1 RECTIFIER DIODES

Rectifier diodes are a major group of semiconductor diodes. Besides rectification, there are other uses to which this type of diode can be applied. In fact, many diodes in this category are referred to as general-purpose diodes. In this section, you will learn that the characteristic curve graphically shows how the diode works, and you will see the diode symbol that is used in circuit schematics. Three diode approximations are presented. Each approximation represents the diode at a different level of accuracy so that you can use the one most appropriate for a given situation. In some cases, the lowest level of accuracy is all that is needed, and additional details only complicate the situation. In other cases, you need the highest level of accuracy so that all factors can be taken into account.

DIODE CHARACTERISTIC CURVE

As you learned in Chapter 1, a diode conducts current when it is forward-biased if the bias voltage exceeds the barrier potential, and the diode prevents current when it is reverse-biased at less than the breakdown voltage. Figure 2–1 shows a diode **characteristic curve,** which is a graph of diode current versus voltage. The upper right quadrant of the graph represents the forward-biased condition. As you can see, there is essentially no forward current (I_F) for forward voltages (V_F) below the barrier potential. As the forward voltage approaches the value of the barrier potential (typically 0.7 V for silicon and 0.3 V for germanium), the current begins to increase. Once the forward voltage reaches the barrier potential, the current increases drastically and must be limited by a series resistor. The voltage across the forward-biased diode remains approximately equal to the barrier potential but increases slightly with forward current.

The lower left quadrant of the graph represents the reverse-biased condition. As the reverse voltage (V_R) increases to the left, the current remains near zero until the breakdown voltage (V_{BR}) is reached. When breakdown occurs, there is a large reverse current which, if not limited, can destroy the diode. Typically, the breakdown voltage is greater than 50 V for most rectifier diodes. Rectifier diodes should not be operated in reverse breakdown.

FIGURE 2–1
Diode characteristic curve.

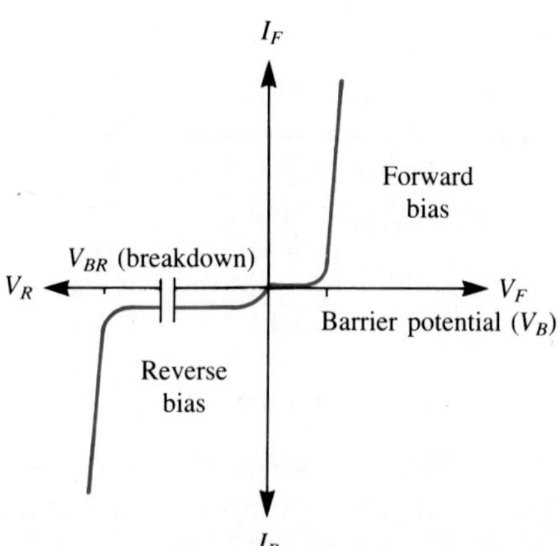

DIODE SYMBOL

Figure 2–2(a) is the standard symbol for a general-purpose diode. The arrow points in the direction of conventional current. The two terminals of the diode are the anode and cathode. When the anode is positive with respect to the cathode, the diode is forward biased and current is from anode to cathode, as shown in Figure 2–2(b). If there is any question about terminal polarities, always check the manufacturer's data book. Remember that when the diode is forward-biased, the barrier potential V_B always appears between anode and cathode, as indicated in the figure. When the anode is negative with respect to the cathode, the diode is reverse-biased, as shown in Figure 2–2(c). The bias battery voltage is designated V_{BB} and is not the same as the barrier potential.

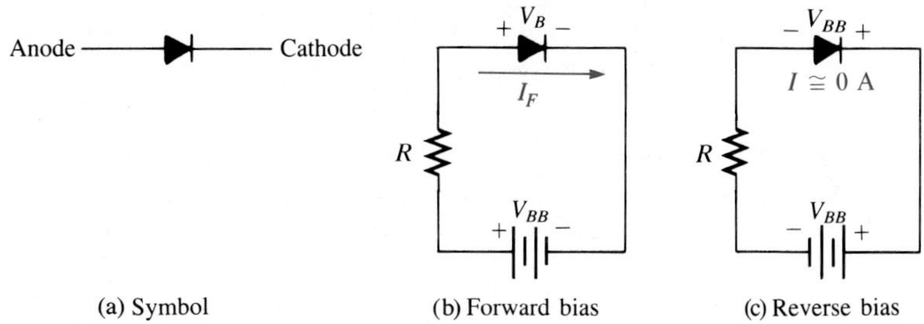

FIGURE 2–2
General-purpose diode and conditions of forward bias and reverse bias. The resistor limits the forward current to a safe value.

Some typical diodes are shown in Figure 2–3(a) to illustrate the variety of physical structures. Figure 2–3(b) illustrates terminal identification.

TESTING THE DIODE WITH A MULTIMETER

The internal battery in analog ohmmeters will forward-bias or reverse-bias a diode, permitting a quick and simple check for proper functioning. Many digital multimeters have a diode test position.

To check the diode in the forward direction, the positive meter lead is connected to the anode and the negative lead to the cathode, as shown in Figure 2–4(a). When a diode is forward-biased, its internal resistance is low (typically around 100 Ω or less). When the meter leads are reversed, as shown in Figure 2–4(b), the internal ohmmeter battery reverse-biases the diode, and a very large resistance value (ideally infinite) is indicated. The pn junction is shorted if a low resistance is indicated in both bias conditions; it is open if a very high resistance is read for both checks. In the diode test position on many digital multimeters, the forward diode voltage is displayed when the diode is good, as shown in Figure 2–4(c).

DIODE APPROXIMATIONS

THE IDEAL MODEL The simplest way to visualize diode operation is to think of it as a switch. When forward-biased, the diode acts as a closed (on) switch; and when reverse-biased, it acts as an open (off) switch, as in Figure 2–5. The characteristic curve

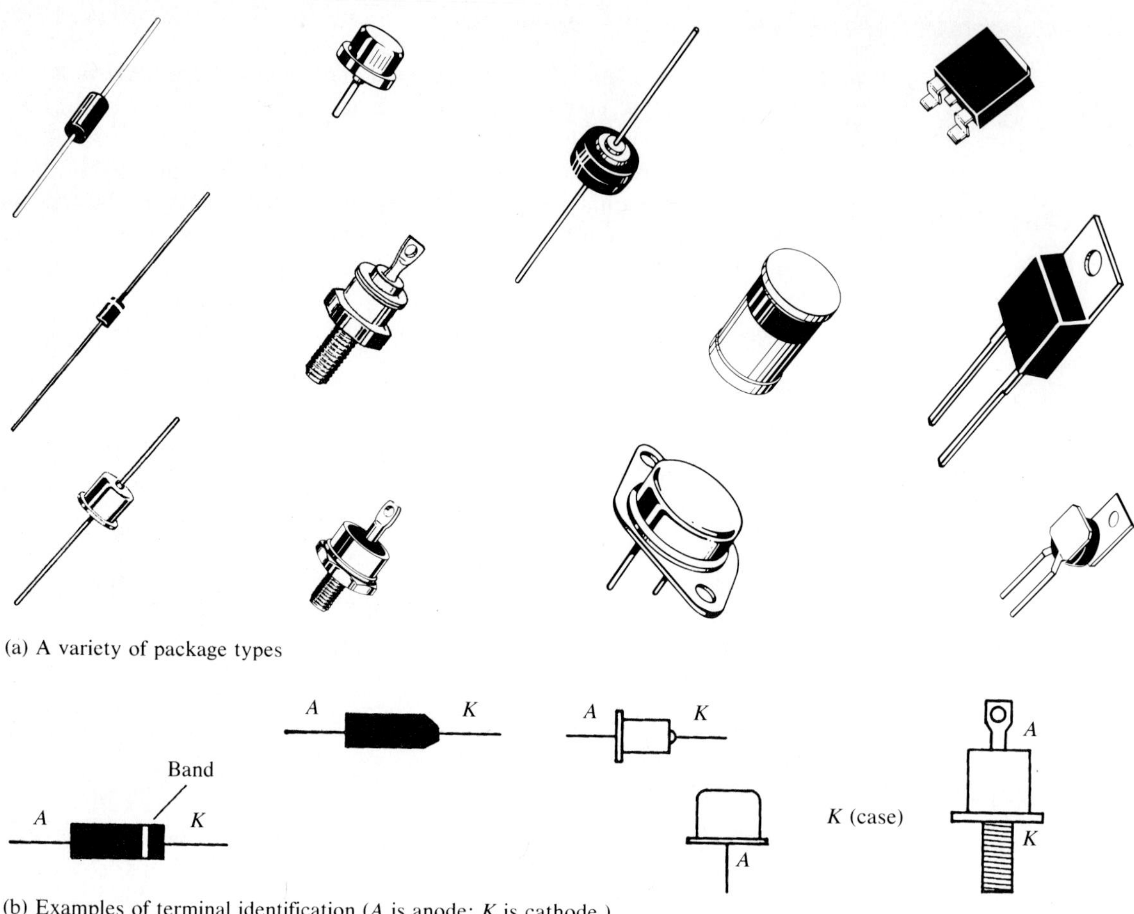

(a) A variety of package types

(b) Examples of terminal identification (A is anode; K is cathode.)

FIGURE 2–3
Typical diodes. Part (a) copyright of Motorola, Inc. Used by permission.

for this approximation is shown in part (c). Note that the forward voltage and the reverse current are always zero. This ideal model, of course, neglects the effect of the barrier potential, the internal resistances, and other parameters. But in many cases it is accurate enough, particularly when the bias voltage is at least ten times greater than the barrier potential.

THE BARRIER POTENTIAL MODEL The next higher level of accuracy is the barrier potential model. In this approximation, the forward-biased diode is represented as a closed switch in series with a small "battery" equal to the barrier potential V_B (0.7 V for Si and 0.3 V for Ge), as shown in Figure 2–6(a). The positive end of the equivalent battery is toward the anode. Keep in mind that the barrier potential cannot actually be measured with a voltmeter, but only has the *effect* of a battery when forward bias is applied. The reverse-biased diode is represented by an open switch, as in the ideal case, because the

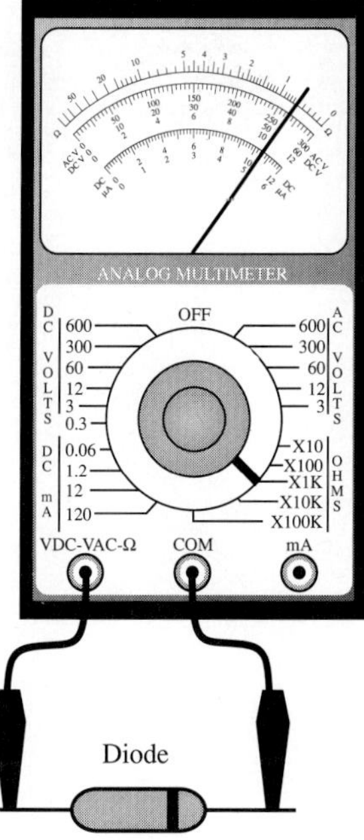

(a) Forward check with analog meter gives very low resistance.

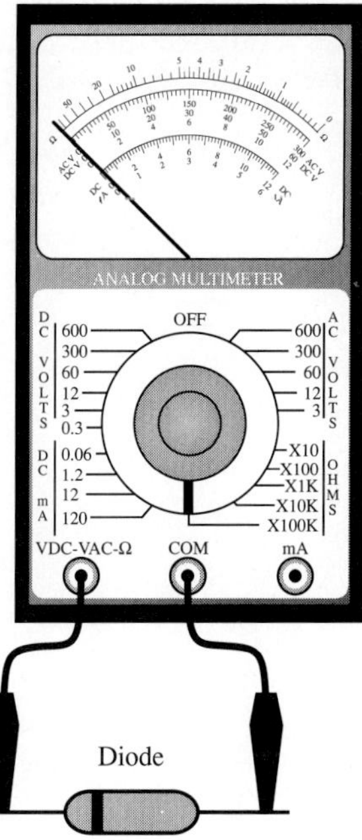

(b) Reverse check with analog meter gives very high resistance.

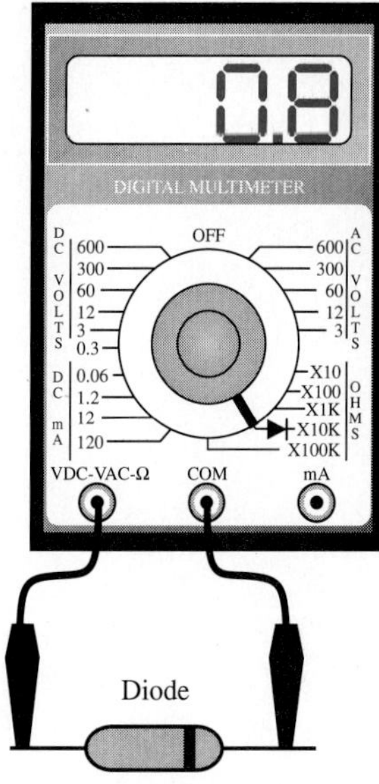

(c) Diode check position on digital meter gives forward voltage (barrier potential plus drop across forward resistance).

FIGURE 2–4
Checking a diode with a multimeter.

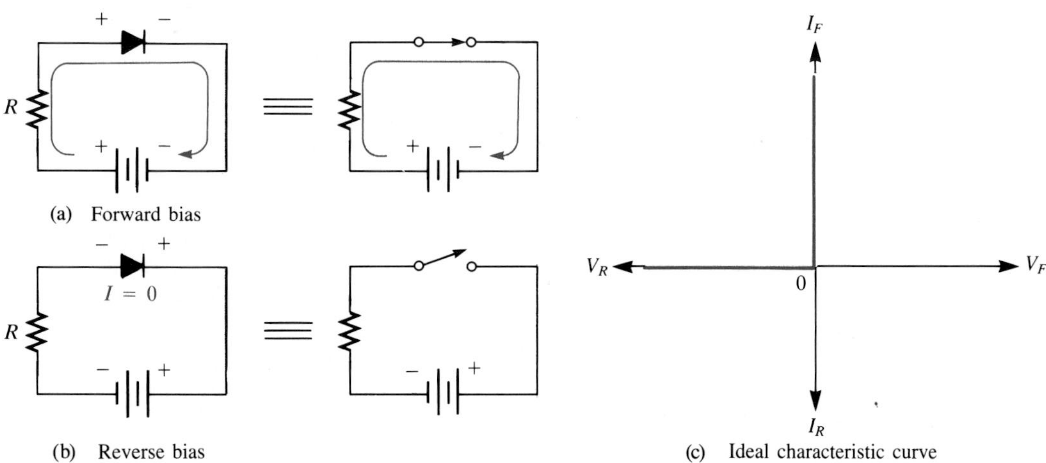

(a) Forward bias

(b) Reverse bias

(c) Ideal characteristic curve

FIGURE 2–5
The ideal switch approximation of the diode.

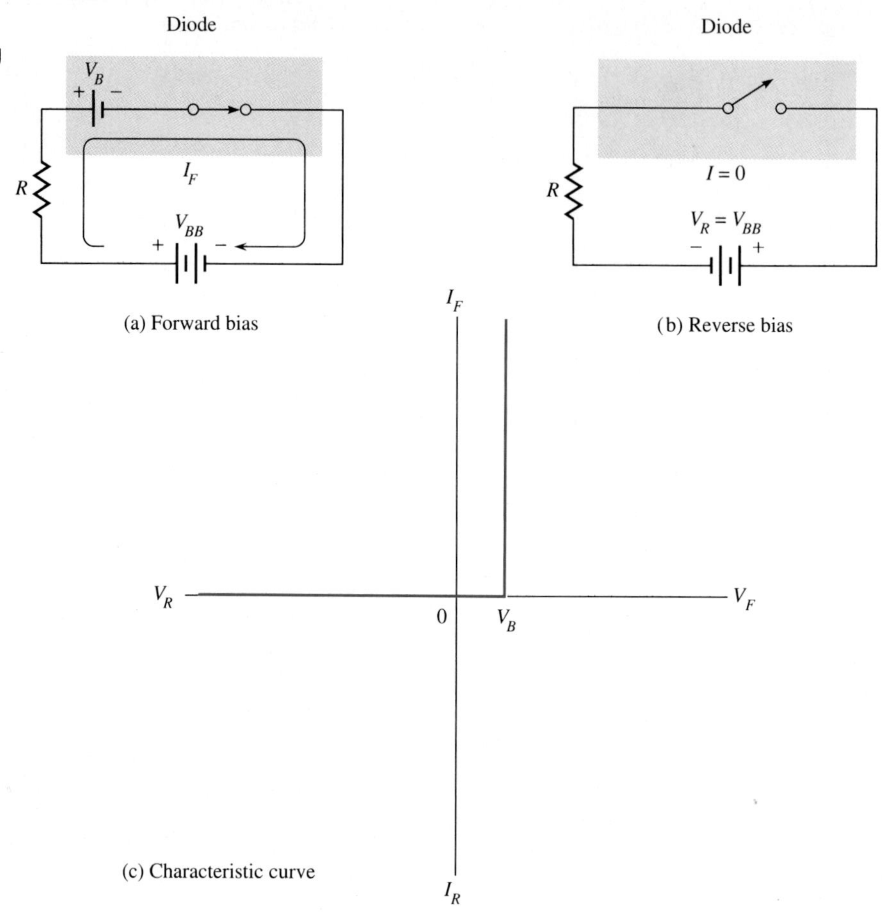

FIGURE 2–6
Diode approximation including the barrier potential.

barrier potential does not affect reverse bias. This is shown in Figure 2–6(b). The characteristic curve for this model is shown in Figure 2–6(c). In this textbook, we use the barrier potential approximation for analysis unless otherwise stated.

THE COMPLETE MODEL One more level of accuracy will be considered at this point. Figure 2–7(a) shows the forward-biased diode model with both the barrier potential and the low forward (bulk) resistance. Figure 2–7(b) shows how the high reverse resistance affects the reverse-biased model. The characteristic curve is shown in Figure 2–7(c). Other parameters such as junction capacitance and breakdown voltage become important only under certain operating conditions and will be considered only where appropriate.

2–1 REVIEW QUESTIONS

1. What are the two conditions under which the rectifier diode is operated?
2. Under what condition is the rectifier diode never intentionally operated?
3. What is the simplest way to visualize a diode?
4. To accurately represent a diode, what factors must be included?
5. Which diode approximation will normally be used in this book?

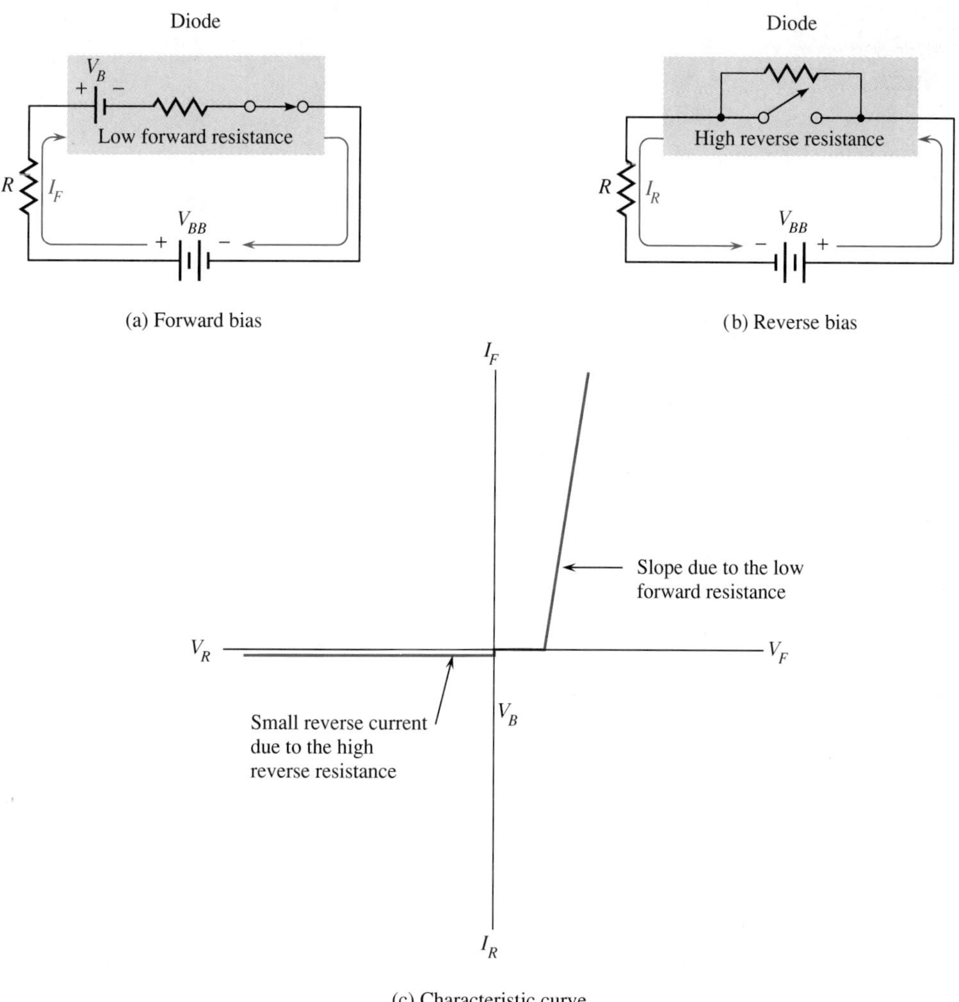

(a) Forward bias

(b) Reverse bias

(c) Characteristic curve

FIGURE 2–7
Diode approximation including barrier potential, forward resistance, and reverse resistance.

2–2 HALF-WAVE RECTIFIERS

Because of their ability to conduct current in one direction and block current in the other direction, diodes are used in circuits called rectifiers that convert ac voltage into dc voltage. Rectifiers are found in all dc power supplies that operate from an ac voltage source. A power supply is an essential part of each electronic system from the simplest to the most complex. In this section, you will study the most basic type of rectifier, the half-wave rectifier. This will prepare you to move into the full-wave rectifier in the next section.

FIGURE 2–8
Half-wave rectifier operation. The diode is considered to be ideal.

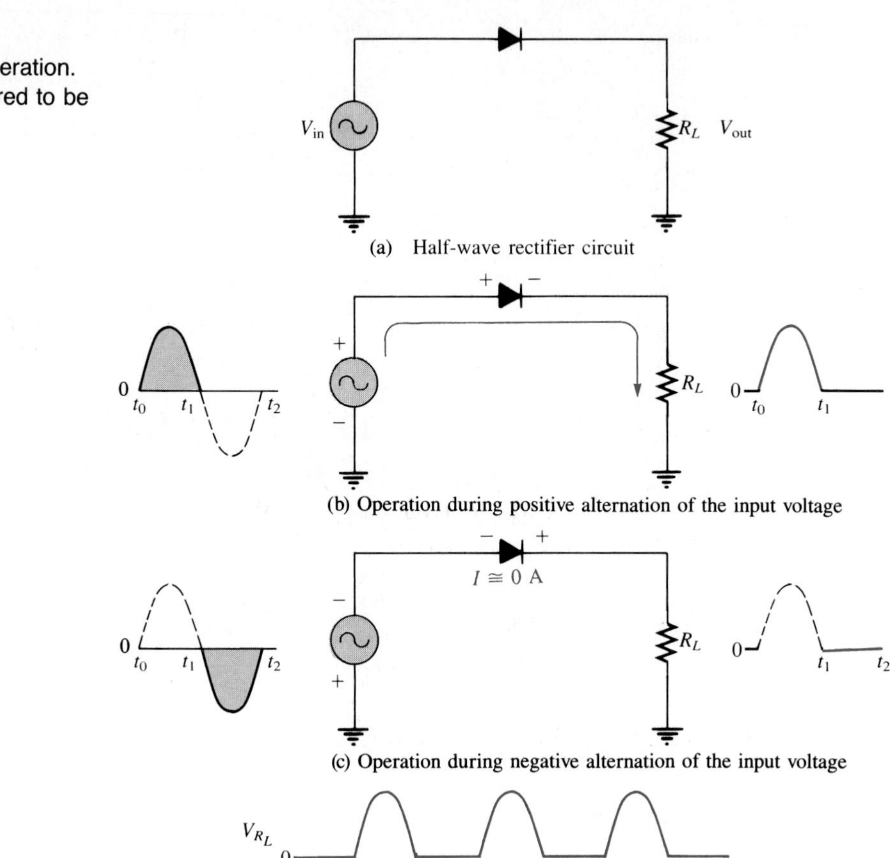

Figure 2–8 illustrates the process called *half-wave rectification*. In part (a), an ac source is connected to a load resistor through an ideal diode. Let's examine what happens during one cycle of the input voltage. When the input sine wave goes positive, the diode is forward-biased and conducts current to the load resistor, as shown in part (b). The current produces a voltage across the load which has the same shape as the positive half-cycle of the input voltage. When the input voltage goes negative during the second half of its cycle, the diode is reverse-biased. There is no current, so the voltage across the load resistor is 0 V, as shown in part (c). The net result is that only the positive half-cycles of the ac input voltage appear across the load, making the output a pulsating dc voltage as shown in part (d).

AVERAGE VALUE OF THE HALF-WAVE OUTPUT VOLTAGE

The average (dc) value of the half-wave rectified output is determined by finding the area under the curve over a full cycle, as illustrated in Figure 2–9. Using calculus to get the area under the half-cycle and then dividing by the period, we get V_{AVG} = area/period =

V_p/π, where V_p is the peak voltage. (See Appendix B for detailed derivation.)

$$V_{AVG} = \frac{V_p}{\pi} \tag{2-1}$$

The average value is the value that would be indicated by a dc voltmeter.

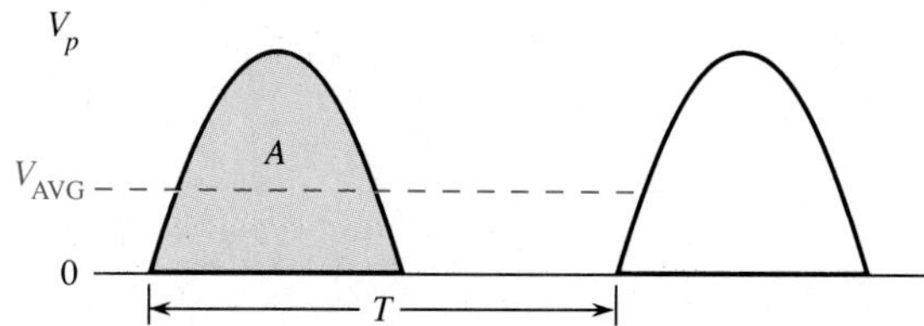

FIGURE 2–9
Average value of half-wave rectified signal.

EXAMPLE 2–1

What is the average (dc) value of the half-wave rectified voltage waveform in Figure 2–10?

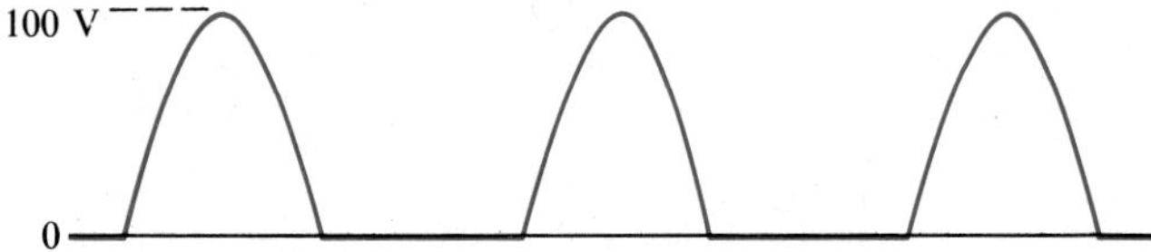

FIGURE 2–10

SOLUTION

$$V_{AVG} = \frac{V_p}{\pi} = \frac{100\text{ V}}{\pi} = 31.83\text{ V}$$

PRACTICE EXERCISE 2–1

Determine the average value of the half-wave voltage if its peak amplitude is 12 V.

EFFECT OF THE BARRIER POTENTIAL ON THE HALF-WAVE RECTIFIER OUTPUT

In the previous discussion, the diode was considered ideal. When the diode barrier potential is taken into account, here is what happens: During the positive half-cycle, the input voltage must overcome the barrier potential before the diode becomes forward-biased. For a silicon diode this results in a half-wave output with a peak value that is 0.7 V less than the peak value of the input (0.3 V less for a germanium diode), as shown in Figure 2–11. The expressions for peak output voltage are, for silicon,

$$V_{p(out)} = V_{p(in)} - 0.7\text{ V} \tag{2-2}$$

and, for germanium,

$$V_{p(out)} = V_{p(in)} - 0.3\text{ V} \tag{2-3}$$

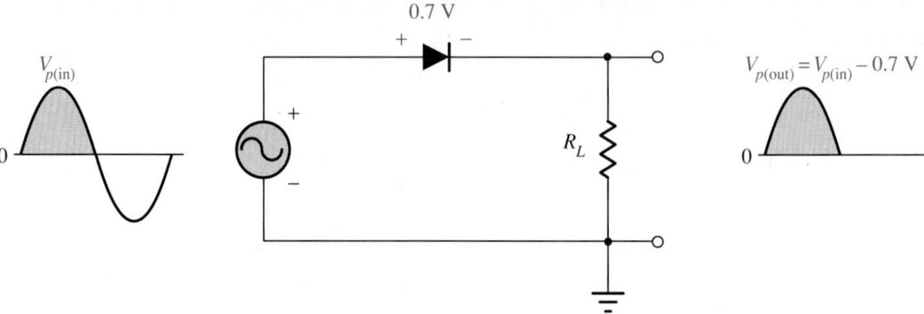

FIGURE 2–11
Effect of barrier potential on half-wave rectified output voltage (silicon diode shown).

In working with diode circuits, it is sometimes practical to neglect the effect of barrier potential when the peak value of the applied voltage is much greater than the barrier potential (at least 10 times. Some people use 100). As mentioned before, we will always use silicon diodes and take the barrier potential into account unless stated otherwise.

EXAMPLE 2–2 Sketch the output voltages of each rectifier circuit for the indicated input voltages, as shown in Figure 2–12. Each diode is silicon, and the barrier potential should be included.

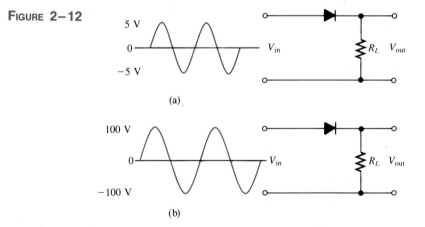

FIGURE 2–12

SOLUTION

The peak output voltage for circuit (a) is

$$V_p = 5\text{ V} - 0.7\text{ V} = 4.3\text{ V}$$

The peak output voltage for circuit (b) is

$$V_{p(out)} = 100\text{ V} - 0.7\text{ V} = 99.3\text{ V}$$

These outputs are shown in Figure 2–13. Note that the barrier potential could have been neglected in circuit (b) with very little error (0.7 percent); but, if it is neglected in circuit (a), a significant error results (14 percent).

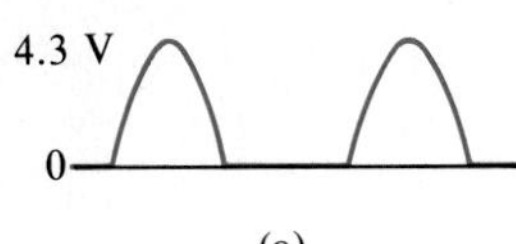

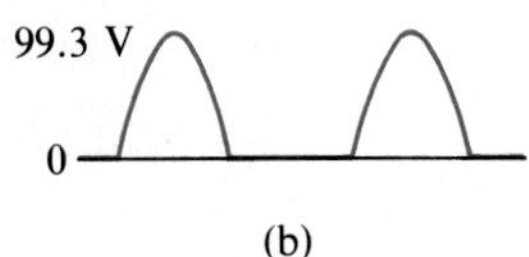

FIGURE 2–13
Outputs of circuits in Figure 2–12.

PRACTICE EXERCISE 2–2

Determine the peak output voltages for the rectifiers in Figure 2–12 if the peak input in part (a) is 3 V and the peak input in part (b) is 210 V.

MAXIMUM REVERSE VOLTAGE

The maximum value of reverse voltage, sometimes designated as *peak inverse voltage* (PIV), occurs at the peak of the negative alternation of the input cycle when the diode is reverse-biased. This condition is illustrated in Figure 2–14. The PIV equals the peak value of the input voltage, and the diode must be capable of withstanding this amount of repetitive reverse voltage.

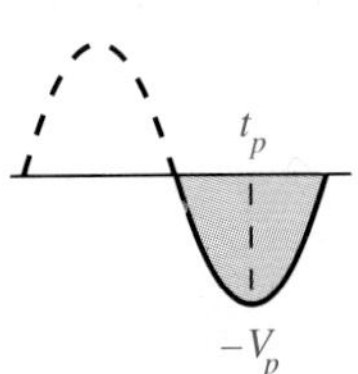

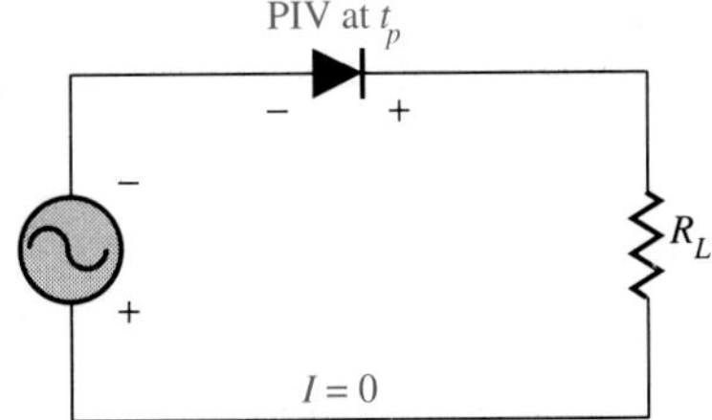

FIGURE 2–14
The PIV occurs at the peak of the half-cycle when the diode is reverse-biased. In this circuit, the PIV occurs at the peak of the negative half-cycle.

HALF-WAVE RECTIFIER WITH TRANSFORMER-COUPLED INPUT

A transformer is often used to couple the ac input voltages from the source to the rectifier circuit, as shown in Figure 2–15. Transformer coupling provides two advantages. First, it allows the source voltage to be stepped up or stepped down as needed. Second, the ac power source is electrically isolated from the rectifier circuit, thus reducing the shock hazard.

From basic ac circuits, the secondary (output) voltage of a transformer equals the turns ratio (N_2/N_1) times the primary (input) voltage, as expressed in Equation (2–4).

$$V_2 = \left(\frac{N_2}{N_1}\right)V_1 \qquad (2\text{–}4)$$

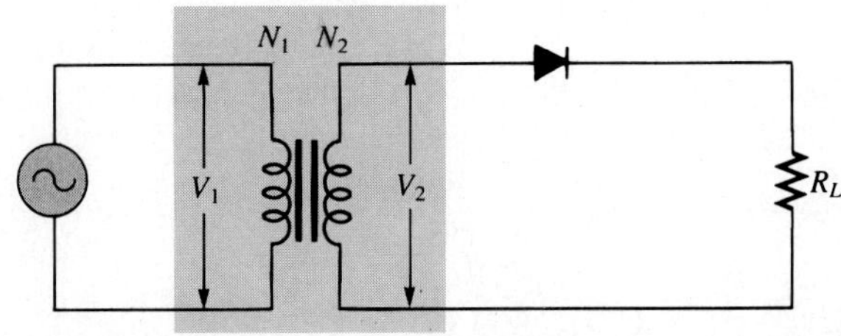

FIGURE 2–15
Half-wave rectifier with transformer-coupled input.

If $N_2 > N_1$, the secondary voltage is greater than the primary voltage. If $N_2 < N_1$, the secondary voltage is less than the primary voltage. If $N_2 = N_1$, then $V_2 = V_1$.

EXAMPLE 2–3

Determine the peak value of the output voltage for Figure 2–16.

FIGURE 2–16

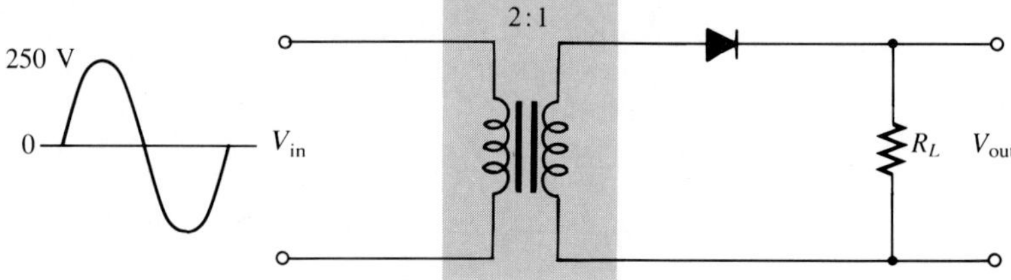

SOLUTION

The turns ratio is

$$\frac{N_2}{N_1} = \frac{1}{2} = 0.5$$

The secondary peak voltage is

$$\begin{aligned} V_{p(2)} &= \left(\frac{N_2}{N_1}\right)V_{p(1)} \\ &= 0.5(250 \text{ V}) \\ &= 125 \text{ V} \end{aligned}$$

The peak rectified output voltage is

$$\begin{aligned} V_{p(out)} &= 125 \text{ V} - 0.7 \text{ V} \\ &= 124.3 \text{ V} \end{aligned}$$

PRACTICE EXERCISE 2–3

(a) Determine the peak value of the output voltage for Figure 2–16 if the turns ratio is 1:2 and $V_{p(in)} = 50$ V.
(b) What is the maximum reverse voltage across the diode?
(c) Describe the output voltage if the diode is turned around.

2–2 REVIEW QUESTIONS

1. At what point on the input cycle does the PIV occur?
2. For a half-wave rectifier, there is current through the load for approximately what percentage of the input cycle?
3. What is the average of a half-wave rectified voltage with a peak value of 10 V?
4. What is the peak value of the output voltage of a half-wave rectifier with a peak sine wave input of 25 V?
5. What PIV rating must a diode have to be used in a rectifier with a peak output voltage of 50 V?

2–3 FULL-WAVE RECTIFIERS

Although half-wave rectifiers have some applications, the full-wave rectifier is the most commonly used type in dc power supplies. In this section, you will use what you learned about half-wave rectification and expand it to full-wave rectifiers. You will learn about two types of full-wave rectifiers: center-tapped and bridge.

The difference between full-wave and half-wave rectification is that a **full-wave rectifier** allows unidirectional current to the load during the entire input cycle, and the **half-wave rectifier** allows this only during one half-cycle. The result of full-wave rectification is a dc output voltage that pulsates every half-cycle of the input, as shown in Figure 2–17.

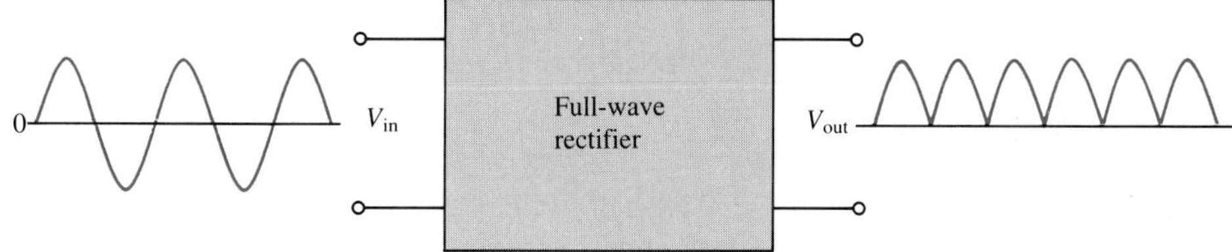

FIGURE 2–17
Full-wave rectification.

Since the number of positive alternations that make up the full-wave rectified voltage is twice that of the half-wave voltage, the average value for a full-wave rectified sine wave voltage is twice that of the half-wave, as expressed in Equation (2–5).

$$V_{AVG} = \frac{2V_p}{\pi} \tag{2–5}$$

EXAMPLE 2–4 Find the average value of the full-wave rectified voltage in Figure 2–18.

FIGURE 2–18

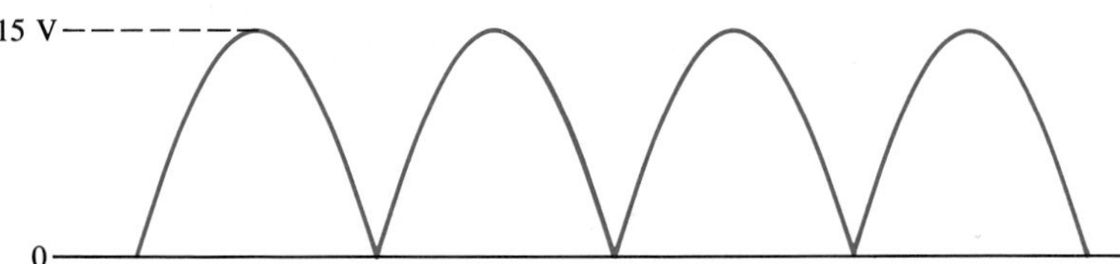

SOLUTION

$$V_{AVG} = \frac{2V_p}{\pi} = \frac{2(15\text{ V})}{\pi} = 9.55\text{ V}$$

PRACTICE EXERCISE 2–4

Find the average value of the full-wave rectified voltage if its peak is 155 V. On what function setting would a multimeter measure this value?

THE CENTER-TAPPED FULL-WAVE RECTIFIER

The center-tapped full-wave rectifier circuit uses two diodes connected to the secondary of a center-tapped transformer, as shown in Figure 2–19. The input signal is coupled through the transformer to the center-tapped secondary. Half of the total secondary voltage appears between the center tap and each end of the secondary winding as shown.

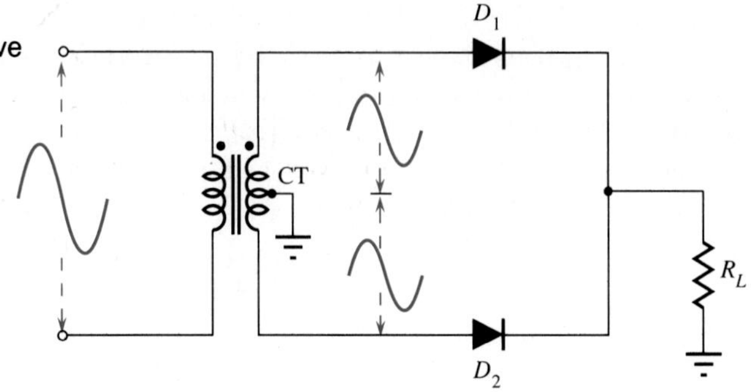

FIGURE 2–19
A center-tapped (CT) full-wave rectifier.

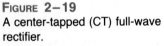

For a positive half-cycle of the input voltage, the polarities of the secondary voltages are shown in Figure 2–20(a). This condition forward-biases the upper diode D_1 and reverse-biases the lower diode D_2. The current path is through D_1 and the load resistor, as indicated. For a negative half-cycle of the input voltage, the voltage polarities on the secondary are shown in Figure 2–20(b). This condition reverse-biases D_1 and forward-biases D_2. The current path is through D_2 and the load resistor, as indicated.

FIGURE 2–20
Basic operation of a center-tapped full-wave rectifier. Note that the current through the load resistor is in the same direction during the entire input cycle.

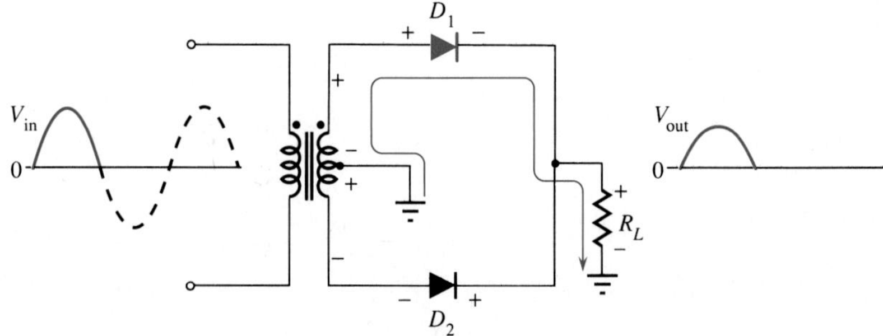

(a) During positive half-cycles, D_1 is forward-biased and D_2 is reverse-biased.

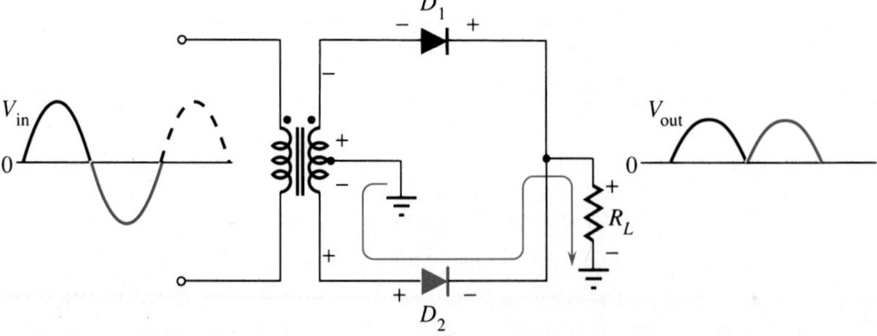

(b) During negative half-cycles, D_2 is forward-biased and D_1 is reverse-biased.

Because the output current during both the positive and negative portions of the input cycle is in the same direction through the load, the output voltage developed across the load resistor is a full-wave rectified dc voltage.

EFFECT OF THE TURNS RATIO ON THE FULL-WAVE OUTPUT VOLTAGE

If the transformer's turns ratio is 1, the peak value of the rectified output voltage equals half the peak value of the primary input voltage less the barrier potential (we will sometimes call it the **diode drop**), as illustrated in Figure 2–21. This is because half of the input voltage appears across each half of the secondary winding.

In order to obtain an output voltage equal to the input (less the barrier potential), a step-up transformer with a turns ratio of 2 must be used, as shown in Figure 2–22. In this case, the total secondary voltage V_2 is twice the primary voltage ($2V_1$), so the voltage across each half of the secondary is equal to V_1.

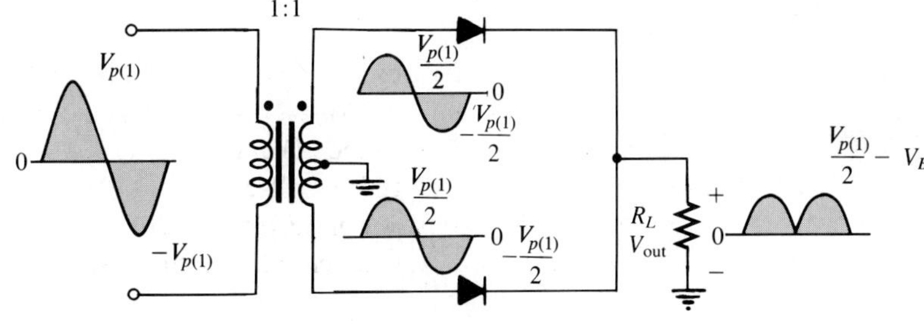

FIGURE 2–21
Center-tapped full-wave rectifier with a transformer turns ratio of 1. $V_{p(1)}$ is the peak value of the primary voltage.

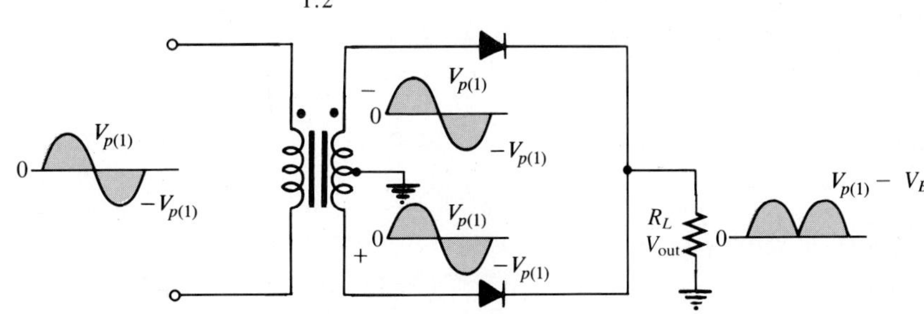

FIGURE 2–22
Center-tapped full-wave rectifier with a transformer turns ratio of 2.

In any case, the output voltage of a center-tapped full-wave rectifier is always one-half of the total secondary voltage, no matter what the turns ratio is.

$$V_{out} = \frac{V_2}{2} \tag{2–6}$$

To include the diode drop V_B, subtract it from $V_2/2$.

$$V_{out} = \frac{V_2}{2} - V_B \tag{2–7}$$

PEAK INVERSE VOLTAGE

Each diode in the full-wave rectifier is alternately forward-biased and then reverse-biased. The maximum reverse voltage that each diode must withstand is the peak secondary voltage $V_{p(2)}$. This can be seen by examining Figure 2–23. When the total secondary voltage has the polarity shown, the maximum anode voltage of D_1 is $+V_{p(2)}/2$ and the maximum anode voltage of D_2 is $-V_{p(2)}/2$. Since D_1 is forward-biased, its cathode is at the same voltage as its anode minus the barrier potential; this is also the voltage on the cathode of D_2. The maximum total reverse voltage across D_2 is therefore

$$\left(\frac{V_{p(2)}}{2} - V_B\right) - \left(-\frac{V_{p(2)}}{2}\right) = \frac{V_{p(2)}}{2} + \frac{V_{p(2)}}{2} - V_B = V_{p(2)} - V_B$$

Since $V_{p(out)} = V_{p(2)}/2 - V_B$, the peak inverse voltage across either diode in the center-tapped full-wave rectifier is

$$\text{PIV} = 2V_{p(out)} + V_B \tag{2–8}$$

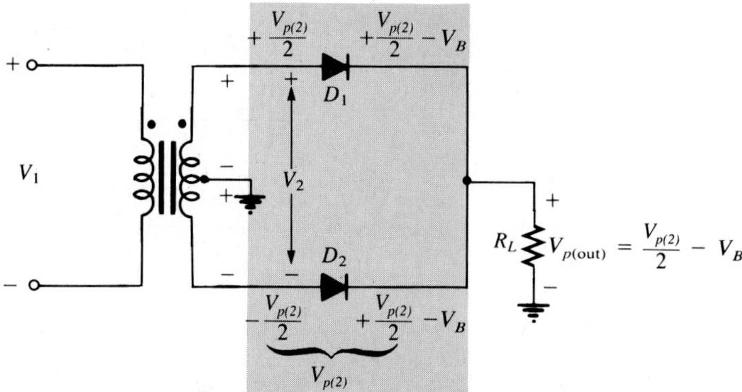

FIGURE 2–23
Diode reverse voltage (D_2 shown reverse-biased). The PIV is twice the peak value of the output voltage.

EXAMPLE 2–5 Show the voltage waveforms across each half of the secondary winding and across R_L when a 25 V peak sine wave is applied to the primary winding in Figure 2–24. Also, what minimum PIV rating must the diodes have?

FIGURE 2–24

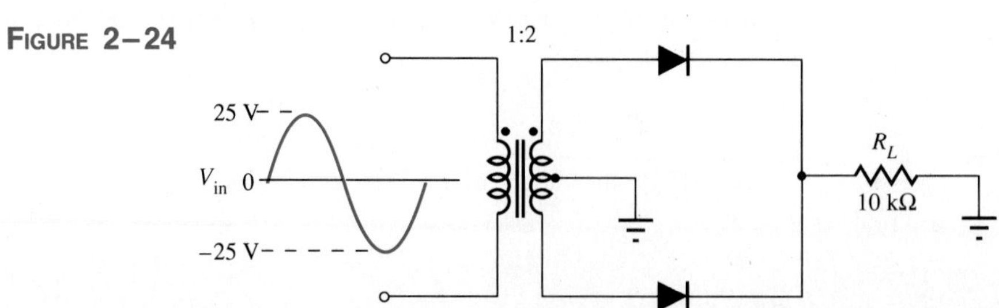

SOLUTION

The total peak secondary voltage is

$$V_{p(2)} = \left(\frac{N_2}{N_1}\right)V_{p(1)} = (2)25\text{ V} = 50\text{ V}$$

There is a 25 V peak across each half of the secondary. The output load voltage has a peak value of 25 V, less the 0.7 V drop across the diode. Each diode must have a minimum PIV rating of 50 V (neglecting diode drop). The waveforms are shown in Figure 2–25.

FIGURE 2–25

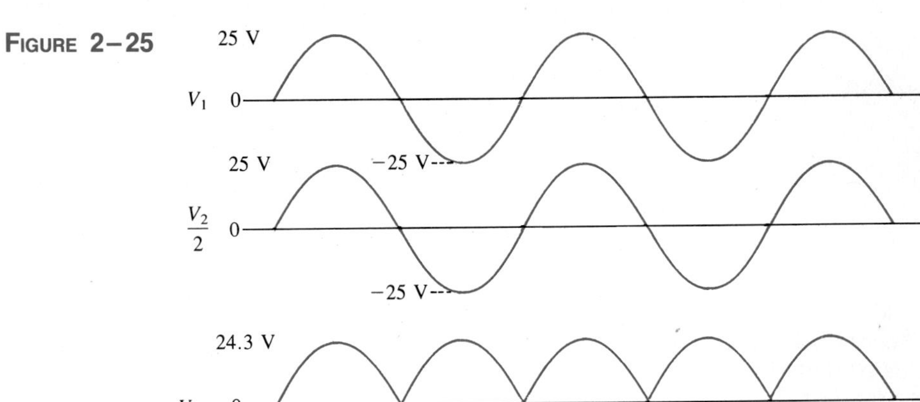

PRACTICE EXERCISE 2–5

What diode PIV rating is required to handle a peak input of 160 V in Figure 2–24?

THE FULL-WAVE BRIDGE RECTIFIER

The full-wave **bridge rectifier** uses four diodes, as shown in Figure 2–26. When the input cycle is positive as in part (a), diodes D_1 and D_2 are forward-biased and conduct current in the direction shown. A voltage is developed across R_L which looks like the positive half of the input cycle. During this time, diodes D_3 and D_4 are reverse-biased. When the input cycle is negative as in part (b), diodes D_3 and D_4 are forward-biased and

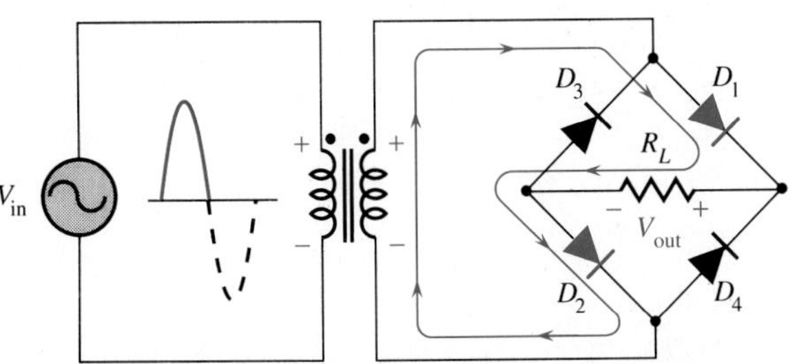

(a) During positive half-cycle of the input, D_1 and D_2 are forward-biased and conduct current. D_3 and D_4 are reverse-biased.

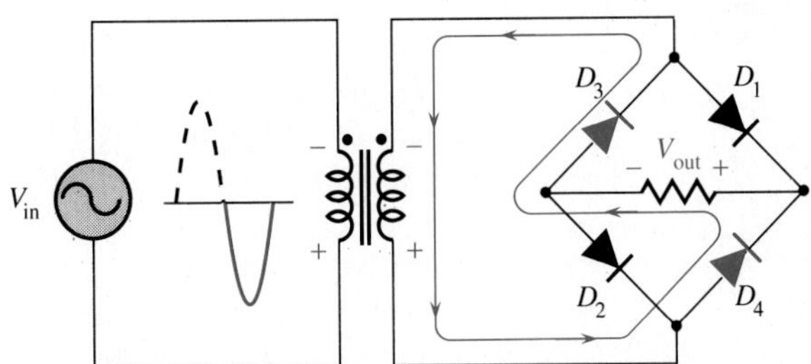

(b) During negative half-cycle, D_3 and D_4 are forward-biased and conduct current. D_1 and D_2 are reverse-biased.

FIGURE 2–26
Operation of full-wave bridge rectifier.

conduct current in the same direction through R_L as during the positive half-cycle. During the negative half-cycle, D_1 and D_2 are reverse-biased. A full-wave rectified output voltage appears across R_L as a result of this action.

BRIDGE OUTPUT VOLTAGE

A bridge rectifier with a transformer-coupled input is shown in Figure 2–27(a). During the positive half-cycle of the total secondary voltage, diodes D_1 and D_2 are forward-biased. Neglecting the diode drops, the secondary voltage V_2 appears across the load resistor. The same is true when D_3 and D_4 are forward-biased during the negative half-cycle.

$$V_{out} = V_2 \qquad (2–9)$$

As you can see in Figure 2–27(b), two diodes are always in series with the load resistor during both the positive and negative half-cycles. If these diode drops are taken into account, the output voltage is

$$V_{out} = V_2 - 2V_B \qquad (2–10)$$

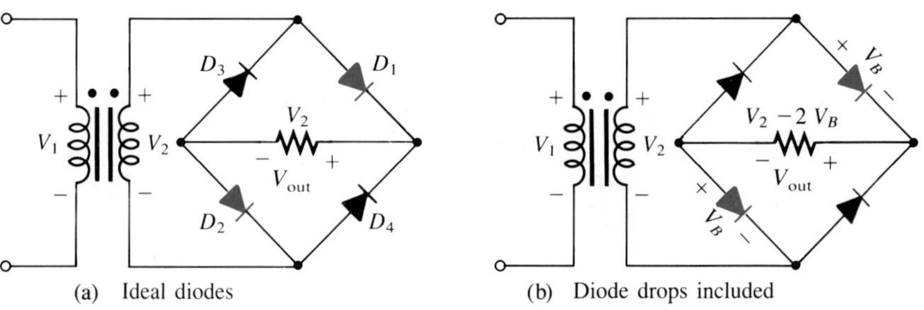

FIGURE 2–27
Bridge output voltage. Color indicates forward bias.

PEAK INVERSE VOLTAGE

Let us assume that D_1 and D_2 are forward-biased and examine the reverse voltage across D_3 and D_4. Visualizing D_1 and D_2 as shorts (ideally), as in Figure 2–28(a), you can see that D_3 and D_4 have a peak inverse voltage equal to the peak secondary voltage. Since the output voltage is *ideally* equal to the secondary voltage, we have

$$\text{PIV} = V_{p(out)}$$

If the diode drops of the forward-biased diodes are included as shown in Figure 2–28(b), the peak inverse voltage across each reverse-biased diode is

$$\text{PIV} = V_{p(out)} + V_B \qquad (2–11)$$

The PIV rating of the bridge diodes is less than that required for the center-tapped configuration. If the diode drop, V_B, is neglected, the bridge rectifier requires diodes with half the PIV rating of those in a center-tapped rectifier for the same output voltage.

FIGURE 2–28
Peak inverse voltage in a bridge rectifier during the positive half-cycle of the input voltage.

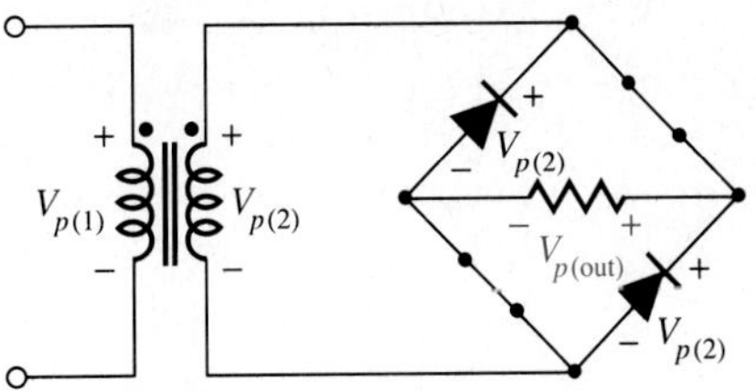

(a) With ideal diodes (forward-biased diodes are shown as shorts)

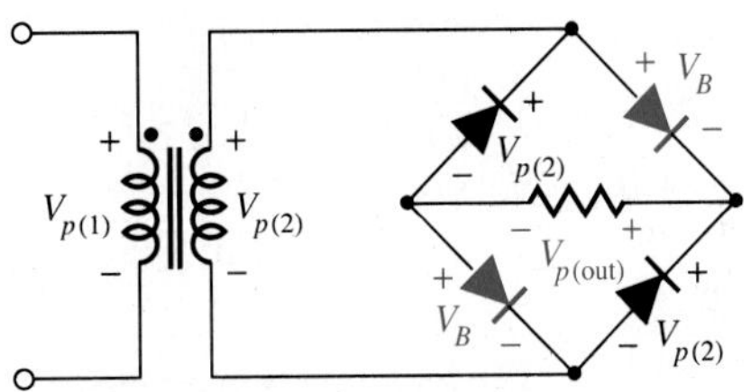

(b) With the barrier potential taken into account (forward-biased diodes are shown in color)

EXAMPLE 2–6

Determine the output voltage for the bridge rectifier in Figure 2–29. What minimum PIV rating is required for the silicon diodes? Notice that the way the load resistor is connected appears different than in previous diagrams, although it is effectively the same except for ground.

FIGURE 2–29

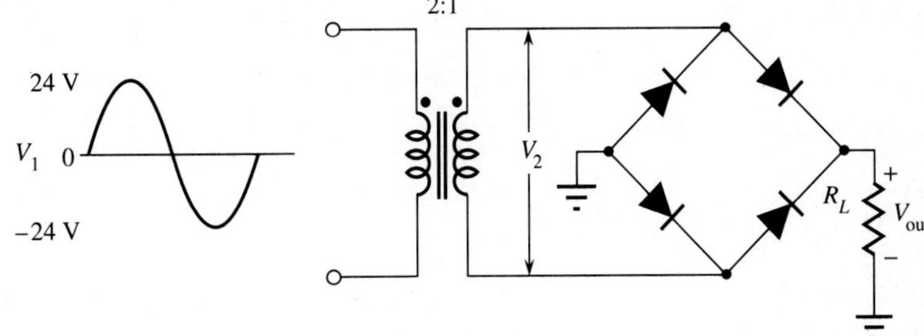

SOLUTION

The peak output voltage is (taking into account the two diode drops)

$$V_{p(out)} = V_{p(2)} - 2V_B$$
$$= \left(\frac{N_2}{N_1}\right)V_{p(in)} - 2V_B$$
$$= \frac{1}{2}V_{p(in)} - 2V_B$$
$$= 12\text{ V} - 1.4\text{ V}$$
$$= 10.6\text{ V}$$

The PIV for each diode is

$$\text{PIV} = V_{p(out)} + V_B = 10.6\text{ V} + 0.7\text{ V} = 11.3\text{ V}$$

PRACTICE EXERCISE 2–6

Determine the peak output voltage for the bridge rectifier in Figure 2–29 if the peak primary voltage is 160 V. What is the PIV rating for the diodes? ■

2–3 REVIEW QUESTIONS

1. How does a full-wave voltage differ from a half-wave voltage?
2. What is the average value of a full-wave rectified voltage with a peak value of 60 V?
3. Which type of full-wave rectifier has the greater output voltage for the same input voltage and transformer turns ratio?
4. For a peak output voltage of 45 V, in which type of rectifier would you use diodes with a PIV rating of 50 V?
5. What minimum PIV rating is required for diodes used in the type of rectifier that was not selected in Question 4?

2–4 RECTIFIER FILTERS

The purpose of a power supply filter is to greatly reduce the fluctuations in the output voltage of a half-wave or full-wave rectifier and produce a nearly constant-level dc voltage. Filtering is necessary because electronic circuits require a constant source of dc voltage and current to provide power and biasing for proper operation. Filtering is done using capacitors, inductors, or combinations of both, as you will see in this section.

In most power supply applications, the standard 60 Hz ac power line voltage must be converted to a sufficiently constant dc voltage. The 60 Hz pulsating dc output of a half-wave rectifier or the 120 Hz pulsating output of a full-wave rectifier must be **filtered** to reduce the large voltage variations. Figure 2–30 illustrates the filtering concept showing a nearly smooth dc output voltage. A full-wave rectifier voltage is applied to the filter's input and, ideally, a constant dc level appears on the output.

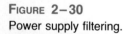

FIGURE 2–30
Power supply filtering.

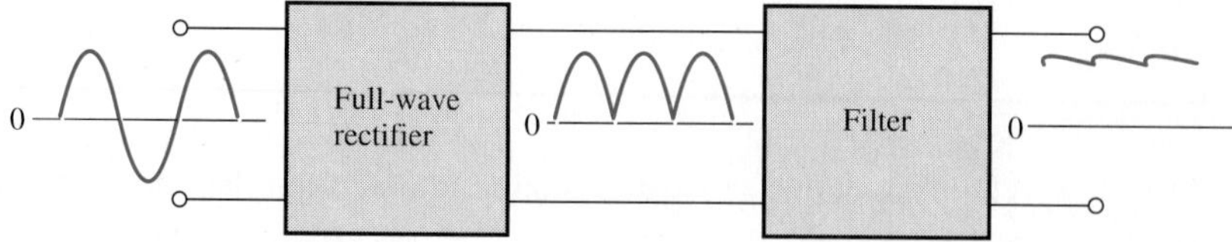

CAPACITOR FILTER

A half-wave rectifier with a capacitor filter is shown in Figure 2–31. R_L represents the load resistance. We will use the half-wave rectifier to illustrate the principle, and then expand the concept to full-wave.

FIGURE 2–31
Operation of a half-wave rectifier with a capacitor filter.

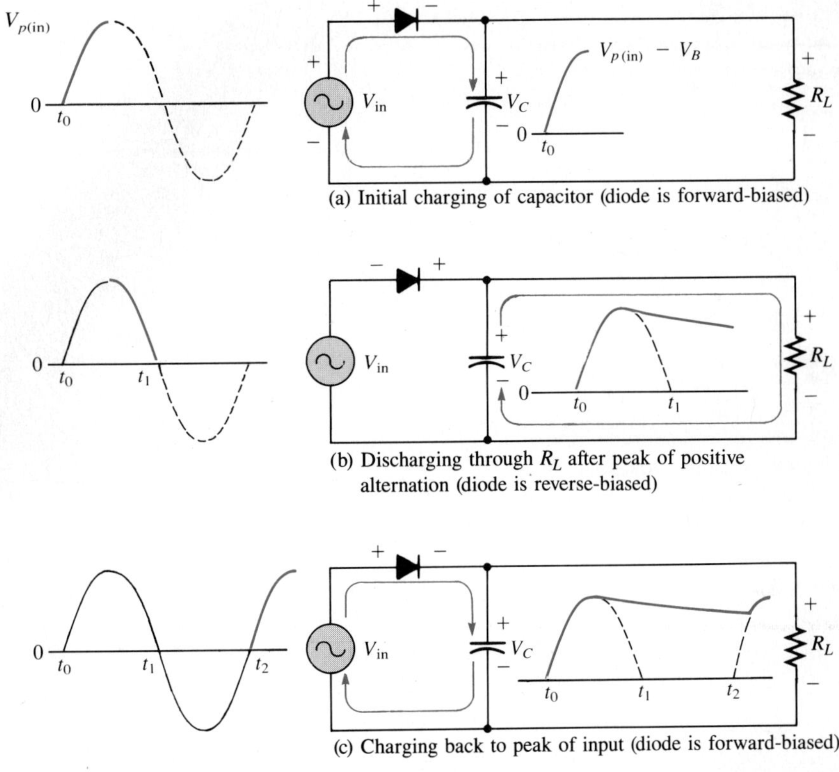

During the positive first quarter-cycle of the input, the diode is forward-biased, allowing the capacitor to charge to within a diode drop of the input peak, as illustrated in Figure 2–31(a). When the input begins to decrease below its peak, as shown in part (b), the capacitor retains its charge and the diode becomes reverse-biased. During the remaining part of the cycle, the capacitor can discharge only through the load resistance at a rate determined by the R_LC time constant. The larger the time constant, the less the capacitor will discharge. During the first quarter of the next cycle, the diode will again become forward-biased when the input voltage exceeds the capacitor voltage by approximately a diode drop. This is illustrated in part (c).

RIPPLE VOLTAGE

As you have seen, the capacitor quickly charges at the beginning of a cycle and slowly discharges after the positive peak (when the diode is reverse-biased). The variation in the output voltage due to the charging and discharging is called the **ripple voltage.** The smaller the ripple, the better the filtering action, as illustrated in Figure 2–32.

For a given input frequency, the output frequency of a full-wave rectifier is twice that of a half-wave rectifier, as illustrated in Figure 2–33. This makes a full-wave rectifier easier to filter. When filtered, the full-wave rectified voltage has less ripple than does a half-wave voltage for the same load resistance and capacitor values. This is because the capacitor discharges less during the shorter interval between full-wave pulses, as shown in Figure 2–34.

FIGURE 2–32
Half-wave ripple voltage (solid waveform).

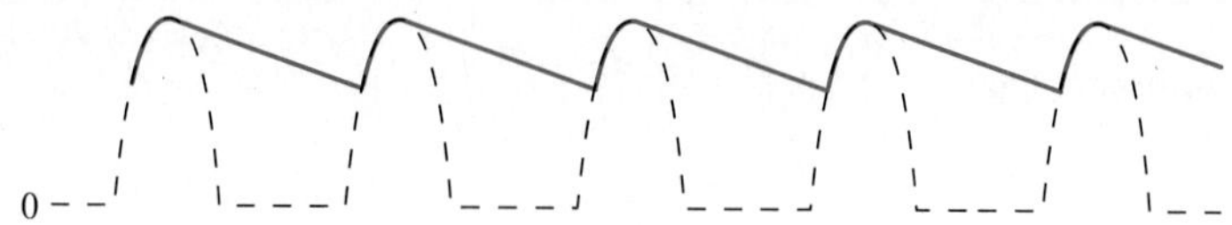

(a) Larger ripple means less effective filtering.

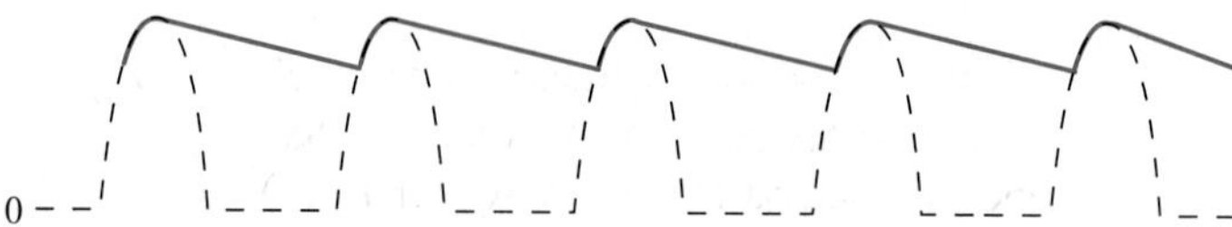

(b) Smaller ripple means more effective filtering.

FIGURE 2–33
Frequencies of half-wave and full-wave rectified voltages derived from a 60 Hz sine wave.

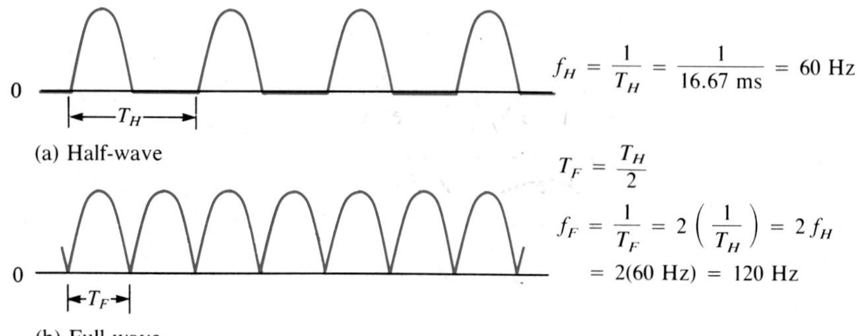

FIGURE 2–34
Comparison of ripple voltages for half-wave and full-wave signals with the same filter capacitor and derived from the same sine wave input.

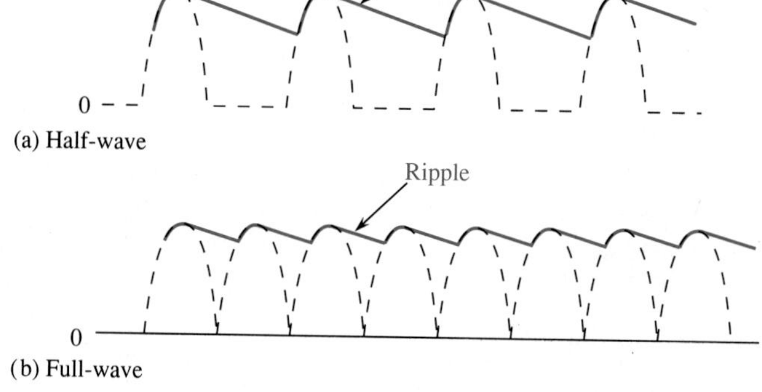

RIPPLE FACTOR

The **ripple factor** is an indication of the effectiveness of the filter and is defined as

$$r = \frac{V_r}{V_{\text{dc}}} \tag{2–12}$$

where V_r is the rms ripple voltage and V_{dc} is the dc (average) value of the filter's output voltage, as illustrated in Figure 2–35. The lower the ripple factor, the better the filter. The ripple factor can be lowered by increasing the value of the filter capacitor.

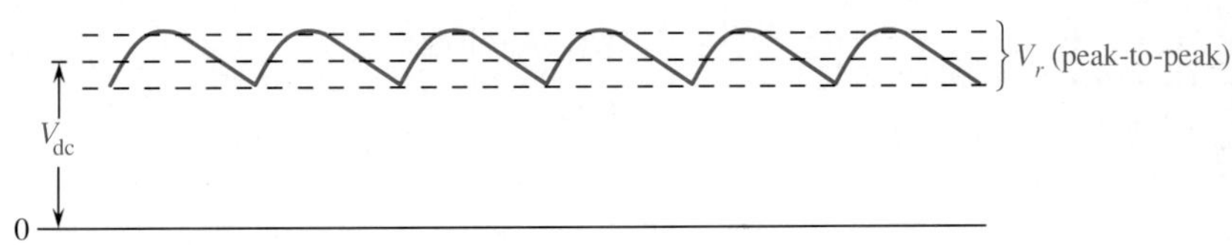

FIGURE 2–35
V_r and V_{dc} determine the ripple factor.

For a full-wave rectifier with a sufficiently high capacitance filter, if V_{dc} is very near in value to the peak rectified input voltage, then the expressions for V_{dc} and V_r are as follows. (Derivations are in Appendix B.)

$$V_{dc} = \left(1 - \frac{0.00417}{R_L C}\right)V_{p(in)} \tag{2–13}$$

$$V_r = \frac{0.0024}{R_L C}V_{p(in)} \tag{2–14}$$

where $V_{p(in)}$ is the peak rectified full-wave voltage applied to the filter.

EXAMPLE 2–7 Determine the ripple factor for the filtered bridge rectifier in Figure 2–36.

FIGURE 2–36

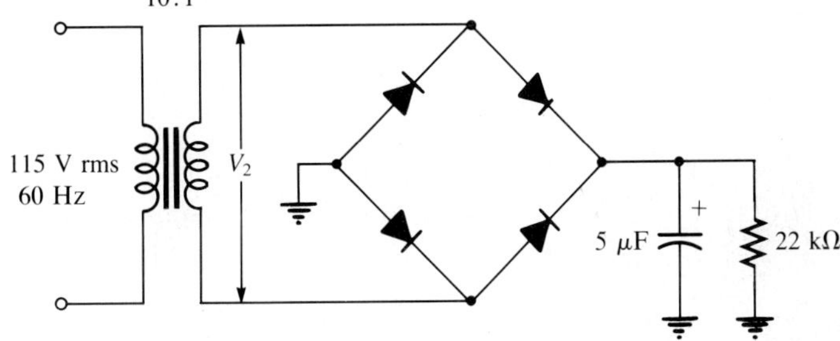

SOLUTION

The peak primary voltage is

$$V_{p(1)} = (1.414)115\text{ V}$$
$$= 162.6\text{ V}$$

The peak secondary voltage is

$$V_{p(2)} = \left(\frac{1}{10}\right)162.6\text{ V}$$
$$= 16.26\text{ V}$$

The peak full-wave rectified voltage at the filter input is

$$V_{p(in)} = V_{p(2)} - 2V_B$$
$$= 16.26\text{ V} - 1.4\text{ V}$$
$$= 14.86\text{ V}$$

The filtered dc output voltage is

$$V_{dc} = \left(1 - \frac{0.00417}{R_L C}\right)V_{p(in)}$$

$$= \left(1 - \frac{0.00417}{(22\ \text{k}\Omega)(5\ \mu\text{F})}\right)14.86\ \text{V}$$

$$= (1 - 0.0379)14.86\ \text{V}$$

$$= 14.3\ \text{V}$$

The rms ripple is

$$V_r = \frac{0.0024V_{p(in)}}{R_L C}$$

$$= \frac{0.0024(14.86\ \text{V})}{(22\ \text{k}\Omega)(5\ \mu\text{F})}$$

$$= 0.324\ \text{V}$$

The ripple factor is

$$r = \frac{V_r}{V_{dc}}$$

$$= \frac{0.324\ \text{V}}{14.3\ \text{V}}$$

$$= 0.0227$$

The percent ripple is 2.27%.

The following BASIC program computes the percent ripple for a bridge rectifier when you input the turns ratio, the load resistance, and the filter capacitance.

```
10 CLS
20 PRINT "THIS PROGRAM COMPUTES THE RIPPLE"
30 PRINT "FACTOR FOR THE BRIDGE RECTIFIER"
40 PRINT "IN FIGURE 2-36 GIVEN N2/N1,"
50 PRINT "RL, AND C."
60 PRINT:PRINT:PRINT
70 INPUT "TO CONTINUE PRESS 'ENTER'";X
80 CLS
90 INPUT "TURNS RATIO (N2/N1)";N
100 INPUT "LOAD RESISTANCE RL IN OHMS";RL
110 INPUT "FILTER CAPACITANCE C IN FARADS";C
120 CLS
130 V2P=N*162.6
140 VIP=V2P-1.4
150 VDC=(1-.00417/(RL*C))*VIP
160 VR=.0024*VIP/(RL*C)
170 PRINT "THE PERCENT RIPPLE IS";(VR/VDC)*100
```

PRACTICE EXERCISE 2–7

Determine the ripple voltage if the filter capacitor in Figure 2–36 is increased to 10 μF and the load resistance changes to 12 kΩ.

SURGE CURRENT IN THE CAPACITOR FILTER

Before the switch in Figure 2–37(a) is closed, the filter capacitor is uncharged. At the instant the switch is closed, voltage is connected to the bridge and the capacitor appears as a short, as shown. This produces an initial surge of current I_S through the two forward-biased diodes. The worst-case situation occurs when the switch is closed at a peak of the secondary voltage and a maximum surge current, $I_{S(max)}$, is produced, as illustrated in the figure.

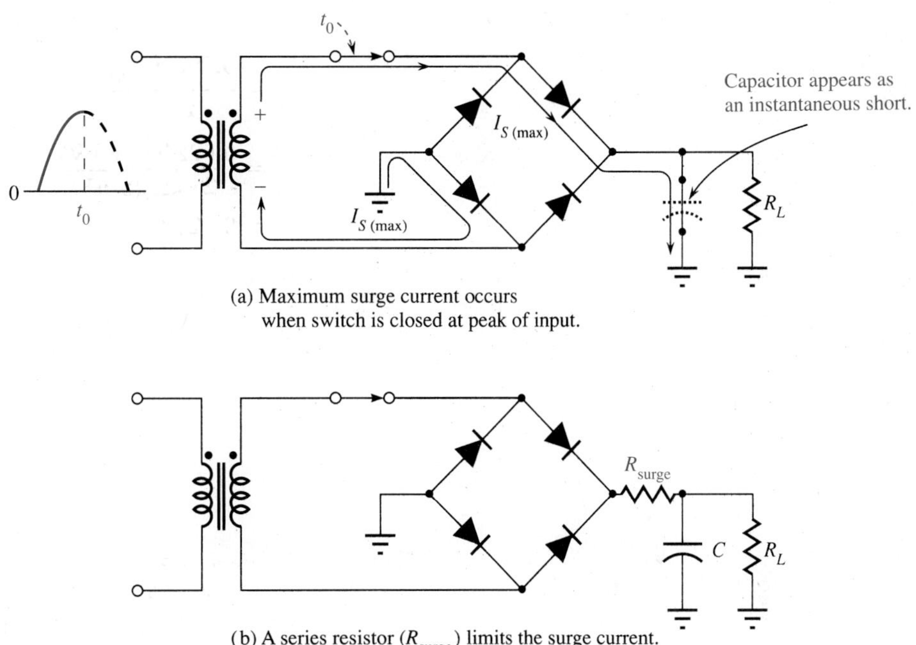

(a) Maximum surge current occurs when switch is closed at peak of input.

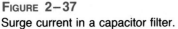

(b) A series resistor (R_{surge}) limits the surge current.

FIGURE 2–37
Surge current in a capacitor filter.

It is possible that the surge current could destroy the diodes, and for this reason a surge-limiting resistor is sometimes connected, as shown in Figure 2–37(b). The value of this resistor must be small compared to R_L. Also, the diodes must have a forward current rating such that they can withstand the momentary surge of current.

THE *LC* FILTER

When a choke is added to the filter, as in Figure 2–38, an additional reduction in the ripple voltage is achieved. The choke has a high reactance at the ripple frequency, and the capacitive reactance is low compared to both X_L and R_L. The two reactances form an ac voltage divider that tends to significantly reduce the ripple from that of a straight capacitor filter, as shown in Figure 2–39.

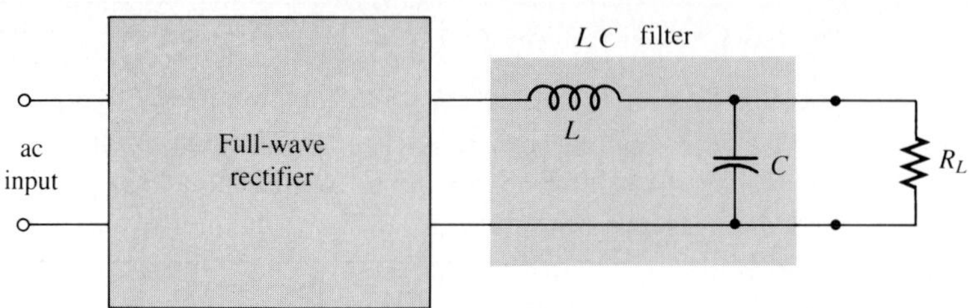

FIGURE 2–38
Rectifier with an LC filter.

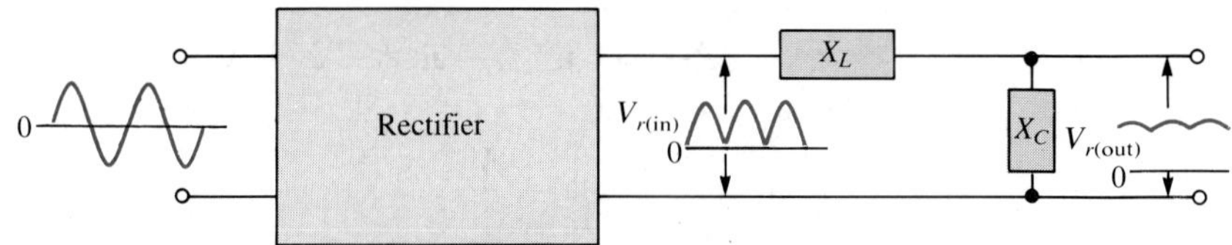

FIGURE 2–39
The LC filter as it looks to the ac component.

The magnitude of the ripple voltage out of the filter is determined with the voltage divider equation.

$$V_{r(out)} = \left(\frac{X_C}{|X_L - X_C|}\right)V_{r(in)} \quad \textbf{(2–15)}$$

To the dc (average) value of the rectified input, the choke presents a winding resistance (R_W) in series with the load resistance, as shown in Figure 2–40. This resistance produces an undesirable reduction of the dc value, and therefore R_W must be small compared to R_L. The dc output voltage is determined as follows.

$$V_{dc(out)} = \left(\frac{R_L}{R_W + R_L}\right)V_{dc(in)} \quad \textbf{(2–16)}$$

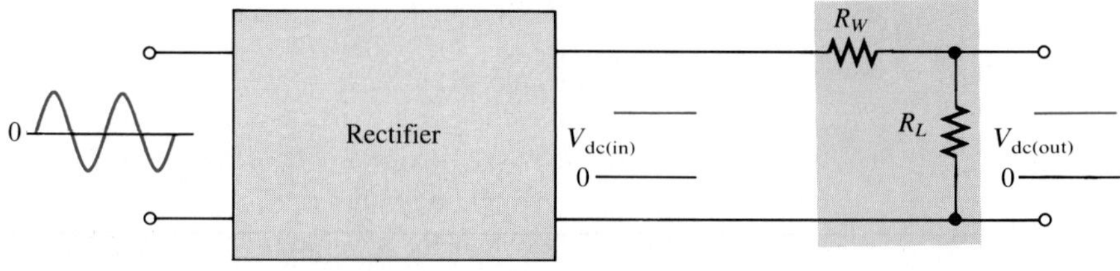

FIGURE 2–40
The LC filter as it looks to the dc component.

EXAMPLE 2–8

A 120 Hz full-wave rectified voltage with a peak value of 162.6 V is applied to the LC filter in Figure 2–41. Determine the filter output in terms of its dc value and the rms ripple voltage. What is the ripple factor? Compare this to the capacitor filter in Example 2–7.

FIGURE 2–41

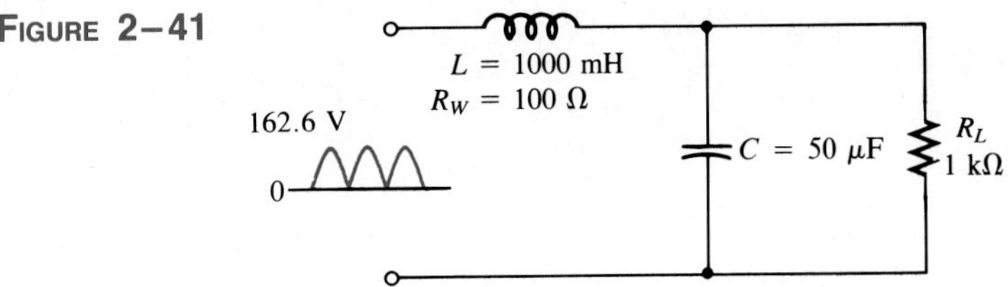

SOLUTION

First, we determine the dc value of the full-wave rectified input using Equation (2–5).

$$V_{dc(in)} = V_{AVG} = \frac{2V_p}{\pi} = \frac{2(162.6\text{ V})}{\pi}$$
$$= 103.5\text{ V}$$

Next, we find the amount of rms input ripple using a formula derived in Appendix B for the rms ripple of an unfiltered full-wave rectified signal.

$$V_{r(in)} = 0.308V_p$$
$$= 0.308(162.6\text{ V})$$
$$= 50.1\text{ V}$$

Now that the input values are known, the output values can be calculated.

$$V_{dc(out)} = \left(\frac{R_L}{R_W + R_L}\right)V_{dc(in)}$$
$$= \left(\frac{1\text{ k}\Omega}{1.1\text{ k}\Omega}\right)103.5\text{ V}$$
$$= 94.1\text{ V}$$

For the ripple calculation we need X_L and X_C.

$$X_L = 2\pi fL$$
$$= 2\pi(120\text{ Hz})(1000\text{ mH})$$
$$= 754\ \Omega$$

$$X_C = \frac{1}{2\pi fC}$$
$$= \frac{1}{2\pi(120\text{ Hz})(50\ \mu\text{F})}$$
$$= 26.5\ \Omega$$

$$V_{r(out)} = \left(\frac{X_C}{|X_L - X_C|}\right)V_{r(in)}$$
$$= \left(\frac{26.5\ \Omega}{|754\ \Omega - 26.5\ \Omega|}\right)50.1\text{ V}$$
$$= 1.82\text{ V rms}$$

The ripple factor is

$$r = \frac{V_{r(out)}}{V_{dc(out)}} = \frac{1.82\ \text{V}}{94.1\ \text{V}} = 0.0193$$

The percent ripple is 1.93%, which is less than that for the capacitor filter in Example 2–7.

PRACTICE EXERCISE 2–8

A 120 Hz full-wave rectified voltage with a peak value of 50 V is applied to the *LC* filter in Figure 2–41 with $L = 300$ mH, $R_W = 50\ \Omega$, $C = 100\ \mu$F, and $R_L = 10$ kΩ. Determine the filter output in terms of its dc value and the rms ripple voltage. What is the ripple factor?

It should be noted at this time that an *LC* rectifier filter produces an output with a dc value approximately equal to the average value of the rectified input. The capacitor filter, however, produces an output with a dc value approximately equal to the peak value of the input. Another point of comparison is that the amount of ripple voltage in the capacitor filter varies inversely with the load resistance. Ripple voltage in the *LC* filter is essentially independent of the load resistance and depends only on X_L and X_C, as long as X_C is sufficiently less than R_L.

π-TYPE AND T-TYPE FILTERS

A one-section π-type filter is shown in Figure 2–42(a). It can be thought of as a capacitor filter followed by an *LC* filter. It is also known as a *capacitor-input* filter. The T-type filter in Figure 2–42(b) is basically an *LC* filter followed by an inductor. It is also known as an *inductor-input* filter.

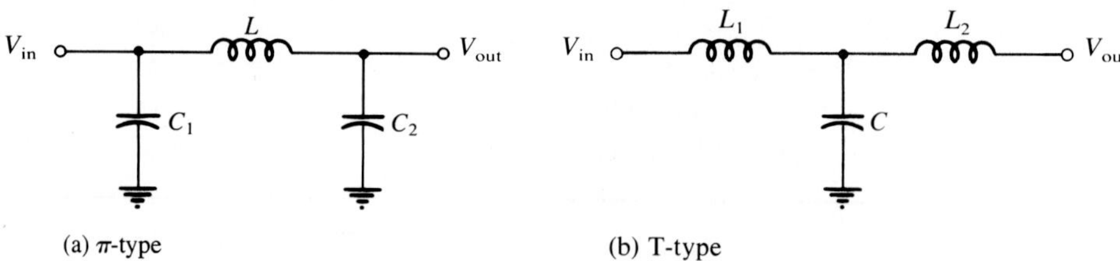

FIGURE 2–42
π-type and T-type *LC* filters.

In the π-type capacitive-input filter, C_1 charges to the peak value of the input voltage and discharges slightly through the load during the remaining portion of the input cycle. The opposing action of the inductor tends to keep the discharge variation to a minimum, resulting in a relatively small amount of ripple at the output. C_2 also helps to keep the output voltage constant.

In the T-type inductor-input filter, the voltage drop across the reactance reduces the output voltage well below the input peak. However, the smoothing action of L_1 and L_2 tends to provide an output with less ripple than the π-type filter. In general, the T-type filter provides less output voltage for a given input voltage than does the π-type, but it provides improved ripple reduction.

2–4 REVIEW QUESTIONS

1. When a 60-Hz sine wave is applied to the input of a half-wave rectifier, what is the output frequency?
2. When a 60-Hz sine wave is applied to the input of a full-wave rectifier, what is the output frequency?
3. What causes the ripple voltage on the output of a capacitor filter?
4. If the load resistance of a power supply is decreased, what happens to the ripple voltage?
5. Define *ripple factor*.
6. Name one advantage of an *LC* filter over a capacitor filter. Name one disadvantage.

2–5 DIODE LIMITING AND CLAMPING CIRCUITS

Diode circuits are sometimes used to clip off portions of signal voltages above or below certain levels; these circuits are called limiters or* clippers. *Another type of diode circuit is used to restore a dc level to an electrical signal; these circuits are called clampers. Both limiter and clamper diode circuits will be examined in this section.

DIODE LIMITERS

Figure 2–43(a) shows a diode **limiter** circuit that limits the positive part of the input voltage. As the input signal goes positive, the diode becomes forward-biased. Since the cathode is at ground potential (0 V), the anode cannot exceed 0.7 V (assuming silicon). So point *A* is limited to +0.7 V when the input exceeds this value. When the input goes back below 0.7 V, the diode reverse-biases and appears as an open. The output voltage looks like the negative part of the input, but with a magnitude determined by the R_s and R_L voltage divider as follows:

$$V_{out} = \left(\frac{R_L}{R_s + R_L}\right)V_{in}$$

If R_s is small compared to R_L, then $V_{out} \cong V_{in}$.

Turn the diode around, as in Figure 2–43(b), and the negative part of the input is clipped off. When the diode is forward-biased during the negative part of the input, point *A* is held at −0.7 V by the diode drop. When the input goes above −0.7 V, the diode is no longer forward-biased; and a voltage appears across R_L proportional to the input.

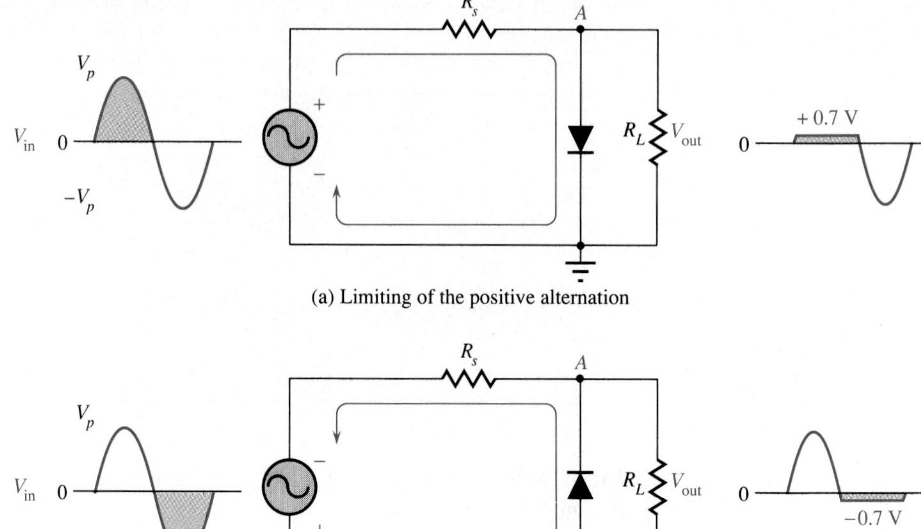

FIGURE 2–43
Diode limiters (clippers).

EXAMPLE 2–9

What would you expect to see displayed on an oscilloscope connected as shown in Figure 2–44?

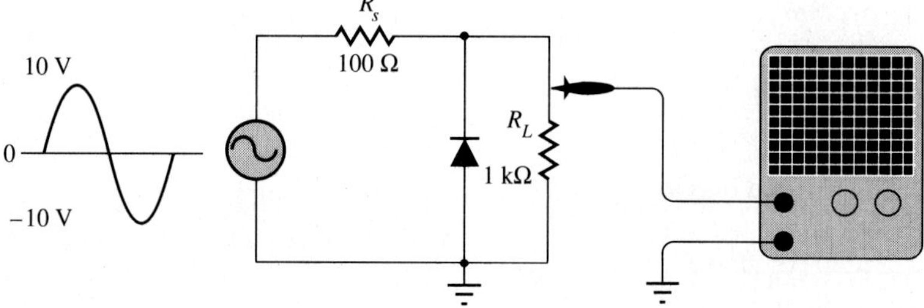

FIGURE 2–44

SOLUTION

The diode conducts when the input voltage goes below −0.7 V. So, we have a negative limiter with a peak output voltage determined by the following equation.

$$V_{p(out)} = \left(\frac{R_L}{R_s + R_L}\right)V_{p(in)}$$
$$= \left(\frac{1\ \text{k}\Omega}{1.1\ \text{k}\Omega}\right)10\ \text{V}$$
$$= 9.09\ \text{V}$$

The scope will display an output waveform as shown in Figure 2–45.

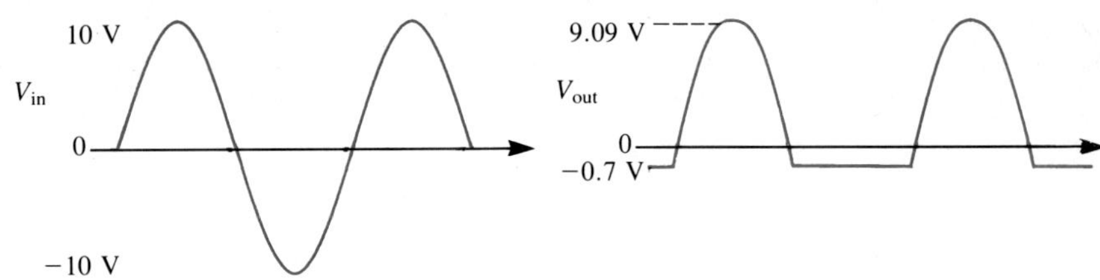

FIGURE 2–45
Waveforms for Figure 2–44.

PRACTICE EXERCISE 2–9

Describe the output waveform for Figure 2–44 if the diode is germanium and R_L is changed to 680 Ω.

ADJUSTMENT OF THE LIMITING LEVEL

The level to which a signal voltage is limited can be adjusted by adding a bias voltage in series with the diode, as shown in Figure 2–46. The voltage at point A must equal V_{BB} + 0.7 V before the diode will conduct. Once the diode begins to conduct, the voltage at point A is limited to V_{BB} + 0.7 V so that all input voltage above this level is clipped off.

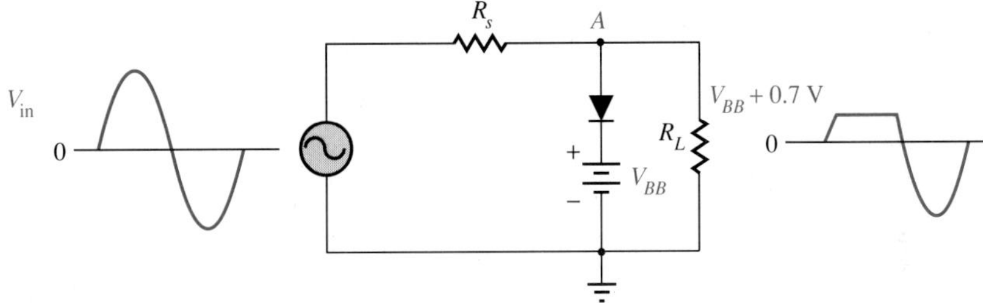

FIGURE 2–46
Positively biased limiter.

If the bias voltage is varied up or down, the limiting level changes correspondingly, as shown in Figure 2–47. If the polarity of the bias voltage is reversed, as in Figure 2–48, voltages above $-V_{BB}$ + 0.7 V are clipped, resulting in an output waveform as shown. The diode is reverse-biased only when the voltage at point A goes below $-V_{BB}$ + 0.7 V.

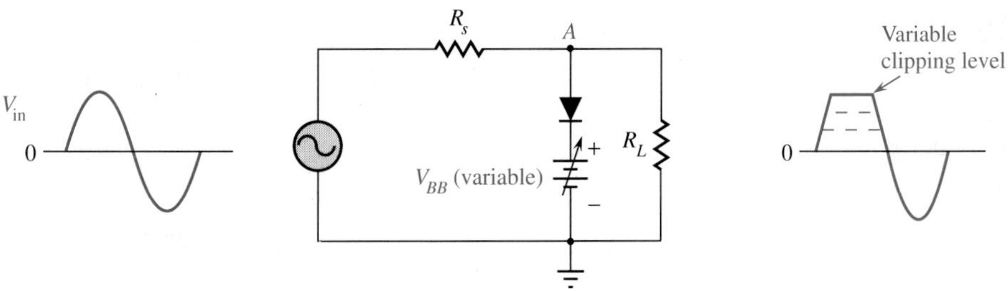

FIGURE 2–47
Positive limiter with variable bias.

FIGURE 2–48

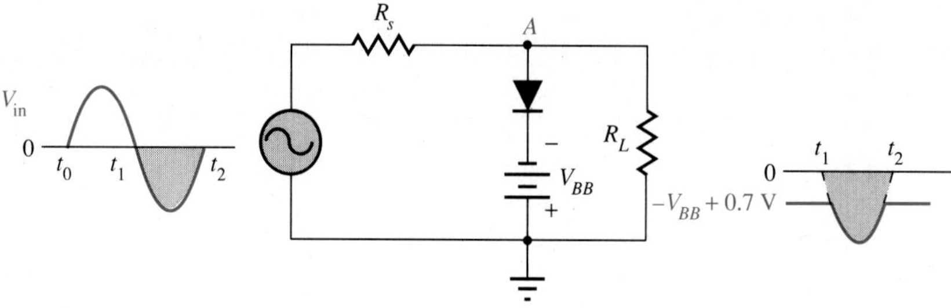

If it is necessary to limit a voltage to a specified negative level, then the diode and bias battery must be connected as in Figure 2–49(a). In this case, the voltage at point A must go below $-V_{BB} - 0.7$ V to forward-bias the diode and initiate limiting action, as shown. Turn the battery around and you get the waveform shown in Figure 2–49(b).

FIGURE 2–49

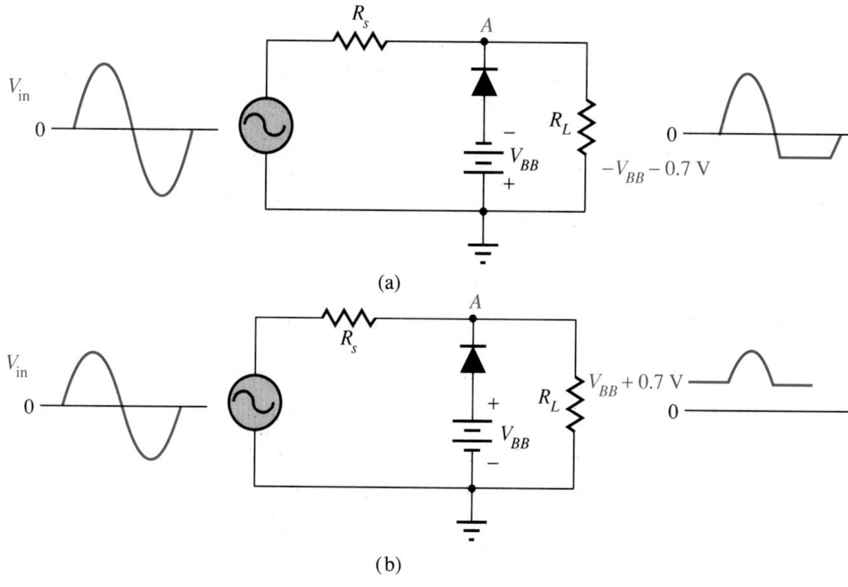

EXAMPLE 2–10

Figure 2–50 shows a circuit combining a positive-biased limiter with a negative-biased limiter. Determine the output waveform. The diodes are silicon.

FIGURE 2–50

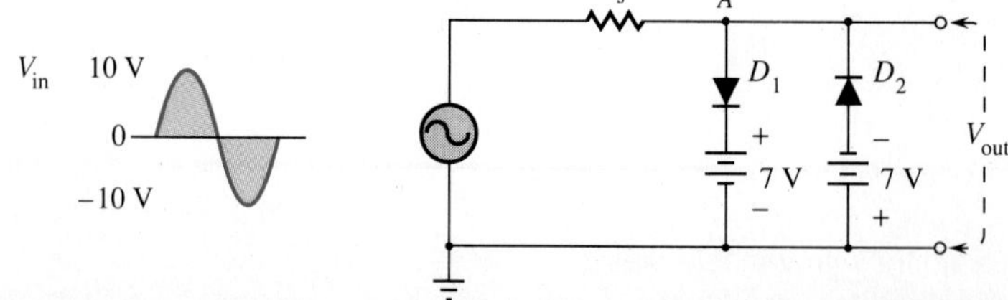

SOLUTION

When the voltage at point A reaches +7.7 V, diode D_1 conducts and limits the waveform to +7.7 V. Diode D_2 does not conduct until the voltage reaches −7.7 V. Therefore, positive voltages above +7.7 V and negative voltages below −7.7 V are clipped off. The resulting output waveform is shown in Figure 2–51.

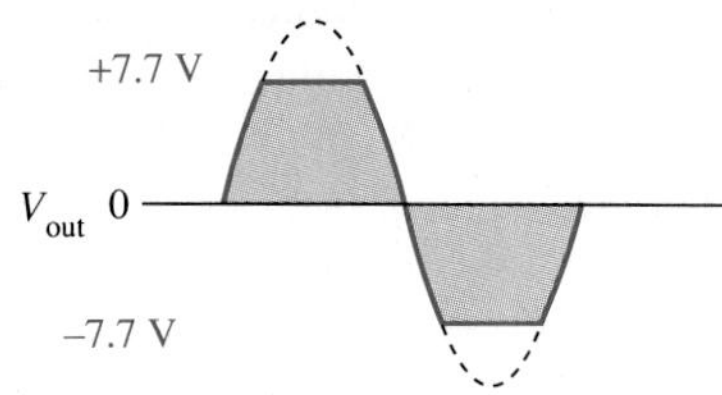

FIGURE 2–51
Output waveform for Figure 2–50.

PRACTICE EXERCISE 2–10

Determine the output waveform in Figure 2–50 if both dc sources are 10 V and the input has a peak value of 20 V.

DIODE CLAMPERS

The purpose of a clamper is to add a dc level to an ac signal. **Clampers** are sometimes known as *dc restorers*. Figure 2–52 shows a diode clamper that inserts a positive dc level. The operation of this circuit can be seen by considering the first negative half-cycle of the input voltage. When the input initially goes negative, the diode is forward-biased, allowing the capacitor to charge to near the peak of the input ($V_{p(in)}$ − 0.7 V), as shown in Figure 2–52(a). Just above the negative peak, the diode is reverse-biased. This is because the cathode is held near $V_{p(in)}$ − 0.7 V by the charge on the capacitor. The

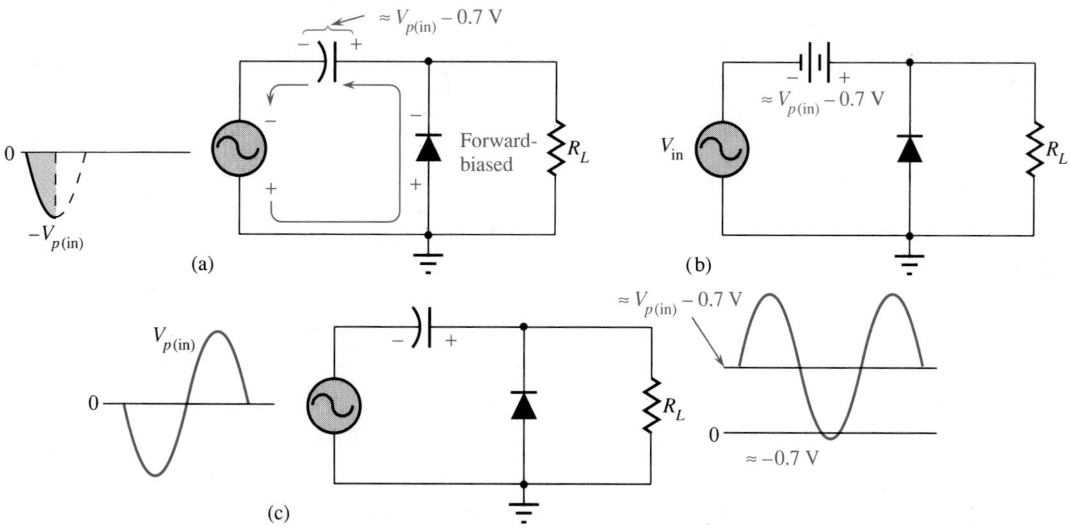

FIGURE 2–52
Positive clamper operation.

capacitor can only discharge through the high resistance of R_L. So, from the peak of one negative half-cycle to the next, the capacitor discharges very little. The amount that is discharged, of course, depends on the value of R_L. For good clamping action, the RC time constant should be at least ten times the period of the input frequency.

The net effect of the clamping action is that the capacitor retains a charge approximately equal to the peak value of the input less the diode drop. The capacitor voltage acts essentially as a battery in series with the input signal, as shown in Figure 2–52(b). The dc voltage of the capacitor adds to the input voltage by superposition, as in Figure 2–52(c). If the diode is turned around, a negative dc voltage is added to the input signal, as shown in Figure 2–53.

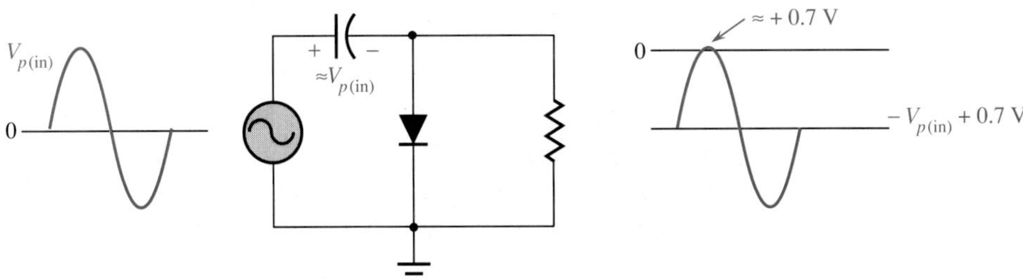

FIGURE 2–53
Negative clamper.

A CLAMPER APPLICATION

A clamping circuit is often used in television receivers as a dc restorer. The incoming composite video signal is normally processed through capacitively coupled amplifiers that eliminate the dc component, thus losing the black and white reference levels and the blanking level. Before being applied to the picture tube, these reference levels must be restored. Figure 2–54 illustrates this process in a general way.

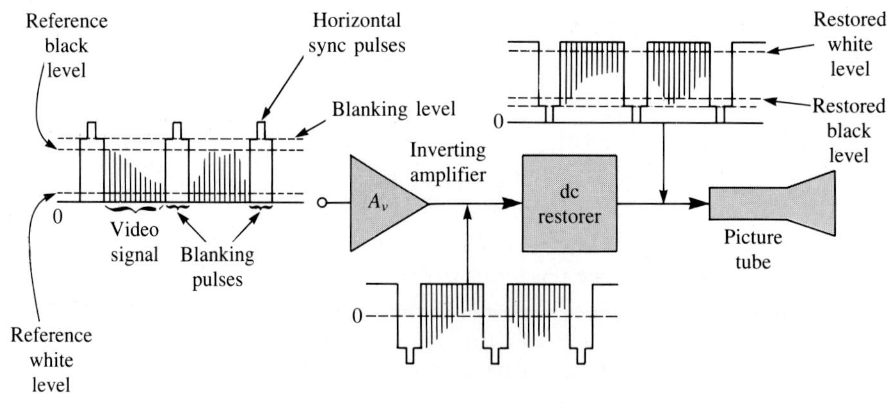

FIGURE 2–54
Clamper (dc restorer) in a TV receiver.

EXAMPLE 2–11 What is the output voltage that you would expect to observe across R_L in the clamper circuit of Figure 2–55? Assume that RC is large enough to prevent significant capacitor discharge.

FIGURE 2–55

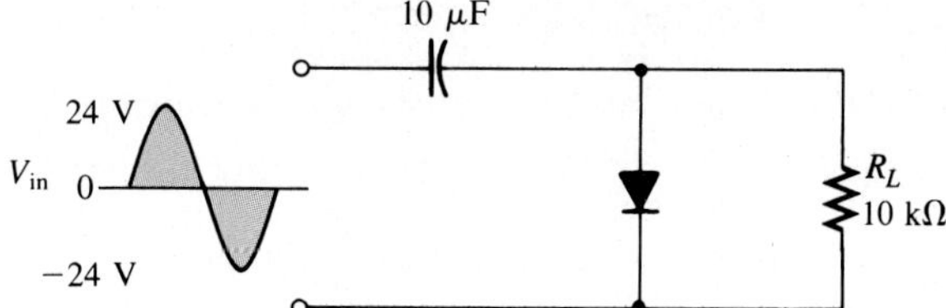

SOLUTION

Ideally, a negative dc value equal to the input peak less the diode drop is inserted by the clamping circuit:

$$\begin{aligned} V_{dc} &\cong V_{p(in)} - 0.7 \text{ V} \\ &= 24 \text{ V} - 0.7 \text{ V} \\ &= 23.3 \text{ V} \end{aligned}$$

Actually, the capacitor will discharge slightly between peaks, and, as a result, the output voltage will have an average value of slightly less than that calculated above. The output waveform goes to approximately 0.7 V above ground, as shown in Figure 2–56.

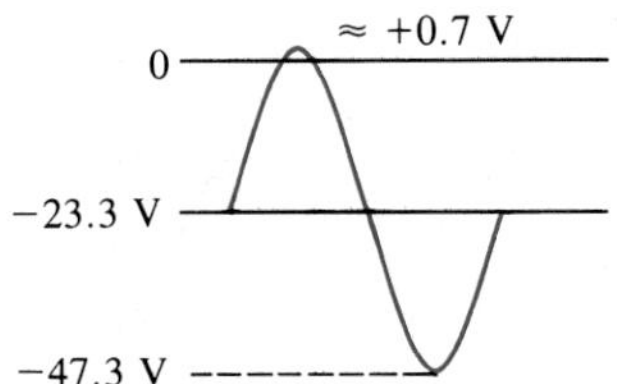

FIGURE 2–56
Output waveform for Figure 2–55.

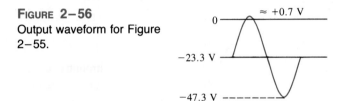

PRACTICE EXERCISE 2–11

What is the output voltage that you would observe across R_L in Figure 2–55 if $C = 22\ \mu\text{F}$ and $R_L = 18\ \text{k}\Omega$? ■

2–5 REVIEW QUESTIONS

1. Discuss how diode limiters and diode clampers differ in terms of their function.
2. What is the difference between a positive limiter and a negative limiter?
3. What is the maximum voltage across an unbiased positive silicon diode limiter during the positive alternation of the input voltage?
4. To limit the output of a positive limiter to 5 V when a 10-V peak input is applied, what value must the bias voltage be?
5. What component in a clamper circuit effectively acts as a battery?

2–6 VOLTAGE MULTIPLIERS

Voltage* multipliers *utilize clamping action to increase peak rectified voltages without the necessity of increasing the input transformer's voltage rating. Multiplication factors of two, three, and four are commonly used. Voltage multipliers are used in high-voltage, low-current applications.

VOLTAGE DOUBLERS

A half-wave voltage doubler is shown in Figure 2–57. During the positive half-cycle of the secondary voltage, diode D_1 is forward-biased and D_2 is reverse-biased. Capacitor C_1 is charged to the peak of the secondary voltage (V_p) less the diode drop with the polarity shown in part (a). During the negative half-cycle, diode D_2 is forward-biased and D_1 is reverse-biased, as shown in part (b). The peak voltage on C_1 adds to the secondary voltage to charge C_2 to approximately $2V_p$. Applying Kirchhoff's law around the loop,

$$V_{C1} - V_{C2} + V_p = 0$$
$$V_{C2} = V_p + V_{C1}$$

Neglecting the diode drop $V_{C1} \cong V_p$,

$$V_{C2} = 2V_p$$

FIGURE 2–57
Half-wave voltage doubler operation.

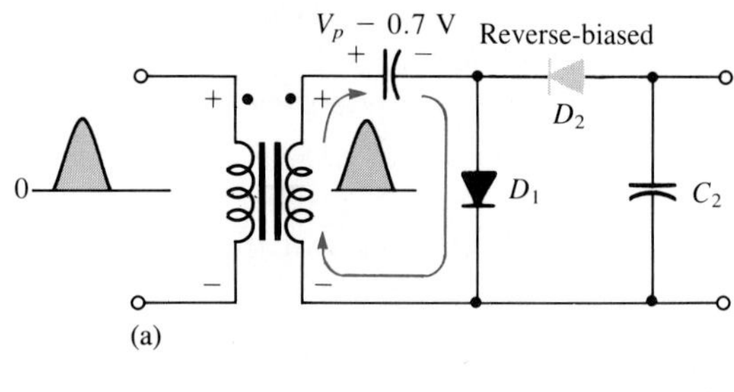

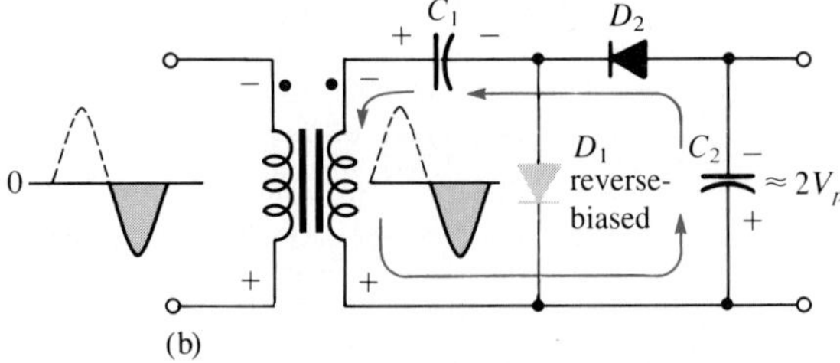

Under a no-load condition, C_2 remains charged to approximately $2V_p$. If a load resistance is connected across the output, C_2 discharges through the load on the next positive half-cycle and is again recharged to $2V_p$ on the following negative half-cycle. The resulting output is a half-wave, capacitor-filtered voltage. The peak inverse voltage across each diode is $2V_p$.

A full-wave doubler is shown in Figure 2–58. When the secondary voltage is positive, D_1 is forward-biased and C_1 charges to approximately V_p, as shown in part (a). During the negative half-cycle, D_2 is forward-biased and C_2 charges to approximately V_p, as shown in part (b). The output voltage, $2V_p$, is taken across the two capacitors in series.

VOLTAGE TRIPLER

The addition of another diode-capacitor section to the half-wave voltage doubler creates a voltage tripler, as shown in Figure 2–59. The operation is as follows: On the positive

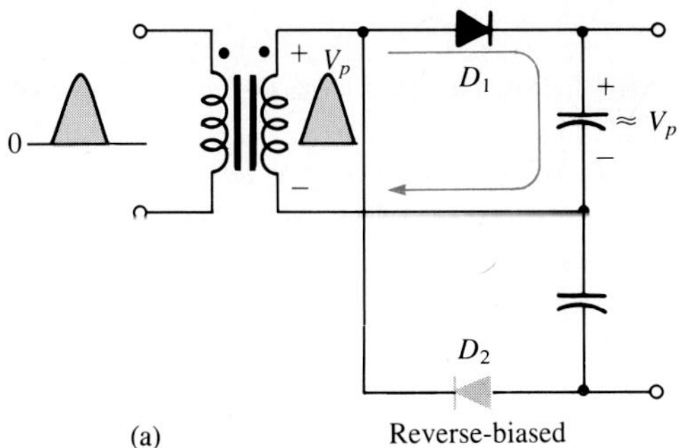

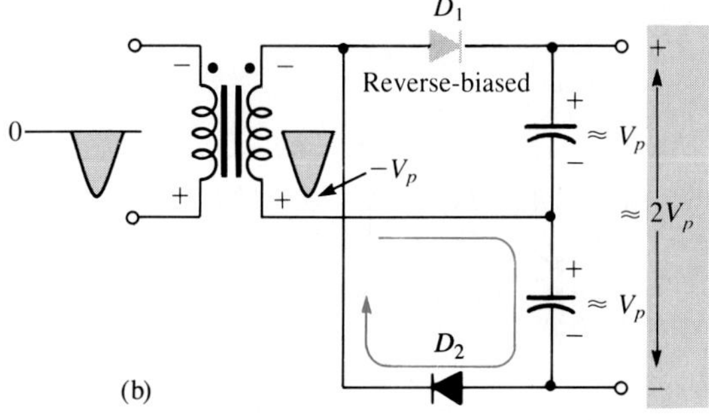

FIGURE 2–58
Full-wave voltage doubler operation.

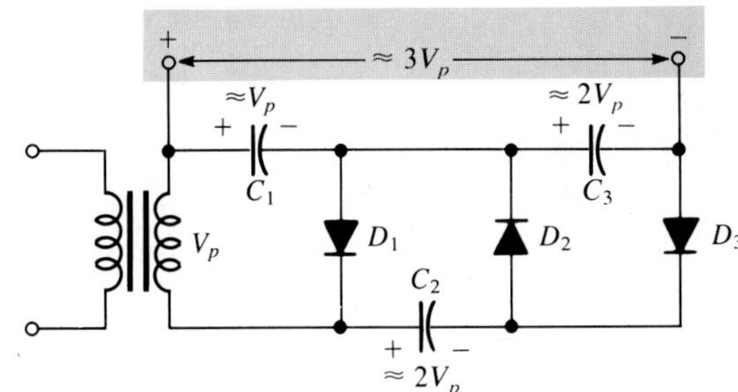

FIGURE 2–59
Voltage tripler.

half-cycle of the secondary voltage, C_1 charges to V_p through D_1. During the negative half-cycle, C_2 charges to $2V_p$ through D_2, as described for the doubler. During the next positive half-cycle, C_3 charges to $2V_p$ through D_3. The tripler output is taken across C_1 and C_3, as shown in the figure.

VOLTAGE QUADRUPLER

The addition of still another diode-capacitor section, as shown in Figure 2–60, produces an output four times the peak secondary voltage. C_4 charges to $2V_p$ through D_4 on a negative half-cycle. The $4V_p$ output is taken across C_2 and C_4, as shown. In both the tripler and quadrupler circuits, the PIV of each diode is $2V_p$.

FIGURE 2–60
Voltage quadrupler.

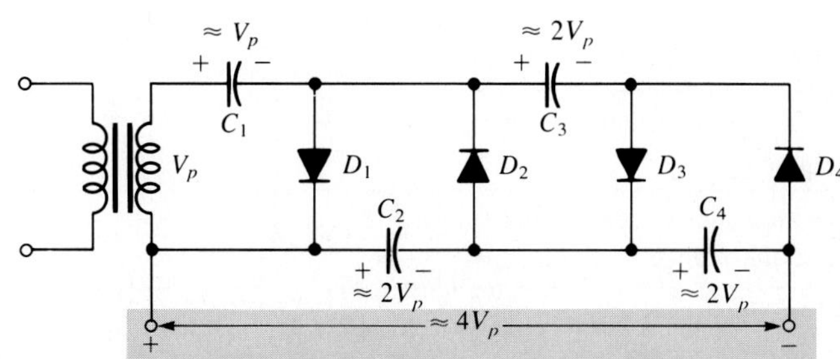

2–6 REVIEW QUESTIONS

1. What must be the peak voltage rating of the transformer secondary for a voltage doubler that produces an output of 200 V?
2. The output voltage of a quadrupler is 622 V. What minimum PIV rating must each diode have?

2–7 THE DIODE DATA SHEET

A manufacturer's data sheet gives detailed information on a device so that it can be used properly in a given application. A typical data sheet provides maximum ratings, electrical characteristics, mechanical data, and graphs of various parameters. In this section, we use a specific example to illustrate a typical data sheet.

Table 2–1 shows the maximum ratings for a certain series of rectifier diodes (1N4001 through 1N4007). These are the absolute maximum values under which the diode can be operated without damage to the device. For greatest reliability and longer life, the diode should always be operated well under these maximums. Generally, the maximum ratings are specified at 25°C and must be adjusted downward for greater temperatures.

TABLE 2–1
Maximum ratings

Rating	Symbol	1N4001	1N4002	1N4003	1N4004	1N4005	1N4006	1N4007	Unit
Peak repetitive reverse voltage Working peak reverse voltage dc blocking voltage	V_{RRM} V_{RWM} V_R	50	100	200	400	600	800	1000	V
Nonrepetitive peak reverse voltage	V_{RSM}	60	120	240	480	720	1000	1200	V
rms reverse voltage	$V_{R(rms)}$	35	70	140	280	420	560	700	V
Average rectified forward current (single-phase, resistive load, 60 Hz, T_A = 75°C)	I_0								A
Nonrepetitive peak surge current (surge applied at rated load conditions)	I_{FSM}	30 (for 1 cycle)							A
Operating and storage junction temperature range	T_j, T_{stg}	−65 to +175							°C

An explanation of some of the parameters from Table 2–1 is as follows:

V_{RRM} The maximum reverse peak voltage that can be applied repetitively across the diode. Notice that in this case, it is 50 V for the 1N4001 and 1 kV for the 1N4007. This is the same as PIV.

V_R The maximum reverse dc voltage that can be applied across the diode.

V_{RSM} The maximum reverse peak value of nonrepetitive voltage that can be applied across the diode.

I_0 The maximum average value of a 60 Hz full-wave rectified forward current.

I_{FSM} The maximum peak value of nonrepetitive (one cycle) forward current. The graph in Figure 2–61 expands on this parameter to show values for more than one cycle at temperatures of 25°C and 175°C. The dashed lines represent values where typical failures occur. Notice what happens on the lower solid line when ten cycles of I_{FSM} are applied. The limit is 15 A rather than the one-cycle value of 30 A.

Table 2–2 lists typical and maximum values of certain electrical characteristics. These items differ from the maximum ratings in that they are not selected by design but

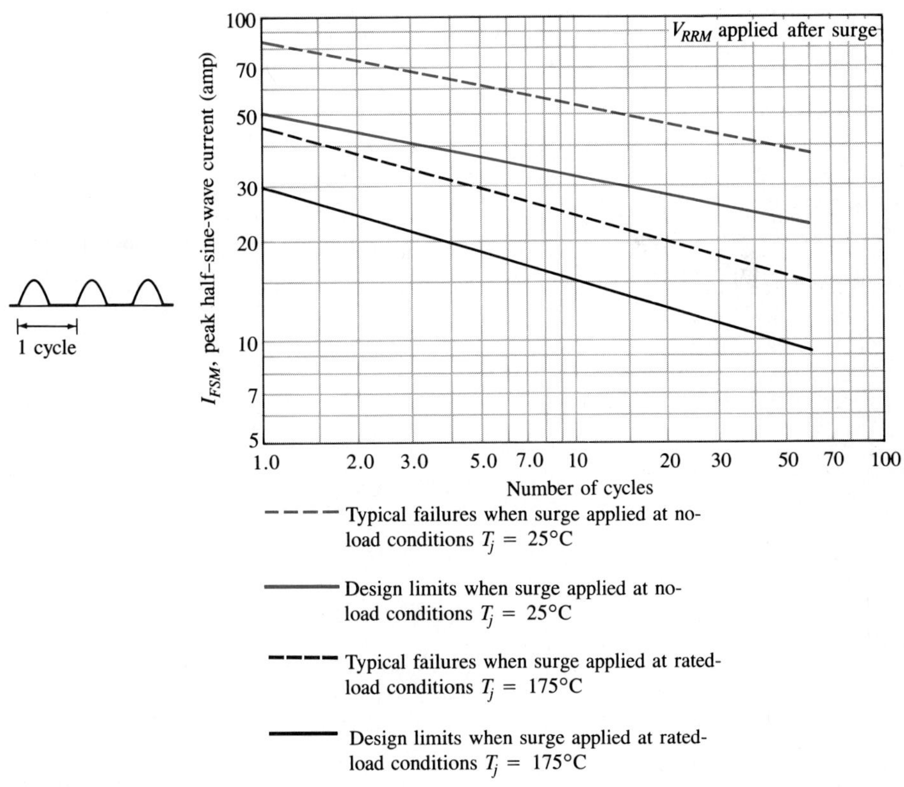

FIGURE 2–61
Nonrepetitive surge capability.

TABLE 2–2
Electrical characteristics

Characteristic and Conditions	Symbol	Typical	Maximum	Unit
Maximum instantaneous forward voltage drop (i_F = 1 A, T_j = 25°C)	v_F	0.93	1.1	V
Maximum full-cycle average forward voltage drop (I_0 = 1 A, T_L = 75°C, 1 inch leads)	$V_{F(AVG)}$	–	0.8	V
Maximum reverse current (rated dc voltage) T_j = 25°C T_j= 100°C	I_R	 0.05 1.0	 10.0 50.0	μA
Maximum full-cycle average reverse current (I_0 = 1 A, T_L = 75°C, 1 inch leads)	$I_{R(AVG)}$	–	30.0	μA

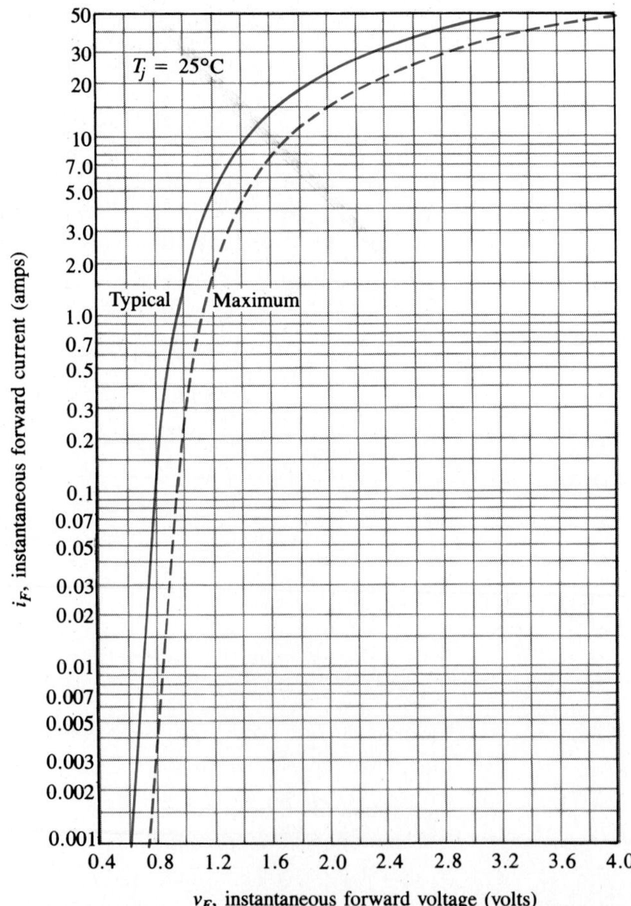

FIGURE 2–62
Forward voltage.

are the result of operating the diode under specified conditions. A brief explanation of these parameters follows:

v_F The instantaneous voltage across the forward-biased diode when the forward current is 1 A at 25°C. Figure 2–62 shows how the forward voltages vary with forward current.

$V_{F(AVG)}$ The maximum forward voltage drop averaged over a full cycle.

I_R The maximum current when the diode is reverse-biased with a dc voltage.

$I_{R(AVG)}$ The maximum reverse current averaged over one cycle (when reverse-biased with an ac voltage).

The mechanical data for these particular diodes as they appear on a typical data sheet are shown in Figure 2–63.

FIGURE 2–63
Mechanical data

	Millimeters		Inches	
DIM	Min	Max	Min	Max
A	5.97	6.60	0.235	0.260
B	2.79	3.05	0.110	0.120
D	0.76	0.86	0.030	0.034
K	27.94	–	1.100	–

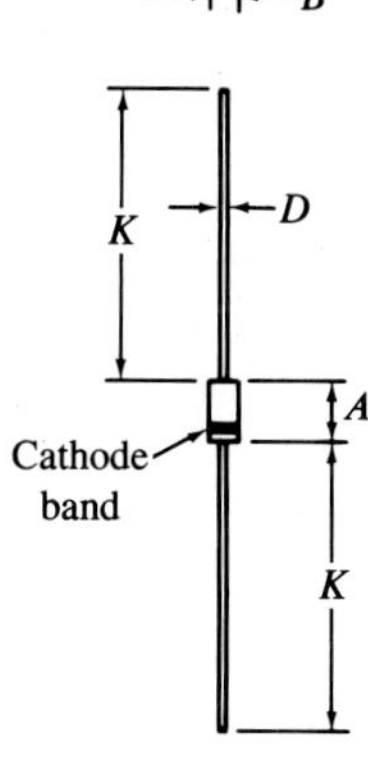

Mechanical characteristics

Case: Transfer Molded Plastic
Maximum lead temperature for soldering purposes: 350°C, 3/8″ from case for 10 seconds at 5 lbs. tension
Finish: All external surfaces are corrosion-resistant, leads are readily solderable
Polarity: Cathode indicated by color band
Weight: 0.40 Grams (approximately)

2–7 REVIEW QUESTIONS

1. List the three rating categories typically given on all diode data sheets.
2. Identify each of the following parameters: **(a)** V_F **(b)** I_R **(c)** I_0.

2–8 TROUBLESHOOTING

Several types of failures can occur in power supply rectifiers. In this section, you will learn how to* troubleshoot *rectifiers for common faults such as open diodes, shorted diodes, open filter capacitors, shorted or leaky filter capacitors, and open or shorted transformer windings.

OPEN DIODE

A half-wave rectifier with a diode that has opened (a common failure mode) is shown in Figure 2–64. In this case, you would measure 0 V dc across the load resistor, as depicted.

FIGURE 2–64
Test for an open diode in a half-wave rectifier shows 0 V at the output.

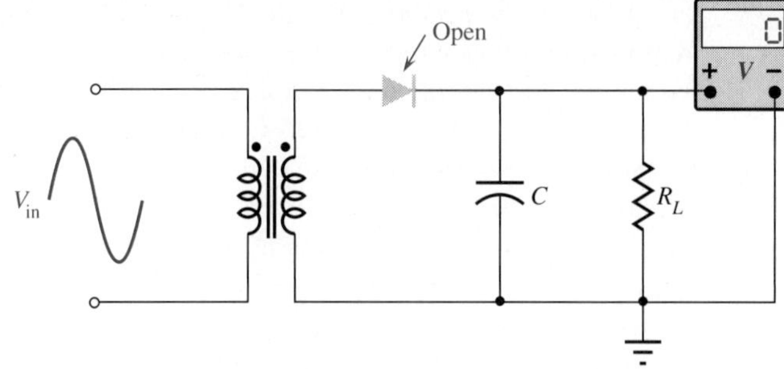

Now consider the center-tapped full-wave rectifier in Figure 2–65. If the circuit is functioning properly, you will see a normal ripple voltage at a frequency of 120 Hz, as shown in part (a). If one of the diodes is open, you will observe a ripple voltage at a frequency of 60 Hz with a larger than normal amplitude, as shown in part (b).

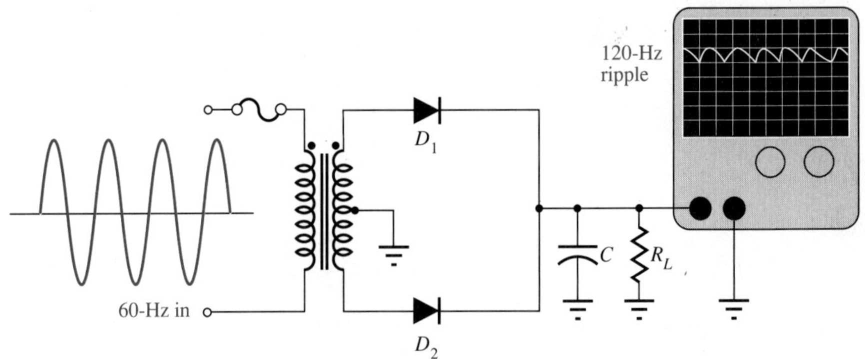

(a) Normally the ripple should have a frequency of 120 Hz.

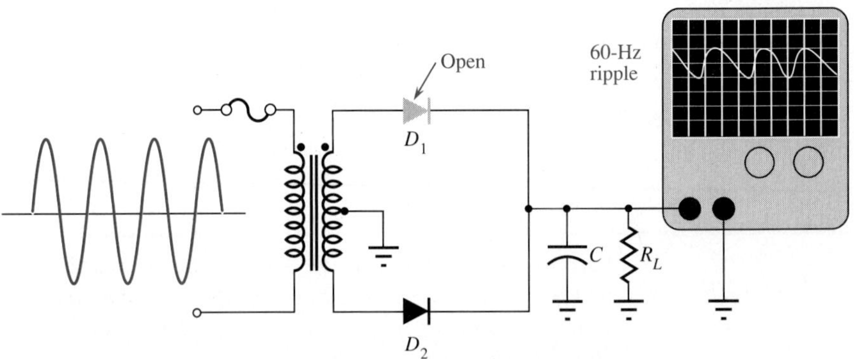

(b) With one diode open the ripple has a frequency of 60 Hz.

FIGURE 2–65
Test for an open diode in a filtered full-wave center-tapped rectifier.

Now let's examine the reason for these observations. If diode D_1 is open, there will be current through R_L only during the negative half-cycle of the input voltage. During the positive half-cycle, the open path caused by the faulty diode prevents current through R_L. The result is a half-wave voltage, as illustrated in Figure 2–66(a). With the filter capacitor in the circuit, the half-wave voltage will allow the capacitor to discharge more than it would with a normal full-wave voltage, resulting in a larger ripple voltage with a frequency of 60 Hz, as shown in Figure 2–66(b). The same observations would be made for an open failure of diode D_2.

An open diode in a bridge rectifier would create symptoms identical to those just discussed for the center-tapped rectifier. As illustrated in Figure 2–67, the open diode would prevent current through R_L during half of the input cycle (in this case, the negative half). This would result in a half-wave output and an increased ripple voltage at 60 Hz, as discussed before.

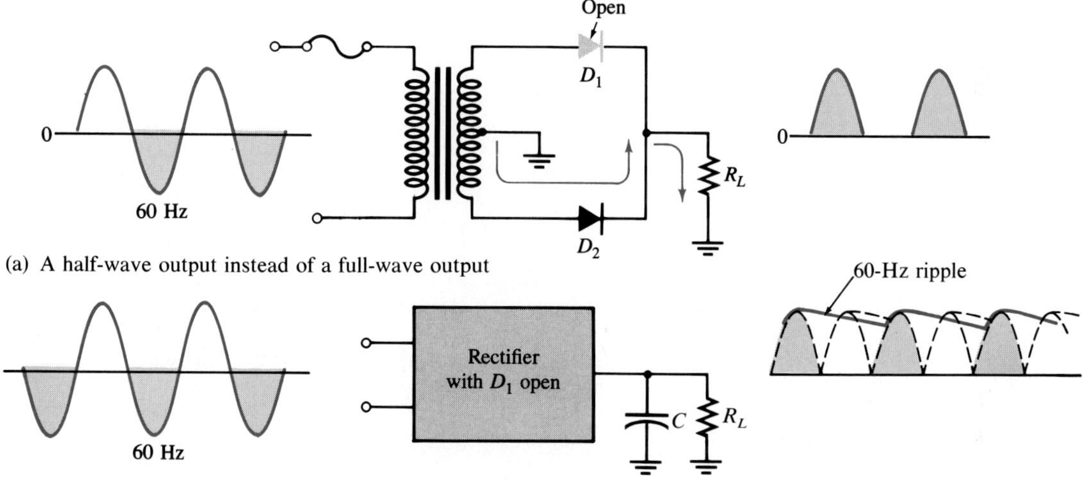

FIGURE 2–66
Effects of an open diode in a center-tapped rectifier.

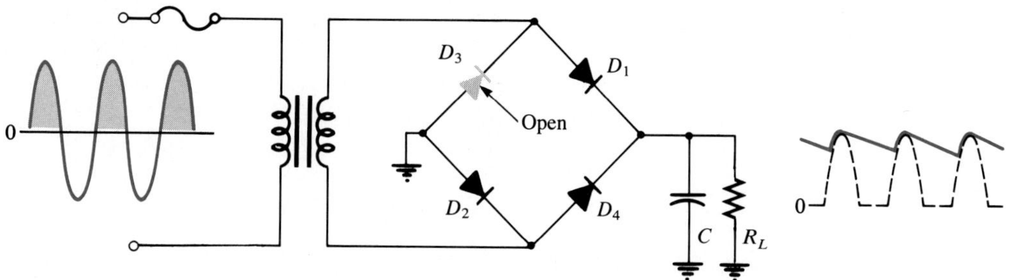

FIGURE 2–67
The effect of an open diode in a bridge rectifier is a half-wave output that results in an increased ripple voltage at 60 Hz.

SHORTED DIODE

A shorted diode is one that has failed such that it has a very low resistance in both directions. If a diode suddenly becomes shorted in a bridge rectifier, an excessively high current will flow during one half of the input cycle possibly burning open one of the diodes. If one of the diodes does not burn open, the transformer can be damaged unless the power supply is properly fused. This condition is illustrated in Figure 2–68 with diode D_1 shorted.

In part (a) of Figure 2–68, current is supplied to the load through the shorted diode during the first positive half-cycle, just as though it were forward-biased. During the negative half-cycle, the current is shorted through D_1 and D_4, as shown in part (b). Again, damage to the transformer is possible unless properly fused. It is likely that this excessive current would burn either or both of the diodes open. If only one of the diodes opened, you would still observe a half-wave voltage on the output. If both diodes (D_1 and D_4 in this case) opened, there would be no voltage developed across the load.

FIGURE 2–68
Effects of a shorted diode in a bridge rectifier circuit.

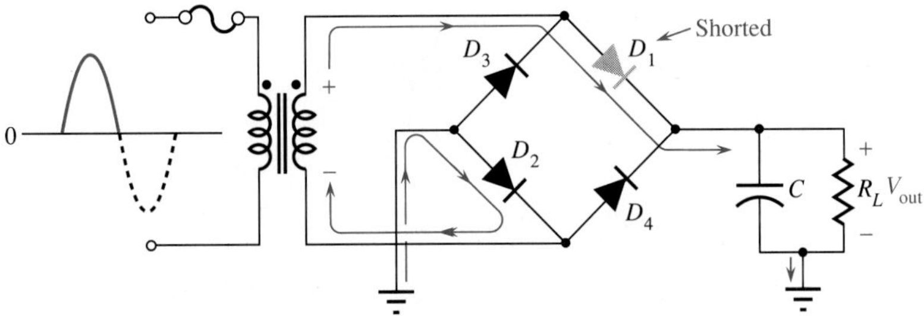

(a) Positive half-cycle: The shorted diode acts as a forward-biased diode, so the load current is normal.

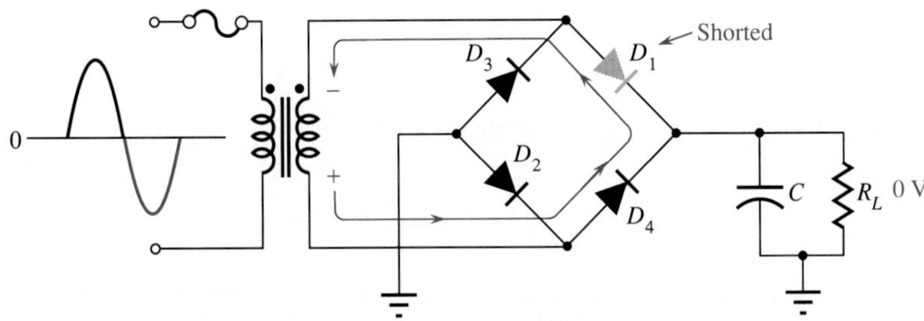

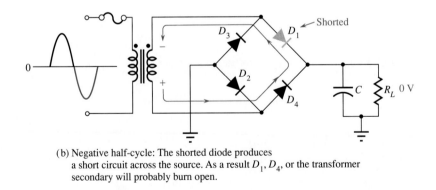

(b) Negative half-cycle: The shorted diode produces a short circuit across the source. As a result D_1, D_4, or the transformer secondary will probably burn open.

OPEN FILTER CAPACITOR

An open filter capacitor results in a full-wave rectified output voltage as shown in Figure 2–69(a).

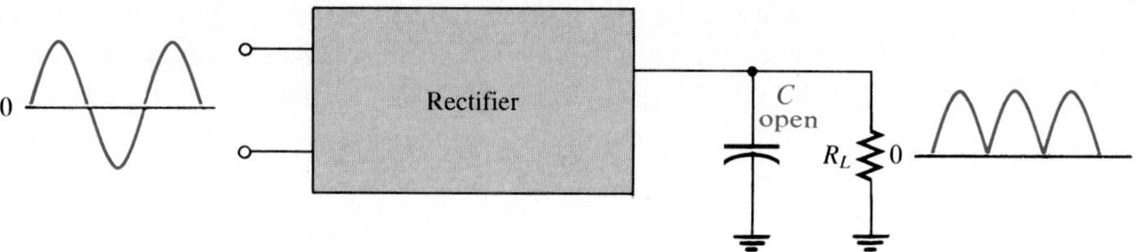

(a) Open filter capacitor produces a full-wave output.

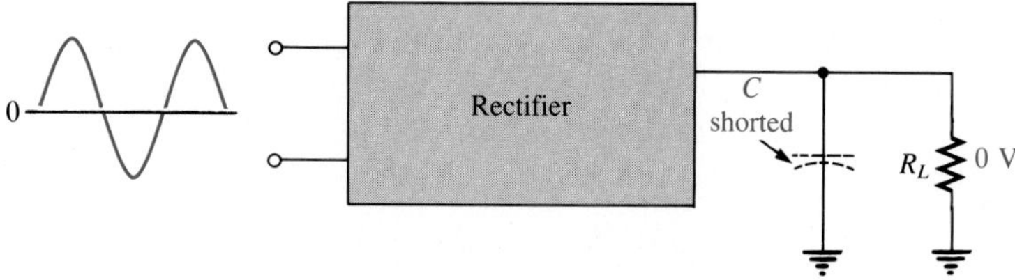

(b) Shorted filter capacitor produces an output of 0 V.

FIGURE 2–69
Symptoms produced by faulty filter capacitors.

SHORTED OR LEAKY FILTER CAPACITOR

A shorted capacitor would most likely cause some or all of the diodes in a full-wave rectifier to open due to excessive current or it would blow the **fuse.** In any event, there would be no dc voltage on the output. Figure 2–69(b) illustrates the condition. A leaky capacitor can be represented by a leakage resistance in parallel with the capacitor, as shown in Figure 2–70(a). The effect of the leakage resistance is to reduce the discharging time constant. This would cause an increase in ripple voltage on the output, as shown in Figure 2–70(b).

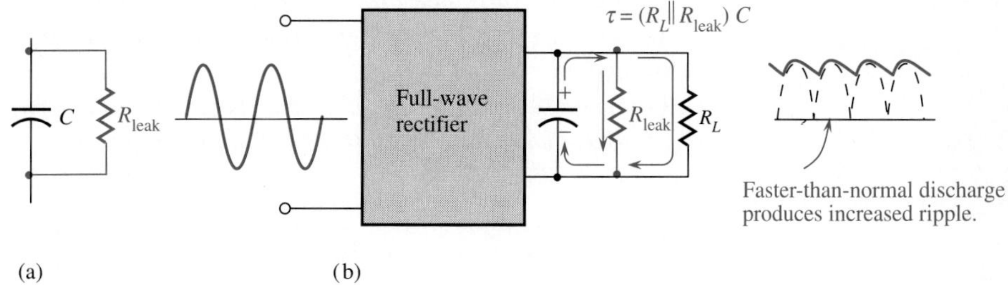

(a) (b)

FIGURE 2–70
Effects of a leaky filter capacitor.

OPEN OR SHORTED TRANSFORMER

Another type of failure that occurs in a power supply circuit involves transformers. If either the primary or the secondary winding of the transformer in a power supply opens, there will be no output voltage. If some of the turns in the primary winding are shorted, the dc output voltage will be greater than it should be, because this effectively increases the turns ratio of the transformer. If some of the turns in the secondary winding are shorted, the dc output voltage will be less than it should be, because this effectively reduces the turns ratio.

2–8 REVIEW QUESTIONS

1. What effect does an open diode have on the output voltage of a half-wave rectifier?
2. What effect does an open diode have on the output voltage of a full-wave rectifier?
3. If one of the diodes in a bridge rectifier shorts, what are some possible consequences?
4. What happens to the output voltage of a rectifier if the filter capacitor becomes very leaky?
5. The primary winding of the transformer in a power supply opens. What will you observe on the rectifier output?
6. The dc output voltage of a filtered rectifier is less than it should be. What might be the problem?

2–9 A SYSTEM APPLICATION

The purpose of the dc power supply in any electronic system is to provide to all parts of the system a constant dc voltage and sufficient dc current from which all the circuits operate. In other words, the dc power supply energizes the system. In this section, you will be dealing with the dc power supply in the radio receiver system shown at the opening of the chapter and you will

- ☐ ***See how a diode rectifier is used in a system application.***
- ☐ ***See how the filter operates in a system application.***
- ☐ ***See how the 110 V ac voltage from a standard outlet is converted to the dc supply voltage.***
- ☐ ***Translate between a printed circuit board and a schematic.***
- ☐ ***Troubleshoot some common power supply failures.***

At the opening of this chapter, you saw the dc power supply printed circuit board with the components assembled on it. This part of the system represents a principal application of the diode and the full-wave bridge rectifier as well as the capacitor filter. An additional component in the power supply is the voltage **regulator.** This device is covered in detail in a later chapter, so we won't spend any time on it here except to state its purpose in the power supply.

Now, so that you can take a closer look at the power supply, let's take it out of the system and put it on the test bench.

ON THE TEST BENCH

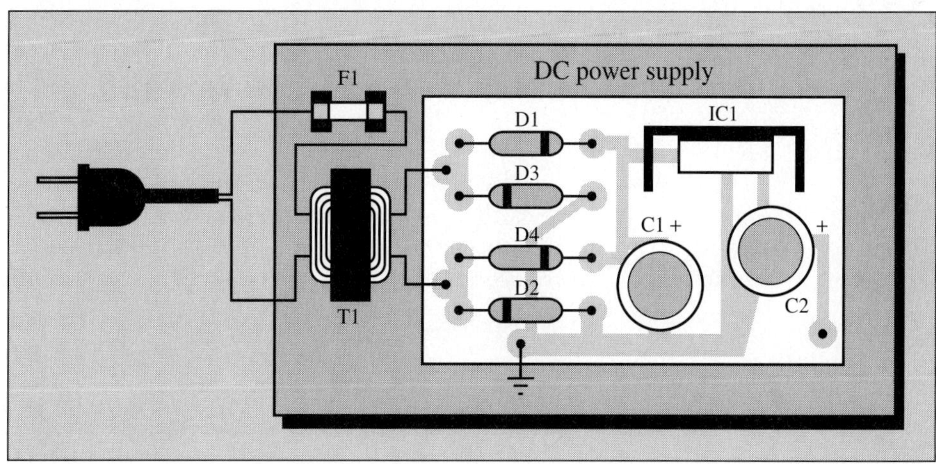

FIGURE 2–71

IDENTIFYING THE COMPONENTS

On the pc board shown in Figure 2–71, the diodes are labeled D1, D2, D3, and D4. The banded end of the diode is the cathode, as indicated in Figure 2–72(a). The filter capacitors are labeled C1 and C2. These are cylindrical electrolytic capacitors as shown in Figure 2–72(b). The 3-terminal voltage regulator is an **integrated circuit** (IC) device labeled IC1. The thick bracket-shaped object is a heat sink for conducting heat away from the device. A pictorial view is shown in Figure 2–72(c) to give you an idea of its shape. Again, we don't get into IC voltage regulators until later in the book. Also mounted in the power supply are the transformer (T1) and the fuse (F1).

FIGURE 2–72

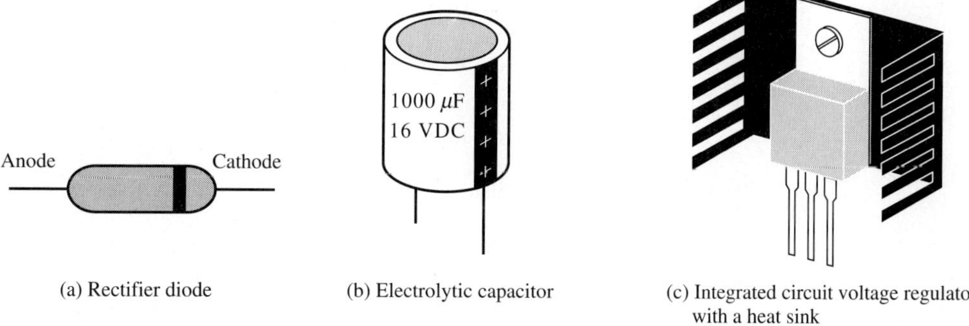

(a) Rectifier diode (b) Electrolytic capacitor (c) Integrated circuit voltage regulator with a heat sink

■ ACTIVITY 1 RELATE THE PC BOARD TO THE SCHEMATIC

Carefully follow the conductive traces on the pc board to see how the components are interconnected. Compare each point (*A* through *J*) on the pc board with the corresponding point on the schematic in Figure 2–73. This exercise will develop your skill in going from a pc board to a schematic or vice versa—*a very important skill for a technician*. Notice that some of the components on the pc board have been made to appear "transparent." This lets you see connections and traces under the components for the purposes of this exercise. Of course, on an actual pc board, you can't see through the components. For each point on the pc board, place the letter on the corresponding point or points on the schematic.

FIGURE 2–73

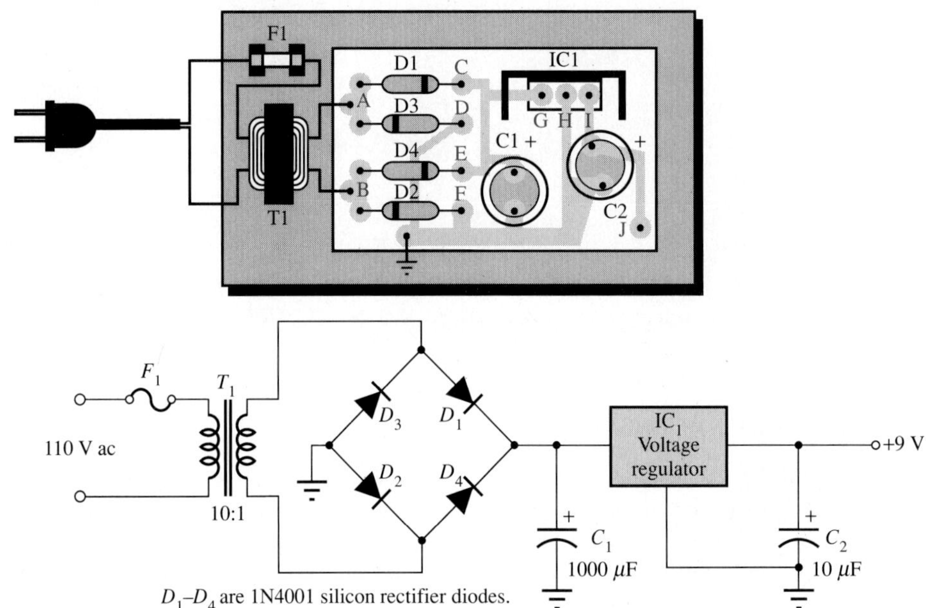

■ ACTIVITY 2 ANALYZE THE POWER SUPPLY CIRCUIT

With 110 V rms applied to the input, determine what the voltages should be at each point indicated (1, 2, 3, and 4) in Figure 2–74 using the oscilloscope or digital multimeter as indicated.

■ ACTIVITY 3 WRITE A TECHNICAL REPORT

Discuss the detailed operation from input to output including voltage values. Tell what each component does and why it is in the circuit. Since we have not studied voltage regulators yet, all you need to know at this point is that the voltage regulator (IC1) takes the filtered output voltage from the rectifier and produces a constant +9 V output under varying load conditions.

■ ACTIVITY 4 TROUBLESHOOT THE POWER SUPPLY FOR EACH OF THE FOLLOWING PROBLEMS BY STATING THE PROBABLE CAUSE OR CAUSES (REFER TO FIGURE 2–74)

1. A full-wave 120-Hz voltage with a peak value of about 14 V at point 2.
2. A large 120-Hz ripple voltage with a peak of about 14 V at point 2.
3. No voltage at point 1.

4. No voltage at point 2.
5. A half-wave 60-Hz voltage with a peak of about 14 V at point 2.
6. No voltage at point 4.

■ ACTIVITY 5 TEST BENCH SPECIAL ASSIGNMENT

Go to Test Bench 1 in the color insert section (which follows page 422) and carry out the assignment that is stated there.

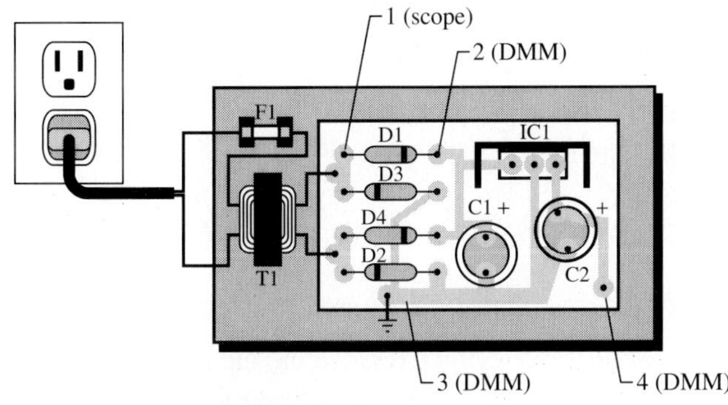

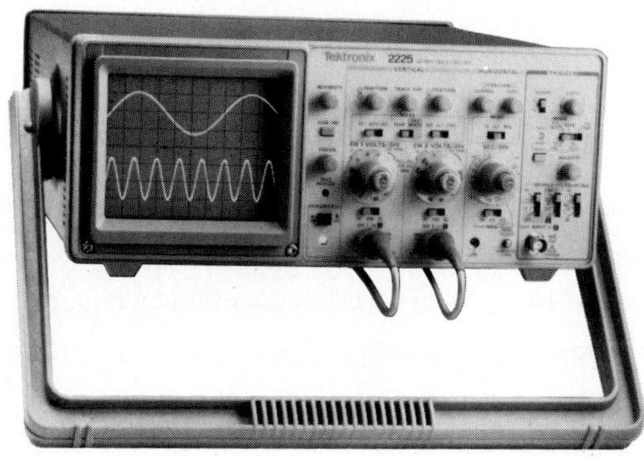

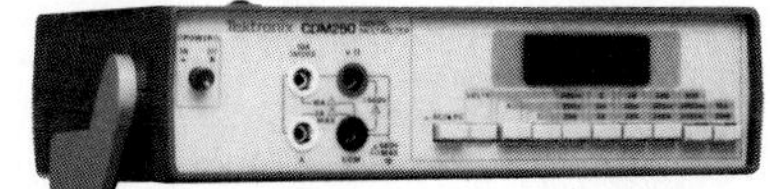

FIGURE 2–74
Test bench set-up. Photos courtesy of Tektronix, Inc.

2–9 REVIEW QUESTIONS

1. List the major parts that make up a basic dc power supply.
2. What is the purpose of a power supply in a system application?
3. What is the input voltage to a typical power supply?
4. From the data sheet for the 1N4001 diode in Appendix A, determine its PIV.
5. Are the diodes used in this power supply operating close to their PIV rating?
6. What is the purpose of capacitor C_1 in the power supply?
7. What is the purpose of capacitor C_2 in the power supply?
8. Why is the voltage regulator connected to a heat sink?
9. What do you think would happen if C_2 were removed?

SUMMARY

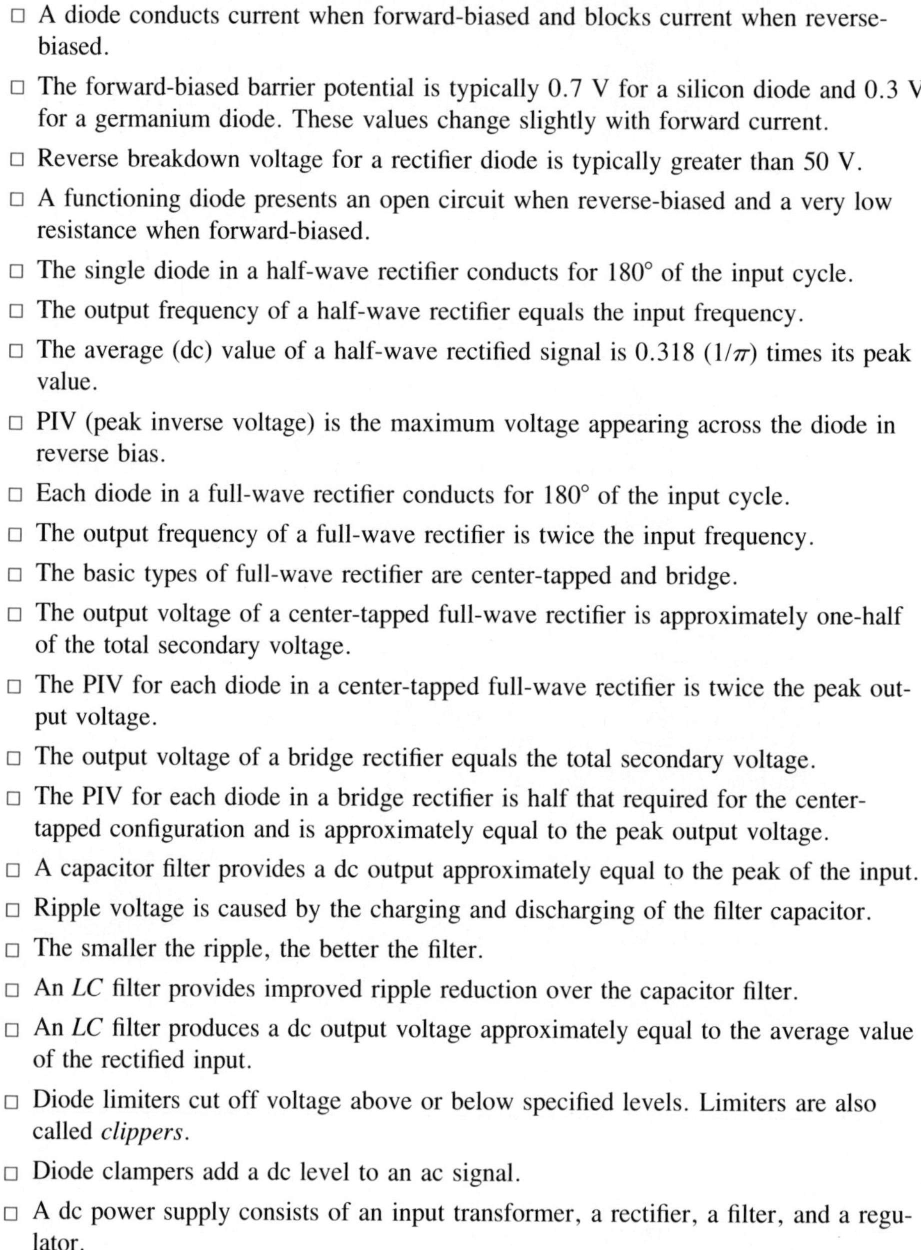

- □ A diode conducts current when forward-biased and blocks current when reverse-biased.
- □ The forward-biased barrier potential is typically 0.7 V for a silicon diode and 0.3 V for a germanium diode. These values change slightly with forward current.
- □ Reverse breakdown voltage for a rectifier diode is typically greater than 50 V.
- □ A functioning diode presents an open circuit when reverse-biased and a very low resistance when forward-biased.
- □ The single diode in a half-wave rectifier conducts for 180° of the input cycle.
- □ The output frequency of a half-wave rectifier equals the input frequency.
- □ The average (dc) value of a half-wave rectified signal is 0.318 ($1/\pi$) times its peak value.
- □ PIV (peak inverse voltage) is the maximum voltage appearing across the diode in reverse bias.
- □ Each diode in a full-wave rectifier conducts for 180° of the input cycle.
- □ The output frequency of a full-wave rectifier is twice the input frequency.
- □ The basic types of full-wave rectifier are center-tapped and bridge.
- □ The output voltage of a center-tapped full-wave rectifier is approximately one-half of the total secondary voltage.
- □ The PIV for each diode in a center-tapped full-wave rectifier is twice the peak output voltage.
- □ The output voltage of a bridge rectifier equals the total secondary voltage.
- □ The PIV for each diode in a bridge rectifier is half that required for the center-tapped configuration and is approximately equal to the peak output voltage.
- □ A capacitor filter provides a dc output approximately equal to the peak of the input.
- □ Ripple voltage is caused by the charging and discharging of the filter capacitor.
- □ The smaller the ripple, the better the filter.
- □ An *LC* filter provides improved ripple reduction over the capacitor filter.
- □ An *LC* filter produces a dc output voltage approximately equal to the average value of the rectified input.
- □ Diode limiters cut off voltage above or below specified levels. Limiters are also called *clippers*.
- □ Diode clampers add a dc level to an ac signal.
- □ A dc power supply consists of an input transformer, a rectifier, a filter, and a regulator.

GLOSSARY

Bridge rectifier A type of full-wave rectifier using four diodes arranged in a bridge configuration.

Center-tapped rectifier A type of full-wave rectifier using a center-tapped transformer and two diodes.

Characteristic curve A graph showing current versus voltage in a diode or other device.

Clamper A circuit that adds a dc level to an ac voltage using a diode and a capacitor.

Clipper See *Limiter*.

Diode drop The voltage across the diode when it is forward-biased. Approximately the same as the barrier potential.

Filter A capacitor or combination of capacitor and inductor used to reduce the variation of the output voltage from a rectifier.

Full-wave rectifier A circuit that converts an ac sine wave input voltage into a pulsating dc voltage with two pulses occurring for each input cycle.

Fuse A protective device that opens when the current exceeds a rated limit.

Half-wave rectifier A circuit that converts an ac sine wave input voltage into a pulsating dc voltage with one pulse occurring for each input cycle.

Integrated circuit (IC) A type of circuit in which all the components are constructed on a single tiny chip of silicon.

Limiter A diode circuit that clips off or removes part of a waveform above and/or below a specified level.

Multiplier A circuit using diodes and capacitors that increases the input voltage by two, three, or four times.

Power supply The circuit that supplies the proper dc voltage and current to operate a system.

Regulator An electronic device or circuit that maintains an essentially constant output voltage for a range of input voltage or load values; one part of a power supply.

Ripple factor A measure of effectiveness of a power supply filter in reducing the ripple voltage.

Ripple voltage The small variation in the dc output voltage of a filtered rectifier caused by the charging and discharging of the filter capacitor.

Troubleshooting The process and technique of identifying and locating faults in an electronic circuit or system.

FORMULAS

(2–1)	$V_{AVG} = \dfrac{V_p}{\pi}$	Half-wave average value
(2–2)	$V_{p(out)} = V_{p(in)} - 0.7\text{ V}$	Peak half-wave rectifier output (silicon)
(2–3)	$V_{p(out)} = V_{p(in)} - 0.3\text{ V}$	Peak half-wave rectifier output (germanium)

(2–4)	$V_2 = \left(\frac{N_2}{N_1}\right)V_1$	Secondary voltage
(2–5)	$V_{AVG} = \frac{2V_p}{\pi}$	Full-wave average value
(2–6)	$V_{out} = \frac{V_2}{2}$	Center-tapped full-wave output
(2–7)	$V_{out} = \frac{V_2}{2} - V_B$	Center-tapped full-wave output (including diode drop, V_B)
(2–8)	$PIV = 2V_{p(out)} + V_B$	Diode peak inverse voltage, center-tapped rectifier
(2–9)	$V_{out} = V_2$	Bridge full-wave output (ideal)
(2–10)	$V_{out} = V_2 - 2V_B$	Bridge full-wave output (including diode drops)
(2–11)	$PIV = V_{p(out)} + V_B$	Diode peak inverse voltage, bridge rectifier
(2–12)	$r = \frac{V_r}{V_{dc}}$	Ripple factor
(2–13)	$V_{dc} = \left(1 - \frac{0.00417}{R_L C}\right)V_{p(in)}$	DC output voltage, capacitor filter
(2–14)	$V_r = \frac{0.0024}{R_L C}V_{p(in)}$	Ripple voltage, capacitor filter
(2–15)	$V_{r(out)} = \left(\frac{X_C}{\lvert X_L - X_C \rvert}\right)V_{r(in)}$	Ripple output, *LC* filter
(2–16)	$V_{dc(out)} = \left(\frac{R_L}{R_W + R_L}\right)V_{dc(in)}$	DC output, *LC* filter

SELF-TEST

1. When forward-biased, a diode
   (a) blocks current **(b)** conducts current
   (c) has a high resistance **(d)** drops a large voltage
2. When a voltmeter is placed across a forward-biased diode, it will read approximately
   (a) the bias battery voltage **(b)** 0 V
   (c) the diode barrier potential **(d)** the total circuit voltage

3. A silicon diode is in series with a 1 kΩ resistor and a 5-V battery. If the anode is connected to the positive source terminal, the cathode voltage with respect to the negative source terminal is

   (a) 0.7 V **(b)** 0.3 V **(c)** 5.7 V **(d)** 4.3 V

4. The positive lead of an analog ohmmeter is connected to the anode of a diode and the negative lead is connected to the cathode. If the needle is at the high end of the ohms scale, the diode is

   (a) shorted **(b)** open **(c)** good **(d)** faulty **(e)** b and d

5. The average value of a half-wave rectified voltage with a peak value of 200 V is

   (a) 63.66 V **(b)** 127.33 V **(c)** 141.4 V **(d)** 0 V

6. When a 60-Hz sine wave is applied to the input of a half-wave rectifier, the output frequency is

   (a) 120 Hz **(b)** 30 Hz **(c)** 60 Hz **(d)** 0 Hz

7. The peak value of the input to a half-wave rectifier is 10 V. The approximate peak value of the output is

   (a) 10 V **(b)** 3.18 V **(c)** 10.7 V **(d)** 9.3 V

8. For the circuit in Question 7, the diode must be able to withstand a reverse voltage of

   (a) 10 V **(b)** 5 V **(c)** 20 V **(d)** 3.18 V

9. The average value of a full-wave rectified voltage with a peak value of 75 V is

   (a) 53 V **(b)** 47.75 V **(c)** 37.5 V **(d)** 23.87 V

10. When a 60-Hz sine wave is applied to the input of a full-wave rectifier, the output frequency is

    (a) 120 Hz **(b)** 60 Hz **(c)** 240 Hz **(d)** 0 Hz

11. The total secondary voltage in a center-tapped full-wave rectifier is 125 V rms. Neglecting the diode drop, the rms output voltage is

    (a) 125 V **(b)** 176.75 V **(c)** 100 V **(d)** 62.5 V

12. When the peak output voltage is 100 V, the PIV for each diode in a center-tapped full-wave rectifier is (neglecting the diode drop)

    (a) 100 V **(b)** 200 V **(c)** 141.14 V **(d)** 50 V

13. When the rms output voltage of a full-wave bridge rectifier is 20 V, the peak inverse voltage across the diodes is (neglecting the diode drop)

    (a) 20 V **(b)** 40 V **(c)** 28.28 V **(d)** 56.56 V

14. The ideal dc output voltage of a capacitor filter is equal to

    (a) the peak value of the rectified voltage

    (b) the average value of the rectified voltage

    (c) the rms value of the rectified voltage

15. A certain power-supply filter produces an output with a ripple of 100 mV and a dc value of 20 V. The ripple factor is

    (a) 0.05 **(b)** 0.0005 **(c)** 0.00005 **(d)** 0.02

16. A 60-V peak full-wave rectified voltage is applied to a capacitor filter. If R_L = 10 kΩ and C = 10 μF, the ripple voltage is

    (a) 0.6 V (b) 6 mV (c) 1.44 V (d) 2.88 V

17. If the load resistance of a capacitor-filtered full-wave rectifier is reduced, the ripple voltage

    (a) increases (b) decreases

    (c) is not affected (d) has a different frequency

18. A 10-V peak-to-peak sine wave voltage is applied across a silicon diode and series resistor. The peak-to-peak voltage across the diode is

    (a) 9.3 V (b) 4.3 V (c) 0.7 V (d) 5.7 V

19. If the input voltage to a voltage tripler has an rms value of 12 V, the dc output voltage is approximately

    (a) 36 V (b) 50.9 V (c) 33.9 V (d) 32.4 V

20. If one of the diodes in a full-wave bridge rectifier opens, the output is

    (a) 0 V (b) one-fourth the amplitude of the input voltage

    (c) a half-wave voltage (d) a 120-Hz voltage

21. If you are checking a 60-Hz full-wave bridge rectifier and observe that the output has a 60 Hz ripple,

    (a) the circuit is working properly (b) there is an open diode

    (c) the transformer secondary is shorted (d) the filter capacitor is leaky

PROBLEMS

SECTION 2–1 RECTIFIER DIODES

1. Determine whether each diode in Figure 2–75 is forward- or reverse-biased.

FIGURE 2–75

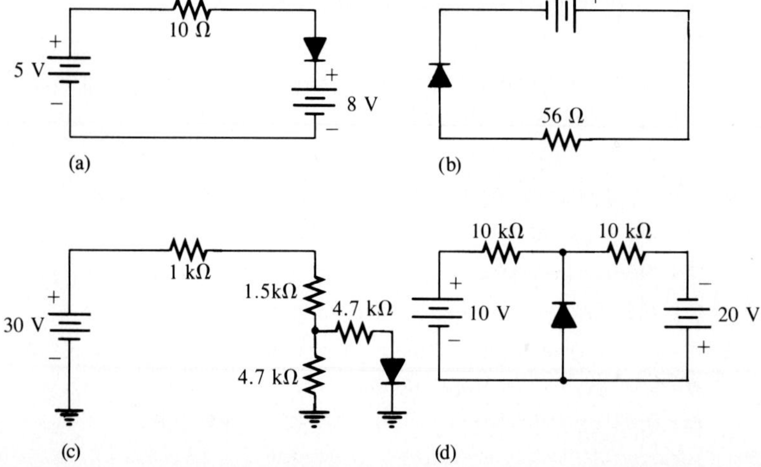

2. Determine the voltage across each diode in Figure 2–75, assuming that they are germanium diodes with a reverse resistance of 50 MΩ.
3. Consider the meter indications in each circuit of Figure 2–76, and determine whether the diode is functioning properly, or whether it is open or shorted.
4. Determine the voltage with respect to ground at each point in Figure 2–77. (The diodes are silicon.)

FIGURE 2–76

FIGURE 2–77

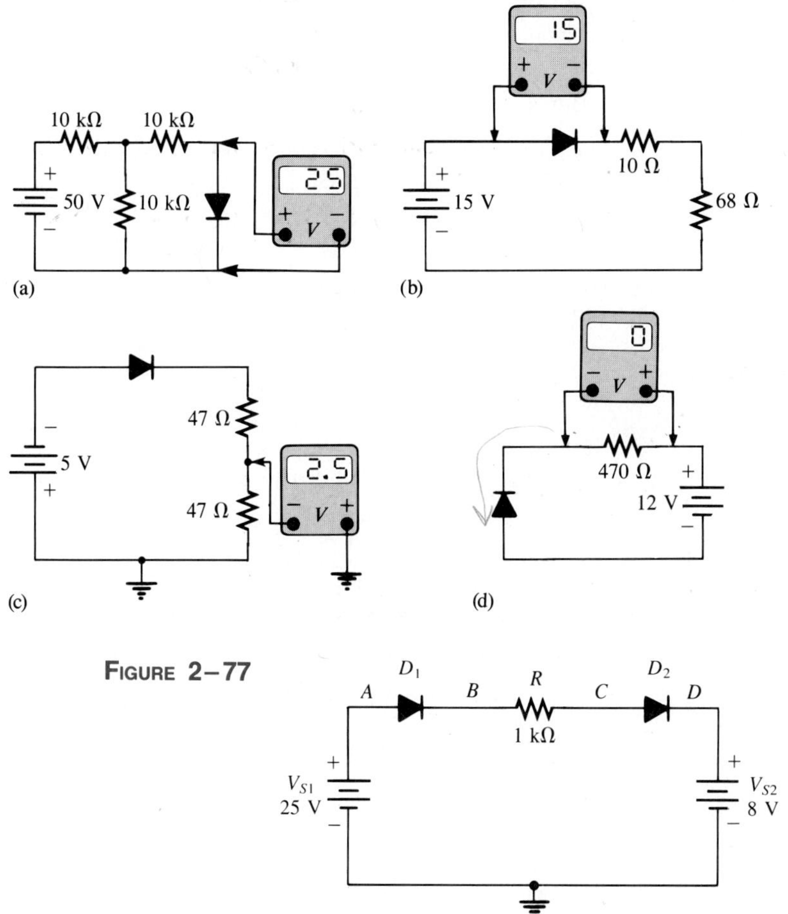

SECTION 2–2 HALF-WAVE RECTIFIERS

5. Draw the output waveform for each circuit in Figure 2–78.

FIGURE 2–78

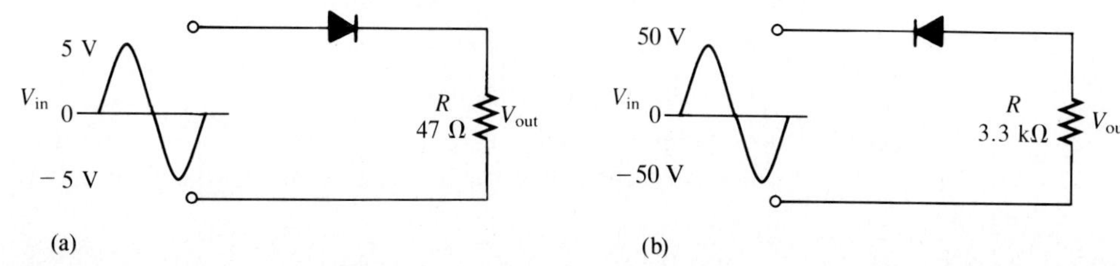

6. What is the peak forward current through each diode in Figure 2–78?
7. A power-supply transformer has a turns ratio of 5:1. What is the secondary voltage if the primary is connected to a 115 V rms source?
8. Determine the peak and average power delivered to R_L in Figure 2–79.

FIGURE 2–79

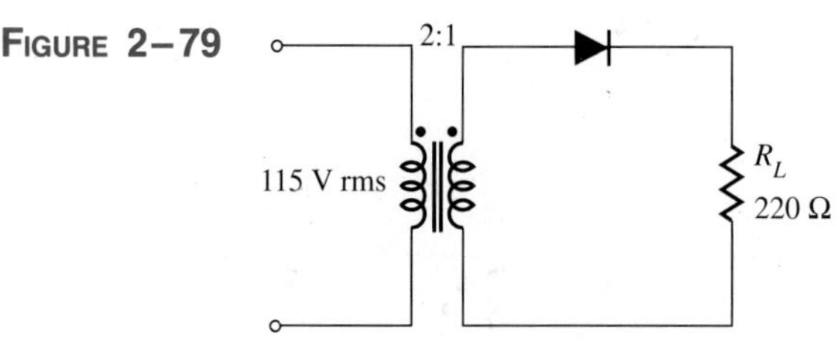

SECTION 2–3 FULL-WAVE RECTIFIERS

9. Find the average value of each voltage in Figure 2–80.

FIGURE 2–80

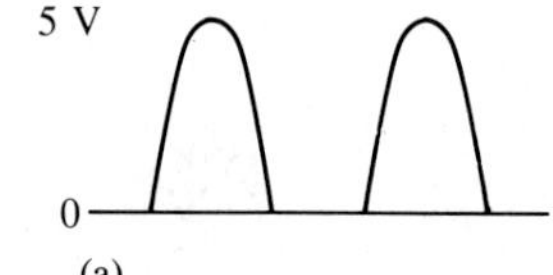

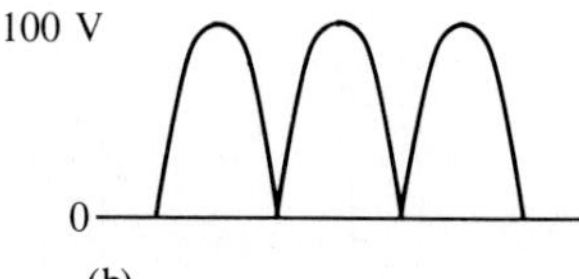

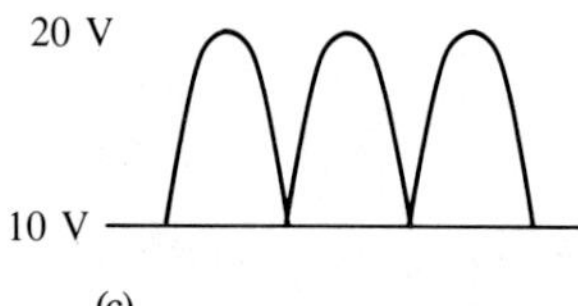

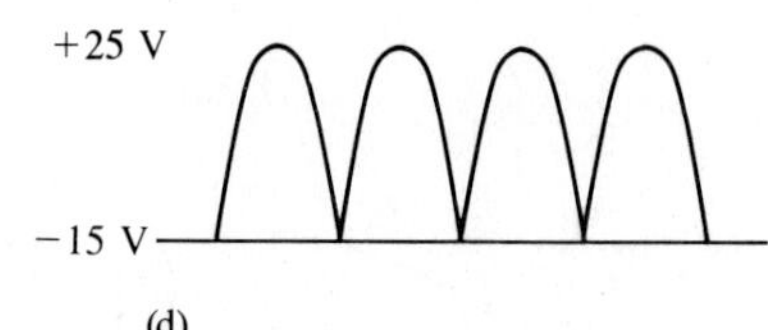

10. Consider the circuit in Figure 2–81.
    (a) What type of circuit is this?
    (b) What is the total peak secondary voltage?
    (c) Find the peak voltage across each half of the secondary.
    (d) Sketch the voltage waveform across R_L.
    (e) What is the peak current through each diode?
    (f) What is the PIV for each diode?
11. Calculate the peak voltage rating of each half of a center-tapped transformer used in a full-wave rectifier that has an average output voltage of 110 V.

12. Show how to connect the diodes in a center-tapped rectifier in order to produce a negative-going full-wave voltage across the load resistor.
13. What PIV rating is required for the diodes in a bridge rectifier that produces an average output voltage of 50 V?
14. The rms output voltage of a bridge rectifier is 20 V. What is the peak inverse voltage across the diodes?
15. Sketch the output voltage of the bridge rectifier in Figure 2–82. Be careful here.

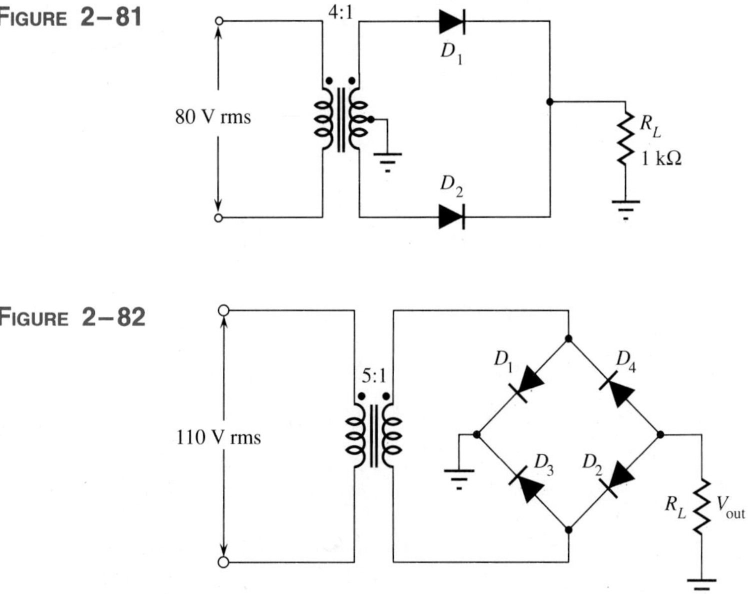

FIGURE 2–81

FIGURE 2–82

SECTION 2–4 RECTIFIER FILTERS

16. A certain rectifier filter produces a dc output voltage of 75 V with an rms ripple of 0.5 V. Calculate the ripple factor.
17. A certain full-wave rectifier has a peak output voltage of 30 V. A 50 μF capacitor filter is connected to the rectifier. Calculate the rms ripple and the dc output voltage developed across a 600 Ω load resistance.
18. What is the percentage of ripple in Problem 17?
19. What value of filter capacitor is required to produce a 1% ripple for a full-wave rectifier having a load resistance of 1.5 kΩ?
20. A full-wave rectifier produces an 80 V peak rectified voltage from a 60 Hz ac source. If a 10 μF filter capacitor is used, determine the ripple factor for a 100 mA peak load current.
21. Determine the ripple and dc output voltages in Figure 2–83. The line voltage has a frequency of 60 Hz, and the winding resistance of the coil is 100 Ω.

22. Refer to Figure 2–83 and sketch the following voltage waveforms in relationship to the input waveforms: V_{AB}, V_{AD}, V_{BD}, and V_{CD}.

FIGURE 2–83

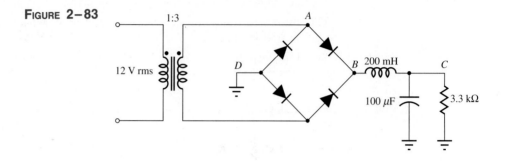

SECTION 2–5 DIODE LIMITING AND CLAMPING CIRCUITS

23. Determine the output waveform for the circuit of Figure 2–84.

FIGURE 2–84

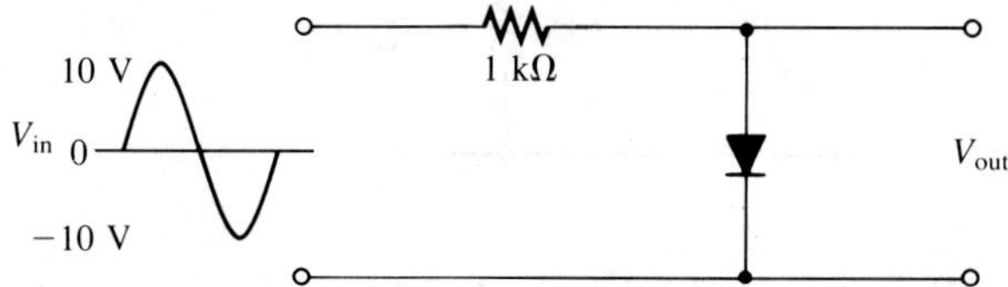

24. Determine the output of the circuit in Figure 2–85(a) for each input in (b), (c), and (d).

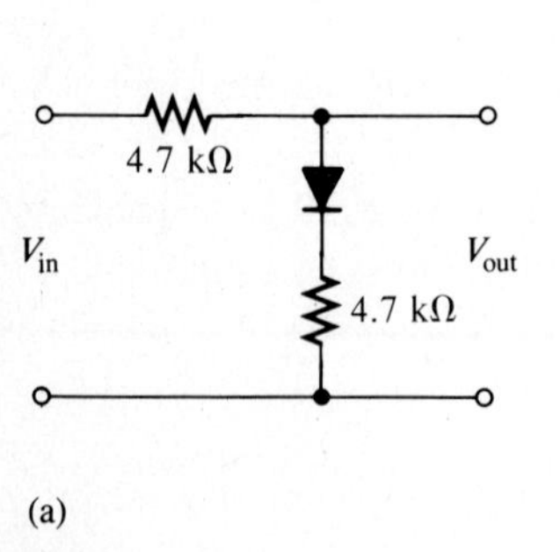

(a)

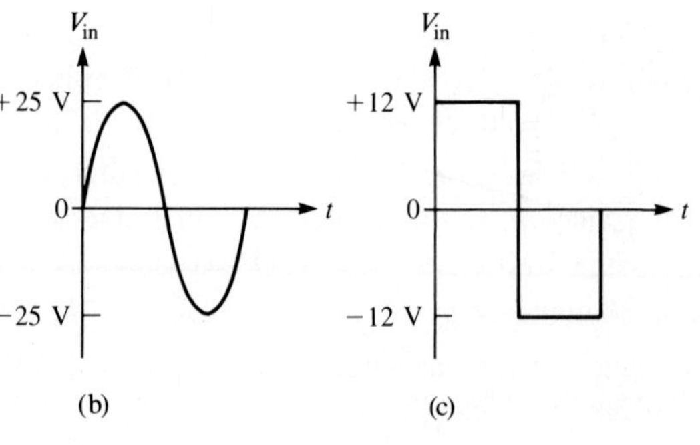

(b) (c)

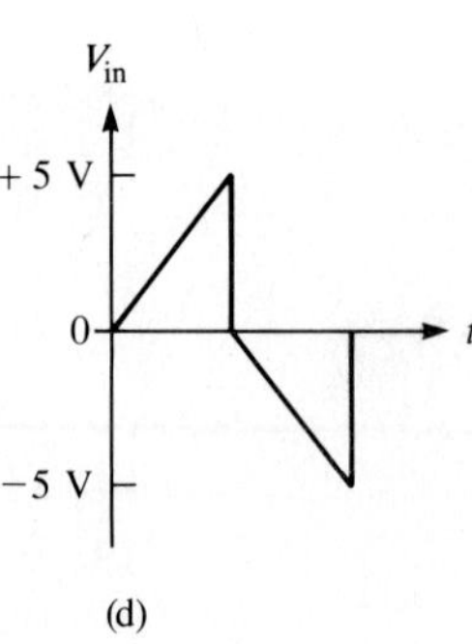

(d)

FIGURE 2–85

25. Determine the output waveform for each circuit in Figure 2–86.

FIGURE 2–86

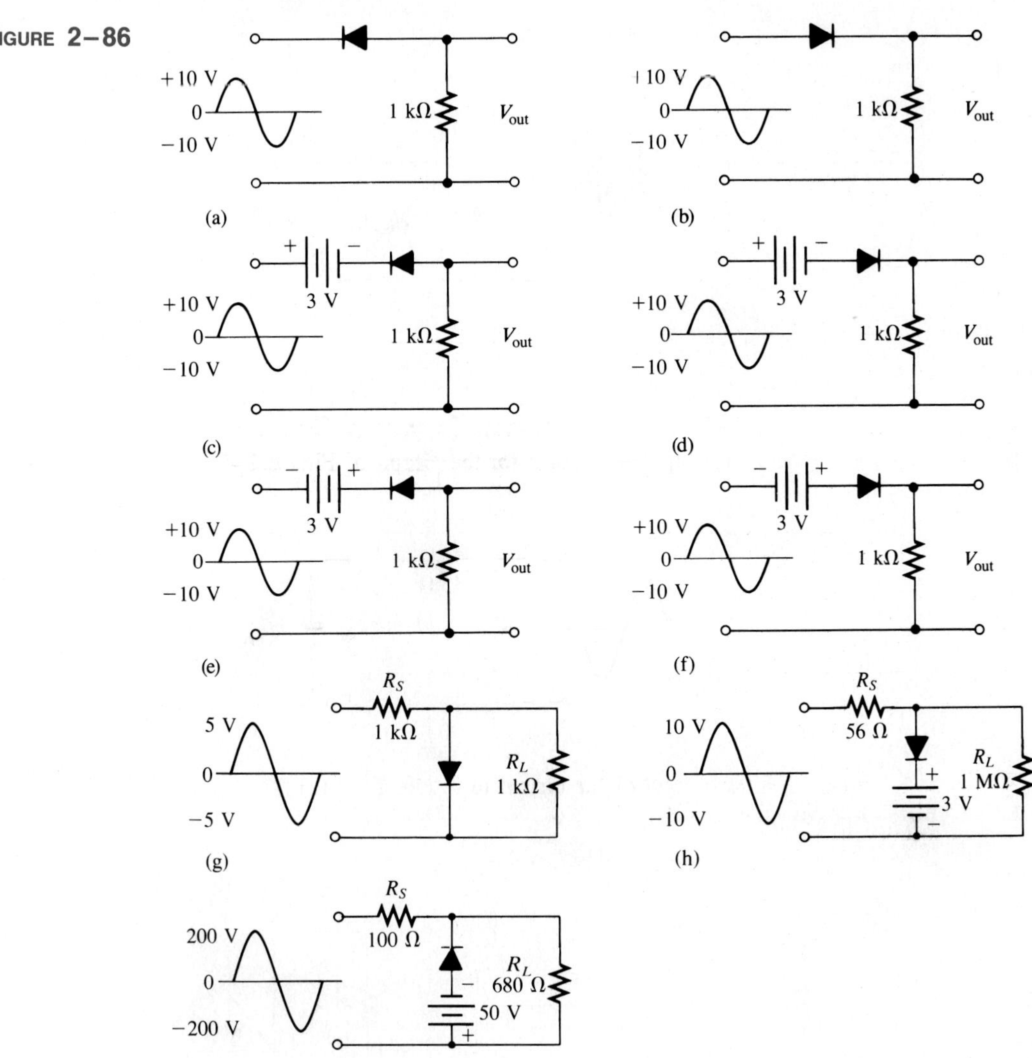

26. Sketch the output waveforms for each circuit in Figure 2–87.

FIGURE 2–87

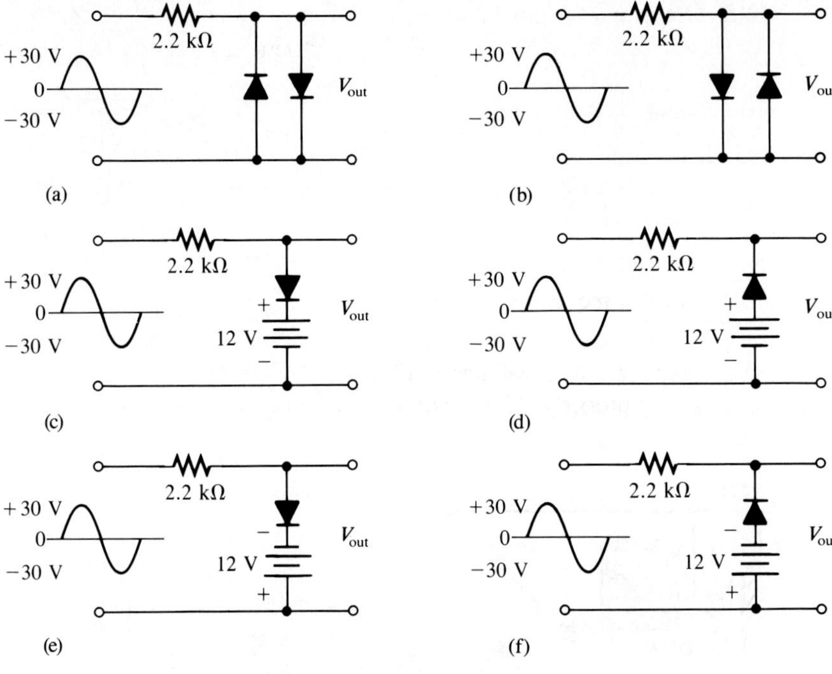

FIGURE 2–88

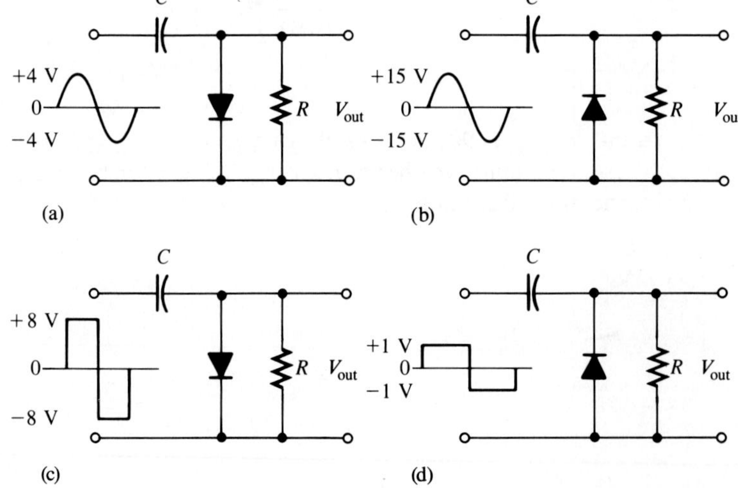

27. Describe the output waveform of each circuit in Figure 2–88. Assume the RC time constant is much greater than the period of the input.

28. Repeat Problem 27 with the diodes turned around.

SECTION 2–6 VOLTAGE MULTIPLIERS

29. A certain voltage doubler has 20 V rms on its input. What is the output voltage? Sketch the circuit, indicating the output terminals and PIV rating for the diode.

30. Repeat Problem 29 for a voltage tripler and quadrupler.

Section 2–7 The Diode Data Sheet

31. From the data sheet in Appendix A, determine how much peak inverse voltage that a 1N1183A diode can withstand.

32. Repeat Problem 31 for a 1N1188A.

33. If the peak output voltage of a full-wave bridge rectifier is 50 V, determine the minimum value of the surge limiting resistor required when 1N1183A diodes are used.

Section 2–8 Troubleshooting

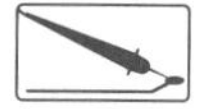

34. If one of the diodes in a bridge rectifier opens, what happens to the output?

35. From the meter readings in Figure 2–89 determine if the rectifier circuit is functioning properly. If it is not, determine the most likely failure(s).

Figure 2–89

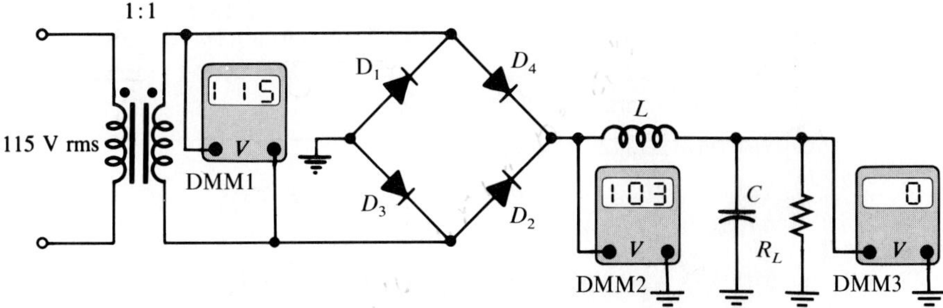

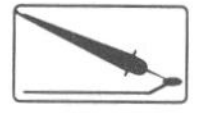

36. Each part of Figure 2–90 shows oscilloscope displays of rectifier output voltages. In each case, determine whether or not the rectifier is functioning properly and if it is not, the most likely failure(s).

Figure 2–90

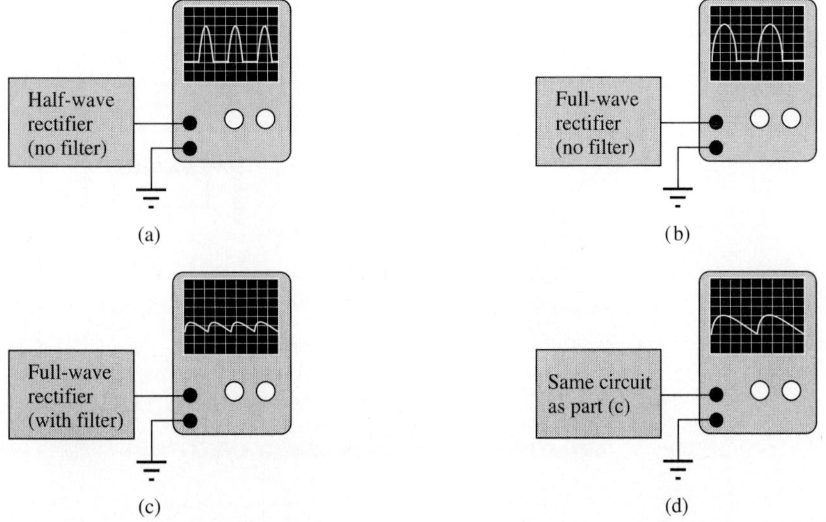

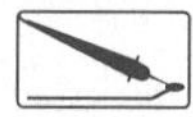

37. Based on the values given, would you expect the circuit in Figure 2–91 to fail? Why?

FIGURE 2–91

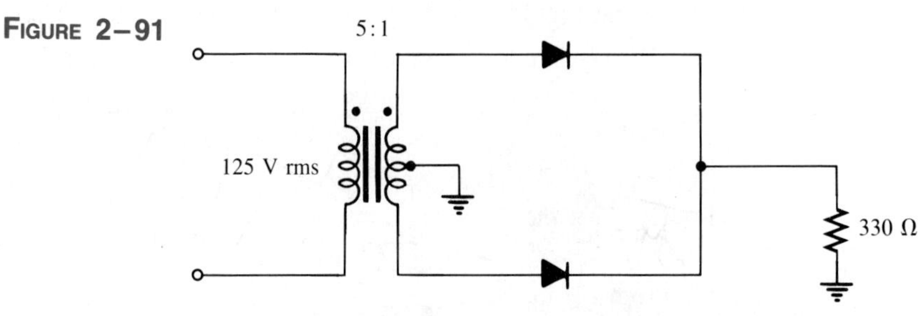

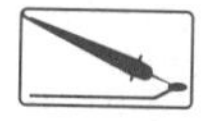

38. Determine the most likely failure in the circuit board of Figure 2–92 for each of the following symptoms. State the corrective action you would take in each case. The transformer has a turns ratio of 1.

(a) No voltage at point 1.

(b) No voltage at point 2. 110 V rms at point 1.

(c) No voltage at point 3. 110 V rms at point 1.

(d) 150 V rms at point 2. Input is correct at 110 V rms.

(e) 68 V rms at point 3. Input is correct at 110 V rms.

(f) A pulsating full-wave rectified voltage with a peak of 155.5 V at point 4.

(g) Excessive 120-Hz ripple voltage at point 5.

(h) Ripple voltage has a frequency of 60 Hz at point 4.

(i) No voltage at point 6.

FIGURE 2–92

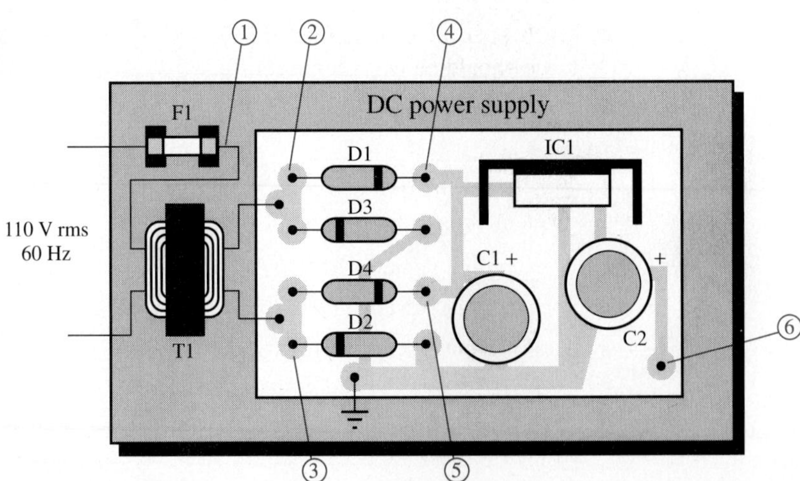

39. In testing the power supply in Figure 2–93, you found the voltage at the positive side of C_1 to have a 60-Hz ripple voltage with a greater than normal amplitude. Just to be sure, you replaced all of the diodes with known good ones. You check the point again to verify proper operation and it *still* has the 60-Hz ripple voltage. What now?

FIGURE 2–93

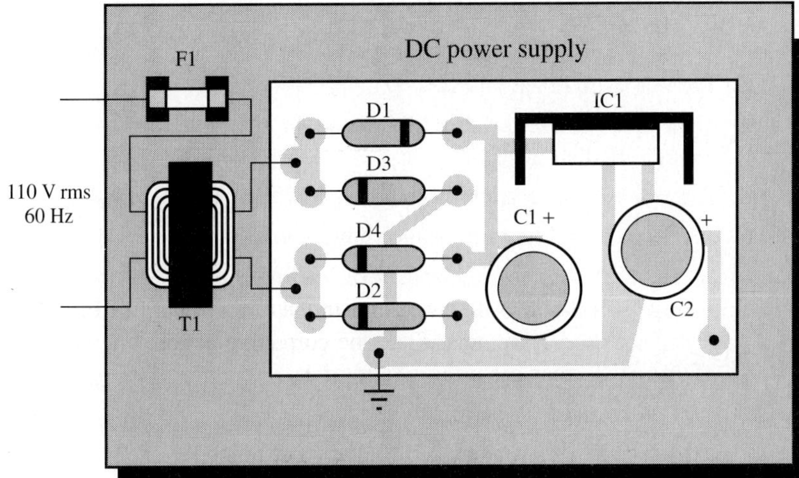

ANSWERS TO REVIEW QUESTIONS

SECTION 2–1

1. The diode is operated in forward bias or in reverse bias.
2. The rectifier diode is never intentionally operated in reverse breakdown.
3. In the simplest approximation, the diode is treated as a closed switch in forward bias and an open switch in reverse bias.
4. For best accuracy, the diode barrier voltage, forward resistance, and reverse resistance should be considered.
5. The barrier potential model will normally be used in this book.

SECTION 2–2

1. PIV across the diode occurs at the peak of the input.
2. There is current through the load for approximately half (50%) of the input cycle.
3. The average value is $10 \text{ V}/\pi = 3.18 \text{ V}$.
4. The peak output voltage is $25 \text{ V} - V_B$.
5. The PIV must be at least 50 V.

SECTION 2–3

1. A full-wave voltage occurs on each half of the input cycle and has a frequency of twice the input frequency. A half-wave voltage occurs once each input cycle and has a frequency equal to the input frequency.

2. The average value of $2(60\text{ V})/\pi = 38.12\text{ V}$.
3. The bridge rectifier has the greater output voltage.
4. The 50 V diodes must be used in the bridge rectifier.
5. In the center-tapped rectifier, diodes with a PIV rating of at least 90 V would be required.

SECTION 2–4

1. The output frequency is 60 Hz.
2. The output frequency is 120 Hz.
3. The ripple voltage is caused by the slight charging and discharging of the capacitor through the load resistor.
4. The ripple voltage amplitude increases when the load resistance decreases.
5. Ripple factor is the ratio of the ripple voltage to the average or dc voltage.
6. *LC* ripple voltage is independent of load resistance. *LC* output voltage equals the average value of the filter input rather than the peak value.

SECTION 2–5

1. Limiters clip off or remove portions of a waveform. Clampers insert a dc level.
2. A positive limiter clips off positive voltages. A negative limiter clips off negative voltages.
3. 0.7 V appears across the diode.
4. The bias voltage must be $5\text{ V} - 0.7\text{ V} = 4.3\text{ V}$.
5. The capacitor acts as a battery.

SECTION 2–6

1. The peak voltage rating must be 100 V.
2. The PIV rating must be at least 311 V.

SECTION 2–7

1. The three rating categories on a diode data sheet are maximum ratings, electrical characteristics, and mechanical data.
2. **(a)** V_F is forward voltage. **(b)** I_R is reverse current. **(c)** I_0 is peak average forward current.

SECTION 2–8

1. An open diode results in no voltage.
2. An open diode produces a half-wave output.
3. The diode will burn open. Transformer will be damaged. Fuse will blow.
4. The amplitude of the ripple voltage increases.
5. There will be no output voltage.
6. The problem might be a partially shorted secondary winding.

SECTION 2–9

1. The transformer, rectifier, filter, and regulator are the major parts of a basic dc power supply.

2. The power supply provides all parts of the system with dc voltage and current to operate the circuits.
3. The input voltage to a typical power supply is 110 V rms (portable power supplies use a battery).
4. The PIV for the 1N4001 diode is 50 V, designated V_{RRM}.
5. No, they experience approximately 15 V PIV and they are rated at 50 V.
6. C_1 filters (smooths) the full-wave output of the rectifier.
7. C_2 provides additional filtering on the regulator output.
8. The voltage regulator is connected to a heat sink to dissipate the heat that it produces.
9. The dc output voltage would not be affected. C_2 removes any noise or transient voltages that may get on the line.

ANSWERS TO PRACTICE EXERCISES

2–1 3.82 V

2–2 **(a)** 2.3 V, 23.3%
(b) 209.3 V, 0.33%; the greater the voltage, the less error

2–3 **(a)** 99.3 V **(b)** 100 V
(c) negative half-cycles rather than positive half cycles

2–4 98.68 V, DC VOLTS function

2–5 320.7 V including diode drop

2–6 78.6 V, 79.3 V

2–7 0.297 V

2–8 $V_{dc} = 31.6$ V, $V_r = 0.96$ V, $r = 0.03$

2–9 A positive peak of 8.72 V and clipped at −0.3 V

2–10 Clipped at +10.7 V and −10.7 V

2–11 Same voltage waveform as Figure 2–56

3

Special Diodes

After completing this chapter, you should be able to

- □ Explain what the zener diode characteristic curve indicates.
- □ Discuss various zener characteristics.
- □ Use a zener diode for voltage regulation.
- □ Define *line regulation* and *load regulation* and explain the difference.
- □ Show how zener diodes can be used as limiters or clippers.
- □ Explain what a varactor is and how it is used.
- □ Discuss the properties of light-emitting diodes.
- □ Discuss the properties of photodiodes.
- □ Describe the features of Schottky, tunnel, PIN, step-recovery, and laser diodes.
- □ Troubleshoot problems in diode circuits.

The last chapter was devoted to general-purpose and rectifier diodes, which are the most widely used types. In this chapter, we will cover several other types of diodes that are designed for specific applications, including the zener, varactor (variable-capacitance), light-emitting, photodiode, Schottky, tunnel, PIN, step-recovery, and laser diodes.

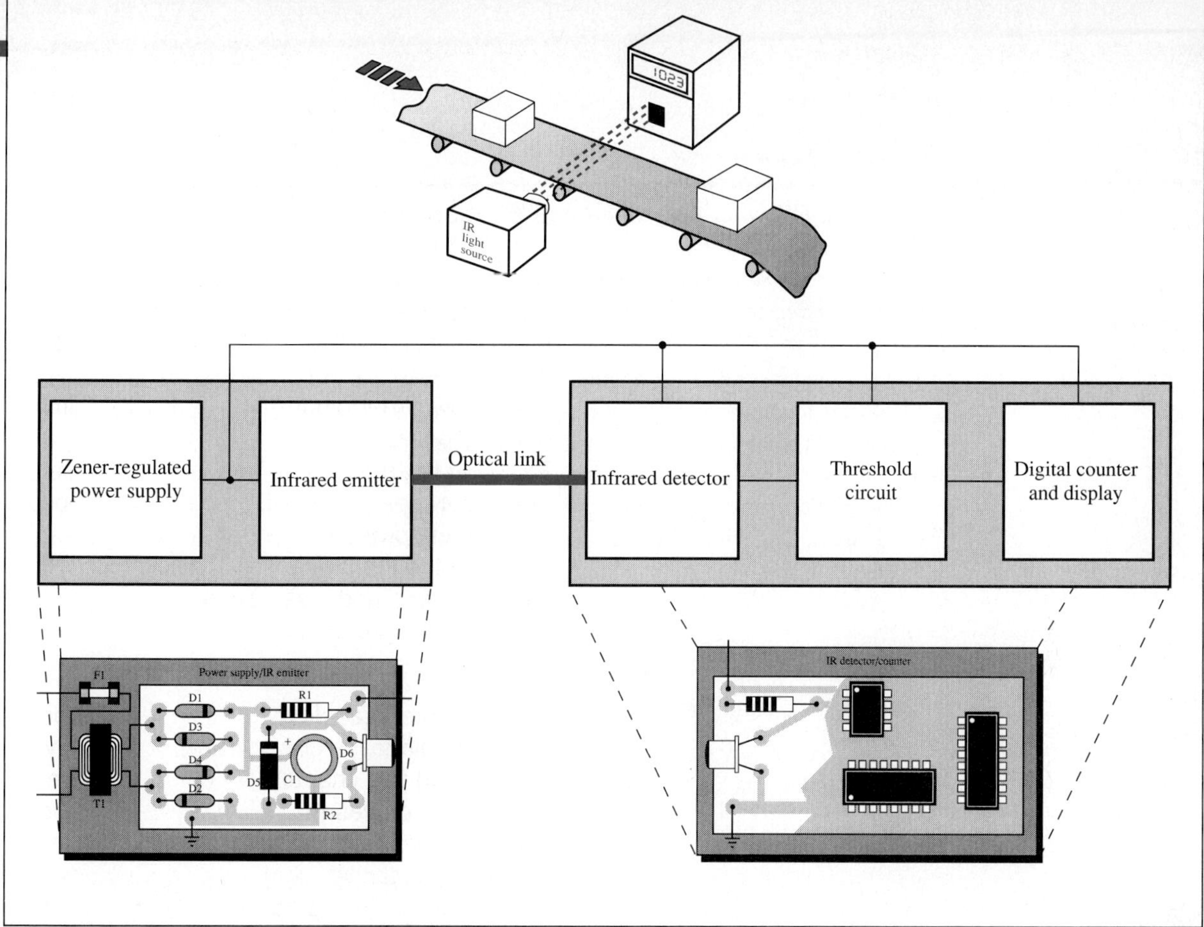

A System Application

The zener diode is sometimes used in dc power supplies to provide simple voltage regulation instead of more sophisticated regulating devices such as the three-terminal regulator that was mentioned in the last chapter's system application. A zener-regulated dc power supply is used in the system shown on this page. In addition to the use of a zener diode in the system application, you will see how two optical devices, the light-emitting diode (LED) and the photodiode, are also used in this system.

The system that we will deal with in this chapter to illustrate device applications is a simple industrial system used to count objects on a conveyor for purposes such as control and inventory. The basic system concept, along with a block diagram and physical components, is shown above.

For the system application in Section 3–7, in addition to the other topics, be sure you understand

- ☐ How a zener diode works.
- ☐ The parameters and ratings of zener diodes.
- ☐ The principles of LEDs (light-emitting diodes).
- ☐ The principles of photodiodes.
- ☐ How to read a data sheet.

3–1 ZENER DIODES

A major application for zener diodes is voltage regulation in dc power supplies. In this section, you will see how the zener maintains a nearly constant dc voltage under the proper operating conditions. You will learn the conditions and limitations for properly using the zener diode and the factors that affect its performance.

FIGURE 3–1
Zener diode symbol.

The schematic symbol for a zener diode is shown in Figure 3–1. The **zener diode** is a silicon pn junction device that differs from the rectifier diode in that it is designed for operation in the reverse breakdown region. The breakdown voltage of a zener diode is set by carefully controlling the doping level during manufacture. Recall, from the discussion of the diode characteristic curve in the last chapter, that when a diode reaches reverse breakdown, its voltage remains almost constant even though the current may change drastically. This volt-ampere characteristic is shown again in Figure 3–2.

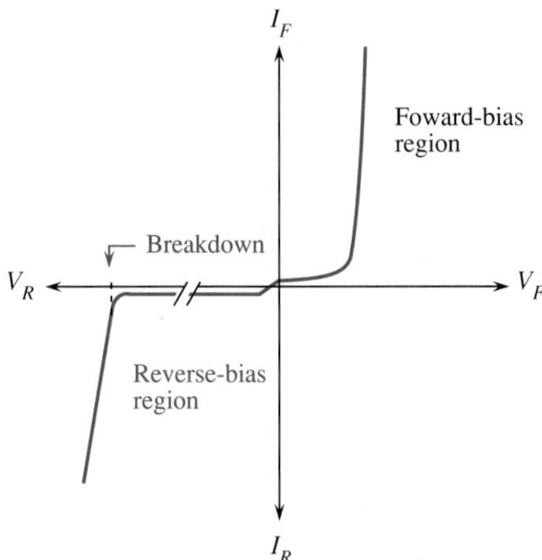

FIGURE 3–2
General diode characteristic.

ZENER BREAKDOWN

There are two types of reverse breakdown in a zener diode. One is the avalanche breakdown that was discussed in Chapter 1; this also occurs in rectifier diodes at a sufficiently high reverse voltage. The other type is zener breakdown, which occurs in a zener diode at low reverse voltages. A zener diode is heavily doped to reduce the breakdown voltage. This causes a very narrow depletion layer. As a result, an intense electric field exists within the depletion layer. Near the breakdown voltage (V_Z), the field is intense enough to pull electrons from their valence bands and create current.

Zener diodes with breakdown voltages of less than approximately 5 V operate predominately in zener breakdown. Those with breakdown voltages greater than approximately 5 V operate predominately in avalanche breakdown. Both types, however, are called *zener diodes*. Zeners are commercially available with breakdown voltages of 1.8 V to 200 V.

BREAKDOWN CHARACTERISTICS

Figure 3–3 shows the reverse portion of a zener diode's characteristic curve. Notice that as the reverse voltage (V_Z) is increased, the reverse current (I_Z) remains extremely small up to the "knee" of the curve. At this point, the breakdown effect begins; the internal zener resistance (R_Z) begins to decrease as the reverse current increases rapidly. From the bottom of the knee, the zener breakdown voltage (V_Z) remains essentially constant although it increases slightly as I_Z increases. This regulating ability is the key feature of the zener diode. *A zener diode maintains a nearly constant voltage across its terminals over a specified range of reverse-current values.*

FIGURE 3–3
Reverse characteristic of a zener diode. V_Z is usually specified at the zener test current, I_{ZT}, and is designated V_{ZT}.

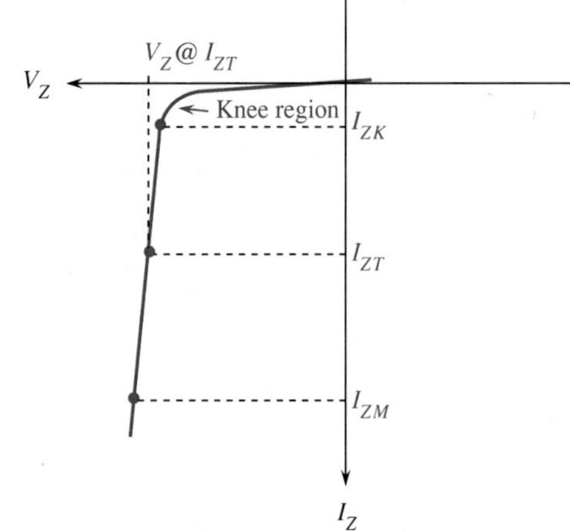

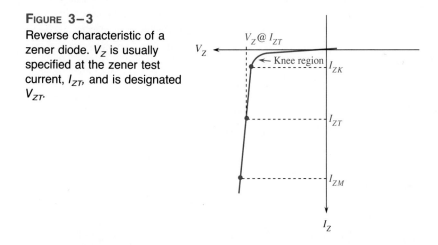

A minimum value of reverse current, I_{ZK}, must be maintained in order to keep the diode in regulation. You can see on the curve that when the reverse current is reduced below the knee of the curve, the voltage changes drastically and regulation is lost. Also, there is a maximum current, I_{ZM}, above which the diode may be damaged. So, basically, the zener diode maintains a nearly constant voltage across its terminals for values of reverse current ranging from I_{ZK} to I_{ZM}. A nominal zener voltage, V_{ZT}, is usually specified on a data sheet at a value of reverse current called the *zener test current,* I_{ZT}.

ZENER EQUIVALENT CIRCUIT

Figure 3–4(a) shows the ideal approximation of a zener diode in reverse breakdown. It acts simply as a battery having a value equal to the nominal zener voltage. Figure 3–4(b) represents the practical equivalent of a zener, where the zener resistance R_Z is included. Since the voltage curve is not ideally vertical, a change in zener current produces a small change in zener voltage, as illustrated in Figure 3–4(c). The ratio of ΔV_Z to ΔI_Z is the resistance, as expressed in the following equation.

$$R_Z = \frac{\Delta V_Z}{\Delta I_Z} \qquad \textbf{(3–1)}$$

Normally, R_Z is specified at I_{ZT}, the zener test current. In most cases, we will assume that R_Z is constant over the full linear range of zener current values.

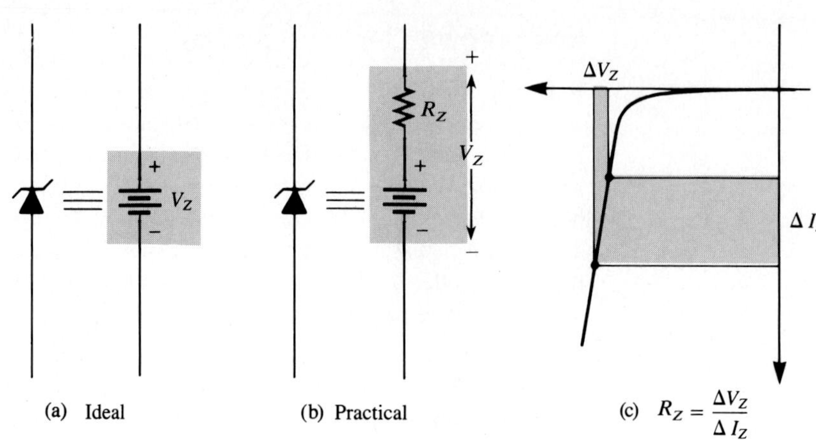

FIGURE 3–4
Zener diode equivalent circuits.

(a) Ideal (b) Practical (c) $R_Z = \dfrac{\Delta V_Z}{\Delta I_Z}$

■ EXAMPLE 3–1

A certain zener diode exhibits a 50 mV change in V_Z for a 2 mA change in I_Z on the linear portion of the characteristic curve between I_{ZK} and I_{ZM}. What is the zener resistance?

SOLUTION

$$R_Z = \frac{\Delta V_Z}{\Delta I_Z} = \frac{50 \text{ mV}}{2 \text{ mA}} = 25\ \Omega$$

PRACTICE EXERCISE 3–1

Calculate the zener resistance if the zener voltage changes 100 mV for a 20 mA change in zener current. ■

■ EXAMPLE 3–2

A certain zener diode has a resistance of 5 Ω. The data sheet gives $V_{ZT} = 6.8$ V at $I_{ZT} =$ 20 mA, $I_{ZK} = 1$ mA, and $I_{ZM} = 50$ mA. What is the voltage across the diode terminals when the current is 30 mA? When the current is 10 mA?

SOLUTION

Figure 3–5 represents the diode. The 30 mA current is a 10 mA increase above $I_{ZT} =$ 20 mA.

$$\Delta I_Z = +10 \text{ mA}$$
$$\Delta V_Z = \Delta I_Z R_Z = (10 \text{ mA})(5\ \Omega) = +50 \text{ mV}$$

The change in voltage due to the increase in current above the I_{ZT} values causes the zener terminal voltage to increase. The zener voltage for $I_Z = 30$ mA is

$$V_Z = 6.8 \text{ V} + \Delta V_Z = 6.8 \text{ V} + 50 \text{ mV} = 6.85 \text{ V}$$

The 10 mA current is a 10 mA decrease below $I_{ZT} = 20$ mA.

$$\Delta I_Z = -10 \text{ mA}$$
$$\Delta V_Z = \Delta I_Z R_Z = (-10 \text{ mA})(5\ \Omega) = -50 \text{ mV}$$

The change in voltage due to the decrease in current below I_{ZT} causes the zener terminal voltage to decrease. The zener voltage for $I_Z = 10$ mA is

$$V_Z = 6.8\text{ V} - \Delta V_Z = 6.8\text{ V} - 50\text{ mV} = 6.75\text{ V}$$

FIGURE 3–5

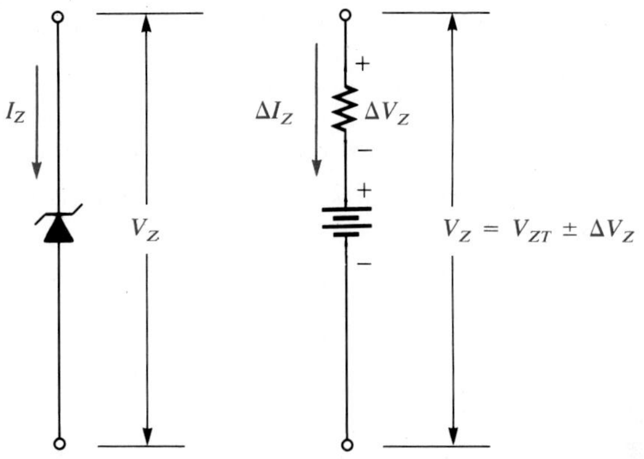

I_Z V_Z ΔI_Z ΔV_Z $V_Z = V_{ZT} \pm \Delta V_Z$

PRACTICE EXERCISE 3–2

Repeat the analysis for a diode with $V_{ZT} = 12$ V at $I_{ZT} = 50$ mA, $I_{ZK} = 0.5$ mA, $I_{ZM} = 100$ mA, and $R_Z = 20\ \Omega$ for a current of 20 mA and a current of 80 mA.

TEMPERATURE COEFFICIENT

The temperature coefficient specifies the percent change in zener voltage for each °C change in temperature. For example, a 12 V zener diode with a temperature coefficient of 0.1%/°C will exhibit a 0.012 V increase in V_Z when the junction temperature increases one Celsius degree. The formula for calculating the change in zener voltage for a given junction temperature change, for a specified temperature coefficient, is as follows:

$$\Delta V_Z = V_Z \times TC \times \Delta T \qquad (3\text{–}2)$$

where V_Z is the nominal zener voltage at 25° C, TC is the temperature coefficient, and ΔT is the change in temperature. A positive TC means that the zener voltage increases with an increase in temperature or decreases with a decrease in temperature. A negative TC means that the zener voltage decreases with an increase in temperature or increases with a decrease in temperature.

EXAMPLE 3–3

An 8.2 V zener diode (8.2 V at 25°C) has a positive temperature coefficient of 0.048%/°C. What is the zener voltage at 60°C?

SOLUTION

$$\begin{aligned}\Delta V_Z &= V_Z \times TC \times \Delta T \\ &= (8.2\text{ V})(0.048\%/°\text{C})(60°\text{C} - 25°\text{C}) \\ &= (8.2\text{ V})(0.00048/°\text{C})(35°\text{C}) \\ &= 0.138\text{ V}\end{aligned}$$

The zener voltage at 60°C is

$$V_Z + \Delta V_Z = 8.2\text{ V} + 0.138\text{ V} = 8.338\text{ V}$$

PRACTICE EXERCISE 3–3

A 12-V zener has a positive temperature coefficient of 0.075%/°C. How much will the zener voltage change when the junction temperature decreases 50°C? ■

3–1 REVIEW QUESTIONS

1. In what region of their characteristic curve are zener diodes operated?
2. At what value of zener current is the zener voltage normally specified?
3. How does the internal zener resistance affect the voltage across the terminals of the device?
4. For a certain zener diode, $V_Z = 10$ V at 30 mA. If $R_Z = 8\ \Omega$, what is the terminal voltage at 50 mA?
5. What does a positive temperature coefficient of 0.05%/°C mean?

3–2 ZENER DIODE APPLICATIONS

As mentioned before, the zener diode is often found as a voltage regulator in dc power supplies. In this section, the concept of voltage regulation is introduced and two types of regulation are examined—line regulation and load regulation. Also, you will see how zeners can be used as simple limiters or clippers.

OUTPUT VOLTAGE REGULATION WITH A VARYING INPUT VOLTAGE

Zener diodes are widely used for voltage regulation. Figure 3–6 illustrates how a zener diode can be used to regulate a varying dc voltage. This is called *input* or **line regulation.** As the input voltage varies (within limits), the zener diode maintains a nearly constant voltage across the output terminals. However, as V_{IN} changes, I_Z will change

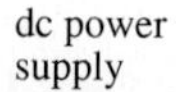

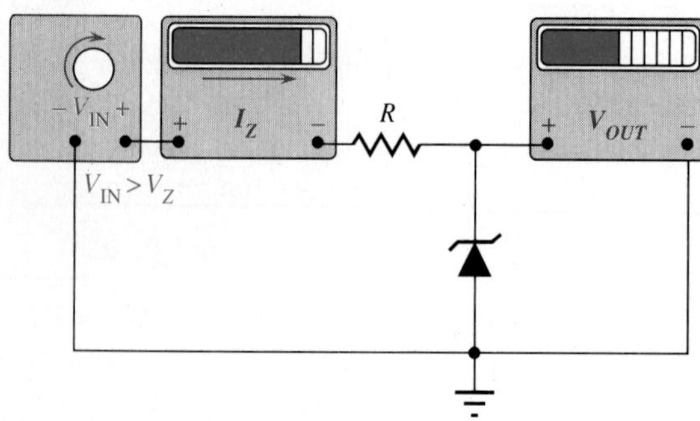

(a) As input voltage increases, V_{OUT} remains constant ($I_{ZK} < I_Z < I_{ZM}$).

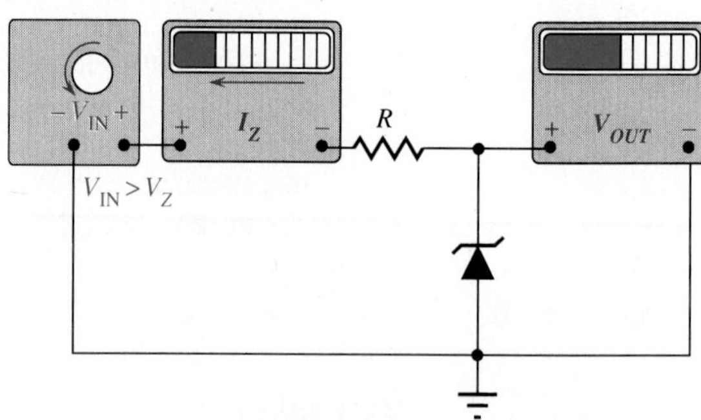

(b) As input voltage decreases, V_{OUT} remains constant ($I_{ZK} < I_Z < I_{ZM}$).

FIGURE 3–6
Zener regulation of a variable input voltage.

proportionally so that the limitations on the input variation are set by the minimum and maximum current values (I_{ZK} and I_{ZM}) with which the zener can operate. R is the series current-limiting resistor. The bar graph on the DMM symbols indicates the relative values and trends. Many modern DMMs provide analog bar graph displays in addition to the digital readout.

For example, suppose that the zener diode in Figure 3–7 can maintain regulation over a range of current values from 4 mA to 40 mA. For the minimum current, the voltage across the 1 kΩ resistor is

$$V_R = (4\text{ mA})(1\text{ k}\Omega) = 4\text{ V}$$

Since

$$V_R = V_{IN} - V_Z$$

then

$$V_{IN} = V_R + V_Z = 4\text{ V} + 10\text{ V} = 14\text{ V}$$

For the maximum current, the voltage across the 1 kΩ resistor is

$$V_R = (40\text{ mA})(1\text{ k}\Omega) = 40\text{ V}$$

Therefore,

$$V_{IN} = 40\text{ V} + 10\text{ V} = 50\text{ V}$$

This shows that this zener diode can regulate an input voltage from 14 V to 50 V and maintain an approximate 10 V output. (The output will vary slightly because of the zener resistance.)

FIGURE 3–7

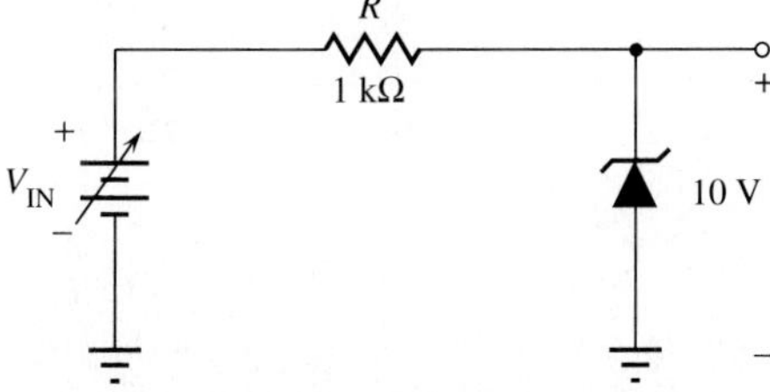

■ **EXAMPLE 3–4** Determine the minimum and the maximum input voltages that can be regulated by the zener diode in Figure 3–8. Assume $I_{ZK} = 1$ mA, $I_{ZM} = 15$ mA, $V_{ZT} = 5.1$ V at $I_{ZT} = 8$ mA, and $R_Z = 10\ \Omega$.

FIGURE 3–8

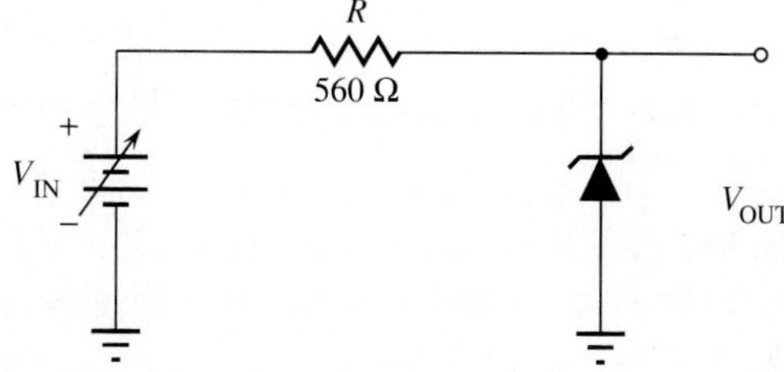

SOLUTION

The equivalent circuit is shown in Figure 3–9. At $I_{ZK} = 1$ mA, the output voltage is

$$\begin{aligned} V_{OUT} &= 5.1\text{ V} - \Delta V_Z = 5.1\text{ V} - \Delta I_Z R_Z \\ &= 5.1\text{ V} - (7\text{ mA})(10\ \Omega) \\ &= 5.1\text{ V} - 0.070\text{ V} \\ &= 5.03\text{ V} \end{aligned}$$

Therefore,

$$\begin{aligned} V_{IN(min)} &= I_{ZK}R + V_{OUT} \\ &= (1\text{ mA})(560\ \Omega) + 5.03\text{ V} \\ &= 5.59\text{ V} \end{aligned}$$

At $I_{ZM} = 15$ mA, the output voltage is

$$\begin{aligned} V_{OUT} &= 5.1\text{ V} + \Delta V_Z \\ &= 5.1\text{ V} + (7\text{mA})(10\ \Omega) \\ &= 5.1\text{ V} + 0.07\text{ V} \\ &= 5.17\text{ V} \end{aligned}$$

Therefore,

$$\begin{aligned} V_{IN(max)} &= I_{ZM}R + V_{OUT} \\ &= (15\text{ mA})(560\ \Omega) + 5.17\text{ V} \\ &= 13.57\text{ V} \end{aligned}$$

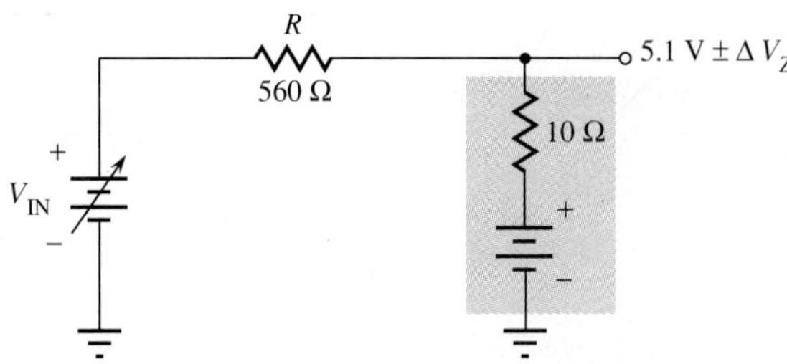

FIGURE 3–9
Equivalent of circuit in Figure 3–8.

PRACTICE EXERCISE 3–4

Determine the minimum and maximum input voltages that can be regulated by the zener diode in Figure 3–8. $I_{ZK} = 0.5$ mA, $I_{ZM} = 60$ mA, $V_{ZT} = 6.8$ V at 30 mA, and $R_Z = 18\ \Omega$.

VOLTAGE REGULATION WITH A VARYING LOAD

Figure 3–10 shows a zener regulator with a variable load resistor across the terminals. The zener diode maintains a nearly constant voltage across R_L as long as the zener current is greater than I_{ZK} and less than I_{ZM}. This is called **load regulation.**

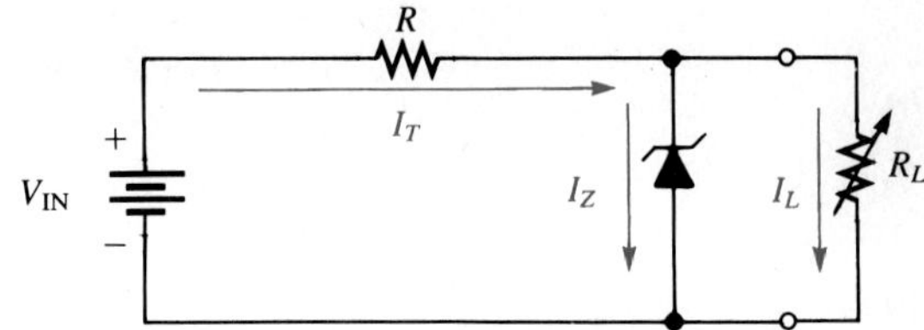

FIGURE 3–10
Zener regulation with a variable load.

NO LOAD TO FULL LOAD

When the output terminals are open ($R_L = \infty$), the load current is zero and *all* of the current is through the zener. When a load resistor is connected, part of the total current is through the zener and part through R_L. As R_L is decreased, the load current, I_L, increases and I_Z decreases. The zener diode continues to regulate until I_Z reaches its minimum value, I_{ZK}. At this point the load current is maximum. The following example will illustrate this.

EXAMPLE 3–5

Determine the minimum and the maximum load currents for which the zener diode in Figure 3–11 will maintain regulation. What is the minimum R_L that can be used? $V_Z = 12$ V, $I_{ZK} = 3$ mA, and $I_{ZM} = 90$ mA. Assume $R_Z = 0\ \Omega$ and V_Z remains a constant 12 V over the range of current values for simplicity.

FIGURE 3–11

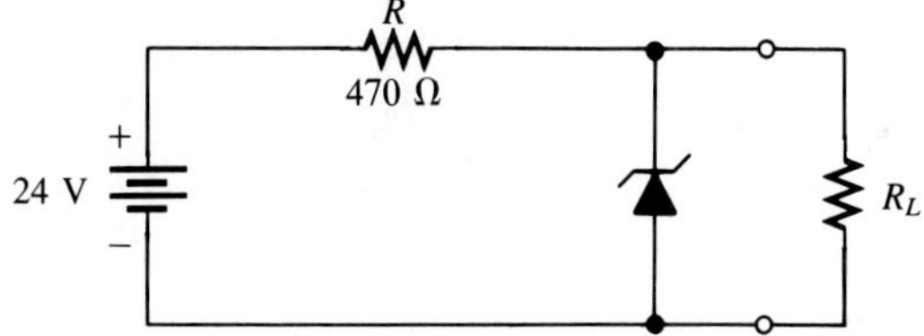

SOLUTION

When $I_L = 0$ A, I_Z is maximum and equal to the total circuit current I_T.

$$I_{Z(\max)} = I_T = \frac{V_{IN} - V_Z}{R} = \frac{24\text{ V} - 12\text{ V}}{470\ \Omega} = 25.53\text{ mA}$$

Since this is much less than I_{ZM}, 0 A is an acceptable minimum value for I_L.

$$I_{L(\min)} = 0\text{ A}$$

The maximum value of I_L occurs when I_Z is minimum ($I_Z = I_{ZK}$), so we can solve for $I_{L(\max)}$ as follows:

$$\begin{aligned} I_{L(\max)} &= I_T - I_{ZK} \\ &= 25.53\text{ mA} - 3\text{ mA} \\ &= 22.53\text{ mA} \end{aligned}$$

The minimum value of R_L is

$$R_{L(\text{min})} = \frac{V_Z}{I_{L(\text{max})}} = \frac{12\text{ V}}{22.53\text{ mA}} = 533\ \Omega$$

PRACTICE EXERCISE 3–5

Find the minimum and maximum load currents for which the circuit in Figure 3–11 will maintain regulation. Determine the minimum R_L that can be used. $V_Z = 3.3$ V (constant), $I_{ZK} = 2$ mA, $I_{ZM} = 75$ mA, and $R_Z = 0\ \Omega$.

In the last example, we made the assumption that R_Z was zero and, therefore, the zener voltage remained constant over the range of currents. This assumption was made in order to simplify the concept of how the regulator works with a varying load. Such a simplifying assumption is often acceptable and in some cases produces results that are accurate enough. In Example 3–6, we will take into account the internal resistance of the zener.

EXAMPLE 3–6

The zener used in the regulator circuit of Figure 3–12 is a 15-V diode. The manufacturer's data sheet gives the following information:

$$V_Z = 15\text{ V @ } I_{ZT}$$
$$I_{ZK} = 1\text{ mA}$$
$$I_{ZT} = 170\text{ mA}$$
$$I_{ZM} = 560\text{ mA}$$
$$R_Z = 3\ \Omega$$

(a) Determine V_{OUT} at I_{ZK} and at I_{ZM}.
(b) Calculate the value of R that should be used.
(c) Determine the minimum value of R_L that can be used.

FIGURE 3–12

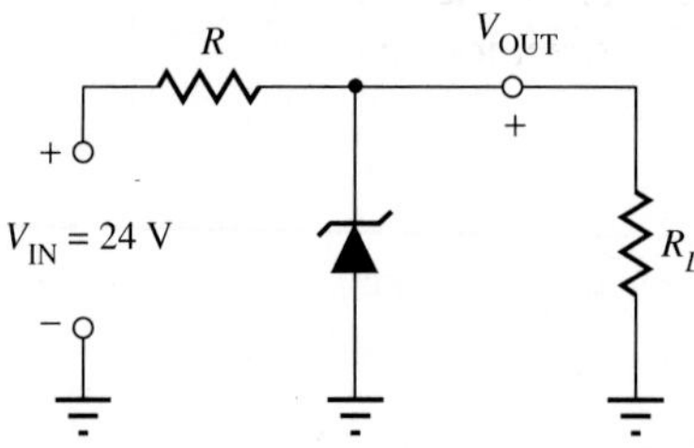

SOLUTION

(a) For I_{ZK},

$$V_Z = 15\text{ V} - (\Delta I_Z)R_Z = 15\text{ V} - (I_{ZT} - I_{ZK})R_Z$$
$$= 15\text{ V} - (169\text{ mA})(3\ \Omega) = 15\text{ V} - 0.507\text{ V} = 14.49\text{ V}$$

For I_{ZM},

$$V_Z = 15\text{ V} + \Delta I_Z R_Z = 15\text{ V} + (390\text{ mA})(3\ \Omega) = 16.17\text{ V}$$

(b) The value of R is calculated for the maximum zener current that occurs when there is no load as shown in Figure 3–13(a).

$$R = \frac{V_{IN} - V_Z}{I_{ZM}} = \frac{24\text{ V} - 16.17\text{ V}}{560\text{ mA}} = 13.98\ \Omega$$

$$R = 15\ \Omega \quad \text{(nearest standard value)}$$

(c) For the minimum load resistance (maximum load current), the zener current is minimum ($I_{ZK} = 1$ mA) as shown in Figure 3–13(b).

$$I_T = \frac{V_{IN} - V_{OUT}}{R} = \frac{24\text{ V} - 14.49\text{ V}}{15\ \Omega} = 634\text{ mA}$$

$$I_L = I_T - I_{ZK} = 634\text{ mA} - 1\text{ mA} = 633\text{ mA}$$

$$R_{L(\min)} = \frac{V_{OUT}}{I_L} = \frac{14.49\text{ V}}{633\text{ mA}} = 22.9\ \Omega$$

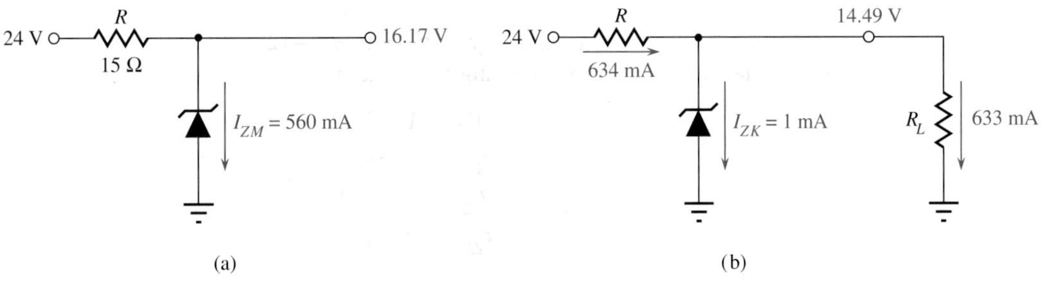

FIGURE 3–13

PRACTICE EXERCISE 3–6

Repeat each part of the preceding analysis if the zener is changed to a 12-V device with the following specifications:

$$V_Z = 12\text{ V @ } I_{ZT}$$
$$I_{ZK} = 1\text{ mA}$$
$$I_{ZT} = 210\text{ mA}$$
$$I_{ZM} = 720\text{ mA}$$
$$R_Z = 2\ \Omega$$

PERCENT REGULATION

The percent regulation is a figure of merit used to specify the performance of a voltage regulator. It can be in terms of input (line) regulation or load regulation. The *percent line regulation* specifies how much change occurs in the output voltage for a given change in input voltage.

$$\text{Percent line regulation} = \frac{\Delta V_{OUT}}{\Delta V_{IN}} \times 100\% \qquad (3\text{–}3)$$

Line regulation is usually expressed as a percent change in V_{OUT} for a one volt change in V_{IN} (%/V).

The *percent load regulation* specifies how much change occurs in the output voltage over a certain range of load current values, usually from minimum current (no load) to maximum current (full load). It is normally expressed as a percentage and can be calculated with the following formula.

$$\text{Percent load regulation} = \frac{V_{NL} - V_{FL}}{V_{FL}} \times 100\% \qquad (3\text{–}4)$$

where V_{NL} is the output voltage with no load, and V_{FL} is the output voltage with full (maximum) load.

EXAMPLE 3–7

A certain regulator has a no-load output voltage of 6 V and a full-load output of 5.82 V. What is the percent load regulation?

SOLUTION

$$\text{Percent load regulation} = \frac{V_{NL} - V_{FL}}{V_{FL}} \times 100\%$$

$$= \frac{6\text{ V} - 5.82\text{ V}}{5.82\text{ V}} \times 100\%$$

$$= 3.09\%$$

PRACTICE EXERCISE 3–7

(a) If the no-load output voltage of a regulator is 24.8 V and the full-load output is 23.9 V, what is the percent load regulation?

(b) If the output voltage changes 0.035 V for a 2 V change in the input voltage, what is the line regulation expressed as %/V?

ZENER LIMITING

Zener diodes can be used in ac applications to limit voltage swings to desired levels. Figure 3–14 shows three basic ways the limiting action of a zener diode can be used. Part (a) shows a zener used to limit the positive peak of a signal voltage to the selected zener voltage. During the negative alternation, the zener acts as a conventional forward-biased diode and limits the negative voltage to −0.7 V. When the zener is turned around, as in part (b), the negative peak is limited by zener action and the positive voltage is limited to 0.7 V. Two back-to-back zeners limit both peaks to the zener voltage plus 0.7 V, as shown in part (c). During the positive alternation, D_2 is functioning as the zener limiter and D_1 is functioning as a conventional forward-biased diode. During the negative alternation, the roles are reversed.

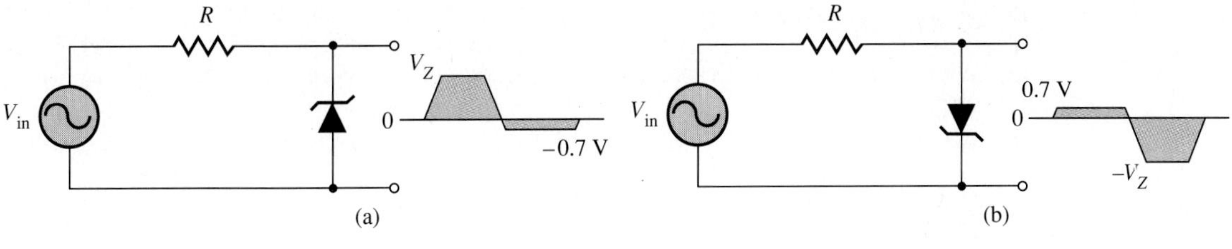

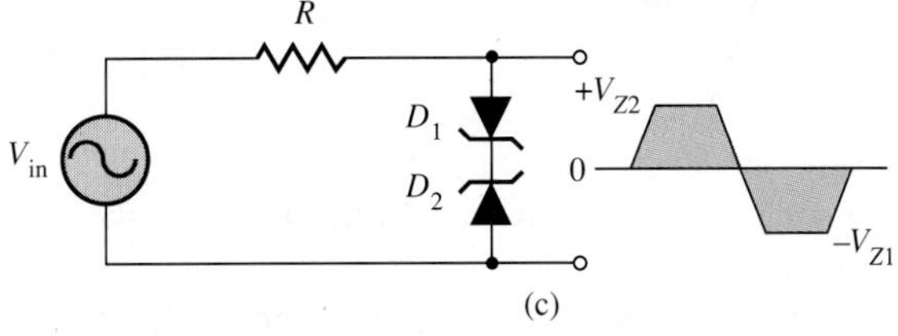

Figure 3–14
Basic zener limiting action with sine wave input.

Example 3–8 Determine the output voltage for each zener limiting circuit in Figure 3–15.

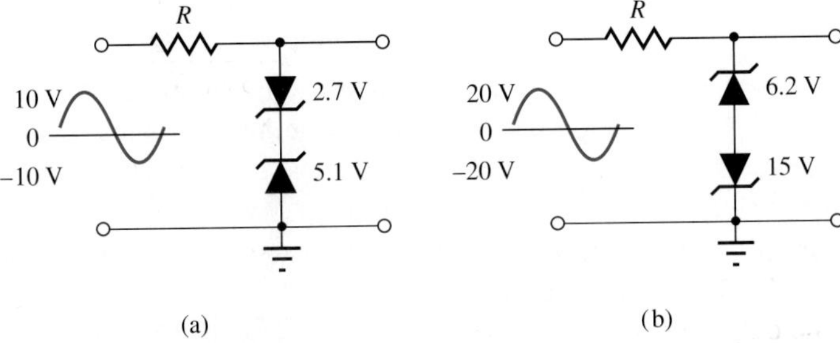

Figure 3–15

Solution

See Figure 3–16 for the resulting output voltages. Remember, when one zener is operating in breakdown, the other one is forward-biased with approximately 0.7 V across it.

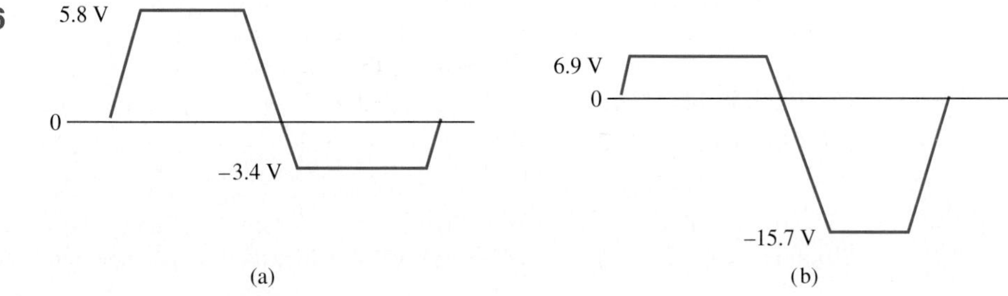

Figure 3–16

PRACTICE EXERCISE 3–8

(a) What is the output in Figure 3–15(a) if the input voltage is increased to a peak value of 20 V?

(b) What is the output in Figure 3–15(b) if the input voltage is decreased to a peak value of 5 V? ■

3–2 REVIEW QUESTIONS

1. Explain the difference between line regulation and load regulation.
2. In a zener diode regulator, what value of load resistance results in the maximum zener current?
3. Define the terms *no-load* and *full-load.*
4. A regulator has an output voltage of 12 V with no load and 11.9 V with full load. What is the percent load regulation?
5. How much voltage appears across a zener diode when it is forward-biased?

3–3 VARACTOR DIODES

Varactor diodes are also known as variable-capacitance diodes because the junction capacitance varies with the amount of reverse-bias voltage. Varactors are specifically designed to take advantage of this variable-capacitance characteristic. The capacitance can be changed by changing the reverse voltage. These devices are commonly used in electronic tuning circuits used in communications systems.

A **varactor** is basically a reverse-biased pn junction diode that utilizes the inherent capacitance of the depletion layer. The depletion layer, created by the reverse bias, acts as a capacitor dielectric because of its nonconductive characteristic. The p and n regions are conductive and act as the capacitor plates, as illustrated in Figure 3–17.

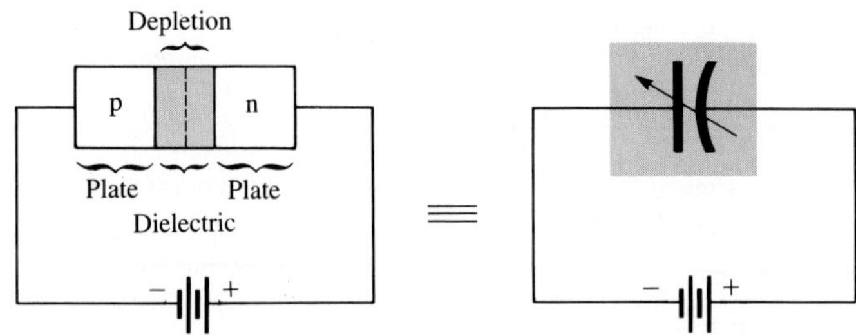

FIGURE 3–17
The reverse-biased varactor diode acts as a variable capacitor.

BASIC OPERATION

When the reverse-bias voltage increases, the depletion layer widens, effectively increasing the dielectric thickness and thus decreasing the capacitance. When the reverse-bias voltage decreases, the depletion layer narrows, thus increasing the capacitance. This action is shown in Figure 3–18(a) and (b). A general curve of capacitance versus voltage

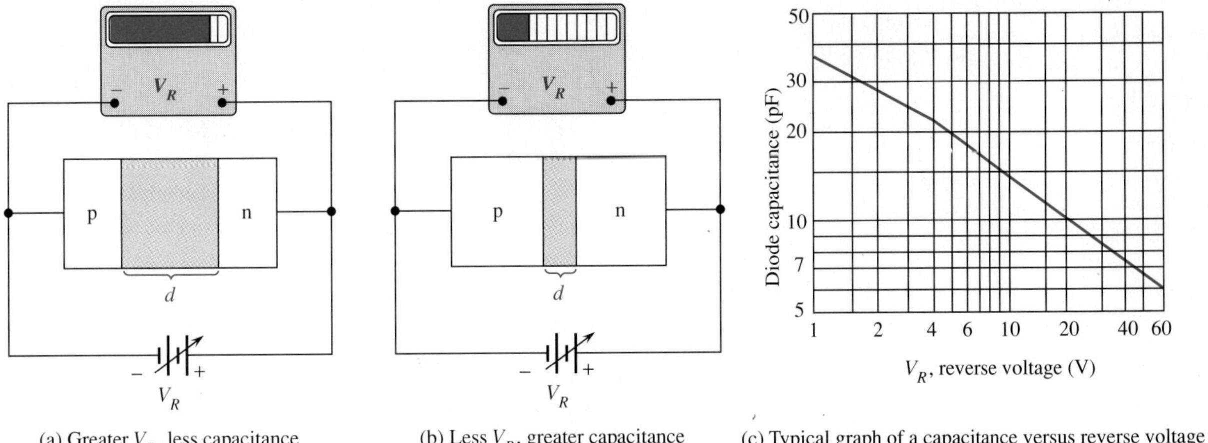

(a) Greater V_R, less capacitance (b) Less V_R, greater capacitance (c) Typical graph of a capacitance versus reverse voltage

FIGURE 3–18
Varactor diode capacitance varies with reverse voltage.

is shown in Figure 3–18(c). Recall that capacitance is determined by the plate area (A), dielectric constant(ϵ), and dielectric thickness (d), as expressed in the following formula.

$$C = \frac{A\epsilon}{d} \qquad (3-5)$$

In a varactor diode, the capacitance parameters are controlled by the method of doping in the depletion layer and the size and geometry of the diode's construction. Varactor capacitances typically range from a few picofarads to a few hundred picofarads. Figure 3–19(a) shows a common symbol for a varactor, and Figure 3–19(b) shows a simplified equivalent circuit. R_s is the reverse series resistance, and C_V is the variable capacitance.

FIGURE 3–19
Varactor diode.

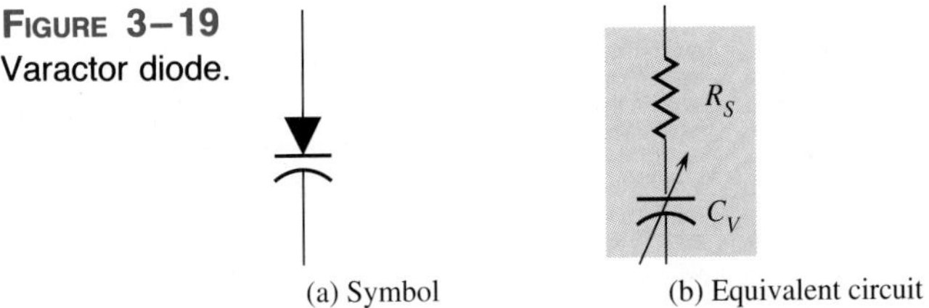

(a) Symbol (b) Equivalent circuit

APPLICATIONS

A major application of varactors is in tuning circuits. For example, electronic tuners in TV and other commercial receivers utilize varactors as one of their elements. When used in a resonant circuit, the varactor acts as a variable capacitor, thus allowing the resonant frequency to be adjusted by a variable voltage level, as illustrated in Figure 3–20 where two varactor diodes provide the total variable capacitance in a parallel resonant (tank) circuit.

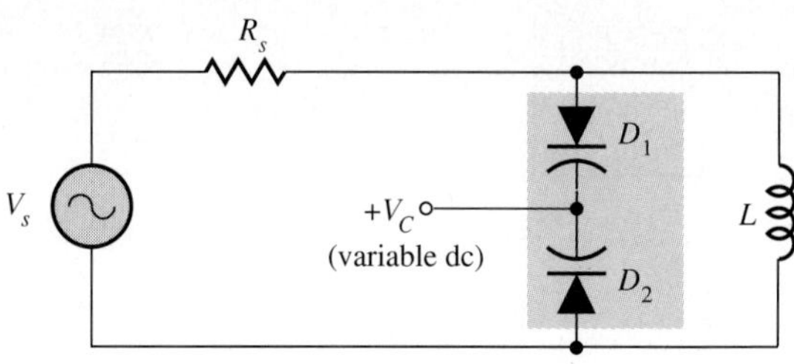

FIGURE 3–20
Varactors in a resonant circuit.

V_C is a variable dc voltage that controls the reverse bias and therefore the capacitance of the diodes. Recall that the resonant frequency of the tank circuit is

$$f_r \cong \frac{1}{2\pi\sqrt{LC}} \tag{3–6}$$

This approximation is valid for $Q \geqq 10$.

EXAMPLE 3–9

The capacitance of a certain varactor can be varied from 5 pF to 50 pF. The diode is used in a tuned circuit similar to that shown in Figure 3–20. Determine the tuning range for the circuit if $L = 10$ mH.

SOLUTION

The equivalent circuit is shown in Figure 3–21. Notice that the varactor capacitances are in series. The minimum total capacitance is

$$C_{T(\text{min})} = \frac{C_{1(\text{min})}C_{2(\text{min})}}{C_{1(\text{min})} + C_{2(\text{min})}}$$
$$= \frac{(5\text{ pF})(5\text{ pF})}{10\text{ pF}}$$
$$= 2.5\text{ pF}$$

The maximum resonant frequency is therefore

$$f_{r(\text{max})} = \frac{1}{2\pi\sqrt{LC}}$$
$$= \frac{1}{2\pi\sqrt{(10\text{ mH})(2.5\text{ pF})}}$$
$$\cong 1\text{ MHz}$$

The maximum total capacitance is

$$C_{T(\text{max})} = \frac{C_{1(\text{max})}C_{2(\text{max})}}{C_{1(\text{max})} + C_{2(\text{max})}}$$
$$= \frac{(50\text{ pF})(50\text{ pF})}{100\text{ pF}}$$
$$= 25\text{ pF}$$

The minimum resonant frequency is therefore

$$f_{r(min)} = \frac{1}{2\pi\sqrt{LC}}$$
$$= \frac{1}{2\pi\sqrt{(10 \text{ mH})(25 \text{ pF})}}$$
$$\cong 318 \text{ kHz}$$

FIGURE 3–21

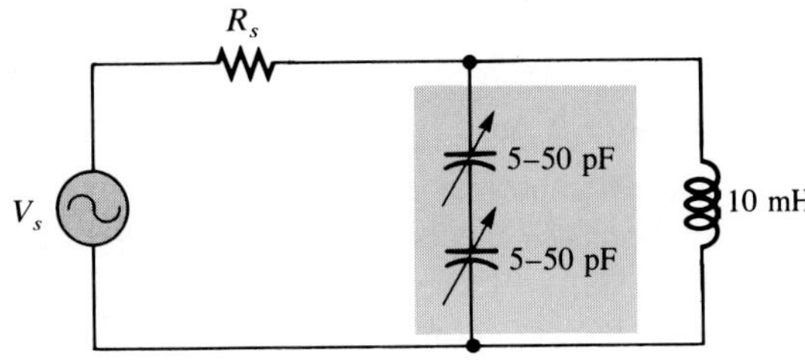

PRACTICE EXERCISE 3–9
Determine the tuning range for Figure 3–21 if L = 100 mH. ■

3–3 REVIEW QUESTIONS

1. What is the key feature of a varactor diode?
2. Under what bias condition is a varactor operated?
3. What part of the varactor produces the capacitance?
4. Based on the graph in Figure 3–18(c), what happens to the diode capacitance when the reverse voltage is increased?

3–4 OPTICAL DIODES

In this section, two types of optoelectronic devices—the light-emitting diode (LED) and the photodiode—are introduced. As the name implies, the LED is a light emitter. The photodiode, on the other hand, is a light detector. We will examine the characteristics of both devices, and you will see an example of their use in a system application at the end of the chapter.

THE LIGHT-EMITTING DIODE (LED)

The basic operation of the **light-emitting diode (LED)** is as follows: When the device is forward-biased, electrons cross the pn junction from the n-type material and recombine with holes in the p-type material. Recall from Chapter 1 that these free electrons are in the conduction band and at a higher energy level than the holes in the valence band. When recombination takes place, the recombining electrons release energy in the form of heat and light. A large exposed surface area on one layer of the semiconductor material permits the photos to be emitted as visible light. Figure 3–22 illustrates this process, called **electroluminescence.**

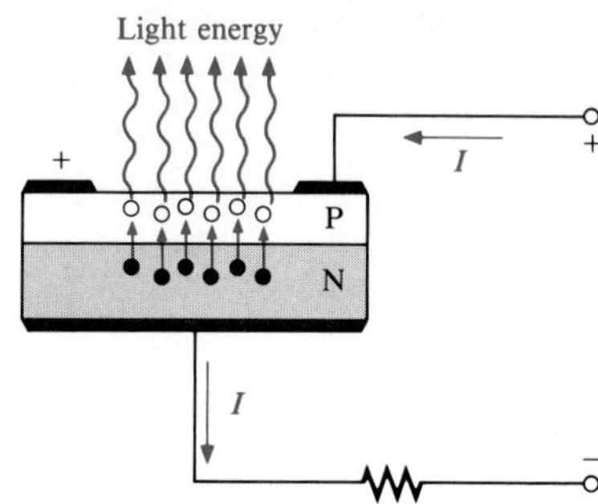

FIGURE 3–22
Electroluminescence in an LED.

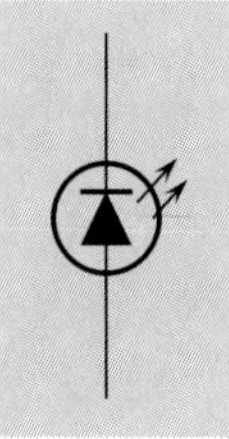

FIGURE 3–23
Symbol for an LED.

The semiconductor materials used in LEDs are gallium arsenide (GaAs), gallium arsenide phosphide (GaAsP), or gallium phosphide (GaP). Silicon and germanium are not used because they are essentially heat-producing materials and are very poor at producing light. GaAs LEDs emit **infrared** (IR) radiation, which is nonvisible, GaAsP produces either red or yellow visible light, and GaP emits red or green visible light. Red is by far the most common. The symbol for an LED is shown in Figure 3–23.

The LED emits light in response to a sufficient forward current, as shown in Figure 3–24(a). The amount of power output translated into light is directly proportional to the forward current, as indicated in Figure 3–24(b).

FIGURE 3–24
Operation of an LED.

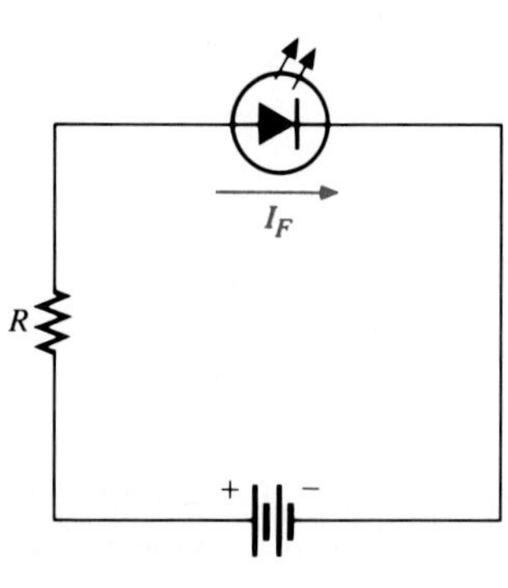

(a) Forward-biased operation

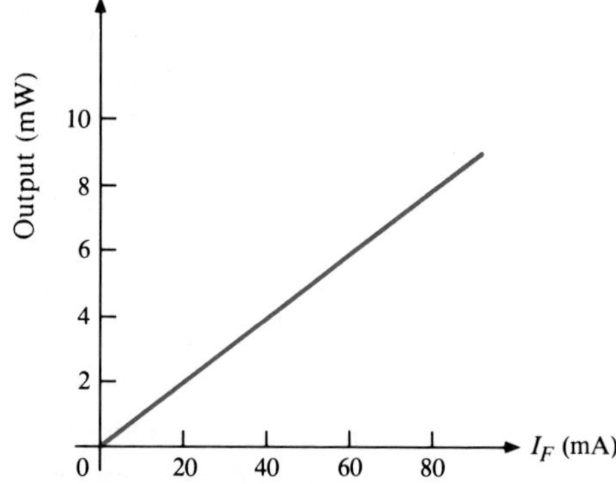

(b) Typical light output versus forward current

The **wavelength** of light determines whether it is visible or infrared. An LED emits light over a specified range of wavelengths as indicated by the **spectral** output curves in Figure 3–25. The curve in part (a) represents the light output versus wavelength for a typical visible red LED and the curve in part (b) is for a typical infrared LED. The wavelength (λ) is expressed in nanometers (nm). The output of the visible red LED peaks at 660 nm, and output for the infrared LED peaks at 940 nm.

The graph in Figure 3–26 is the **radiation** pattern for a typical LED. It shows how directional the emitted light is and that it depends on the type of lens structure of the LED. The narrower the radiation pattern, the more the light is concentrated in a particular direction.

Typical LEDs are shown in Figure 3–27. Photodiodes, to be studied next, generally have the same appearance.

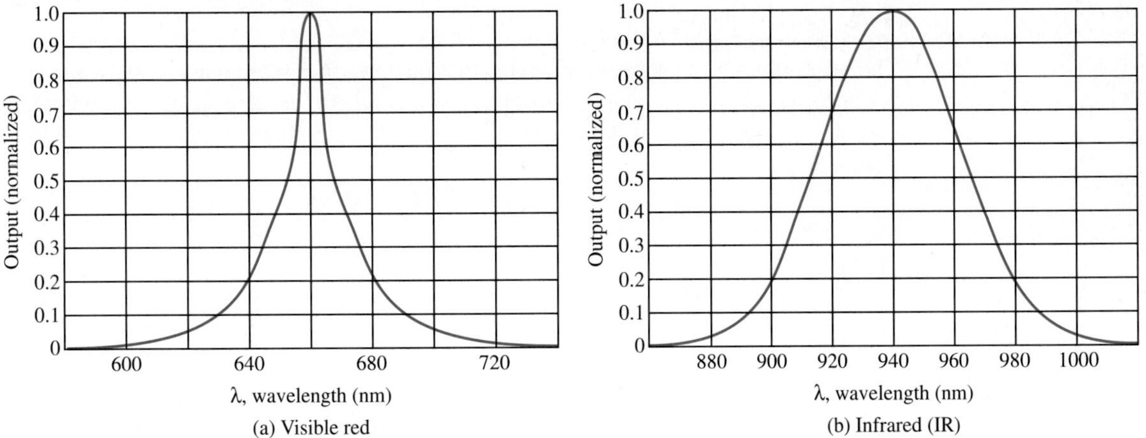

FIGURE 3–25
Typical spectral output curves for LEDs.

FIGURE 3–26
Typical radiation pattern of an LED.

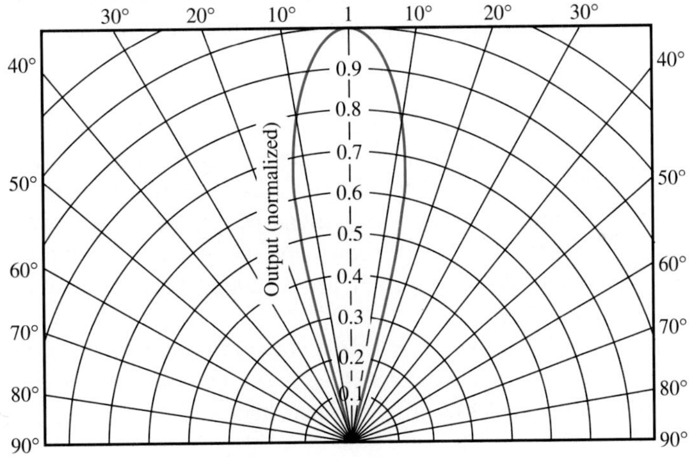

FIGURE 3–27
LEDs (copyright of Motorola, Inc. Used by permission).

APPLICATIONS LEDs are used for indicator lamps and readout displays on a wide variety of instruments, ranging from consumer appliances to scientific apparatus. A common type of display device using LEDs is the seven-segment display. Combinations of the segments form the ten decimal digits. Also, IR-emitting diodes are employed in optical coupling applications, often in conjunction with fiber optics.

THE PHOTODIODE

The **photodiode** is a pn junction device that operates in reverse bias, as shown in Figure 3–28(a),where I_λ is the reverse current. The photodiode has a small transparent window that allows light to strike the pn junction. An alternate photodiode symbol is shown in Figure 3–28(b).

FIGURE 3–28
Photodiode.

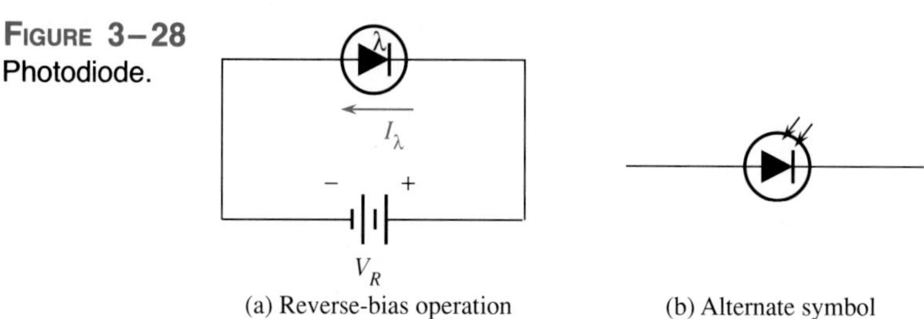

(a) Reverse-bias operation (b) Alternate symbol

Recall that when reverse-biased, a rectifer diode has a very small reverse leakage current. The same is true for the photodiode. The reverse-biased current is produced by thermally generated electron hole pairs in the depletion layer, which are swept across the junction by the electric field created by the reverse voltage. In a rectifier diode, the reverse leakage current increases with temperature due to an increase in the number of electron hole pairs.

A photodiode differs from a rectifier diode in that the reverse current increases with the light intensity at the pn junction. When there is no incident light, the reverse current I_λ is almost negligible and is called the **dark current.** An increase in the amount of light intensity, expressed in **lumens** per square meter (lm/m^2), produces an increase in the reverse current, as shown by the graph in Figure 3–29(a). For a given value of reverse-bias voltage, Figure 3–29(b) shows a set of characteristic curves for a typical photodiode.

From the characteristic curve in Figure 3–29(b), the dark current for this particular device is approximately 25 μA at a reverse-bias voltage of 3 V. Therefore, the resistance of the device with no incident light is

$$R_R = \frac{V_R}{I_\lambda} = \frac{3\text{ V}}{25\ \mu\text{A}} = 120\text{ k}\Omega$$

At 25,000 lm/m^2, the current is approximately 375 μA at −3 V. The resistance under this condition is

$$R_R = \frac{V_R}{I_\lambda} = \frac{3\text{ V}}{375\ \mu\text{A}} = 8\text{ k}\Omega$$

These calculations show that the photodiode can be used as a variable-resistance device controlled by light intensity.

Figure 3–30 illustrates that the photodiode allows essentially no reverse current (except for a very small dark current) when there is no incident light. When a light beam strikes the photodiode, it conducts an amount of reverse current that is proportional to the light intensity.

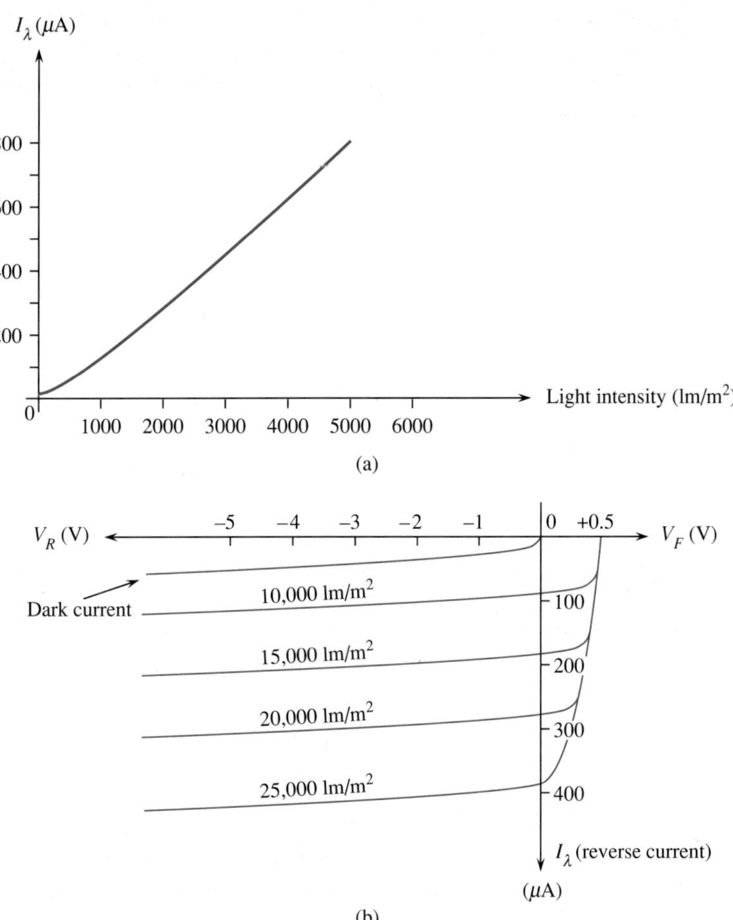

FIGURE 3–29 Typical photodiode characteristics.

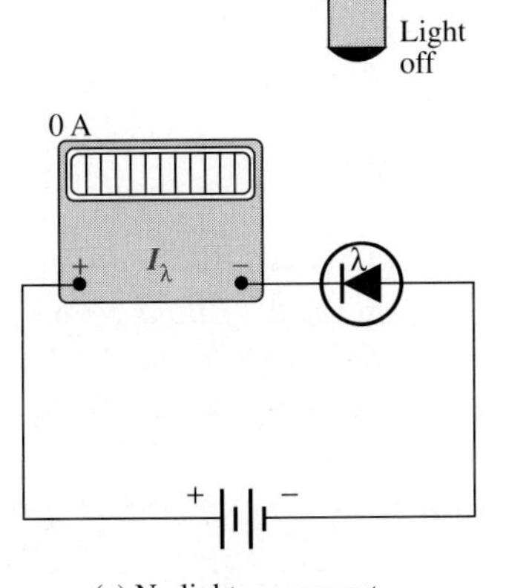

(a) No light, no current

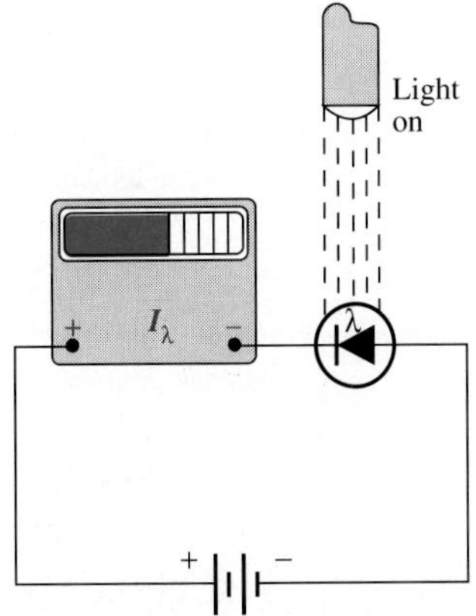

(b) When there is incident light, resistance decreases and reverse current flows.

FIGURE 3–30 Operation of a photodiode.

A typical spectral response curve is shown in Figure 3–31(a). Also, the response of a photodiode to incident light is dependent on the angle at which the light strikes its optical surface as indicated in the graph of Figure 3–31(b). The maximum response occurs when the light strikes head-on at an angle of 0° and falls off as the angle increases in either direction.

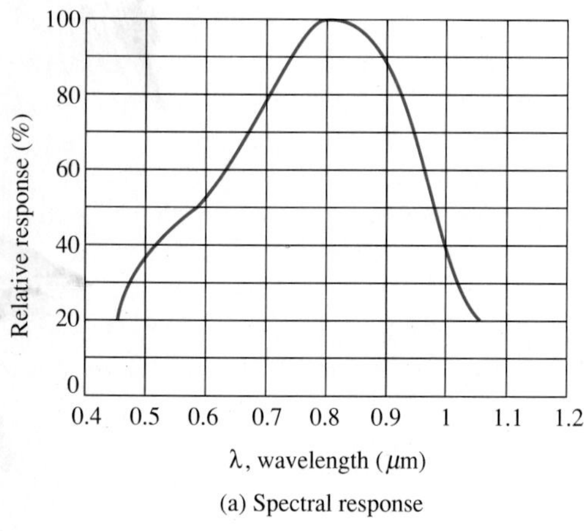

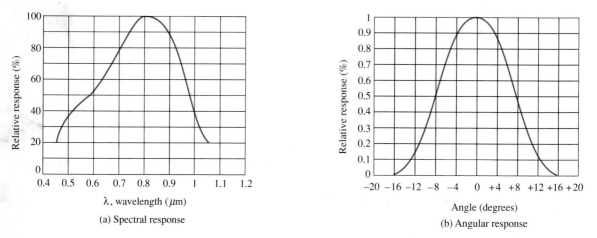

(a) Spectral response

1
0.9
0.8
0.7
0.6
0.5
0.4
0.3
0.2
0.1
0
Relative response (%)
−20 −16 −12 −8 −4 0 +4 +8 +12 +16 +20
Angle (degrees)

(b) Angular response

FIGURE 3–31
Response curves for a typical photodiode.

3–4 REVIEW QUESTIONS

1. Name two types of LEDs in terms of their light-emission spectrum.
2. Which has the greatest wavelength, visible light or infrared?
3. In what bias condition is an LED normally operated?
4. What happens to the light emission of an LED as the forward current increases?
5. To what ranges of the light spectrum do photodiodes typically respond?
6. In what bias condition is a photodiode normally operated?
7. When the intensity of the incident light on a photodiode increases, what happens to its internal reverse resistance?
8. What is *dark current?*

3–5 OTHER TYPES OF DIODES

In this section, we introduce several types of diodes that you are less likely to run into as a technician but are nevertheless important. These are the Schottky diode, the tunnel diode, the PIN diode, the step-recovery diode, and the laser diode.

THE SCHOTTKY DIODE

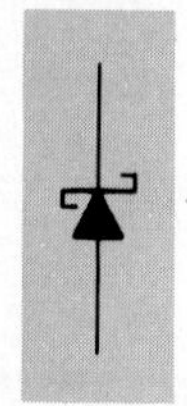

FIGURE 3–32
Schottky diode symbol.

Schottky diodes are used primarily in high-frequency and fast-switching applications. They are also known as *hot-carrier diodes*. A Schottky diode symbol is shown in Figure 3–32. A Schottky diode is formed by joining a doped semiconductor region (usually n-type) with a metal such as gold, silver, or platinum. So, rather than a pn junction, there is a metal-to-semiconductor junction, as shown in Figure 3–33.

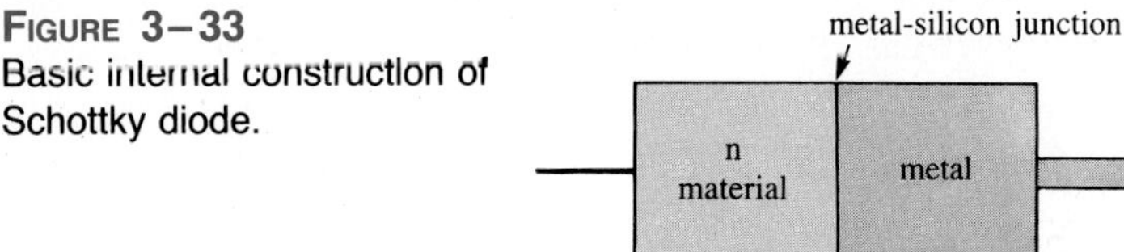

FIGURE 3–33
Basic internal construction of Schottky diode.

The Schottky diode operates only with majority carriers. There are no minority carriers as in other types of diodes. The metal region is heavily occupied with conduction-band electrons, and the n-type semiconductor region is lightly doped. When forward-biased, the higher-energy electrons in the n region are injected into the metal region where they give up their excess energy very rapidly. Since there are no minority carriers, as in a conventional diode, there is a very rapid response to a change in bias. The Schottky is a very fast-switching diode, and most of its applications make use of this property. It can be used to rectify very high-frequency signals, for example. Also, it is used in many digital circuits to decrease switching times.

THE TUNNEL DIODE

Tunnel diodes exhibit a special characteristic known as *negative resistance*. This feature makes them useful in oscillator and microwave amplifier applications. Three alternate symbols are shown in Figure 3–34. Tunnel diodes are constructed with germanium or gallium arsenide by doping the p and n regions much more heavily than in a conventional rectifier diode. This heavy doping results in a very narrow depletion layer. The heavy doping allows conduction for all reverse voltages so that there is no breakdown effect as with the conventional diode. This is shown in Figure 3–35.

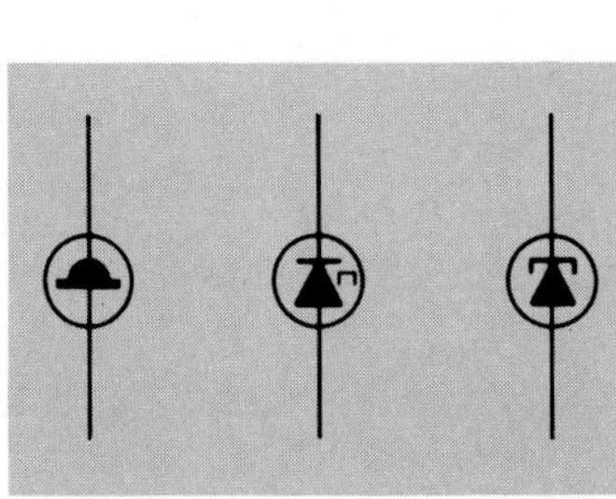

FIGURE 3–34
Tunnel diode symbols.

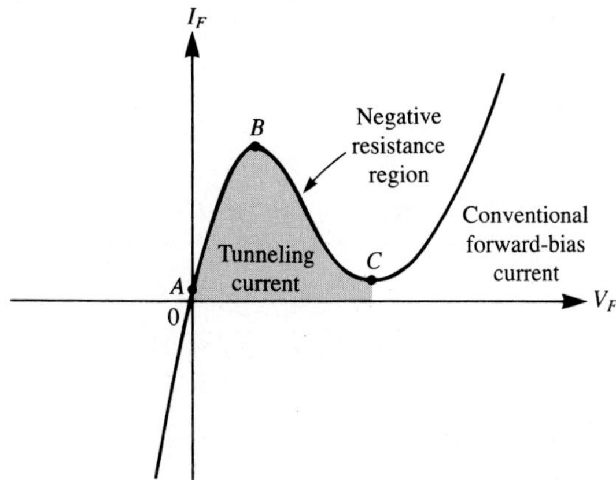

FIGURE 3–35
Tunnel diode characteristic.

Also, the extremely narrow depletion layer permits electrons to ''tunnel'' across the pn junction at very low forward-biased voltages, and the diode acts as a conductor. This is shown in Figure 3–35 between points A and B. At point B, the forward voltage begins to develop a barrier, and the current begins to decrease as the forward voltage continues to increase. This is the *negative-resistance region*.

$$R_F = \frac{\Delta V_F}{\Delta I_F} \tag{3–7}$$

This effect is opposite to that described by Ohm's law, where an increase in voltage results in an increase in current. At point C, the diode begins to act as a conventional forward-biased diode.

An Application A parallel resonant circuit can be represented by a capacitance, inductance, and resistance in parallel, as in Figure 3–36(a). R_p is the parallel equivalent of the series winding resistance of the coil. When the tank circuit is ''shocked'' into oscillation as in Figure 3–36(b), a damped sinusoidal output results. The damping is due to the resistance of the tank, which prevents a sustained oscillation because energy is lost when there is current through the resistance.

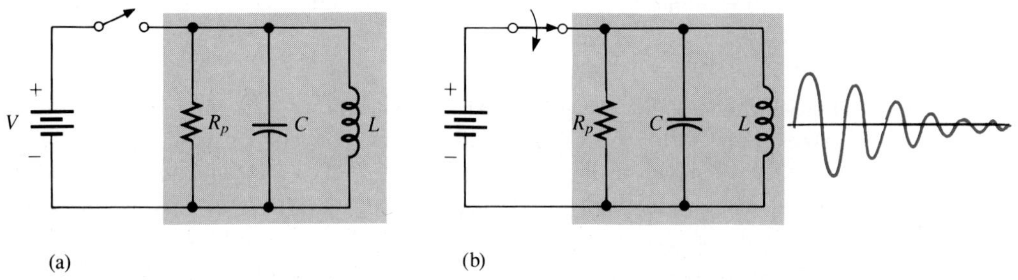

Figure 3–36
Parallel resonant circuit.

If a tunnel diode is placed in series with the tank circuit and biased at the center of the negative-resistance portion of its characteristic curve, as shown in Figure 3–37, a

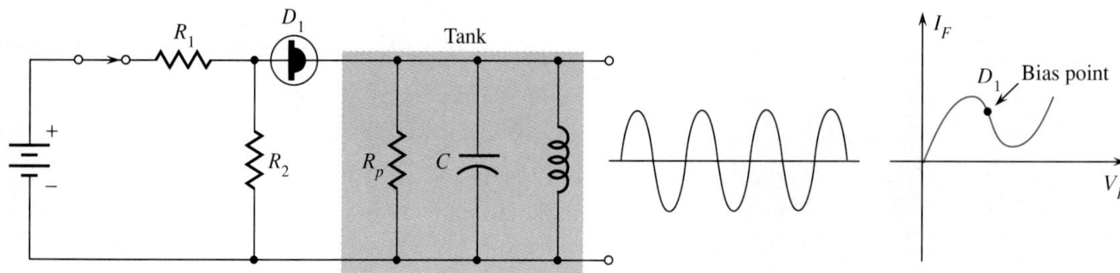

Figure 3–37
Tunnel diode oscillator.

sustained oscillation (constant sine wave) will result on the output. This is because the negative-resistance characteristic of the tunnel diode counteracts the positive-resistance characteristic of the tank resistance.

THE PIN DIODE

The PIN diode consists of heavily doped p and n regions separated by an intrinsic (undoped) region, as shown in Figure 3–38(a). When reverse-biased, the PIN diode acts like an almost constant capacitance. When forward-biased, it acts like a variable resistance, as shown in Figure 3–38(b) and (c). The forward resistance of the intrinsic region decreases with increasing current.

FIGURE 3–38
PIN diode.

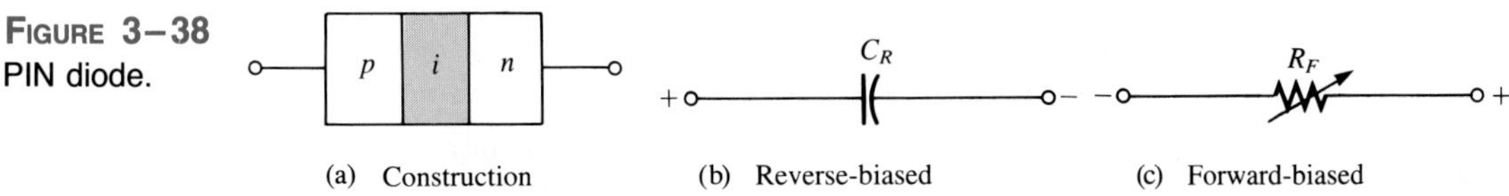

The PIN diode is used as a dc-controlled microwave switch operated by rapid changes in bias or as a modulating device that takes advantage of the variable forward-resistance characteristic. Since no rectification occurs at the pn junction, a high-frequency signal can be modulated (varied) by a lower-frequency bias variation. A PIN diode can also be used in attenuator applications because its resistance can be controlled by the amount of current.

THE STEP-RECOVERY DIODE

The step-recovery diode employs graded doping where the doping level of the semiconductor materials is reduced as the pn junction is approached. This produces an abrupt turn-off time by allowing a very fast release of stored charge when switching from forward to reverse bias. It also allows a rapid re-establishment of forward current when switching from reverse to forward bias. This diode is used in very fast-switching applications.

THE LASER DIODE

The term **laser** stands for *l*ight *a*mplification by *s*timulated *e*mission of *r*adiation. Laser light is **monochromatic,** meaning that it consists of a single color and not a mixture of colors. Laser light is also called **coherent light,** meaning that it is a single wavelength or a very narrow band of wavelengths, as opposed to incoherent light, which consists of a wide band of wavelengths. The laser diode normally emits coherent light, whereas the LED emits incoherent light.

The basic construction of a laser diode is shown in Figure 3–39(a). A pn junction is formed by two layers of doped gallium arsenide. The length of the pn junction bears a precise relationship with the wavelength of the light to be emitted. There is a highly reflective surface at one end of the junction and a partially reflective surface at the other end. External leads provide the anode and cathode connections.

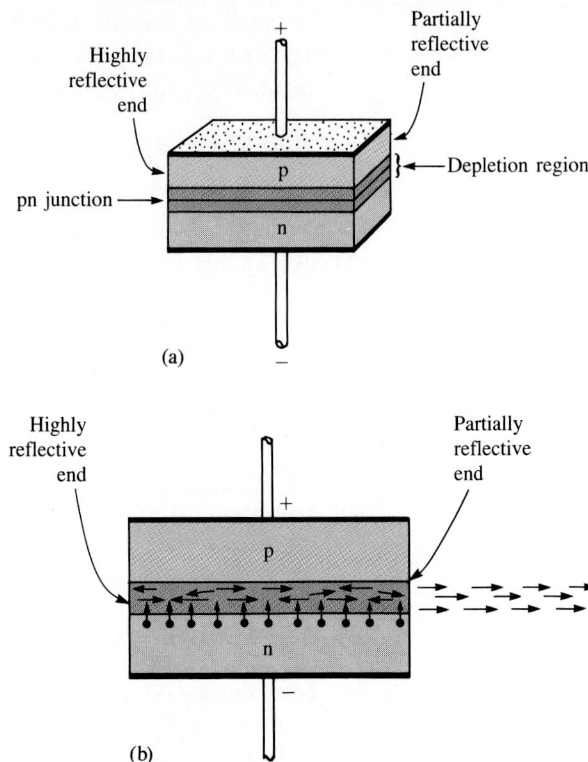

FIGURE 3–39
Basic laser diode construction and operation.

The basic operation is as follows. The pn junction is forward-biased by an external voltage source. As electrons move across the junction, recombination occurs in the depletion region, just as in an ordinary diode. As electrons fall into holes to recombine, **photons** are released. A released photon can strike an atom, causing another photon to be released. As the forward current is increased, more electrons enter the depletion region and cause more photons to be emitted. Eventually some of the photons that are randomly drifting within the depletion region strike the reflected surfaces perpendicularly. These reflected photons move along the depletion region, striking atoms and releasing additional photons. This back-and-forth movement of photons increases as the generation of photons "snowballs" until a very intense beam of laser light is formed by the photons that pass through the partially reflective end of the pn junction.

Each photon produced in this process is identical to the other photons in energy level, phase relationship, and frequency. So a single wavelength of intense light emerges from the laser diode, as indicated in Figure 3–39(b). Laser diodes have a threshold level of current above which the laser action occurs and below which the diode behaves essentially as an LED, emitting incoherent light.

AN APPLICATION Laser diodes and photodiodes are used in the pick-up system of compact disk (CD) players. Audio information (sound) is digitally recorded in stereo on the surface of a compact disk in the form of microscopic "pits" and "flats." A lens arrangement focuses the laser beam from the diode onto the CD surface. As the CD rotates, the lens and beam follow the track under control of a servomotor. The laser light, which is altered by the pits and flats along the recorded track, is reflected back from the track through a lens and optical system to infrared photodiodes. The signal from the photodiodes is then used to reproduce the digitally recorded sound.

3–5 REVIEW QUESTIONS

1. What are the primary application areas for Schottky diodes?
2. What is a hot-carrier diode?
3. What is the key characteristic of a tunnel diode?
4. What is one application for a tunnel diode?
5. Name the three regions of a PIN diode.
6. What does *laser* means?
7. What is the difference between incoherent and coherent light?

3–6 TROUBLESHOOTING

In this section, you will see how a faulty zener diode can affect the output of a regulated dc power supply. Like other diodes, the zener can fail open, it can exhibit degraded performance in which its internal resistance increases significantly, or it can short out. Generally, however, when a zener shorts internally, its current suddenly becomes excessive; and you end up with an open diode.

A ZENER-REGULATED DC POWER SUPPLY

Figure 3–40 shows a filtered dc power supply that produces a constant 30 V before it is regulated down to 15 V. The zener diode is the same as the one in Example 3–6. A no-load check of the regulated output voltage shows approximately 16 V as indicated in part (a). The voltage expected at maximum zener current (I_{ZM}) for this particular diode is 16.17 V. In part (b), a potentiometer is connected to provide a variable load resistance. It is adjusted to a minimum value for a full-load test as determined by the following calculations. The full-load test is at minimum zener current (I_{ZK}). The expected output voltage of 14.49 V is indicated in the figure.

$$I_T = \frac{30 \text{ V} - 14.49 \text{ V}}{27 \ \Omega} = 574 \text{ mA}$$

$$I_L = 574 \text{ mA} - 1 \text{ mA} = 573 \text{ mA}$$

$$R_{L(\text{min})} = \frac{14.49 \text{ V}}{573 \text{ mA}} \cong 25 \ \Omega$$

CASE 1: ZENER DIODE OPEN

If the zener diode fails open, the power supply test gives the approximate results indicated in Figure 3–41. In the no-load check shown in part (a), the output voltage is 30 V because there is no voltage dropped between the filtered output of the power supply and the output terminal. This definitely indicates an open between the output terminal and ground. In the full-load check, the voltage of 14.42 V results from the voltage-divider action of the 27-Ω series resistor and the 25-Ω load. In this case, this is too close to the normal reading to be a reliable fault indication but the no-load check will verify the problem.

FIGURE 3–40
Regulated power supply test.

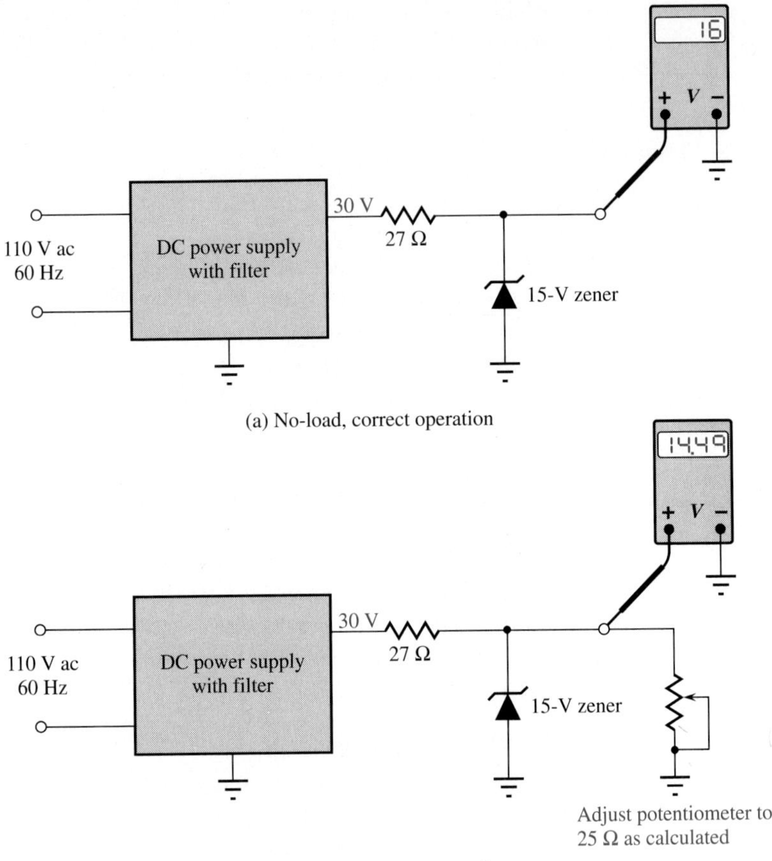

CASE 2: EXCESSIVE ZENER RESISTANCE

As indicated in Figure 3–42, a no-load check that results in an output voltage greater than the maximum zener voltage but less than the power supply output voltage indicates that the zener has failed such that its internal resistance is more than it should be. The 22-V output in this case is 5.83 V higher than the expected value of 16.17 V. That additional voltage is caused by the drop across the excessive internal resistance of the zener.

3–6 REVIEW QUESTIONS

1. In a filtered power supply with a zener regulator, what are the symptoms of an open zener diode?
2. If a zener regulator fails so that the zener resistance is greater than the specified value, is the output voltage more or less than it should be?
3. If you measure 0 V at the output of a zener-regulated power supply, what is the most likely fault(s)?

FIGURE 3–41
Indications of an open zener.

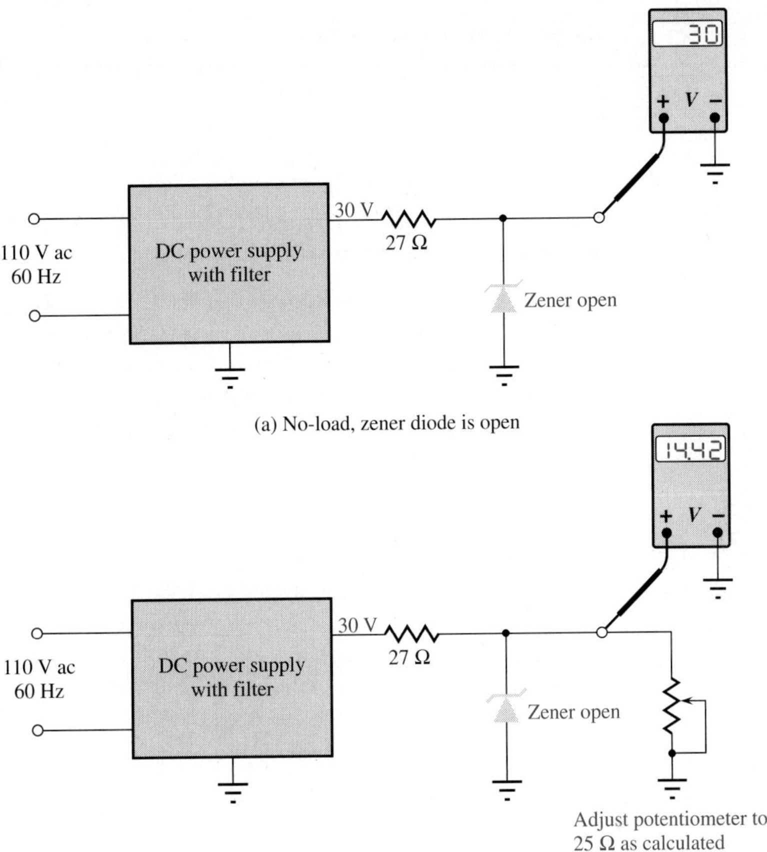

FIGURE 3–42
Indication of excessive zener resistance.

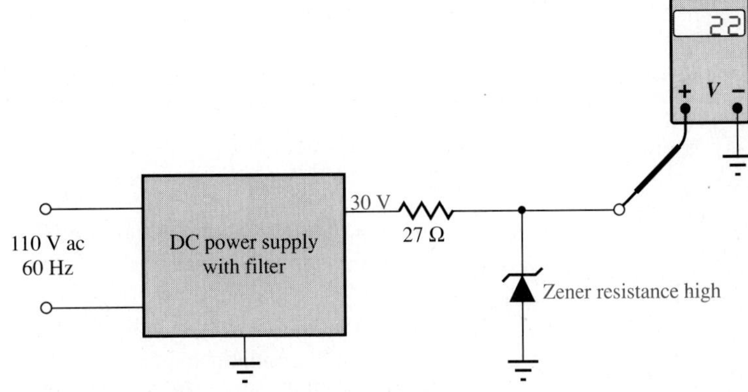

3–7 A SYSTEM APPLICATION

The optical counting system presented at the opening of this chapter incorporates the zener diode, the LED, and the photodiode, as well as many other types of devices. Of course, at this point, we are interested in only how the zener and the optical diodes are used in this system application. The system power supply is similar to the one you worked with in the last chapter except that there is a zener regulator in this one rather than the three-terminal regulator. In this section, you will

- ☐ ***See how a zener diode regulator is used in a specific system application.***
- ☐ ***See how the LED operates in a system application.***
- ☐ ***See how the photodiode operates in a system application.***
- ☐ ***Translate between a printed circuit board and a schematic.***
- ☐ ***Troubleshoot some common system problems.***

A BRIEF DESCRIPTION OF THE SYSTEM

At the opening of this chapter you saw an illustration of the optical-counting system and its basic configuration. The system is used to count objects as they pass on a moving conveyor and can be used in production line monitoring, loading, shipping, inventory control, and the like. The system consists of the power supply, the infrared emitter circuit, the infrared detector, threshold circuit, and counter/display circuits as shown in Figure 3–43 in block diagram form.

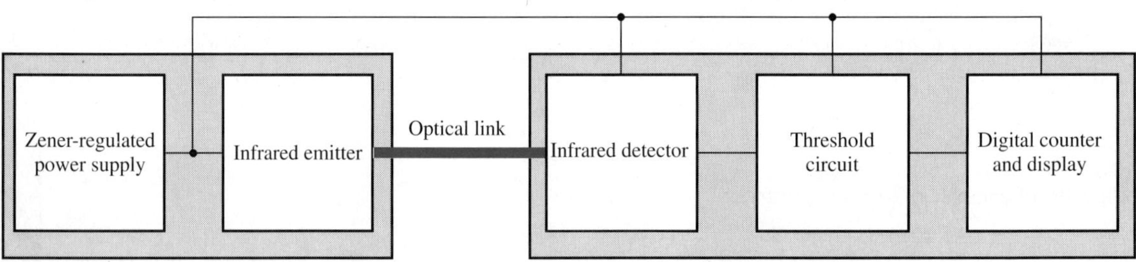

FIGURE 3–43

The power supply and infrared emitter are housed in a single unit and positioned on one side of the conveyor. The detector/counter unit is positioned directly opposite on the other side. The dc power supply provides power to both units. The emitter produces a continuous pattern of infrared light directed toward the detector. As long as the detector is receiving light, nothing happens. When an object passes between the two units and blocks the infrared light, the detector reacts. This reaction is sensed by the threshold circuit, which sends a signal to a digital counter, advancing its count by one. The total count can be displayed and/or sent to a central computer.

In this section, we are not concerned with the details of the threshold circuit or the digital counter and display circuit. In fact, they are beyond the scope of this book and perhaps beyond your background at this time. They are included only to make a complete system and to show that the circuits we are concerned with must **interface** with other types of circuits in order to perform the required system function. At this point, we are strictly interested in how some of the devices covered in this chapter function in the system.

Physically, the emitter unit consists of one printed circuit board that contains the regulated dc power supply and the infrared emitter circuit. The detector/counter unit consists of one printed circuit board that contains the infrared detector circuit, a threshold circuit, and the digital counter and display circuits. All of the circuits on this board except the infrared detector are covered in another course, and you do not need to know the details of how they work for the purposes of this exercise.

Now, so that you can take a closer look at the circuits, let's take them out of the system and put them on the test bench.

ON THE TEST BENCH

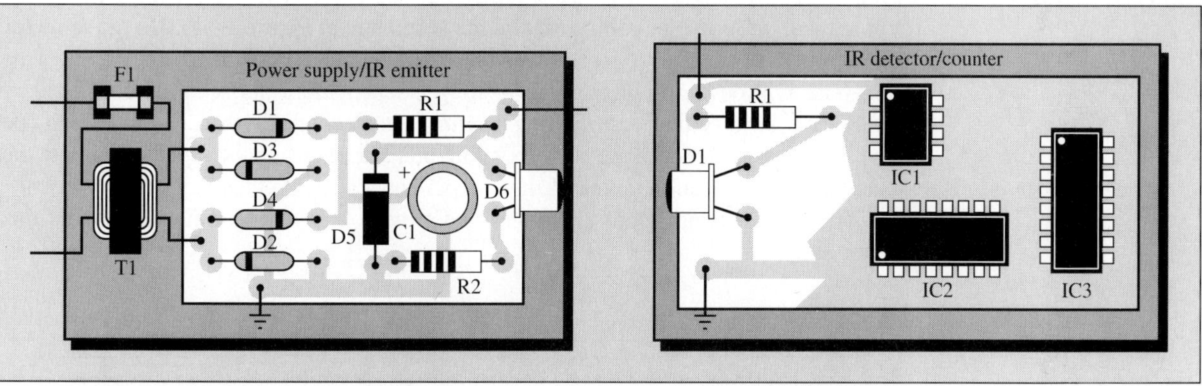

FIGURE 3–44

IDENTIFYING THE COMPONENTS

As you can see in Figure 3–44, the power supply/IR emitter circuit board is very similar to the power supply board that you worked with in the last chapter except for the regulator and the infrared emitter circuit. The dc voltage regulator is made up of a resistor R1 and zener diode D5. The resistor and zener are shown in Figure 3–45(a). The infrared emitter circuit consists of R2 and D6, which is an LED. The infrared detector circuit on the second board is formed by R1 and photodiode D1. Both the LED and the photodiode are shown in Figure 3–45(b). As you can see, they are identical in appearance and can only be distinguished by a part number stamped on the case or on the package from which they are taken. The digital portion of this board is shown in the shaded area without detail. The digital integrated circuits, IC1, IC2, and IC3, are in SOIC (Small Outline Integrated Circuit) packages. SOIC is a surface-mount technology packaging configuration. Although digital circuits are studied in other courses, you see here an example of a common situation in many systems where linear (analog) and digital circuits are interfaced.

FIGURE 3–45

Circuit board components. Part (b) copyright of Motorola, Inc. Used by permission.

ACTIVITY 1 RELATE THE PC BOARD TO THE SCHEMATIC

Carefully follow the conductive traces on each pc board to see how the components are interconnected. Compare each point (*A* through *H*) on the power supply/IR emitter board with the corresponding point on the schematic in Figure 3–46. Do the same for each point (*A* through *C*) on the IR detector/counter board in Figure 3–47. This exercise will develop your skill in relating a schematic to an actual circuit—*a very important skill for an electronics technician or technologist*. For each point on the pc board, place the letter of the corresponding point or points on the schematic. As in the last system application, some of the components on the pc boards are "transparent" so you can see connections and traces under the components for the purposes of this exercise.

FIGURE 3–46

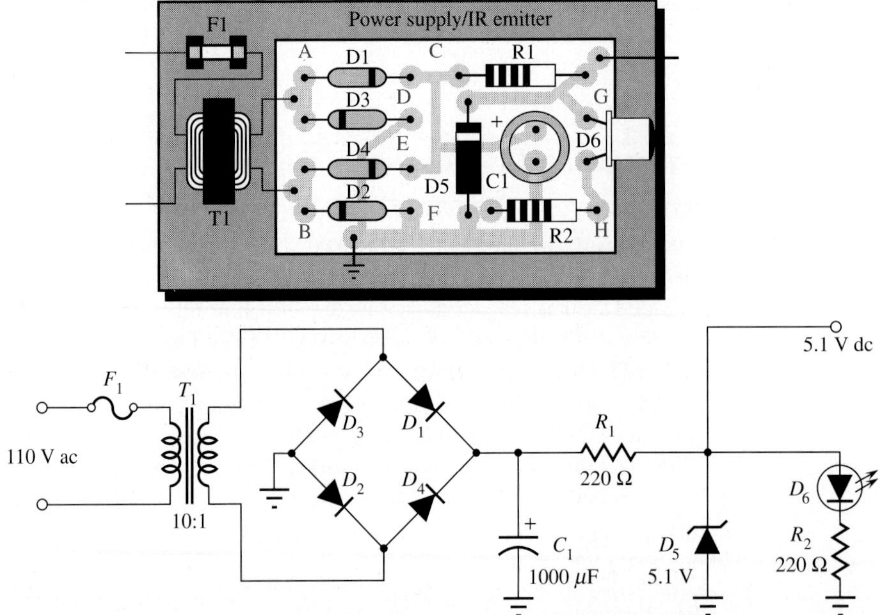

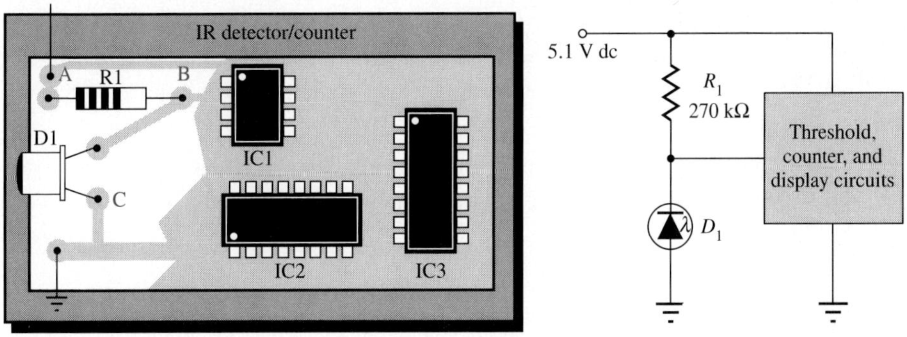

FIGURE 3–47

■ ACTIVITY 2 ANALYZE THE CIRCUITS

STEP 1 With the power supply/IR emitter board plugged into an ac outlet, determine the dc voltages at each point indicated in Figure 3–48. Assume that the circuit is operating properly.

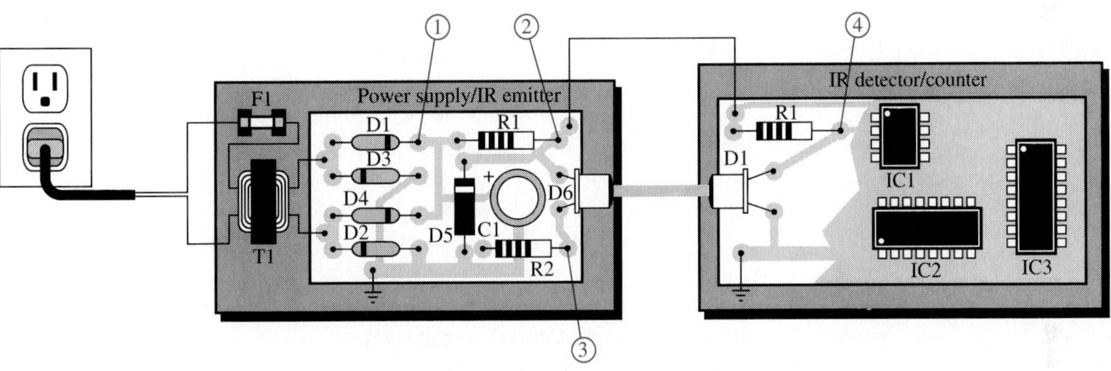

FIGURE 3–48

STEP 2 With the dc supply voltage from the power supply connected to the IR detector/counter board and with light blocked from the photodiode lens, determine the voltage at point 4 in Figure 3–48. Assume that the dark current is 0.025 nA.

STEP 3 With the two units optically aligned and with the power on, determine the voltage at point 4 in the detector circuit. With the IR beam alternately turned on and off by an opaque object passed back and forth in front of the lens, what is the voltage at point 4? Assume that a reverse current of 10 μA flows through the photodiode with incident light. Neglect any current drawn by the digital circuits. Assuming there is a digital readout, what should you see on the readout in the system as you turn the IR beam on and off?

■ ACTIVITY 3 WRITE A TECHNICAL REPORT

Discuss the detailed operation of both the power supply/IR emitter unit and the IR detector/counter unit. You worked with a similar power supply in the system application in the last chapter, so you can utilize that information. Your emphasis should be on the voltage regulation and the operation of the IR optical system. Of course, you do not need to worry about the digital part at this time.

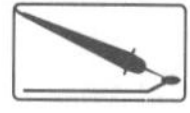

■ ACTIVITY 4 TROUBLESHOOT THE SYSTEM FOR EACH OF THE FOLLOWING PROBLEMS BY STATING THE PROBABLE CAUSE OR CAUSES (REFER TO FIGURE 3–48)

1. No voltage at point 1.
2. No voltage at point 2.
3. A voltage at point 2 that varies from 0 V to about 5 V with a frequency of 120 Hz.
4. A constant dc voltage of approximately 6.73 V at point 2.
5. A constant dc voltage of approximately 10 V at point 2.
6. A floating level at point 4.
7. Approximately 5.1 V at point 4 with no light obstruction between the emitter and the detector.
8. A voltage of 0 V at point 4.

3–7 REVIEW QUESTIONS

1. List all of the types of diodes used in this system.
2. What is the purpose of resistor R2 on the power supply/IR emitter board?
3. What would you do to change the dc voltage to 8.2 V in this system?
4. Refer to the data sheets in Appendix A to specify the appropriate device for Question 3.
5. Does the LED emit continuous IR or does it turn on and off as objects pass by?
6. What is the purpose of the photodiode in this system?
7. Why are two physical units required in this system?
8. Refer to the data sheets in Appendix A to determine the dark current for the MRD500 photodiode if the reverse voltage were increased to 20 V.

SUMMARY

- □ The zener diode operates in reverse breakdown.
- □ There are two breakdown mechanisms in a zener diode: avalanche breakdown and zener breakdown.
- □ When $V_Z < 5$ V, zener breakdown is predominant.
- □ When $V_Z > 5$ V, avalanche breakdown is predominant.
- □ A zener diode maintains a nearly constant voltage across its terminals over a specified range of zener currents.
- □ Zener diodes are used as voltage regulators and limiters.
- □ Zener diodes are available for many voltages ranging from 1.8 V to 200 V.
- □ Regulation of output voltage over a range of input voltages is called *input* or *line regulation*.

- □ Regulation of output voltage over a range of load currents is called *load regulation*.
- □ The smaller the percent regulation, the better.
- □ A varactor diode acts as a variable capacitor under reverse-biased conditions.
- □ The capacitance of a varactor varies inversely with reverse-biased voltage.
- □ The Schottky diode has a metal-to-semiconductor junction. It is used in fast-switching applications.
- □ The tunnel diode is used in oscillator circuits.
- □ An LED emits light when forward-biased.
- □ LEDs are available for either infrared or visible light.
- □ The photodiode exhibits an increase in reverse current with light intensity.
- □ The PIN diode has a p region, an n region, and an intrinsic region, and displays a variable resistance characteristic when forward-biased and a constant capacitance when reverse-biased.
- □ A laser diode is similar to an LED except that it emits coherent (single wavelength) light when the forward current exceeds a threshold value.

GLOSSARY

Coherent light Light having only one wavelength.

Dark current The amount of thermally generated reverse current in a photodiode in the absence of light.

Electroluminescence The process of releasing light energy by the recombination of electrons in a semiconductor.

Infrared Light that has a range of wavelengths greater than visible light.

Interfacing The process of making the output of one type of circuit compatible with the input of another so that they can operate properly together.

Laser *L*ight *A*mplification by *S*timulated *E*mission of *R*adiation.

Light-emitting diode (LED) A type of diode that emits light when forward current flows.

Line regulation The percent change in output voltage for a given change in line (input) voltage.

Load regulation The percent change in output voltage for a given change in load current.

Lumen Unit of light measurement.

Monochromatic Light of a single frequency; one color.

Photodiode A diode in which the reverse current varies directly with the amount of light.

Photon A particle of light energy.

Radiation The process of emitting electromagnetic or light energy.

Schottky diode A diode using only majority carriers and intended for high-frequency operation.

Spectral Pertaining to a range of frequencies.

Tunnel diode A diode exhibiting a negative resistance characteristic.

Varactor A variable capacitance diode.

Wavelength The distance in space occupied by one cycle of an electromagnetic or light wave.

Zener diode A diode designed for limiting the voltage across its terminals in reverse bias.

FORMULAS

(3–1)	$R_Z = \dfrac{\Delta V_Z}{\Delta I_Z}$	Zener resistance
(3–2)	$\Delta V_Z = V_Z \times TC \times \Delta T$	V_Z temperature change
(3–3)	$\text{Percent line regulation} = \dfrac{\Delta V_{\text{OUT}}}{\Delta V_{\text{IN}}} \times 100\%$	Line regulation
(3–4)	$\text{Percent load regulation} = \dfrac{V_{\text{NL}} - V_{\text{FL}}}{V_{\text{FL}}} \times 100\%$	Load regulation
(3–5)	$C = \dfrac{A\epsilon}{d}$	Capacitance
(3–6)	$f_r \cong \dfrac{1}{2\pi\sqrt{LC}}$	Resonant frequency
(3–7)	$R_F = \dfrac{\Delta V_F}{\Delta I_F}$	Negative resistance

SELF-TEST

1. The cathode of a zener diode in a voltage regulator is normally
   (a) more positive than the anode **(b)** more negative than the anode
   (c) at +0.7 V **(d)** grounded
2. If a certain zener diode has a zener voltage of 3.6 V, it operates in
   (a) regulated breakdown
   (b) zener breakdown
   (c) forward conduction
   (d) avalanche breakdown
3. For a certain 12-V zener diode, a 10-mA change in zener current produces a 0.1 V change in zener voltage. The zener resistance for this current range is
   (a) 1 Ω **(b)** 100 Ω **(c)** 10 Ω **(d)** 0.1 Ω

4. The data sheet for a particular zener gives $V_Z = 10$ V at $I_{ZT} = 500$ mA. R_Z for these conditions is

   (a) 50 Ω (b) 20 Ω (c) 10 Ω (d) unknown

5. Line regulation is determined by
   (a) load current
   (b) zener current and load current
   (c) changes in load resistance and output voltage
   (d) changes in output voltage and input voltage

6. Load regulation is determined by
   (a) changes in load current and input voltage
   (b) changes in load current and output voltage
   (c) changes in load resistance and input voltage
   (d) changes in zener current and load current

7. A no-load condition means that
   (a) the load has infinite resistance
   (b) the load has zero resistance
   (c) the output terminals are open
   (d) a and c

8. A varactor diode exhibits
   (a) a variable capacitance that depends on reverse voltage
   (b) a variable resistance that depends on reverse voltage
   (c) a variable capacitance that depends on forward current
   (d) a constant capacitance over a range of reverse voltages

9. An LED
   (a) emits light when reverse-biased
   (b) senses light when reverse-biased
   (c) emits light when forward-biased
   (d) acts as a variable resistance

10. Compared to a visible-red LED, an infrared LED
    (a) produces light with shorter wavelengths
    (b) produces light of all wavelengths
    (c) produces only one color of light
    (d) produces light with longer wavelengths

11. The internal resistance of a photodiode
    (a) increases with light intensity when reverse-biased
    (b) decreases with light intensity when reverse-biased
    (c) increases with light intensity when forward-biased
    (d) decreases with light intensity when forward-biased

12. A diode that has a negative resistance characteristic is the
    (a) Schottky diode (b) tunnel diode
    (c) laser diode (d) hot-carrier diode

13. An infrared LED is optically coupled to a photodiode. When the LED is turned off, the reading on an ammeter in series with the reverse-biased photodiode will
    (a) not change (b) decrease (c) increase (d) fluctuate

14. In order for a system to function properly, the various types of circuits that make up the system must be
    (a) properly biased (b) properly interconnected (c) properly interfaced
    (d) all of the above (e) a and b

PROBLEMS

SECTION 3–1 ZENER DIODES

1. A certain zener diode has a $V_Z = 7.5$ V and an $R_Z = 5\ \Omega$ at a certain current. Sketch the equivalent circuit.

2. From the characteristic curve in Figure 3–49, what is the approximate minimum zener current (I_{ZK}) and the approximate zener voltage I_{ZK}?

FIGURE 3–49

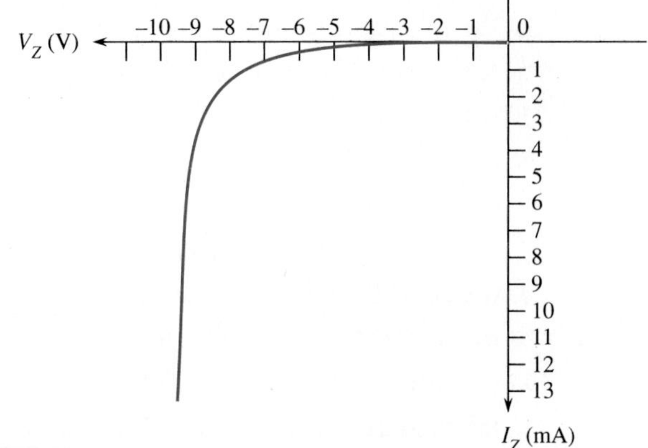

3. When the reverse current in a particular zener diode increases from 20 mA to 30 mA, the zener voltage changes from 5.6 V to 5.65 V. What is the resistance of this device?

4. A zener has a resistance of 15 Ω. What is its terminal voltage at 50 mA if V_{ZT} = 4.7 V at I_{ZT} = 25 mA?

5. A certain zener diode has the following specifications: V_Z = 6.8 V at 25°C and TC = +0.04%/°C. Determine the zener voltage at 70°C.

SECTION 3–2 ZENER DIODE APPLICATIONS

6. Determine the minimum input voltage required for regulation to be established in Figure 3–50. Assume an ideal zener diode with $I_{ZK} = 1.5$ mA and $V_Z = 14$ V.

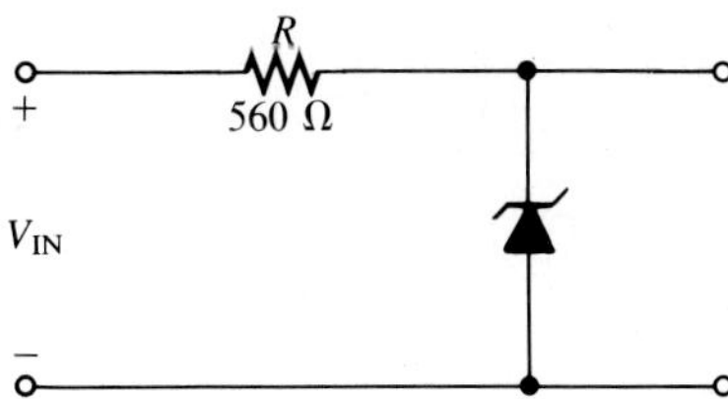

FIGURE 3–50

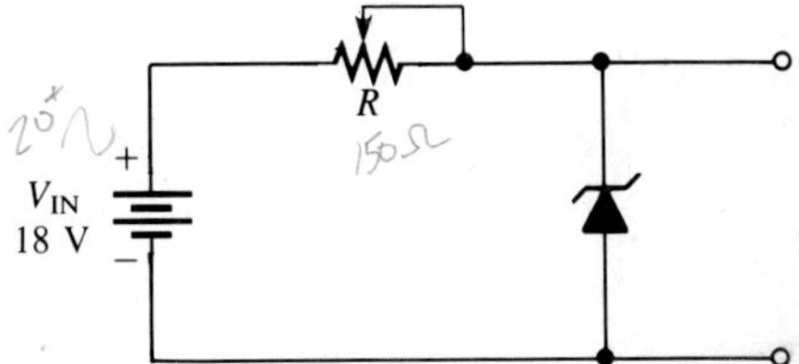

FIGURE 3–51

7. Repeat Problem 6 with $R_Z = 20\ \Omega$ and $V_{ZT} = 14$ V at 30 mA.

8. To what value must R be adjusted in Figure 3–51 to make $I_Z = 40$ mA? Assume $V_Z = 12$ V at 30 mA and $R_Z = 30\ \Omega$.

9. A 20 V peak sine wave voltage is applied to the circuit in Figure 3–51 in place of the dc source. Sketch the output waveform. Use the parameter values established in Problem 8.

10. A loaded zener regulator is shown in Figure 3–52. $V_Z = 5.1$ V at 35 mA, $I_{ZK} = 1$ mA, $R_Z = 12\ \Omega$, and $I_{ZM} = 70$ mA. Determine the minimum and maximum permissible load currents.

FIGURE 3–52

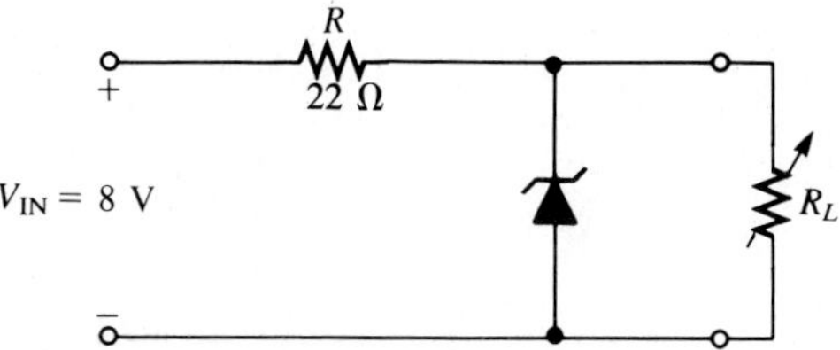

11. Find the percent load regulation in Problem 10.

12. For the circuit of Problem 10, assume the input voltage is varied from 6 V to 12 V. Determine the approximate percent input regulation with no load.

13. The no-load output voltage of a certain zener regulator is 8.23 V, and the full-load output is 7.98 V. Calculate the percent load regulation.

14. In a certain zener regulator, the output voltage changes 0.2 V when the input voltage goes from 5 V to 10 V. What is the percent input regulation?

15. The output voltage of a zener regulator is 3.6 V at no load and 3.4 V at full load. Determine the percent load regulation.

SECTION 3–3 VARACTOR DIODES

16. Figure 3–53 is a curve of reverse voltage versus capacitance for a certain varactor. Determine the change in capacitance if V_R varies from 5 V to 20 V.

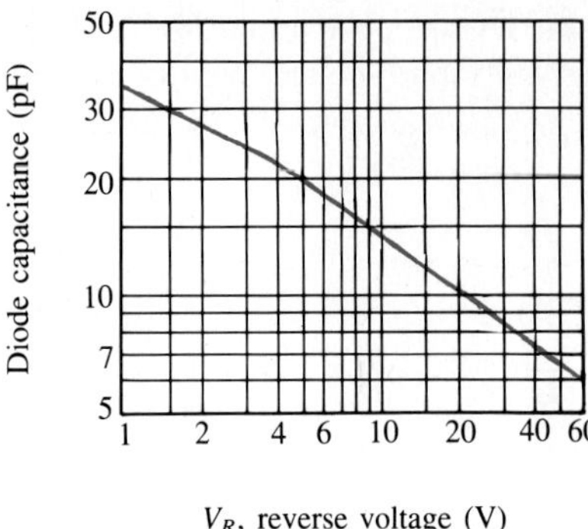

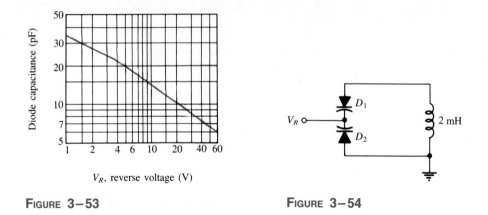

FIGURE 3–53

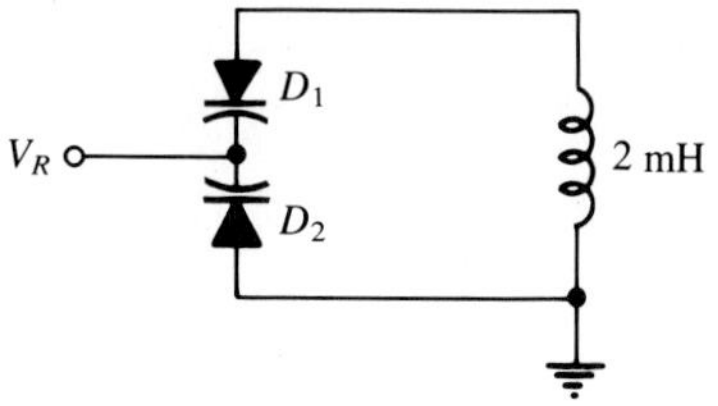

FIGURE 3–54

17. Refer to Figure 3–53 and determine the value of V_R that produces 25 pF.
18. What capacitance value is required for each of the varactors in Figure 3–54 to produce a resonant frequency of 1 MHz?
19. At what value must the control voltage be set in Problem 18 if the varactors have the characteristic curve in Figure 3–53?

SECTION 3–4 OPTICAL DIODES

20. The LED in Figure 3–55(a) has a light-producing characteristic as shown in 3–55(b). Determine the amount of radiant power produced in mW.

FIGURE 3–55

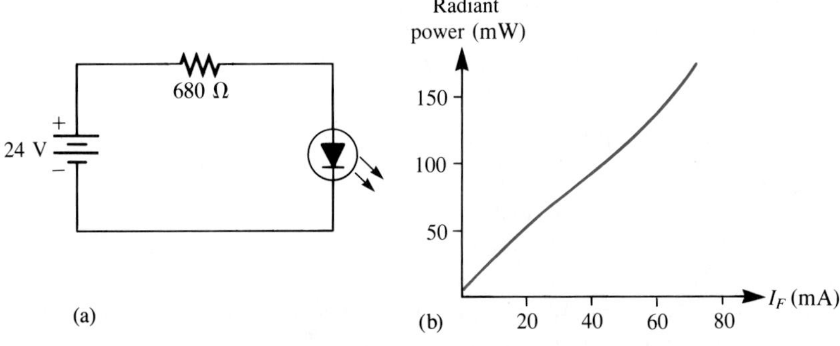

21. Refer to the data sheet for the MLED930 in Appendix A.
    (a) At what wavelength does the peak output occur?
    (b) What is the range of light wavelengths for which the LED produces at least half of its peak output?
    (c) What is the radiant output power when the forward current is 10 mA?
22. For a certain photodiode at a given light intensity, the reverse resistance is 200 kΩ and the reverse voltage is 10 V. What is the current through the device?
23. What is the resistance of each photodiode in Figure 3–56?

FIGURE 3–56

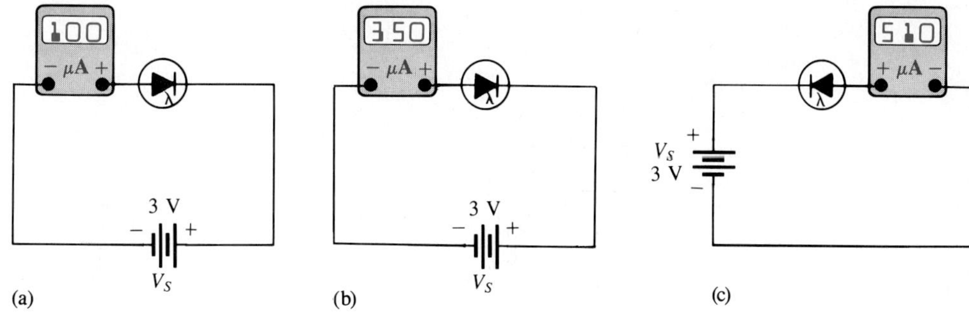

24. When the switch in Figure 3–57 is closed, will the microammeter reading increase or decrease? Assume D_1 and D_2 are optically coupled.

FIGURE 3–57

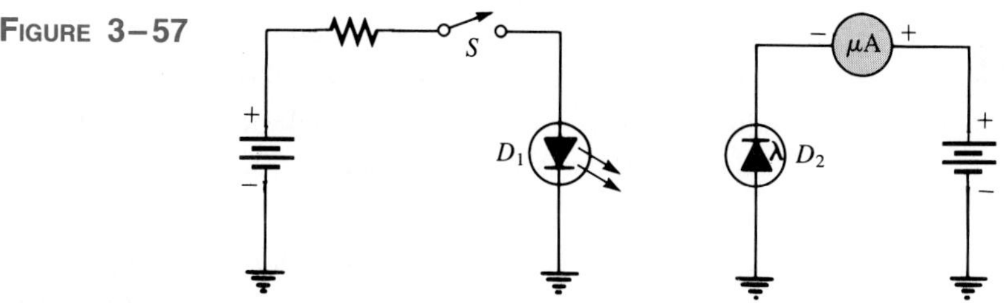

SECTION 3–5 OTHER TYPES OF DIODES

25. The current in a certain tunnel diode changes from 0.25 mA to 0.15 mA when the forward voltage is adjusted from 125 mV to 200 mV. What is the resistance in this region?

26. In what type of circuit are tunnel diodes commonly used?

27. What purpose do the reflective surfaces in the laser diode serve? Why is one end only partially reflective?

SECTION 3–6 TROUBLESHOOTING

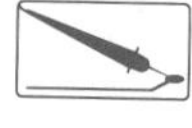
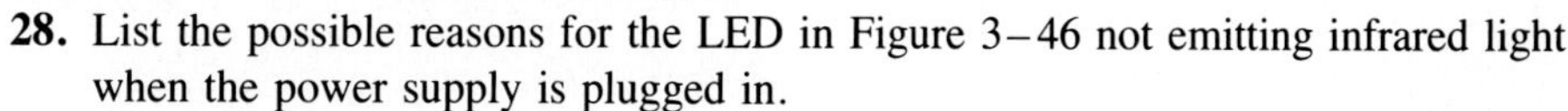

28. List the possible reasons for the LED in Figure 3–46 not emitting infrared light when the power supply is plugged in.

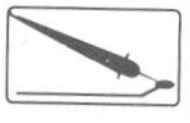

29. List the possible reasons for the photodiode in Figure 3–47 not responding to the infrared light from the LED. List the steps in sequence that you would take to isolate the problem.

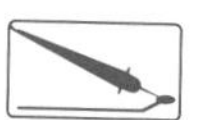

30. For each set of measured voltages at the points (1, 2, and 3) indicated in Figure 3–58, determine if they are correct and if not, identify the most likely fault(s). State what you would do to correct the problem once it is isolated. Refer to the data sheet for information on the 1N963A.

(a) V_1 = 110 V rms, $V_2 \cong$ 30 V dc, $V_3 \cong$ 12 V dc

(b) V_1 = 110 V rms, $V_2 \cong$ 30 V dc, $V_3 \cong$ 30 V dc

FIGURE 3–58

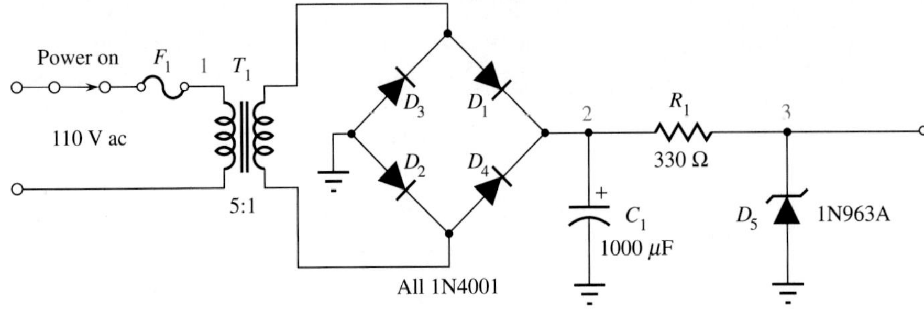

(c) $V_1 = 0$ V, $V_2 = 0$ V, $V_3 = 0$ V

(d) $V_1 = 110$ V rms, $V_2 \cong 30$ V peak full-wave 120-Hz, $V_3 = 12$ V 120-Hz pulsating voltage

(e) $V_1 = 110$ V rms, $V_2 = 0$ V, $V_3 = 0$ V

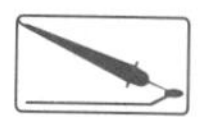

31. Draw the schematic for the circuit board in Figure 3–59 and determine what the correct output voltages should be. Refer to the data sheets for information on the 1N754 and the 1N970A.

FIGURE 3–59

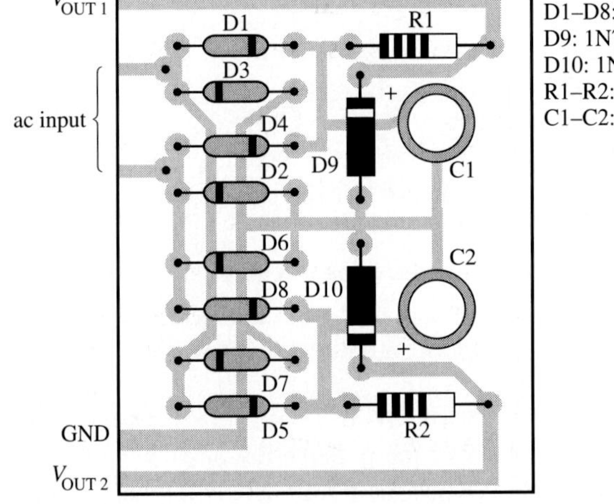

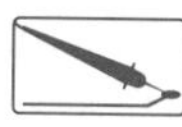

32. The ac input of the circuit board in Figure 3–59 is connected to the secondary of a transformer with a turns ratio of 1 that is operating from the 110 V ac source. When you measure the output voltages, both are zero. What do you think has failed, and what is the reason for this failure?

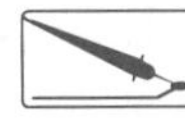

33. Select a transformer turns ratio that will provide the greatest possible secondary voltage compatible with the circuit board in Figure 3–59.

34. In Figure 3–59, V_{OUT2} is zero and V_{OUT1} is correct. What are possible reasons for this?

ANSWERS TO REVIEW QUESTIONS

SECTION 3–1

1. Zener diodes are operated in the reverse-bias region.
2. The test current, I_{ZT}
3. The zener resistance causes the voltage to vary slightly with current.
4. $V_Z = 10\text{ V} + (20\text{ mA})(8\ \Omega) = 10.16\text{ V}$.
5. The zener voltage increases (or decreases) 0.05% for each centrigrade degree increase (or decrease).

SECTION 3–2

1. Line regulation is a change in output voltage for a given change in input voltage. Load regulation is a change in output voltage for a given change in load current.
2. An infinite resistance (open)
3. With no load, there is no current to a load. With full load, there is maximum current to the load.
4. Percent load regulation = ((12 V − 11.9 V)/11.9 V)100% = 0.84%
5. Approximately 0.7 V, just like a rectifier diode

SECTION 3–3

1. A varactor exhibits variable capacitance.
2. A varactor is operated in reverse bias.
3. The depletion region
4. Capacitance decreases with more reverse bias.

SECTION 3–4

1. Infrared and visible light
2. Infrared has the greater wavelengths.
3. An LED operates in forward bias.
4. Light emission increases with forward current.
5. Portions of both the visible and infrared spectra
6. A photodiode operates in reverse bias.
7. The internal resistance decreases.
8. Dark current is the reverse photodiode current when there is no light.

SECTION 3–5

1. High-frequency and fast-switching circuits
2. *Hot carrier* is another name for Schottky diodes.
3. Tunnel diodes have negative resistance.
4. Oscillators
5. *P* region, *N* region, and intrinsic (*I*) region
6. *l*ight *a*mplification by *s*timulated *e*mission of *r*adiation

7. Coherent light has only a single wavelength, but incoherent light has a wide band of wavelengths.

SECTION 3–6

1. The output voltage is too high and equal to the rectifier output.
2. More
3. Series limiting resistor open, fuse blown

SECTION 3–7

1. Rectifier, zener, light-emitting, photo
2. R_2 limits the LED current.
3. Use an 8.2-V zener diode.
4. 1N756 or 1N959A
5. The LED emits a continuous IR beam.
6. To sense the absence of the IR beam
7. An object must be able to pass between the LED and the photodiode.
8. Approximately 0.06 nA

ANSWERS TO PRACTICE EXERCISES

3–1 5 Ω

3–2 $V_Z = 12.6$ V at 80 mA, $V_Z = 11.4$ V at 20 mA

3–3 0.45 V to 11.55 V

3–4 $V_{IN(min)} = 6.55$ V, $V_{IN(max)} = 40.9$ V

3–5 $I_{L(min)} = 0$ A, $I_{L(max)} = 42$ mA, $R_{L(min)} = 79\ \Omega$

3–6 **(a)** 11.58 V at I_{ZK}, 13.02 V at I_{ZM} **(b)** 15.25 Ω **(c)** 14.25 Ω

3–7 **(a)** 3.77% **(b)** 1.75%/V

3–8 **(a)** A waveform identical to Figure 3–16(a)
(b) A sine wave with a peak value of 5 V

3–9 100.66 kHz to 318.31 kHz

4

Bipolar Junction Transistors

After completing this chapter, you should be able to

- □ Describe the basic construction of a bipolar junction transistor (BJT).
- □ Discuss the internal operation of a BJT.
- □ Identify npn and pnp transistors and explain the differences between them.
- □ Define the transistor currents and tell how they are related.
- □ Interpret the characteristic curves of a transistor.
- □ Explain how a transistor is biased for use as an amplifier.
- □ Explain how a transistor produces gain in amplifier circuits.
- □ Use important transistor parameters in calculations.
- □ Explain how the transistor is used as a switch.
- □ Identify various transistor packaging configurations and their terminals.
- □ Troubleshoot basic transistor bias circuits.
- □ Test transistors in-circuit or out-of-circuit.

The transistor was invented by a team of three men at Bell Laboratories in 1947. Although this first transistor was not a bipolar junction device, it was the beginning of a technological revolution that is still continuing. All of the complex electronic devices and systems today are an outgrowth of early developments in semiconductor transistors.

There are two basic types of transistors—the bipolar junction transistor (BJT), which we will begin to study in this chapter, and the field-effect transistor (FET), which we will cover in later chapters. The BJT is used in two broad areas—as a linear amplifier to boost or amplify an electrical signal and as an electronic switch. Both of these applications are introduced in this chapter.

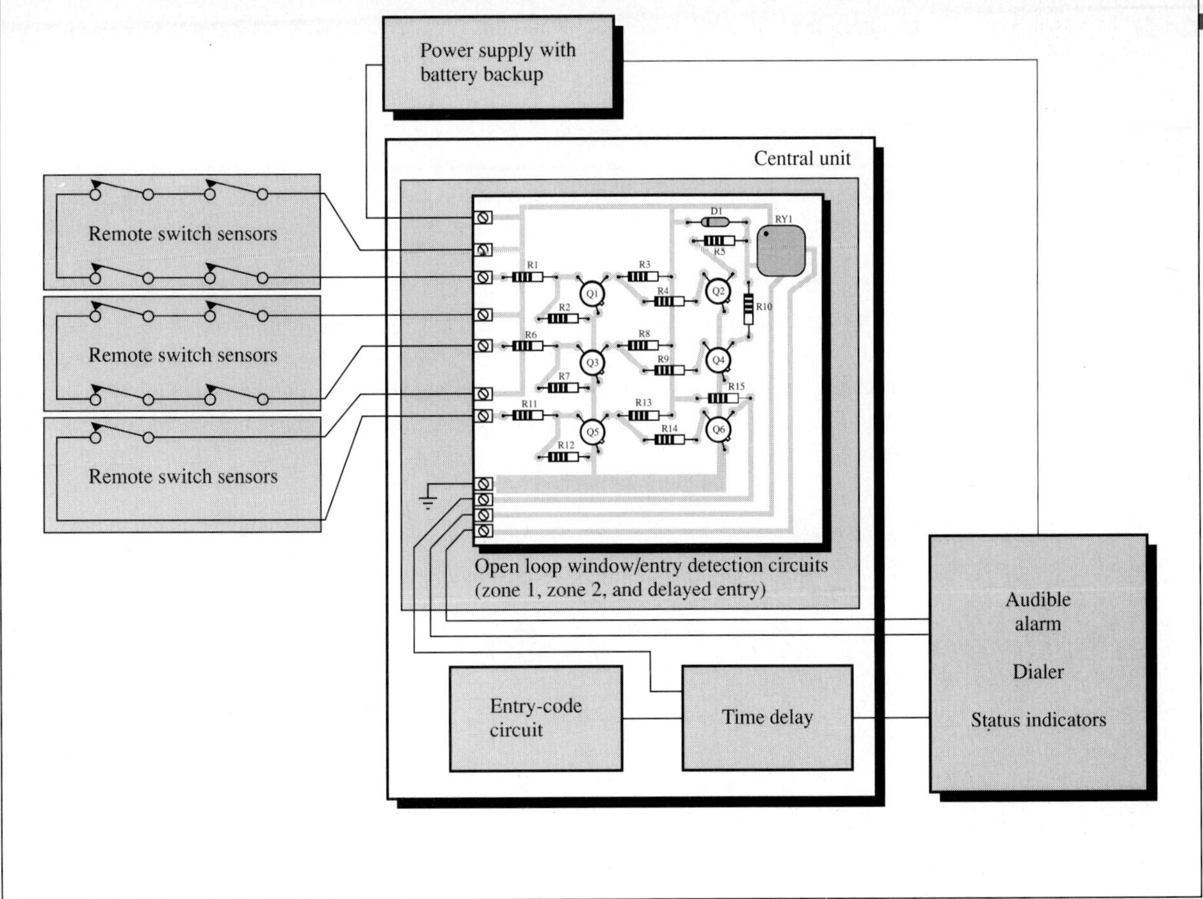

A System Application

Transistors are used in one form or another in just about every electronic system imaginable. A simple electronic security system is the focus of this chapter's system application section. You will see how transistors, operating in their switching modes, are used to perform a very basic function in the system shown in the diagram above. Also, you will see how the transistor is used to operate nonelectronic devices such as, in this case, the electromechanical relay.

For the system application in Section 4–8, in addition to the other topics, be sure you understand

- ☐ How a bipolar junction transistor works.
- ☐ How the currents in a transistor are calculated.
- ☐ The meaning of *saturation* and *cutoff*.
- ☐ The principles of biasing a transistor.
- ☐ How to read a data sheet.

4–1 TRANSISTOR CONSTRUCTION

The basic structure of the bipolar junction transistor (BJT) determines its operating characteristics. In this section, you will see how semiconductor materials are joined to form a transistor, and you will learn the symbols used in schematic diagrams.

The **bipolar junction transistor** (BJT) is constructed with three doped semiconductor regions separated by two pn junctions. The three regions are called **emitter, base,** and **collector.** The two types of bipolar transistors are shown in Figure 4–1. One type consists of two n regions separated by a p region (npn), and the other consists of two p regions separated by an n region (pnp).

FIGURE 4–1
Basic bipolar transistor construction.

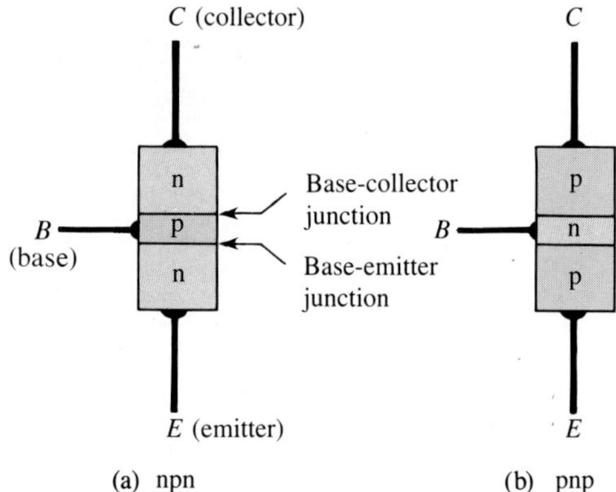

The pn junction joining the base region and the emitter region is called the *base-emitter junction*. The junction joining the base region and the collector region is called the *base-collector junction,* as indicated in Figure 4–1. A wire lead connects to each of the three regions, as shown. These leads are labeled *E, B,* and *C* for emitter, base, and collector, respectively.

The base region is lightly doped and very narrow compared to the heavily doped emitter and collector regions. (The reason for this is discussed in the next section.) Figure 4–2 shows the schematic symbols for the npn and pnp bipolar transistors. The term **bipolar** refers to the use of both holes and electrons as carriers in the transistor structure.

FIGURE 4–2
Standard bipolar junction transistor (BJT) symbols.

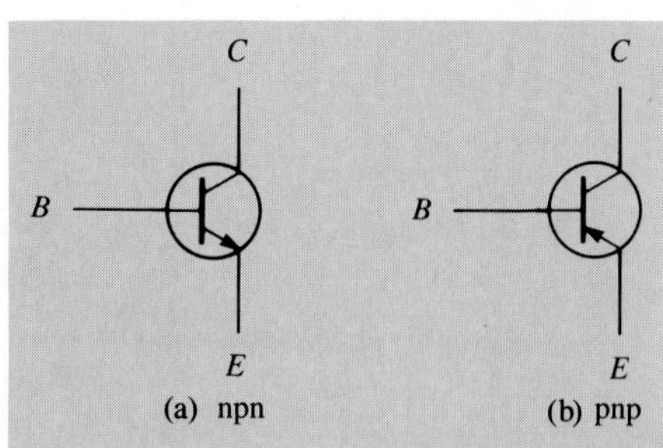

4–1 REVIEW QUESTIONS

1. Name the two types of BJTs according to their construction.
2. The BJT is a three-terminal device. Name the three terminals.
3. What separates the three regions in a BJT?

4–2 BASIC TRANSISTOR OPERATION

In order for the transistor to operate properly as an amplifier, the two pn junctions must be correctly biased with external voltages. In this section, we use the npn transistor for illustration. The operation of the pnp is the same as for the npn except that the roles of the electrons and holes, the bias voltage polarities, and the current directions are all reversed.

Figure 4–3 shows the proper bias arrangement for both npn and pnp transistors. Notice that in both cases the base-emitter *(BE)* junction is forward-biased and the base-collector *(BC)* junction is reverse-biased.

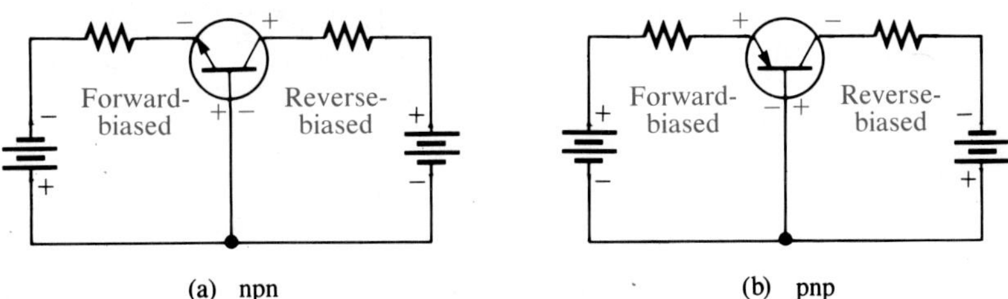

FIGURE 4–3
Forward-reverse bias of a bipolar transistor.

Now, let's examine what happens inside the transistor when it is forward-reverse biased. The forward bias from base to emitter narrows the *BE* depletion layer, and the reverse bias from base to collector widens the *BC* depletion layer, as depicted in Figure 4–4(a). The n-type emitter region is teeming with conduction-band (free) electrons that easily diffuse across the forward-biased *BE* junction into the p-type base region, just as in a forward-biased diode. The base region is lightly doped and very narrow so that it has a very limited number of holes. Thus, only a small percentage of all the electrons flowing across the *BE* junction can combine with the available holes. These relatively few recombined electrons flow out of the base lead as valence electrons, forming the small base current I_B, as shown in Figure 4–4(b).

Most of the electrons flowing from the emitter into the narrow base region do not recombine and diffuse into the *BC* depletion layer. Once in this layer they are pulled across the reverse-biased *BC* junction by the depletion layer field set up by the force of attraction between the positive and negative ions. Actually, you can think of the electrons as being pulled across the reverse-biased *BC* junction by the attraction of the positive ions on the other side. This is illustrated in Figure 4–4(c). The electrons now move through the collector region, out through the collector lead, and into the positive terminal of the external dc source. This forms the collector current, I_C, as shown. The amount of collector current depends directly on the amount of base current and is essentially independent of the dc collector voltage.

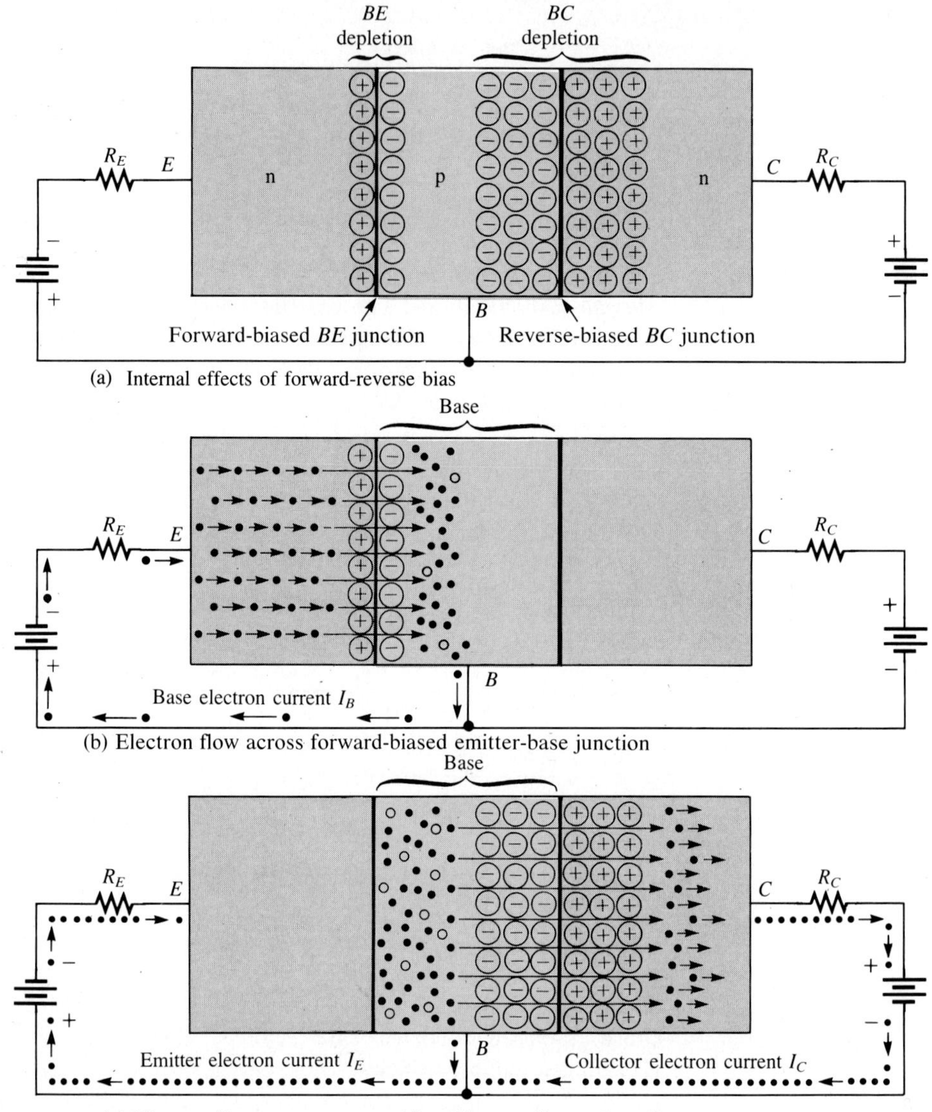

FIGURE 4–4
Illustration of BJT action. The base region is very narrow, but it is shown wider here for clarity.

TRANSISTOR CURRENTS

The directions of the currents in an npn transistor are as shown in Figure 4–5(a), and those for a pnp are shown in Figure 4–5(b). The currents are indicated on the corresponding schematic symbols in parts (c) and (d) of the figure. Notice that the arrow on the emitter of the transistor symbols points in the direction of conventional current. An examination of these diagrams shows that the emitter current is the sum of the collector and base currents, expressed as follows.

$$I_E = I_C + I_B \tag{4-1}$$

As mentioned before, I_B is very small compared to I_E or I_C. The capital-letter subscripts indicate dc values.

FIGURE 4–5
Transistor current directions.

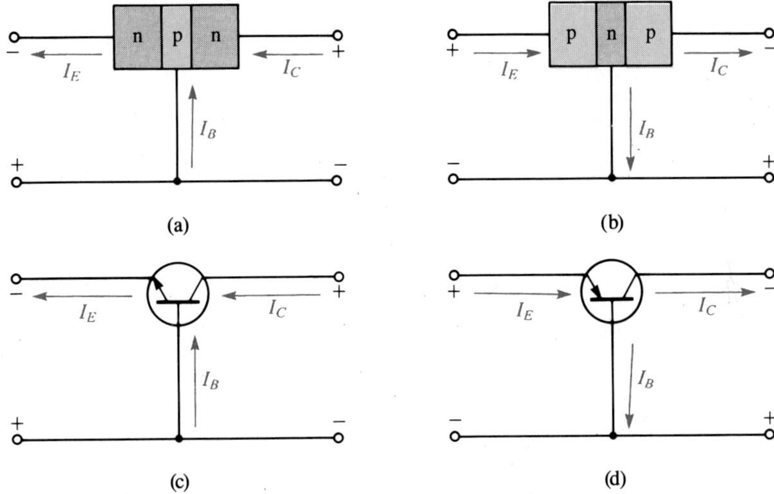

4–2 REVIEW QUESTIONS

1. What are the bias conditions of the base-emitter and base-collector junctions for a transistor to operate as an amplifier?
2. Which is the largest of the three transistor currents?
3. Is the base current smaller or larger than the emitter current?
4. Is the base region much narrower or much wider than the collector and emitter regions?
5. If the collector current is 1 mA and the base current is 1 μA, what is the emitter current?

4–3 TRANSISTOR PARAMETERS AND RATINGS

In this section, you will first learn how to set up a dc circuit to properly bias a transistor. Two important parameters, β_{dc} (dc current gain) and α_{dc} are introduced and used to analyze a transistor circuit. Also, transistor characteristic curves are covered, and you will learn how a transistor's operation can be determined from these curves. Finally, maximum ratings of a transistor are discussed.

When a transistor is connected to bias voltages, as shown in Figure 4–6 for both npn and pnp types, V_{BB} forward-biases the base-emitter junction, and V_{CC} reverse-biases the base-collector junction.

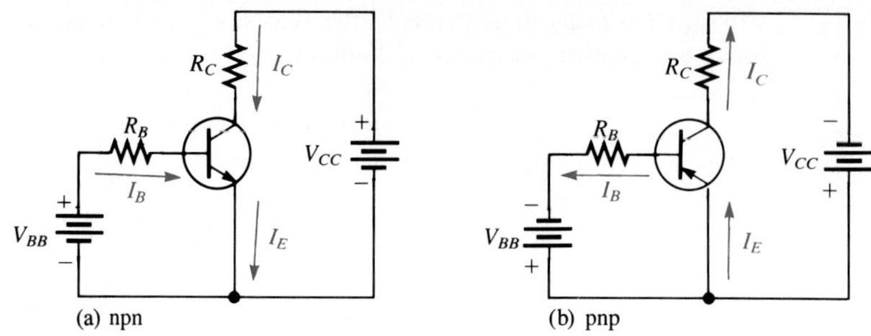

FIGURE 4–6
Transistor bias circuits.

DC BETA (β_{dc}) AND DC ALPHA (α_{dc})

The ratio of the collector current I_C to the base current I_B is the **beta** or the dc current **gain** (β_{dc}) of a transistor.

$$\beta_{dc} = \frac{I_C}{I_B} \quad (4\text{–}2)$$

Typical values of β_{dc} range from 20 to 200 or higher. β_{dc} is usually designated as h_{FE} on transistor data sheets. The ratio of the collector current to the emitter current I_E is the dc **alpha** (α_{dc}).

$$\alpha_{dc} = \frac{I_C}{I_E} \quad (4\text{–}3)$$

Typically, values of α_{dc} range from 0.95 to 0.99 or greater.

RELATIONSHIP OF α_{dc} AND β_{dc}

Starting with the current formula $I_E = I_C + I_B$ and dividing by I_C, we get

$$\frac{I_E}{I_C} = \frac{I_C}{I_C} + \frac{I_B}{I_C}$$

$$= 1 + \frac{I_B}{I_C}$$

Since $\beta_{dc} = I_C/I_B$ and $\alpha_{dc} = I_C/I_E$, the equation becomes

$$\frac{1}{\alpha_{dc}} = 1 + \frac{1}{\alpha_{dc}}$$

Rearranging, we get

$$\frac{1}{\alpha_{dc}} = \frac{\beta_{dc} + 1}{\beta_{dc}}$$

$$\alpha_{dc} = \frac{\beta_{dc}}{\beta_{dc} + 1} \quad (4\text{–}4)$$

Equation (4–4) allows us to calculate α_{dc} if we know β_{dc}. By simple algebra, a formula for β_{dc} in terms of α_{dc} is derived as follows from Equation (4–4).

$$\alpha_{dc}(\beta_{dc} + 1) = \beta_{dc}$$
$$\alpha_{dc}\beta_{dc} + \alpha_{dc} = \beta_{dc}$$
$$\alpha_{dc} = \beta_{dc} - \alpha_{dc}\beta_{dc}$$
$$\beta_{dc}(1 - \alpha_{dc}) = \alpha_{dc}$$

$$\beta_{dc} = \frac{\alpha_{dc}}{1 - \alpha_{dc}} \tag{4–5}$$

EXAMPLE 4–1 Determine β_{dc} and α_{dc} for a transistor where $I_B = 50\ \mu A$ and $I_C = 3.65$ mA.

SOLUTION

$$\beta_{dc} = \frac{I_C}{I_B} = \frac{3.65 \text{ mA}}{50\ \mu\text{A}} = 73$$

$$\alpha_{dc} = \frac{\beta_{dc}}{\beta_{dc} + 1} = \frac{73}{74} = 0.986$$

PRACTICE EXERCISE 4–1

A certain transistor has a β_{dc} of 200. When the base current is 50 μA, determine the collector current. What is α_{dc}?

DC ANALYSIS

Consider the circuit configuration in Figure 4–7. There are three transistor currents and three voltages: I_B, I_E, I_C, V_{BE}, V_{CB}, and V_{CE}.

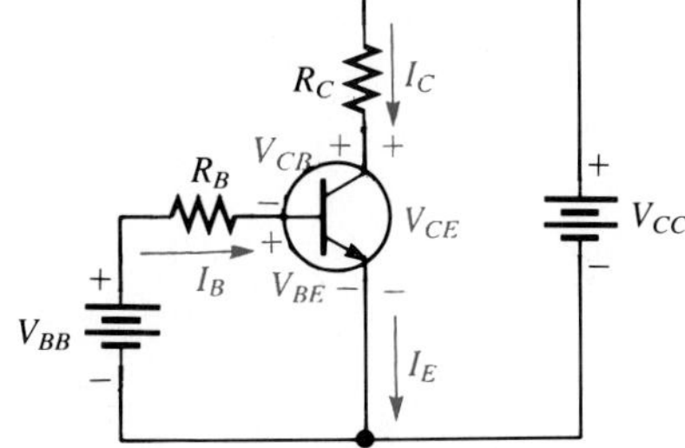

FIGURE 4–7 Transistor currents and voltages.

As you know, V_{BB} forward-biases the base-emitter junction and V_{CC} reverse-biases the base-collector junction. When the base-emitter junction is forward-biased, it is like a forward-biased diode and has a forward voltage drop of

$$V_{BE} \cong 0.7 \text{ V} \tag{4–6}$$

The voltage across R_B is

$$V_{R_B} = V_{BB} - V_{BE}$$

and

$$V_{R_B} = I_B R_B$$

Substituting,

$$I_B R_B = V_{BB} - V_{BE}$$

and

$$I_B = \frac{V_{BB} - V_{BE}}{R_B} \tag{4–7}$$

From Equation (4–3), the expression for I_E is

$$I_E = \frac{I_C}{\alpha_{dc}}$$

From Equation (4–2), the collector current is

$$I_C = \beta_{dc} I_B$$

The drop across R_C is

$$V_{R_C} = I_C R_C$$

The voltage at the collector with respect to the emitter (ground) is

$$V_{CE} = V_{CC} - I_C R_C \tag{4–8}$$

The voltage between the base and collector is

$$V_{CB} = V_{CE} - V_{BE} \tag{4–9}$$

EXAMPLE 4–2 Determine I_B, I_C, I_E, α_{dc}, V_{CE}, and V_{CB} in the circuit of Figure 4–8. The transistor has a $\beta_{dc} = 150$.

SOLUTION

$$I_B = \frac{V_{BB} - V_{BE}}{R_B} = \frac{5\text{ V} - 0.7\text{ V}}{10\text{ k}\Omega} = 430\ \mu\text{A}$$

$$I_C = \beta_{dc} I_B = (150)(430\ \mu\text{A}) = 64.5\text{ mA}$$

$$\alpha_{dc} = \frac{\beta_{dc}}{\beta_{dc} + 1} = \frac{150}{151} = 0.993$$

$$I_E = \frac{I_C}{\alpha_{dc}} = \frac{64.5\text{ mA}}{0.993} = 64.95\text{ mA}$$

$$V_{CE} = V_{CC} - I_C R_C = 10\text{ V} - (64.5\text{ mA})(100\ \Omega)$$
$$= 10\text{ V} - 6.45\text{ V} = 3.55\text{ V}$$

$$V_{CB} = V_{CE} - V_{BE} = 3.55\text{ V} - 0.7\text{ V} = 2.85\text{ V}$$

Since the collector is at a higher voltage than the base, the collector-base junction is reverse-biased.

FIGURE 4–8

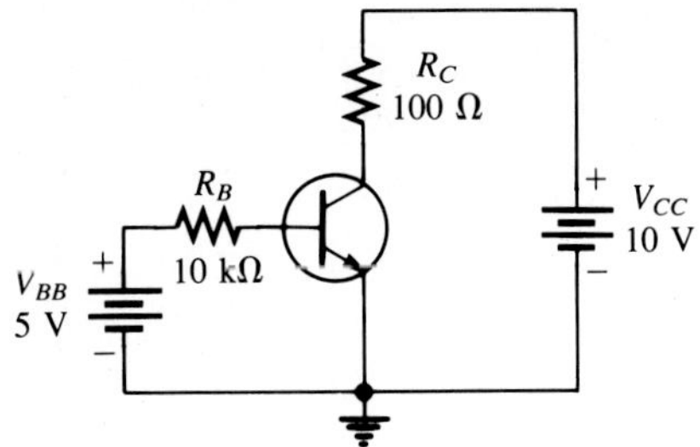

The following **BASIC** computer program will compute the transistor dc currents and voltages when you input V_{BB}, V_{CC}, R_B, R_C, and β_{dc}.

```
10  CLS
20  PRINT "THIS PROGRAM COMPUTES THE DIRECT CURRENTS"
30  PRINT "AND DC VOLTAGES FOR THE CIRCUIT IN FIGURE"
40  PRINT "4-8 GIVEN VALUES FOR VBB, VCC, RB, RC,"
50  PRINT "AND DC BETA."
60  PRINT:PRINT:PRINT
70  INPUT "TO CONTINUE PRESS 'ENTER'";X
80  CLS
90  INPUT "VBB IN VOLTS";VBB
100 INPUT "VCC IN VOLTS";VCC
110 INPUT "RB IN OHMS";RB
120 INPUT "RC IN OHMS";RC
130 INPUT "DC BETA";B
140 CLS
150 IB=(VBB-.7)/RB
160 IC=B*IB
170 ALPHA=B/(B+1)
180 IE=IC/ALPHA
190 VCE=VCC-IC*RC
200 IF VCE<=0 THEN PRINT "TRANSISTOR IS SATURATED" ELSE 220
210 END
220 VBC=.7-VCE
230 PRINT "IB =";IB;"A"
240 PRINT "IC =";IC;"A"
250 PRINT "IE =";IE;"A"
260 PRINT "VCE =";VCE;"V"
270 PRINT "VCB =";VCB;"V"
```

PRACTICE EXERCISE 4–2

Determine I_B, I_C, I_E, V_{CE}, and V_{CB} in Figure 4–8 for the following values: $R_B = 22\ \text{k}\Omega$, $R_C = 220\ \Omega$, $V_{BB} = 6$ V, $V_{CC} = 9$ V, and $\beta_{dc} = 90$.

COLLECTOR CHARACTERISTIC CURVES

With a circuit like Figure 4–9(a), a set of curves can be generated to show how I_C varies with V_{CE} for various values of I_B. These are the *collector characteristic curves*. Notice that both V_{BB} and V_{CC} are adjustable. If V_{BB} is set to produce a specific value of I_B and $V_{CC} = 0$, then $I_C = 0$ and $V_{CE} = 0$. Now, as V_{CC} is gradually increased, V_{CE} will increase and so will I_C. This is indicated on the color-shaded portion of the curve between points *A* and *B* in Figure 4–9(b).

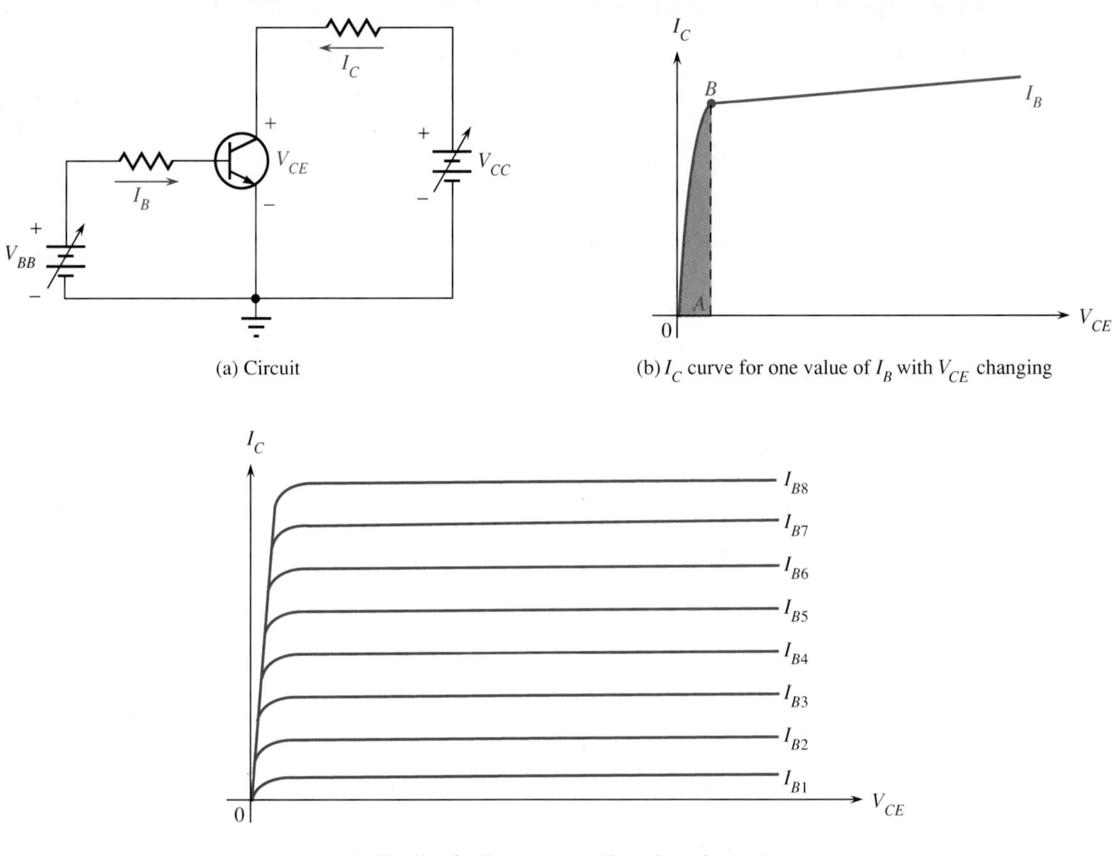

(a) Circuit

(b) I_C curve for one value of I_B with V_{CE} changing

(c) Family of collector curves ($I_{B1} < I_{B2} < I_{B3}$, etc.)

FIGURE 4–9
Collector characteristic curves.

When V_{CE} reaches approximately 0.7 V, the base-collector junction becomes reverse-biased and I_C reaches its full value determined by the relationship $I_C = \beta_{dc} I_B$. At this point, the I_C levels off to an almost constant value as V_{CE} continues to increase. This action appears to the right of point B on the curve. Actually, I_C increases slightly as V_{CE} increases due to widening of the base-collector depletion layer, which results in fewer holes for recombination in the base region. By using other values of I_B, additional I_C-versus-V_{CE} curves can be produced, as shown in Figure 4–9(c). These curves constitute a *family* of collector curves for a given transistor.

EXAMPLE 4–3

Sketch the family of collector curves for the circuit in Figure 4–10 for $I_B = 5\ \mu$A to 25 μA in 5 μA increments. Assume $\beta_{dc} = 100$.

SOLUTION

Using the relationship $I_C = \beta_{dc} I_B$, values of I_C are calculated and tabulated in Table 4–1. The resulting curves are plotted in Figure 4–11.

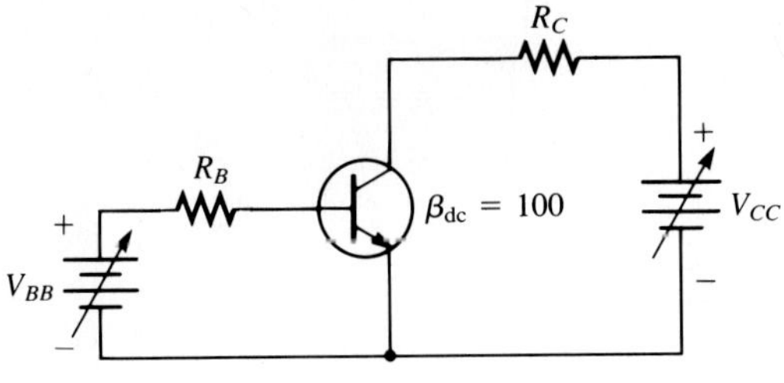

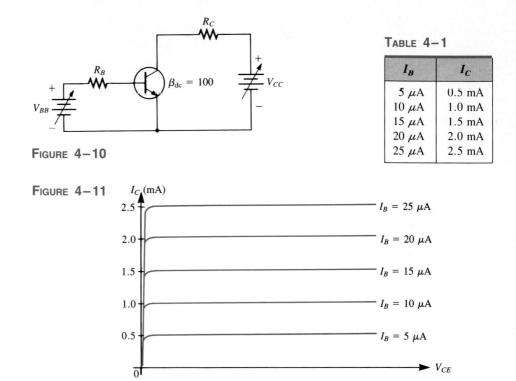

FIGURE 4–10

TABLE 4–1

I_B	I_C
5 μA	0.5 mA
10 μA	1.0 mA
15 μA	1.5 mA
20 μA	2.0 mA
25 μA	2.5 mA

FIGURE 4–11

PRACTICE EXERCISE 4–3

Where would the curve for $I_B = 0$ appear on the graph in Figure 4–11, neglecting leakage?

CUTOFF AND SATURATION

When $I_B = 0$, the transistor is cut off. This is shown in Figure 4–12 with the base lead open, resulting in a base current of zero. Under this condition, there is a very small amount of collector leakage current, I_{CEO}, due mainly to thermally produced carriers. Because I_{CEO} is extremely small, it will usually be neglected in our circuit calculations. In **cutoff,** both the base-emitter and the base-collector junctions are reverse-biased.

FIGURE 4–12
Collector leakage current (I_{CEO}) in cutoff.

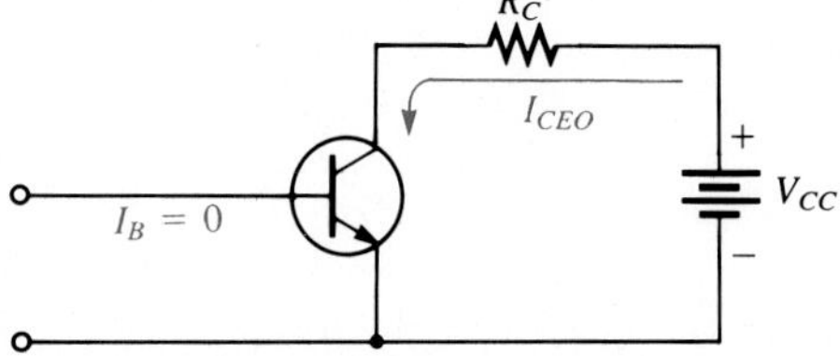

Now let's consider the condition known as **saturation.** When the base current in Figure 4–13(a) is increased, the collector current also increases and V_{CE} decreases as a result of more drop across R_C. When V_{CE} reaches its saturation value, $V_{CE(sat)}$, the

FIGURE 4–13
Saturation.

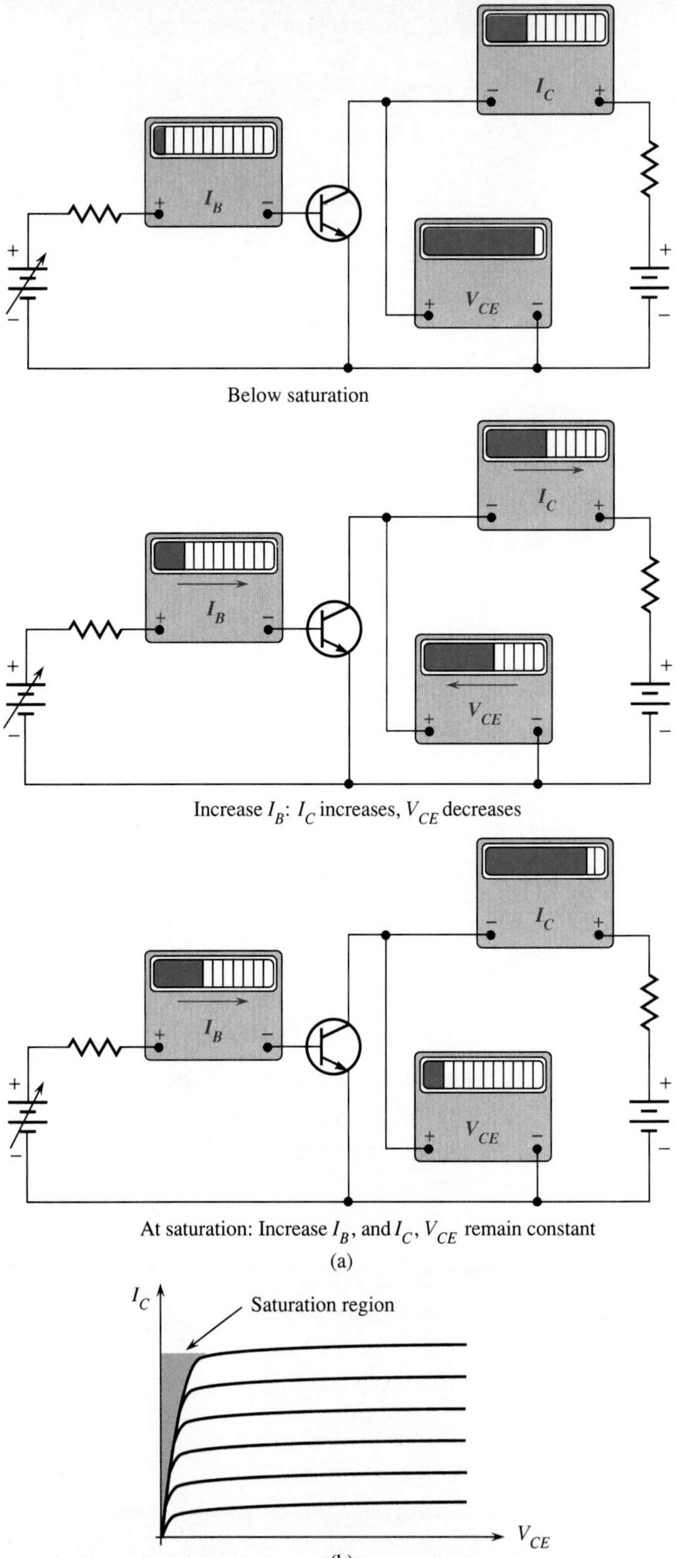

Below saturation

Increase I_B: I_C increases, V_{CE} decreases

At saturation: Increase I_B, and I_C, V_{CE} remain constant

(a)

I_C

Saturation region

V_{CE}

(b)

base-collector junction becomes forward-biased and I_C can increase no further even with a continued increase in I_B. At the point of saturation, the relation $I_C = \beta_{dc}I_B$ is no longer valid. $V_{CE(sat)}$ for a transistor occurs somewhere below the knee of the collector curves, as shown in Figure 4–13(b), and it is usually only a few tenths of a volt for silicon transistors.

EXAMPLE 4–4 Determine whether or not the transistor in Figure 4–14 is in saturation. Assume $V_{CE(sat)} = 0.2$ V.

FIGURE 4–14

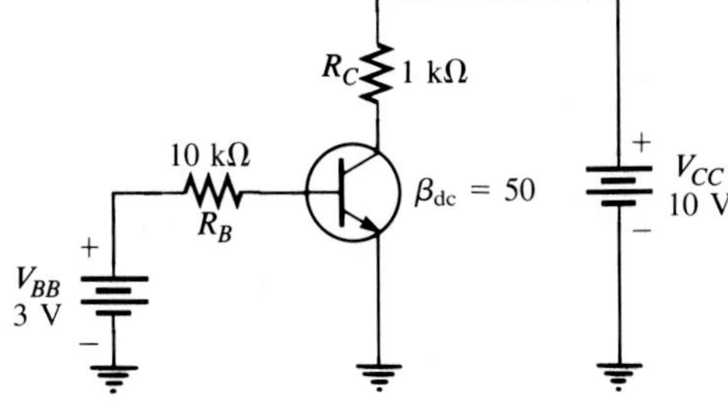

SOLUTION

First, determine $I_{C(sat)}$.

$$I_{C(sat)} = \frac{V_{CC} - V_{CE(sat)}}{R_C} \cong \frac{10\text{ V} - 0.2\text{ V}}{1\text{ k}\Omega} = \frac{9.8\text{ V}}{1\text{ k}\Omega} = 9.8\text{ mA}$$

Now, let's see if I_B is large enough to produce $I_{C(sat)}$.

$$I_B = \frac{V_{BB} - V_{BE}}{R_B} = \frac{3\text{ V} - 0.7\text{ V}}{10\text{ k}\Omega} = \frac{2.3\text{ V}}{10\text{ k}\Omega} = 0.23\text{ mA}$$

$$I_C = \beta_{dc}I_B = (50)(0.23\text{ mA}) = 11.5\text{ mA}$$

This shows that with the specified β_{dc}, this base current is capable of producing an I_C greater than $I_{C(sat)}$. Therefore, the transistor is saturated, and the collector current value of 11.5 mA is never reached. If you further increase I_B, the collector current remains at its saturation value.

PRACTICE EXERCISE 4–4

Determine whether or not the transistor in Figure 4–14 is saturated for the following values: $\beta_{dc} = 25$, $V_{BB} = 1.5$ V, $R_B = 6.8$ kΩ, $R_C = 1.8$ kΩ, and $V_{CC} = 12$ V.

MORE ABOUT β_{dc}

The β_{dc} is a very important bipolar transistor parameter that we need to examine further. β_{dc} is not really a constant but varies with both collector current and temperature. Keeping the junction temperature constant and increasing I_C causes β_{dc} to increase to a maximum. A further increase in I_C beyond this maximum point causes β_{dc} to decrease. If I_C is held constant and the temperature is varied, β_{dc} changes directly with the temperature. If the

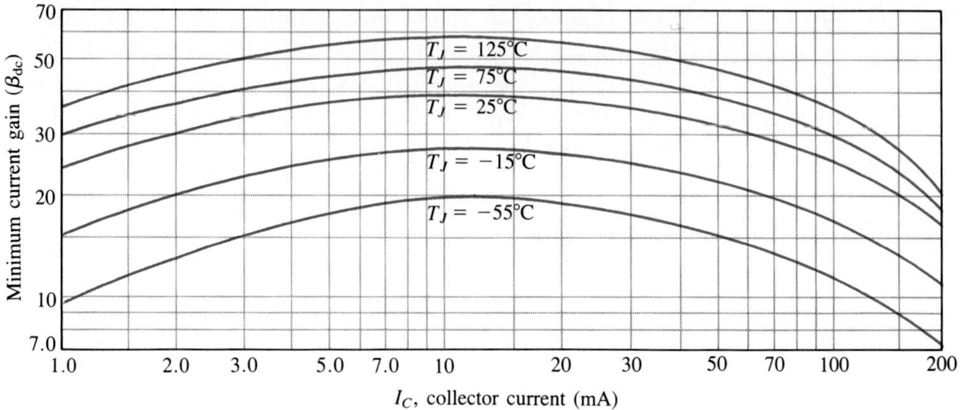

FIGURE 4–15
Variation of β_{dc} with I_C for several temperatures.

temperature goes up, β_{dc} goes up and vice versa. Figure 4–15 shows the variation of β_{dc} with I_C and junction temperature (T_J) for a typical transistor.

A transistor data sheet usually specifies β_{dc} (h_{FE}) at specific I_C values. Even at fixed values of I_C and temperature, β_{dc} varies from device to device for a given transistor. The β_{dc} specified at a certain value of I_C is usually the minimum value, $\beta_{dc(min)}$, although the maximum and typical values are also sometimes specified.

MAXIMUM RATINGS

The transistor, like any other electronic device, has limitations on its operation. These limitations are stated in the form of maximum ratings and are normally specified on the manufacturer's data sheet. Typically, maximum ratings are given for collector-to-base voltage, collector-to-emitter voltage, emitter-to-base voltage, collector current, and power dissipation.

The product of V_{CE} and I_C must not exceed the maximum power dissipation. Both V_{CE} and I_C cannot be maximum at the same time. If V_{CE} is maximum, I_C can be calculated as

$$I_C = \frac{P_{D(max)}}{V_{CE}} \tag{4–10}$$

If I_C is maximum, V_{CE} can be calculated by rearranging Equation (4–10) as follows:

$$V_{CE} = \frac{P_{D(max)}}{I_C} \tag{4–11}$$

For a given transistor, a maximum power dissipation curve can be plotted on the collector characteristic curves, as shown in Figure 4–16(a). These values are tabulated in Figure 4–16(b). The $P_{D(max)}$ for this transistor is 0.5 W, $V_{CE(max)}$ is 20 V, and $I_{C(max)}$ is 50 mA. The curve shows that this particular transistor cannot be operated in the shaded portion of the graph. $I_{C(max)}$ is the limiting rating between points A and B, $P_{D(max)}$ is the

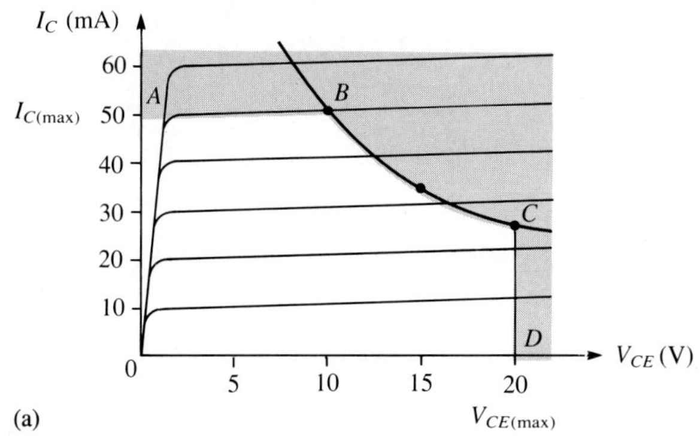

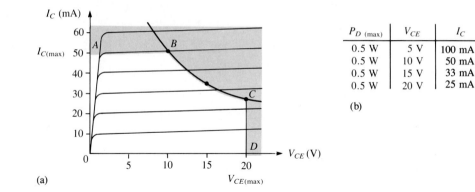

$P_{D\,(max)}$	V_{CE}	I_C
0.5 W	5 V	100 mA
0.5 W	10 V	50 mA
0.5 W	15 V	33 mA
0.5 W	20 V	25 mA

(b)

FIGURE 4–16
Maximum power dissipation curve.

limiting rating between points B and C, and $V_{CE(max)}$ is the limiting rating between points C and D.

EXAMPLE 4–5

A certain transistor is to be operated with $V_{CE} = 6$ V. If its maximum power rating is 0.25 W, what is the most collector current that it can withstand?

SOLUTION

$$I_C = \frac{P_{D(max)}}{V_{CE}} = \frac{0.25\text{ W}}{6\text{ V}} = 41.67\text{ mA}$$

Remember that this is not necessarily the maximum I_C. The transistor can handle more collector current if V_{CE} is reduced, as long as $P_{D(max)}$ is not exceeded.

PRACTICE EXERCISE 4–5

If $P_{D(max)} = 1$ W, how much voltage is allowed from collector to emitter if the transistor is operating with $I_C = 100$ mA?

EXAMPLE 4–6

The silicon transistor in Figure 4–17 has the following maximum ratings: $P_{D(max)} = 0.8$ W, $V_{CE(max)} = 15$ V, $I_{C(max)} = 100$ mA, $V_{CB(max)} = 20$ V, and $V_{EB(max)} = 10$ V. Determine the maximum value to which V_{CC} can be adjusted without exceeding a rating. Which rating would be exceeded first?

FIGURE 4–17

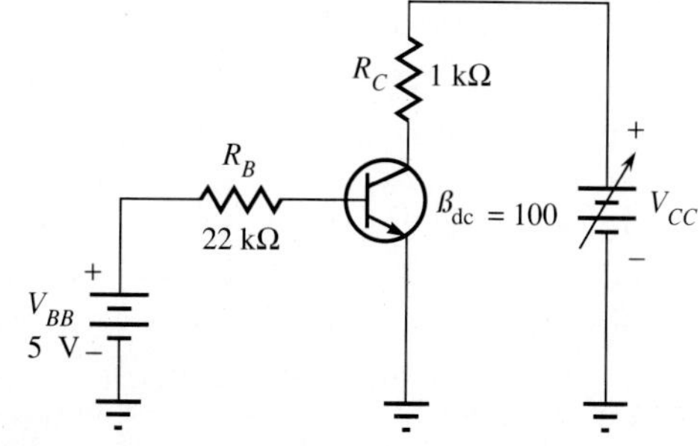

SOLUTION

First find I_B so that I_C can be determined.

$$I_B = \frac{V_{BB} - V_{BE}}{R_B} = \frac{5\text{ V} - 0.7\text{ V}}{22\text{ k}\Omega} = 195.5\ \mu\text{A}$$

$$I_C = \beta_{dc}I_B = (100)(195.5\ \mu\text{A}) = 19.55\text{ mA}$$

I_C is much less than $I_{C(max)}$ and will not change with V_{CC}. It is determined only by I_B and β_{dc}.

The voltage drop across R_C is

$$V_{R_C} = I_C R_C = (19.55\text{ mA})(1\text{ k}\Omega) = 19.55\text{ V}$$

Now we can determine the value of V_{CC} when $V_{CE} = V_{CE(max)} = 15$ V.

$$V_{R_C} = V_{CC} - V_{CE}$$

So,

$$\begin{aligned} V_{CC(max)} &= V_{CE(max)} + V_{R_C} \\ &= 15\text{ V} + 19.55\text{ V} \\ &= 34.55\text{ V} \end{aligned}$$

V_{CC} can be increased to 34.55 V, under the existing conditions, before $V_{CE(max)}$ is exceeded. However, we do not know whether or not $P_{D(max)}$ has been exceeded at this point. Let's find out.

$$P_D = V_{CE(max)}I_C = (15\text{ V})(19.55\text{ mA}) = 0.293\text{ W}$$

Since $P_{D(max)}$ is 0.8 W, it is *not* exceeded when $V_{CE} = 34.55$ V. So, $V_{CE(max)}$ is the limiting rating in this case. If the base current is removed causing the transistor to turn off, $V_{CE(max)}$ will be exceeded because the entire supply voltage, V_{CC}, will be dropped across the transistor.

PRACTICE EXERCISE 4–6

The transistor in Figure 4–17 has the following maximum ratings: $P_{D(max)} = 0.5$ W, $V_{CE(max)} = 25$ V, $I_{C(max)} = 200$ mA, $V_{CB(max)} = 30$ V, and $V_{EB(max)} = 15$ V. Determine the maximum value to which V_{CC} can be adjusted without exceeding a rating. Which rating would be exceeded first?

DERATING $P_{D(max)}$

$P_{D(max)}$ is usually specified at 25°C. For higher temperatures, $P_{D(max)}$ is less. Data sheets often give derating factors for determining $P_{D(max)}$ at any temperature above 25°C. For example, a derating factor of 2 mW/°C indicates that the maximum power dissipation is reduced 2 mW for each degree Centigrade increase in temperature.

EXAMPLE 4–7

A certain transistor has a $P_{D(max)}$ of 1 W at 25°C. The derating factor is 5 mW/°C. What is the $P_{D(max)}$ at a temperature of 70°C?

SOLUTION

The change (reduction) in $P_{D(\max)}$ is

$$\Delta P_{D(\max)} = (5 \text{ mW/°C})(70°\text{C} - 25°\text{C}) = (5 \text{ mW/°C})(45°\text{C}) = 225 \text{ mW}$$

Therefore, the $P_{D(\max)}$ at 70°C is

$$1 \text{ W} - 225 \text{ mW} = 775 \text{ mW} = 0.775 \text{ W}$$

PRACTICE EXERCISE 4–7

A transistor has a $P_{D(\max)} = 5$ W at 25°C. The derating factor is 10 mW/°C. What is the $P_{D(\max)}$ at 70°C? ■

4–3 REVIEW QUESTIONS

1. Define β_{dc} and α_{dc}.
2. If the dc current gain of a transistor is 100, determine β_{dc} and α_{dc}.
3. What two variables are plotted on a collector characteristic curve?
4. What bias conditions must exist for a transistor to operate as an amplifier?
5. Does β_{dc} increase or decrease with temperature?
6. For a given type of transistor, can β_{dc} be considered to be a constant?

4–4 THE TRANSISTOR AS A VOLTAGE AMPLIFIER

***Amplification** is the process of **linearly** increasing the amplitude of an electrical signal and is one of the major properties of a transistor. As you learned, the transistor exhibits current gain (called β). When a transistor is biased, as previously described, the BE junction has a low resistance due to forward bias and the BC junction has a high resistance due to reverse bias.*

Since I_B is extremely small, I_C is approximately equal to I_E. Actually, I_C is always slightly less than I_E. Therefore, Equation (4–1) can be restated as an approximation:

$$I_E \cong I_C$$

With this in mind, let's consider the transistor in Figure 4–18(a) with an ac input voltage, V_{in}, applied in series with the *BE* bias voltage and with an external resistor, R_C, connected in series with the *BC* bias voltage. Because the dc sources appear ideally as shorts to the ac voltage, the ac equivalent circuit is as shown in Figure 4–18(b). The forward-biased base-emitter junction appears as a low resistance to the ac signal. This internal ac emitter resistance of the transistor is called r_e. The ac emitter current in Figure 4–18(b) is

$$I_e = \frac{V_{in}}{r_e}$$

Since $I_c \cong I_e$, the output voltage developed across R_C is

$$V_{out} \cong I_e R_C$$

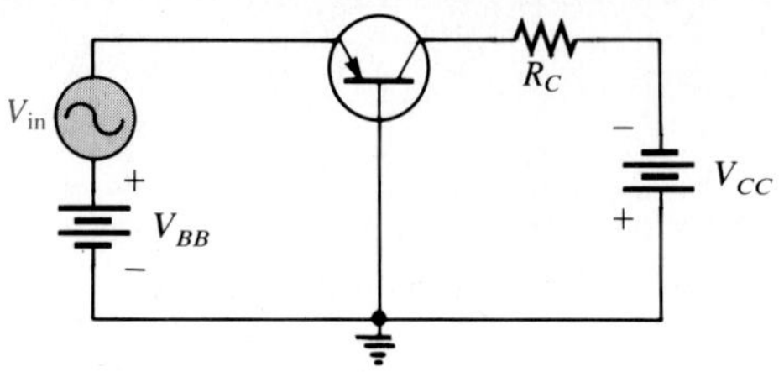

(a) ac input and dc bias

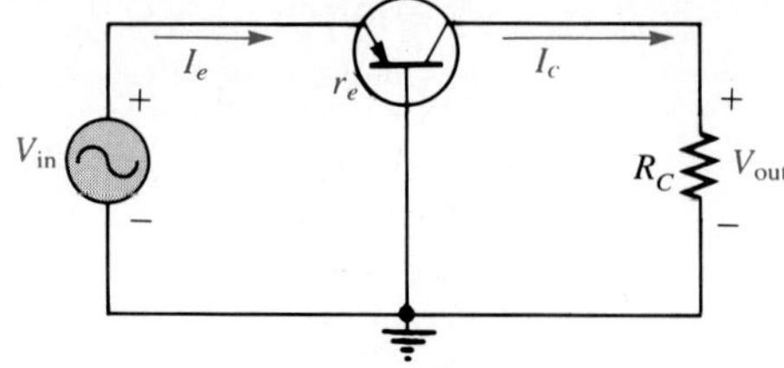

(b) ac equivalent circuit

FIGURE 4–18

Biased transistor with ac input signal.

The ratio of V_{out} to V_{in} is called the ac *voltage gain* (A_v) and is expressed as follows:

$$A_v = \frac{V_{out}}{V_{in}} \tag{4–12}$$

Substituting, we get

$$A_v = \frac{V_{out}}{V_{in}} \cong \frac{I_e R_C}{I_e r_e}$$

$$A_v \cong \frac{R_C}{r_e} \tag{4–13}$$

This shows that the transistor circuit in Figure 4–18 provides amplification or voltage gain, dependent on the value of R_C and r_e. Remember that *lowercase subscripts indicate ac quantities,* and the lowercase r is an internal transistor parameter. Various types of amplifiers are covered thoroughly in later chapters.

EXAMPLE 4–8

Determine the voltage gain and output voltage in Figure 4–19 if $r_e = 50\ \Omega$.

FIGURE 4–19

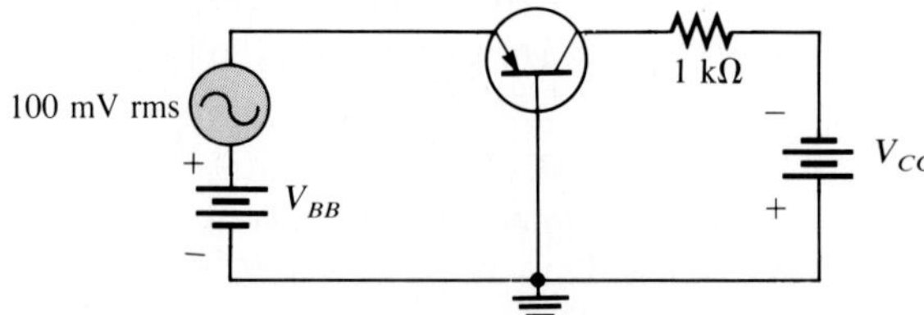

SOLUTION

The voltage gain is

$$A_v \cong \frac{R_C}{r_e} = \frac{1\text{ k}\Omega}{50\ \Omega} = 20$$

Therefore, the output voltage is

$$V_{out} = A_v V_{in} = (20)(100\text{ mV}) = 2\text{ V rms}$$

PRACTICE EXERCISE 4–8

What value of R_C in Figure 4–19 will it take to have a voltage gain of 50?

4–4 REVIEW QUESTIONS

1. What is amplification?
2. How is voltage gain defined?
3. Name two factors that determine the voltage gain of an amplifier.
4. What is the voltage gain of a transistor amplifier that has an output of 5 V rms and an input of 250 mV rms?
5. A transistor connected as in Figure 4–19 has an $r_e = 20\ \Omega$. If R_C is 1200 Ω, what is the voltage gain?

4–5 THE TRANSISTOR AS A SWITCH

In the previous section, we discussed the transistor as a linear amplifier. The second major application area is switching applications. When used as an electronic switch, a transistor is normally operated alternately in cutoff and saturation.

Figure 4–20 illustrates the basic operation of the transistor as a switching device. In part (a), the transistor is cut off because the base-emitter junction is not forward-biased. In this condition, there is, ideally, an open between collector and emitter as indicated by the switch equivalent. In part (b), the transistor is saturated because the base-emitter junction is forward-biased and the base current is large enough to cause the collector current to reach its saturated value. In this condition, there is, ideally, a short between collector and emitter as indicated by the switch equivalent. Actually, a voltage drop of a few tenths of a volt normally occurs, which is the saturation voltage.

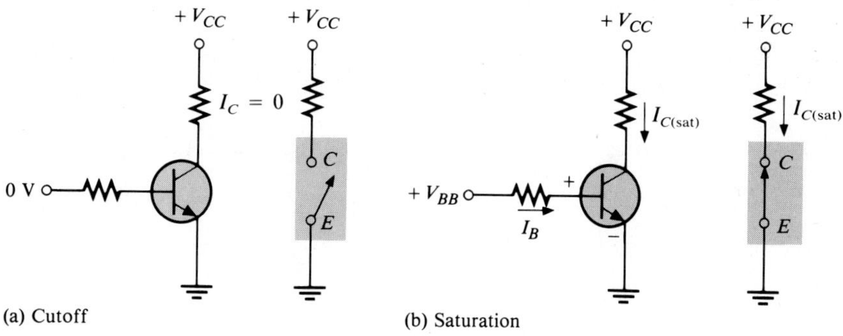

FIGURE 4–20
Ideal switching action of a transistor.

CONDITIONS IN CUTOFF

As mentioned before, a transistor is in cutoff when the base-emitter junction is *not* forward-biased. Neglecting leakage current, all of the currents are zero, and V_{CE} is equal to V_{CC}.

$$V_{CE(\text{cutoff})} = V_{CC} \tag{4–14}$$

CONDITIONS IN SATURATION

As you have learned, when the emitter junction is forward-biased and there is enough base current to produce a maximum collector current, the transistor is saturated. The formula for collector saturation current is

$$I_{C(\text{sat})} = \frac{V_{CC} - V_{CE(\text{sat})}}{R_C} \qquad (4\text{–}15)$$

Since V_{CE} is very small at saturation, an approximation for the collector saturation current is

$$I_{C(\text{sat})} \cong \frac{V_{CC}}{R_C} \qquad (4\text{–}16)$$

The minimum value of base current needed to produce saturation is

$$I_{B(\text{min})} = \frac{I_{C(\text{sat})}}{\beta_{\text{dc}}} \qquad (4\text{–}17)$$

I_B should be significantly greater than $I_{B(\text{min})}$ to keep the transistor well into saturation.

EXAMPLE 4–9

(a) For the transistor-switching circuit in Figure 4–21, what is V_{CE} when $V_{\text{IN}} = 0$ V?
(b) What minimum value of I_B is required to saturate this transistor if β_{dc} is 200? Assume $V_{CE(\text{sat})} = 0$ V.
(c) Calculate the maximum value of R_B when $V_{\text{IN}} = 5$ V.

FIGURE 4–21

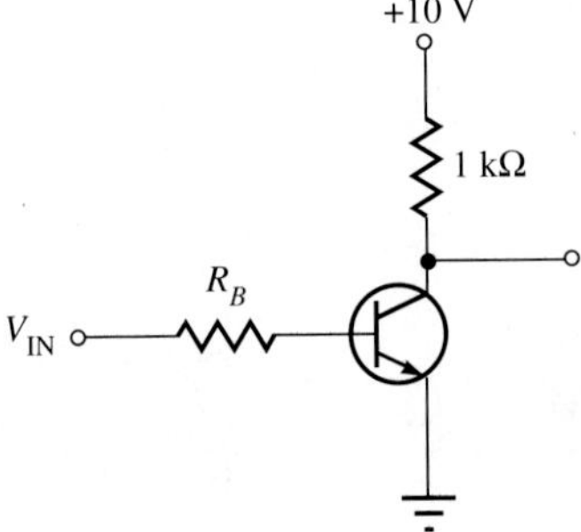

SOLUTION

(a) When $V_{\text{IN}} = 0$ V, the transistor is off and $V_{CE} = V_{CC} = 10$ V.
(b) Since $V_{CE(\text{sat})} = 0$ V,

$$I_{C(\text{sat})} = \frac{V_{CC}}{R_C} = \frac{10\text{ V}}{1\text{ k}\Omega} = 10\text{ mA}$$

$$I_B = \frac{I_{C(\text{sat})}}{\beta_{\text{dc}}} = \frac{10\text{ mA}}{200} = 0.05\text{ mA}$$

This is the value of I_B necessary to drive the transistor to the point of saturation. *Any further increase in I_B will drive the transistor deeper into saturation but will not increase I_C.*

(c) When the transistor is saturated, $V_{BE} = 0.7$ V. The voltage across R_B is

$$V_{IN} - 0.7 \text{ V} = 4.3 \text{ V}$$

The maximum value of R_B needed to allow a minimum I_B of 0.05 mA is calculated by Ohm's law as follows:

$$R_B = \frac{V_{IN} - 0.7 \text{ V}}{I_B} = \frac{4.3 \text{ V}}{0.05 \text{ mA}} = 86 \text{ k}\Omega$$

PRACTICE EXERCISE 4–9

Determine the minimum value of I_B required to saturate the transistor in Figure 4–21 if β_{dc} is 125 and $V_{CE(sat)}$ is 0.2 V.

A SIMPLE APPLICATION

The transistor in Figure 4–22 is used as a switch to turn the visible-light LED on and off. For example, a square wave input voltage with a period of 2 s is applied to the input as indicated. When the square wave is at 0 V, the transistor is in cutoff and, since there is no collector current, the LED does not emit light. When the square wave goes to its high level, the transistor saturates. This forward-biases the LED, and the collector current through the LED causes it to emit light. So, we have a blinking LED that is on for 1 s and off for 1 s.

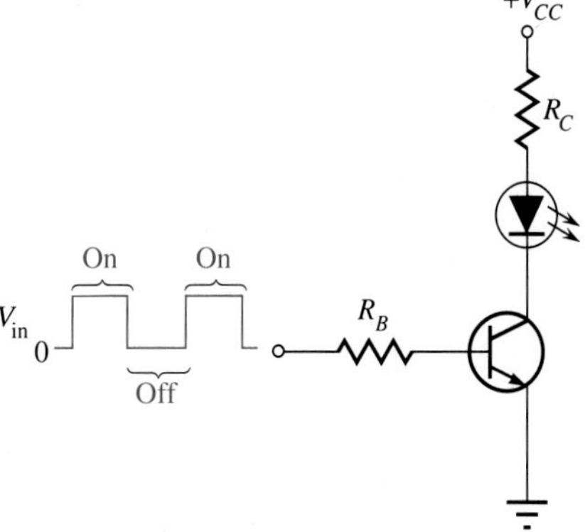

FIGURE 4–22
A transistor used to switch an LED on and off.

EXAMPLE 4–10

The LED in Figure 4–22 requires 30 mA to emit a sufficient level of light. Therefore, we want the collector current to be approximately 30 mA. For the following circuit values, determine the amplitude of the square wave input voltage necessary to make sure that the transistor saturates. Use double the minimum value of base current to ensure saturation. $V_{CC} = 9$ V, $V_{CE(sat)} = 0.3$ V, $R_C = 270\ \Omega$, $R_B = 3.3$ kΩ, $\beta_{dc} = 50$.

SOLUTION

$$I_{C(sat)} = \frac{V_{CC} - V_{CE(sat)}}{R_C} = \frac{9\text{ V} - 0.3\text{ V}}{270\ \Omega} = 32\text{ mA}$$

$$I_{B(min)} = \frac{I_{C(sat)}}{\beta_{dc}} = \frac{32\text{ mA}}{50} = 0.64\text{ mA}$$

To ensure saturation, use twice the value of $I_{B(min)}$, which gives 1.28 mA. Then

$$I_B = \frac{V_{RB}}{R_B} = \frac{V_{in} - V_{BE}}{R_B} = \frac{V_{in} - 0.7\text{ V}}{3.3\text{ k}\Omega}$$

Solving for V_{in},

$$V_{in} - 0.7\text{ V} = (3.3\text{ k}\Omega)(1.28\text{ mA})$$
$$V_{in} = (3.3\text{ k}\Omega)(1.28\text{ mA}) + 0.7\text{ V} = 4.92\text{ V}$$

PRACTICE EXERCISE 4–10

If you change the LED in Figure 4–22 to one that requires 50 mA for a specified light emission and you can't increase the input amplitude above 5 V or V_{CC} above 9 V, how would you modify the circuit? Specify the component(s) to be changed and the value(s). ■

4–5 REVIEW QUESTIONS

1. When a transistor is used as a switch, in what two states is it operated?
2. When is the collector current maximum?
3. When is the collector current approximately zero?
4. Under what condition is $V_{CE} = V_{CC}$?
5. When is V_{CE} minimum?

4–6 TRANSISTOR PACKAGES AND TERMINAL IDENTIFICATION

Transistors are available in a wide range of package types for various applications. Those with mounting studs or heat sinks are usually power transistors. Low and medium power transistors are usually found in smaller metal or plastic cases. Still another package classification is for high-frequency devices. You should be familiar with common transistor packages and be able to identify the emitter, base, and collector terminals. This section is about transistor packages and terminal identification.

TRANSISTOR CATEGORIES

As mentioned, manufacturers generally classify their bipolar junction transistors into three broad categories—general purpose/small-signal devices, power devices, and RF (radio frequency/microwave) devices. Although each of these categories, to a large degree, has its own unique package types, you will find certain types of packages used in more than one device category. While keeping in mind there is some overlap, we will look at transistor packages for each of the three categories, so that you will be able to recognize a transistor when you see one on a circuit board and have a good idea of what general category it is in.

General-Purpose/Small-Signal Transistors General-purpose/small-signal transistors are generally used for low or medium power amplifiers or switching circuits. The packages are either plastic or metal cases. Certain types of packages contain multiple transistors. Figure 4–23 illustrates common plastic cases, Figure 4–24 shows packages called *metal cans,* and Figure 4–25 shows multiple-transistor packages. Some of the multiple-transistor packages such as the dual in-line (DIP) and the small-outline (SO) are the same as those used for many integrated circuits. Pin connections are shown so you can identify the emitter, base, and collector.

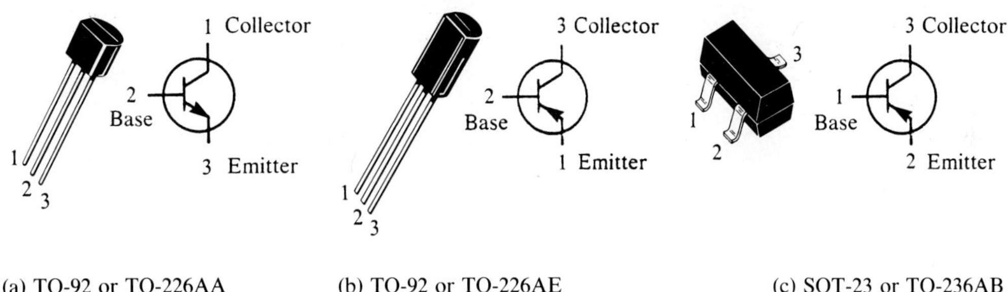

(a) TO-92 or TO-226AA (b) TO-92 or TO-226AE (c) SOT-23 or TO-236AB

Figure 4–23
Plastic cases for general-purpose/small-signal transistors. Both old and new JEDEC TO numbers are given. (Copyright of Motorola, Inc. Used by permission.)

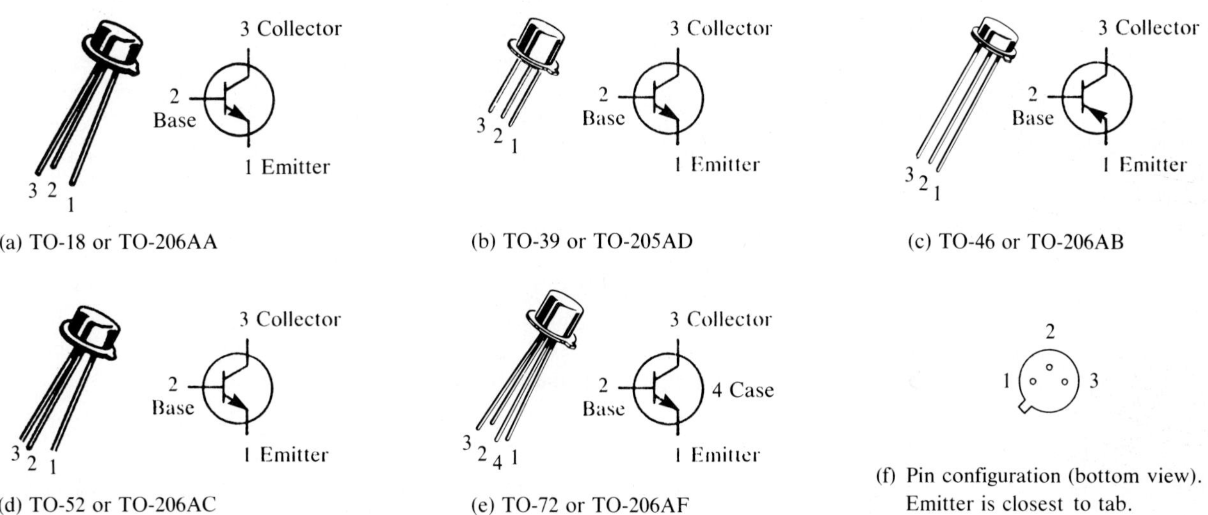

(a) TO-18 or TO-206AA (b) TO-39 or TO-205AD (c) TO-46 or TO-206AB

(d) TO-52 or TO-206AC (e) TO-72 or TO-206AF (f) Pin configuration (bottom view). Emitter is closest to tab.

Figure 4–24
Metal cases for general-purpose/small-signal transistors (copyright of Motorola, Inc. Used by permission).

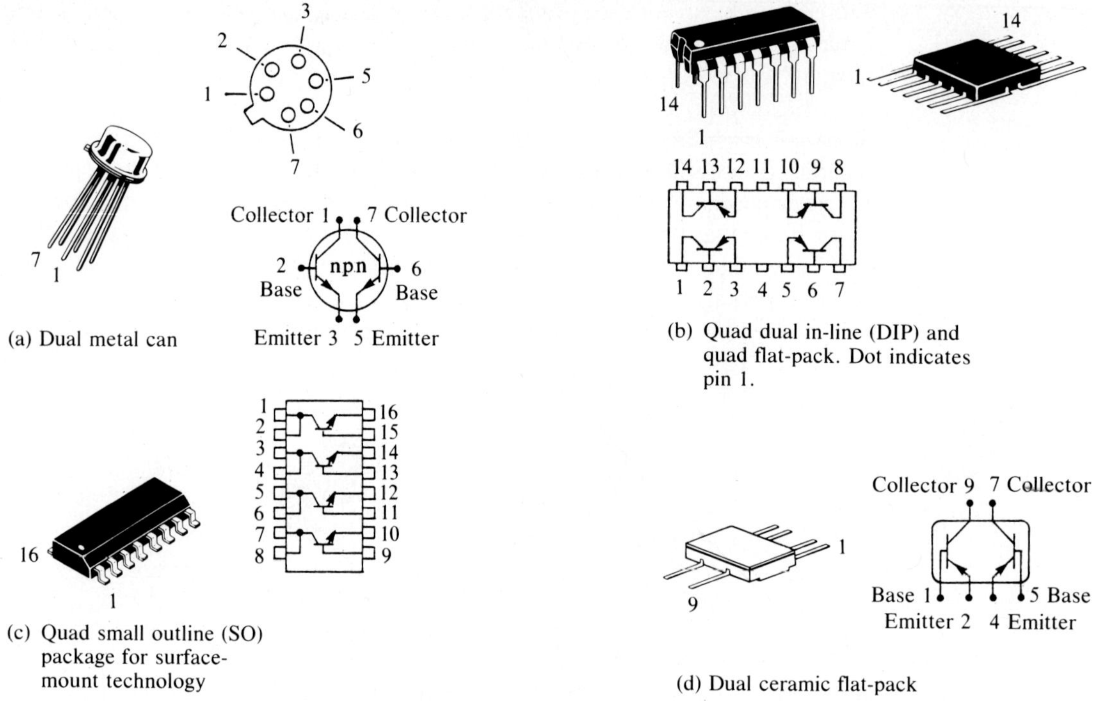

FIGURE 4–25
Typical multiple-transistor packages (copyright of Motorola, Inc. Used by permission).

POWER TRANSISTORS Power transistors are used to handle large currents (typically more than 1 A) and/or large voltages. For example, the final audio stage in a stereo system uses a power transistor amplifier to drive the speakers. Figure 4–26 shows some common package configurations. In most applications, the metal tab or the metal case is common to the collector and is thermally connected to a heat sink for heat dissipation. Notice in part (g) how the small transistor chip is mounted inside the much larger package.

RF TRANSISTORS RF transistors are designed to operate at extremely high frequencies and are commonly used for various purposes in communications systems and other high-frequency applications. Their unusual shapes and lead configurations are designed to optimize certain high-frequency parameters. Figure 4–27 shows some examples.

4–6 REVIEW QUESTIONS

1. List the three broad categories of bipolar junction transistors.
2. In a single-transistor metal case, how do you identify the leads?
3. In power transistors, the metal mounting tab or case is connected to which transistor region?

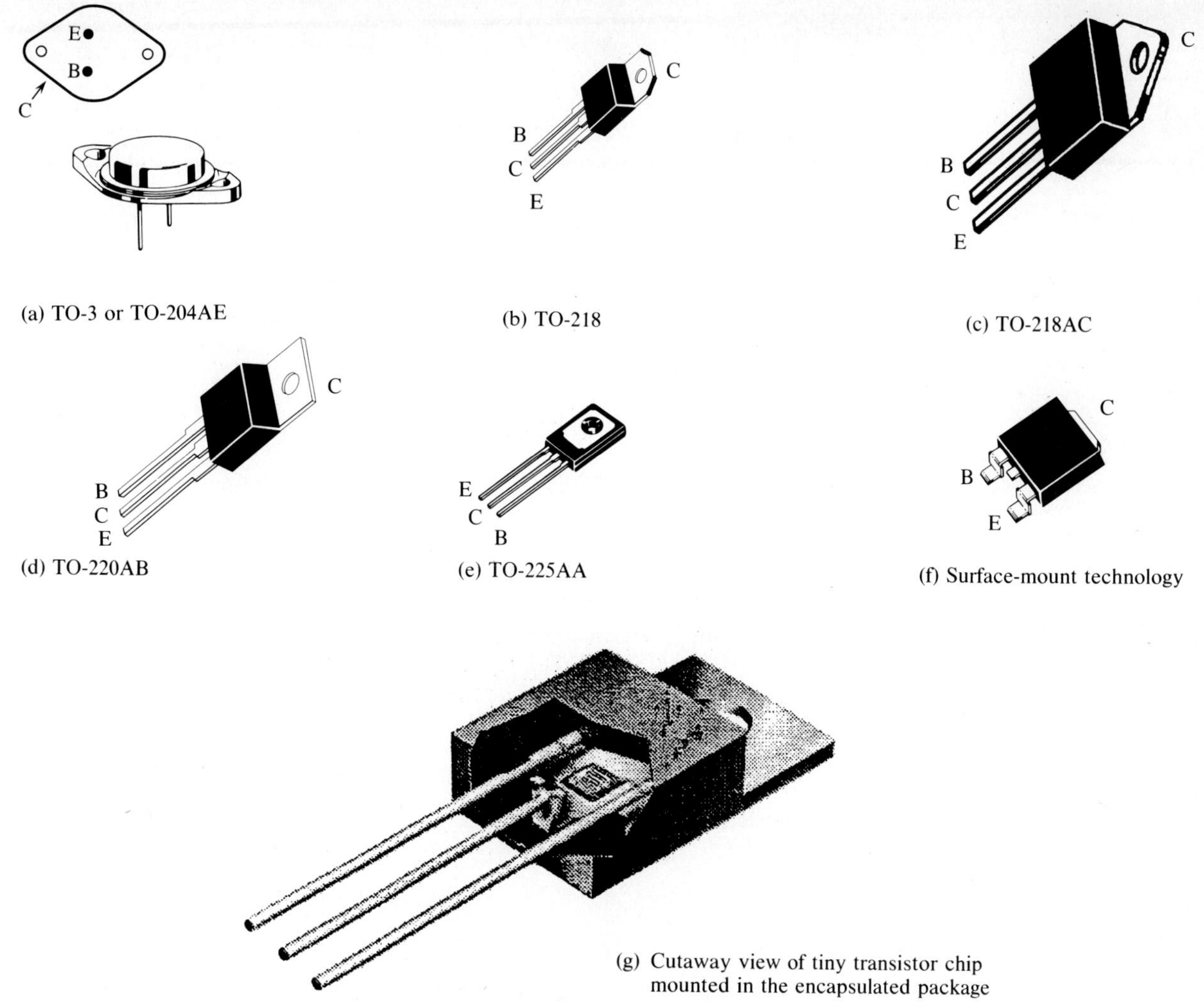

E
B
C
(a) TO-3 or TO-204AE
C
B
C
E
(b) TO-218
C
B
C
E
(c) TO-218AC
C
B
C
E
(d) TO-220AB
E
C
B
(e) TO-225AA
C
B
E
(f) Surface-mount technology
(g) Cutaway view of tiny transistor chip mounted in the encapsulated package

FIGURE 4–26
Typical power transistors (copyright of Motorola, Inc. Used by permission).

FIGURE 4–27
Examples of RF transistors (copyright of Motorola, Inc. Used by permission).

C
E
E
B
(a)
C
E
E
B
(b)
B
E
C
E
(c)
(d)

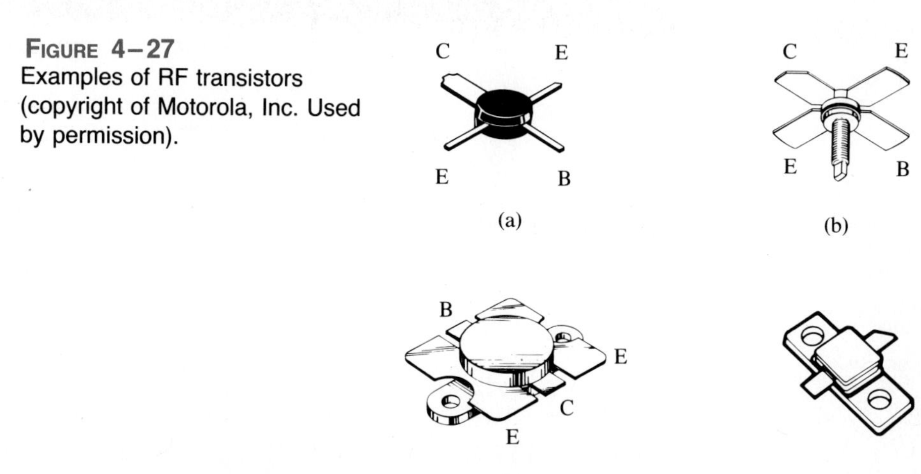

4–7 TROUBLESHOOTING

As you already know, a critical skill in electronics is the ability to identify a circuit malfunction and to isolate the failure to a single component if possible. In this section, the basics of troubleshooting transistor bias circuits and testing individual transistors are covered.

TROUBLESHOOTING A TRANSISTOR BIAS CIRCUIT

Several things can cause a failure in a transistor bias circuit. The most common faults are open base or collector resistor, loss of base or collector bias voltage, open transistor junctions, and open contacts or interconnections on the printed circuit board. Figure 4–28 illustrates a basic four-point check that will help isolate one of these faults in the circuit. The term *floating* refers to a point that is effectively disconnected from a ''solid'' voltage or ground. A very small, erratically fluctuating voltage is sometimes observed.

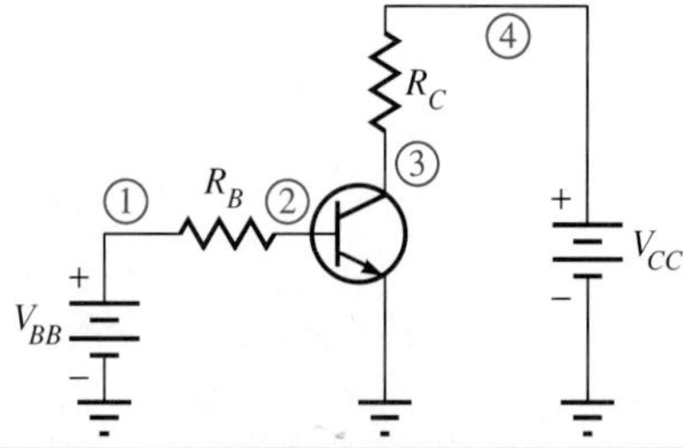

FIGURE 4–28
Test point troubleshooting guide for a basic transistor bias circuit.

	Test point voltages			
Symptom	1	2	3	4
A. No V_{BB} Source shorted Source open	 0 V Floating	 0 V Floating	 V_{CC} V_{CC}	 V_{CC} V_{CC}
B. No V_{CC} Source shorted Source open	 V_{BB} V_{BB}	 0.7 V 0.7 V	 0 V Floating	 0 V Floating
C. R_{BN} open	V_{BB}	Floating	V_{CC}	V_{CC}
D. R_C open	V_{BB}	0.7 V	Floating	V_{CC}
E. Base-to-emitter junction open	V_{BB}	V_{BB}	V_{CC}	V_{CC}
F. Base-to-collector junction open	V_{BB}	0.7 V	V_{CC}	V_{CC}
G. Low β	V_{BB}	0.7 V	Higher than normal	V_{CC}

EXAMPLE 4–11 From the voltage measurements in Figure 4–29, what is the most likely problem?

FIGURE 4–29

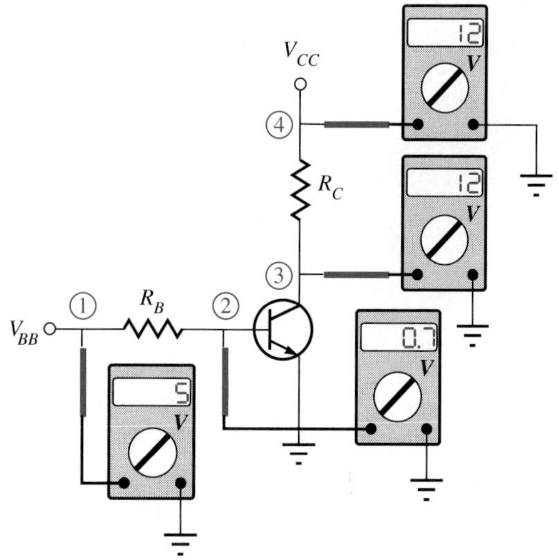

SOLUTION

Since $V_{CE} = 12$ V, the transistor is obviously in cutoff. The V_{BB} measurement appears OK and the voltage at the base is correct. Therefore, the collector-base junction of the transistor must be open. Replace the transistor and check the voltages again.

PRACTICE EXERCISE 4–11

What is the voltage at point 2 if the *BE* junction opens in Figure 4–29?

TRANSISTOR TESTING

An individual transistor can be tested either in-circuit or out-of-circuit. For example, let's say that an amplifier on a particular printed circuit (pc) board has malfunctioned. You perform a check and find that the symptoms are the same as Item F in the table of Figure 4–28, indicating an open from base to collector. This can mean either a bad transistor or an open in an external circuit connection. Your next step is to find out which one.

Good troubleshooting practice dictates that you do not remove a component from a circuit board unless you are reasonably sure that it is bad or you simply cannot isolate the problem down to a single component. When components are removed, there is a risk of damage to the pc board contacts and traces.

The next step is to do an in-circuit check of the transistor using a transistor tester similar to the ones shown in Figure 4–30. The three clip-leads are connected to the transistor terminals and the tester gives a GOOD/BAD indication.

CASE 1

If the transistor tests BAD, it should be carefully removed and replaced by a known good one. An out-of circuit check of the replacement device is usually a good idea, just to make sure it is OK. The transistor is plugged into the socket on the transistor tester for out-of-circuit tests.

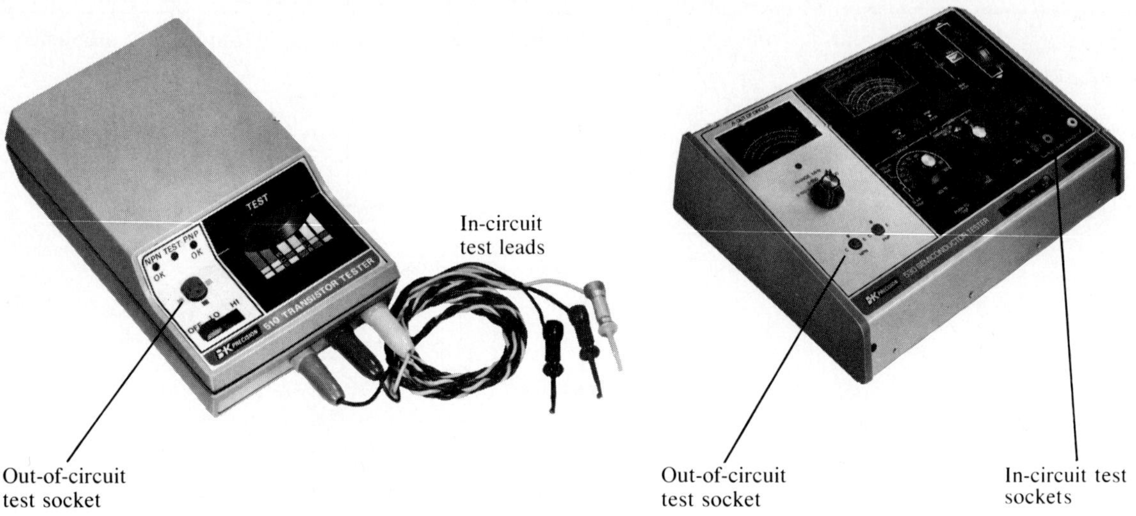

FIGURE 4–30
Transistor testors (courtesy of B & K Precision).

CASE 2

If the transistor tests GOOD in-circuit, examine the circuit board for a poor connection at the collector pad or for a break in the connecting trace. A poor solder joint often results in an open or a highly resistive contact. The physical point at which you actually measure the voltage is very important in this case. If you measure on the collector side of an external open, you will get the indication listed in Item D (point 3) of the table in Figure 4–28. If you measure on the side of the external open closest to R_C, you will get the indication in Item F (point 3) of the table. This situation is illustrated in Figure 4–31.

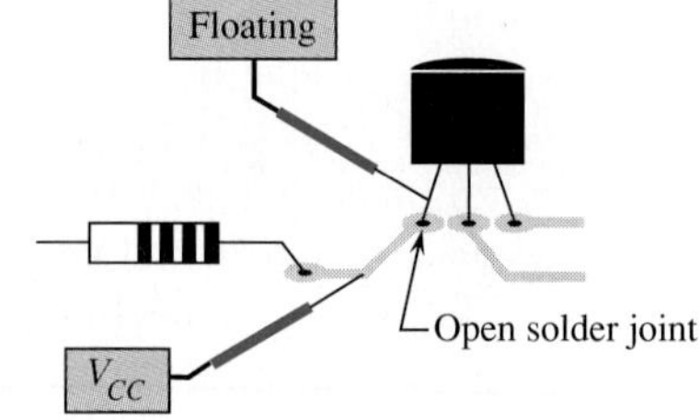

FIGURE 4–31
The indication of an open, when it is in the external circuit, depends on where you measure.

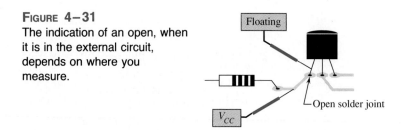

IMPORTANCE OF POINT-OF-MEASUREMENT IN TROUBLESHOOTING

It is interesting to look back and notice that if you had taken the initial measurement on the transistor lead itself and the open were internal to the transistor, you would have measured V_{CC}. This would have indicated a bad transistor even before the tester was used. This simple concept emphasizes the importance of point-of-measurement in certain troubleshooting situations as further illustrated in Figure 4–32 for the case just discussed. This is also valid for opens at the base and emitter of a transistor.

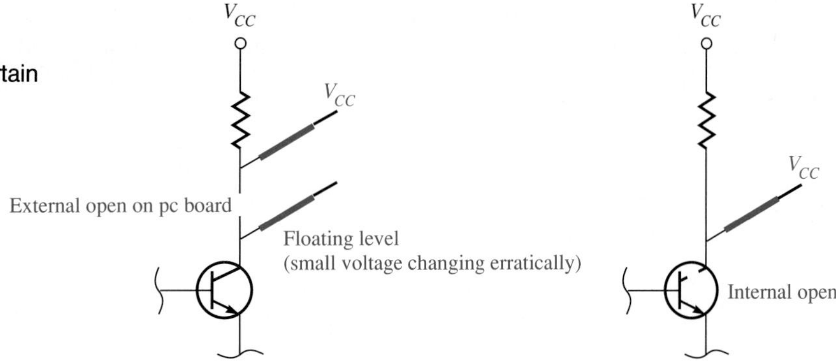

FIGURE 4–32 Importance of the point-of-measurement for certain open failures in transistor circuits.

EXAMPLE 4–12

What fault do the measurements in Figure 4–33 indicate?

FIGURE 4–33

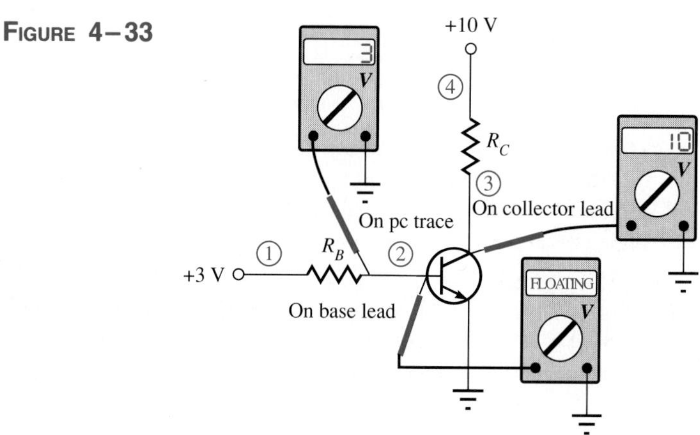

SOLUTION

The transistor is in cutoff, as indicated by the 10-V measurement on the collector lead. The base bias voltage of 3 V appears on the pc board contacts but not on the transistor lead. This indicates that there is an open external to the transistor between the two measured points. Check the solder joint at the base contact on the pc board. If the open were internal, there would be 3 V on the base lead.

PRACTICE EXERCISE 4–12

If the meter in Figure 4–33 that now reads 3 V indicates a floating point, what is the most likely fault?

OUT-OF-CIRCUIT MULTIMETER TEST

If a transistor tester is not available, an analog multimeter can be used to perform a basic diode check on the transistor junctions for opens or shorts. For this test, a transistor can be viewed as two back-to-back diodes as shown in Figure 4–34. The *BC* junction is one diode and the *BE* junction is the other. Also, many DMMs have a transistor test feature as Figure 4–35 indicates.

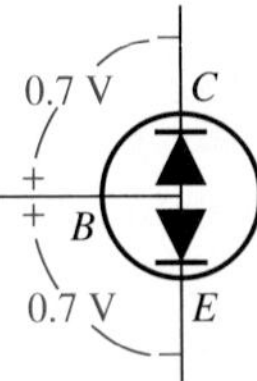

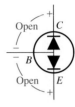

(a) Both junctions should read approximately 0.7 V when forward-biased.

(b) Both junctions should read open (ideally) when reverse-biased.

FIGURE 4–34
The transistor is viewed as two diodes for a junction check with a multimeter.

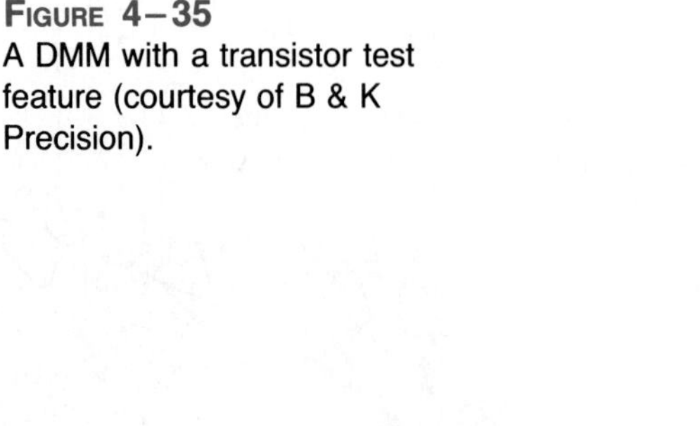

FIGURE 4–35
A DMM with a transistor test feature (courtesy of B & K Precision).

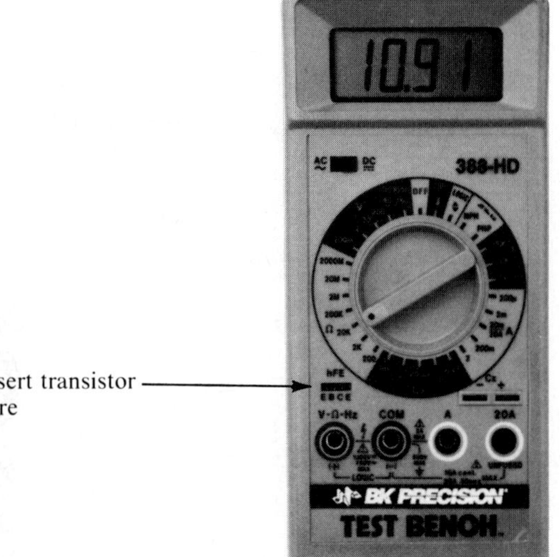

LEAKAGE MEASUREMENT

Very small leakage currents exist in all transistors and in most cases are small enough to neglect (usually nA). When a transistor is connected as shown in Figure 4–36(a) with the base open ($I_B = 0$), it is in cutoff. Ideally $I_C = 0$; but actually there is a small current from collector to emitter, as mentioned earlier, called I_{CEO} (collector-to-emitter current with base open). This leakage current is usually in the nA range for silicon. A faulty transistor will often have excessive leakage current and can be checked in a transistor tester, which connects an ammeter as shown in part (a). Another leakage current in transistors is the reverse collector-to-base current, I_{CBO}. This is measured with the emitter open, as shown in Figure 4–36(b). If it is excessive, a shorted collector-base junction is likely.

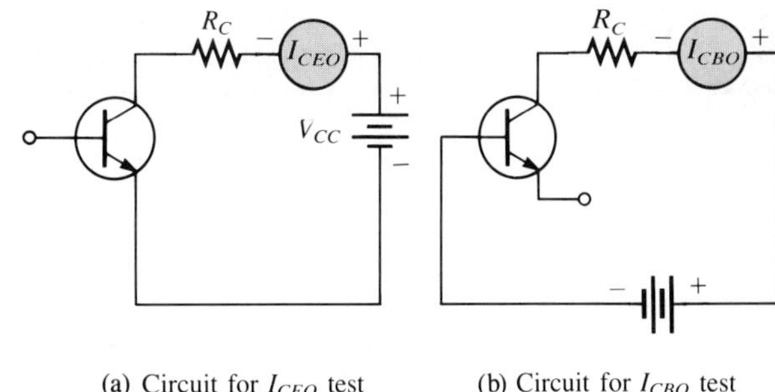

FIGURE 4–36
Leakage current test circuits.

GAIN MEASUREMENT

In addition to leakage tests, the typical transistor tester also checks the β_{dc}. A known value of I_B is applied and the resulting I_C is measured. The reading will indicate the value of the I_C/I_B ratio, although in some units only a relative indication is given. Most testers provide for an in-circuit β_{dc} check, so that a suspected device does not have to be removed from the circuit for testing.

CURVE TRACERS

The *curve tracer* is an oscilloscope type of instrument that can display transistor characteristics such as a family of collector curves. In addition to the measurement and display of various transistor characteristics, diode curves can also be displayed, as well as the β_{dc}. A typical curve tracer is shown in Figure 4–37.

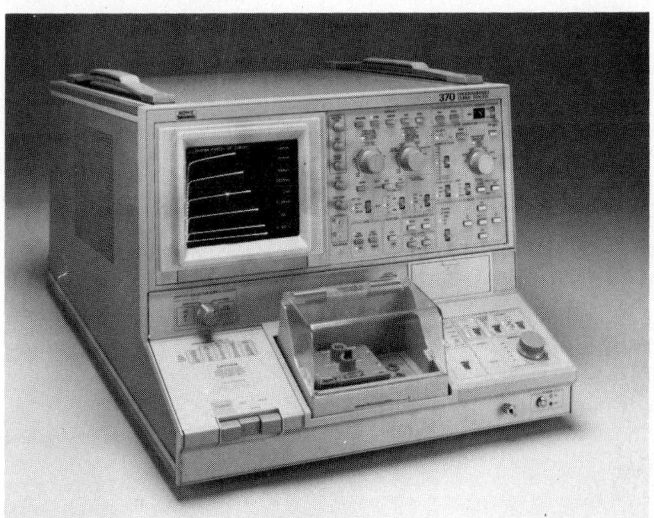

FIGURE 4–37
Curve tracer (courtesy of Tektronix, Inc.).

4–7 REVIEW QUESTIONS

1. If a transistor on a circuit board is suspected of being faulty, what should you do?
2. In a transistor bias circuit, such as the one in Figure 4–28, what happens if R_B opens?
3. In a circuit such as the one in Figure 4–28, what are the base and collector voltages if there is an open between the emitter and ground?

4–8 A SYSTEM APPLICATION

At the opening of this chapter, you saw a block diagram for a basic electronic security system. The system has several parts, but in this section we are concerned with the transistor circuits that detect an open in the loops containing remote sensors. In this particular application, the transistors are used as switching devices, and you will

- ☐ *See how transistors are used in a switching application.*
- ☐ *See how a transistor is used to activate a relay.*
- ☐ *See how one transistor is used to drive another transistor.*
- ☐ *Translate between a printed circuit board and a schematic.*
- ☐ *Troubleshoot some common transistor circuit failures.*

A BRIEF DESCRIPTION OF THE SYSTEM

The circuit board shown in the system diagram in Figure 4–38 contains the transistor circuits for detecting when one of the remote switch sensors is open. There are three remote loops in this particular system. One loop protects all windows/doors in one area of the structure (Zone 1), and another loop protects the windows/doors in a second area

FIGURE 4–38

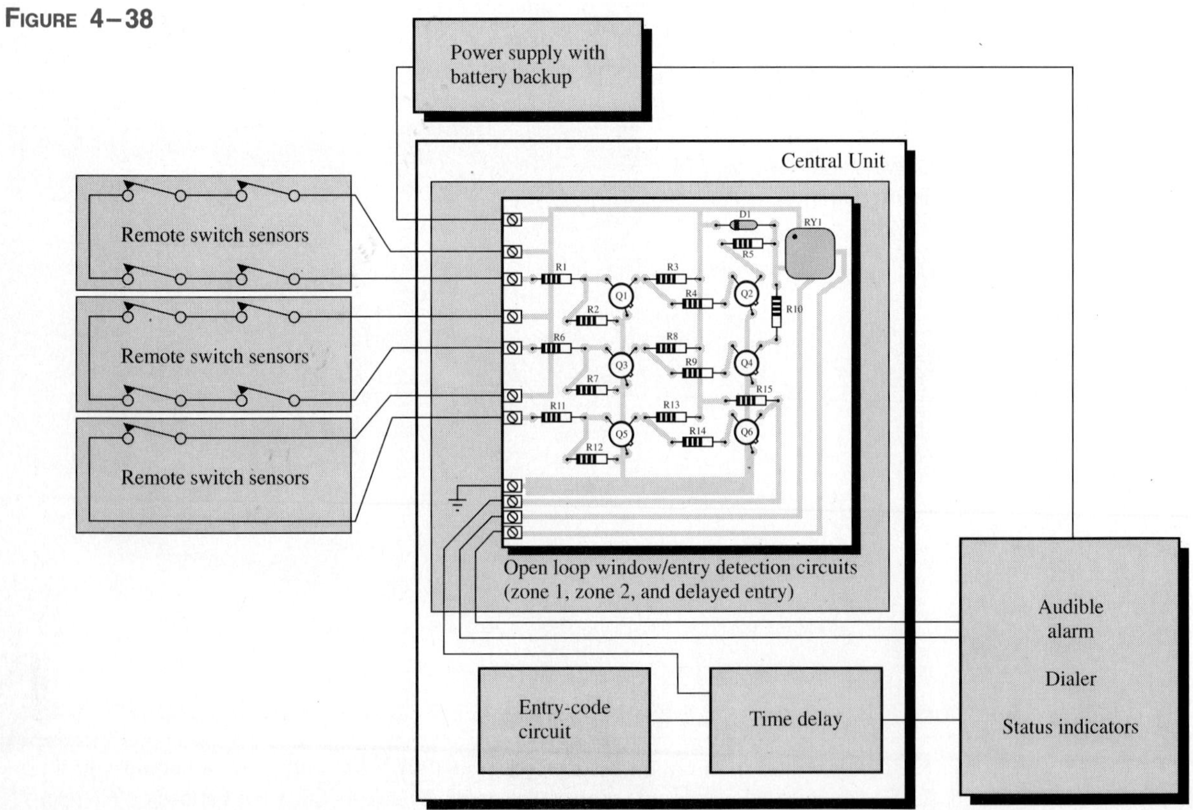

(Zone 2). The third loop protects the main entry door. When an intrusion occurs, a switch sensor at the point of intrusion breaks contact and opens the loop. This causes the input to the detection circuit for that loop to go to a 0-V level and activate the circuit which, in turn, energizes the relay causing it to set off an alarm and/or initiate an automatic telephone dialing sequence. Either of the detection circuits for Zone 1 or Zone 2 can energize the common relay. The output of the detection circuit for the main entry goes to a time delay circuit to allow time for keying in an entry code that will disarm the system. If, after a sufficient time, no code or an improper code has been entered, the delay circuit will initiate an alarm.

Although there are many aspects to a security system and a system can range from very simple to very complex, in this application we are concentrating only on the detection circuits board.

Now, so you can take a closer look at the detection circuits, let's take the board out of the system and put it on the test bench.

ON THE TEST BENCH

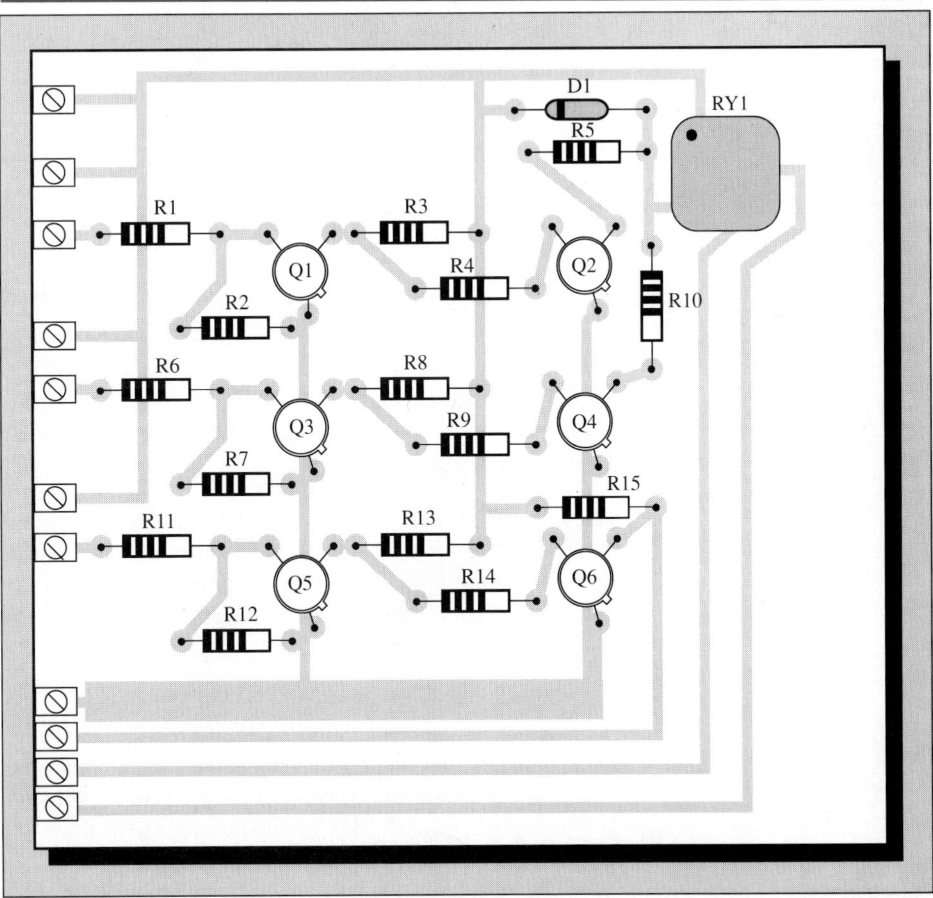

FIGURE 4–39

IDENTIFYING THE COMPONENTS

There are three separate detection circuits on the board in Figure 4–39, each consisting of two transistors and their associated resistors. The transistors are labeled Q1, Q2, and so forth. There is one relay labeled RY1 and a diode D1 for suppression of negative transients across the relay coil. As shown in Figure 4–40(a), the transistors are housed in TO-18 (TO-206AA) metal cans (remember, the emitter is closest to the tab). A detail of the relay is shown in Figure 4–40 (b). You are already familiar with resistors and diodes.

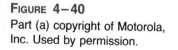

FIGURE 4–40
Part (a) copyright of Motorola, Inc. Used by permission.

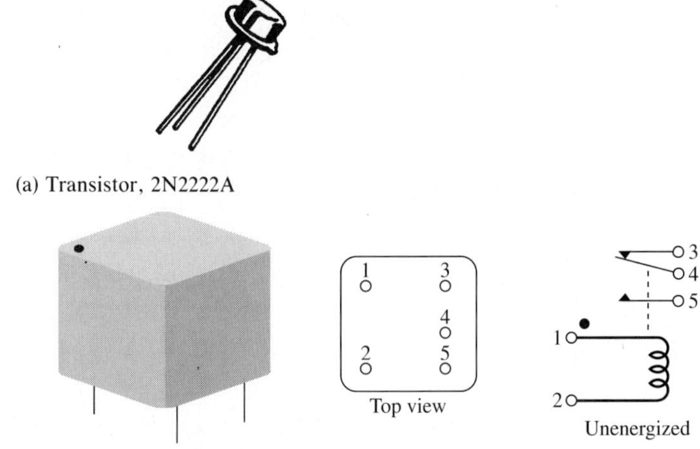

(a) Transistor, 2N2222A

(b) Relay, pin numbering, and schematic

ACTIVITY 1 RELATE THE PC BOARD TO THE SCHEMATIC

Carefully follow the traces on the pc board to see how the components are interconnected. Compare each point (*A* through *X*) on the circuit board with the corresponding point on the schematic in Figure 4–41. For each point on the pc board, place the letter on the corresponding point or points on the schematic.

ACTIVITY 2 ANALYZE THE CIRCUITS

Calculate the amount of base current and collector current for each transistor when it is saturated. Refer to the β_{dc} (h_{FE}) information on the data sheet for the 2N2222 in Appendix A to make sure there is sufficient base current to keep the transistor well in saturation. Assume that the relay has a coil resistance of 15 Ω and requires a minimum of 400 mA to operate.

ACTIVITY 3 WRITE A TECHNICAL REPORT

Discuss the detailed operation of the transistor detection circuits on the pc board. Also, describe the inputs and outputs indicating what other areas of the system they connect to and describe the purpose served by each one.

ACTIVITY 4 DEVELOP A COMPLETE TEST PROCEDURE

Someone with no knowledge of how this circuitry works should be able to use your test procedure and verify that all of the circuits operate properly. The circuit board test

FIGURE 4–41

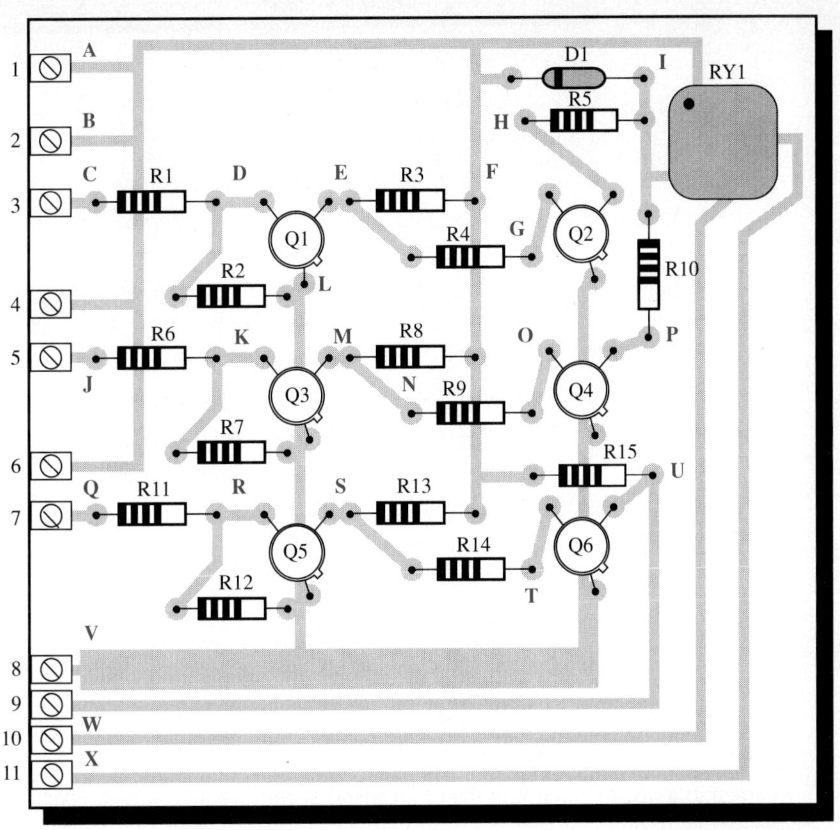

A
B
C
D
E
F
G
H
I
J
K
L
M
N
O
P
Q
R
S
T
U
V
W
X
D1
R5
RY1
R1
R2
R3
R4
R6
R7
R8
R9
R10
R11
R12
R13
R14
R15
Q1
Q2
Q3
Q4
Q5
Q6
1
2
3
4
5
6
7
8
9
10
11

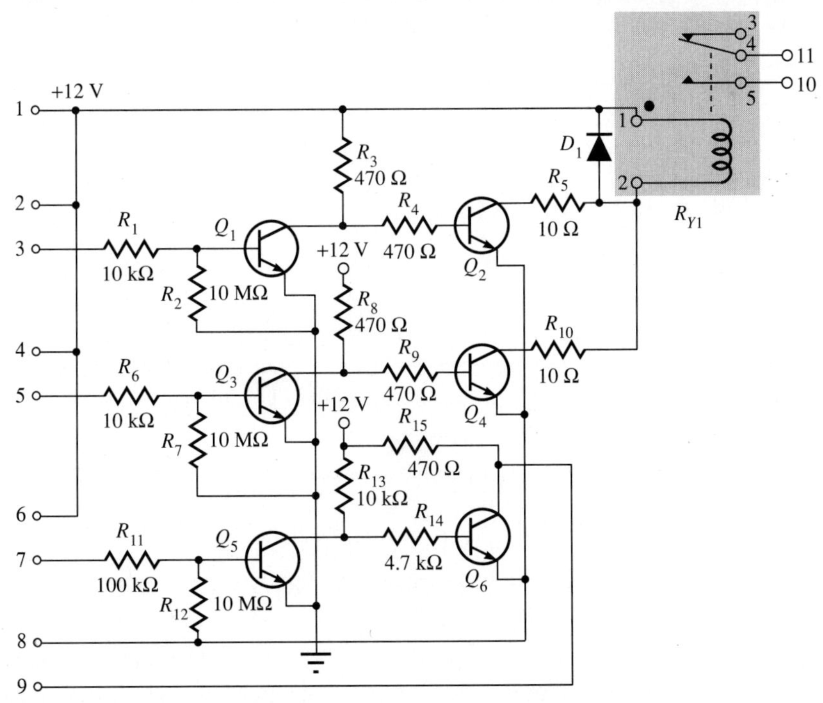

+12 V
R_1 10 kΩ
R_2 10 MΩ
R_3 470 Ω
R_4 470 Ω
R_5 10 Ω
R_6 10 kΩ
R_7 10 MΩ
R_8 470 Ω
R_9 470 Ω
R_{10} 10 Ω
R_{11} 100 kΩ
R_{12} 10 MΩ
R_{13} 10 kΩ
R_{14} 4.7 kΩ
R_{15} 470 Ω
Q_1
Q_2
Q_3
Q_4
Q_5
Q_6
D_1
R_{Y1}

is to be done on the test bench and not in the functional system, so you must simulate operational conditions.

STEP 1 Begin with the most basic and necessary thing—power.

STEP 2 Develop a detailed step-by-step procedure for simulating the external remote loops.

STEP 3 For each loop detector circuit, specify the points at which to measure voltage, the value of the voltage, and under what conditions it is to be measured. Detail each step required to fully check out all the circuits.

STEP 4 Specify in detail any other tests in addition to voltage measurements.

■ ACTIVITY 5 TROUBLESHOOT THE CIRCUIT BOARD FOR EACH OF THE FOLLOWING PROBLEMS BY STATING THE PROBABLE CAUSE OR CAUSES IN EACH CASE

1. Approximately 5.65 V at the collector of Q1 with the remote loop switch closed.
2. Relay is energized (contact from 10 to 11) continuously and independent of remote switches.
3. Collector lead of Q4 floating.
4. Pin 9 is always at 12 V. Q6 collector lead is also at 12 V.
5. When 12 V is applied to pin 5 with the closed loop switch, the base of Q3 is 0 V.
6. Approximately 3.6 V at the collector of Q5 when 12 V is applied to pin 7.

■ ACTIVITY 6 TEST BENCH SPECIAL ASSIGNMENT

Go to Test Bench 2 in the color insert section (which follows page 422) and carry out the assignment that is stated there.

4–8 REVIEW QUESTIONS

1. State the basic purpose of the detector circuit board in the security system.
2. How many of the circuits on the board are identical?
3. When is there a contact closure between pins 10 and 11 on the pc board?
4. Based on your knowledge of coils and diodes, why do you think D1 is in the circuit?

SUMMARY

- ☐ The bipolar junction transistor (BJT) is constructed with three regions—base, collector, and emitter.
- ☐ The BJT has two pn junctions, the base-emitter junction and the base-collector junction.
- ☐ Current in a BJT consists of both free electrons and holes, thus the term *bipolar*.
- ☐ The base region is very narrow and lightly doped compared to the collector and emitter regions.
- ☐ The two types of bipolar junction transistor are the npn and the pnp.
- ☐ To operate as an amplifier, the base-emitter junction must be forward-biased and the base-collector junction must be reverse-biased. This is called *forward-reverse bias*.

- □ The three currents in the transistor are the base current (I_B), emitter current (I_E), and collector current (I_C).
- □ I_B is very small compared to I_C and I_E.
- □ The dc current gain of a transistor is the ratio of I_C to I_B and is designated β_{dc}. Values typically range from less than 20 to several hundred.
- □ β_{dc} is usually referred to as h_{FE} on transistor data sheets.
- □ The ratio of I_C to I_E is called α_{dc}. Values typically range from 0.95 to 0.99.
- □ When a transistor is forward-reverse biased, the voltage gain depends on the internal emitter resistance and the external collector resistance.
- □ A transistor can be operated as an electronic switch in cutoff and saturation.
- □ In cutoff, both pn junctions are reverse-biased and there is essentially no collector current. The transistor ideally behaves like an open switch between collector and emitter.
- □ In saturation, both pn junctions are forward-biased and the collector current is maximum. The transistor ideally behaves like a closed switch between collector and emitter.
- □ There is a variation in β_{dc} over temperature and also from one transistor to another of the same type.
- □ There are many types of transistor packages using plastic, metal, or ceramic.
- □ It is best to check a transistor in-circuit before removing it.
- □ Common faults are open junctions, low β_{dc}, excessive leakage currents, and external opens and shorts on the circuit board.

GLOSSARY

Alpha (α) The ratio of collector current to emitter current in a bipolar junction transistor.

Amplification The process of increasing the power, voltage, or current by electronic means.

Amplifier An electronic circuit having the capability to amplify power, voltage, or current.

Base One of the semiconductor regions in a bipolar junction transistor. The base is very narrow compared to the other regions.

BASIC A computer programming language: *B*eginner's *A*ll-purpose *S*ymbolic *I*nstruction *C*ode.

Beta (β) The ratio of collector current to base current in a bipolar junction transistor. Current gain.

Bipolar Characterized by both free electrons and holes as current carriers.

Bipolar junction transistor (BJT) A transistor constructed with three doped semiconductor regions separated by two pn junctions.

Collector One of the three semiconductor regions of a BJT.

Cutoff The nonconducting state of a transistor.

Emitter One of the three semiconductor regions of a BJT.

Gain The amount by which an electrical signal is increased or amplified.

Linear Characterized by a straight-line relationship.

Saturation The state of a BJT in which the collector current has reached a maximum and is independent of the base current.

Transistor A semiconductor device used for amplification and switching applications.

FORMULAS

(4–1)	$I_E = I_C + I_B$	Transistor currents
(4–2)	$\beta_{dc} = \frac{I_C}{I_B}$	DC current gain
(4–3)	$\alpha_{dc} = \frac{I_C}{I_E}$	DC alpha
(4–4)	$\alpha_{dc} = \frac{\beta_{dc}}{\beta_{dc} + 1}$	β_{dc}-to-α_{dc} conversion
(4–5)	$\beta_{dc} = \frac{\alpha_{dc}}{1 - \alpha_{dc}}$	α_{dc}-to-β_{dc} conversion
(4–6)	$V_{BE} \cong 0.7\text{ V}$	Base-to-emitter voltage (silicon)
(4–7)	$I_B = \frac{V_{BB} - V_{BE}}{R_B}$	Base current
(4–8)	$V_{CE} = V_{CC} - I_C R_C$	Collector-to-emitter voltage (common-emitter)
(4–9)	$V_{CB} = V_{CE} - V_{BE}$	Collector-to-base voltage
(4–10)	$I_C = \frac{P_{D(max)}}{V_{CE}}$	Maximum I_C for given V_{CE}
(4–11)	$V_{CE} = \frac{P_{D(max)}}{I_C}$	Maximum V_{CE} for given I_C
(4–12)	$A_v = \frac{V_{out}}{V_{in}}$	AC voltage gain
(4–13)	$A_v \cong \frac{R_C}{r_e}$	Approximate ac voltage gain
(4–14)	$V_{CE(cutoff)} = V_{CC}$	Cutoff condition

(4–15) $$I_{C(\text{sat})} = \frac{V_{CC} - V_{CE(\text{sat})}}{R_C}$$ Collector saturation current

(4–16) $$I_{C(\text{sat})} \cong \frac{V_{CC}}{R_C}$$ Approximate collector saturation current

(4–17) $$I_{B(\text{min})} = \frac{I_{C(\text{sat})}}{\beta_{\text{dc}}}$$ Minimum base current for saturation

SELF-TEST

1. The three terminals of a bipolar junction transistor are called
(a) p, n, p **(b)** n, p, n
(c) input, output, ground **(d)** base, emitter, collector

2. In a pnp transistor, the p-regions are
(a) base and emitter **(b)** base and collector
(c) emitter and collector

3. For operation as an amplifier, the base of an npn transistor must be
(a) positive with respect to the emitter
(b) negative with respect to the emitter
(c) negative with respect to the collector
(d) 0 V

4. The emitter current is always
(a) greater than the base current
(b) less than the collector current
(c) greater than the collector current
(d) a and c

5. The β_{dc} of a transistor is its
(a) current gain **(b)** voltage gain
(c) power gain **(d)** internal resistance

6. If I_C is 50 times larger than I_B, then β_{dc} is
(a) 0.02 **(b)** 100 **(c)** 50 **(d)** 500

7. If β_{dc} is 100, the value of α_{dc} is
(a) 99 **(b)** 0.99 **(c)** 101 **(d)** 0.01

8. The approximate voltage across the forward-biased base-emitter junction of a silicon BJT is
(a) 0 V **(b)** 0.7 V **(c)** 0.3 V **(d)** V_{BB}

9. The bias condition for a transistor to be used as a linear amplifier is called
(a) forward-reverse **(b)** forward-forward
(c) reverse-reverse **(d)** collector bias

10. If the output of a transistor amplifier is 5 V rms and the input is 100 mV rms, the voltage gain is

(a) 5 **(b)** 500 **(c)** 50 **(d)** 100

11. When operated in cutoff and saturation, the transistor acts like

(a) a linear amplifier **(b)** a switch

(c) a variable capacitor **(d)** a variable resistor

12. In cutoff, V_{CE} is

(a) 0 V **(b)** minimum **(c)** maximum

(d) equal to V_{CC} **(e)** a and b **(f)** c and d

13. In saturation, V_{CE} is

(a) 0.7 V **(b)** equal to V_{CC}

(c) minimum **(d)** maximum

14. To saturate a BJT,

(a) $I_B = I_{C(sat)}$ **(b)** $I_B > I_{C(sat)}/\beta_{dc}$

(c) V_{CC} must be at least 10 V **(d)** the emitter must be grounded

15. Once in saturation, a further increase in base current will

(a) cause the collector current to increase

(b) not affect the collector current

(c) cause the collector current to decrease

(d) turn the transistor off

16. If the base-emitter junction is open, the collector voltage is

(a) V_{CC} **(b)** 0 V **(c)** floating **(d)** 0.2 V

PROBLEMS

SECTION 4–1 TRANSISTOR CONSTRUCTION

1. The majority carriers in the base region of an npn transistor are ________.
2. Explain the purpose of a thin, lightly doped base.

SECTION 4–2 BASIC TRANSISTOR OPERATION

3. Why is the base current in a transistor so much less than the collector current?
4. In a certain transistor circuit, the base current is 2 percent of the 30 mA emitter current. Determine the collector current.
5. For normal operation of a pnp transistor, the base must be (+ or −) with respect to the emitter, and (+ or −) with respect to the collector.
6. What is the value of I_C for $I_E = 5.34$ mA and $I_B = 475$ μA?

SECTION 4–3 TRANSISTOR PARAMETERS AND RATINGS

7. What is the α_{dc} when $I_C = 8.23$ mA and $I_E = 8.69$ mA?
8. A certain transistor has an $I_C = 25$ mA and an $I_B = 200$ μA. Determine the β_{dc}.

9. What is the β_{dc} of a transistor if α_{dc} is 0.96?
10. What is the α_{dc} if β_{dc} is 30?
11. A certain transistor exhibits an α_{dc} of 0.96. Determine I_C when $I_E = 9.35$ mA.
12. A base current of 50 μA is applied to the transistor in Figure 4–42, and a voltage of 5 V is dropped across R_C. Determine the β_{dc} of the transistor.

FIGURE 4–42

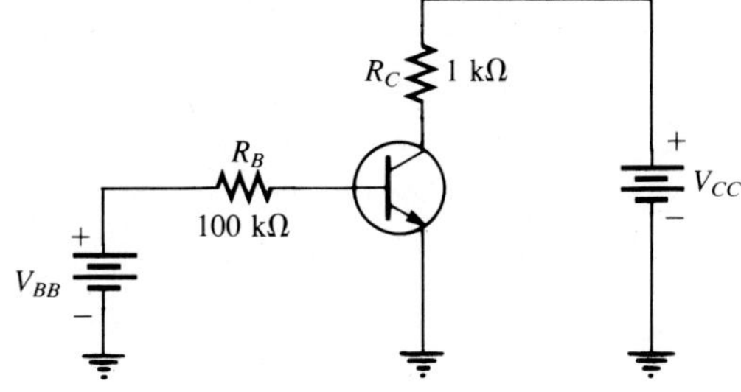

13. Calculate α_{dc} for the transistor in Problem 12.
14. Determine each current in Figure 4–43? What is the β_{dc}?

FIGURE 4–43

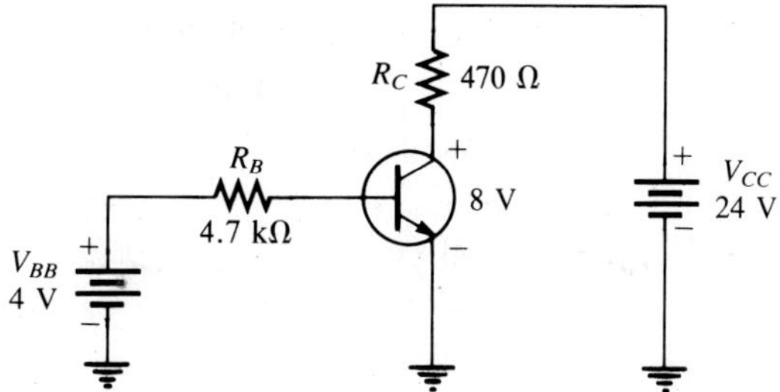

15. Find V_{CE}, V_{BE}, and V_{CB} in both circuits of Figure 4–44.

FIGURE 4–44

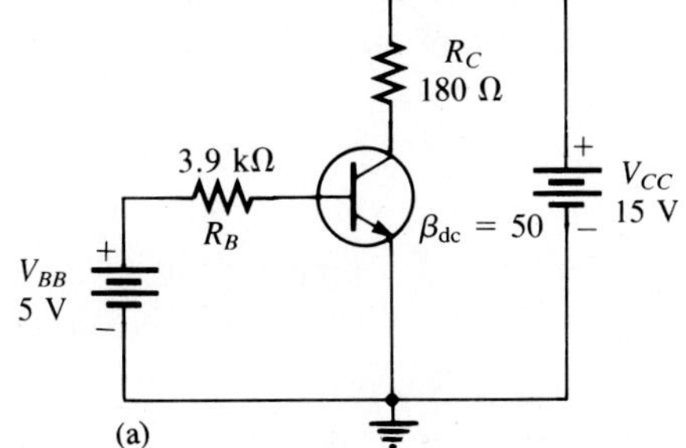

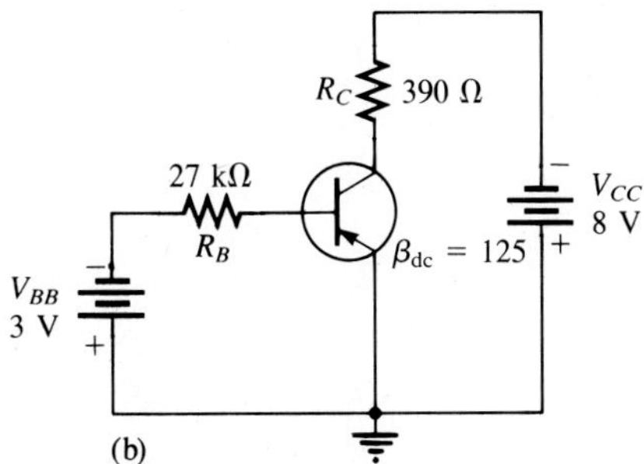

16. Determine whether or not the transistors in Figure 4–44 are saturated.
17. Find I_B, I_E, and I_C in Figure 4–45. $\alpha_{dc} = 0.98$.

FIGURE 4–45

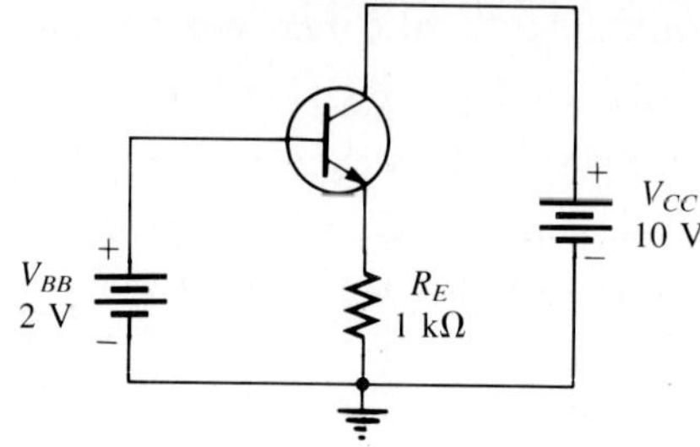

18. Determine the terminal voltages of each transistor with respect to ground for each circuit in Figure 4–46. Also determine V_{CE}, V_{BE}, and V_{BC}.

FIGURE 4–46

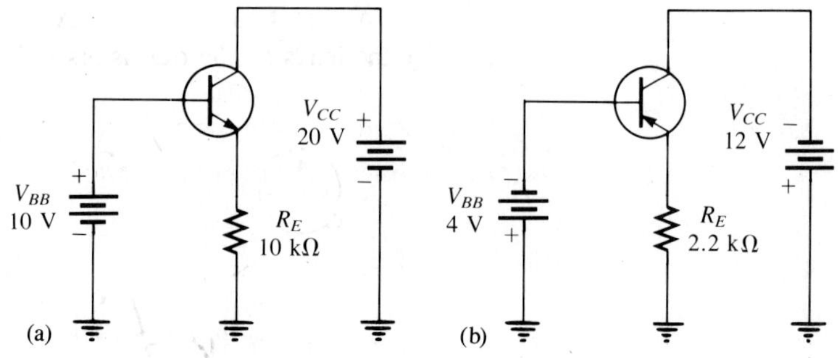

19. If the β_{dc} in Figure 4–46(a) changes from 100 to 150 due to a temperature increase, what is the change in collector current?

20. A certain transistor is to be operated at a collector current of 50 mA. How high can V_{CE} go without exceeding a $P_{D(max)}$ of 1.2 W?

21. The power dissipation derating factor for a certain transistor is 1 mW/°C. The $P_{D(max)}$ is 0.5 W at 25°C. What is $P_{D(max)}$ at 100°C?

SECTION 4–4 THE TRANSISTOR AS A VOLTAGE AMPLIFIER

22. A transistor amplifier has a voltage gain of 50. What is the output voltage when the input voltage is 100 mV?

23. To achieve an output of 10 V with an input of 300 mV, what voltage gain is required?

24. A 50 mV signal is applied to the base of a properly biased transistor with r_e = 10 Ω and R_C = 560 Ω. Determine the signal voltage at the collector.

SECTION 4–5 THE TRANSISTOR AS A SWITCH

25. Determine $I_{C(sat)}$ for the transistor in Figure 4–47. What is the value of I_B necessary to produce saturation? What minimum value of V_{IN} is necessary for saturation?

26. The transistor in Figure 4–48 has a β_{dc} of 50. Determine the value of R_B required to ensure saturation when V_{IN} is 5 V. What must V_{IN} be to cut off the transistor?

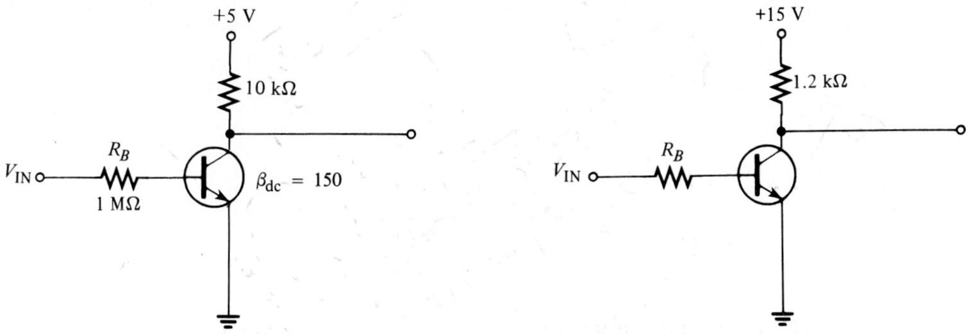

FIGURE 4–47

FIGURE 4–48

SECTION 4–6 TRANSISTOR PACKAGES AND TERMINAL IDENTIFICATION

27. Identify the leads on the transistors in Figure 4–49. Bottom views are shown.

FIGURE 4–49

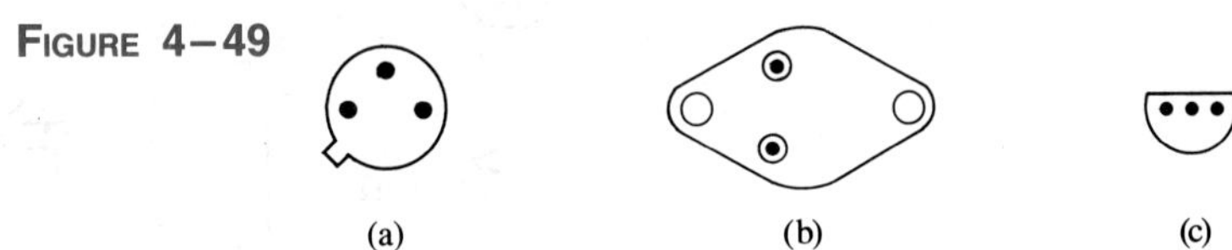

28. What is the most probable category of each transistor in Figure 4–50?

FIGURE 4–50
(Copyright of Motorola, Inc. Used by permission.)

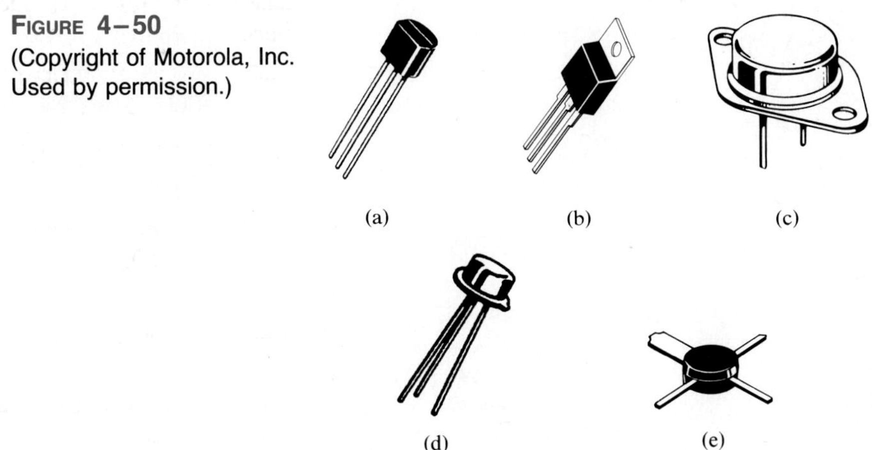

SECTION 4–7 TROUBLESHOOTING

29. In an out-of-circuit test of a good npn transistor, what should an analog ohmmeter indicate when its positive probe is touching the emitter and the negative probe is touching the base? When its positive probe is touching the base and the negative probe is touching the collector?

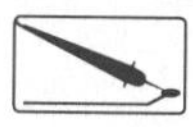

30. What is the most likely problem, if any, in each circuit of Figure 4–51? Assume a β_{dc} of 75.

FIGURE 4–51

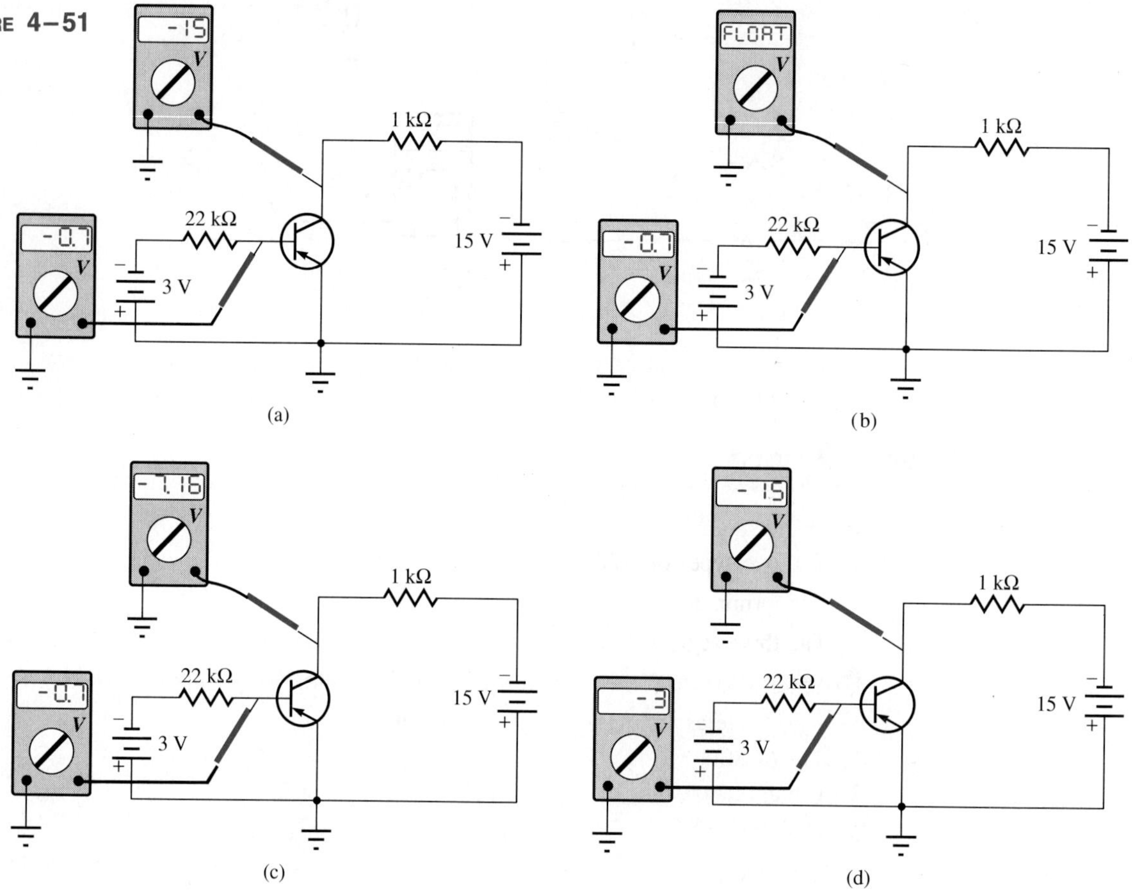

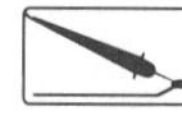

31. What is the value of the β_{dc} of each transistor in Figure 4–52?

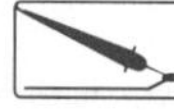

32. This problem relates to the circuit board and schematic in Figure 4–41. A remote switch loop is connected between pins 2 and 3. When the remote switches are closed, the relay (RY_1) contacts between pin 10 and pin 11 are normally open. When a remote switch is opened, the relay contacts do not close. Determine the possible causes of this malfunction.

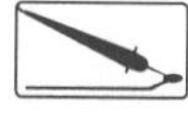

33. This problem relates to Figure 4–41. The relay contacts remain closed between pins 10 and 11 no matter what any of the inputs are. This means that the relay is energized continuously. What are the possible faults?

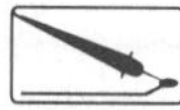

34. This problem relates to Figure 4–41. Pin 9 stays at approximately 0.1 V, regardless of the input at pin 7. What do you think is wrong? What would you check first?

FIGURE 4–52

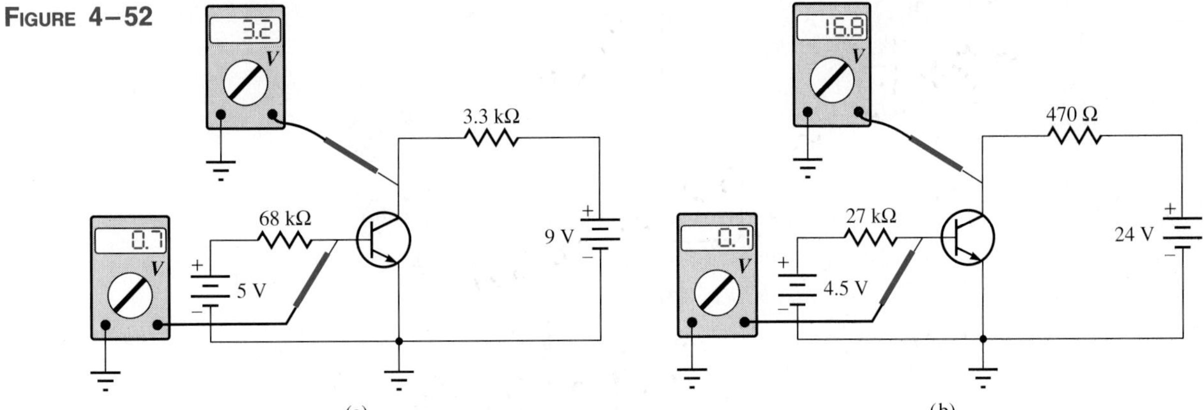

ANSWERS TO REVIEW QUESTIONS

SECTION 4–1

1. The two types of BJTs are npn and pnp.
2. The terminals of a BJT are base, collector, and emitter.
3. The three regions of a BJT are separated by two pn junctions.

SECTION 4–2

1. To operate as an amplifier, the base-emitter is forward-biased and the base-collector is reverse-biased.
2. The emitter current is the largest.
3. The base current is much smaller than the emitter current.
4. The base region is very narrow compared to the other two regions.
5. $I_E = 1\text{ mA} + 1\ \mu\text{A} = 1.001\text{ mA}$

SECTION 4–3

1. $\beta_{dc} = I_C/I_B$; $\alpha_{dc} = I_C/I_E$
2. $\beta_{dc} = 100$; $\alpha_{dc} = 100/(100 + 1) = 0.99$
3. I_C is plotted versus V_{CE}.
4. Forward-reverse bias is required for amplifier operation.
5. β_{dc} increases with temperature.
6. β_{dc} generally varies some from one device to the next for a given type.

SECTION 4–4

1. Amplification is the process where a smaller signal is used to produce a larger identical signal.
2. Voltage gain is the ratio of output voltage to input voltage.
3. R_C and r_e determine the voltage gain.

4. A_v = 5 V/250 mV = 20
5. A_v = 1200 Ω/20 Ω = 60

SECTION 4–5

1. A transistor switch operates in cutoff and saturation.
2. The collector current is maximum in saturation.
3. The collector current is approximately zero in cutoff.
4. $V_{CE} = V_{CC}$ in cutoff.
5. V_{CE} is minimum in saturation.

SECTION 4–6

1. Three categories of BJT are small signal/general purpose, power, and RF.
2. Going clockwise from tab: emitter, base, and collector (bottom view).
3. The metal mounting tab or case in power transistors is the collector.

SECTION 4–7

1. First, test it in-circuit.
2. If R_B opens, the transistor is in cutoff.
3. The base voltage is V_{BB} and the collector voltage is V_{CC}.

SECTION 4–8

1. The detector board detects when there is an open in one of the remote loops.
2. The Zone 1 and Zone 2 circuits are identical.
3. There is contact closure between pin 10 and pin 11 when the relay is activated as a result of an open switch in Zone 1 or Zone 2 loops.
4. The diode clips off any negative voltage induced in the coil to prevent possible damage to a transistor.

ANSWERS TO PRACTICE EXERCISES

4–1 10 mA, 0.995

4–2 $I_B = 241\ \mu A$, $I_C = 21.7$ mA, $I_E = 21.94$ mA, $V_{CE} = 4.23$ V, $V_{CB} = 3.53$ V

4–3 Along the horizontal axis

4–4 Not saturated

4–5 10 V

4–6 $V_{CC(max)} = 44.55$ V, $V_{CE(max)}$ is exceeded first.

4–7 4.55 W

4–8 2.5 kΩ

4–9 78.4 μA

4–10 Reduce R_C to 160 Ω and R_B to 2.2 kΩ.

4–11 5 V

4–12 R_B open

5

Bipolar Transistor Biasing

After completing this chapter, you should be able to

- □ Explain the purpose of dc bias in transistor circuits.
- □ Show how bias affects the operation of a bipolar junction transistor.
- □ Define *linear operation* in relation to transistor characteristic curves and load lines.
- □ Analyze the bias of a transistor that operates as a linear amplifier.
- □ Explain how the dc bias point (*Q* point) affects the linear operation of a transistor.
- □ Determine by graphical analysis how much input signal a given biased transistor can take without being driven into either cutoff or saturation.
- □ Discuss various types of dc bias circuits and their characteristics.
- □ Explain the meaning of bias stability and describe how the type of bias affects the stability of a transistor circuit.
- □ Troubleshoot biased transistor circuits.

As you saw in the last chapter, a transistor must be properly biased in order to operate as an amplifier. The purpose of dc biasing is to establish a steady level of transistor current and voltage, called the *dc operating point* or *quiescent (Q) point.* In this chapter, several types of bias circuits are studied. This material lays the groundwork for the study of amplifiers, oscillators, and other circuits that cannot function without proper biasing.

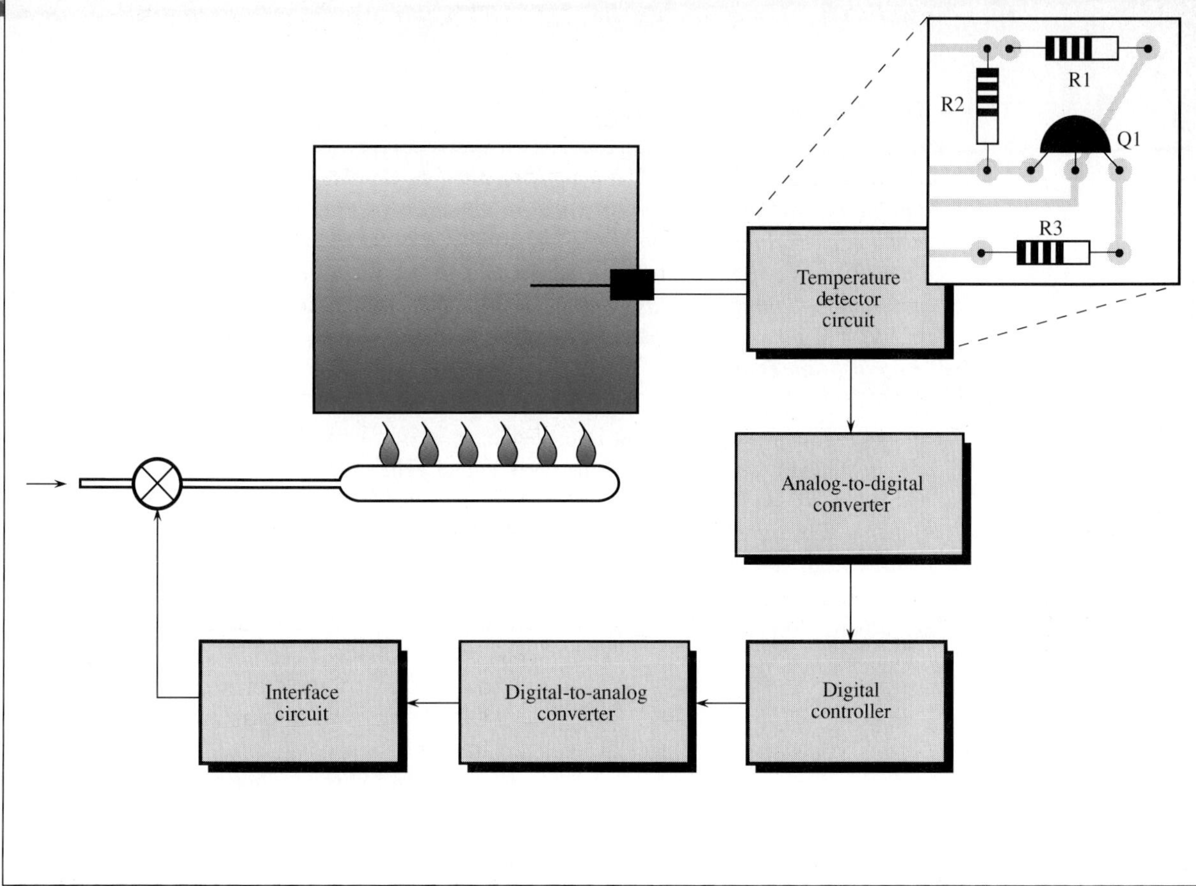

A System Application

A proportional heat control system that maintains a liquid at a precise temperature for a certain industrial manufacturing process is depicted above. The system application for this chapter deals with a temperature detection circuit that produces a dc output voltage proportional to the temperature of the substance in the tank. This circuit uses the variable resistance of a temperature sensor as part of its bias circuit. The output of the temperature detector circuit is converted to digital form by an analog-to-digital converter (covered in Chapter 14) for precise control of the burner. If the temperature of the substance tends to increase, the voltage fed back to the control valve decreases the fuel supply to the burner and thus reduces the temperature. If the temperature of the substance tends to decrease, the voltage fed back to the valve increases the fuel supply to the burner and thus raises the heat.

This type of closed-loop system maintains a nearly constant temperature using this feedback method. This basic concept is typical of many industrial process control systems.

In this system application, our attention is focused on the temperature detection circuit and how it interfaces with the rest of the system. Although coverage of the digital portions of the system are beyond the scope of this book, it is very important to realize that all sorts of devices must work together in real-world systems.

For the system application in Section 5–7, in addition to the other topics, be sure you understand

- □ How transistor bias circuits work.
- □ How voltages and currents are affected by bias resistance values.
- □ The effects of loading on transistor bias.

5–1 THE DC OPERATING POINT

As mentioned in Chapter 4, a transistor must be dc-biased in order to operate as an amplifier. A dc operating point must be set so that signal variations at the input terminal are amplified and accurately reproduced at the output terminal. As you learned in Chapter 4, when you bias a transistor, you establish a certain current and voltage condition. This means, for example, that at the dc operating point, I_C and V_{CE} have specified values. The dc operating point is often referred to as the Q point (quiescent point).

Figure 5–1 shows proper and improper biasing of an amplifier. In part (a), the output signal is an amplified replica of the input signal. The output signal swings equally above and below the dc level of the output. Improper biasing can cause distortion in the output signal, as illustrated in parts (b) and (c). Part (b) illustrates limiting of the positive portion of the output voltage as a result of a dc operating point being too close to cutoff. Part (c) shows limiting of the negative portion of the output voltage as a result of a dc operating point being too close to saturation.

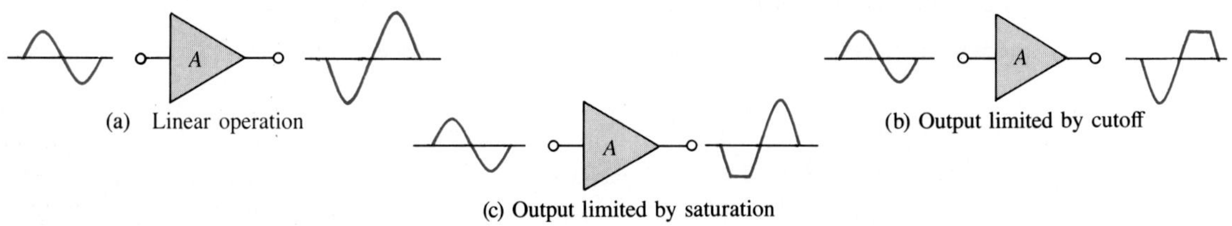

FIGURE 5–1
Examples of linear and nonlinear operation of an inverting amplifier.

GRAPHICAL ANALYSIS

The transistor in Figure 5–2(a) is biased with variables V_{CC} and V_{BB} to obtain certain values of I_B, I_C, I_E, and V_{CE}. The collector characteristic curves for this particular transistor are shown in Figure 5–2(b), and we will use these to graphically illustrate the effects of dc bias. To do this, as illustrated in Figure 5–3, we will set various values of I_B and observe what happens to I_C and V_{CE}.

To start, V_{BB} is adjusted to produce an I_B of 200 μA, as shown in Figure 5–3(a). Since $I_C = \beta_{dc}I_B$, the collector current is 20 mA, as indicated, and

$$V_{CE} = V_{CC} - I_C R_C = 10\text{ V} - 4\text{ V} = 6\text{ V}$$

This Q point is shown on the graph of Figure 5–3(d) as Q_1.

Next, V_{BB} is increased to produce an I_B of 300 μA, an I_C of 30 mA, and a resulting V_{CE} of 4 V, as shown in Figure 5–3(b). The Q point for this condition is indicated by Q_2

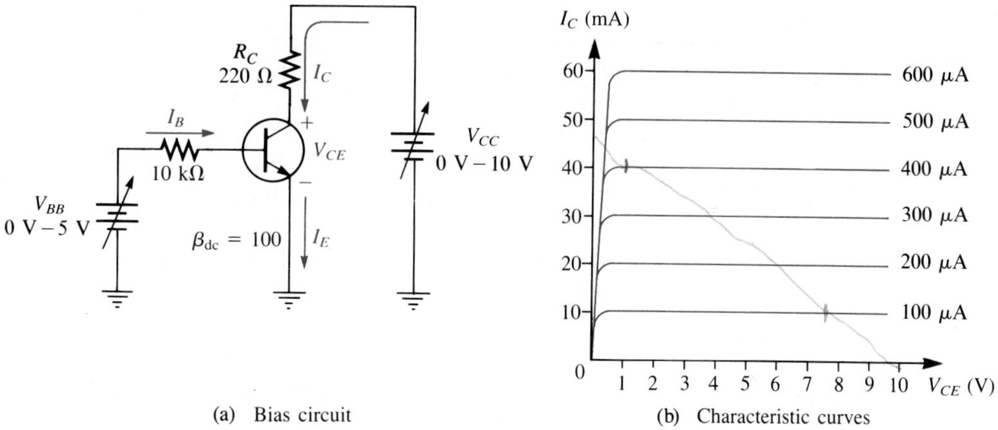

FIGURE 5–2
A transistor circuit with variable bias voltages.

on the graph. Finally, V_{BB} is increased to give an I_B of 400 μA, an I_C of 40 mA, and a V_{CE} of 2 V, as in Figure 5–3(c). Q_3 is the corresponding Q point on the graph.

DC LOAD LINE

Notice that when I_B increases, I_C increases and V_{CE} decreases. When I_B decreases, I_C decreases and V_{CE} increases. So, as V_{BB} is adjusted up or down, the dc operating point of the transistor moves along a sloping straight line, called the dc **load line,** connecting each Q point. At any point along the line, values of I_B, I_C, and V_{CE} can be picked off the graph, as shown in Figure 5–4.

The dc load line intersects the V_{CE} axis at 10 V, the point where $V_{CE} = V_{CC}$. This is the transistor cutoff point because I_B and I_C are zero (ideally). Actually, there is a small leakage current, I_{CBO}, at cutoff as indicated, and therefore V_{CE} is slightly less than 10 V.

The dc load line intersects the I_C axis at 50 mA ideally. This is the transistor saturation point because I_C is maximum (ideally 50 mA) at the point where $V_{CE} = 0$ V and $I_C = V_{CC}/R_C$. Actually, there is a small voltage ($V_{CE(\text{sat})}$) across the transistor, and $I_{C(\text{sat})}$ is slightly less than 50 mA, as indicated in Figure 5–4.

LINEAR OPERATION

The region along the load line including all points between saturation and cutoff is generally known as the *linear region* of the transistor's operation. As long as the transistor is operated in this region, the output voltage is ideally a linear reproduction of the input. For example, assume a sine wave voltage is superimposed on V_{BB}, causing the base current to vary 100 μA above and below the Q-point value of 300 μA. This, in turn, causes the collector current to vary 10 mA above and below its Q-point value of 30 mA.

FIGURE 5–3
Q-point adjustments.

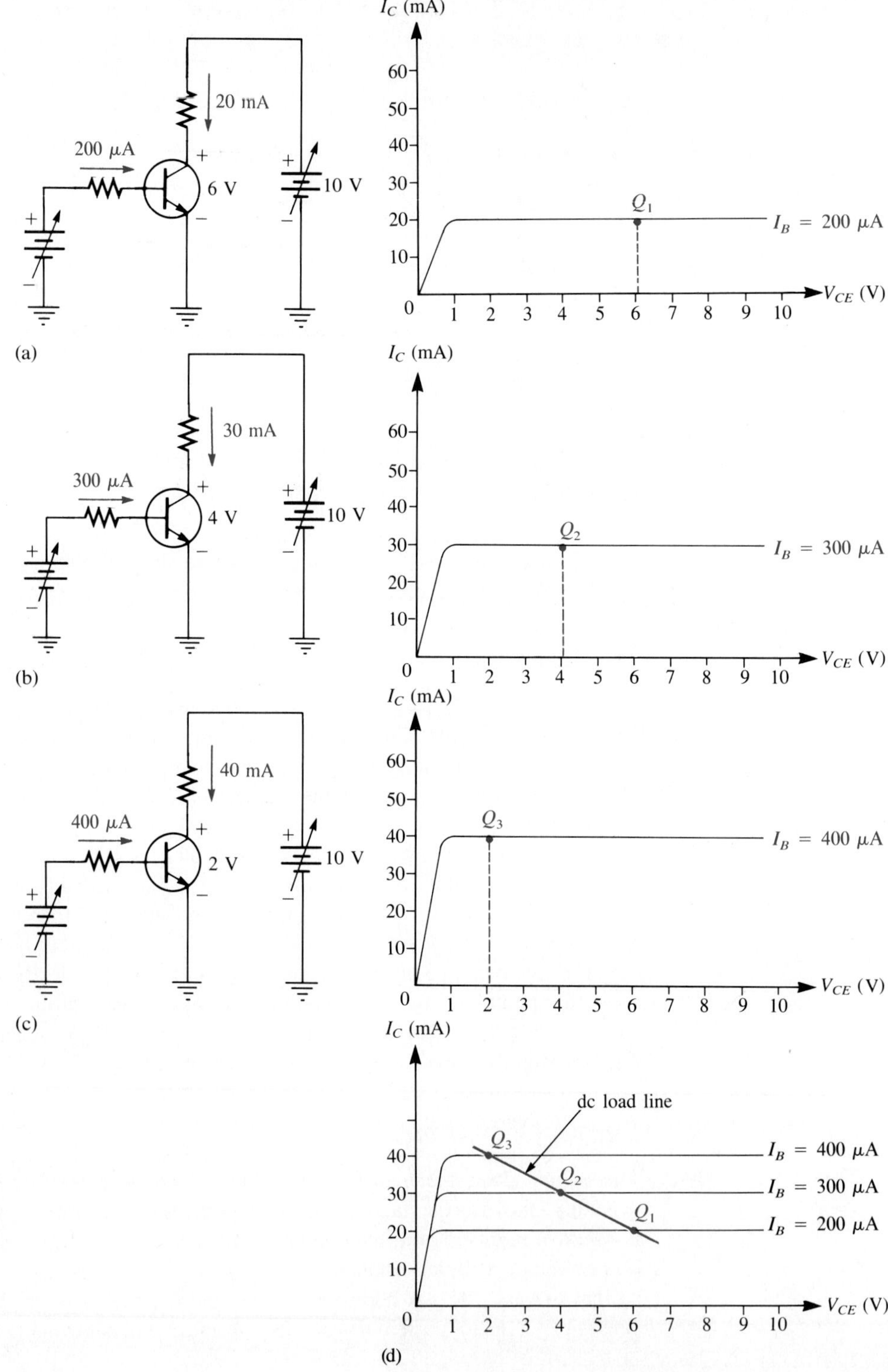

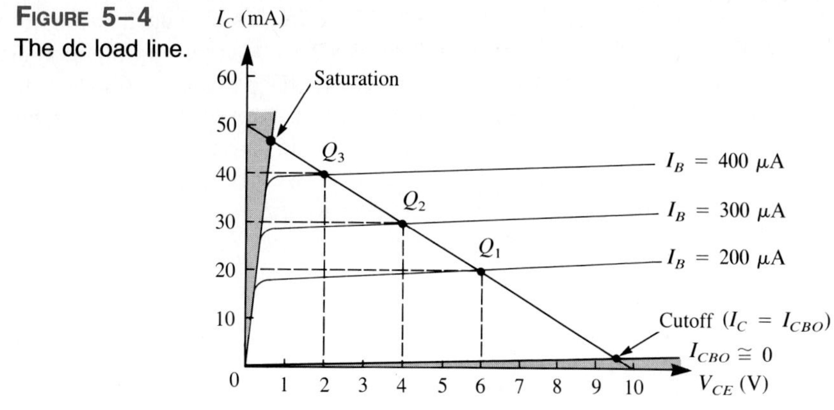

Figure 5–4
The dc load line.

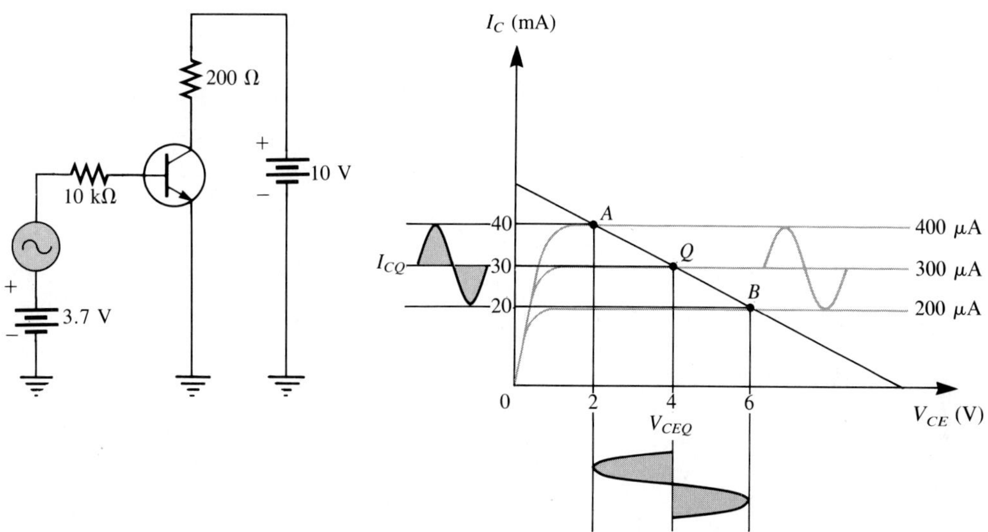

Figure 5–5
Variations in collector current and collector-to-emitter voltage as a result of a variation in base current.

As a result of the variation in collector current, the collector-to-emitter voltage varies 2 V above and below its *Q*-point value of 4 V. This action is shown in Figure 5–5.

Point *A* on the load line corresponds to the positive peak of the input sine wave. Point *B* corresponds to the negative peak, and the *Q* point corresponds to the zero value of the sine wave, as indicated. V_{CEQ} and I_{CQ} are *Q*-point values with no signal applied.

Distortion of the Output

As previously mentioned, under certain input signal conditions the location of the *Q* point on the load line can cause one peak of the output signal to be limited or clipped, as shown in parts (a) and (b) of Figure 5–6. In each case the input signal is too large for the Q point location and is driving the transistor into cutoff or saturation during a portion of the input cycle. When both peaks are limited as in Figure 5–6(c), the transistor is being driven into both saturation and cutoff by an excessively large input signal. When only the positive peak is limited, the transistor is being driven into cutoff but not saturation. When only the negative peak is limited, the transistor is being driven into saturation but not cutoff. A good rule-of-thumb for a distortion-free output is to limit the maximum peak V_{CE} swing to $0.95V_{CC}$ and the minimum peak to $0.05V_{CC}$.

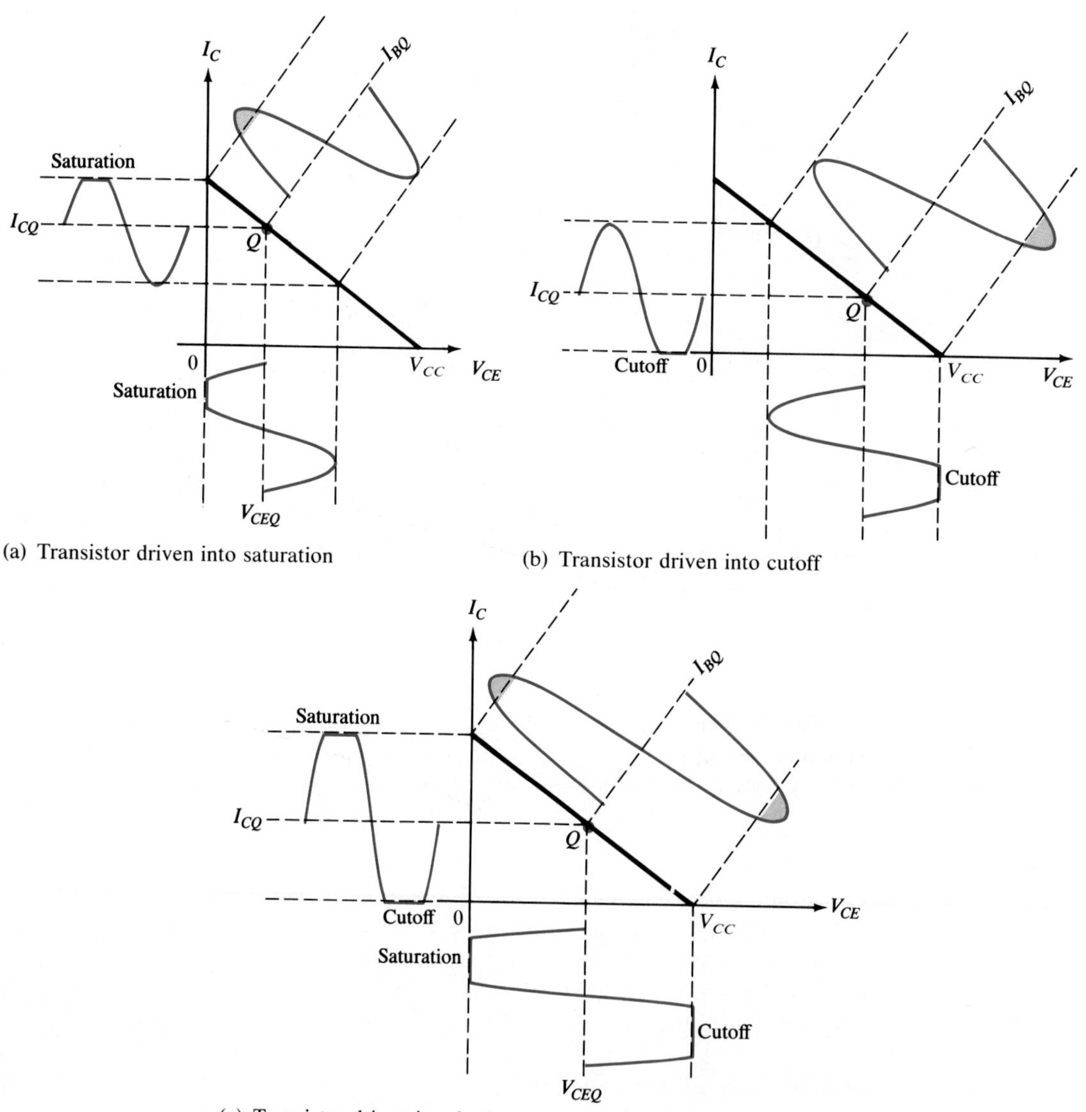

FIGURE 5–6
Graphical illustration of saturation and cutoff.

EXAMPLE 5–1

Determine the Q point in Figure 5–7, and find the maximum peak value of base current for linear operation ($\beta_{dc} = 200$).

FIGURE 5–7

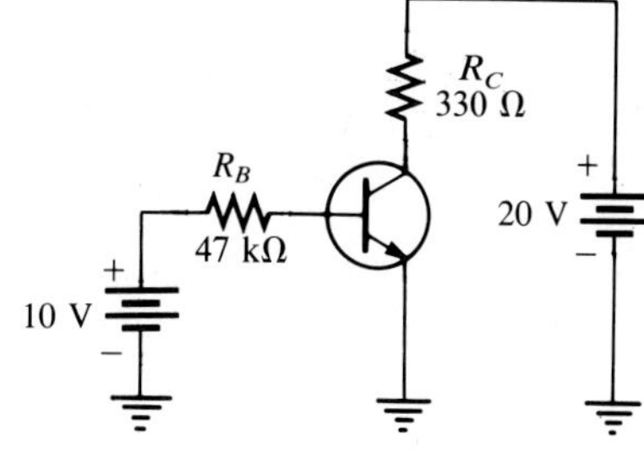

SOLUTION

The Q point is defined by I_C and V_{CE}.

$$I_B = \frac{V_{BB} - 0.7}{R_B} = \frac{9.3\ \text{V}}{47\ \text{k}\Omega} = 198\ \mu\text{A}$$

$$I_C = \beta_{dc} I_B = (200)(198\ \mu\text{A}) = 39.6\ \text{mA}$$

$$\begin{aligned} V_{CE} &= V_{CC} - I_C R_C \\ &= 20\ \text{V} - 13.07\ \text{V} \\ &= 6.93\ \text{V} \end{aligned}$$

The Q point is at $I_C = 39.6$ mA and at $V_{CE} = 6.93$ V. Since $I_{C(\text{cutoff})} = 0$, we need to know $I_{C(\text{sat})}$ to determine how much variation in collector current can occur and still maintain linear operation of the transistor.

$$I_{C(\text{sat})} = \frac{V_{CC}}{R_C} = \frac{20\ \text{V}}{330\ \Omega} = 60.6\ \text{mA}$$

The dc load line is graphically illustrated in Figure 5–8, showing that before saturation is reached, I_C can increase an amount ideally equal to

$$I_{C(\text{sat})} - I_{CQ} = 6.06\ \text{mA} - 39.6\ \text{mA} = 21\ \text{mA}$$

FIGURE 5–8

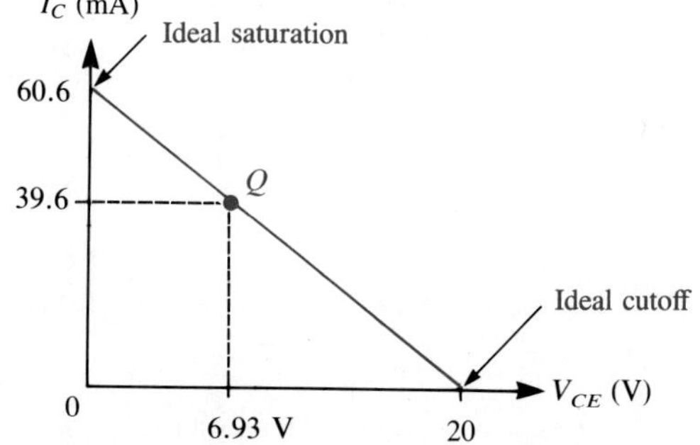

However, I_C can decrease by 39.6 mA before cutoff is reached. Therefore, the limiting excursion is 21 mA because the Q point is closer to saturation than to cutoff. The 21 mA is the maximum peak variation of the collector current. Actually, it would be slightly less in practice because $V_{CE(\text{sat})}$ is not quite zero. The maximum peak variation of the base current is determined as follows:

$$I_{b(\text{peak})} = \frac{I_{c(\text{peak})}}{\beta_{\text{dc}}} = \frac{21\text{ mA}}{200} = 105\ \mu\text{A}$$

PRACTICE EXERCISE 5–1

Find the Q point in Figure 5–7, and determine the maximum peak value of base current for linear operation for the following circuit values: $\beta_{\text{dc}} = 100$, $R_C = 1\text{ k}\Omega$, and $V_{CC} = 24$ V. ■

5–1 REVIEW QUESTIONS

1. What are the upper and lower limits on a dc load line in terms of V_{CE} and I_C?
2. Define *Q point*.
3. At what point on the load line does saturation begin? At what point does cutoff occur?

5–2 BASE BIAS

In this section and several following sections, we will be looking at various methods for dc biasing a transistor circuit without using a separate base bias source. You will learn the advantages and disadvantages of each method.

In the previous discussions, a separate battery, V_{BB}, was used to bias the base-emitter junction. A more practical method is to use V_{CC} as a single bias source, as shown in Figure 5–9(a). To simplify the schematics, the battery symbol can be omitted and replaced by a line termination with a voltage indicator, shown in Figure 5–9(b) as we have done before. The analysis of the circuit in Figure 5–9 for the linear region is as follows.

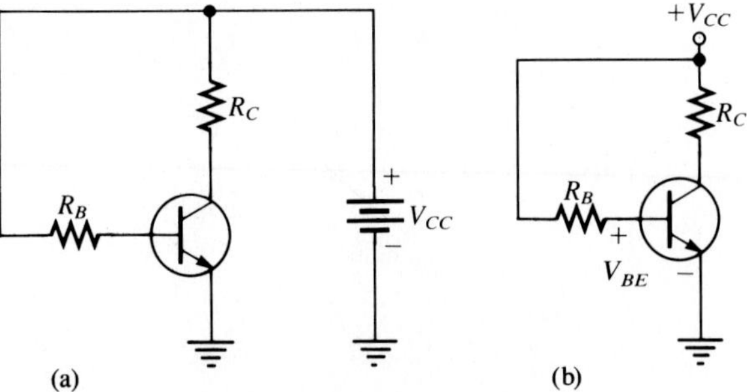

FIGURE 5–9
Base bias.

The drop across R_B is $V_{CC} - V_{BE}$. Therefore,

$$I_B = \frac{V_{CC} - V_{BE}}{R_B} \tag{5–1}$$

and, neglecting leakage current, I_{CBO},

$$I_C = \beta_{dc} I_B \tag{5-2}$$

The collector-to-emitter voltage equals the collector supply voltage minus the drop across R_C.

$$V_{CE} = V_{CC} - I_C R_C \tag{5-3}$$

Substituting $\beta_{dc} I_B$ for I_C,

$$V_{CE} = V_{CC} - \beta_{dc} I_B R_C \tag{5-4}$$

EFFECT OF β_{dc} ON THE Q POINT

Notice that Equations (5–2) and (5–4) include β_{dc}. The disadvantage of this is that a variation in β_{dc} causes both I_C and V_{CE} to change, thus changing the Q point of the transistor and making the base-biased circuit beta-dependent. Recall that β_{dc} varies with temperature. In addition, there is a large spread of values from one device to another of the same type.

EXAMPLE 5–2

The base-biased circuit in Figure 5–10 is subjected to an increase in junction temperature from 25°C to 75°C. If $\beta_{dc} = 100$ at 25°C and 150 at 75°C, determine the percent change in Q-point values (I_C and V_{CE}) over the temperature range. Neglect any change in V_{BE} and the effects of any leakage current. The transistor is silicon.

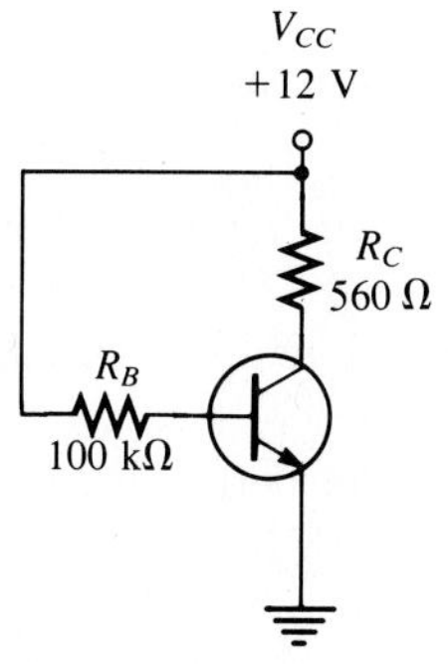

FIGURE 5–10

SOLUTION

At 25°C,

$$I_B = \frac{V_{CC} - V_{BE}}{R_B} = \frac{12\text{ V} - 0.7\text{ V}}{100\text{ k}\Omega} = 113\ \mu\text{A}$$

$$I_C = \beta_{dc} I_B = (100)(113\ \mu\text{A}) = 11.3\text{ mA}$$

$$\begin{aligned} V_{CE} &= V_{CC} - I_C R_C \\ &= 12\text{ V} - (11.3\text{ mA})(560\ \Omega) \\ &= 5.67\text{ V} \end{aligned}$$

At 75°C,

$$I_B = 113\ \mu\text{A}$$

$$I_C = \beta_{dc} I_B = (150)(113\ \mu\text{A}) = 16.95\text{ mA}$$

$$\begin{aligned} V_{CE} &= V_{CC} - I_C R_C \\ &= 12\text{ V} - (16.95\text{ mA})(560\ \Omega) \\ &= 2.51\text{ V} \end{aligned}$$

The percent change in I_C is

$$\begin{aligned} \%\ \Delta I_C &= \frac{I_{C(75°)} - I_{C(25°)}}{I_{C(25°)}} \times 100\% \\ &= \frac{16.95\text{ mA} - 11.3\text{ mA}}{11.3\text{ mA}} \times 100\% \\ &= 50\%\ \text{(an increase)} \end{aligned}$$

The percent change in V_{CE} is

$$\% \Delta V_{CE} = \frac{V_{CE(75°)} - V_{CE(25°)}}{V_{CE(25°)}} \times 100\%$$
$$= \frac{2.51\text{ V} - 5.67\text{ V}}{5.67\text{ V}} \times 100\%$$
$$= -55.7\% \text{ (a decrease)}$$

As you can see, the Q point is extremely dependent on β_{dc} in this circuit and therefore makes the base bias arrangement very unstable.

PRACTICE EXERCISE 5–2

If $\beta_{dc} = 50$ at 0°C and 125 at 100°C for the circuit in Figure 5–10, determine the percent change in the Q-point values over the temperature range.

OTHER FACTORS INFLUENCING BIAS STABILITY

In addition to being affected by a change in β_{dc}, the bias point can be affected by changes in V_{BE} and I_{CBO}. The base-to-emitter voltage, V_{BE}, decreases with an increase in temperature. As you can see from the following equation, a decrease in V_{BE} tends to increase I_B.

$$I_B = \frac{V_{CC} - V_{BE}}{R_B}$$

The effect of a change in V_{BE} is negligible if $V_{CC} >> V_{BE}$ (V_{CC} at least 10 times greater than V_{BE}).

The reverse leakage current, I_{CBO}, has the effect of decreasing the net base current and thus increasing the base voltage because it creates a voltage drop across R_B with a polarity that adds to the base bias voltage, as shown in Figure 5–11. In modern transistors I_{CBO} is extremely small (nA), and its effect on the bias is negligible if $V_{BB} >> I_{CBO}R_B$.

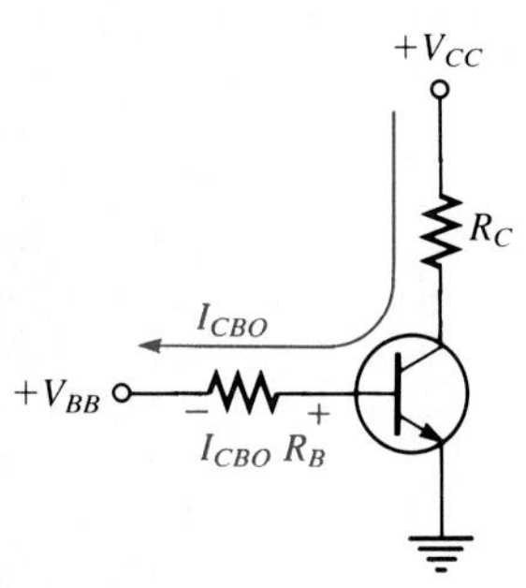

FIGURE 5–11
Effect of I_{CBO}.

5–2 REVIEW QUESTIONS

1. What is an advantage of base bias over using two separate dc supplies?
2. What is the main disadvantage of the base bias method?

5–3 EMITTER BIAS

Although this method of biasing requires two separate dc voltage sources, one positive and the other negative, it does have an important advantage as you will learn.

This type of bias circuit uses both a positive and a negative supply voltage, as shown in Figure 5–12. In this circuit, the base is at approximately zero volts, and the $-V_{EE}$ supply

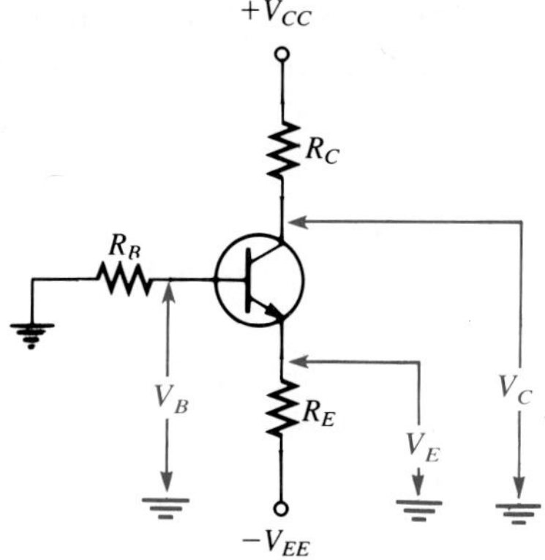

FIGURE 5–12
An npn transistor with emitter bias.

forward-biases the base-emitter junction. Expressions for the currents and voltages are as follows. The single-variable subscripts indicate voltages with respect to ground.

$$V_B \cong 0 \tag{5–5}$$

$$V_E \cong -V_{BE} \tag{5–6}$$

$$I_E = \frac{V_E - V_{EE}}{R_E} \tag{5–7}$$

$$I_C \cong I_E \tag{5–8}$$

$$V_C = V_{CC} - I_C R_C \tag{5–9}$$

$$V_{CE} = V_C - V_E \tag{5–10}$$

EXAMPLE 5–3 Find I_E, I_C, and V_{CE} in Figure 5–13.

SOLUTION

$$V_E \cong -V_{BE} = -0.7 \text{ V}$$

$$I_E = \frac{V_{BE} - V_{EE}}{R_E} = \frac{-0.7 \text{ V} - (-10 \text{ V})}{4.7 \text{ k}\Omega} = \frac{9.3 \text{ V}}{4.7 \text{ k}\Omega} = 1.98 \text{ mA}$$

$$I_C \cong I_E = 1.98 \text{ mA}$$

$$V_C = V_{CC} - I_C R_C = 10 \text{ V} - (1.98 \text{ mA})(1 \text{ k}\Omega) = 8.02 \text{ V}$$

$$V_{CE} = V_C - V_E = 8.02 \text{ V} - (-0.7 \text{ V}) = 8.72 \text{ V}$$

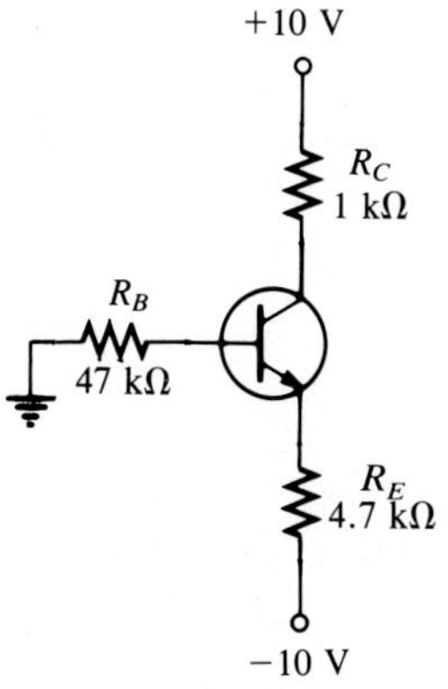

FIGURE 5–13

These are the Q-point values for the circuit. The dc load line is graphically illustrated in Figure 5–14. The approximate collector saturation current is determined as follows.

$$I_{C(\text{sat})} = \frac{V_{CC} - (-V_{EE})}{R_C + R_E} = \frac{20 \text{ V}}{5.7 \text{ k}\Omega} = 3.5 \text{ mA}$$

The collector-to-emitter voltage at cutoff is

$$V_{CC} - (-V_{EE}) = 10 \text{ V} - (-10 \text{ V}) = 20 \text{ V}$$

The dc load line, illustrated in Figure 5–14, shows that I_C can increase an amount ideally equal to

$$I_{C(\text{sat})} - I_{CQ} = 3.5 \text{ mA} - 1.98 \text{ mA} = 1.52 \text{ mA}$$

before saturation is reached. I_C can decrease by 1.98 mA before cutoff is reached. As you can see, this circuit is biased closer to saturation than to cutoff.

FIGURE 5–14

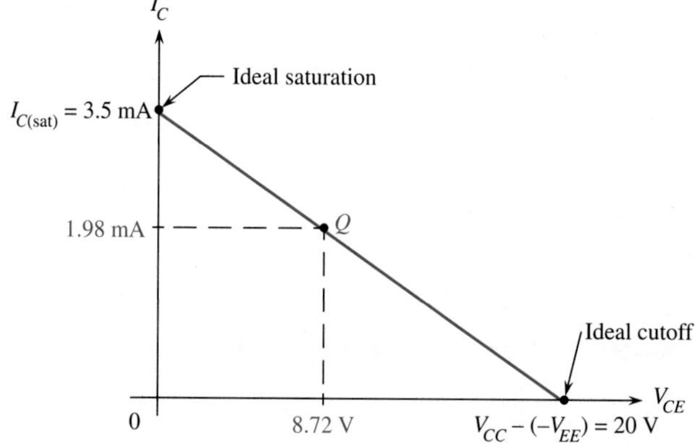

PRACTICE EXERCISE 5–3

Find I_E, I_C, and V_{CE} for the circuit in Figure 5–13 with the following component values: $R_B = 100 \text{ k}\Omega$, $R_C = 680\ \Omega$, $R_E = 3.3 \text{ k}\Omega$, $V_{CC} = +15 \text{ V}$, and $V_{EE} = -15 \text{ V}$.

STABILITY OF EMITTER BIAS

Applying Kirchhoff's law around the base-emitter circuit in Figure 5–12, which has been redrawn in Figure 5–15 for analysis, we get the following equation:

$$V_{EE} - I_B R_B - V_{BE} - I_E R_E = 0$$

$$V_{EE} = I_B R_B + V_{BE} + I_E R_E$$

Since

$$I_B \cong \frac{I_E}{\beta_{\text{dc}}}$$

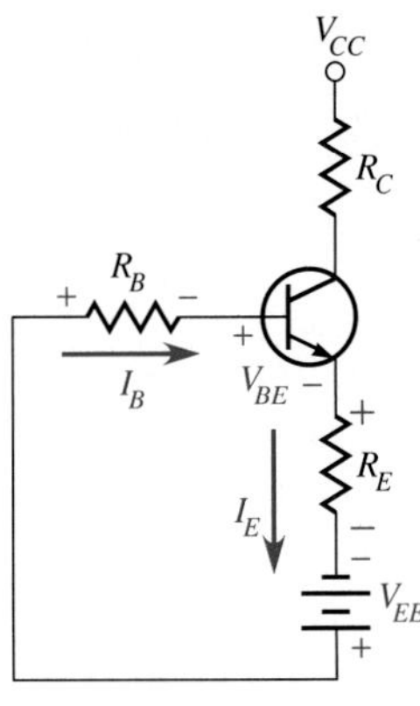

FIGURE 5–15
Kirchhoff's voltage law applies around the base-emitter loop.

then

$$\left(\frac{I_E}{\beta_{dc}}\right)R_B + I_E R_E + V_{BE} = V_{EE}$$

Factoring out I_E,

$$I_E\left(\frac{R_B}{\beta_{dc}} + R_E\right) + V_{BE} = V_{EE}$$

Transposing V_{BE} and then solving for I_E,

$$I_E = \frac{V_{EE} - V_{BE}}{R_E + R_B/\beta_{dc}} \qquad \textbf{(5–11)}$$

If $R_E >> R_B/\beta_{dc}$, the equation becomes

$$I_E \cong \frac{V_{EE} - V_{BE}}{R_E} \qquad \textbf{(5–12)}$$

A further simplification can be made if $V_{EE} >> V_{BE}$.

$$I_E \cong \frac{V_{EE}}{R_E} \qquad \textbf{(5–13)}$$

This shows that the emitter current can be essentially independent of β_{dc} and V_{BE}, as long as the conditions mentioned above are satisfied. Of course, if I_E is independent of β_{dc} and V_{BE}, then the Q point is not affected appreciably by variations in these parameters. Thus, emitter bias can provide a stable bias point.

■ EXAMPLE 5–4

Determine how much the Q point in Figure 5–16 will change over a temperature range where β_{dc} increases from 50 to 100 and V_{BE} decreases from 0.7 V to 0.6 V.

SOLUTION

For $\beta_{dc} = 50$ and $V_{BE} = 0.7$ V,

$$I_C \cong I_E = \frac{|V_{EE}| - V_{BE}}{R_E + R_B/\beta_{dc}}$$
$$= \frac{20\text{ V} - 0.7\text{ V}}{10\text{ k}\Omega + 10\text{ k}\Omega/50}$$
$$= 1.892\text{ mA}$$

$$V_C = V_{CC} - I_C R_C$$
$$= 20\text{ V} - (1.892\text{ mA})(4.7\text{ k}\Omega)$$
$$= 11.11\text{ V}$$

$$V_E = V_{EE} + I_E R_E$$
$$= -20\text{ V} + (1.892\text{ mA})(10\text{ k}\Omega)$$
$$= -1.08\text{ V}$$

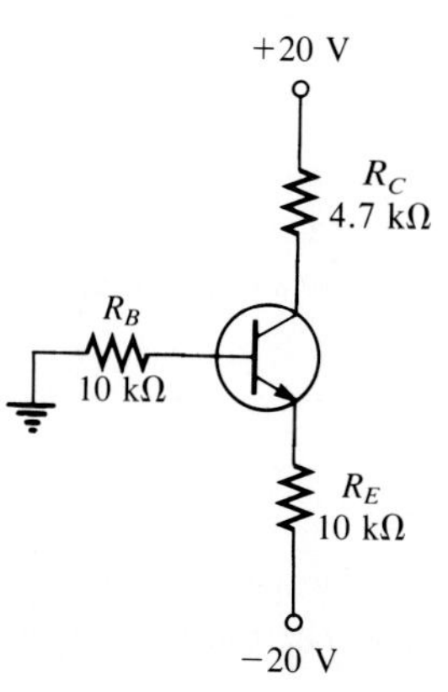

FIGURE 5–16

Therefore,

$$\begin{aligned} V_{CE} &= V_C - V_E \\ &= 11.11\ \text{V} - (-1.08\ \text{V}) \\ &= 12.19\ \text{V} \end{aligned}$$

For $\beta_{dc} = 100$ and $V_{BE} = 0.6$ V,

$$\begin{aligned} I_C \cong I_E &= \frac{|V_{EE}| - V_{BE}}{R_E + R_B/\beta_{dc}} \\ &= \frac{20\ \text{V} - 0.6\ \text{V}}{10\ \text{k}\Omega + 10\ \text{k}\Omega/100} \\ &= 1.921\ \text{mA} \end{aligned}$$

$$\begin{aligned} V_C &= V_{CC} - I_C R_C \\ &= 20\ \text{V} - (1.921\ \text{mA})(4.7\ \text{k}\Omega) \\ &= 10.97\ \text{V} \end{aligned}$$

$$\begin{aligned} V_E &= V_{EE} + I_E R_E \\ &= -20\ \text{V} + (1.921\ \text{mA})(10\ \text{k}\Omega) \\ &= -0.79\ \text{V} \end{aligned}$$

Therefore,

$$\begin{aligned} V_{CE} &= 10.97\ \text{V} - (-0.79\ \text{V}) \\ &= 11.76\ \text{V} \end{aligned}$$

The percent change in I_C as β_{dc} changes from 50 to 100 is

$$\frac{1.921\ \text{mA} - 1.892\ \text{mA}}{1.892\ \text{mA}} \times 100\% = 1.53\%$$

The percent change in V_{CE} is

$$\frac{11.76\ \text{V} - 12.19\ \text{V}}{12.19\ \text{V}} \times 100\% = -3.53\%$$

Keep in mind that the smal! change in the Q point in this example resulted from a doubling of β_{dc} and a 0.1 V change in V_{BE}.

PRACTICE EXERCISE 5–4

Determine how much the Q point in Figure 5–16 changes over a temperature range where β_{dc} increases from 20 to 75 and V_{BE} decreases from 0.75 V to 0.59 V. The supply voltages are ±10 V.

EMITTER-BIASED PNP

Figure 5–17 shows a pnp transistor with emitter bias. The basic difference is that the polarities of the supply voltages are reversed from those of the npn. The operation and analysis are basically the same, as Example 5–5 illustrates.

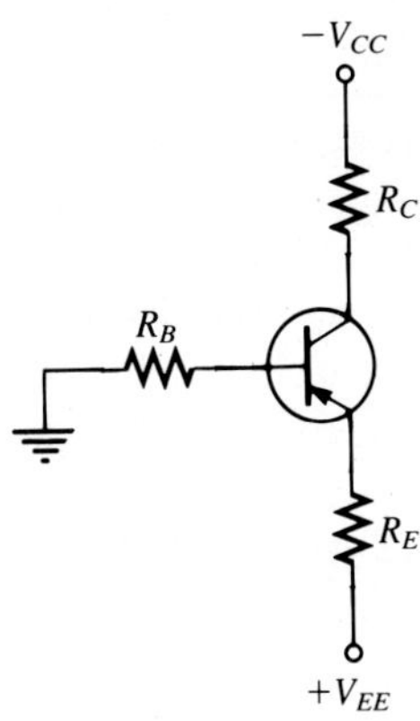

FIGURE 5–17
Emitter-biased pnp transistor.

EXAMPLE 5–5 Determine V_C, V_E, and V_{CE} in the pnp circuit of Figure 5–18. Assume $\beta_{dc} = 100$. Notice how the transistor is oriented in this diagram.

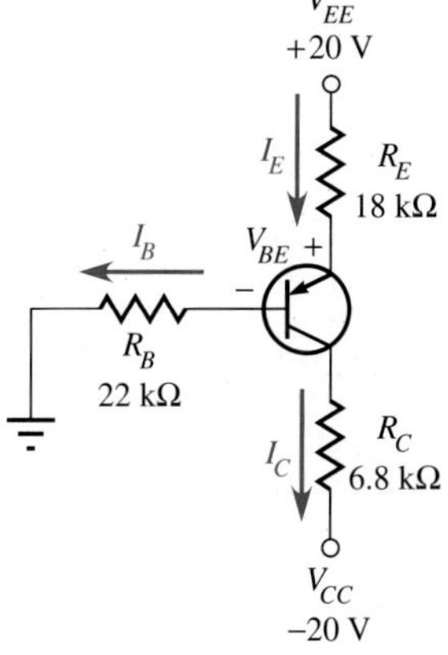

FIGURE 5–18

SOLUTION

$$I_C \cong I_E \cong \frac{V_{EE} - V_{BE}}{R_E + R_B/\beta_{dc}} = \frac{20\text{ V} - 0.7\text{ V}}{18\text{ k}\Omega + 22\text{ k}\Omega/100} = 1.06\text{ mA}$$

$$V_C = V_{CC} + I_C R_C = -20\text{ V} + (1.06\text{ mA})(6.8\text{ k}\Omega) = -12.8\text{ V}$$

$$V_E = V_{EE} - I_E R_E = 20\text{ V} - (1.06\text{ mA})(18\text{ k}\Omega) = 0.92\text{ V}$$

Therefore,

$$V_{CE} = V_C - V_E = -12.8\text{ V} - 0.92\text{ V} = -13.72\text{ V}$$

PRACTICE EXERCISE 5–5

Does V_C become more or less negative if R_C is increased? Will an increase in R_C change the collector current significantly? Calculate the collector voltage if R_C is 8.2 kΩ. ■

5–3 REVIEW QUESTIONS

1. Why is emitter bias more stable than base bias?
2. For an emitter-biased npn transistor, what is the approximate relationship of V_B and V_E? For a pnp transistor?
3. What is the main disadvantage of emitter bias?
4. An emitter-biased pnp transistor has dc supply voltages of ± 15 V. If R_E is 10 kΩ, what is the approximate emitter current?

5–4 VOLTAGE-DIVIDER BIAS

Next, we will study a method of biasing a transistor for linear operation using a resistive voltage-divider. This is the most widely used biasing method, for reasons that you will discover in this section.

A dc bias voltage at the base of the transistor is developed by a resistive voltage-divider consisting of R_1 and R_2 as shown in Figure 5–19. At point A, there are two current paths to ground: one through R_2 and the other through the base-emitter junction of the transistor.

If the base current is much smaller than the current through R_2, the bias circuit can be viewed as a simplified voltage-divider consisting of R_1 and R_2, as indicated in Figure 5–20(a). If I_B is *not* small enough to neglect compared to I_2, then the dc input resistance, $R_{IN(base)}$, looking in at the base of the transistor must be considered. $R_{IN(base)}$ appears in parallel with R_2, as shown in Figure 5–20(b).

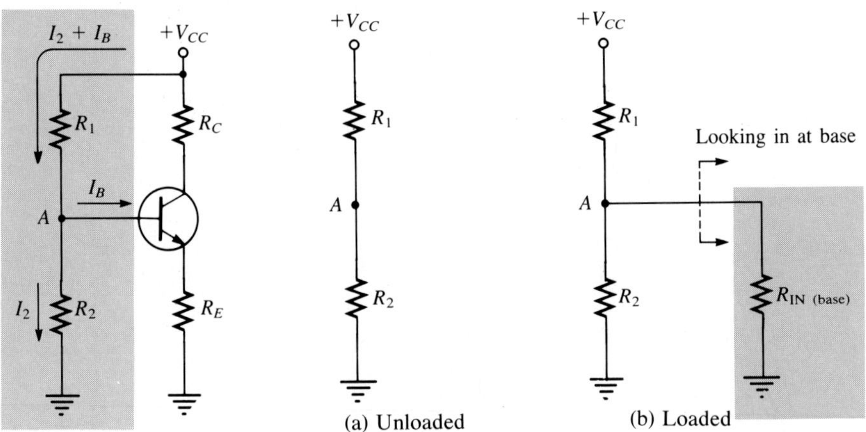

FIGURE 5–19
Voltage-divider bias.

FIGURE 5–20
Simplified voltage-divider.

INPUT RESISTANCE AT THE BASE

To develop an expression for the dc input resistance at the base of a transistor, we will use the diagram in Figure 5–21. V_{IN} is applied between base and ground, and I_{IN} is the

FIGURE 5–21
DC input resistance is V_{IN}/I_{IN}.

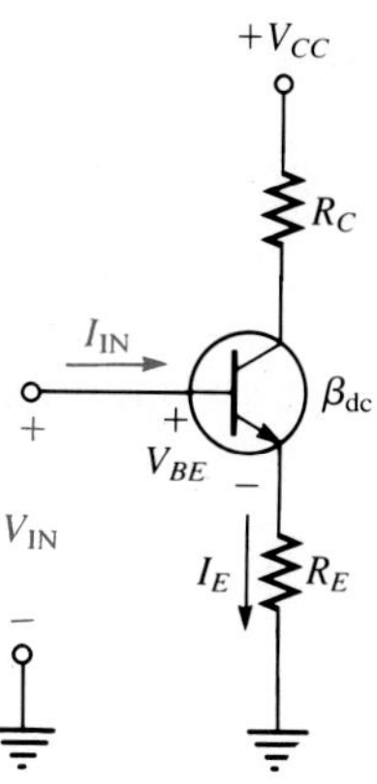

current into the base as shown. By Ohm's law,

$$R_{IN(base)} = \frac{V_{IN}}{I_{IN}} \qquad (5\text{–}14)$$

Kirchhoff's law applied to the base-emitter circuit yields

$$V_{IN} = V_{BE} + I_E R_E$$

With the assumption that $V_{BE} << I_E R_E$, the equation reduces to

$$V_{IN} \cong I_E R_E$$

Now, since $I_E \cong I_C = \beta_{dc} I_B$,

$$V_{IN} \cong \beta_{dc} I_B R_E$$

Since the input current is the base current ($I_{IN} = I_B$), substitution into Equation (5–14) gives

$$R_{IN(base)} \cong \frac{\beta_{dc} I_B R_E}{I_B}$$

So,

$$R_{IN(base)} \cong \beta_{dc} R_E \qquad (5\text{–}15)$$

EXAMPLE 5–6 Determine the dc input resistance at the base of the transistor in Figure 5–22. $\beta_{dc} = 125$.

SOLUTION

$$\begin{aligned} R_{IN(base)} &\cong \beta_{dc} R_E \\ &= (125)(1\ \text{k}\Omega) \\ &= 125\ \text{k}\Omega \end{aligned}$$

FIGURE 5–22

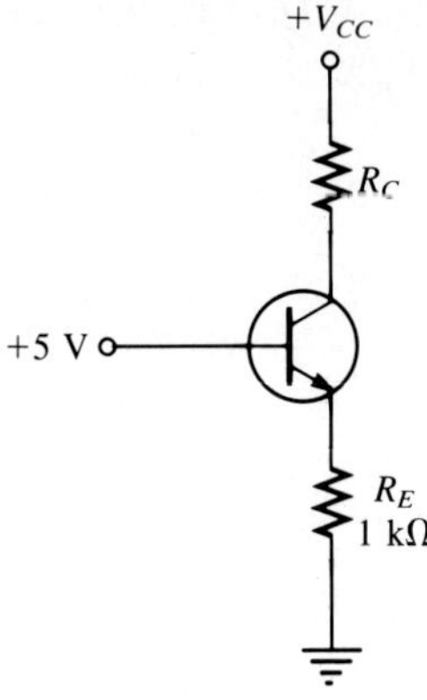

PRACTICE EXERCISE 5–6

What is $R_{IN(base)}$ if $\beta_{dc} = 60$ and $R_E = 910\ \Omega$ in Figure 5–22?

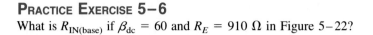

ANALYSIS OF A VOLTAGE-DIVIDER BIAS CIRCUIT

A voltage-divider biased npn transistor is shown in Figure 5–23. We begin the analysis by determining the voltage at the base using the voltage-divider formula, which is developed as follows.

$$R_{IN(base)} = \beta_{dc} R_E$$

The resistance from base to ground is

$$R_2 \| \beta_{dc} R_E$$

A voltage-divider is formed by the resistance from base to ground and R_2. So, applying the voltage-divider principle, we get

$$V_B = \frac{R_2 \| \beta_{dc} R_E}{R_1 + (R_2 \| \beta_{dc} R_E)} V_{CC} \tag{5–16}$$

If $\beta_{dc} R_E >> R_2$, then the formula simplifies to

$$V_B \cong \left(\frac{R_2}{R_1 + R_2}\right) V_{CC} \tag{5–17}$$

Once the base voltage is known, the emitter voltage is V_B less the value of the V_{BE} drop.

$$V_E = V_B - V_{BE} \tag{5–18}$$

Now the emitter current is found using Ohm's law.

$$I_E = \frac{V_E}{R_E} \tag{5–19}$$

Once I_E is known, all other circuit values can be found.

$$I_C \cong I_E$$
$$V_C = V_{CC} - I_C R_C$$
$$V_{CE} = V_C - V_E$$

V_{CE} can be expressed in terms of I_E as follows, since $I_C \cong I_E$.

$$V_{CE} \cong V_{CC} - I_E R_C - I_E R_E$$
$$V_{CE} \cong V_{CC} - I_E(R_C + R_E) \tag{5-20}$$

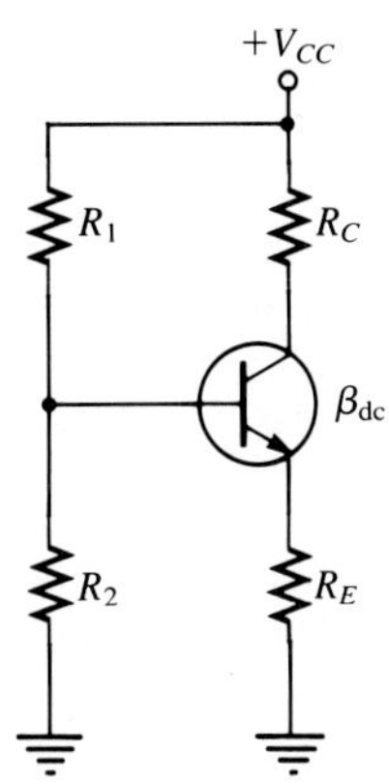

FIGURE 5–23
An npn transistor with voltage-divider bias.

EXAMPLE 5–7

Determine V_{CE} and I_C in Figure 5–24, where $\beta_{dc} = 100$ for the silicon transistor.

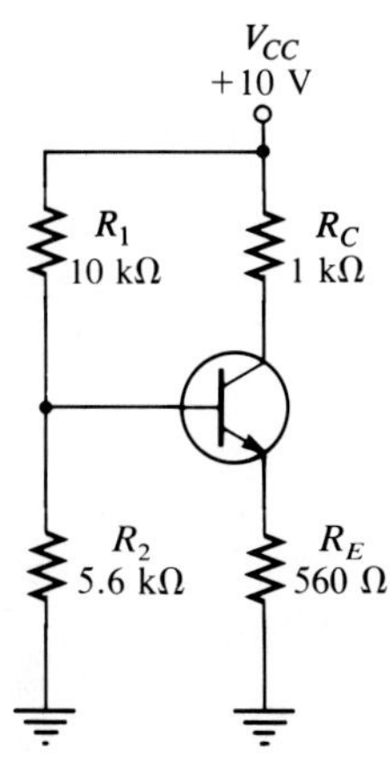

FIGURE 5–24

SOLUTION

$$R_{IN(base)} = \beta_{dc} R_E = (100)(560\ \Omega) = 56\ \text{k}\Omega$$

A common rule-of-thumb is that if two resistors are in parallel and one is at least ten times the other, the total resistance is taken to be approximately equal to the smallest value although, in some cases, this may result in unacceptable inaccuracy.

In this case, $R_{IN(base)} = 10R_2$, so we will choose to neglect $R_{IN(base)}$, although, of course, some accuracy is lost. In the practice exercise, you will rework this example taking $R_{IN(base)}$ into account and compare the difference. Proceeding with the analysis, we get

$$V_B \cong \left(\frac{R_2}{R_1 + R_2}\right)V_{CC} = \left(\frac{5.6\ \text{k}\Omega}{15.6\ \text{k}\Omega}\right)10\ \text{V} = 3.59\ \text{V}$$

So,

$$V_E = V_B - V_{BE} = 3.59\ \text{V} - 0.7\ \text{V} = 2.89\ \text{V}$$

and

$$I_E = \frac{V_E}{R_E} = \frac{2.89\ \text{V}}{560\ \Omega} = 5.16\ \text{mA}$$

Therefore,

$$I_C \cong 5.16 \text{ mA}$$

and

$$\begin{aligned} V_{CE} &\cong V_{CC} - I_E(R_C + R_E) \\ &= 10 \text{ V} - 5.16 \text{ mA}(1.56 \text{ k}\Omega) \\ &= 1.95 \text{ V} \end{aligned}$$

Since $V_{CE} > 0$ V (or a few tenths of a volt), we know that the transistor is not in saturation.

The following program can be used to compute the values of the voltages and currents in a circuit similar to that in Figure 5–24 for various values of β_{dc}, V_{CC}, and resistances.

```
10  CLS
20  PRINT "THIS PROGRAM COMPUTES VB, VE, VCE, IE, AND IC
30  PRINT "FOR THE CIRCUIT IN FIGURE 5-24 GIVEN R1, R2,
40  PRINT "RC, RE, VCC, AND DC BETA."
50  PRINT:PRINT:PRINT
60  INPUT "TO CONTINUE PRESS 'ENTER'";X
70  CLS
80  INPUT "R1 IN OHMS"R1
90  INPUT "R2 IN OHMS";R2
100 INPUT "RC IN OHMS";RC
110 INPUT "RE IN OHMS";RE
120 INPUT "VCC IN VOLTS";VCC
130 INPUT "DC BETA";B
140 CLS
150 RIN=B*RE
160 IF RIN>=10*R2 THEN R=R2 ELSE R=R2*RIN/(R+RIN)
170 VB=(R/(R+R1))*VCC
180 VE=VB-.7
190 IE=VE/RE
200 IC=IE
210 VCE=VCC-IC*(RC+RE)
220 IF VE<=0 THEN PRINT "TRANSISTOR IS CUTOFF"
230 IF VE<=0 THEN GOTO 250
240 IF VCE<=0 THEN PRINT "TRANSISTOR IS SATURATED" ELSE 260
250 END
260 PRINT "VB =";VB;"V"
270 PRINT "VE =";VE;"V"
280 PRINT "VCE =";VCE;"V"
290 PRINT "IC =";IC;"A"
300 PRINT "IE =";IE;"A"
```

PRACTICE EXERCISE 5–7

Rework this example taking into account $R_{IN(base)}$ and compare the results.

STABILITY OF VOLTAGE-DIVIDER BIAS

First, we will get an equivalent base-emitter circuit for Figure 5–23 using Thevenin's theorem. Looking out from the base terminal, the bias circuit can be redrawn as shown in

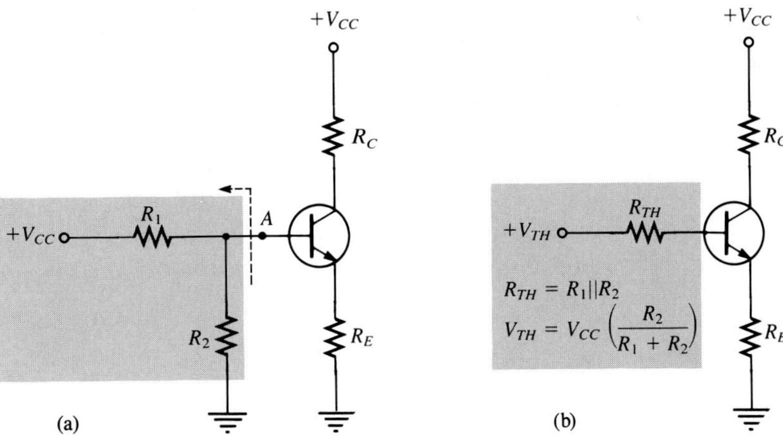

FIGURE 5–25
Thevenizing the bias circuit.

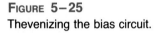

Figure 5–25(a). Applying Thevenin's theorem to the circuit left of point A, we get the following results:

$$R_{TH} = \frac{R_1R_2}{R_1 + R_2}$$

and

$$V_{TH} = \left(\frac{R_2}{R_1 + R_2}\right)V_{CC}$$

The Thevenin equivalent is shown in Figure 5–25(b). Writing Kirchhoff's equation around the equivalent base-emitter loop gives

$$V_{TH} = I_BR_{TH} + V_{BE} + I_ER_E$$

Substituting I_E/β_{dc} for I_B,

$$V_{TH} = I_E(R_E + R_{TH}/\beta_{dc}) + V_{BE}$$

or

$$I_E = \frac{V_{TH} - V_{BE}}{R_E + R_{TH}/\beta_{dc}}$$

If $R_E >> R_{TH}/\beta_{dc}$, then

$$I_E \cong \frac{V_{TH} - V_{BE}}{R_E} \tag{5–21}$$

This shows that I_E is essentially independent of β_{dc} (notice that β_{dc} does not appear in the equation) under the condition indicated. This is easy to achieve in practice by selecting a value for R_E that is at least ten times the resistance of the parallel combination of the voltage-divider resistors divided by the minimum β_{dc}. The voltage-divider bias is popular because good stability is achieved with a single-polarity supply voltage.

VOLTAGE-DIVIDER BIASED PNP

As you know, a pnp transistor requires bias polarities opposite to the npn. This can be accomplished with a negative collector supply voltage, as in Figure 5–26(a), or with a positive emitter supply voltage, as in Figure 5–26(b). In a schematic, the pnp is often drawn upside down so that the supply voltage line can be drawn across the top of the schematic and ground at the bottom, as in Figure 5–27. The analysis procedure is basically the same as for an npn circuit, as demonstrated in the following steps with reference to Figure 5–27. Assuming that $\beta_{dc}R_E >> R_2$, the base voltage is

$$V_B = \left(\frac{R_1}{R_1 + R_2}\right)V_{EE} \quad (5\text{–}22)$$

and

$$V_E = V_B + V_{BE} \quad (5\text{–}23)$$

So,

$$I_E = \frac{V_{EE} - V_E}{R_E} \quad (5\text{–}24)$$

and

$$V_C = I_C R_C \quad (5\text{–}25)$$

$$V_{EC} = V_E - V_C \quad (5\text{–}26)$$

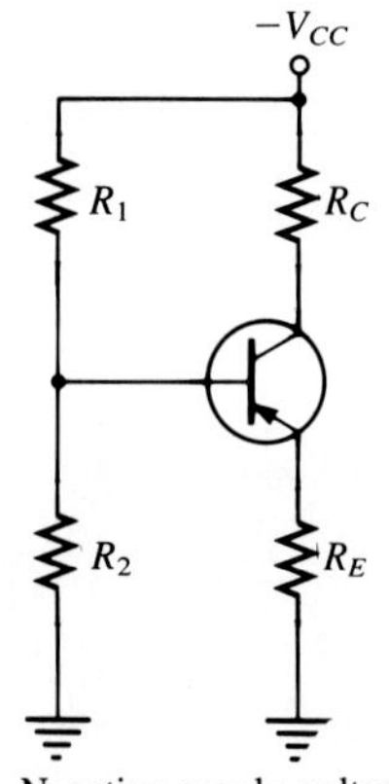

(a) Negative supply voltage

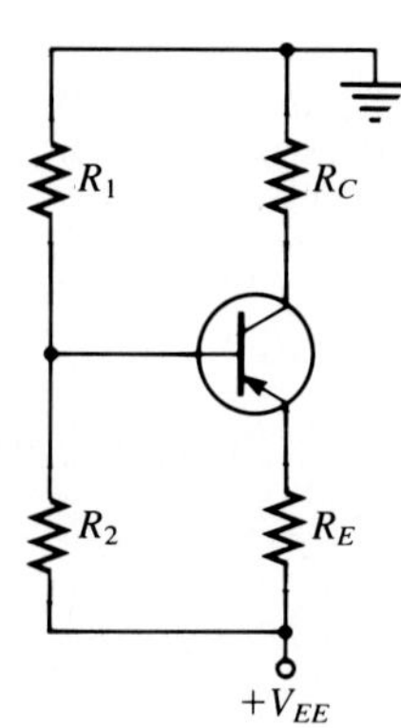

(b) Positive supply voltage

FIGURE 5–26
Voltage-divider biased pnp transistor.

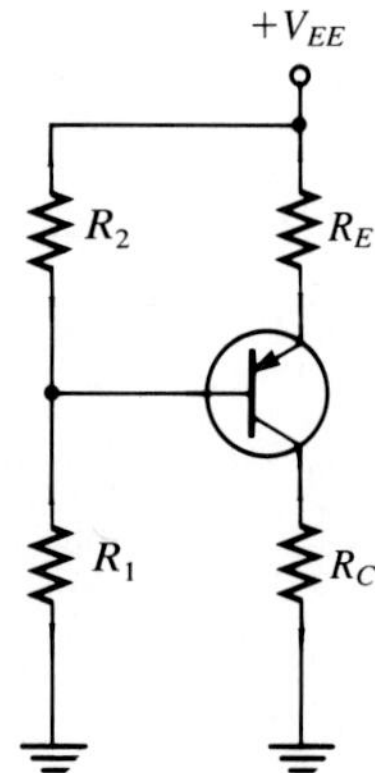

FIGURE 5–27

EXAMPLE 5–8 Find I_C and V_{EC} in Figure 5–28.

SOLUTION

Let's check to see if $R_{IN(base)}$ can be neglected.

$$R_{IN(base)} = \beta_{dc}R_E = (150)(1\text{ k}\Omega) = 150\text{ k}\Omega$$

FIGURE 5–28

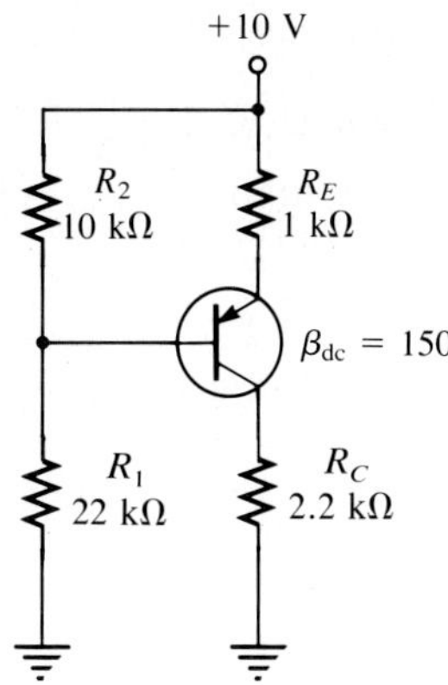

Since 150 kΩ is more than ten times R_2, the condition $\beta_{dc}R_E >> R_2$ is met and $R_{IN(base)}$ can be neglected. First calculate V_B.

$$V_B = \left(\frac{R_1}{R_1 + R_2}\right)V_{EE} = \left(\frac{22\text{ k}\Omega}{32\text{ k}\Omega}\right)10\text{ V} = 6.88\text{ V}$$

Then

$$V_E = 6.88\text{ V} + 0.7\text{ V} = 7.58\text{ V}$$

and

$$I_E = \frac{V_{EE} - V_E}{R_E} = \frac{10\text{ V} - 7.58\text{ V}}{1\text{ k}\Omega} = 2.42\text{ mA}$$

From I_E, we can get I_C and V_{CE} as follows:

$$I_C \cong I_E = 2.42\text{ mA}$$

and

$$V_C = I_C R_C = (2.42\text{ mA})(2.2\text{ k}\Omega) = 5.32\text{ V}$$

Therefore,

$$\begin{aligned} V_{EC} &= V_E - V_C \\ &= 7.58\text{ V} - 5.32\text{ V} \\ &= 2.26\text{ V} \end{aligned}$$

PRACTICE EXERCISE 5–8

Determine I_C and V_{EC} in Figure 5–28 with $R_{IN(base)}$ taken into account. ■

■ EXAMPLE 5–9

Find I_C and V_{CE} in Figure 5–26(a) for R_1 = 68 kΩ, R_2 = 47 kΩ, R_C = 1.8 kΩ, R_E = 2.2 kΩ, V_{CC} = −6 V, and β_{dc} = 75.

SOLUTION

$$R_{IN(base)} = \beta_{dc}R_E = 75(2.2\text{ k}\Omega) = 165\text{ k}\Omega$$

Since $R_{IN(base)}$ is not ten times greater than R_2, it must be taken into account.

$$V_B = \left(\frac{R_2 \| R_{IN(base)}}{R_1 + R_2 \| R_{IN(base)}}\right)V_{CC} = \left(\frac{47\text{ k}\Omega \| 165\text{ k}\Omega}{68\text{ k}\Omega + 47\text{ k}\Omega \| 165\text{ k}\Omega}\right)(-6\text{ V})$$

$$= \left(\frac{36.6\text{ k}\Omega}{68\text{ k}\Omega + 36.6\text{ k}\Omega}\right)(-6\text{ V}) = -2.1\text{ V}$$

$$V_E = V_B + V_{BE} = -2.1\text{ V} + 0.7\text{ V} = -1.4\text{ V}$$

$$I_E = \frac{V_E}{R_E} = \frac{-1.4\text{ V}}{2.2\text{ k}\Omega} = -0.636\text{ mA}$$

$$I_C \cong I_E = -0.636\text{ mA}$$

$$V_C = V_{CC} - I_C R_C = -6\text{ V} - (-0.636\text{ mA})(1.8\text{ k}\Omega) = -4.86\text{ V}$$

$$V_{CE} = -4.86\text{ V} - (-1.4\text{ V}) = -3.46\text{ V}$$

PRACTICE EXERCISE 5–9

What value of β_{dc} is required in this example in order to neglect $R_{IN(base)}$ in keeping with our basic ten-times rule? ■

5–4 REVIEW QUESTIONS

1. If the voltage at the base of a transistor is 5 V and the base current is 5 μA, what is the dc input resistance at the base?
2. If a transistor has a dc beta of 190 and its emitter resistor is 1 kΩ, what is the dc input resistance at the base?
3. What bias voltage is developed at the base of a transistor if both resistors in the voltage divider are equal and $V_{CC} = +10$ V? Assume the input resistance at the base is large enough to neglect.
4. What are two advantages of voltage-divider bias?

5–5 COLLECTOR-FEEDBACK BIAS

Another type of bias arrangement is the collector-feedback circuit. This is a negative feedback connection that provides a relatively stable Q point by reducing the effect of variations in β_{dc}. It is also a simple circuit in terms of the components required.

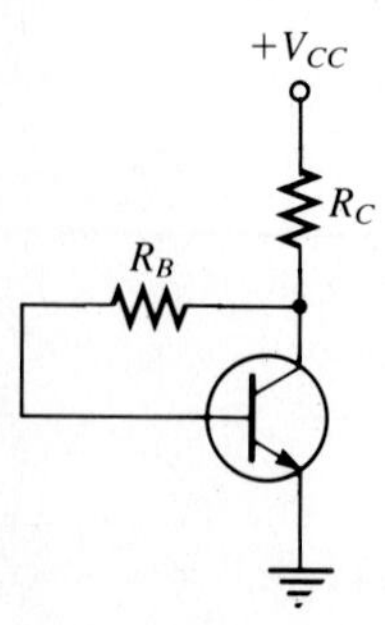

FIGURE 5–29
Collector-feedback bias.

In Figure 5–29, the base resistor R_B is connected to the collector rather than to V_{CC}, as in the base bias arrangement discussed earlier. The collector voltage provides the bias for the base-emitter junction.

The operation of the circuit is as follows. As you know, β_{dc} increases with temperature, which causes I_C to increase. An increase in I_C produces more voltage drop across R_C, lowering the collector voltage. This, in turn, reduces the I_B, which tends to offset the original increase in I_C. The result is that the circuit tends to maintain a stable value of collector current, keeping the Q point fixed. The reverse action occurs when the temperature decreases. Figure 5–30 illustrates this **feedback** stabilizing action.

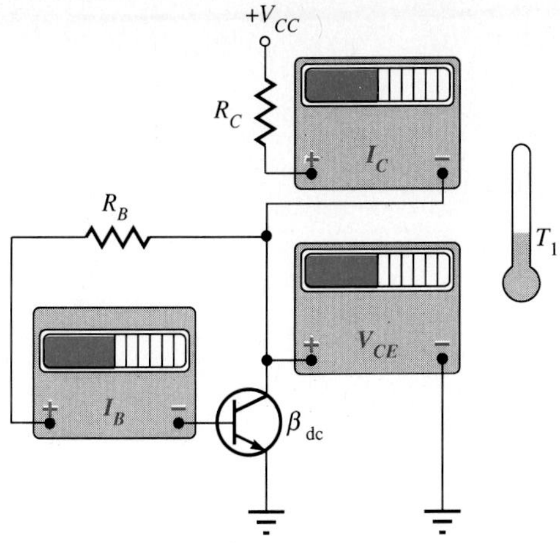

(a) Stabilized at initial temperature

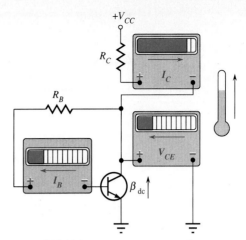

(b) Initial response to temperature rise

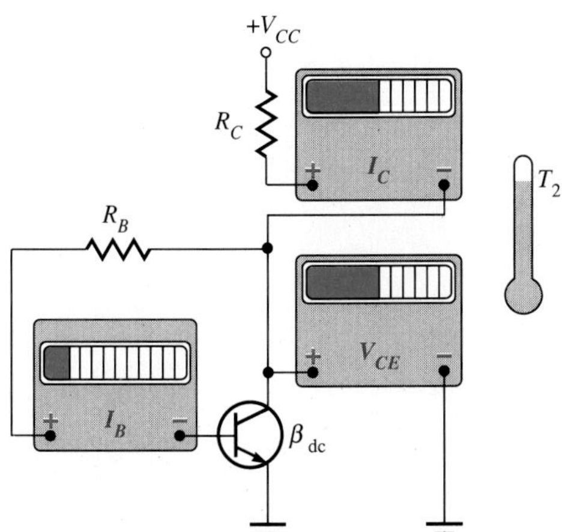

(c) Stabilized at higher temperature

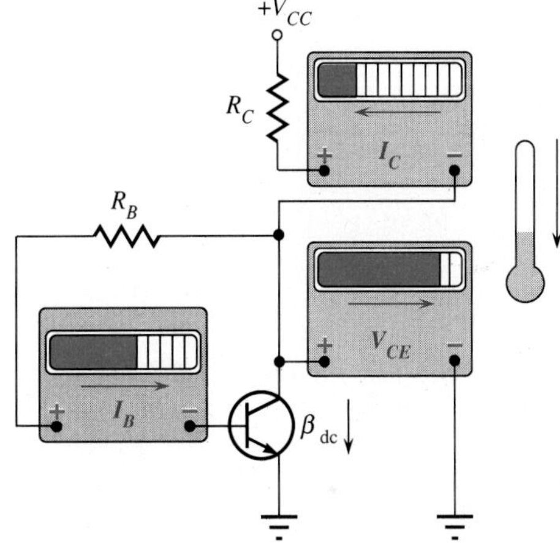

(d) Initial response to temperature drop

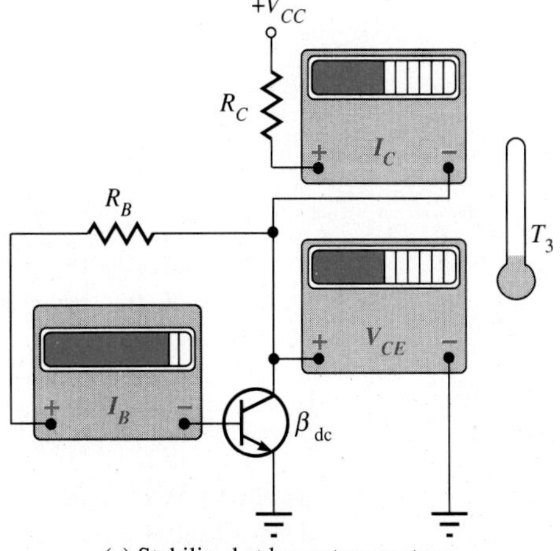

(e) Stabilized at lower temperature

FIGURE 5–30
Collector-feedback stabilization over temperature changes.

ANALYSIS OF COLLECTOR FEEDBACK

By Ohm's law, the base current can be expressed as

$$I_B = \frac{V_C - V_{BE}}{R_B} \tag{5-27}$$

The collector voltage is

$$V_C \cong V_{CC} - I_C R_C$$

Also,

$$I_B = \frac{I_C}{\beta_{dc}}$$

Substituting in Equation (5–27), we get

$$\frac{I_C}{\beta_{dc}} = \frac{V_{CC} - I_C R_C - V_{BE}}{R_B}$$

Rearranging,

$$\frac{I_C R_B}{\beta_{dc}} + I_C R_{CC} = V_{CC} - V_{BE}$$

Solving for I_C,

$$I_C(R_C + R_B/\beta_{dc}) = V_{CC} - V_{BE}$$

$$I_C = \frac{V_{CC} - V_{BE}}{R_C + R_B/\beta_{dc}} \tag{5-28}$$

EXAMPLE 5–10 Calculate the Q-point values (I_C and V_{CE}) for the circuit in Figure 5–31.

FIGURE 5–31

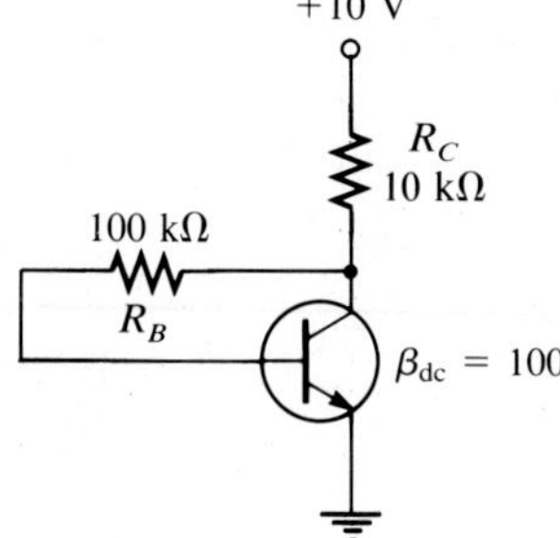

SOLUTION
Using Equation (5–28),

$$I_C = \frac{V_{CC} - V_{BE}}{R_C + R_B/\beta_{dc}} = \frac{10\text{ V} - 0.7\text{ V}}{10\text{ k}\Omega + 100\text{ k}\Omega/100} = 0.845\text{ mA}$$

Since the emitter is grounded,

$$\begin{aligned} V_{CE} &= V_C \cong V_{CC} - I_C R_C \\ &= 10\text{ V} - (0.845\text{ mA})(10\text{ k}\Omega) \\ &= 1.55\text{ V} \end{aligned}$$

PRACTICE EXERCISE 5–10

Calculate the Q-point values in Figure 5–31 if β_{dc} is 250.

5–5 REVIEW QUESTIONS

1. Explain how an increase in β_{dc} causes a reduction in base current in a collector-feedback circuit.
2. In a certain collector-feedback circuit, $R_B = 47\text{ k}\Omega$, $R_C = 2.2\text{ k}\Omega$, and $V_{CC} = 15\text{ V}$. If $I_C = 5\text{ mA}$, what is I_B?

5–6 TROUBLESHOOTING

In a biased transistor circuit, either the transistor can fail or one or more resistors in the bias circuit can fail. We will examine several possibilities in this section using the voltage-divider bias arrangement for illustration. Most circuit failures result from open resistors, internally open transistor leads and junctions, or shorted junctions. In general, these failures will produce an apparent cutoff or saturation condition when voltage is measured at the collector.

Figure 5–32 indicates certain failures that will produce a collector voltage equal to V_{CC}, thus making it appear that the transistor is in cutoff. (It actually may be.)

Now, let's examine these failures in more detail using specific values. If the resistor R_1 opens, the base is at zero volts because of R_2 to ground, and the transistor is in cutoff since $I_B = 0$. These voltage levels are shown in Figure 5–33(a).

An open emitter resistor, of course, prevents current in the transistor except for a small I_{CBO}. Therefore, the collector voltage is approximately V_{CC}, the base voltage of 3.2 V is determined by the voltage-divider, and the emitter voltage is 0.7 V below V_B, as indicated in Figure 5–33(b).

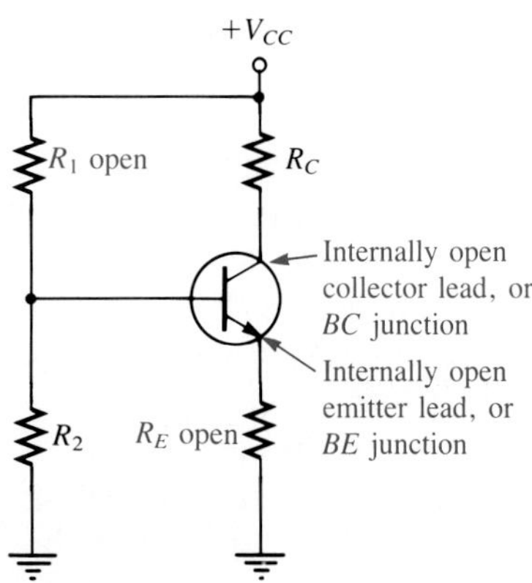

FIGURE 5–32
Typical failures that produce an apparent *off* condition ($V_C = V_{CC}$).

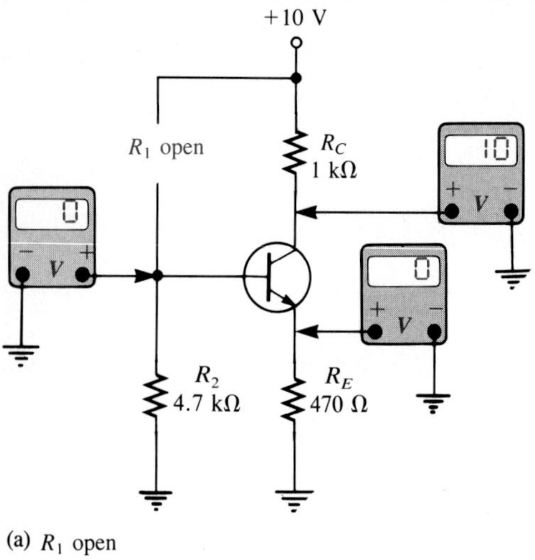

(a) R_1 open

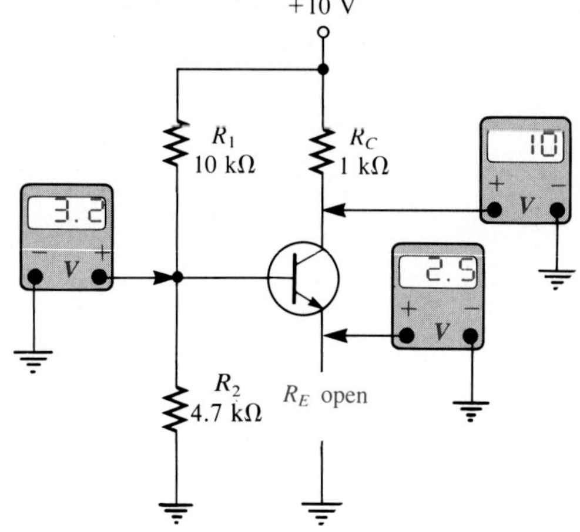

(b) R_E open

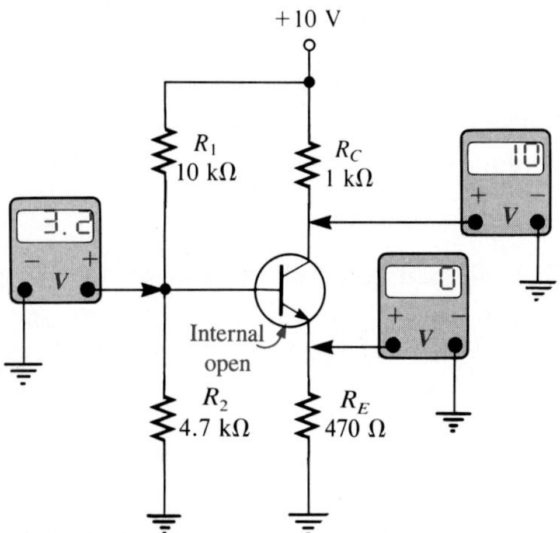

(c) Internal open at emitter

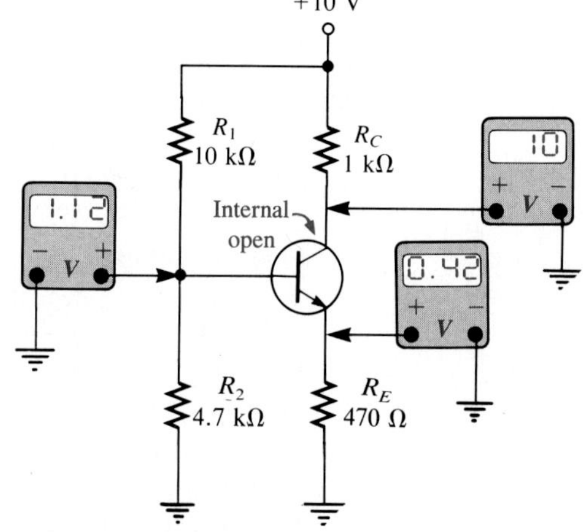

(d) Internal open at collector

FIGURE 5–33
Open component failures and indications.

An internally open emitter lead or open base-emitter junction will cause the same voltage readings at the collector and base as did the open R_E. However, an internally open emitter causes the external emitter terminal to be at ground potential because there is no current through R_E. This condition is shown in Figure 5–33(c). With an internally open collector lead or open base-collector junction, the collector terminal is at V_{CC}. The base is at approximately 1.12 V because $R_{IN(base)}$ is not $\beta_{dc}R_E$ but simply R_E. The emitter is 0.7 V less than the base, as shown in Figure 5–33(d).

Next, we will consider the two open-resistor conditions shown in Figure 5–34. An open collector resistor will cause the collector terminal of the transistor to be

FIGURE 5–34

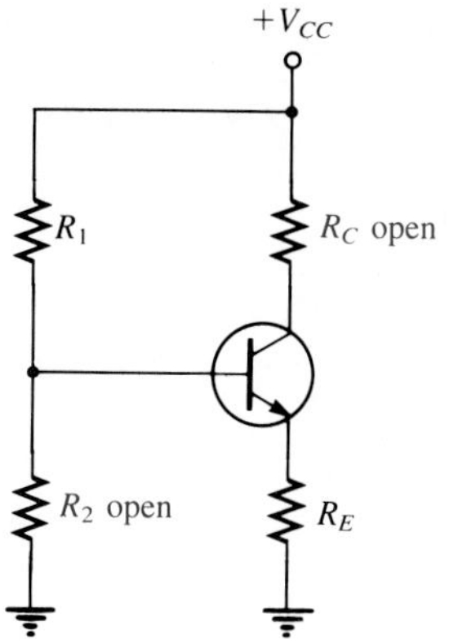

approximately 0.7 V below the base. You may think that the collector should be at zero volts, but the positive base voltage of 1.12 V forward-biases the base-collector junction so that, when an instrument is connected to the collector, a voltage reading is obtained, as indicated in Figure 5–35(a), which is 0.7 V below the base voltage.

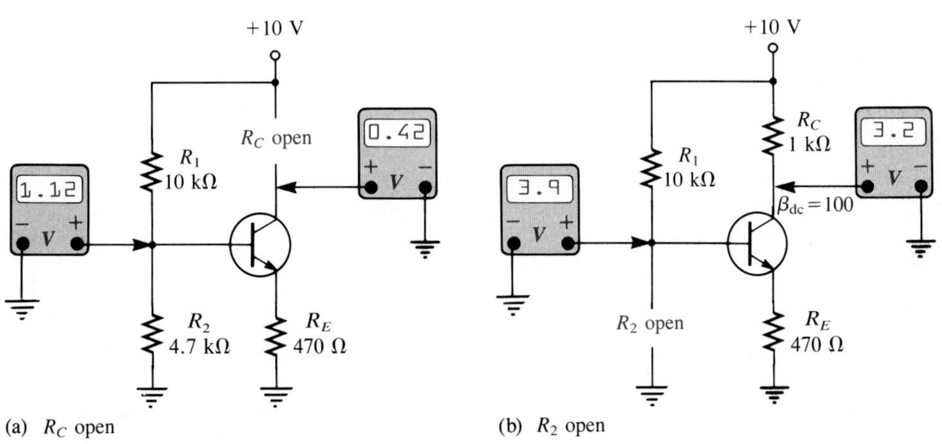

(a) R_C open

(b) R_2 open

FIGURE 5–35
Examples of open component failures.

An open R_2 increases the bias voltage on the base because of the voltage-divider action of R_1 and $R_{IN(base)}$. The maximum V_B possible, if saturation does not occur, is developed as follows.

$$R_{IN(base)} = \beta_{dc}R_E = (100)(470\ \Omega) = 47\ \text{k}\Omega$$

$$V_{B(max)} = \left(\frac{R_{IN(base)}}{R_1 + R_{IN(base)}}\right)V_{CC}$$

$$= \left(\frac{47\ \text{k}\Omega}{57\ \text{k}\Omega}\right)10\ \text{V}$$

$$= 8.25\ \text{V}$$

This excessive base bias *tries* to produce an emitter current of

$$\frac{V_E}{R_E} = \frac{7.55\text{ V}}{470\ \Omega} = 16.06\text{ mA}$$

Since

$$I_{C(\text{sat})} = \frac{V_{CC}}{R_C + R_E} = \frac{10\text{ V}}{1.47\text{ k}\Omega} = 6.8\text{ mA}$$

the transistor saturates well before the 16.06 mA value. Thus, V_B is limited to 4.04 V, which is 0.7 V greater than the saturation value of V_E, as shown by the following steps.

$$\begin{aligned} V_{E(\text{sat})} &\cong I_{C(\text{sat})}R_E \\ &= (6.8\text{ mA})(470\ \Omega) \\ &= 3.2\text{ V} \\ V_B &= 3.2\text{ V} + 0.7\text{ V} \\ &= 3.9\text{ V} \end{aligned}$$

This condition is illustrated in Figure 5–35(b).

5–6 REVIEW QUESTIONS

1. How do you determine when a transistor is saturated? Cut off?
2. In a voltage-divider biased npn transistor circuit, you measure V_{CC} at the collector and an emitter voltage 0.7 V less than the base voltage. Is the transistor functioning in cutoff, or is R_E open?
3. What symptoms does an open R_C produce?

5–7 A SYSTEM APPLICATION

At the opening of this chapter, you saw a block diagram for a basic industrial temperature control system. The system has several parts, but in this section we are concerned with the transistor circuit that monitors temperature changes in a tank and provides a proportional output that is used to precisely control the temperature. In this particular application, the transistor is used as a linear dc device and you will

- ***See how a bias circuit can be used to control a transistor's output voltage.***
- ***See how a linear transistor dc circuit is used in a simple system application.***
- ***Translate between a printed circuit board and a schematic.***
- ***Troubleshoot some common transistor circuit failures.***

A BRIEF DESCRIPTION OF THE SYSTEM

The system in Figure 5–36 is an example of a process control system that uses the principle of closed-loop feedback to maintain a substance at a contant temperature of 50°C + 1°C. The temperature sensor in the tank is a **thermistor,** which is a

FIGURE 5–36

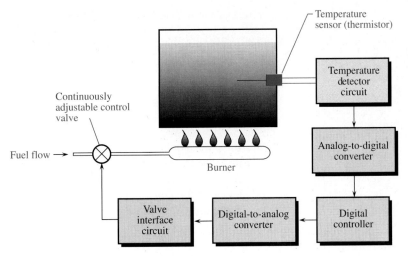

temperature-sensitive resistor whose resistance has a negative temperature coefficient. The thermistor forms part of the bias circuit for the transistor detector.

A small change in temperature causes the resistance of the thermistor to change, which results in a proportional change in the output voltage from the transistor detector. Since thermistors are typically nonlinear devices, the purpose of the digital portions of the system is to precisely compensate for the nonlinear characteristic of the temperature detector in order to provide a continuous linear adjustment of the fuel flow to the burner to offset any small shift in temperature away from the desired value.

The temperature detector board that we are concerned with in this system application is obviously only a small part of the total system. It is very interesting, however, to see how the variations in a transistor's bias point can be used in a situation such as this. Although the digital portions of the system are not covered, they serve to point out that in the real world you will have to deal with systems with various types of elements. Analog-to-digital and digital-to-analog converters are introduced in Chapter 14.

Now, so you can take a closer look at the temperature detector circuit, let's take it out of the system and put it on the test bench.

ON THE TEST BENCH

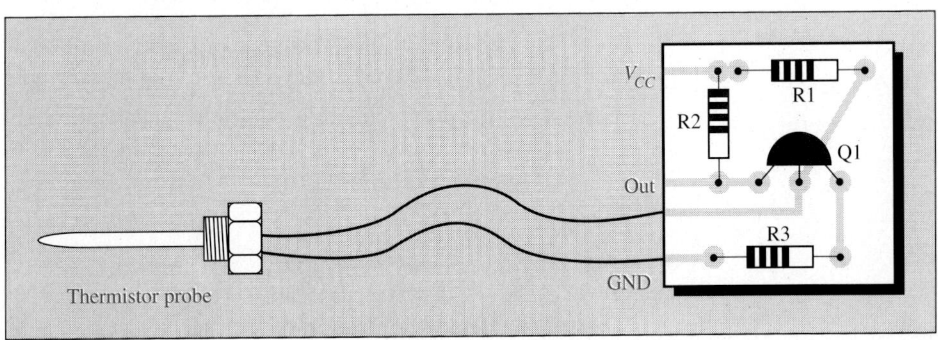

FIGURE 5–37

IDENTIFYING THE COMPONENTS

The 2N2222A transistor, Q1, is in a TO-92 (TO-226AA) plastic case. Refer back to Section 4–6 for transistor package information. The thermistor is in the form of a probe that mounts into the side of the tank.

ACTIVITY 1 RELATE THE PC BOARD TO THE SCHEMATIC

Develop a schematic diagram of the circuit board in Figure 5–37 including the thermistor. Arrange the schematic into a form that is familiar to you and identify the type of biasing used. The symbol for a thermistor is provided in Figure 5–38(a) and the resistor color codes are given in Figure 5–38(b). As mentioned, the thermistor has a nonlinear negative temperature coefficient.

FIGURE 5–38

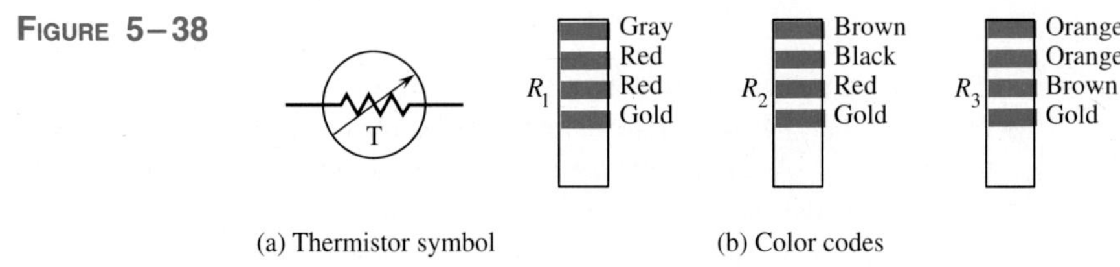

(a) Thermistor symbol (b) Color codes

ACTIVITY 2 ANALYZE THE CIRCUIT

STEP 1 Roughly estimate the collector current and pick the most appropriate value available for the β_{dc} (h_{FE}) from the transistor data sheet in Appendix A.

STEP 2 With a dc supply of +15 V, analyze the temperature detector circuit to determine the output voltage at 50°C and at 1°C above and below 50°C. Use 0.7 V for the base-emitter voltage. Since accuracy on the graph is difficult to achieve for this small range, assume that the thermistor resistances for the controlled temperature range are as follows:

At T = 50°C, R = 2.75 kΩ
At T = 49°C, R = 3.1 kΩ
At T = 51°C, R = 2.5 kΩ

Will your values change significantly if you take into account the 1-MΩ input resistance of the analog-to-digital converter?

STEP 3 With the circuit in a controlled temperature environment, determine the output voltage over a range from 30°C to 110°C at 20° intervals. Use the graph in Figure 5–39.

ACTIVITY 3 WRITE A TECHNICAL REPORT

Discuss the detailed operation of the detector circuit including the thermistor. You may wish to refer to the resistance versus temperature curve for the thermistor in Figure 5–39.

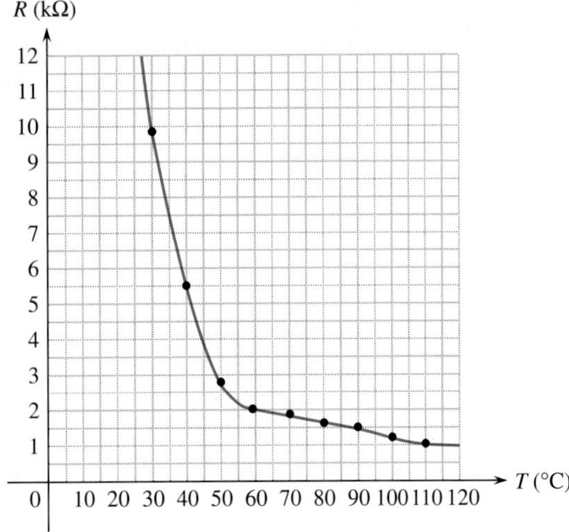

FIGURE 5–39
Thermistor resistance versus temperature curve.

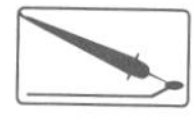

■ **ACTIVITY 4 TROUBLESHOOT THE CIRCUIT BOARD FOR EACH OF THE FOLLOWING PROBLEMS BY STATING THE PROBABLE CAUSE OR CAUSES IN EACH CASE**

1. Collector of Q1 remains at approximately 0.1 V.
2. Collector of Q1 remains at 15 V.
3. For each of the above conditions, if there is more than one possible fault, describe how you will go about isolating the problem.

5–7 REVIEW QUESTIONS

1. What is the basic purpose of the detector circuit board in the process control system?
2. In Step 3 of Activity 2, does the transistor go into saturation and/or cutoff?
3. If you found that the transistor goes into saturation at some point, determine the temperature.
4. If you found that the transistor goes into cutoff at some point, determine the temperature.

SUMMARY

- The purpose of biasing is to establish a proper dc operating point (Q point).
- The Q point of a circuit is defined by specific values for I_C and V_{CE}. These values are sometimes called the *coordinates* of the Q point.
- A dc load line passes through the Q point on a transistor's collector curves and intersects the vertical axis at approximately $I_{C(sat)}$ and the horizontal axis at $V_{CE(off)}$.
- The linear operating region of a transistor lies along the load line below saturation and above cutoff.
- The dc input resistance looking in at the base of a bipolar transistor is approximately $\beta_{dc}R_E$.
- The base bias circuit arrangement has poor stability because its Q point varies widely with β_{dc}.

- Emitter bias generally provides good Q-point stability but requires both positive and negative supply voltages.
- Voltage-divider bias provides good Q-point stability with a single-polarity supply voltage. It is the most common bias circuit.
- Collector-feedback bias provides good stability using negative feedback from collector to base.

GLOSSARY

Feedback The process of returning a portion of a circuit's output back to the input in such a way as to oppose a change in the output.

Load line A straight line on the characteristic curves of an amplifier that represents the operating range of the amplifier's voltages and currents.

***Q* point** The dc operating (bias) point of an amplifier specified by voltage and current values.

Thermistor A temperature-sensitive resistor with a negative temperature coefficient.

FORMULAS

BASE BIAS

(5–1) $$I_B = \frac{V_{CC} - V_{BE}}{R_B}$$

(5–2) $$I_C = \beta_{dc} I_B$$

(5–3) $$V_{CE} = V_{CC} - I_C R_C$$

(5–4) $$V_{CE} = V_{CC} - \beta_{dc} I_B R_C$$

EMITTER BIAS

(5–5) $$V_B \cong 0$$

(5–6) $$V_E \cong -V_{BE}$$

(5–7) $$I_E = \frac{V_E - V_{EE}}{R_E}$$

(5–8) $$I_C \cong I_E$$

(5–9) $$V_C = V_{CC} - I_C R_C$$

(5–10) $$V_{CE} = V_C - V_E$$

(5–11) $$I_E = \frac{V_{EE} - V_{BE}}{R_E + R_B/\beta_{dc}}$$

(5–12) $$I_E \cong \frac{V_{EE} - V_{BE}}{R_E}, \quad R_E >> R_B/\beta_{dc}$$

(5–13) $$I_E \cong \frac{V_{EE}}{R_E}, \quad V_{EE} >> V_{BE}$$

VOLTAGE-DIVIDER BIAS (NPN)

(5–14) $$R_{\text{IN(base)}} = \frac{V_{\text{IN}}}{I_{\text{IN}}}$$

(5–15) $$R_{\text{IN(base)}} \cong \beta_{\text{dc}}R_E$$

(5–16) $$V_B = \frac{R_2 \| \beta_{\text{dc}}R_E}{R_1 + (R_2 \| \beta_{\text{dc}}R_E)} V_{CC}$$

(5–17) $$V_B \cong \left(\frac{R_2}{R_1 + R_2}\right) V_{CC}, \ \beta_{\text{dc}}R_E >> R_2$$

(5–18) $$V_E = V_B - V_{BE}$$

(5–19) $$I_E = \frac{V_E}{R_E}$$

(5–20) $$V_{CE} \cong V_{CC} - I_E(R_C + R_E)$$

(5–21) $$I_E \cong \frac{V_{TH} - V_{BE}}{R_E}, \ R_E >> R_{TH}/\beta_{\text{dc}}$$

VOLTAGE-DIVIDER BIAS (PNP)

(5–22) $$V_B = \left(\frac{R_1}{R_1 + R_2}\right) V_{EE}, \ \beta_{\text{dc}}R_E >> R_2$$

(5–23) $$V_E = V_B + V_{BE}$$

(5–24) $$I_E = \frac{V_{EE} - V_E}{R_E}$$

(5–25) $$V_C = I_C R_C$$

(5–26) $$V_{EC} = V_E - V_C$$

COLLECTOR-FEEDBACK BIAS

(5–27) $$I_B = \frac{V_C - V_{BE}}{R_B}$$

(5–28) $$I_C = \frac{V_{CC} - V_{BE}}{R_C + R_B/\beta_{\text{dc}}}$$

SELF-TEST

1. The maximum value of collector current in a biased transistor is

(a) $\beta_{\text{dc}}I_B$ **(b)** $I_{C(\text{sat})}$ **(c)** greater than I_E **(d)** $I_E - I_B$

2. Ideally, a dc load line is a straight line drawn on the collector characteristic curves between

(a) the Q point and cutoff **(b)** the Q point and saturation

(c) $V_{CE(\text{cutoff})}$ and $I_{C(\text{sat})}$ **(d)** $I_B = 0$ and $I_B = I_C/\beta_{\text{dc}}$

3. If a sine wave voltage is applied to the base of a biased npn transistor and the resulting sine wave collector voltage is clipped near zero volts, the transistor is

(a) being driven into saturation **(b)** being driven into cutoff

(c) operating nonlinearly **(d)** a and c **(e)** b and c

4. The dc beta for a given type of transistor

(a) varies with temperature **(b)** is a fixed constant

(c) varies from device to device **(d)** a and c

5. The disadvantage of base bias is that

(a) it is very complex **(b)** it produces low gain

(c) it is too beta dependent **(d)** it produces high leakage current

6. Emitter bias is

(a) essentially independent of β_{dc} **(b)** very dependent on β_{dc}

(c) provides a stable bias point **(d)** a and c

7. In an emitter-biased circuit, $R_E = 2.7\ \text{k}\Omega$ and $V_{EE} = 15$ V. The emitter current

(a) is 5.3 mA **(b)** is 2.7 mA **(c)** is 180 mA **(d)** cannot be determined

8. The input resistance at the base of a biased transistor depends mainly on

(a) β_{dc} **(b)** R_B **(c)** R_E **(d)** β_{dc} and R_E

9. In a voltage-divider biased transistor circuit such as in Figure 5–23, $R_{IN(base)}$ can generally be neglected in calculations when

(a) $R_{IN(base)} > R_2$ **(b)** $R_2 > 10R_{IN(base)}$

(c) $R_{IN(base)} > 10R_2$ **(d)** $R_1 << R_2$

10. In a certain voltage-divider biased npn transistor, V_B is 2.95 V. The dc emitter voltage is approximately

(a) 2.25 V **(b)** 2.95 V **(c)** 3.65 V **(d)** 0.7 V

11. Voltage-divider bias

(a) cannot be independent of β_{dc}

(b) can be essentially independent of β_{dc}

(c) is not widely used

(d) requires fewer components than all the other methods

12. Collector-feedback bias is

(a) based on the principle of positive feedback

(b) based on beta multiplication

(c) based on the principle of negative feedback

(d) not very stable

13. In a voltage-divider biased npn transistor, if the upper voltage-divider resistor (the one connected to V_{CC}) opens,

(a) the transistor goes into cutoff **(b)** the transistor goes into saturation

(c) the transistor burns out **(d)** the supply voltage is too high

14. In a voltage-divider biased npn transistor, if the lower voltage-divider resistor (the one connected to ground) opens,
    (a) the transistor is not affected
    (b) the transistor may be driven into cutoff
    (c) the transistor may be driven into saturation
    (d) the collector current will decrease
15. In a voltage-divider biased pnp transistor, there is no base current, but the base voltage is approximately correct. The most likely problem(s) is
    (a) a bias resistor is open
    (b) the collector resistor is open
    (c) the base-emitter junction is open
    (d) the emitter resistor is open
    (e) a and c
    (f) c and d

PROBLEMS

SECTION 5–1 THE DC OPERATING POINT

1. The output (collector voltage) of a biased transistor amplifier is shown in Figure 5–40. Is the transistor biased too close to cutoff or too close to saturation?
2. What is the Q point for a biased transistor as in Figure 5–2 with $I_B = 150\ \mu\text{A}$, $\beta_{dc} = 75$, $V_{CC} = 18$ V, and $R_C = 1\ \text{k}\Omega$?
3. What is the saturation value of collector current in Problem 2?
4. What is the cutoff value of V_{CE} in Problem 2?
5. Determine the intercept points of the dc load line on the vertical and horizontal axes of the collector-characteristic curves for the circuit in Figure 5–41.
6. Assume that you wish to bias the transistor in Figure 5–41 with $I_B = 20\ \mu\text{A}$. To what voltage must you change the V_{BB} supply? What are I_C and V_{CE} at the Q point, given that $\beta_{dc} = 50$?
7. Design a biased-transistor circuit using $V_{BB} = V_{CC} = 10$ V, for a Q point of $I_C = 5$ mA and $V_{CE} = 4$ V. Assume $\beta_{dc} = 100$. The design involves finding R_B, R_C, and the *minimum* power rating of the transistor. (The actual power rating should be greater.) Sketch the circuit.

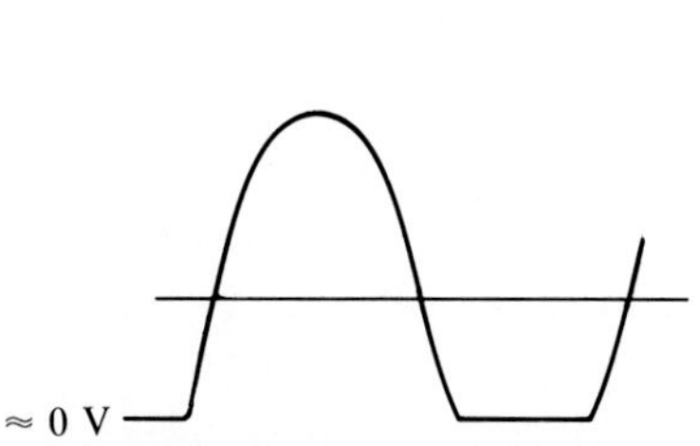

FIGURE 5–40

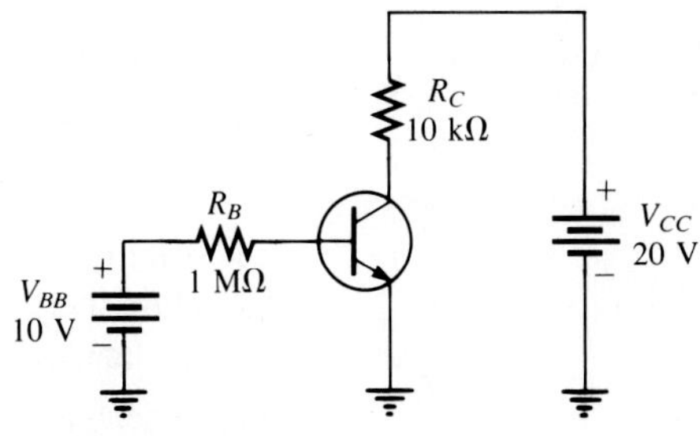

FIGURE 5–41

FIGURE 5–42

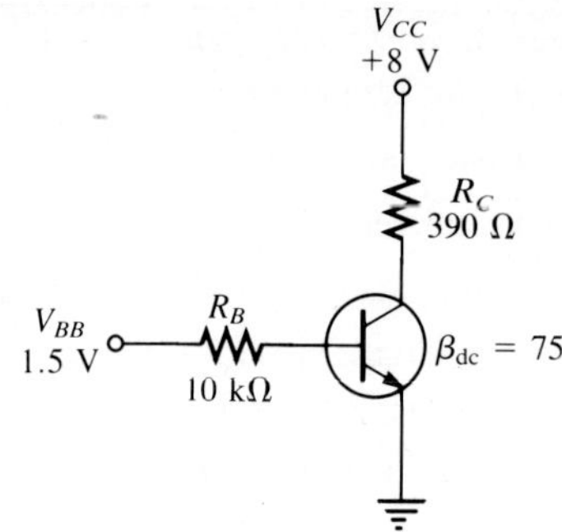

8. Determine whether the transistor in Figure 5–42 is biased in cutoff, saturation, or the linear region. Keep in mind that $I_C = \beta_{dc} I_B$ is valid only in the linear region.

SECTION 5–2 BASE BIAS

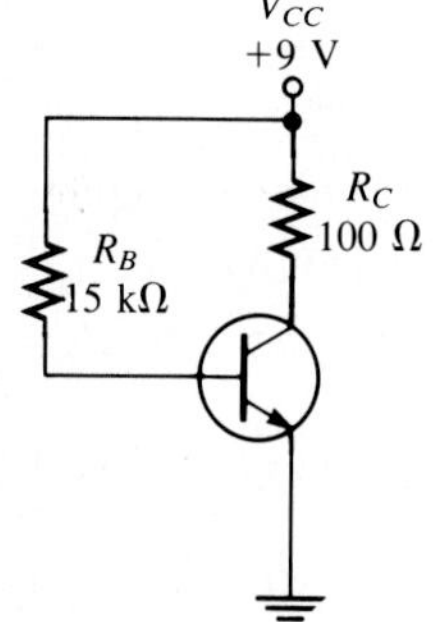

FIGURE 5–43

9. Determine I_B, I_C, and V_{CE} for a base-biased transistor circuit with the following values: β_{dc} = 90, V_{CC} = 12 V, R_B = 22 kΩ, and R_C = 100 Ω.
10. If β_{dc} in Problem 9 doubles over temperature, what are the Q-point values?
11. You have two base-biased circuits connected for testing. They are identical except that one is biased with a separate V_{BB} source and the other is biased with the base resistor connected to V_{CC}. Ammeters are connected to measure collector current in each circuit. You vary the V_{CC} supply voltage and observe that the collector current varies in one circuit, but not in the other. In which circuit does the collector current change? Explain your observation.
12. The data sheet for a particular transistor specifies a minimum β_{dc} of 50 and a maximum β_{dc} of 125. What range of Q-point values can be expected if an attempt is made to mass-produce the circuit in Figure 5–43? Is this range acceptable if the Q point must remain in the transistor's linear region?
13. The base-biased circuit in Figure 5–43 is subjected to a temperature variation from 0°C to 70°C. The β_{dc} decreases by 50 percent at 0°C and increases by 75 percent at 70°C from its nominal value of 110 at 25°C. What are the changes in I_C and V_{CE} over the temperature range of 0°C to 70°C?

SECTION 5–3 EMITTER BIAS

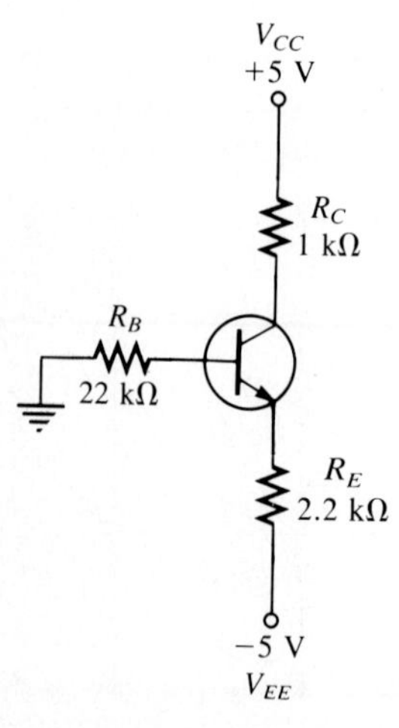

FIGURE 5–44

14. Analyze the circuit in Figure 5–44 to determine the correct voltage at the transistor terminals with respect to ground.
15. To what value can R_E in Figure 5–44 be reduced without the transistor going into saturation?
16. Taking V_{BE} into account in Figure 5–44, how much will I_E change with a temperature increase from 25°C to 100°C? The V_{BE} is 0.7 V at 25°C and decreases 2.5 mV per degree Celsius. Assume β_{dc} has no effect.
17. When can the effect of a change in β_{dc} be neglected in the emitter-biased circuit?
18. Determine I_C and V_{CE} in the pnp emitter-biased circuit of Figure 5–45.

FIGURE 5–45

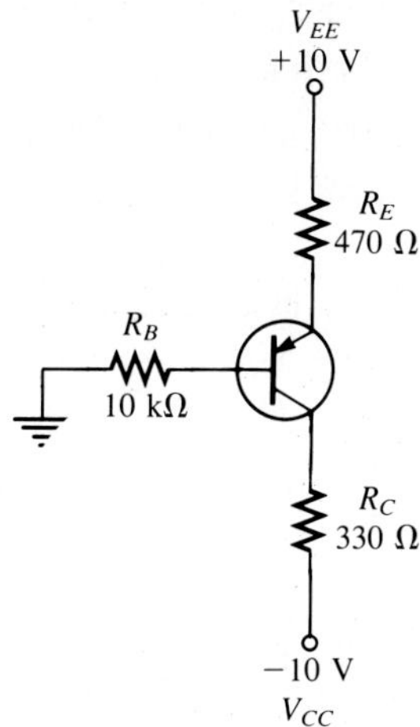

SECTION 5–4 VOLTAGE-DIVIDER BIAS

19. What is the minimum value of β_{dc} in Figure 5–46 that makes $R_{IN(base)} \geqq 10R_2$?

20. The bias resistor R_2 in Figure 5–46 is replaced by a 15 kΩ potentiometer. What minimum resistance setting causes saturation?

21. If the potentiometer described in Problem 20 is set at 2 kΩ, what are the values for I_C and V_{CE}?

22. Determine all transistor terminal voltages with respect to ground in Figure 5–47. Do not neglect the input resistance at the base or V_{BE}.

23. Show the connections required to replace the transistor in Figure 5–47 with a pnp device.

24. **(a)** Determine V_B in Figure 5–48.

(b) If R_E is doubled, what is the value of V_B?

25. **(a)** Find the Q-point values for Figure 5–48.

(b) Find the minimum power rating of the transistor in Figure 5–48.

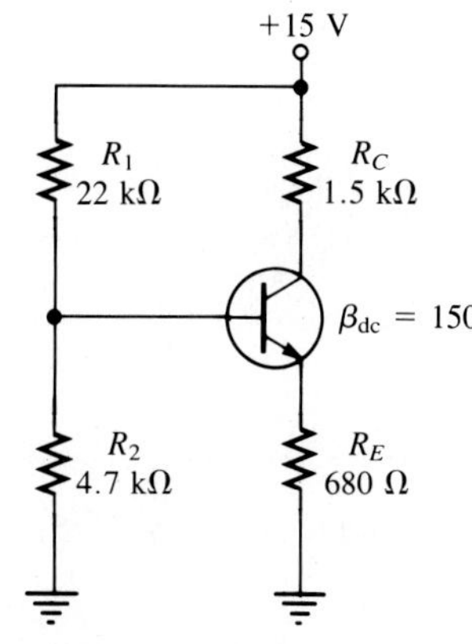

FIGURE 5–46

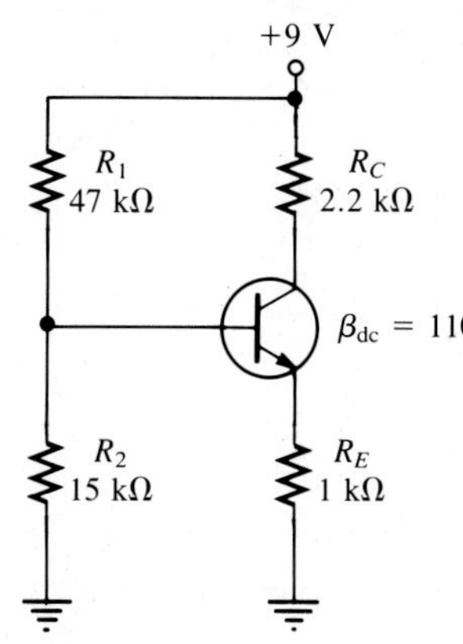

FIGURE 5–47

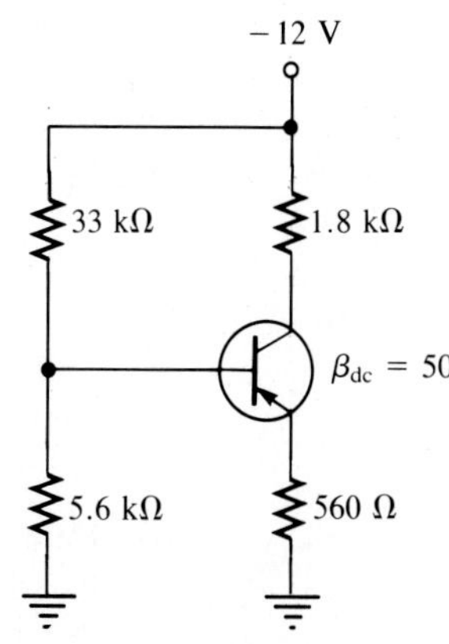

FIGURE 5–48

SECTION 5–5 COLLECTOR-FEEDBACK BIAS

26. Determine V_B, V_C, and I_C in Figure 5–49.
27. What value of R_C can be used to decrease I_C in Problem 26 by 25 percent?
28. What is the minimum power rating for the transistor in Problem 27?
29. A collector-feedback circuit uses an npn transistor with $V_{CC} = 12$ V, $R_C = 1.2$ kΩ, and $R_B = 47$ kΩ. Determine the collector voltage and the collector current if $\beta_{dc} = 200$.

FIGURE 5–49

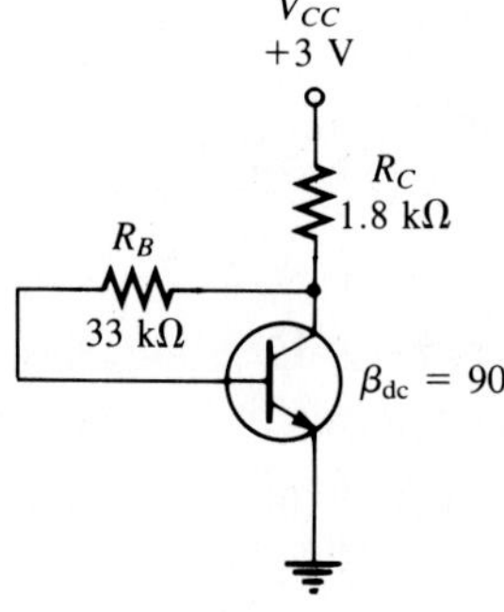

SECTION 5–6 TROUBLESHOOTING

30. Assume the emitter becomes shorted to ground in Figure 5–50 by a solder splash or stray wire clipping. What do the meters read?

FIGURE 5–50

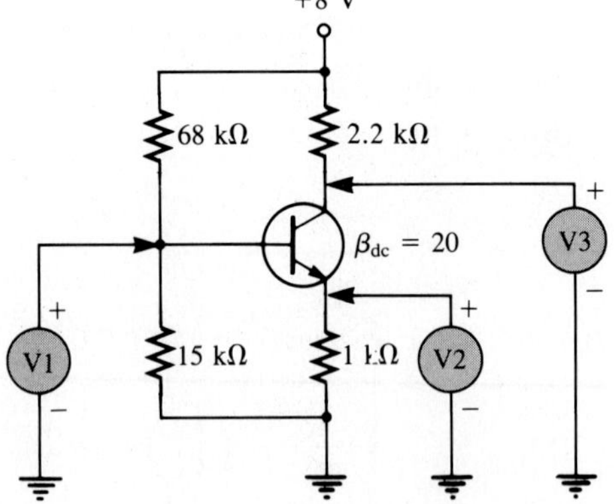

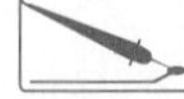

31. Determine the most probable failures, if any, in each circuit of Figure 5–51, based on the indicated measurements.

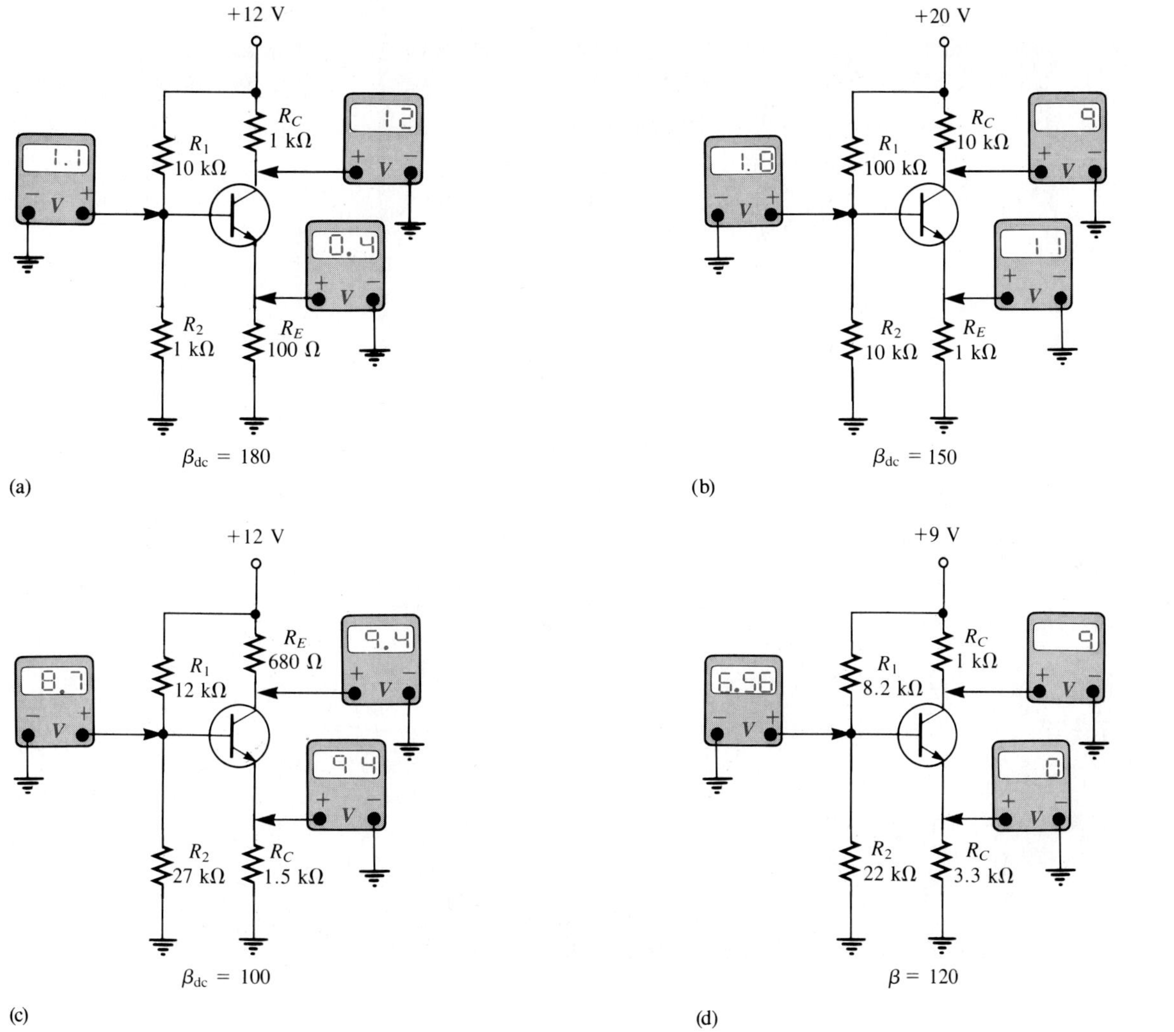

(a) $\beta_{dc} = 180$

(b) $\beta_{dc} = 150$

(c) $\beta_{dc} = 100$

(d) $\beta = 120$

FIGURE 5–51

32. Determine if the readings in the breadboard circuit of Figure 5–52 are correct. If they are not, isolate the problem(s). The transistor is a pnp device with a specified dc beta range of 35 to 100.

33. Determine each meter reading and range setting in Figure 5–52 for each of the following faults:

(a) The 680 kΩ resistor open

(b) The 5.6 kΩ resistor open

(c) The 10 kΩ resistor open

(d) The 1 kΩ resistor open

(e) A short from emitter to ground

(f) An open base-emitter junction

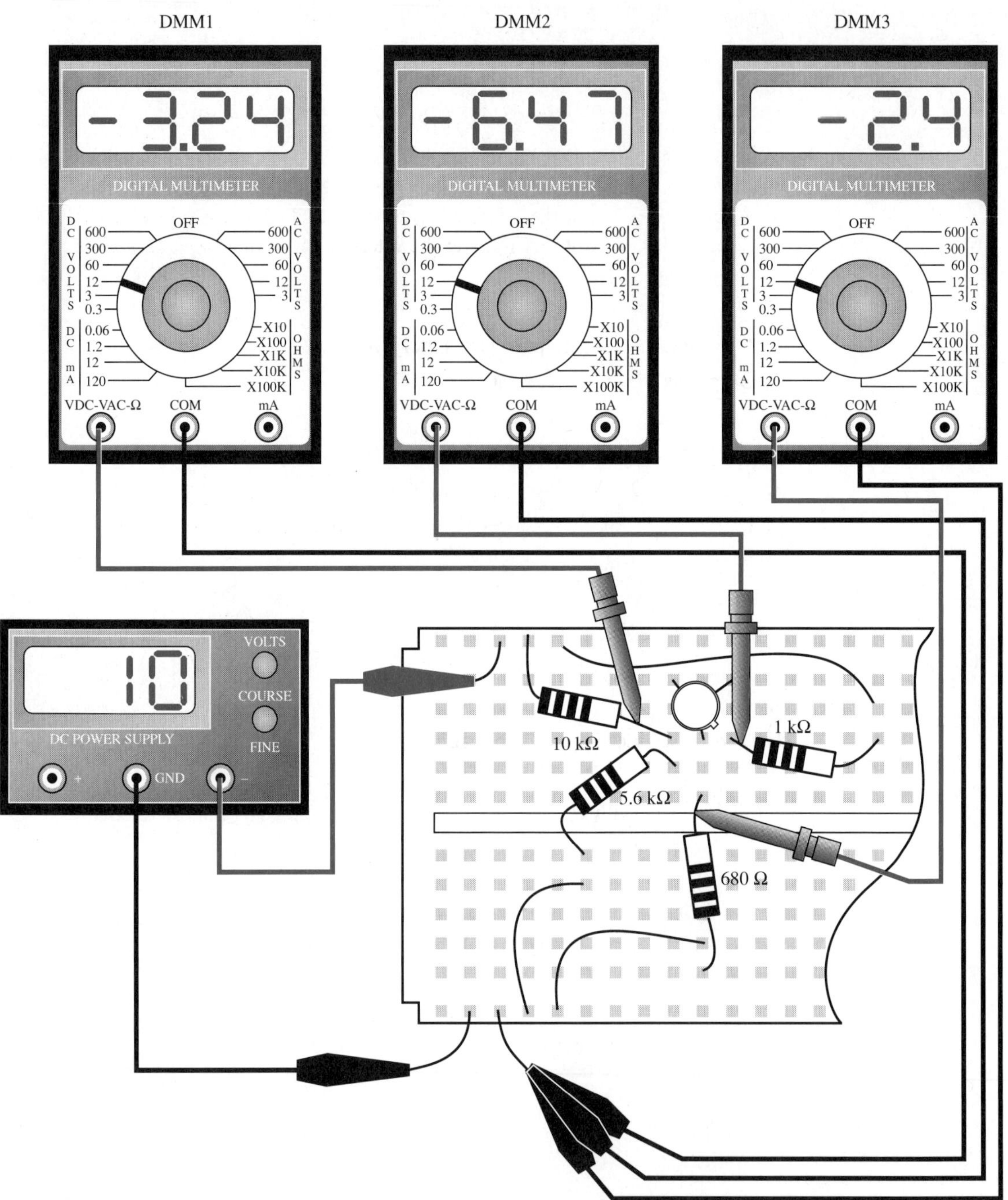

DMM1
DMM2
DMM3
-3.24
-6.47
-2.4
DIGITAL MULTIMETER
OFF
DC VOLTS
600
300
60
12
3
0.3
AC VOLTS
DC mA
0.06
1.2
12
120
OHMS
X10
X100
X1K
X10K
X100K
VDC-VAC-Ω
COM
mA
10
VOLTS
COURSE
FINE
DC POWER SUPPLY
+
GND
–
10 kΩ
1 kΩ
5.6 kΩ
680 Ω

FIGURE 5–52

ANSWERS TO REVIEW QUESTIONS

SECTION 5–1

1. The upper load line limit is $I_{C(\text{sat})}$ and $V_{CE(\text{sat})}$. The lower limit is $I_C = 0$ and $V_{CE(\text{cutoff})}$.
2. The Q point is the dc point at which a transistor is biased specified by V_{CE} and I_C.
3. Saturation begins at the intersection of the load line and the vertical portion of the collector curve. Cutoff occurs at the intersection of the load line and the $I_B = 0$ curve.

SECTION 5–2

1. Base bias does not require two supply voltages.
2. Base bias is beta-dependent.

SECTION 5–3

1. Emitter bias is much less dependent on the value of beta than is base bias.
2. $V_E = V_B - 0.7$ V, $V_E = V_B + 0.7$ V
3. Emitter bias requires two separate supply voltages.
4. $I_E = 14.3 \text{ V}/10 \text{ k}\Omega = 1.43$ mA

SECTION 5–4

1. $R_{\text{IN(base)}} = V_{\text{IN}}/I_{\text{IN}} = 5 \text{ V}/5\ \mu\text{A} = 1 \text{ M}\Omega$
2. $R_{\text{IN(base)}} = \beta_{\text{dc}}R_E = 190(1 \text{ k}\Omega) = 190 \text{ k}\Omega$
3. $V_B = 5$ V
4. Voltage-divider bias is stable and requires only one supply voltage.

SECTION 5–5

1. I_C increases with β_{dc}, causing a reduction in V_C and, therefore, less voltage across R_B, thus less I_B.
2. $I_B = (V_C - V_{BE})/R_B = (4 \text{ V} - 0.7 \text{ V})/47 \text{ k}\Omega = 70.2\ \mu\text{A}$

SECTION 5–6

1. A transistor is saturated when $V_{CE} = 0$ V. A transistor is in cutoff when $V_{CE} = V_{CC}$.
2. R_E is open because the BE junction of the transistor is still forward-biased.
3. If R_C is open, V_C is about 0.7 V less than V_B.

SECTION 5–7

1. The detector circuit senses a change in temperature and produces a proportional change in output voltage.
2. Yes, it goes into saturation.
3. The transistor saturates when the thermistor resistance is approximately 5.3 kΩ. From the graph in Figure 5–39, the temperature is approximately 42°C.
4. Not applicable

ANSWERS TO PRACTICE EXERCISES

5–1 $I_{CQ} = 19.8$ mA, $V_{CEQ} = 4.2$ V, $I_{b(\text{peak})} = 42\ \mu$A

5–2 $\%\ \Delta I_C = 150\%$, $\%\ \Delta V_{CE} = 53.7\%$

5–3 $I_E = 4.33$ mA, $I_C \cong 4.33$ mA, $V_{CE} = 12.7$ V

5–4 $\%\ \Delta I_C = 5.45\%$, $\%\ \Delta V_{CE} = -11.2\%$

5–5 Less negative, No, -11.3 V

5–6 54.6 kΩ

5–7 $V_{CE} = 2.56$ V, $I_C = 4.77$ mA

5–8 $I_C \cong 2.29$ mA, $V_{EC} = 2.67$ V

5–9 214

5–10 $I_{CQ} = 0.894$ mA, $V_{CEQ} = 1.06$ V

6

Small-Signal Bipolar Amplifiers

After completing this chapter, you should be able to

- □ Describe how a transistor circuit operates as a small-signal amplifier.
- □ Represent a bipolar junction transistor by an equivalent circuit.
- □ Define the hybrid (*h*) parameters of a transistor.
- □ Define the *r* parameters of a transistor.
- □ List the characteristics of common-emitter, common-collector, and common-base amplifiers.
- □ Analyze an amplifier by reducing it to its equivalent circuit.
- □ Explain what determines the voltage gain, input resistance, and output resistance of an amplifier.
- □ Explain how an emitter bypass capacitor affects the voltage gain.
- □ Discuss how loading affects the voltage gain of an amplifier.
- □ Determine resistor values necessary to set a specified voltage gain.
- □ Show how voltage gain is increased by connecting amplifiers in cascade.
- □ Express voltage gain and power gain in decibels.
- □ Distinguish between direct-coupled, transformer-coupled, and capacitively coupled amplifiers.
- □ Troubleshoot multistage amplifiers.

The things you learned about biasing a transistor in Chapter 5 are carried forward into this chapter where transistor circuits are used as small-signal amplifiers. The term *small-signal* refers to the use of signals that take up a relatively small percentage of an amplifier's operational range; that is, signals that use only a small portion of the load line. Additionally, you will learn how to reduce an amplifier to an equivalent dc and ac circuit for easier analysis, and you will learn about multistage amplifiers.

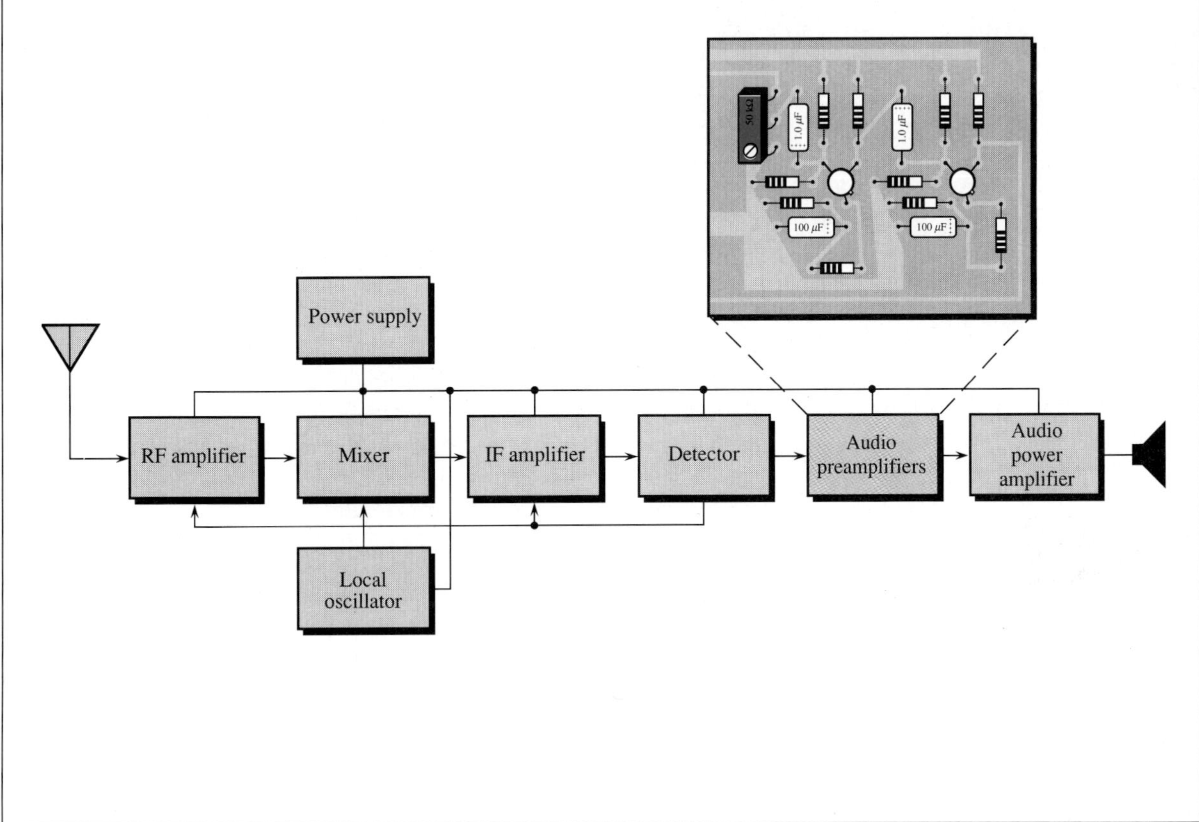

A System Application

The radio receiver is a system that most of us use every day. The block diagram for an AM superheterodyne receiver is shown above with the audio preamplifier as our focus in this chapter's system application. You do not need to understand the complete system in order to apply what you will learn in this chapter to the audio amplifier. Parts of this system use circuits that have not been covered yet. We will focus on some of these at appropriate points throughout the book.

The receiver selects **amplitude-modulated** (AM) frequencies between 535 kHz and 1605 kHz in the broadcast band and extracts the audio signal. The audio preamplifier amplifies the audio signal coming from the detector before it goes to the power amplifier. The power amplifier boosts the audio power sufficiently to drive the speaker.

For the system application in Section 6–8, in addition to the other topics, be sure you understand

- □ The principles of common-emitter amplifier operation.
- □ How loading affects an amplifier.
- □ The concept of multistage amplifiers.
- □ The use of capacitors for coupling and bypass functions.

6–1 SMALL-SIGNAL AMPLIFIER OPERATION

The biasing of a transistor is purely a dc operation. The purpose, however, is to establish a Q point about which variations in current and voltage can occur in response to an ac signal. In applications where very small signal voltages must be amplified, such as from an antenna, variations about the Q point are relatively small. Amplifiers designed to handle these small ac signals are called small-signal amplifiers.

A biased transistor with an ac source capacitively coupled to the base and a load capacitively coupled to the collector is shown in Figure 6–1. The coupling capacitors block dc and thus prevent the source resistance and the load resistance from changing the bias voltages at the base and collector. The signal voltage causes the base voltage to vary above and below its dc bias level. The resulting variation in base current produces a larger variation in collector current because of the current gain of the transistor.

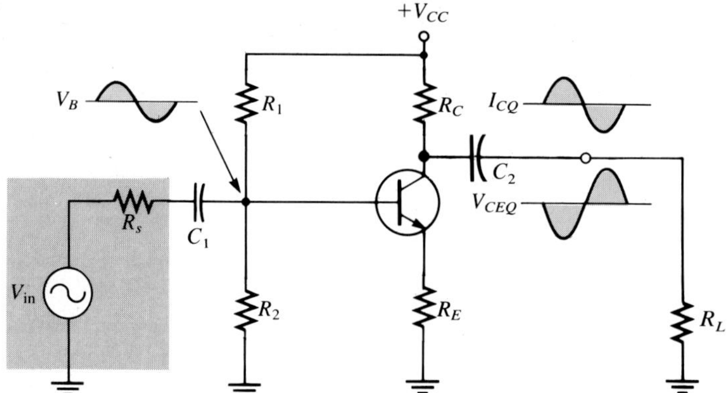

FIGURE 6–1
An amplifier with voltage-divider bias driven by an ac source with an internal resistance, R_s.

As the collector current increases, the collector voltage decreases. The collector current varies above and below its Q-point value in-phase with the base current, and the collector-to-emitter voltage varies above and below its Q-point value 180° out-of-phase with the base voltage, as illustrated in Figure 6–1.

A GRAPHICAL PICTURE

The operation just described can be illustrated graphically on the collector-characteristic curves, as shown in Figure 6–2. The signal at the base drives the base current equally above and below the Q point on the ac load line, as shown by the arrows. Lines projected from the peaks of the base current, across to the I_C axis, and down to the V_{CE} axis, indicate the peak-to-peak variations of the collector current and collector-to-emitter voltage, as shown. The ac load line differs from the dc load line because the effective collector resistance is less than for the dc circuit due to the capacitively coupled load. This difference is covered in Chapter 9 in relation to power amplifiers.

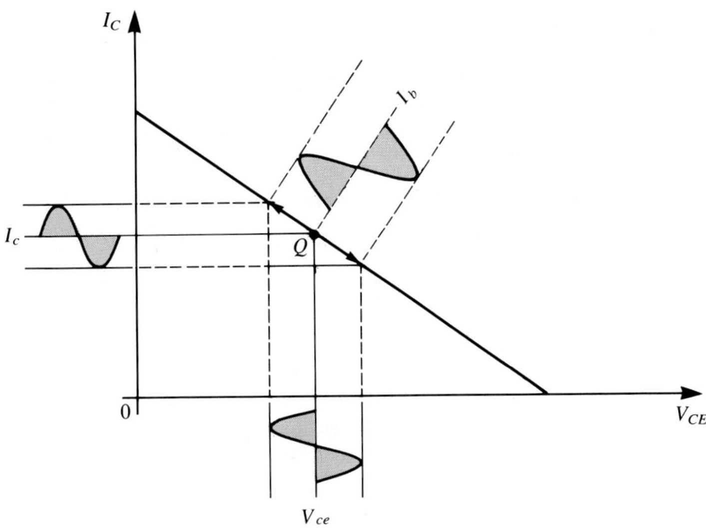

FIGURE 6–2
Graphical operation of the amplifier showing the base current, collector current, and collector-to-emitter voltage swings.

EXAMPLE 6–1

The ac load line operation of a certain amplifier extends 10 μA above and below the Q-point base current value of 50 μA, as shown in Figure 6–3. Determine the resulting peak-to-peak values of collector current and collector-to-emitter voltage from the graph.

FIGURE 6–3

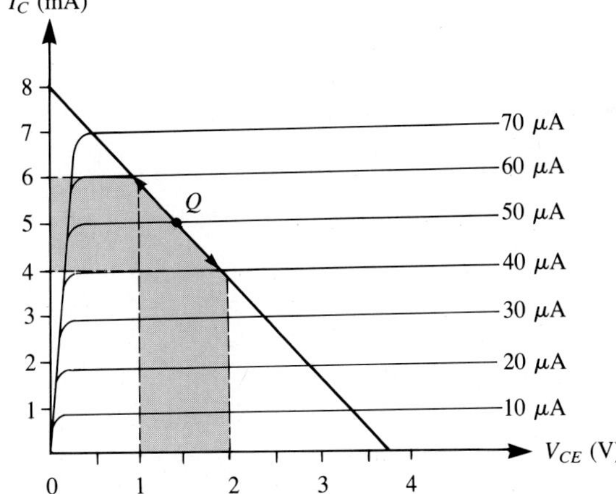

SOLUTION

Projections on the graph of Figure 6–3 show a peak-to-peak collector current of 2 mA and a peak-to-peak collector-to-emitter voltage of 1 V.

PRACTICE EXERCISE 6–1

What are the Q-point values of I_C and V_{CE} in Figure 6–3?

LABELS

The labels used for dc quantities in the previous chapters were identified by uppercase (capital) subscripts such as I_C, I_E, V_C, V_{CE}, and so forth. We will use lowercase subscripts to indicate ac quantities of rms, peak, and peak-to-peak currents and voltages, such as I_c, I_e, I_b, V_c, V_{ce}, and so forth. These symbols will represent rms values unless otherwise stated. Also, instantaneous quantities are represented by both lowercase letters and subscripts, such as i_c, i_e, i_b, v_e, v_{ce}, and so forth.

In addition to currents and voltages, resistances often have different values when a circuit is analyzed from an ac viewpoint as opposed to a dc viewpoint. Lowercase subscripts are used to identify ac resistance values. For example, R_c is the ac collector resistance, and R_C is the dc collector resistance. You will see the need for this distinction later. Resistance values *internal* to the transistor use a lowercase r. An example is the internal ac emitter resistance, r_e.

6–1 REVIEW QUESTIONS

1. When I_b is at its positive peak, I_c is at its ________ peak, and V_{ce} is at its ________ peak.
2. What is the difference between V_{CE} and V_{ce}?

6–2 TRANSISTOR AC EQUIVALENT CIRCUITS

In order to better visualize the operation of a transistor in an amplifier circuit, it is often useful to represent the device by an equivalent circuit. An equivalent circuit uses various internal transistor parameters to represent the transistor's operation. Two types of equivalent circuit representations are described in this section. One is based on h parameters and the other on r parameters.

h PARAMETERS

Because they are typically specified on a manufacturer's data sheet, h (hybrid) parameters are important. These parameters are usually specified because they are relatively easy to measure. The four basic ac h parameters and their descriptions are given in Table 6–1. Each of the four h parameters carries a second subscript letter to designate the common-emitter (e), common-base (b), or common-collector (c) configuration. These are listed in Table 6–2.

We can describe the three amplifier circuit configurations this way:

- When the emitter is connected to ac ground, the input signal is applied to the base, and the output signal is on the collector, the circuit is a **common-emitter** (CE) type. (The term *common* refers to the ground point.)
- When the collector is connected to ac ground, the input signal is applied to the base, and the output signal is on the emitter, the circuit is a **common-collector** (CC) type.
- When the base is connected to ac ground, the input signal is applied to the emitter, and the output signal is on the collector, the circuit is a **common-base** (CB) type.

TABLE 6–1
Basic ac h parameters

h Parameter	Description	Condition
h_i	Input resistance	Output shorted
h_r	Voltage feedback ratio	Input open
h_f	Forward current gain	Output shorted
h_o	Output conductance	Input open

TABLE 6–2
Subscripts of h parameters

Configuration	h Parameters
Common-Emitter	h_{ie}, h_{re}, h_{fe}, h_{oe}
Common-Base	h_{ib}, h_{rb}, h_{fb}, h_{ob}
Common-Collector	h_{ic}, h_{rc}, h_{fc}, h_{oc}

We will examine the characteristics of each of these bipolar amplifier configurations later in this chapter.

DEFINITIONS OF h PARAMETERS

Each of the h parameters is derived from ac measurements taken from operating transistor characteristic curves. h_i is the impedance (ac resistance) looking in at the input terminal of the transistor with the output shorted, as indicated in Figure 6–4(a), for a common-emitter connection. It is the ratio of input voltage to input current, expressed as follows:

$$h_{ie} = \frac{V_b}{I_b} \tag{6–1}$$

h_r is a measure of the amount of output voltage that is reflected (fed back) to the input with the input open. The common-emitter measurement circuit is shown in Figure 6–4(b). It is the ratio of input voltage to output voltage.

$$h_{re} = \frac{V_b}{V_c} \tag{6–2}$$

FIGURE 6–4
AC equivalent circuits for defining h parameters using the common-emitter configuration.

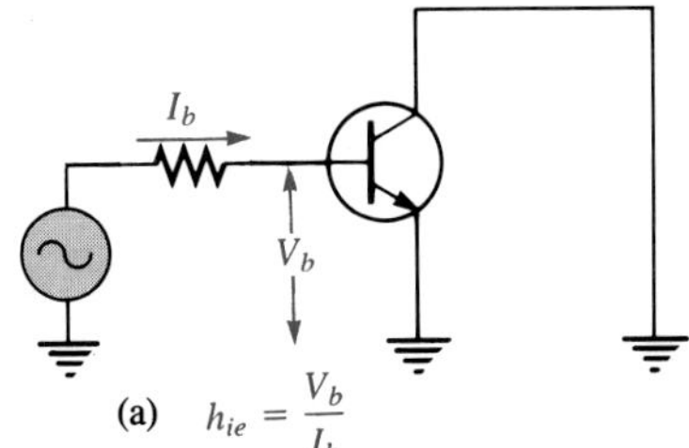

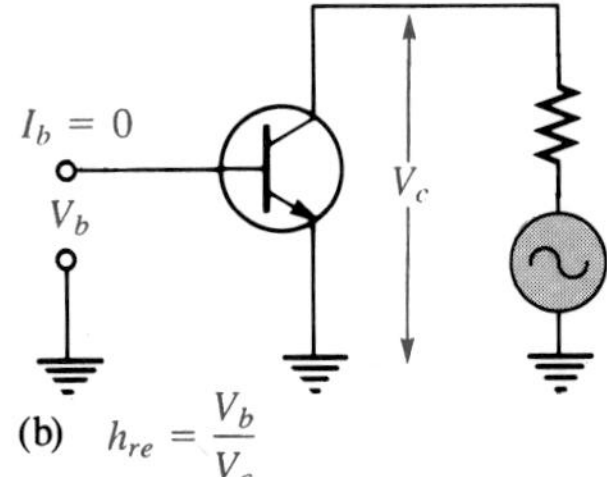

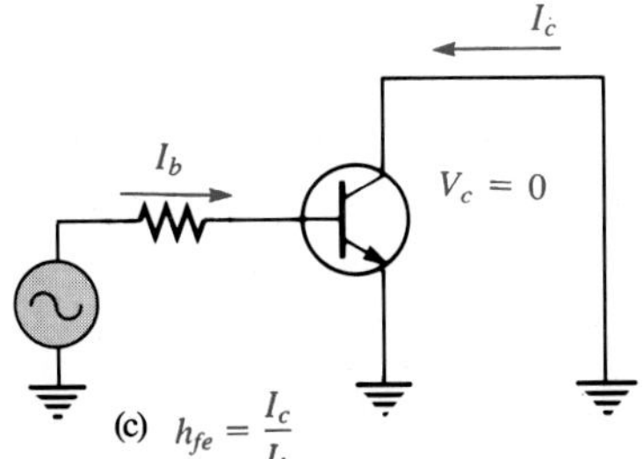

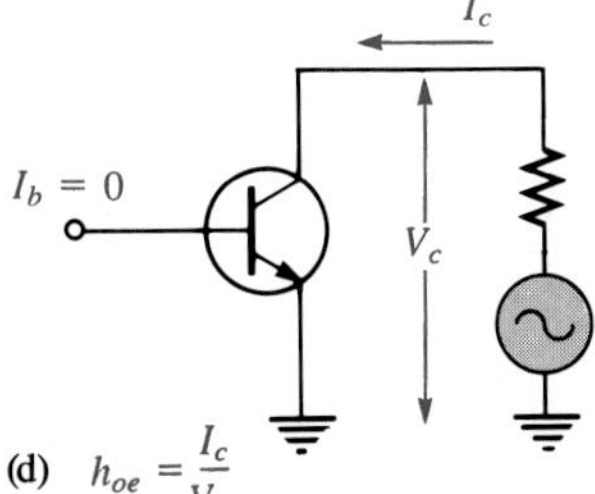

h_f is the forward current gain measured with the output shorted, as shown in Figure 6–4(c). For the common-emitter configuration, h_{fe} is expressed as

$$h_{fe} = \frac{I_c}{I_b} \quad (6\text{–}3)$$

Finally, h_o is the admittance (conductance) looking in at the output terminal with the input open, as shown in Figure 6–4(d), and it is expressed as

$$h_{oe} = \frac{I_c}{V_c} \quad (6\text{–}4)$$

The units of h_{oe} are siemens (S). Table 6–3 summarizes the h-parameter ratios for each amplifier configuration.

TABLE 6–3
h-parameter ratios for the three amplifier configurations

Common-Emitter	Common-Base	Common-Collector
$h_{ie} = V_b/I_b$	$h_{ib} = V_e/I_b$	$h_{ic} = V_b/I_b$
$h_{re} = V_b/V_c$	$h_{rb} = V_e/V_c$	$h_{rc} = V_b/V_e$
$h_{fe} = I_c/I_b$	$h_{fb} = I_c/I_e$	$h_{fc} = I_e/I_b$
$h_{oe} = I_c/V_c$	$h_{ob} = I_c/V_c$	$h_{oc} = I_e/V_e$

HYBRID EQUIVALENT CIRCUITS

The general form of the h-parameter equivalent circuit is shown in Figure 6–5. The input impedance, h_i, appears in series at the input. The reverse voltage ratio, h_r, is multiplied by the output voltage (h_rV_{out}) to produce an equivalent voltage source in series with the input. The forward current gain, h_f, is multiplied by the input current (h_fI_{in}) and appears as an equivalent current source in the output. The output admittance, h_o, appears across the output terminals. Specifically, there are three hybrid equivalent circuits—one for the common-emitter, one for the common-base, and one for the common-collector configuration, as shown in Figure 6–6.

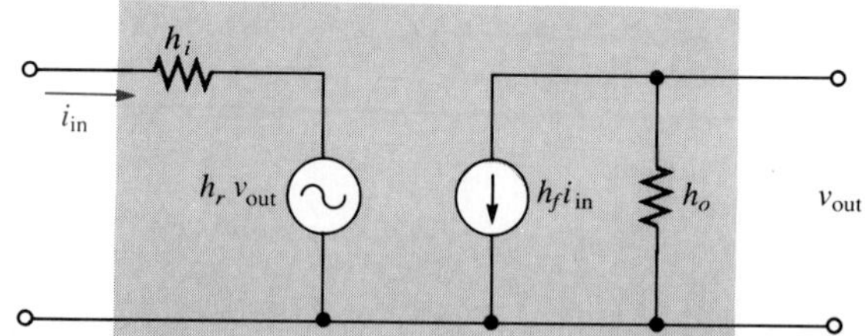

FIGURE 6–5
Generalized h parameter equivalent circuit for a bipolar junction transistor.

r PARAMETERS

The h parameters are important because they are given on data sheets, so you need to know what they mean. There is, however, another widely used set of parameters that are

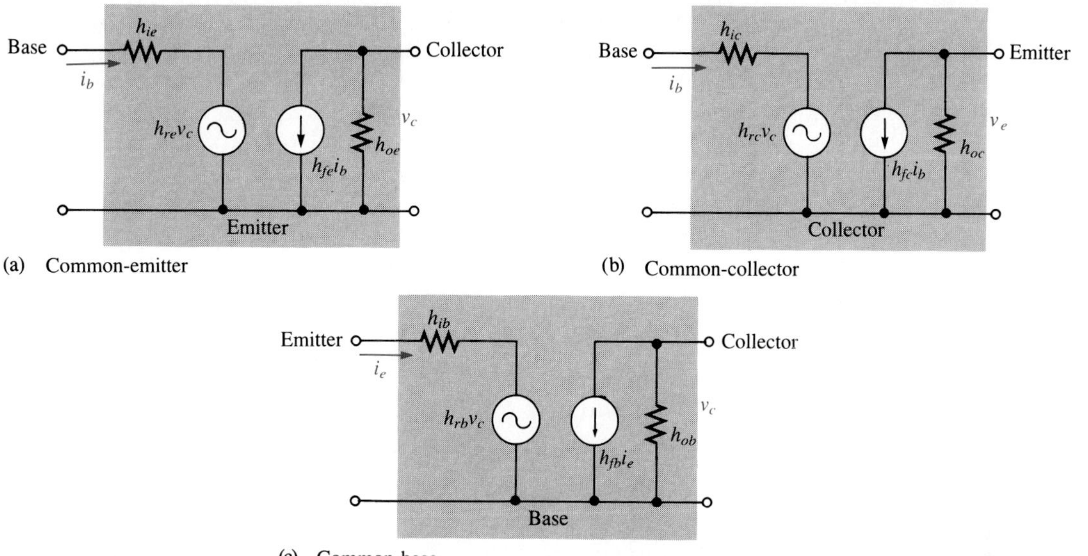

FIGURE 6–6
h parameter equivalent circuits for the three amplifier configurations.

perhaps easier to work with than the *h* parameters. These are the *r* parameters. The five *r* parameters are given in Table 6–4.

TABLE 6–4
r parameters

r Parameter	Description
α_{ac}	ac alpha (I_c/I_e)
β_{ac}	ac beta (I_c/I_b)
r_e	ac emitter resistance
r_b	ac base resistance
r_c	ac collector resistance

RELATIONSHIPS OF *h* PARAMETERS AND *r* PARAMETERS

The ac current gain parameters, α_{ac} and β_{ac}, convert directly from *h* parameters as follows:

$$\alpha_{ac} = h_{fb} \quad (6\text{–}5)$$

$$\beta_{ac} = h_{fe} \quad (6\text{–}6)$$

Recall that we used α_{dc} and β_{dc} in previous chapters. These are dc parameters and sometimes have values different from those of the ac parameters. The difference between β_{dc} and β_{ac} will be discussed later.

Because data sheets often provide only common-emitter h parameters, the following formulas show how to convert them to the remaining r parameters.

$$r_e = \frac{h_{re}}{h_{oe}} \tag{6–7}$$

$$r_c = \frac{h_{re} + 1}{h_{oe}} \tag{6–8}$$

$$r_b = h_{ie} - \frac{h_{re}}{h_{oe}}(1 + h_{fe}) \tag{6–9}$$

We will use r parameters throughout the text.

r-PARAMETER EQUIVALENT CIRCUITS

An r-parameter equivalent circuit is shown in Figure 6–7. For most general analysis work, Figure 6–7 can be simplified as follows: The effect of the ac base resistance r_b is usually small enough to neglect, so that it can be replaced by a short. The ac collector resistance is usually several megohms and can be replaced by an open. The resulting simplified r-parameter equivalent circuit is shown in Figure 6–8.

The interpretation of this equivalent circuit in terms of a transistor's ac operation is as follows: A resistance r_e appears between the emitter and base terminals. This is the resistance "seen" by an ac signal when applied to the base or emitter terminals of a forward-biased transistor. The collector effectively acts as a current source of $\alpha_{ac}I_e$ or, equivalently, $\beta_{ac}I_b$. These factors are shown with a standard transistor symbol in Figure 6–9.

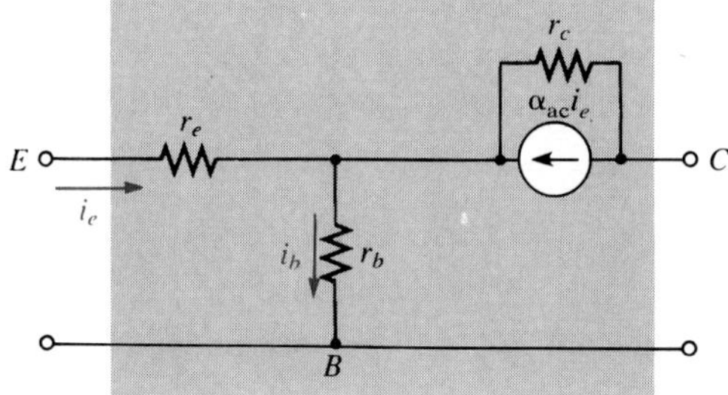

FIGURE 6–7
Generalized r-parameter equivalent circuit for a bipolar junction transistor.

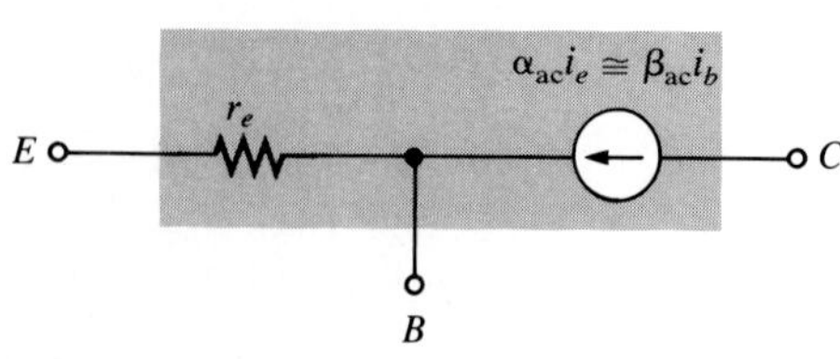

FIGURE 6–8
Simplified r-parameter equivalent circuit for a transistor.

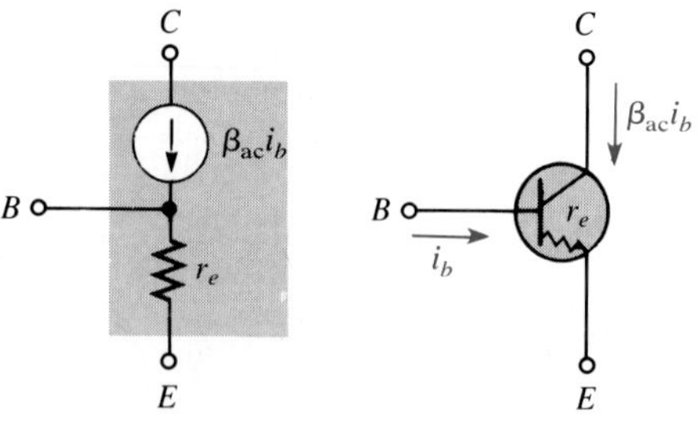

FIGURE 6–9
Relation of transistor symbol to r-parameter equivalent.

DETERMINING r_e BY FORMULA

Instead of using h parameters to get r_e, we can also use the simple formula of Equation (6–10) to calculate an approximate value.

$$r_e \cong \frac{25 \text{ mV}}{I_E} \tag{6–10}$$

Although the formula is simple, its derivation is not and is therefore reserved for Appendix B for those who are interested.

EXAMPLE 6–2 Determine the r_e of a transistor that is operating with a dc emitter current of 2 mA.

SOLUTION

$$I_E = 2 \text{ mA}$$

$$r_e \cong \frac{25 \text{ mV}}{I_E} = \frac{25 \text{ mV}}{2 \text{ mA}} = 12.5 \ \Omega$$

PRACTICE EXERCISE 6–2

What is I_E if $r_e = 8 \ \Omega$?

COMPARISON OF THE AC BETA (β_{ac}) TO THE DC BETA (β_{dc})

For a typical transistor, a graph of I_C versus I_B is nonlinear, as shown in Figure 6–10(a). If we pick a Q point on the curve and cause the base current to vary an amount ΔI_B, then the collector current will vary an amount ΔI_C as shown in part (b). At different points on the curve, the ratio $\Delta I_C/\Delta I_B$ will be different, and it may also differ from the I_C/I_B ratio at the Q point. Since $\beta_{dc} = I_C/I_B$ and $\beta_{ac} = \Delta I_C/\Delta I_B$, the values of these two quantities can differ. Remember that $\beta_{dc} = h_{FE}$ and $\beta_{ac} = h_{fe}$.

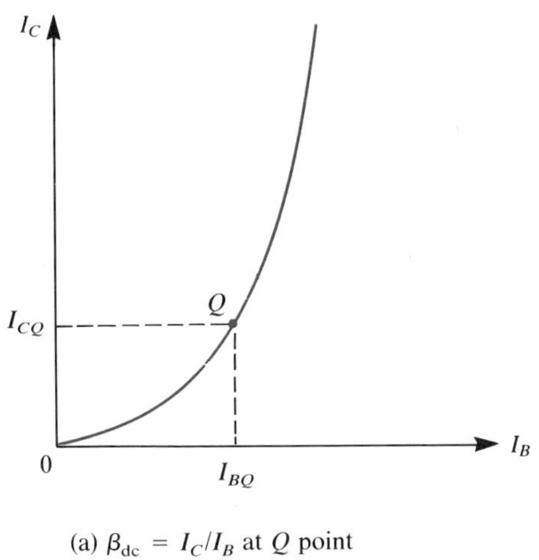

(a) $\beta_{dc} = I_C/I_B$ at Q point

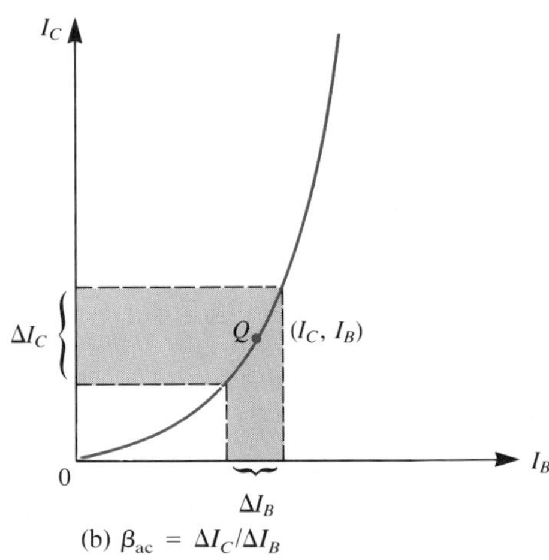

(b) $\beta_{ac} = \Delta I_C/\Delta I_B$

FIGURE 6–10
I_C-versus-I_B curve illustrates the difference between $\beta_{dc} = I_C / I_B$ and $\beta_{ac} = \Delta I_C / \Delta I_B$.

6–2 REVIEW QUESTIONS

1. Define each of the parameters: h_{ie}, h_{re}, h_{fe}, and h_{oe}.
2. Which h parameter is equivalent to β_{ac}?
3. If $I_E = 15$ mA, what is the approximate value of r_e?

6–3 COMMON-EMITTER AMPLIFIERS

Now that you have an idea of how a transistor can be represented in an ac circuit, a complete amplifier circuit will be examined. The common-emitter (CE) configuration is covered in this section. The common-collector and common-base configurations are covered in the following sections.

Figure 6–11 shows a common-emitter amplifier with voltage-divider bias and coupling capacitors, C_1 and C_2, on the input and output and a bypass capacitor C_3 from emitter to ground. The circuit has a combination of dc and ac operation, both of which must be considered.

DC ANALYSIS

To analyze the amplifier in Figure 6–11, the dc bias values must first be determined. To do this, a dc equivalent circuit is developed by simply replacing the coupling and bypass capacitors with opens, as shown in Figure 6–12. The analysis for this dc circuit was covered in Chapter 5, but we will go through it again here. Recall that the dc input resistance at the base is

$$R_{IN(base)} \cong \beta_{dc}R_E$$
$$R_{IN(base)} = (150)(560\ \Omega) = 84\ \text{k}\Omega$$

Since this is more than ten times R_2, it will be neglected when calculating the base voltage. Keep in mind that some accuracy is sacrificed.

$$V_B = \left(\frac{R_2}{R_1 + R_2}\right)V_{CC} = \left(\frac{4.7\ \text{k}\Omega}{26.7\ \text{k}\Omega}\right)12\ \text{V} = 2.11\ \text{V}$$

and

$$V_E = 2.11\ \text{V} - 0.7\ \text{V} = 1.41\ \text{V}$$

Therefore,

$$I_E = \frac{V_E}{R_E} = \frac{1.41\ \text{V}}{560\ \Omega} = 2.5\ \text{mA}$$

Since $I_C \cong I_E$, then

$$V_C = V_{CC} - I_CR_C = 12\ \text{V} - 2.5\ \text{V} = 9.5\ \text{V}$$

Finally,

$$V_{CE} = V_C - V_E = 9.5\ \text{V} - 1.41\ \text{V} = 8.09\ \text{V}$$

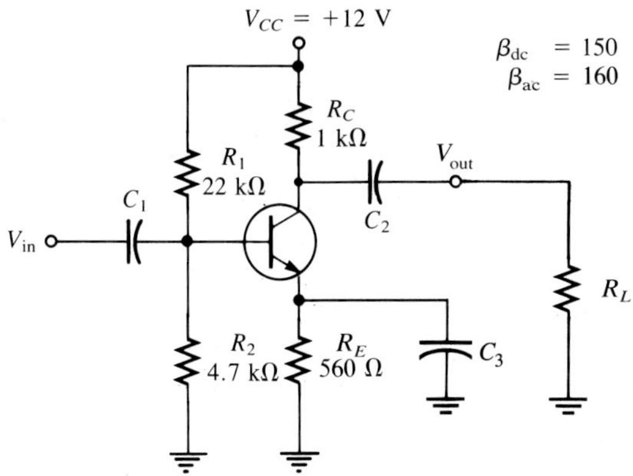

FIGURE 6–11
A common-emitter amplifier.

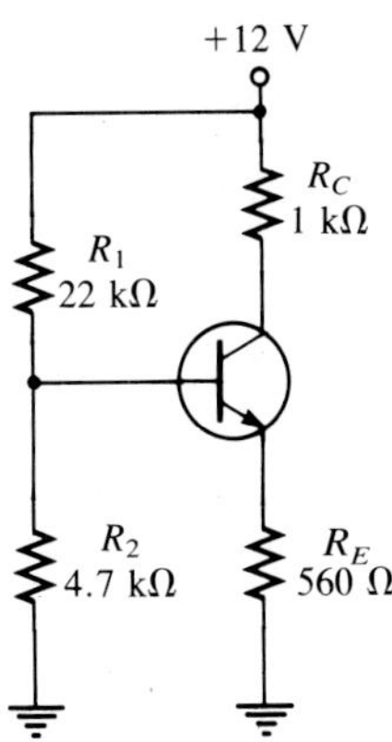

FIGURE 6–12
DC equivalent circuit for the amplifier in Figure 6–11.

AC EQUIVALENT CIRCUIT

To analyze the signal (ac) operation of the amplifier in Figure 6–11, an ac equivalent circuit is developed as follows: The coupling capacitors C_1 and C_2 are replaced by effective shorts. This is based on the simplifying assumption that $X_C \cong 0$ at the signal frequency. The bypass capacitor C_3 is omitted for this initial analysis, but we will consider it later.

The dc source is replaced by a ground. This is based on the assumption that the voltage source has an internal resistance of approximately 0 Ω, so that no ac voltage is developed across the source terminals. Therefore, the V_{CC} terminal is at a zero-volt ac potential and is called *ac ground*.

The ac equivalent circuit is shown in Figure 6–13(a). Notice in the figure that both R_C and R_1 have one end connected to ac ground because, in the actual circuit, they are connected to V_{CC} (ac ground). In ac analysis, the ac ground and the actual ground are treated as the same point.

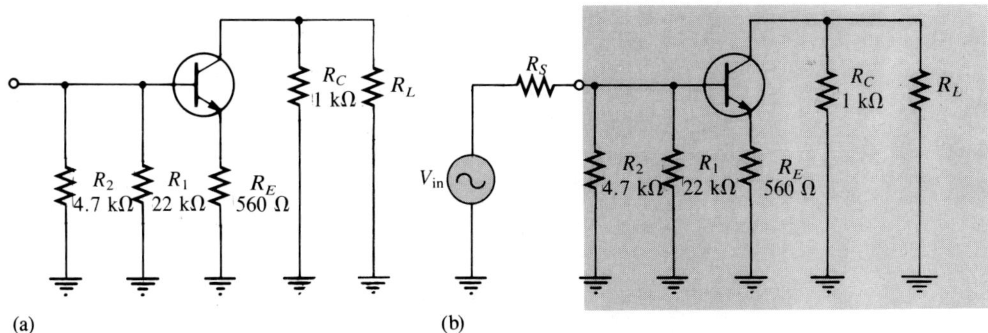

FIGURE 6–13
AC equivalent circuit for the amplifier in Figure 6–11 with the emitter bypass capacitor C_3 removed for initial analysis.

SIGNAL (AC) VOLTAGE AT THE BASE

An ac voltage source is shown connected to the input in Figure 6–13(b). If the internal resistance of the ac source is 0 Ω, then all of the input signal voltage appears at the base terminal. If, however, the ac source has a nonzero internal resistance, then three factors must be taken into account in determining the actual signal voltage at the base. These are the source resistance, the bias resistance, and the input resistance at the base. This is illustrated in Figure 6–14(a) and is simplified by combining R_1, R_2, and $R_{in(base)}$ in parallel to get the total input resistance, R_{in}, as shown in Figure 6–14(b). As you can see, the input voltage V_{in} is divided down by R_s (source resistance) and R_{in}, so that the signal voltage at the base of the transistor is

$$V_b = \left(\frac{R_{in}}{R_s + R_{in}}\right)V_{in} \tag{6–11}$$

Of course, if $R_s << R_{in}$, then $V_b \cong V_{in}$.

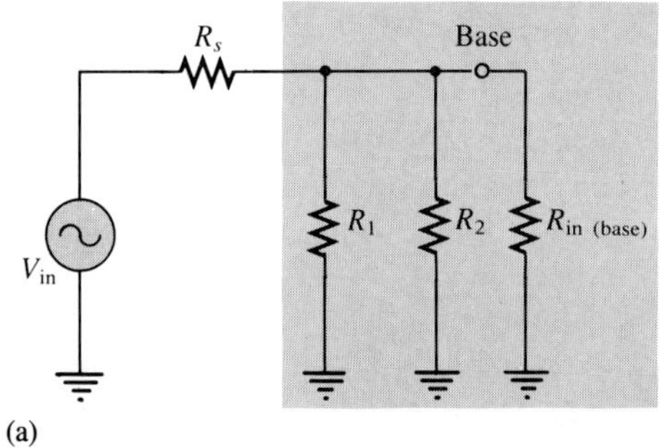

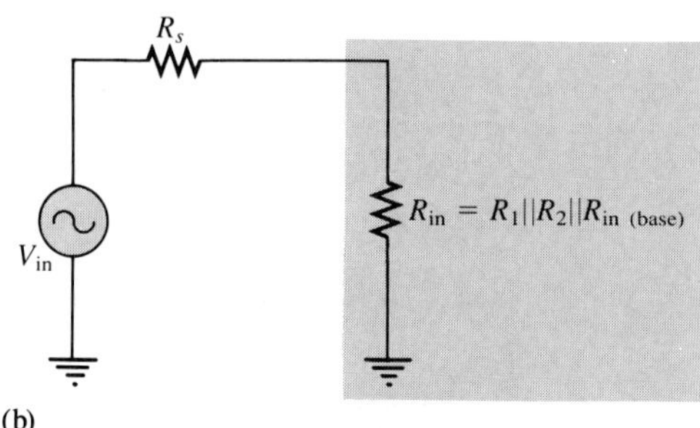

FIGURE 6–14
AC equivalent base circuit.

INPUT RESISTANCE

To develop an expression for the input resistance as seen by an ac source looking in at the base, we will use the simplified r-parameter model of the transistor. Figure 6–15 shows it connected with external resistors R_E and R_C. The input resistance looking in at the base with no emitter bypass capacitor is

$$R_{in(base)} = \frac{V_b}{I_b} \tag{6–12}$$

$$V_b = I_e(r_e + R_E)$$

and

$$I_b \cong \frac{I_e}{\beta_{ac}}$$

Substituting and cancelling I_e, we get

$$R_{in(base)} = \beta_{ac}(r_e + R_E) \tag{6–13}$$

The total input resistance seen by the source is the parallel combination of R_1, R_2, and $R_{in(base)}$.

$$R_{in} = R_1 \| R_2 \| R_{in(base)} \tag{6–14}$$

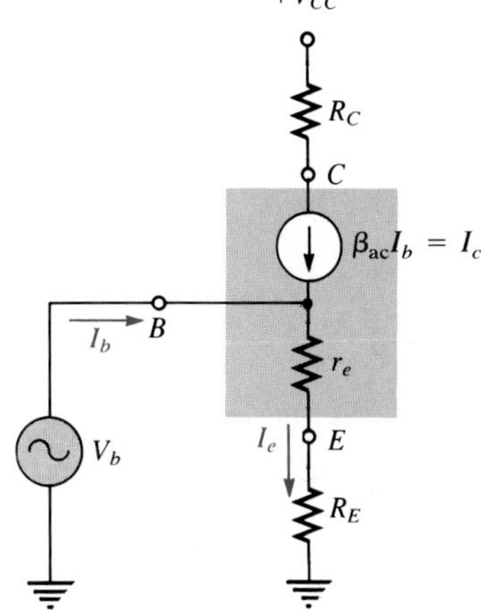

FIGURE 6–15
r-parameter transistor model (inside shaded block) connected to external circuit.

OUTPUT RESISTANCE

With the load removed, the output resistance looking in at the collector is approximately equal to the collector resistor.

$$R_{out} \cong R_C \tag{6–15}$$

Actually, $R_{out} = R_C \| r_c$, but since the internal ac collector resistance of the transistor, r_c, is typically much larger than R_C, the approximation is usually valid.

EXAMPLE 6–3 Determine the signal voltage at the base in Figure 6–16. This is the ac equivalent of the amplifier in Figure 6–11 without C_3, and with a 10 mV rms, 300-Ω signal source. I_E was previously found to be 2.5 mA.

FIGURE 6–16

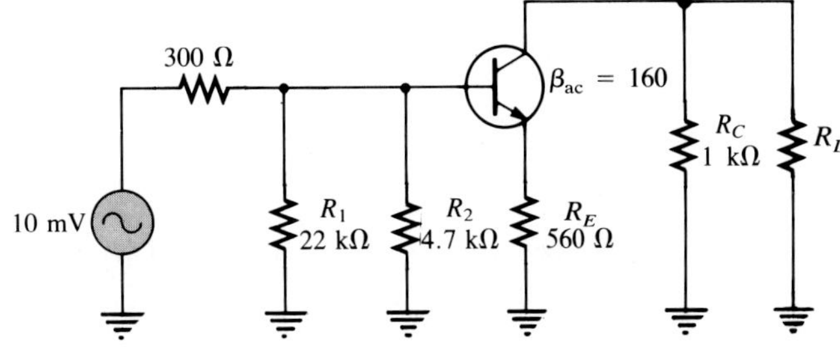

SOLUTION

First,

$$r_e = \frac{25\text{ mV}}{I_E} = \frac{25\text{ mV}}{2.5\text{ mA}} = 10\ \Omega$$

Then,

$$R_{\text{in(base)}} = \beta_{\text{ac}}(r_e + R_E) = 160(570\ \Omega) = 91.2\text{ k}\Omega$$

Next, determine the total input resistance viewed from the source:

$$\frac{1}{R_{\text{in}}} = \frac{1}{R_1} + \frac{1}{R_2} + \frac{1}{R_{\text{in(base)}}}$$
$$= \frac{1}{22\text{ k}\Omega} + \frac{1}{4.7\text{ k}\Omega} + \frac{1}{91.2\text{ k}\Omega}$$

and

$$R_{\text{in}} = 3.7\text{ k}\Omega$$

The input signal voltage is divided down by R_s and R_{in}, so the signal voltage at the base is the voltage across R_{in}.

$$V_b = \left(\frac{R_{\text{in}}}{R_s + R_{\text{in}}}\right)V_{\text{in}}$$
$$= \left(\frac{3.7\text{ k}\Omega}{4\text{ k}\Omega}\right)10\text{ mV}$$
$$= 9.25\text{ mV}$$

As you can see, there is some attenuation (reduction) of the input signal due to the source resistance and amplifier input resistance.

PRACTICE EXERCISE 6–3

Determine the signal voltage at the base in Figure 6–16 if the source resistance is 75 Ω and another transistor with an ac beta of 200 is inserted.

VOLTAGE GAIN

The ac voltage gain expression is developed using the equivalent circuit in Figure 6–17 with no collector load resistance. The gain is the ratio of ac output voltage (V_c) to ac input voltage at the base (V_b).

$$A_v = \frac{V_c}{V_b} \qquad (6\text{–}16)$$

Notice in the figure that $V_c = \alpha_{\text{ac}} I_e R_C \cong I_e R_C$ and $V_b = I_e(r_e + R_E)$. Therefore,

$$A_v = \frac{I_e R_C}{I_e(r_e + R_E)}$$

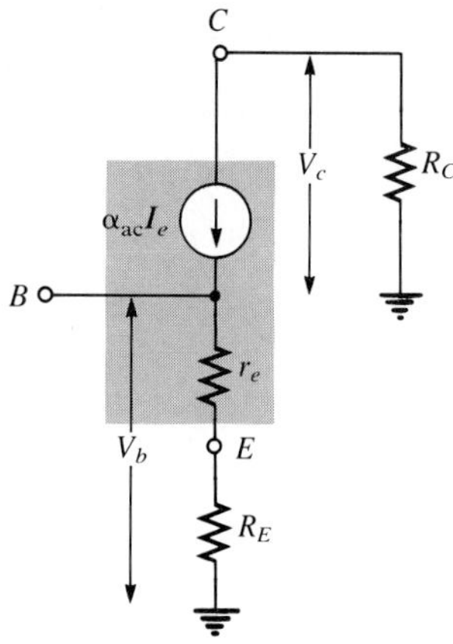

FIGURE 6–17
Equivalent circuit for obtaining ac voltage gain.

Cancelling the I_e terms, we get

$$A_v = \frac{R_C}{r_e + R_E} \tag{6–17}$$

Equation (6–17) is the voltage gain from base to collector. To get the overall gain of the amplifier from signal input to collector, the attenuation of the input circuit must be included. The **attenuation** (signal reduction) from input to base multiplied by the gain from base to collector is the overall amplifier gain. You can see this better with an example. Suppose a 10 mV signal is applied to the input, and the input network is such that the base voltage is 5 mV. The attenuation is therefore 5 mV/10 mV = 0.5. Now assume the amplifier has a voltage gain from base to collector of 20. The output voltage is 5 mV × 20 = 100 mV. Therefore, the overall gain is 100 mV/10 mV = 10 and is equal to the attenuation times the gain (0.5 × 20 = 10). This is illustrated in Figure 6–18.

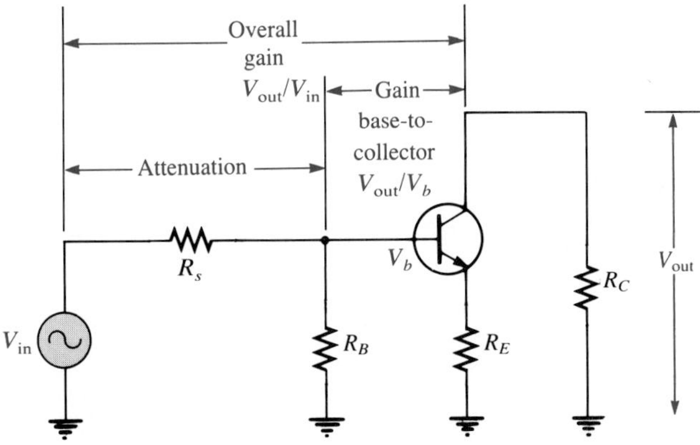

FIGURE 6–18
Base circuit attenuation and overall gain.

The expression for the attenuation in the base circuit where R_s and R_{in} act as a voltage divider is

$$\frac{V_b}{V_{in}} = \frac{R_{in}}{R_s + R_{in}} \qquad (6\text{–}18)$$

The overall gain A'_v is the product of A_v and V_b/V_{in}.

$$A'_v = A_v\left(\frac{V_b}{V_{in}}\right) \qquad (6\text{–}19)$$

EMITTER BYPASS CAPACITOR INCREASES GAIN

When the bypass capacitor is connected across R_E, as shown in Figure 6–19(a), the emitter is effectively at ac ground, as shown in Figure 6–19(b). The capacitor is selected large enough so that X_C is much less than R_E and effectively appears as a short at the signal frequency. It is important to recognize that the bypass capacitor does not alter the dc bias of the transistor because it is open to dc. Since the emitter resistance R_E is assumed to be shorted by the capacitor at the signal frequency, the ac voltage gain expression becomes

$$A_v = \frac{R_C}{r_e} \qquad (6\text{–}20)$$

FIGURE 6–19
Bypassing R_E with C_3.

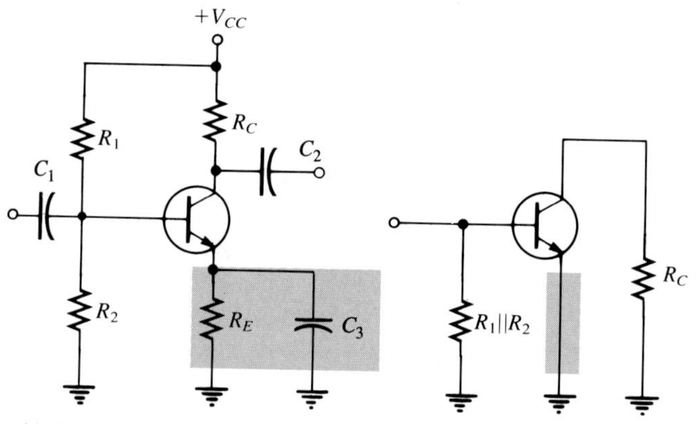

EXAMPLE 6–4 Calculate the base-to-collector voltage gain of the amplifier in Figure 6–11 with and without an emitter bypass capacitor and with no load resistor.

SOLUTION

Recall that $r_e = 10\ \Omega$ for this particular amplifier. Without C_3, the gain is

$$A_v = \frac{R_C}{r_e + R_E} = \frac{1\ \text{k}\Omega}{570\ \Omega} = 1.75$$

With C_3, the gain is

$$A_v = \frac{R_C}{r_e} = \frac{1\ \text{k}\Omega}{10\ \Omega} = 100$$

What a difference the bypass capacitor makes!

PRACTICE EXERCISE 6–4

Determine the base-to-collector voltage gain (unloaded) in Figure 6–11 with R_E bypassed, for the following circuit values: $R_C = 1.8\ \text{k}\Omega$, $R_E = 1\ \text{k}\Omega$, $R_1 = 33\ \text{k}\Omega$, and $R_2 = 6.8\ \text{k}\Omega$.

EFFECT OF AN AC LOAD ON VOLTAGE GAIN

When the load R_L is connected to the output through the coupling capacitor C_2, as shown in Figure 6–20(a), the collector resistance at the signal frequency is effectively R_C in parallel with R_L. Remember, the upper end of R_C is effectively at ac ground. The ac equivalent circuit is shown in Figure 6–20(b). The total ac collector resistance is

$$R_c = \frac{R_C R_L}{R_C + R_L} \tag{6–21}$$

Replacing R_C with R_c in the voltage gain expression gives

$$A_v = \frac{R_c}{r_e} \tag{6–22}$$

Since $R_c < R_C$, the voltage gain is reduced. Of course, if $R_L >> R_C$, then $R_c \cong R_C$. So the load has very little effect on the gain.

FIGURE 6–20
A common-emitter amplifier with an ac (capacitively) coupled load.

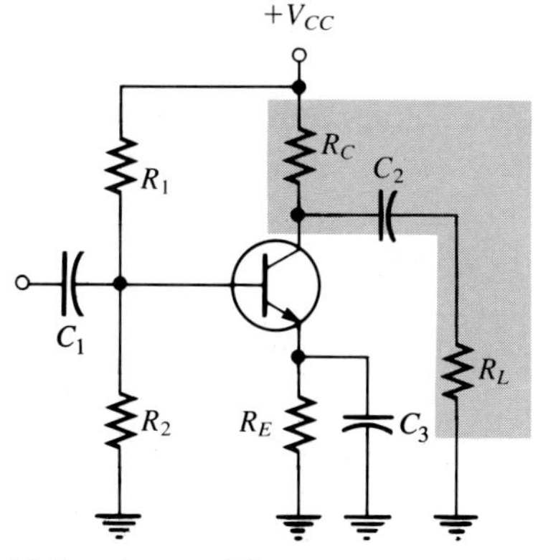

(a) Complete amplifier

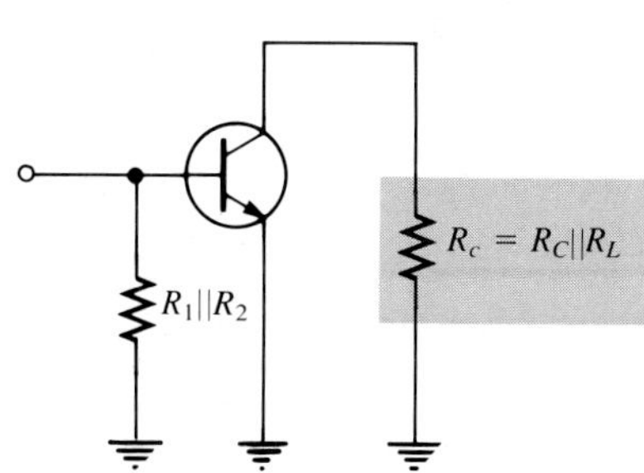

(b) ac equivalent ($X_{C2} = 0$)

EXAMPLE 6–5

Calculate the base-to-collector voltage gain of the amplifier in Figure 6–11 when the load resistance is 5.1 kΩ. Assume the emitter is effectively bypassed. Recall that $r_e = 10\ \Omega$.

SOLUTION

The ac collector resistance is

$$R_c = \frac{R_C R_L}{R_C + R_L} = \frac{(1\ \text{k}\Omega)(5.1\ \text{k}\Omega)}{6.1\ \text{k}\Omega} = 836\ \Omega$$

Therefore,

$$A_v = \frac{R_c}{r_e} = \frac{836\ \Omega}{10\ \Omega} = 83.6$$

The unloaded gain was 100 in Example 6–4.

PRACTICE EXERCISE 6–5

Determine the base-to-collector voltage gain in Figure 6–11 when a 10 kΩ load resistance is connected from collector to ground. Change the resistance values as follows: $R_C = 1.8\ \text{k}\Omega$, $R_E = 1\ \text{k}\Omega$, $R_1 = 33\ \text{k}\Omega$, and $R_2 = 6.8\ \text{k}\Omega$. The emitter resistor is bypassed.

GAIN STABILITY

The internal emitter resistance of a transistor, r_e, varies considerably with temperature, and since $A_v = R_c/r_e$, the voltage gain also varies. To achieve gain **stability,** a technique of *partially* bypassing R_E is often used to minimize the dependency on r_e and thus on temperature. Figure 6–21 shows that only a portion of the total emitter resistor is bypassed by C_3. The effect of this method is to reduce the gain but make it essentially independent of variations in r_e. The voltage gain for the circuit in Figure 6–21 is

$$A_v = \frac{R_c}{r_e + R_{E1}} \tag{6–23}$$

If $R_{E1} >> r_e$, then $A_v \cong R_c/R_{E1}$.

The total emitter resistance $R_{E1} + R_{E2}$ is used in dc biasing, but only R_{E1} is used in the ac operation to "swamp out" r_e, thus making its effect on circuit operation and particularly the voltage gain negligible.

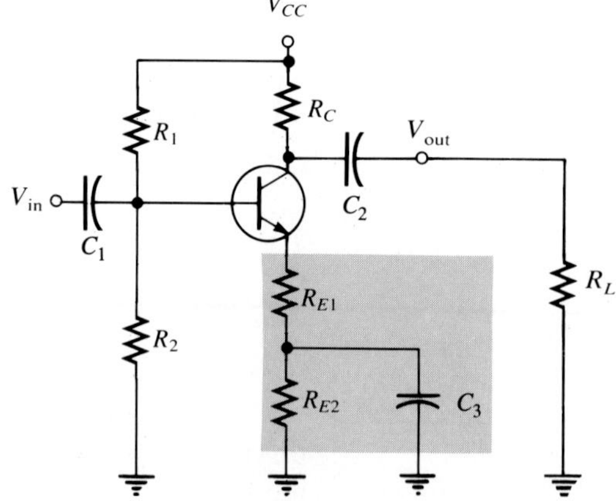

FIGURE 6–21
Partially bypassing the emitter resistance for gain stability.

PHASE INVERSION

The output voltage at the collector of a common-emitter amplifier is 180° out of phase with the input voltage at the base. The phase inversion is sometimes indicated by a negative sign in front of voltage gain, $-A_v$. The next example pulls together the concepts covered so far as they relate to the common-emitter amplifier.

EXAMPLE 6–6

For the amplifier in Figure 6–22, determine the total collector voltage (dc and ac).

SOLUTION

We will first determine the dc bias values. Refer to the dc equivalent circuit in Figure 6–23.

$$R_{IN(base)} = \beta_{dc}(R_{E1} + R_{E2}) = 150(940\ \Omega) = 141\ k\Omega$$

Since $R_{IN(base)}$ is approximately fourteen times larger than R_2, we will neglect it in the dc base voltage calculation.

$$V_B = \left(\frac{10\ k\Omega}{47\ k\Omega + 10\ k\Omega}\right)10\ V = 1.75\ V$$

$$V_E = V_B - 0.7\ V = 1.75\ V - 0.7\ V = 1.05\ V$$

$$I_E = \frac{V_E}{R_{E1} + R_{E2}} = \frac{1.05}{940\ \Omega} = 1.12\ mA$$

$$V_C = V_{CC} - I_C R_C = 10\ V - (1.12\ mA)(4.7\ k\Omega) = 5.26\ V$$

Next is the ac analysis, based on the ac equivalent circuit in Figure 6–24. The first thing to do in the ac analysis is calculate r_e.

$$r_e \cong \frac{25\ mV}{I_E} = \frac{25\ mV}{1.12\ mA} = 22.32\ \Omega$$

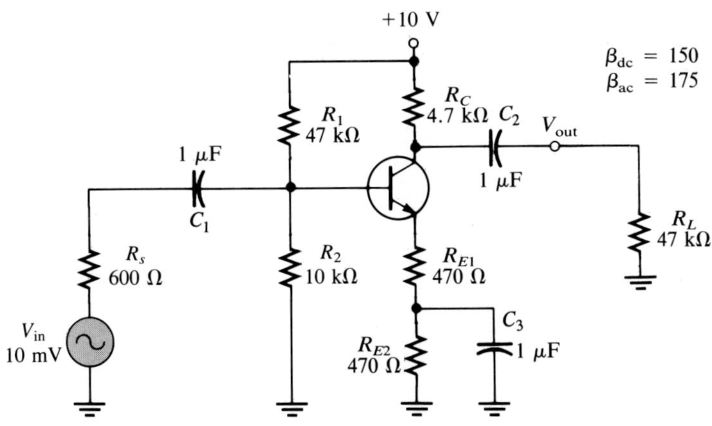

FIGURE 6–22

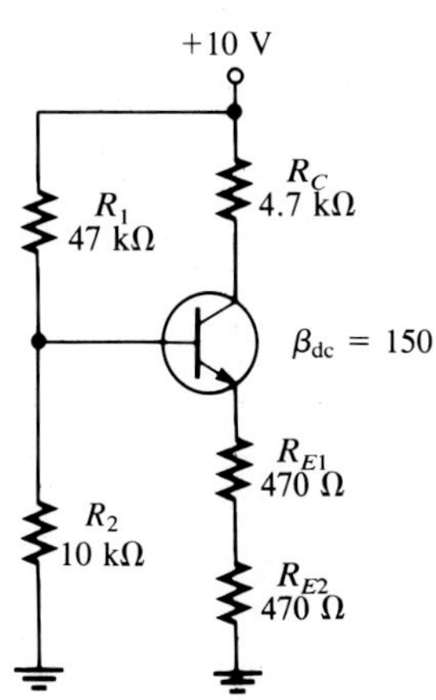

FIGURE 6–23
DC equivalent for the circuit in Figure 6–22.

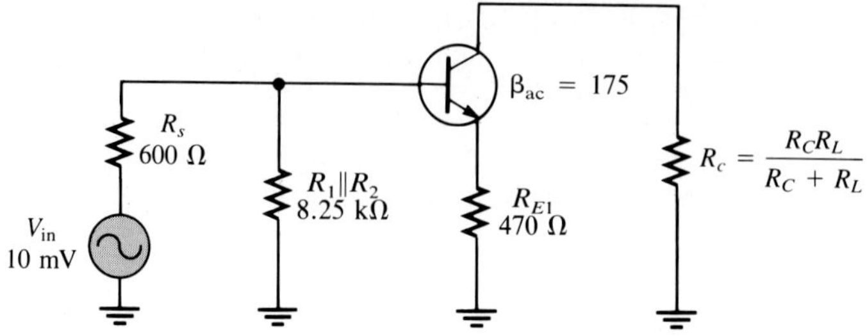

FIGURE 6–24
AC equivalent for the circuit in Figure 6–22.

Next, we determine the attenuation in the base circuit. Looking from the 600-Ω source, the total R_{in} is

$$R_{in} = R_1 \| R_2 \| R_{in(base)}$$

$$R_{in(base)} = \beta_{ac}(r_e + R_{E1}) = 175(492.32\ \Omega) \cong 86.16\ \text{k}\Omega$$

Therefore,

$$R_{in} = 47\ \text{k}\Omega \| 10\ \text{k}\Omega \| 86.16\ \text{k}\Omega = 7.53\ \text{k}\Omega$$

The attenuation from input to base is

$$\frac{V_b}{V_{in}} = \frac{R_{in}}{R_s + R_{in}} = \frac{7.53\ \text{k}\Omega}{600\ \Omega + 7.53\ \text{k}\Omega} = 0.93$$

Before A_v can be determined, we must know R_c.

$$R_c = \frac{R_C R_L}{R_C + R_L} = \frac{(4.7\ \text{k}\Omega)(47\ \text{k}\Omega)}{4.7\ \text{k}\Omega + 47\ \text{k}\Omega} = 4.27\ \text{k}\Omega$$

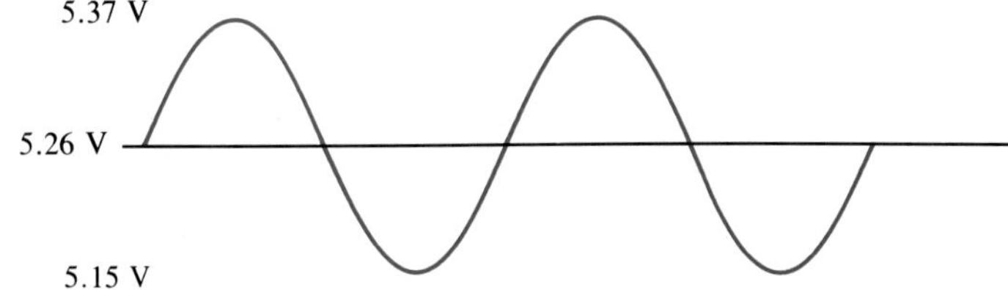

(a) Collector voltage

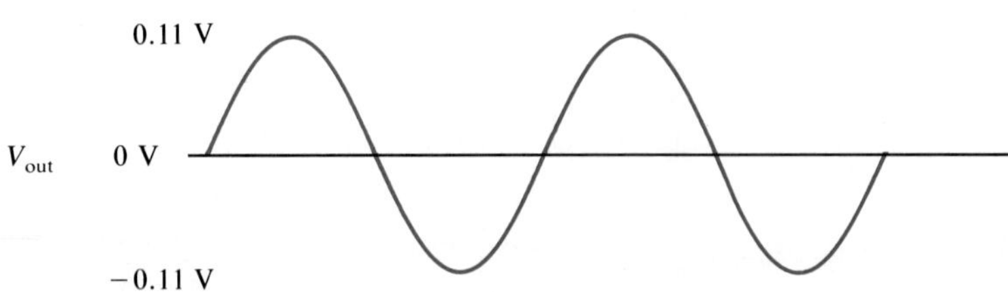

(b) Output voltage

FIGURE 6–25
Voltages for Figure 6–22.

The voltage gain from base to collector is

$$A_v = \frac{R_c}{r_e + R_{E1}} = \frac{4.27\ \text{k}\Omega}{492.32\ \Omega} = 8.67$$

The overall gain is

$$A'_v = (0.93)(8.67) = 8.06$$

The source produces 10 mV, so the ac voltage at the collector is

$$V_c = A'_v V_{in} = (8.06)(10\ \text{mV}) = 80.6\ \text{mV}$$

The total collector voltage is the signal of 80.6 mV rms riding on a dc level of 5.26 V, as shown in Figure 6–25(a), where the peak values are shown. The coupling capacitor C_2 keeps the dc level from getting to the output. So, V_{out} is equal to the ac portion of the collector voltage, as indicated in Figure 6–25(b).

PRACTICE EXERCISE 6–6

What is A_v in Figure 6–22 with R_L removed?

CURRENT GAIN

The current gain from base to collector is I_c/I_b or β_{ac}. However, the overall current gain of the amplifier is

$$A_i = \frac{I_c}{I_{in}} \qquad (6\text{–}24)$$

I_{in} is the total current from the source, part of which is base current and part of which is in the bias network ($R_1 \| R_2$), as shown in Figure 6–26. The total input current is

$$I_{in} = \frac{V_{in}}{R_{in}} \qquad (6\text{–}25)$$

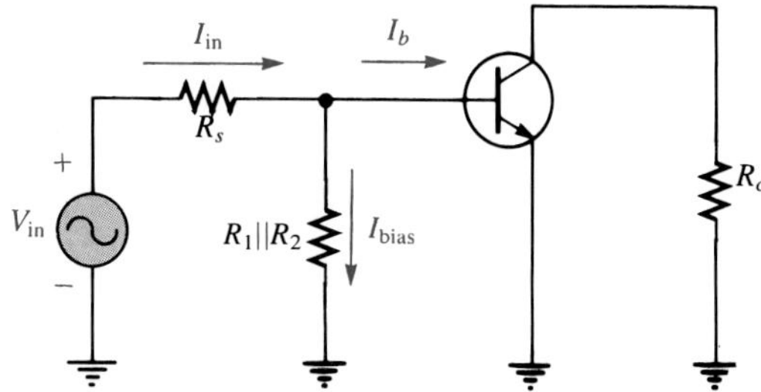

FIGURE 6–26
Total ac input current (directions shown are for the positive half-cycle of V_{in}).

POWER GAIN

The power gain is the product of the overall voltage gain and the current gain.

$$A_p = A'_v A_i \qquad (6\text{–}26)$$

6–3 REVIEW QUESTIONS

1. In the dc equivalent circuit of an amplifier, how are the capacitors treated?
2. When the emitter resistor is bypassed with a capacitor, how is the gain of the amplifier affected?
3. What is the purpose of partially bypassing the emitter resistor?
4. List the elements included in the total input resistance of a common-emitter amplifier.
5. What elements determine the overall voltage gain of a common-emitter amplifier?
6. When a load resistor is capacitively coupled to the collector of a CE amplifier, is the voltage gain increased or decreased?
7. What is the phase relationship of the input and output voltages of a CE amplifier?

6–4 COMMON-COLLECTOR AMPLIFIERS

The common-collector (CC) amplifier is usually referred to as an emitter-follower. The input is applied to the base through a coupling capacitor, and the output is at the emitter. There is no collector resistor. The voltage gain of a CC amplifier is approximately 1, and its main advantage is its high input resistance, as you will see.

Figure 6–27 shows an emitter-follower circuit with voltage-divider bias.

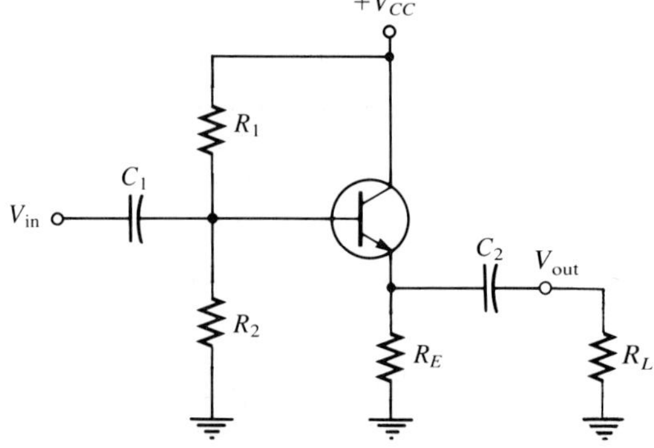

FIGURE 6–27
Emitter-follower with voltage-divider bias.

VOLTAGE GAIN

As in all amplifiers, the voltage gain is $A_v = V_{out}/V_{in}$. For the emitter-follower, V_{out} is I_eR_e and V_{in} is $I_e(r_e + R_e)$, as shown in Figure 6–28. Therefore, the voltage gain is $I_eR_e/I_e(r_e + R_e)$. The currents cancel, and the base-to-emitter voltage gain expression simplifies to

$$A_v = \frac{R_e}{r_e + R_e} \tag{6–27}$$

where R_e is the parallel combination of R_E and R_L. Notice here that the gain is always slightly less than 1. If $R_e >> r_e$, then a good approximation is $A_v \cong 1$. Since the output voltage is at the emitter, it is in phase with the base or input voltage, so there is no inversion. Because of this and because the voltage gain is approximately 1, the output voltage closely follows the input voltage; thus the term **emitter-follower.**

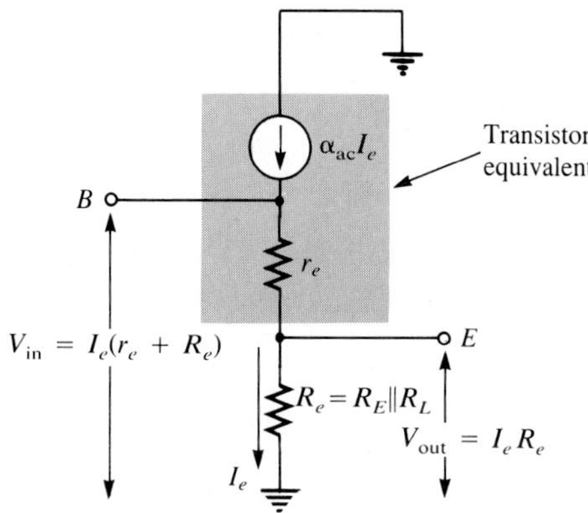

FIGURE 6–28
Emitter-follower model for voltage gain derivation.

INPUT RESISTANCE

The emitter-follower is characterized by a high input resistance; this is what makes it a very useful circuit. Because of the high input resistance, it can be used as a buffer to minimize loading effects when one circuit is driving another. The derivation of the input resistance looking in at the base is similar to that for the common-emitter amplifier. In a common-collector circuit, however, the emitter resistor is *never* bypassed because the output is taken across R_e.

$$R_{in(base)} = \frac{V_b}{I_b} = \frac{I_e(r_e + R_e)}{I_b}$$

$$\cong \frac{\beta_{ac} I_b(r_e + R_e)}{I_b}$$

$$R_{in(base)} \cong \beta_{ac}(r_e + R_e) \quad \textbf{(6–28)}$$

If $R_e >> r_e$, then the input resistance at the base is

$$R_{in(base)} \cong \beta_{ac} R_e \quad \textbf{(6–29)}$$

The bias resistors in Figure 6–27 appear in parallel with $R_{in(base)}$ looking from the input source, just as in the common-emitter circuit.

$$R_{in} = R_1 \| R_2 \| R_{in(base)} \quad \textbf{(6–30)}$$

OUTPUT RESISTANCE

With the load removed, the output resistance for the emitter-follower is approximated as follows:

$$R_{out} \cong \left(\frac{R_s}{\beta_{ac}}\right)\|R_E \qquad \textbf{(6–31)}$$

The derivation of this expression is relatively involved and several simplifying assumptions have been made, as shown in Appendix B. The output resistance is very low, making the emitter-follower useful for driving low resistance loads.

CURRENT GAIN

The overall gain for the emitter-follower is I_e/I_{in}. I_{in} can be calculated as V_{in}/R_{in}. If the parallel combination of the bias resistors R_1 and R_2 is much greater than $R_{in(base)}$, then most of the input current goes into the base; thus the current gain of the amplifier approaches the current gain of the transistor, β_{ac}. This is because very little signal current is in the bias resistors. Stated concisely, if

$$R_1\|R_2 >> \beta_{ac}R_e$$

then

$$A_i \cong \beta_{ac} \qquad \textbf{(6–32)}$$

Otherwise,

$$A_i = \frac{I_e}{I_{in}} \qquad \textbf{(6–33)}$$

β_{ac} is the maximum achievable current gain in both common-collector and common-emitter amplifiers.

POWER GAIN

The common-collector power gain is the product of the voltage gain and the current gain. For the emitter-follower, the power gain is approximately equal to the current gain because the voltage gain is approximately one.

$$A_p = A_vA_i$$

Since $A_v \cong 1$,

$$A_p \cong A_i \qquad \textbf{(6–34)}$$

EXAMPLE 6–7 Determine the total input resistance of the emitter-follower in Figure 6–29. Also find the voltage gain, current gain, and power gain. Assume $\beta_{ac} = 175$.

SOLUTION

$$R_e = R_E\|R_L = 1\ \text{k}\Omega\|10\ \text{k}\Omega = 909\ \Omega$$

FIGURE 6–29

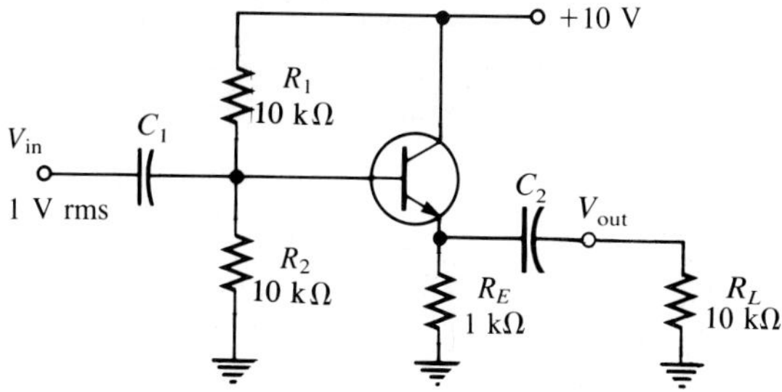

The approximate resistance looking in at the base is

$$R_{in(base)} \cong \beta_{ac}R_e = (175)(909\ \Omega) = 159\ \text{k}\Omega$$

The total input resistance is

$$\begin{aligned} R_{in} &= R_1 \| R_2 \| R_{in(base)} \\ &= 10\ \text{k}\Omega \| 10\ \text{k}\Omega \| 159\ \text{k}\Omega = 4.85\ \text{k}\Omega \end{aligned}$$

The voltage gain is

$$A_v \cong 1$$

By using r_e, a more precise value of A_v can be determined if necessary.

$$I_E = \frac{V_E}{R_E} = \frac{4.3\ \text{V}}{1\ \text{k}\Omega} = 4.3\ \text{mA}$$

and

$$r_e \cong \frac{25\ \text{mV}}{I_E} = \frac{25\ \text{mV}}{4.3\ \text{mA}} = 5.8\ \Omega$$

So,

$$A_v = \frac{R_e}{r_e + R_e} = \frac{909\ \Omega}{914.8\ \Omega} = 0.994$$

The difference is hardly worth the trouble in most cases. The current gain is

$$A_i = \frac{I_e}{I_{in}}$$

$$I_e = \frac{V_e}{R_e} = \frac{A_v V_b}{R_e} \cong \frac{1\ \text{V}}{909\ \Omega} = 1.1\ \text{mA}$$

$$I_{in} = \frac{V_{in}}{R_{in}} = \frac{1\ \text{V}}{4.85\ \text{k}\Omega} = 0.21\ \text{mA}$$

$$A_i = \frac{I_e}{I_{in}} = \frac{1.1\ \text{mA}}{0.21\ \text{mA}} = 5.24$$

The power gain is

$$A_p \cong A_i = 5.24$$

PRACTICE EXERCISE 6–7

Find the input resistance, voltage gain, current gain, and power gain for the circuit in Figure 6–29 with the following values: $V_{CC} = 12$ V, $R_1 = R_2 = 22$ kΩ, $R_E = 1.8$ kΩ, $R_L = 10$ kΩ, and $\beta_{ac} = 100$.

THE DARLINGTON PAIR

As you have seen, β_{ac} is a major factor in determining the input resistance. The β_{ac} of the transistor limits the maximum achievable input resistance you can get from a given emitter-follower circuit.

One way to boost input resistance is to use a **Darlington pair,** as shown in Figure 6–30. The collectors of two transistors are connected, and the emitter of the first drives the base of the second. This configuration achieves β_{ac} multiplication as shown in the following steps. The emitter current of the first transistor is

$$I_{e1} \cong \beta_{ac1} I_b$$

This emitter current becomes the base current for the second transistor, producing a second emitter current of

$$I_{e2} \cong \beta_{ac2} I_{e1}$$
$$I_{e2} \cong \beta_{ac1} \beta_{ac2} I_b$$

Therefore, the effective current gain of the Darlington pair is

$$\beta_{ac} = \beta_{ac1} \beta_{ac2} \quad (6\text{–}35)$$

The input resistance is $\beta_{ac1}\beta_{ac2}R_E$.

The Darlington pair is widely used in audio power amplifiers, high-current motor switches, and other power switching applications.

FIGURE 6–30
A Darlington pair multiplies β_{ac}.

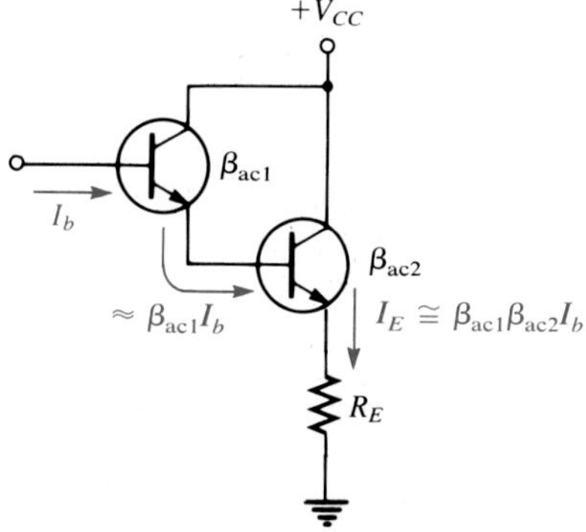

6–4 REVIEW QUESTIONS

1. What is a common-collector amplifier called?
2. What is the ideal maximum voltage gain of a common-collector amplifier?
3. What characteristic of the common-collector amplifier makes it a very useful circuit?

6–5 COMMON-BASE AMPLIFIERS

The common-base (CB) amplifier is the least used of the three basic amplifier configurations. It provides high-voltage gain with no current gain. Since it has a low-input resistance, the CB amplifier is the most appropriate type for certain high-frequency applications where sources tend to have very low-resistance outputs.

A typical common-base amplifier is shown in Figure 6–31. The base is the common terminal and is at ac ground because of capacitor C_2, and the input signal is applied at the emitter. The output is capacitively coupled from the *collector*.

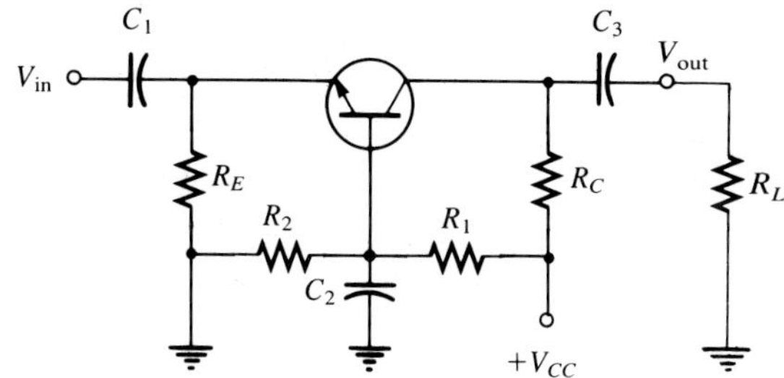

FIGURE 6–31
Common-base amplifier with voltage-divider bias.

VOLTAGE GAIN

The voltage gain from emitter to collector is developed as follows.

$$A_v = \frac{V_c}{V_e} = \frac{I_c R_c}{I_e r_e} \cong \frac{I_e R_c}{I_e r_e}$$

$$A_v \cong \frac{R_c}{r_e} \qquad \textbf{(6–36)}$$

where $R_c = R_C \| R_L$. Notice that the gain expression is the same as for the common-emitter amplifier. There is no phase inversion from emitter to collector.

INPUT RESISTANCE

The resistance looking in at the emitter is

$$R_{in(emitter)} = \frac{V_{in}}{I_{in}} = \frac{V_e}{I_e} = \frac{I_e r_e}{I_e}$$

$$R_{in(emitter)} = r_e \qquad \textbf{(6–37)}$$

R_E, of course, appears in parallel with $R_{in(emitter)}$; however, r_e is normally so small compared to R_E that the expression in Equation (6–37) is valid for the total input resistance also.

OUTPUT RESISTANCE

Looking into the collector and base terminals, the ac collector resistance, r_c, appears in parallel with R_C. Since r_c is typically much larger than R_C, a good approximation for the output resistance is

$$R_{out} \cong R_C \tag{6-38}$$

CURRENT GAIN

The current gain is the output current divided by the input current. I_c is the ac output current, and I_e is the ac input current. Since $I_c \cong I_e$, the current gain is approximately 1.

$$A_i \cong 1 \tag{6-39}$$

POWER GAIN

Since current gain is approximately 1 for the common-base amplifier, the power gain is approximately equal to the voltage gain.

$$A_p \cong A_v \tag{6-40}$$

EXAMPLE 6–8 Find the input resistance, voltage gain, current gain, and power gain for the amplifier in Figure 6–32. $\beta_{dc} = 250$.

FIGURE 6–32

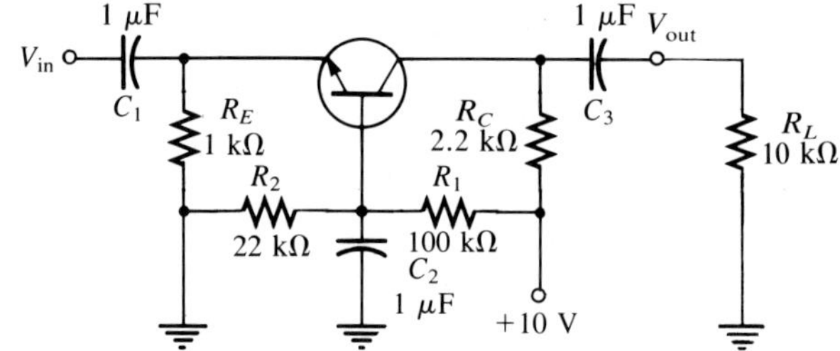

SOLUTION

First, we find I_E so that r_e can be found. Then $R_{in} \cong r_e$. If $\beta_{dc}R_E >> R_2$, then

$$V_B \cong \left(\frac{R_2}{R_1 + R_2}\right)V_{CC} = \left(\frac{22\text{ k}\Omega}{122\text{ k}\Omega}\right)10\text{ V} = 1.8\text{ V}$$

$$V_E = V_B - 0.7\text{ V} = 1.8\text{ V} - 0.7\text{ V} = 1.1\text{ V}$$

$$I_E = \frac{V_E}{R_E} = \frac{1.1\text{ V}}{1\text{ k}\Omega} = 1.1\text{ mA}$$

Therefore,

$$R_{in} \cong r_e = \frac{25\text{ mV}}{1.1\text{ mA}} = 22.73\ \Omega$$

The ac voltage gain is

$$R_c = R_C \| R_L = 2.2\ \text{k}\Omega \| 10\ \text{k}\Omega = 1.8\ \text{k}\Omega$$

$$A_v = \frac{R_c}{r_e} = \frac{1.8\ \text{k}\Omega}{22.73\ \Omega} = 79.2$$

Also, $A_i = 1$ and $A_p \cong A_v = 79.2$.

PRACTICE EXERCISE 6–8

Find A_v in Figure 6–32 if $\beta_{dc} = 50$.

SUMMARY OF THE THREE AMPLIFIER CONFIGURATIONS

Table 6–5 summarizes the important characteristics of each amplifier. Relative values are also indicated for a general comparison.

TABLE 6–5

Comparison of amplifier configurations. The current gains and input/output resistances, with bias resistors neglected, indicate maximum achievable values.

	Common-Emitter (CE)	Common-Collector (CC)	Common-Base (CB)
Voltage gain A_v	R_c/r_e High	1 Low	R_c/r_e High
Maximum current gain $A_{i(max)}$	β_{ac} High	β_{ac} High	1 Low
Power gain A_p	A_iA_v Very high	A_i High	A_v High
Input resistance R_{in}	$\beta_{ac}r_e$ Low	$\beta_{ac}R_e$ High	r_e Very low
Output resistance R_{out}	R_C High	$\left(\frac{R_s}{\beta_{ac}}\right) \| R_E$ Very low	R_C High
Phase relationship of V_{in} and V_{out}	180° Out-of-phase	0° In-phase	0° In-phase

6–5 REVIEW QUESTIONS

1. Can the same voltage gain be achieved with a common-base as with a common-emitter amplifier?
2. Does the common-base amplifier have a very low or a very high input resistance?

6–6 MULTISTAGE AMPLIFIERS

Several amplifiers can be connected in a* cascaded *arrangement with the output of one amplifier driving the input of the next. Each amplifier in the cascaded arrangement is known as a* stage. *The purpose of a* multistage *arrangement is to increase the overall gain.

MULTISTAGE GAIN

The overall gain A'_v of cascaded amplifiers as in Figure 6–33 is the product of the individual gains.

$$A'_v = A_{v1}A_{v2}A_{v3} \cdots A_{vn} \tag{6–41}$$

where n is the number of stages.

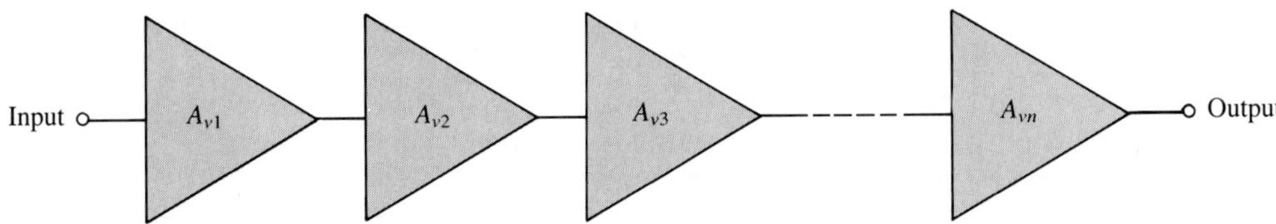

FIGURE 6–33
Cascaded amplifiers. Each triangular symbol represents a separate amplifier.

DECIBEL VOLTAGE GAIN

Amplifier voltage gain is often expressed in **decibels** (dB) as follows:

$$A_v \text{ (dB)} = 20 \log A_v \tag{6–42}$$

This is particularly useful in multistage systems because the overall dB voltage gain is the sum of the individual dB gains.

$$A'_v \text{ (dB)} = A_{v1} \text{ (dB)} + A_{v2} \text{ (dB)} + \cdots + A_{vn} \text{ (dB)} \tag{6–43}$$

EXAMPLE 6–9

A given cascaded amplifier arrangement has the following voltage gains: $A_{v1} = 10$, $A_{v2} = 15$, and $A_{v3} = 20$. What is the overall gain? Also express each gain in dB and determine the total dB voltage gain.

SOLUTION

$$A'_v = A_{v1}A_{v2}A_{v3} = (10)(15)(20) = 3000$$
$$A_{v1} \text{ (dB)} = 20 \log 10 = 20 \text{ dB}$$
$$A_{v2} \text{ (dB)} = 20 \log 15 = 23.52 \text{ dB}$$
$$A_{v3} \text{ (dB)} = 20 \log 20 = 26.02 \text{ dB}$$
$$A'_v \text{ (dB)} = 20 \text{ dB} + 23.52 \text{ dB} + 26.02 \text{ dB} = 69.54 \text{ dB}$$

PRACTICE EXERCISE 6–9

In a certain multistage amplifier, the individual stages have the following voltage gains:

$A_{v1} = 25$, $A_{v2} = 5$, and $A_{v3} = 12$. What is the overall gain? Express each gain in dB and determine the total dB voltage gain.

MULTISTAGE ANALYSIS

For purposes of illustration, the two-stage capacitively coupled amplifier in Figure 6–34 is used. Notice that both stages are identical common-emitter amplifiers with the output of the first stage capacitively coupled to the input of the second stage. Capacitive coupling prevents the dc bias of one stage from affecting that of the other. Notice, also, that the transistors are designated Q_1 and Q_2.

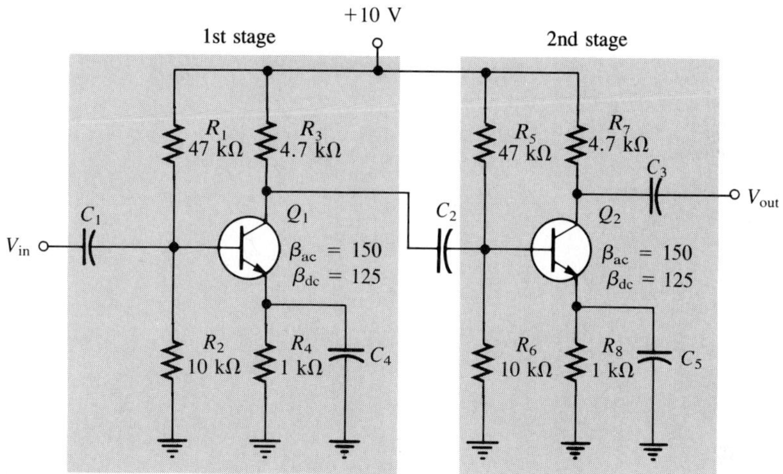

FIGURE 6–34
A two-stage common-emitter amplifier.

LOADING EFFECTS

In determining the gain of the first stage, we must consider the **loading** effect of the second stage. Because the coupling capacitor C_2 appears as a short to the signal frequency, the total input resistance of the second stage presents an ac load to the first stage. Looking from the collector of Q_1, the two biasing resistors, R_5 and R_6, appear in parallel with the input resistance at the base of Q_2. In other words, the signal at the collector of Q_1 "sees" R_3 and R_5, R_6, and $R_{in(base)}$ of the second stage all in parallel to ac ground. Thus the effective ac collector resistance of Q_1 is the total of all these in parallel, as Figure 6–35 illustrates. The voltage gain of the first stage is reduced by the loading of the second stage, because the effective ac collector resistance of the first stage is less than the actual value of its collector resistor, R_3. Remember that $A_v = R_c/r_e$.

VOLTAGE GAIN OF THE FIRST STAGE

The ac collector resistance of the first stage is

$$R_{c1} = R_3 \| R_5 \| R_6 \| R_{in(base2)}$$

Keep in mind that lowercase subscripts denote ac quantities such as for R_c.

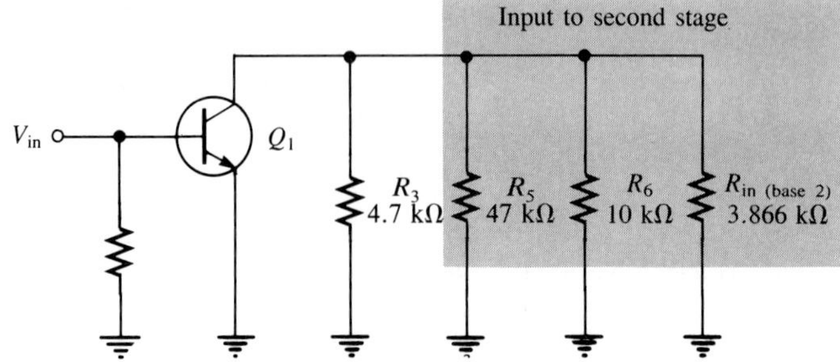

FIGURE 6–35
AC equivalent of first stage in Figure 6–34, showing loading from second stage.

You can verify that $I_E = 1.05$ mA, $r_e = 23.8\ \Omega$, and $R_{in(base2)} = 3.57\ k\Omega$. The effective ac collector resistance of the first stage is as follows:

$$R_{c1} = 4.7\ k\Omega \| 47\ k\Omega \| 10\ k\Omega \| 3.57\ k\Omega$$
$$= 1.63\ k\Omega$$

Therefore, the base-to-collector voltage gain of the first stage is

$$A_{v1} = \frac{R_{c1}}{r_e} = \frac{1.63\ k\Omega}{23.8\ \Omega} = 68.5$$

VOLTAGE GAIN OF THE SECOND STAGE

The second stage has no load resistor, so the ac collector resistance is R_7, and the gain is

$$A_{v2} = \frac{R_7}{r_e} = \frac{4.7\ k\Omega}{23.8\ \Omega} = 197.5$$

Compare this to the gain of the first stage, and notice how much the loading effect of the second stage reduced the gain.

OVERALL VOLTAGE GAIN

The overall amplifier gain with no load on the output is

$$A'_v = A_{v1}A_{v2} = (68.5)(197.5) \cong 13{,}529$$

If an input signal of, say, 100 μV is applied to the first stage and if the attenuation of the input base circuit is neglected, an output from the second stage of (100 μV)(13,529) = 1.3529 V will result. The overall gain can be expressed in dB as follows:

$$A'_v\ (\text{dB}) = 20 \log (13{,}529) = 82.63\ \text{dB}$$

DC VOLTAGE LEVELS IN THE CAPACITIVELY COUPLED MULTISTAGE AMPLIFIER

Since both stages in Figure 6–34 are identical, the dc voltages for Q_1 and Q_2 are the same. The dc base voltage for Q_1 and Q_2 is

$$V_B \cong \left(\frac{R_2}{R_1 + R_2}\right)10\ \text{V} = \left(\frac{10\ k\Omega}{57\ k\Omega}\right)10\ \text{V} = 1.75\ \text{V}$$

The dc emitter and collector voltages are as follows:

$$V_E = V_B - 0.7\text{ V} = 1.05\text{ V}$$

$$I_E = \frac{V_E}{R_4} = \frac{1.05\text{ V}}{1\text{ k}\Omega} = 1.05\text{ mA}$$

$$I_C \cong I_E = 1.05\text{ mA}$$

$$V_C = V_{CC} - I_C R_3 = 10\text{ V} - (1.05\text{ mA})(4.7\text{ k}\Omega) = 5.07\text{ V}$$

DIRECT-COUPLED MULTISTAGE AMPLIFIERS

A basic two-stage, direct-coupled amplifier is shown in Figure 6–36. Notice that there are no coupling or bypass capacitors in this circuit. The dc collector voltage of the first stage provides the base-bias voltage for the second stage. Because of the direct coupling, this type of amplifier has a better low-frequency response than the capacitively coupled type because the reactance of coupling and bypass capacitors at very low frequencies becomes excessive for practical capacitor values. The increased reactance produces signal loss and gain reduction.

FIGURE 6–36
A basic two-stage direct-coupled amplifier.

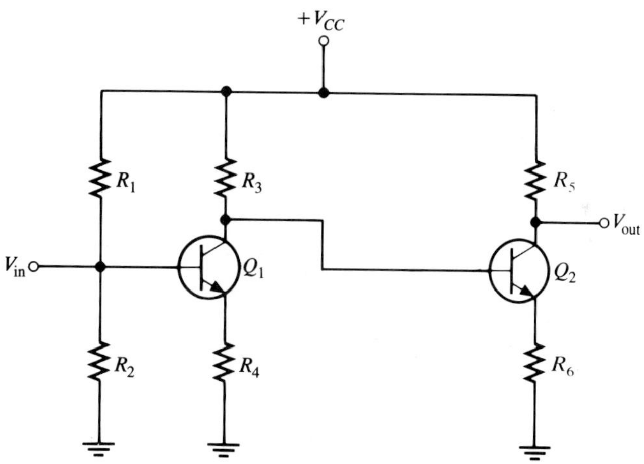

Direct-coupled amplifiers, on the other hand, can be used to amplify low frequencies all the way down to dc (0 Hz) without loss of gain because there are no capacitive reactances in the circuit. The disadvantage of direct-coupled amplifiers is that small changes in the dc bias voltages from temperature effects or power-supply variation are amplified by the succeeding stages, which can result in a significant drift in the dc levels throughout the circuit.

TRANSFORMER-COUPLED MULTISTAGE AMPLIFIERS

A basic transformer-coupled two-stage amplifier is shown in Figure 6–37. Transformer coupling is often used in high-frequency amplifiers such as those in the **RF** (radio frequency) and IF (intermediate frequency) sections of radio and TV receivers. At lower

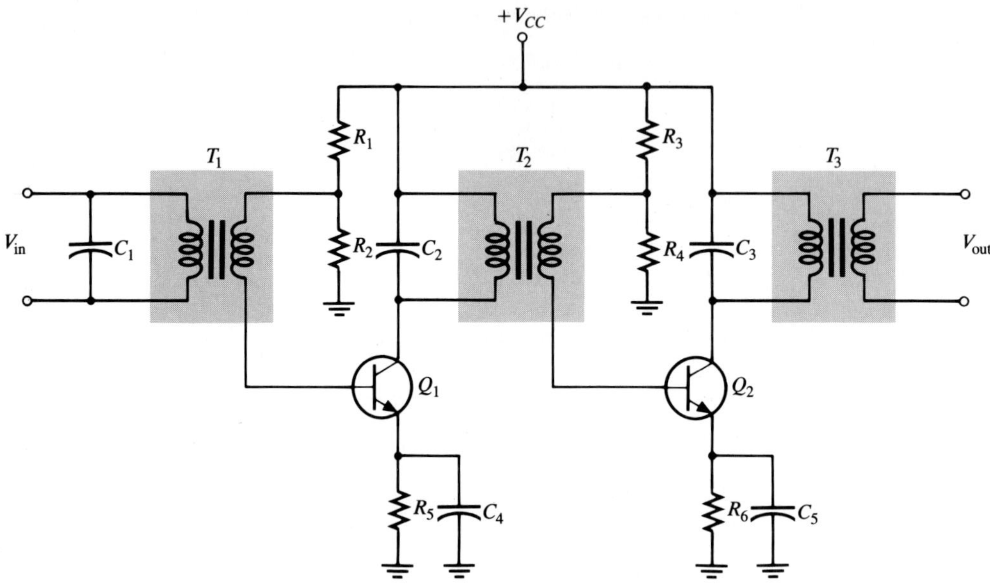

FIGURE 6–37
A basic two-stage transformer-coupled amplifier.

frequency ranges such as **audio,** the size of transformers is usually prohibitive. Capacitors are usually connected across the primary windings of the transformers to obtain resonance and increased selectivity for the band of frequencies to be amplified.

6–6 REVIEW QUESTIONS

1. What does the term *stage* mean?
2. How is the overall gain of a multistage amplifier determined?
3. Express a voltage gain of 500 in dB.

6–7 TROUBLESHOOTING

In working with any circuit, you must first know how it is supposed to work before you can troubleshoot it for a failure. The two-stage capacitively coupled amplifier discussed in Section 6-6 is used to illustrate a typical troubleshooting procedure.

The proper signal levels and dc voltage levels for the capacitively coupled two-stage amplifier (determined in the previous section) are shown in Figure 6–38.

TROUBLESHOOTING PROCEDURE

A basic procedure for troubleshooting called *signal tracing,* which is usually done with an oscilloscope, is illustrated in Figure 6–39 using the two-stage amplifier as an example. This general procedure can be expanded to any number of stages.

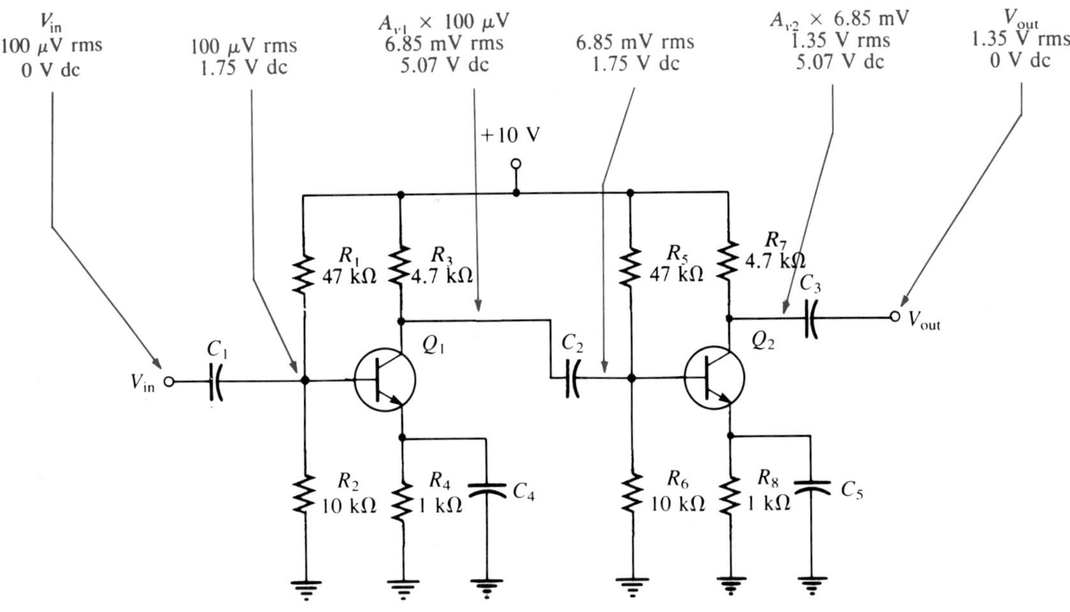

FIGURE 6–38
Two-stage amplifier with proper ac and dc voltage levels indicated.

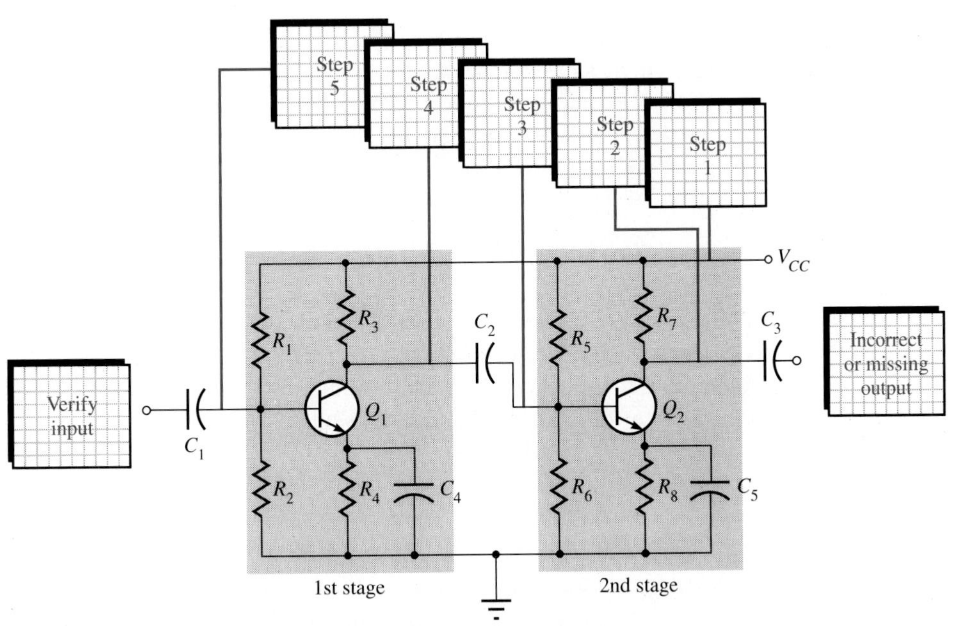

FIGURE 6–39
Basic troubleshooting procedure for a two-stage amplifier with no output signal.

We will begin by assuming that the final output signal of the amplifier has been determined to be missing. We will also assume that there is an input signal and that this has been verified. When troubleshooting, you generally start at the point where the signal is missing and work back point-by-point toward the input until a correct voltage is found. The fault then lies somewhere between the point of the first good voltage check and the missing or incorrect voltage point.

Before you begin checking the voltages, it is usually a good idea to visually check the circuit board or assembly for obvious problems such as broken or poor connections, solder splashes, wire clippings, or burned components.

STEP 1 Check the dc supply voltage. Often something as simple as a blown fuse or the power switch being in the *off* position may be the problem. The circuit will certainly not work without power.

STEP 2 Check the voltage at the collector of Q_2. If the correct signal is present at this point, the coupling capacitor, C_3, is open. If there is no signal at this point, proceed to Step 3.

STEP 3 Check the signal at the base of Q_2. If the signal is present at this point, the fault is in the second stage of the amplifier. First, do an in-circuit check of the transistor. If it is OK, then one of the biasing resistors may be open; and if this is the case, the dc voltages will be incorrect. If there is no signal at this point, proceed to Step 4.

STEP 4 Check the signal at the collector of Q_1. If the signal is present at this point, the coupling capacitor, C_2, is open. If there is no signal at the point, proceed to Step 5.

STEP 5 Check the signal at the base of Q_1. If the signal is present at this point, the fault is in the first stage of the amplifier. First, do an in-circuit check of the transistor. If it is OK, then one of the biasing resistors may be open, and if this is the case, the dc voltages will be incorrect. If there is no signal at this point, the coupling capacitor, C_1, is open because we started out by verifying that there is a correct signal at the input.

6–7 REVIEW QUESTIONS

1. If C_5 in Figure 6–39 were open, how would the output signal be affected? How would the dc level at the collector of Q_2 be affected?
2. If R_5 in Figure 6–39 were open, how would the output signal be affected?
3. If the coupling capacitor C_2 shorted out, would any of the dc voltages in the amplifier be changed? If so, which ones?

6–8 A SYSTEM APPLICATION

The purpose of the audio preamplifier in the receiver system that you saw at the opening of this chapter is to accept a very small audio signal out of the detector circuit and amplify it to a level for input to the power amplifier that provides audio

power to the speaker. In this application, the transistors are used as common-emitter small-signal linear amplifiers. In this section, you will

- ☐ ***See how a small-signal amplifier is used in a system application.***
- ☐ ***See how a two-stage amplifier is used in a system application.***
- ☐ ***See how a potentiometer is used for audio volume control.***
- ☐ ***Translate between a printed circuit board and a schematic.***
- ☐ ***Troubleshoot some common amplifier failures.***

A BRIEF DESCRIPTION OF THE SYSTEM

The block diagram for the AM radio receiver is shown in Figure 6–40. A very basic system description follows. The antenna picks up all radiated signals that pass by and feeds them into the RF amplifier. The voltages induced in the antenna by the electromagnetic radiation are extremely small. The RF amplifier is tuned to select and amplify a desired frequency within the AM broadcast band. Since it is a frequency-selective circuit, the RF amplifier eliminates essentially all but the selected frequency band. The output of the RF amplifier goes to the mixer where it is combined with the output of the local oscillator which is 455 kHz above the selected RF frequency. In the mixer, a nonlinear process called *heterodyning* takes place and produces one frequency that is the sum of the RF and local oscillator frequencies and another frequency that is the difference (always 455 kHz). The sum frequency is filtered out and only the 455 kHz difference frequency is used. This frequency is amplitude modulated just like the much higher RF frequency and, therefore, contains the audio signal. The IF (intermediate frequency) amplifier is tuned to 455 kHz and amplifies the mixer output. The detector takes the amplified 455 kHz AM signal and recovers the audio from it while eliminating the intermediate frequency. The output of the detector is a small audio signal that goes to the audio preamplifier and then to the power amplifier, which drives the speaker to convert the electrical audio signal into sound.

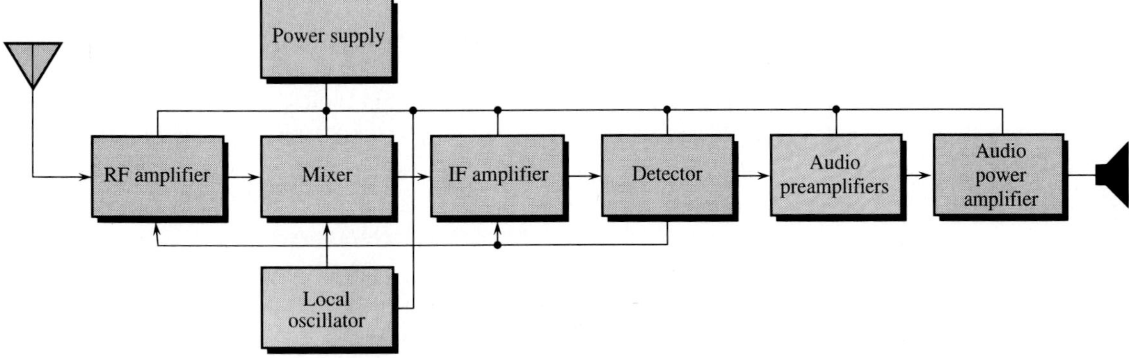

FIGURE 6–40

Now, so that you can take a closer look at the audio preamplifier, let's take it out of the system and put it on the test bench.

ON THE TEST BENCH

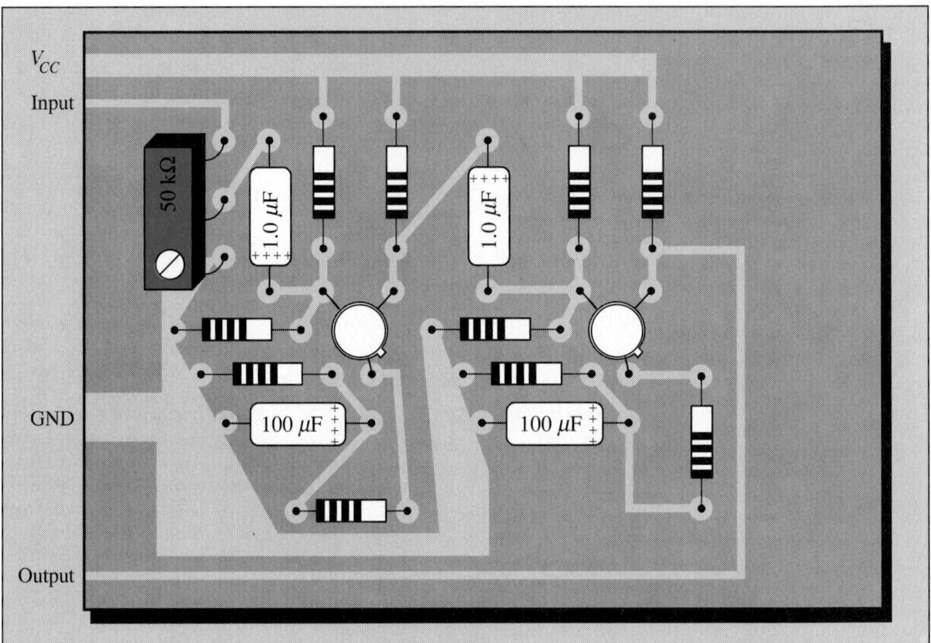

FIGURE 6–41

■ ACTIVITY 1 RELATE THE PC BOARD TO THE SCHEMATIC

Notice that there are no component labels on this pc board in Figure 6–41. Label the components using the schematic in Figure 6–42 as a guide. That is, find and label R_1, R_2, and so forth.

■ ACTIVITY 2 ANALYZE THE AMPLIFIER

Refer to the 2N3947 data sheet in Appendix A and use minimum values of β when available.

STEP 1 Calculate all of the dc voltages in the audio preamp.

STEP 2 Determine the voltage gain of each stage and the overall voltage gain of the two-stage amplifier. Don't forget to take into account the loading effect of the input resistance of the second stage. Assume that the second stage is unloaded and all capacitive reactances are zero.

STEP 3 Specify what you would expect to happen as the frequency is reduced to a sufficiently low value. Why?

STEP 4 Repeat Step 2 with a 15 kΩ load resistor connected to the output of the second stage through a coupling capacitor.

FIGURE 6–42

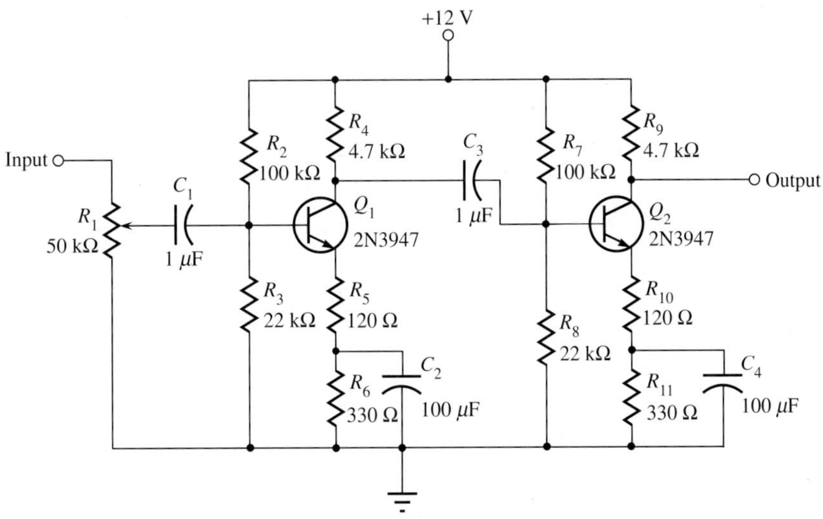

■ **ACTIVITY 3 WRITE A TECHNICAL REPORT**

Discuss the overall operation of the amplifier circuit. State the purpose of each component on the board. Make sure to explain how the potentiometer R_1 is used. Use the results of Activity 2 as appropriate.

■ **ACTIVITY 4 TROUBLESHOOT THE CIRCUIT BOARD FOR EACH OF THE FOLLOWING PROBLEMS BY STATING THE PROBABLE CAUSE OR CAUSES IN EACH CASE**

1. No output signal when there is a verified input signal.
2. Proper voltages at base of Q_2, but signal voltage at the collector is less than it should be.
3. Proper signal at collector of Q_1, but no signal at base of Q_2.
4. Amplitude of output signal much less than it should be.
5. Collector of Q_2 at +12 V dc and no signal voltage with a verified signal at the base.

■ **ACTIVITY 5 TEST BENCH SPECIAL ASSIGNMENT**

Go to Test Bench 3 in the color insert section (which follows page 422) and carry out the assignment that is stated there.

6–8 REVIEW QUESTIONS

1. Explain why R_1 is in the circuit of Figure 6–42.
2. What happens if C_2 opens?
3. What happens if C_3 opens?
4. How can you reduce the gain of each stage without changing the dc voltages?

SUMMARY

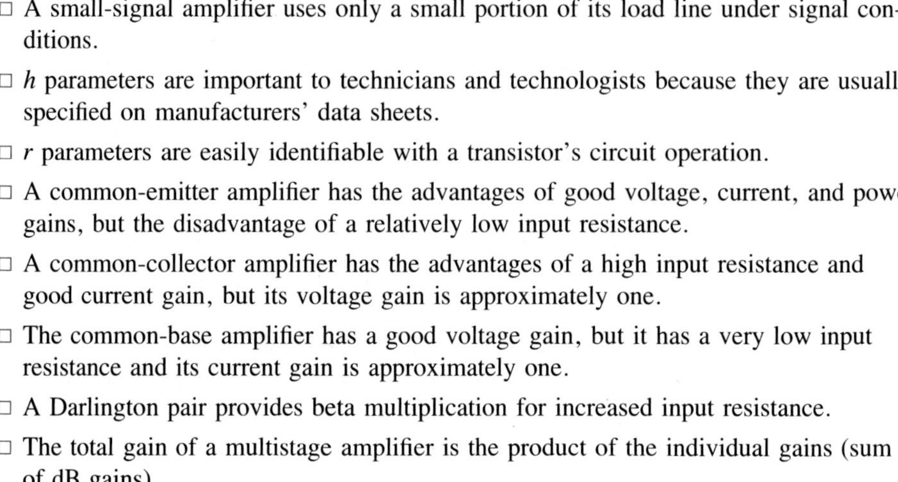

- □ A small-signal amplifier uses only a small portion of its load line under signal conditions.
- □ *h* parameters are important to technicians and technologists because they are usually specified on manufacturers' data sheets.
- □ *r* parameters are easily identifiable with a transistor's circuit operation.
- □ A common-emitter amplifier has the advantages of good voltage, current, and power gains, but the disadvantage of a relatively low input resistance.
- □ A common-collector amplifier has the advantages of a high input resistance and good current gain, but its voltage gain is approximately one.
- □ The common-base amplifier has a good voltage gain, but it has a very low input resistance and its current gain is approximately one.
- □ A Darlington pair provides beta multiplication for increased input resistance.
- □ The total gain of a multistage amplifier is the product of the individual gains (sum of dB gains).

GLOSSARY

Amplitude modulation (AM) A communication method in which a lower frequency signal modulates (varies) the amplitude of a higher frequency signal (carrier).

Attenuation The reduction in the level of power, current, or voltage.

Audio Related to the range of sound waves that can be heard by the human ear.

Cascade An arrangement of circuits in which the output of one circuit becomes the input to the next.

Common-base (CB) A BJT amplifier configuration in which the base is the common (grounded) terminal.

Common-collector (CC) A BJT amplifier configuration in which the collector is the common (grounded) terminal.

Common-emitter (CE) A BJT amplifier configuration in which the emitter is the common (grounded) terminal.

Darlington A configuration of two transistors in which the collectors are connected and the emitter of the first drives the base of the second to achieve beta multiplication.

Decibel (dB) The unit of the logarithmic expression of a ratio, such as power or voltage.

Emitter-follower A popular term for a common-emitter amplifier.

Loading The amount of current drawn from the output of a circuit through a load resistance.

Multistage Characterized by having more than one stage. A cascaded arrangement of two or more amplifiers.

RF Radio frequency.

Stability A measure of how well an amplifier maintains its design values (Q point, gain, etc.) over changes in beta and temperature.

Stage One of the amplifier circuits in a multistage configuration.

FORMULAS

h PARAMETERS

(6–1)	$h_{ie} = \dfrac{V_b}{I_b}$	Input impedance, common-emitter configuration
(6–2)	$h_{re} = \dfrac{V_b}{V_c}$	Reverse voltage ratio, common-emitter configuration
(6–3)	$h_{fe} = \dfrac{I_c}{I_b}$	Current gain, common-emitter configuration
(6–4)	$h_{oe} = \dfrac{I_c}{V_c}$	Output admittance, common-emitter configuration

CONVERSIONS FROM *h* PARAMETERS TO *r* PARAMETERS

(6–5)	$\alpha_{ac} = h_{fb}$	AC alpha
(6–6)	$\beta_{ac} = h_{fe}$	AC beta
(6–7)	$r_e = \dfrac{h_{re}}{h_{oe}}$	AC emitter resistance
(6–8)	$r_c = \dfrac{h_{re} + 1}{h_{oe}}$	AC collector resistance
(6–9)	$r_b = h_{ie} - \dfrac{h_{re}}{h_{oe}}(1 + h_{fe})$	AC base resistance
(6–10)	$r_e \cong \dfrac{25 \text{ mV}}{I_E}$	AC emitter resistance

COMMON-EMITTER

(6–11)	$V_b = \left(\dfrac{R_{in}}{R_s + R_{in}}\right)V_{in}$	Base signal voltage
(6–12)	$R_{in(base)} = \dfrac{V_b}{I_b}$	Input resistance at base
(6–13)	$R_{in(base)} = \beta_{ac}(r_e + R_E)$	Input resistance at base
(6–14)	$R_{in} = R_1 \| R_2 \| R_{in(base)}$	Total input resistance, voltage-divider bias

(6–15)	$R_{out} \cong R_C$	Output resistance
(6–16)	$A_v = \dfrac{V_c}{V_b}$	Voltage gain, base-to-collector
(6–17)	$A_v = \dfrac{R_C}{r_e + R_E}$	Voltage gain, base-to-collector, unloaded, unbypassed R_E
(6–18)	$\dfrac{V_b}{V_{in}} = \dfrac{R_{in}}{R_s + R_{in}}$	Attenuation, base circuit
(6–19)	$A'_v = A_v\left(\dfrac{V_b}{V_{in}}\right)$	Voltage gain, input-to-collector
(6–20)	$A_v = \dfrac{R_C}{r_e}$	Voltage gain, base-to-collector, unloaded, bypassed R_E
(6–21)	$R_c = \dfrac{R_C R_L}{R_C + R_L}$	AC collector resistance
(6–22)	$A_v = \dfrac{R_c}{r_e}$	Voltage gain, base-to-collector, loaded, bypassed R_E
(6–23)	$A_v = \dfrac{R_c}{r_e + R_{E1}}$	Voltage gain, base-to-collector, partially bypassed R_E (swamping)
(6–24)	$A_i = \dfrac{I_c}{I_{in}}$	Current gain
(6–25)	$I_{in} = \dfrac{V_{in}}{R_{in}}$	Input signal current
(6–26)	$A_p = A'_v A_i$	Power gain

COMMON-COLLECTOR

(6–27)	$A_v = \dfrac{R_e}{r_e + R_e}$	Voltage gain, base-to-emitter
(6–28)	$R_{in(base)} \cong \beta_{ac}(r_e + R_e)$	Input resistance at base
(6–29)	$R_{in(base)} \cong \beta_{ac} R_e$	Input resistance at base when $r_e << R_e$
(6–30)	$R_{in} = R_1 \| R_2 \| R_{in(base)}$	Total input resistance, voltage-divider biase
(6–31)	$R_{out} \cong \left(\dfrac{R_s}{\beta_{ac}}\right) \| R_E$	Output resistance
(6–32)	$A_i = \beta_{ac}$	Current gain when $R_1 \| R_2 >> \beta R_e$
(6–33)	$A_i = \dfrac{I_e}{I_{in}}$	Current gain

(6–34)	$A_p \cong A_i$	Power gain
(6–35)	$\beta_{ac} = \beta_{ac1}\beta_{ac2}$	Darlington pair

COMMON-BASE

(6–36)	$A_v \cong \dfrac{R_c}{r_e}$	Voltage gain, emitter-to-collector
(6–37)	$R_{in(emitter)} = r_e$	Input resistance at emitter
(6–38)	$R_{out} \cong R_C$	Output resistance
(6–39)	$A_i \cong 1$	Current gain
(6–40)	$A_p \cong A_v$	Power gain

MULTISTAGE AMPLIFIER

(6–41)	$A'_v = A_{v1}A_{v2}A_{v3} \cdots A_{vn}$	Overall gain
(6–42)	$A_v\ (\text{dB}) = 20 \log A_v$	dB voltage gain
(6–43)	$A'_v\ (\text{dB}) = A_{v1}\ (\text{dB}) + A_{v2}\ (\text{dB}) + \cdots + A_{vn}\ (\text{dB})$	Overall dB voltage again

SELF-TEST

1. A small-signal amplifier

(a) uses only a small portion of its load line

(b) always has an output signal in the mV range

(c) goes into saturation once on each input cycle

(d) is always a common-emitter amplifier

2. The parameter h_{fe} corresponds to

(a) β_{dc} **(b)** β_{ac} **(c)** r_e **(d)** r_c

3. If the dc emitter current in a certain transistor amplifier is 3 mA, the approximate value of r_e is

(a) 3 kΩ **(b)** 3 Ω **(c)** 8.33 Ω **(d)** 0.33 kΩ

4. A certain common-emitter amplifier has a voltage gain of 100. If the emitter bypass capacitor is removed,

(a) the circuit will become unstable **(b)** the voltage gain will decrease

(c) the voltage gain will increase **(d)** the Q point will shift

5. For a common-collector amplifier, $R_E = 100\ \Omega$, $r_e = 10\ \Omega$, and $\beta_{ac} = 150$. The ac input resistance at the base is

(a) 1500 Ω **(b)** 15 kΩ **(c)** 110 Ω **(d)** 16.5 kΩ

6. If a 10 mV signal is applied to the base of the common-collector circuit in Question 5, the output signal is approximately

(a) 100 mV **(b)** 150 mV **(c)** 1.5 V **(d)** 10 mV

7. For a common-emitter amplifier, $R_C = 1\ \text{k}\Omega$, $R_E = 390\ \Omega$, $r_e = 15\ \Omega$, and $\beta_{ac} = 75$. Assuming that R_E is completely bypassed at the operating frequency, the voltage gain is

   (a) 66.67 **(b)** 2.56 **(c)** 2.47 **(d)** 75

8. In the circuit of Question 7, if the frequency is reduced to the point where $X_{C(bypass)} = R_E$, the voltage gain

   (a) remains the same **(b)** is less **(c)** is greater

9. In a certain common-collector circuit, the current gain is 50. The power gain is approximately

   (a) $50A_v$ **(b)** 50 **(c)** 1 **(d)** a and b

10. In a Darlington configuration, each transistor has an ac beta of 125. If R_E is 560 Ω, the input resistance is

    (a) 560 Ω **(b)** 70 kΩ **(c)** 8.75 MΩ **(d)** 140 kΩ

11. The input resistance of a common-base amplifier is

    (a) very low **(b)** very high

    (c) the same as a CE **(d)** the same as a CC

12. In a common-emitter amplifier with voltage-divider bias, $R_{in(base)} = 68\ \text{k}\Omega$, $R_1 = 33\ \text{k}\Omega$, and $R_2 = 15\ \text{k}\Omega$. The total input resistance is

    (a) 68 kΩ **(b)** 8.95 kΩ **(c)** 22.2 kΩ **(d)** 12.29 kΩ

13. A CE amplifier is driving a 10 kΩ load. If $R_C = 2.2\ \text{k}\Omega$ and $r_e = 10\ \Omega$, the voltage gain is approximately

    (a) 220 **(b)** 1000 **(c)** 10 **(d)** 180

14. Each stage of a four-stage amplifier has a voltage gain of 15. The overall voltage gain is

    (a) 60 **(b)** 15 **(c)** 50,625 **(d)** 3078

15. The overall gain found in Question 14 can be expressed in decibels as

    (a) 94.09 dB **(b)** 47.04 dB **(c)** 35.56 dB **(d)** 69.77 dB

PROBLEMS

SECTION 6–1 SMALL-SIGNAL AMPLIFIER OPERATION

1. What is the lowest value of dc collector current to which a transistor having the characteristic curves in Figure 6–3 can be biased and still retain linear operation with a peak-to-peak base current swing of 20 μA?
2. What is the highest value of I_C under the conditions described in Problem 1?

SECTION 6–2 TRANSISTOR AC EQUIVALENT CIRCUITS

3. What are the hybrid parameters that can be measured with each of the test circuits in Figure 6–43, and what is the value of each?

FIGURE 6–43

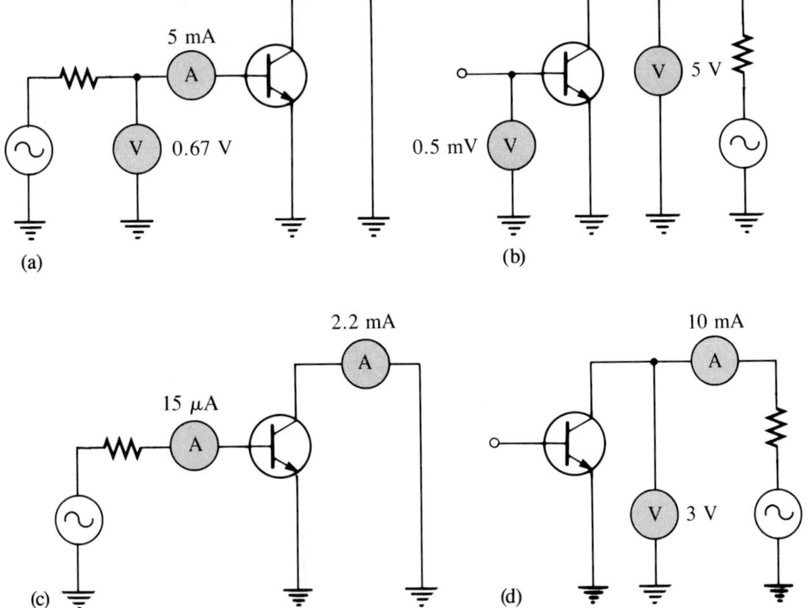

4. Use the h-parameter values obtained in Problem 3 to determine the r parameters β_{ac}, α_{ac}, r_e, r_b, and r_c.
5. A certain transistor has a dc beta (h_{FE}) of 130. If the dc base current is 10 μA, determine r_e. $\alpha_{dc} = 0.99$.
6. At the dc bias point of a certain transistor circuit, $I_B = 15\ \mu$A and $I_C = 2$ mA. Also, a variation in I_B of 3 μA about the Q point produces a variation in I_C of 0.35 mA about the Q point. Determine β_{dc} and β_{ac}.

SECTION 6–3 COMMON-EMITTER AMPLIFIERS

7. Draw the dc equivalent circuit and the ac equivalent circuit for the unloaded amplifier in Figure 6–44.

FIGURE 6–44

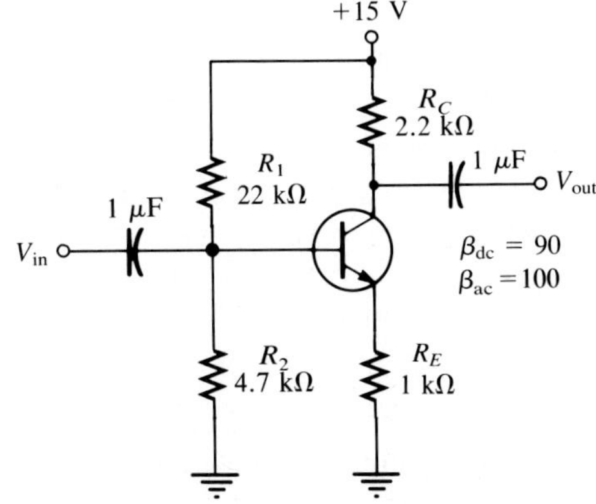

8. Determine the following values for the amplifier in Figure 6–44:
   (a) $R_{in(base)}$ (b) R_{in} (c) A_v
9. Connect a bypass capacitor across R_E in Figure 6–44, and repeat Problem 8.
10. Connect a 10 kΩ load resistor to the output in Figure 6–44, and repeat Problem 9.
11. Determine the following dc values for the amplifier in Figure 6–45.
    (a) V_B (b) V_E
    (c) I_E (d) I_C
    (e) V_E (f) V_{CE}

FIGURE 6–45

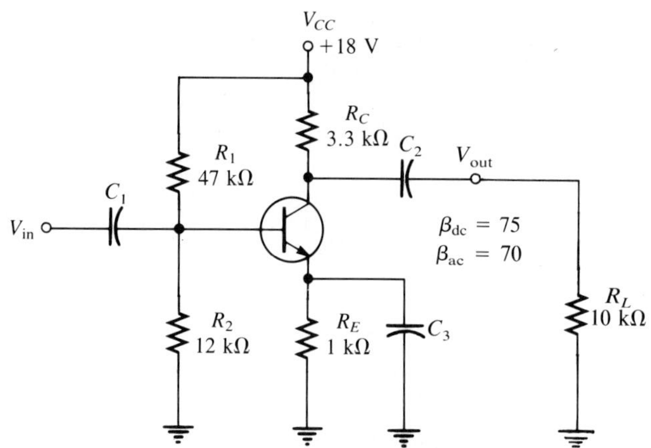

12. Determine the following ac values for the amplifier in Figure 6–45.
    (a) $R_{in(base)}$ (b) R_{in} (c) A_v
    (d) A_i (e) A_p
13. Assume that a 600 Ω, 12 μV rms voltage source is driving the amplifier in Figure 6–45. Determine the overall gain by taking into account the attenuation in the base circuit, and find the *total* output voltage (ac and dc). What is the phase relationship of the collector signal voltage to the base signal voltage?
14. The amplifier in Figure 6–46 has a variable gain control, using a 100 Ω potentiometer for R_E with the wiper ac-grounded. As the potentiometer is adjusted, more or less of R_E is bypassed to ground, thus varying the gain. The total R_E remains constant to dc, keeping the bias fixed. Determine the maximum and minimum gains for this unloaded amplifier.
15. If a load resistance of 600 Ω is placed on the output of the amplifier in Figure 6–46, what are the maximum and minimum gains?
16. Find the overall maximum voltage gain for the amplifier in Figure 6–46 with a 1 kΩ load, if it is being driven by a 300 Ω source.

FIGURE 6–46

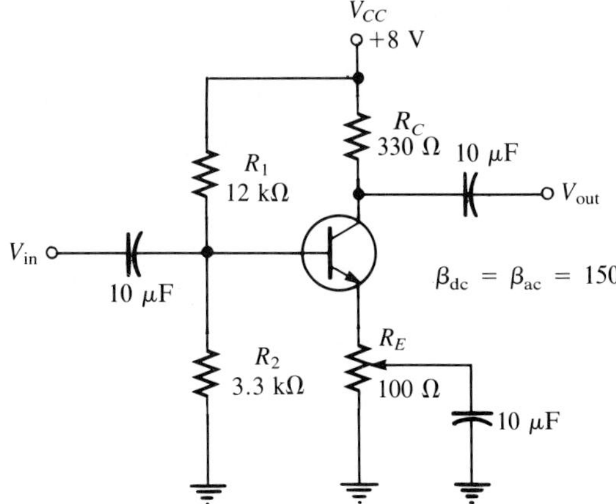

17. Modify the schematic to show how you would "swamp out" the temperature effects of r_e in Figure 6–45 by making R_e at least ten times larger than r_e. Keep the same total R_E. How does this affect the voltage gain?

SECTION 6–4 COMMON-COLLECTOR AMPLIFIERS

18. Determine the *exact* voltage gain for the unloaded emitter-follower in Figure 6–47.

FIGURE 6–47

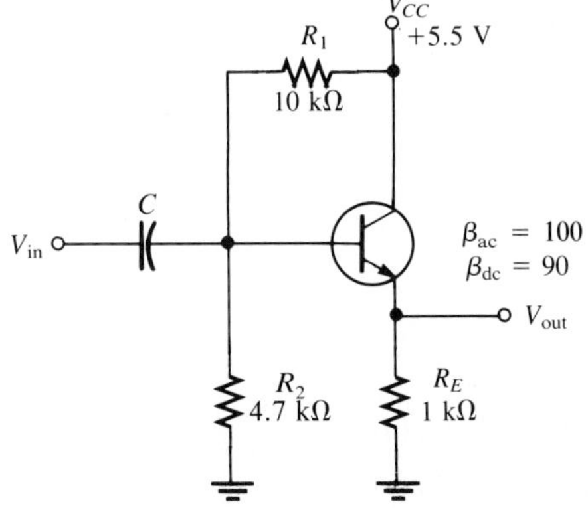

19. What is the total input resistance in Figure 6–47? What is the dc output voltage?
20. A load resistance is capacitively coupled to the emitter in Figure 6–47. In terms of signal operation, the load appears in parallel with R_E and reduces the effective emitter resistance. How does this affect the voltage gain?
21. In problem 20, what value of R_L will cause the voltage gain to drop to 0.9?
22. For the circuit in Figure 6–48, determine the following:
    (a) Q_1 and Q_2 dc terminal voltages (b) Overall β_{ac}
    (c) r_e for each transistor (d) Total input resistance
23. Find the overall current gain A_i in Figure 6–48.

FIGURE 6–48

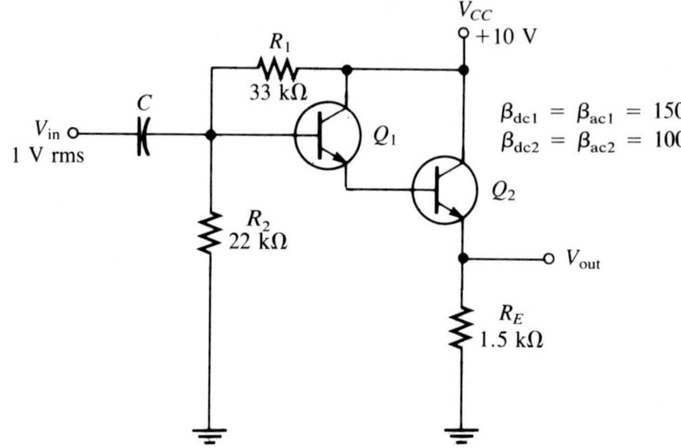

SECTION 6–5 COMMON-BASE AMPLIFIERS

24. What is the main disadvantage of the common-base amplifier compared to the common-emitter and the emitter-follower amplifiers?

25. Find $R_{in(emitter)}$, A_v, A_i, and A_p for the unloaded amplifier in Figure 6–49.

FIGURE 6–49

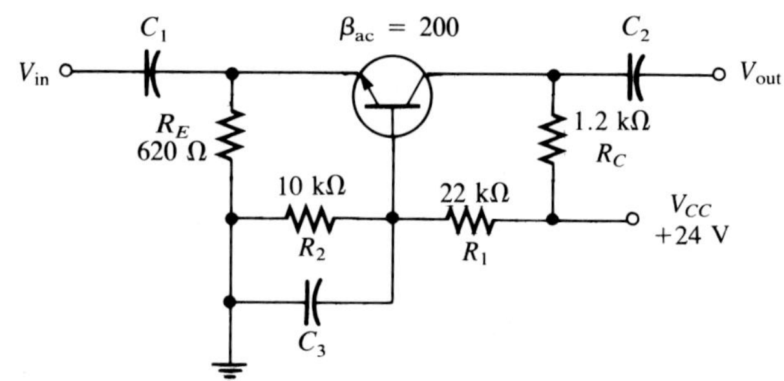

26. Match the following generalized characteristics with the appropriate amplifier configuration.

(a) Unity current gain, good voltage gain, very low input resistance

(b) Good current gain, good voltage gain, low input resistance

(c) Good current gain, unity voltage gain, high input resistance

SECTION 6–6 MULTISTAGE AMPLIFIERS

27. Each of two cascaded amplifier stages has an $A_v = 20$. What is the overall gain?

28. Each of three cascaded amplifier stages has a dB voltage gain of 10 dB. What is the overall dB voltage gain? What is the actual overall voltage gain?

29. For the two-stage, capacitively coupled amplifier in Figure 6–50, find the following values:

(a) Voltage gain of each stage **(b)** Overall voltage gain

(c) Express the gains found in (a) and (b) in dB.

FIGURE 6–50

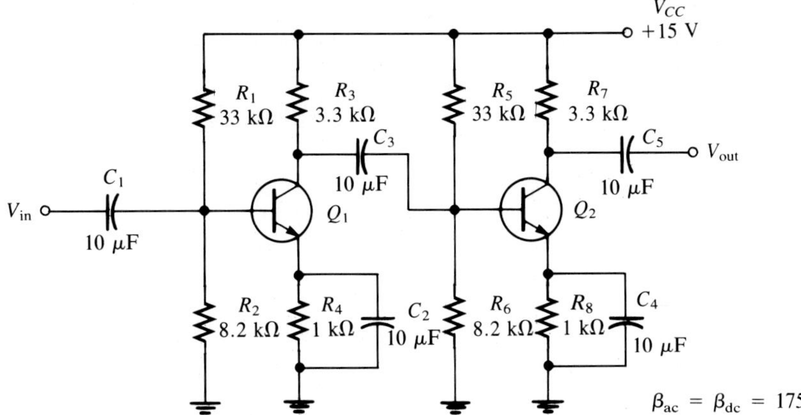

30. If the multistage amplifier in Figure 6–50 is driven by a 75 Ω, 50 μV source and the second stage is loaded with an R_L = 18 kΩ, determine

(a) Voltage gain of each stage **(b)** Overall voltage gain

(c) Express the gains found in (a) and (b) in dB.

31. Figure 6–51 shows a direct-coupled (that is, with no coupling capacitors between stages) two-stage amplifier. The dc bias of the first stage sets the dc bias of the second. Determine all dc voltages for both stages and the overall ac voltage gain.

FIGURE 6–51

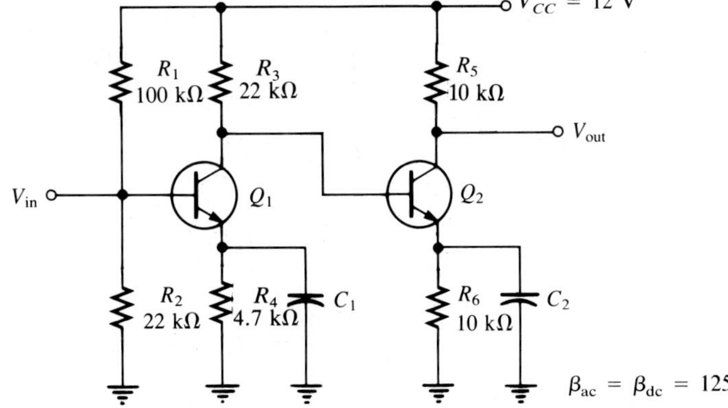

32. Express the following voltage gains in dB:

(a) 12 **(b)** 50

(c) 100 **(d)** 2500

33. Express the following dB voltage gains as actual voltage ratios:

(a) 3 dB **(b)** 6 dB

(c) 10 dB **(d)** 20 dB

(e) 40 dB

SECTION 6–7 TROUBLESHOOTING

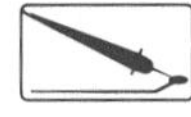

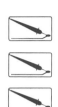

34. Assume that the coupling capacitor C_2 is shorted in Figure 6–34. What dc voltage will appear at the collector of Q_1?

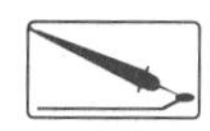

35. Assume that R_5 opens in Figure 6–34. Will Q_2 be in cutoff or in conduction? What dc voltage will you observe at the Q_2 collector?

36. Refer to Figure 6–50 and determine the general effect of each of the following failures:

(a) C_2 opens **(b)** C_3 opens

(c) C_4 opens **(d)** C_2 shorts

(e) Base-collector junction of Q_1 opens **(f)** Base-emitter junction of Q opens

37. Assume that you must troubleshoot the amplifier in Figure 6–50. Set up a table of test point values, input, output, and all transistor terminals that include both dc and rms values that you expect to observe when a 300 Ω test signal source with a 25 μV rms output is used.

38. Determine if the output voltage in Figure 6–52 is correct. If not, what is the most likely problem? This board was used in the system application.

FIGURE 6–52

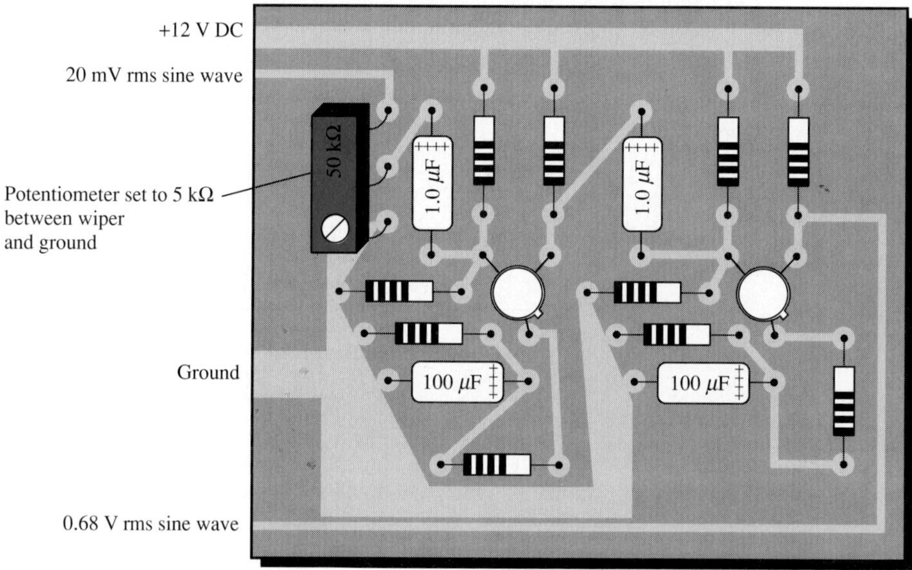

ANSWERS TO REVIEW QUESTIONS

SECTION 6–1

1. Positive, negative

2. V_{CE} is a dc quantity and V_{ce} is an ac quantity.

SECTION 6–2

1. h_{ie}—input resistance common-emitter configuration, h_{re}—voltage feedback ratio common-emitter configuration, h_{fe}—forward current gain common-emitter configuration, h_{oe}—output conductance common-emitter configuration
2. h_{fe} is equivalent to β_{ac}.
3. $r_e = 25\ \text{mV}/15\ \text{mA} = 1.67\ \Omega$

SECTION 6–3

1. The capacitors are treated as opens.
2. The gain increases with a bypass capacitor.
3. Partial bypassing swamps out the effects of r_e.
4. Total input resistance includes the bias resistors, r_e, and any unbypassed R_E.
5. The gain is determined by R_c, r_e, and any unbypassed R_E.
6. The voltage gain decreases with a load.
7. The input and output are 180 degrees out-of-phase.

SECTION 6–4

1. A common-collector amplifier is an emitter-follower.
2. The maximum voltage gain of a common-collector amplifier is one.
3. A common-collector amplifier has a high input resistance.

SECTION 6–5

1. Yes
2. The common-base amplifier has a very low input resistance.

SECTION 6–6

1. A stage is one amplifier in a cascaded arrangement.
2. The overall gain is the product of the individual gains.
3. $20 \log(500) = 53.98$ dB

SECTION 6–7

1. If C_5 opens, the gain drops. The dc level would not be affected.
2. Q_2 would be biased in cutoff.
3. The collector voltage of Q_1 and the base voltage of Q_2 would change.

SECTION 6–8

1. R_1 adjusts the volume by increasing or decreasing the input voltage.
2. The gain of the first stage decreases if C_2 opens.
3. There will be no output if C_3 opens because the signal is not coupled through to the second stage.
4. The gain can be reduced by keeping the same total R_E, but bypassing less of the total—for example, interchanging the two existing emitter resistors in each stage.

ANSWERS TO PRACTICE EXERCISES

6–1 $I_C = 5$ mA, $V_{CE} = 1.5$ V

6–2 3.125 mA

6–3 9.8 mV

6–4 97.3

6–5 82.5

6–6 9.56

6–7 $R_{\text{in}} = 10.37$ kΩ, $A_v = 0.995$, $A_i = 6.7$, $A_p \cong 6.7$

6–8 45.4

6–9 $A_v' = 1500$, A_{v1} (dB) = 27.96 dB, A_{v2} (dB) = 13.98 dB, A_{v3} (dB) = 21.58 dB, A_v' (dB) = 63.52 dB

7

Field-Effect Transistors and Biasing

After completing this chapter, you should be able to

- □ Describe the basic structure and operation of junction field-effect transistors.
- □ Explain why FETs are voltage-controlled devices.
- □ Compare FETs to BJTs, which are current-controlled devices.
- □ Define JFET data sheet parameters.
- □ Distinguish between pinch-off and cutoff in a JFET.
- □ Interpret JFET transconductance curves.
- □ Analyze JFET self-bias and voltage-divider bias circuits.
- □ Describe the basic structure of MOSFETs.
- □ Explain the difference between JFETs and MOSFETs.
- □ Explain the difference between a D-MOSFET and an E-MOSFET.
- □ Discuss V-MOS technology.
- □ Interpret MOSFET data sheet parameters.
- □ Analyze MOSFET zero-bias, drain-feedback bias, and voltage-divider bias circuits.
- □ Troubleshoot FET circuits.

Bipolar junction transistors (BJTs) were covered in previous chapters. Now we turn our attention to the second major type of transistor, the field-effect transistor (FET). FETs are unipolar devices because, unlike bipolar transistors that use both electron and hole current, they operate only with one type of charge carrier. The two main types of FETs are the junction field-effect transistor (JFET) and the metal oxide semiconductor field-effect transistor (MOSFET). You will learn about both types in this chapter.

Recall that the bipolar transistor is a current-controlled device; that is, the base current controls the amount of collector current. The FET is different. It is a voltage-controlled device, where the voltage between two of the terminals (gate and source) controls the current through the device. As you will learn, a major feature of FETs is their very high input resistance.

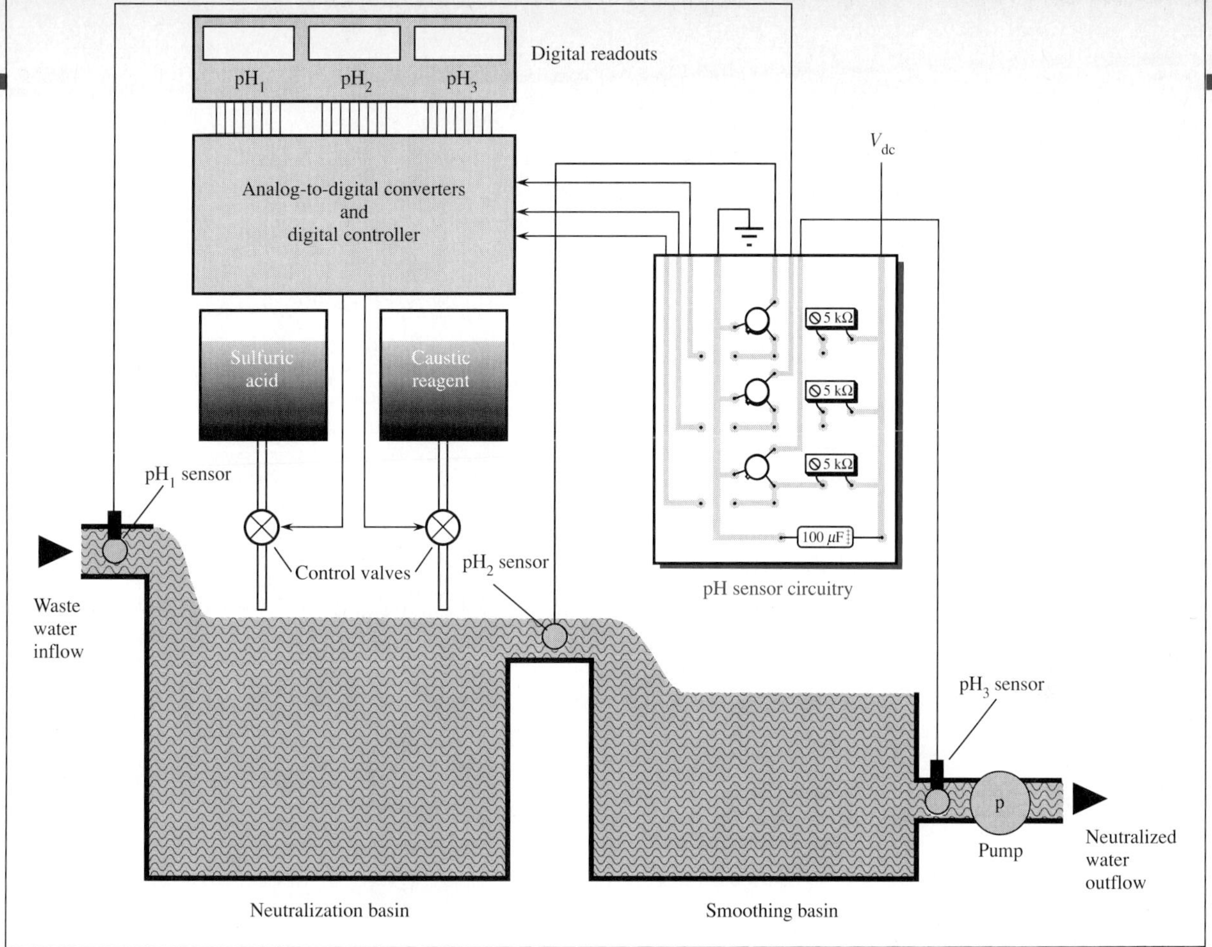

A System Application

The diagram shown here represents a basic waste water neutralization system that is part of an overall waste treatment facility. The pH of the water is adjusted to a neutral value of 7. Solutions with lower pHs are acidic and those with higher pHs are bases. pH sensors are located at three points, and the sensor outputs are modified by the FET sensor circuits and converted to digital form by the analog-to-digital (A/D) converters. The digital data are used by the controller to neutralize the water by controlling the amount of either acid or base that is added, and the pH values are displayed on the digital panel readouts. Since the pH sensors have a high output resistance, FETs are used in the pH sensor circuitry because of their high input resistance. Each FET output is proportional to the measured pH value and is changed to a digital number by the A/D converters. Since our focus will be on the FET sensor circuits, you do not need to know how A/D converters work at this point. They are covered in a later chapter.

For the system application in Section 7–8, in addition to the other topics, be sure you understand

- ☐ Why FET input resistance is very high.
- ☐ Why FETs are voltage-controlled devices.
- ☐ How to interpret a transfer characteristic curve.

7–1 THE JUNCTION FIELD-EFFECT TRANSISTOR (JFET)

Recall that the bipolar junction transistor (BJT) is a current-controlled device; that is, the base current controls the amount of collector current. The* field-effect transistor, *(FET) is different; it is a voltage-controlled device in which the voltage at the gate terminal controls the amount of current through the device. Also, compared to the BJT, the FET has a very high input resistance, which makes it superior in certain applications.

The **junction field-effect transistor (JFET)** is a type of FET that operates with a reverse-biased junction to control current in a channel. Depending on their structure, JFETs fall into either of two categories, n-channel or p-channel. Figure 7–1(a) shows the basic structure of an n-channel JFET. Wire leads are connected to each end of the n-channel; the **drain** is at the upper end and the **source** is at the lower end. Two p-type regions are diffused in the n-type material to form a **channel,** and both p-type regions are connected to the **gate** lead. For simplicity, the gate lead is shown connected to only one of the p regions. A p-channel JFET is shown in Figure 7–1(b).

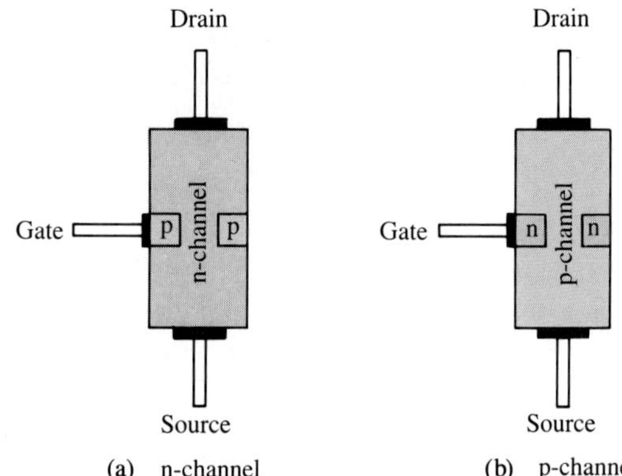

FIGURE 7–1
Basic structure of the two types of JFET.

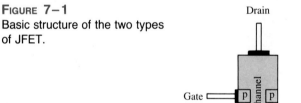

BASIC OPERATION

To illustrate the operation of a JFET, Figure 7–2 shows bias voltages applied to an n-channel device. V_{DD} provides a drain-to-source voltage and supplies current from drain to source. V_{GG} sets the reverse-bias voltage between the gate and the source, as shown.

The JFET is *always* operated with the gate-source pn junction reverse-biased. Reverse-biasing of the gate-source junction with a negative gate voltage produces a *depletion region* in the n-channel and thus increases its resistance. The channel width can be controlled by varying the gate voltage, whereby the amount of drain current, I_D, can also be controlled. This concept is illustrated in Figure 7–3. The color-shaded areas represent the depletion region created by the reverse bias. It is wider toward the drain end of the channel because the reverse-bias voltage between the gate and the drain is greater than that between the gate and the source. We will discuss JFET characteristic curves and some important parameters in the next section.

FIGURE 7–2
Biased n-channel JFET.

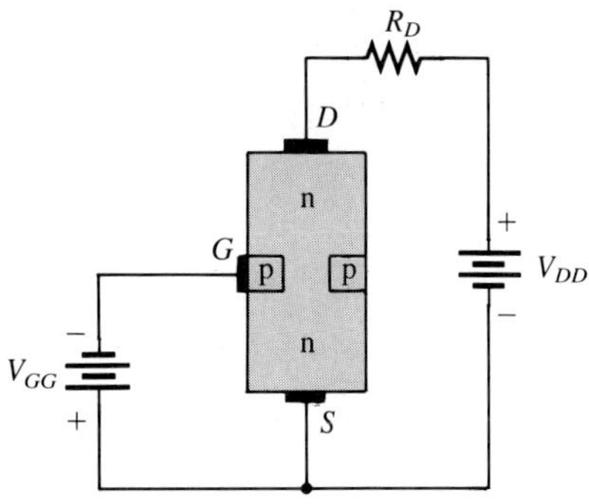

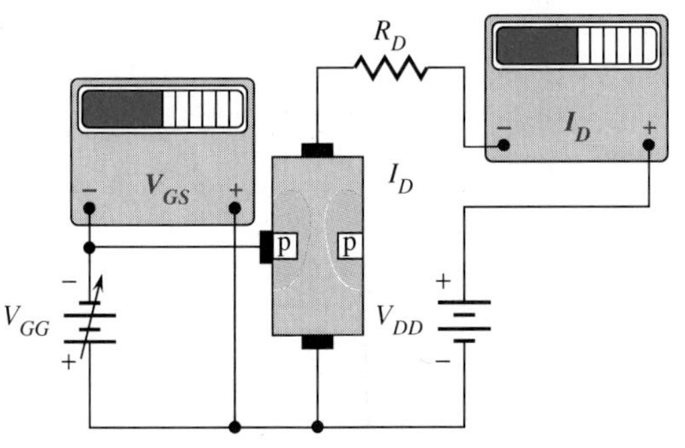

(a) JFET biased for conduction

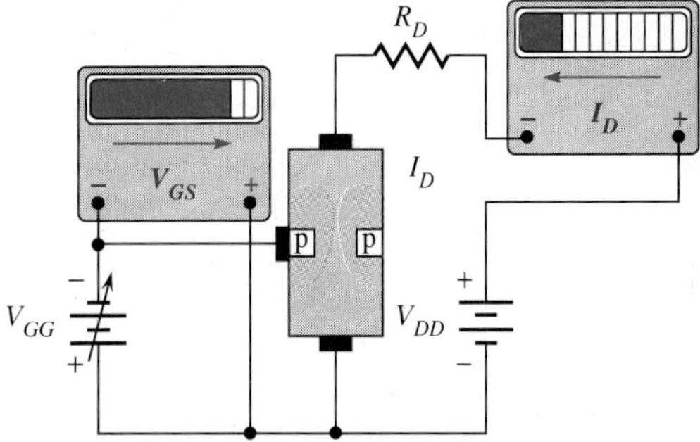

(b) Greater V_{GS} narrows the channel, thus decreasing I_D

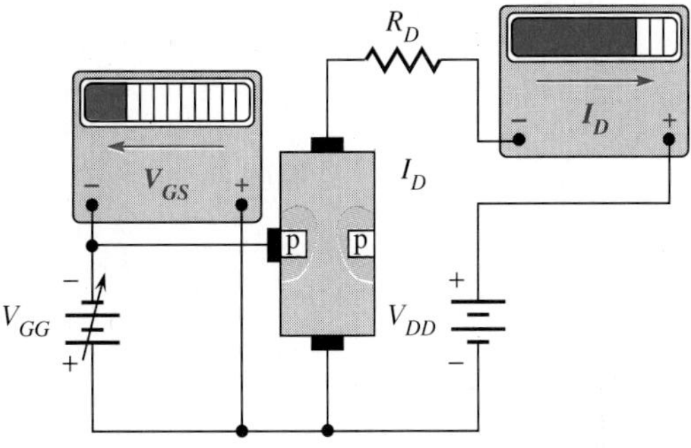

(c) Less V_{GS} widens channel and increases I_D

FIGURE 7–3
Effects of V_{GS} on channel width and on drain current.

JFET SYMBOLS

The schematic symbols for both n-channel and p-channel JFETs are shown in Figure 7–4. Notice that the arrow on the gate points "in" for n-channel and "out" for p-channel.

FIGURE 7–4
JFET schematic symbols.

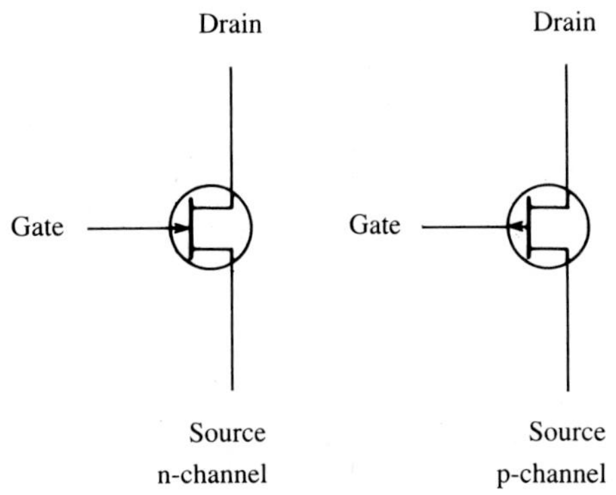

7–1 REVIEW QUESTIONS

1. Name the three terminals of a JFET.
2. Does an n-channel JFET require a positive or negative value for V_{GS}?
3. How is the drain current controlled in a JFET?

7–2 JFET CHARACTERISTICS AND PARAMETERS

In this section, you will see how the JFET operates as a voltage-controlled, constant current device. You will also learn about cutoff and pinch-off as well as JFET transfer characteristics.

First, let's consider the case where the gate-to-source voltage is zero ($V_{GS} = 0$ V). This is produced by shorting the gate to the source, as in Figure 7–5(a) where both are grounded. As V_{DD} (and thus V_{DS}) is increased from zero, I_D will increase proportionally, as shown in the graph of Figure 7–5(b) between points A and B. In this region, the channel resistance is essentially constant because the depletion region is not large enough to have significant effect. This is called the *ohmic region* because V_{DS} and I_D are related by Ohm's law.

At point B in Figure 7–5(b), the curve levels off and I_D becomes essentially constant. As V_{DS} increases from point B to point C, the reverse-bias voltage from gate to drain (V_{GD}) produces a depletion region large enough to offset the increase in V_{DS}, thus keeping I_D relatively constant.

PINCH-OFF

For $V_{GS} = 0$ V, the value of V_{DS} at which I_D becomes essentially constant (point B on the curve in Figure 7–5(b)) is the **pinch-off voltage,** V_P. For a given JFET, V_P has a fixed value. As you can see, a continued increase in V_{DS} above the pinch-off voltage produces

FIGURE 7–5
The drain characteristic curve for $V_{GS} = 0$ showing pinch-off.

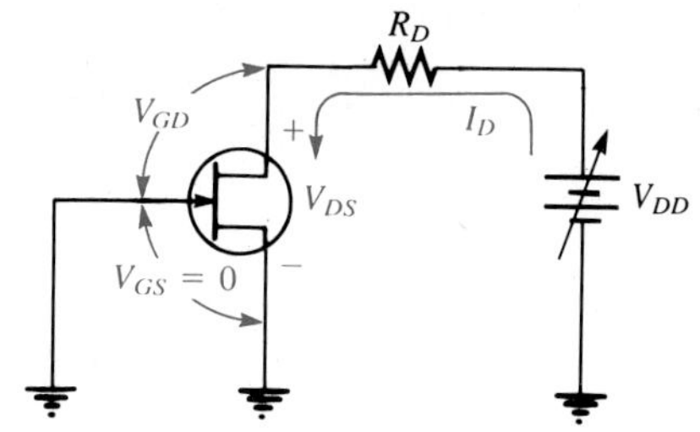

(a) JFET with $V_{GS} = 0$ V and a variable V_{DS}

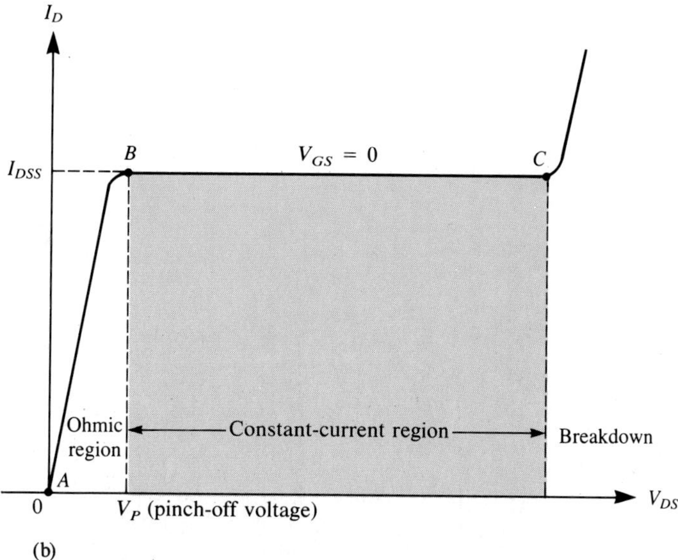

(b)

an almost constant drain current. This value of drain current is I_{DSS} (*D*rain to *S*ource current with gate *S*horted) and is always specified on JFET data sheets. I_{DSS} is the *maximum* drain current that a specific JFET can produce regardless of the external circuit, and it is always specified for the condition, $V_{GS} = 0$ V.

Continuing along the graph in Figure 7–5(b), breakdown occurs at point C when I_D begins to increase very rapidly with any further increase in V_{DS}. Breakdown can result in irreversible damage to the device, so JFETs are always operated below breakdown and within the *constant-current region* (between points B and C on the graph). The JFET action that produces the drain characteristic curve for $V_{GS} = 0$ V is illustrated in Figure 7–6.

V_{GS} CONTROLS I_D

Let's connect a bias voltage, V_{GG}, from gate to source as shown in Figure 7–7(a). As V_{GS} is set to increasingly more negative values by adjusting V_{GG}, a family of drain characteristic curves is produced as shown in Figure 7–7(b). Notice that I_D decreases as

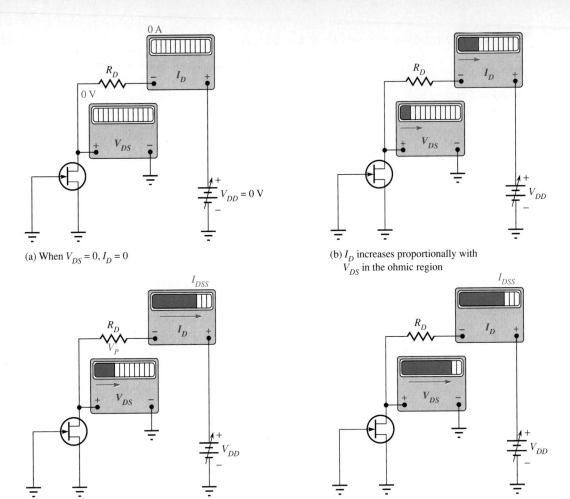

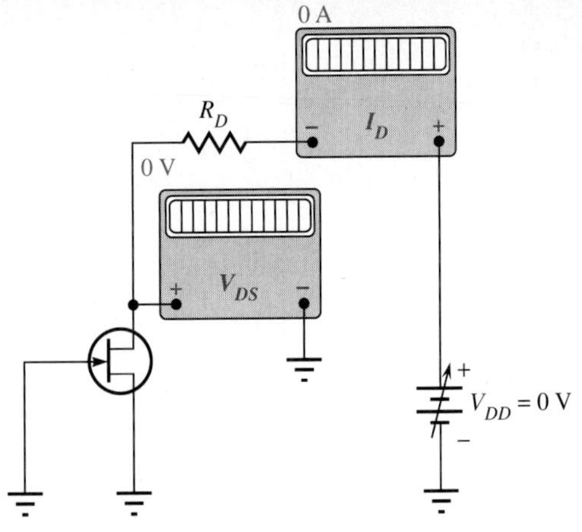

(a) When $V_{DS} = 0, I_D = 0$

R_D I_D V_{DS} V_{DD}

(b) I_D increases proportionally with V_{DS} in the ohmic region

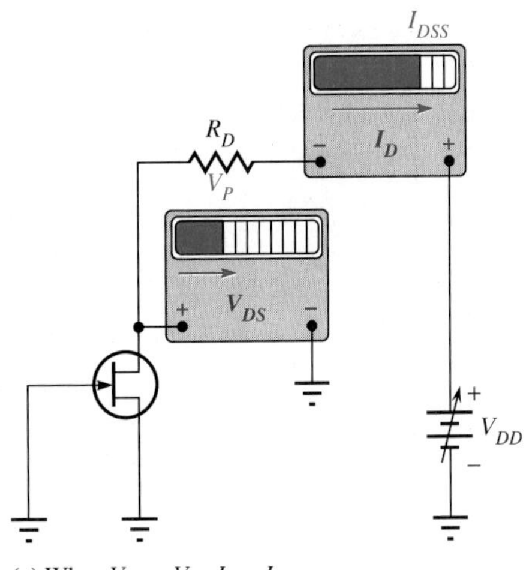

(c) When $V_{DS} = V_P, I_D = I_{DSS}$

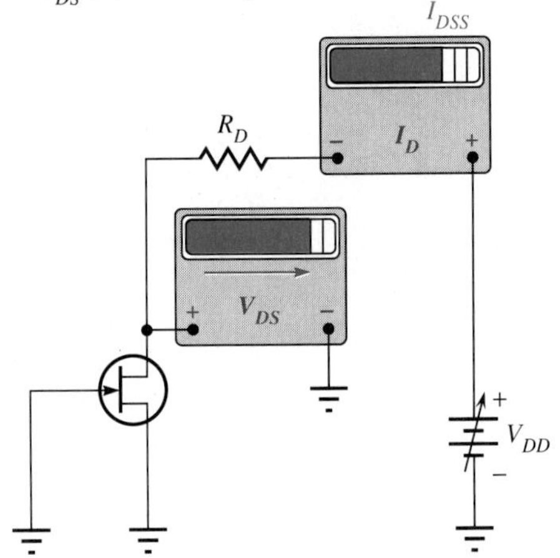

(d) As V_{DS} increases further, I_D remains at I_{DSS}

FIGURE 7–6
JFET action that produces the characteristic curve for $V_{GS} = 0$ V.

FIGURE 7–7
Pinch-off occurs at a lower V_{DS} as V_{GS} is increased to more negative values.

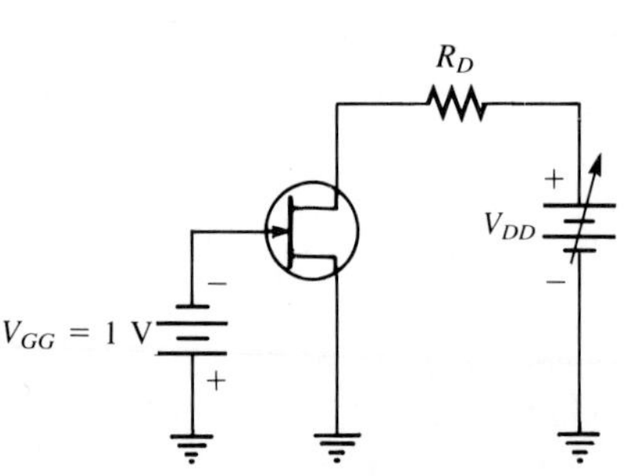

(a) JFET biased at $V_{GS} = -1$ V

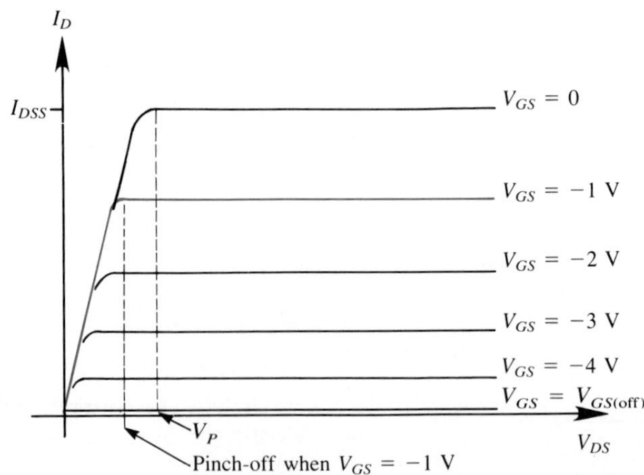

(b) Family of drain characteristic curves

the magnitude of V_{GS} is increased to larger negative values. Also notice that, for each increase in V_{GS}, the JFET reaches pinch-off (where constant current begins) at values of V_{DS} less than V_P. So, the amount of drain current is controlled by V_{GS}, as illustrated in Figure 7–8.

CUTOFF

The value of V_{GS} that makes I_D approximately zero is the cutoff value, $V_{GS(\text{off})}$. The JFET must be operated between $V_{GS} = 0$ V and $V_{GS(\text{off})}$. For this range of gate-to-source voltages, I_D will vary from a maximum of I_{DSS} to a minimum of almost zero.

As you have seen, for an n-channel JFET, the more negative V_{GS} is, the smaller I_D becomes in the constant-current region. When V_{GS} has a sufficiently large negative value, I_D is reduced to zero. This cutoff effect is caused by the widening of the depletion region to a point where it completely closes the channel as shown in Figure 7–9.

The basic operation of a p-channel JFET is the same as for an n-channel device except that it requires a negative V_{DD} and a positive V_{GS} as illustrated in Figure 7–10.

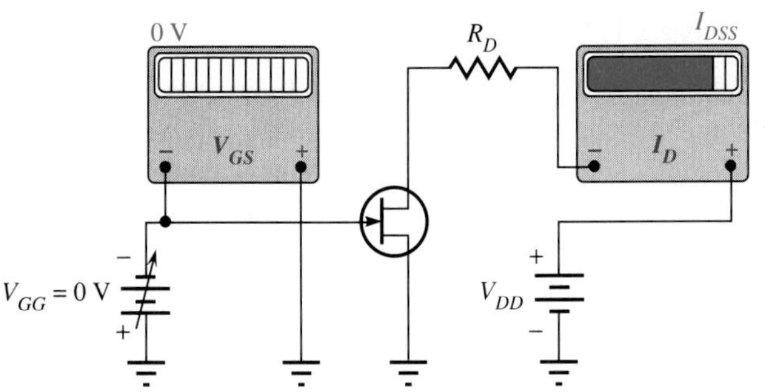

(a) $V_{GS} = 0$ V, $I_D = I_{DSS}$

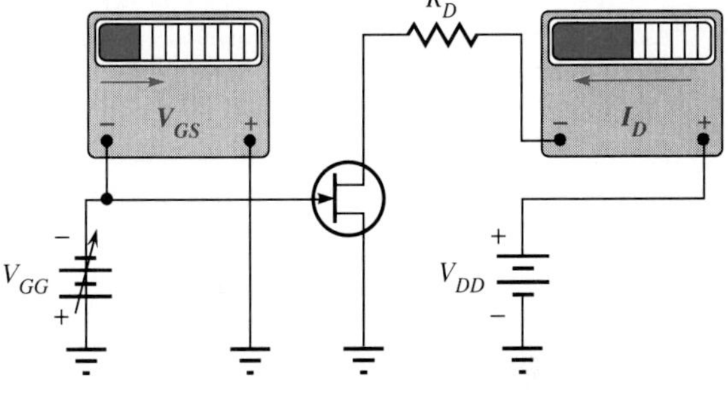

(b) V_{GS} negative, I_D decreases

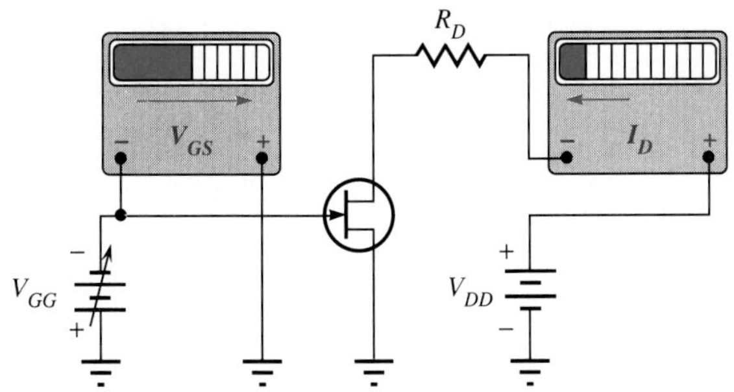

(c) I_D continues to decrease as V_{GS} is made more negative

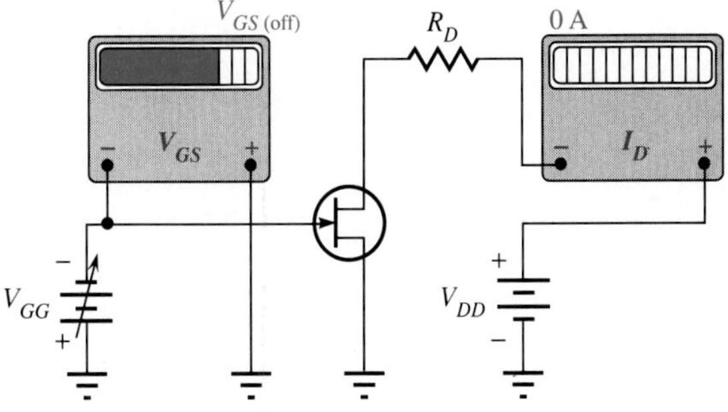

(d) I_D continues to decrease until $V_{GS} = -V_{GS(\text{off})}$

FIGURE 7–8
V_{GS} controls I_D.

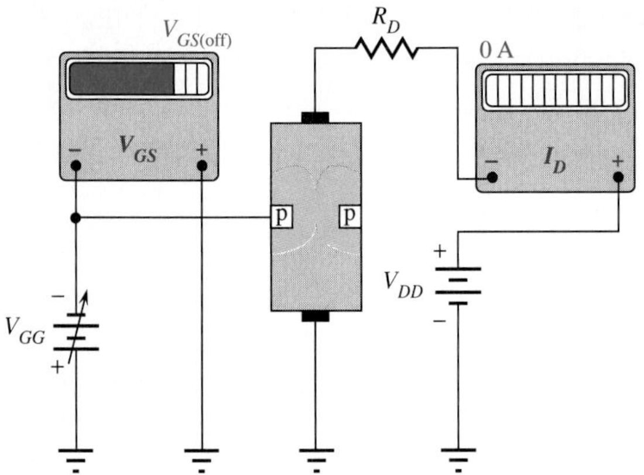

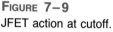

FIGURE 7–9
JFET action at cutoff.

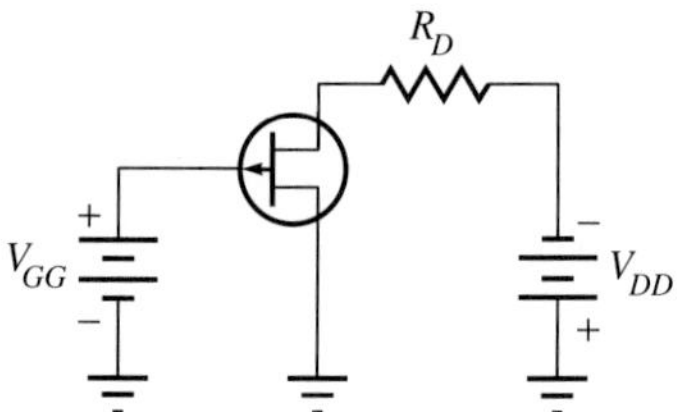

FIGURE 7–10
A biased p-channel JFET.

COMPARISON OF PINCH-OFF AND CUTOFF

As you have seen, there is definitely a difference between pinch-off and cutoff. There is also a connection. V_P is the value of V_{DS} at which the drain current becomes constant and is always measured at $V_{GS} = 0$ V. However, pinch-off occurs for V_{DS} values less than V_P when V_{GS} is nonzero. So, although V_P is a constant, the minimum value of V_{DS} at which I_D becomes constant varies with V_{GS}.

$V_{GS(off)}$ and V_P are always equal in magnitude but opposite in sign. A data sheet usually will give either $V_{GS(off)}$ or V_P, but not both. However, when you know one, you have the other. For example, if $V_{GS(off)} = -5$ V, then $V_P = +5$ V.

EXAMPLE 7–1

For the JFET in Figure 7–11, $V_{GS(off)} = -4$ V and $I_{DSS} = 12$ mA. Determine the *minimum* value of V_{DD} required to put the device in the constant-current region of operation.

FIGURE 7–11

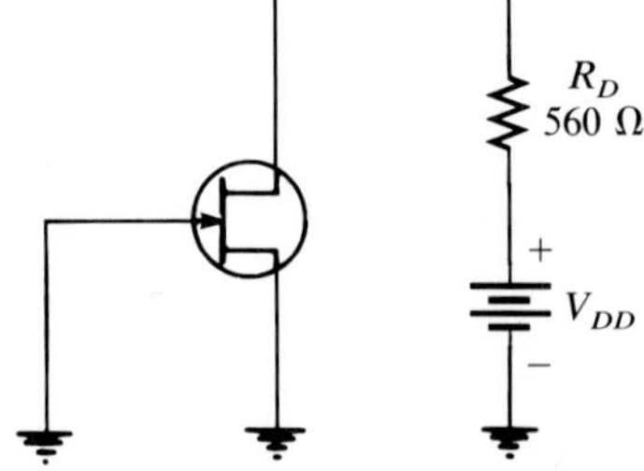

SOLUTION

Since $V_{GS(off)} = -4$ V, $V_P = 4$ V. The minimum value of V_{DS} for the JFET to be in its constant-current region is

$$V_{DS} = V_P = 4 \text{ V}$$

In the constant-current region with $V_{GS} = 0$ V,

$$I_D = I_{DSS} = 12 \text{ mA}$$

The drop across the drain resistor is

$$V_{RD} = (12 \text{ mA})(560 \ \Omega) = 6.72 \text{ V}$$

Applying Kirchhoff's law around the drain circuit gives

$$\begin{aligned} V_{DD} &= V_{DS} + V_{RD} \\ &= 4 \text{ V} + 6.72 \text{ V} = 10.72 \text{ V} \end{aligned}$$

This is the value of V_{DD} to make $V_{DS} = V_P$ and put the device in the constant-current region.

PRACTICE EXERCISE 7–1

If V_{DD} is increased to 15 V, what is the drain current? ■

■ EXAMPLE 7–2

A particular p-channel JFET has a $V_{GS(\text{off})} = +4$ V. What is I_D when $V_{GS} = +6$ V?

SOLUTION

The p-channel JFET requires a positive gate-to-source voltage. The more positive the voltage, the less the drain current. When $V_{GS} = 4$ V, I_D is 0. Any further increase in V_{GS} keeps the JFET cut off, so I_D remains 0.

PRACTICE EXERCISE 7–2

What is V_P for the JFET described in this example? ■

JFET TRANSFER CHARACTERISTIC

You have learned that a range of V_{GS} values from zero to $V_{GS(\text{off})}$ controls the amount of drain current. For an n-channel JFET, $V_{GS(\text{off})}$ is negative, and for a p-channel JFET, $V_{GS(\text{off})}$ is positive. Because V_{GS} does control I_D, the relationship between these two quantities is very important. Figure 7–12 is a typical transfer characteristic curve that illustrates graphically the relationship between V_{GS} and I_D.

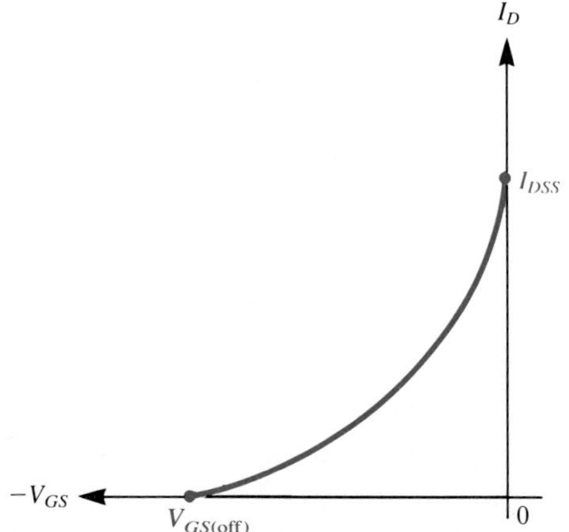

FIGURE 7–12
JFET transfer characteristic curve (n-channel).

Notice that the bottom end of the curve is at a point on the V_{GS} axis equal to $V_{GS(\text{off})}$, and the top end of the curve is at a point on the I_D axis equal to I_{DSS}. This curve, of course, shows that the operating limits of a JFET are

$$I_D = 0 \text{ when } V_{GS} = V_{GS(\text{off})}$$
$$I_D = I_{DSS} \text{ when } V_{GS} = 0$$

The transfer characteristic curve can be developed from the drain characteristic curves by plotting values of I_D for the values of V_{GS} taken from the family of drain curves in the pinch-off region, as illustrated in Figure 7–13 for a specific set of curves. Notice that each point on the transfer characteristic curve corresponds to specific values of V_{GS} and I_D on the drain curves. For example, when $V_{GS} = -2$ V, $I_D = 4.32$ mA. Also, for this specific JFET, $V_{GS(\text{off})} = -5$ V and $I_{DSS} = 12$ mA.

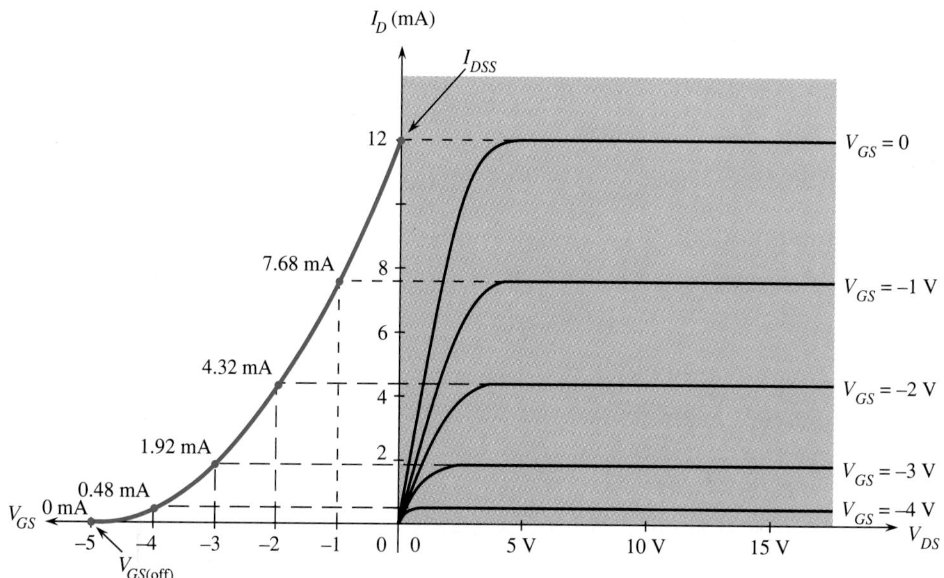

FIGURE 7–13
Example of the development of an n-channel JFET transfer characteristic curve (left) from the JFET drain characteristic curves (right).

A JFET transfer characteristic curve is nearly parabolic in shape and can therefore be expressed approximately as

$$I_D = I_{DSS}\left[1 - \frac{V_{GS}}{V_{GS(\text{off})}}\right]^2 \qquad \textbf{(7–1)}$$

With Equation (7–1), I_D can be determined for any V_{GS} if $V_{GS(\text{off})}$ and I_{DSS} are known. These quantities are usually available from the data sheet for a given JFET. Notice the squared term in the equation. Because of its form, a parabolic relationship is known as a *square law,* and therefore, JFETs and MOSFETs are often referred to as *square-law devices*.

EXAMPLE 7–3

The data sheet for a certain JFET indicates that $I_{DSS} = 15$ mA and $V_{GS(\text{off})} = -5$ V. Determine the drain current for $V_{GS} = 0$ V, -1 V, and -4 V.

SOLUTION

For $V_{GS} = 0$ V, $I_D = I_{DSS} = 15$ mA. For $V_{GS} = -1$ V, we use Equation (7–1).

$$I_D = I_{DSS}\left[1 - \frac{V_{GS}}{V_{GS(\text{off})}}\right]^2 = (15\text{ mA})\left(1 - \frac{-1\text{ V}}{-5\text{ V}}\right)^2$$
$$= (15\text{ mA})(1 - 0.2)^2 = (15\text{ mA})(0.64)$$
$$= 9.6\text{ mA}$$

For $V_{GS} = -4$ V,

$$I_D = (15\text{ mA})\left(1 - \frac{-4\text{ V}}{-5\text{ V}}\right)^2$$
$$= (15\text{ mA})(1 - 0.8)^2$$
$$= (15\text{ mA})(0.04)$$
$$= 0.6\text{ mA}$$

The following program computes I_D for JFETs, given the data sheet parameters I_{DSS} and $V_{GS(\text{off})}$ and a value of V_{GS}. You can input any number of V_{GS} values; the computer will tabulate the corresponding I_D values.

```
10  DIM VBIAS(100)
20  CLS
30  PRINT "COMPUTATION OF DRAIN CURRENT FOR JFETS"
40  PRINT
50  PRINT "THE FOLLOWING INPUTS ARE REQUIRED:"
60  PRINT "(1) IDSS FROM THE DATA SHEET"
70  PRINT "(2) VGS(OFF) FROM THE DATA SHEET"
80  PRINT "(3) VGS AS REQUIRED"
90  PRINT:PRINT:PRINT
100 INPUT "TO CONTINUE PRESS 'ENTER' ";X
110 CLS
120 INPUT "VALUE OF IDSS IN MILLIAMPS";IDSS
130 INPUT "VALUE OF VGS(OFF) IN VOLTS";VGSOFF
140 INPUT "FOR HOW MANY VALUES OF VGS DO YOU WANT ID COMPUTED";N
150 CLS
160 FOR A=1 TO N
170 INPUT "VALUE OF VGS";VBIAS(A)
180 NEXT
190 CLS
200 PRINT "VGS (VOLTS)", "ID (MA)":PRINT
210 FOR A=1 TO N
220 ID=IDSS*(1-VBIAS(A)/VGSOFF)[2
230 PRINT VBIAS(A),ID
240 NEXT
```

PRACTICE EXERCISE 7–3

Determine I_D for $V_{GS} = -3$ V in this example.

JFET FORWARD TRANSCONDUCTANCE

The forward **transconductance,** g_m, is the change in drain current (ΔI_D) for a given change in gate-to-source (ΔV_{GS}) voltage with the drain-to-source voltage constant. It is expressed as a ratio and has the unit of siemens (S).

$$g_m = \frac{\Delta I_D}{\Delta V_{GS}} \quad (7\text{–}2)$$

Other common designations for this parameter are g_{fs} and y_{fs} (forward transfer admittance). As you will see in the next chapter, g_m is very important in FET amplifiers as a major factor in determining the voltage gain.

Because the transfer characteristic (transconductance) curve for a JFET is nonlinear, g_m varies in value depending on the location on the curve as set by V_{GS}. The value for g_m is greater near the top of the curve (near $V_{GS} = 0$) than it is near the bottom (near $V_{GS(\text{off})}$), as illustrated in Figure 7–14. A data sheet normally gives the value of g_m measured at $V_{GS} = 0$ V (g_{m0}). For example, the data sheet for the 2N3823 JFET specifies a minimum g_{m0} of 3500 μS with $V_{DS} = 15$ V.

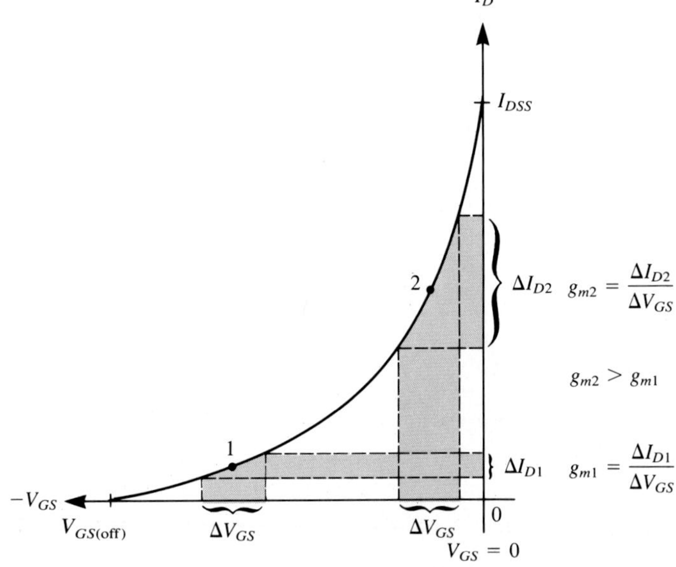

FIGURE 7–14
g_m varies depending on the bias point (V_{GS}).

Given g_{m0}, you can calculate an approximate value for g_m at any point on the transfer characteristic curve using the following formula.

$$g_m = g_{m0}\left[1 - \frac{V_{GS}}{V_{GS(\text{off})}}\right] \quad (7\text{–}3)$$

When a value for g_{m0} is not available, it can be calculated using values of I_{DSS} and $V_{GS(\text{off})}$. The vertical lines indicate an absolute value (no sign).

$$g_{m0} = \frac{2I_{DSS}}{|V_{GS(\text{off})}|} \tag{7-4}$$

EXAMPLE 7–4

The following information is included on the data sheet for a certain JFET: $I_{DSS} = 20$ mA, $V_{GS(\text{off})} = -8$ V, and $g_{m0} = 4000$ μS. Determine the forward transconductance for $V_{GS} = -4$ V, and find I_D at this point.

SOLUTION

First, g_m is found using Equation (7–3).

$$\begin{aligned} g_m &= g_{m0}\left[1 - \frac{V_{GS}}{V_{GS(\text{off})}}\right] \\ &= (4000\ \mu\text{S})\left(1 - \frac{-4\text{ V}}{-8\text{ V}}\right) \\ &= 2000\ \mu\text{S} \end{aligned}$$

Next, using Equation (7–1), I_D at $V_{GS} = -4$ V is calculated.

$$\begin{aligned} I_D &= I_{DSS}\left[1 - \frac{V_{GS}}{V_{GS(\text{off})}}\right]^2 \\ &= (20\text{ mA})\left(1 - \frac{-4\text{ V}}{-8\text{ V}}\right)^2 \\ &= 5\text{ mA} \end{aligned}$$

PRACTICE EXERCISE 7–4

A given JFET has the following characteristics: $I_{DSS} = 12$ mA, $V_{GS(\text{off})} = -5$ V, and $g_{m0} = 3000$ μS. Find g_m and I_D when $V_{GS} = -2$ V.

INPUT RESISTANCE AND CAPACITANCE

A JFET operates with its gate-source junction reverse-biased. Therefore, the input resistance at the gate is very high. This high input resistance is one advantage of the JFET over the bipolar transistor. (Recall that a bipolar transistor operates with a forward-biased base-emitter junction.) JFET data sheets often specify the input resistance by giving a value for the gate reverse current I_{GSS} at a certain gate-to-source voltage. The input resistance can then be determined using the following equation.

$$R_{\text{IN}} = \left|\frac{V_{GS}}{I_{GSS}}\right| \tag{7-5}$$

For example, the 2N3970 data sheet lists a maximum I_{GSS} of 250 pA for $V_{GS} = -20$ V at 25°C. I_{GSS} increases with temperature, so the input resistance decreases.

The input capacitance C_{iss} of a JFET is considerably greater than that of a bipolar transistor because the JFET operates with a reverse-biased pn junction. Recall that a reverse-biased pn junction acts as a capacitor whose capacitance depends on the amount of reverse voltage. For example, the 2N3970 has a maximum C_{iss} of 25 pF for $V_{GS} = 0$.

EXAMPLE 7–5 A certain JFET has an I_{GSS} of 1 nA for $V_{GS} = -20$ V. Determine the input resistance.

SOLUTION

$$R_{IN} = \left|\frac{V_{GS}}{I_{GSS}}\right| = \frac{20 \text{ V}}{1 \text{ nA}} = 20{,}000 \text{ M}\Omega$$

PRACTICE EXERCISE 7–5

Determine the minimum input resistance for the 2N3970.

DRAIN-TO-SOURCE RESISTANCE

You learned from the drain characteristic curve that, above pinch-off, the drain current is relatively constant over a range of drain-to-source voltages. Therefore, a large change in V_{DS} produces only a very small change in I_D. The ratio of these changes is the drain-to-source resistance of the device, r_{ds}.

$$r_{ds} = \frac{\Delta V_{DS}}{\Delta I_D} \tag{7–6}$$

Data sheets often specify this parameter as output conductance, g_{os}, or output admittance, y_{os}. Typical values for r_{ds} are on the order of several thousand ohms.

7–2 REVIEW QUESTIONS

1. The drain-to-source voltage at the pinch-off point of a particular JFET is 7 V. If the gate-to-source voltage is zero, what is V_P?
2. The V_{GS} of a certain n-channel JFET is increased negatively. Does the drain current increase or decrease?
3. What value must V_{GS} have to produce cutoff in a p-channel JFET with a $V_P = -3$ V?

7–3 JFET BIASING

Using some of the FET parameters discussed in the previous sections, we will now see how to dc-bias JFETs. The purpose of biasing is to select the proper dc gate-to-source voltage to establish a desired value of drain current. You will learn about two major types of bias circuits, self-bias and voltage-divider bias.

SELF-BIAS

Recall that a JFET must be operated such that the gate-source junction is always reverse-biased. This condition requires a negative V_{GS} for an n-channel JFET and a positive V_{GS} for a p-channel JFET. This can be achieved using the self-bias arrangements shown in Figure 7–15. The gate is biased at approximately 0 V by resistor R_G connected to ground. The reverse leakage current I_{GSS} does produce a very small voltage across R_G as indicated, but this can be neglected in most cases, and it can be assumed that R_G has no voltage drop across it as we will do.

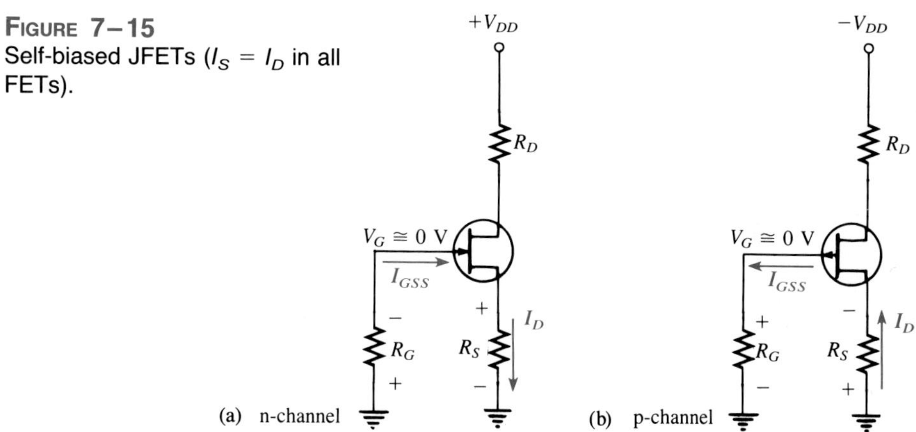

FIGURE 7–15
Self-biased JFETs ($I_S = I_D$ in all FETs).

For the n-channel JFET in Figure 7–15(a), I_D produces a voltage drop across R_S and makes the source positive with respect to ground. Since $V_G = 0$ and $V_S = I_D R_S$, the gate-to-source voltage is

$$V_{GS} = V_G - V_S = 0 - I_D R_S$$

so

$$V_{GS} = -I_D R_S \tag{7–7}$$

For the p-channel JFET shown in Figure 7–15(b), the current through R_S produces a negative voltage at the source and therefore

$$V_{GS} = +I_D R_S \tag{7–8}$$

In the following analysis, the n-channel JFET is used for illustration. Keep in mind that analysis of the p-channel JFET is the same except for opposite-polarity voltages. The drain voltage with respect to ground is determined as follows.

$$V_D = V_{DD} - I_D R_D \tag{7–9}$$

Since $V_S = I_D R_S$, the drain-to-source voltage is

$$V_{DS} = V_D - V_S$$
$$V_{DS} = V_{DD} - I_D(R_D + R_S) \tag{7–10}$$

EXAMPLE 7–6 Find V_{DS} and V_{GS} in Figure 7–16, given that $I_D = 5$ mA.

FIGURE 7–16

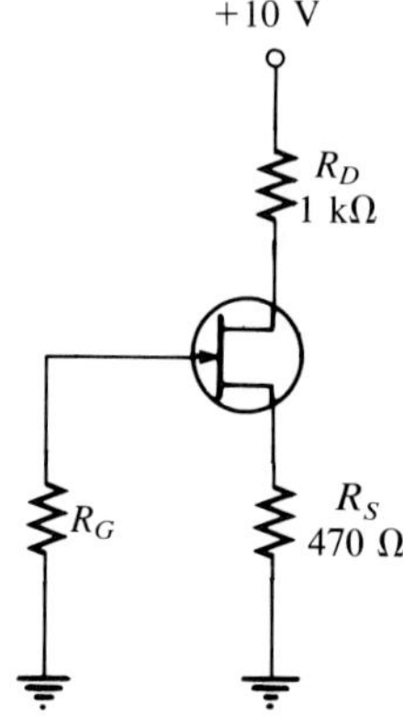

SOLUTION

$$V_S = I_D R_S = (5 \text{ mA})(470 \ \Omega) = 2.35 \text{ V}$$
$$V_D = V_{DD} - I_D R_D = 10 \text{ V} - (5 \text{ mA})(1 \text{ k}\Omega)$$
$$= 10 \text{ V} - 5 \text{ V} = 5 \text{ V}$$

Therefore,

$$V_{DS} = V_D - V_S = 5 \text{ V} - 2.35 \text{ V} = 2.65 \text{ V}$$

Since $V_G = 0$ V,

$$V_{GS} = V_G - V_S = -2.35 \text{ V}$$

PRACTICE EXERCISE 7–6

Determine V_{DS} and V_{GS} in Figure 7–16 when $I_D = 8$ mA. Assume that $R_D = 860 \ \Omega$, $R_S = 390 \ \Omega$, and $V_{DD} = 12$ V.

SETTING THE Q POINT OF A SELF-BIASED JFET

The basic approach to establishing a JFET bias point is to determine I_D for a desired value of V_{GS}. Then calculate the required value of R_S using the relationship derived from Equation (7–7) and stated in Equation (7–11). The vertical lines indicate an absolute value.

$$R_S = \left| \frac{V_{GS}}{I_D} \right| \qquad \textbf{(7–11)}$$

For a desired value of V_{GS}, I_D can be determined in either of two ways: from the transfer characteristic curve for the particular JFET, or from Equation (7–1) using I_{DSS} and $V_{GS(\text{off})}$ from the JFET data sheet. The next two examples illustrate these procedures.

EXAMPLE 7–7

Determine the value of R_S required to self-bias an n-channel JFET having the transfer characteristic curve shown in Figure 7–17 at $V_{GS} = -5$ V.

FIGURE 7–17

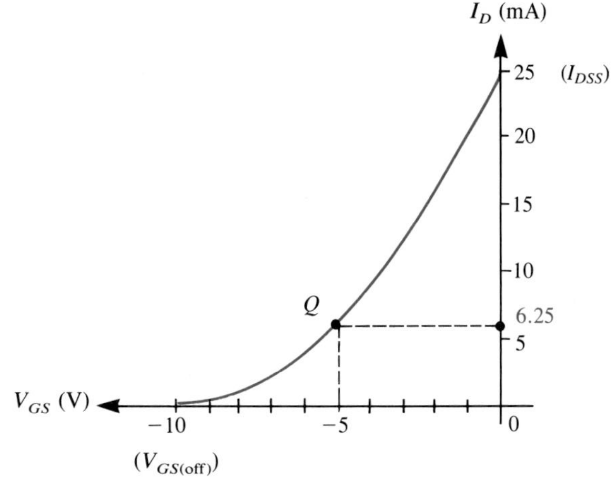

SOLUTION

From the graph, $I_D = 6.25$ mA when $V_{GS} = -5$ V. Calculating R_S, we get

$$R_S = \left|\frac{V_{GS}}{I_D}\right| = \frac{5 \text{ V}}{6.25 \text{ mA}} = 800 \ \Omega$$

PRACTICE EXERCISE 7–7

Find R_S for $V_{GS} = -3$ V.

EXAMPLE 7–8

Determine the value of R_S required to self-bias a p-channel JFET with $I_{DSS} = 25$ mA and $V_{GS(off)} = 15$ V. V_{GS} is to be 5 V.

SOLUTION

I_D is calculated using Equation (7–1).

$$I_D = I_{DSS}\left[1 - \frac{V_{GS}}{V_{GS(off)}}\right]^2$$
$$= (25 \text{ mA})\left(1 - \frac{5 \text{ V}}{15 \text{ V}}\right)^2$$
$$= (25 \text{ mA})(1 - 0.333)^2$$
$$= 11.12 \text{ mA}$$

Now, R_S can be determined.

$$R_S = \left|\frac{V_{GS}}{I_D}\right| = \frac{5 \text{ V}}{11.12 \text{ mA}} = 450 \ \Omega$$

Since 450 Ω is not a standard value, use a 470-Ω resistor.

PRACTICE EXERCISE 7–8

Find the value of R_S required to self-bias a p-channel JFET with $I_{DSS} = 18$ mA and $V_{GS(\text{off})} = 8$ V. $V_{GS} = 4$ V.

MIDPOINT BIAS

It is often desirable to bias a JFET near the midpoint of its transfer characteristic curve where $I_D = I_{DSS}/2$. Under signal conditions, midpoint bias allows a maximum amount of drain current swing between I_{DSS} and 0. Using Equation (7–1), it is shown in Appendix B that I_D is approximately one-half of I_{DSS} when $V_{GS} = V_{GS(\text{off})}/3.414$.

$$I_D = I_{DSS}\left[1 - \frac{V_{GS}}{V_{GS(\text{off})}}\right]^2$$

$$= I_{DSS}\left[1 - \frac{V_{GS(\text{off})}/3.414}{V_{GS(\text{off})}}\right]^2$$

$$= I_{DSS}(1 - 0.2929)^2 = I_{DSS}(0.707)^2 \cong 0.5I_{DSS}$$

So, by selecting $V_{GS} = V_{GS(\text{off})}/3.414$, we get a midpoint bias in terms of I_D.

To set the drain voltage at midpoint ($V_D = V_{DD}/2$), a value of R_D is selected to produce the desired voltage drop. R_G is chosen arbitrarily large to prevent loading on the driving stage in a cascaded amplifier arrangement. Example 7–9 illustrates these concepts.

EXAMPLE 7–9

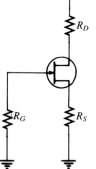

Select resistor values in Figure 7–18 to set up an approximate midpoint bias. The JFET parameters are $I_{DSS} = 15$ mA and $V_{GS(\text{off})} = -8$ V. V_D should be 6 V (one-half of V_{DD}).

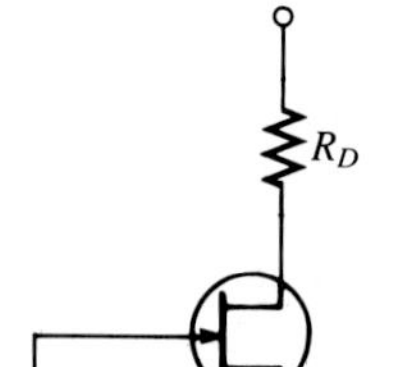

FIGURE 7–18

SOLUTION

For midpoint bias, $I_D \cong I_{DSS}/2 = 7.5$ mA and $V_{GS} \cong V_{GS(\text{off})}/3.414 = -8\text{ V}/3.414 = -2.343$ V.

$$R_S = \left|\frac{V_{GS}}{I_D}\right| = \frac{2.343\text{ V}}{7.5\text{ mA}}$$

$$= 312\ \Omega$$

$$V_D = V_{DD} - I_D R_D$$

$$I_D R_D = V_{DD} - V_D$$

$$R_D = \frac{V_{DD} - V_D}{I_D} = \frac{12\text{ V} - 6\text{ V}}{7.5\text{ mA}}$$

$$= 800\ \Omega$$

Use the nearest standard values of 330 Ω and 820 Ω. R_G can be arbitrarily large if I_{GSS} is assumed to be negligible. R_G essentially establishes the input resistance to the JFET stage because it appears in parallel with the very high resistance of the reverse-biased pn junction.

PRACTICE EXERCISE 7–9

Select resistor values in Figure 7–18 to set up an approximate midpoint bias. The JFET parameters are $I_{DSS} = 10$ mA and $V_{GS(\text{off})} = -10$ V. $V_{DD} = 15$ V.

GRAPHICAL ANALYSIS OF A SELF-BIASED JFET

You can use the transfer characteristic curve of a JFET and certain parameters to determine the Q point (I_D and V_{GS}) of a self-biased circuit. A circuit is shown in Figure 7–19(a) and a transfer characteristic curve is shown in Figure 7–19(b). If a curve is not available from the data sheet, you can plot it from Equation (7–1) using data sheet values for I_{DSS} and $V_{GS(\text{off})}$.

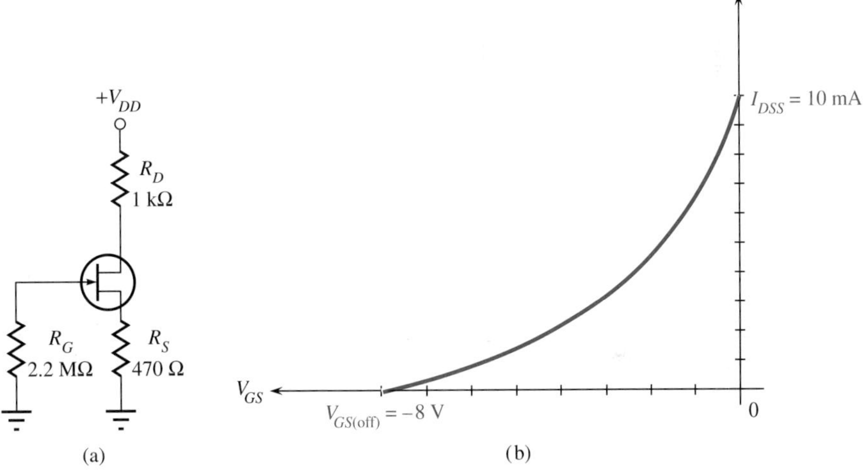

FIGURE 7–19 A self-biased JFET and its transfer characteristic curve.

To determine the Q point, a self-bias dc load line is established as follows. First, calculate V_{GS} when I_D is zero.

$$V_{GS} = -I_D R_S = (0)(470\ \Omega) = 0\ \text{V}$$

This establishes a point at the origin on the graph ($I_D = 0$, $V_{GS} = 0$). Next, get I_{DSS} from the data sheet and calculate V_{GS} when $I_D = I_{DSS}$. From the curve, $I_{DSS} = 10$ mA for the JFET in Figure 7–19.

$$V_{GS} = -I_D R_S = -(10\ \text{mA})(470\ \Omega) = -4.7\ \text{V}$$

This establishes a second point on the graph ($I_D = 10$ mA, $V_{GS} = -4.7$ V). Now, with two points, the load line can be drawn on the graph of the transfer characteristic as shown in Figure 7–20. The point where the line intersects the transfer characteristic curve is the Q point of the circuit.

FIGURE 7–20
The intersection of the self-bias dc load line and the transfer characteristic curve is the Q point.

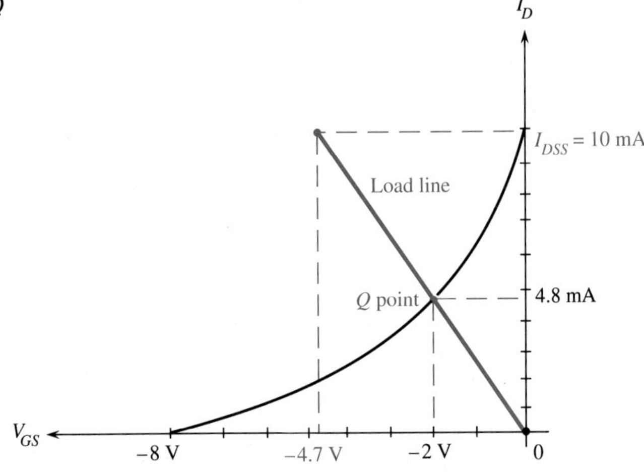

EXAMPLE 7–10 Determine the Q point for the JFET circuit in Figure 7–21(a). The transfer characteristic curve is given in Figure 7–21(b).

FIGURE 7–21

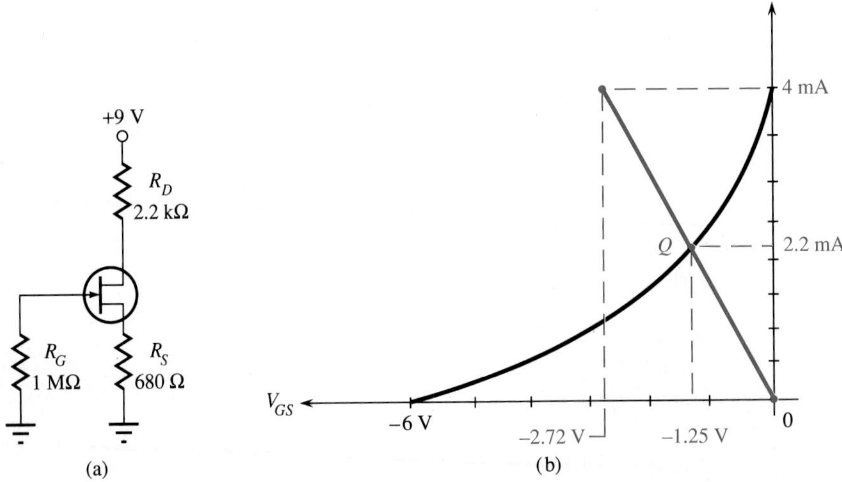

SOLUTION

For $I_D = 0$,

$$V_{GS} = -I_D R_S = (0)(680\ \Omega) = 0\ \text{V}$$

This gives a point at the origin. From the curve, $I_{DSS} = 4$ mA. So for $I_D = I_{DSS} = 4$ mA,

$$V_{GS} = -I_D R_S = -(4\ \text{mA})(680\ \Omega) = -2.72\ \text{V}$$

This gives a second point at 4 mA and −2.72 V. A line is now drawn between the two points and the values of I_D and V_{GS} at the intersection of the line, and the curve is taken from the graph as illustrated in Figure 7–21(b). The Q-point values from the graph are $I_D = 2.2$ mA and $V_{GS} = -1.25$ V.

PRACTICE EXERCISE 7–10

If R_S is increased to 1 kΩ in Figure 7–21(a), what is the new Q point?

VOLTAGE-DIVIDER BIAS

An n-channel JFET with voltage-divider bias is shown in Figure 7–22. The voltage at the source of the JFET must be more positive than the voltage at the gate in order to keep the gate-source junction reverse-biased.

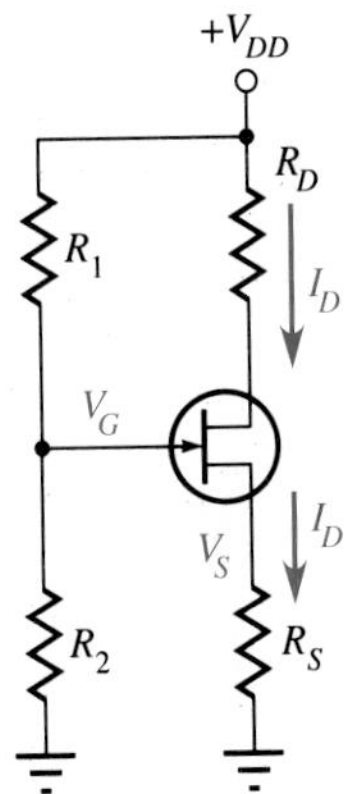

FIGURE 7–22
An n-channel JFET with voltage-divider bias.

The source voltage is

$$V_S = I_D R_S$$

The gate voltage is set by resistors R_1 and R_2 as expressed by the following equation:

$$V_G = \left(\frac{R_2}{R_1 + R_2}\right) V_{DD} \tag{7–12}$$

The gate-to-source voltage is

$$V_{GS} = V_G - V_S$$

and the source voltage is

$$V_S = V_G - V_{GS}$$

The drain current can be expressed as

$$I_D = \frac{V_S}{R_S} = \frac{V_G - V_{GS}}{R_S} \tag{7–13}$$

EXAMPLE 7–11 Determine I_D and V_{GS} for the JFET with voltage-divider bias in Figure 7–23, given that $V_D = 7$ V.

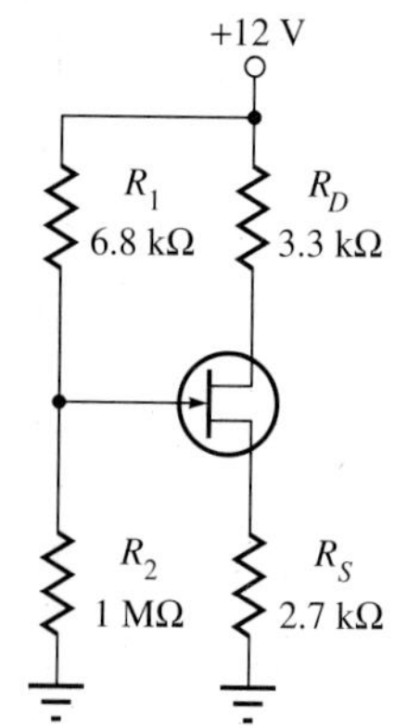

FIGURE 7–23

SOLUTION

$$I_D = \frac{V_{DD} - V_D}{R_D} = \frac{12\text{ V} - 7\text{ V}}{3.3\text{ k}\Omega} = \frac{5\text{ V}}{3.3\text{ k}\Omega} = 1.52\text{ mA}$$

$$V_S = I_D R_S = (1.52\text{ mA})(2.7\text{ k}\Omega) = 4.1\text{ V}$$

$$V_G = \left(\frac{R_2}{R_1 + R_2}\right)V_{DD} = \left(\frac{1\text{ M}\Omega}{7.8\text{ M}\Omega}\right)12\text{ V} = 1.54\text{ V}$$

$$V_{GS} = V_G - V_S = 1.54\text{ V} - 4.1\text{ V} = -2.56\text{ V}$$

If V_D had not been given in this example, the Q-point values could not have been found.

PRACTICE EXERCISE 7–11
Given that $V_D = 9$ V in Figure 7–23, determine the Q point.

GRAPHICAL ANALYSIS OF A JFET WITH VOLTAGE-DIVIDER BIAS

An approach similar to the one used for self-bias can be used with voltage-divider bias to graphically determine the Q point of a circuit on the transfer characteristic curve.

In a JFET with voltage-divider bias when $I_D = 0$, V_{GS} is not zero, as in the self-biased case, because the voltage divider produces a voltage at the gate independent of the drain current. The voltage-divider dc load line is determined as follows.

For $I_D = 0$,

$$V_S = I_D R_S = (0)R_S = 0\text{ V}$$

$$V_{GS} = V_G - V_S = V_G - 0\text{ V} = V_G$$

Therefore, one point on the line is at $I_D = 0$ and $V_{GS} = V_G$.

For $V_{GS} = 0$,

$$I_D = \frac{V_G - V_{GS}}{R_S} = \frac{V_G}{R_S}$$

A second point on the line is at $I_D = V_G/R_S$ and $V_{GS} = 0$. The dc load line is shown in Figure 7–24.

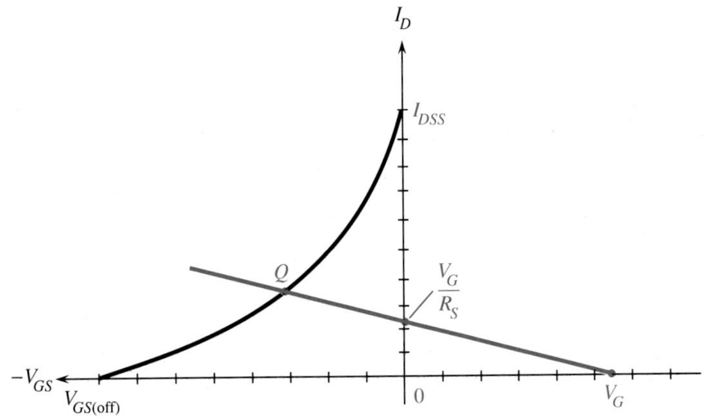

FIGURE 7–24
DC load line for a JFET with voltage-divider bias.

EXAMPLE 7–12 Determine the Q point for the JFET with voltage-divider bias in Figure 7–25(a), given the transfer characteristic curve in Figure 7–25(b).

FIGURE 7–25

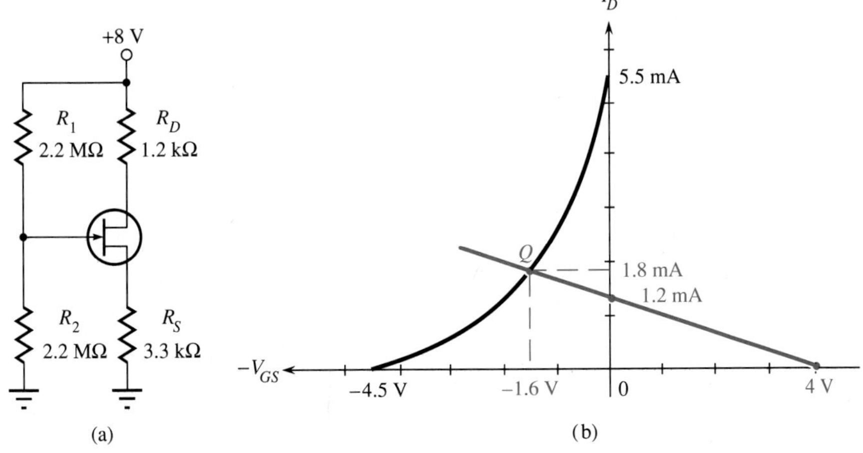

SOLUTION

First, we establish the two points for the bias line.

For $I_D = 0$,

$$V_{GS} = V_G = \left(\frac{R_2}{R_1 + R_2}\right)V_{DD} = \left(\frac{2.2\ \text{M}\Omega}{4.4\ \text{M}\Omega}\right)8\ \text{V} = 4\ \text{V}$$

The first point is at $I_D = 0$ and $V_{GS} = 4$ V.

For $V_{GS} = 0$,

$$I_D = \frac{V_G - V_{GS}}{R_S} = \frac{V_G}{R_S} = \frac{4\text{ V}}{3.3\text{ k}\Omega} = 1.2\text{ mA}$$

The second point is at $I_D = 1.2$ mA and $V_{GS} = 0$.

The load line is drawn in Figure 7–25(b) and the Q-point values of $I_D = 1.8$ mA and $V_{GS} = -1.6$ V are picked off the graph, as indicated.

PRACTICE EXERCISE 7–12

Change R_S to 4.7 kΩ and determine the Q point for the circuit in Figure 7–25(a).

Q-POINT STABILITY

Unfortunately, the transfer characteristic of a JFET can differ considerably from one device to another of the same type. If, for example, a 2N5459 JFET is replaced in a given bias circuit with another 2N5459, the transfer characteristic curve can vary greatly as illustrated in Figure 7–26(a). In this case, the maximum I_{DSS} is 16 mA and the minimum I_{DSS} is 4 mA. Likewise, the maximum $V_{GS(off)}$ is −8 V and the minimum $V_{GS(off)}$ is −2 V. This means that if you have a pile of 2N5459s and you arbitrarily pick one out, it can have values anywhere within these ranges.

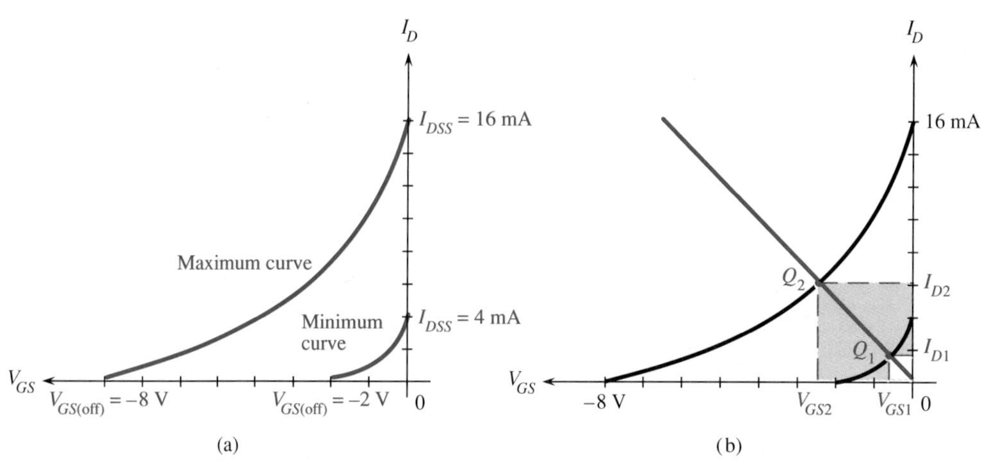

FIGURE 7–26
Variation in the transfer characteristic of 2N5459 JFETs and the effect on the Q point.

If a self-bias dc load line is drawn as illustrated in Figure 7–26(b), the *same* circuit using a 2N5459 can have a Q point anywhere along the line from Q_1, the minimum bias point, to Q_2, the maximum bias point. Accordingly, the drain current can be any value between I_{D1} and I_{D2}, as shown. This means that the dc voltage at the drain can have a range of values depending on I_D. Also, the gate-to-source voltage can be any value between V_{GS1} and V_{GS2}, as indicated.

With voltage-divider bias, the dependency of I_D on the range of Q points is reduced because the slope of the bias line is less than for self-bias. Although V_{GS} varies quite a bit for both self-bias and voltage-divider bias, I_D is much more stable with voltage-divider bias, as illustrated in Figure 7–27.

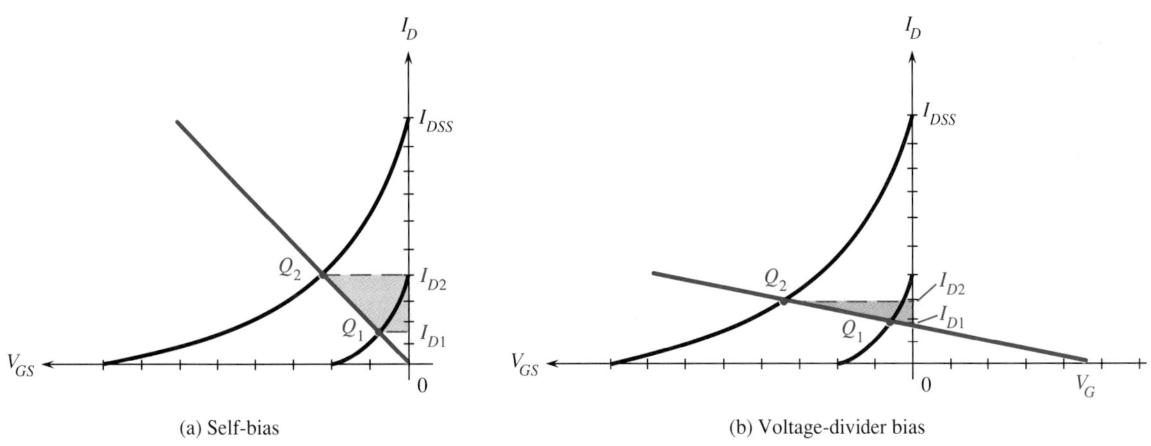

FIGURE 7–27
The change in I_D between the minimum and the maximum Q points is much less for a JFET with voltage-divider bias than for a self-biased JFET.

7–3 REVIEW QUESTIONS

1. Should a p-channel JFET have a positive or a negative V_{GS}?
2. In a certain self-biased n-channel JFET circuit, $I_D = 8$ mA and $R_S = 1$ kΩ. Determine V_{GS}.
3. An n-channel JFET with voltage-divider bias has a gate voltage of 3 V and a source voltage of 5 V. What is V_{GS}?

7–4 THE METAL OXIDE SEMICONDUCTOR FET (MOSFET)

The metal oxide semiconductor field-effect transistor (MOSFET) is the second category of field-effect transistor. The* MOSFET *differs from the JFET in that it has no pn junction structure; instead, the gate of the MOSFET is insulated from the channel by a silicon dioxide (SIO_2) layer. The two basic types of MOSFETs are* depletion *(D) and* enhancement *(E). Because of the insulated gate, these devices are sometimes called IGFETs.

DEPLETION MOSFET (D-MOSFET)

Figure 7–28 illustrates the basic structure of D-MOSFETs. The drain and source are diffused into the substrate material and then connected by a narrow channel adjacent to the insulated gate. Both n-channel and p-channel devices are shown in the figure. We will use the n-channel device to describe the basic operation. The p-channel operation is the same, except the voltage polarities are opposite those of the n-channel.

FIGURE 7–28
Basic structure of D-MOSFETs.

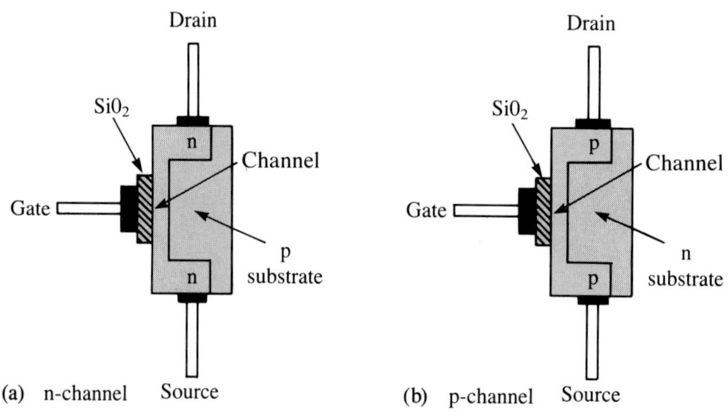

The D-MOSFET can be operated in either of two modes—the *depletion mode* or the *enhancement mode*—and is sometimes called a *depletion/enhancement MOSFET*. Since the gate is insulated from the channel, either a positive or a negative gate voltage can be applied. The MOSFET operates in the depletion mode when a negative gate-to-source voltage is applied and in the enhancement mode when a positive gate-to-source voltage is applied. These devices are generally operated in the depletion mode.

DEPLETION MODE Visualize the gate as one plate of a parallel-plate capacitor and the channel as the other plate. The silicon dioxide insulating layer is the dielectric. With a negative gate voltage, the negative charges on the gate repel conduction electrons from the channel, leaving positive ions in their place. Thereby, the n-channel is depleted of some of its electrons, thus decreasing the channel conductivity. The greater the negative voltage on the gate, the greater the depletion of n-channel electrons. At a sufficiently negative gate-to-source voltage, $V_{GS(\text{off})}$, the channel is totally depleted and the drain current is zero. This depletion mode is illustrated in Figure 7–29(a). Like the n-channel JFET, the n-channel D-MOSFET conducts drain current for gate-to-source voltages between $V_{GS(\text{off})}$ and zero. In addition, the D-MOSFET conducts for values of V_{GS} above zero.

ENHANCEMENT MODE With a positive gate voltage, more conduction electrons are attracted into the channel, thus increasing (enhancing) the channel conductivity, as illustrated in Figure 7–29(b).

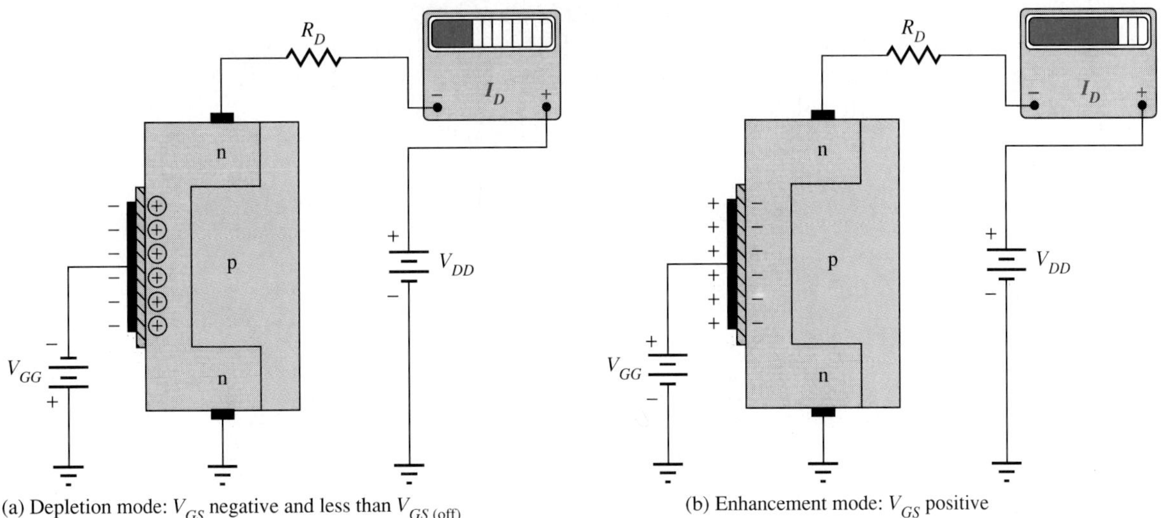

FIGURE 7–29
Operation of n-channel D-MOSFET.

D-MOSFET SYMBOLS The schematic symbols for both the n-channel and the p-channel depletion MOSFETs are shown in Figure 7–30. The substrate, indicated by the arrow, is normally (but not always) connected internally to the source. Sometimes, there is a separate substrate pin. An inward substrate arrow is for n-channel, and an outward arrow is for p-channel.

FIGURE 7–30
D-MOSFET schematic symbols.

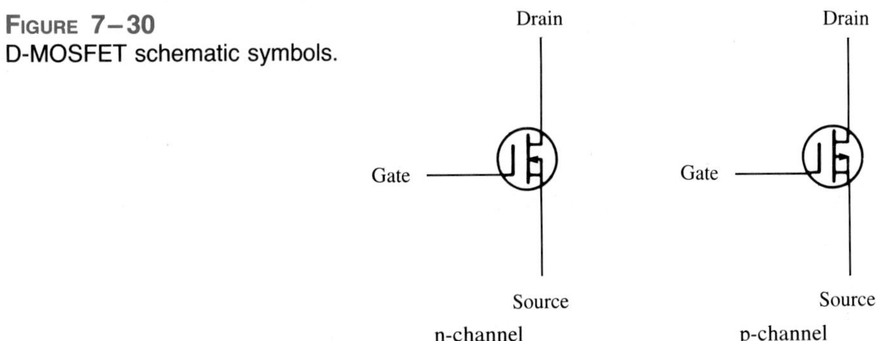

ENHANCEMENT MOSFET (E-MOSFET)

The E-MOSFET operates *only* in the enhancement mode and has no depletion mode. It differs in construction from the D-MOSFET in that it has no physical channel. Notice in Figure 7–31(a) that the substrate extends completely to the SiO_2 layer. For an n-channel device, a positive gate voltage above a threshold value induces a channel by creating a thin layer of negative charges in the substrate region adjacent to the SiO_2 layer, as shown in Figure 7–31(b). The conductivity of the channel is enhanced by increasing the gate-to-source voltage and thus pulling more electrons into the channel. For any gate voltage below the threshold value, there is no channel.

FIGURE 7–31
E-MOSFET construction and operation (n-channel).

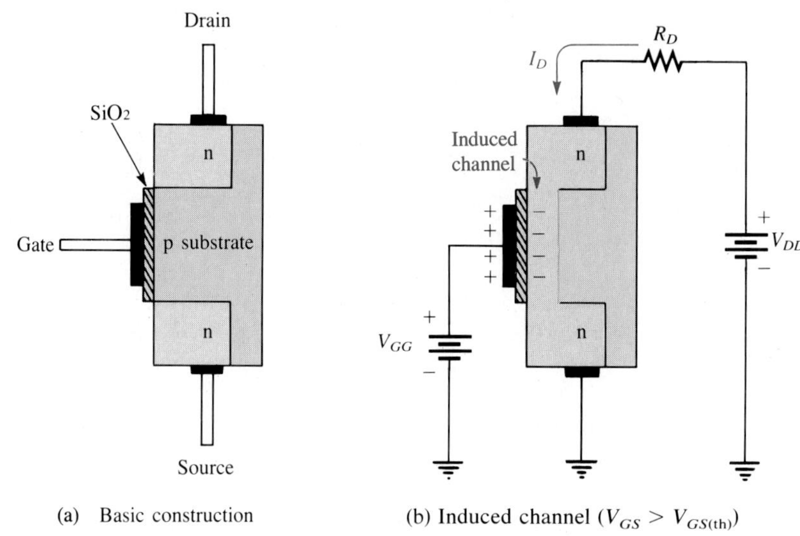

(a) Basic construction (b) Induced channel ($V_{GS} > V_{GS(th)}$)

The schematic symbols for the n-channel and p-channel E-MOSFETs are shown in Figure 7–32. The broken lines symbolize the absence of a physical channel. Like the D-MOSFET, some devices have a separate substrate connection.

FIGURE 7–32
E-MOSFET schematic symbols.

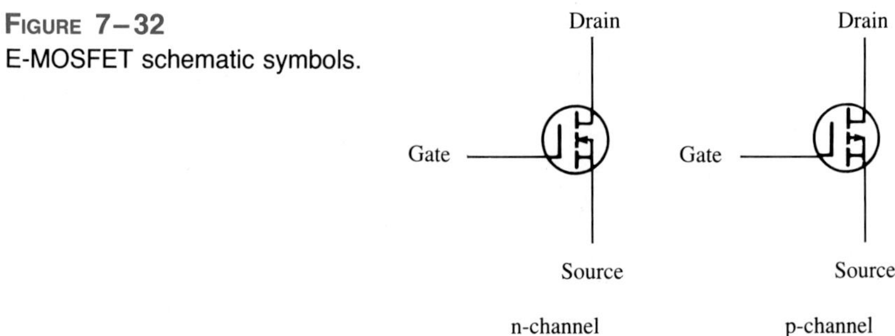

n-channel p-channel

V-MOSFET

The V-MOSFET is a special type of E-MOSFET with a physical structure that permits higher current operation and greater drain-to-source voltage than the conventional E-MOSFET. Thus, the V-MOSFET is particularly suited for high power use and competes successfully with bipolar power transistors in many amplifier and switching applications. The structures of a conventional E-MOSFET and a V-MOSFET are compared in Figure 7–33. Notice that they are quite different in structure but similar in operation. Both the V-MOSFET and the E-MOSFET are enhancement devices and have no channel between drain and source until the gate is made positive with respect to the source (for n-channel devices).

FIGURE 7–33
Basic structures of a conventional E-MOSFET and a V-MOSFET.

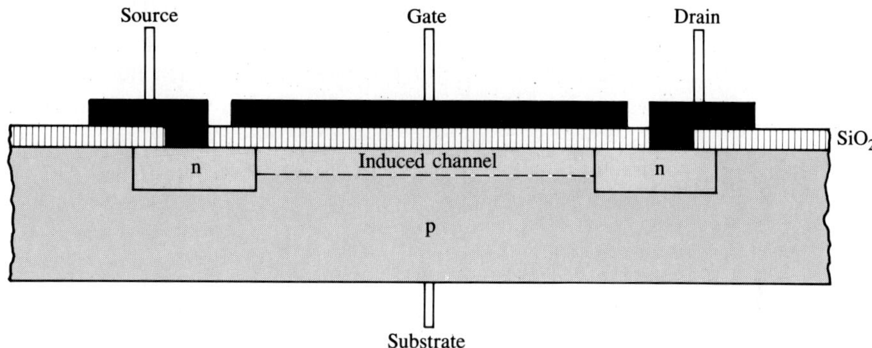

(a) Conventional E-MOSFET

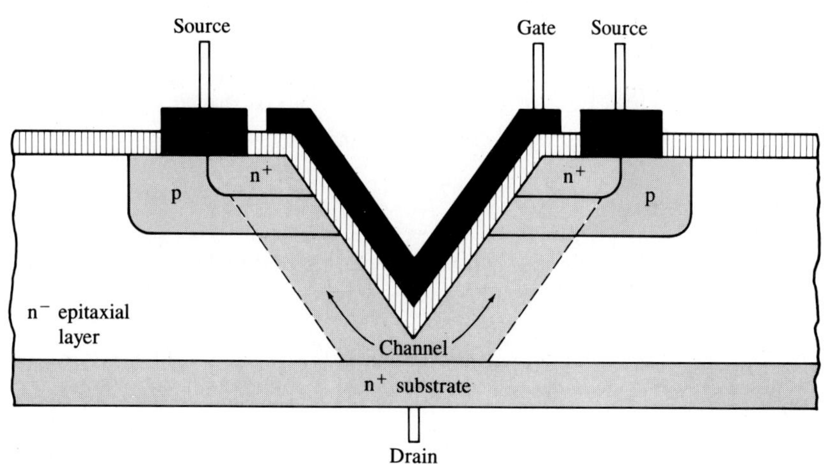

(b) V-MOSFET

When the gate is positive, an n-channel is formed close to the gate in the E-MOSFET, parallel with the surface. The length of the E-MOSFET channel depends on size limitations of the photographic masks used in the diffusion process during manufacture, and the channel is very narrow. The V-MOSFET has two source connections, and the channel is induced vertically along both sides of the V-shaped cut between the drain (n^+ substrate where n^+ indicates a higher doping level than n^-) and the source connections, creating a relatively wide channel. The channel length is set by the thickness of the layers, which is controlled by doping densities and diffusion time rather than by mask dimensions. The V-geometry makes possible much shorter and wider channels than in the conventional E-MOSFET. The shorter, wider channels in the V-MOSFET allow for higher currents and, thus, greater power dissipation. Frequency response is also greatly improved.

DUAL-GATE MOSFETS

The dual-gate MOSFET can be either a depletion or an enhancement type. The only difference is that it has two gates, as shown in Figure 7–34. As previously mentioned, one drawback of an FET is its high input capacitance, which restricts its use at higher frequencies. By using a dual-gate device, the input capacitance is reduced, thus making the device very useful in high-frequency RF amplifier applications. Another advantage of the dual gate arrangement is that it allows for an automatic gain control (AGC) input in certain RF amplifiers.

FIGURE 7–34
Dual-gate n-channel MOSFET symbols.

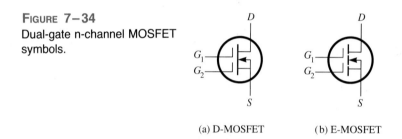

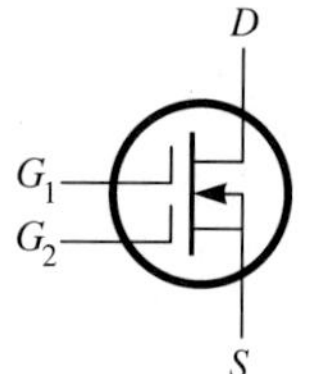

(a) D-MOSFET

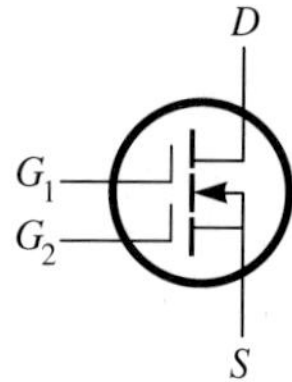

(b) E-MOSFET

7–4 REVIEW QUESTIONS

1. Name the two basic types of MOSFETs.
2. If the gate-to-source voltage in a depletion MOSFET is zero, what is the current from drain to source?
3. If the gate-to-source voltage in an E-MOSFET is zero, what is the current from drain to source?

7–5 MOSFET CHARACTERISTICS AND PARAMETERS

Much of the discussion concerning JFET characteristics and parameters applies equally to MOSFETs. In this section, the differences are highlighted.

D-MOSFET TRANSFER CHARACTERISTIC

As previously discussed, the D-MOSFET can operate with either positive or negative gate voltages. This is indicated on the transfer characteristic curves in Figure 7–35 for both n-channel and p-channel MOSFETs. The point on the curves where $V_{GS} = 0$ corresponds to I_{DSS}. The point where $I_D = 0$ corresponds to $V_{GS(\text{off})}$. As with the JFET, $V_{GS(\text{off})} = -V_p$.

The square-law expression in Equation (7–1) for the JFET curve also applies to the D-MOSFET curve, as the following example demonstrates.

FIGURE 7–35
D-MOSFET transfer characteristic curves.

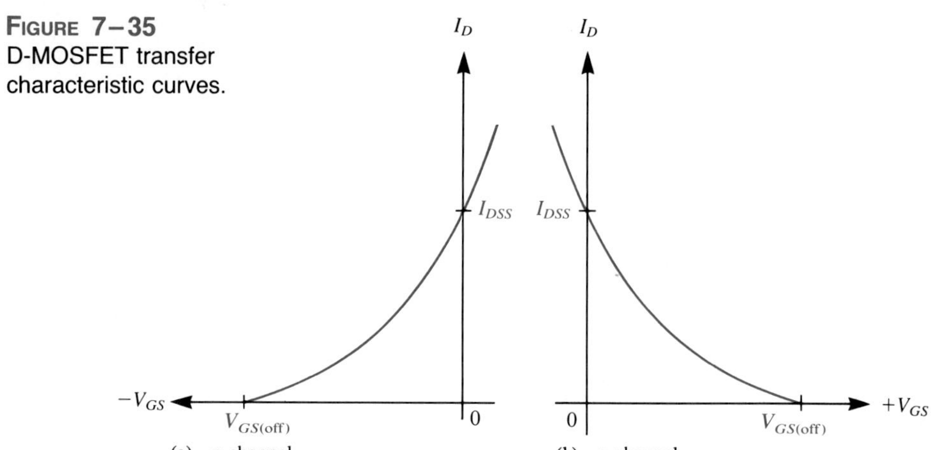

EXAMPLE 7–13 For a certain D-MOSFET, $I_{DSS} = 10$ mA and $V_{GS(\text{off})} = -8$ V.

(a) Is this an n-channel or a p-channel?
(b) Calculate I_D at $V_{GS} = -3$ V.
(c) Calculate I_D at $V_{GS} = +3$ V.

SOLUTION

(a) The device has a negative $V_{GS(\text{off})}$; therefore, it is an n-channel MOSFET.

(b)
$$I_D = I_{DSS}\left[1 - \frac{V_{GS}}{V_{GS(\text{off})}}\right]^2$$
$$= (10 \text{ mA})\left(1 - \frac{-3 \text{ V}}{-8 \text{ V}}\right)^2$$
$$= 3.9 \text{ mA}$$

(c)
$$I_D = (10 \text{ mA})\left(1 - \frac{+3 \text{ V}}{-8 \text{ V}}\right)^2$$
$$= 18.9 \text{ mA}$$

PRACTICE EXERCISE 7–13
For a certain D-MOSFET, $I_{DSS} = 18$ mA and $V_{GS(\text{off})} = +10$ V.

(a) Is this an n-channel or a p-channel?
(b) Determine I_D at $V_{GS} = +4$ V.
(c) Determine I_D at $V_{GS} = -4$ V.

E-MOSFET TRANSFER CHARACTERISTIC

The E-MOSFET uses only channel *enhancement*. Therefore, an n-channel device requires a positive gate-to-source voltage, and a p-channel device requires a negative gate-to-source voltage. Figure 7–36 shows the transfer characteristic curves for both types of

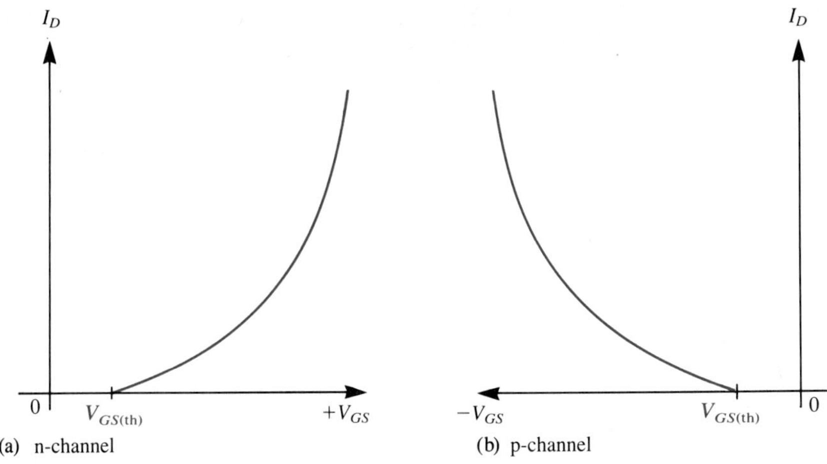

FIGURE 7–36
E-MOSFET transfer characteristic curves.

E-MOSFETs. As you can see, there is no drain current when $V_{GS} = 0$. Therefore, the E-MOSFET does not have an I_{DSS} parameter, as do the JFET and the D-MOSFET. Notice also that there is no drain current until V_{GS} reaches a certain nonzero value called the *threshold voltage* $V_{GS(\text{th})}$.

The equation for the parabolic transfer characteristic curve of the E-MOSFET differs from that of the JFET and the D-MOSFET because the curve starts at $V_{GS(\text{th})}$ rather than $V_{GS(\text{off})}$ on the horizontal axis and never intersects the vertical axis. The equation for the E-MOSFET transfer characteristic curve is

$$I_D = K[V_{GS} - V_{GS(\text{th})}]^2 \tag{7–14}$$

The constant K depends on the particular MOSFET and can be determined from the data sheet by taking the specified value of I_D, called $I_{D(\text{on})}$, at the given value of V_{GS} and substituting the values into Equation (7–14).

EXAMPLE 7–14

The data sheet for a certain E-MOSFET gives $I_{D(\text{on})} = 3$ mA at $V_{GS} = 10$ V and $V_{GS(\text{th})} = 5$ V. Determine the drain current for $V_{GS} = 8$ V.

SOLUTION

First, solve for K using Equation (7–14).

$$K = \frac{I_{D(\text{on})}}{[V_{GS} - V_{GS(\text{th})}]^2}$$

$$= \frac{3\text{ mA}}{(10\text{ V} - 5\text{ V})^2}$$

$$= \frac{3\text{ mA}}{25\text{ V}^2}$$

$$= 0.12\text{ mA/V}^2$$

Next, using the value of K, I_D is calculated for $V_{GS} = 8$ V.

$$I_D = K[V_{GS} - V_{GS(\text{th})}]^2$$
$$= 0.12 \text{ mA/V}^2(8 \text{ V} - 5 \text{ V})^2$$
$$= 1.08 \text{ mA}$$

PRACTICE EXERCISE 7–14

The data sheet for an E-MOSFET gives $I_{D(\text{on})} = 5$ mA at $V_{GS} = 8$ V and $V_{GS(\text{th})} = 4$ V. Find I_D when $V_{GS} = 6$ V.

HANDLING PRECAUTIONS

All MOS devices are subject to damage from electrostatic discharge (ESD). Because the gate of a MOSFET is insulated from the channel, the input resistance is extremely high (ideally infinite). The gate leakage current I_{GSS} for a typical MOSFET is in the pA range, whereas the gate reverse current for a typical JFET is in the nA range. The input capacitance results from the insulated gate structure. Excess static charge can be accumulated because the input capacitance combines with the very high input resistance and can result in damage to the device. To avoid damage from ESD, certain precautions should be taken when handling MOSFETs:

1. MOS devices should be shipped and stored in conductive foam.
2. All instruments and metal benches used in assembly or test should be connected to earth ground (round or third prong of 110-V wall outlets).
3. The assembler's or handler's wrist should be connected to earth ground with a length of wire and a high-value series resistor.
4. Never remove an MOS device (or any other device, for that matter) from the circuit while the power is on.
5. Do not apply signals while the dc power supply is off.

7–5 REVIEW QUESTIONS

1. What is the major difference in construction of the D-MOSFET and the E-MOSFET?
2. What two parameters do not exist for an E-MOSFET?
3. What is ESD?

7–6 MOSFET BIASING

In this section, we will look at three ways to bias a MOSFET—zero-bias, voltage-divider bias, and drain-feedback bias. As you know, biasing is very important in amplifier applications, which you will study in the next chapter.

D-MOSFET BIAS

Recall that D-MOSFETs can be operated with either positive or negative values of V_{GS}. A simple bias method is to set $V_{GS} = 0$ so that an ac signal at the gate varies the gate-to-source voltage above and below this bias point. A MOSFET with zero bias is shown in Figure 7–37. Since $V_{GS} = 0$, $I_D = I_{DSS}$ as indicated. The drain-to-source voltage is expressed as follows.

$$V_{DS} = V_{DD} - I_{DSS}R_D \qquad (7\text{–}15)$$

FIGURE 7–37
A zero-biased D-MOSFET.

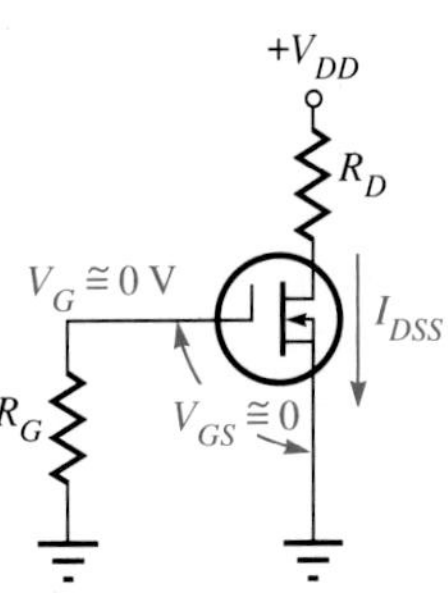

■ EXAMPLE 7–15

Determine the drain-to-source voltage in the circuit of Figure 7–38. The MOSFET data sheet gives $V_{GS(\text{off})} = -8$ V and $I_{DSS} = 12$ mA.

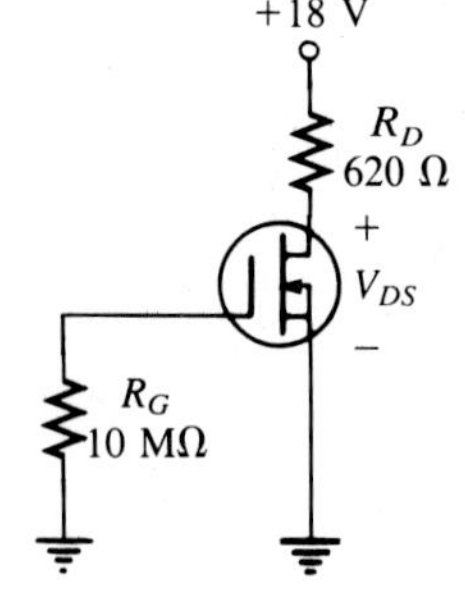

FIGURE 7–38

SOLUTION

Since $I_D = I_{DSS} = 12$ mA, the drain-to-source voltage is calculated as follows.

$$\begin{aligned} V_{DS} &= V_{DD} - I_{DSS}R_D \\ &= 18\text{ V} - (12\text{ mA})(620\ \Omega) \\ &= 10.56\text{ V} \end{aligned}$$

PRACTICE EXERCISE 7–15

Find V_{DS} in Figure 7–38 when $V_{GS(\text{off})} = -10$ V and $I_{DSS} = 20$ mA.

■

E-MOSFET BIAS

Recall that E-MOSFETs must have a V_{GS} greater than the threshold value, $V_{GS(\text{th})}$. Figure 7–39 shows two ways to bias an E-MOSFET (D-MOSFETs can also be biased using these methods). An n-channel device is used for purposes of illustration. In either the voltage-divider or drain-feedback bias arrangement, the purpose is to make the gate

FIGURE 7–39
E-MOSFET biasing arrangements.

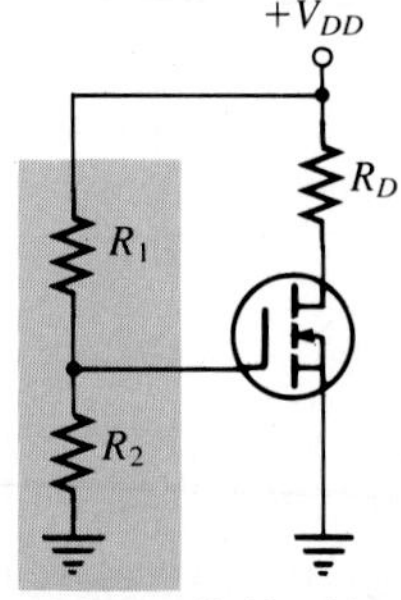

(a) Voltage-divider bias

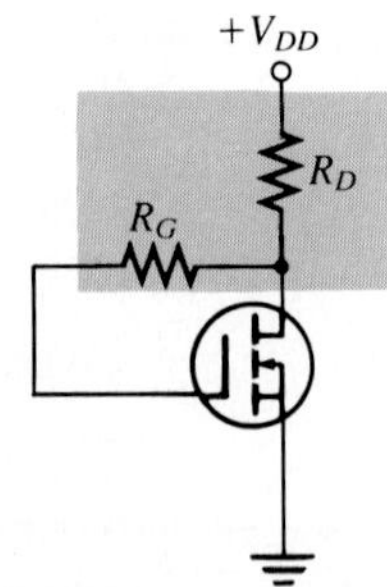

(b) Drain-feedback bias

voltage more positive than the source by an amount exceeding $V_{GS(th)}$. Equations for the analysis of the voltage-divider bias in Figure 7–39(a) are as follows:

$$V_{GS} = \left(\frac{R_2}{R_1 + R_2}\right)V_{DD}$$

$$V_{DS} = V_{DD} - I_D R_D$$

where $I_D = K[V_{GS} - V_{GS(th)}]^2$ from Equation (7–14).

EXAMPLE 7–16 Determine V_{GS} and V_{DS} for the E-MOSFET circuit in Figure 7–40. The data sheet for this particular MOSFET gives $I_{D(on)} = 3$ mA at $V_{GS} = 10$ V and $V_{GS(th)} = 5$ V.

FIGURE 7–40

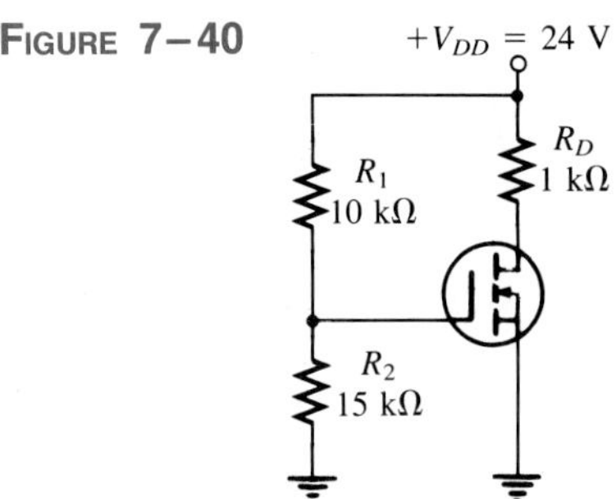

SOLUTION

$$V_{GS} = \left(\frac{R_2}{R_1 + R_2}\right)V_{DD} = \left(\frac{15\text{ k}\Omega}{25\text{ k}\Omega}\right)24\text{ V} = 14.4\text{ V}$$

To find I_D, the constant K must first be determined using the parameters given.

$$K = \frac{I_{D(on)}}{[V_{GS} - V_{GS(th)}]^2} = \frac{3\text{ mA}}{(10\text{ V} - 5\text{ V})^2} = \frac{3\text{ mA}}{25\text{ V}^2} = 0.12\text{ mA/V}^2$$

Now I_D can be calculated for $V_{GS} = 14.4$ V.

$$\begin{aligned} I_D &= K[V_{GS} - V_{GS(th)}]^2 \\ &= (0.12\text{ mA/V}^2)(14.4\text{ V} - 5\text{ V})^2 \\ &= (0.12\text{ mA/V}^2)(9.4\text{ V})^2 \\ &= 10.6\text{ mA} \end{aligned}$$

Finally, V_{DS} can be calculated.

$$\begin{aligned} V_{DS} &= V_{DD} - I_D R_D \\ &= 24\text{ V} - (10.6\text{ mA})(1\text{ k}\Omega) \\ &= 13.4\text{ V} \end{aligned}$$

PRACTICE EXERCISE 7–16

Determine V_{GS} and V_{DS} for the E-MOSFET in Figure 7–40 when $I_{D(on)} = 4$ mA at $V_{GS} = 8$ V and $V_{GS(th)} = 3$ V.

In the drain-feedback bias circuit in Figure 7–39(b), there is negligible gate current and, therefore, no voltage drop across R_G. This makes $V_{GS} = V_{DS}$.

EXAMPLE 7–17

Determine the amount of drain current in Figure 7–41. The MOSFET has a $V_{GS(th)} = 3$ V.

FIGURE 7–41

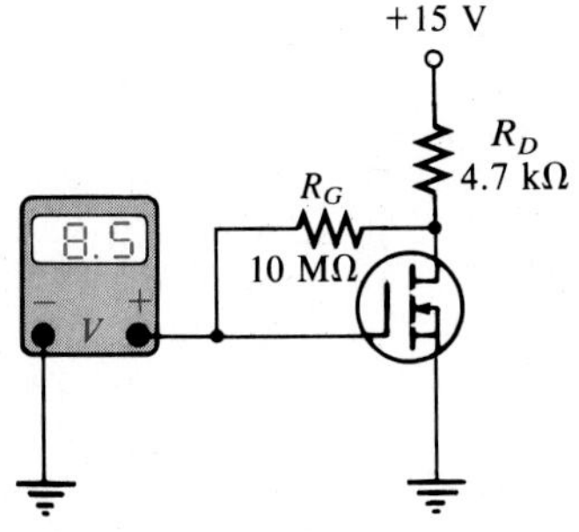

SOLUTION

The meter indicates $V_{GS} = 8.5$ V. Since this is a drain-feedback configuration, $V_{DS} = V_{GS} = 8.5$ V.

$$I_D = \frac{V_{DD} - V_{DS}}{R_D} = \frac{15\text{ V} - 8.5\text{ V}}{4.7\text{ k}\Omega} = 1.38\text{ mA}$$

PRACTICE EXERCISE 7–17

Determine I_D if the meter in Figure 7–41 reads 5 V.

7–6 REVIEW QUESTIONS

1. For a D-MOSFET biased at $V_{GS} = 0$, is the drain current zero, I_{GSS}, or I_{DSS}?
2. For an n-channel E-MOSFET with $V_{GS(th)} = 2$ V, V_{GS} must be in excess of what value in order to conduct?

7–7 TROUBLESHOOTING

In this section, we discuss some common faults that may be encountered in several FET circuits and give probable causes for each fault.

CASE 1: WHEN $V_D = V_{DD}$ IN A SELF-BIASED JFET

For this condition, the drain current must be zero as illustrated in Figure 7–42(a) because there is no voltage drop across R_D. As in any circuit, it is good troubleshooting practice to first check for obvious problems such as an open or poor ground connection, as well as charred resistors. Next, disconnect power and measure suspected resistors for opens. If these are OK, the JFET is probably bad. A list of faults that can produce this symptom is as follows:

1. No ground connection at bottom of R_S.
2. R_D open.
3. R_S open.
4. Open drain lead connection.
5. Open source lead connection.
6. Internal open between drain and source.

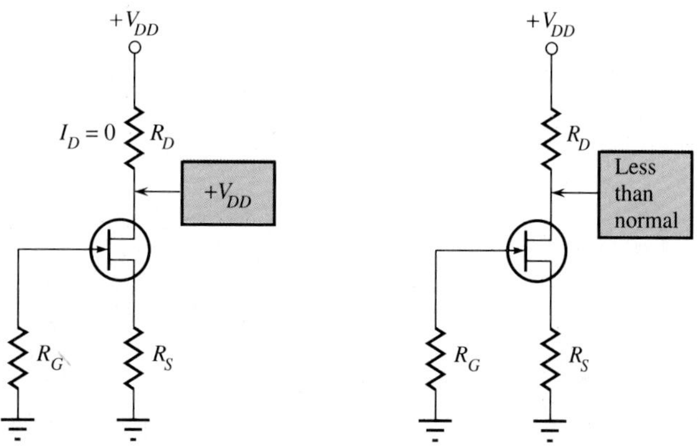

(a) Drain voltage equal to supply voltage (b) Drain voltage less than normal

FIGURE 7–42
Two symptoms in a self-biased JFET circuit.

CASE 2: WHEN V_D IS SIGNIFICANTLY LESS THAN NORMAL IN A SELF-BIASED JFET

For this condition, unless the supply voltage is lower than it should be, the drain current must be larger than normal because the drop across R_D is too much. Figure 7–42(b) indicates this situation. This symptom can be caused by any one of the following.

1. Open R_G
2. Open gate lead
3. Open pn junction

Any one of these faults will cause the depletion region in the JFET to disappear and the channel to widen, so that the drain current is limited only by R_D, R_S, and the small channel resistance.

FAULTS IN D-MOSFET AND E-MOSFET CIRCUITS

One fault that is difficult to detect is when either the gate or R_G opens in a zero-biased D-MOSFET or an E-MOSFET with drain-feedback bias. In a zero-biased D-MOSFET, the gate-to-source voltage remains zero when an open occurs in the gate circuit; thus, the drain current doesn't change, and the bias appears normal as indicated in Figure 7–43.

In an E-MOSFET circuit with voltage-divider bias, an open R_1 makes the gate voltage zero. This causes the transistor to be off and act like an open switch because a gate-to-source threshold voltage greater than zero is required to turn the device on. This condition is illustrated in Figure 7–44(b).

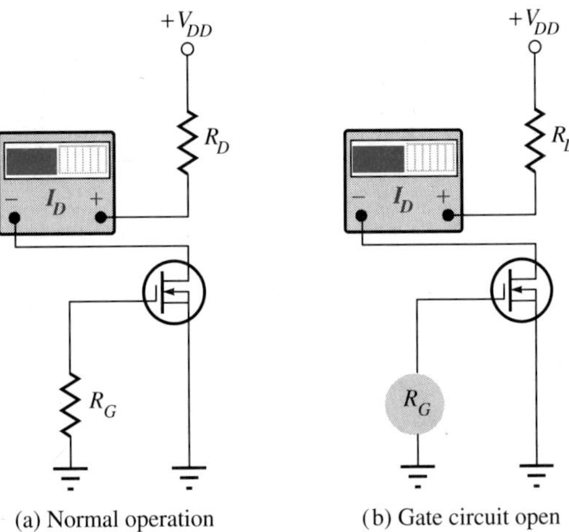

(a) Normal operation (b) Gate circuit open

FIGURE 7–43 An open fault in the gate circuit of a D-MOSFET causes no change in I_D.

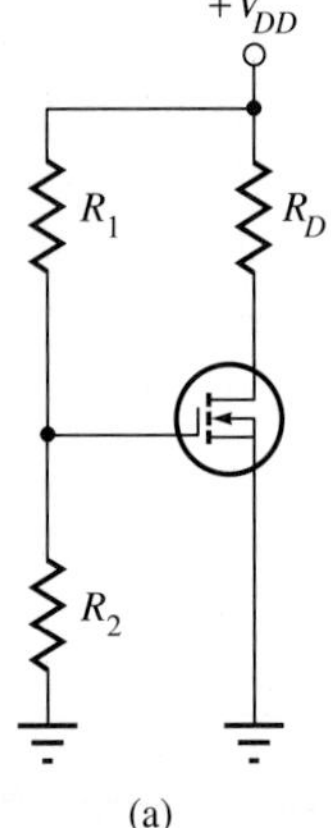

(a)

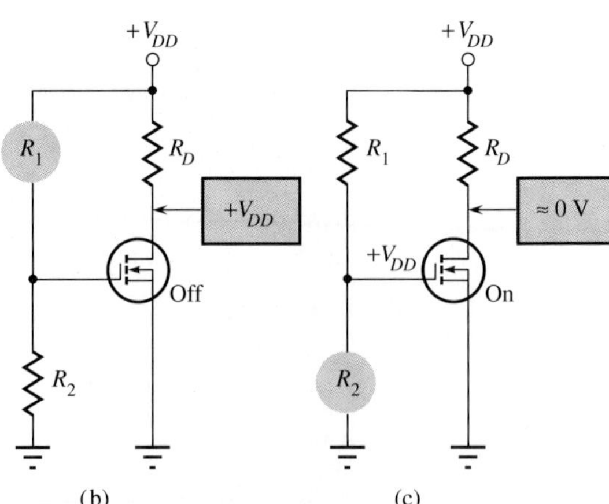

(b) (c)

FIGURE 7–44 Failures in an E-MOSFET circuit with voltage-divider bias.

If R_2 opens, the gate is at $+V_{DD}$ and the channel resistance is very low so the device approximates a closed switch. The drain current is limited only by R_D. This condition is illustrated in Figure 7–44(c).

7–7 REVIEW QUESTIONS

1. In a self-biased JFET circuit, the drain voltage equals V_{DD}. If the JFET is OK, what are other possible faults?
2. Why doesn't the drain current change when an open occurs in the gate circuit of a zero-biased D-MOSFET circuit?
3. If the gate of an E-MOSFET becomes shorted to ground in a circuit with voltage-divider bias, what is the drain voltage?

7–8 A SYSTEM APPLICATION

The purpose of the FET circuits in the waste water neutralization system introduced at the beginning of the chapter is to accept the voltages from the pH sensors and to provide output voltages to the A/D converters that are proportional to the value of the pH of the water at each measuring point. The key things we are interested in are the FET's high input resistance and the variation of drain current with gate-to-source voltage or transfer characteristic. In this section, you will

- ☐ ***See how the high input resistance of the D-MOSFET prevents loading of the sensor output.***
- ☐ ***Use the data sheet to get information from the transconductance curve.***
- ☐ ***Translate between a printed circuit board and a schematic.***
- ☐ ***Troubleshoot some FET failures.***

A BRIEF DESCRIPTION OF THE SYSTEM

The block diagram for the waste water neutralization system is shown in Figure 7–45. The purpose of the system is to measure and control the pH of the waste water being processed. The pH is an index of the chemical hydrogen activity of a solution. Values range from 1 to 2 for strong acids, through 7 for neutral solutions such as very pure water, on up to 11 or more for strong bases. Typical waste water is not a strong acid or a strong base, so the range of pH values that we must deal with is somewhat smaller than the values mentioned.

The pH of the water is measured at the inlet and outlet of the neutralization basin and at the outlet of the smoothing basin where the pH should have a value of 7, indicating pure water. The output voltage from each sensor is fed into the pH sensor circuitry, which consists of three identical FET circuits. The output voltages from these FET circuits are sent to the A/D converters and digital controller. Here, the voltage levels are changed to digital information for processing. Based on the digitized pH values, the controller

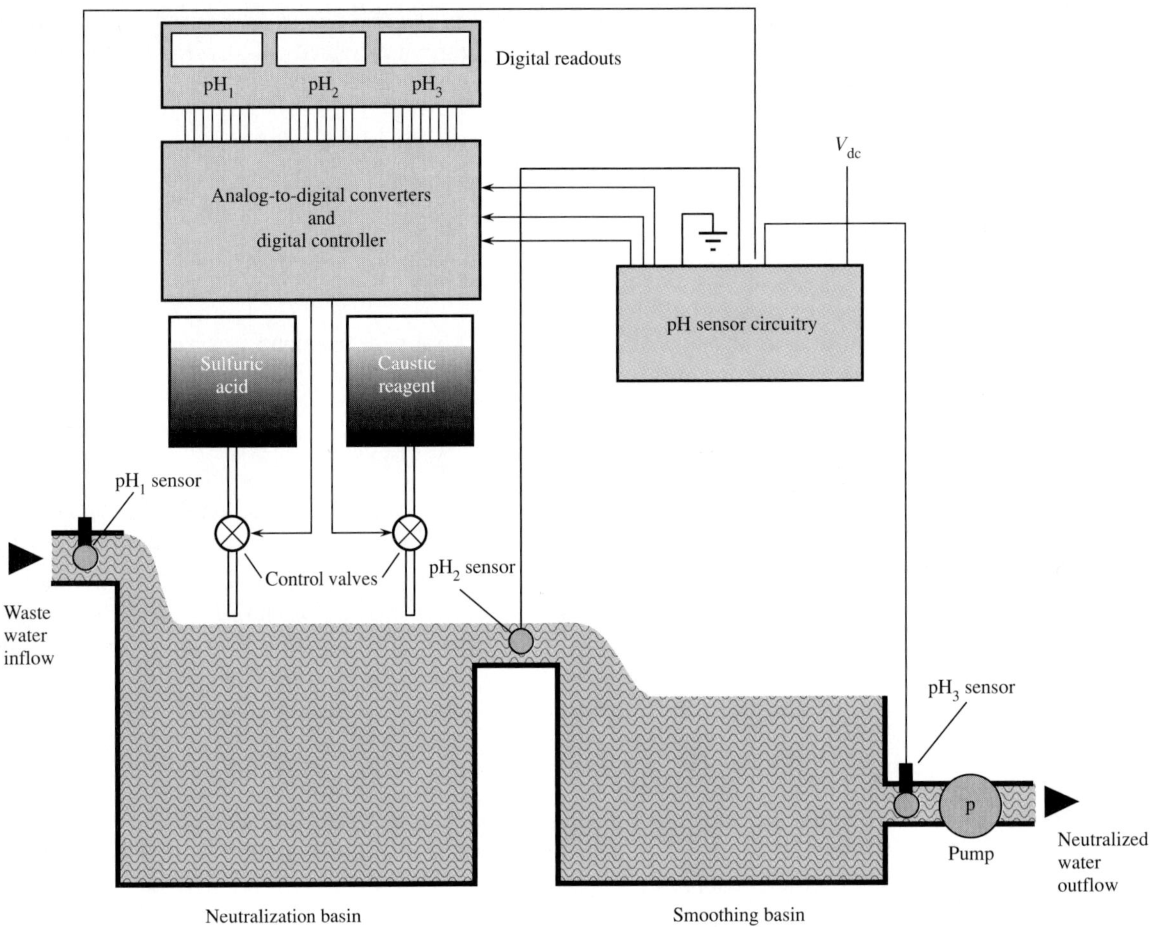

FIGURE 7–45

decides whether to add sulfuric acid or caustic reagent to the water and determines the quantity to be added. The digital controller operates the control valves for just the right amount of chemical to properly adjust the pH level. Also, the digitized pH values are sent to a display panel for visual monitoring.

Although the overall system operation is very interesting, in this system application we will concentrate only on the pH sensor circuitry.

Now, so that you can take a closer look at the pH sensor circuits, let's take the board out of the system and put it on the test bench.

ON THE TEST BENCH

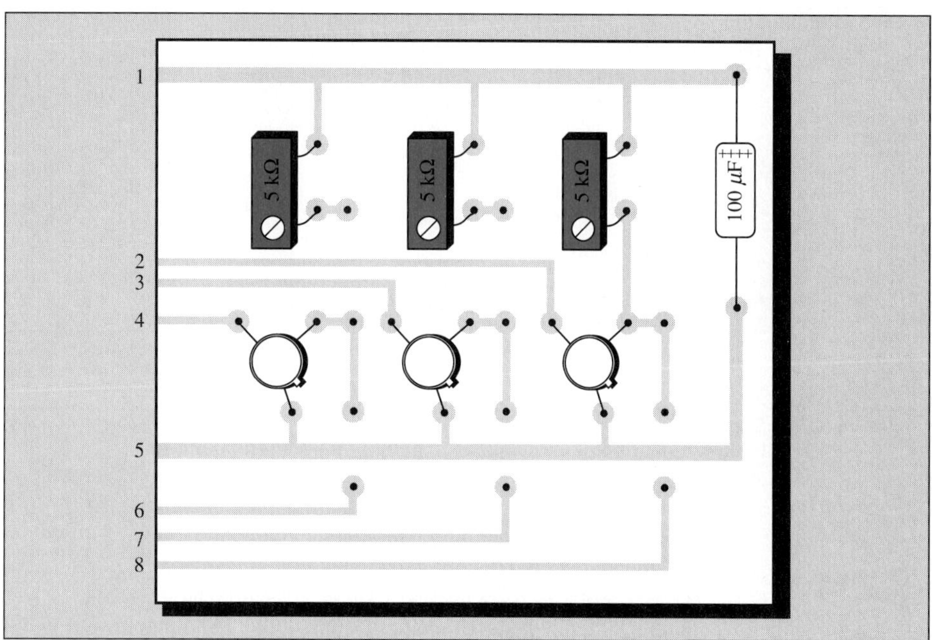

FIGURE 7–46

ACTIVITY 1 RELATE THE PC BOARD TO THE SCHEMATIC

Develop a circuit schematic for the pc board in Figure 7–46. You will also have to assign component labels for reference, since these are not provided on the pc board. Notice that this is a two-sided pc board, where some of the conductive traces are on the bottom side. Feedthroughs, indicated by empty pads, connect the top and bottom sides. In this particular case, you should be able to determine the connections that are on the bottom side without seeing that side.

ACTIVITY 2 ANALYZE THE CIRCUIT

STEP 1 Determine the input resistance of the 2N3797 D-MOSFET that is used in this circuit from information given on the data sheet in Appendix A. Use typical values and/or values specified at 25°C, if there is a choice.

STEP 2 Using the information provided in Figure 7–47 on the pH sensor element and data from the MOSFET data sheet, determine the value to which the potentiometers must be adjusted to provide a dc voltage of 7 V at the drain for a neutral solution (pH = 7). The dc supply voltage is +15 V.

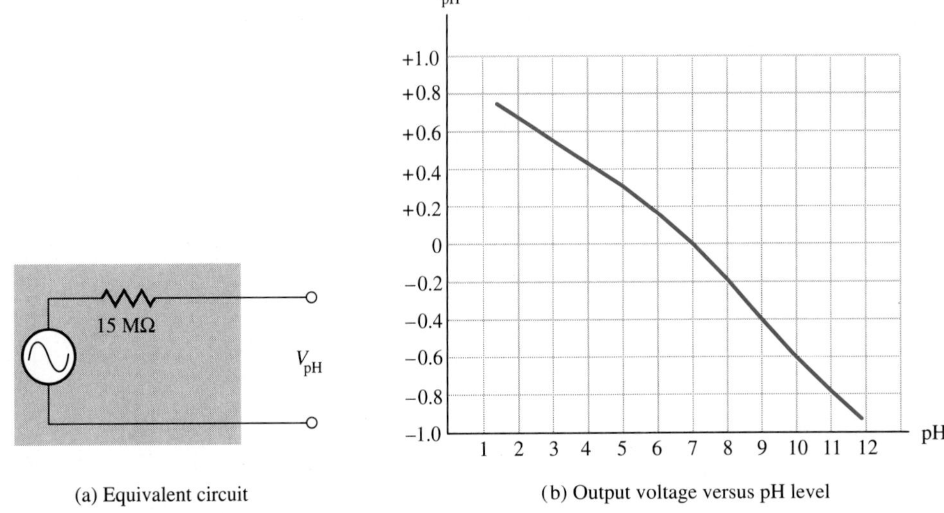

(a) Equivalent circuit (b) Output voltage versus pH level

FIGURE 7–47
Equivalent circuit and output characteristic curve for the pH sensor used in this system.

STEP 3 Using the transfer characteristic curve from the MOSFET data sheet, determine the range of output voltage (drain voltage) for a +500 mV change from the neutral output of the pH sensor. What is the range of pH values represented?

■ ACTIVITY 3 WRITE A TECHNICAL REPORT

Discuss the overall operation of the pH sensor circuit. State the purpose of each component on the board. Use the results of Activity 2 where appropriate.

■ ACTIVITY 4 TROUBLESHOOT THE CIRCUIT BOARD FOR EACH OF THE FOLLOWING PROBLEMS BY STATING THE PROBABLE CAUSE OR CAUSES IN EACH CASE

1. +15 V at pin 6.
2. Voltage at pin 8 doesn't change when you vary the voltage at pin 2 above and below 0 V.
3. Floating level at pin 7.

7–8 REVIEW QUESTIONS

1. What do you think is the purpose of the 100 μF capacitor on the pc board?
2. How much larger is the input resistance of the MOSFET than the internal resistance of the pH sensor element? Is there any significant loading effect?
3. How do you compensate for variations in the specified value of I_{DSS}?
4. Can you directly substitute JFETs for MOSFETs in this application? If not, why?

SUMMARY

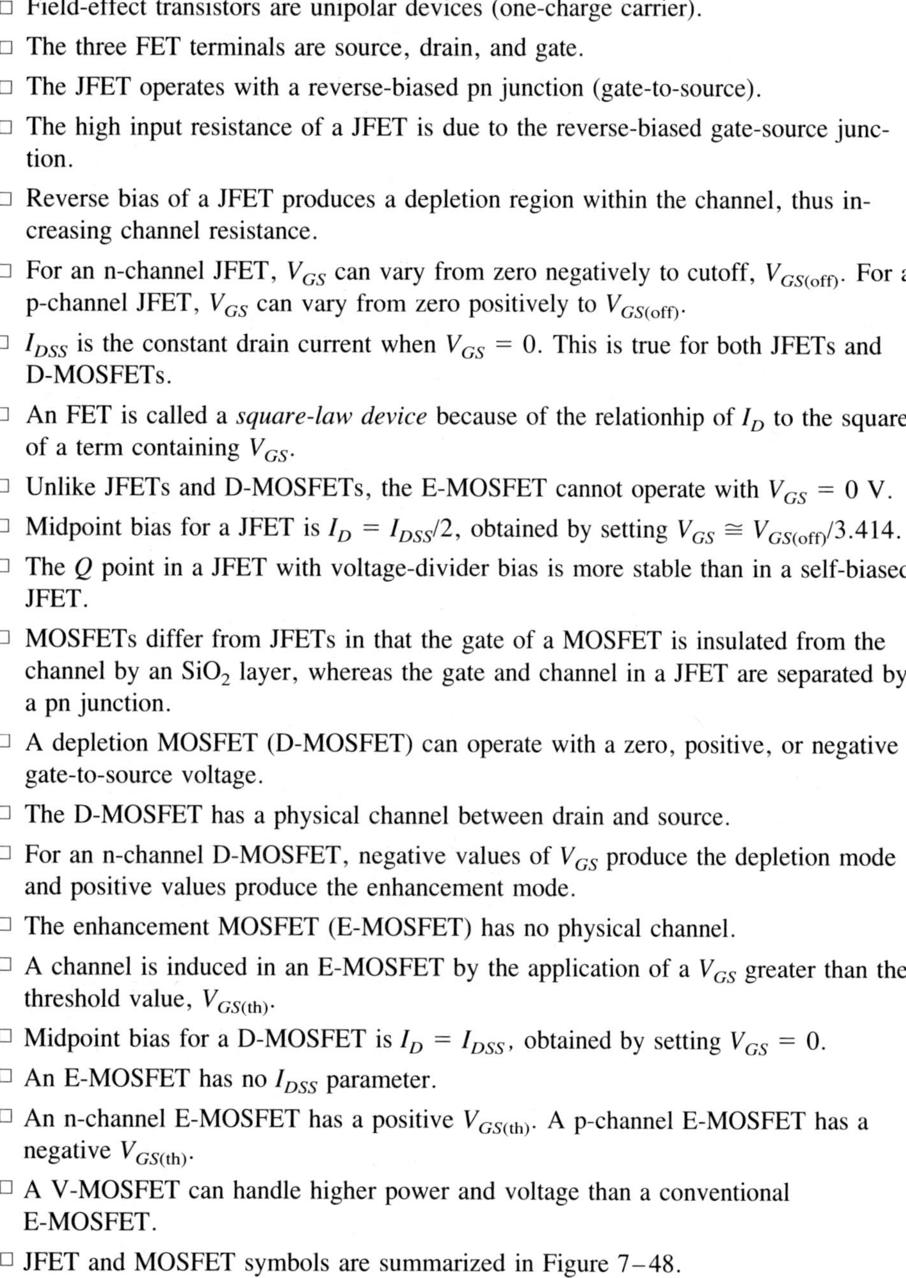

- □ Field-effect transistors are unipolar devices (one-charge carrier).
- □ The three FET terminals are source, drain, and gate.
- □ The JFET operates with a reverse-biased pn junction (gate-to-source).
- □ The high input resistance of a JFET is due to the reverse-biased gate-source junction.
- □ Reverse bias of a JFET produces a depletion region within the channel, thus increasing channel resistance.
- □ For an n-channel JFET, V_{GS} can vary from zero negatively to cutoff, $V_{GS(\text{off})}$. For a p-channel JFET, V_{GS} can vary from zero positively to $V_{GS(\text{off})}$.
- □ I_{DSS} is the constant drain current when $V_{GS} = 0$. This is true for both JFETs and D-MOSFETs.
- □ An FET is called a *square-law device* because of the relationhip of I_D to the square of a term containing V_{GS}.
- □ Unlike JFETs and D-MOSFETs, the E-MOSFET cannot operate with $V_{GS} = 0$ V.
- □ Midpoint bias for a JFET is $I_D = I_{DSS}/2$, obtained by setting $V_{GS} \cong V_{GS(\text{off})}/3.414$.
- □ The Q point in a JFET with voltage-divider bias is more stable than in a self-biased JFET.
- □ MOSFETs differ from JFETs in that the gate of a MOSFET is insulated from the channel by an SiO_2 layer, whereas the gate and channel in a JFET are separated by a pn junction.
- □ A depletion MOSFET (D-MOSFET) can operate with a zero, positive, or negative gate-to-source voltage.
- □ The D-MOSFET has a physical channel between drain and source.
- □ For an n-channel D-MOSFET, negative values of V_{GS} produce the depletion mode and positive values produce the enhancement mode.
- □ The enhancement MOSFET (E-MOSFET) has no physical channel.
- □ A channel is induced in an E-MOSFET by the application of a V_{GS} greater than the threshold value, $V_{GS(\text{th})}$.
- □ Midpoint bias for a D-MOSFET is $I_D = I_{DSS}$, obtained by setting $V_{GS} = 0$.
- □ An E-MOSFET has no I_{DSS} parameter.
- □ An n-channel E-MOSFET has a positive $V_{GS(\text{th})}$. A p-channel E-MOSFET has a negative $V_{GS(\text{th})}$.
- □ A V-MOSFET can handle higher power and voltage than a conventional E-MOSFET.
- □ JFET and MOSFET symbols are summarized in Figure 7–48.

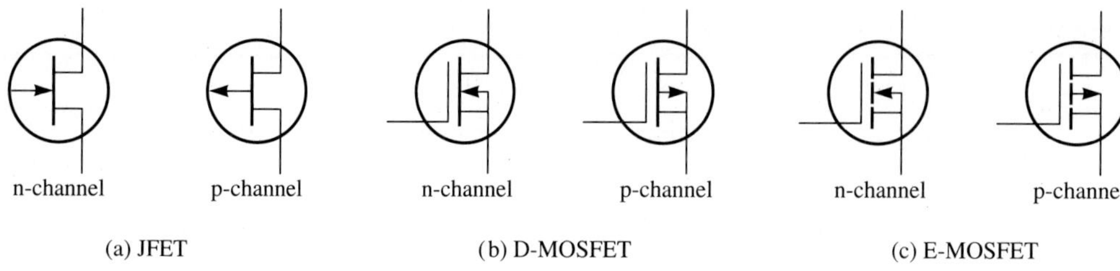

FIGURE 7–48

GLOSSARY

Channel The conductive path between the drain and source in an FET.

Depletion In a MOSFET, the process of removing or depleting the channel of charge carriers and thus decreasing the channel conductivity.

Drain One of the three terminals of an FET.

Enhancement In a MOSFET, the process of creating a channel or increasing the conductivity of the channel by the addition of charge carriers.

Field-Effect Transistor (FET) A type of unipolar, voltage-controlled transistor that uses an induced electric field to control current.

Gate One of the three terminals of an FET.

Junction Field-Effect Transistor (JFET) One of two major types of field-effect transistor.

MOSFET Metal oxide semiconductor field-effect transistor. One of two major types of FET.

Pinch-off voltage The value of the drain-to-source voltage of an FET at which the drain current becomes constant when the gate-to-source voltage is zero.

Source One of the three terminals of an FET.

Transconductance The ratio of a change in drain current for a change in gate-to-source voltage in an FET.

FORMULAS

(7–1) $$I_D = I_{DSS}\left[1 - \frac{V_{GS}}{V_{GS(\text{off})}}\right]^2$$ JFET/D-MOSFET transfer characteristic

(7–2) $$g_m = \frac{\Delta I_D}{\Delta V_{GS}}$$ Transconductance

(7–3) $$g_m = g_{m0}\left[1 - \frac{V_{GS}}{V_{GS(\text{off})}}\right]$$ Transconductance

(7–4)	$g_{m0} = \dfrac{2I_{DSS}}{\lvert V_{GS(\text{off})}\rvert}$	Transconductance
(7–5)	$R_{\text{IN}} = \left\lvert \dfrac{V_{GS}}{I_{GSS}} \right\rvert$	Input resistance
(7–6)	$r_{ds} = \dfrac{\Delta V_{DS}}{\Delta I_D}$	Drain-to-source resistance
(7–7)	$V_{GS} = -I_D R_S$	N-channel JFET
(7–8)	$V_{GS} = +I_D R_S$	P-channel JFET
(7–9)	$V_D = V_{DD} - I_D R_D$	Drain-to-ground voltage (self-bias)
(7–10)	$V_{DS} = V_{DD} - I_D(R_D + R_S)$	Drain-to-source voltage (self-bias)
(7–11)	$R_S = \left\lvert \dfrac{V_{GS}}{I_D} \right\rvert$	Source resistance
(7–12)	$V_G = \left(\dfrac{R_2}{R_1 + R_2}\right)V_{DD}$	Voltage-divider bias gate voltage
(7–13)	$I_D = \dfrac{V_G - V_{GS}}{R_S}$	Voltage-divider bias drain current
(7–14)	$I_D = K[V_{GS} - V_{GS(\text{th})}]^2$	E-MOSFET transfer characteristic
(7–15)	$V_{DS} = V_{DD} - I_{DSS}R_D$	Drain-to-source voltage (zero bias)

SELF-TEST

1. The JFET is

(a) a unipolar device **(b)** a voltage-controlled device

(c) a current-controlled device **(d)** a and c

(e) a and b

2. The channel of a JFET is between the

(a) gate and drain **(b)** drain and source

(c) gate and source **(d)** input and output

3. A JFET always operates with

(a) the gate-to-source pn junction reverse-biased

(b) the gate-to-source pn junction forward-biased

(c) the drain connected to ground

(d) the gate connected to the source

4. For $V_{GS} = 0$ V, the drain current becomes constant when V_{DS} exceeds

(a) cutoff **(b)** V_{DD}

(c) V_P **(d)** 0 V

5. The constant-current region of an FET lies between
   (a) cutoff and saturation **(b)** cutoff and pinch-off
   (c) 0 and I_{DSS} **(d)** pinch-off and breakdown
6. I_{DSS} is
   (a) the drain current with the source shorted
   (b) the drain current at cutoff
   (c) the maximum possible drain current
   (d) the midpoint drain current
7. Drain current in the constant-current region increases when
   (a) the gate-to-source bias voltage decreases
   (b) the gate-to-source bias voltage increases
   (c) the drain-to-source voltage increases
   (d) the drain-to-source voltage decreases
8. In a certain FET circuit, $V_{GS} = 0$ V, $V_{DD} = 15$ V, $I_{DSS} = 15$ mA, and $R_D = 470\ \Omega$. If R_D is decreased to 330 Ω, I_{DSS} is
   (a) 19.5 mA **(b)** 10.5 mA **(c)** 15 mA **(d)** 1 mA
9. At cutoff, the JFET channel is
   (a) as its widest point **(b)** completely closed by the depletion region
   (c) extremely narrow **(d)** reverse-biased
10. A certain JFET data sheet gives $V_{GS(\text{off})} = -4$ V. The pinch-off voltage, V_P,
    (a) cannot be determined **(b)** is −4 V
    (c) depends on V_{GS} **(d)** is +4 V
11. The JFET in Question 10
    (a) is an n-channel **(b)** is a p-channel
    (c) can be either
12. For a certain JFET, $I_{GSS} = 10$ nA at $V_{GS} = 10$ V. The input resistance is
    (a) 100 MΩ **(b)** 1 MΩ **(c)** 1000 MΩ **(d)** 1000 mΩ
13. For a certain p-channel JFET, $V_{GS(\text{off})} = 8$ V. The value of V_{GS} for an approximate midpoint bias is
    (a) 4 V **(b)** 0 V **(c)** 1.25 V **(d)** 2.34 V
14. A MOSFET differs from a JFET mainly because
    (a) of the power rating **(b)** the MOSFET has two gates
    (c) the JFET has a pn junction **(d)** MOSFETs do not have a physical channel
15. A certain D-MOSFET is biased at $V_{GS} = 0$ V. Its data sheet specifies $I_{DSS} = 20$ mA and $V_{GS(\text{off})} = -5$ V. The value of the drain current
    (a) is 0 A **(b)** cannot be determined **(c)** is 20 mA
16. An n-channel D-MOSFET with a positive V_{GS} is operating in
    (a) the depletion mode **(b)** the enhancement mode
    (c) cutoff **(d)** saturation

17. A certain p-channel E-MOSFET has a $V_{GS(th)} = -2$ V. If $V_{GS} = 0$ V, the drain current is
(a) 0 A **(b)** $I_{D(on)}$ **(c)** maximum **(d)** I_{DSS}

18. A V-MOSFET is a special type of
(a) D-MOSFET **(b)** JFET **(c)** E-MOSFET **(d)** a and c

PROBLEMS

SECTION 7–1 THE JUNCTION FIELD-EFFECT TRANSISTOR (JFET)

1. The V_{GS} of a p-channel JFET is increased from 1 V to 3 V.
(a) Does the depletion region narrow or widen?
(b) Does the resistance of the channel increase or decrease?

2. Why must the gate-to-source voltage of an n-channel JFET always be either 0 or negative?

3. Sketch the schematic diagrams for a p-channel and an n-channel JFET. Label the terminals.

4. Show how to connect bias batteries between the gate and source of the JFETs in Figure 7–49.

FIGURE 7–49

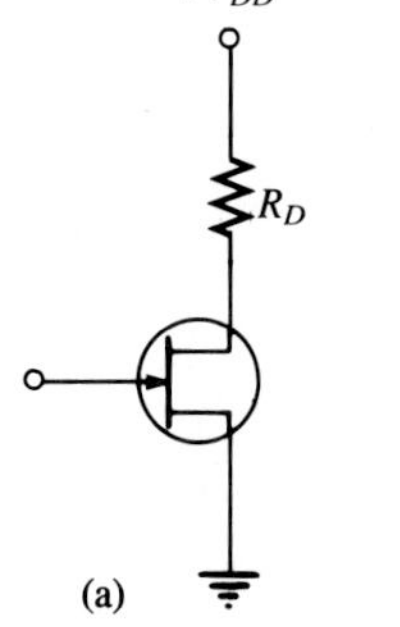

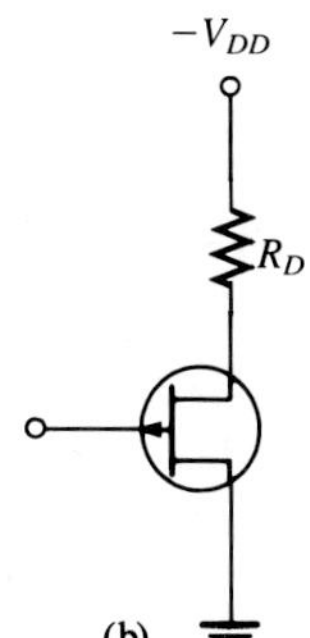

SECTION 7–2 JFET CHARACTERISTICS AND PARAMETERS

5. A JFET has a specified pinch-off voltage of 5 V. When $V_{GS} = 0$, what is V_{DS} at the point where the drain current becomes constant?

6. A certain n-channel JFET is biased such that $V_{GS} = -2$ V. What is the value of $V_{GS(off)}$ if V_P is specified to be 6 V? Is the device on?

7. A certain JFET data sheet gives $V_{GS(off)} = -8$ V and $I_{DSS} = 10$ mA. When $V_{GS} = 0$, what is I_D for values of V_{DS} above pinch-off? $V_{DD} = 15$ V.

8. A certain p-channel JFET has a $V_{GS(off)} = 6$ V. What is I_D when $V_{GS} = 8$ V?

9. The JFET in Figure 7–50 has a $V_{GS(off)} = -4$ V. Assume that you increase the supply voltage, V_{DD}, beginning at zero until the ammeter reaches a steady value. What does the voltmeter read at this point?

FIGURE 7–50

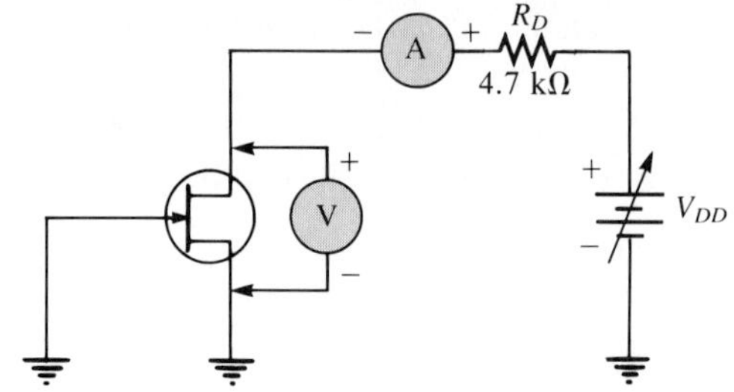

10. The following parameters are obtained from a certain JFET data sheet: $V_{GS(off)} = -8$ V and $I_{DSS} = 5$ mA. Determine the values of I_D for each value of V_{GS} ranging from 0 V to −8 V in 1 V steps. Plot the transfer characteristic curve from these data.

11. For the JFET in Problem 10, what value of V_{GS} is required to set up a drain current of 2.25 mA?

12. For a particular JFET, $g_{m0} = 3200\ \mu$S. What is g_m when $V_{GS} = -4$ V, given that $V_{GS(off)} = -8$ V?

13. Determine the forward transconductance of a JFET biased at $V_{GS} = -2$ V. From the data sheet, $V_{GS(off)} = -7$ V and $g_m = 2000\ \mu$S at $V_{GS} = 0$ V. Also determine the forward transfer admittance, y_{fs}.

14. A p-channel JFET data sheet shows that $I_{GSS} = 5$ nA at $V_{GS} = 10$ V. Determine the input resistance.

15. Using Equation (7–1), plot the transfer characteristic curve for a JFET with $I_{DSS} = 8$ mA and $V_{GS(off)} = -5$ V using at least four points.

SECTION 7–3 JFET BIASING

16. An n-channel self-biased JFET has a drain current of 12 mA and a 100 Ω source resistor. What is the value of V_{GS}?

17. Determine the value of R_S required for a self-biased JFET to produce a V_{GS} of −4 V when $I_D = 5$ mA.

18. Determine the value of R_S required for a self-biased JFET to produce $I_D = 2.5$ mA when $V_{GS} = -3$ V.

19. $I_{DSS} = 20$ mA and $V_{GS(off)} = -6$ V for a particular JFET.
    (a) What is I_D when $V_{GS} = 0$ V?
    (b) What is I_D when $V_{GS} = V_{GS(off)}$?
    (c) If V_{GS} is increased from −4 V to −1 V, does I_D increase or decrease?

20. For each circuit in Figure 7–51, determine V_{DS} and V_{GS}.

21. Using the curve in Figure 7–52, determine the value of R_S required for a 9.5 mA drain current.

22. Set up a midpoint bias for a JFET with $I_{DSS} = 14$ mA and $V_{GS(off)} = -10$ V. Use a 24 V dc source as the supply voltage. Show the circuit and resistor values. Indicate the values of I_D, V_{GS}, and V_{DS}.

23. Determine the total input resistance in Figure 7–53. $I_{GSS} = 20$ nA at $V_{GS} = -10$ V.

FIGURE 7–51

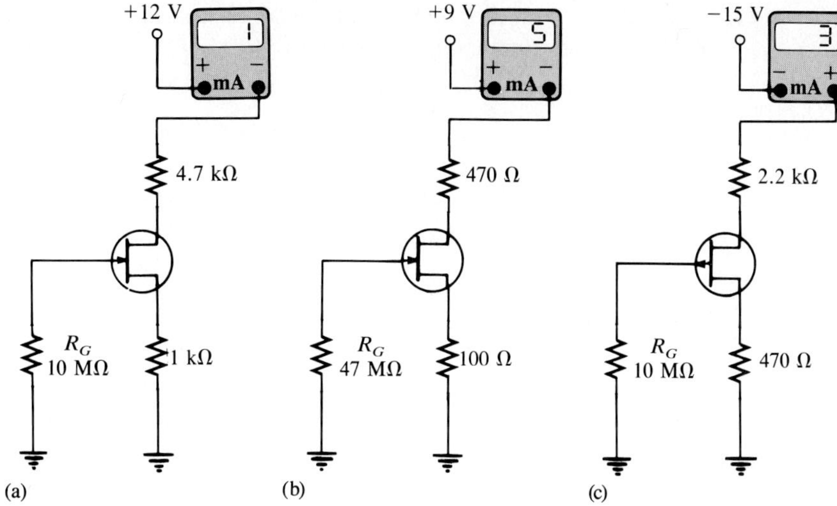

FIGURE 7–52

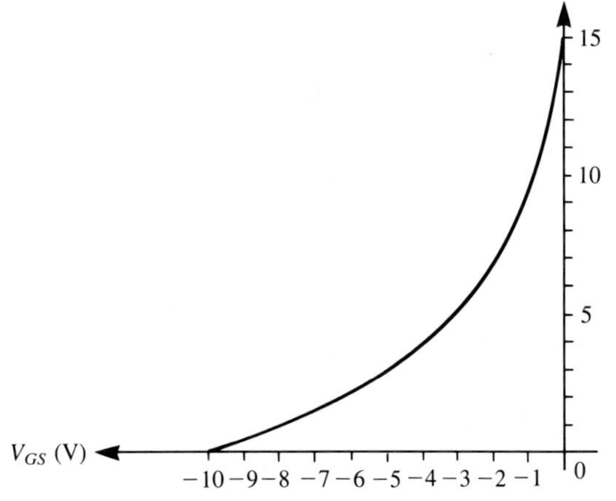

FIGURE 7–53

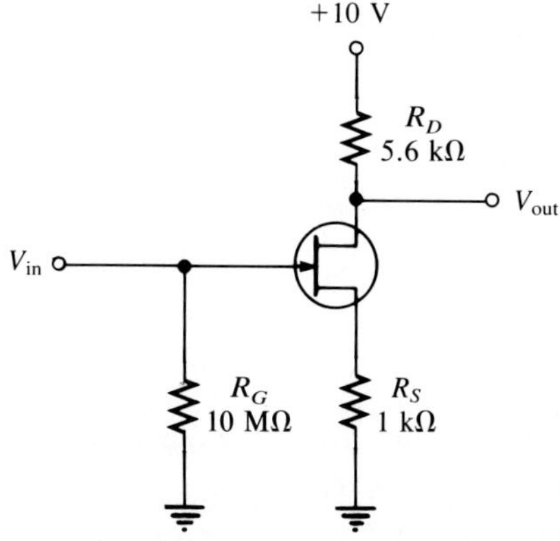

24. Graphically determine the Q point for the circuit in Figure 7–54(a) using the transfer characteristic curve in Figure 7–54(b).

FIGURE 7–54

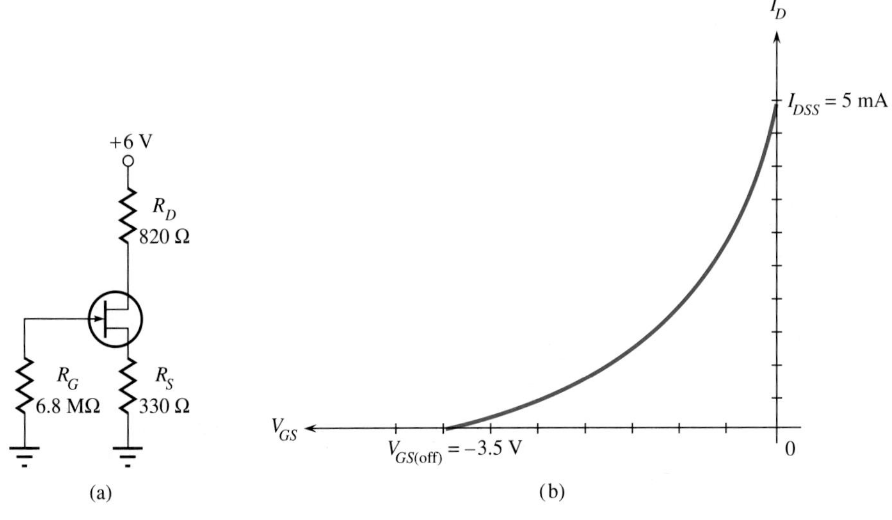

25. Find the Q point for the p-channel JFET circuit in Figure 7–55.

FIGURE 7–55

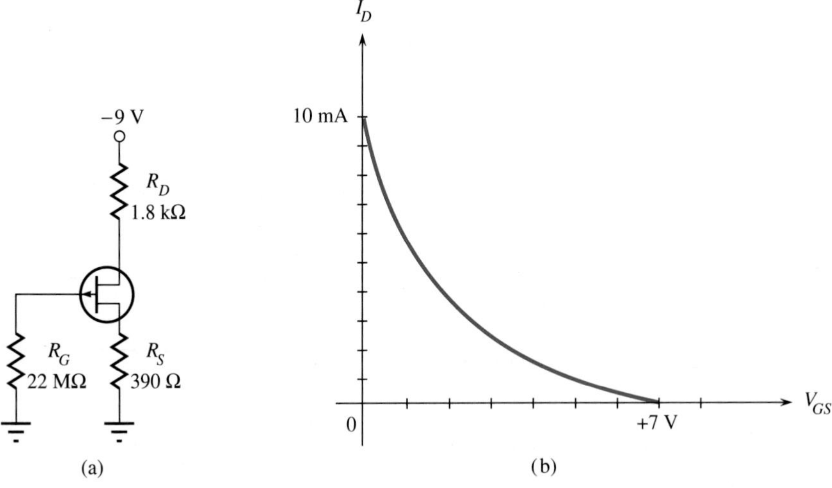

26. Given that the drain-to-ground voltage in Figure 7–56 is 5 V, determine the Q point of the circuit.
27. Find the Q-point values for the JFET with voltage-divider bias in Figure 7–57.

FIGURE 7–56

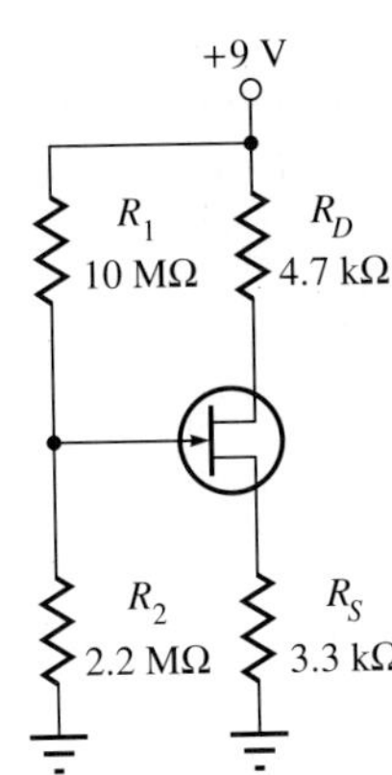

FIGURE 7–57

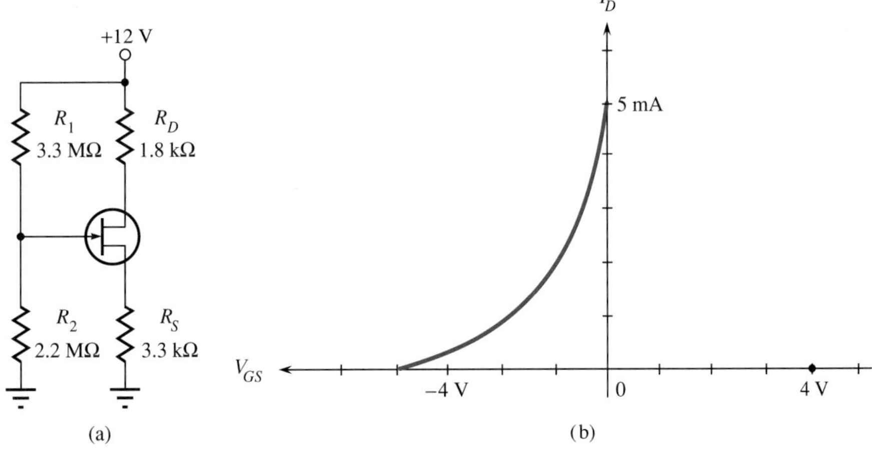

SECTION 7–4 THE METAL OXIDE SEMICONDUCTOR FET (MOSFET)

28. Sketch the schematic symbols for n-channel and p-channel D-MOSFETs and E-MOSFETs. Label the terminals.

29. In what mode is an n-channel D-MOSFET with a positive V_{GS} operating?

30. Describe the basic difference between a D-MOSFET and an E-MOSFET.

31. Explain why both types of MOSFETs have an extremely high input resistance at the gate.

SECTION 7–5 MOSFET CHARACTERISTICS AND PARAMETERS

32. The data sheet for a certain D-MOSFET gives $V_{GS(\text{off})} = -5$ V and $I_{DSS} = 8$ mA.

(a) Is this device p-channel or n-channel?

(b) Determine I_D for values of V_{GS} ranging from -5 V to $+ 5$ V in increments of 1 V.

(c) Plot the transfer characteristic curve using the data from part (b).

33. Determine I_{DSS}, given $I_D = 3$ mA, $V_{GS} = -2$ V, and $V_{GS(off)} = -10$ V.

34. The data sheet for an E-MOSFET reveals that $I_{D(on)} = 10$ mA at $V_{GS} = -12$ V and $V_{GS(th)} = -3$ V. Find I_D when $V_{GS} = -6$ V.

SECTION 7–6 MOSFET BIASING

35. Determine in which mode (depletion or enhancement) each D-MOSFET in Figure 7–58 is biased.

FIGURE 7–58

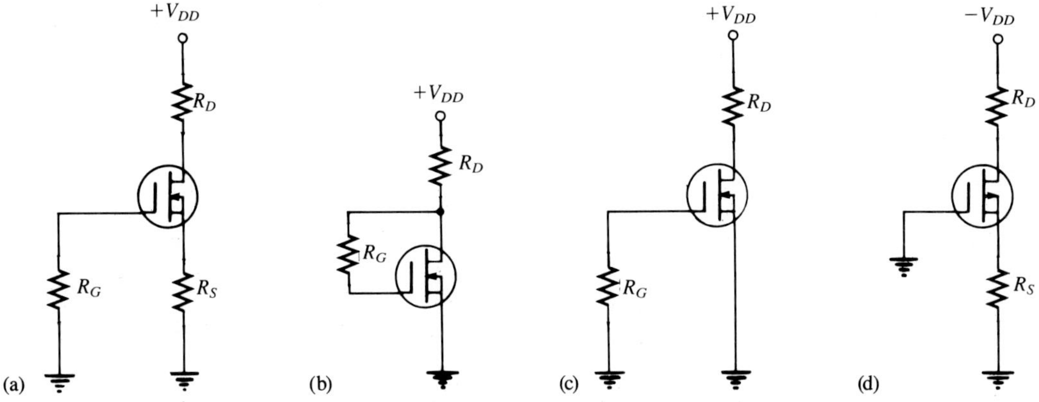

36. Each E-MOSFET in Figure 7–59 has a $V_{GS(th)}$ of +5 V or −5 V, depending on whether it is an n-channel or a p-channel device. Determine whether each MOSFET is on or off.

FIGURE 7–59

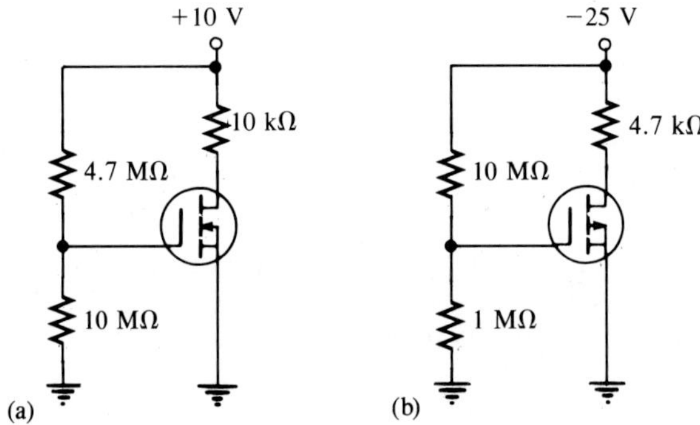

37. Determine V_{DS} for each circuit in Figure 7–60. $I_{DSS} = 8$ mA.

38. Find V_{GS} and V_{DS} for the E-MOSFETs in Figure 7–61. Data sheet information is listed with each circuit.

39. Based on the identical V_{GS} measurements, determine the drain current and drain-to-source voltage for each circuit in Figure 7–62.

40. Determine the actual gate-to-source voltage in Figure 7–63 by taking into account the gate leakage current I_{GSS}. Assume that I_{GSS} is 50 pA and I_D is 1 mA under the existing bias conditions.

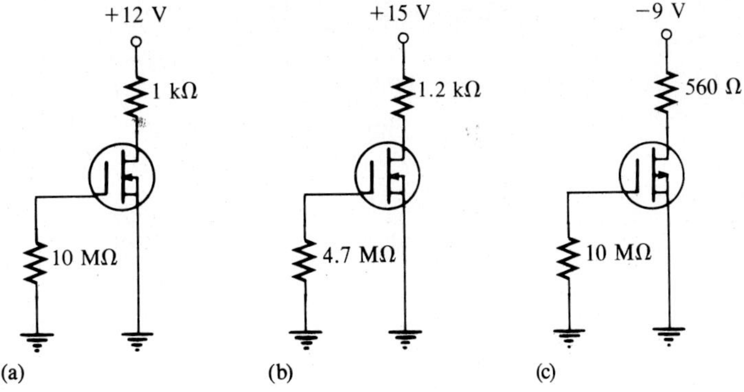

FIGURE 7–60

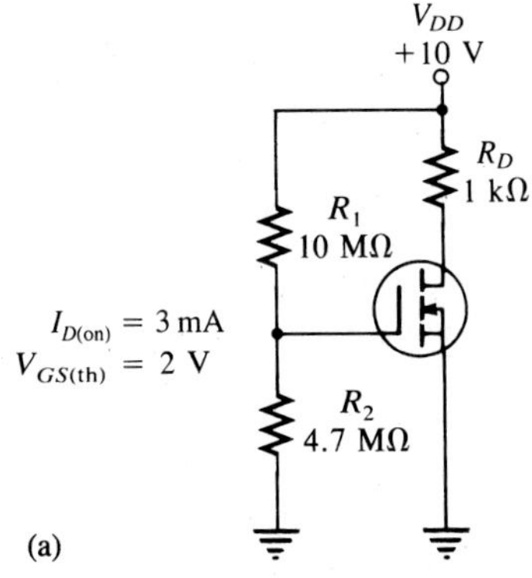

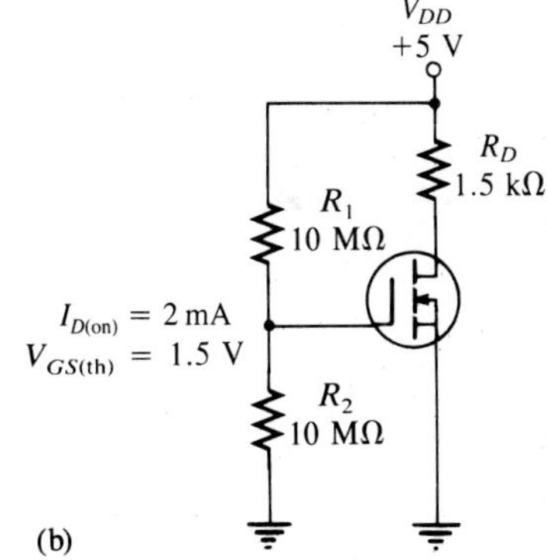

FIGURE 7–61

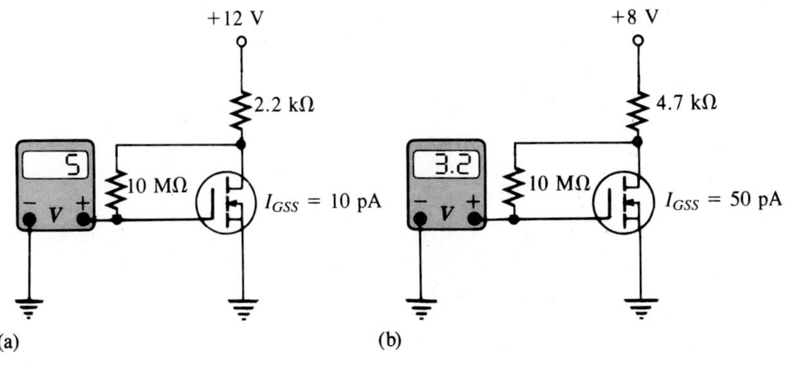

FIGURE 7–62

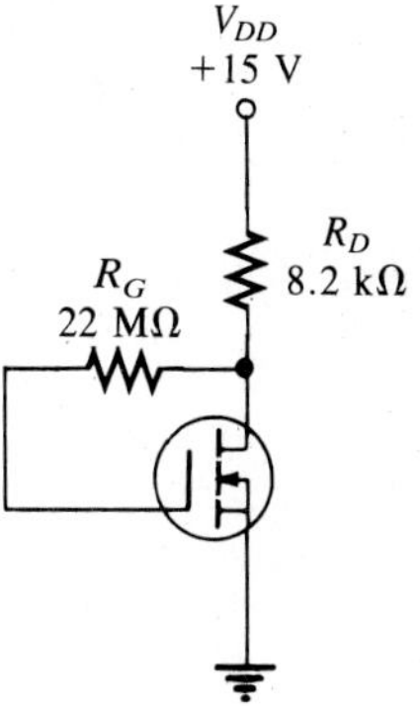

FIGURE 7–63

SECTION 7–7 TROUBLESHOOTING

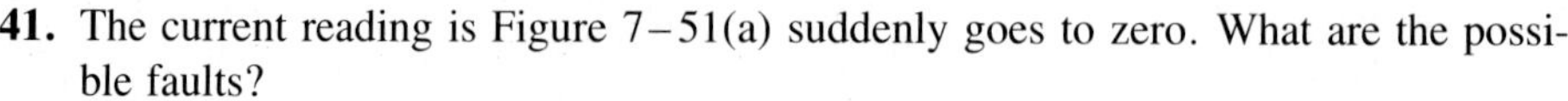

41. The current reading is Figure 7–51(a) suddenly goes to zero. What are the possible faults?

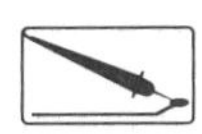

42. The current reading in Figure 7–51(b) suddenly jumps to approximately 16 mA. What are the possible faults?

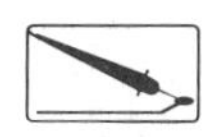

43. If the supply voltage in Figure 7–51(c) is accidentally changed to −20 V, what would you see on the ammeter?

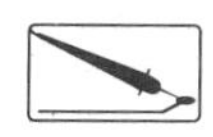

44. You measure +10 V at the drain of the MOSFET in Figure 7–59(a). The transistor checks good and the ground connections are OK. What can be the problem?

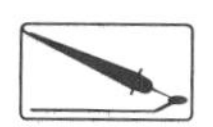

45. You measure approximately 0 V at the drain of the MOSFET in Figure 7–59(b). You can find no shorts and the transistor checks good. What is the most likely problem?

ANSWERS TO REVIEW QUESTIONS

SECTION 7–1

1. Drain, source, and gate
2. An n-channel JFET requires a negative V_{GS}.
3. I_D is controlled by V_{GS}.

SECTION 7–2

1. When $V_{DS} = 7$ V at pinch-off and $V_{GS} = 0$ V, $V_P = -7$ V.
2. As V_{GS} increases negatively, I_D decreases.
3. For $V_P = -3$ V, $V_{GS(off)} = +3$ V.

SECTION 7–3

1. A p-channel JFET requires a positive V_{GS}.
2. $V_{GS} = V_G - V_S = 0\text{ V} - (8\text{ mA})(1\text{ k}\Omega) = -8\text{ V}$
3. $V_{GS} = V_G - V_S = 3\text{ V} - 5\text{ V} = -2\text{ V}$

SECTION 7–4

1. Depletion MOSFET (D-MOSFET) and enhancement MOSFET (E-MOSFET)
2. When $V_{GS} = 0$ V for a D-MOSFET, $I_D = I_{DSS}$.
3. When $V_{GS} = 0$ V for an E-MOSFET, $I_D = 0$ A.

SECTION 7–5

1. The D-MOSFET has a physical channel; the E-MOSFET does not.
2. The E-MOSFET does not have I_{DSS} or $V_{GS(off)}$.
3. ESD is electrostatic discharge.

SECTION 7–6

1. When $V_{GS} = 0$ V, the drain current is equal to I_{DSS}.
2. V_{GS} must exceed $V_{GS(th)} = 2$ V for conduction to occur.

SECTION 7–7

1. R_D or R_S open, no ground connection
2. Because V_{GS} remains at approximately zero
3. The device is off and $V_D = V_{DD}$.

SECTION 7–8

1. This is a decoupling capacitor that removes variations from the dc voltage.
2. The input resistance of the MOSFET is $V_{GS}/I_{GSS} = 10 \times 10^{12}\ \Omega$. There is no significant loading.
3. The potentiometers are used to adjust for variations in I_{DSS}.
4. No, a JFET cannot operate with both positive and negative values of V_{GS}.

ANSWERS TO PRACTICE EXERCISES

7–1 I_D remains at approximately 12 mA.

7–2 −4 V

7–3 2.4 mA

7–4 $g_m = 1800\ \mu S$, $I_D = 4.32$ mA

7–5 80,000 MΩ

7–6 $V_{DS} = 2$ V, $V_{GS} = -3.12$ V

7–7 231 Ω

7–8 889 Ω

7–9 $R_S = 586\ \Omega$, $R_D = 1500\ \Omega$

7–10 $V_{GS} \cong -1.75$ V, $I_D \cong 1.8$ mA

7–11 $I_D = 0.91$ mA, $V_{GS} = -\ 0.92$ V

7–12 $I_D \cong 1.05$ mA, $V_{GS} \cong -2.25$ V

7–13 (a) p-channel (b) 6.48 mA (c) 35.28 mA

7–14 1.24 mA

7–15 5.6 V

7–16 $V_{GS} = 14.4$ V, $V_{DS} = 3.2$ V

7–17 2.13 mA

8

Small-Signal FET Amplifiers

After completing this chapter, you should be able to

- □ Use FETs in amplifier circuits.
- □ Represent an FET by a simple equivalent circuit.
- □ Discuss the factors that affect the voltage gain of an FET amplifier.
- □ Describe the characteristics of common-source, common-drain, and common-gate amplifiers.
- □ Troubleshoot an FET amplifier.

The things you learned about biasing FETs in the last chapter are carried forward into this chapter where FET circuits are used as small-signal amplifiers. Because of their high input resistance and other factors, FETs are often preferred over bipolar transistors for certain applications in communications systems, such as the FM receiver. Many of the concepts that relate to amplifiers using BJTs apply as well to FET amplifiers. The three FET amplifier configurations are common-source, common-drain, and common-gate. These are analogous to the common-emitter, common-collector, and common-base configurations in bipolar amplifiers.

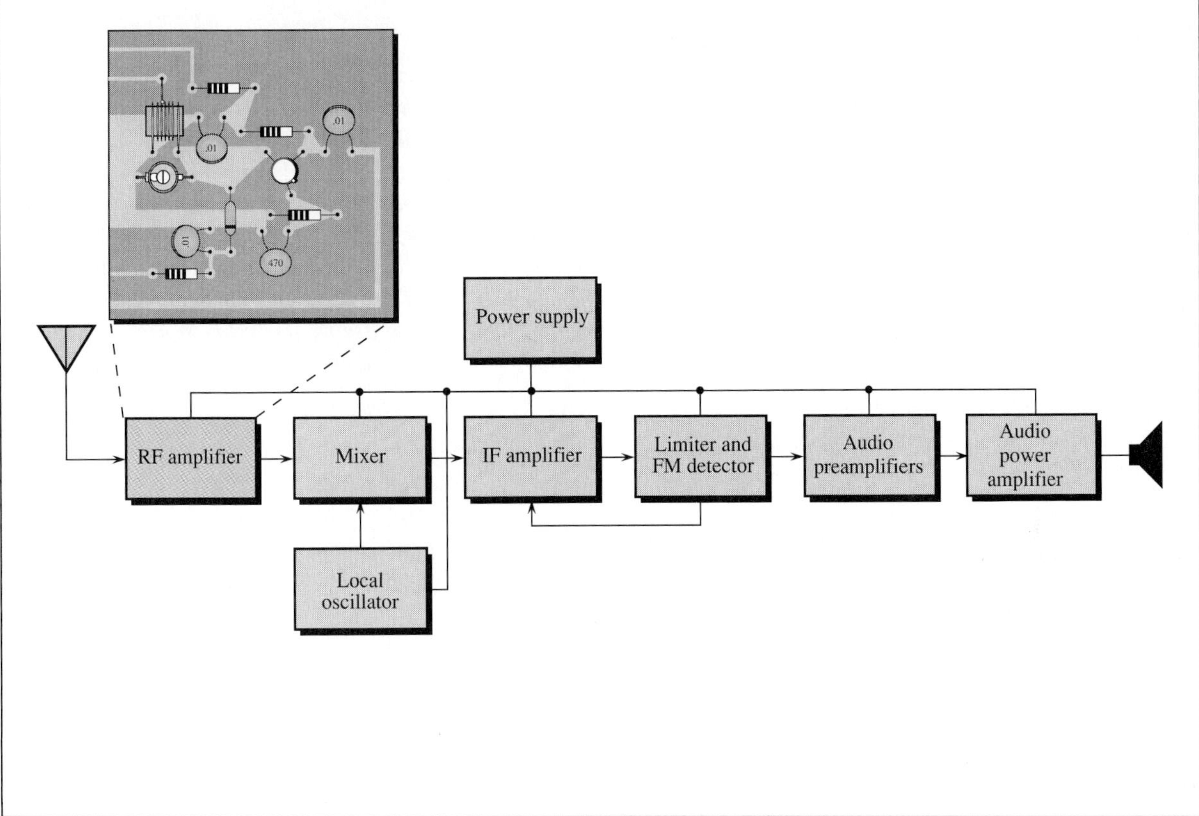

A System Application

You have already been introduced to the basic radio receiver. In this chapter, we come back to the receiver for our system application. This time we will look at an FM receiver, which is represented by the block diagram above, and focus on the RF amplifier. As before, you do not need to understand the complete system in order to apply what you will learn in this chapter to the RF amplifier. Parts of this system use circuits that have not been covered yet, and we will focus on some of these later.

The FM receiver selects frequencies from 88 MHz to 108 MHz in the broadcast band and extracts the audio signal. The purpose of the RF amplifier is to accept a small signal from the antenna using a tuned *LC* circuit to select the desired frequency from the wide range of transmitted frequencies and to provide initial amplification. After the selected signal has been amplified by the RF amplifier, it goes to the mixer where it is combined with a local oscillator frequency to produce the intermediate frequency (IF). The RF amplifier, mixer, and local oscillator together are often referred to as the receiver "front end."

For the system application in Section 8–7, in addition to the other topics, be sure you understand

- □ The principles of FET amplifier operation.
- □ The principles of parallel resonant circuits (go back to your dc/ac text if necessary).

8–1 SMALL-SIGNAL FET AMPLIFIER OPERATION

The concept of small-signal amplifiers was covered in Chapter 6 in relation to bipolar transistors. The concept applies also to small-signal amplifiers using FETs. As you have learned, significant differences exist between BJTs and FETs in terms of their parameters and characteristics, but when used in amplifier circuits, the purpose and final result are the same regardless of the type of transistor: A small input signal is amplified by a desired amount. Because of their high input resistance, FETs are better suited in certain applications. The CE, CC, and CB bipolar amplifier configurations have FET counterparts. These are the common-source (CS), the common-drain (CD), and the common-gate (CG) amplifiers.

JFET AMPLIFIER

A self-biased n-channel JFET with an ac source capacitively coupled to the gate is shown in Figure 8–1(a). The resistor, R_G, serves two purposes: It keeps the gate at approximately 0 V dc (because I_{GSS} is extremely small), and its large value (usually several megohms) prevents loading of the ac signal source. The bias voltage is created by the drop across R_S. The bypass capacitor, C_2, keeps the source of the FET effectively at ac ground.

The signal voltage causes the gate-to-source voltage to swing above and below its Q-point value, causing a swing in drain current. As the drain current increases, the voltage drop across R_D also increases, causing the drain voltage to decrease. The drain current swings above and below its Q-point value in-phase with the gate-to-source voltage. The drain-to-source voltage swings above and below its Q-point value 180° out-of-phase with the gate-to-source voltage, as illustrated in Figure 8–1(b).

FIGURE 8–1
JFET common-source amplifier.

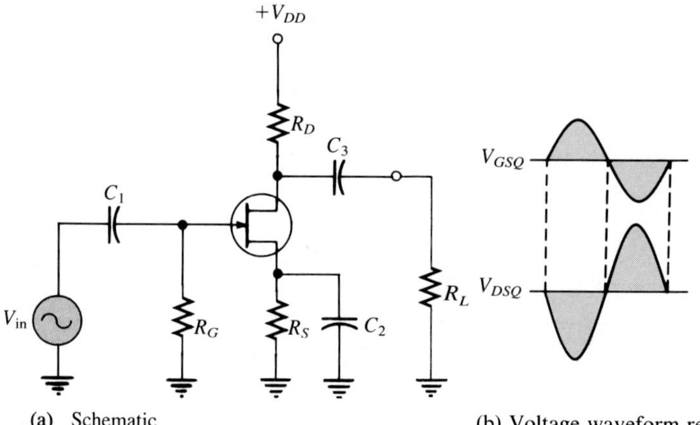

A GRAPHICAL PICTURE

The operation just described for an n-channel JFET can be illustrated graphically on both the transfer characteristic curve and the drain characteristic curve in Figure 8–2. Figure

8–2(a) shows how a sinusoidal variation, V_{gs}, produces a corresponding variation in I_d. As V_{gs} swings from the Q point to a more negative value, I_d decreases from its Q-point value. As V_{gs} swings to a less negative value, I_d increases. Figure 8–2(b) shows a view of the same operation using the drain curves. The signal at the gate drives the drain current equally above and below the Q point on the load line, as indicated by the arrows. Lines projected from the peaks of the gate voltage across to the I_D axis and down to the V_{DS} axis indicate the peak-to-peak variations of the drain current and drain-to-source voltage, as shown.

FIGURE 8–2
JFET characteristic curves.

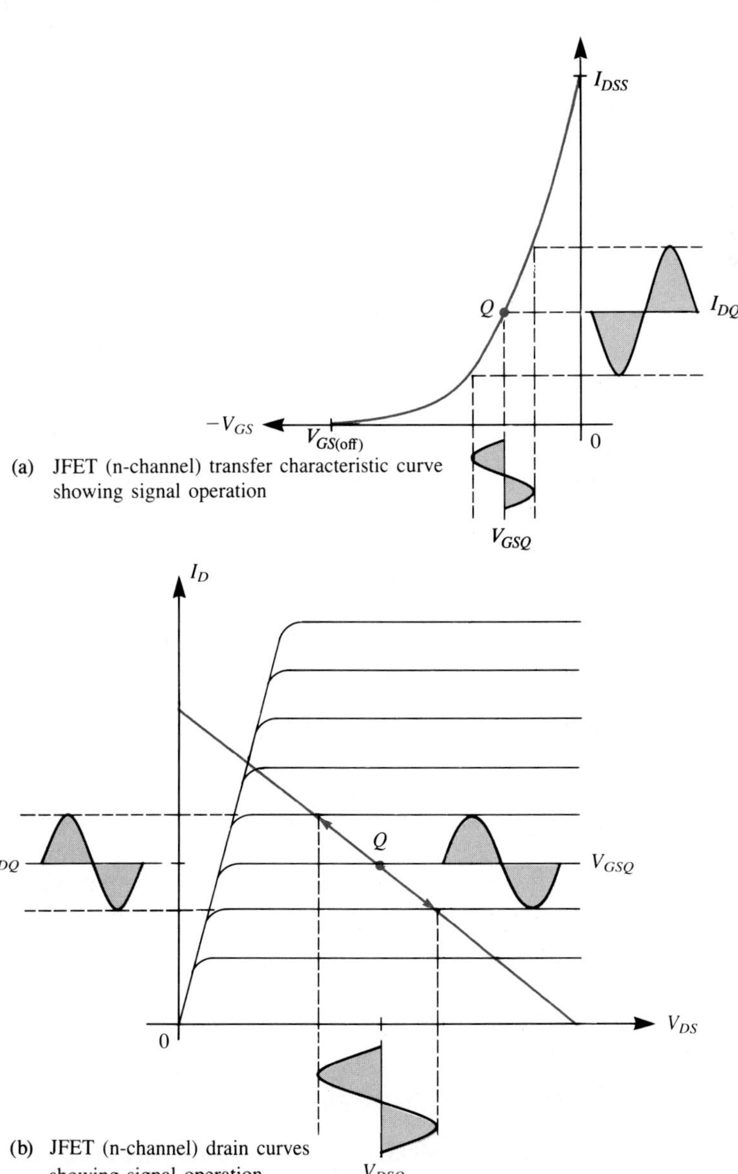

D-MOSFET AMPLIFIER

A zero-biased n-channel D-MOSFET with an ac source capacitively coupled to the gate is shown in Figure 8–3. The gate is at approximately 0 V dc and the source terminal is at ground, thus making $V_{GS} = 0$ V.

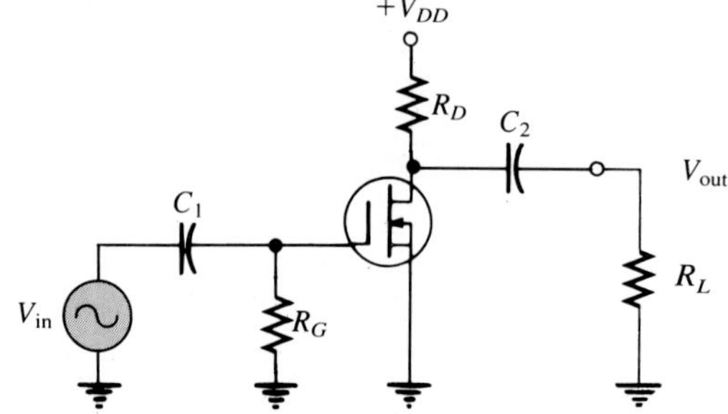

FIGURE 8–3
Zero-biased D-MOSFET common-source amplifier.

The signal voltage causes V_{gs} to swing above and below its zero value, producing a swing in I_d as shown in Figure 8–4. The negative swing in V_{gs} produces the depletion mode, and I_d decreases. The positive swing in V_{gs} produces the enhancement mode, and I_d increases. Note that the enhancement mode is to the right of the vertical axis ($V_{GS} = 0$), and the depletion mode is to the left.

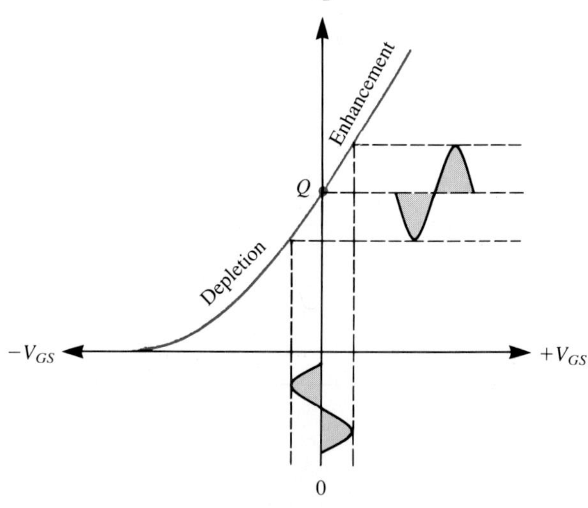

FIGURE 8–4
Depletion-enhancement operation of D-MOSFET shown on transfer characteristic curve.

E-MOSFET AMPLIFIER

An n-channel E-MOSFET with voltage-divider bias with an ac signal source capacitively coupled to the gate is shown in Figure 8–5. The gate is biased with a positive voltage such that $V_{GS} > V_{GS(th)}$.

FIGURE 8–5
Common-source E-MOSFET amplifier with voltage-divider bias.

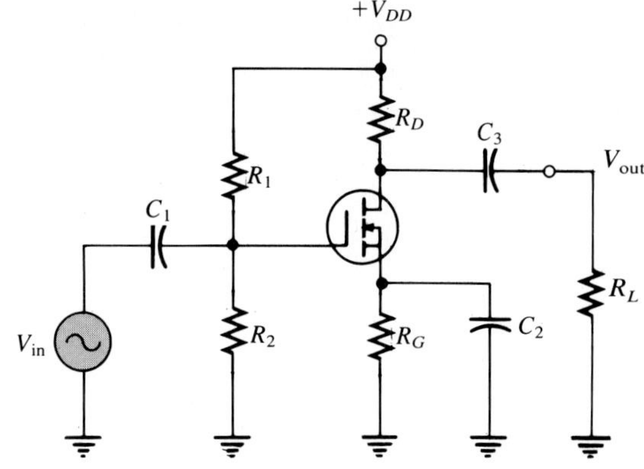

As with the JFET and D-MOSFET, the signal voltage produces a swing in V_{gs} above and below its Q-point value. This, in turn, causes a swing in I_d, as illustrated in Figure 8–6. Operation is entirely in the enhancement mode.

FIGURE 8–6
E-MOSFET (n-channel) operation shown on transfer characteristic curve.

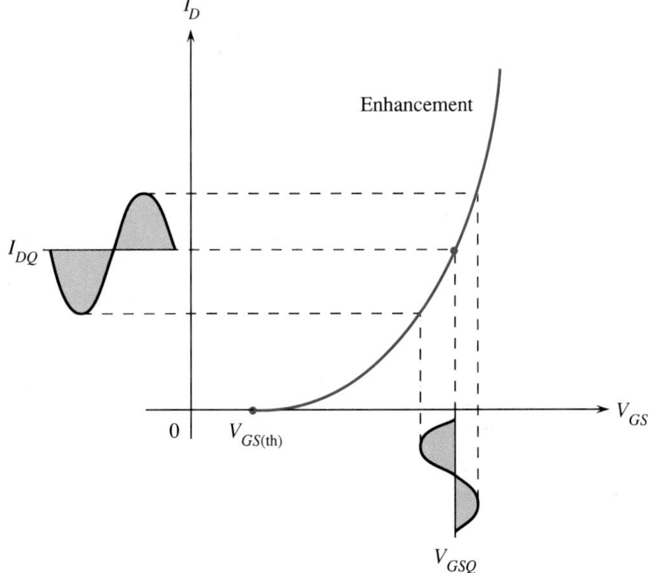

EXAMPLE 8–1 Transfer characteristic curves for an n-channel JFET, D-MOSFET, and E-MOSFET are shown in Figure 8–7. Determine the peak-to-peak variation in I_d when V_{gs} is varied ± 1 V about its Q-point value for each curve.

SOLUTION

(a) The JFET Q point is at $V_{GS} = -2$ V and $I_D = 2.3$ mA. From the graph in Figure 8–7(a), $I_D = 3.3$ mA when $V_{GS} = -1$ V, and $I_D = 1.5$ mA when $V_{GS} = -3$ V. The peak-to-peak drain current is therefore 1.8 mA.

FIGURE 8–7

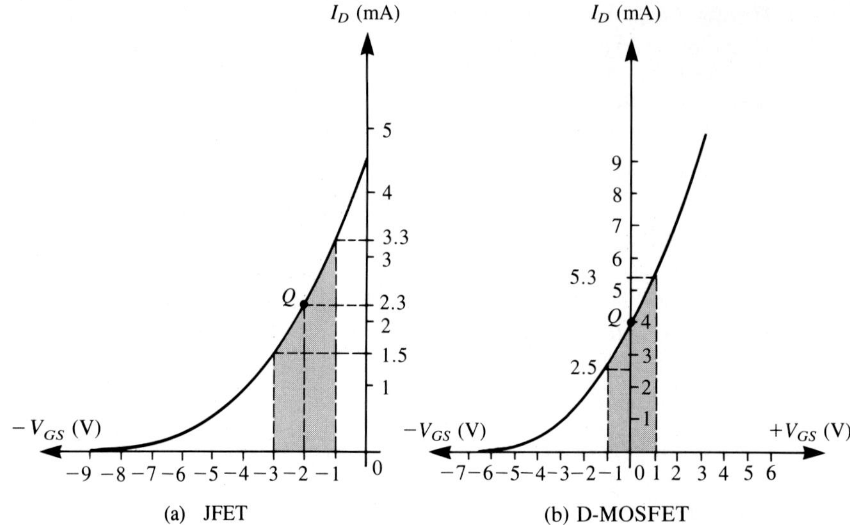

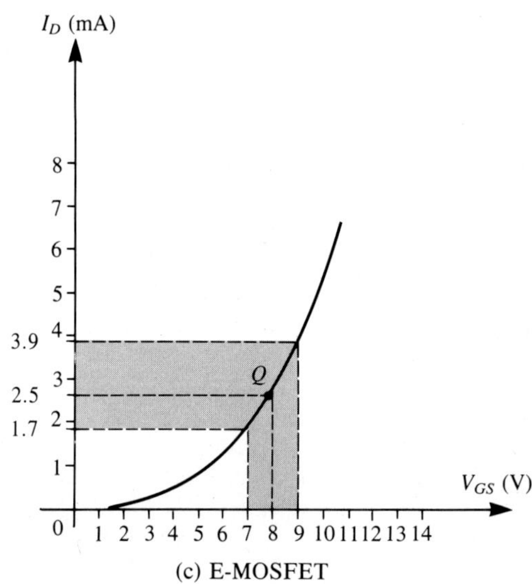

(b) The D-MOSFET Q point is at $V_{GS} = 0$ V and $I_D = I_{DSS} = 4$ mA. From the graph in Figure 8–7(b), $I_D = 2.5$ mA when $V_{GS} = -1$ V, and $I_D = 5.3$ mA when $V_{GS} = +1$ V. The peak-to-peak drain current is therefore 2.8 mA.

(c) The E-MOSFET Q point is at $V_{GS} = +8$ V and $I_D = 2.5$ mA. From the graph in Figure 8–7(c), $I_D = 3.9$ mA when $V_{GS} = +9$ V, and $I_D = 1.7$ mA when $V_{GS} = +7$ V. The peak-to-peak drain current is therefore 2.2 mA.

PRACTICE EXERCISE 8–1

As the Q point is moved toward the bottom end of the curves in Figure 8–7, does the variation in I_D increase or decrease for the same ± 1 V variation in V_{GS}? In addition to the change in the amount that I_D varies, what else will happen? ■

8–1 REVIEW QUESTIONS

1. When V_{gs} is at its positive peak, at what points are I_d and V_{ds}?
2. What is the difference between V_{gs} and V_{GS}?
3. Which of the three types of FETs can operate with a gate-to-source Q-point value of 0 V?

8–2 FET AMPLIFICATION

In this section, you will learn about the amplification properties of FETs and how the gain is affected by certain parameters and circuit components. We will boil the FET down to a simple equivalent circuit to get to the essence of its operation.

The transconductance was defined by Equation (7–2) as $g_m = \Delta I_D/\Delta V_{GS}$. In terms of ac quantities, $g_m = I_d/V_{gs}$. By rearranging, we get

$$I_d = g_m V_{gs} \tag{8–1}$$

This equation says that the output current I_d equals the input voltage V_{gs} multipled by the transconductance g_m.

EQUIVALENT CIRCUIT

An FET equivalent circuit representing the relationship in Equation (8–1) is shown in Figure 8–8. In part (a), the internal resistance, r_{gs}, appears between gate and source, and a current source equal to $g_m V_{gs}$ appears between drain and source. Also, the internal drain-to-source resistance, r_{ds}, is included. In part (b), a simplified ideal model is shown. The resistance, r_{gs}, is assumed to be infinitely large so that there is an open circuit between gate and source. Also, r_{ds} is assumed large enough to neglect.

FIGURE 8–8
Internal FET equivalent circuits.

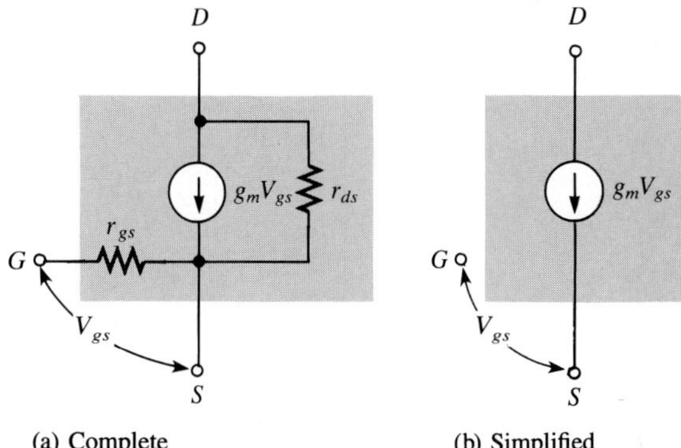

VOLTAGE GAIN

An FET ideal equivalent circuit with an external ac drain resistance is shown in Figure 8–9. The ac voltage gain of this circuit is V_{out}/V_{in}, where $V_{in} = V_{gs}$. The voltage gain expression is therefore

$$A_v = \frac{V_{ds}}{V_{gs}} \tag{8–2}$$

From the equivalent circuit, we see that

$$V_{ds} = I_d R_d$$

From the definition of transconductance in Chapter 7,

$$V_{gs} = \frac{I_d}{g_m}$$

Substituting these two expressions into Equation (8–2),

$$A_v = \frac{I_d R_d}{I_d/g_m} = \frac{g_m I_d R_d}{I_d}$$

$$A_v = g_m R_d \tag{8–3}$$

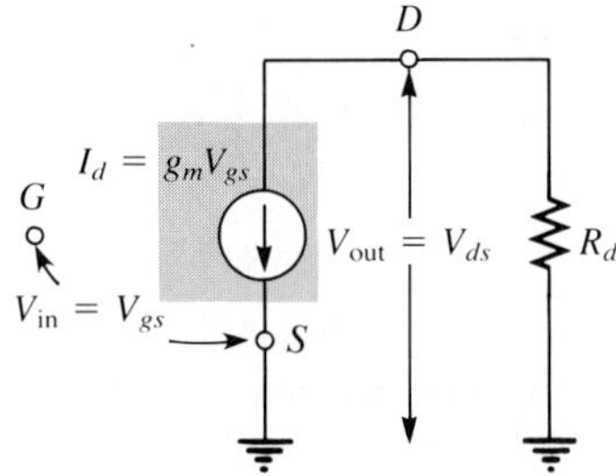

FIGURE 8–9
Simplified FET equivalent circuit with external drain resistor.

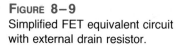

EXAMPLE 8–2 A certain JFET has a g_m = 4 mS. With an external ac drain resistance of 1.5 kΩ, what is the ideal voltage gain?

SOLUTION

$$A_v = g_m R_d = (4 \text{ mS})(1.5 \text{ k}\Omega) = 6$$

PRACTICE EXERCISE 8–2

What is the ideal voltage gain when g_m = 6000 μS and R_d = 2.2 kΩ?

EFFECT OF r_{ds} ON GAIN

If the internal drain-to-source resistance of the FET is taken into account, it appears in parallel with R_d, as indicated in Figure 8–10. The resulting gain expression is stated in Equation (8–4). As you can see, if r_{ds} is not sufficiently greater than $R_d (r_{ds} \geqq 10R_d)$, the gain is reduced from the ideal case of Equation (8–3) to the following.

$$A_v = g_m\left(\frac{R_d r_{ds}}{R_d + r_{ds}}\right) \tag{8–4}$$

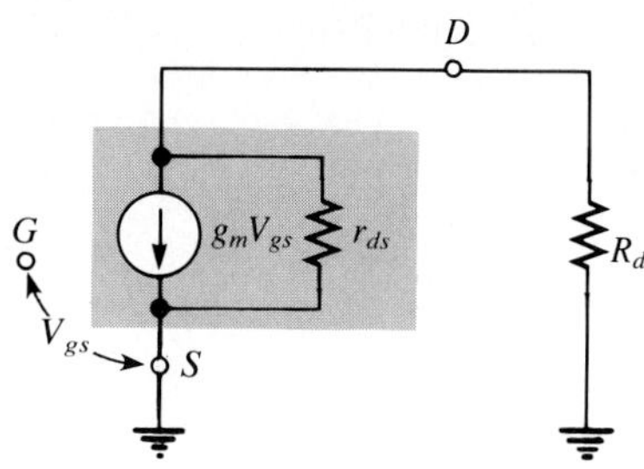

FIGURE 8–10
FET equivalent circuit including the internal drain-to-source resistance, r_{ds}.

EXAMPLE 8–3 The JFET in Example 8–2 has an r_{ds} = 10 kΩ. Determine the voltage gain when r_{ds} is taken into account.

SOLUTION

The r_{ds} is effectively in parallel with the external ac drain resistance R_d. Therefore, Equation (8–4) is used.

$$A_v = g_m\left(\frac{R_d r_{ds}}{R_d + r_{ds}}\right)$$

$$= (4\text{ mS})\left[\frac{(1.5\text{ k}\Omega)(10\text{ k}\Omega)}{1.5\text{ k}\Omega + 10\text{ k}\Omega}\right]$$

$$= (4\text{ mS})(1.3\text{ k}\Omega) = 5.2$$

The gain is reduced from a value of 6 (Example 8–2) because r_{ds} is in parallel with R_d.

PRACTICE EXERCISE 8–3

A JFET has a g_m = 6 mS, an r_{ds} = 5 kΩ, and an external ac drain resistance of 1 kΩ. What is the voltage gain?

EFFECT OF EXTERNAL SOURCE RESISTANCE ON GAIN

Including an external resistance from the FET's source terminal to ground results in the equivalent circuit of Figure 8–11. Examination of this circuit shows that the total input voltage between the gate and ground is

$$V_{in} = V_{gs} + I_d R_s$$

The output voltage taken across R_d is

$$V_{out} = I_d R_d$$

Therefore, the voltage gain is developed as follows.

$$A_v = \frac{V_{out}}{V_{in}} = \frac{I_d R_d}{V_{gs} + I_d R_s}$$

$$= \frac{g_m V_{gs} R_d}{V_{gs} + g_m V_{gs} R_s} = \frac{g_m V_{gs} R_d}{V_{gs}(1 + g_m R_s)}$$

$$A_v = \frac{g_m R_d}{1 + g_m R_s} \qquad \textbf{(8–5)}$$

FIGURE 8–11

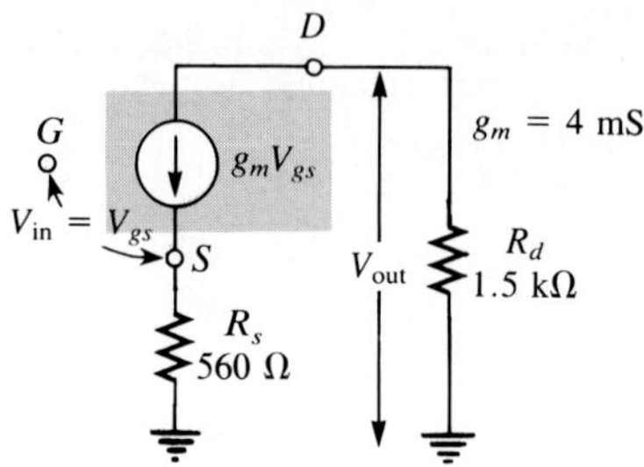

■ **EXAMPLE 8–4**

An FET equivalent circuit is shown in Figure 8–11. Determine the voltage gain when the output is taken across R_d. Neglect r_{ds}.

SOLUTION

There is an external source resistor, so the voltage gain is

$$A_v = \frac{g_mR_d}{1 + g_mR_s} = \frac{(4 \text{ mS})(1.5 \text{ k}\Omega)}{1 + (4 \text{ mS})(560 \text{ }\Omega)} = \frac{6}{1 + 2.24} = \frac{6}{3.24} = 2.68$$

This is the same circuit as in Example 8–2 except for R_s. As you can see, R_s reduces the voltage gain. The voltage gain was ideally 6 in Example 8–2.

PRACTICE EXERCISE 8–4

For the circuit in Figure 8–11, g_m = 3.5 mS, R_s = 330 Ω, and R_d = 1.8 kΩ. Find the voltage gain when the output is taken across R_d. Neglect r_{ds}. ■

8–2 REVIEW QUESTIONS

1. One FET has a transconductance of 3000 μS and another has a transconductance of 3.5 mS. Which one can produce the highest voltage gain, with all other circuit components the same?
2. An FET circuit has a g_m = 2500 μS and an R_d = 10 kΩ. Ideally, what voltage gain can it produce?
3. Two FETs have the same g_m. One has an r_{ds} = 50 kΩ and the other has an r_{ds} = 100 kΩ under the same conditions. Which FET can produce the higher voltage gain when used in a circuit with R_d = 10 kΩ?

8–3 COMMON-SOURCE AMPLIFIERS

Now that you have an idea of how an FET functions as an amplifying device, we will look at a complete amplifier circuit. The common-source (CS) amplifier covered in this section is comparable to the common-emitter BJT amplifier that you studied in Chapter 6.

COMMON-SOURCE JFET AMPLIFIERS

Figure 8–12 shows a **common-source** amplifier with a self-biased n-channel JFET. There are coupling capacitors on the input and output in addition to the source bypass capacitor. The circuit has a combination of dc and ac operations.

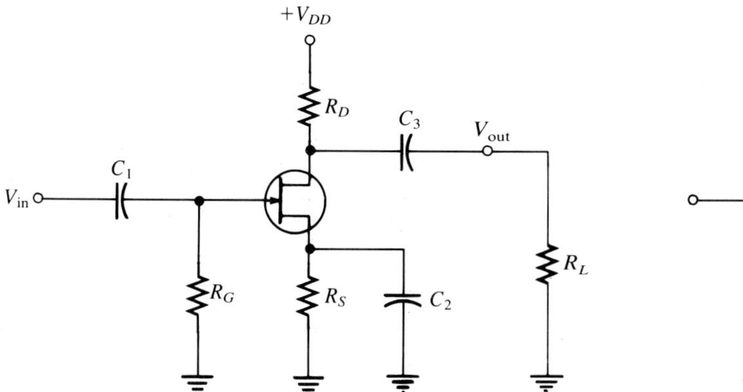

FIGURE 8–12
JFET common-source amplifier.

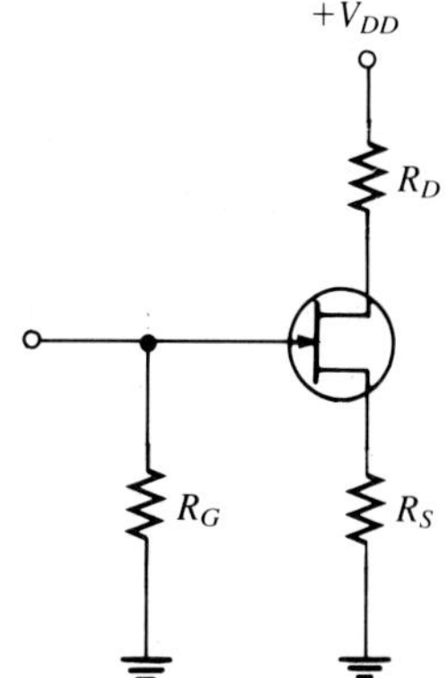

FIGURE 8–13
DC equivalent circuit for the amplifier in Figure 8–12.

DC ANALYSIS

To analyze the amplifier in Figure 8–12, the dc bias values must first be determined. To do this, a dc equivalent circuit is developed by replacing all capacitors with opens, as shown in Figure 8–13. Analysis of dc bias was covered in Chapter 7, but we will go through it again here. First, I_D must be determined before any analysis can be done. If the circuit is biased at the midpoint of the load line, I_D can be calculated using I_{DSS} from the FET data sheet.

$$I_D = \frac{I_{DSS}}{2} \quad \textbf{(8–6)}$$

Otherwise, I_D must be known before any other dc calculations can be made. Determination of I_D from circuit parameter values is tedious because Equation (8–7) must be solved for I_D. (This equation is derived by substitution of $V_{GS} = I_D R_S$ into Equation (8–2).) Solution of the equation for I_D involves expanding it into a quadratic form and then finding the root of the quadratic. This is developed in Appendix B.

$$I_D = I_{DSS}\left[1 - \frac{I_D R_S}{V_{GS(\text{off})}}\right]^2 \quad \textbf{(8–7)}$$

To make the solution of Equation (8–7) easier, refer to the following BASIC computer program. The required inputs are the data sheet values of I_{DSS} and $V_{GS(\text{off})}$ and the value of R_S taken from the circuit. Once I_D is determined, the dc analysis can proceed using the following relationships:

$$V_S = V_{GS} = I_D R_S$$
$$V_D = V_{DD} - I_D R_D$$
$$V_{DS} = V_D - V_S$$

The following program computes I_D from Equation (8–7) for self-biased JFETs:

```
10  CLS
20  PRINT "THIS PROGRAM COMPUTES JFET DRAIN CURRENT"
30  PRINT
40  PRINT "THE REQUIRED INPUTS ARE AS FOLLOWS"
50  PRINT "(1) IDSS FROM DATA SHEET
60  PRINT "(2) VGS(OFF) FROM DATA SHEET"
70  PRINT "(3) RS FROM CIRCUIT DIAGRAM"
80  PRINT:PRINT:PRINT
90  INPUT "TO CONTINUE PRESS 'ENTER'";X
100 CLS
110 INPUT "VALUE OF IDSS IN AMPS";IDSS
120 INPUT "VALUE OF VGS(OFF) IN VOLTS";VGSOFF
130 INPUT "VALUE OF RS IN OHMS";RS
140 CLS
150 A=RS[2*IDSS/VGSOFF[2
160 B=-(1+2*RS*IDSS/ABS(VGSOFF))
170 C=IDSS
180 DI=(-B-SQR(B[2-4*A*C))/(2*A)
190 PRINT "ID=";ABS(DI);"A"
```

AC Equivalent Circuit

To analyze the signal operation of the amplifier in Figure 8–12, an ac equivalent circuit is developed as follows. The capacitors are replaced by effective shorts, based on the simplifying assumption that $X_C \cong 0$ at the signal frequency. The dc source is replaced by a ground, based on the assumption that the voltage source has a zero internal resistance. The V_{DD} terminal is at a zero-volt ac potential and therefore acts as an ac ground.

The ac equivalent circuit is shown in Figure 8–14(a). Notice that the $+V_{DD}$ end of R_d and the source terminal are both effectively at ac ground. Recall that in ac analysis, the ac ground and the actual circuit ground are treated as the same point.

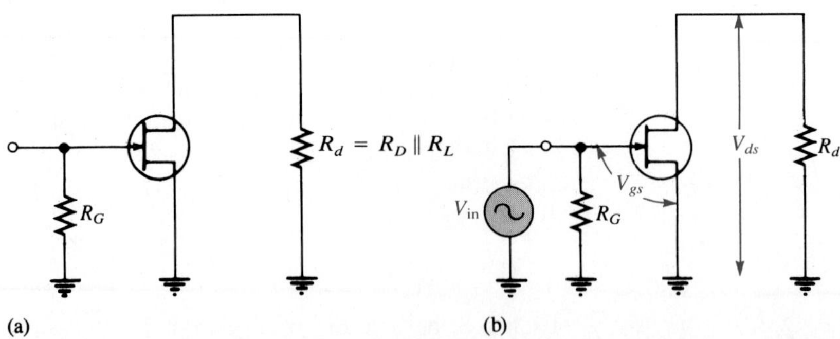

FIGURE 8–14
AC equivalent for the amplifier in Figure 8–12.

SIGNAL VOLTAGE AT THE GATE

An ac voltage source is shown connected to the input in Figure 8–14(b). Since the input resistance to the FET is extremely high, practically all of the input voltage from the signal source appears at the gate with very little voltage dropped across the internal source resistance.

$$V_{gs} = V_{in}$$

THE OUTPUT VOLTAGE

The expression for voltage gain that was developed in Section 8–2 applies to the common-source amplifier.

$$A_v = g_m R_d \qquad (8\text{–}8)$$

The output signal voltage V_{ds} at the drain is

$$V_{out} = V_{ds} = A_v V_{gs}$$

or

$$V_{out} = g_m R_d V_{in} \qquad (8\text{–}9)$$

where $R_d = R_D \| R_L$.

EXAMPLE 8–5 What is the total output voltage of the unloaded amplifier in Figure 8–15? The g_m is 4500 μS, I_{DSS} is 8 mA, and $V_{GS(off)}$ is −10 V.

FIGURE 8–15

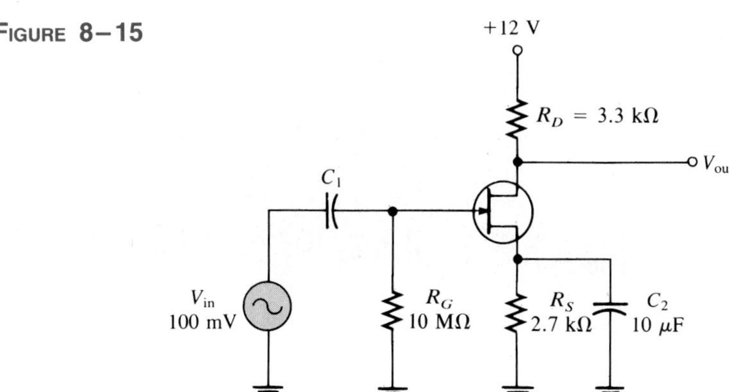

SOLUTION

First, let's find the dc output voltage using the computer program. When executed with the parameter values given, the computer determines that I_D = 2 mA. From this, V_D is calculated.

$$\begin{aligned} V_D &= V_{DD} - I_D R_D \\ &= 12\text{ V} - (2\text{ mA})(3.3\text{ k}\Omega) \\ &= 5.4\text{ V} \end{aligned}$$

Next, the ac output voltage is found.

$$\begin{aligned} V_{out} &= g_m R_D V_{in} \\ &= (4500\ \mu\text{S})(3.3\ \text{k}\Omega)(100\ \text{mV}) \\ &= 1.485\ \text{V rms} \end{aligned}$$

The total output voltage is an ac signal with a peak-to-peak value of 1.485 V × 2.828 = 4.2 V, riding on a dc level of 5.4 V.

PRACTICE EXERCISE 8–5

What will happen in the amplifier of Figure 8–15 if a transistor with $V_{GS(off)} = -2$ V is used? Assume the other parameters are the same.

EFFECT OF AN AC LOAD ON GAIN

When the load is connected to the amplifier's output through a coupling capacitor, as shown in Figure 8–16(a), the drain resistance at the signal frequency is effectively R_D in parallel with R_L. Remember that the upper end of R_D is at ac ground. The ac equivalent circuit is shown in Figure 8–16(b). The total ac drain resistance is

$$R_d = \frac{R_D R_L}{R_D + R_L} \tag{8–10}$$

The effect of R_L is to reduce the unloaded voltage gain, as the next example illustrates.

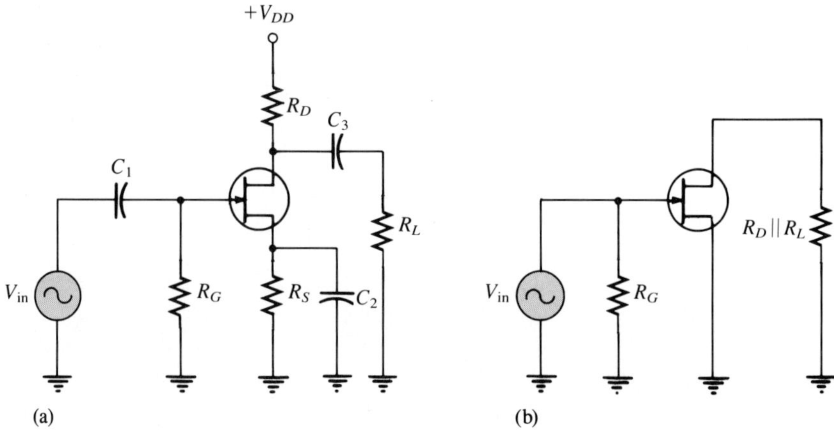

FIGURE 8–16
JFET amplifier and its ac equivalent.

EXAMPLE 8–6

If a 4.7 kΩ load resistor is ac coupled to the output of the amplifier in Example 8–5, what is the resulting rms output voltage?

SOLUTION

The ac drain resistance is

$$R_d = \frac{R_D R_L}{R_D + R_L} = \frac{(3.3\ \text{k}\Omega)(4.7\ \text{k}\Omega)}{8\ \text{k}\Omega} = 1.94\ \text{k}\Omega$$

Calculation of V_{out} yields

$$V_{out} = g_m R_d V_{in} = (4500\ \mu\text{S})(1.94\ \text{k}\Omega)(100\ \text{mV}) = 873\ \text{mV rms}$$

The unloaded ac output voltage was 1.485 V rms in Example 8–5.

PRACTICE EXERCISE 8–6

If a 10 kΩ load resistor is ac coupled to the output of the amplifier in Example 8–5 and the JFET is replaced with one having a $g_m = 3000\ \mu$S, what is the resulting rms output voltage?

PHASE INVERSION

The output voltage (at the drain) is 180° out of phase with the input voltage (at the gate). The phase inversion is sometimes denoted by a negative voltage gain, $-A_v$. Recall that the common-emitter bipolar amplifier also exhibited a phase inversion.

INPUT RESISTANCE

Because the input to a common-source amplifier is at the gate, the input resistance is extremely high. Ideally, it approaches infinity and can be neglected. As you know, the high input resistance is produced by the reverse-biased pn junction in a JFET and by the insulated gate structure in a MOSFET. The actual input resistance seen by the signal source is the gate-to-ground resistor R_G in parallel with the FET's input resistance, V_{GS}/I_{GSS}. The reverse leakage current I_{GSS} is typically given on the data sheet for a specific value of V_{GS} so that the input resistance of the device can be calculated.

EXAMPLE 8–7

What input resistance is seen by the signal source in Figure 8–17? $I_{GSS} = 30$ nA at $V_{GS} = 10$ V.

FIGURE 8–17

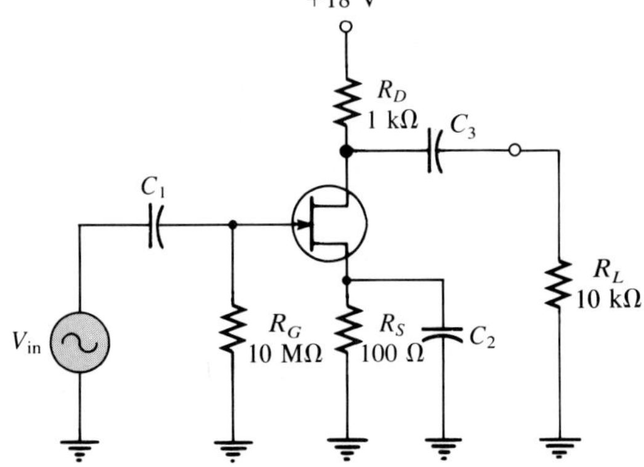

SOLUTION

The input resistance at the gate of the JFET is

$$R_{IN(gate)} = \frac{V_{GS}}{I_{GSS}} = \frac{10\ \text{V}}{30\ \text{nA}} = 333\ \text{M}\Omega$$

The input resistance seen by the signal source is

$$R_{\text{in}} = R_G \| R_{\text{IN(gate)}} = 10\ \text{M}\Omega \| 333\ \text{M}\Omega = 9.7\ \text{M}\Omega$$

PRACTICE EXERCISE 8–7

How much does the total input resistance change if $I_{GSS} = 1$ nA at $V_{GS} = 10$ V?

COMMON-SOURCE MOSFET AMPLIFIERS

In the previous discussion, the JFET was used to illustrate common-source amplifier analysis. The considerations for MOSFET amplifiers are similar, with the exception of the biasing arrangements required. A D-MOSFET is usually biased at $V_{GS} = 0$ and an E-MOSFET at a V_{GS} greater than threshold.

Figure 8–18 shows a common-source amplifier using a D-MOSFET. The dc analysis of this amplifier is somewhat easier than for a JFET because $I_D = I_{DSS}$ at $V_{GS} = 0$. Once I_D is known, the analysis involves calculating only V_D.

$$V_D = V_{DD} - I_D R_D$$

The ac analysis is the same as for the JFET amplifier.

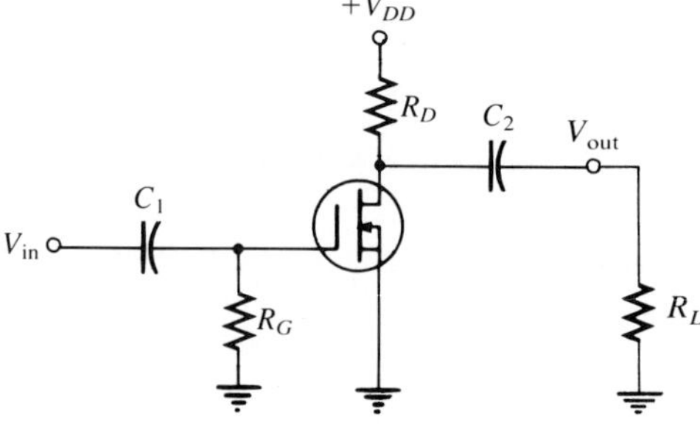

FIGURE 8–18
D-MOSFET common-source amplifier.

EXAMPLE 8–8 The D-MOSFET used in the amplifier of Figure 8–19 has an I_{DSS} of 12 mA and a g_m of 3.2 mS. Determine both the dc drain voltage and ac output voltage. $V_{\text{in}} = 500$ mV.

FIGURE 8–19

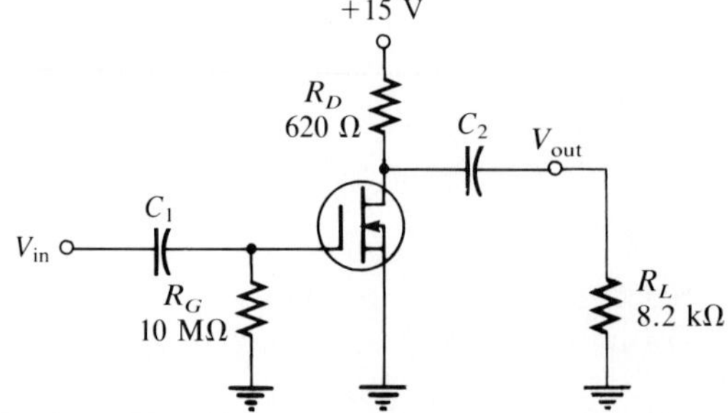

SOLUTION

Since the amplifier is zero-biased,

$$I_D = I_{DSS} = 12 \text{ mA}$$

and, therefore,

$$V_D = V_{DD} - I_D R_D = 15 \text{ V} - (12 \text{ mA})(620 \ \Omega) = 7.56 \text{ V}$$
$$R_d = R_D \| R_L = 620 \ \Omega \| 8.2 \text{ k}\Omega = 576 \ \Omega$$

The signal voltage output is

$$V_{out} = g_m R_d V_{in} = (3.2 \text{ mS})(576 \ \Omega)(500 \text{ mV}) = 922 \text{ mV}$$

PRACTICE EXERCISE 8–8

If a D-MOSFET with $g_m = 5$ mS and $I_{DSS} = 10$ mA replaces the one in this example, what is the ac output voltage when $V_{in} = 500$ mV?

Figure 8–20 shows a common-source amplifier using an E-MOSFET. This circuit uses voltage-divider bias to achieve a V_{GS} above threshold. The general dc analysis proceeds as follows.

$$V_{GS} = \left(\frac{R_2}{R_1 + R_2}\right) V_{DD}$$
$$I_D = K[V_{GS} - V_{GS(th)}]^2$$
$$V_{DS} = V_{DD} - I_D R_D$$

The ac analysis is the same as for the JFET and D-MOSFET circuits.

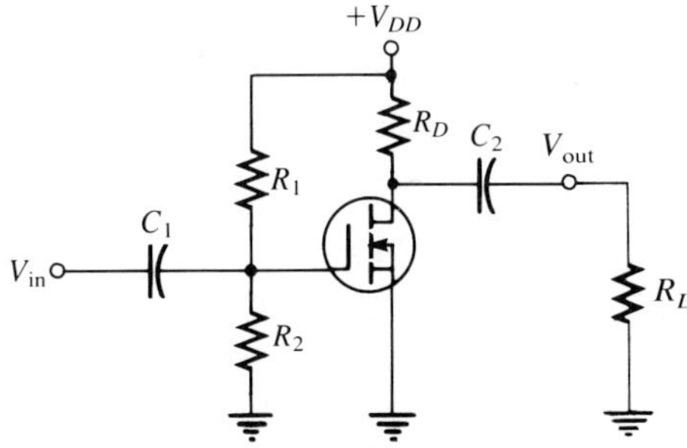

FIGURE 8–20
E-MOSFET common-source amplifier.

EXAMPLE 8–9

A common-source amplifier using an E-MOSFET is shown in Figure 8–21. Find V_{GS}, I_D, V_{DS}, and the ac output voltage. $I_{D(on)} = 5$ mA at $V_{GS} = 10$ V, $V_{GS(th)} = 4$ V, and $g_m = 5.5$ mS. $V_{in} = 50$ mV.

SOLUTION

$$V_{GS} = \left(\frac{R_2}{R_1 + R_2}\right) V_{DD} = \left(\frac{33 \text{ k}\Omega}{80 \text{ k}\Omega}\right) 15 \text{ V} = 6.19 \text{ V}$$

Using $V_{GS} = 10\text{ V}$,

$$K = \frac{I_{D(\text{on})}}{[V_{GS} - V_{GS(\text{th})}]^2} = \frac{5\text{ mA}}{(10\text{ V} - 4\text{ V})^2} = 0.139\text{ mA/V}^2$$

Therefore,

$$I_D = K[V_{GS} - V_{GS(\text{th})}]^2 = 0.139\text{ mA/V}^2(6.19\text{ V} - 4\text{ V})^2 = 0.667\text{ mA}$$
$$V_{DS} = V_{DD} - I_D R_D = 15\text{ V} - (0.667\text{ mA})(3.3\text{ k}\Omega) = 12.8\text{ V}$$
$$R_d = R_D \| R_L = 3.3\text{ k}\Omega \| 33\text{ k}\Omega = 3\text{ k}\Omega$$

The ac output voltage is

$$\begin{aligned} V_{out} &= g_m R_d V_{in} \\ &= (5.5\text{ mS})(3\text{ k}\Omega)(50\text{ mV}) \\ &= 0.825\text{ V} \end{aligned}$$

FIGURE 8–21

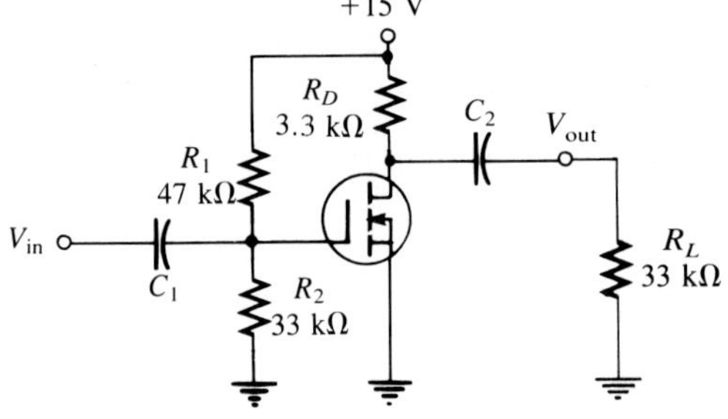

PRACTICE EXERCISE 8–9

In Figure 8–21, $I_{D(\text{on})} = 8\text{ mA}$ at $V_{GS} = 12\text{ V}$, $V_{GS(\text{th})} = 3\text{ V}$, and $g_m = 3.5\text{ mS}$. Find V_{GS}, I_D, V_{DS}, and the ac output voltage. $V_{in} = 50\text{ mV}$. ■

8–3 REVIEW QUESTIONS

1. What factors determine the voltage gain of a common-source FET amplifier?
2. A certain amplifier has an $R_D = 1\text{ k}\Omega$. When a load resistance of $1\text{ k}\Omega$ is capacitively coupled to the drain, how much does the gain change?

8–4 COMMON-DRAIN AMPLIFIERS

The common-drain (CD) amplifier covered in this section is comparable to the common-collector bipolar amplifier. Recall that the CC amplifier is called an emitter-follower. The common-drain amplifier is sometimes called a* source-follower *because the voltage at the source is approximately the same amplitude as the input (gate) voltage and in phase with it. In other words, the source voltage follows the input voltage.

A **common-drain** JFET amplifier is shown in Figure 8–22. Self-biasing is used in this circuit. The input signal is applied to the gate through a coupling capacitor, and the output is at the source terminal. There is no drain resistor.

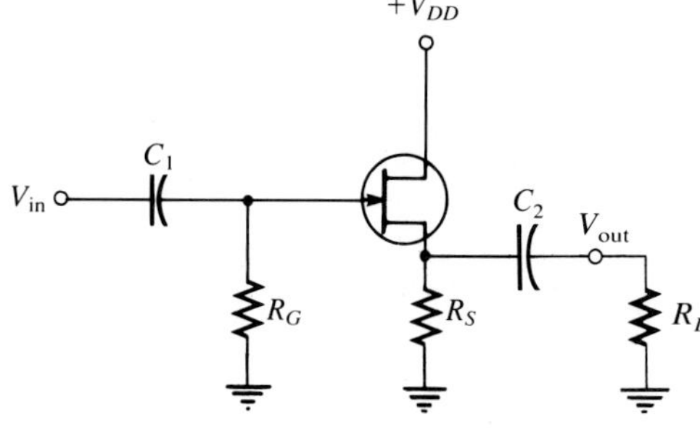

FIGURE 8–22
JFET common-drain amplifier (source-follower).

VOLTAGE GAIN

As in all amplifiers, the voltage gain is $A_v = V_{out}/V_{in}$. For the source-follower, V_{out} is I_dR_s and V_{in} is $V_{gs} + I_dR_s$, as shown in Figure 8–23. Therefore, the gate-to-source voltage gain is $I_dR_s/(V_{gs} + I_dR_s)$. Substituting $I_d = g_mV_{gs}$ into the expression gives the following result.

$$A_v = \frac{g_mV_{gs}R_s}{V_{gs} + g_mV_{gs}R_s}$$

Cancelling V_{gs}, we get

$$A_v = \frac{g_mR_s}{1 + g_mR_s} \tag{8–11}$$

Notice here that the gain is always slightly less than one. If $g_mR_s >> 1$, then a good approximation is $A_v \cong 1$. Since the output voltage is at the source, it is in phase with the gate (input) voltage.

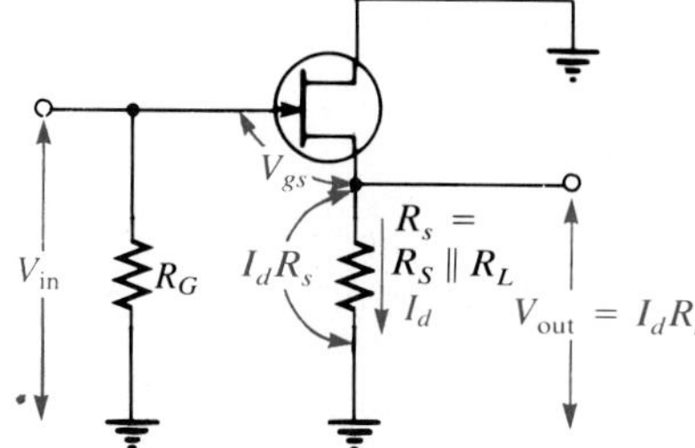

FIGURE 8–23
Voltages in a common-drain amplifier.

INPUT RESISTANCE

Because the input signal is applied to the gate, the input resistance seen by the input signal source is extremely high, just as in the common-source amplifier configuration. The gate resistor R_G, in parallel with the input resistance looking in at the gate, is the total input resistance.

EXAMPLE 8–10 Determine the voltage gain of the amplifier in Figure 8–24 using the data sheet information in Appendix A. Also, determine the input resistance. Use minimum data sheet values where available.

FIGURE 8–24

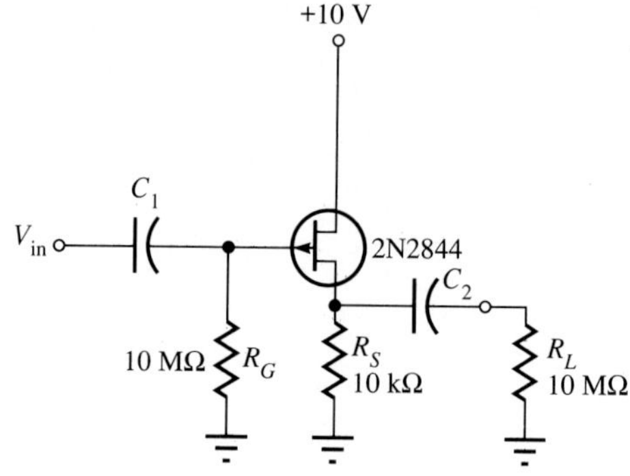

SOLUTION

Since $R_L >> R_S$, $R_s \cong R_S$. From the data sheet in Appendix A, $g_m = y_{fs} = 1400\ \mu S$ (minimum). The gain is

$$A_v = \frac{g_m R_S}{1 + g_m R_S} = \frac{(1400\ \mu S)(10\ k\Omega)}{1 + (1400\ \mu S)(10\ k\Omega)} = 0.933$$

From the data sheet, $I_{GSS} = 10$ nA (maximum) at $V_{GS} = 5$ V. Therefore,

$$R_{IN(gate)} = \frac{5\ V}{10\ nA} = 500\ M\Omega$$

$$R_{IN} = R_G \| R_{IN(gate)} = 10\ M\Omega \| 500\ M\Omega = 9.8\ M\Omega$$

PRACTICE EXERCISE 8–10

If the g_m of the JFET in the source follower of Figure 8–24 is doubled, what is the voltage gain?

8–4 REVIEW QUESTIONS

1. What is the maximum voltage gain of a common-drain amplifier?
2. What factors influence the voltage gain?

8–5 COMMON-GATE AMPLIFIERS

The common-gate FET amplifier configuration introduced in this section is comparable to the common-base BJT amplifier. Like the CB, the common-gate (CG) amplifier has a low input resistance. This is different from the CS and CD configurations, which have very high input resistances.

A typical **common-gate** amplifier is shown in Figure 8–25. The gate is connected directly to ground. The input signal is applied at the source terminal through C_1. The output is coupled through C_2 from the drain terminal.

FIGURE 8–25
JFET common-gate amplifier.

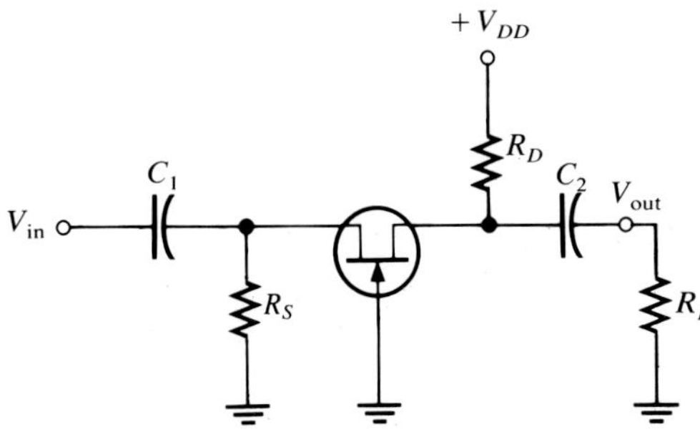

VOLTAGE GAIN

The voltage gain from source to drain is developed as follows.

$$A_v = \frac{V_{out}}{V_{in}} = \frac{V_d}{V_{gs}} = \frac{I_d R_d}{V_{gs}} = \frac{g_m V_{gs} R_d}{V_{gs}}$$

$$A_v = g_m R_d \qquad (8\text{–}12)$$

where $R_d = R_D \| R_L$. Notice that the gain expression is the same as for the common-source JFET amplifier.

INPUT RESISTANCE

As you have seen, both the common-source and common-drain configurations have extremely high input resistances because the gate is the input terminal. In contrast, the common-gate configuration has a low input resistance, as shown in the following steps.

First, the input current (source current) is equal to the drain current.

$$I_{in} = I_s = I_d = g_m V_{gs}$$

Second, the input voltage equals V_{gs}.

$$V_{in} = V_{gs}$$

Therefore, the input resistance at the source terminal is

$$R_{in(source)} = \frac{V_{in}}{I_{in}} = \frac{V_{gs}}{g_m V_{gs}}$$

$$R_{in(source)} = \frac{1}{g_m} \qquad (8\text{–}13)$$

If, for example, g_m has a value of 4000 μS, then

$$R_{in(source)} = \frac{1}{4000\ \mu S} = 250\ \Omega$$

EXAMPLE 8–11 Determine the voltage gain and input resistance of the amplifier in Figure 8–26.

FIGURE 8–26

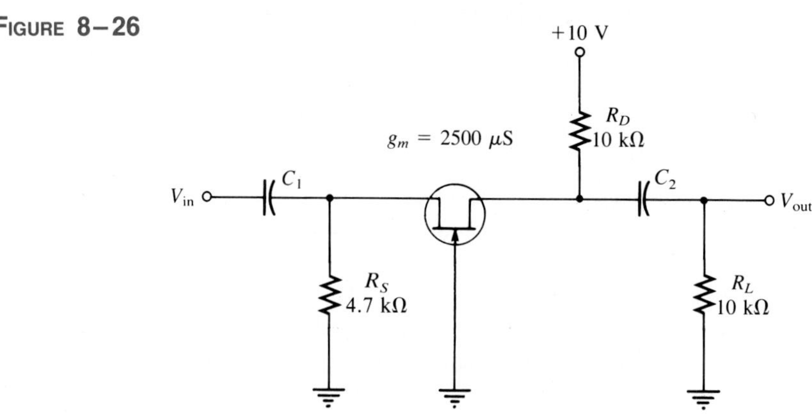

SOLUTION

This common-gate amplifier has a load resistor, so the effective drain resistance is $R_D \| R_L$ and the gain is

$$A_v = g_m(R_D \| R_L) = (2500\ \mu S)(10\ k\Omega \| 10\ k\Omega) = 12.5$$

The input resistance at the source terminal is

$$R_{in(source)} = \frac{1}{g_m} = \frac{1}{2500\ \mu S} = 400\ \Omega$$

The signal source actually sees R_S in parallel with $R_{in(source)}$, so the total input resistance is

$$R_{in} = 400\ \Omega \| 4.7\ k\Omega = 369\ \Omega$$

PRACTICE EXERCISE 8–11

How much does the input resistance change in Figure 8–26 if R_S is changed to 10 kΩ?

SUMMARY OF GAIN AND INPUT RESISTANCE CHARACTERISTICS

A summary of the gain and input resistance characteristics for the three FET amplifier configurations is given in Table 8–1.

8–5 REVIEW QUESTIONS

1. What is a major difference between a common-gate amplifier and the other two configurations?
2. What common factor determines the voltage gain and the input resistance of a common-gate amplifier?

TABLE 8–1
FET amplifier gain and input resistance formulas

	Common-Source	Common-Drain	Common-Gate
Voltage gain A_v	$g_m R_d$	$\dfrac{g_m R_s}{1 + g_m R_s}$	$g_m R_d$
Input resistance	$\left(\dfrac{V_{GS}}{I_{GSS}}\right) \| R_G$	$\left(\dfrac{V_{GS}}{I_{GSS}}\right) \| R_G$	$\left(\dfrac{1}{g_m}\right) \| R_S$

8–6 TROUBLESHOOTING

The technician who understands the basics of circuit operation and who can, if necessary, perform basic analysis on a given circuit is much more valuable than the individual who is limited to carrying out routine test procedures. In this section, we will discuss how to test a circuit board that has only a schematic with no specified test procedure or voltage levels. Basic knowledge of how the circuit operates and the ability to do a quick circuit analysis are very useful in this case.

Assume that you are given a circuit board pulled from the audio amplifier section of a sound system and told simply that it is not working properly. The first step is to obtain the system schematic and locate this particular circuit on it. Notice that the circuit is a two-stage FET amplifier, as shown in Figure 8–27. The problem is approached in the following sequence.

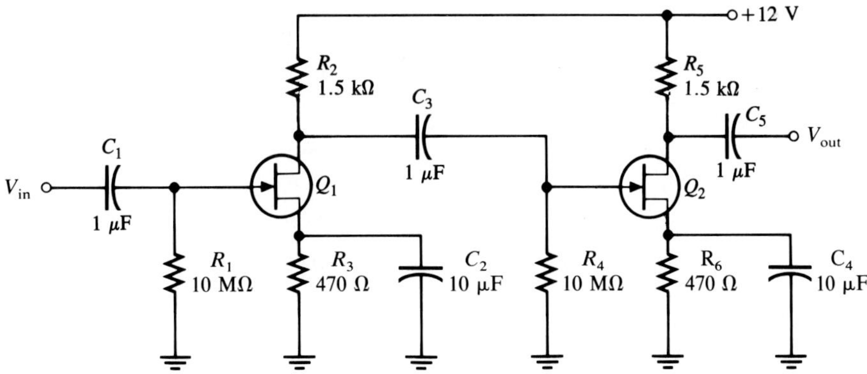

FIGURE 8–27
A two-stage FET amplifier circuit.

STEP 1 Determine what the voltage levels in the circuit should be so that you know what to look for. First, pull a data sheet on the particular transistor (both Q_1 and Q_2 are found to be the same type of transistor from the label on the case) and determine the g_m so that you can calculate the voltage gain. Assume that for this particular device, a typical g_m of 5000 μS is specified. Now, calculate the expected voltage gain of each stage (notice they are identical). Because input resistance is very high, the second stage does not significantly load the first stage, as in a bipolar amplifier.

$$A_v = g_m R_2 = (5000\ \mu S)(1.5\ k\Omega) = 7.5$$

Since the stages are identical, the overall gain should be

$$A'_v = (7.5)(7.5) = 56.25$$

We will ignore dc levels at this time and concentrate on signal tracing.

STEP 2 Arrange a test set-up to permit connection of an input test signal, a dc supply voltage, and ground to the circuit board. The schematic shows that the dc supply voltage must be +12 V. Choose 10 mV rms as an input test signal. This value is arbitrary (although the capability of your signal source is a factor), but small enough that the expected output signal voltage is well below the absolute peak-to-peak limit of 12 V set by the supply voltage and ground (we know that the output voltage swing cannot go higher than 12 V or lower than 0 V).

STEP 3 The test set-up is shown in Figure 8–28. Adjust the dc supply voltage to +12 V and the sine wave source to 10 mV rms (we will assume it is capable of going that low). Set the frequency of the sine wave signal source to an arbitrary value in the audio range (say 10 kHz), since you know this is an audio amplifier. (By the way, the audio frequency range is generally accepted as 20 Hz to 20 kHz.)

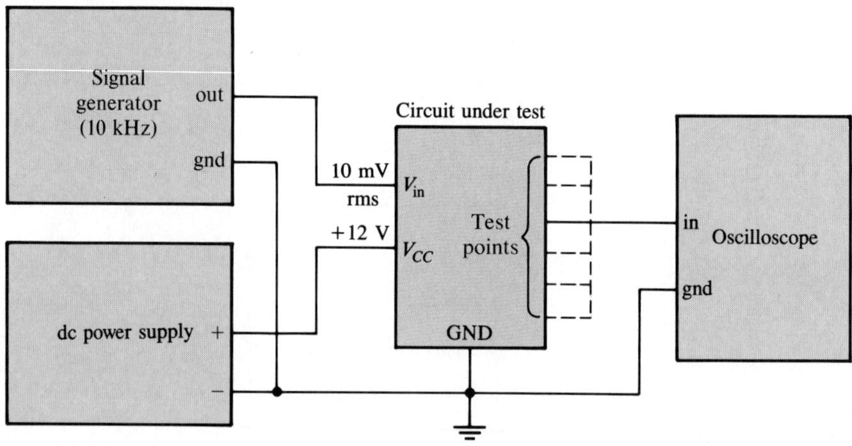

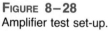

FIGURE 8–28
Amplifier test set-up.

STEP 4 Check the input signal at the gate of Q_1 and the output signal at the drain of Q_2 with an oscilloscope. The results are shown in Figure 8–29(a) and (b). The measured output voltage has a peak value of 226 mV. The expected peak output voltage is

$$V_{out} = V_{in}A'_v = (14.14\ mV)(56.25) = 795\ mV$$

The output is much less than it should be.

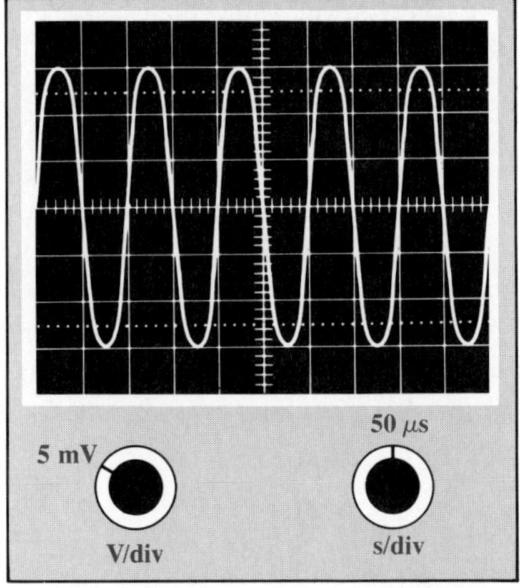

(a) Q_1 gate: 10 mV rms (14.14 mV peak)

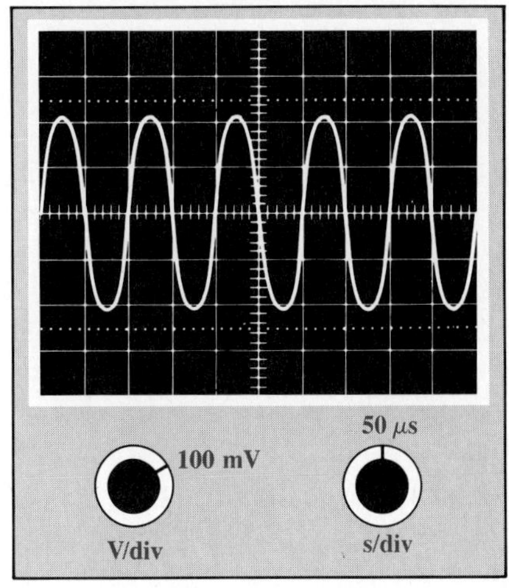

(b) V_{out}: 226 mV peak

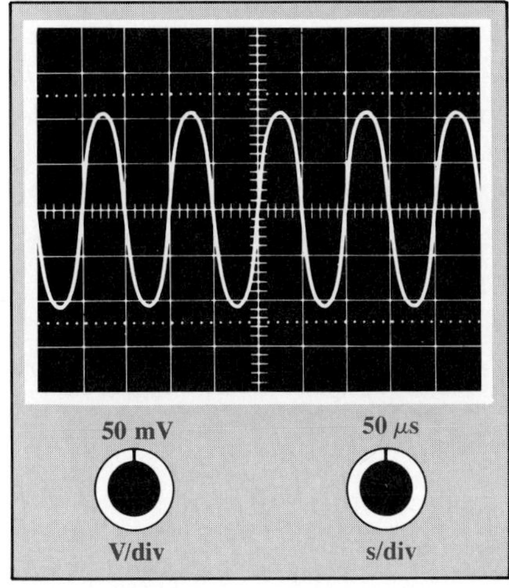

(c) Q_2 drain: 106 mV peak

FIGURE 8–29
Oscilloscope displays of signals in the two-stage amplifier of Figure 8–27.

STEP 5 Trace the signal from the output back toward the input to determine the fault. Figure 8–29 shows the oscilloscope displays of the measured signal voltages. Part (c) indicates that the voltage at the gate of Q_2 is 106 mV peak, as expected (14.14 mV × 7.5 = 106 mV). This signal is properly coupled from the drain of Q_1. Therefore, the problem lies in the second stage. The gain of Q_2 is much lower than it should be (2.13 instead of 7.5).

STEP 6 Consider the possible causes of the observed malfunction. There are three possible reasons the gain is low: (1) Q_2 has a lower transconductance (g_m) than specified; (2) R_5 has a lower value than shown on the schematic; or (3) the bypass capacitor C_4 is open.

The only way to check the g_m is by replacing Q_2 with a new transistor of the same type and rechecking the output signal. You can make certain that R_5 is the proper value by removing one end of the resistor from the circuit board and measuring the resistance with an ohmmeter. To avoid having to remove a component, the best way to start isolating the fault is by checking the signal voltage at the source of Q_2. If the capacitor is working properly, there will be only a dc voltage at the source. The presence of a signal voltage at the source indicates that C_4 is open. With R_6 unbypassed, the gain expression is $g_mR_d/(1 + g_mR_d)$ rather than simply g_mR_d, thus resulting in less gain.

8–6 REVIEW QUESTIONS

1. What is the prerequisite to effective troubleshooting?
2. Assume that C_2 in Figure 8–27 opened. What symptoms would indicate this failure?
3. If C_3 opened, would the gain of the first stage be affected?

8–7 A SYSTEM APPLICATION

The RF amplifier in the receiver system that you saw at the opening of this chapter is a tuned amplifier. It uses a resonant circuit to eliminate all frequencies except those in a narrow band at a selected frequency. The amplifier takes the small input signals from the antenna, filters out all but the selected frequency and amplifies it. The output of the amplifier then goes to the mixer circuit. In this section, you will

- ☐ ***See how a tuned FET amplifier is used in a system application.***
- ☐ ***See how a varactor diode (studied earlier) is used.***
- ☐ ***Translate between a printed circuit board and a schematic.***
- ☐ ***Troubleshoot some amplifier failures.***

A BRIEF DESCRIPTION OF THE SYSTEM

The block diagram for the FM receiver is shown in Figure 8–30. A very basic system description is as follows. The antenna picks up all radiated signals that pass across it and feeds them into the RF amplifier. The voltages induced in the antenna by the electromagnetic radiation are extremely small. The RF amplifier is tuned to select and amplify a desired frequency within the FM broadcast band (88 MHz to 108 MHz). Since it is a frequency-selective circuit, the RF amplifier eliminates all but a narrow band centered at the selected frequency.

The output of the RF amplifier goes to the mixer where it is combined with the output of the local oscillator, which is 10.7 MHz above the selected RF frequency. In the mixer, a nonlinear process called *heterodyning* takes place and produces one frequency that is the sum of the RF and local oscillator frequencies and another frequency that is the difference (always 10.7 MHz). The sum frequency is filtered out and only the 10.7 MHz difference frequency is used. This difference frequency contains the same audio information as the higher RF frequency. The IF (intermediate frequency) amplifier is tuned to 10.7 MHz and

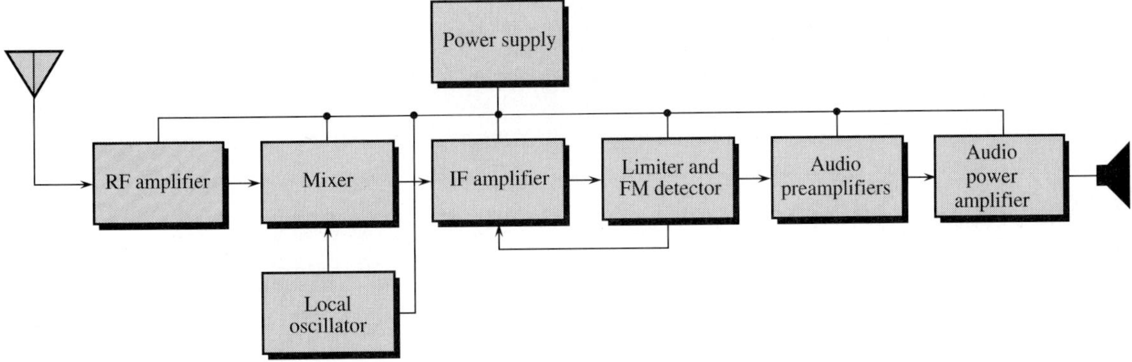

FIGURE 8–30

amplifies the mixer output. The limiter and FM detector or discriminator takes the amplified 10.7 MHz FM signal and recovers the audio from it while eliminating the intermediate frequency. The output of the FM detector is a small audio signal that goes to the audio preamplifier and then to the power amplifier, which drives the speaker.

Now, so that you can take a closer look at the tuned RF amplifier, let's take it out of the system and put it on the test bench.

ON THE TEST BENCH

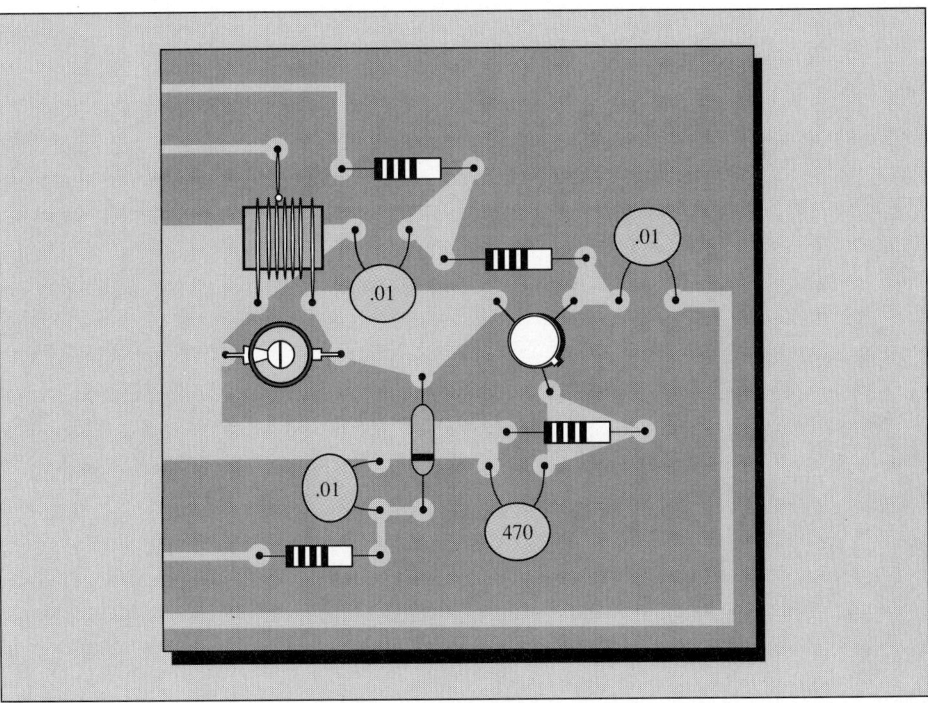

FIGURE 8–31

ACTIVITY 1 RELATE THE PC BOARD TO THE SCHEMATIC

The pc board is shown in Figure 8–31. Using the schematic in Figure 8–32 as reference, identify and label the components. Also label the inputs and outputs.

FIGURE 8–32

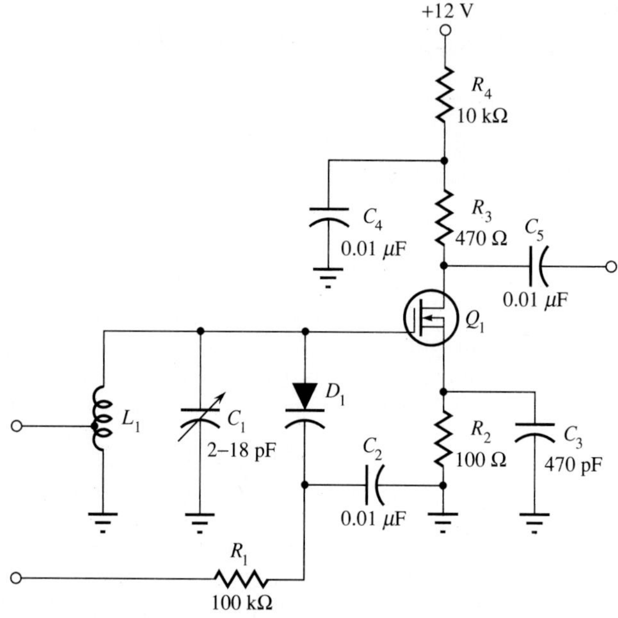

ACTIVITY 2 ANALYZE THE CIRCUIT

STEP 1 If the voltage applied to the varactor diode can be varied from 1.5 V to 8 V, determine the approximate capacitance range. Use the graph in Chapter 3 (Figure 3–18(c)).

STEP 2 What is the total capacitance range of the resonant circuit if the FET has an input capacitance of 5 pF and there is 1 pF of stray capacitance between the gate and ground due to the conductors on the pc board?

STEP 3 Determine an approximate value for the inductor that will allow a tuning range from 88 MHz to 108 MHz.

STEP 4 Determine the value to which C_1 should be adjusted to ensure that the varactor can provide the necessary tuning range.

ACTIVITY 3 WRITE A TECHNICAL REPORT

Discuss the overall operation of the RF amplifier circuit. State the purpose of each component on the board. Explain how the varactor diode is used. Use the results of Activity 2 as appropriate.

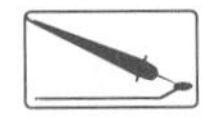

ACTIVITY 4 TROUBLESHOOT THE CIRCUIT BOARD FOR EACH OF THE FOLLOWING PROBLEMS BY STATING THE PROBABLE CAUSE OR CAUSES IN EACH CASE

1. No output signal when there is a verified input signal to L_1.
2. Output signal amplitude does not change when the tuning voltage is varied over its range.
3. The drain of Q_1 is at a constant +12 V.
4. Proper signal at drain but no signal on final output.
5. DC voltage at drain is much less than it should be.
6. The frequency range over which there is a proper output signal does not cover the FM band from 88 MHz to 108 MHz.

8–7 REVIEW QUESTIONS

1. If the dc voltage on the right side of R_1 is 5 V in Figure 8–32, what is the voltage at the cathode of D_1? Explain.
2. What is the purpose of capacitors C_2 and C_4 in Figure 8–32?
3. What is the purpose of capacitor C_3? C_5?
4. Which components make up the resonant circuit in Figure 8–32?
5. What is the dc voltage on the gate of Q_1?
6. Is D_1 reverse-biased or forward-biased?

SUMMARY

- □ The transconductance g_m of an FET relates the output current I_d to the input voltage V_{gs}.
- □ The voltage gain of a common-source amplifier is determined largely by the transconductance g_m and the drain resistance R_d.
- □ The internal drain-to-source resistance r_{ds} of the FET influences (reduces) the gain if it is not sufficiently greater than R_d so that it can be neglected.
- □ An unbypassed resistance between source and ground (R_S) reduces the voltage gain of an FET amplifier.
- □ A load resistance connected to the drain of a common-source amplifier reduces the voltage gain.
- □ There is a 180° phase inversion between gate and drain voltages.
- □ The input resistance at the gate of FETs is extremely high.
- □ The voltage gain of a common-drain amplifier (source-follower) is always slightly less than 1.
- □ There is no phase inversion between gate and source.
- □ The input resistance of a common-gate amplifier is the reciprocal of g_m because the input is at the source terminal.
- □ The total voltage gain of a multistage amplifier is the product of the individual voltage gains (sum of dB gains).

GLOSSARY

Common-drain An FET amplifier configuration in which the drain is the grounded terminal.

Common-gate An FET amplifier configuration in which the gate is the grounded terminal.

Common-source An FET amplifier configuration in which the source is the grounded terminal.

Source-follower The common-drain amplifier.

FORMULAS

FET AMPLIFICATION

(8–1)	$I_d = g_m V_{gs}$	Drain current
(8–2)	$A_v = \dfrac{V_{ds}}{V_{gs}}$	General voltage gain
(8–3)	$A_v = g_m R_d$	Voltage gain
(8–4)	$A_v = g_m\left(\dfrac{R_d r_{ds}}{R_d + r_{ds}}\right)$	Gain considering r_{ds}
(8–5)	$A_v = \dfrac{g_m R_d}{1 + g_m R_s}$	Gain with R_s

COMMON-SOURCE AMPLIFIER

(8–6)	$I_D = \dfrac{I_{DSS}}{2}$	For centered Q point
(8–7)	$I_D = I_{DSS}\left[1 - \dfrac{I_D R_S}{V_{GS(\text{off})}}\right]^2$	Self-biased JFET current
(8–8)	$A_v = g_m R_d$	Voltage gain
(8–9)	$V_{\text{out}} = g_m R_d V_{\text{in}}$	Output voltage
(8–10)	$R_d = \dfrac{R_D R_L}{R_D + R_L}$	AC drain resistance with load

COMMON-DRAIN AMPLIFIER

(8–11)	$A_v = \dfrac{g_m R_s}{1 + g_m R_s}$	Voltage gain

COMMON-GATE AMPLIFIER

(8–12)	$A_v = g_m R_d$	Voltage gain
(8–13)	$R_{\text{in(source)}} = \dfrac{1}{g_m}$	Input resistance

SELF-TEST

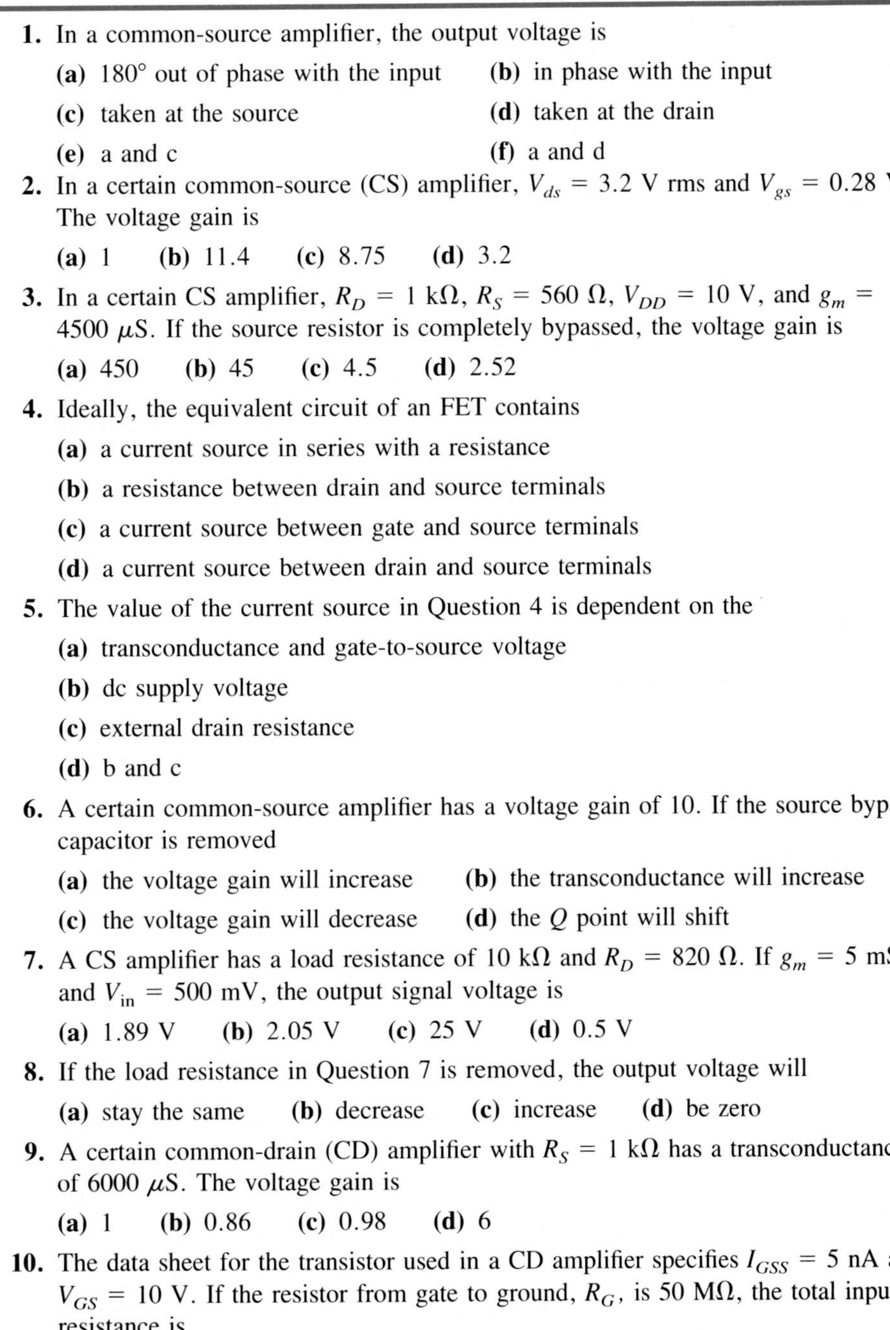

1. In a common-source amplifier, the output voltage is
   (a) 180° out of phase with the input (b) in phase with the input
   (c) taken at the source (d) taken at the drain
   (e) a and c (f) a and d
2. In a certain common-source (CS) amplifier, $V_{ds} = 3.2$ V rms and $V_{gs} = 0.28$ V. The voltage gain is
   (a) 1 (b) 11.4 (c) 8.75 (d) 3.2
3. In a certain CS amplifier, $R_D = 1$ kΩ, $R_S = 560$ Ω, $V_{DD} = 10$ V, and $g_m =$ 4500 μS. If the source resistor is completely bypassed, the voltage gain is
   (a) 450 (b) 45 (c) 4.5 (d) 2.52
4. Ideally, the equivalent circuit of an FET contains
   (a) a current source in series with a resistance
   (b) a resistance between drain and source terminals
   (c) a current source between gate and source terminals
   (d) a current source between drain and source terminals
5. The value of the current source in Question 4 is dependent on the
   (a) transconductance and gate-to-source voltage
   (b) dc supply voltage
   (c) external drain resistance
   (d) b and c
6. A certain common-source amplifier has a voltage gain of 10. If the source bypass capacitor is removed
   (a) the voltage gain will increase (b) the transconductance will increase
   (c) the voltage gain will decrease (d) the Q point will shift
7. A CS amplifier has a load resistance of 10 kΩ and $R_D = 820$ Ω. If $g_m = 5$ mS and $V_{\text{in}} = 500$ mV, the output signal voltage is
   (a) 1.89 V (b) 2.05 V (c) 25 V (d) 0.5 V
8. If the load resistance in Question 7 is removed, the output voltage will
   (a) stay the same (b) decrease (c) increase (d) be zero
9. A certain common-drain (CD) amplifier with $R_S = 1$ kΩ has a transconductance of 6000 μS. The voltage gain is
   (a) 1 (b) 0.86 (c) 0.98 (d) 6
10. The data sheet for the transistor used in a CD amplifier specifies $I_{GSS} = 5$ nA at $V_{GS} = 10$ V. If the resistor from gate to ground, R_G, is 50 MΩ, the total input resistance is
    (a) 50 MΩ (b) 200 MΩ (c) 40 MΩ (d) 20.5 MΩ

11. The common-gate (CG) amplifier differs from both the CS and CD configurations in that it has a
    (a) much higher voltage gain **(b)** much lower voltage gain
    (c) much higher input resistance **(d)** much lower input resistance
12. If you are looking for both good voltage gain and high input resistance, you must use a
    (a) CS amplifier **(b)** CD amplifier **(c)** CG amplifier
13. For small-signal operation, an n-channel JFET must be biased at
    (a) $V_{GS} = 0$ V **(b)** $V_{GS} = V_{GS(off)}$
    (c) $-V_{GS(off)} < V_{GS} < 0$ V **(d)** $0 \text{ V} < V_{GS} < +V_{GS(off)}$
14. Two FET amplifiers are cascaded. The first stage has a voltage gain of 5 and the second stage has a voltage gain of 7. The overall voltage gain is
    (a) 35 **(b)** 12 **(c)** dependent on the second stage loading
15. If there is an internal open between the drain and source in a CS amplifier, the drain voltage is equal to
    (a) 0 V **(b)** V_{DD} **(c)** a value less than normal **(d)** V_{GS}

PROBLEMS

SECTION 8–1 SMALL-SIGNAL FET AMPLIFIER OPERATION

1. Identify the type of FET and its bias arrangement in Figure 8–33. Ideally, what is V_{GS}?

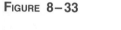

FIGURE 8–33

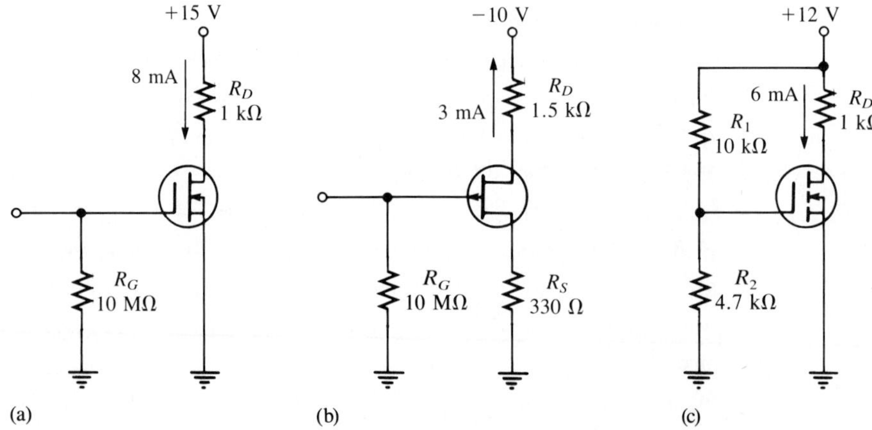

2. Calculate the dc voltages from each terminal to ground for the FETs in Figure 8–33.
3. Identify each characteristic curve in Figure 8–34 by the type of FET that it represents.

FIGURE 8–34

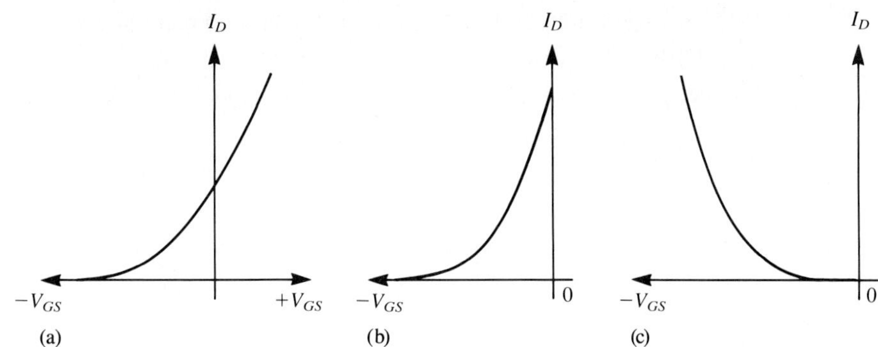

4. Refer to the JFET transfer characteristic curve in Figure 8–7(a) and determine the peak-to-peak value of I_d when V_{gs} is varied ± 1.5 V about its Q-point value.
5. Repeat Problem 4 for the curves in Figure 8–7(b) and Figure 8–7(c).

SECTION 8–2 FET AMPLIFICATION

6. An FET has a $g_m = 6000\ \mu$S. Determine the rms drain current for each of the following values of V_{gs}.
   (a) 10 mV **(b)** 150 mV
   (c) 0.6 V **(d)** 1 V
7. The gain of a certain JFET amplifier with a source resistance of zero is 20. Determine the drain resistance if the g_m is 3500 μS.
8. A certain FET amplifier has a g_m of 4.2 mS, $r_{ds} = 12$ kΩ, and $R_D = 4.7$ kΩ. What is the voltage gain? Assume the source resistance is 0 Ω.
9. What is the gain for the amplifier in Problem 8 if the source resistance is 1 kΩ?

SECTION 8–3 COMMON-SOURCE AMPLIFIERS

10. Given that $I_D = 3.5$ mA in Figure 8–35, find V_{DS} and V_{GS}. $V_{GS(\text{off})} = -7$ V and $I_{DSS} = 8$ mA.
11. If a 50 mV rms input signal is applied to the amplifier in Figure 8–35, what is the peak-to-peak output voltage? $g_m = 5000\ \mu$S.

FIGURE 8–35

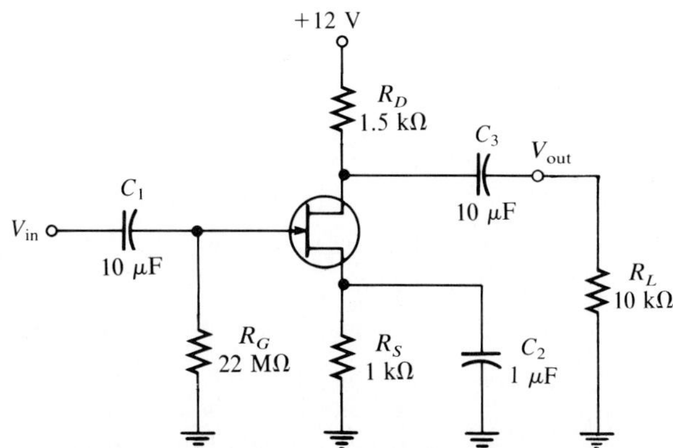

12. If a 1500 Ω load is ac coupled to the output in Figure 8–35, what is the resulting output voltage (rms) when a 50 mV rms input is applied? $g_m = 5000\ \mu S$.
13. Determine the voltage gain of each common-source amplifier in Figure 8–36.

FIGURE 8–36

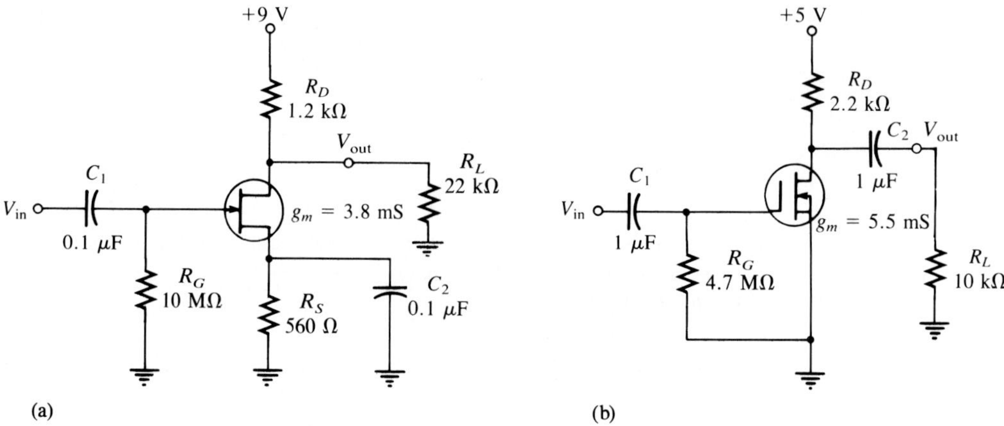

14. Draw the dc and ac equivalent circuits for the amplifier in Figure 8–37.

FIGURE 8–37

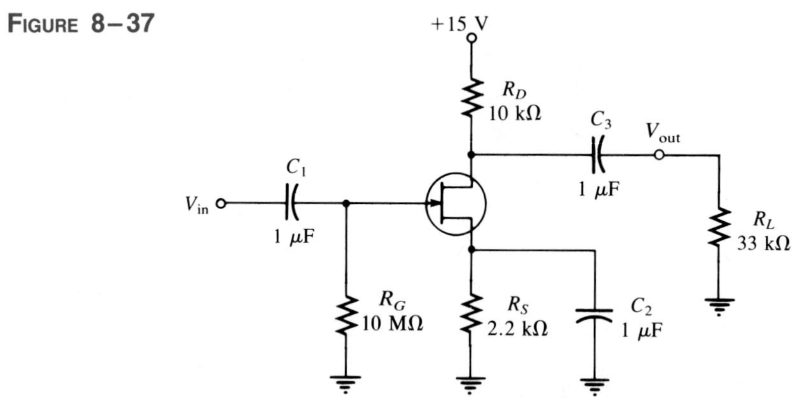

15. Determine the drain current in Figure 8–37 given that $I_{DSS} = 15$ mA, $V_{GS(off)} = -4$ V, and $g_m = 5000\ \mu S$. The Q point is centered.
16. What is the gain of the amplifier in Figure 8–37 if C_2 is removed?
17. A 4.7 kΩ resistor is connected in parallel with R_L in Figure 8–37. What is the voltage gain?
18. For the common-source amplifier in Figure 8–38 determine I_D, V_{GS}, and V_{DS} for a centered Q point. $I_{DSS} = 9$ mA, $V_{GS(off)} = -3$ V, and $g_m = 2500\ \mu S$.
19. If a 10 mV rms signal is applied to the input of the amplifier in Figure 8–38, what is the rms value of the output signal?

FIGURE 8–38

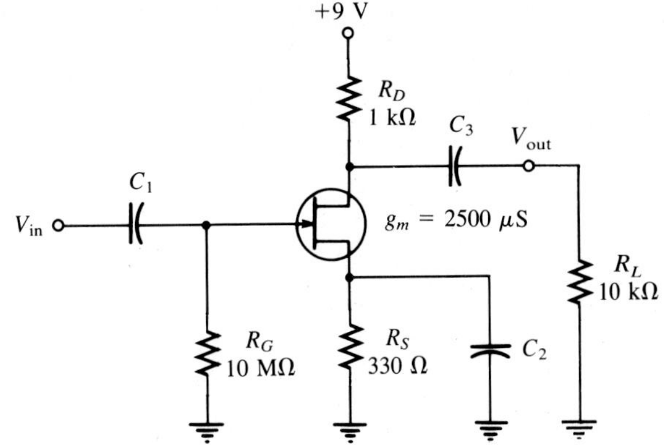

20. Determine V_{GS}, I_D, and V_{DS} for the amplifier in Figure 8–39. $I_{D(on)} = 18$ mA at $V_{GS} = 10$ V, $V_{GS(th)} = 2.5$ V, and $g_m = 3000\ \mu$S.
21. Determine the input resistance seen by the signal source in Figure 8–40. Assume $I_{GSS} = 25$ nA at $V_{GS} = -15$ V.

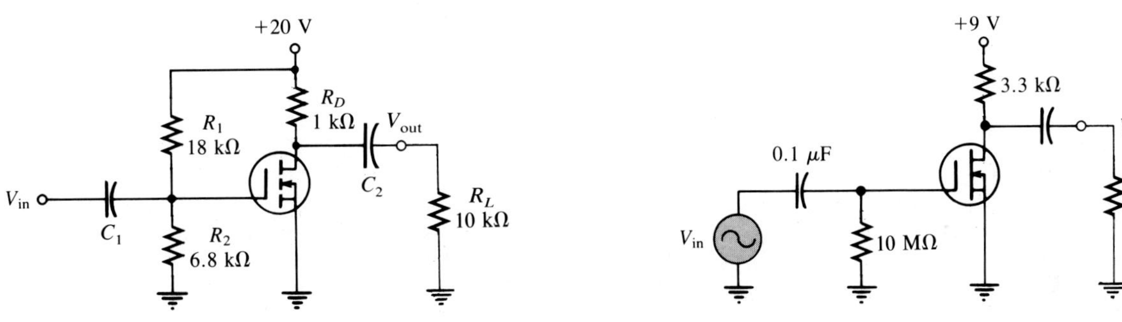

FIGURE 8–39

FIGURE 8–40

22. Determine the total drain voltage waveform (dc and ac) and the V_{out} waveform in Figure 8–41. $g_m = 4.8$ mS and $I_{DSS} = 15$ mA. Observe that $V_{GS} = 0$.

FIGURE 8–41

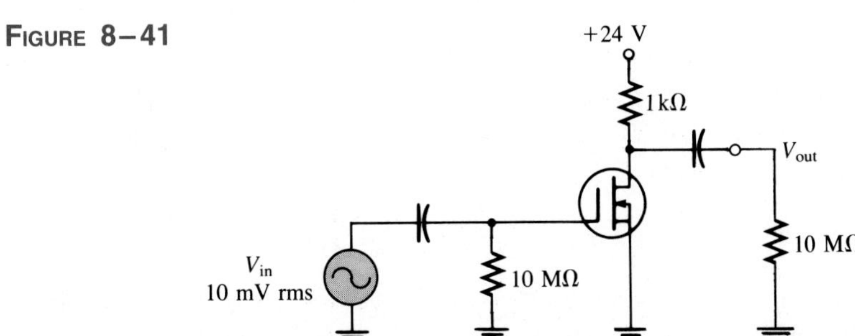

FIGURE 8–42

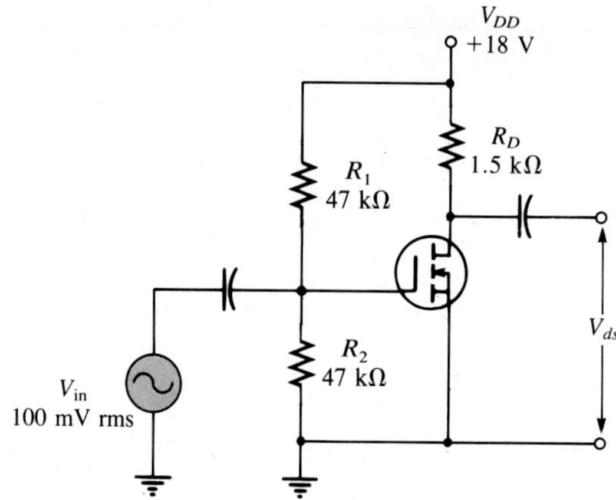

23. For the unloaded amplifier in Figure 8–42, find V_{GS}, I_D, V_{DS}, and the rms output voltage V_{ds}. $I_{D(on)} = 8$ mA at $V_{GS} = 12$ V, $V_{GS(th)} = 4$ V, and $g_m = 4500\ \mu$S.

SECTION 8–4 COMMON-DRAIN AMPLIFIERS

24. For the source-follower in Figure 8–43, determine the voltage gain and input resistance. $I_{GSS} = 50$ pA at $V_{GS} = -15$ V and $g_m = 5500\ \mu$S.

FIGURE 8–43

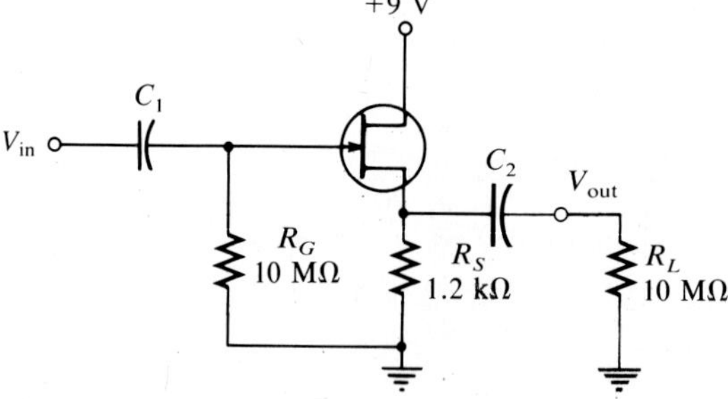

25. If the JFET in Figure 8–43 is replaced with one having a g_m of 3000 μS, what are the gain and the input resistance with all other conditions the same?

26. Find the gain of each amplifier in Figure 8–44.

FIGURE 8–44

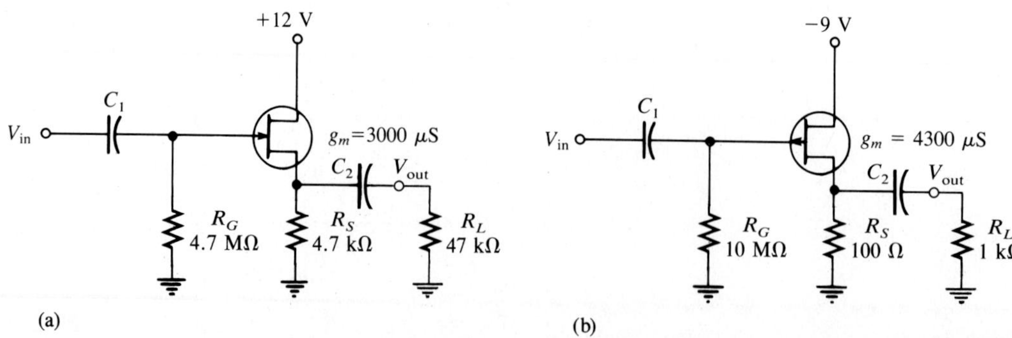

27. Determine the voltage gain of each amplifier in Figure 8–44 when the capacitively coupled load is changed to 10 kΩ.

SECTION 8–5 COMMON-GATE AMPLIFIERS

28. A common-gate amplifier has a g_m = 4000 μS and R_d = 1.5 kΩ. What is its gain?

29. What is the input resistance of the amplifier in Problem 28?

30. Determine the voltage gain and input resistance of the common-gate amplifier in Figure 8–45.

FIGURE 8–45

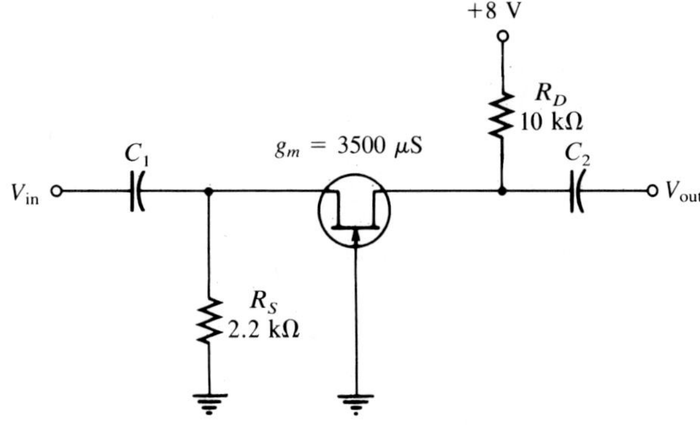

SECTION 8–6 TROUBLESHOOTING

31. What symptom(s) would indicate each of the following failures when a signal voltage is applied to the input in Figure 8–46?

(a) Q_1 open from drain to source **(b)** R_3 open **(c)** C_2 shorted
(d) C_3 open **(e)** Q_2 open from drain to source

FIGURE 8–46

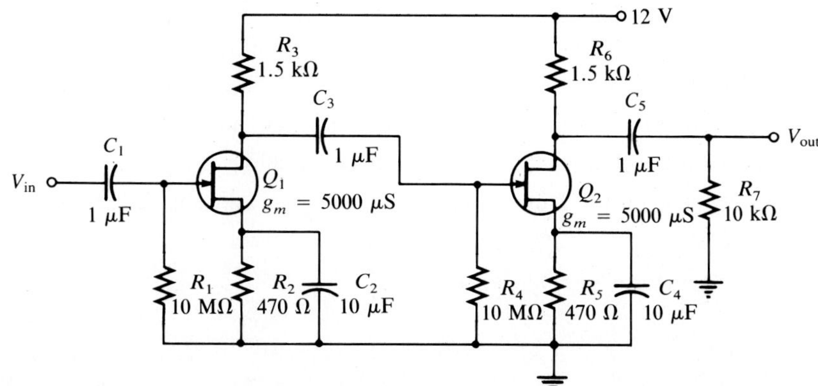

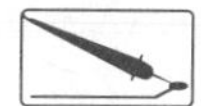

32. If V_{in} = 10 mV rms in Figure 8–46, what is V_{out} for each of the following faults?

(a) C_1 open **(b)** C_4 open

(c) a short from the source of Q_2 to ground **(d)** Q_2 has open gate

ANSWERS TO REVIEW QUESTIONS

SECTION 8–1

1. I_d is at its positive peak and V_{ds} is at its negative peak.
2. V_{gs} is an ac quantity, V_{GS} is a dc quantity.
3. The D-MOSFET can operate with V_{GS} = 0 V at the Q point.

SECTION 8–2

1. The FET with g_m = 3.5 mS can produce the highest gain.
2. $A_v = g_m R_d$ = (2500 μS)(10 kΩ) = 25
3. The one with R_{ds} = 100 kΩ can produce the highest gain.

SECTION 8–3

1. Voltage gain is determined by g_m and R_d.
2. The gain is halved because $R_d = R_D/2$.

SECTION 8–4

1. The maximum voltage gain of a CD amplifier is 1.
2. The voltage gain is determined by g_m and R_s.

SECTION 8–5

1. The CG amplifier has a low input resistance.
2. g_m affects both voltage gain and input resistance.

SECTION 8–6

1. To be a good troubleshooter, you must understand the circuit.
2. There would be a lower than normal first stage gain.
3. Yes, but there would be no signal to the second stage.

SECTION 8–7

1. It is 5 V also because direct current cannot flow through D_1 or C_2, so there is no drop across R_1.
2. C_2 and C_4 are decoupling capacitors to shunt any ac ripple to ground.
3. C_3 is a source bypass. C_5 is for signal coupling.
4. L_1, C_1, D_1 (FET input capacitance and stray capacitance are also factors.)
5. 0 V
6. D_1 operates in reverse bias.

ANSWERS TO PRACTICE EXERCISES

8–1 Decreases; distortion and clipping at cutoff

8–2 13.2

8–3 5

8–4 2.92

8–5 I_D will be 584 μA. V_D will increase to 10.1 V.

8–6 744 mV

8–7 $R_{\text{in}} = 99\ \text{M}\Omega$; $\Delta R_{\text{in}} = 22.1\ \text{M}\Omega$

8–8 1.55 V

8–9 $V_{GS} = 6.19$ V, $I_D = 1$ mA, $V_{DS} = 11.7$ V, $V_{\text{out}} = 0.578$ V

8–10 0.966

8–11 $R_{\text{in}} = 384.6\ \Omega$; $\Delta R_{\text{in}} = 15.6\ \Omega$

9

Power Amplifiers

After completing this chapter, you should be able to

- □ Distinguish between large-signal and small-signal operation.
- □ Define class A, class B, and class C operation.
- □ Discuss the factors that contribute to nonlinear distortion.
- □ Describe how the classification of an amplifier is determined by its bias point.
- □ Explain the operation of a class B push-pull amplifier.
- □ Explain crossover distortion.
- □ Define amplifier efficiency.
- □ Compare the efficiencies of class A, class B, and class C amplifiers.
- □ Discuss the basic operation of class C tuned amplifiers.
- □ Troubleshoot an amplifier.

Power amplifiers are large-signal amplifiers. This generally means that a much larger portion of the load line is used during signal operation than in a small-signal amplifier. In this chapter, we will cover the three general classes of large-signal amplifiers—class A, class B, and class C. These amplifier classifications are based on the percentage of the input cycle for which the amplifier operates in its linear region. Each class has a unique circuit configuration because of the way it must be operated. In this chapter, BJT amplifiers are used to illustrate the basic concepts, but each class of amplifier can also be implemented with FETs.

The emphasis in large-signal amplifiers is on power amplification. Power amplifiers are normally used as the final stage of a communications receiver or transmitter to provide signal power to speakers or to a transmitting antenna.

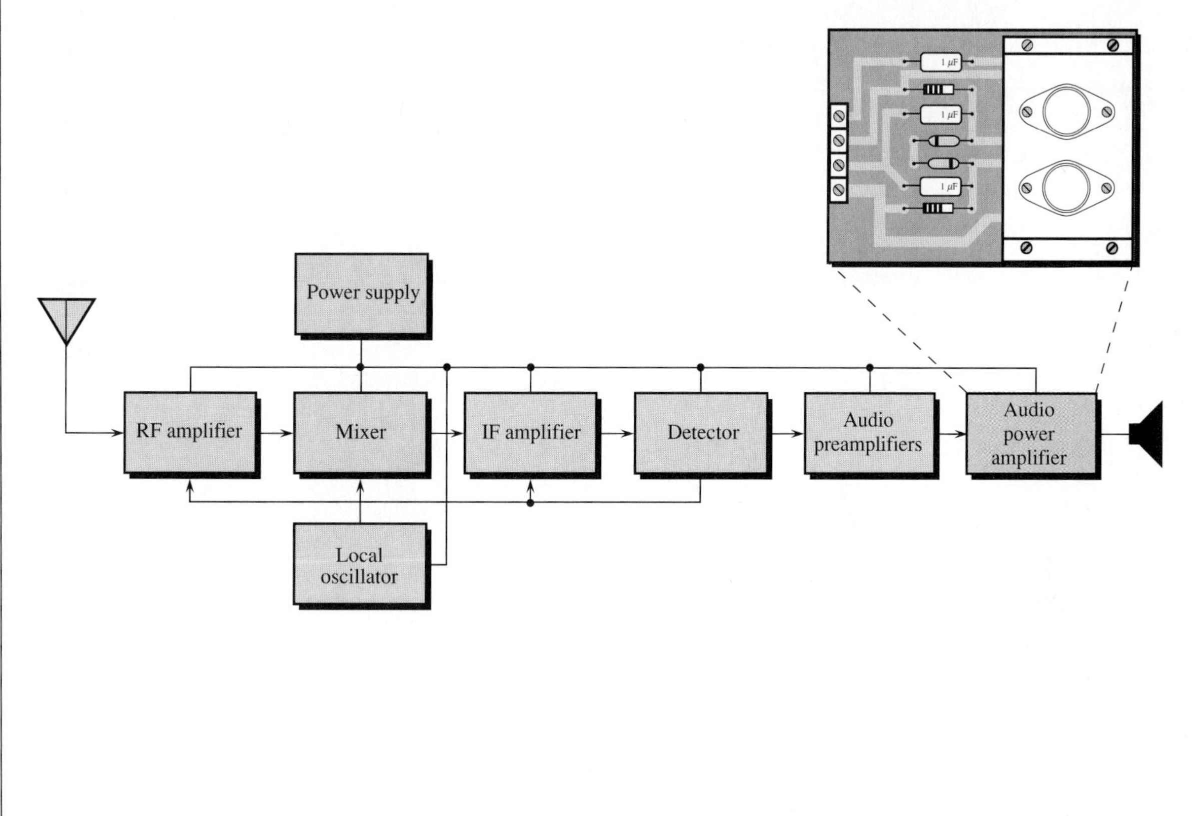

A System Application

In this chapter, we come back to the receiver system that you worked with in some previous chapters. In Chapter 6, we focused on the audio amplifier. Recall that the purpose of the audio amplifier is to increase the small audio signal voltage coming from the detector circuit. The output of the audio amplifier goes to the power amplifier, which increases the signal power in order to drive the speaker. In this system application, our focus will be on the class B push-pull power amplifier, and you will see how the audio amplifier and the power amplifier work together to produce an audio output voltage that is converted into sound by the speaker.

For the system application in Section 9–5, in addition to the other topics, be sure you understand

- □ The principles of large-signal amplification.
- □ How class B push-pull amplifiers work.
- □ The difference between power amplification and voltage amplification.

9–1 CLASS A AMPLIFIERS

When a common-emitter, common-collector, or common-base amplifier is biased so that it operates in the linear region for the full 360° of the input cycle, it is a class A amplifier. In this mode of operation, the amplifier does not go into either cutoff or saturation; therefore, the output voltage waveform has the same shape as the input waveform. A class A amplifier can be either inverting or noninverting. All of the small-signal amplifiers that you studied in Chapter 8 were class A. In this section, you will learn about* large-signal *class A amplifiers.

Class A operation is illustrated in Figure 9–1, where the output waveform is an amplified replica of the input and may be either in phase or 180° out of phase with the input.

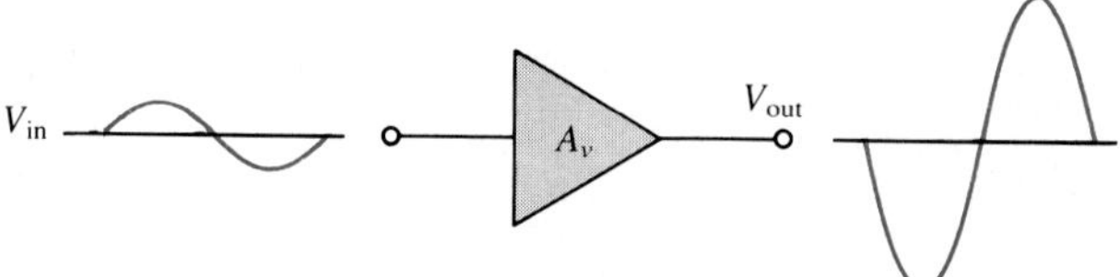

FIGURE 9–1
Class A amplifier operation (inverting).

Q POINT MUST BE CENTERED FOR MAXIMUM OUTPUT SIGNAL

When the Q point is at the center of the ac load line (midway between saturation and cutoff), a maximum class A signal can be obtained. This is graphically illustrated in the ac load line in Figure 9–2(a). Ideally, the collector current can vary from its Q-point value, I_{CQ}, up to its saturation value $I_{c(sat)}$, and down to its cutoff value of zero. This operation is indicated in Figure 9–2(b).

FIGURE 9–2
Maximum class A output (centered Q point).

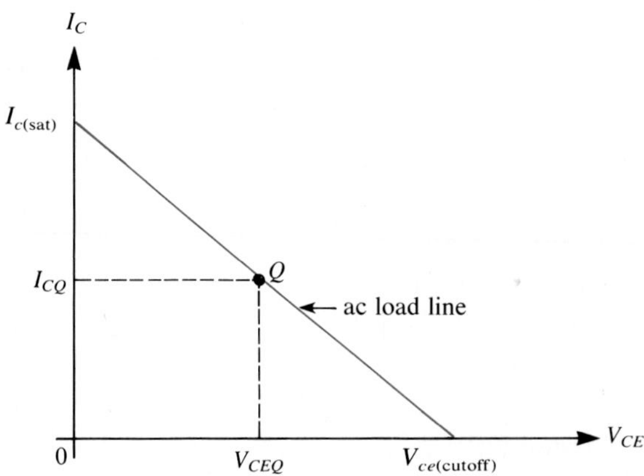

(a)

FIGURE 9–2
(continued)

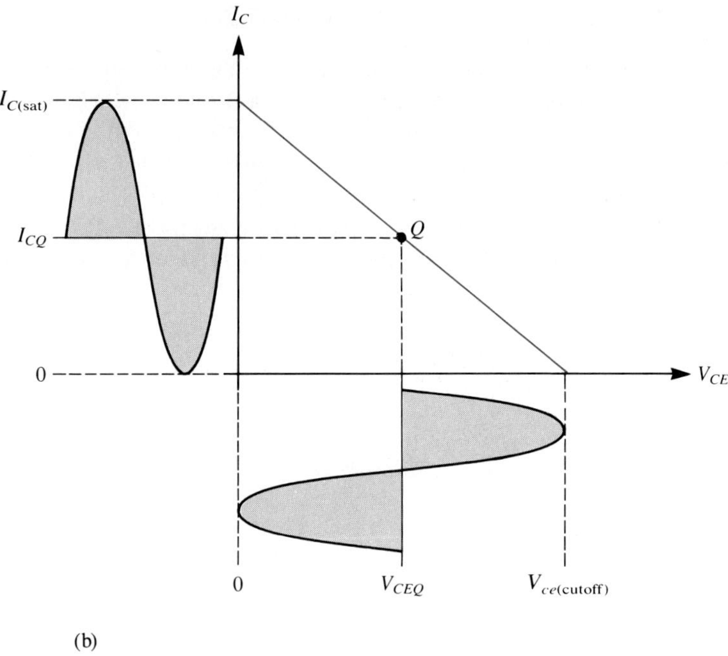

(b)

As you can see in Figure 9–2(b), the peak value of the collector current equals I_{CQ}, and the peak value of the collector-to-emitter voltage equals V_{CEQ}. This is the largest signal possible from a class A amplifier. When the input signal is too large, the amplifier is driven into cutoff and saturation, as illustrated in Figure 9–3.

FIGURE 9–3
Waveforms are clipped off at cutoff and saturation because amplifier is overdriven (too large an input signal) about a centered Q point.

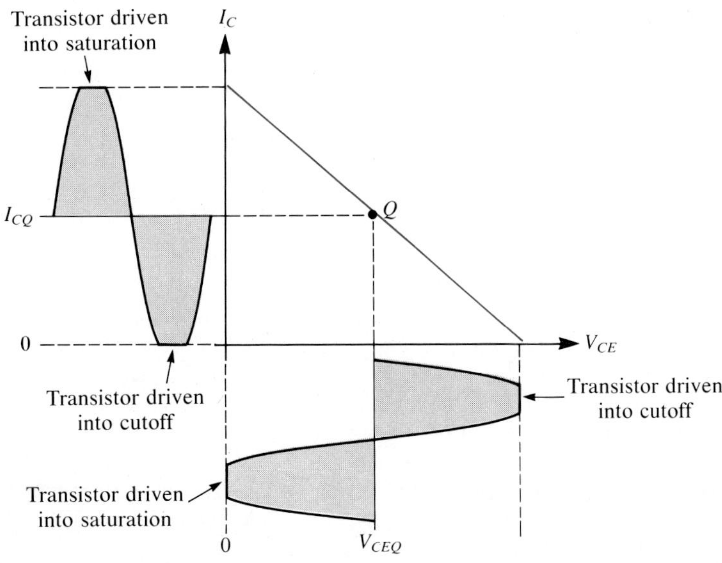

NONCENTERED Q POINT LIMITS OUTPUT SWING

If the Q point is not centered on the ac load line, the output signal is limited to less than the possible maximum. Figure 9–4(a) shows an ac load line with the Q point moved away from center toward cutoff. The output swing is limited by cutoff in this case. The collector current can swing only down to near zero and an equal amount above I_{CQ}. The collector-to-emitter voltage can swing only up to its cutoff value and an equal amount below V_{CEQ}. If the amplifier is driven any further than this, it will go into cutoff, as shown in Figure 9–4(b), and the waveforms will appear to be clipped off on one peak.

FIGURE 9–4
Q point closer to cutoff.

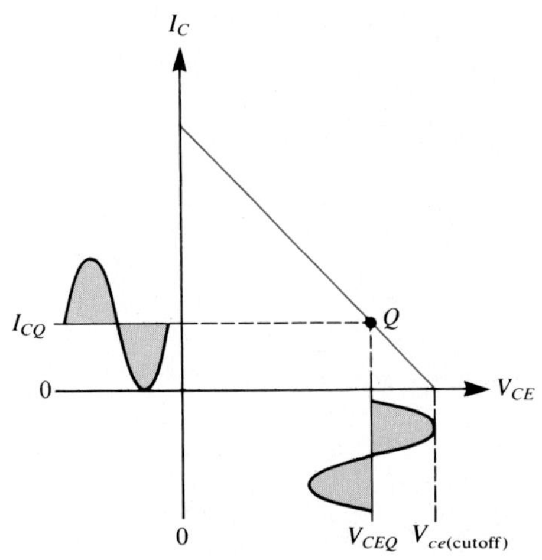

(a) Amplitude limited by cutoff

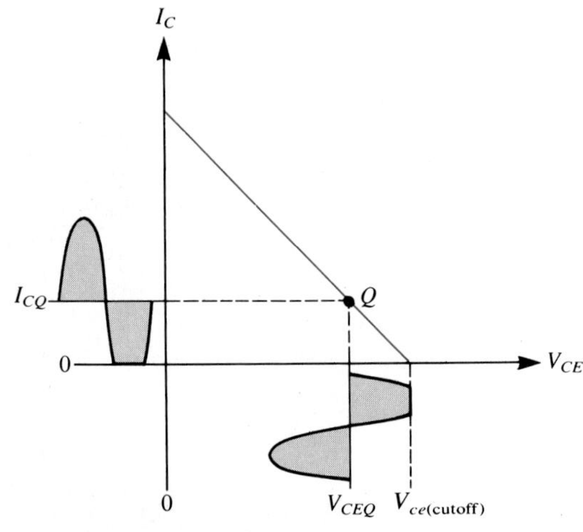

(b) Driven into cutoff by an increase in input amplitude

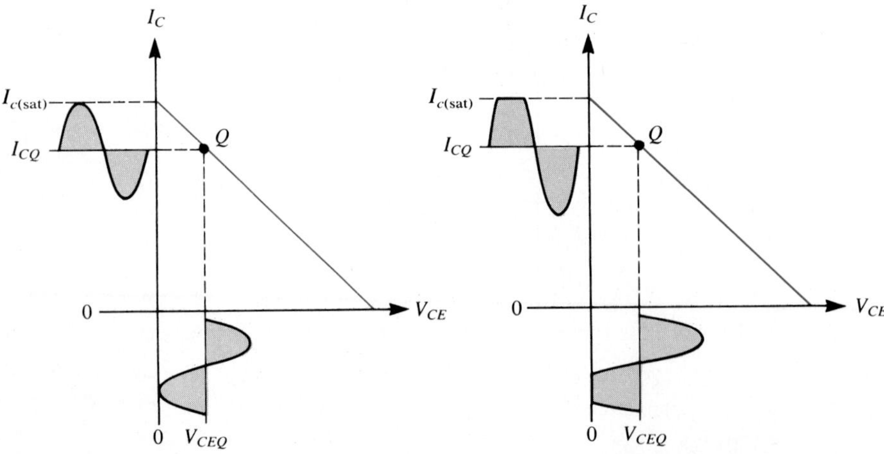

(a) Amplitude limited by saturation

(b) Driven into saturation by an increase in input amplitude

FIGURE 9–5
Q point closer to saturation.

Figure 9–5(a) shows an ac load line with the Q point moved away from center toward saturation. In this case, the output swing is limited by saturation. The collector current can swing only up to near saturation and an equal amount below I_{CQ}. The collector-to-emitter voltage can swing only down to near its saturation value and an equal amount above V_{CEQ}. If the amplifier is driven any further than this, it will go into saturation, as shown in Figure 9–5(b).

LARGE-SIGNAL LOAD LINE OPERATION

Recall that an amplifier such as that shown in Figure 9–6 can be represented in terms of either its dc or its ac equivalent.

FIGURE 9–6
Common-emitter amplifier with signal source.

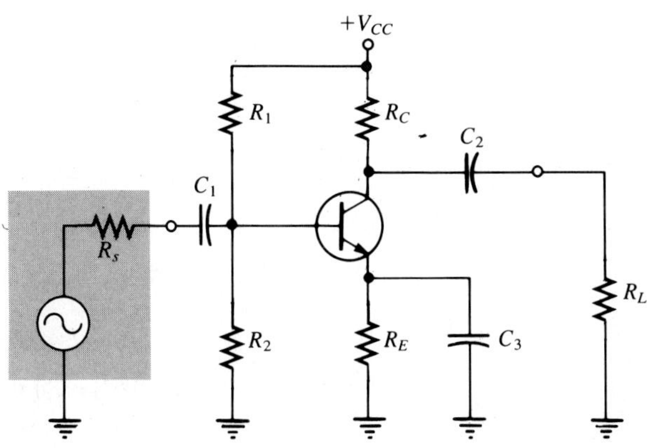

FIGURE 9–7
DC equivalent circuit and dc load line for the amplifier in Figure 9–6.

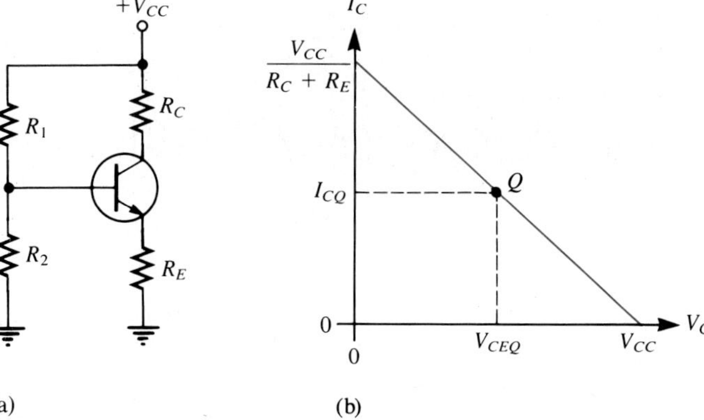

DC LOAD LINE Using the dc equivalent in Figure 9–7(a), we can determine the dc load line as follows: $I_{C(sat)}$ occurs when $V_{CE} \cong 0$, so

$$I_{C(sat)} \cong \frac{V_{CC}}{R_C + R_E}$$

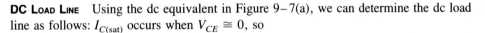

$V_{CE(\text{cutoff})}$ occurs when $I_C \cong 0$, so

$$V_{CE(\text{cutoff})} \cong V_{CC}$$

The dc load line is shown in Figure 9–7(b).

AC LOAD LINE From the ac viewpoint, the circuit in Figure 9–6 looks different than it does from the dc viewpoint. The collector resistance is different because R_L is in parallel with R_C due to the coupling capacitor C_2, and the emitter resistance is zero due to the bypass capacitor C_3; therefore, the ac load line is different from the dc load line. How much collector current can there be under ac conditions before saturation occurs? To answer this question, we will refer to the ac equivalent circuit and ac load line in Figure 9–8. Note that a lower-case subscript indicates an ac quantity and an upper-case subscript indicates a dc quantity. For example, R_c is the ac collector resistance and R_C is the dc collector resistance. I_{CQ} and V_{CEQ} are the dc Q-point coordinates. Going from the Q point to the saturation point, the collector-to-emitter voltage changes from V_{CEQ} to zero; that is, $\Delta V_{CE} = V_{CEQ}$. The swing in collector current going from the Q point to saturation is therefore

$$\Delta I_C = \frac{\Delta V_{CE}}{R_C \| R_L} = \frac{V_{CEQ}}{R_c}$$

where $R_c = R_C \| R_L$ is the ac collector resistance. The maximum (saturation) ac collector current is

$$I_{c(\text{sat})} = I_{CQ} + \Delta I_C$$

Thus,

$$I_{c(\text{sat})} = I_{CQ} + \frac{V_{CEQ}}{R_c} \tag{9–1}$$

Going from the Q point to the cutoff point, the collector current swings from I_{CQ} to near 0; that is, $\Delta I_C = I_{CQ}$. The swing in collector-to-emitter voltage going from the Q

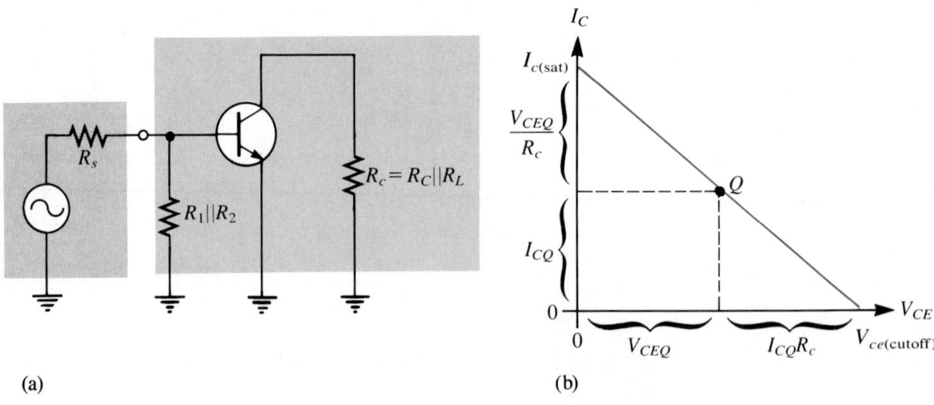

FIGURE 9–8
AC equivalent circuit and ac load line for the amplifier in Figure 9–6.

point to cutoff is therefore

$$\Delta V_{CE} = (\Delta I_C)R_c = I_{CQ}R_c$$

The cutoff value of ac collector-to-emitter voltage is

$$V_{ce(\text{cutoff})} = V_{CEQ} + I_{CQ}R_c \qquad (9\text{–}2)$$

These results are shown on the ac load line of Figure 9–9. The corresponding dc load line is shown for comparison.

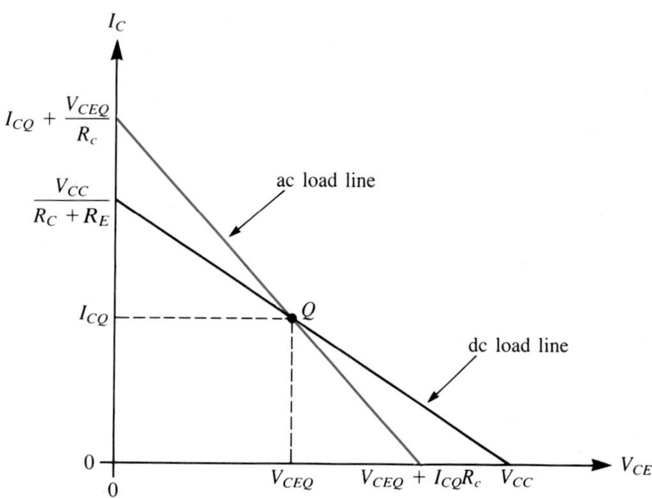

FIGURE 9–9
DC and ac load lines.

EXAMPLE 9–1 Determine the collector current and the collector-to-emitter voltage at the points of saturation and cutoff in Figure 9–10 under signal (ac) operation. Assume $X_{C1} = X_{C2} = X_{C3} \cong 0$.

SOLUTION

The Q-point values for this amplifier are determined as follows:

$$V_{BQ} = \left(\frac{4.7\text{ k}\Omega}{14.7\text{ k}\Omega}\right)10\text{ V} = 3.2\text{ V}$$

neglecting $R_{\text{IN(base)}}$ because it is sufficiently greater than R_2. Then,

$$I_{EQ} = \frac{3.2\text{ V} - 0.7\text{ V}}{470\ \Omega} = 5.3\text{ mA}$$

Therefore,

$$I_{CQ} \cong 5.3\text{ mA}$$

and

$$V_{CQ} \cong V_{CC} - I_{CQ}R_C = 10\text{ V} - (5.3\text{ mA})(1\text{ k}\Omega) = 4.7\text{ V}$$

Therefore,

$$V_{CEQ} = V_{CQ} - I_{EQ}R_E = 4.7\text{ V} - 2.5\text{ V} = 2.2\text{ V}$$

The point of saturation under ac conditions is determined as follows:

$$V_{ce(\text{sat})} \cong 0\text{ V}$$

$$I_{c(\text{sat})} = I_{CQ} + \frac{V_{CEQ}}{R_c}$$

Remember, lower-case subscripts indicate ac quantities. The ac collector resistance is

$$R_c = R_C \| R_L = \frac{R_C R_L}{R_C + R_L} = \frac{(1\text{ k}\Omega)(1.5\text{ k}\Omega)}{2.5\text{ k}\Omega} = 600\ \Omega$$

Thus,

$$\begin{aligned} I_{c(\text{sat})} &= 5.3\text{ mA} + \frac{2.2\text{ V}}{600\ \Omega} \\ &= 5.3\text{ mA} + 3.67\text{ mA} \\ &= 8.97\text{ mA} \end{aligned}$$

The point of cutoff under ac conditions is determined as follows:

$$\begin{aligned} I_{c(\text{cutoff})} &= 0\text{ A} \\ V_{ce(\text{cutoff})} &= V_{CEQ} + I_{CQ}R_c \\ &= 2.2\text{ V} + (5.3\text{ mA})(600\ \Omega) \\ &= 5.38\text{ V} \end{aligned}$$

These results show that the collector current can swing up to almost 8.97 mA or the collector-to-emitter voltage can swing up to almost 5.38 V without the peaks being clipped.

FIGURE 9–10

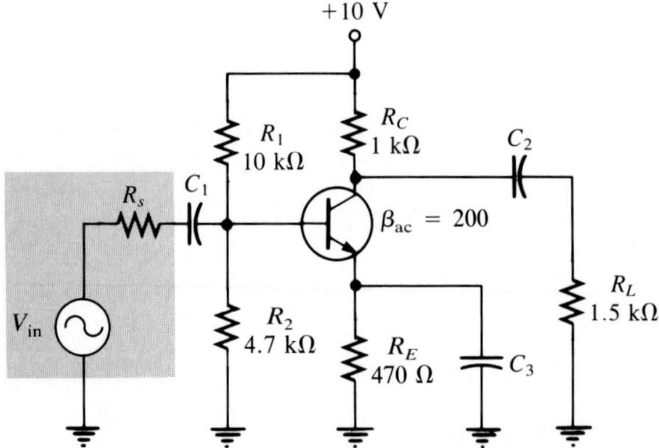

PRACTICE EXERCISE 9–1

Determine I_c and V_{ce} at the points of saturation and cutoff in Figure 9–10 for the following circuit values: $V_{CC} = 15$ V and $\beta_{ac} = 150$.

As described earlier, a centered Q point allows a maximum unclipped output swing. A closer look at Example 9–1 shows that the Q point is not centered, and therefore, the unclipped output swing is somewhat less than it could be if the Q point were centered on the ac load line. This is examined further in Example 9–2.

EXAMPLE 9–2

Figure 9–11 shows the ac load line for the circuit in Example 9–1. Notice that the Q point is *not* centered. The maximum output swings are

$$\begin{aligned}\Delta I_C &= I_{c(sat)} - I_{CQ} \\ &= 8.97\text{ mA} - 5.3\text{ mA} \\ &= 3.67\text{ mA}\end{aligned}$$

and

$$\Delta V_{CE} = V_{CEQ} = 2.2\text{ V}$$

Determine the maximum output swings for collector current and collector-to-emitter voltage when the Q point is centered, assuming the same load line is maintained.

FIGURE 9–11

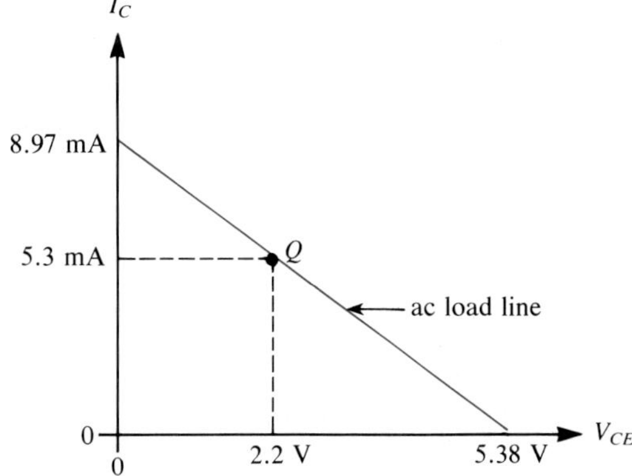

SOLUTION

The maximum output swings for a centered Q point are

$$\Delta I_C = I_{CQ} = \frac{I_{c(sat)}}{2} = \frac{8.97\text{ mA}}{2} = 4.49\text{ mA}$$

$$\Delta V_{CE} = V_{CEQ} = \frac{V_{ce(cutoff)}}{2} = \frac{5.38\text{ V}}{2} = 2.69\text{ V}$$

PRACTICE EXERCISE 9–2

Explain what happens to the output voltage if the Q point is shifted to $V_{CE} = 3$ V.

CENTERING THE Q POINT ON THE AC LOAD LINE

To have a centered Q point on the ac load line in Figure 9–9, I_{CQ} must be midway between zero and the upper (saturation) end of the ac load line and V_{CEQ} must be midway between zero and the lower (cutoff) end of the ac load line.

$$I_{CQ} = \frac{I_{CQ} + V_{CEQ}/R_c}{2} \tag{9–3}$$

$$V_{CEQ} = \frac{V_{CEQ} + I_{CQ}R_c}{2} \tag{9–4}$$

The condition for a centered Q point is developed as follows using Equation (9–4). Developing Equation (9–3) will produce the same result.

$$2V_{CEQ} = V_{CEQ} + I_{CQ}R_c$$

$$2V_{CEQ} - V_{CEQ} = I_{CQ}R_c$$

$$V_{CEQ} = I_{CQ}R_c \tag{9–5}$$

The Q point can be moved to an approximate center position on the load line by changing I_{CQ} until both sides of Equation (9–5) are approximately equal.

To move the Q point toward cutoff on the ac load line without affecting the load line itself, I_{CQ} should be decreased by increasing R_E. To move the Q point toward saturation on the load line, I_{CQ} should be increased by decreasing R_E.

■ EXAMPLE 9–3

As you saw in Example 9–2, the circuit in Figure 9–10 does *not* have a centered Q point. Select a value for R_E that will produce a Q point that is approximately centered on the ac load line.

SOLUTION

For a centered Q point,

$$V_{CEQ} = I_{CQ}R_c$$

We know from Example 9–2 that the collector current for a centered Q point should be 4.49 mA. $R_C = 600\ \Omega$ from Example 9–1.

$$I_{CQ}R_c = (4.49\ \text{mA})(600\ \Omega) = 2.69\ \text{V}$$

$$V_{CEQ} = V_{CC} - I_{CQ}(R_C - R_E) = 2.69\ \text{V}$$

$$10\ \text{V} - 4.49\ \text{mA}(1\ \text{k}\Omega - R_E) = 2.69\ \text{V}$$

$$10\ \text{V} - 4.49\ \text{V} - (4.49\ \text{mA})R_E = 2.69\ \text{V}$$

$$(4.49\ \text{mA})R_E = 2.82\ \text{V}$$

$$R_E = \frac{2.82\ \text{V}}{4.49\ \text{mA}} = 628\ \Omega$$

Choosing the nearest standard value,

$$R_E = 620\ \Omega$$

This will produce an approximately centered Q point.

PRACTICE EXERCISE 9–3

If R_C in Figure 9–10 is increased to 1.2 kΩ, what should you do to R_E to keep a centered Q point?

■

LARGE-SIGNAL VOLTAGE GAIN

The voltage gain of a class A large-signal amplifier is determined in the same way as for a small-signal amplifier with the exception that the formula $r_e \cong 25\text{ mV}/I_E$ is not valid for the large-signal amplifier. This is because the signal swings over a large portion of the transconductance curve. Since $r_e = \Delta V_{BE}/\Delta I_C$, the value is different for large-signal operation than it is for small-signal conditions because of the nonlinearity of the curve.

The large-signal ac emitter resistance, r_e', can be determined graphically from the transconductance curve, as shown in Figure 9–12, using the relationship in Equation (9–6). The prime mark (′) distinguishes the large-signal parameter from the small-signal r_e.

$$r_e' = \frac{\Delta V_{BE}}{\Delta I_C} \tag{9–6}$$

The voltage-gain formula for a common-emitter, large-signal amplifier is therefore

$$A_v = \frac{R_c}{r_e'} \tag{9–7}$$

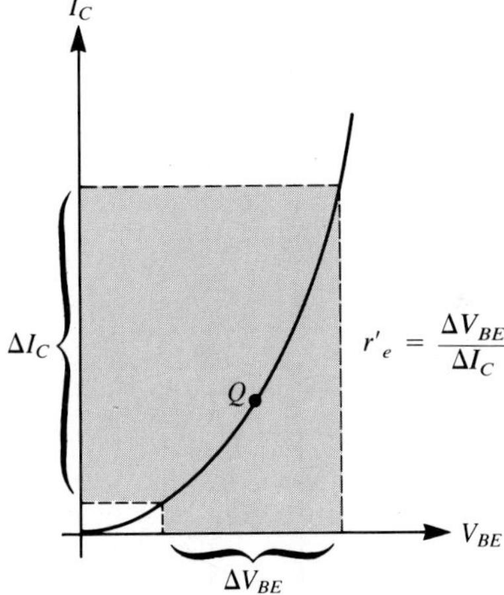

FIGURE 9–12
Determination of r_e' from the transconductance curve.

EXAMPLE 9–4 Find the large-signal voltage gain of the amplifier in Figure 9–13. Assume that r_e' has been found to be 8 Ω from graphical data.

SOLUTION

$$R_c = \frac{(1\text{ k}\Omega)(1.5\text{ k}\Omega)}{2.5\text{ k}\Omega} = 600\ \Omega$$

$$A_v = \frac{R_c}{r_e'} = \frac{600\ \Omega}{8\ \Omega} = 75$$

FIGURE 9–13

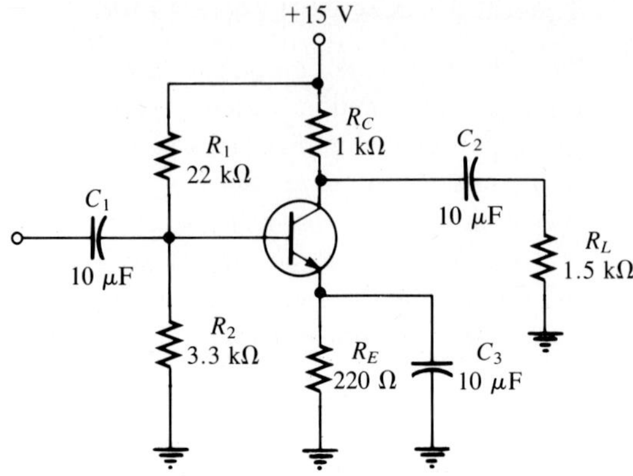

PRACTICE EXERCISE 9–4

Find the large-signal voltage gain of the amplifier in Figure 9–13 if the supply voltage is changed to 9 V, R_L is changed to 1 kΩ, and r'_e is 10 Ω.

NONLINEAR DISTORTION

When the collector current swings over a large portion of the transconductance curve, distortion can occur on the negative half-cycle. This is caused by the greater nonlinearity on the lower end of the curve, as shown in Figure 9–14. This distortion can be sufficiently

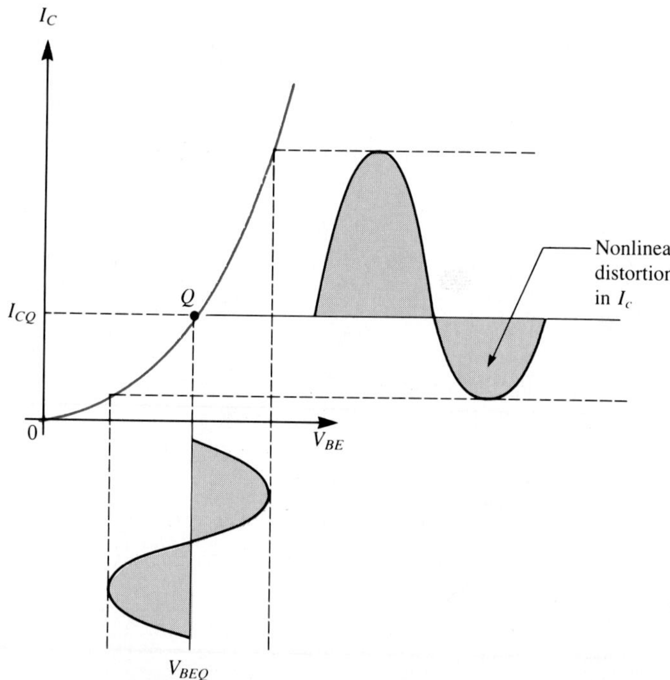

FIGURE 9–14
Example of nonlinear distortion.

reduced by keeping the collector current on the more linear part of the curve (at higher values of I_{CQ} and V_{BEQ}). This can be done by increasing the base bias voltage, which will result in more collector current and an increase in V_{BE} due to more voltage drop across r_e.

POWER GAIN

The main purpose of a large-signal amplifier is to achieve power gain. If we assume that the large-signal current gain A_i is approximately equal to β_{dc}, then the power gain, A_p, for a common-emitter amplifier is

$$A_p = A_i A_v = \beta_{dc} A_v$$

$$A_p = \beta_{dc}\left(\frac{R_c}{r_e'}\right) \qquad (9\text{–}8)$$

QUIESCENT POWER

The power dissipation of a transistor with no signal input is the product of its Q-point current and voltage.

$$P_{DQ} = I_{CQ}V_{CEQ} \qquad (9\text{–}9)$$

The quiescent power is the maximum power that the class A transistor must handle; therefore, its power rating should exceed this value.

OUTPUT POWER

In general, for any Q-point location on the ac load line, the output power of a common-emitter amplifier is the product of the rms collector current and the rms collector-to-emitter voltage.

$$P_{out} = V_{ce}I_c \qquad (9\text{–}10)$$

Let's now consider the output power for three cases of Q-point location.

Q POINT CLOSER TO SATURATION When the Q point is closer to saturation, the maximum collector-to-emitter voltage swing is V_{CEQ}, and the maximum collector current swing is V_{CEQ}/R_c, as shown in Figure 9–15(a). The output power is therefore

$$P_{out} = (0.707V_{CEQ}/R_c)(0.707V_{CEQ})$$

$$P_{out} = \frac{0.5V_{CEQ}^2}{R_c} \qquad (9\text{–}11)$$

where $R_c = R_C \| R_L$.

Q POINT CLOSER TO CUTOFF When the Q point is closer to cutoff, the maximum collector current swing is I_{CQ}, and the collector-to-emitter voltage swing is $I_{CQ}R_c$, as shown in Figure 9–15(b). The output power is therefore

$$P_{out} = (0.707I_{CQ})(0.707I_{CQ}R_c)$$

$$P_{out} = 0.5I_{CQ}^2R_c \qquad (9\text{–}12)$$

Q POINT CENTERED When the Q point is centered, the maximum collector current swing is I_{CQ}, and the maximum collector-to-emitter voltage swing is V_{CEQ}, as shown in Figure 9–15(c). The output power is therefore

$$P_{out} = (0.707V_{CEQ})(0.707I_{CQ})$$

$$P_{out} = 0.5V_{CEQ}I_{CQ} \tag{9–13}$$

This is the maximum ac output power from a class A amplifier under signal conditions. Notice that it is one-half the quiescent power dissipation.

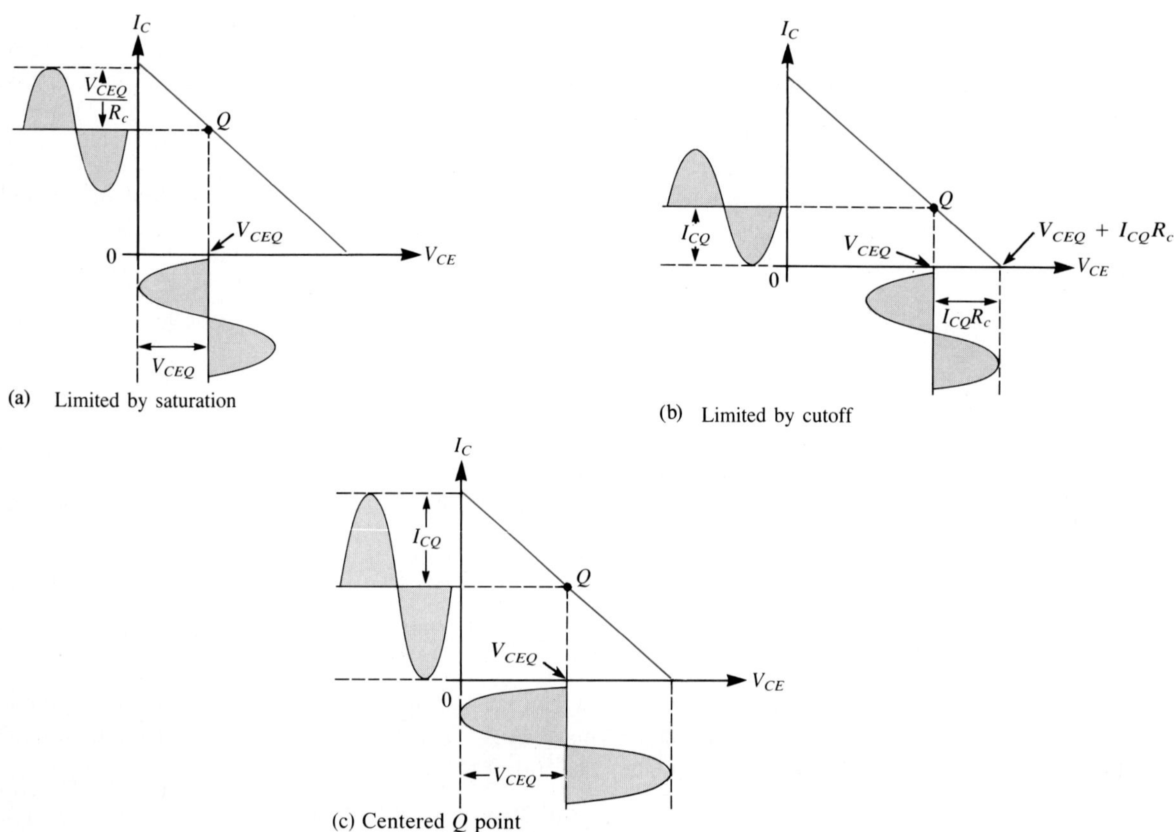

FIGURE 9–15
AC load line operation showing limitations of output voltage swings.

EFFICIENCY

Efficiency of an amplifier is the ratio of ac output power to dc input power. The dc input power is the dc supply voltage times the current drawn from the supply.

$$P_{DC} = V_{CC}I_{CC} \tag{9–14}$$

The average supply current I_{CC} equals I_{CQ}, and the supply voltage V_{CC} is twice V_{CEQ} when the Q point is centered. The maximum efficiency is therefore

$$\eta_{max} = \frac{P_{out}}{P_{DC}} = \frac{0.5V_{CEQ}I_{CQ}}{V_{CC}I_{CC}}$$

$$= \frac{0.5V_{CEQ}I_{CQ}}{2V_{CEQ}I_{CQ}} = \frac{0.5}{2}$$

$$\eta_{max} = 0.25 \quad (9\text{–}15)$$

Thus, 25 percent is the highest possible efficiency available from a class A amplifier and is approached only when the Q point is at the center of the ac load line.

EXAMPLE 9–5 Determine the following values for the amplifier in Figure 9–16 when operated with the maximum possible output signal.

(a) Minimum transistor power rating
(b) AC output power
(c) Efficiency

FIGURE 9–16

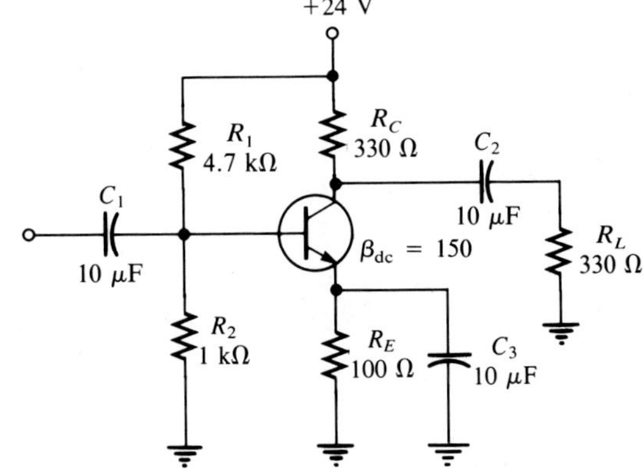

SOLUTION

The dc values are first determined. Neglecting $R_{IN(base)}$,

$$V_B = \left(\frac{R_2}{R_1 + R_2}\right)V_{CC} = \left(\frac{1\text{ k}\Omega}{5.7\text{ k}\Omega}\right)24\text{ V} = 4.2\text{ V}$$

Then

$$V_E = V_B - V_{BE} = 4.2\text{ V} - 0.7\text{ V} = 3.5\text{ V}$$

$$I_E = \frac{V_E}{R_E} = \frac{3.5\text{ V}}{100\ \Omega} = 35\text{ mA}$$

Therefore,

$$I_{CQ} \cong 35\text{ mA}$$

and

$$V_C = V_{CC} - I_{CQ}R_C = 24\text{ V} - (35\text{ mA})(330\ \Omega)$$
$$= 12.45\text{ V}$$
$$V_{CEQ} = V_C - V_E = 12.45\text{ V} - 3.5\text{ V}$$
$$= 8.95\text{ V}$$

(a) The transistor power rating must be greater than

$$P_D = V_{CEQ}I_{CQ} = (8.95\text{ V})(35\text{ mA}) = 0.313\text{ W}$$

(b) To make a calculation of ac output power under a *maximum* signal condition, we must know the location of the Q point relative to center. This will tell us whether I_{CQ} or V_{CEQ} is the limiting factor if the Q point is not centered. The ac load line values are as follows:

$$R_c = R_C \| R_L = 330\ \Omega \| 330\ \Omega = 165\ \Omega$$

$$I_{c(\text{sat})} = I_{CQ} + \frac{V_{CEQ}}{R_c}$$
$$= 35\text{ mA} + \frac{8.95\text{ V}}{165\ \Omega}$$
$$= 89.2\text{ mA}$$

and

$$V_{ce(\text{cutoff})} = V_{CEQ} + I_{CQ}R_c$$
$$= 8.95\text{ V} + (35\text{ mA})(165\ \Omega)$$
$$= 14.73\text{ V}$$

A *centered* Q point is at

$$I_{CQ} = \frac{89.2\text{ mA}}{2} = 44.6\text{ mA}$$

and

$$V_{CEQ} = \frac{14.73\text{ V}}{2} \cong 7.37\text{ V}$$

These are shown on the ac load line in Figure 9–17. The *actual* Q point for this amplifier is closer to cutoff, as shown in the figure. Therefore, the maximum collector current swing is I_{CQ}, and the ac output power is

$$P_{\text{out}} = 0.5 I_{CQ}^2 R_c$$
$$= 0.5(35\text{ mA})^2(165\ \Omega)$$
$$= 101\text{ mW}$$

(c) The efficiency is

$$\eta = \frac{P_{\text{out}}}{P_{\text{DC}}} = \frac{P_{\text{out}}}{V_{CC}I_{CC}} = \frac{P_{\text{out}}}{V_{CC}I_{CQ}}$$
$$= \frac{101\text{ mW}}{(24\text{ V})(35\text{ mA})} = 0.12$$

This efficiency of 12 percent is considerably less than the maximum possible efficiency (25 percent) because the actual Q point is not centered.

FIGURE 9–17

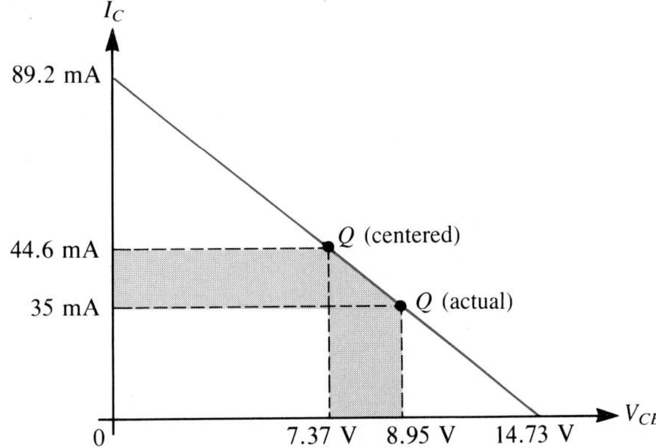

The following program computes efficiency.

```
10  DIM RL(100):CLS
20  PRINT "THIS PROGRAM COMPUTES CLASS A EFFICIENCY FOR VARIOUS
30  PRINT "LOADS IN AN AMPLIFIER OF THE TYPE IN FIGURE 9-16."
40  PRINT:PRINT
50  PRINT "ENTER RESISTOR VALUE IN OHMS."
60  INPUT "R1";R1
70  INPUT "R2";R2
80  INPUT "RC";RC
90  INPUT "RE";RE
100 CLS
110 INPUT "DC SUPPLY VOLTAGE, VCC";VCC
120 INPUT "FOR HOW MANY VALUES OF RL DO YOU WANT THE
    EFFICIENCY COMPUTED";N
130 FOR A=1 to N
140 INPUT "RL";RL(A)
150 NEXT
160 CLS
170 VB=R2/(R1+R2)*VCC
180 VE=VB-.7
190 ICQ=VE/RE
200 PRINT "RL", "EFFICIENCY":PRINT
210 FOR A=1 TO N
220 TRC=RC*RL(A)/(RC+RL(A))
230 P=.5*ICQ[2*TRC
240 EFF=P/(VCC*ICQ)
250 IF EFF>.25 THEN PRINT "AMPLIFIER IS NOT OPERATING CLASS A"
    ELSE PRINT RL(A), EFF
260 NEXT
```

PRACTICE EXERCISE 9–5

Suggest ways to increase the efficiency of the amplifier in Figure 9–16.

MAXIMUM LOAD POWER

The maximum power to the load in a class A amplifier occurs when the Q point is centered, as shown in Figure 9–15(c). The maximum peak load voltage equals V_{CEQ}, assuming a negligible voltage drop across the output coupling capacitor.

$$P_L = \frac{V_L^2}{R_L} = \frac{(0.707V_{CEQ})^2}{R_L}$$

$$P_L = \frac{0.5V_{CEQ}^2}{R_L} \qquad (9\text{–}16)$$

Sometimes the load power is defined as the output power.

EXAMPLE 9–6

Determine the maximum load power for the amplifier in Example 9–5.

SOLUTION

When the Q point is centered, $V_{CEQ} = 7.37$ V.

$$P_{L(\max)} = \frac{0.5V_{CEQ}^2}{R_L} = \frac{0.5(7.37\text{ V})^2}{330\ \Omega} = 82.3\text{ mW}$$

PRACTICE EXERCISE 9–6

If the Q point in the amplifier is shifted to $V_{CEQ} = 5$ V, what is the load power? ■

9–1 REVIEW QUESTIONS

1. Basically, why does the ac load line differ from the dc load line?
2. What is the optimum Q-point location for class A amplifiers?
3. What is the maximum efficiency of a class A amplifier?
4. How do you approach maximum efficiency in a class A amplifier?

9–2 CLASS B PUSH-PULL AMPLIFIERS

When an amplifier is biased such that it operates in the linear region for 180° of the input cycle and is in cutoff for 180°, it is a class B amplifier. The primary advantage of a class B amplifier over a class A is that the class B is more efficient; you can get more output power for a given amount of input power. A disadvantage of class B is that it is more difficult to implement the circuit in order to get a linear reproduction of the input waveform. As you will see in this section, the term push-pull refers to a common type of class B amplifier circuit in which the input wave shape is approximately reproduced at the output.

The class B operation is illustrated in Figure 9–18, where the output waveform is shown relative to the input.

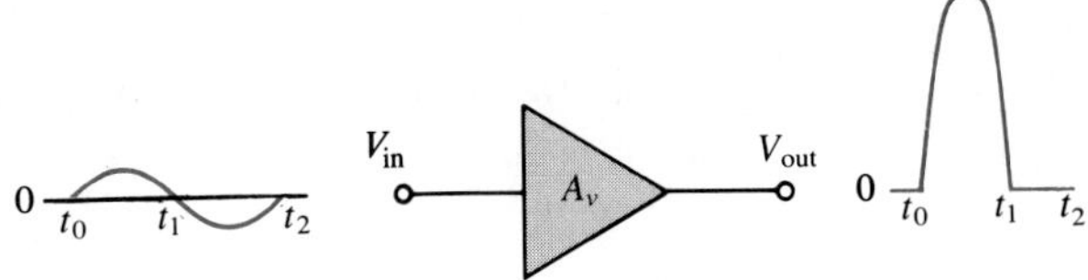

FIGURE 9–18
Class B amplifier operation (noninverting).

Q POINT IS AT CUTOFF

The class B amplifier is biased at cutoff so that $I_{CQ} = 0$ and $V_{CEQ} = V_{CE(\text{cutoff})}$. It is brought out of cutoff and operates in its linear region when the input signal drives it into conduction. This is illustrated in Figure 9–19 with an emitter-follower circuit. Obviously, the output is not a replica of the input. Therefore, a two-transistor configuration, known as a **push-pull** amplifier, is necessary to get a sufficiently good reproduction of the input waveform.

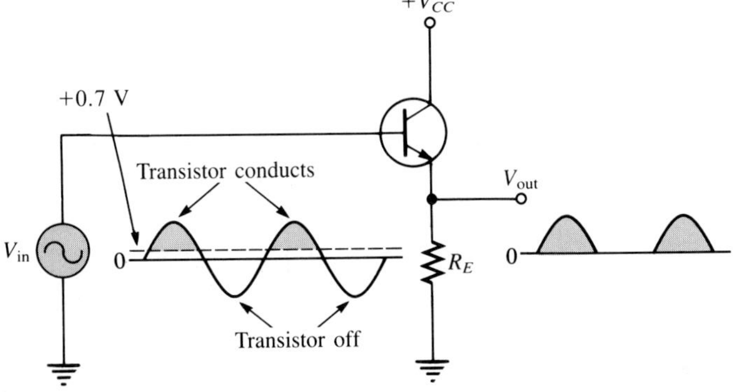

FIGURE 9–19
Common-collector class B amplifier.

PUSH-PULL OPERATION

Figure 9–20 shows one type of push-pull class B amplifier using two emitter-followers. This is a **complementary** amplifier because one emitter-follower uses an npn transistor and the other a pnp, which conduct on *opposite* alternations of the input cycle. Notice that there is no dc base bias voltage ($V_B = 0$); thus, only the signal voltage drives the transistors into conduction. Q_1 conducts during the positive half of the input cycle, and Q_2 conducts during the negative half.

CROSSOVER DISTORTION

When the dc base voltage is zero, the input signal voltage must exceed V_{BE} before a transistor conducts. Because of this, there is a time interval between the positive and negative alternations of the input when neither transistor is conducting, as shown in Figure 9–21. The resulting distortion in the output waveform is quite common and is called **crossover distortion.**

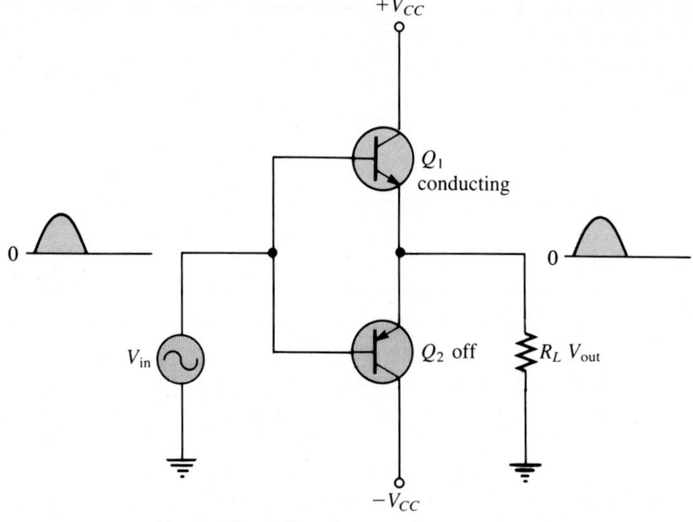

(a) Positive half-cycle

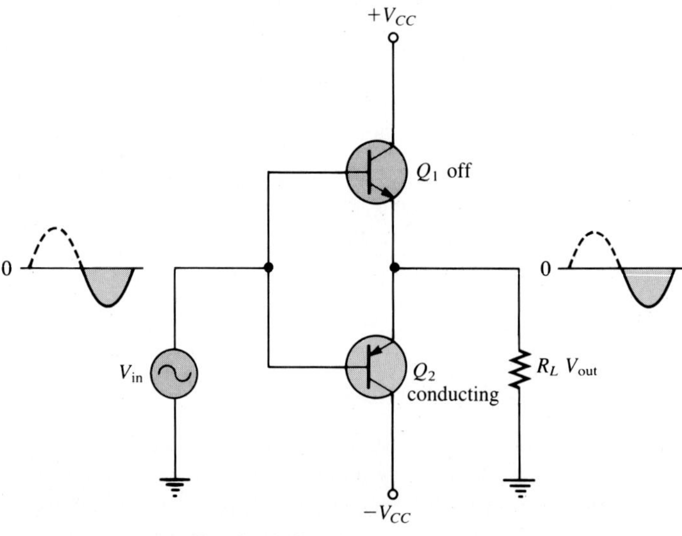

(b) Negative half-cycle

FIGURE 9–20
Class B push-pull operation.

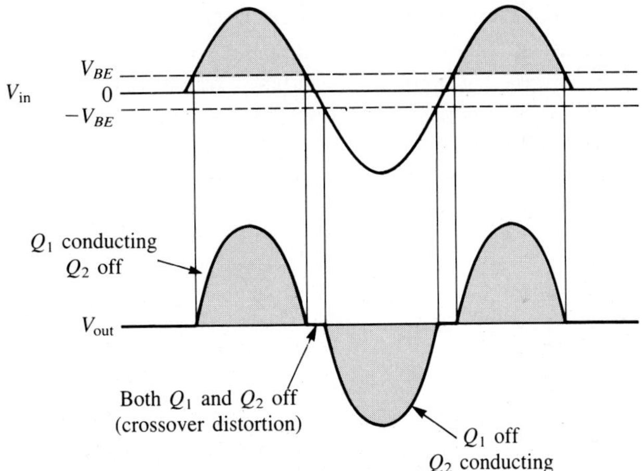

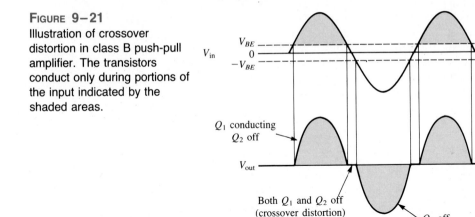

FIGURE 9–21
Illustration of crossover distortion in class B push-pull amplifier. The transistors conduct only during portions of the input indicated by the shaded areas.

BIASING THE PUSH-PULL AMPLIFIER

To eliminate crossover distortion, both transistors in the push-pull arrangement must be slightly above cutoff when there is no signal. This can be done with a voltage-divider arrangement, as shown in Figure 9–22(a). It is, however, difficult to maintain a stable bias point with this circuit due to changes in V_{BE} over temperature changes. (The requirement for dual-polarity power supplies is eliminated when R_L is capacitively coupled.) A more suitable arrangement is shown in Figure 9–22(b). When the diode characteristics of D_1 and D_2 are closely matched to the transconductance characteristics of the transistors, a stable bias is maintained. This can also be accomplished by using the base-emitter junctions of two additional transistors instead of D_1 and D_2.

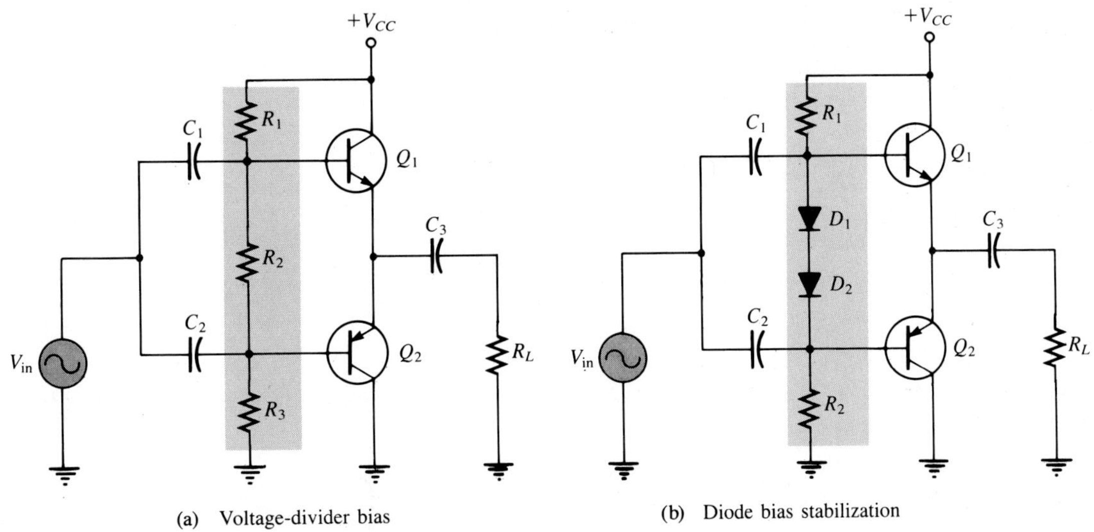

FIGURE 9–22
Biasing the push-pull amplifier to eliminate crossover distortion.

The dc equivalent circuit of the push-pull amplifier is shown in Figure 9–23. R_1 and R_2 are of equal value; therefore the voltage at point A between the two diodes is $V_{CC}/2$. Assuming that both diodes and both transistors are identical, the drop across D_1 equals the V_{BE} of Q_1, and the drop across D_2 equals the V_{BE} of Q_2. As a result, the voltage at the emitters is also $V_{CC}/2$, and therefore, $V_{CEQ1} = V_{CEQ2} = V_{CC}/2$, as indicated. Because both transistors are biased near cutoff, $I_{CQ} \cong 0$.

FIGURE 9–23
DC equivalent of push-pull amplifier.

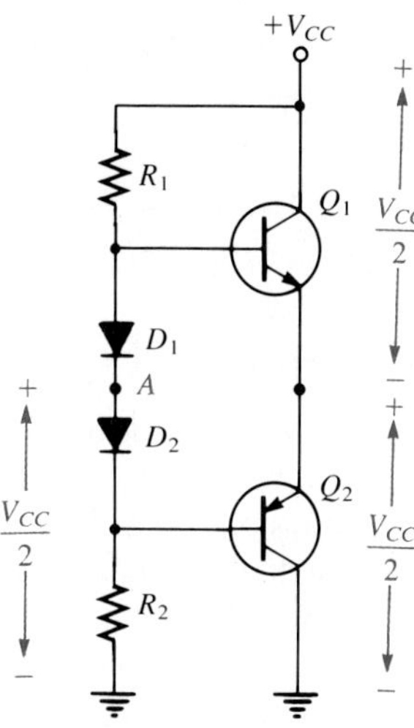

EXAMPLE 9–7

Determine the dc voltages at the bases and emitters of Q_1 and Q_2 in Figure 9–24. Also determine V_{CEQ} for each transistor. Assume $V_{D1} = V_{D2} = V_{BE} = 0.7$ V.

FIGURE 9–24

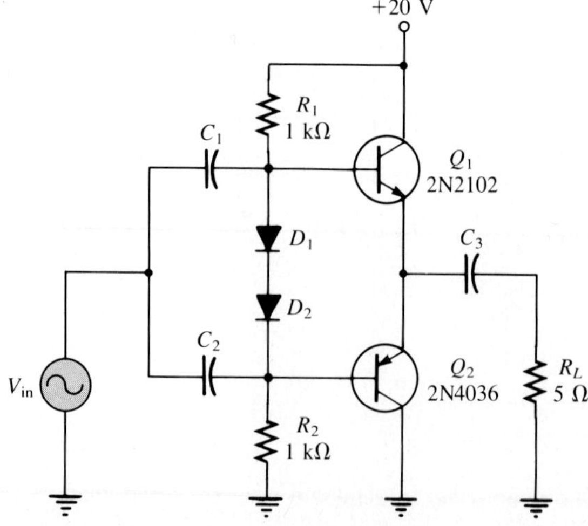

SOLUTION

The equivalent circuit for the bias network is shown in Figure 9–25(a).

$$I_T = \frac{V_{CC} - V_{D1} - V_{D2}}{R_1 + R_2} = \frac{20\text{ V} - 1.4\text{ V}}{2\text{ k}\Omega} = 9.3\text{ mA}$$

$$V_{B1} = V_{CC} - I_T R_1 = 20\text{ V} - (9.3\text{ mA})(1\text{ k}\Omega) = 10.7\text{ V}$$

and

$$V_{B2} = V_{B1} - V_{D1} - V_{D2} = 10.7\text{ V} - 1.4\text{ V} = 9.3\text{ V}$$

$$V_{E1} = V_{E2} = 10.7\text{ V} - 0.7\text{ V} = 10\text{ V}$$

Therefore,

$$V_{CEQ1} = V_{CEQ2} = \frac{V_{CC}}{2} - \frac{20\text{ V}}{2}$$

$$= 10\text{ V}$$

These values are shown in Figure 9–25(b).

FIGURE 9–25

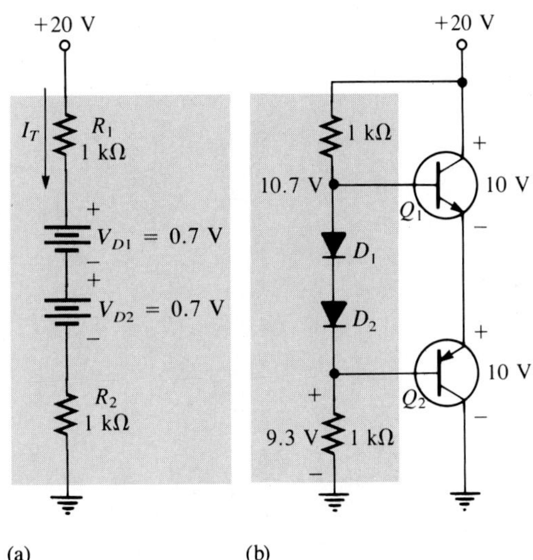

PRACTICE EXERCISE 9–7

If R_1 and R_2 are changed to 2.2 kΩ, what are the base and emitter voltages in Figure 9–24?

AC OPERATION

Under maximum conditions, transistors Q_1 and Q_2 are alternately driven from near cutoff to near saturation. During the positive alternation of the input signal, the Q_1 emitter is driven from its Q-point value of $V_{CC}/2$ to near V_{CC}, producing a positive peak voltage

FIGURE 9–26
Ideal ac push-pull operation.

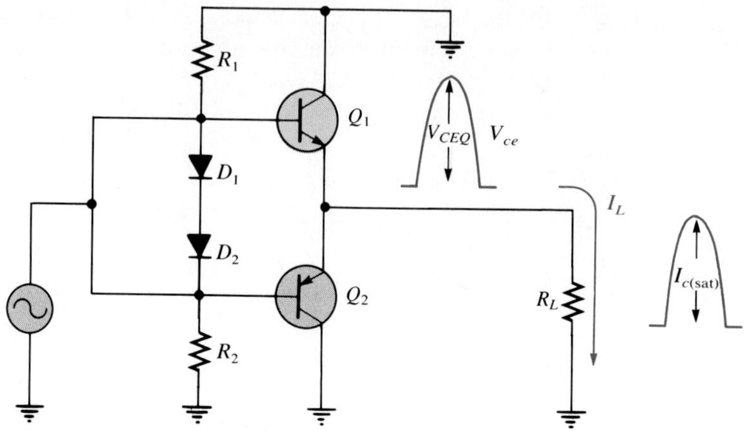

(a) Q_1 conducting with maximum signal output

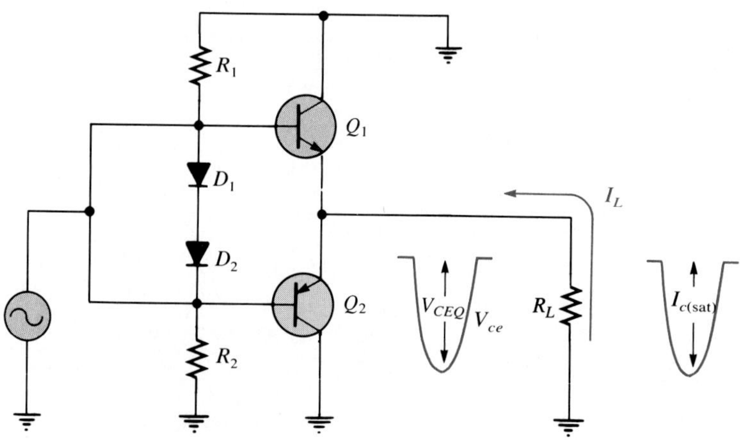

(b) Q_2 conducting with maximum signal output

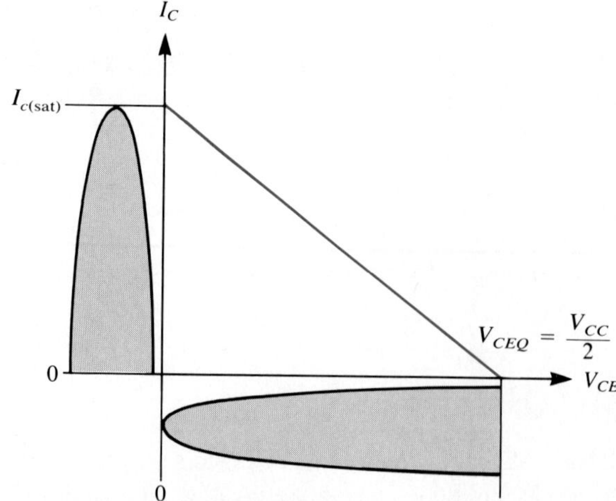

(c) AC load line for each transistor

approximately equal to V_{CEQ}. At the same time, the Q_1 current swings from its Q-point value near zero to near-saturation value, as shown in Figure 9–26(a).

During the negative alternation of the input signal, the Q_2 emitter is driven from its Q point value of $V_{CC}/2$ to near zero, producing a negative peak voltage approximately equal to V_{CEQ}. Also, the Q_2 current swings from near zero to near-saturation value, as shown in Figure 9–26(b).

In terms of the ac load line operation, the V_{ce} of both transistors swings from $V_{CC}/2$ to near zero, and the current swings from zero to $I_{c(sat)}$, as shown in Figure 9–26(c). Because the peak voltage across each transistor is V_{CEQ}, the ac saturation current is

$$I_{c(sat)} = \frac{V_{CEQ}}{R_L} \tag{9–17}$$

Since $I_e \cong I_c$ and the output current is the emitter current, the peak output current is also V_{CEQ}/R_L.

EXAMPLE 9–8 Determine the maximum peak values for the output voltage and current in Figure 9–27.

FIGURE 9–27

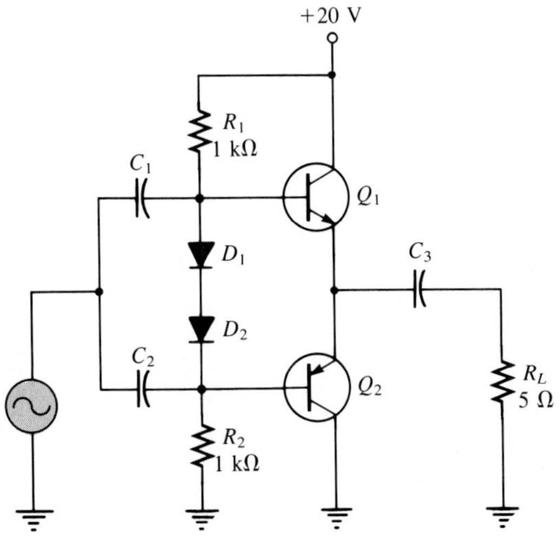

+20 V
R_1 1 kΩ
C_1
Q_1
D_1
C_3
D_2
C_2
Q_2
R_L 5 Ω
R_2 1 kΩ

SOLUTION

The maximum peak output voltage is

$$V_{out(peak)} \cong V_{CEQ} = \frac{V_{CC}}{2} = \frac{20\text{ V}}{2} = 10\text{ V}$$

The maximum peak output current is

$$I_{out(peak)} \cong I_{c(sat)} = \frac{V_{CEQ}}{R_L} = \frac{10\text{ V}}{5\ \Omega} = 2\text{ A}$$

PRACTICE EXERCISE 9–8

Find the maximum peak values for the output voltage and current in Figure 9–27 if V_{CC} is lowered to 15 V and the load resistance is changed to 10 Ω.

MAXIMUM OUTPUT POWER

It has been shown that the maximum peak output current is approximately $I_{c(sat)}$, and the maximum peak output voltage is approximately V_{CEQ}. The maximum average output power is therefore

$$P_{out} = V_{out(rms)} I_{out(rms)}$$

Since

$$V_{out(rms)} = 0.707 V_{out(peak)} = 0.707 V_{CEQ}$$

and

$$I_{out(rms)} = 0.707 I_{out(peak)} = 0.707 I_{c(sat)}$$

then

$$P_{out} = 0.5 V_{CEQ} I_{c(sat)} \quad \textbf{(9–18)}$$

Substituting $V_{CC}/2$ for V_{CEQ}, we get

$$P_{out} = 0.25 V_{CC} I_{c(sat)} \quad \textbf{(9–19)}$$

DC INPUT POWER

The input power comes from the V_{CC} supply and is

$$P_{DC} = V_{CC} I_{CC} \quad \textbf{(9–20)}$$

Since each transistor draws current for a half-cycle, the current is a half-wave signal with an average of

$$I_{CC} = \frac{I_{c(sat)}}{\pi} \quad \textbf{(9–21)}$$

So,

$$P_{DC} = \frac{V_{CC} I_{c(sat)}}{\pi} \quad \textbf{(9–22)}$$

EFFICIENCY

The great advantage of push-pull class B amplifiers over class A is a much higher efficiency. This advantage usually overrides the difficulty of biasing the class B push-pull amplifier to eliminate crossover distortion. The efficiency is again defined as the ratio of ac output power to dc input power.

$$\text{Efficiency} = \frac{P_{out}}{P_{DC}}$$

The maximum efficiency for a class B amplifier is designated η_{max} and is developed as follows, starting with Equation (9–19).

$$P_{out} = 0.25V_{CC}I_{c(sat)}$$

$$\eta_{max} = \frac{P_{out}}{P_{DC}} = \frac{0.25V_{CC}I_{c(sat)}}{V_{CC}I_{c(sat)}/\pi} = 0.25\pi$$

$$\eta_{max} = 0.785 \qquad (9\text{–}23)$$

or, as a percentage

$$\eta_{max} = 78.5\%$$

Recall that the maximum efficiency for class A is 0.25 (25 percent).

EXAMPLE 9–9 Find the maximum ac output power and the dc input power of the amplifier in Figure 9–28.

FIGURE 9–28

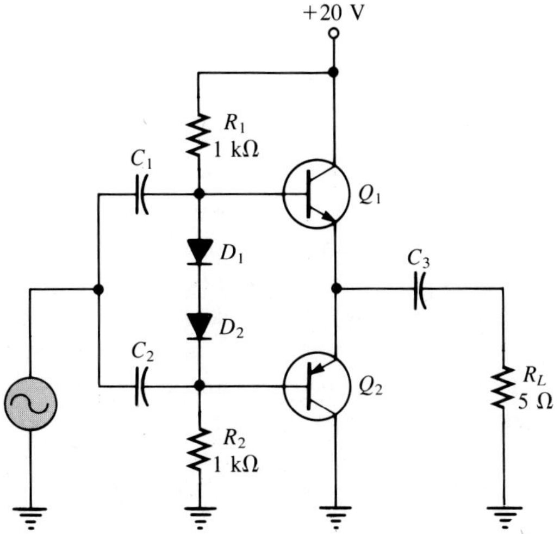

SOLUTION

$I_{c(sat)}$ for this same circuit was found to be 2 A in Example 9–8.

$$\begin{aligned} P_{out} &= 0.25V_{CC}I_{c(sat)} \\ &= 0.25(20\text{ V})(2\text{ A}) \\ &= 10\text{ W} \end{aligned}$$

$$\begin{aligned} P_{DC} &= \frac{V_{CC}I_{c(sat)}}{\pi} \\ &= \frac{(20\text{ V})(2\text{ A})}{\pi} \\ &= 12.73\text{ W} \end{aligned}$$

PRACTICE EXERCISE 9–9

Determine the maximum ac output power and the dc input power in Figure 9–28 for $V_{CC} = 15$ V and $R_L = 10\ \Omega$. ■

9–2 REVIEW QUESTIONS

1. Where is the Q point for a class B amplifier?
2. What causes crossover distortion?
3. What is the maximum efficiency of a push-pull class B amplifier?
4. Explain the purpose of the push-pull configuration for class B.

9–3 CLASS C AMPLIFIERS

Class C amplifiers are biased so that conduction occurs for much less than 180°. Class C amplifiers are more efficient than either class A or push-pull class B, which means that more output power can be obtained from class C operation. Because the output waveform is severely distorted, class C amplifiers are normally limited to applications as tuned amplifiers at radio frequencies (RF) as you will see in this section.

BASIC OPERATION

Class C operation is illustrated in Figure 9–29. A basic common-emitter class C amplifier with a resistive load is shown in Figure 9–30(a). It is biased below cutoff with the $-V_{BB}$ supply. The ac source voltage has a peak value that is slightly greater than $V_{BB} + V_{BE}$ so that the base voltage exceeds the barrier potential of the base-emitter junction for a short time near the positive peak of each cycle, as illustrated in Figure 9–30(b). During this short interval, the transistor is turned on. When the entire ac load line is used, as shown in Figure 9–30(c), the maximum collector current is approximately $I_{C(sat)}$, and the minimum collector voltage is approximately $V_{CE(sat)}$.

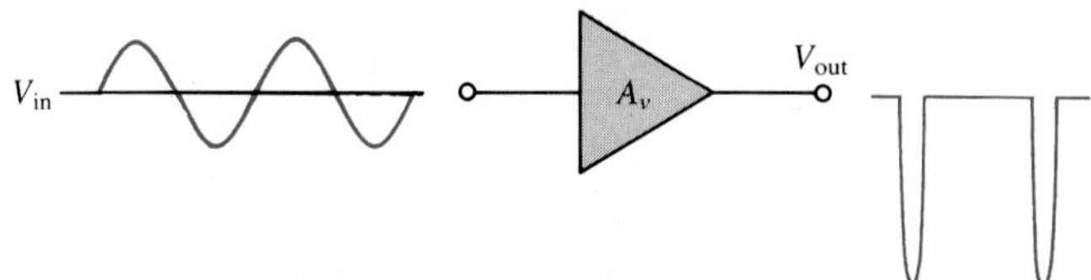

FIGURE 9–29
Class C amplifier operation.

POWER DISSIPATION

The power dissipation of the transistor in a class C amplifier is low because it is on for only a small percentage of the input cycle. Figure 9–31(a) shows the collector current pulses. The time between the pulses is the *period* (T) of the ac input voltage. To avoid complex mathematics, we will use ideal pulse approximations for the collector current and the collector voltage during the on-time of the transistor, as shown in Figure 9–31(b). Using this simplification, the current amplitude is $I_{C(sat)}$ and the voltage amplitude is

FIGURE 9–30
Class C operation.

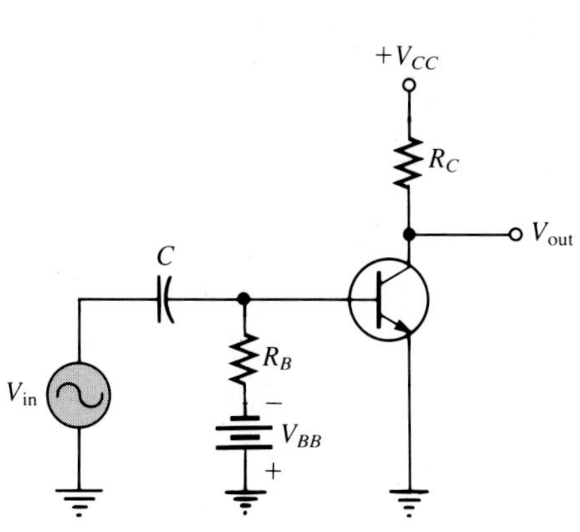

(a) Basic class C amplifier circuit

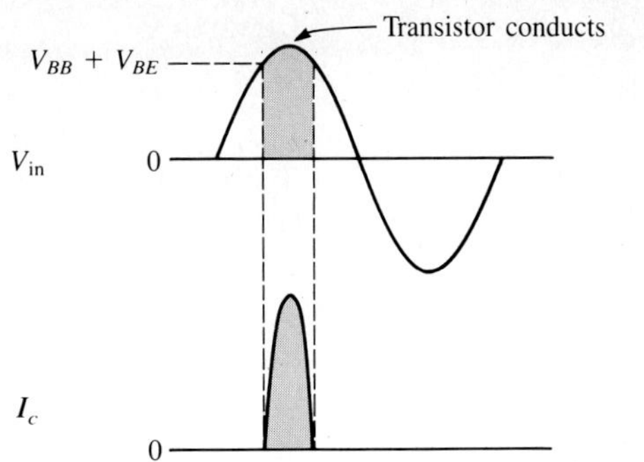

(b) Input voltage and output current waveforms

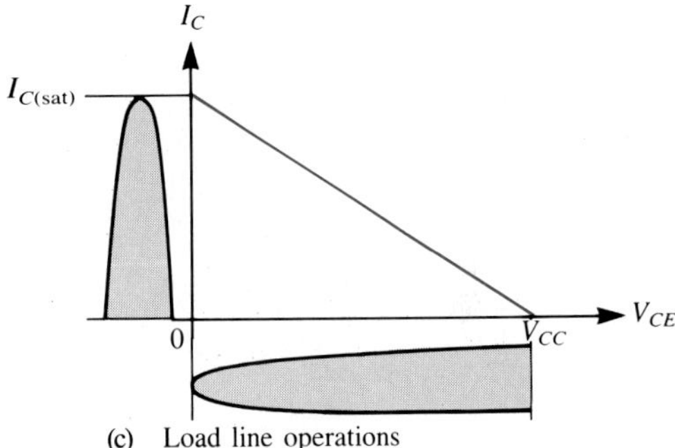

(c) Load line operations

FIGURE 9–31
Class C waveforms.

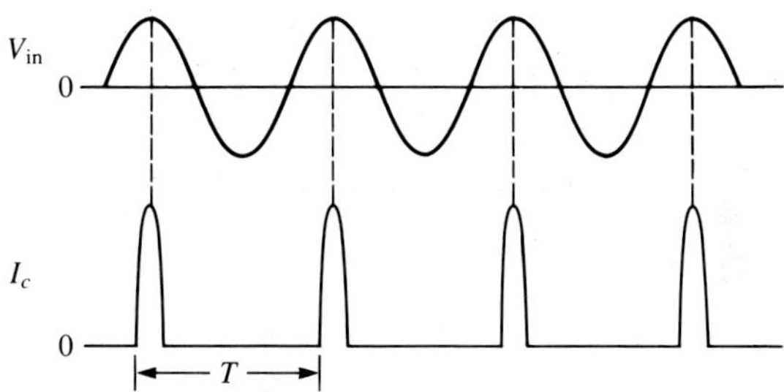

(a) Collector current pulses

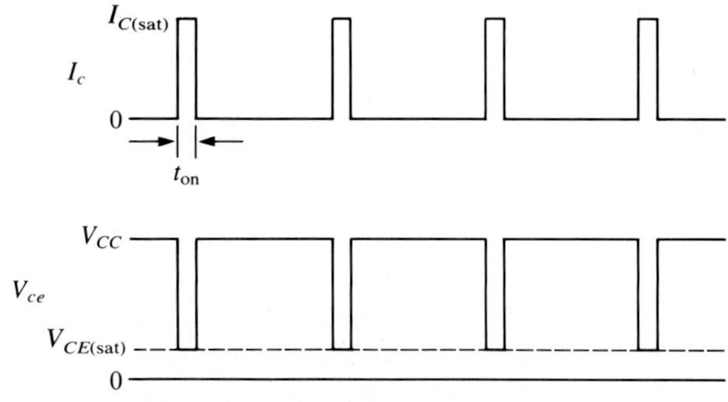

(b) Ideal class C waveforms

$V_{CE(sat)}$ during the time the transistor is on, if the output swings over the entire load line. The power dissipation during the on-time is therefore

$$P_{D(on)} = V_{CE(sat)}I_{C(sat)}$$

The transistor is on for a short time, t_{on}, and off for the rest of the input cycle. Therefore, the power dissipation averaged over the entire cycle is

$$P_{D(avg)} = \left(\frac{t_{on}}{T}\right)P_{D(on)}$$

$$P_{D(avg)} = \left(\frac{t_{on}}{T}\right)V_{CE(sat)}I_{C(sat)} \qquad (9\text{–}24)$$

EXAMPLE 9–10

A class C amplifier is driven by a 200 kHz signal. The transistor is on for 1 μs, and the amplifier is operating over 100 percent of its load line. If $I_{C(sat)} = 100$ mA and $V_{CE(sat)} = 0.2$ V, what is the average power dissipation?

SOLUTION

$$T = \frac{1}{200\text{ kHz}} = 5\ \mu\text{s}$$

Therefore,

$$P_{D(avg)} = \left(\frac{t_{on}}{T}\right)V_{CE(sat)}I_{C(sat)}$$
$$= (0.2)(0.2\text{ V})(100\text{ mA})$$
$$= 4\text{ mW}$$

PRACTICE EXERCISE 9–10

If the frequency is reduced from 200 kHz to 150 kHz, what is the average power dissipation?

TUNED OPERATION

Because the collector voltage (output) is not a replica of the input, the resistively loaded class C amplifier is of no value in linear applications. It is therefore necessary to use a class C amplifier with a parallel resonant circuit (tank), as shown in Figure 9–32(a). The resonant frequency of the tank circuit is determined by the formula $f_r = 1/(2\pi\sqrt{LC})$. The short pulse of collector current on each cycle of the input initiates and sustains the oscillation of the tank circuit so that an output sine wave voltage is produced, as illustrated in Figure 9–32(b).

The current pulse charges the capacitor to approximately $+V_{CC}$, as shown in Figure 9–33(a). After the pulse, the capacitor quickly discharges, thus charging the inductor. Then, after the capacitor completely discharges, the inductor's magnetic field collapses and then quickly recharges C to near V_{CC} in a direction opposite to the previous charge. This completes one half-cycle of the oscillation, as shown in parts (b) and (c) of Figure 9–33. Next, the capacitor discharges again, quickly increasing the inductor's magnetic field. The inductor then quickly recharges the capacitor back to a positive peak less than

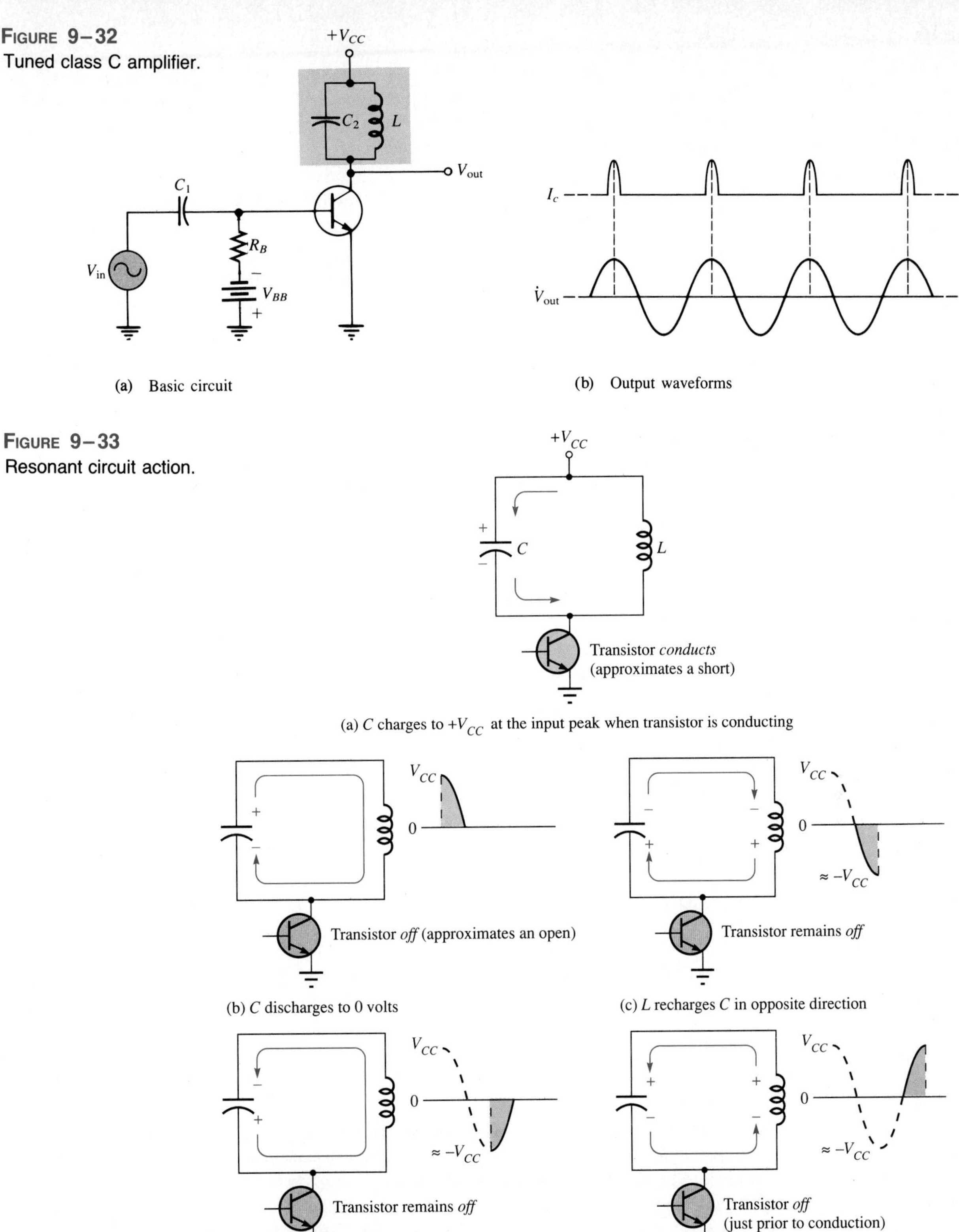

FIGURE 9–32
Tuned class C amplifier.
$+V_{CC}$
C_2
L
V_{out}
C_1
R_B
V_{in}
V_{BB}
I_c
V_{out}
(a) Basic circuit
(b) Output waveforms
FIGURE 9–33
Resonant circuit action.
$+V_{CC}$
C
L
Transistor conducts (approximates a short)
(a) C charges to $+V_{CC}$ at the input peak when transistor is conducting
V_{CC}
0
Transistor off (approximates an open)
(b) C discharges to 0 volts
$\approx -V_{CC}$
Transistor remains off
(c) L recharges C in opposite direction
Transistor remains off
(d) C discharges to 0 volts
Transistor off (just prior to conduction)
(e) L recharges C

the previous one, due to energy loss in the winding resistance. This completes the second half-cycle, as shown in parts (d) and (e) of Figure 9–33. The peak-to-peak output voltage is therefore approximately equal to $2V_{CC}$.

The amplitude of each successive cycle of the oscillation will be less than that of the previous cycle because of energy loss in the resistance of the tank circuit, as shown in Figure 9–34(a), and the oscillation will eventually die out. However, the regular recurrences of the collector current pulse re-energizes the resonant circuit and sustains the oscillations at a constant amplitude. When the tank circuit is tuned to the frequency of the input signal, re-energizing occurs on each cycle of the tank voltage V_r, as shown in Figure 9–34(b).

FIGURE 9–34
Tank circuit oscillations. V_r is the voltage across the tank circuit.

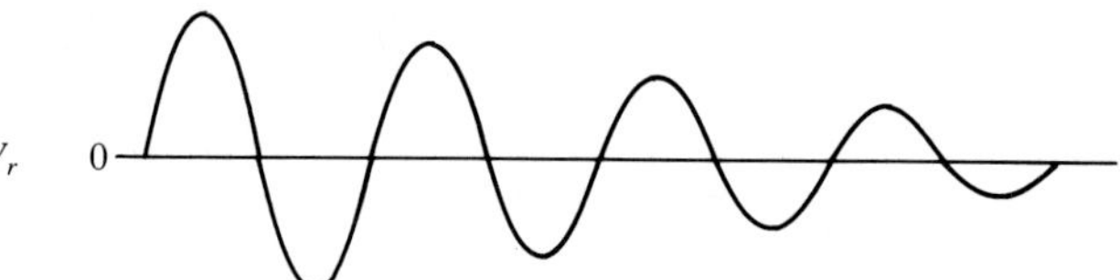

(a) Oscillation dies out due to energy loss

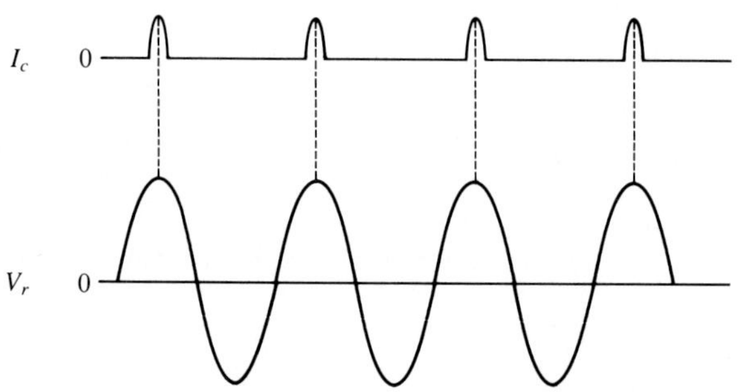

(b) Oscillation is sustained by short pulses of collector current

MAXIMUM OUTPUT POWER

Since the voltage developed across the tank circuit has a peak-to-peak value of approximately $2V_{CC}$, the maximum output power can be expressed as

$$P_{out} = \frac{V_{rms}^2}{R_c} = \frac{(0.707V_{CC})^2}{R_c}$$

$$P_{out} = \frac{0.5V_{CC}^2}{R_c} \qquad (9\text{–}25)$$

R_c is the equivalent parallel resistance of the collector tank circuit and represents the parallel combination of the coil resistance and the load resistance. It usually has a low value. The total power that must be supplied to the amplifier is

$$P_T = P_{out} + P_{D(avg)}$$

Therefore, the efficiency is

$$\eta = \frac{P_{out}}{P_{out} + P_{D(avg)}} \quad (9\text{–}26)$$

When $P_{out} >> P_{D(avg)}$, the class C efficiency closely approaches 1 (100 percent).

EXAMPLE 9–11 Suppose the class C amplifier described in Example 9–10 has a V_{CC} equal to 24 V, and the R_C is 100 Ω. Determine the efficiency.

SOLUTION

From Example 9–10, $P_{D(avg)} = 4$ mW.

$$P_{out} = \frac{0.5V_{CC}^2}{R_c} = \frac{0.5(24\text{ V})^2}{100\ \Omega} = 2.88\text{ W}$$

Therefore,

$$\eta = \frac{P_{out}}{P_{out} + P_{D(avg)}} = \frac{2.88\text{ W}}{2.88\text{ W} + 4\text{ mW}} = 0.9986$$

or

$$\eta \times 100\% = 99.86\%$$

PRACTICE EXERCISE 9–11

What happens to the efficiency of the amplifier if R_c is increased?

CLAMPER BIAS FOR A CLASS C AMPLIFIER

Figure 9–35 shows a class C amplifier with a base bias clamping circuit. The base-emitter junction functions as a diode. When the input signal goes positive, capacitor C_1 is charged to the peak value with the polarity shown in Figure 9–36(a). This action produces an average voltage at the base of approximately $-V_p$. This places the transistor in cutoff except at the positive peaks, when the transistor conducts for a short interval. For good

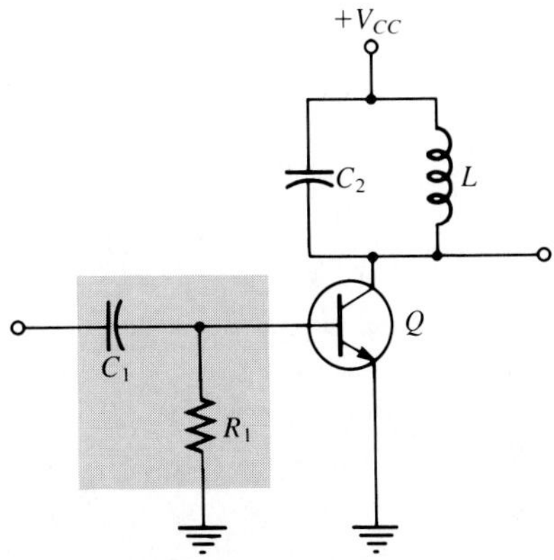

FIGURE 9–35
Tuned class C amplifier with clamper bias.

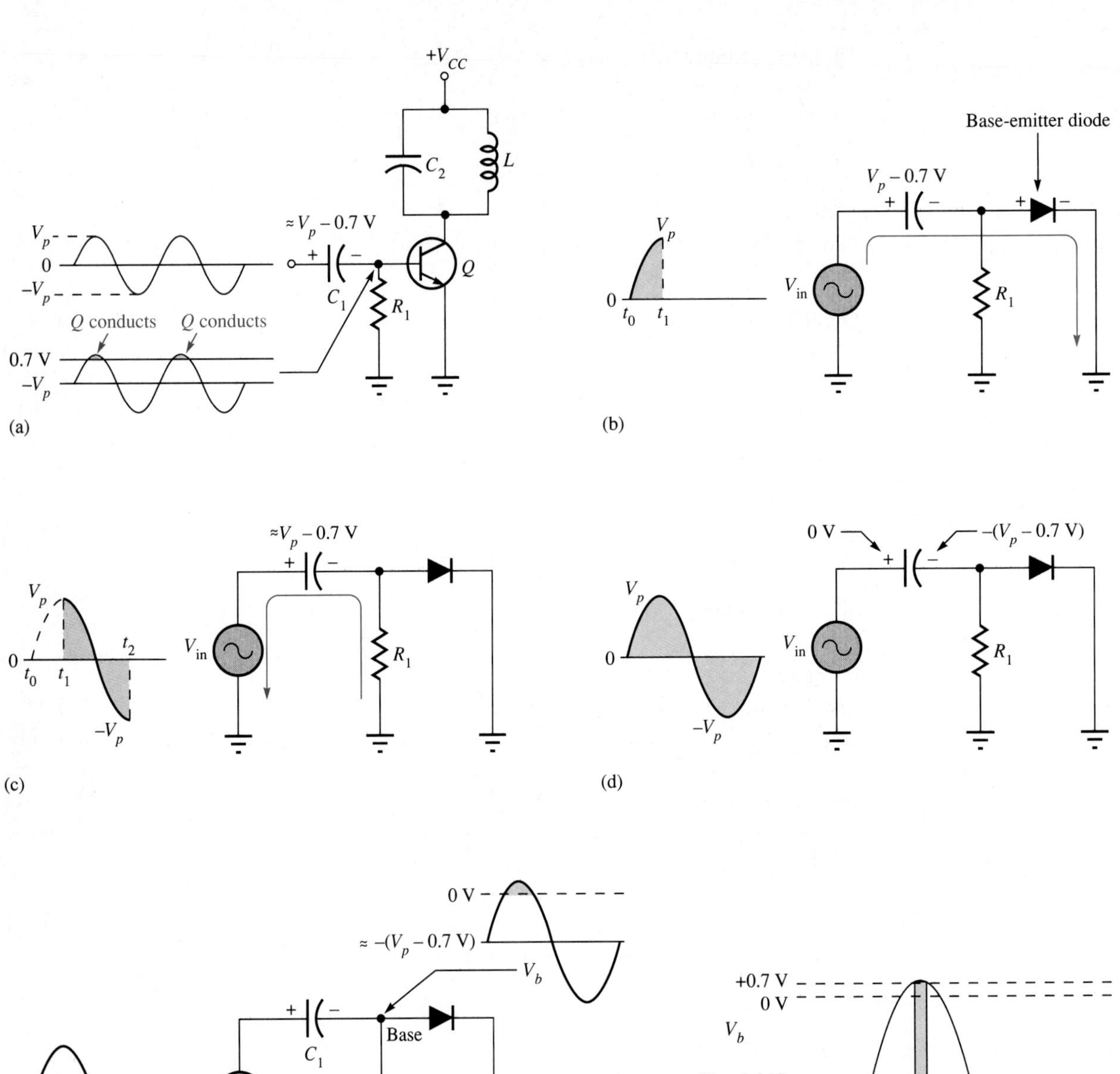

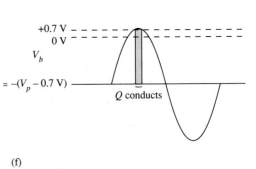

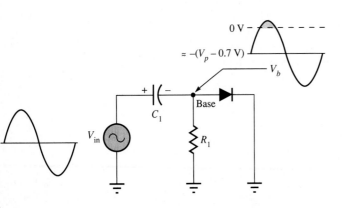

FIGURE 9–36
Clamper bias action.

clamping action, the R_1C_1 time constant of the clamping circuit must be much greater than the period of the input signal. Parts (b) through (f) of Figure 9–36 illustrate the bias clamping action in more detail. During the time up to the positive peak of the input (t_0 to t_1), the capacitor charges to $V_p - 0.7$ V through the base-emitter diode, as shown in part (b). During the time from t_1 to t_2, as shown in part (c), the capacitor discharges very little because of the large RC time constant. The capacitor, therefore, maintains an average charge slightly less than $V_p - 0.7$ V.

Since the dc value of the input signal is zero (positive side of C_1), the dc voltage at the base (negative side of C_1) is slightly more positive than $-(V_p - 0.7 \text{ V})$, as indicated in Figure 9–36(d). As shown in Figure 9–36(e), the capacitor couples the ac input signal through to the base so that the voltage at the transistor's base is the ac signal riding on a dc level slightly more positive than $-(V_p - 0.7 \text{ V})$. Near the positive peaks of the input voltage, the base voltage goes slightly above 0.7 V and causes the transistor to conduct for a short time, as shown in Figure 9–36(f).

9–3 REVIEW QUESTIONS

1. At what point is a class C amplifier normally biased?
2. What is the purpose of the tuned circuit in a class C amplifier?
3. A certain class C amplifier has a power dissipation of 100 mW and an output power of 1 W. What is its efficiency?

9–4 TROUBLESHOOTING

In this section, an example of isolating a component failure in a circuit is presented. We will use a class A amplifier with the output voltage monitored by an oscilloscope. Several incorrect output waveforms will be examined and the most likely faults will be discussed.

As shown in Figure 9–37, the class A amplifier should have a normal sine wave output when a sinusoidal input signal is applied. Now we will consider several incorrect output waveforms and the most likely causes in each case. In Figure 9–38(a), the scope displays

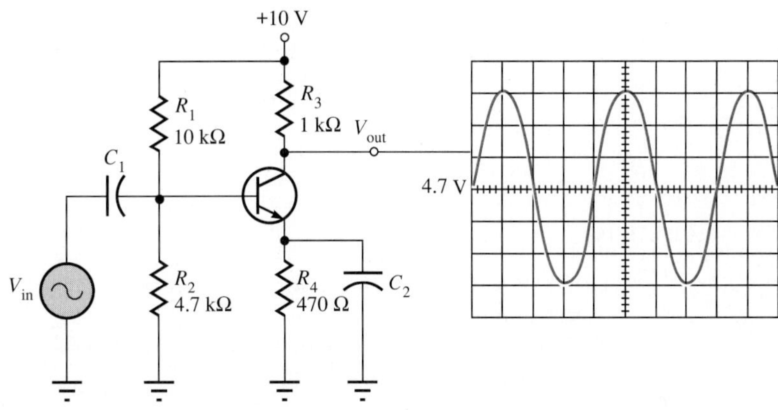

FIGURE 9–37
Class A amplifier with proper output voltage swing.

a dc level equal to the dc supply voltage, indicating that the transistor is in cutoff. The two most likely causes of this condition are (1) the transistor has an open pn junction, or (2) R_4 is open, preventing collector and emitter current.

In Figure 9–38(b), the scope displays a dc level at the collector approximately equal to the dc emitter voltage. The two probable causes of this indication are (1) the transistor is shorted from collector to emitter, or (2) R_2 is open, causing the transistor to be biased in saturation. In the second case, a sufficiently large input signal can bring the transistor out of saturation on its negative peaks, resulting in short pulses on the output. In Figure 9–38(c), the scope displays an output waveform that indicates the transistor is cut off except during a small portion of the input cycle. Possible causes of this indication are (1) the Q point has shifted down due to a drastic out-of-tolerance change in a resistor value, or (2) R_1 is open, biasing the transistor in cutoff. The display shows that the input signal is sufficient to bring it out of cutoff for a small portion of the cycle. In Figure 9–38(d), the scope displays an output waveform that indicates the transistor is saturated except during a small portion of the input cycle. Again, it is possible that a resistance change has caused a drastic shift in the Q point up toward saturation, or R_2 is open, causing the transistor to be biased in saturation, and the input signal is bringing it out of saturation for a small portion of the cycle.

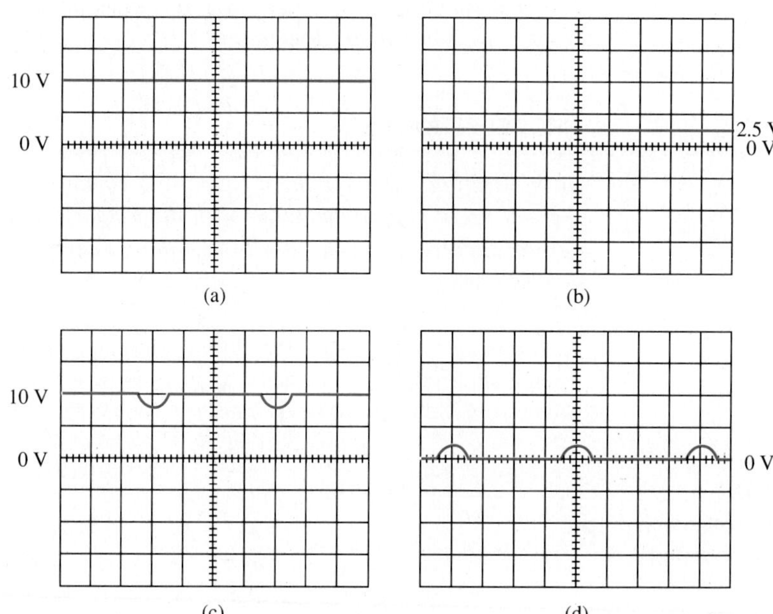

FIGURE 9–38
Oscilloscope displays of output voltage for the amplifier in Figure 9–37, illustrating several types of failures.

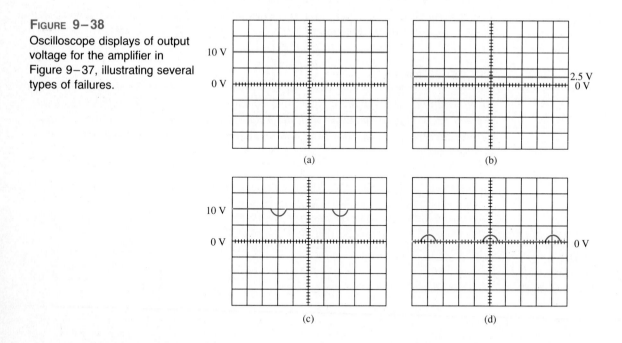

9–4 REVIEW QUESTIONS

1. What would you check for it you noticed clipping at both peaks of the output waveform?
2. A significant loss of gain in the amplifier of Figure 9–37 would most likely be caused by what type of failure?

TEST BENCH
SPECIAL ASSIGNMENTS

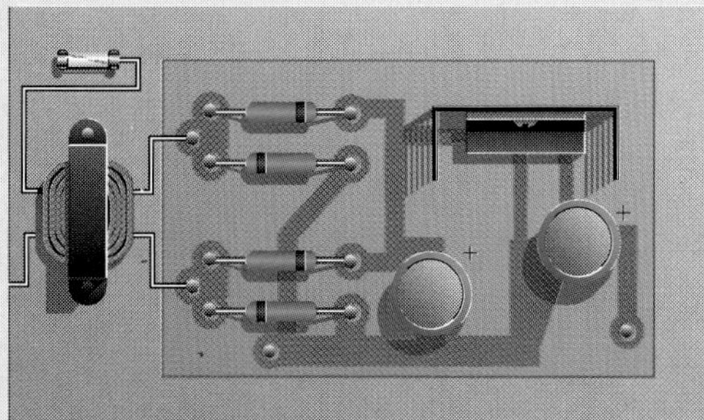

TEST BENCH 1 *Chapter* 2
DC Power Supply Board

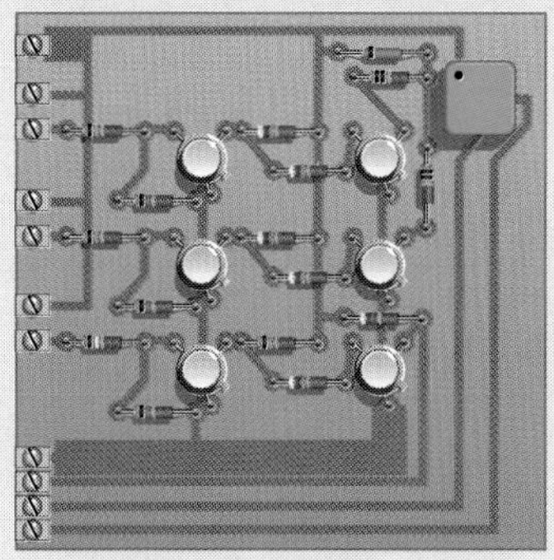

TEST BENCH 2 *Chapter* 4
Intrusion Alarm Detection Board

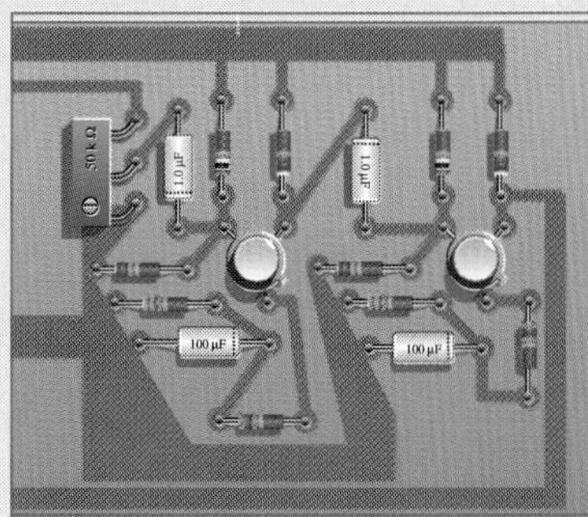

TEST BENCH 3 *Chapter* 6
Audio Preamplifier Board

The seven Test Bench assignments in this color section are each related to a different System Application as indicated in these graphics. These are representative of the circuit boards that you will be working with. Each Test Bench in this section contains a circuit board page and an instrumentation page. The assignment is stated on the circuit board page of each Test Bench.

Use this section only when directed by the special assigment activity in the System Application section of a chapter. You may use the special worksheets available from your instructor to facilitate performing these assignments as well as the other activities in the System Application.

Refer to the back page of this section for useful information.

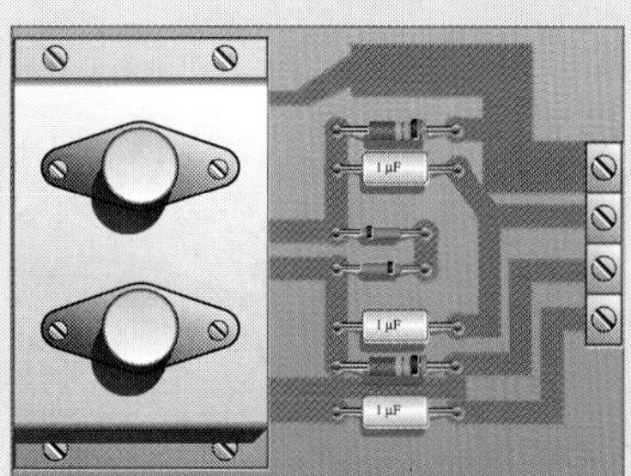

TEST BENCH 4 *Chapter* 9
Audio Power Amplifier Board

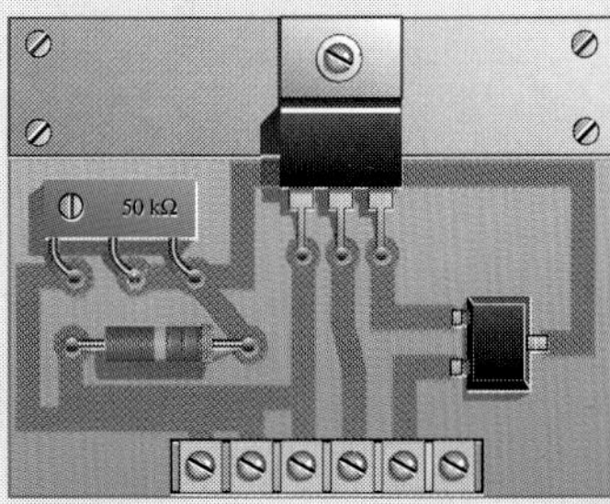

TEST BENCH 5 *Chapter* 11
Motor Speed Control Board

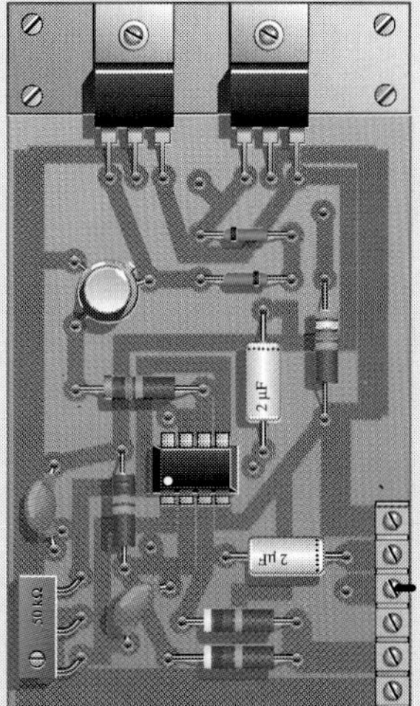

TEST BENCH 6 *Chapter* 13
Stereo Amplifer Board

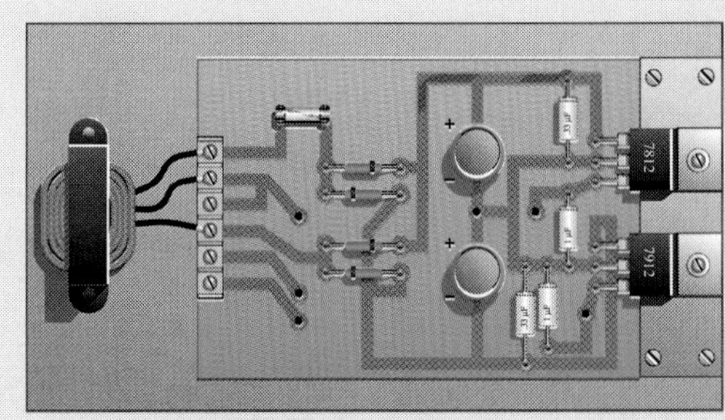

TEST BENCH 7 *Chapter* 17
Dual Power Supply Board

Assignment:
Evaluate the instrument settings and readings for this test setup and determine if the circuit is operating properly. If it is not operating properly, isolate the fault(s). The colored circled numbers indicate connections between the pc board and the instruments: 1(blue) goes to 1(blue), 2 (red) goes to 2 (red), etc. The scope probe is X1.

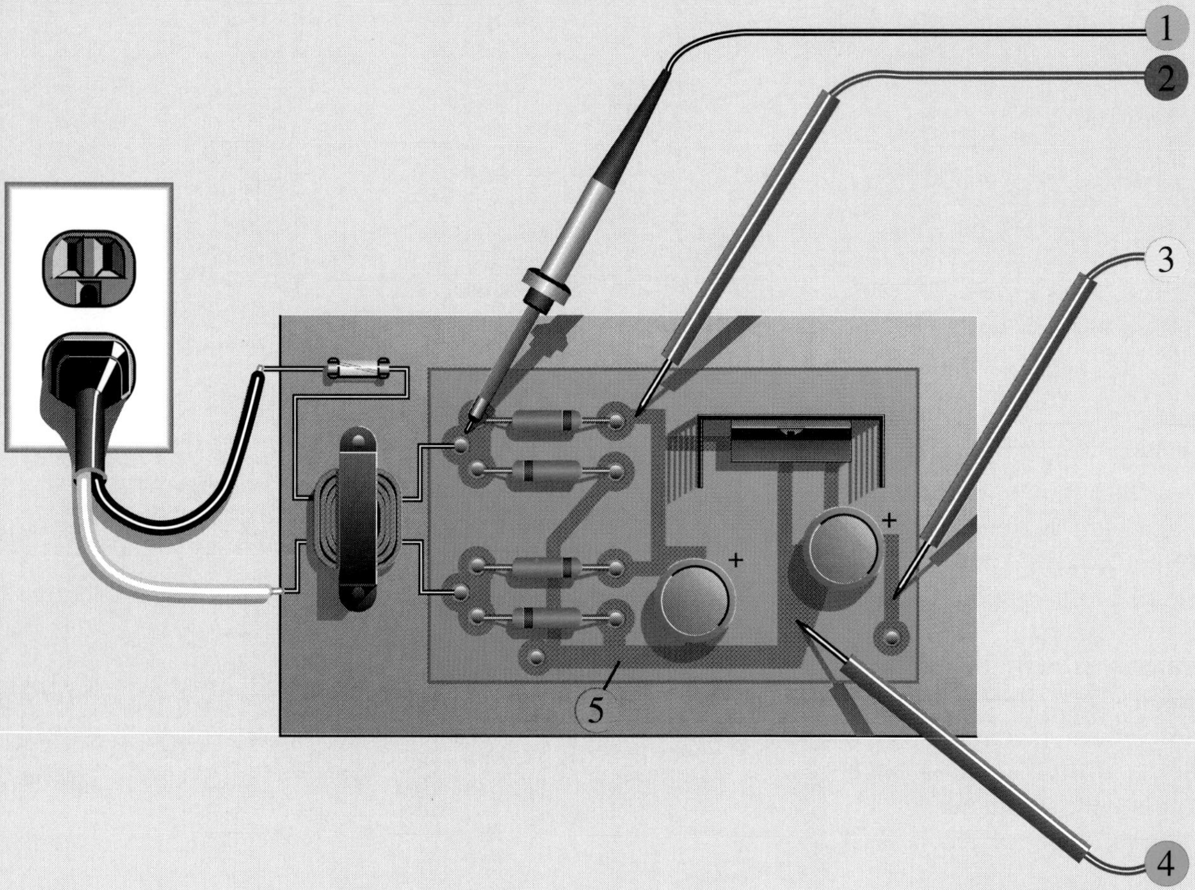

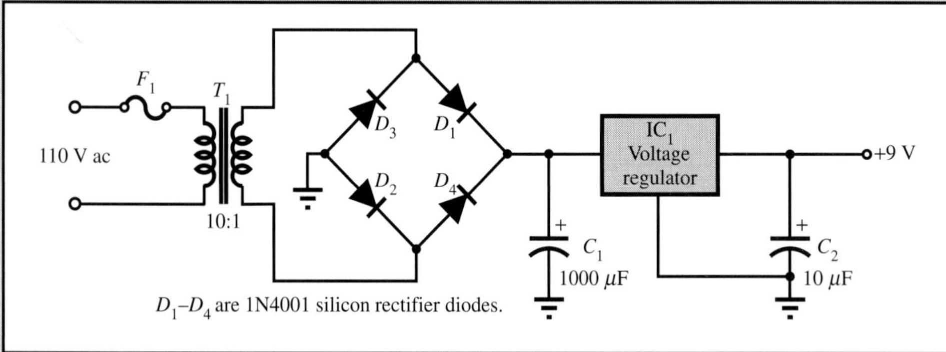

TEST BENCH 1

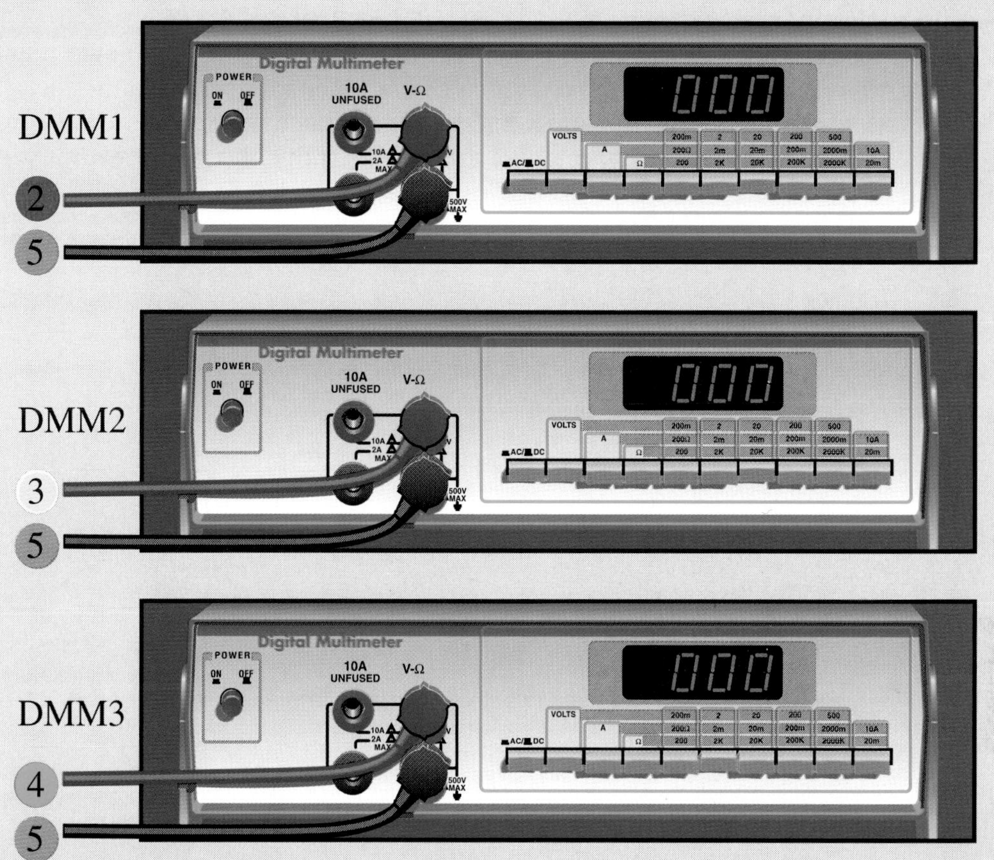

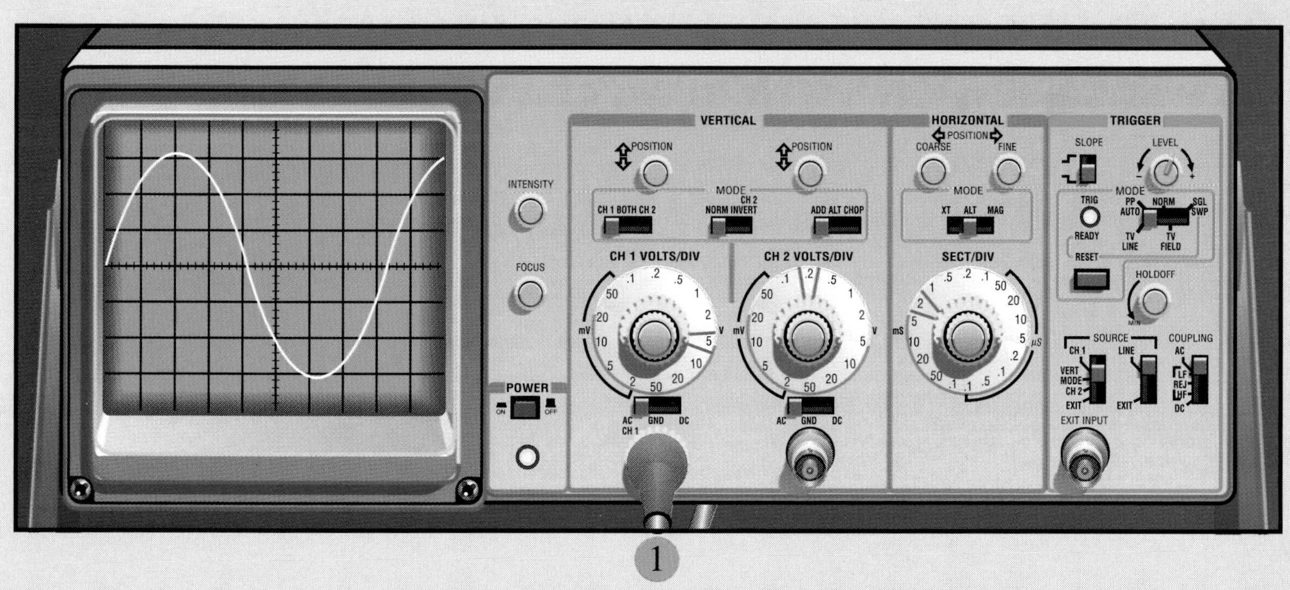

For Chapter 2
System Application

Assignment:

1. Create a point-to-point wire list to specify how the switch box and the pc board are interconnected for testing each circuit. Each switch terminal is designated by a letter and the board terminals are numbered. The toggle switches are single-pole-single-throw.
2. Select the proper function and range setting for each meter.
3. With the instruments connected as shown, determine the meter readings for all possible combinations of switch positions.

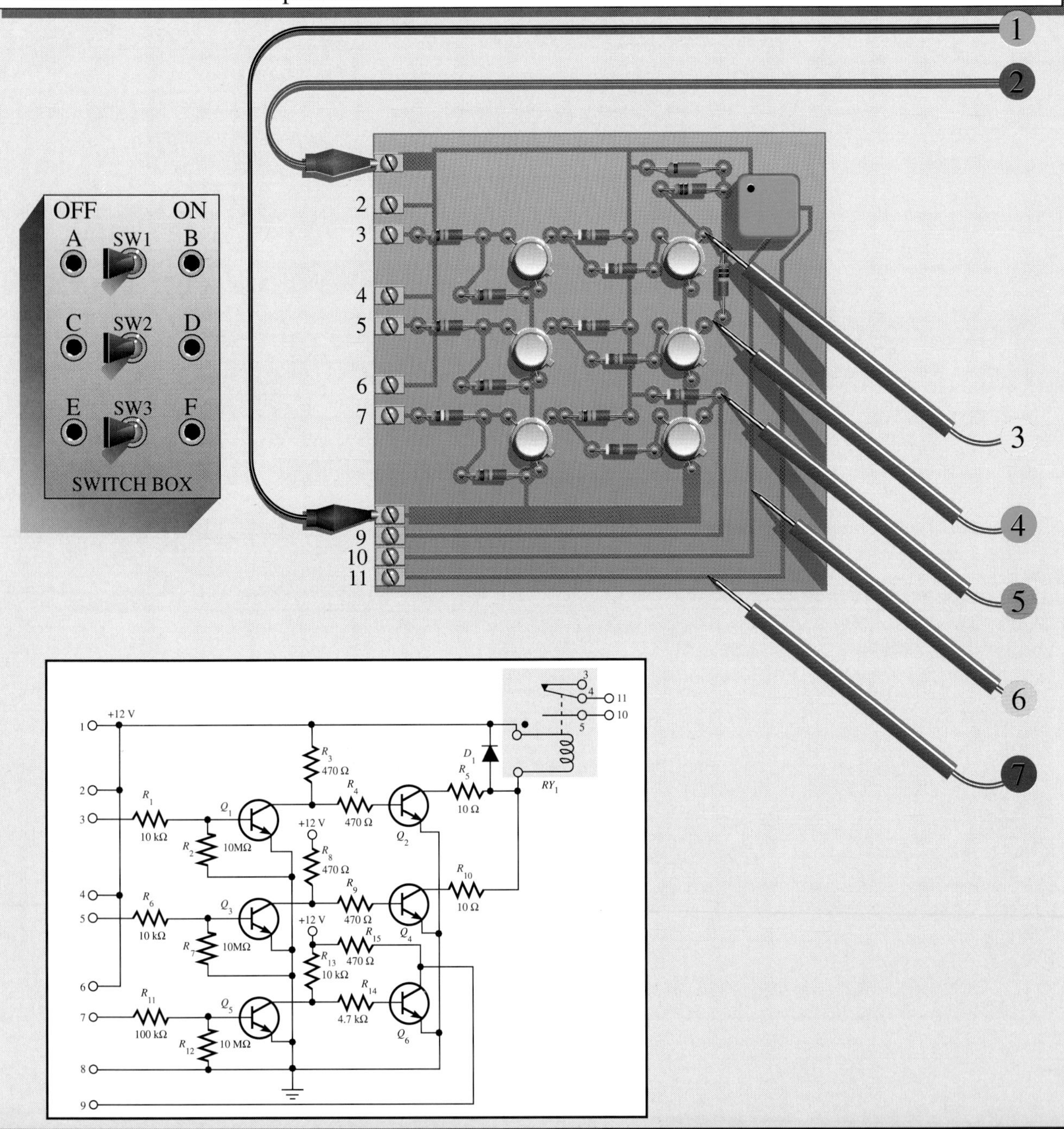

TEST BENCH 2

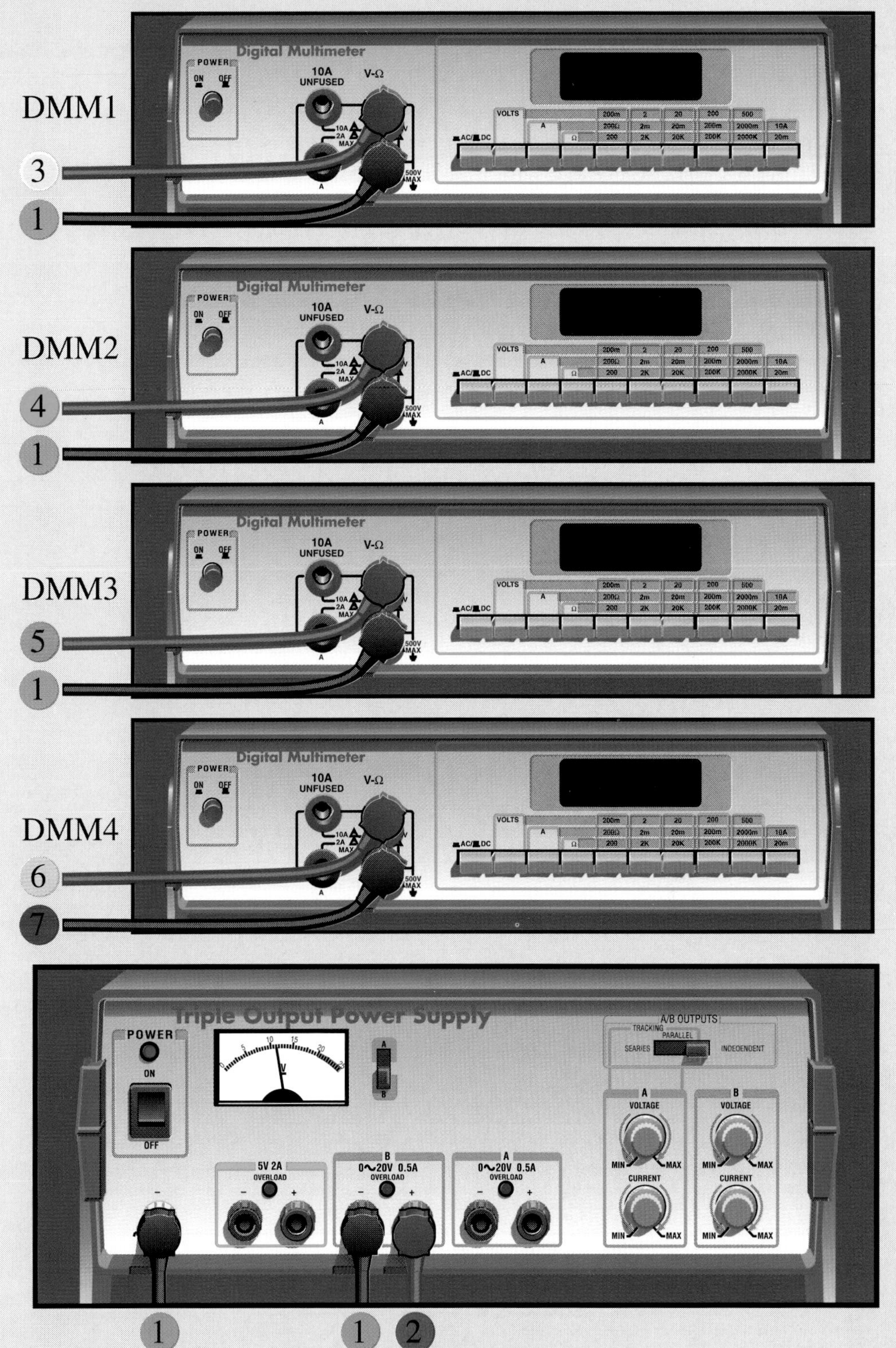

For Chapter 4
System Application

Assignment:

1. Create a point-to-point cabling list to specify how the instruments and boards are interconnected to perform steps 2 and 3. Use the numbers and letters in the blue circles to specify each point. It may or may not be necessary to use all indicated points.
2. Indicate on the DMM the dc voltage at the collector of Q1 with no input signal.
3. Apply a 5 kHz sine wave with a peak-to-peak value of 100 mV to the input and adjust the potentiometer for a 10 mV peak-to-peak at the base of Q1. To what value must the potentiometer be adjusted?
4. Indicate the waveforms of the input and output voltages on the oscilloscope.
5. In each case, specify all the appropriate instrument settings required to generate and/or display the proper voltages. Settings shown may not be correct.

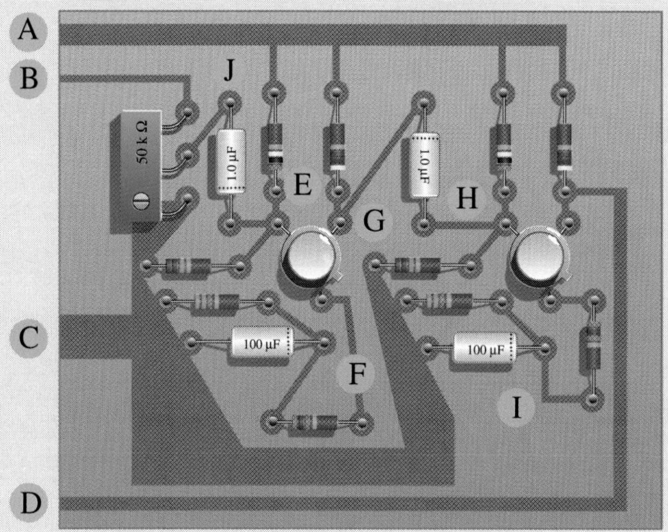

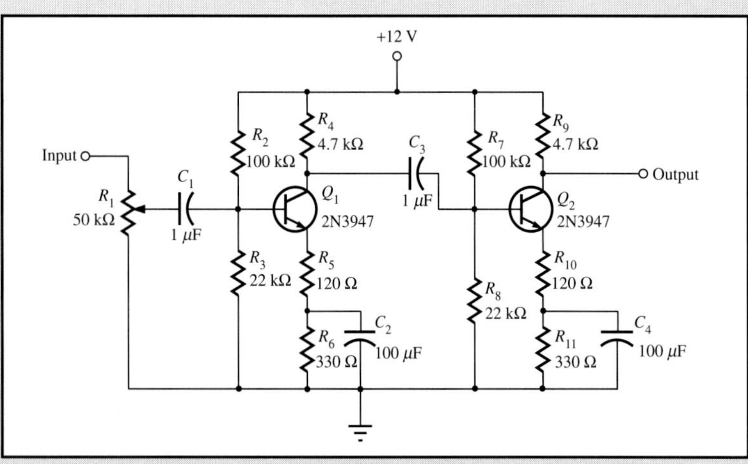

TEST BENCH 3

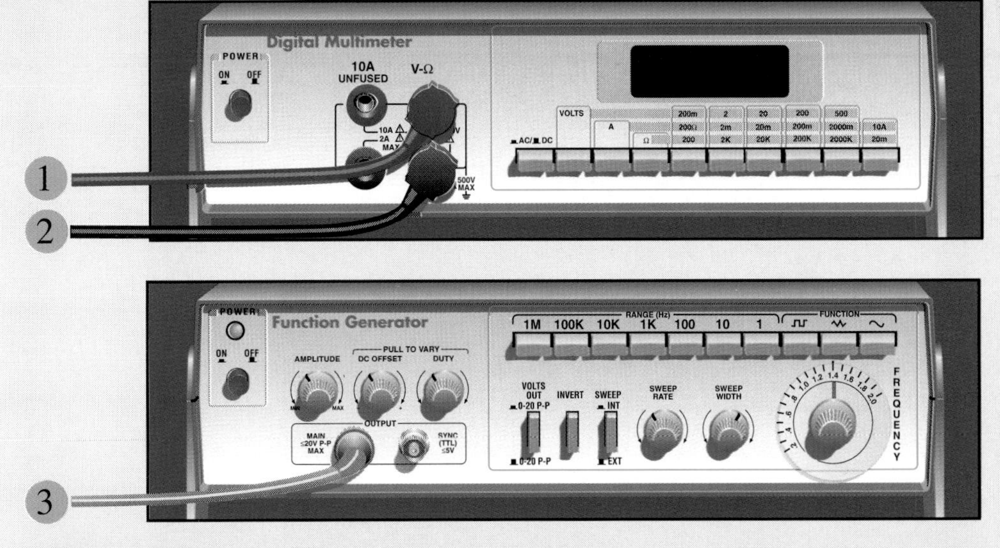

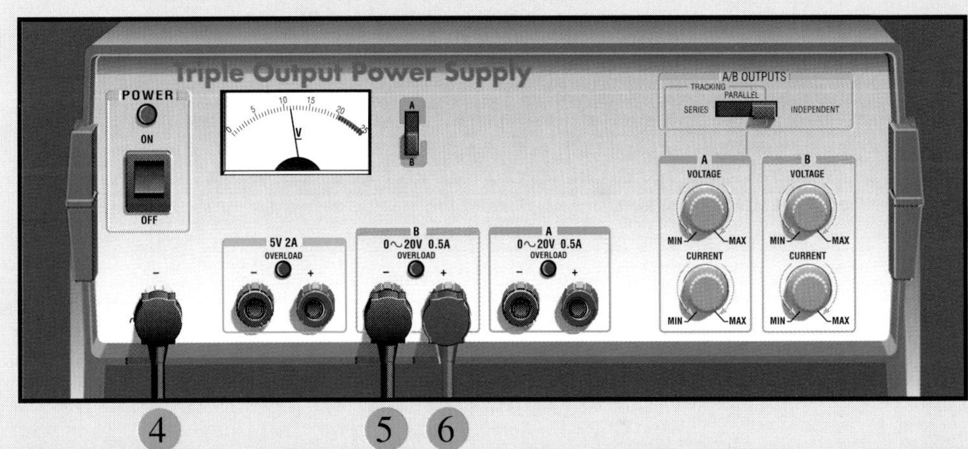

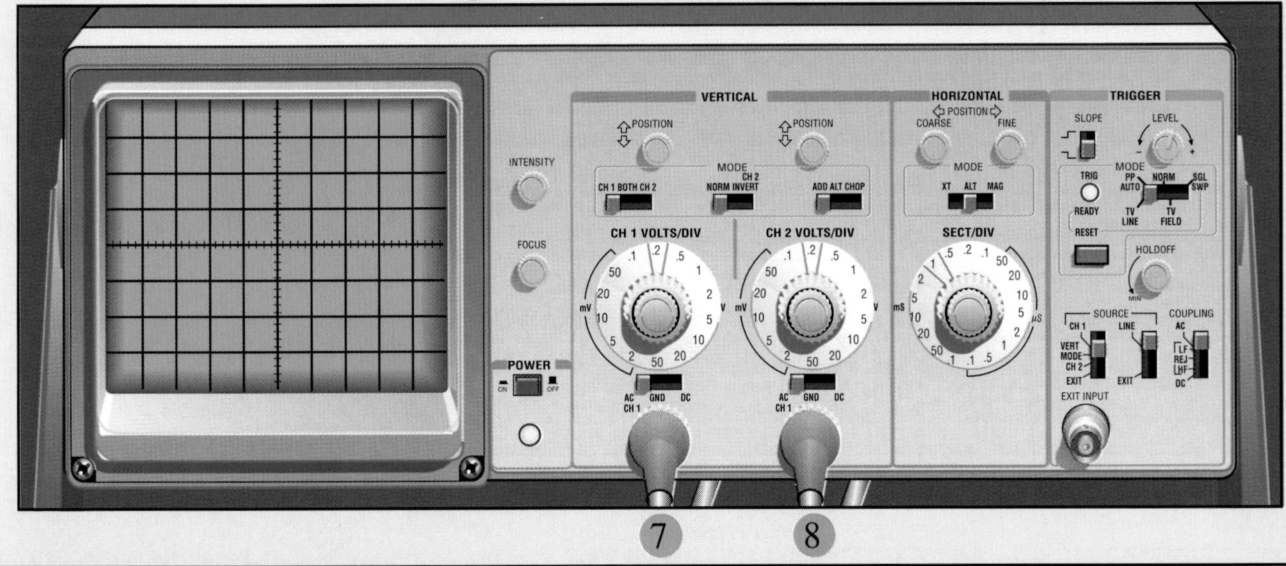

For Chapter 6
System Applicatio

***Assignment*:**
Evaluate the instrument settings and readings for this test set-up and determine if the circuit is operating properly. If it is not operating properly, determine the most probable fault(s). The color circled numbers indicate connections between the pc board and the instruments: 1 (blue) goes to 1 (blue), 2 (red) goes to 2 (red), etc. Top waveform on the scope is CH 1.

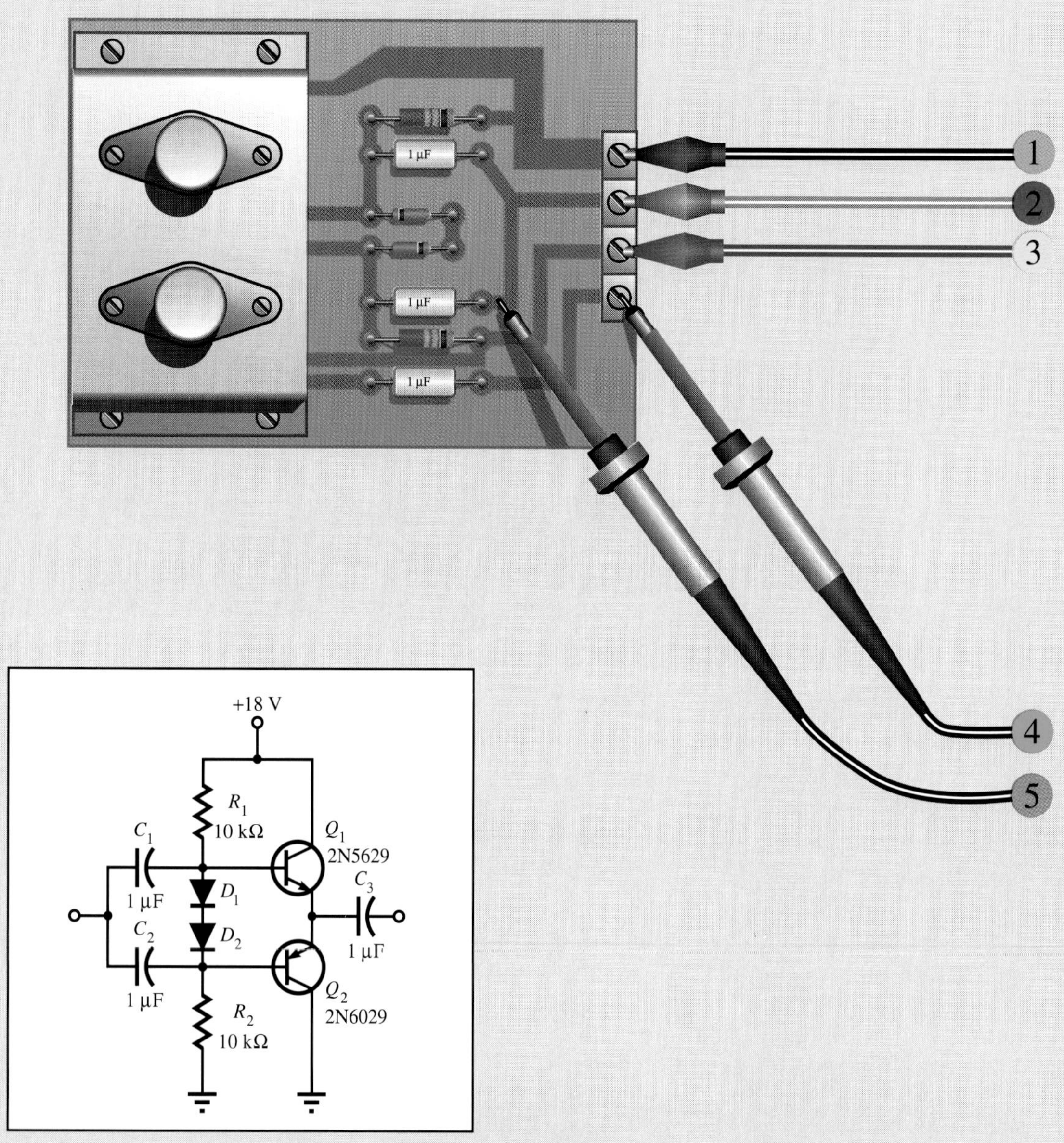

TEST BENCH 4

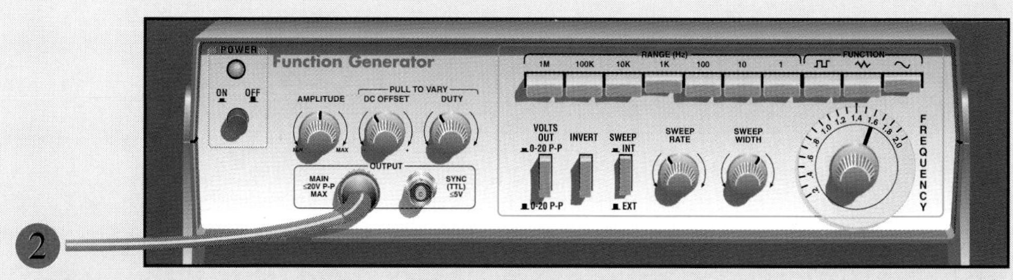

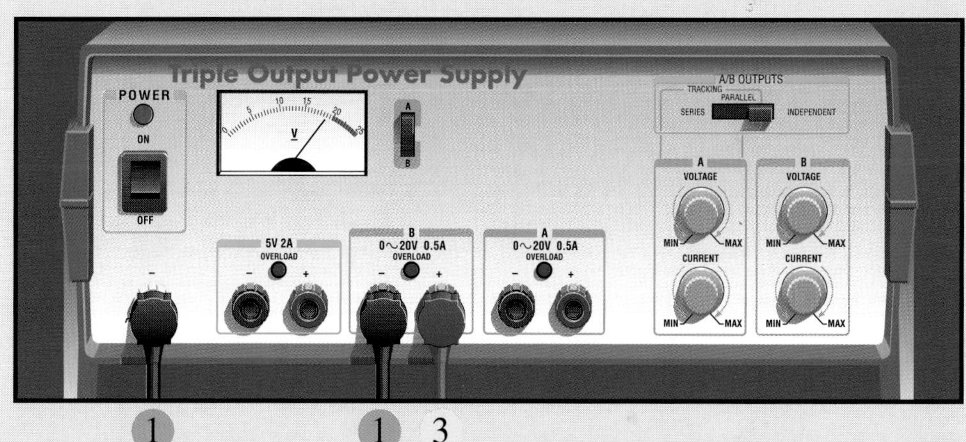

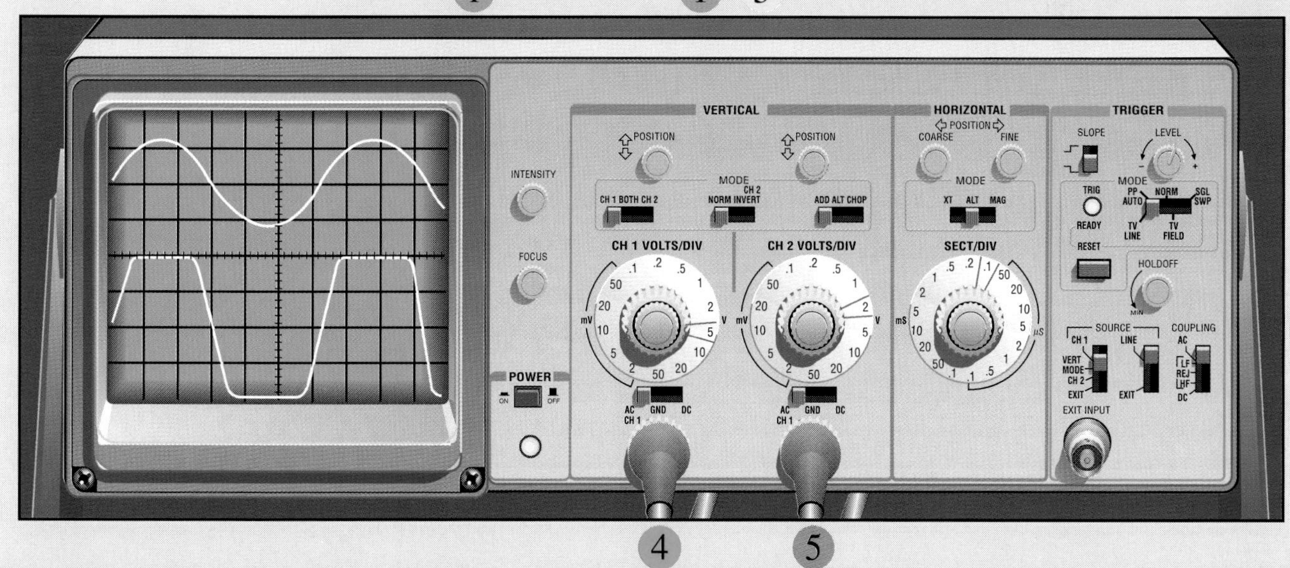

For Chapter 9
System Application

Assignment:
Evaluate the instrument settings and readings for this test set-up and determine if the circuit is operating properly. If it is not operating properly, determine the most probable fault(s). The color circled numbers indicate connections between the pc board and the instruments: 1 (blue) goes to 1 (blue), 2 (red) goes to 2 (red), etc. Top waveform on the scope is CH 1.

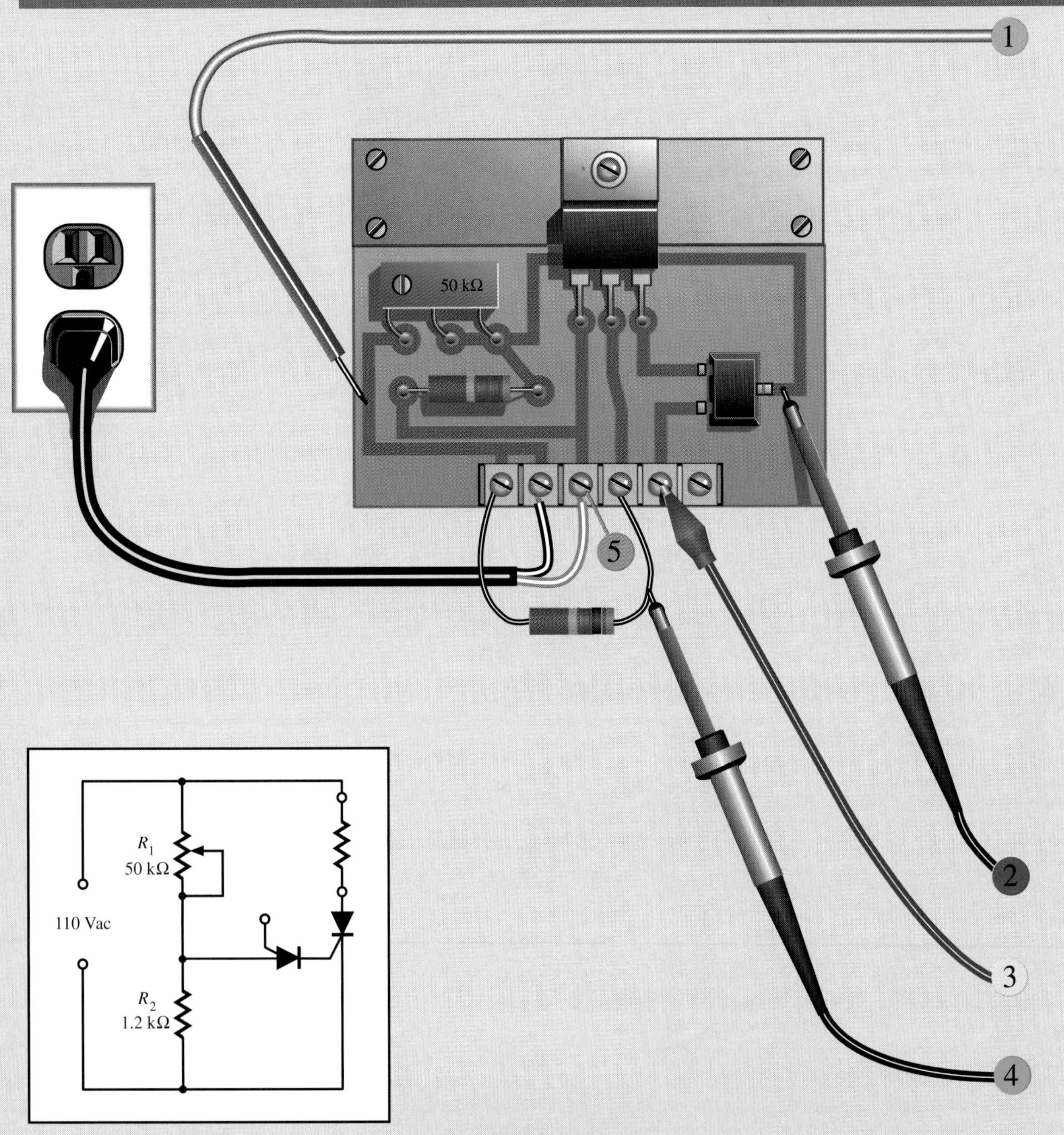

TEST BENCH 5

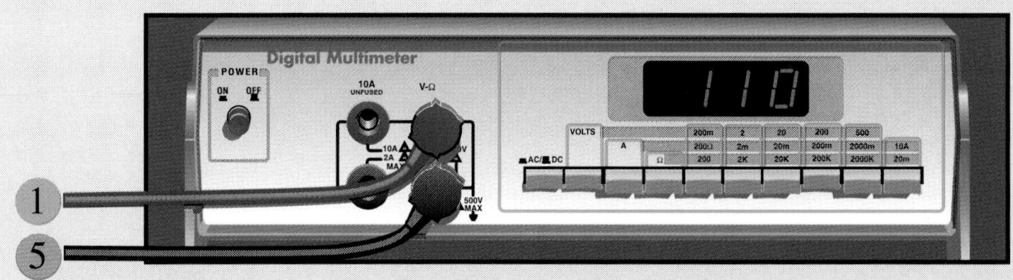

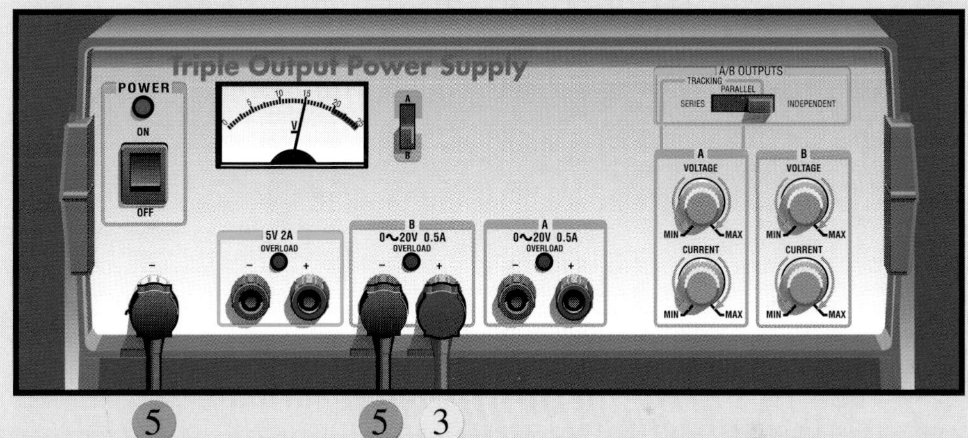

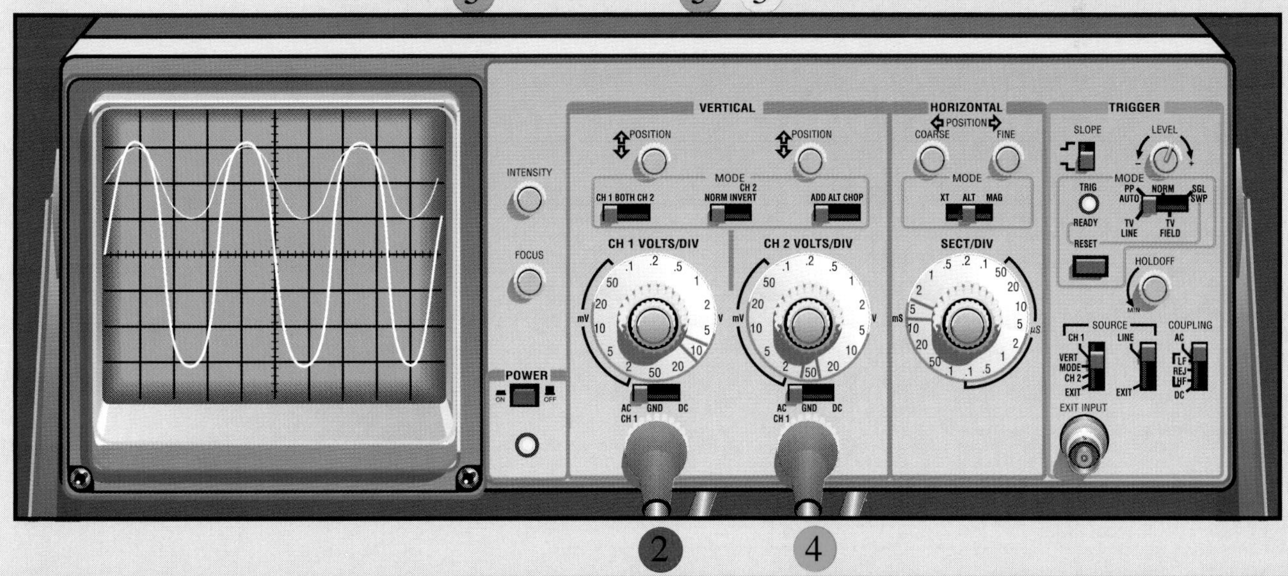

For Chapter 11
System Application

Assignment:
Evaluate the instrument settings and readings for this test setup and determine if the circuit is operating properly. If it is not operating properly, determine the most probable fault(s). The color circled numbers indicate connections between the pc board and the instruments: 1(blue) goes to 1(blue), 2 (red) goes to 2 (red), etc. Top waveform on the scope is CH 1.

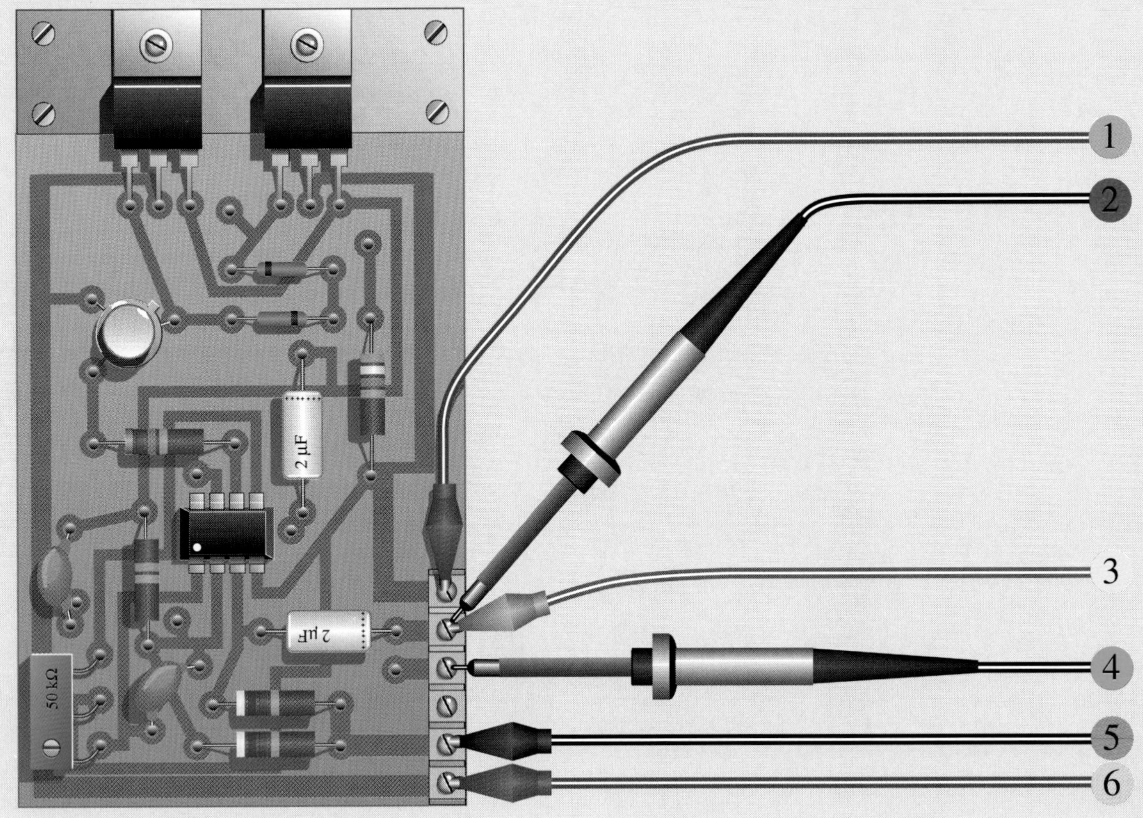

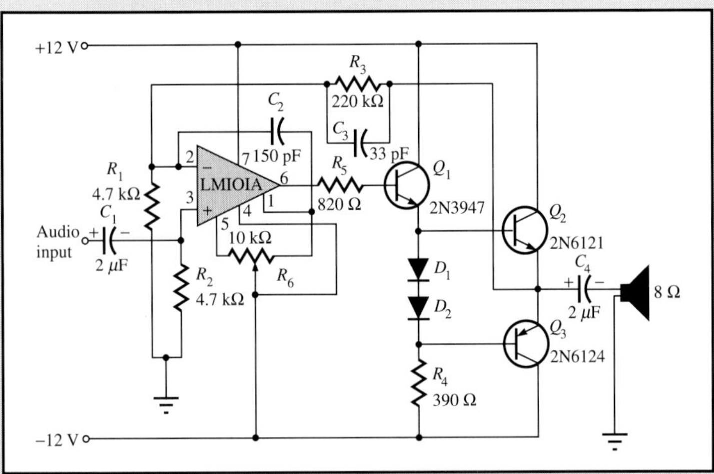

TEST BENCH 6

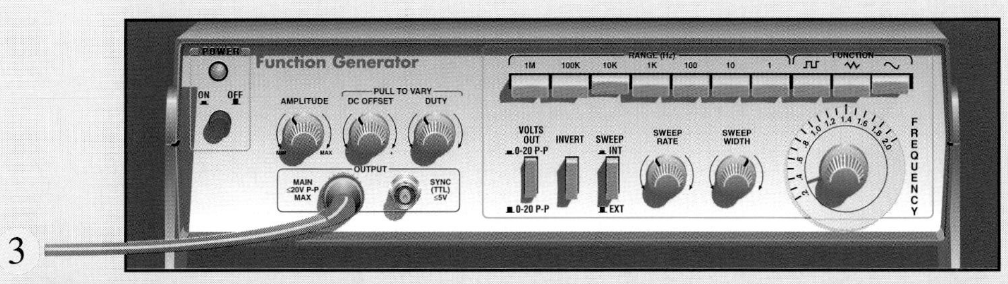

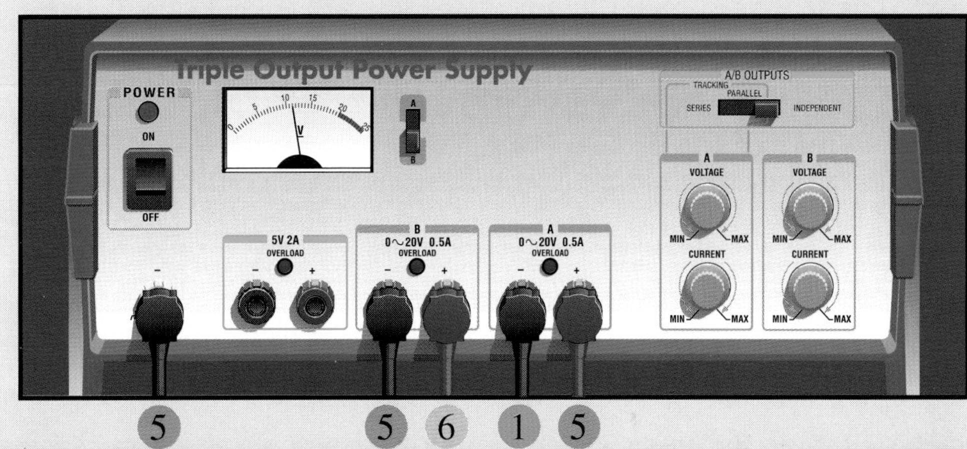

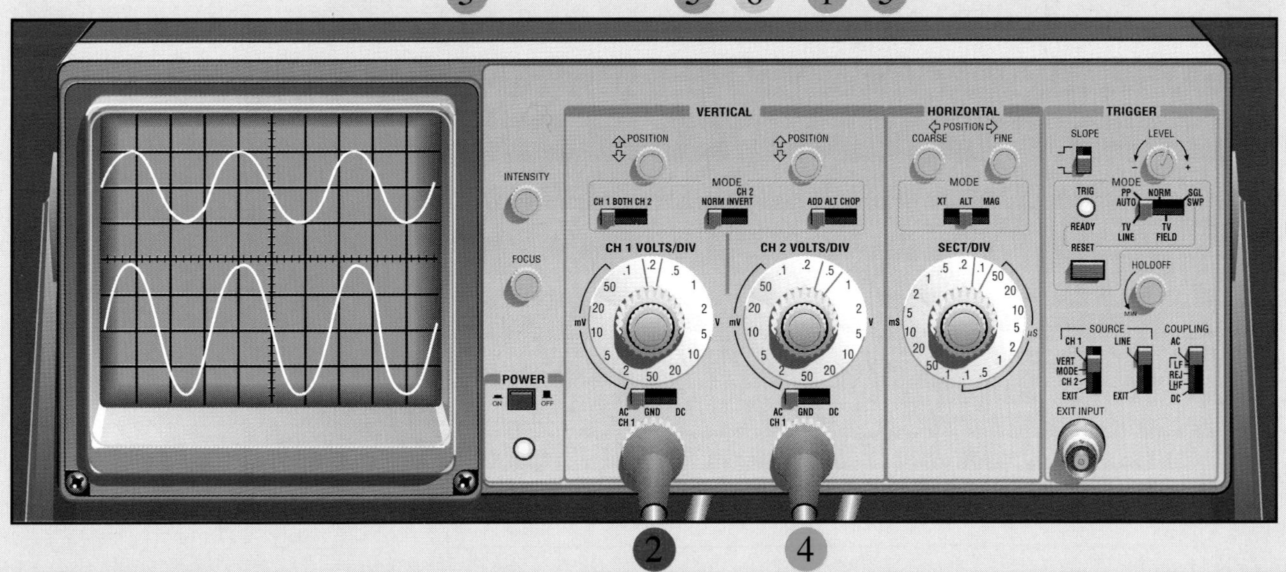

For Chapter 13
System Application

***Assignment*:**

1. Set up the range and function controls on the instruments for observing the voltages at the probed points. Existing settings may not be correct.
2. Indicate the readings and displays you should observe if the power supply is working properly.
3. Indicate the readings and displays you should observe if diode D1 is open.
4. Indicate the readings and displays you should observe if capacitor C2 is open.

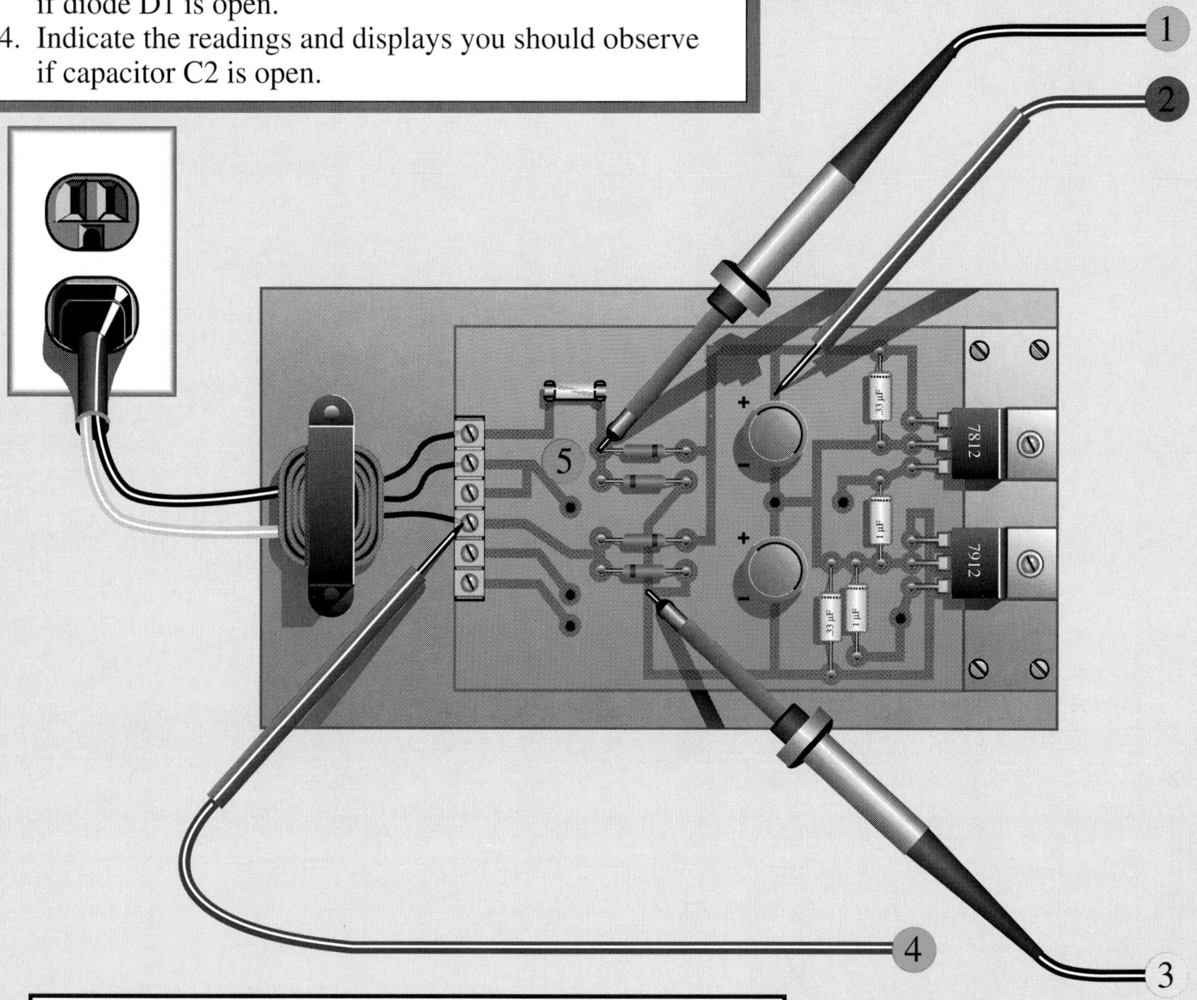

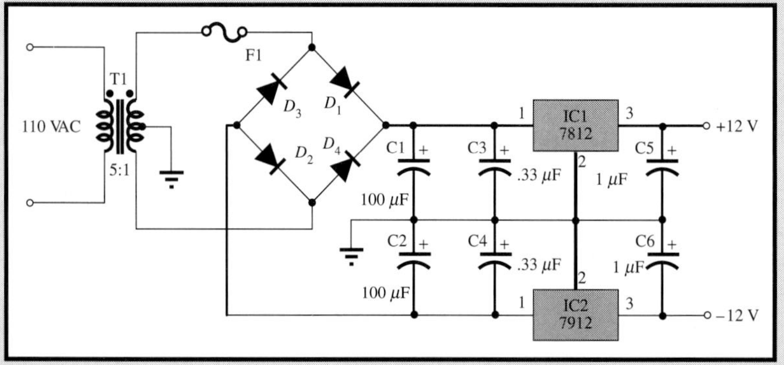

TEST BENCH 7

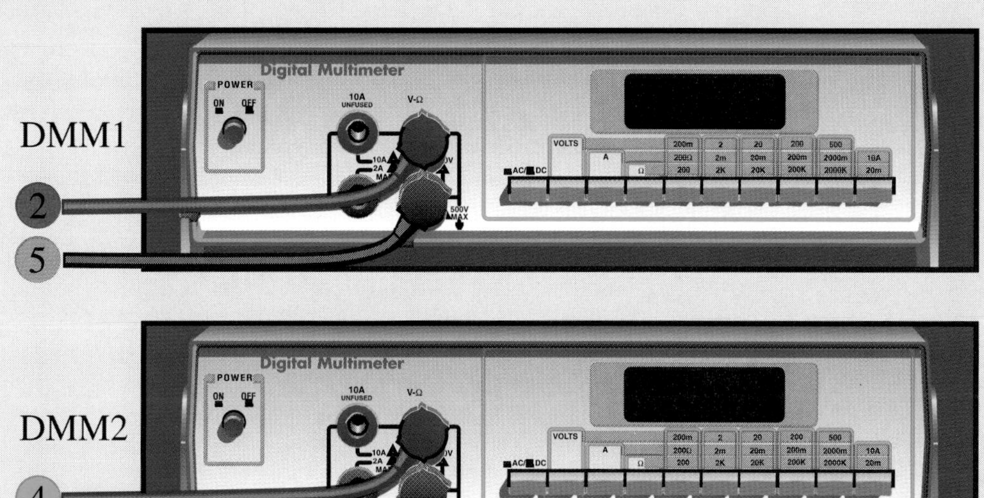

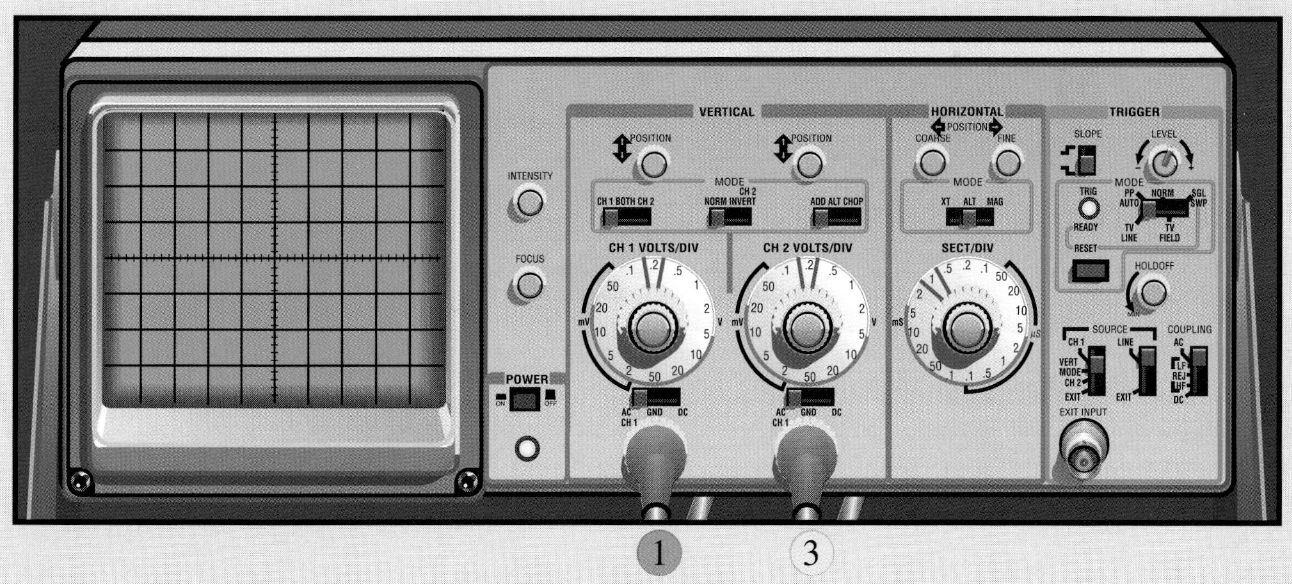

For Chapter 17
System Application

TEST BENCH INFORMATION

There are seven Test Bench assignments in this color section. Each Test Bench contains a two-page spread and is related to one of the System Application sections in the text. The left page contains the assignment, the circuit board with connections and/or labeled points, and a schematic of the board. The right page contains certain instruments that are connected to the board or are to be connected to the board.

Connections between the circuit board and the instruments are indicated by corresponding numbers in color-coded circles. For example, a probe attached to the board with a *red circle 2* goes to the instrument input(s) or output(s) also labeled with a *red circle 2*. In the Test Bench where no cables are shown, you are to properly connect the instruments by matching the circled letters on the board to the appropriate circled numbers on the instruments.

The assignments in this section are of three basic types:

1. Evaluation of indicated instrument settings and readings for the purpose of troubleshooting the board.
2. Selection of instrument setting and proper readings for specified inputs to circuit boards.
3. Hookup of instruments and test fixtures for a test setup to properly check out a board.

Instrument readings that involve circular dial, rotary switch, or slide switch settings should be evident. Examples are the *Time/Div* switch on the oscilloscope and the *frequency* dial on the function generator.

Resistor color code

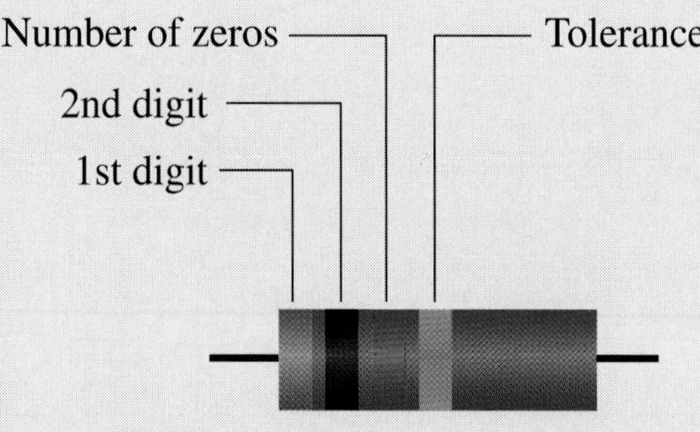

9–5 A SYSTEM APPLICATION

The purpose of the audio power amplifier in the receiver system that you saw at the opening of this chapter is to take the audio signal on the output of the preamplifiers and increase the power to a level that is sufficient for driving the speaker. In this application, a complementary pair of power transistors is used in a class B push-pull configuration. In this section, you will

- ***See how a large-signal amplifier is used in a system application.***
- ***See how a typical power transistor is mounted on a heat sink.***
- ***Translate between a printed circuit board and a schematic.***
- ***Become familiar with the mechanical assembly of a power transistor on a heat sink.***
- ***Troubleshoot some common amplifier failures.***

A BRIEF DESCRIPTION OF THE SYSTEM

The block diagram for the AM receiver is shown in Figure 9–39. A very basic system description was given in Chapter 6 and is repeated here for review. The antenna picks up all radiated electromagnetic signals that pass by and feeds them into the RF amplifier. The voltages induced in the antenna by the electromagnetic radiation are extremely small. The RF amplifier is tuned to select and amplify a desired frequency within the AM broadcast band. Since it is a frequency-selective circuit, the RF amplifier eliminates essentially all but a narrow band centered at the selected frequency. The output of the RF amplifier goes to the mixer where it is combined with the output of the local oscillator, which is 455 kHz above the selected RF frequency. In the mixer, a nonlinear process called *heterodyning* takes place and produces one frequency that is the sum of the RF and local oscillator frequencies and another frequency that is the difference (always 455 kHz). The sum frequency is filtered out and only the 455 kHz difference frequency is used. This frequency is amplitude modulated just like the much higher RF frequency and, therefore, contains the audio signal. The IF (intermediate frequency) amplifier is tuned to 455 kHz and amplifies the mixer output. The detector takes the amplified 455 kHz AM signal and

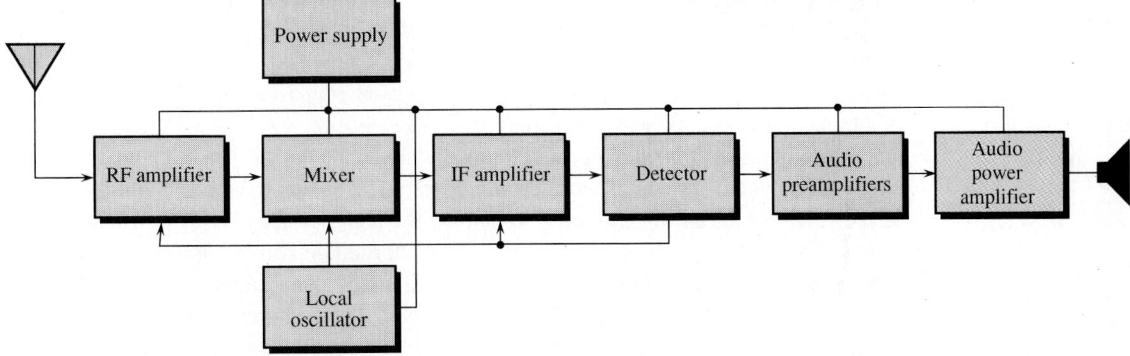

FIGURE 9–39

recovers the audio from it while eliminating the intermediate frequency. The output of the detector is a small audio signal that goes to the audio preamplifier and then to the power amplifier, which drives the speaker to convert the electrical audio signal into sound.

Now, so that you can take a closer look at the power amplifier, let's take it out of the system and put it on the test bench.

ON THE TEST BENCH

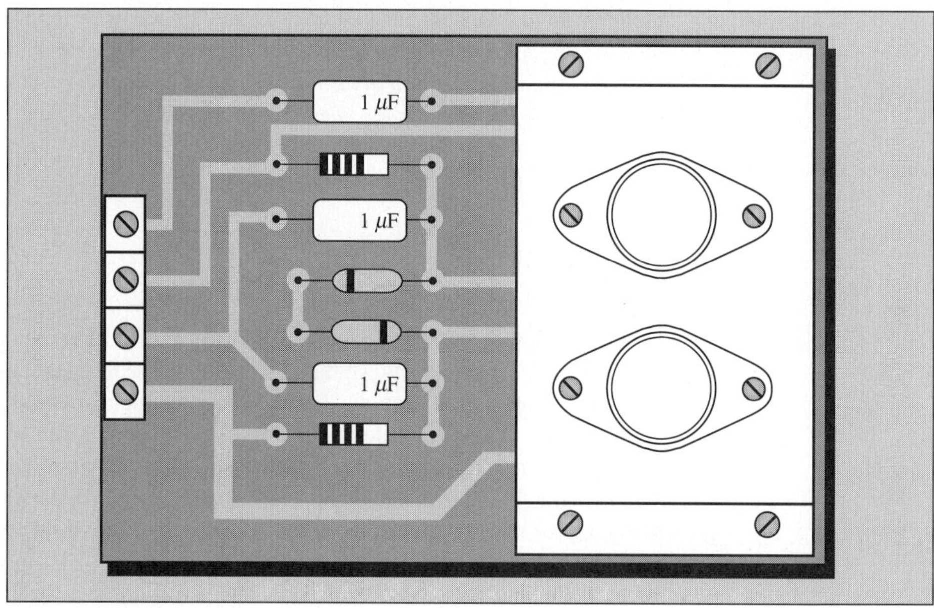

FIGURE 9–40

THE POWER TRANSISTOR HEAT SINK ASSEMBLY

The two power transistors are mounted on a heat sink, which in this case is simply a raised metal bracket that is mounted onto the pc board in Figure 9–40. The heat sink provides additional surface area to conduct the heat away from the transistors more quickly to prevent overheating. Heat sinks come in a variety of shapes and sizes from simple plates or brackets to more elaborate designs.

The conductors on the pc board run under the heat sink bracket and terminate in connection pads from which short jumper wires interconnect the circuit with the transistor sockets. The diagram in Figure 9–41 shows the TO-204AA (TO-3) transistor case and how it is typically assembled on a heat sink. A thermal compound applied to the insulator conducts heat from the transistor cases to the heat sink. The jumper wires are soldered to the socket pins and connected to the pc board pads. The complementary transistors are the 2N5629 (npn) and the 2N6029 (pnp).

FIGURE 9–41

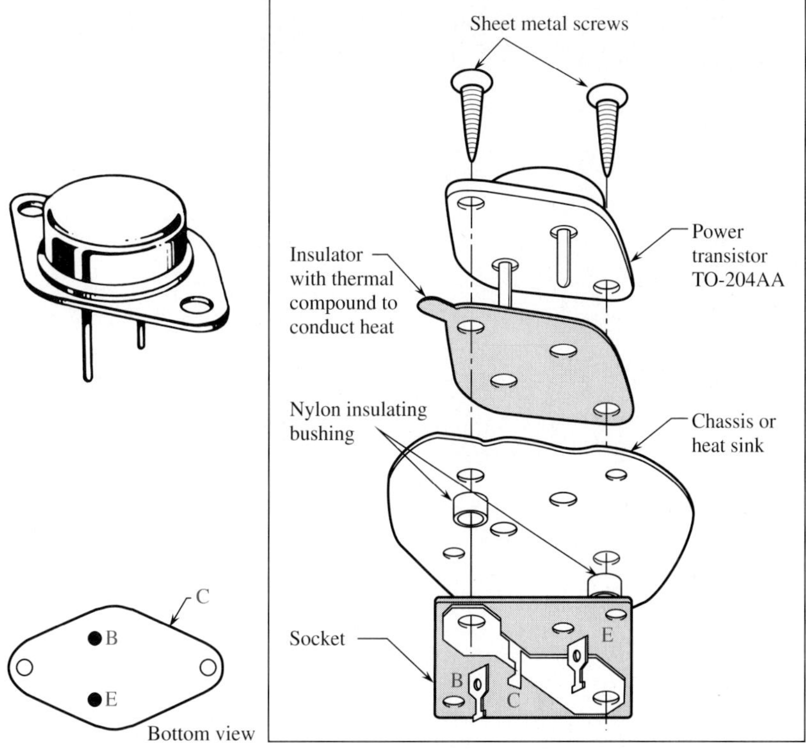

(a) TO-204AA case (b) Typical transistor/heat sink assembly

■ ACTIVITY 1 INTERCONNECT PC BOARD AND HEAT SINK ASSEMBLY

Figure 9–42 shows the pc board and the power transistor heat sink before they are combined. You are looking at a bottom view of the heat sink assembly. The pins on the transistor sockets will be wired to the appropriate pads on the board and then the heat sink bracket will be flipped over and attached to the board. Notice that the sockets are not identical to the one in Figure 9–41, but they are very similar. Develop a wire list that shows which socket pin and pc board pad connect together using the appropriate numbers and letters.

■ ACTIVITY 2 RELATE THE PC BOARD ASSEMBLY TO THE SCHEMATIC

First assign labels to all of the components (R_1, C_1, Q_1, and so forth) and then determine the function of each input and output and properly assign labels to those. Next draw the circuit schematic. The resistors are both 10 kΩ and the dc supply voltage is +18 V.

■ ACTIVITY 3 ANALYZE THE CIRCUIT

STEP 1 Determine the dc voltages on the terminals of each transistor and between the diodes. Assume the forward-biased base-emitter voltages and the diode voltages are 0.7 V.

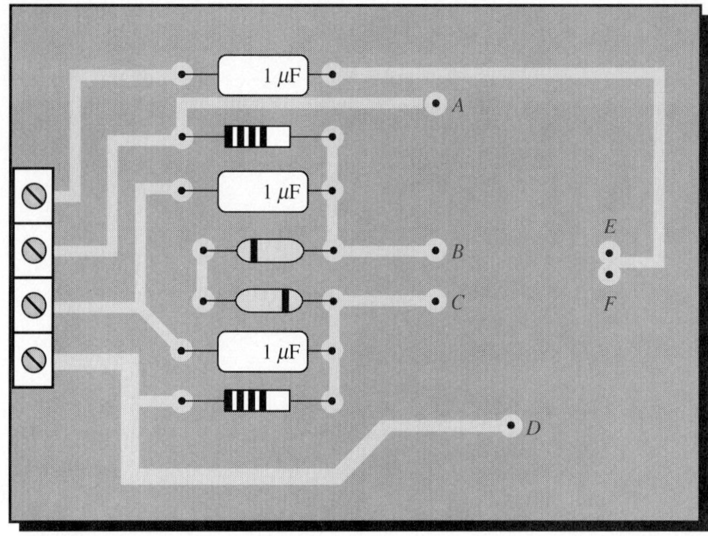

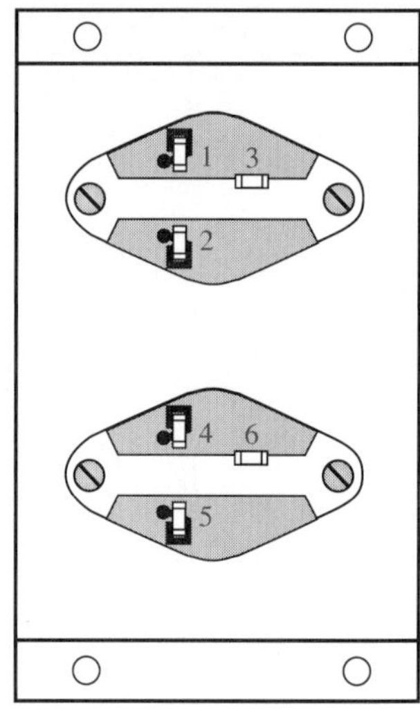

FIGURE 9–42

STEP 2 What is the approximate voltage gain of the amplifier?

STEP 3 Determine the maximum average output power if the amplifier is driving an 8-Ω speaker.

ACTIVITY 4 WRITE A TECHNICAL REPORT

Discuss the overall operation of the amplifier circuit. State the purpose of each component on the board. Use the results of Activity 3 when appropriate.

ACTIVITY 5 TROUBLESHOOT THE CIRCUIT BOARD FOR EACH OF THE FOLLOWING PROBLEMS BY STATING THE PROBABLE CAUSE OR CAUSES IN EACH CASE

1. The dc voltage at the base of the pnp transistor is zero.
2. There is no signal output when there is a signal input and the dc bias voltages are correct.
3. A signal is getting to the base of the pnp transistor, but there is no signal at the base of the npn transistor.

ACTIVITY 6 TEST BENCH SPECIAL ASSIGNMENT

Go to Test Bench 4 in the color insert section (which follows page 422) and carry out the assignment that is stated there.

9–5 REVIEW QUESTIONS

1. Why is a heat sink required in this type of application?
2. What happens to the audio output if one of the input coupling capacitors opens?
3. What might cause both transistors to be on at the same time?
4. Can you significantly increase the voltage gain of this power amplifier?

SUMMARY

- □ A class A amplifier operates entirely in the linear region of the transistor's characteristic curves. The transistor conducts during the entire input cycle.
- □ The Q point must be centered on the load line for maximum class A output signal swing.
- □ The maximum efficiency of a class A amplifier is 25 percent.
- □ A class B amplifier operates in the linear region for half of the input cycle (180°), and it is in cutoff for the other half.
- □ The Q point is at cutoff for class B operation.
- □ Class B amplifiers are normally operated in a push-pull configuration in order to produce an output that is a replica of the input.
- □ The maximum efficiency of a class B amplifier is 78.5 percent.
- □ A class C amplifier operates in the linear region for only a small part of the input cycle.
- □ The class C amplifier is biased below cutoff.
- □ Class C amplifiers are normally operated as tuned amplifiers to produce a sinusoidal output.
- □ The maximum efficiency of a class C amplifier is higher than that of either class A or class B amplifiers. Under conditions of low power dissipation and high output power, the efficiency can approach 100 percent.

GLOSSARY

Complementary pair Two transistors, one npn and one pnp, having matched characteristics.

Crossover distortion Distortion in the output of a class B push-pull amplifier at the point where each transistor changes from the cutoff state to the *on* state.

Large-signal A signal that operates an amplifier over a significant portion of its load line.

Push-Pull A type of class B amplifier with two transistors in which one transistor conducts for one half-cycle and the other conducts for the other half-cycle.

FORMULAS

LARGE-SIGNAL CLASS A AMPLIFIERS

(9–1) $$I_{c(\text{sat})} = I_{CQ} + \frac{V_{CEQ}}{R_c}$$ Saturation current

(9–2) $$V_{ce(\text{cutoff})} = V_{CEQ} + I_{CQ}R_c$$ Cutoff voltage

(9–3) $$I_{CQ} = \frac{I_{CQ} + V_{CEQ}/R_c}{2}$$ Centered Q point

(9–4) $$V_{CEQ} = \frac{V_{CEQ} + I_{CQ}R_c}{2}$$ Centered Q point

(9–5)	$V_{CEQ} = I_{CQ}R_c$	Centered Q point
(9–6)	$r'_e = \dfrac{\Delta V_{BE}}{\Delta I_C}$	AC emitter resistance
(9–7)	$A_v = \dfrac{R_c}{r'_e}$	Voltage gain
(9–8)	$A_p = \beta_{\text{dc}}\left(\dfrac{R_c}{r'_e}\right)$	Power gain
(9–9)	$P_{DQ} = I_{CQ}V_{CEQ}$	Quiescent power
(9–10)	$P_{\text{out}} = V_{ce}I_c$	Output power
(9–11)	$P_{\text{out}} = \dfrac{0.5V_{CEQ}^2}{R_c}$	Output power (Q point closer to saturation)
(9–12)	$P_{\text{out}} = 0.5I_{CQ}^2R_c$	Output power (Q point closer to cutoff)
(9–13)	$P_{\text{out}} = 0.5V_{CEQ}I_{CQ}$	Output power (Q point centered)
(9–14)	$P_{\text{DC}} = V_{CC}I_{CC}$	DC input power
(9–15)	$\eta_{\max} = 0.25$	Maximum efficiency
(9–16)	$P_L = \dfrac{0.5V_{CEQ}^2}{R_L}$	Load power

Large-Signal Class B Push-Pull Amplifiers

(9–17)	$I_{c(\text{sat})} = \dfrac{V_{CEQ}}{R_L}$	Saturation current
(9–18)	$P_{\text{out}} = 0.5V_{CEQ}I_{c(\text{sat})}$	Output power
(9–19)	$P_{\text{out}} = 0.25V_{CC}I_{c(\text{sat})}$	Output power
(9–20)	$P_{\text{DC}} = V_{CC}I_{CC}$	DC input power
(9–21)	$I_{CC} = \dfrac{I_{c(\text{sat})}}{\pi}$	DC supply current
(9–22)	$P_{\text{dc}} = \dfrac{V_{CC}I_{c(\text{sat})}}{\pi}$	DC input power
(9–23)	$\eta_{\max} = 0.785$	Maximum efficiency

Large-Signal Class C Amplifiers

(9–24)	$P_{D(\text{avg})} = \left(\dfrac{t_{\text{on}}}{T}\right)V_{CE(\text{sat})}I_{C(\text{sat})}$	Average power
(9–25)	$P_{\text{out}} = \dfrac{0.5V_{CC}^2}{R_c}$	Output power
(9–26)	$\eta = \dfrac{P_{\text{out}}}{P_{\text{out}} + P_{D(\text{avg})}}$	Efficiency

SELF-TEST

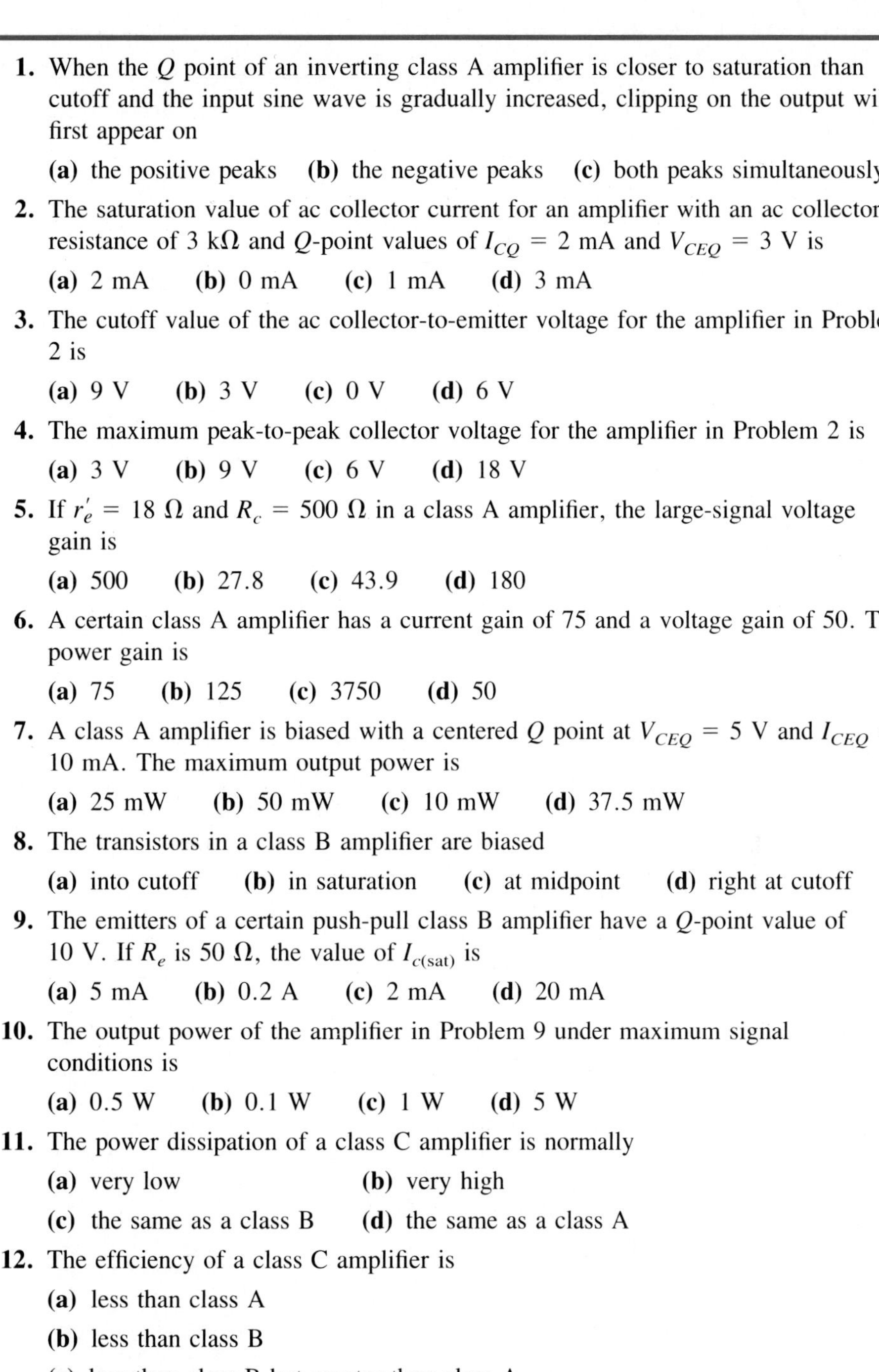

1. When the Q point of an inverting class A amplifier is closer to saturation than cutoff and the input sine wave is gradually increased, clipping on the output will first appear on

   (a) the positive peaks **(b)** the negative peaks **(c)** both peaks simultaneously

2. The saturation value of ac collector current for an amplifier with an ac collector resistance of 3 kΩ and Q-point values of $I_{CQ} = 2$ mA and $V_{CEQ} = 3$ V is

   (a) 2 mA **(b)** 0 mA **(c)** 1 mA **(d)** 3 mA

3. The cutoff value of the ac collector-to-emitter voltage for the amplifier in Problem 2 is

   (a) 9 V **(b)** 3 V **(c)** 0 V **(d)** 6 V

4. The maximum peak-to-peak collector voltage for the amplifier in Problem 2 is

   (a) 3 V **(b)** 9 V **(c)** 6 V **(d)** 18 V

5. If $r'_e = 18\ \Omega$ and $R_c = 500\ \Omega$ in a class A amplifier, the large-signal voltage gain is

   (a) 500 **(b)** 27.8 **(c)** 43.9 **(d)** 180

6. A certain class A amplifier has a current gain of 75 and a voltage gain of 50. The power gain is

   (a) 75 **(b)** 125 **(c)** 3750 **(d)** 50

7. A class A amplifier is biased with a centered Q point at $V_{CEQ} = 5$ V and $I_{CEQ} =$ 10 mA. The maximum output power is

   (a) 25 mW **(b)** 50 mW **(c)** 10 mW **(d)** 37.5 mW

8. The transistors in a class B amplifier are biased

   (a) into cutoff **(b)** in saturation **(c)** at midpoint **(d)** right at cutoff

9. The emitters of a certain push-pull class B amplifier have a Q-point value of 10 V. If R_e is 50 Ω, the value of $I_{c(sat)}$ is

   (a) 5 mA **(b)** 0.2 A **(c)** 2 mA **(d)** 20 mA

10. The output power of the amplifier in Problem 9 under maximum signal conditions is

    (a) 0.5 W **(b)** 0.1 W **(c)** 1 W **(d)** 5 W

11. The power dissipation of a class C amplifier is normally

    (a) very low **(b)** very high
    (c) the same as a class B **(d)** the same as a class A

12. The efficiency of a class C amplifier is

    (a) less than class A
    (b) less than class B
    (c) less than class B but greater than class A
    (d) greater than class A or class B

PROBLEMS

SECTION 9–1 CLASS A AMPLIFIERS

1. Determine the approximate values $I_{C(sat)}$ and $V_{CE(cutoff)}$ in Figure 9–43. Assume $V_{CE(sat)} \cong 0$ V.

FIGURE 9–43

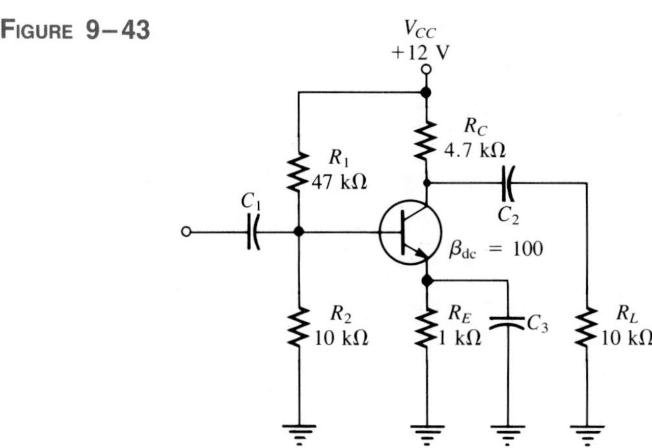

2. Sketch the ac equivalent circuit for the amplifier in Figure 9–43. Calculate the saturation value of the collector current under signal conditions.
3. Find the cutoff value of the ac collector-to-emitter voltage in Figure 9–43.
4. In Figure 9–44 assume β_{dc} for each transistor increases by 50 percent with a certain temperature rise. How is the maximum output voltage affected in each circuit?

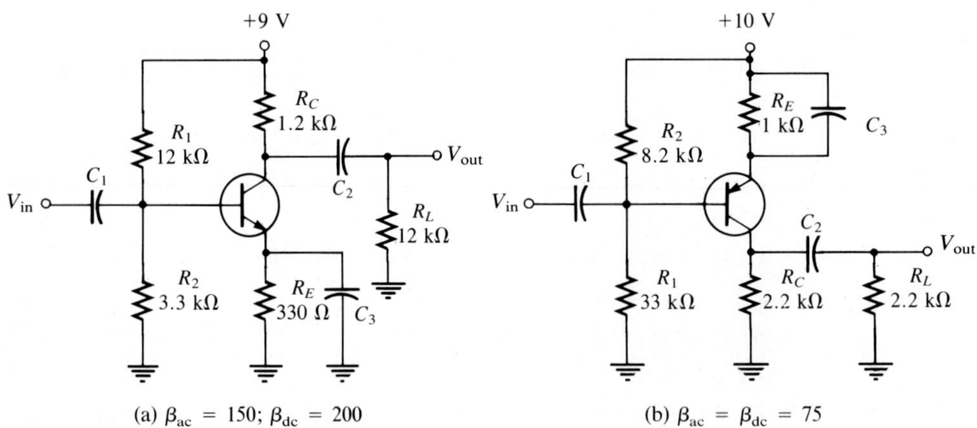

FIGURE 9–44

5. What is the maximum peak value of collector current that can be realized in each circuit of Figure 9–45? What is the maximum peak value of output voltage?

FIGURE 9–45

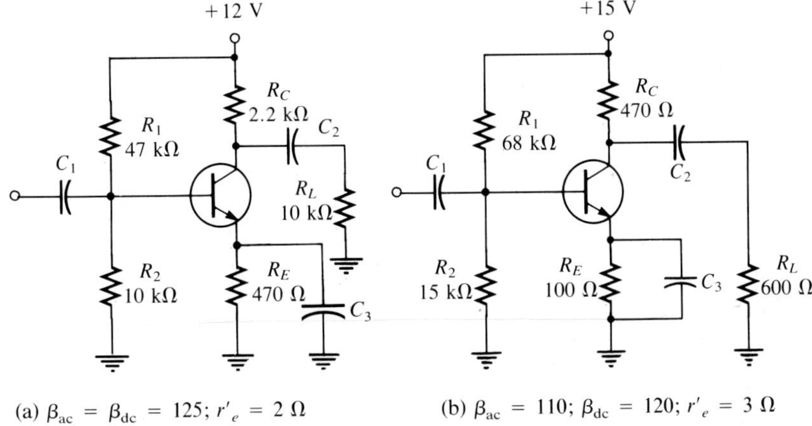

(a) $\beta_{ac} = \beta_{dc} = 125$; $r'_e = 2\ \Omega$ (b) $\beta_{ac} = 110$; $\beta_{dc} = 120$; $r'_e = 3\ \Omega$

6. Find the large-signal voltage gain for each circuit in Figure 9–45.
7. Determine the maximum rms value of input voltage that can be applied to each amplifier in Figure 9–45 without producing a clipped output voltage.
8. Determine the minimum power rating for each of the transistors in Figure 9–46.

FIGURE 9–46

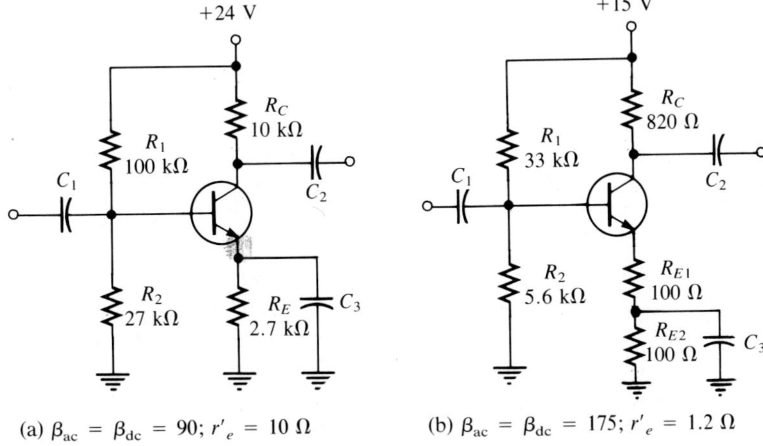

(a) $\beta_{ac} = \beta_{dc} = 90$; $r'_e = 10\ \Omega$ (b) $\beta_{ac} = \beta_{dc} = 175$; $r'_e = 1.2\ \Omega$

9. Find the maximum output signal power and efficiency for each amplifier in Figure 9–46 with a 10 kΩ load resistor.

SECTION 9–2 CLASS B PUSH-PULL AMPLIFIERS

10. What dc voltage would you expect to measure with each dc meter in Figure 9–47?

FIGURE 9–47

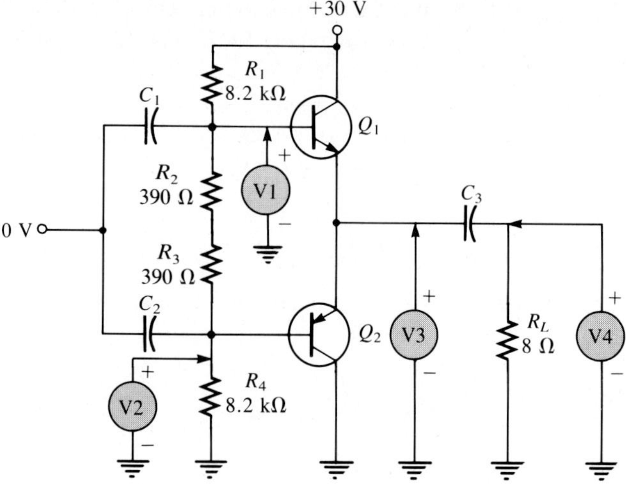

11. Determine the dc voltages at the bases and emitters of Q_1 and Q_2 in Figure 9–48. Also determine V_{CEQ} for each transistor.

FIGURE 9–48

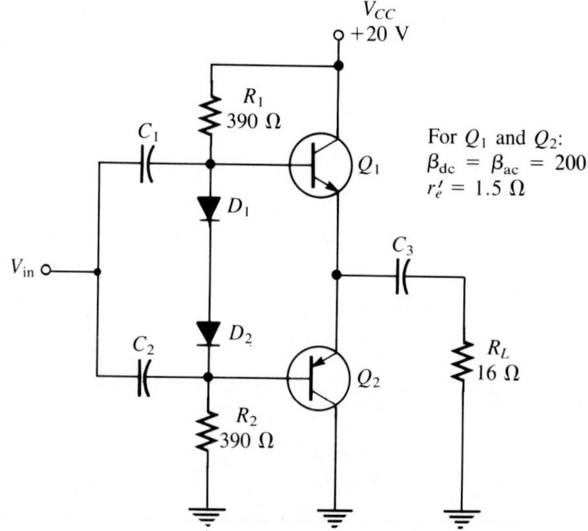

12. Determine the maximum peak output voltage and peak load current for the circuit in Figure 9–48.

13. Find the maximum signal power achievable with the class B push-pull amplifier in Figure 9–48. Find the dc input power.

14. The efficiency of a certain class B push-pull amplifier is 0.71, and the dc input power is 16.25 W. What is the ac output power?

15. A certain class B push-pull amplifier has an R_e of 8 Ω and a V_{CEQ} of 12 V. Determine I_{CC}, P_{dc}, and P_{out}. Assuming this amplifier is operating under maximum output conditions, what is V_{CC}?

16. In Figure 9–49, what rms input voltage will produce a maximum output voltage swing? Use an ideal voltage gain of one for the emitter-followers and assume the maximum peak output voltage equals $V_{CC}/2$.

FIGURE 9–49

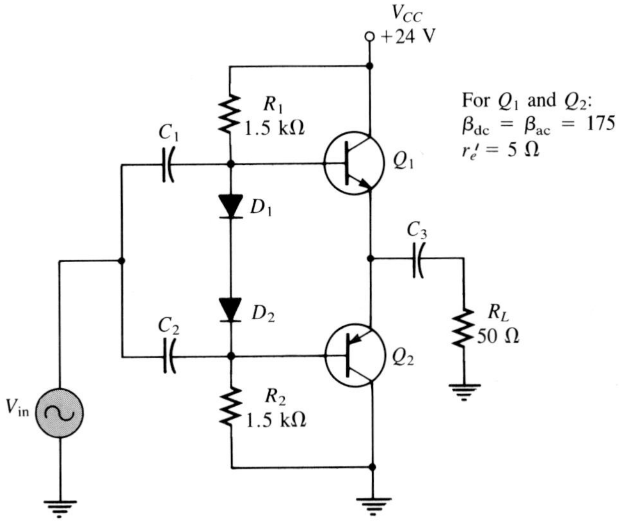

SECTION 9–3 CLASS C AMPLIFIERS

17. A certain class C amplifier transistor is on for 10 percent of the input cycle. If $V_{CE(sat)} = 0.18$ V and $I_{C(sat)} = 25$ mA, what is the average power dissipation for maximum output?

18. What is the resonant frequency of a tank circuit with $L = 10$ mH and $C = 0.001$ μF?

19. What is the maximum peak-to-peak output voltage of a tuned class C amplifier with $V_{CC} = 12$ V?

20. Determine the efficiency of the class C amplifier described in Problem 17 if $V_{CC} = 15$ V and the equivalent parallel resistance in the collector tank circuit is 50 Ω.

SECTION 9–4 TROUBLESHOOTING

21. Refer to Figure 9–49. What would you expect to observe across R_L if C_1 opened?

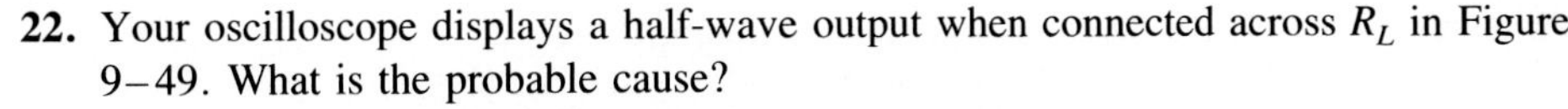

22. Your oscilloscope displays a half-wave output when connected across R_L in Figure 9–49. What is the probable cause?

23. Determine the possible fault or faults, if any, for each circuit in Figure 9–50 based on the indicated dc voltage measurements.

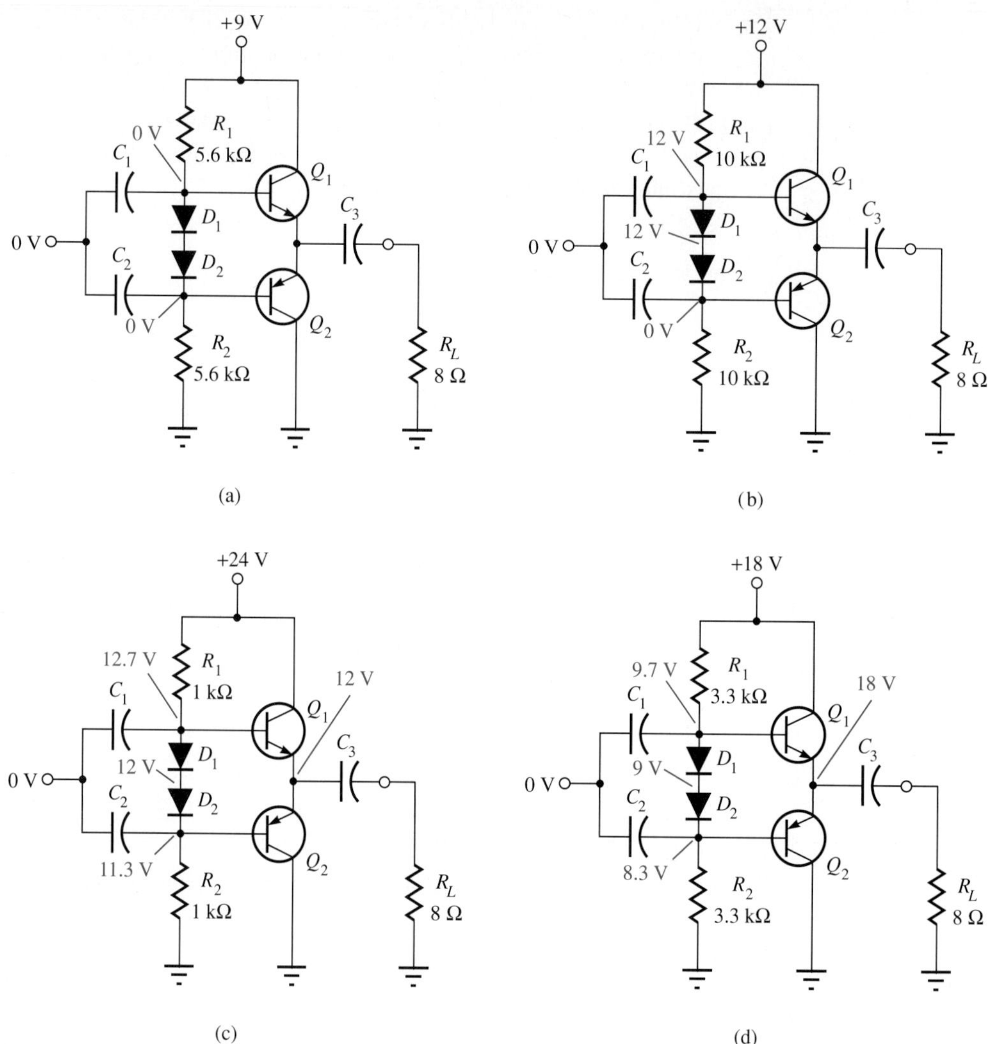

FIGURE 9–50

ANSWERS TO REVIEW QUESTIONS

SECTION 9–1

1. The ac load resistance has a value different from that of the dc load resistance.
2. Centered on the load line
3. Maximum efficiency of class A is 25%.
4. Center the Q point.

SECTION 9–2

1. Cutoff
2. The barrier potential of the base-emitter junction
3. Maximum efficiency of class B is 78.5%.
4. To reproduce both positive and negative alternations of the input signal with greater efficiency

SECTION 9–3

1. Class C is biased well into cutoff.
2. The purpose of the tuned circuit is to produce a sine wave output.
3. $\eta = [1\text{ W}/(1\text{ W} + 0.1\text{ W})]100 = 90.9\%$

SECTION 9–4

1. Excess input signal voltage
2. Open bypass capacitor, C_2

SECTION 9–5

1. To dissipate heat produced by the power transistors
2. You will get only the upper or lower half of the audio signal on the output.
3. An open bias diode
4. No, because each transistor is connected as an emitter follower with a maximum voltage gain of one.

ANSWERS TO PRACTICE EXERCISES

9–1 $V_{ce(sat)} = 0\text{ V}$, $I_{c(sat)} = 12.35\text{ mA}$, $V_{ce(cutoff)} = 7.41\text{ V}$, $I_{c(cutoff)} = 0\text{ A}$

9–2 The maximum output swing increases to 3 V.

9–3 Decrease R_E to 360 Ω.

9–4 $A_v = 50$

9–5 Shift the Q point toward the center of the load line by increasing I_{CQ} and decreasing V_{CEQ}.

9–6 37.9 mW

9–7 The base and emitter voltages will remain the same.

9–8 7.5 V, 0.75 A

9–9 $P_{out} = 2.8\text{ W}$, $P_{DC} = 3.58\text{ W}$

9–10 3 mW

9–11 The efficiency decreases.

10

Amplifier Frequency Response

After completing this chapter, you should be able to

- □ Define what is meant by *frequency response.*
- □ Define and apply Miller's theorem.
- □ Determine the input and output capacitances of an amplifier.
- □ Discuss the characteristics of an amplifier's frequency response.
- □ Identify the critical frequencies in an amplifier's response.
- □ Analyze the effects of coupling and bypass capacitors at low frequencies.
- □ Discuss how the transistor's internal capacitances limit the response of an amplifier at high frequencies.
- □ Express amplifier gain in decibels.
- □ Define the midrange gain of an amplifier.
- □ Define the term *roll-off* and explain what factors determine it.
- □ Define and explain a Bode plot.
- □ Compare direct-coupled and capacitively coupled amplifiers in terms of their low-frequency response.
- □ Measure an amplifier's frequency response by two methods: Amplitude vs. frequency and step response.

In the previous chapters on amplifiers, the effects of the input frequency on the amplifier's operation were neglected in order to focus on other concepts. We considered the coupling and bypass capacitors to be ideal shorts and the internal transistor capacitances to be ideal opens. This is valid when the frequency is in the amplifier's midrange. As you know, capacitive reactance decreases with increasing frequency and vice versa. When the frequency is low enough, the coupling and bypass capacitors can no longer be considered as shorts because their reactances are large enough to have a significant effect. Also, when the frequency is high enough, the internal device capacitances can no longer be considered as opens because their reactances become small enough to have a significant effect. A complete picture of an amplifier's response must take into account the full

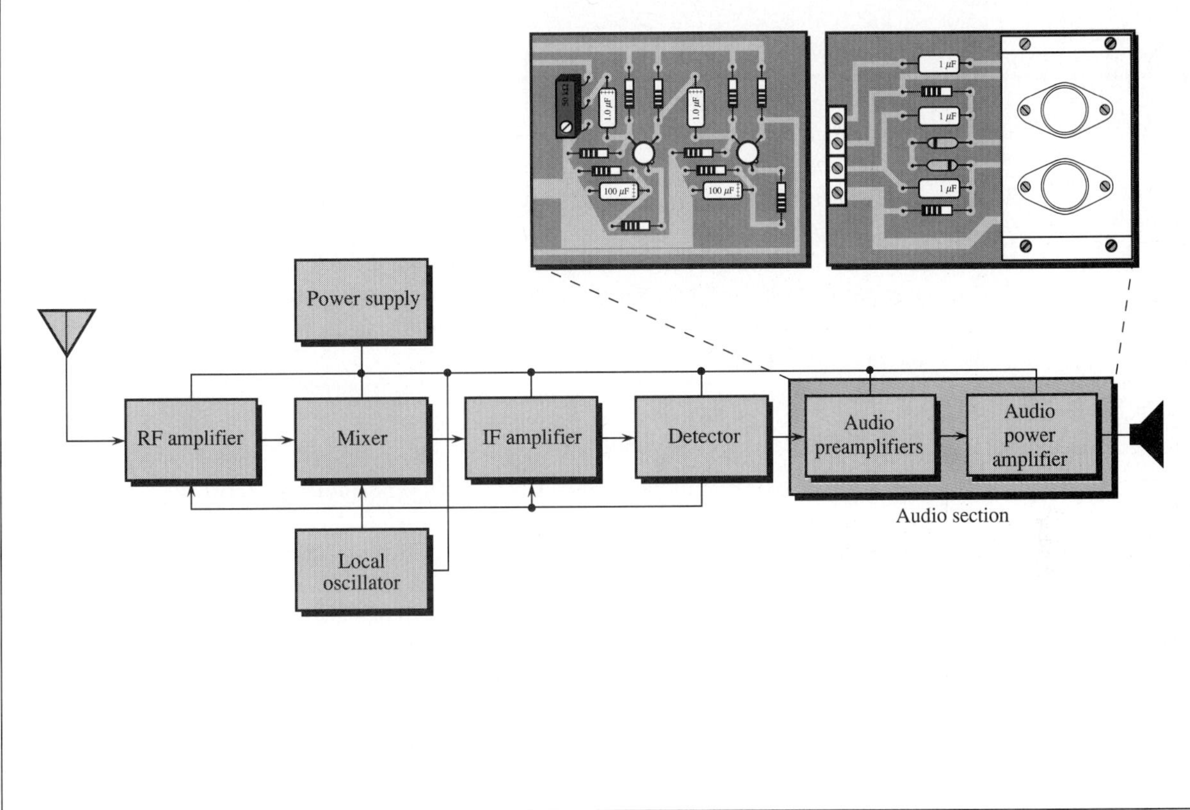

range of frequencies over which the amplifier may operate.

In this chapter, we examine the frequency effects on amplifier gain and phase shift. The coverage applies to both BJT and FET amplifiers, and a mix of both are included to illustrate the concepts.

A System Application

In this chapter, we continue to look at the receiver system and focus on two of the circuits you have worked with in previous chapters—the audio preamplifier and the audio power amplifier. These two circuits combine to form the audio section of the receiver, and you will be focusing on its frequency response characteristics. The general frequency range that the human ear can hear is from about 20 Hz to about 20 kHz. Everyone cannot hear the full range and each person's hearing ability differs. In order to reproduce sound properly, an audio amplifier must be able to uniformly amplify signals within this general range.

For the system application in Section 10–8, in addition to the other topics, be sure you understand

- ☐ The principles of amplifier frequency response.
- ☐ How to determine an amplifier's critical frequencies.
- ☐ The concept of bandwidth.
- ☐ The factors in an amplifier that affect the bandwidth.

10–1 GENERAL CONCEPTS

In our previous studies of amplifiers, the capacitive reactance of the coupling and bypass capacitors was assumed to be 0 Ω at the signal frequency and, therefore, had no effect on the amplifier's gain or phase shift. Also, the internal transistor capacitances were assumed to be small enough to neglect at the signal frequency. All of these simplifying assumptions are valid and necessary for studying amplifier theory. However, they do give a limited picture of an amplifier's total operation, so in this section we begin to study the frequency effects of these capacitances. The frequency response of an amplifier is the change in gain or phase shift over a specified range of signal frequencies.

EFFECT OF COUPLING CAPACITORS

Recall from basic circuit theory that $X_C = 1/2\pi fC$. This formula shows that the capacitive reactance varies inversely with frequency. At lower frequencies the reactance is greater, and it decreases as the frequency increases. At lower frequencies—for example, audio frequencies below 10 Hz—capacitively coupled amplifiers such as those in Figure 10–1 have less voltage gain than they have at higher frequencies. The reason is that at lower frequencies more signal voltage is dropped across C_1, C_2, and C_3 because their reactances are higher. This is the result of the limitation imposed by availability and physical size of large values of capacitance required in these applications. Also, a phase shift is introduced by the coupling capacitors because C_1 forms a lead network with the R_{in} of the amplifier, and C_2 forms a lead network with R_L and R_C or R_D. A *lead network* is an *RC* circuit in which the output voltage across *R* leads the input voltage.

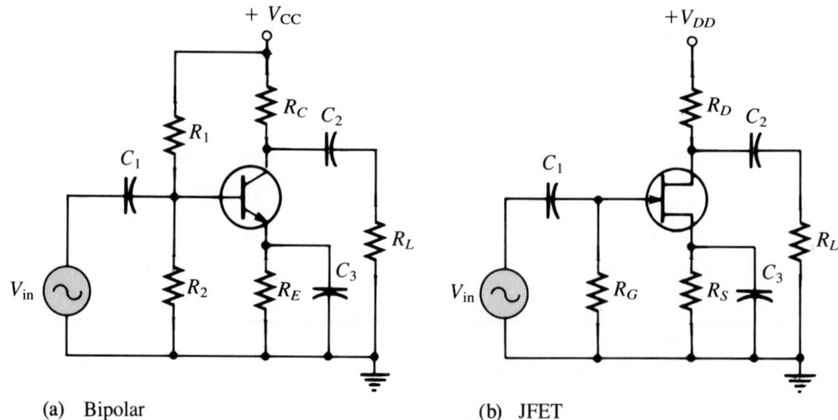

FIGURE 10–1
Typical capacitively coupled amplifiers.

EFFECT OF BYPASS CAPACITORS

At lower frequencies, the bypass capacitor C_3 is not a short. So, the emitter (or FET source terminal) is not at ac ground. X_{C3} in parallel with R_E (or R_S) creates an impedance that reduces the gain. This is illustrated in Figure 10–2.

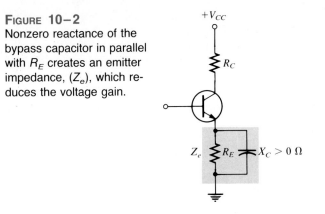

FIGURE 10–2
Nonzero reactance of the bypass capacitor in parallel with R_E creates an emitter impedance, (Z_e), which reduces the voltage gain.

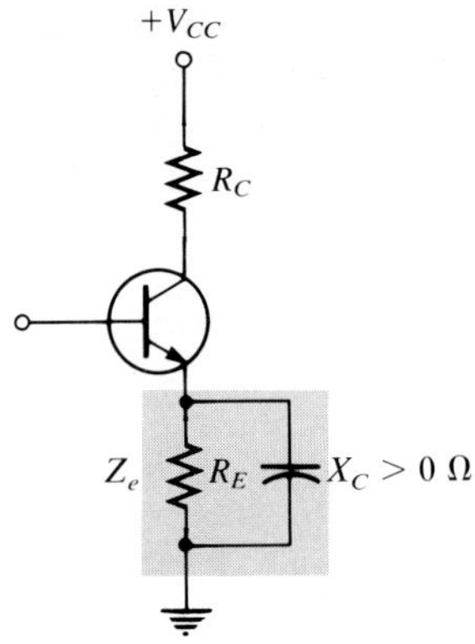

EFFECT OF INTERNAL TRANSISTOR CAPACITANCES

At high frequencies, the coupling and bypass capacitors become effective shorts and do not affect the amplifier's response. Undesired internal device capacitances, however, do come into play, reducing the amplifier's gain and introducing phase shift as the signal frequency increases.

Figure 10–3 shows the internal capacitances for both a bipolar transistor and a JFET. In the case of the bipolar transistor, C_{be} is the base-emitter junction capacitance, C_{bc} is the base-collector junction capacitance, and C_{ce} is the capacitance from collector to emitter. In the case of the JFET, C_{gs} is the internal capacitance between gate and source, C_{gd} is the internal capacitance between gate and drain, and C_{ds} is the capacitance from drain to source.

FIGURE 10–3
Internal transistor capacitances.

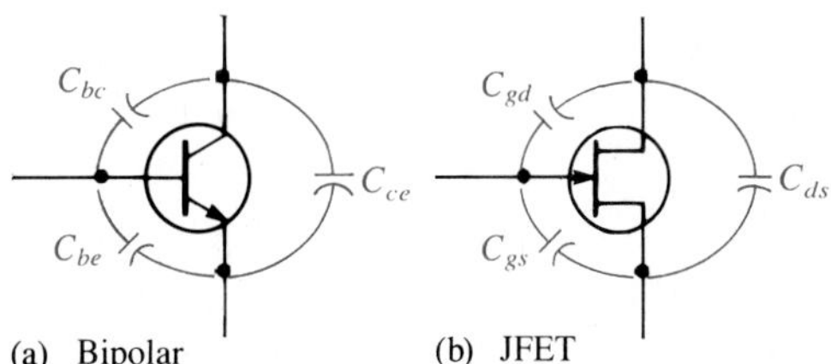

Data sheets often refer to the bipolar transistor capacitance C_{bc} as the output capacitance, often designated C_{ob}. The capacitance C_{be} is often designated as the input capacitance, C_{ib}. Data sheets for FETs normally specify input capacitance C_{iss} and reverse transfer capacitance, C_{rss}. From these, C_{gs} and C_{gd} can be calculated, as you will see later.

At lower frequencies, the internal capacitances have a very high reactance because of their low capacitance value (usually only a few picofarads). Therefore, they look like opens and have no effect on the transistor's performance. As the frequency goes up, the internal capacitive reactances go down, and at some point they begin to have a significant effect on the transistor's gain. When the reactance of C_{be} (or C_{gs}) becomes small enough, a significant amount of the signal voltage is lost due to a voltage-divider effect of the source resistance and the capacitive reactance of C_{be}, as illustrated in Figure 10–4(a). When the reactance of C_{bc} (or C_{gd}) becomes small enough, a significant amount of output signal voltage is fed back out-of-phase with the input (negative feedback), thus effectively reducing the voltage gain. This is illustrated in Figure 10–4(b).

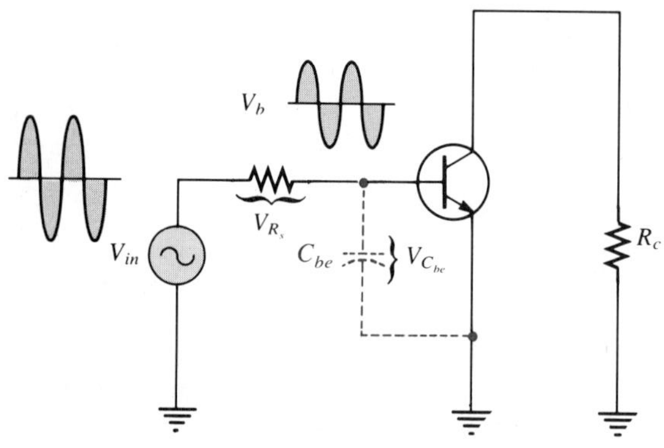

(a) Effect of C_{be}, where V_b is reduced by the voltage-divider action of R_s and $X_{C_{be}}$

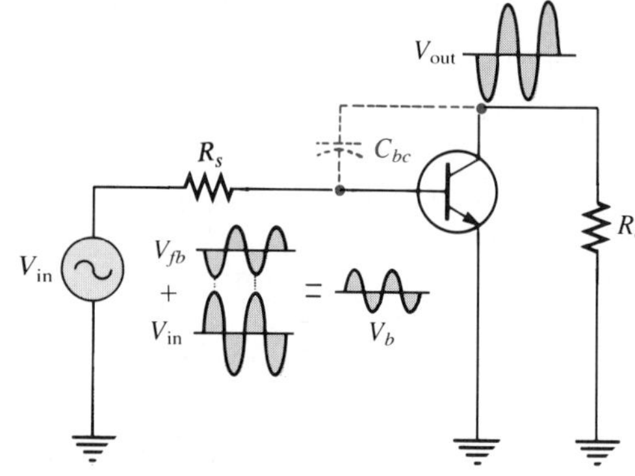

(b) Effect of C_{bc}, where part of V_{out} (V_{fb}) goes back through C_{bc} to the base and reduces the input signal because it is approximately 180° out of phase with V_{in}

FIGURE 10–4
AC equivalent circuit for a bipolar amplifier showing effects of the internal capacitances C_{be} and C_{bc}.

10–1 REVIEW QUESTIONS

1. In an ac amplifier, which capacitors affect the low-frequency gain?
2. How is the high-frequency gain of an amplifier limited?

10–2 MILLER'S THEOREM AND DECIBELS

Before we get into the analysis of amplifier frequency response, two important topics must be considered—Miller's theorem and the concept of decibels. Miller's theorem deals with the relationships of the capacitance and voltage gain in an amplifier, and it will help you understand high-frequency response. Decibels are a common form of gain measurement and are used in expressing amplifier response.

MILLER'S THEOREM

Miller's theorem can be used to simplify the analysis of inverting amplifiers at high frequencies where the internal transistor capacitances are important. The capacitance C_{bc} in bipolar transistors (C_{gd} in FETs) between the input (base or gate) and the output (collector or drain) is shown in Figure 10–5(a) in a generalized form. A_v is the voltage gain of the amplifier at midrange frequencies, and C represents either C_{bc} or C_{gd}.

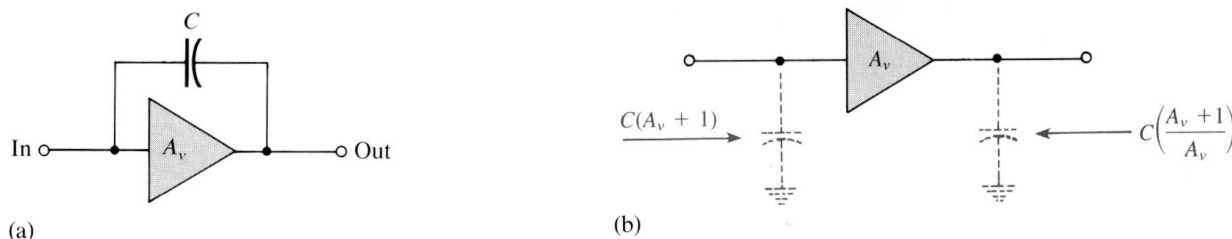

FIGURE 10–5
General case of Miller input and output capacitances. C is C_{bc} or C_{gd}.

Miller's theorem states that C effectively appears as a capacitance from input to ground, as shown in Figure 10–5(b), that can be expressed as follows.

$$C_{\text{in(Miller)}} = C(A_v + 1) \tag{10–1}$$

This shows that C_{bc} (or C_{gd}) has a much greater impact on input capacitance than its actual value. For example, if $C_{bc} = 6$ pF and the amplifier gain is 50, then $C_{\text{in(Miller)}} = 306$ pF. Figure 10–6 shows how this effective input capacitance appears in the actual ac equivalent circuit in parallel with C_{be} (or C_{gs}).

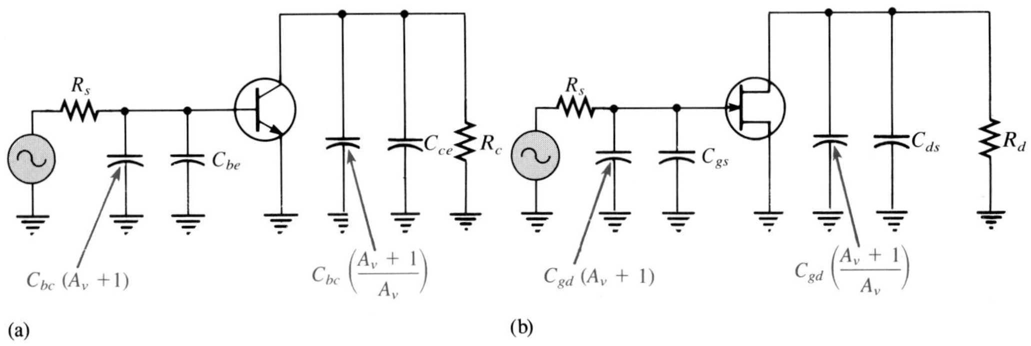

FIGURE 10–6
AC equivalent circuits showing effective Miller capacitances.

Miller's theorem also states that C effectively appears as a capacitance from output to ground, as shown in Figure 10–5(b), that can be expressed as follows.

$$C_{\text{out(Miller)}} = C\left(\frac{A_v + 1}{A_v}\right) \tag{10–2}$$

Equation (10–2) shows that if the voltage gain is 10 or greater, $C_{\text{out(Miller)}}$ is approximately equal to C_{bc} or C_{gd} because $(A_v + 1)/A_v$ is approximately equal to 1. Figure 10–6 also shows how this effective output capacitance appears in the ac equivalent circuit for bipolar transistors and FETs. Equations (10–1) and (10–2) are derived in Appendix B.

EXAMPLE 10–1 Apply Miller's theorem to the small-signal amplifier in Figure 10–7 to determine a high-frequency equivalent circuit, given that $C_{bc} = 3$ pF and $C_{be} = 1$ pF.

FIGURE 10–7

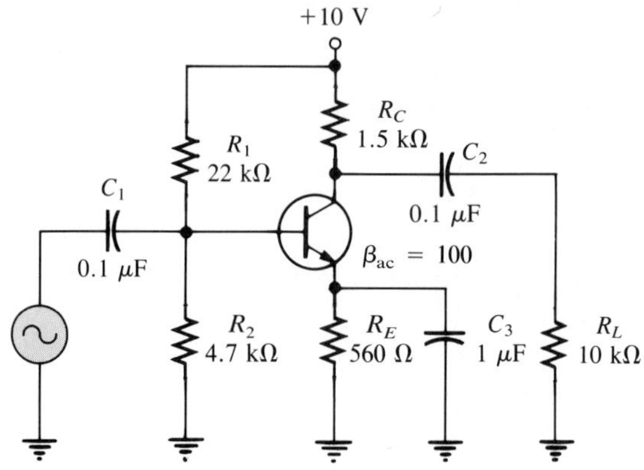

SOLUTION

The voltage gain must first be found before Miller's theorem can be applied. Recall that the voltage gain is expressed as

$$A_v = \frac{R_c}{r_e}$$

The ac collector resistance, R_c, equals R_C in parallel with R_L.

$$R_c = 1.5\ \text{k}\Omega \| 10\ \text{k}\Omega = 1.3\ \text{k}\Omega$$

To find r_e, I_E must be known.

$$V_B = \left(\frac{R_2}{R_1 + R_2}\right)V_{CC} = \left(\frac{4.7\ \text{k}\Omega}{26.7\ \text{k}\Omega}\right)10\ \text{V} = 1.76\ \text{V}$$

$$I_E = \frac{V_E}{R_E} = \frac{1.76\ \text{V} - 0.7\ \text{V}}{560\ \Omega} = 1.89\ \text{mA}$$

$$r_e = \frac{25\ \text{mV}}{I_E} = \frac{25\ \text{mV}}{1.89\ \text{mA}} \cong 13.23\ \Omega$$

Therefore,

$$A_v = \frac{1.3\ \text{k}\Omega}{13.23\ \Omega} = 98$$

Applying Miller's theorem, we get

$$C_{in(Miller)} = C_{bc}(A_v + 1) = (3\ \text{pF})(99) = 297\ \text{pF}$$

and

$$C_{out(Miller)} = C_{bc}\left(\frac{A_v + 1}{A_v}\right) \cong 3\ \text{pF}$$

The high-frequency equivalent circuit for the amplifier in Figure 10–7 is shown in Figure 10–8.

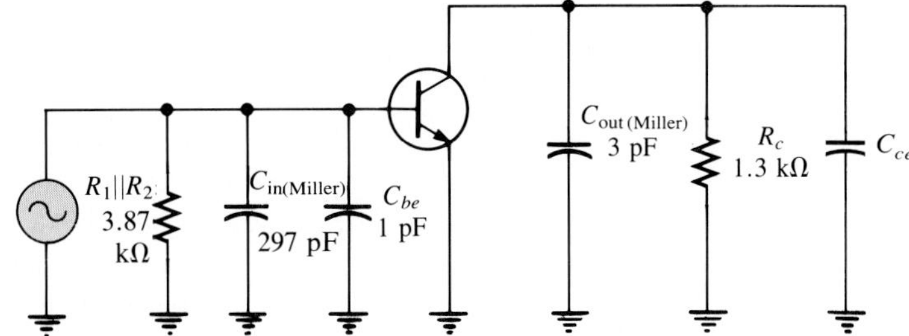

FIGURE 10–8
High-frequency equivalent circuit for the amplifier in Figure 10–7.

PRACTICE EXERCISE 10–1
If the voltage gain of the amplifier in Figure 10–7 is reduced to 50, what are the input and output capacitances?

DECIBELS

Chapter 6 introduced the use of decibels in expressing gain. Because of the importance of the decibel unit in amplifier measurements, additional coverage of this topic is necessary before going any further.

The basis for the decibel unit stems from the logarithmic response of the human ear to the intensity of sound. The *decibel* is a measurement of the ratio of one power to another or one voltage to another. Power gain is expressed in decibels (dB) by the following formula.

$$A_p \text{ (dB)} = 10 \log A_p \qquad \textbf{(10–3)}$$

where A_p is the actual power gain, P_{out}/P_{in}. Voltage gain is expressed in decibels by the following formula.

$$A_v \text{ (dB)} = 20 \log A_v \qquad \textbf{(10–4)}$$

If A_v is greater than one, the dB gain is positive. If A_v is less than one, the dB gain is negative and is usually called *attenuation*.

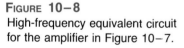

EXAMPLE 10–2 Express each of the following ratios in dB.

(a) $\dfrac{P_{out}}{P_{in}} = 250$ **(b)** $\dfrac{P_{out}}{P_{in}} = 100$ **(c)** $A_v = 10$

(d) $A_p = 0.5$ **(e)** $\dfrac{V_{out}}{V_{in}} = 0.707$

SOLUTION

(a) $A_p \text{ (dB)} = 10 \log(250) = 24 \text{ dB}$
(b) $A_p \text{ (dB)} = 10 \log(100) = 20 \text{ dB}$
(c) $A_v \text{ (dB)} = 20 \log(10) = 20 \text{ dB}$
(d) $A_p \text{ (dB)} = 10 \log(0.5) = -3 \text{ dB}$
(e) $A_v \text{ (dB)} = 20 \log(0.707) = -3 \text{ dB}$

PRACTICE EXERCISE 10–2

Express each of the following gains in dB: **(a)** $A_v = 1200$, **(b)** $A_p = 50$, **(c)** $A_v = 125{,}000$.

0 dB REFERENCE

It is often convenient in amplifier analysis to assign a certain value of gain as the 0 dB reference. This does not mean that the actual voltage gain is 1 (which is 0 dB); it means that the reference gain, no matter what its actual value, is used as a reference with which to compare other values of gain and is therefore assigned a 0 dB value.

Many amplifiers exhibit a maximum gain over a certain range of frequencies and a reduced gain at frequencies below and above this range. The maximum gain is called the *midrange gain* in this case and is assigned a 0 dB value. Any value of gain below **midrange** can be referenced to 0 dB and expressed as a negative dB value. For example, if the midrange voltage gain of a certain amplifier is 100 and the gain at a certain frequency below midrange is 50, then this reduced gain can be expressed as $20 \log(50/100) = 20 \log(0.5) = -6$ dB. This indicates that it is 6 dB *below* the 0 dB reference. So a halving of the output voltage for a steady input voltage is a 6 dB reduction in the gain. Correspondingly, a doubling of the output voltage is a 6 dB increase in the gain.

FIGURE 10–9
Normalized gain-versus-frequency curve.

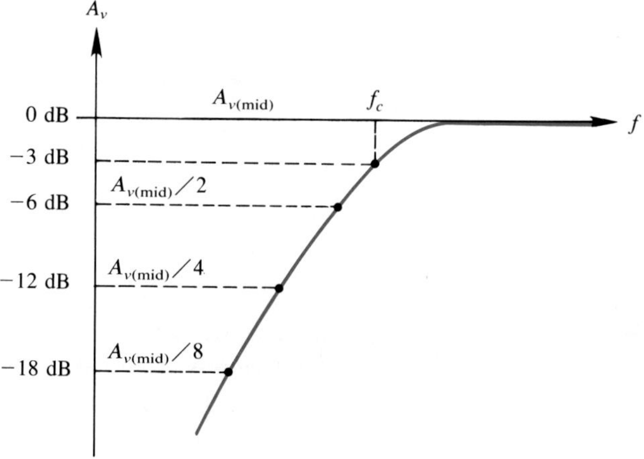

Figure 10–9 illustrates a gain-versus-frequency curve showing several dB points. Table 10–1 shows how doubling or halving voltage ratios translates into dB values. Notice that each step in the doubling or halving of the voltage ratio increases or decreases the dB value by 6 dB. Notice in Table 10–1 that every time the voltage ratio is doubled, the dB value increases by 6 dB, and every time the ratio is halved, the dB value decreases by 6 dB.

TABLE 10–1
dB values corresponding to doubling and halving of the voltage ratio

Voltage Gain (A_v)	dB (with Respect to Zero Reference)
32	20 log(32) = 30 dB
16	20 log(16) = 24 dB
8	20 log(8) = 18 dB
4	20 log(4) = 12 dB
2	20 log(2) = 6 dB
1	20 log(1) = 0 dB
0.707	20 log(0.707) = −3 dB
0.5	20 log(0.5) = −6 dB
0.25	20 log(0.25) = −12 dB
0.125	20 log(0.125) = −18 dB
0.0625	20 log(0.0625) = −24 dB
0.03125	20 log(0.03125) = −30 dB

THE CRITICAL FREQUENCY

The **critical** or *corner* **frequency** is the frequency at which the output power drops to one-half of its midrange value. This corresponds to a 3 dB reduction in the power gain, as expressed by the following formula.

$$A_p\ (\text{dB}) = 10\log(0.5) = -3\ \text{dB}$$

Also, when the output voltage is 70.7 percent of its midrange value, it is expressed in dB as

$$A_v\ (\text{dB}) = 20\log(0.707) = -3\ \text{dB}$$

This tells us that at the critical frequency, the voltage gain is down 3 dB or 70.7% of its midrange value. At this same frequency, the power is one-half of its midrange value.

EXAMPLE 10–3 A certain amplifier has a midrange rms output voltage of 10 V. What is the rms output voltage for each of the following dB gain reductions with a constant rms input voltage?

(a) −3 dB **(b)** −6 dB **(c)** −12 dB **(d)** −24 dB

SOLUTION
(a) At −3 dB, $V_{out} = 0.707(10\text{ V}) = 7.07\text{ V}$
(b) At −6 dB, $V_{out} = 0.5(10\text{ V}) = 5\text{ V}$
(c) At −12 dB, $V_{out} = 0.25(10\text{ V}) = 2.5\text{ V}$
(d) At −24 dB, $V_{out} = 0.0625(10\text{ V}) = 0.625\text{ V}$

PRACTICE EXERCISE 10–3
Determine the output voltage at the following dB levels for a midrange value of 50 V: **(a)** 0 dB, **(b)** −18 dB, **(c)** −30 dB.

dBm MEASUREMENT

A unit that is often used in measuring power is the dBm. The term *dBm* means decibels referenced to 1 mW of power. When dBm is used, all power measurements are relative to a reference level of 1 mW. A 3-dBm increase corresponds to doubling of the power, and a 3-dBm decrease corresponds to a halving of the power. For example, +3 dBm corresponds to 2 mW (twice 1 mW), and −3 dBm corresponds to 0.5 mW (half of 1 mW). Table 10–2 shows several dBm values.

TABLE 10–2
Power in terms of dBm

Power	dBm
0.03125 mW	−15 dBm
0.0625 mW	−12 dBm
0.125 mW	−9 dBm
0.25 mW	−6 dBm
0.5 mW	−3 dBm
1 mW	0 dBm
2 mW	3 dBm
4 mW	6 dBm
8 mW	9 dBm
16 mW	12 dBm
32 mW	15 dBm

10–2 REVIEW QUESTIONS

1. How much increase in actual voltage gain corresponds to +12 dB?
2. Convert a power gain of 25 to decibels.
3. What power corresponds to 0 dBm?

10–3 LOW-FREQUENCY AMPLIFIER RESPONSE

In this section, we will examine how the voltage gain and phase shift of a capacitively coupled amplifier are affected by frequencies below which the capacitive reactance becomes significant. At the end of the section, a comparison is made to direct-coupled amplifiers.

A typical capacitively coupled small-signal common-emitter amplifier is shown in Figure 10–10 (the approach for FETs is similar). Assuming that the coupling and bypass capacitors are ideal shorts at the midrange signal frequency, the voltage gain can be calculated as you have done before. Once this value of gain is determined, you have the midrange gain of the amplifier as restated in Equation (10–5), where $R_c = R_C \| R_L$.

$$A_{v(\text{mid})} = \frac{R_c}{r_e} \tag{10–5}$$

The amplifier in Figure 10–10 has three high-pass *RC* networks that affect its gain as the frequency is reduced below midrange. These are shown in the low-frequency

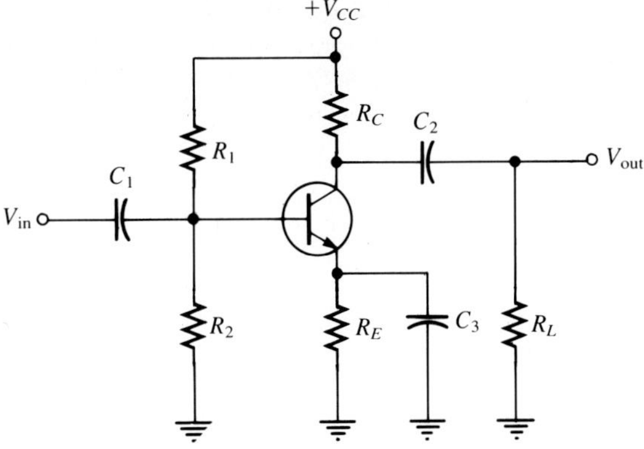

FIGURE 10–10
Typical capacitively coupled amplifier.

equivalent circuit in Figure 10–11. Unlike the ac equivalent circuit used in previous chapters, which represented midrange response ($X_C \cong 0$), the low-frequency equivalent retains the coupling and bypass capacitors because X_C is not small enough to neglect.

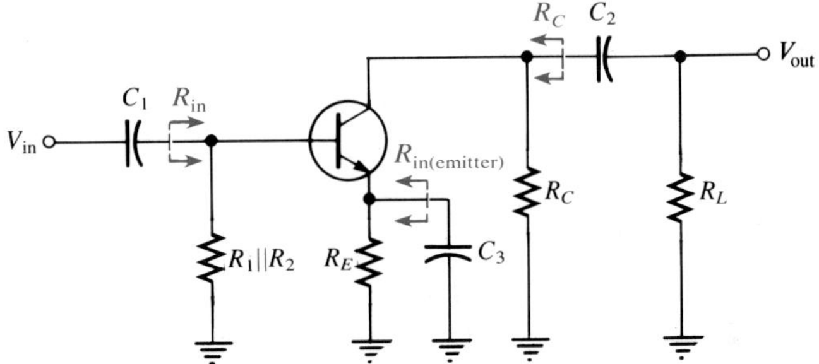

FIGURE 10–11
Low-frequency equivalent of the amplifier in Figure 10–10.

One RC network is formed by the input coupling capacitor C_1 and the input resistance of the amplifier. The second RC network is formed by the output coupling capacitor C_2, the resistance looking in at the collector, and the load resistance. The third RC network that affects the low-frequency response is formed by the emitter-bypass capacitor C_3 and the resistance looking in at the emitter.

THE INPUT *RC* NETWORK

The RC network for the amplifier in Figure 10–10 that is formed by C_1 and the amplifier's input resistance is shown in Figure 10–12. (Input resistance was discussed in Chapter 6.) As the signal frequency decreases, X_{C1} increases. This causes less voltage to be applied across the input resistance of the amplifier because more is dropped across X_{C1}. As you can see, this reduces the overall voltage gain of the amplifier. Recall from basic ac circuit theory that the input/output relationship for the RC network in Figure 10–12 (neglecting the internal resistance of the input signal source) can be stated as

$$V_{out} = \left(\frac{R_{in}}{\sqrt{R_{in}^2 + X_{C1}^2}}\right)V_{in} \qquad (10\text{–}6)$$

FIGURE 10–12
RC network formed by the input coupling capacitor and the amplifier's input resistance.

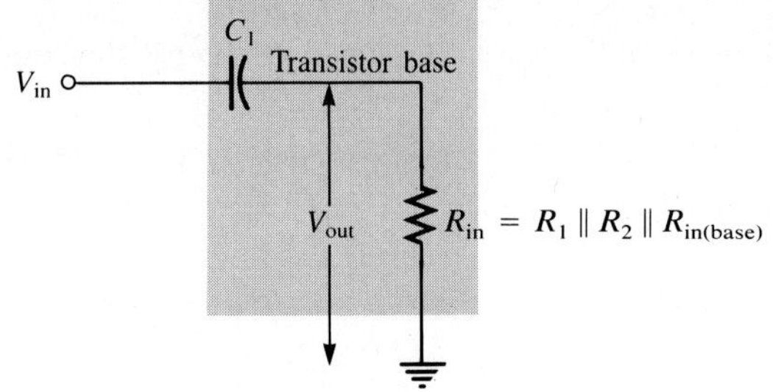

As previously mentioned, a critical point in the amplifier's response is generally accepted to occur when the output voltage is 70.7 percent of its midrange value. This condition occurs in the input *RC* network when $X_{C1} = R_{in}$, as shown in the following steps using Equation (10–6).

$$V_{out} = \left(\frac{R_{in}}{\sqrt{R_{in}^2 + R_{in}^2}}\right)V_{in}$$

$$= \left(\frac{R_{in}}{\sqrt{2R_{in}^2}}\right)V_{in}$$

$$= \left(\frac{R_{in}}{\sqrt{2}\,R_{in}}\right)V_{in}$$

$$= \left(\frac{1}{\sqrt{2}}\right)V_{in}$$

$$V_{out} = 0.707V_{in}$$

In terms of dB measurement,

$$20\log\left(\frac{V_{out}}{V_{in}}\right) = 20\log(0.707) = -3\text{ dB}$$

This particular condition is often called the *−3 dB point* of the amplifier response; the overall gain is 3 dB less than at midrange frequencies because of the attenuation of the input *RC* network. The frequency f_c at which this condition occurs is called the *lower critical frequency* (also known as the *lower corner, lower break,* or **lower cutoff frequency**) and can be calculated as follows.

$$X_{C1} = R_{in}$$

$$\frac{1}{2\pi f_c C_1} = R_{in}$$

$$f_c = \frac{1}{2\pi R_{in} C_1} \qquad \textbf{(10–7)}$$

If the resistance of the input source is taken into account, Equation (10–7) becomes

$$f_c = \frac{1}{2\pi(R_s + R_{in})C_1} \qquad \textbf{(10–8)}$$

EXAMPLE 10–4

For an input *RC* network in a certain amplifier, $R_{in} = 1\ \text{k}\Omega$ and $C_1 = 1\ \mu\text{F}$. Neglect the source resistance.

(a) Determine the lower critical frequency.
(b) What is the attenuation of the *RC* network at the lower critical frequency?
(c) If the midrange voltage gain of the amplifier is 100, what is the gain at the lower critical frequency?

SOLUTION

(a) $$f_c = \frac{1}{2\pi R_{in}C_1} = \frac{1}{2\pi(1\ \text{k}\Omega)(1\ \mu\text{F})} = 159\ \text{Hz}$$

(b) At f_c, $X_{C1} = R_{in}$. Therefore, $V_{out}/V_{in} = 0.707$ from Equation (10–6).
(c) $A_v = 0.707A_{v(mid)} = 0.707(100) = 70.7$

PRACTICE EXERCISE 10–4

For an input *RC* network in a certain amplifier, $R_{in} = 10\ \text{k}\Omega$ and $C_1 = 2.2\ \mu\text{F}$.

(a) What is f_c?
(b) What is the attenuation at f_c?
(c) If $A_{v(mid)} = 500$, what is A_v at f_c?

GAIN ROLL-OFF AT LOW FREQUENCIES

As you have seen, the input *RC* network reduces the overall gain of an amplifier by 3 dB when the frequency is reduced to the critical value f_c. As the frequency continues to decrease, the overall gain also continues to decrease. This decrease in gain with frequency is called **roll-off.** *For each time the frequency is reduced by ten below* f_c, *there is a 20 dB reduction in gain.*

Let's take a frequency that is one-tenth of the critical frequency ($f = 0.1 f_c$). Since $X_{C1} = R_{in}$ at f_c, then $X_{C1} = 10R_{in}$ at $0.1 f_c$ because of the inverse relationship of X_{C1} and f. The attenuation of the *RC* network is therefore

$$\frac{V_{out}}{V_{in}} = \frac{R_{in}}{\sqrt{R_{in}^2 + X_{C1}^2}} = \frac{R_{in}}{\sqrt{R_{in}^2 + (10R_{in})^2}}$$
$$= \frac{R_{in}}{\sqrt{R_{in}^2 + 100R_{in}^2}} = \frac{R_{in}}{\sqrt{R_{in}^2(1 + 100)}}$$
$$= \frac{R_{in}}{R_{in}\sqrt{101}} = \frac{1}{\sqrt{101}} \cong \frac{1}{10} = 0.1$$

The dB attenuation is

$$20\log(V_{out}/V_{in}) = 20\log(0.1) = -20\ \text{dB}$$

dB/DECADE A ten-times change in frequency is called a **decade.** So, for the input *RC* network, the attenuation is reduced by 20 dB for each decade decrease in frequency below the critical frequency. This causes the overall gain to drop 20 dB per decade. For example, if the frequency is reduced to one-hundredth of f_c (a two-decade decrease), the amplifier gain drops 20 dB for each decade, giving a total decrease in gain of -20 dB + (-20 dB) = -40 dB. This is illustrated in Figure 10–13, which is a graph of dB gain versus frequency. It is the low-frequency response curve for the amplifier showing the effect of the input *RC* network on the voltage gain.

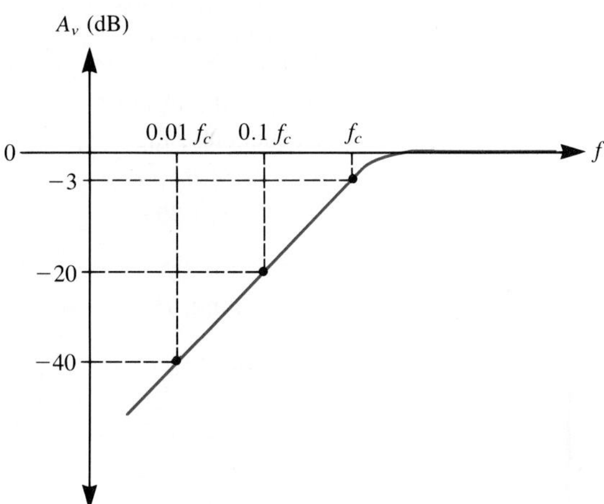

FIGURE 10–13
dB gain versus frequency for an *RC* lead network.

dB/OCTAVE Sometimes, the gain roll-off of an amplifier is expressed in dB/octave rather than dB/decade. An **octave** corresponds to a doubling or halving of the frequency. For example, an increase in frequency from 100 Hz to 200 Hz is an octave. Likewise, a decrease in frequency from 100 kHz to 50 kHz is also an octave. A rate of -20 dB/decade is approximately -6 dB/octave, a rate of -40 dB/decade is approximately -12 dB/octave, and so on.

EXAMPLE 10–5

The midrange voltage gain of a certain amplifier is 100. The input *RC* network has a lower critical frequency of 1 kHz. Determine the actual voltage gain at $f = 1$ kHz, $f = 100$ Hz, and $f = 10$ Hz.

SOLUTION

When $f = 1$ kHz, the gain is 3 dB less than at midrange. At -3 dB, the gain is reduced by a factor of 0.707.

$$A_v = (0.707)(100) = 70.7$$

When $f = 100$ Hz, $f = 0.1\, f_c$, so the gain is 17 dB less than at -3 dB. The gain at -20 dB is one-tenth of that at the midrange frequencies.

$$A_v = (0.1)(100) = 10$$

When $f = 10$ Hz, $f = 0.01\, f_c$, so the gain is 20 dB less than at -20 dB. The gain at -40 dB is one-tenth of that at -20 dB or one-hundredth that at the midrange frequencies.

$$A_v = (0.1)(10) = 1$$

PRACTICE EXERCISE 10–5

The midrange gain of an amplifier is 300. The lower critical frequency of the input *RC* network is 400 Hz. Determine the actual voltage gain at 400 Hz, 40 Hz, and 4 Hz.

PHASE SHIFT IN THE INPUT *RC* NETWORK

In addition to reducing the voltage gain, the input *RC* network also causes an increasing phase shift through the amplifier as the frequency decreases. At midrange frequencies, the phase shift through the *RC* network is approximately zero because $X_{C1} \cong 0$. At lower frequencies, higher values of X_{C1} cause a phase shift to be introduced, and the output voltage of the *RC* network leads the input voltage. As you learned in ac circuit theory, the phase angle is expressed as

$$\theta = \tan^{-1}\left(\frac{X_{C1}}{R_{in}}\right) \tag{10–9}$$

For midrange frequencies, $X_{C1} \cong 0$, so

$$\theta = \tan^{-1}\left(\frac{0}{R_{in}}\right) = \tan^{-1}(0) = 0°$$

At the critical frequency, $X_{C1} = R_{in}$, so

$$\theta = \tan^{-1}\left(\frac{R_{in}}{R_{in}}\right) = \tan^{-1}(1) = 45°$$

A decade below the critical frequency, $X_{C1} = 10R_{in}$, so

$$\theta = \tan^{-1}\left(\frac{10R_{in}}{R_{in}}\right) = \tan^{-1}(10) = 84.29°$$

This analysis shows that the phase shift through the input *RC* network approaches 90° as the frequency is reduced a few decades below midrange. A plot of phase angle versus frequency is shown in Figure 10–14. The net result is that the voltage at the base of the transistor leads the input signal voltage below midrange, as shown in Figure 10–15.

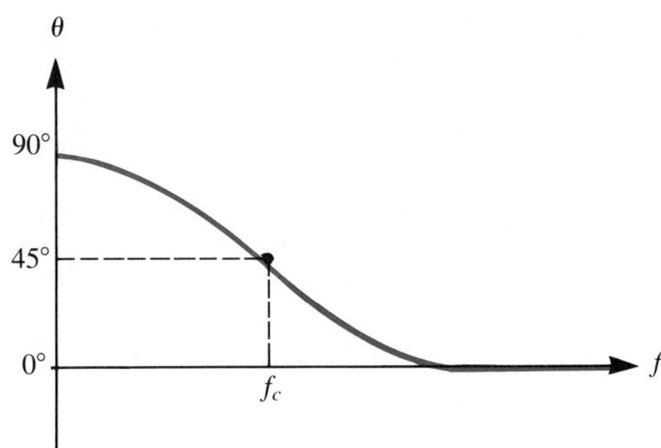

FIGURE 10–14
Phase angle versus frequency for an *RC* lead network.

FIGURE 10–15
Input *RC* network causes base voltage to lead the input voltage (below midrange) by an amount equal to the circuit phase angle.

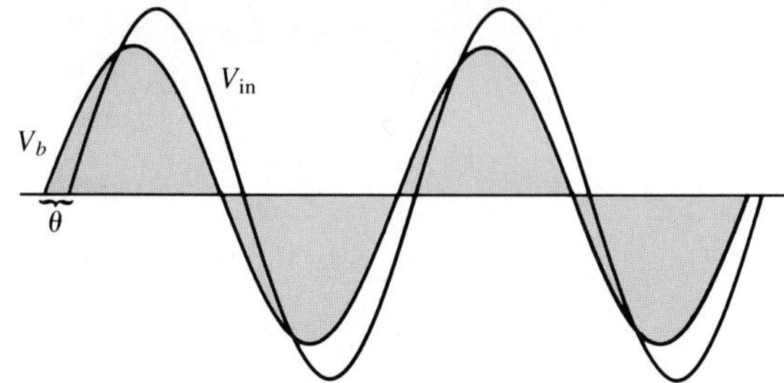

THE OUTPUT *RC* NETWORK

The second high-pass *RC* network in the amplifier of Figure 10–10 is formed by the coupling capacitor C_2 and the resistance looking in at the collector plus R_L, as shown in Figure 10–16(a). In determining the output resistance, the transistor is treated as an ideal current source (with infinite internal resistance), and the upper end of R_C is effectively ac ground, as shown in Figure 10–16(b). Therefore, Thevenizing the circuit to the left of the capacitor produces an equivalent voltage source and series resistance, as shown in Figure 10–16(c). The critical frequency of this *RC* network is

$$f_c = \frac{1}{2\pi(R_C + R_L)C_2} \tag{10–10}$$

The effect of the output *RC* network on the amplifier gain is similar to that of the input *RC* network. As the signal frequency decreases, X_{C2} increases. This causes less voltage across the load resistance because more is dropped across X_{C2}. The signal voltage is reduced by a factor of 0.707 when the frequency is reduced to the lower critical value, f_c. This corresponds to a 3 dB reduction in voltage gain.

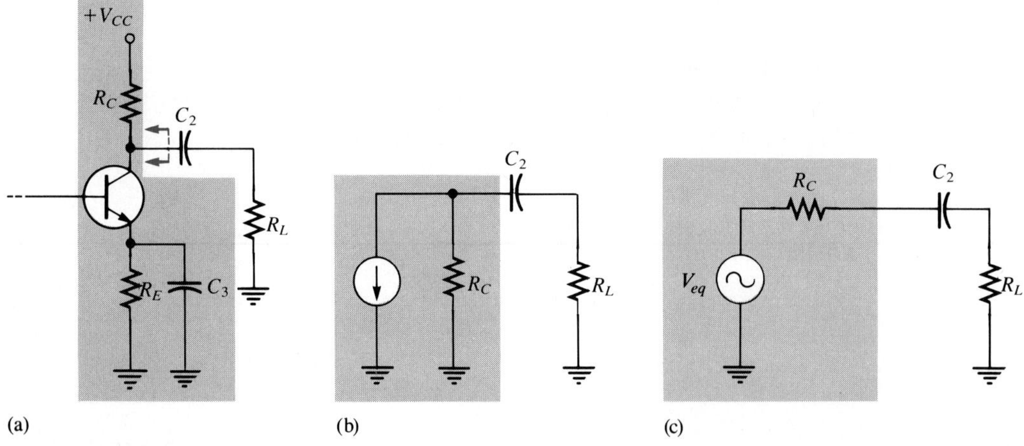

FIGURE 10–16
Development of low-frequency output *RC* network.

EXAMPLE 10–6

For an output *RC* network in a certain amplifier, $R_C = 10\ \text{k}\Omega$, $C_2 = 0.1\ \mu\text{F}$, and $R_L = 10\ \text{k}\Omega$.

(a) Determine the critical frequency.
(b) What is the attenuation of the *RC* network at the critical frequency?
(c) If the midrange voltage gain of the amplifier is 50, what is the gain at the critical frequency?

SOLUTION

(a) $f_c = \dfrac{1}{2\pi(R_C + R_L)C_2} = \dfrac{1}{2\pi(20\ \text{k}\Omega)(0.1\ \mu\text{F})} = 79.58\ \text{Hz}$

(b) For the midrange frequencies, $X_{C2} = 0$, thus the attenuation of the network is

$$\frac{V_{out}}{V_{in}} = \frac{R_L}{R_C + R_L} = \frac{10\ \text{k}\Omega}{20\ \text{k}\Omega} = 0.5$$

or, in dB, 20 log(0.5) = −6 dB. This shows that, in this case, the midrange gain is reduced by 6 dB because of the load resistor. At the critical frequency, $X_{C2} = R_C + R_L$.

$$\frac{V_{out}}{V_{in}} = \frac{R_L}{\sqrt{(R_C + R_L)^2 + X_{C2}^2}} = \frac{10\ \text{k}\Omega}{\sqrt{(20\ \text{k}\Omega)^2 + (20\ \text{k}\Omega)^2}} = 0.354$$

or, in dB, 20 log(0.354) = −9 dB. As you can see, the gain at f_c is 3 dB less than the gain at midrange.

(c) $A_v = 0.707A_{v(mid)} = 0.707(50) = 35.35$

PRACTICE EXERCISE 10–6

The output *RC* network in a certain amplifier has the following values: $R_C = 3.9\ \text{k}\Omega$, $C_2 = 1\ \mu\text{F}$, and $R_L = 8.2\ \text{k}\Omega$.

(a) Find the critical frequency.
(b) What is the attenuation at f_c?
(c) If $A_{v(mid)}$ is 100, what is the gain at f_c?

PHASE SHIFT IN THE OUTPUT *RC* NETWORK

The phase shift in the output *RC* network is

$$\theta = \tan^{-1}\left(\frac{X_{C2}}{R_C + R_L}\right) \qquad \textbf{(10–11)}$$

As previously discussed, $\theta \cong 0°$ in the midrange of frequencies and approaches 90° as the frequency approaches zero (X_{C2} approaches infinity). At f_c, the phase shift is 45°.

THE BYPASS NETWORK

The third *RC* network that affects the low-frequency gain of the amplifier in Figure 10–10 includes the bypass capacitor C_3. For midrange frequencies, it is assumed that $X_{C3} \cong 0$

and effectively shorts the emitter to ground so that the amplifier gain is R_c/r_e, as you already know. As the frequency is reduced, X_{C3} increases and no longer provides a sufficiently low reactance to effectively place the emitter at ac ground. This is illustrated in Figure 10–17. Because the impedance from emitter to ground increases, the gain decreases; recall the formula $A_v = R_c/(r_e + R_e)$. In this case, R_e becomes Z_e, an impedance formed by R_E in parallel with X_{C3}.

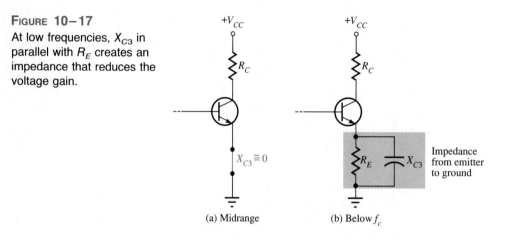

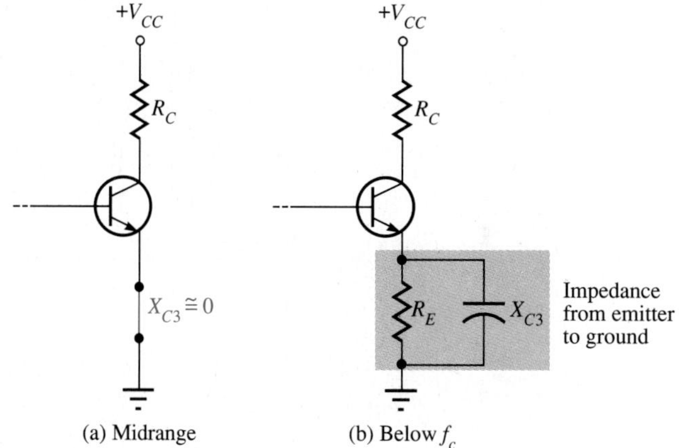

FIGURE 10–17
At low frequencies, X_{C3} in parallel with R_E creates an impedance that reduces the voltage gain.

The bypass *RC* network is formed by C_3 and the resistance looking in at the emitter $R_{in(emitter)}$, as shown in Figure 10–18(a). The resistance looking in at the emitter is derived as follows. First, Thevenin's theorem is applied looking from the base of the transistor toward the input V_{in}, as shown in Figure 10–18(b). This results in an equivalent resistance R_{TH} in series with the base, as shown in Figure 10–18(c). The output resistance looking in at the emitter is measured with the input grounded, as shown in Figure 10–18(d), and is expressed as follows.

$$R_{in(emitter)} = \frac{V_e}{I_e} + r_e \cong \frac{V_b}{\beta_{ac}I_b} + r_e = \frac{I_bR_{TH}}{\beta_{ac}I_b} + r_e$$

$$R_{in(emitter)} = \frac{R_{TH}}{\beta_{ac}} + r_e \qquad \textbf{(10–12)}$$

Looking from the capacitor C_3, $R_{TH}/\beta_{ac} + r_e$ is in parallel with R_E, as shown in Figure 10–18(e). Thevenizing again, we get the equivalent *RC* network shown in Figure 10–18(f). The critical frequency for the bypass network is

$$f_c = \frac{1}{2\pi[(r_e + R_{TH}/\beta_{ac})\|R_E]C_3} \qquad \textbf{(10–13)}$$

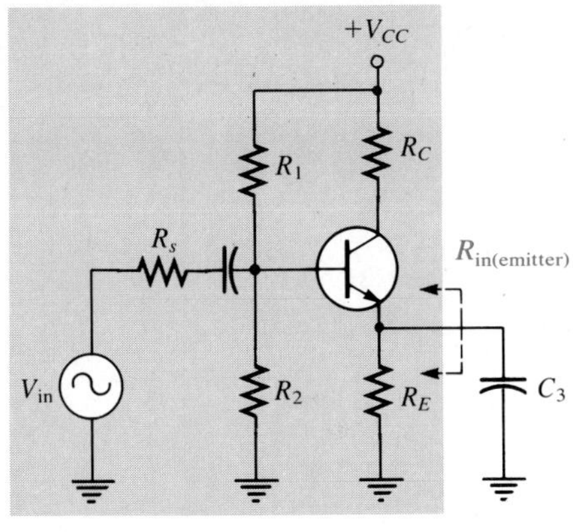

(a)

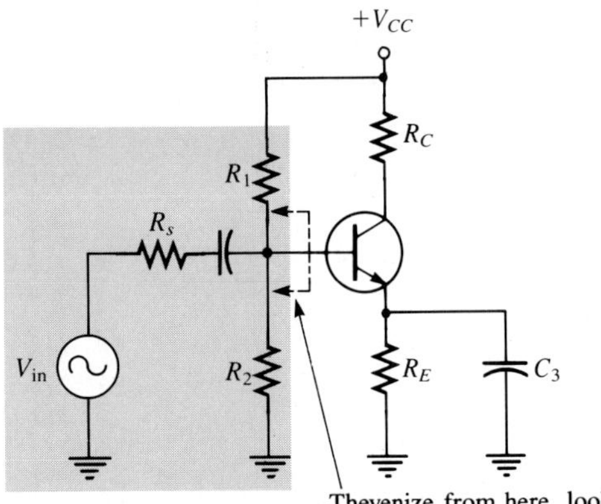

(b)

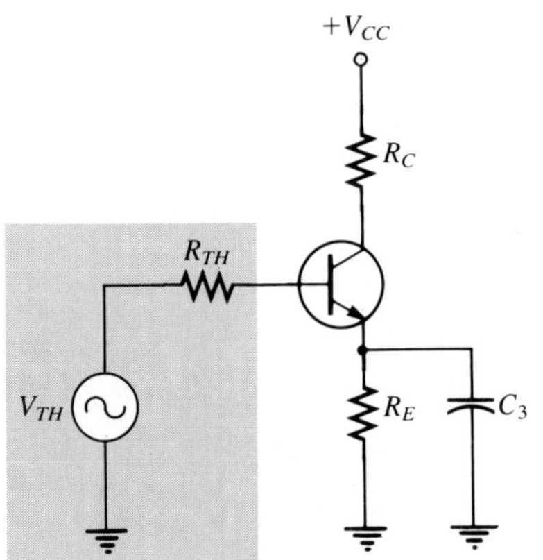

(c)

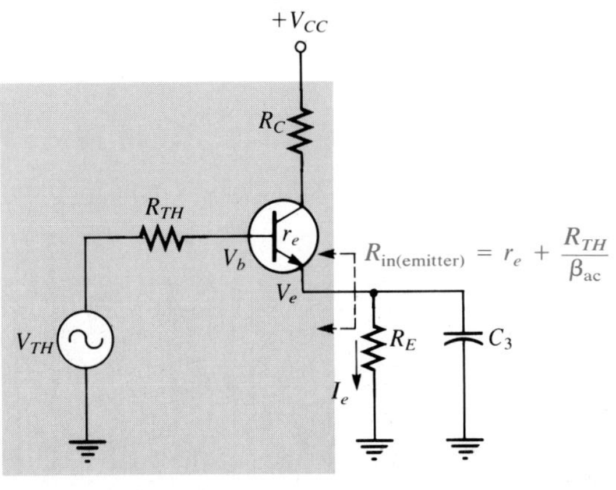

(d)

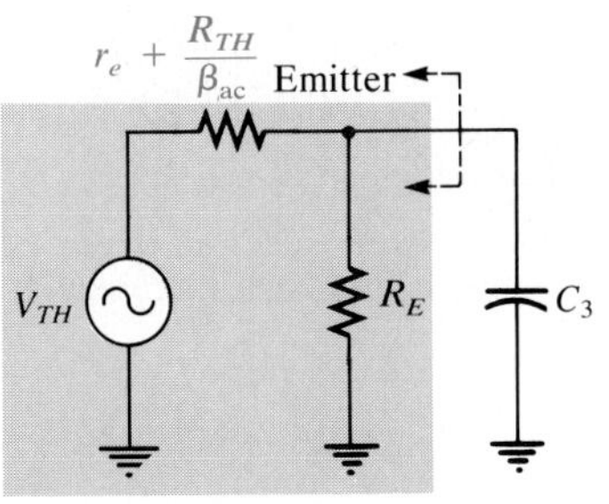

(e)

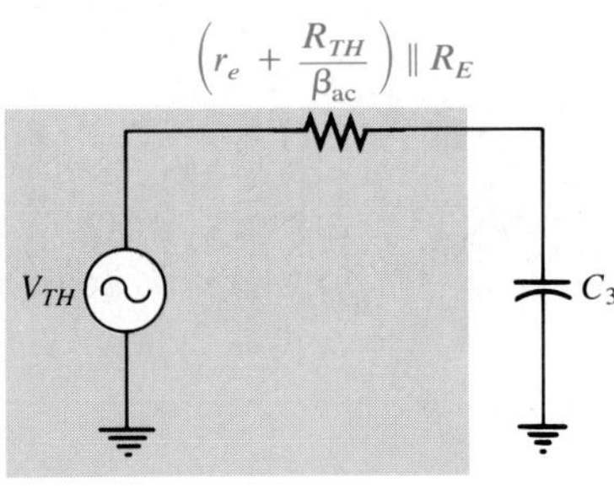

(f)

FIGURE 10–18
Development of the bypass *RC* network.

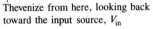

EXAMPLE 10–7 Determine the critical frequency of the bypass network in Figure 10–19 ($r_e = 17\ \Omega$).

FIGURE 10–19

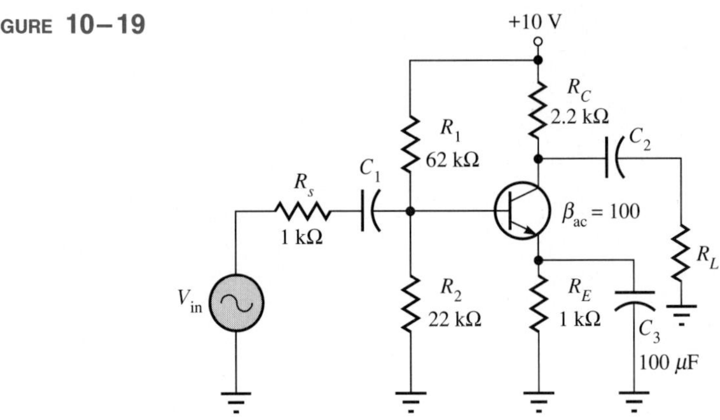

SOLUTION

Thevenize the base circuit (looking from the base toward the input source).

$$R_{TH} = R_1 \| R_2 \| R_s = 62\text{ k}\Omega \| 22\text{ k}\Omega \| 1\text{ k}\Omega \cong 942\ \Omega$$

The resistance looking in at the emitter is

$$R_{in(emitter)} = r_e + \frac{R_{TH}}{\beta_{ac}} = 17\ \Omega + 9.42\ \Omega = 26.42\ \Omega$$

The resistance of the equivalent RC network is $R_{in(emitter)} \| R_E$.

$$R_{in(emitter)} \| R_E = 26.42\ \Omega \| 1000\ \Omega = 25.74\ \Omega$$

The critical frequency of the bypass network is

$$f_c = \frac{1}{2\pi(R_{in(emitter)} \| R_E)C_3} = \frac{1}{2\pi(25.74\ \Omega)(100\ \mu\text{F})} = 61.8\text{ Hz}$$

PRACTICE EXERCISE 10–7

In Figure 10–19, the source resistance, R_s, is 50 Ω, and the transistor ac beta is 150. Determine the critical frequency of the bypass network.

BODE PLOT

A plot of dB gain versus frequency on semilog graph paper is called a **Bode plot.** A generalized Bode plot for an RC network like that shown in Figure 10–20(a) appears in part (b) of the figure. The ideal response curve is drawn with a solid line. Notice that it is flat (0 dB) down to the critical frequency, at which point the gain drops at −20 dB/decade as shown. Above f_c are the midrange frequencies. The actual response curve is shown with the dashed line. Notice that it decreases gradually in midrange and is down to −3 dB at the critical frequency. Often, the ideal response is used to simplify

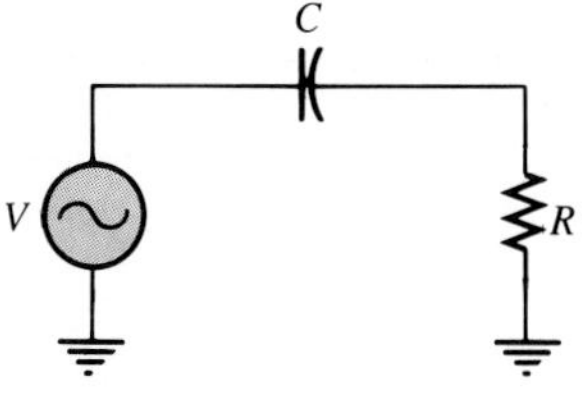

(a) *RC* lead network

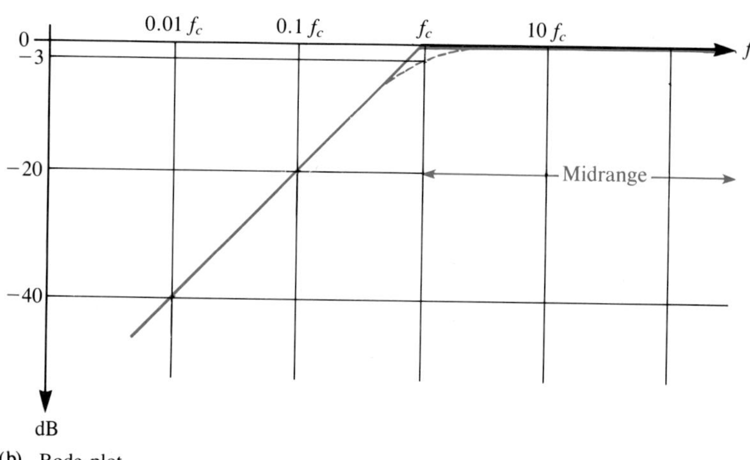

(b) Bode plot

FIGURE 10–20 *RC* network and its low-frequency response.

amplifier analysis. As previously mentioned, the critical frequency at which the curve "breaks" into a −20 dB/decade drop is often called the *lower break frequency*.

TOTAL LOW-FREQUENCY RESPONSE

Now that we have individually examined the three high-pass *RC* networks that affect the amplifier's gain at low frequencies, we will look at the combined effect of all three networks. Each network has a critical frequency determined by the *RC* values. The critical frequencies are not necessarily all equal. If one of the *RC* networks has a critical (break) frequency higher than the other two, then it is the *dominant* network. The dominant network determines the frequency at which the overall gain of the amplifier begins to drop at −20 dB/decade. The other networks cause an additional −20 dB/decade roll-off below their respective critical (break) frequencies.

To get a better picture of what happens at low frequencies, refer to the Bode plot in Figure 10–21, which shows the superimposed ideal responses for the three *RC* networks (dashed lines). In this example, each *RC* network has a different critical frequency. The input *RC* network is dominant (highest f_c), and the bypass network has the lowest f_c. The overall response is the solid color line.

Here is what happens. As the frequency is reduced from midrange, the first "break point" occurs at $f_{c(\text{input})}$ and the gain begins to drop at −20 dB/decade. This constant roll-off rate continues until $f_{c(\text{output})}$ is reached. At this break point, the output *RC* network adds another −20 dB/decade to make a total roll-off of −40 dB/decade. This constant roll-off continues until $f_{c(\text{bypass})}$ is reached. At this break point, the bypass *RC* network adds still another −20 dB/decade, making the gain roll-off at −60 dB/decade.

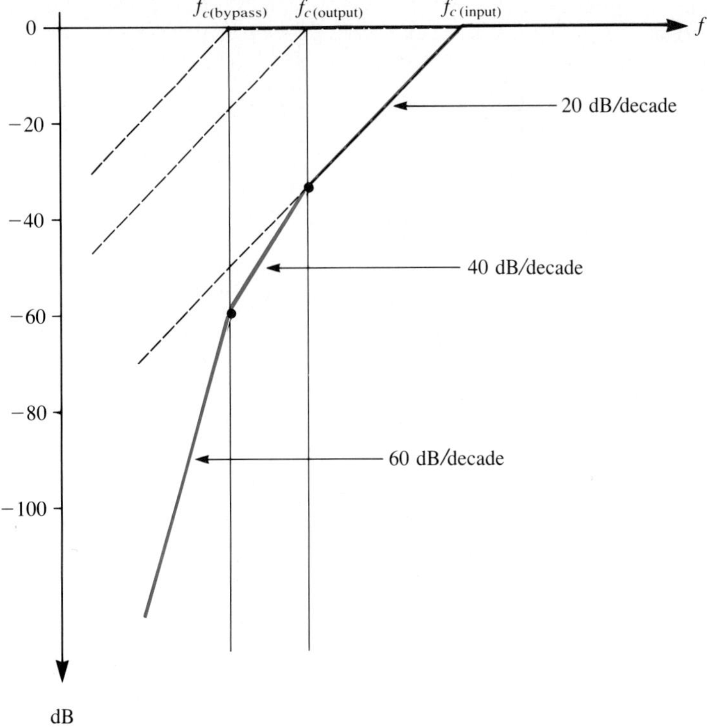

FIGURE 10–21
Composite Bode plot for three low-frequency *RC* networks with different critical frequencies.

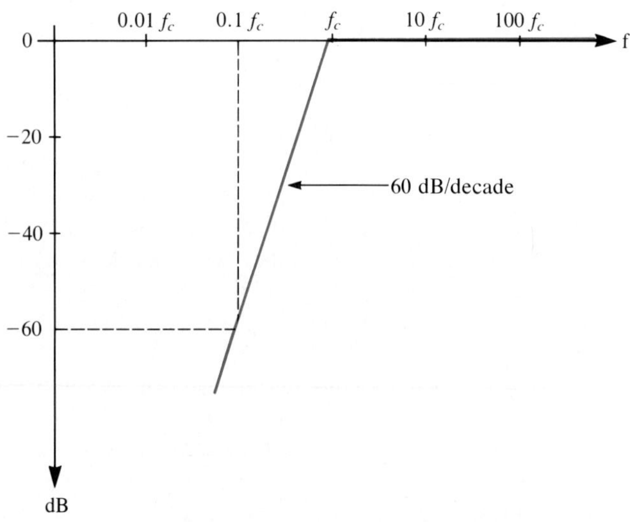

FIGURE 10–22
Composite Bode plot where all three networks have same f_c.

If all three *RC* networks have the same critical frequency, the response curve has one break point at that value of f_c, and the gain rolls off at -60 dB/decade, as shown in Figure 10–22. Keep in mind that the ideal response curves have been used. Actually, the midrange gain does not extend down to the dominant critical frequency but is really at -9 dB below the midrange gain at that point (-3 dB for each *RC* network).

EXAMPLE 10–8 Determine the total low-frequency response of the amplifier in Figure 10–23. $\beta_{ac} = 100$ and $r_e = 13.9\ \Omega$.

FIGURE 10–23

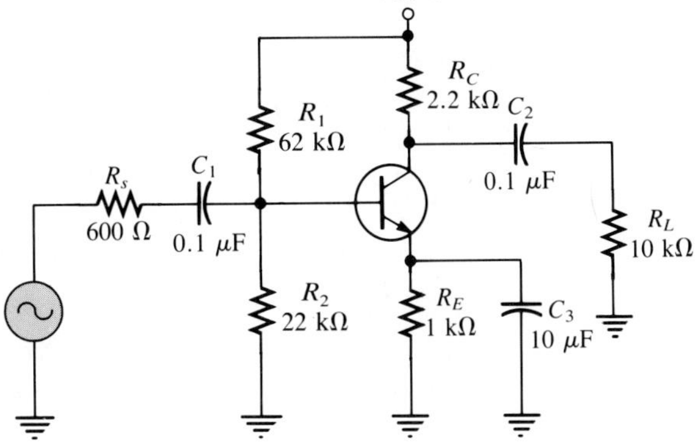

SOLUTION

Each RC network is analyzed to determine its critical frequency. For the input RC network with the source resistance R_s taken into account,

$$R_{in} = R_1 \| R_2 \| \beta_{ac} r_e = 62\ \text{k}\Omega \| 22\ \text{k}\Omega \| 1.39\ \text{k}\Omega = 1.28\ \text{k}\Omega$$

$$f_{c(input)} = \frac{1}{2\pi(R_s + R_{in})C_1}$$

$$= \frac{1}{2\pi(600\ \Omega + 1.28\ \text{k}\Omega)(0.1\ \mu\text{F})}$$

$$= 847\ \text{Hz}$$

For the bypass RC network,

$$R_{TH} = R_1 \| R_2 \| R_s = 62\ \text{k}\Omega \| 22\ \text{k}\Omega \| 600\ \Omega \cong 600\ \Omega$$

$$R_{in(emitter)} = \frac{R_{TH}}{\beta_{ac}} + r_e = \frac{600\ \Omega}{100} + 13.9\ \Omega = 19.9\ \Omega$$

$$f_{c(bypass)} = \frac{1}{2\pi(R_{in(emitter)} \| R_E)C_3}$$

$$= \frac{1}{2\pi(19.9\ \Omega \| 1\ \text{k}\Omega)(10\ \mu\text{F})}$$

$$= \frac{1}{2\pi(19.5\ \Omega)(10\ \mu\text{F})} = 816\ \text{Hz}$$

For the output RC network,

$$f_{c(output)} = \frac{1}{2\pi(R_C + R_L)C_2}$$

$$= \frac{1}{2\pi(2.2\ \text{k}\Omega + 10\ \text{k}\Omega)(0.1\ \mu\text{F})}$$

$$= 130.5\ \text{Hz}$$

The above analysis shows that the input network produces the dominant (highest) lower critical frequency. The midrange gain of the amplifier is

$$A_{v(\text{mid})} = \frac{R_c}{r_e} = \frac{2.2\ \text{k}\Omega \| 10\ \text{k}\Omega}{13.9\ \Omega} = 129.7$$

The midrange attenuation of the input network is

$$\frac{R_1 \| R_2 \| \beta_{ac} r_e}{R_s + R_1 \| R_2 \| \beta_{ac} r_e} = \frac{62\ \text{k}\Omega \| 22\ \text{k}\Omega \| 1390\ \Omega}{600\ \Omega + 62\ \text{k}\Omega \| 22\ \text{k}\Omega \| 1390\ \Omega}$$

$$= \frac{1280}{1880} = 0.68$$

The overall gain is

$$A'_{v(\text{mid})} = 0.68(129.7) = 88.2$$
$$A'_{v(\text{mid})} = 20 \log(88.2) = 38.9\ \text{dB}$$

The Bode plot of the low-frequency response of this amplifier is shown in Figure 10–24.

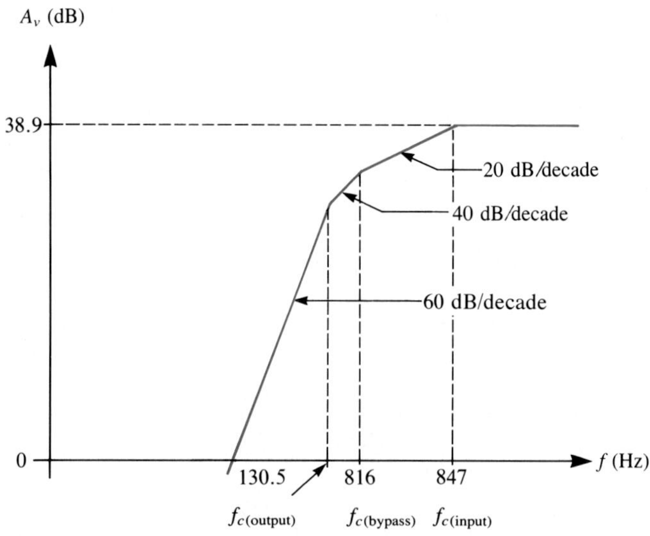

FIGURE 10–24
Bode plot for the low-frequency response of the amplifier in Figure 10–23.

PRACTICE EXERCISE 10–8

If the overall gain of the amplifier is reduced, how are the critical frequencies affected?

LOW-FREQUENCY RESPONSE OF DIRECT-COUPLED AMPLIFIERS

Recall from Chapter 6 that direct-coupled amplifiers have no coupling or bypass capacitors, which allows the frequency response to extend down to dc (0 Hz). Direct-coupled amplifiers are popular and commonly used in linear ICs because of their simplicity and because they can effectively amplify signals with frequencies less than

10 Hz (the frequency to which capacitively coupled amplifiers are generally limited because of the size of the capacitors required).

A direct-coupled amplifier has the same gain for a dc input voltage as it does for a signal with a midrange frequency. The gain of a direct-coupled amplifier such as the one in Figure 10–25 is determined by the ratio R_C/R_E and remains constant at lower frequencies because there are no frequency-dependent components.

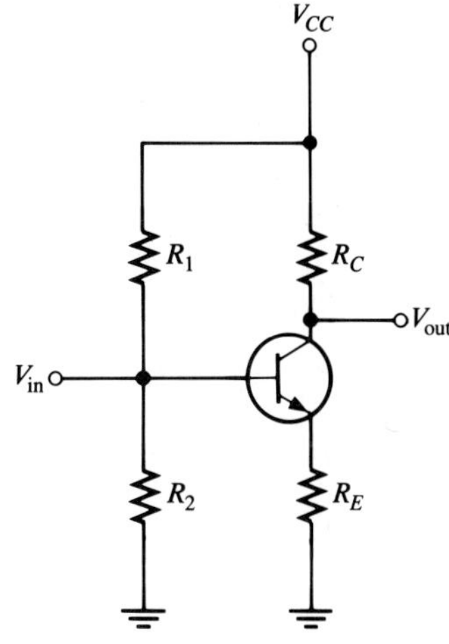

FIGURE 10–25
Basic direct-coupled amplifier stage.

10–3 REVIEW QUESTIONS

1. A certain amplifier exhibits three critical frequencies in its low-frequency response: f_{c1} = 130 Hz, f_{c2} = 167 Hz, and f_{c3} = 75 Hz. Which is the dominant critical frequency?
2. If the midrange gain of the amplifier in Question 1 is 50 dB, what is the gain at the dominant f_c?
3. A certain *RC* network has an f_c = 235 Hz, above which the attenuation is 0 dB. What is the dB attenuation at 23.5 Hz?

10–4 HIGH-FREQUENCY AMPLIFIER RESPONSE

You have seen how the coupling and bypass capacitors affect the voltage gain of an amplifier at lower frequencies where the capacitive reactances are significant. In the midrange of the amplifier, the effects of the capacitors are minimal and can be neglected. If the frequency is increased sufficiently, a point is reached where the transistor's internal capacitances begin to have a significant effect on the gain. (These capacitances were discussed in Section 10–1. In this section, we use a bipolar transistor amplifier to illustrate the principles, but the approach for FET amplifiers is similar. The basic differences are the specifications of the internal capacitances and the input resistance. This coverage applies to both capacitively coupled and direct-coupled amplifiers.

A high-frequency ac equivalent circuit for the amplifier in Figure 10–26(a) is shown in Figure 10–26(b). Notice that the coupling and bypass capacitors are treated effectively as shorts, and the three internal capacitances, C_{be}, C_{bc}, and C_{ce} which are significant only at high frequencies, appear in the diagram. As previously mentioned, C_{be} is sometimes called the input capacitance and C_{bc} the output capacitance. C_{be} (C_{ib}) is specified on data sheets at a certain value of V_{BE}. For example, a 2N2222A has a C_{be} of 25 pF at V_{BE} = 0.5 V dc, $I_C = 0$, and $f = 100$ kHz. Also, C_{bc} (C_{ob}) is specified at a certain value of V_{CB}. The 2N2222A has a maximum C_{bc} of 8 pF at V_{CB} = 10 V dc. C_{ce} is usually not specified on data sheets; and since it is very small, we will neglect it.

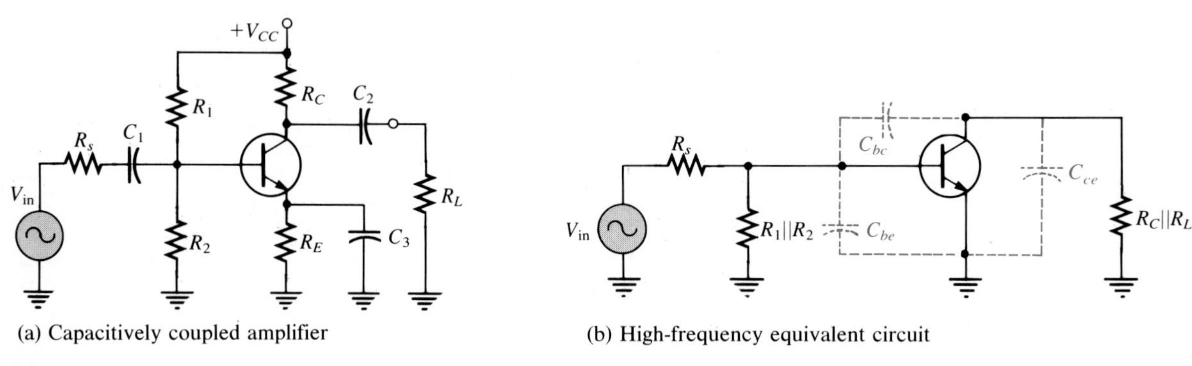

(a) Capacitively coupled amplifier

(b) High-frequency equivalent circuit

FIGURE 10–26
Capacitively coupled amplifier and high-frequency equivalent circuit.

USING MILLER'S THEOREM FOR HIGH-FREQUENCY ANALYSIS

By applying Miller's theorem to the circuit in Figure 10–26(b) and using the midrange gain, we can get a circuit that can be analyzed for high-frequency response. Looking in from the signal source, the capacitance C_{bc} effectively appears in the Miller input capacitance from base to ground.

$$C_{in(Miller)} = C_{bc}(A_v + 1) \tag{10–14}$$

C_{be} simply appears as a capacitance to ac ground, as shown in Figure 10–27, in parallel with $C_{in(Miller)}$. Looking in at the collector, C_{bc} effectively appears in the Miller output capacitance from collector to ground. As shown in Figure 10–27, it appears in parallel with R_c.

$$C_{out(Miller)} = C_{bc}\left(\frac{A_v + 1}{A_v}\right) \tag{10–15}$$

These two Miller capacitances create a high-frequency input RC network and a high-frequency output RC network. These two networks differ from the low-frequency input and output lead networks in that the capacitances go to ground and therefore act as lag networks (low-pass filters).

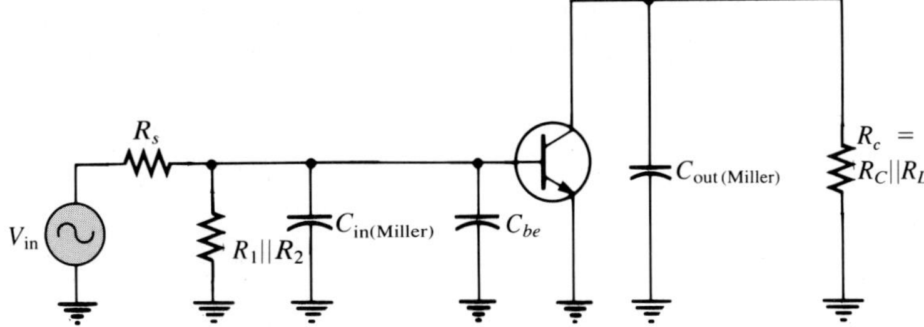

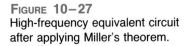

FIGURE 10–27
High-frequency equivalent circuit after applying Miller's theorem.

THE INPUT *RC* NETWORK

At high frequencies, the input network appears as in Figure 10–28(a), where $\beta_{ac}r_e$ is the input resistance at the base of the transistor because the bypass capacitor effectively shorts the emitter to ground. By combining C_{be} and $C_{in(Miller)}$ in parallel and repositioning, we get the simplified network shown in Figure 10–28(b). Next, by Thevenizing the circuit to the left of the capacitor, as indicated, the input *RC* network is reduced to the equivalent form in Figure 10–28(c).

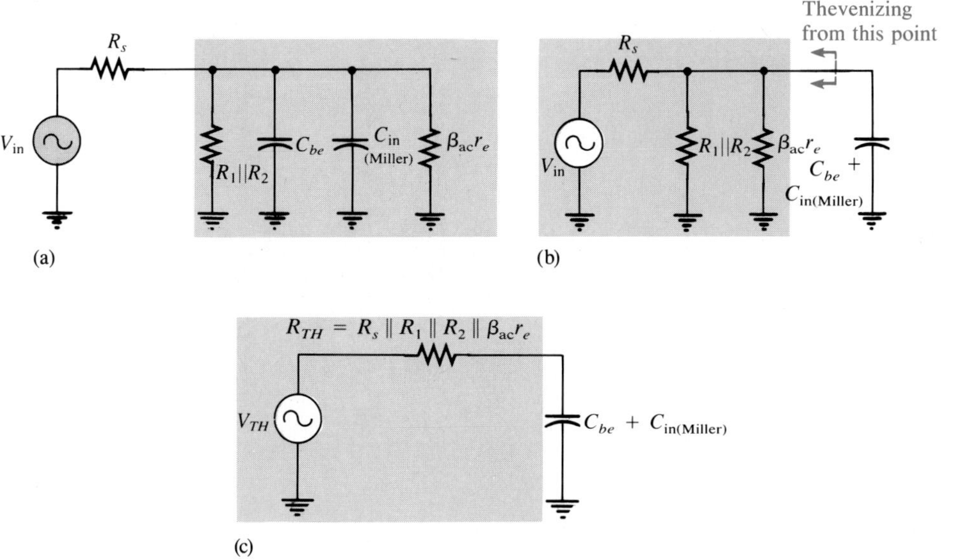

FIGURE 10–28
Development of the high-frequency input *RC* network.

As the frequency increases, the capacitive reactance becomes smaller. This causes the signal voltage at the base to decrease; thus the amplifier's gain decreases. The reason for this is that the capacitance and resistance act as a voltage divider and, as the frequency increases, more voltage is dropped across the resistance and less across the capacitance. At the critical frequency, the gain is 3 dB less than its midrange value. Just as with the low-frequency response, the critical frequency f_c is the frequency at which the capacitive reactance is equal to the resistance.

$$X_{C_T} = R_s \| R_1 \| R_2 \| \beta_{ac} r_e \qquad (10\text{–}16)$$

Therefore,

$$\frac{1}{2\pi f_c C_T} = R_s \| R_1 \| R_2 \| \beta_{ac} r_e$$

and

$$f_c = \frac{1}{2\pi (R_s \| R_1 \| R_2 \| \beta_{ac} r_e) C_T} \qquad (10\text{–}17)$$

where R_s is the resistance of the signal source and $C_T = C_{be} + C_{in(Miller)}$. As the frequency goes above f_c, the input RC network rolls off the gain at a rate of -20 dB/decade just as in the low-frequency response.

EXAMPLE 10–9

Derive the high-frequency input RC network for the amplifier in Figure 10–29. Also determine the critical frequency. The transistor's data sheet provides the following: $\beta_{ac} = 125$, $C_{be} = 20$ pF, and $C_{bc} = 3$ pF.

FIGURE 10–29

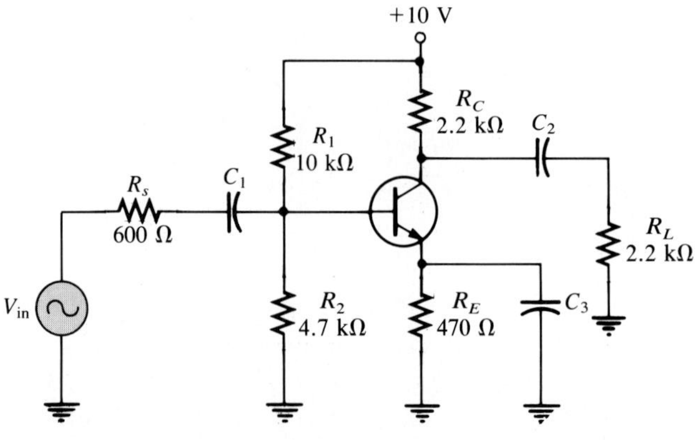

SOLUTION

The first thing to do is find r_e.

$$V_B = \left(\frac{4.7 \text{ k}\Omega}{14.7 \text{ k}\Omega}\right) 10 \text{ V} = 3.2 \text{ V}$$

$$V_E = V_B - 0.7 \text{ V} = 2.5 \text{ V}$$

$$I_E = \frac{V_E}{R_E} = \frac{2.5 \text{ V}}{470 \ \Omega} = 5.3 \text{ mA}$$

$$r_e = \frac{25 \text{ mV}}{I_E} = 4.72 \ \Omega$$

The resistance of the input network is

$$R_s \| R_1 \| R_2 \| \beta_{ac} r_e = 600\ \Omega \| 10\ \text{k}\Omega \| 4.7\ \text{k}\Omega \| 125(4.72\ \Omega) = 272\ \Omega$$

Next, in order to determine the capacitance, the midrange gain of the amplifier must be determined so that Miller's theorem can be applied.

$$A_{v(\text{mid})} = \frac{R_c}{r_e} = \frac{1.1\ \text{k}\Omega}{4.72\ \Omega} = 233$$

Applying Miller's theorem, we get

$$C_{\text{in(Miller)}} = C_{bc}(A_{v(\text{mid})} + 1) = (3\ \text{pF})(234) = 702\ \text{pF}$$

The total input capacitance is $C_{\text{in(Miller)}}$ in parallel with C_{be}.

$$C_T = C_{\text{in(Miller)}} + C_{be} = 702\ \text{pF} + 20\ \text{pF} = 722\ \text{pF}$$

The resulting high-frequency input RC network is shown in Figure 10–30. The critical frequency is

$$f_c = \frac{1}{2\pi(272\ \Omega)(722\ \text{pF})} = 810.4\ \text{kHz}$$

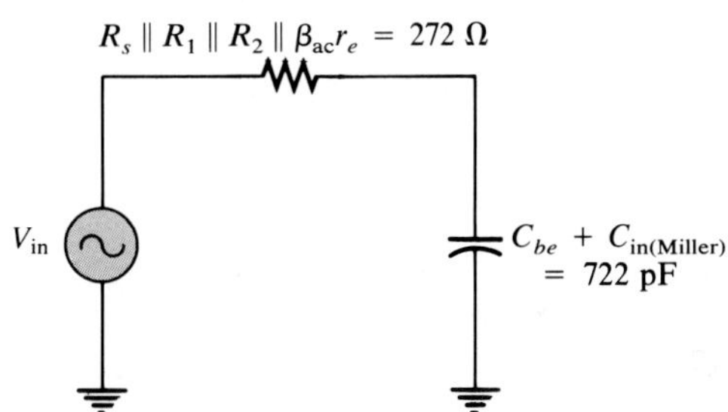

FIGURE 10–30
High-frequency input RC network for the amplifier in Figure 10–29.

PRACTICE EXERCISE 10–9

Determine the high-frequency input RC network for Figure 10–29 and find its critical frequency if a transistor with the following specifications is used: $\beta_{ac} = 75$, $C_{be} = 15$ pF, $C_{bc} = 2$ pF.

PHASE SHIFT OF THE INPUT *RC* NETWORK

Because the output voltage of a high-frequency input RC network is across the capacitor, it acts as a lag network. That is, the output of the network lags the input. The phase angle is expressed as

$$\phi = \tan^{-1}\left(\frac{R_s \| R_1 \| R_2 \| \beta_{ac} r_e}{X_{C_T}}\right) \tag{10–18}$$

At the critical frequency, the phase angle is 45° with the signal voltage at the base of the transistor lagging the input signal. As the frequency increases above f_c, the phase angle increases above 45° and approaches 90° when the frequency is sufficiently high.

THE OUTPUT *RC* NETWORK

The high-frequency output *RC* network is formed by the Miller output capacitance and the resistance looking in at the collector, as shown in Figure 10–31(a). In determining the output resistance, the transistor is treated as a current source (open) and one end of R_C is effectively ac ground, as shown in Figure 10–31(b). By rearranging the position of the capacitance in the diagram and Thevenizing the circuit to the left, as shown in Figure 10–31(c), we get the equivalent circuit in Figure 10–31(d). The equivalent output *RC* network consists of a resistance equal to R_C and R_L in parallel and a capacitance as follows.

$$C_{out(Miller)} = C_{bc}\left(\frac{A_v + 1}{A_v}\right)$$

If the voltage gain is at least 10, this formula can be approximated as

$$C_{out(Miller)} \cong C_{bc} \qquad \textbf{(10–19)}$$

The critical frequency is determined with the following equation, where $R_c = R_C \| R_L$.

$$f_c = \frac{1}{2\pi R_c C_{out(Miller)}} \qquad \textbf{(10–20)}$$

Just as in the input *RC* network, the output network reduces the gain by 3 dB at the critical frequency. When the frequency goes above the critical value, the gain drops at a

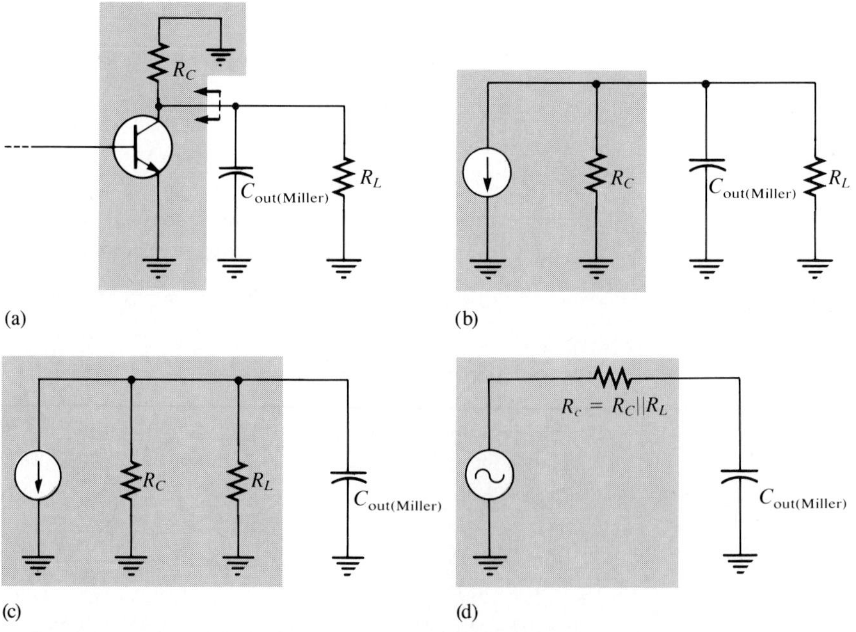

FIGURE 10–31
Development of high-frequency output *RC* network.

−20 dB/decade rate. The phase shift introduced by the output *RC* network is

$$\phi = \tan^{-1}\left(\frac{R_c}{X_{C_{\text{out(Miller)}}}}\right) \tag{10–21}$$

EXAMPLE 10–10

Determine the critical frequency of the amplifier in Example 10–9 (Figure 10–29) due to the output *RC* network.

SOLUTION

The Miller output capacitance is as follows.

$$C_{\text{out(Miller)}} = C_{bc}\left(\frac{A_v + 1}{A_v}\right) = (3\text{ pF})\left(\frac{233 + 1}{233}\right) \cong 3\text{ pF}$$

The equivalent resistance is

$$R_c = R_C \| R_L = 2.2\text{ k}\Omega \| 2.2\text{ k}\Omega = 1.1\text{ k}\Omega$$

The equivalent network is shown in Figure 10–32, and the critical frequency is as follows ($C_{\text{out(Miller)}} \cong C_{bc}$).

$$f_c = \frac{1}{2\pi R_c C_{bc}} = \frac{1}{2\pi(1.1\text{ k}\Omega)(3\text{ pF})} = 48.2\text{ MHz}$$

FIGURE 10–32

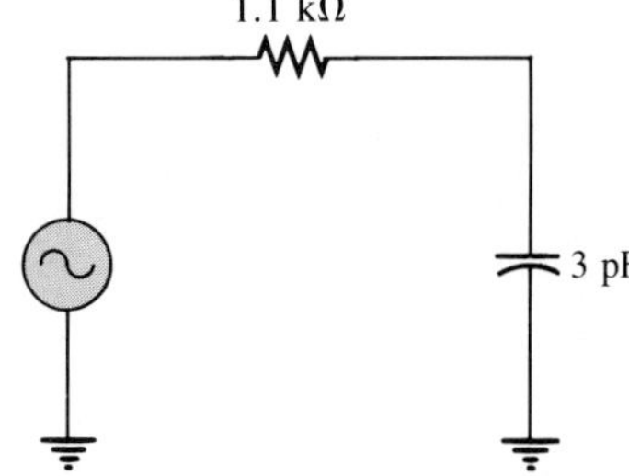

PRACTICE EXERCISE 10–10

Although we have been neglecting C_{ce}, let's assume that its value is given as 1 pF in this case and we want to take it into account. What is f_c?

TOTAL HIGH-FREQUENCY RESPONSE

As you have seen, two *RC* networks created by the internal transistor capacitances influence the high-frequency response of an amplifier. As the frequency increases and reaches the high end of its midrange values, one of the *RC* networks will cause the amplifiers' gain to begin dropping off. The frequency at which this occurs is the dominant critical frequency; it is the lower of the two critical frequencies. An ideal high-frequency Bode plot is shown in Figure 10–33(a). It shows the first break point at $f_{c(\text{input})}$ where the voltage gain begins to roll off at −20 dB/decade. At $f_{c(\text{output})}$, the gain begins dropping at

−40 dB/decade because each *RC* network is providing a −20 dB/decade roll-off. Figure 10–33(b) shows a nonideal Bode plot where the gain is actually −3 dB below midrange at $f_{c(\text{input})}$. Other possibilities are that the output *RC* network is dominant or that both networks have the same critical frequency.

FIGURE 10–33
High-frequency Bode plots.

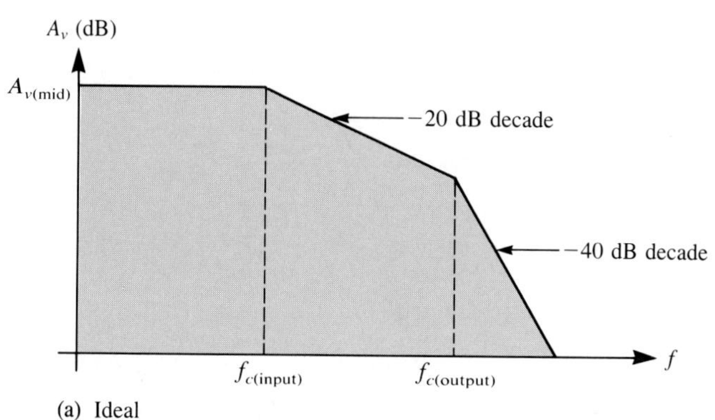

(a) Ideal

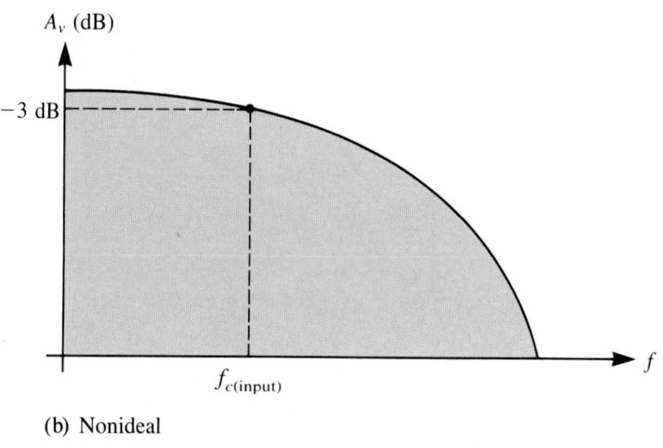

(b) Nonideal

10–4 REVIEW QUESTIONS

1. What determines the high-frequency response of an amplifier?
2. If an amplifier has a midrange voltage gain of 80 and the transistor's C_{bc} is 4 pF, what is the Miller input capacitance?
3. A certain amplifier has $f_{c(\text{input})} = 3.5$ MHz and $f_{c(\text{output})} = 8.2$ MHz. Which network dominates the high-frequency response?

10–5 TOTAL AMPLIFIER RESPONSE

In the previous sections, you learned how each RC network in an amplifier affects the frequency response. In this section, we will bring these concepts together and examine the total response of typical amplifiers and the specifications relating to their performance.

Figure 10–34(b) shows a generalized response curve (Bode plot) for an amplifier of the type shown in Figure 10–34(a). As previously discussed, the three break points at the lower critical frequencies—f_{c1}, f_{c2}, and f_{c3}—are produced by the three low-frequency *RC* lead networks formed by the coupling and bypass capacitors. The break points at the upper critical frequencies, f_{c4} and f_{c5}, are produced by the two high-frequency *RC* lag networks formed by the transistor's internal capacitances.

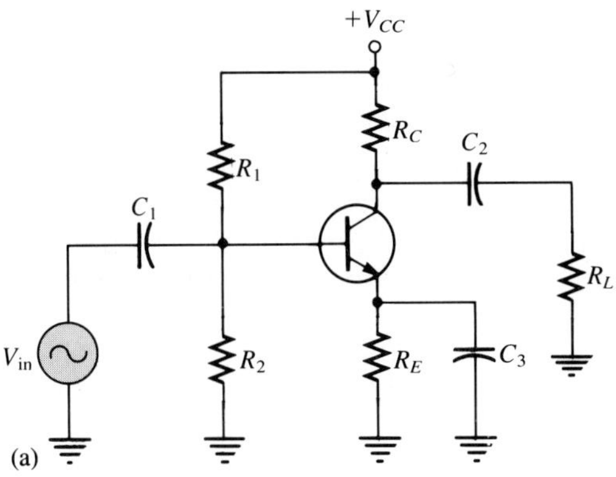

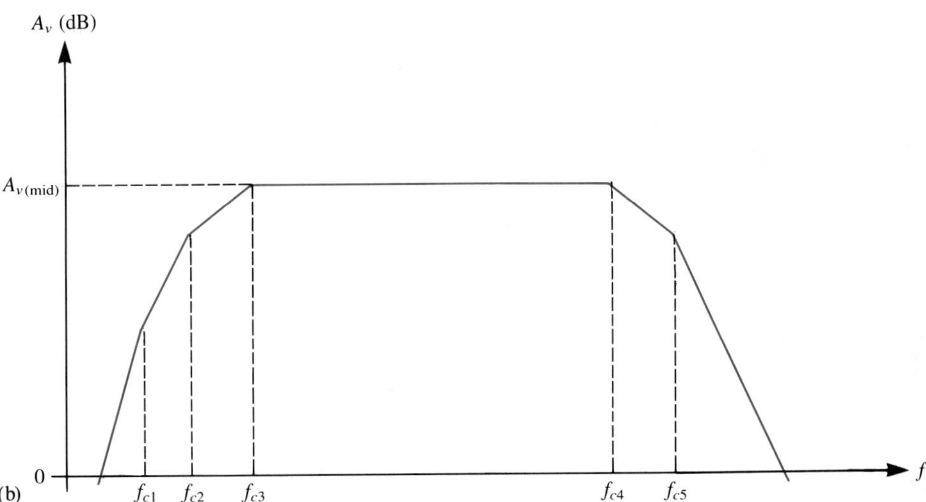

FIGURE 10–34
Amplifier and its generalized ideal response curve (Bode plot).

Of particular interest are the two dominant critical frequencies f_{c3} and f_{c4} in Figure 10–34(b). These two frequencies are where the gain of the amplifier is 3 dB below its midrange value. From now on, these frequencies are referred to as the *lower critical frequency* (f_{cl}) and the *upper critical frequency* (f_{ch}).

BANDWIDTH

An amplifier normally operates with frequencies between f_{cl} and f_{ch}. As you know, when the input signal frequency is at f_{cl} or f_{ch}, the output signal voltage level is 70.7 percent of its midrange value or -3 dB. If the signal frequency drops below f_{cl}, the gain and thus the output signal level drops at 20 dB/decade until the next critical frequency is reached. The same is true when the signal frequency goes above f_{ch}.

The range (band) of frequencies lying between f_{cl} and f_{ch} constitute the **bandwidth** of the amplifier, as illustrated in Figure 10–35. Only the dominant critical frequencies appear in the response curve because they determine the bandwidth. Also, sometimes the other critical frequencies are far enough away from the dominant frequencies that they play no significant role in the total amplifier response and can be neglected for simplicity. The bandwidth is expressed as

$$BW = f_{ch} - f_{cl} \tag{10–22}$$

Ideally, all signal frequencies lying in an amplifier's bandwidth are amplified equally. For example, if a 10 mV rms signal is applied to an amplifier with a gain of 20, it is amplified to 200 mV rms. Likewise, a 50 mV rms signal is amplified to 1 V rms.

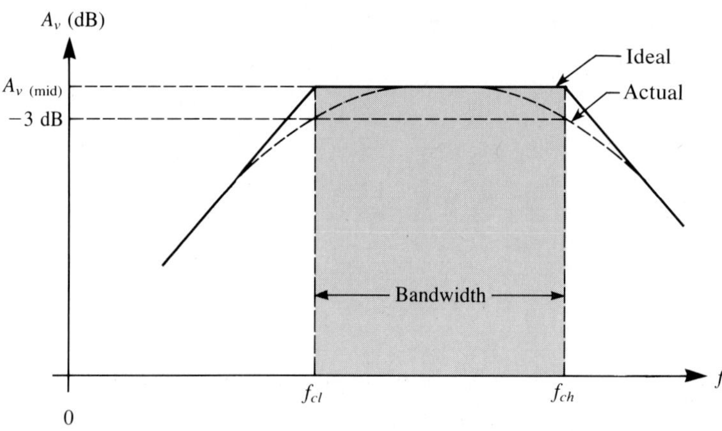

FIGURE 10–35 Response curve illustrating bandwidth of an amplifier.

EXAMPLE 10–11 What is the bandwidth of an amplifier having an f_{cl} of 200 Hz and an f_{ch} of 2 kHz?

SOLUTION

$$BW = f_{ch} - f_{cl} = 2000 \text{ Hz} - 200 \text{ Hz} = 1800 \text{ Hz}$$

Notice that bandwidth has the unit of hertz.

PRACTICE EXERCISE 10–11

If f_{cl} is increased, does the bandwidth increase or decrease? If f_{ch} is increased, does the bandwidth increase or decrease?

GAIN-BANDWIDTH PRODUCT

A characteristic of amplifiers whereby the product of the voltage gain and the bandwidth is always a constant is called the **gain-bandwidth product.** Let's assume that the lower critical frequency of a particular amplifier is much less than the upper critical frequency.

$$f_{cl} << f_{ch}$$

The bandwidth can then be approximated as

$$BW = f_{ch} - f_{cl} \cong f_{ch}$$

The simplified Bode plot for this condition is shown in Figure 10–36. Notice that f_{cl} is neglected, and the bandwidth equals f_{ch}. Beginning at f_{ch}, the gain rolls off until unity gain (0 dB) is reached. The frequency at which the amplifier's gain is one is called the *unity-gain frequency*, f_T. The significance of f_T is that it always equals the product of the midrange voltage gain times the bandwidth and is a constant for a given transistor.

$$f_T = A_{v(\text{mid})}BW \quad (10\text{–}23)$$

FIGURE 10–36
Simplified response curve where f_{cl} is negligible (assumed to be zero) compared to f_{ch}.

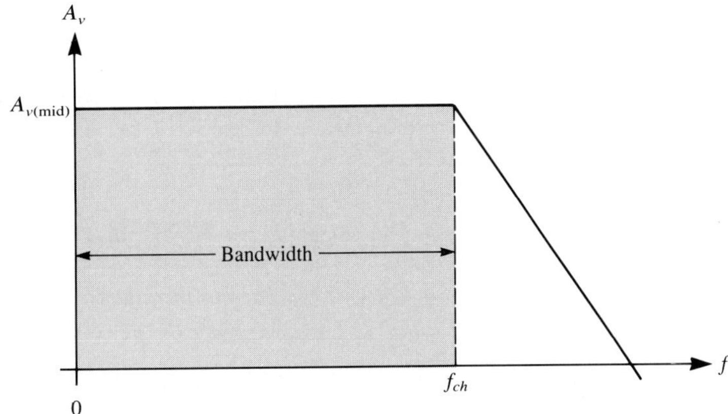

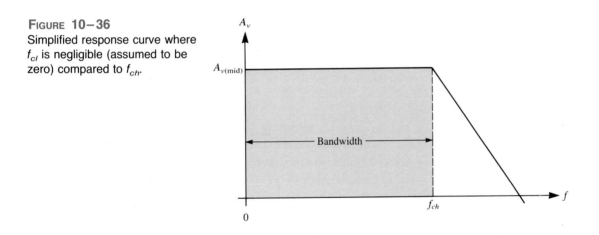

For the case shown in Figure 10–36, $f_T = A_{v(\text{mid})}f_{ch}$. For example, if a transistor data sheet specifies $f_T = 100$ MHz, this means that the transistor is capable of producing a voltage gain of one up to 100 MHz, or a gain of 100 up to 1 MHz, or any combination of gain and bandwidth that produces a product of 100 MHz.

EXAMPLE 10–12 A certain transistor has an f_T of 175 MHz. When this transistor is used in an amplifier with a midrange voltage gain of 50, what bandwidth can be achieved?

SOLUTION

$$f_T = A_{v(\text{mid})}BW$$

$$BW = \frac{f_T}{A_{v(\text{mid})}} = \frac{175 \text{ MHz}}{50} = 3.5 \text{ MHz}$$

PRACTICE EXERCISE 10–12
Am amplifier has a midrange voltage gain of 20 and a bandwidth of 1 MHz. What is the f_T of the transistor?

HALF-POWER POINTS

The upper and lower critical frequencies are sometimes called the *half-power frequencies*. This term is derived from the fact that the output power of an amplifier at its critical frequencies is one-half of its midrange power, as previously mentioned. This can be shown as follows, starting with the fact that the output voltage is 0.707 of its midrange value at the critical frequencies.

$$V_{out(f_c)} = 0.707V_{out(mid)}$$

$$P_{out(f_c)} = \frac{V^2_{out(f_c)}}{R_{out}} = \frac{(0.707V_{out(mid)})^2}{R_{out}}$$

$$= \frac{0.5V^2_{out(mid)}}{R_{out}} = 0.5P_{out(mid)}$$

10–5 REVIEW QUESTIONS

1. What is the gain of an amplifier at f_T?
2. What is the bandwidth of an amplifier when $f_{ch} = 25$ kHz and $f_{cl} = 100$ Hz?
3. The f_T of a certain transistor is 130 MHz. What gain can be achieved with a bandwidth of 50 MHz?

10–6 FREQUENCY RESPONSE OF FET AMPLIFIERS

In this section, we will look at the frequency response of FET amplifiers using both D-MOSFETs and JFETs. As you will see, the approach is basically the same as for BJTs.

A zero-biased amplifier with capacitive coupling on the input and output is shown in Figure 10–37.

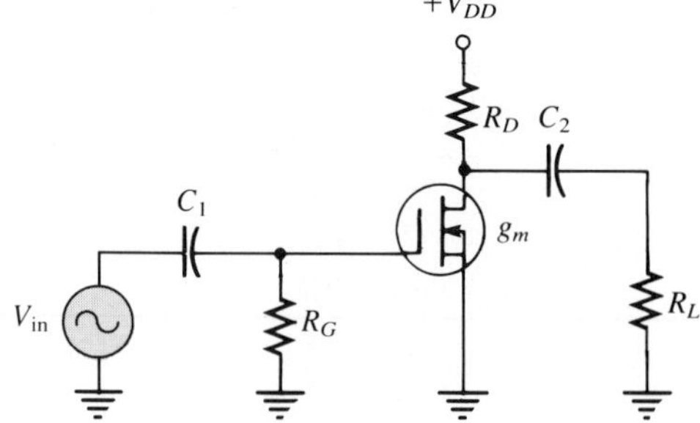

FIGURE 10–37
Zero-biased D-MOSFET amplifier.

LOW-FREQUENCY RESPONSE

The midrange voltage gain of a zero-biased amplifier is determined as follows.

$$A_{v(mid)} = g_m R_d \quad (10\text{–}24)$$

As you know, this is the gain at frequencies high enough so that the capacitive reactances are approximately zero. The amplifier in Figure 10–37 has only two *RC* networks that influence its low-frequency response. One network is formed by the input coupling capacitor C_1 and the input resistance, as shown in Figure 10–38. The other network is formed by the output coupling capacitor C_2 and the output resistance looking in at the drain.

FIGURE 10–38
Input *RC* network for the amplifier in Figure 10–37.

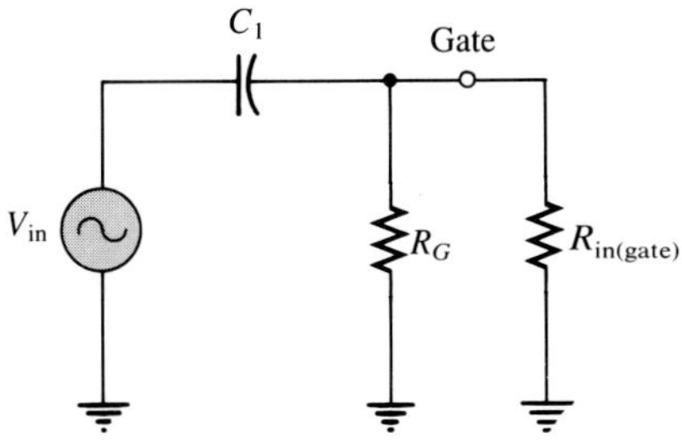

THE LOW-FREQUENCY INPUT *RC* NETWORK Just as in the previous case for the bipolar transistor amplifier, the reactance of the input coupling capacitor increases as the frequency decreases. When $X_{C1} = R_{in}$, the gain is down 3 dB below its midrange value. The lower critical frequency is

$$f_c = \frac{1}{2\pi R_{in} C_1}$$

The input resistance is

$$R_{in} = R_G \| R_{in(gate)}$$

where $R_{in(gate)}$ is determined from data sheet information as follows.

$$R_{in(gate)} = \left| \frac{V_{GS}}{I_{GSS}} \right|$$

So,

$$f_c = \frac{1}{2\pi(R_G \| R_{in(gate)})C_1} \quad \textbf{(10–25)}$$

The gain roll-off below f_c is 20 dB/decade, as previously shown, and the phase shift is $\theta = \tan^{-1}(X_{C1}/R_{in})$.

EXAMPLE 10–13 What is the critical frequency of the input *RC* network in Figure 10–39?

FIGURE 10–39

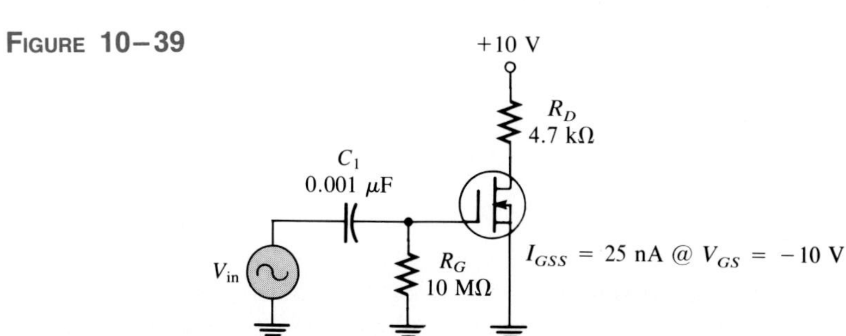

SOLUTION

First determine R_{in}.

$$R_{in(gate)} = \left|\frac{V_{GS}}{I_{GSS}}\right| = \frac{10\ \text{V}}{25\ \text{nA}} = 400\ \text{M}\Omega$$

$$R_{in} = R_G \| R_{in(gate)} = 10\ \text{M}\Omega \| 400\ \text{M}\Omega = 9.8\ \text{M}\Omega$$

$$f_c = \frac{1}{2\pi R_{in} C_1} = \frac{1}{2\pi\,(9.8\ \text{M}\Omega)(0.001\ \mu\text{F})} = 16.2\ \text{Hz}$$

The critical frequency of the input *RC* network of an FET amplifier is usually very low because of the very high input resistance.

PRACTICE EXERCISE 10–13

How much does the critical frequency of the input *RC* network change if the FET is replaced by one with $I_{GSS} = 10$ nA @ $V_{GS} = -8$ V?

THE LOW-FREQUENCY OUTPUT *RC* NETWORK The second *RC* network that affects the low-frequency response of the amplifier in Figure 10–37 is formed by the coupling capacitor C_2 and the output resistance looking in at the drain, as shown in Figure 10–40(a). R_L is also included. As in the case of the bipolar transistor, the FET is treated as a current source, and the upper end of R_D is effectively ac ground, as shown in Figure 10–40(b). The Thevenin equivalent of the circuit to the left of C_2 is shown in Figure 10–40(c). The critical frequency for this network is determined as follows.

$$f_c = \frac{1}{2\pi(R_D + R_L)C_2} \tag{10–26}$$

The effect of the output *RC* network on the amplifier's gain below the midrange is similar to that of the input *RC* network. The network with the higher critical frequency

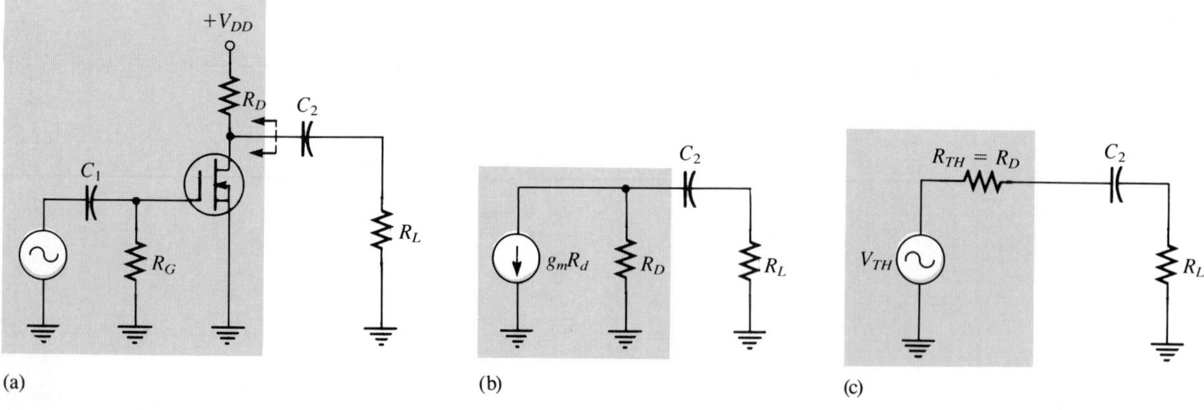

FIGURE 10–40
Development of the output *RC* network.

dominates because it is the one that first causes the gain to roll off as the frequency drops below its midrange values. The phase shift in the output RC network is

$$\theta = \tan^{-1}\left(\frac{X_{C2}}{R_D + R_L}\right) \tag{10-27}$$

Again, at the critical frequency, the phase angle is 45° and approaches 90° as the frequency approaches zero. However, starting at the critical frequency, the phase angle decreases from 45° and becomes very small as the frequency goes higher.

EXAMPLE 10–14

Determine the total low-frequency response of the FET amplifier in Figure 10–41. Assume that the load is another identical amplifier with the same R_{in}. $I_{GSS} = 100$ nA at $V_{GS} = -12$ V.

FIGURE 10–41

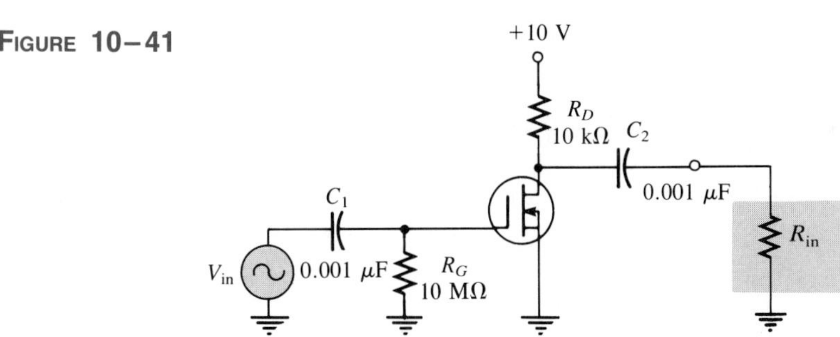

SOLUTION

First, we find the critical frequency for the input RC network.

$$R_{in(gate)} = \left|\frac{V_{GS}}{I_{GSS}}\right| = \frac{12 \text{ V}}{100 \text{ nA}} = 120 \text{ M}\Omega$$

$$R_{in} = R_G \| R_{in(gate)} = 10 \text{ M}\Omega \| 120 \text{ M}\Omega = 9.2 \text{ M}\Omega$$

$$f_{c(input)} = \frac{1}{2\pi R_{in} C_1} = \frac{1}{2\pi(9.2 \text{ M}\Omega)(0.001 \ \mu\text{F})} = 17.3 \text{ Hz}$$

The output RC network has a critical frequency of

$$f_{c(output)} = \frac{1}{2\pi(R_D + R_L)C_2}$$

$$= \frac{1}{2\pi(9.21 \text{ M}\Omega)(0.001 \ \mu\text{F})}$$

$$= 17.3 \text{ Hz}$$

PRACTICE EXERCISE 10–14

If the circuit in Figure 10–41 were operated with no load, how is the low-frequency response affected?

HIGH-FREQUENCY RESPONSE

The approach to the high-frequency analysis of an FET amplifier is very similar to that of a bipolar amplifier. The basic differences are the specifications of the internal FET capacitances and the determination of the input resistance.

Figure 10–42(a) shows a JFET common-source amplifier that we will use to illustrate high-frequency analysis. An equivalent circuit for high-frequency analysis of the amplifier is shown in Figure 10–42(b). Notice that the coupling and bypass capacitors are assumed to have negligible reactances and are thus considered to be shorts. The internal capacitances C_{gs} and C_{gd} appear in the equivalent circuit because their reactances are significant at high frequencies, and thus they influence the amplifier's response.

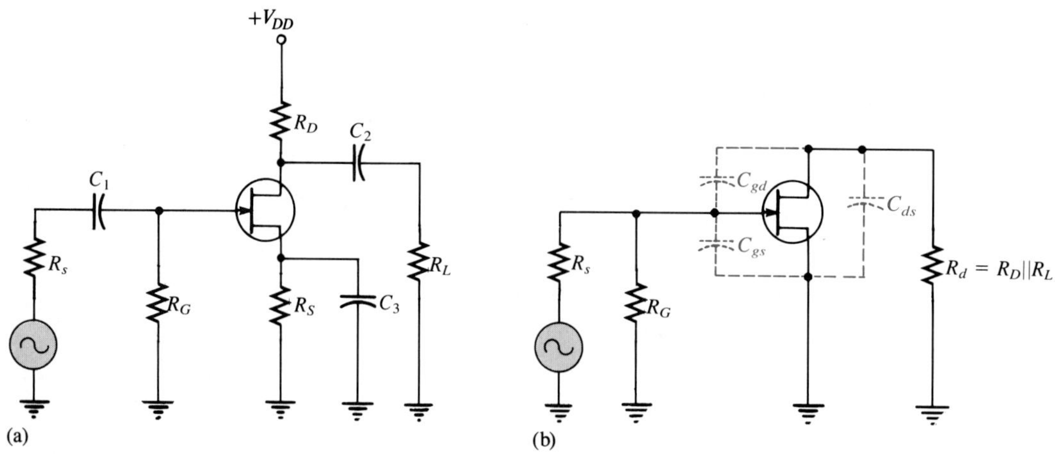

FIGURE 10–42
JFET amplifier and its high-frequency equivalent circuit.

GETTING THE VALUES OF C_{gs}, C_{gd}, AND C_{ds} FET data sheets do not normally provide values for C_{gs}, C_{gd}, or C_{ds}. Instead, three other values are specified because they are easier to measure. These are C_{iss}, the input capacitance, C_{rss}, the reverse transfer capacitance, and C_{oss}, the output capacitance. Because of the manufacturer's method of measurement, the following relationships allow us to determine the capacitor values needed for analysis.

$$C_{gd} = C_{rss} \quad (10\text{–}28)$$

$$C_{gs} = C_{iss} - C_{rss} \quad (10\text{–}29)$$

$$C_{ds} = C_{oss} - C_{rss} \quad (10\text{–}30)$$

C_{oss} is not specified as often as the other values on data sheets. Sometimes, it is designated as $C_{d(\text{sub})}$, the drain-to-substrate capacitance. In cases where a value is not available, we must either assume a value or neglect it.

EXAMPLE 10–15 The data sheet for a 2N3823 JFET gives C_{iss} = 6 pF and C_{rss} = 2 pF. Determine C_{gd} and C_{gs}.

SOLUTION

$$C_{gd} = C_{rss} = 2 \text{ pF}$$
$$C_{gs} = C_{iss} - C_{rss} = 6 \text{ pF} - 2 \text{ pF} = 4 \text{ pF}$$

PRACTICE EXERCISE 10–15

Although C_{oss} is not specified on the data sheet for the 2N3823 JFET, assume a value of 3 pF and determine C_{ds}.

USING MILLER'S THEOREM Miller's theorem is applied the same way in FET amplifier high-frequency analysis as was done in bipolar transistor amplifiers. Looking in from the signal source in Figure 10–42(b), C_{gd} effectively appears in the Miller input capacitance as follows.

$$C_{in(Miller)} = C_{gd}(A_v + 1) \tag{10–31}$$

C_{gs} simply appears as a capacitance to ac ground in parallel with $C_{in(Miller)}$, as shown in Figure 10–43. Looking in at the drain, C_{gd} effectively appears in the Miller output capacitance from drain to ground in parallel with R_d, as shown in Figure 10–43.

$$C_{out(Miller)} = C_{gd}\left(\frac{A_v + 1}{A_v}\right) \tag{10–32}$$

These two Miller capacitances contribute to a high-frequency input RC network and a high-frequency output RC network. Both act as lag networks (low-pass filters).

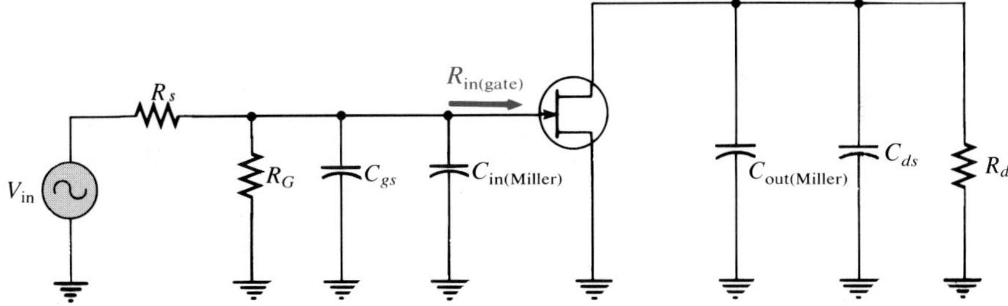

FIGURE 10–43 High-frequency equivalent circuit after applying Miller's theorem.

THE HIGH-FREQUENCY INPUT *RC* NETWORK The high-frequency input network forms a low-pass type of filter and is shown in Figure 10–44(a). Because both R_G and the input resistance at the gate of FETs are extremely high, the controlling resistance for the input network is the resistance of the signal source as long as $R_s << R_{in}$. This is because R_s appears in parallel with R_{in} when Thevenin's theorem is applied. The simplified input RC network appears in Figure 10–44(b). The critical frequency is

$$f_c = \frac{1}{2\pi R_s C_T} \tag{10–33}$$

where $C_T = C_{gs} + C_{in(Miller)}$. The input network produces a phase shift of

$$\phi = \tan^{-1}\left(\frac{R_s}{X_{C_T}}\right) \tag{10–34}$$

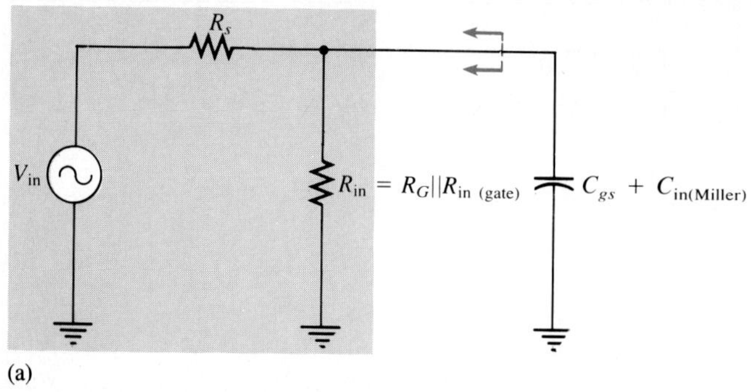

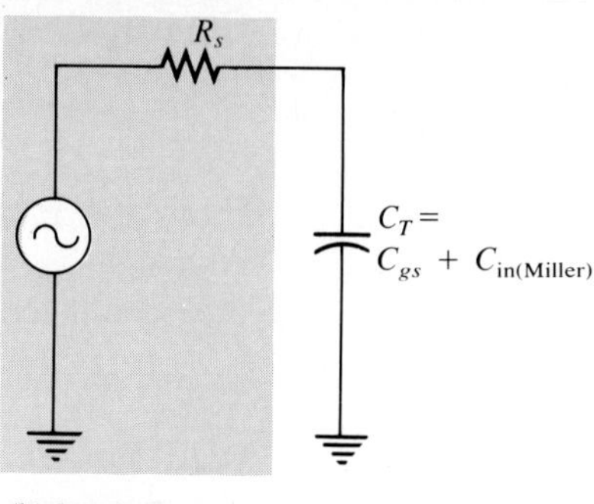

(b) Neglecting R_{in}

FIGURE 10–44
High-frequency input *RC* network.

The effect of the input *RC* network is to reduce the midrange gain of the amplifier by 3 dB at the critical frequency and to cause the gain to decrease at −20 dB/decade beyond f_c.

EXAMPLE 10–16 Find the critical frequency of the input *RC* network for the amplifier in Figure 10–45. $C_{iss} = 8$ pF and $C_{rss} = 3$ pF. $g_m = 6500\ \mu$S.

FIGURE 10–45

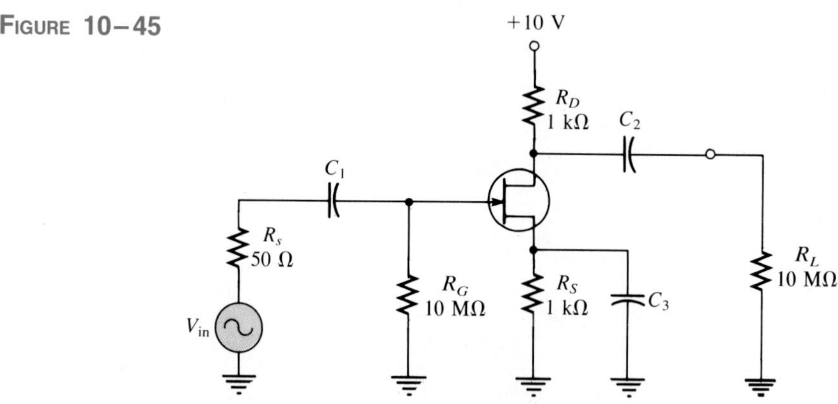

SOLUTION

$$C_{gd} = C_{rss} = 3\text{ pF}$$
$$C_{gs} = C_{iss} - C_{rss} = 8\text{ pF} - 3\text{ pF} = 5\text{ pF}$$

The input *RC* network is derived as follows.

$$A_v = g_m R_d = g_m(R_D \| R_L) \cong (6500\ \mu\text{S})(1\text{ k}\Omega) = 6.5$$
$$C_{in(Miller)} = C_{gd}(A_v + 1) = (3\text{ pF})(7.5) = 22.5\text{ pF}$$

The total input capacitance is

$$C_T = C_{gs} + C_{in(Miller)} = 5\text{ pF} + 22.5\text{ pF} = 27.5\text{ pF}$$

The critical frequency is

$$f_c = \frac{1}{2\pi R_s C_T} = \frac{1}{2\pi(50\ \Omega)(27.5\ \text{pF})} = 115.75\ \text{MHz}$$

PRACTICE EXERCISE 10–16
If the gain of the amplifier in Figure 10–45 is increased to 10, what happens to f_c?

THE HIGH-FREQUENCY OUTPUT *RC* NETWORK The high-frequency output *RC* network is formed by the Miller output capacitance and the output resistance looking in at the drain, as shown in Figure 10–46(a). As in the case of the bipolar transistor, the FET is treated as a current source. When we apply Thevenin's theorem, we get an equivalent output *RC* network consisting of R_D in parallel with R_L and an equivalent output capacitance as follows.

$$C_{\text{out(Miller)}} = C_{gd}\left(\frac{A_v + 1}{A_v}\right)$$

This equivalent output network is shown in Figure 10–46(b). The critical frequency of the output *RC* lag network is

$$f_c = \frac{1}{2\pi R_d C_{\text{out(Miller)}}} \tag{10–35}$$

The output network produces a phase shift of

$$\phi = \tan^{-1}\left(\frac{R_d}{X_{C_{\text{out(Miller)}}}}\right) \tag{10–36}$$

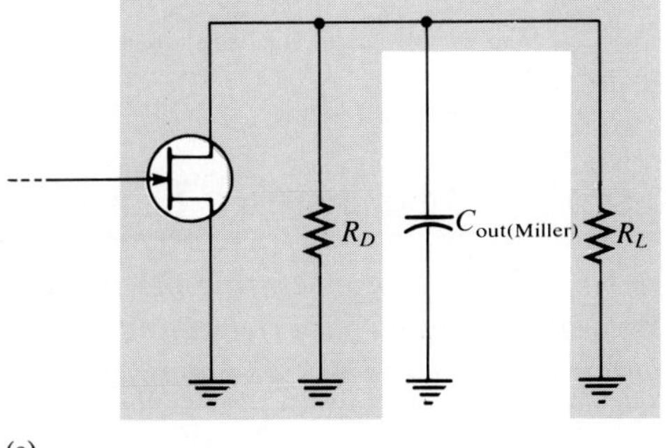

(a)

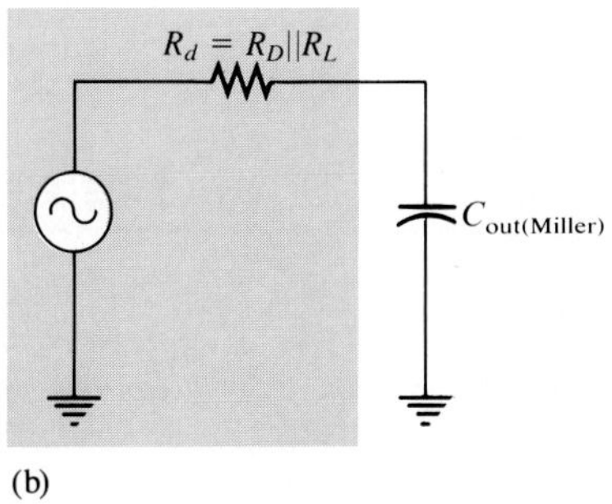

(b)

FIGURE 10–46
High-frequency output *RC* network.

EXAMPLE 10–17 Determine the critical frequency of the output *RC* network for the amplifier in Figure 10–45. What is the phase shift introduced by this network at the critical frequency?

SOLUTION

Since R_L is very large compared to R_D, it can be neglected, and the equivalent output resistance is

$$R_d = R_D = 1 \text{ k}\Omega$$

The equivalent output capacitance is

$$C_{out(Miller)} = C_{gd}\left(\frac{A_v + 1}{A_v}\right) = 3 \text{ pF}\left(\frac{7.5}{6.5}\right) = 3.46 \text{ pF}$$

Therefore, the critical frequency is

$$f_c = \frac{1}{2\pi R_d C_{out(Miller)}} = \frac{1}{2\pi(1 \text{ k}\Omega)(3.46 \text{ pF})} = 46 \text{ MHz}$$

The phase angle is always 45° at f_c for one RC network. At high frequencies, the output network causes a phase lag.

In Example 10–16, the critical frequency of the input RC network was found to be 115.75 MHz. Therefore, the critical frequency for the output network is dominant since it is the lesser of the two.

PRACTICE EXERCISE 10–17

If A_v of the amplifier in Figure 10–45 is increased to 10, what is the f_c of the output network? ■

10–6 REVIEW QUESTIONS

1. What is the amount of phase shift contributed by an input network when $X_C = 0.5R_{in}$ at a certain frequency below f_{cl}?
2. What is f_c when $R_D = 1.5 \text{ k}\Omega$, $R_L = 5 \text{ k}\Omega$, and $C_2 = 0.002\ \mu\text{F}$?
3. What are the capacitances that are usually specified on an FET data sheet?
4. If $C_{gs} = 4$ pF and $C_{gd} = 3$ pF, what is the total input capacitance of an FET amplifier whose voltage gain is 25?

10–7 FREQUENCY RESPONSE MEASUREMENT TECHNIQUES

In this section, two basic methods of measuring the frequency response of an amplifier are covered. We will concentrate on determining the two dominant critical frequencies. From these values, we can get the bandwidth.

FREQUENCY AND AMPLITUDE MEASUREMENT

Figure 10–47 shows an amplifier driven by a sine wave voltage source and with a dual-channel oscilloscope connected to the input and to the output. The input frequency is set to a midrange value, and its amplitude is adjusted to establish an output signal reference level, as shown in Figure 10–47(a). This output voltage reference level for midrange should be set at a convenient value within the linear operation of the amplifier,

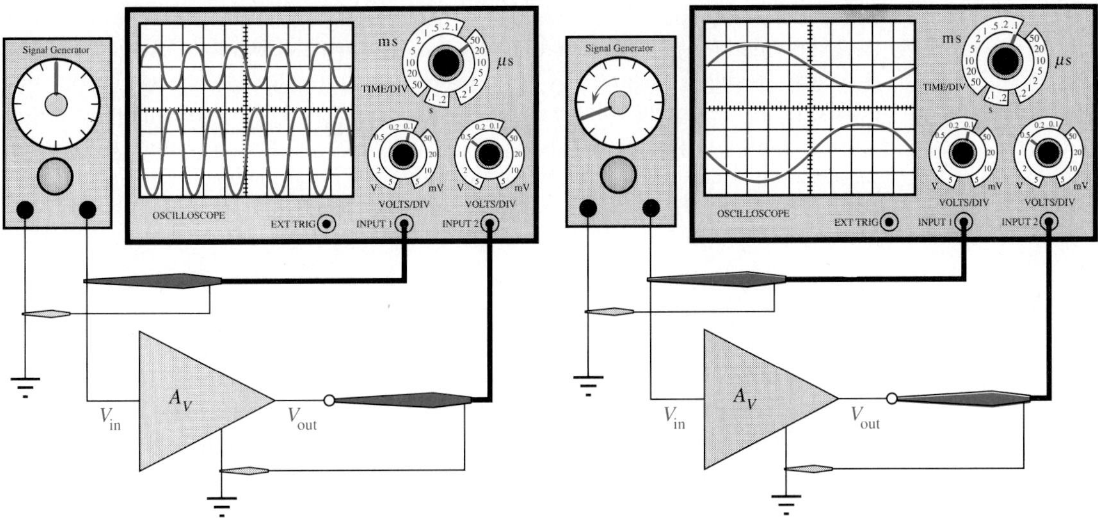

(a) At a midrange frequency setting (10 kHz in this case), the input voltage is adjusted for an output voltage of 1 V peak.

(b) Frequency is reduced until the output voltage is 0.707 V peak. This is the lower critical frequency.

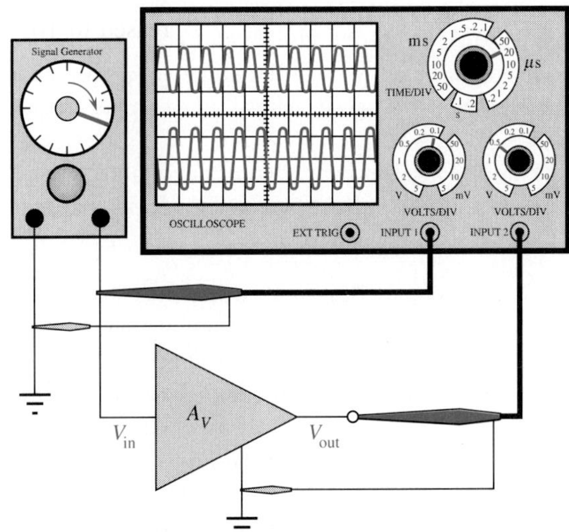

(c) Frequency is increased until the output voltage is again 0.707 V peak. This is the upper critical frequency.

FIGURE 10–47
A general procedure for measuring an amplifier's frequency response.

for example, 100 mV, 1 V, 10 V, and so on. In this case, we set the output signal to a peak value of 1 V.

Next, the frequency of the input voltage is decreased until the peak value of the output drops to 0.707 V. The amplitude of the input must be kept constant as the frequency is reduced. Readjustment may be necessary because of changes in loading of the voltage source with frequency. When the output is 0.707 V, the frequency is measured, and we have the value for f_{cl} as indicated in Figure 10–47(b).

Next, the input frequency is increased back up through midrange and beyond until the peak value of the output voltage again drops to 0.707 V. Again, the amplitude of the input must be kept constant as the frequency is increased. When the output is 0.707 V, the frequency is measured and we have the value for f_{ch} as indicated in Figure 10–47(c). From these two frequency measurements, the bandwidth is found by the formula $BW = f_{ch} - f_{cl}$.

RESPONSE OF LINEAR AMPLIFIERS TO PULSE INPUTS

As you know, pulse waveforms contain **harmonics** and are a composite of frequencies from dc (0 Hz) on up to very high frequencies. The flat parts of an ideal pulse waveform are constant, nonchanging voltages during the pulse interval and between pulses and represent the dc component of the waveform. The rising and falling edges of the pulses, because they are fast-changing, contain the high-frequency components. A pulse consists of two voltage **steps:** one at the rising edge when the pulse goes from low to high and the other at the falling edge when it goes from high to low. When we apply a pulse waveform to the input of an amplifier, we are actually applying a dc voltage plus all the harmonic frequencies that make up the pulse waveform. Suppose an amplifier has a bandwidth that extends from dc (0 Hz) to an infinitely high frequency. The output of the amplifier, in this case, would have the same shape as the input waveform (except it would be amplified) as illustrated in Figure 10–48(a) because all of the input frequency components would pass through.

Practical amplifiers, however, do not have infinite bandwidths and the limitation of the frequency response will affect the shape of the output pulse waveform as you will see.

EFFECT OF THE LOW-FREQUENCY RESPONSE ON THE PULSE SHAPE If a pulse waveform is applied to a noninverting capacitively coupled amplifier with a low-frequency response that does not extend down to dc, the flat portions of the pulse waveform are affected, as shown in Figure 10–48(b). The limitation of the high-frequency response is neglected for the time being. The "sloping" of the flat portions is caused by the coupling capacitors in the amplifier. Let's look at the input *RC* network. The input coupling capacitor and the input resistance form a differentiator. The time constant of this differentiator determines how fast the capacitor charges and discharges with a pulse input and thereby adds a "slope" to the flat portions of the input pulse. The "slope" is actually an exponential curve. The coupling capacitor determines at what frequency the sloping effect becomes significant.

EFFECT OF THE HIGH-FREQUENCY RESPONSE ON THE PULSE SHAPE At high frequencies, the rising and falling edges of the pulses are affected as shown in Figure 10–48(c). This assumes a low-frequency response down to dc. The rounding off of the edges of the pulses is caused by the charging and discharging of the internal input and output capacitances that come into play only at higher frequencies. These capacitances form integrators with the circuit resistances. The value of these capacitances determine at what frequency the rounding off becomes significant.

FIGURE 10–48 Illustration of the general response of a linear noninverting amplifier to a pulse input.

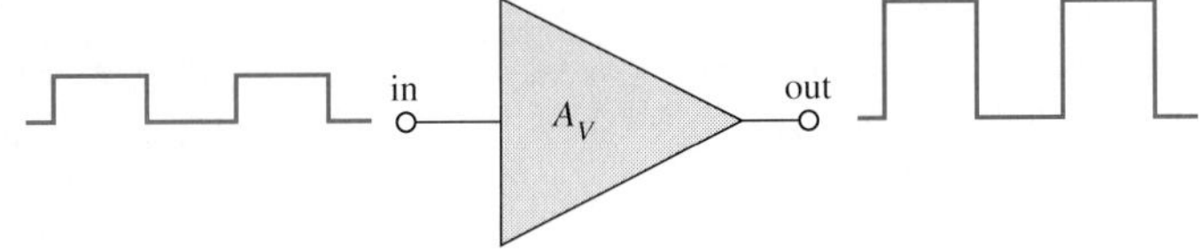

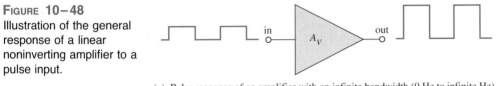

(a) Pulse response of an amplifier with an infinite bandwidth (0 Hz to infinite Hz)

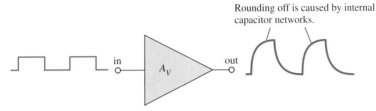

Sloping portions are caused by coupling capacitor networks.

(b) Low-frequency pulse response of amplifier with limited bandwidth

Rounding off is caused by internal capacitor networks.

(c) High-frequency pulse response of amplifier with limited bandwidth

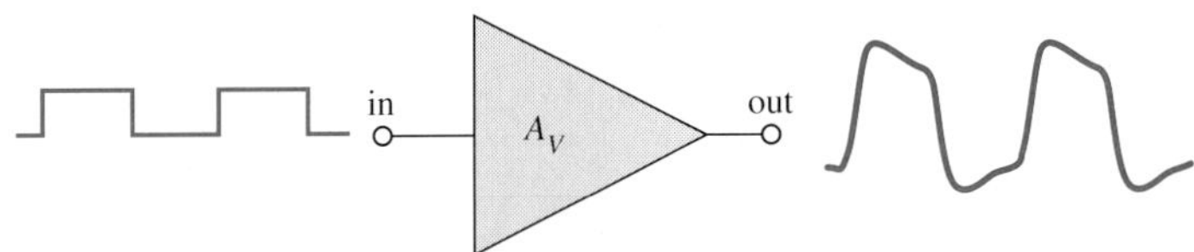

(d) Overall pulse response of amplifier with limited bandwidth

The combined low- and high-frequency responses of an amplifier cause both rounding of the pulse edges and a sloping of the flat portions, as indicated in Figure 10–48(d).

STEP RESPONSE MEASUREMENT

The low and high critical frequencies of an amplifier can be determined using the *step response method* by applying a voltage step to the input of the amplifier and measuring the rise and fall times of the resulting output voltage. A basic test setup is shown in Figure 10–49 where a capacitively coupled, noninverting amplifier is used for illustration. The input step is created by the rising edge of a pulse that has a long duration compared to the rise and fall times to be measured. A pulse generator is the input signal source.

HIGH-FREQUENCY MEASUREMENT When a step input is applied, the amplifier's high-frequency *RC* networks (internal capacitances) prevent the output from responding immediately to

FIGURE 10–49
Basic test setup for measuring the step response of an amplifier.

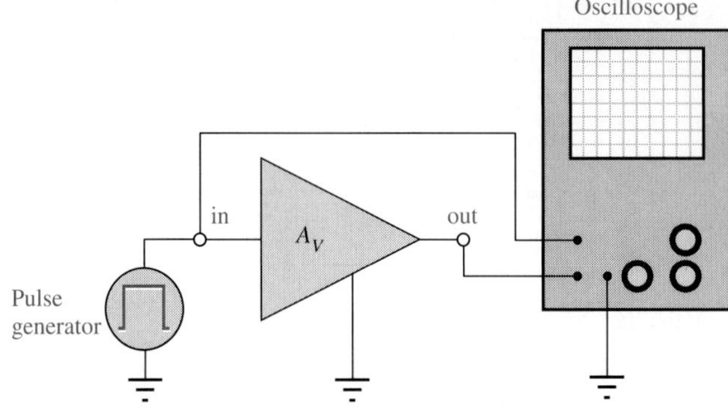

the step input. As a result, the output voltage has a rise-time (t_r) associated with it, as shown in Figure 10–50(a). In fact, the rise time is inversely related to the high critical frequency (f_{ch}) of the amplifier. As f_{ch} becomes lower, the rise time of the output becomes greater. The oscilloscope display illustrates how the rise time is measured from the 10% amplitude point to the 90% amplitude point. The scope must be set on a short time base so the relatively short interval of the rise time can be accurately observed. Once this measurement is made, f_{ch} can be calculated with the following formula.

$$f_{ch} = \frac{0.35}{t_r} \tag{10–37}$$

LOW-FREQUENCY MEASUREMENT To determine the low critical frequency (f_{cl}) of the amplifier, the step input must be of sufficiently long duration to observe the full charging time of the low-frequency *RC* networks (coupling capacitances), which cause the "sloping" of the output and which we will refer to as the fall time (t_f). This is illustrated in Figure 10–50(b). The fall time is inversely related to the low critical frequency of the amplifier. As f_{cl} becomes higher, the fall time of the output becomes less. The scope display illustrates how the fall time is measured from the 90% point to the 10% point. The scope must be set on a long time base so the complete interval of the fall time can be observed. Once this measurement is made, f_{cl} can be determined with the following formula. The derivations of Equations 10–37 and 10–38 can be found in Appendix B.

$$f_{cl} = \frac{0.35}{t_f} \tag{10–38}$$

10–7 REVIEW QUESTIONS

1. The rise time and the fall time of the amplifier's output voltage are measured between what two points on the voltage transition?
2. Which waveform is the input in each part of Figure 10–47?
3. In Figure 10–47, what are the lower and upper critical frequencies?
4. In Figure 10–50, what is the rise time in this case?
5. In Figure 10–50, what is the fall time in this case?
6. What is the bandwidth of the amplifier whose step response is measured in Figure 10–50?

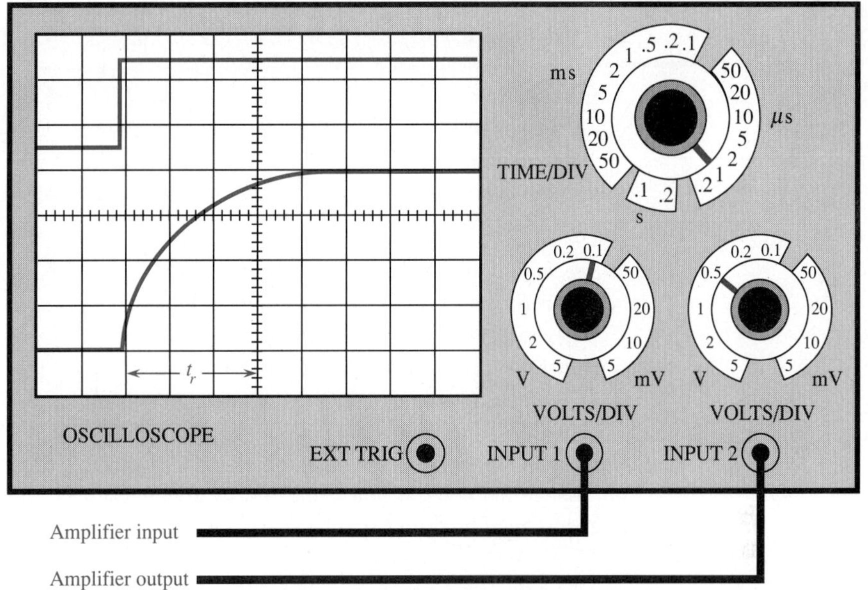

(a) Measurement of output rise time for determination of upper critical frequency

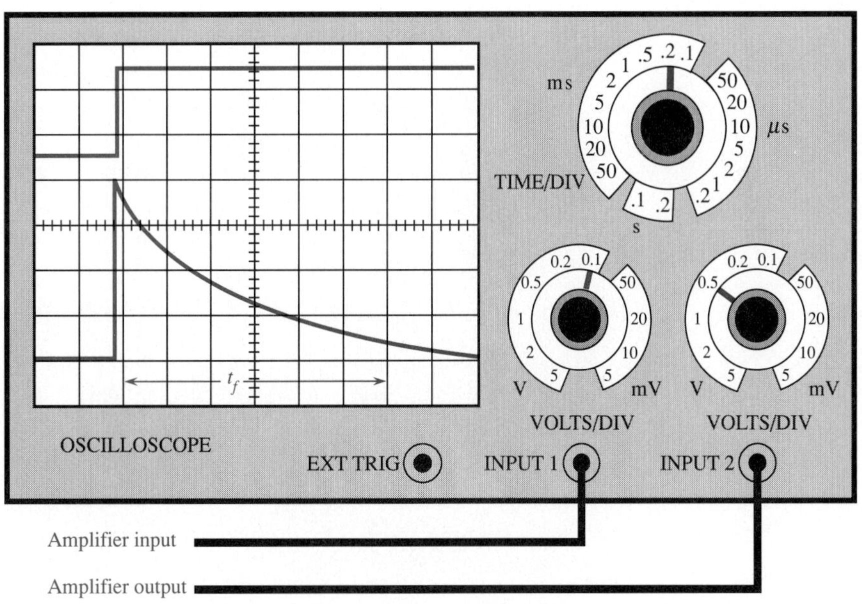

(b) Measurement of output fall time for determination of lower critical frequency

FIGURE 10–50
Measurement of the rise and fall times associated with an amplifier's step response.

10–8 A SYSTEM APPLICATION

You worked with the audio preamplifier board in Chapter 6 and with the power amplifier board in Chapter 9. Now, you will be concentrating on both circuits working together as the receiver's audio section. The focus of this system application is to analyze and check out the audio section for the proper frequency response. In this section, you will

- ☐ ***See how two types of audio amplifiers work as a single system function.***
- ☐ ***Determine the bandwidth of each amplifier individually.***
- ☐ ***Determine the overall bandwidth of the audio section.***
- ☐ ***Translate between a printed circuit board and a schematic.***
- ☐ ***Troubleshoot some common amplifier failures.***

In the previous chapters, you got a basic idea of the overall receiver system, so we will not repeat that here. If you want to review the system or details of the individual boards, refer to Chapters 6 and 9.

Now, so that you can take a closer look at the audio section, let's take it out of the receiver system and put it on the test bench.

ON THE TEST BENCH

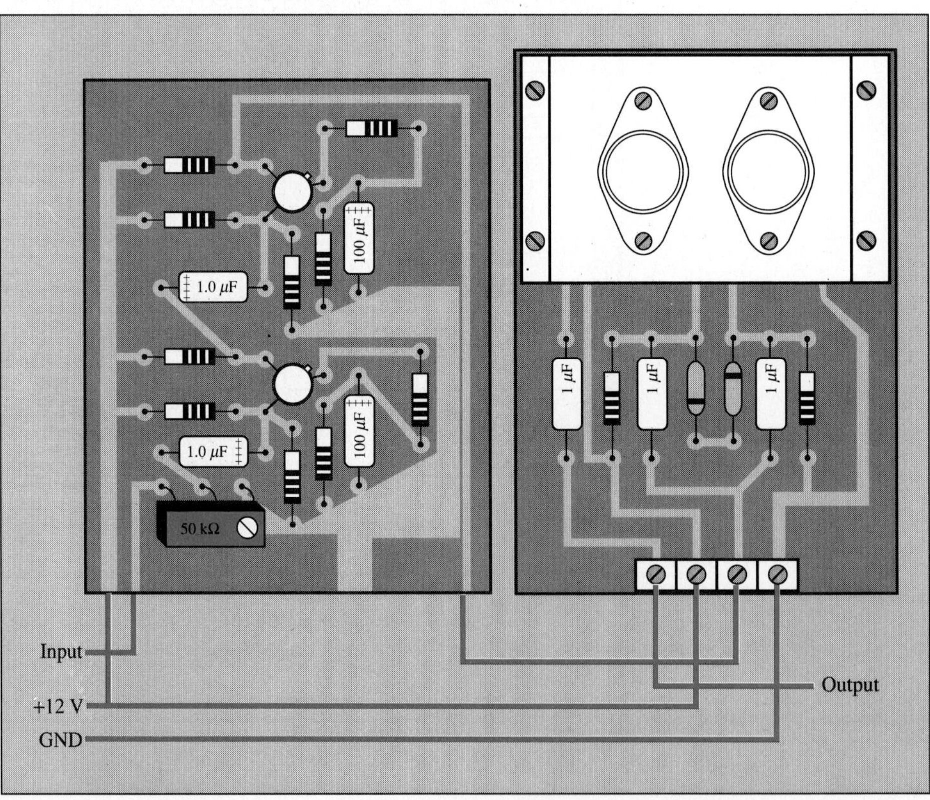

FIGURE 10–51

ACTIVITY 1 RELATE THE PC BOARDS TO THE SCHEMATIC

The schematic diagrams for the two circuit boards in Figure 10–51 were developed in Chapters 6 and 9. Now, take the two schematics shown in Figure 10–52 and combine them into one, showing how the boards interface. Relabel all the components and their values.

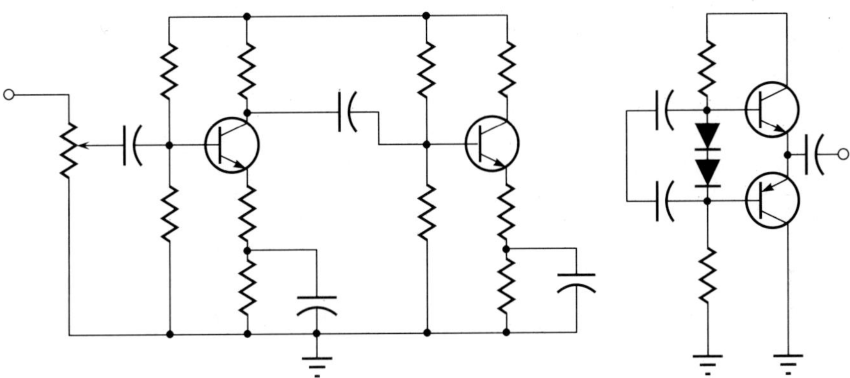

FIGURE 10–52

ACTIVITY 2 ANALYZE THE CIRCUITS

Working from your newly developed schematic, calculate the low and high dominant critical frequencies for the audio section. Use minimum data sheet values when available and neglect any parameter that is not specified. Assume an 8-Ω speaker load. Once the critical frequencies are known, determine the bandwidth. Is the overall response within the audio range?

ACTIVITY 3 WRITE A TECHNICAL REPORT

Using the results from Activity 2, discuss the overall operation of the combined audio section. State the purpose of each circuit board as part of the overall system function.

ACTIVITY 4 TROUBLESHOOT THE AUDIO SECTION BOARDS FOR EACH OF THE FOLLOWING PROBLEMS BY STATING THE PROBABLE CAUSE OR CAUSES IN EACH CASE

1. No final output signal when there is a verified input signal.
2. Proper voltages at base of the second preamp transistor, but no signal voltages at the collector.
3. Proper signal at collector of the first preamp transistor, but no signal at base of the second one.
4. Proper signal at collector of second preamp transistor, but no signal at final output.
5. Amplitude of output signal much less than it should be.

10–8 REVIEW QUESTIONS

1. How can the lower critical frequency of the audio section be reduced?
2. If the internal capacitances of the transistor were greater, would the upper critical frequency increase or decrease?
3. How can the upper critical frequency be increased?

SUMMARY

- The coupling and bypass capacitors of an amplifier affect the low-frequency response.
- The internal transistor capacitances affect the high-frequency response.
- Critical frequencies are values of frequency at which the *RC* networks start to cause a reduction in voltage gain.
- Each *RC* network causes the gain to drop at a rate of 20 dB/decade.
- For the low-frequency networks, the *highest* critical frequency is the dominant critical frequency.
- For the high-frequency networks, the *lowest* critical frequency is the dominant critical frequency.
- A decade of frequency change is a ten-times change (increase or decrease).
- An octave of frequency change is a two-times change (increase or decrease).
- The bandwidth of an amplifier is the range of frequencies between the lower critical frequency and the upper critical frequency.
- The gain-bandwidth product is a transistor parameter that is a constant and equal to the unity-gain frequency.

GLOSSARY

Bandwidth The characteristic of certain electronic circuits that specifies the usable range of frequencies that pass from input to output.

Bode plot An idealized graph of the gain in dB versus frequency used to graphically illustrate the response of an amplifier or filter.

Critical frequency The frequency at which the response of an amplifier or filter is 3 dB less than at midrange.

Cutoff frequency Another term for critical frequency.

Decade A ten-times increase or decrease in the value of a quantity such as frequency.

Gain-bandwidth product A characteristic of amplifiers whereby the product of the gain and the bandwidth is always constant.

Harmonics The frequencies contained in a composite waveform that are integer multiples of the repetition frequency (fundamental).

Midrange The frequency range of an amplifier lying between the lower and upper critical frequencies.

Octave A two-times increase or decrease in the value of a quantity such as frequency.

Roll-off The decrease in the gain of an amplifier above and below the critical frequencies.

Step A fast voltage transition from one level to another.

FORMULAS

MILLER'S THEOREM AND DECIBELS

(10–1)	$C_{\text{in(Miller)}} = C(A_v + 1)$	Miller input capacitance, where $C = C_{bc}$ or C_{gd}
(10–2)	$C_{\text{out(Miller)}} = C\left(\dfrac{A_v + 1}{A_v}\right)$	Miller output capacitance, where $C = C_{bc}$ or C_{gd}
(10–3)	$A_p\ (\text{dB}) = 10 \log A_p$	dB power gain
(10–4)	$A_v\ (\text{dB}) = 20 \log A_v$	dB voltage gain

LOW-FREQUENCY ANALYSIS

(10–5)	$A_{v(\text{mid})} = \dfrac{R_c}{r_e}$	Midrange gain (bipolar)
(10–6)	$V_{\text{out}} = \left(\dfrac{R_{\text{in}}}{\sqrt{R_{\text{in}}^2 + X_{C1}^2}}\right)V_{\text{in}}$	*RC* lead network
(10–7)	$f_c = \dfrac{1}{2\pi R_{\text{in}} C_1}$	Critical low frequency (input)
(10–8)	$f_c = \dfrac{1}{2\pi(R_s + R_{\text{in}})C_1}$	With source resistance considered
(10–9)	$\theta = \tan^{-1}\left(\dfrac{X_{C1}}{R_{\text{in}}}\right)$	Phase angle, input network
(10–10)	$f_c = \dfrac{1}{2\pi(R_C + R_L)C_2}$	Critical low frequency (output)
(10–11)	$\theta = \tan^{-1}\left(\dfrac{X_{C2}}{R_C + R_L}\right)$	Phase angle, output network
(10–12)	$R_{\text{in(emitter)}} = \dfrac{R_{TH}}{\beta_{\text{ac}}} + r_e$	Output resistance (emitter)
(10–13)	$f_c = \dfrac{1}{2\pi[(r_e + R_{TH}/\beta_{\text{ac}}) \parallel R_E]C_3}$	Critical low frequency (bypass)

HIGH-FREQUENCY ANALYSIS

(10–14) $C_{in(Miller)} = C_{bc}(A_v + 1)$ Miller input capacitance

(10–15) $C_{out(Miller)} = C_{bc}\left(\frac{A_v + 1}{A_v}\right)$ Miller output capacitance

(10–16) $X_{C_T} = R_s \| R_1 \| R_2 \beta_{ac} r_e$ Condition at f_c

(10–17) $f_c = \frac{1}{2\pi(R_s \| R_1 \| R_2 \| \beta_{ac} r_e) C_T}$ Critical high frequency (input)

(10–18) $\phi = \tan^{-1}\left(\frac{R_s \| R_1 \| R_2 \| \beta_{ac} r_e}{X_{C_T}}\right)$ Phase angle, intput network

(10–19) $C_{out(Miller)} \cong C_{bc}$ When $A_v > 10$

(10–20) $f_c = \frac{1}{2\pi R_c C_{out(Miller)}}$ Critical high frequency (output)

(10–21) $\phi = \tan^{-1}\left(\frac{R_c}{X_{C_{out(Miller)}}}\right)$ Phase angle, output network

TOTAL RESPONSE

(10–22) $BW = f_{ch} = f_{cl}$ Bandwidth

(10–23) $f_T = A_{v(mid)} BW$ Gain-bandwidth product

FET FREQUENCY RESPONSE

(10–24) $A_{v(mid)} = g_m R_d$ Midrange gain

(10–25) $f_c = \frac{1}{2\pi(R_G \| R_{in(gate)}) C_1}$ Critical low frequency (input)

(10–26) $f_c = \frac{1}{2\pi(R_D + R_L) C_2}$ Critical low frequency (output)

(10–27) $\theta = \tan^{-1}\left(\frac{X_{C2}}{R_D + R_L}\right)$ Phase angle, output network

(10–28) $C_{gd} = C_{rss}$ Gate-to-drain capacitance

(10–29) $C_{gs} = C_{iss} - C_{rss}$ Gate-to-source capacitance

(10–30) $C_{ds} = C_{oss} - C_{rss}$ Drain-to-Source capacitance

(10–31) $C_{in(Miller)} = C_{gd}(A_v + 1)$ Miller input capacitance

(10–32) $C_{out(Miller)} = C_{gd}\left(\frac{A_v + 1}{A_v}\right)$ Miller output capacitance

(10–33) $f_c = \frac{1}{2\pi R_s C_T}$ Critical high frequency (input)

(10–34) $\phi = \tan^{-1}\left(\frac{R_s}{X_{C_T}}\right)$ Phase angle, input network

(10–35) $f_c = \frac{1}{2\pi R_d C_{out(Miller)}}$ Critical high frequency (output)

(10–36) $\phi = \tan^{-1}\left(\frac{R_d}{X_{C_{out(Miller)}}}\right)$ Phase angle, output network

MEASUREMENT TECHNIQUES

(10–37) $f_{ch} = \dfrac{0.35}{t_r}$ Upper critical frequency

(10–38) $f_{cl} = \dfrac{0.35}{t_f}$ Lower critical frequency

SELF-TEST

1. The low-frequency response of an amplifier is determined in part by

(a) the voltage gain **(b)** the type of transistor

(c) the supply voltage **(d)** the coupling capacitors

2. The high-frequency response of an amplifier is determined in part by

(a) the gain-bandwidth product **(b)** the bypass capacitor

(c) the internal transistor capacitances **(d)** the roll-off

3. The bandwidth of an amplifier is determined by

(a) the midrange gain **(b)** the critical frequencies

(c) the roll-off rate **(d)** the input capacitance

4. The gain of a certain amplifier decreases by 6 dB when the frequency is reduced from 1 kHz to 10 Hz. The roll-off is

(a) 3 dB/decade **(b)** 6 dB/decade

(c) 3 dB/octave **(d)** 6 dB/octave

5. The gain of a particular amplifier at a given frequency decreases by 6 dB when the frequency is doubled. The roll-off is

(a) 12 dB/decade **(b)** 20 dB/decade

(c) 6 dB/octave **(d)** b and c

6. The Miller input capacitance of an amplifier is dependent, in part, on

(a) the input coupling capacitor **(b)** the voltage gain

(c) the bypass capacitor **(d)** none of these

7. An amplifier has the following critical frequencies: 1.2 kHz, 950 Hz, 8 kHz, and 8.5 kHz. The bandwidth is

(a) 7550 Hz **(b)** 7300 Hz **(c)** 6800 Hz **(d)** 7050 Hz

8. Ideally, the midrange gain of an amplifier

(a) increases with frequency **(b)** decreases with frequency

(c) remains constant with frequency **(d)** depends on the coupling capacitors

9. The upper frequency at which an amplifier's gain is one is called the

(a) unity-gain frequency **(b)** midrange frequency

(c) corner frequency **(d)** break frequency

10. When the voltage gain of an amplifier is increased, the bandwidth
    (a) is not affected (b) increases
    (c) decreases (d) becomes distorted
11. If the f_T of the transistor used in a certain amplifier is 75 MHz and the bandwidth is 10 MHz, the voltage gain must be
    (a) 750 (b) 7.5 (c) 10 (d) 1
12. In the midrange of an amplifier's bandwidth, the peak output voltage is 6 V. At the lower critical frequency, the peak output voltage is
    (a) 3 V (b) 3.82 V (c) 8.48 V (d) 4.24 V
13. At the upper critical frequency, the peak output voltage of a certain amplifier is 10 V. The peak voltage in the midrange of the amplifier is
    (a) 7.07 V (b) 6.37 V (c) 14.14 V (d) 10 V
14. In the step response of a noninverting amplifier, a longer rise time means
    (a) a narrower bandwidth (b) a lower f_{cl}
    (c) a higher f_{ch} (d) a and b
15. The lower critical frequency of a direct-coupled amplifier with no bypass capacitor is
    (a) variable (b) 0 Hz
    (c) dependent on the bias (d) none of these

PROBLEMS

SECTION 10–1 GENERAL CONCEPTS

1. In a capacitively coupled amplifier, the input coupling capacitor and the output coupling capacitor form two of the networks (along with the respective resistances) that determine the low-frequency response. Assuming that the input and output impedances are the same and neglecting the bypass network, which network will first cause the gain to drop from its midrange value as the frequency is lowered?
2. Explain why the coupling capacitors do not have a significant effect on gain at sufficiently high-signal frequencies.
3. List the capacitances that affect high-frequency gain in both bipolar and FET amplifiers.

SECTION 10–2 MILLER'S THEOREM AND DECIBELS

4. In the amplifier of Figure 10–53, list the capacitors that affect the low-frequency response of the amplifier and those that affect the high-frequency response.
5. Determine the Miller input capacitance in Figure 10–53.
6. Determine the Miller output capacitance in Figure 10–53.
7. Express the midrange gain in dB for the amplifier in Figure 10–53.

FIGURE 10–53

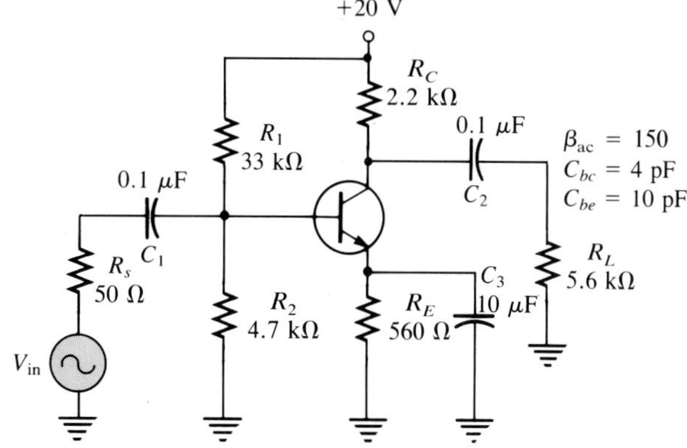

8. Determine the Miller input and output capacitances for the amplifier in Figure 10–54. Necessary data sheet information is given in the diagram.

FIGURE 10–54

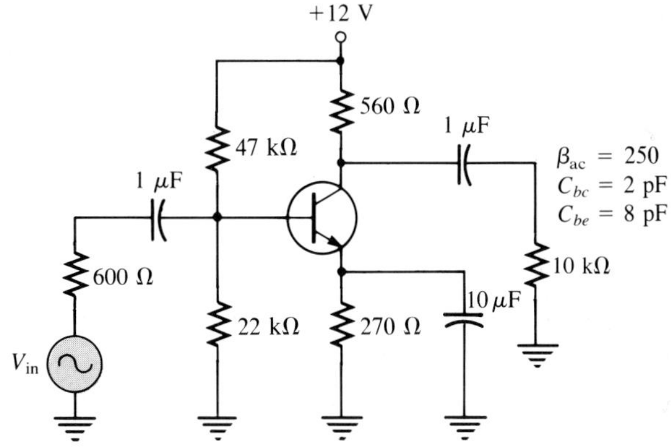

9. Determine the high-frequency equivalent circuit for the amplifier in Figure 10–54.

10. Repeat Problems 8 and 9 for the amplifier in Figure 10–55.

FIGURE 10–55

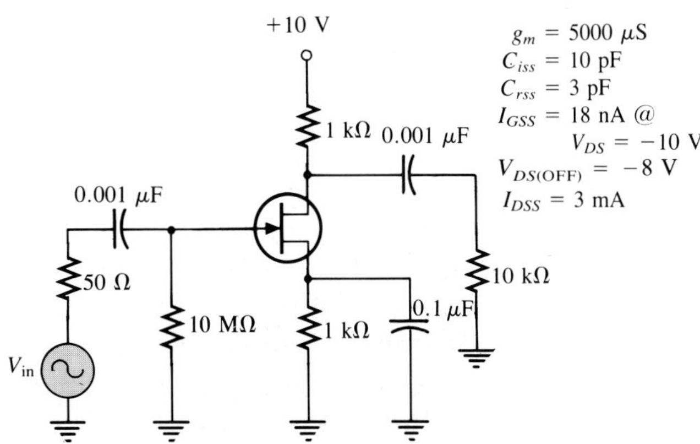

11. A certain amplifier exhibits an output power of 5 W with an input power of 0.5 W. What is the power gain in dB?

12. If the output voltage of an amplifier is 1.2 V rms and its voltage gain is 50, what is the rms input voltage? What is the gain in dB?

13. The midrange voltage gain of a certain amplifier is 65. At a certain frequency beyond midrange, the gain drops to 25. What is the gain reduction in dB?

14. What are the dBm values corresponding to the following power values?

(a) 2 mW **(b)** 1 mW **(c)** 4 mW **(d)** 0.25 mW

SECTION 10–3 LOW-FREQUENCY AMPLIFIER RESPONSE

15. Determine the critical frequencies of each *RC* network in Figure 10–56.

FIGURE 10–56

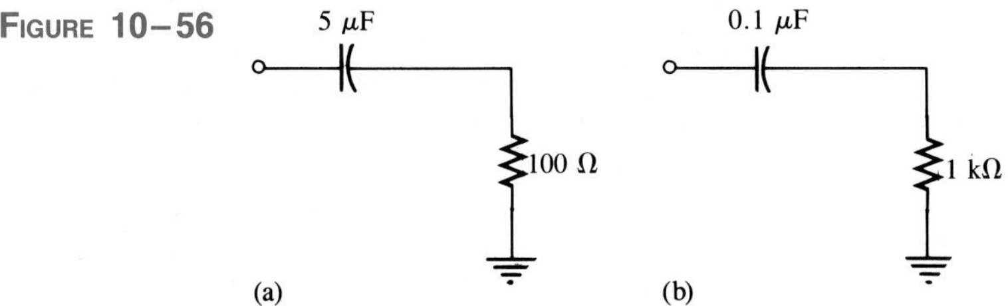

16. Determine the critical frequencies associated with the low-frequency response of the amplifier in Figure 10–57. Which is the dominant critical frequency? Sketch the Bode plot.

FIGURE 10–57

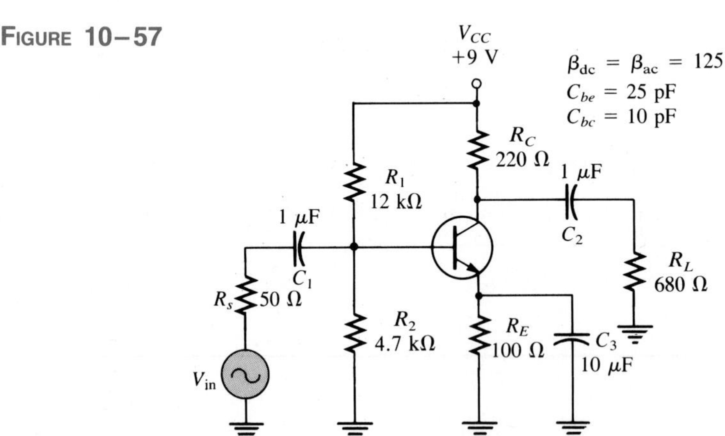

17. Determine the voltage gain of the amplifier in Figure 10–57 at one-tenth of the dominant critical frequency, at the dominant critical frequency, and at ten times the dominant critical frequency for the low-frequency response.

18. Determine the phase shift at each of the frequencies used in Problem 17.

SECTION 10–4 HIGH-FREQUENCY AMPLIFIER RESPONSE

19. Determine the critical frequencies associated with the high-frequency response of the amplifier in Figure 10–57. Identify the dominant critical frequency and sketch the Bode plot.

20. Determine the voltage gain of the amplifier in Figure 10–57 at the following frequencies: 0.1 f_c, f_c, 10 f_c, and 100 f_c, where f_c is the dominant critical frequency in the high-frequency response.

SECTION 10–5 TOTAL AMPLIFIER RESPONSE

21. A particular amplifier has the following low critical frequencies: 25 Hz, 42 Hz, and 136 Hz. It also has high critical frequencies of 8 kHz and 20 kHz. Determine the upper and lower critical frequencies.

22. Determine the bandwidth of the amplifier in Figure 10–57.

23. f_T = 200 MHz is taken from the data sheet of a transistor used in a certain amplifier. If the midrange gain is determined to be 38 and if f_{cl} is low enough to be neglected compared to f_{ch}, what bandwidth would you expect? What value of f_{ch} would you expect?

24. If the midrange gain of a given amplifier is 50 dB and therefore 47 dB at f_{ch}, how much gain is there at 2 f_{ch}? At 4 f_{ch}? At 10 f_{ch}?

SECTION 10–6 FREQUENCY RESPONSE OF FET AMPLIFIERS

25. Determine the critical frequencies associated with the low-frequency response of the amplifier in Figure 10–58. Indicate the dominant critical frequency and draw the Bode plot.

FIGURE 10–58

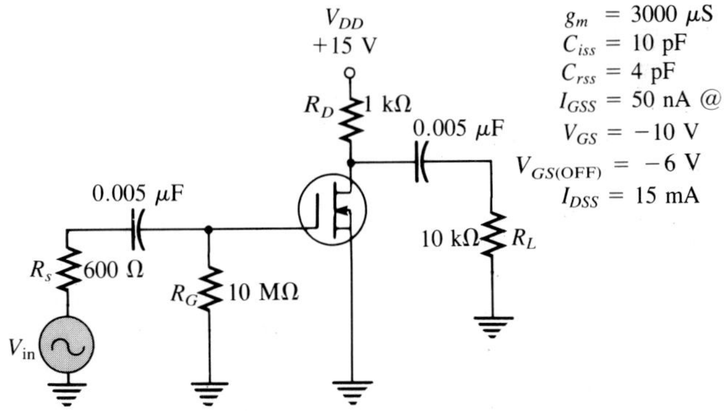

26. **(a)** Find the voltage gain of the amplifier in Figure 10–58 at the following frequencies: f_c, 0.1 f_c, and 10 f_c, where f_c is the dominant critical frequency.
(b) Find the phase shift at each frequency in part (a).

27. The data sheet for the FET in Figure 10–58 gives C_{rss} = 4 pF and C_{iss} = 10 pF. Determine the critical frequencies associated with the high-frequency response of the amplifier, and indicate the dominant frequency.

28. Determine the voltage gain in dB and the phase shift at each of the following multiples of the dominant critical frequency in Figure 10–58 for the high-frequency response: $0.1\,f_c$, f_c, $10\,f_c$, and $100\,f_c$.

SECTION 10–7 FREQUENCY RESPONSE MEASUREMENT TECHNIQUES

29. In a step response test of a certain amplifier, $t_r = 20$ ns and $t_f = 1$ ms. Determine f_{cl} and f_{ch}.

30. Suppose you are measuring the frequency response of an amplifier with a signal source and an oscilloscope. Assume that the signal level and frequency are set such that the oscilloscope indicates an output voltage level of 5 V rms in the midrange of the amplifier's response. If you wish to determine the upper critical frequency, indicate what you would do and what scope indication you would look for.

31. Determine the bandwidth of the amplifier in Figure 10–59 from the indicated results of the step-response test.

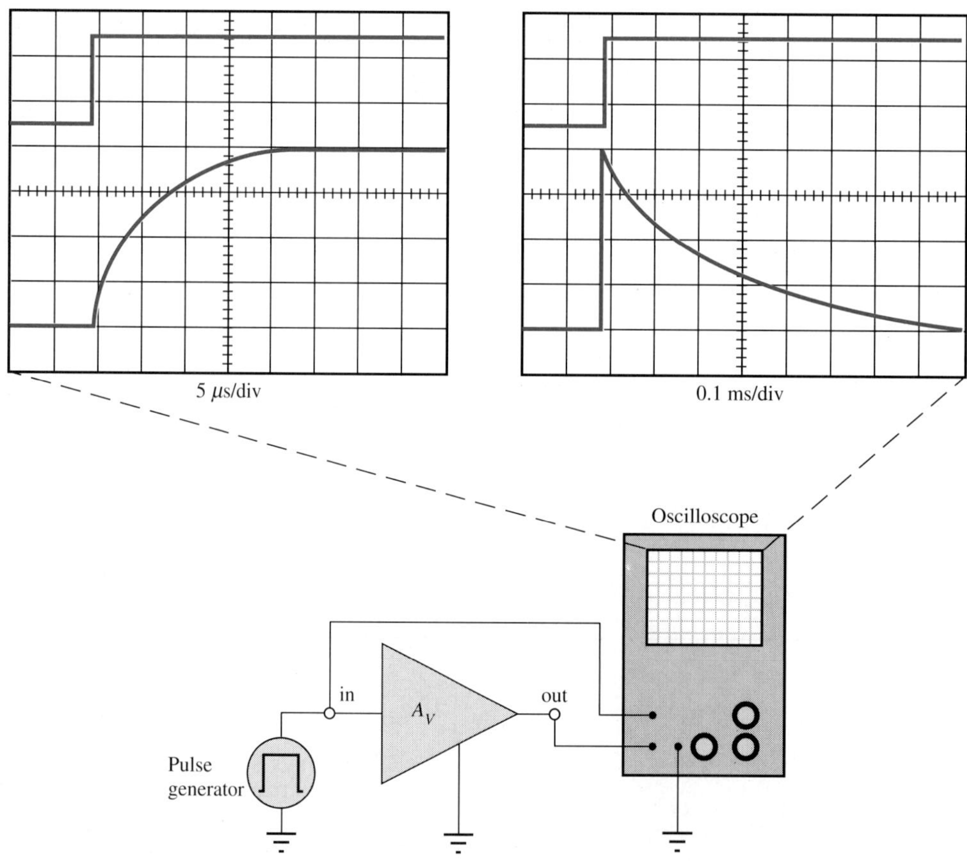

FIGURE 10–59

ANSWERS TO REVIEW QUESTIONS

SECTION 10–1

1. The coupling and bypass capacitors affect the low-frequency gain.
2. The high-frequency gain is limited by internal capacitances.

SECTION 10–2

1. +12 dB corresponds to a voltage gain of approximately 4.
2. $A_p = 10 \log(25) = 13.98$ dB
3. 0 dBm corresponds to 1 mW.

SECTION 10–3

1. $f_{c2} = 167$ Hz is dominant.
2. 50 dB − 3 dB = 47 dB
3. −20 dB at one decade below f_c.

SECTION 10–4

1. The internal transistor capacitances determine the high-frequency response.
2. $C_{in(Miller)} = C_{bc}(A_v + 1) = 4\text{ pF}(81) = 324\text{ pF}$
3. The input network dominates.

SECTION 10–5

1. The gain is 1 at f_T.
2. $BW = 25\text{ kHz} - 100\text{ Hz} = 24.9\text{ kHz}$
3. $A_v = 130\text{ MHz}/50\text{ MHz} = 2.6$

SECTION 10–6

1. $\theta = \tan^{-1}(0.5) = 26.6°$
2. $f_c = 1/(2\pi 6500\ \Omega)(0.002\ \mu\text{F}) = 12.24\text{ kHz}$
3. C_{iss} and C_{rss} are usually specified on the data sheet.
4. $C_{in(total)} = 3\text{ pF}(26) + 4\text{ pF} = 82\text{ pF}$

SECTION 10–7

1. Rise time is between the 10% and 90% points and fall time is between the 90% and 10% points.
2. Input is the top trace.
3. $f_{cl} = 1$ kHz, $f_{ch} = 50$ kHz
4. $t_r = 3\ \mu$s
5. $t_f = 1.2$ ms
6. $BW = 0.35/3\ \mu\text{s} - 0.35/1.2\text{ ms} = 116.67\text{ kHz} - 291.67\text{ Hz} = 116{,}378.33\text{ Hz}$

SECTION 10–8

1. Reduce f_{cl} by increasing the value of the coupling capacitors.
2. Higher internal capacitances reduce the upper critical frequency.
3. A reduction in gain will increase the upper critical frequency because the gain-bandwidth product is a constant for a given transistor.

ANSWERS TO PRACTICE EXERCISE

10–1 154 pF, 3.06 pF

10–2 **(a)** 61.58 dB **(b)** 16.99 dB **(c)** 101.9 dB

10–3 **(a)** 50 V **(b)** 6.25 V **(c)** 1.5625 V

10–4 **(a)** 7.23 Hz **(b)** 0.707 **(c)** 353.5

10–5 212.1 @ 400 Hz, 30 @ 40 Hz, 3 @ 4 Hz

10–6 **(a)** 13.15 Hz **(b)** 0.479 **(c)** 70.7

10–7 93.4 Hz

10–8 No effect

10–9 208 Ω in series with 483 pF, f_c = 1.58 MHz

10–10 36.17 MHz

10–11 Decreases, increases

10–12 20 MHz

10–13 From 16.2 Hz to 16.1 Hz

10–14 Ideally, the low-frequency response is not affected because the input network is unchanged.

10–15 1 pF

10–16 f_c decreases to 83.77 MHz.

10–17 48.2 MHz

11

Thyristors and Special Devices

After completing this chapter, you should be able to

- □ Describe what a thyristor is and name some important thyristor devices.
- □ Explain the basic operation of the Shockley diode.
- □ Describe how an SCR works and discuss some of its applications.
- □ Show how an SCS differs from an SCR.
- □ Describe the basic operation of diacs and triacs.
- □ Explain the principles of the UJT.
- □ Use a UJT in a relaxation oscillator circuit.
- □ Explain the operation of the PUT and discuss its similarities to both the SCR and the UJT.
- □ Discuss phototransistors and how they are used in optical-coupling devices.

In this chapter, we examine several devices. Some or all of the topics in this chapter may be treated as optional in those electronics programs where these devices would be studied in more depth in a later industrial electronics course. First, we discuss a family of devices that are constructed of four semiconductor layers (pnpn) and are known as thyristors. Thyristors include the Shockley diode, the silicon-controlled rectifier (SCR), the silicon-controlled switch (SCS), the diac, and the triac. These various types of thyristors share certain characteristics in addition to their four-layer construction. They act as open circuits capable of withstanding a certain rated voltage until they are triggered. When triggered on, they become low-resistance current paths and remain so, even after the trigger is removed, until the current is reduced to a certain level or until they are triggered off, depending on the type of device. Other devices described in this chapter include the unijunction transistor (UJT), the programmable unijunction transistor (PUT), and several optoelectronic devices.

Thyristors can be used to control the amount of ac power to a load and are used in lamp dimmers, motor speed controls, ignition systems,

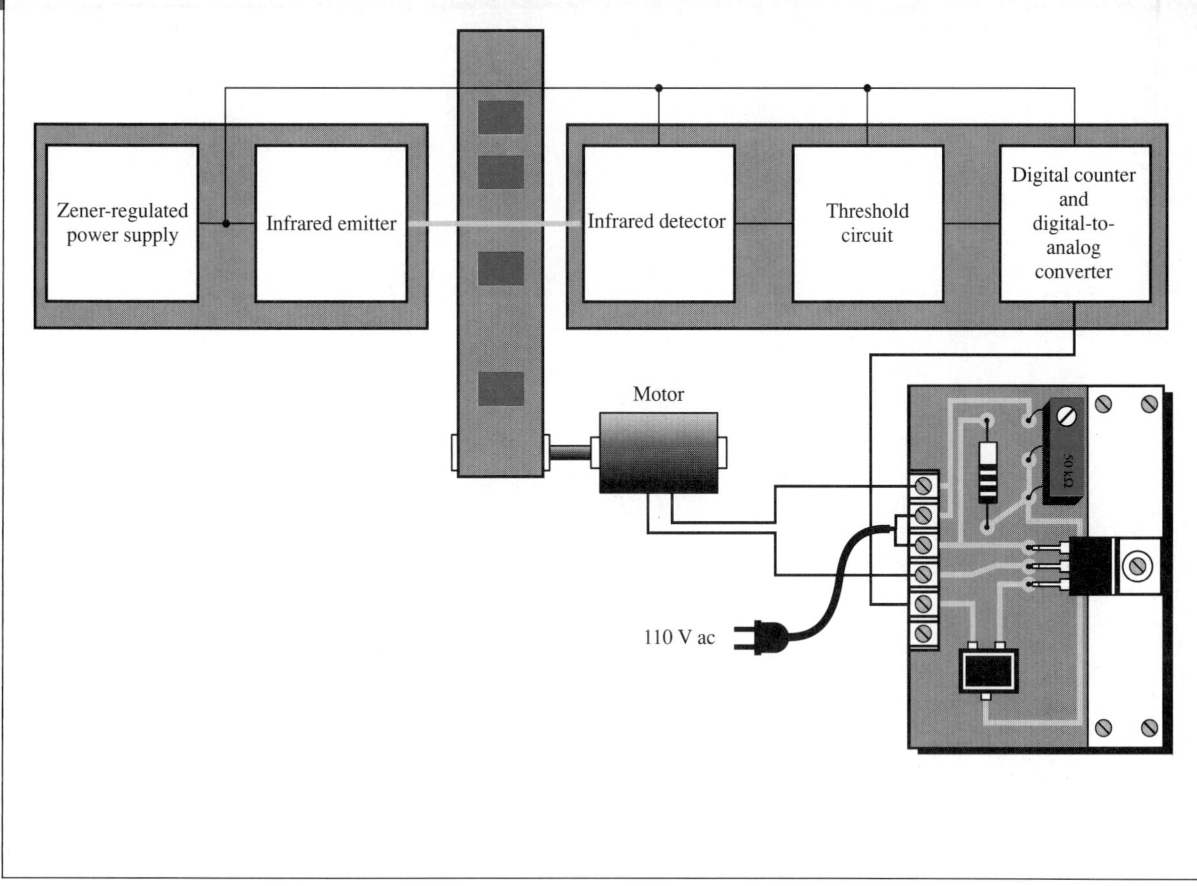

and charging circuits, to name a few. UJTs and PUTs are used as trigger devices for thyristors and also in oscillators and timing circuits.

A System Application

In Chapter 3, you worked with an optical counting system. In this system application, that basic system is modified and expanded to include a speed control for the conveyor track. The optical detector and counter portion of the system counts the number of objects passing by on the moving track in a specified interval of time. The resulting digital number is converted to a proportional analog voltage that is used by the motor speed-control circuit to adjust the speed of the conveyor accordingly. If too many objects pass the detector in the specified time interval, the speed of the electric motor that drives the conveyor is reduced. If too few objects pass the detector in the specified time interval, the motor speed is increased. This system can be used in an industrial setting to regulate the rate at which parts on an assembly line, for example, reach their destination. The focus of the system application in this chapter is the motor speed-control circuit.

For the system application in Section 11–11, in addition to the other topics, be sure you understand

- □ The operation of an SCR.
- □ The operation of a PUT.

11–1 THE SHOCKLEY DIODE

The* Shockley diode *is a* thyristor *with two terminals, the anode and the cathode. It is constructed of four semiconductor layers that form a pnpn structure. The device acts as a switch and remains off until the forward voltage reaches a certain value; then it turns on and conducts. Conduction continues until the current is reduced below a specified value.

The basic construction of a Shockley diode and its schematic symbol are shown in Figure 11–1. The pnpn structure can be represented by an equivalent circuit consisting of a pnp transistor and an npn transistor, as shown in Figure 11–2(a). The upper pnp layers form Q_1 and the lower npn layers form Q_2, with the two middle layers shared by both equivalent transistors. Notice that the base-emitter junction of Q_1 corresponds to pn junction 1 in Figure 11–1, the base-emitter junction of Q_2 corresponds to pn junction 3, and the base-collector junctions of both Q_1 and Q_2 correspond to pn junction 2.

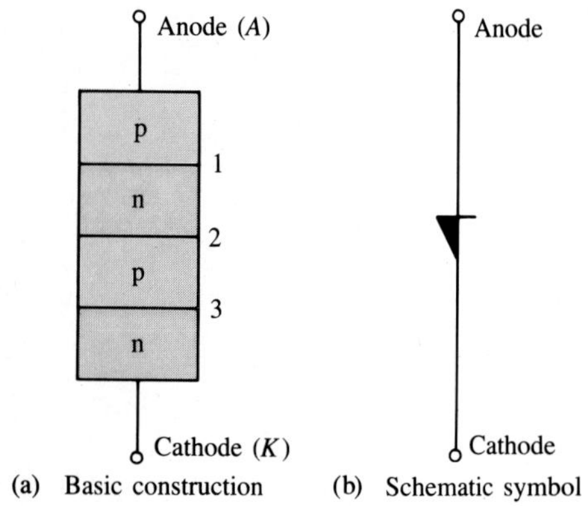

FIGURE 11–1
The Shockley diode.

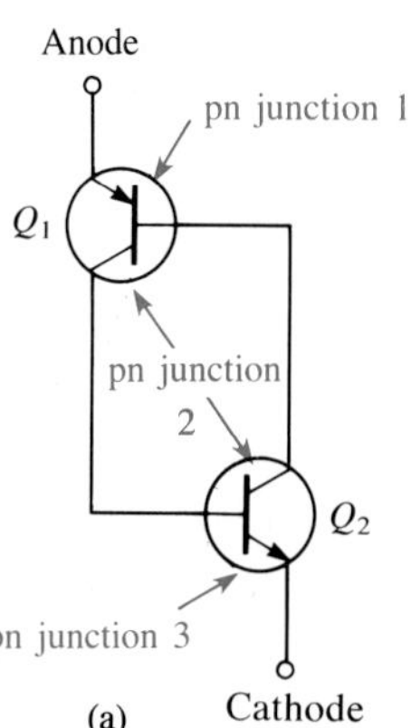

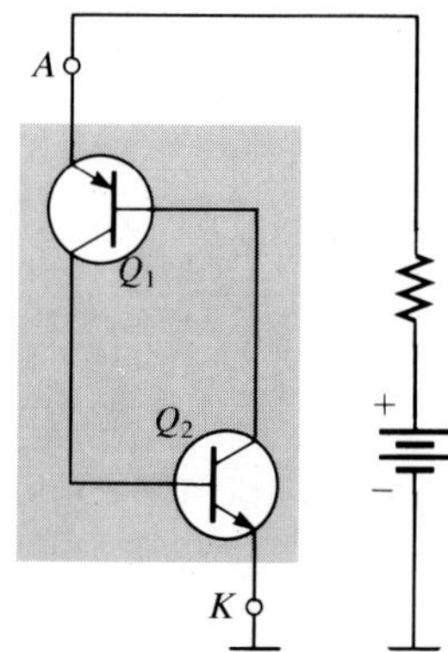

FIGURE 11–2
Shockley diode equivalent circuit.

BASIC OPERATION

When a positive bias voltage is applied to the anode with respect to the cathode, as shown in Figure 11–2(b), the base-emitter junctions of Q_1 and Q_2 (pn junctions 1 and 3 in Figure 11–1[a]) are forward-biased, and the common base-collector junction (pn junction 2 in Figure 11–1[a]) is reverse-biased. Therefore, both equivalent transistors are in the linear region.

At low values of forward-bias voltage, an expression for the anode current is developed as follows, using the standard transistor relationships and Figure 11–3. In this analysis, the component of leakage current, I_{CBO}, is taken into account.

$$I_{B1} = I_{E1} - I_{C1} - I_{CBO1}$$

FIGURE 11–3
Currents in a basic Shockley diode equivalent circuit.

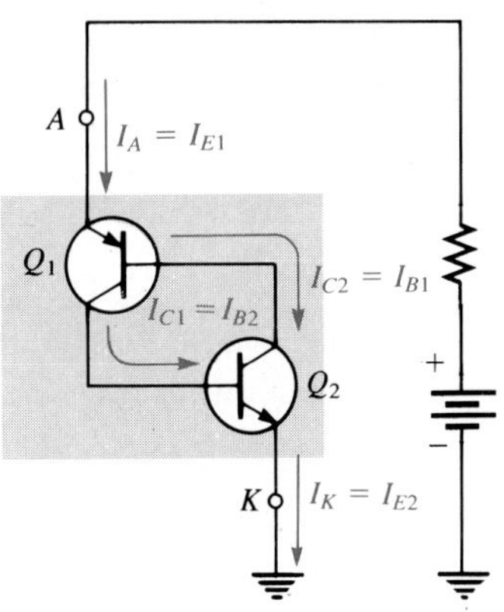

Since $I_C = \alpha_{dc} I_E$,

$$\begin{aligned} I_{B1} &= I_{E1} - \alpha_{dc} I_{E1} - I_{CBO1} \\ &= (1 - \alpha_{dc1}) I_{E1} - I_{CBO1} \end{aligned}$$

The anode current, I_A, is the same as I_{E1}, and therefore

$$I_{B1} = (1 - \alpha_{dc1}) I_A - I_{CBO1} \tag{11–1}$$

$$I_{C2} = \alpha_{dc2} I_{E2} + I_{CBO2}$$

The cathode current, I_K, is the same as I_{E2}, and therefore

$$I_{C2} = \alpha_{dc2} I_K + I_{CBO2} \tag{11–2}$$

Since $I_{C2} = I_{B1}$,

$$\alpha_{dc2} I_K + I_{CBO2} = (1 - \alpha_{dc1}) I_A - I_{CBO1}$$

I_A and I_K are equal. Substituting I_A for I_K in the above equation and solving for I_A, we get

$$\begin{aligned} \alpha_{dc2} I_A + I_{CBO2} &= (1 - \alpha_{dc1}) I_A - I_{CBO1} \\ \alpha_{dc2} I_A - (1 - \alpha_{dc1}) I_A &= -I_{CBO1} - I_{CBO2} \\ I_A[(1 - \alpha_{dc1}) - \alpha_{dc2}] &= I_{CBO1} + I_{CBO2} \end{aligned}$$

$$I_A = \frac{I_{CBO1} + I_{CBO2}}{1 - (\alpha_{dc1} + \alpha_{dc2})} \tag{11–3}$$

At low-current levels, the transistor alpha (α_{dc}) is very small. Therefore, at low-bias levels, there is very little anode current in the Shockley diode as Equation (11–3) demonstrates, and thus it is in the *off* state or forward-blocking region.

EXAMPLE 11–1 A certain Shockley diode is biased in the forward-blocking region with an anode-to-cathode voltage of 20 V. Under this bias condition, $\alpha_{dc1} = 0.35$ and $\alpha_{dc2} = 0.45$. The leakage currents are 100 nA. Determine the anode current and the forward resistance of the diode.

SOLUTION

$$I_A = \frac{I_{CBO1} + I_{CBO2}}{1 - (\alpha_{dc1} + \alpha_{dc2})}$$

$$= \frac{200 \text{ nA}}{1 - 0.8} = \frac{200 \text{ nA}}{0.2} = 1\ \mu\text{A}$$

This is the forward current when the device is off, but forward-biased with V_{AK} = +20 V. The forward resistance is therefore

$$R_{AK} = \frac{V_{AK}}{I_A} = \frac{20 \text{ V}}{1\ \mu\text{A}} = 20 \text{ M}\Omega$$

PRACTICE EXERCISE 11–1

If the anode current is 2 μA, what is the Shockley diode's forward resistance in the forward-blocking region?

FORWARD-BREAKOVER VOLTAGE

At this time, the operation of the Shockley diode may seem very strange because it is forward-biased, yet it acts essentially as an open switch. As previously mentioned, there is a region of forward bias, called the *forward-blocking region,* in which the device has a very high forward resistance (ideally an open). The region exists from $V_{AK} = 0$ up to a value of V_{AK} called the *forward-breakover* voltage, $V_{BR(F)}$. This is indicated on the Shockley diode characteristic curve in Figure 11–4.

As V_{AK} is increased, the anode current I_A gradually increases, as shown on the graph. As I_A increases, so do α_{dc1} and α_{dc2} increase. At the point where $\alpha_{dc1} + \alpha_{dc2} = 1$, the denominator in Equation (11–3) becomes zero, and I_A would become infinitely large, if not limited by an external series resistor.

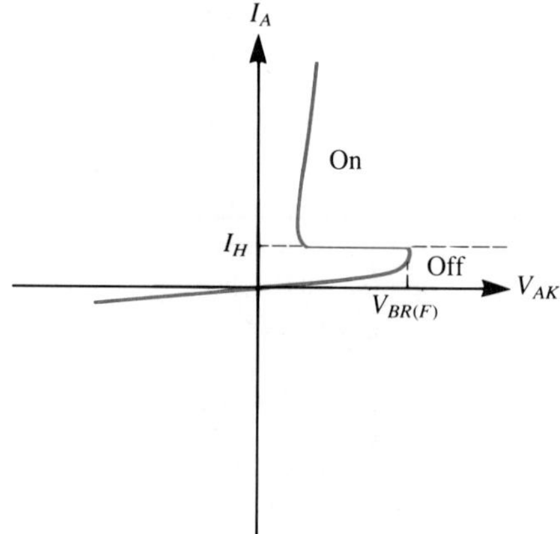

FIGURE 11–4
Shockley diode characteristic curve.

At this point, $V_{AK} = V_{BR(F)}$, and the internal transistor structures become saturated. When this happens, the forward voltage drop V_{AK} suddenly decreases to a low value approximately equal to $V_{BE} + V_{CE(\text{sat})}$ as I_A increases, and the Shockley diode enters the forward-conduction region as indicated in Figure 11–4. Now, the device is in the *on* state and acts as a closed switch. The *on/off* states of the Shockley diode are illustrated in Figure 11–5.

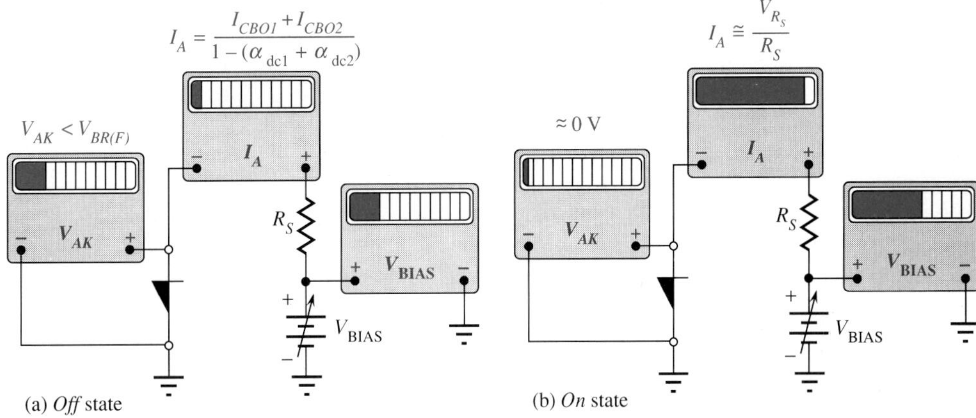

FIGURE 11–5 *On/off* states of the Shockley diode.

EXAMPLE 11–2

(a) Determine the value of anode current in Figure 11–6(a), when the device is off and the dc alphas are 0.4. The leakage currents are 0.08 μA each.

(b) Determine the value of anode current in Figure 11–6(b) when the device is on. $V_{BR(F)} = 40$ V. Assume $V_{BE} = 0.7$ V and $V_{CE(\text{sat})} = 0.1$ V for the internal transistor structure.

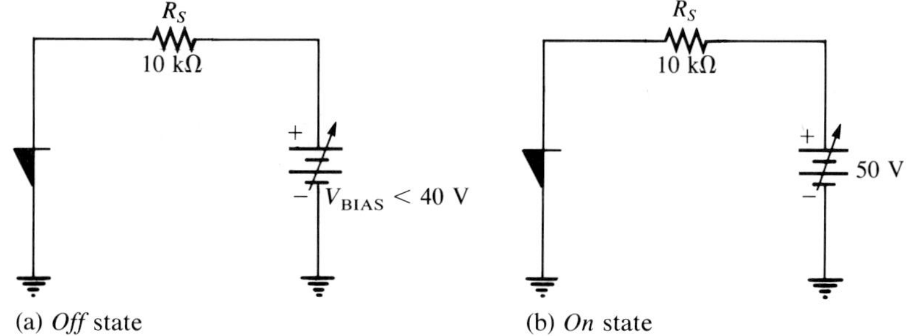

FIGURE 11–6

SOLUTION

(a) $$I_A = \frac{I_{CBO1} + I_{CBO2}}{1 - (\alpha_{dc1} + \alpha_{dc2})} = \frac{0.16\ \mu\text{A}}{1 - 0.8} = 0.8\ \mu\text{A}$$

(b) The voltage at the anode is

$$V_{BE} + V_{CE(\text{sat})} = 0.7\text{ V} + 0.1\text{ V} = 0.8\text{ V}$$

The voltage across R_S is

$$V_{R_S} = V_{\text{BIAS}} - V_A = 50\text{ V} - 0.8\text{ V} = 49.2\text{ V}$$

The anode current is

$$I_A = \frac{V_{R_S}}{R_S} = \frac{49.2\text{ V}}{10\text{ k}\Omega} = 4.92\text{ mA}$$

PRACTICE EXERCISE 11–2

What is the forward resistance of the Shockley diode in Figure 11–6(b)?

HOLDING CURRENT

Once the Shockley diode is conducting (in the *on* state), it will continue to conduct until the anode current is reduced below a specified level, called the *holding current,* I_H. This parameter is also indicated on the characteristic curve in Figure 11–4. *When I_A falls below I_H, the device rapidly switches back to the off state and enters the forward-blocking region.*

SWITCHING CURRENT

The value of the anode current at the point where the device switches from the forward-blocking region (off) to the forward-conduction region (on) is called the *switching current,* I_S. This value of current is always less than the holding current, I_H, which is indicated on the characteristic curve of Figure 11–4.

AN APPLICATION

The circuit in Figure 11–7(a) is a relaxation **oscillator.** The operation is as follows. When the switch is closed, the capacitor charges through R until its voltage reaches the forward-breakover voltage of the Shockley diode. At this point the diode switches into conduction, and the capacitor rapidly discharges through the diode. Discharging continues until the current through the diode falls below the holding value. At this point, the diode switches back to the *off* state, and the capacitor begins to charge again. The result of this action is a voltage waveform across C like that shown in Figure 11–7(b).

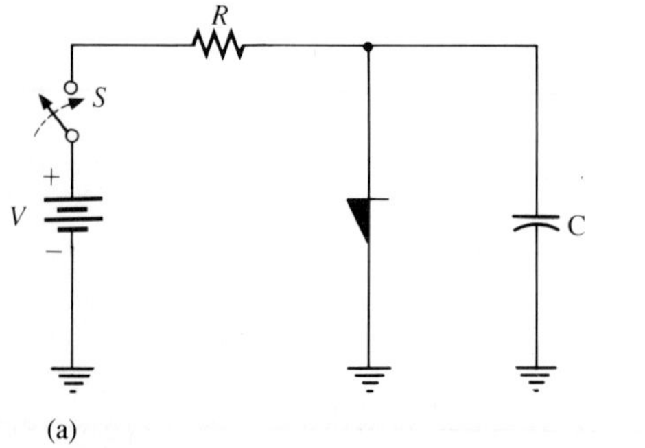

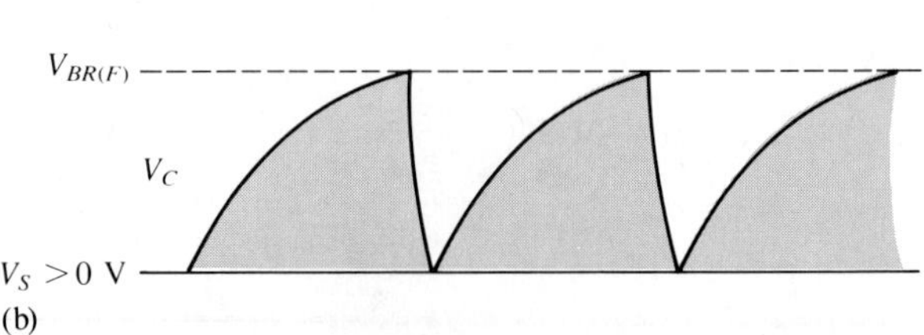

FIGURE 11–7
Shockley diode relaxation oscillator.

11–1 REVIEW QUESTIONS

1. Why is the Shockley diode classified as a thyristor?
2. What is the forward-blocking region?
3. What happens when the anode-to-cathode voltage exceeds the forward-breakover voltage?
4. Once it is on, how can the Shockley diode be turned off?

11–2 THE SILICON-CONTROLLED RECTIFIER (SCR)

The* silicon-controlled rectifer (SCR) *is another four-layer pnpn device similar to the Shockley diode except with three terminals: anode, cathode, and gate. Like the Shockley diode, the SCR has two possible states of operation. In the off state, it acts ideally as an open circuit between the anode and the cathode; actually, rather than an open, there is a very high resistance. In the on state, the SCR acts ideally as a short from the anode to the cathode; actually there is a small on (forward) resistance. The SCR is used in many applications, including motor controls, time delay circuits, heater controls, phase controls, and relay controls, to name a few.

The basic structure of an SCR is shown in Figure 11–8(a) and the schematic symbol is shown in Figure 11–8(b). Typical SCR packages are shown in Figure 11–8(c). Other types of thyristors are found in the same or similar packages.

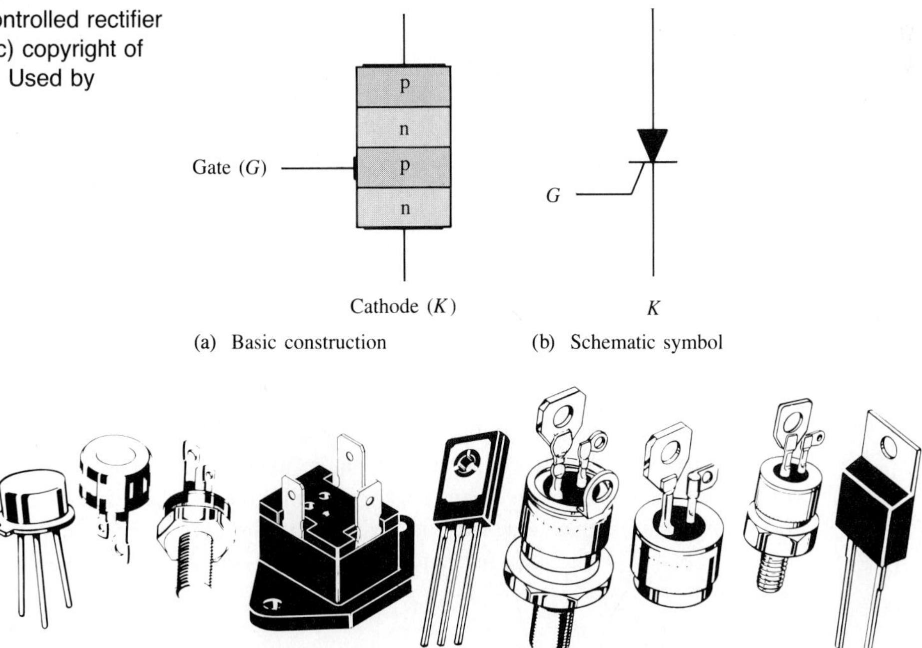

(a) Basic construction (b) Schematic symbol

(c) Typical packages

FIGURE 11–8
The silicon-controlled rectifier (SCR). Part (c) copyright of Motorola, Inc. Used by permission.

SCR Equivalent Circuit

As was done with the Shockley diode, the SCR operation can best be understood by thinking of its internal pnpn structure as a two-transistor arrangement, as shown in Figure 11–9. This structure is like that of the Shockley diode except for the gate connection. The upper pnp layers act as a transistor, Q_1, and the lower npn layers act as a transistor, Q_2. Again, notice that the two middle layers are ''shared.''

Turning the SCR On

When the gate current I_G is zero, as shown in Figure 11–10(a), the device acts as a Shockley diode in the *off* state. In this state, the very high resistance between the anode and cathode can be approximated by an open switch, as indicated. When a positive pulse of current **(trigger)** is applied to the gate, both transistors turn on (the anode must be more positive than the cathode). This action is shown in Figure 11–10(b). I_{B2} turns on Q_2, providing a path for I_{B1} into the Q_2 collector, thus turning on Q_1. The collector current of Q_1 provides additional base current for Q_2 so that Q_2 stays in conduction after the trigger pulse is removed from the gate. By this regenerative action, Q_2 sustains the saturated conduction of Q_1 by providing a path for I_{B1}; in turn, Q_1 sustains the saturated conduction of Q_2 by providing I_{B2}. Thus, the device stays on (latches) once it is triggered on, as shown in Figure 11–10(c). In this state, the very low resistance between the anode and cathode can be approximated by a closed switch, as indicated.

Like the Shockley diode, an SCR can also be turned on without gate triggering by increasing anode-to-cathode voltage to a value exceeding the forward-breakover voltage $V_{BR(F)}$, as shown on the characteristic curve in Figure 11–11(a). The forward-breakover voltage decreases as I_G is increased above 0 V, as shown by the set of curves in Figure 11–11(b). Eventually, a value of I_G is reached at which the SCR turns on at a very low anode-to-cathode voltage. So, as you can see, the gate current controls the value of forward voltage $V_{BR(F)}$ required for turn-on.

Although anode-to-cathode voltages in excess of $V_{BR(F)}$ will not damage the device if current is limited, this situation should be avoided because the normal control of the SCR is lost. It should always be triggered on only with a pulse at the gate.

Turning the SCR Off

When the gate returns to 0 V after the trigger pulse is removed, the SCR cannot turn off; it stays in the forward-conduction region. The anode current must drop below the value of the holding current, I_H, in order for turn-off to occur. The holding current is indicated in Figure 11–11.

There are two basic methods for turning off an SCR—*anode current interruption* and *forced commutation*. The anode current can be interrupted by either a momentary series or parallel switching arrangement, as shown in Figure 11–12. The series switch in part (a) simply reduces the anode current to zero and causes the SCR to turn off. The parallel switch in part (b) routes part of the total current away from the SCR, thereby reducing the anode current to a value less than I_H.

FIGURE 11–9
SCR equivalent circuit.

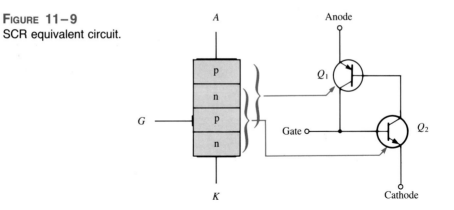

FIGURE 11–10
The SCR turn-on process with the switch equivalents shown.

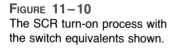

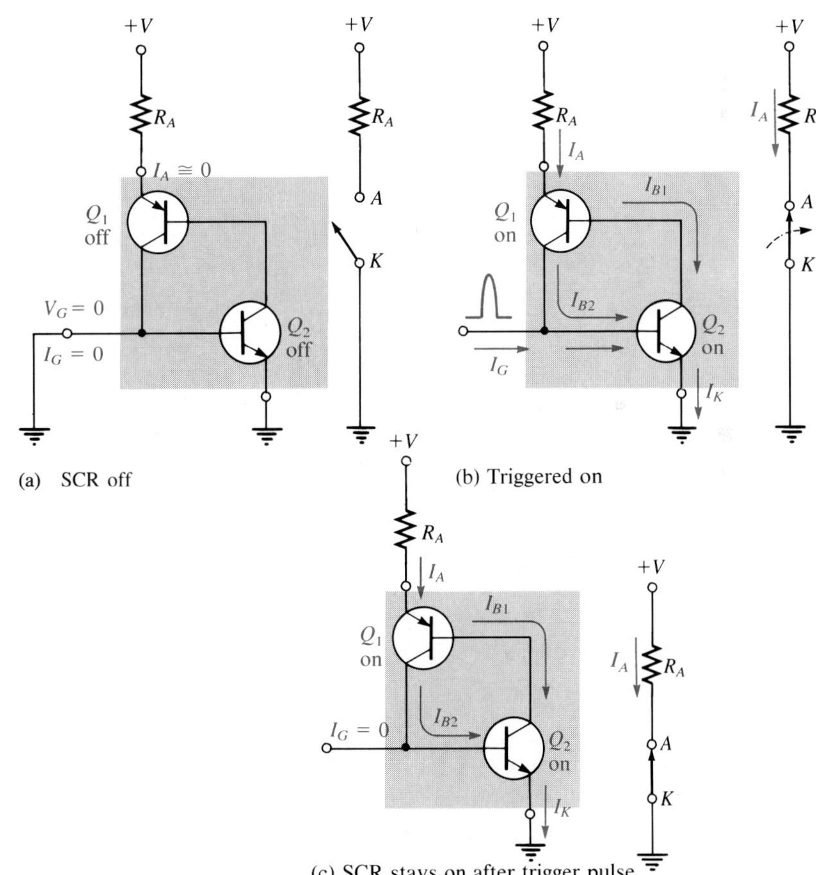

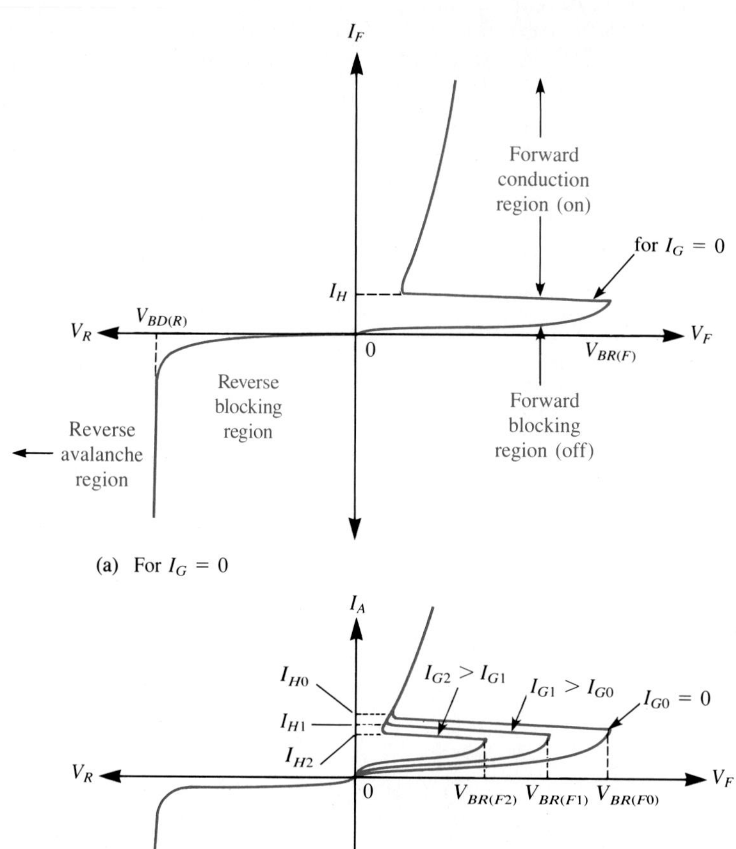

FIGURE 11–11
SCR characteristic curves.

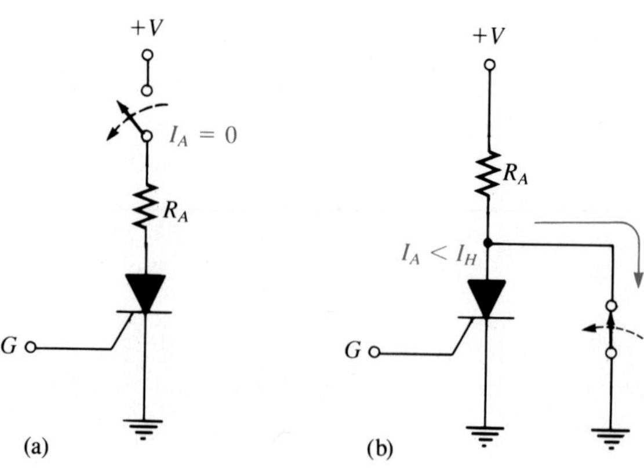

FIGURE 11–12
SCR turn-off by anode current interruption.

The **forced commutation** method basically requires momentarily forcing current through the SCR in the direction opposite to the forward conduction so that the net forward current is reduced below the holding value. The basic circuit, as shown in Figure 11–13, consists of a switch (normally a transistor switch) and a battery in parallel with the SCR. While the SCR is conducting, the switch is open, as shown in part (a). To turn off the SCR, the switch is closed, placing the battery across the SCR and forcing current through it opposite to the forward current, as shown in part (b). Typically, turn-off times for SCRs range from a few microseconds up to about 30 μs.

FIGURE 11–13
SCR turn-off by forced commutation.

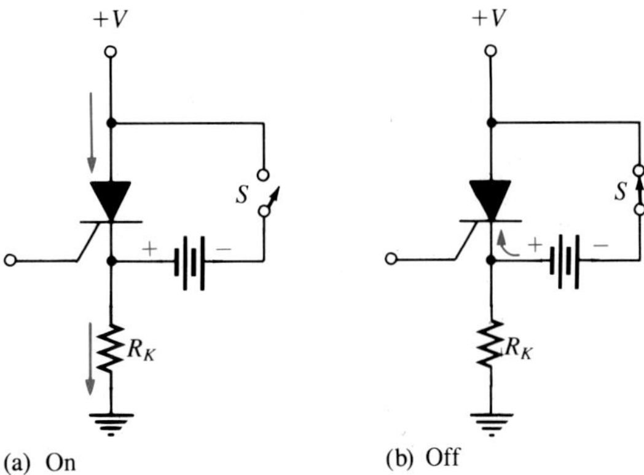

SCR CHARACTERISTICS AND RATINGS

Several of the most important SCR characteristics and ratings are defined as follows. Use the curve in Figure 11–11(a) for reference where appropriate.

FORWARD-BREAKOVER VOLTAGE, $V_{BR(F)}$ This is the voltage at which the SCR enters the forward-conduction region. The value of $V_{BR(F)}$ is maximum when $I_G = 0$ and is designated $V_{BR(F0)}$. When the gate current is increased, $V_{BR(F)}$ decreases and is designated $V_{BR(F1)}$, $V_{BR(F2)}$, and so on, for increasing steps in gate current (I_{G1}, I_{G2}, and so on).

HOLDING CURRENT, I_H This is the value of anode current below which the SCR switches from the forward-conduction region to the forward-blocking region. The value increases with decreasing values of I_G and is maximum for $I_G = 0$.

GATE TRIGGER CURRENT, I_{GT} This is the value of gate current necessary to switch the SCR from the forward-blocking region to the forward-conduction region under specified conditions.

AVERAGE FORWARD CURRENT, $I_{F(AVG)}$ This is the maximum continuous anode current (dc) that the device can withstand in the conduction state under specified conditions.

FORWARD-CONDUCTION REGION This region corresponds to the *on* condition of the SCR where there is forward current from anode to cathode through the very low resistance (approximate short) of the SCR.

FORWARD- AND REVERSE-BLOCKING REGIONS These regions correspond to the *off* condition of the SCR where the forward current from anode to cathode is blocked by the effective open circuit of the SCR.

REVERSE-BREAKDOWN VOLTAGE, $V_{BD(R)}$ This parameter specifies the value of reverse voltage from cathode to anode at which the device breaks into the avalanche region and begins to conduct heavily (the same as in a pn junction diode).

11–2 REVIEW QUESTIONS

1. What is an SCR?
2. Name the SCR terminals.
3. How can an SCR be turned on (made to conduct)?
4. How can an SCR be turned off?

11–3 SCR APPLICATIONS

The SCR has many uses in the areas of power control and switching applications. A few of the basic applications are described in this section.

ON-OFF CONTROL OF CURRENT

Figure 11–14 shows an SCR circuit that permits current to be switched to a load by the momentary closure of switch S_1 and removed from the load by the momentary closure of switch S_2.

Assuming the SCR is initially off, momentary closure of S_1 provides a pulse of current into the gate, thus triggering the SCR on so that it conducts current through R_L. The SCR remains in conduction even after the momentary contact of S_1 is removed. When S_2 is momentarily closed, current is shunted around the SCR, thus reducing its anode current below the holding value I_H. This turns the SCR off and thus reduces the load current to zero.

HALF-WAVE POWER CONTROL

A common application of SCRs is in the control of ac power for lamp dimmers, electric heaters, and electric motors.

A half-wave, variable-resistance, phase-control circuit is shown in Figure 11–15; 120 V ac are applied across terminals A and B; R_L represents the resistance of the load (for example, a heating element or lamp filament). R_1 is a current-limiting resistor, and potentiometer R_2 sets the trigger level for the SCR. By adjusting R_2, the SCR can be made to trigger at any point on the positive half-cycle of the ac waveform between 0° and 90°, as shown in Figure 11–16.

When the SCR triggers near the beginning of the cycle (approximately 0°), as in Figure 11–16(a), it conducts for approximately 180° and maximum power is delivered to the load. When it fires near the peak of the positive half-cycle (90°), as in Figure

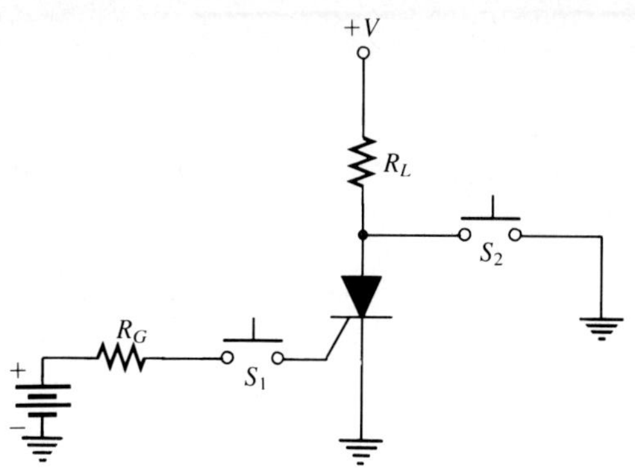

FIGURE 11–14
On-off SCR control circuit.

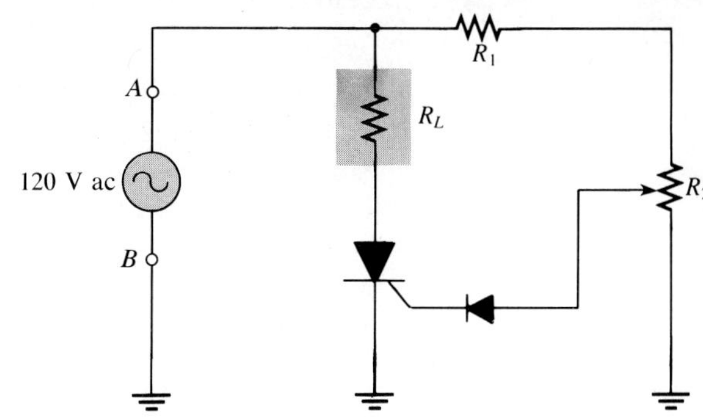

FIGURE 11–15
Half-wave, variable-resistance, phase-control circuit.

FIGURE 11–16
Operation of the phase-control circuit.

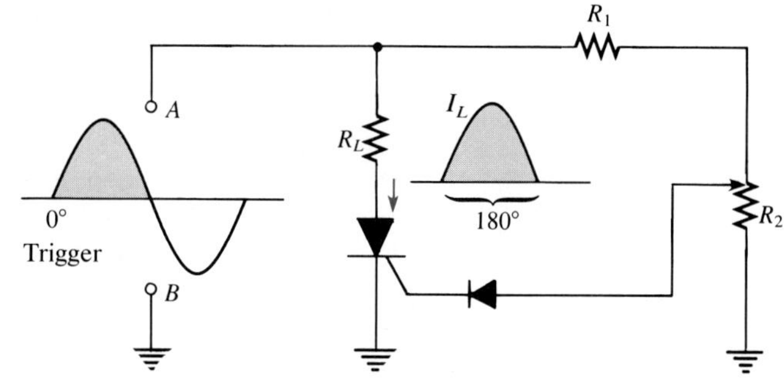

(a) 180° conduction

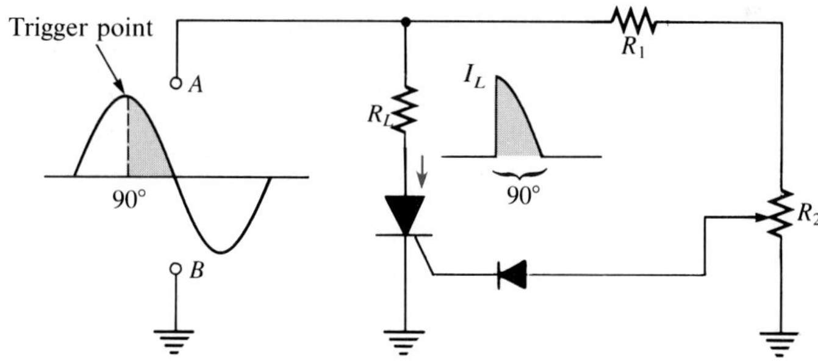

(b) 90° conduction

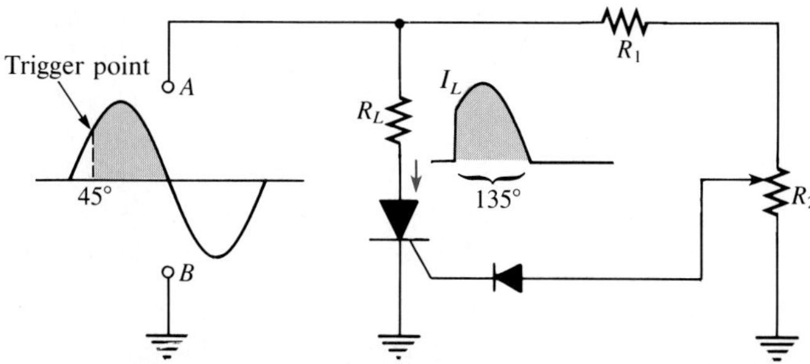

(c) 135° conduction

11–16(b), the SCR conducts for approximately 90° and less power is delivered to the load. By adjusting R_2, triggering can be made to occur anywhere between these two extremes, and therefore, a variable amount of power can be delivered to the load. Figure 11–16(c) shows triggering at the 45° point as an example. When the ac input goes negative, the SCR turns off and does not conduct again until the trigger point on the next positive half-cycle. The diode prevents the negative ac voltage from being applied to the gate of the SCR.

Lighting System for Power Interruptions

As another example of SCR applications, we will examine a circuit that will maintain lighting by using a backup battery when there is an ac power failure. Figure 11–17 shows a center-tapped full-wave rectifier used for providing ac power to a low-voltage lamp. As long as the ac power is available, the battery charges through diode D_3 and R_1.

The SCR's cathode voltage is established when the capacitor charges to the peak value of the full-wave rectified ac (6.3 V rms less the drops across R_2 and D_1). The anode is at the 6 V battery voltage, making it less positive than the cathode, thus preventing

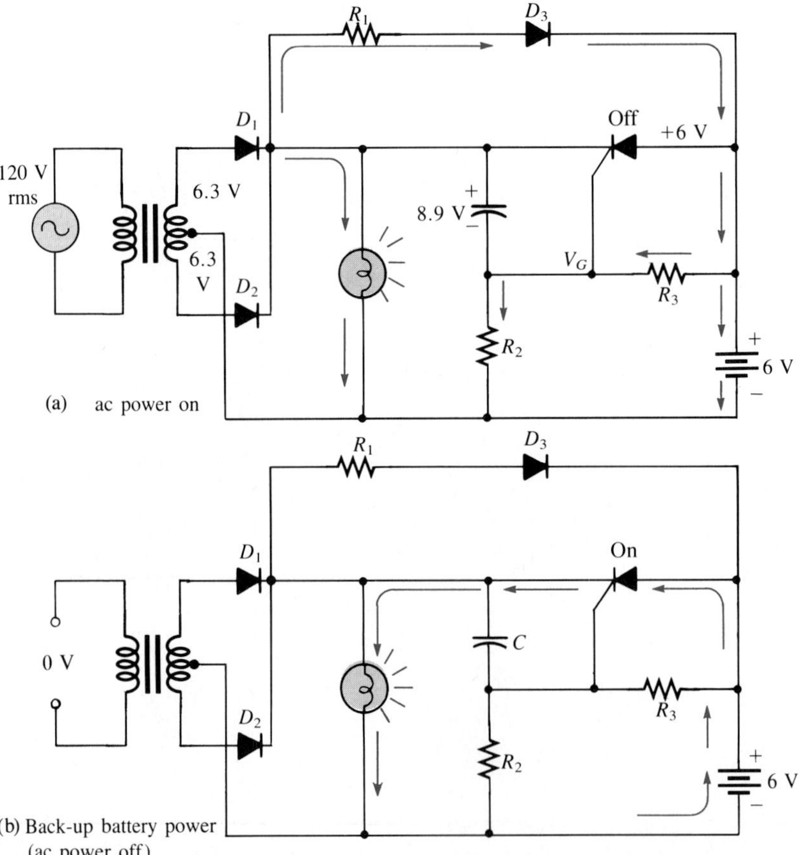

FIGURE 11–17 Automatic back-up lighting circuit.

conduction. The SCR's gate is at a voltage established by the voltage divider made up of R_2 and R_3. Under these conditions the lamp is illuminated by the ac input power and the SCR is off, as shown in Figure 11–17(a).

When there is an interruption of ac power, the capacitor discharges through the closed path D_3, R_1, and R_3, making the cathode less positive than the anode or the gate. This action establishes a triggering condition, and the SCR begins to conduct. Current from the battery is through the SCR and the lamp, thus maintaining illumination, as shown in Figure 11–17(b). When ac power is restored, the capacitor recharges and the SCR turns off. The battery begins recharging.

AN OVER-VOLTAGE PROTECTION CIRCUIT

Figure 11–18 shows a simple over-voltage protection circuit, sometimes called a "crowbar" circuit, in a dc power supply. The dc output voltage from the regulator is monitored by the zener diode D_1 and the resistive voltage divider (R_1 and R_2). The upper limit of the output voltage is set by the zener voltage. If this voltage is exceeded, the zener conducts and the voltage divider produces an SCR trigger voltage. The trigger voltage fires the SCR, which is connected across the line voltage. The SCR current causes the fuse to blow, thus disconnecting the line voltage from the power supply.

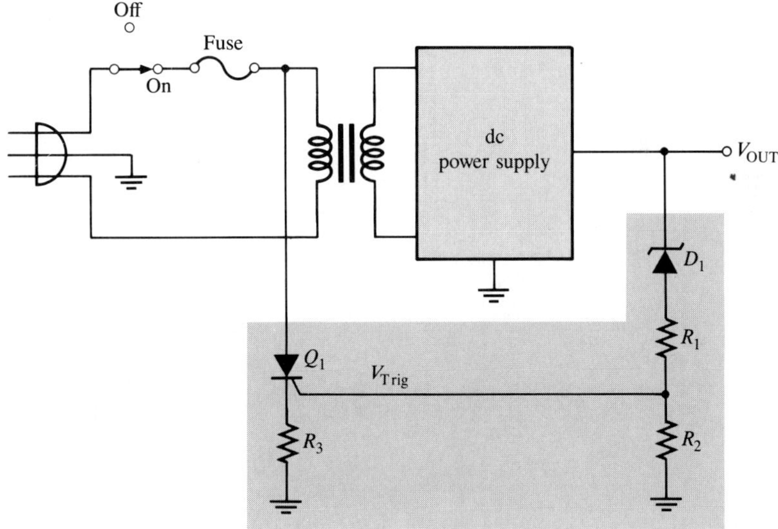

FIGURE 11–18
A basic SCR over-voltage protection circuit.

11–3 REVIEW QUESTIONS

1. If the potentiometer in Figure 11–16 is set at its midpoint, the SCR will conduct for more than ________ degrees but less than ________ degrees of the input cycle.
2. In Figure 11–17, what is the purpose of diode D_3?

11–4 THE SILICON-CONTROLLED SWITCH (SCS)

The* silicon-controlled switch (SCS) *is similar in construction to the SCR. The SCS, however, has two gate terminals, the cathode gate and the anode gate. The SCS can be turned on and off using either gate terminal. Remember that the SCR can be only turned on using its gate terminal. Normally, the SCS is available in power ratings lower than those of the SCR.

The SCS symbol and terminal identification are shown in Figure 11–19. As with the previous four-layer devices, the basic operation of the SCS can be understood by referring to the transistor equivalent, shown in Figure 11–20. To start, assume that both Q_1 and Q_2 are off, and therefore that the SCS is not conducting. A positive pulse on the cathode gate drives Q_2 into conduction and thus provides a path for Q_1 base current. When Q_1 turns on, its collector current provides base drive to Q_2, thus sustaining the *on* state of the device. This regenerative action is the same as in the turn-on process of the SCR and the Shockley diode and is illustrated in Figure 11–20(a).

FIGURE 11–19
The silicon-controlled switch (SCS).

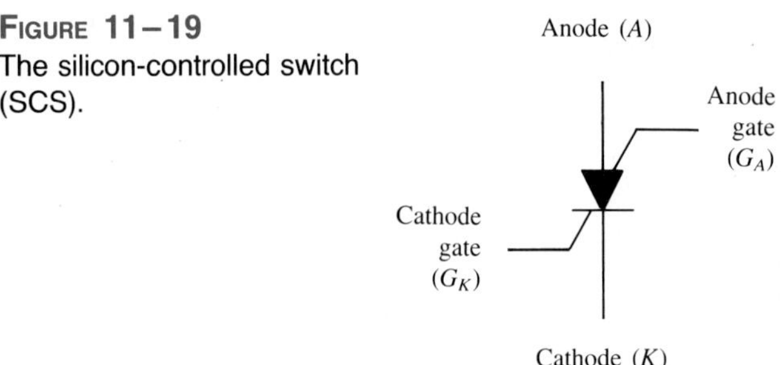

FIGURE 11–20
SCS operation.

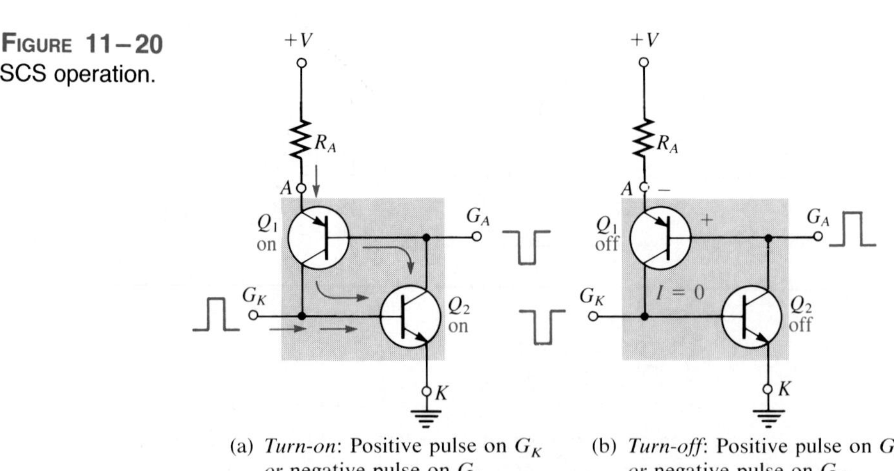

(a) *Turn-on*: Positive pulse on G_K *or* negative pulse on G_A

(b) *Turn-off*: Positive pulse on G_A *or* negative pulse on G_K

The SCS can also be turned on with a negative pulse on the anode gate, as indicated in Figure 11–20(a). This drives Q_1 into conduction which, in turn, provides base current for Q_2. Once Q_2 is on, it provides a path for Q_1 base current, thus sustaining the *on* state.

To turn the SCS off, a positive pulse is applied to the anode gate. This reverse-biases the base-emitter junction of Q_1 and turns it off. Q_2, in turn, cuts off and the SCS ceases conduction, as shown in Figure 11–20(b). The device can also be turned off with a negative pulse on the cathode gate, as indicated in part (b). The SCS typically has a faster turn-off time than the SCR.

In addition to the positive pulse on the anode gate or the negative pulse on the cathode gate, there are other methods for turning off an SCS. Figure 11–21(a) and (b) shows two switching methods to reduce the anode current below the holding value. In each case, the bipolar transistor acts as a switch.

FIGURE 11–21
Transistor switches reduce I_A below I_H and turn off the SCS.

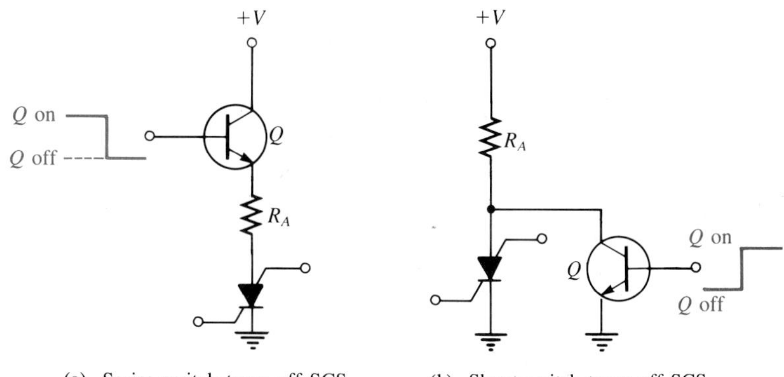

APPLICATIONS

The SCS and SCR are used in similar applications. The SCS has the advantage of faster turn-off with pulses on either gate terminal; however, it is more limited in terms of maximum current and voltage ratings. Also, the SCS is sometimes used in digital applications such as counters, registers, and timing circuits.

11–4 REVIEW QUESTIONS

1. Explain the difference between an SCS and an SCR.
2. How can an SCS be turned on?
3. Describe four ways an SCS can be turned off.

11–5 THE DIAC AND TRIAC

Both the diac and the triac are types of thyristors that can conduct current in both directions (bilateral). The difference between the two is that the diac has two terminals, the anode and the cathode, while the triac has a third terminal, which is the gate. The diac functions basically like two parallel Shockley diodes turned in opposite directions. The triac functions basically like two parallel SCRs turned in opposite directions with a common gate terminal.

THE DIAC

The **diac** basic construction and schematic symbol are shown in Figure 11–22. Notice that there are two terminals, labelled A_1 and A_2. Conduction occurs in the diac when the breakover voltage is reached with either polarity across the two terminals. The curve in Figure 11–23 illustrates this characteristic. Once breakover occurs, current is in a direction depending on the polarity of the voltage across the terminals. The device turns off when the current drops below the holding value.

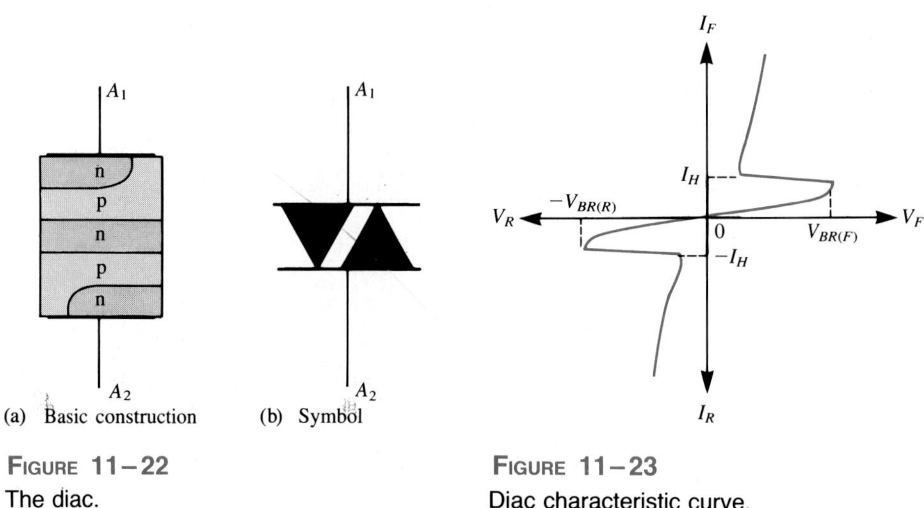

FIGURE 11–22
The diac.

FIGURE 11–23
Diac characteristic curve.

The equivalent circuit of a diac consists of four transistors arranged as shown in Figure 11–24(a). When the diac is biased as in Figure 11–24(b), the pnpn structure from A_1 to A_2 provides the four-layer device operation as was described for the Shockley diode. In the equivalent circuit, Q_1 and Q_2 are forward-biased, and Q_3 and Q_4 are reverse-biased. The device operates on the upper right portion of the characteristic curve in Figure 11–23 under this bias condition. When the diac is biased as shown in Figure 11–24(c), the pnpn structure from A_2 to A_1 is used. In the equivalent circuit, Q_3 and Q_4 are forward-biased, and Q_1 and Q_2 are reverse-biased. The device operates on the lower left portion of the characteristic curve, as shown in Figure 11–23.

THE TRIAC

The **triac** is like a diac with a gate terminal. The triac can be turned on by a pulse of gate current and does not require the breakover voltage to initiate conduction, as does the diac. Basically, the triac can be thought of simply as two SCRs connected in parallel and in opposite directions with a common gate terminal. Unlike the SCR, the triac can conduct current in either direction when it is triggered on, depending on the polarity of the voltage across its A_1 and A_2 terminals. Figure 11–25(a) and (b) shows the basic construction and

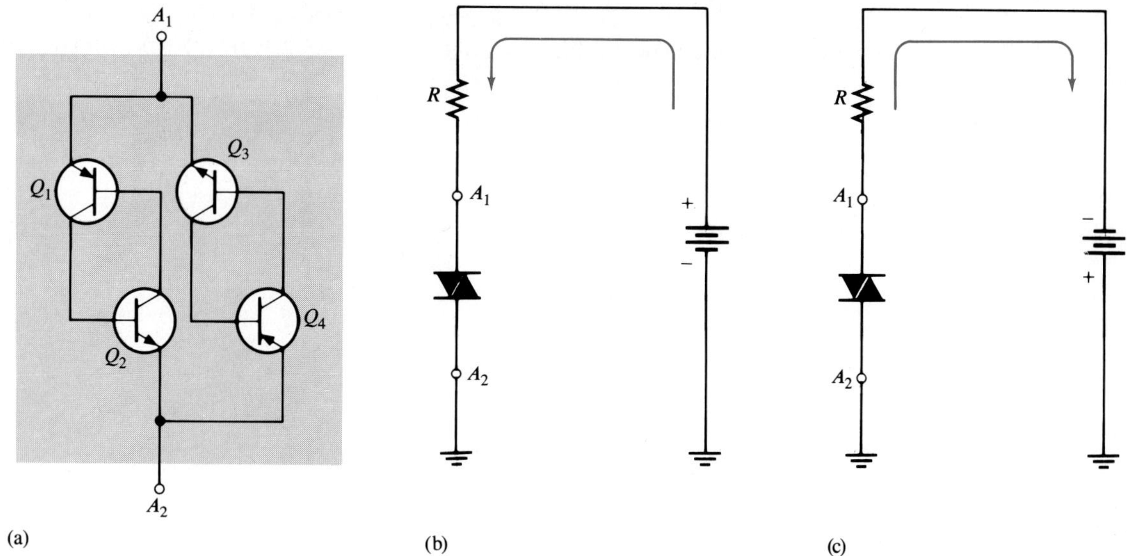

FIGURE 11–24
Diac equivalent circuit and bias conditions.

FIGURE 11–25
The triac.

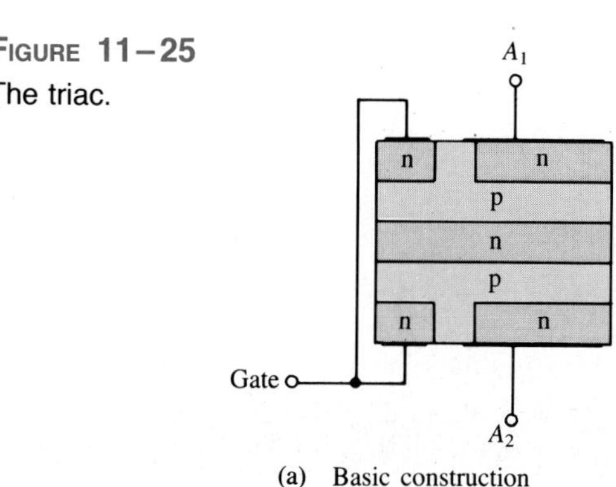

(a) Basic construction

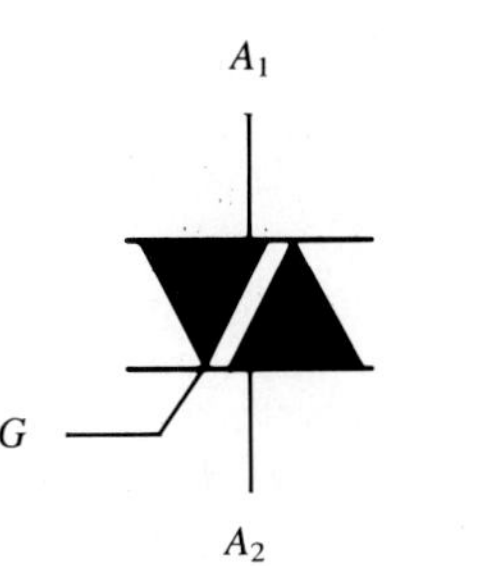

(b) Symbol

schematic symbol for the triac. The characteristic curve is shown in Figure 11–26. Notice that the breakover potential decreases as the gate current increases, just as with the SCR.

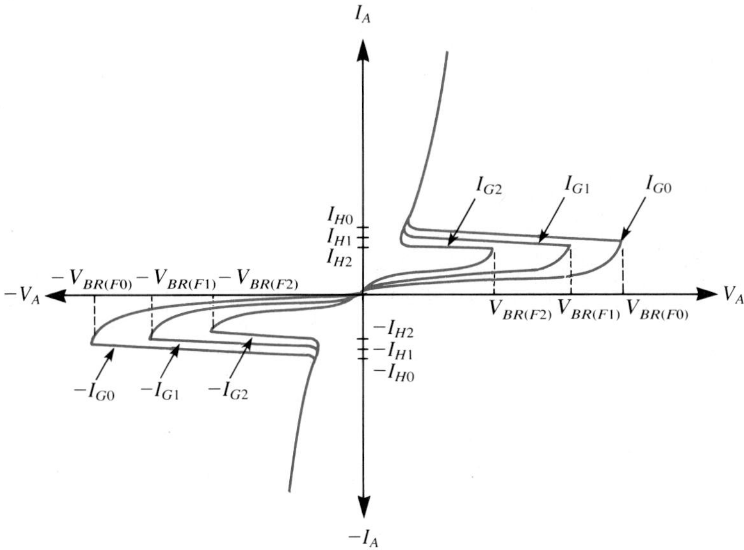

FIGURE 11–26
Triac characteristic curves.

As with other four-layer devices, the triac ceases to conduct when the anode current drops below the specified value of the holding current, I_H. The only way to turn off the triac is to reduce the current to a sufficiently low level.

Figure 11–27 shows the triac being triggered into both directions of conduction. In part (a), terminal A_1 is biased positive with respect to A_2, so the triac conducts as shown when triggered by a positive pulse at the gate terminal. The transistor equivalent circuit in part (b) shows that Q_1 and Q_2 conduct when a positive trigger pulse is applied. In part (c), terminal A_2 is biased positive with respect to A_1, so the triac conducts as shown. In this case, Q_3 and Q_4 conduct as indicated in part (d) upon application of a positive trigger pulse.

APPLICATIONS

Like the SCR, triacs are also used to control average power to a load by the method of phase control. The triac can be triggered such that the ac power is supplied to the load for a controlled portion of each half-cycle. During each positive half-cycle of the ac, the triac is off for a certain interval, called the *delay angle* (measured in degrees), and then it is triggered on and conducts current through the load for the remaining portion of the positive half-cycle, called the *conduction angle*. Similar action occurs on the negative half-cycle except that, of course, current is conducted in the opposite direction through the load. Figure 11–28 illustrates this action.

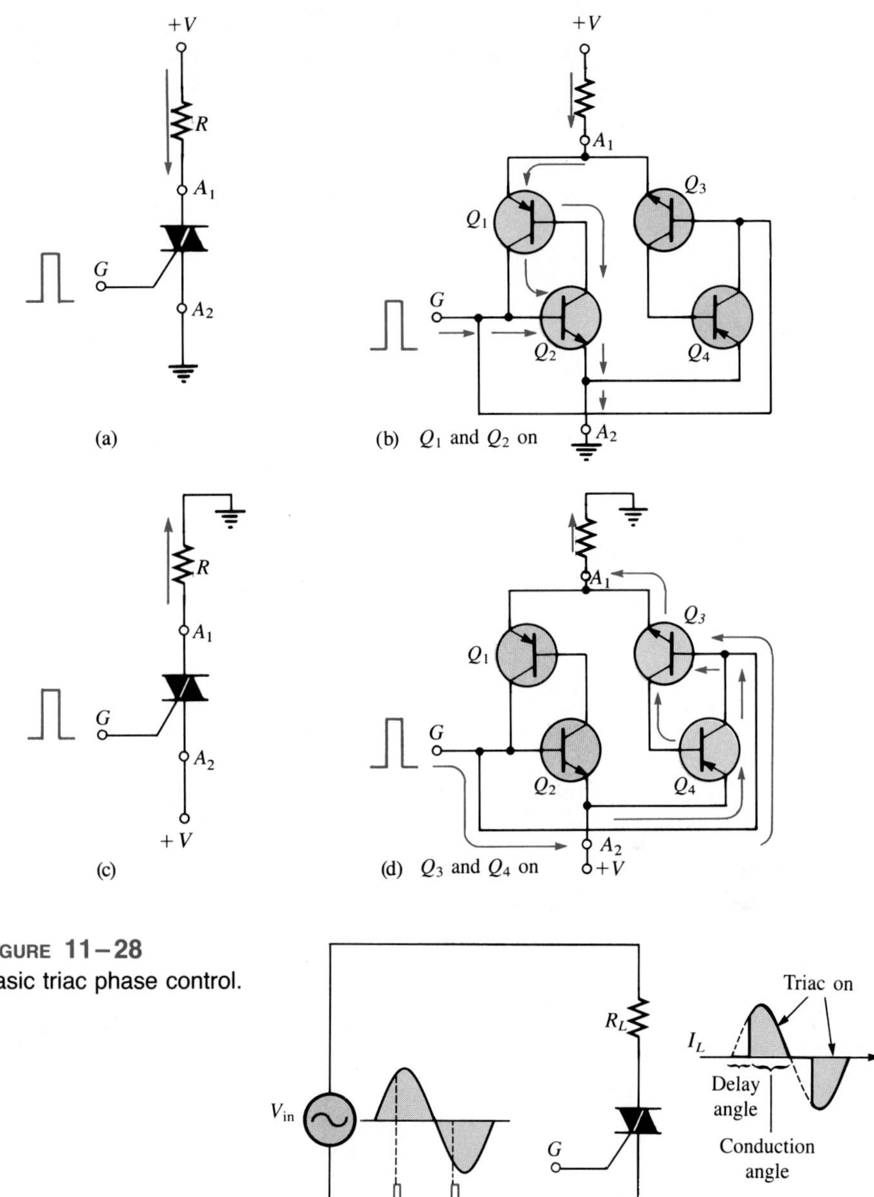

FIGURE 11–27
Bilateral operation of a triac.

FIGURE 11–28
Basic triac phase control.

One example of phase control using a triac is illustrated in Figure 11–29(a). Diodes are used to provide trigger pulses to the gate of the triac. Diode D_1 conducts during the positive half-cycle. The value of R_1 sets the point on the positive half-cycle at which the triac triggers. Notice that during this portion of the ac cycle, A_1 and G are positive with respect to A_2.

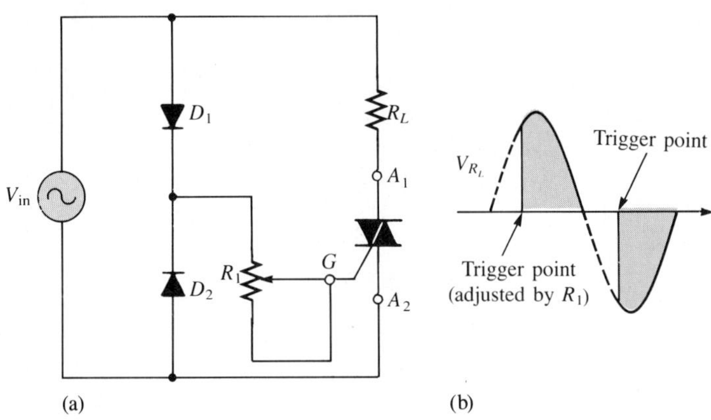

FIGURE 11–29
Triac phase-control circuit.

Diode D_2 conducts during the negative half-cycle, and R_1 sets the trigger point. Notice that during this portion of the ac cycle, A_2 and G are positive with respect to A_1. The resulting waveform across R_L is shown in Figure 11–29(b).

In the phase-control circuit, it is necessary that the triac turn off at the end of each positive and each negative alternation of the ac. Figure 11–30 illustrates that there is an interval near each 0 crossing where the triac current drops below the holding value, thus turning the device off.

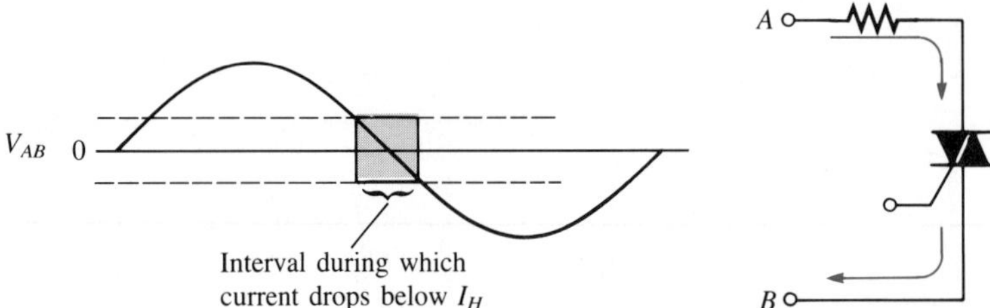

FIGURE 11–30
Triac turn-off interval.

11–5 REVIEW QUESTIONS

1. Compare the diac to the Shockley diode in terms of basic operation.
2. Compare the triac with the SCR in terms of basic operation.
3. How does a triac differ from a diac?

11–6 THE UNIJUNCTION TRANSISTOR (UJT)

The unijunction transistor is not a member of the thyristor family because it does not have a four-layer type of construction. The term unijunction refers to the fact that the UJT has a single pn junction. As you will see in this section, the UJT is useful in certain oscillator applications and as a triggering device in thyristor circuits.

The **unijunction transistor (UJT)** is a three-terminal device whose basic construction is shown in Figure 11–31(a); the schematic symbol appears in Figure 11–31(b). Notice the terminals are labelled *emitter, base 1* (*B*1), and *base 2* (*B*2). Do no confuse this symbol with that of a JFET; the difference is that *the arrow is at an angle for the UJT*. The UJT has only one pn junction, and therefore, the characteristics of this device are very different from those of either the bipolar junction transistor or the FET, as you will see.

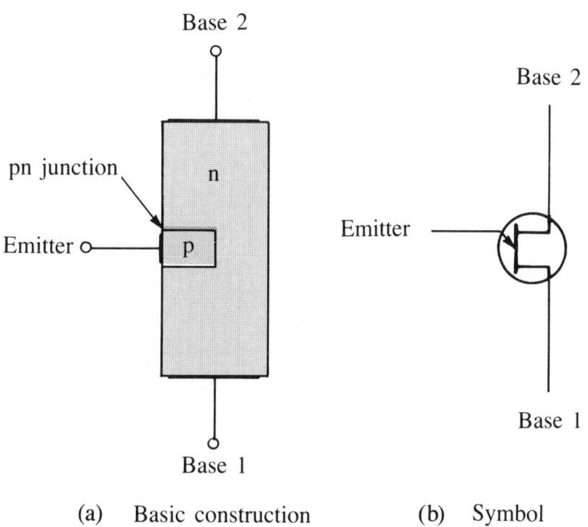

FIGURE 11–31
The unijunction transistor (UJT).

EQUIVALENT CIRCUIT

The equivalent circuit for the UJT, shown in Figure 11–32(a), will aid in understanding the basic operation. The diode shown in the figure represents the pn junction, r_{B1} represents the internal bulk resistance of the silicon bar between the emitter and base 1, and r_{B2} represents the bulk resistance between the emitter and base 2. The sum $r_{B1} + r_{B2}$ is the total resistance between the base terminals and is called the *interbase resistance*, r_{BB}.

$$r_{BB} = r_{B1} + r_{B2} \qquad (11\text{–}4)$$

The value of r_{B1} varies inversely with emitter current I_E, and therefore, it is shown as a variable resistor. Depending on I_E, the value of r_{B1} can vary from several thousand

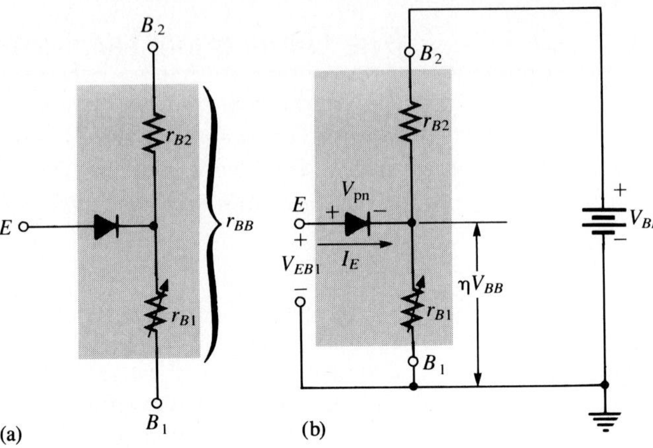

FIGURE 11–32
UJT equivalent circuit.

ohms down to tens of ohms. r_{B1} and r_{B2} form a voltage divider when the device is biased as shown in Figure 11–32(b). The voltage across the resistance r_{B1} can be expressed as

$$V_{r_{B1}} = \left(\frac{r_{B1}}{r_{BB}}\right)V_{BB} \tag{11–5}$$

STANDOFF RATIO

The ratio r_{B1}/r_{BB} is a UJT characteristic called the intrinsic **standoff ratio** and designated by η (Greek *eta*).

$$\eta = \frac{r_{B1}}{r_{BB}} \tag{11–6}$$

As long as the applied emitter voltage V_{EB1} is less than $V_{r_{B1}} + V_{\text{pn}}$, there is no emitter current because the pn junction is not forward-biased (V_{pn} is the barrier potential of the pn junction). The value of emitter voltage that causes the pn junction to become forward-biased is called V_P (peak-point voltage) and is expressed as

$$V_P = \eta V_{BB} + V_{\text{pn}} \tag{11–7}$$

When V_{EB1} reaches V_P, the pn junction becomes forward-biased and I_E begins. Holes are injected into the n-type bar from the p-type emitter. This increase in holes causes an increase in free electrons, thus increasing the conductivity between emitter and $B1$ (decreasing r_{B1}).

After turn-on, the UJT operates in a negative resistance region up to a certain value of I_E, as shown by the characteristic curve in Figure 11–33. As you can see, after the peak point ($V_E = V_P$ and $I_E = I_P$), V_E decreases as I_E continues to increase, thus producing the negative resistance characteristic. Beyond the valley point ($V_E = V_V$ and $I_E = I_V$), the device is in saturation, and V_E increases very little with an increasing I_E.

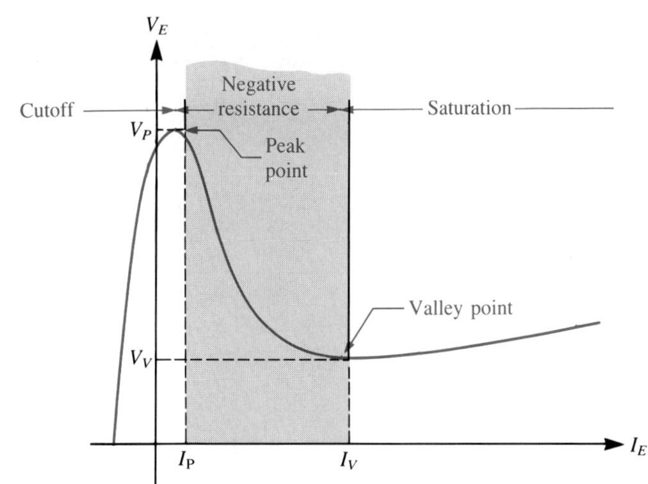

FIGURE 11–33
UJT characteristic curve for a fixed value of V_{BB}.

■ **EXAMPLE 11–3** The data sheet of a certain UJT gives $\eta = 0.6$. Determine the peak-point emitter voltage V_P if $V_{BB} = 20$ V.

SOLUTION

$$\begin{aligned} V_P &= \eta V_{BB} + V_{pn} \\ &= 0.6(20\text{ V}) + 0.7\text{ V} \\ &= 12.7\text{ V} \end{aligned}$$

PRACTICE EXERCISE 11–3

How can the peak-point emitter voltage of a UJT be increased? ■

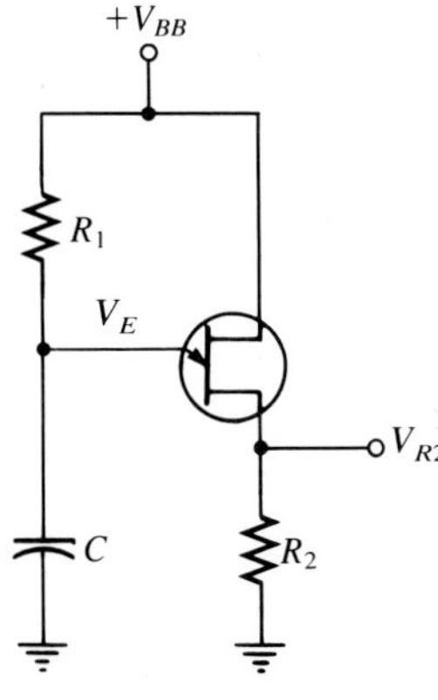

FIGURE 11–34
Relaxation oscillator.

UJT APPLICATIONS

The UJT is often used as a trigger device for SCRs and triacs. Other applications include nonsinusoidal oscillators, sawtooth generators, phase control, and timing circuits. Figure 11–34 shows a UJT relaxation oscillator as an example of one application. Also, this type of circuit is basic to other timing and trigger circuits.

The operation is as follows. When dc power is applied, the capacitor C charges exponentially through R_1 until it reaches the peak-point voltage V_P. At this point, the pn junction becomes forward-biased, and the emitter characteristic goes into the negative resistance region (V_E decreases and I_E increases). The capacitor then quickly discharges through the forward-biased junction, r_B, and R_2. When the capacitor voltage decreases to the valley-point voltage V_V, the UJT turns off, the capacitor begins to charge again, and the cycle is repeated, as shown in the emitter voltage waveform in Figure 11–35 (top). During the discharge time of the capacitor, the UJT is conducting. Therefore, a voltage is developed across R_2, as shown in the waveform diagram in Figure 11–35 (bottom).

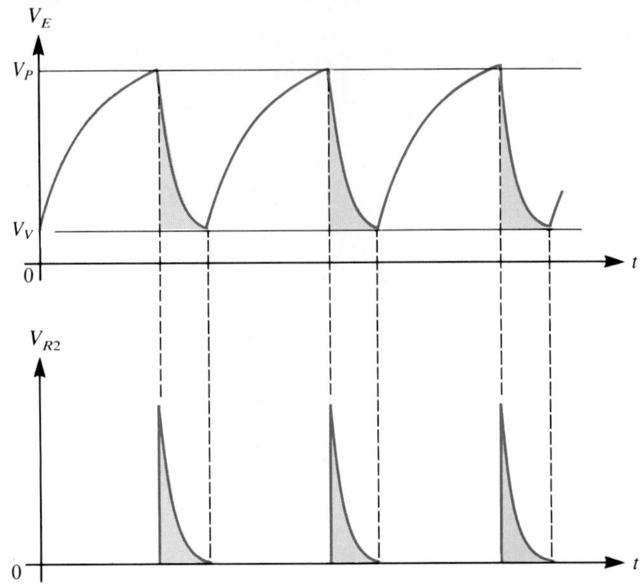

FIGURE 11–35
Waveforms for UJT relaxation oscillator.

CONDITIONS FOR TURN-ON AND TURN-OFF

In the relaxation oscillator of Figure 11–34, certain conditions must be met for the UJT to reliably turn on and turn off. First, to ensure turn-on, R_1 must not limit I_E at the peak point to less than I_P. To ensure this, the voltage drop across R_1 at the peak point should be greater than $I_P R_1$. Thus, the condition for turn-on is

$$V_{BB} - V_P > I_P R_1$$

or

$$R_1 < \frac{V_{BB} - V_P}{I_P} \tag{11–8}$$

To ensure turn-off of the UJT at the valley point, R_1 must be large enough that I_E (at the valley point) can decrease below the specified value of I_V. This means that the voltage across R_1 at the valley point must be less than $I_V R_1$. Thus, the condition for turn-off is

$$V_{BB} - V_V < I_V R_1$$

or

$$R_1 > \frac{V_{BB} - V_V}{I_V} \tag{11–9}$$

So, for a proper turn-on and turn-off, R_1 must be in the range

$$\frac{V_{BB} - V_P}{I_P} > R_1 > \frac{V_{BB} - V_V}{I_V}$$

EXAMPLE 11–4 Determine a value of R_1 in Figure 11–36 that will ensure proper turn-on and turn-off of the UJT. The characteristic of the UJT exhibits the following values: $\eta = 0.5$, $V_V = 1$ V, $I_V = 10$ mA, $I_P = 20$ μA, and $V_P = 14$ V.

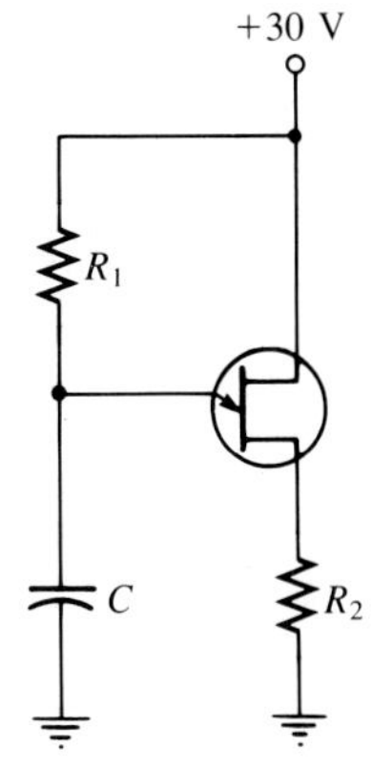

FIGURE 11–36

SOLUTION

$$\frac{V_{BB} - V_P}{I_P} > R_1 > \frac{V_{BB} - V_V}{I_V}$$

$$\frac{30\text{ V} - 14\text{ V}}{20\ \mu\text{A}} > R_1 > \frac{30\text{ V} - 1\text{ V}}{10\text{ mA}}$$

$$800\text{ k}\Omega > R_1 > 2.9\text{ k}\Omega$$

As you can see, R_1 has quite a wide range of possible values.

PRACTICE EXERCISE 11–4

Determine a value of R_1 in Figure 11–36 that will ensure proper turn-on and turn-off for the following values: $\eta = 0.33$, $V_V = 0.8$ V, $I_V = 15$ mA, $I_P = 35$ μA, and $V_P = 18$ V.

Another example of a thyristor application includes a UJT, triac, bipolar transistor, and diodes. It is the temperature-sensitive heater control circuit shown in Figure 11–37. The basic operation is as follows. The ac line voltage is full-wave rectified by the bridge formed by D_1 through D_4. The output of the bridge is applied to the control circuit through R_1 and clamped to a fixed voltage by the zener diode. The thermistor (temperature-sensitive resistor) R_T and the variable resistor R_2 control the bias for Q_1. R_2 is adjusted so that transistor Q_1 is off at a predetermined temperature. When Q_1 is off, the capacitor C_1 is uncharged, and therefore both the UJT (Q_2) and the triac are off.

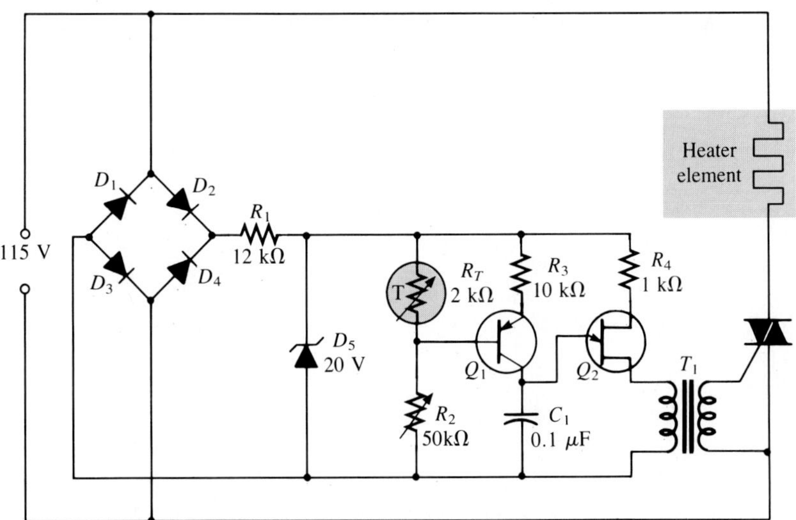

FIGURE 11–37
Temperature-sensitive heater control.

When the temperature decreases below the set temperature, the resistance of R_T increases, thus causing Q_1 to conduct. This allows C_1 to charge up to a voltage sufficient to trigger the UJT. The resulting UJT output pulse is coupled through the transformer and triggers the triac, which conducts current through the heater element (load). As the element heats up, its resistance increases until the current decreases below the holding level of the triac, thus turning it off. If the temperature continues to decrease, the resistance of R_T increases more and causes Q_1 to conduct harder, thus charging C_1 faster. This triggers the triac sooner in the ac cycle, so that more power is delivered to the load. When the temperature increases, the resistance of R_T decreases, causing Q_1 to conduct less. As a result, C_1 takes longer to charge and the triac is triggered later in the ac cycle, so that less power is delivered to the load. When the desired temperature is reached, Q_1 is off, the triac is off, and no more power is delivered to the load.

11–6 REVIEW QUESTIONS

1. Name the UJT terminals.
2. What is the intrinsic standoff ratio?
3. In a basic UJT relaxation oscillator such as in Figure 11–34, what three factors determine the period of oscillation?

11–7 THE PROGRAMMABLE UNIJUNCTION TRANSISTOR (PUT)

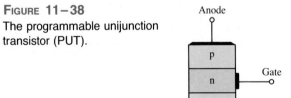

The* programmable unijunction transistor (PUT) *is actually a type of thyristor and not like the UJT at all in terms of structure. The only similarity to a UJT is that the PUT can be used in some oscillator applications to replace the UJT. The PUT is more similar to an SCR except that its anode-to-gate voltage can be used to both turn on and turn off the device.

The structure of the PUT is similar to that of an SCR (four-layer) except that the gate is brought out as shown in Figure 11–38. Notice that the gate is connected to the n region adjacent to the anode. This pn junction controls the *on* and *off* states of the device. The gate is always biased positive with respect to the cathode. When the anode voltage exceeds the gate voltage by approximately 0.7 V, the pn junction is forward-biased and the PUT turns on. The PUT stays on until the anode voltage falls back below this level, then the PUT turns off.

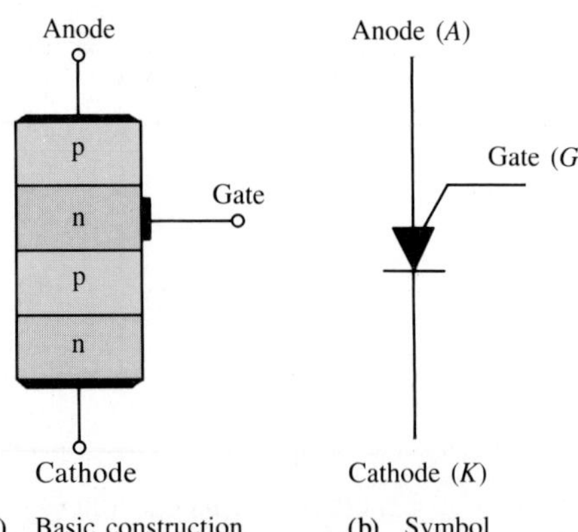

FIGURE 11–38
The programmable unijunction transistor (PUT).

SETTING THE TRIGGER VOLTAGE

The gate can be biased to a desired voltage with an external voltage divider, as shown in Figure 11–39(a), so that when the anode voltage exceeds this "programmed" level, the PUT turns off.

FIGURE 11–39
PUT biasing.

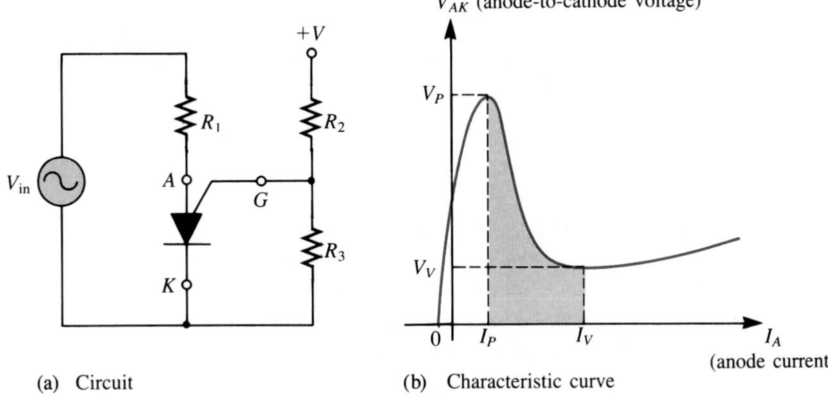

(a) Circuit (b) Characteristic curve

AN APPLICATION

A plot of the anode-to-cathode voltage V_{AK} versus anode current I_A in Figure 11–39(b) reveals a characteristic curve similar to that of the UJT. Therefore, the PUT replaces the UJT in many applications. One such application is the relaxation oscillator in Figure 11–40(a). Its basic operation is as follows.

The gate is biased at +9 V by the voltage divider consisting of resistors R_2 and R_3. When dc power is applied, the PUT is off and the capacitor charges toward +18 V

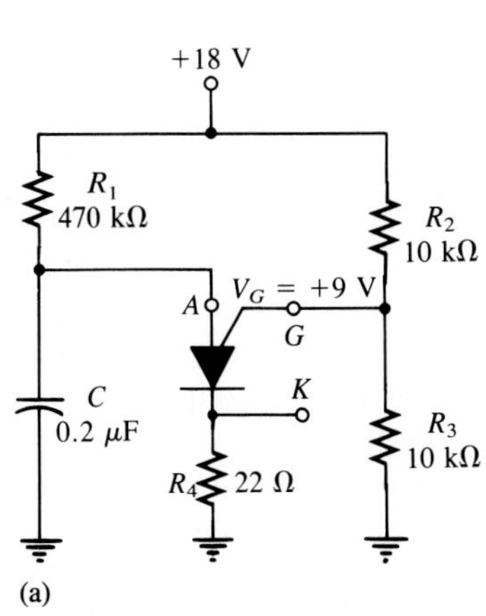

(a)

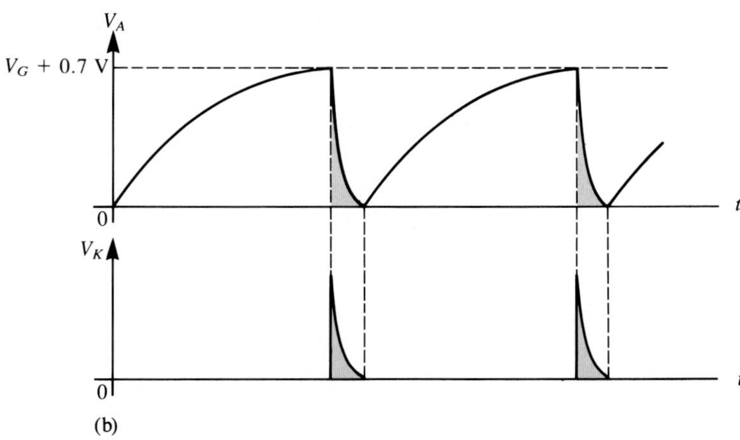

(b)

FIGURE 11–40
PUT relaxation oscillator.

through R_1. When the capacitor reaches $V_G + 0.7$ V, the PUT turns on and the capacitor rapidly discharges through the low *on* resistance of the PUT and R_4. A voltage spike is developed across R_4 during the discharge. As soon as the capacitor discharges, the PUT turns off and the charging cycle starts over, as shown by the waveforms in Figure 11–40(b).

11–7 REVIEW QUESTIONS

1. What does the term *programmable* mean as used in programmable unijunction transistor (PUT)?
2. Compare the structure and the operation of a PUT to those of other devices such as the UJT and SCR.

11–8 THE PHOTOTRANSISTOR

The phototransistor has a light-sensitive, collector-base pn junction. It is exposed to incident light through a lens opening in the transistor package. When there is no incident light, there is a small thermally generated collector-to-emitter leakage current, I_{CEO}; this is called the dark current and is typically in the nA range. When light strikes the collector-base pn junction, a base current I_λ is produced that is directly proportional to the light intensity. This action produces a collector current that increases with I_λ. Except for the way base current is generated, the phototransistor behaves as a conventional bipolar transistor. In many cases, there is no electrical connection to the base.

The relationship between the collector current and the light-generated base current in a phototransistor is

$$I_C = \beta_{dc} I_\lambda \tag{11–10}$$

The schematic symbol and some typical phototransistors are shown in Figure 11–41. Since the actual photogeneration of base current occurs in the collector-base region, the larger the physical area of this region, the more base current is generated. Thus, a typical phototransistor is designed to offer a large area to the incident light, as the simplified structure diagram in Figure 11–42 illustrates.

A **phototransistor** can be either a two-lead or a three-lead device. In the three-lead configuration, the base lead is brought out so that the device can be used as a conventional bipolar transistor with or without the additional light-sensitivity feature. In the two-lead configuration, the base is not electrically available, and the device can be used only with light as the input. In many applications, the phototransistor is used in the two-lead version. Figure 11–43 shows a phototransistor with a biasing circuit and typical collector characteristic curves. Notice that each individual curve on the graph corresponds to a certain value of light intensity (in this case, the units are mW/cm^2) and that the collector current increases with light intensity.

Phototransistors are not sensitive to all light but only to light within a certain range of wavelengths. They are most sensitive to particular wavelengths, as shown by the peak of the spectral response curve in Figure 11–44.

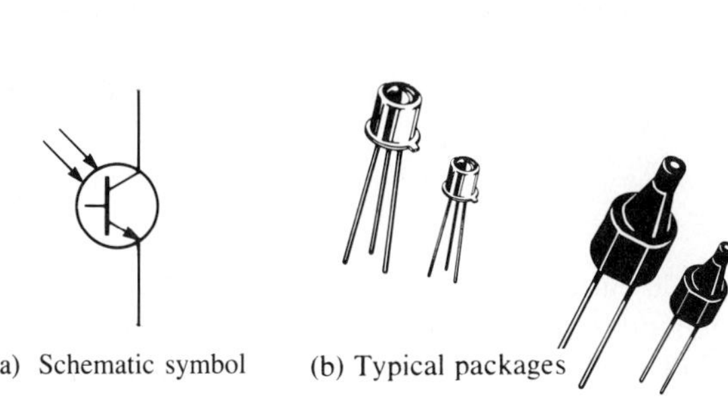

FIGURE 11–41
Phototransistor. Part (b) copyright of Motorola, Inc. Used by permission.

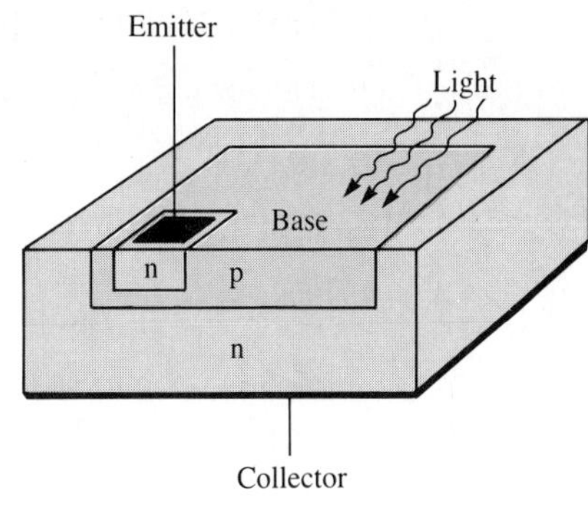

FIGURE 11–42
Typical phototransistor chip structure.

FIGURE 11–43
Phototransistor bias circuit and typical collector characteristic curves.

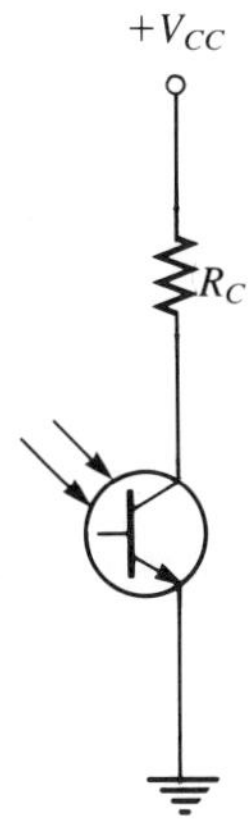

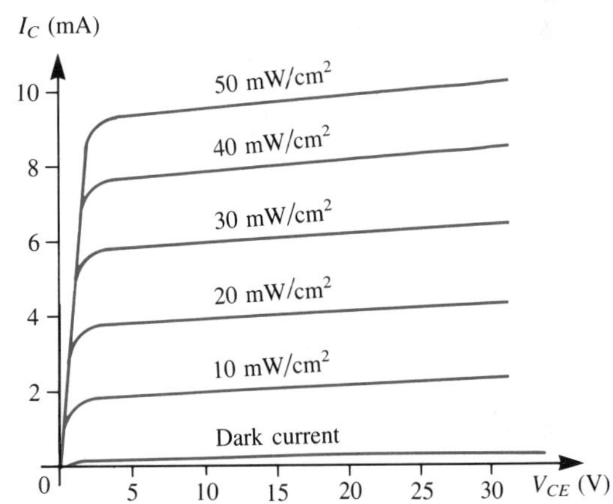

FIGURE 11–44
Typical phototransistor spectral response.

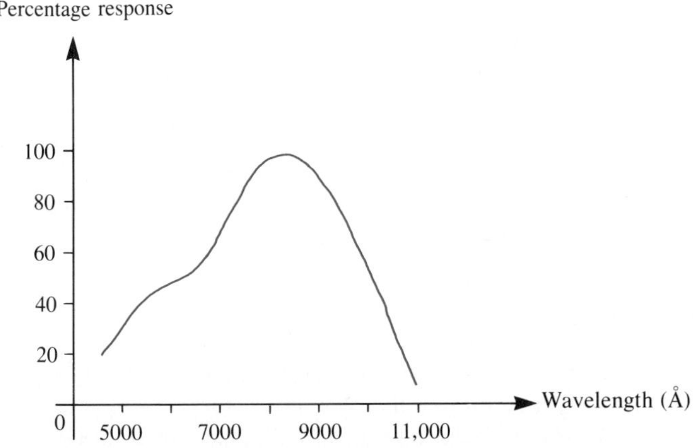

PHOTODARLINGTON

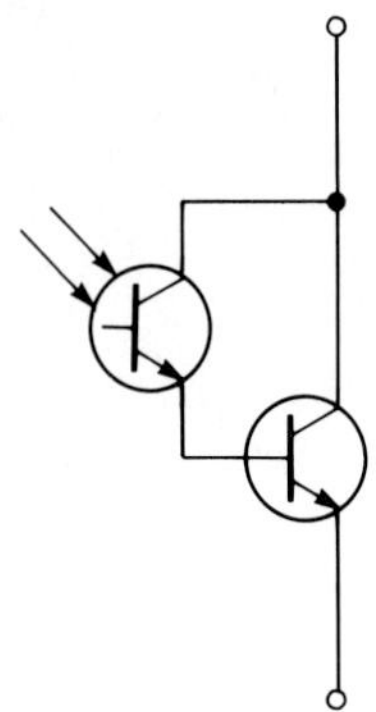

FIGURE 11–45
Photodarlington.

The photodarlington consists of a phototransistor connected in a darlington arrangement with a conventional transistor, as shown in Figure 11–45. Because of the higher current gain, this device has a much higher collector current and exhibits a greater light sensitivity than does a regular phototransistor.

APPLICATIONS

Phototransistors are used in a wide variety of applications. A few are presented in this section. A light-operated relay circuit is shown in Figure 11–46. The phototransistor Q_1 drives the bipolar transistor Q_2. When there is sufficient incident light on Q_1, transistor Q_2 is driven into saturation, and collector current through the relay coil energizes the relay.

Figure 11–47 shows a circuit in which a relay is de-energized by incident light on the phototransistor. When there is insufficient light, transistor Q_2 is biased on, keeping the relay energized. When there is sufficient light, phototransistor Q_1 turns on; this pulls the base of Q_2 low, thus turning Q_2 off and de-energizing the relay.

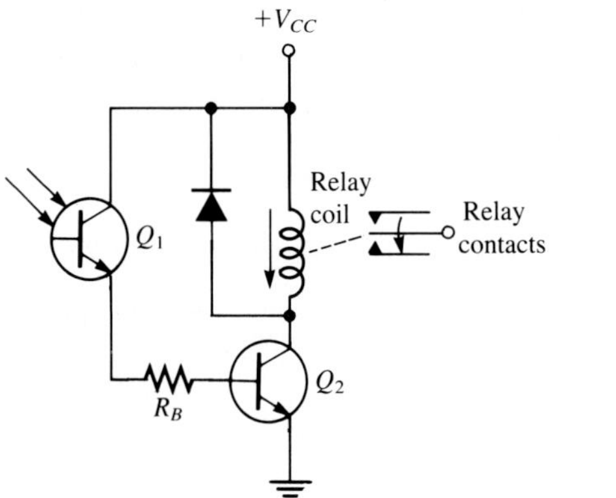

FIGURE 11–46
Light-operated relay circuit.

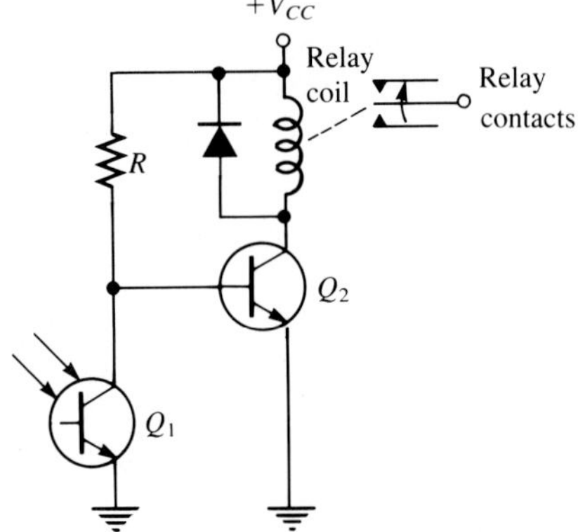

FIGURE 11–47
Darkness-operated relay circuit.

These relay circuits can be used in a variety of applications such as automatic door activators, process counters, and various alarm systems. Another simple application is illustrated in Figure 11–48. The phototransistor is normally on, holding the gate of the SCR low. When the light is interrupted, the phototransistor turns off. The high-going transition on the collector triggers the SCR and sets off the alarm mechanism. The momentary contact switch S_1 provides for resetting the alarm. Smoke and intrusion detection are possible uses for this circuit.

11–8 REVIEW QUESTIONS

1. How does a phototransistor differ from a conventional bipolar transistor?
2. Some, but not all, phototransistors have external ________ leads.
3. The collector current in a phototransistor circuit depends on what two factors?

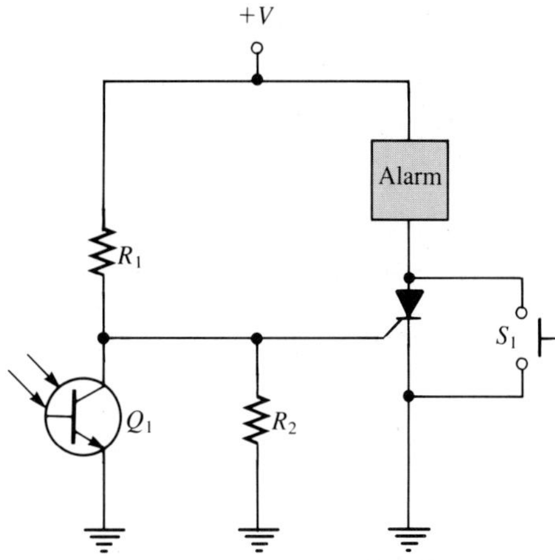

FIGURE 11–48
Light-interruption alarm.

11–9 THE LIGHT-ACTIVATED SCR (LASCR)

The* light-activated silicon-controlled rectifier (LASCR) *operates essentially as does the conventional SCR except that it can also be light-triggered. Most LASCRs have an available gate terminal so that the device can also be triggered by an electrical pulse just as a conventional SCR.

Figure 11–49 shows an LASCR schematic symbol and typical packages. The LASCR is most sensitive to light when the gate terminal is open. If necessary, a resistor from the gate to the cathode can be used to reduce the sensitivity. Figure 11–50 shows an LASCR used to energize a latching relay. The input source turns on the lamp; and the resulting incident light triggers the LASCR. The anode current energizes the relay and closes the contact. Notice that the input source is electrically isolated from the rest of the circuit.

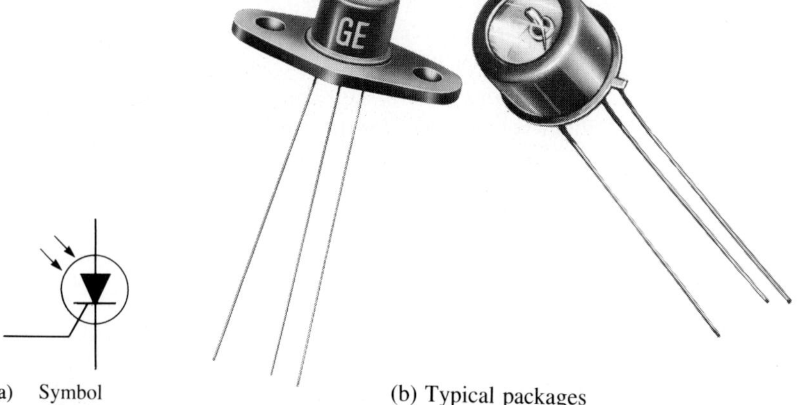

FIGURE 11–49
Light-activated SCRs. Part (b) courtesy of General Electric.

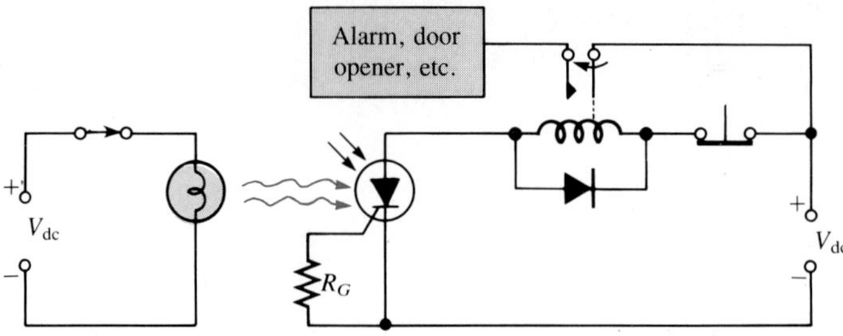

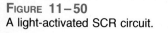

FIGURE 11–50
A light-activated SCR circuit.

11–9 REVIEW QUESTIONS

1. Can most LASCRs be operated as conventional SCRs?
2. What is required in Figure 11–50 to turn off the LASCR and de-energize the relay?

11–10 OPTICAL COUPLERS

Optical couplers are designed to provide complete electrical isolation between an input circuit and an output circuit. The usual purpose of isolation is to provide protection from high-voltage transients, surge voltage, or low-level noise that could possibly result in an erroneous output or damage to the device. Optical couplers also allow interfacing circuits with different voltage levels, different grounds, and so on.

The input circuit is typically an LED, but the output circuit can take several forms, such as the phototransistor shown in Figure 11–51(a). When the input voltage forward-biases the LED, light transmitted to the phototransistor turns it on, producing current through the external load, as shown in Figure 11–51(b). Some typical devices are shown in Figure 11–51(c).

Several other types of couplers are shown in Figure 11–52. The darlington transistor coupler in Figure 11–52(a) can be used when increased output current capability is needed beyond that provided by the phototransistor output. The disadvantage is that the photodarlington has a switching speed less than that of the phototransistor.

An LASCR output coupler is shown in Figure 11–52(b). This device can be used in applications where, for example, a low-level input voltage is required to latch a high-voltage relay for the purpose of activating some type of electromechanical device.

A phototriac output coupler is illustrated in Figure 11–52(c). This device is designed for applications requiring isolated triac triggering, such as switching a 110 V ac line from a low-level input.

Figure 11–52(d) shows an optically isolated ac linear coupler. This device converts an input current variation to an output voltage variation. The output circuit consists of an

FIGURE 11–52 (OPPOSITE)
Common types of optical-coupling devices.

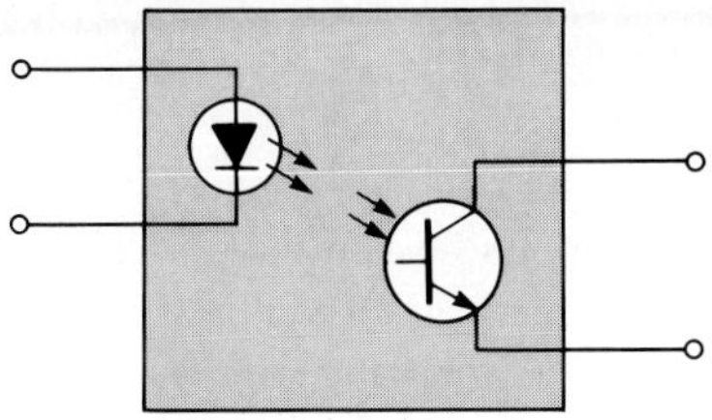

V_{IN} + − $+V_{CC}$ R_L V_{OUT}

(a) Basic device

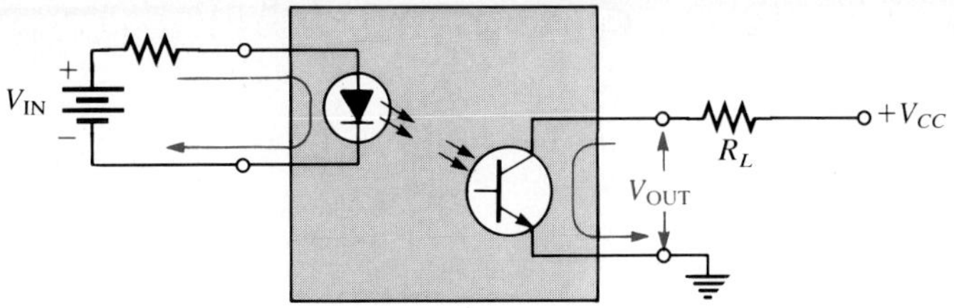

(b) Device with external connections

(c) Typical packages

FIGURE 11–51
Phototransistor couplers. Part (c) copyright of Motorola, Inc. Used by permission.

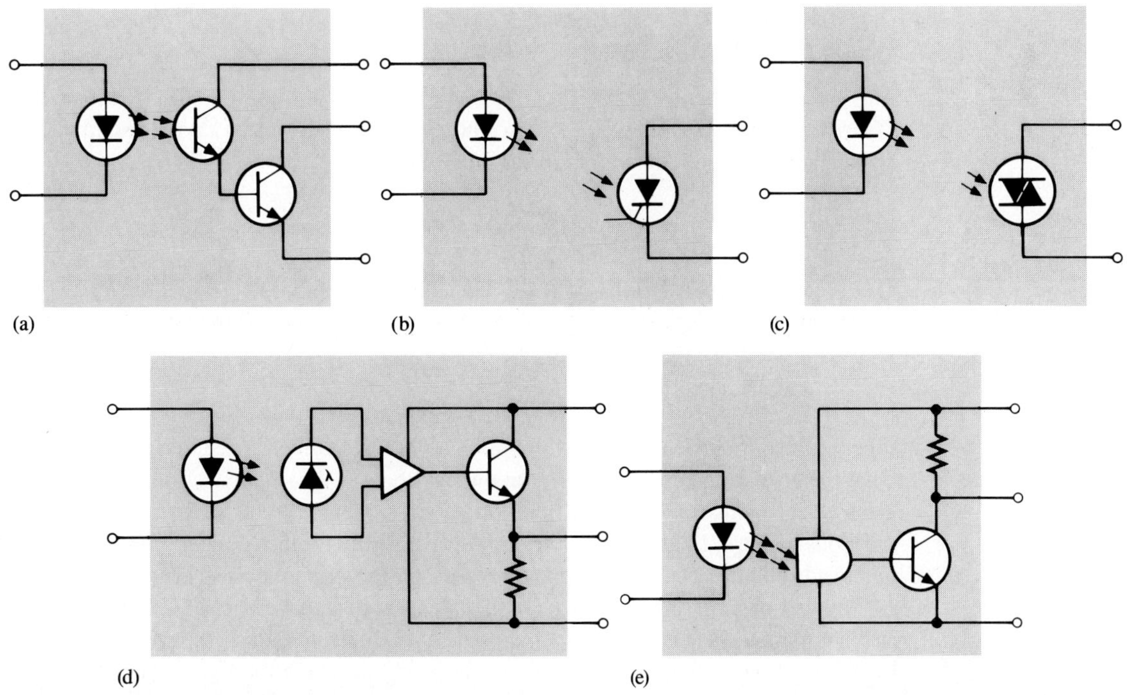

amplifier with a photodiode across its input terminals. Light variations emitted from the LED are picked up by the photodiode, providing an input signal to the amplifier. The output of the amplifier is buffered with the emitter-follower stage. The optically isolated ac linear coupler can be used for telephone line coupling, peripheral equipment isolation, and audio applications, among many others.

Part (e) of Figure 11–52 shows a digital output coupler. This device consists of a high-speed detector circuit followed by a transistor buffer stage. When there is current through the input LED, the detector is light-activated and turns on the output transistor, so that the collector switches to a low-voltage level. When there is no current through the LED, the output is at the high-voltage level. The digital output coupler can be used in applications requiring compatibility with digital circuits, such as interfacing computer terminals with peripheral devices.

Isolation Voltage

The *isolation voltage* of an optical coupler is the maximum voltage that can exist between the input and output terminals without dielectric breakdown occurring. Typical values are about 7500 V ac peak.

DC Current Transfer Ratio

This parameter is the ratio of the output current to the input current through the LED. It is usually expressed as a percentage. For a phototransistor output, typical values range from 2 percent to 100 percent. For a photodarlington output, typical values range from 50 percent to 500 percent.

LED Trigger Current

This parameter applies to the LASCR output coupler and the phototriac output device. The *trigger current* is the value of current required to trigger the thyristor output device. Typically, the trigger current is in the mA range.

Transfer Gain

This parameter applies to the optically isolated ac linear coupler. The *transfer gain* is the ratio of output voltage to input current. A typical value is 200 mV/mA.

Fiber Optics

Fiber optics provides a means for coupling a photo-emitting device to a photodetector via a light-transmitting cable consisting of optical fibers. Typical applications include medical electronics, industrial controls, microprocessor systems, security systems, and communications. Glass fiber cables are used to maximize the optical coupling between optoelectronic devices. Fiber optics is based on the principle of internal reflection.

Every material that conducts light has an **index of refraction,** n. A ray of light striking the interface between two materials with different indices of refraction will either be reflected or refracted, depending on the angle at which the light strikes the interface surface, as illustrated in Figure 11–53. If the incident angle, θ_i, is equal to or greater than a certain value called the *critical angle,* θ_c, the light is reflected. If θ_i is less than the critical angle, the light is refracted.

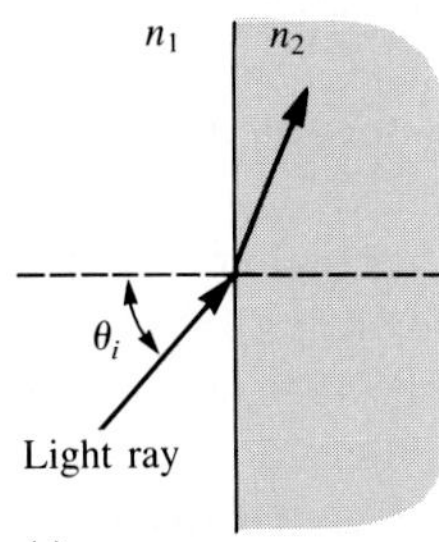

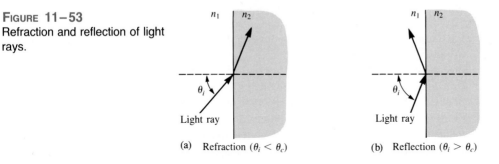

(a) Refraction ($\theta_i < \theta_c$)

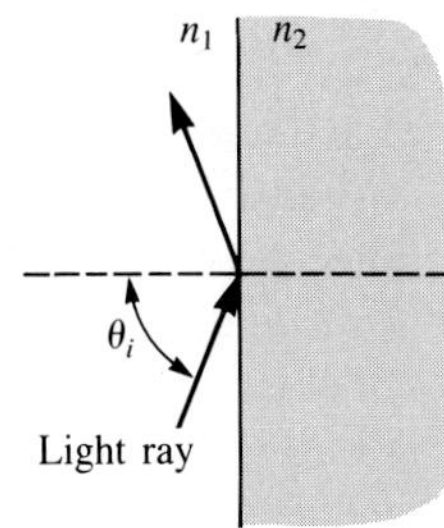

(b) Reflection ($\theta_i > \theta_c$)

FIGURE 11–53
Refraction and reflection of light rays.

A glass fiber is clad with a layer of glass with a lower index of refraction than the fiber. A ray of light entering the end of the cable will be refracted as shown in Figure 11–54. If, after refraction, it approaches and hits the glass interface at an angle greater than the critical angle, it is reflected within the fiber as shown. Since, according to a law of physics, the angle of reflection must equal the angle of incidence, the light ray will bounce down the fiber and emerge, refracted, at the other end, as shown in the figure.

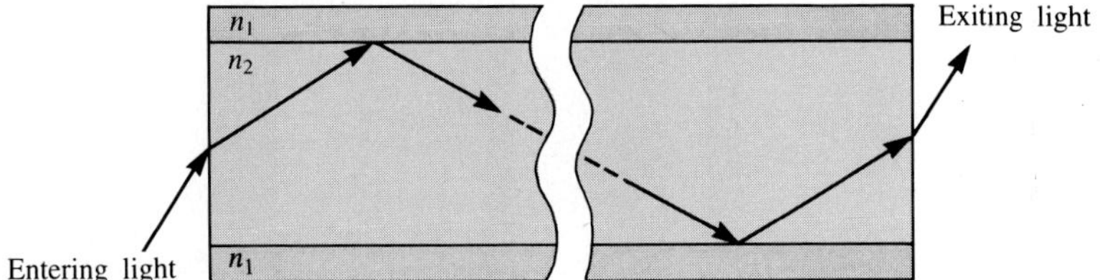

FIGURE 11–54
A light ray in optical fiber.

11–10 REVIEW QUESTIONS

1. What type of input device is normally used in an optical coupler?
2. List five types of output devices used in optical couplers.

11–11 A SYSTEM APPLICATION

The conveyor speed-control system presented at the beginning of this chapter includes the optical counting system that originally appeared in Chapter 3. This system has been modified to interface with a motor speed-control circuit that adjusts the rate at which the conveyor track moves. The motor speed-control circuit utilizes an SCR and a PUT. In this section, you will

- ☐ ***See how an SCR is used in a common system application.***
- ☐ ***See how a PUT can be used in a system application.***

- See how an electronic circuit interfaces with a mechanical device.
- Translate between a printed circuit board and a schematic.
- Troubleshoot some common system problems.

A BRIEF DESCRIPTION OF THE SYSTEM

Each time an object on the moving conveyor passes the optical detector and interrupts the infrared light beam, the digital counter is advanced by one. The count of the passing objects is accumulated over a specified period of time and converted to an analog voltage by a digital-to-analog converter. This analog voltage is directly proportional to the number of objects that pass the beam in the specified time period and is applied directly to the gate of a PUT. The purpose of the motor speed-control circuit shown in Figure 11–55 is to maintain a constant number of objects in the given time period by adjusting the speed of an electric motor that drives the conveyor track. The analog voltage on the gate of the PUT determines the point on the ac cycle that the SCR is triggered on. For a higher gate voltage, the SCR turns on later in the half-cycle and thus delivers less average power to the motor causing its speed to decrease. For a lower gate voltage, the SCR turns on earlier and causes the motor speed to increase.

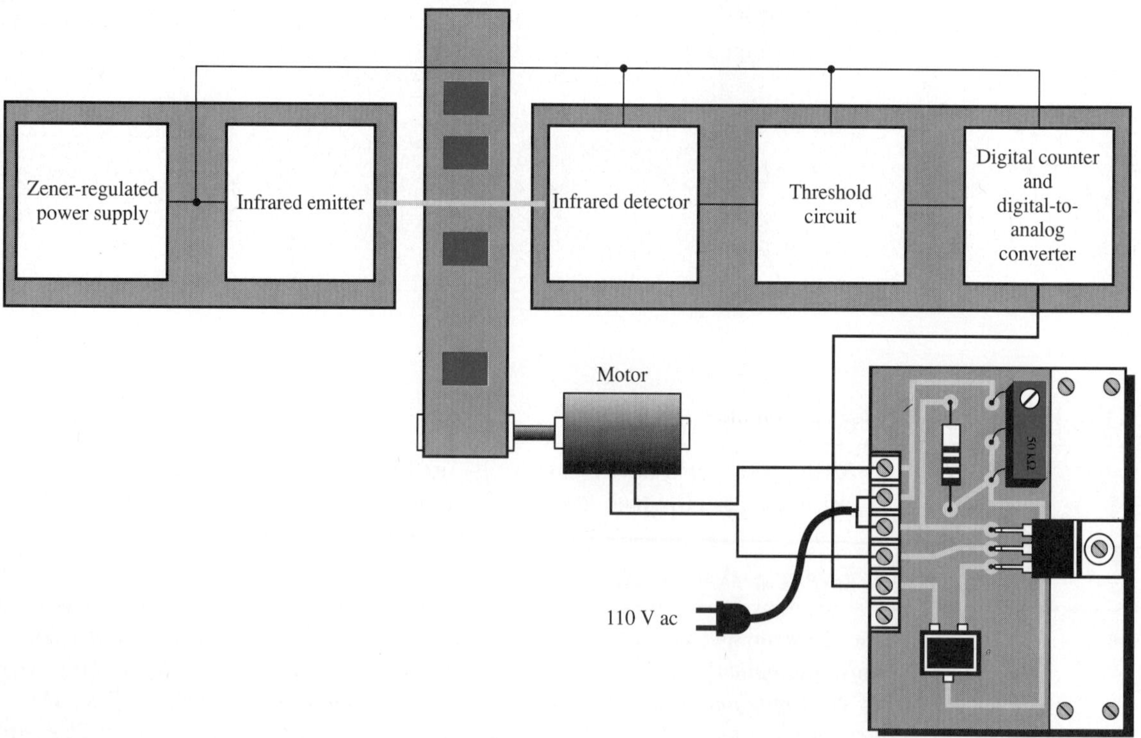

FIGURE 11–55

Now, so that you can take a closer look at the motor speed control circuit, let's take it out of the system and put it on the test bench.

ON THE TEST BENCH

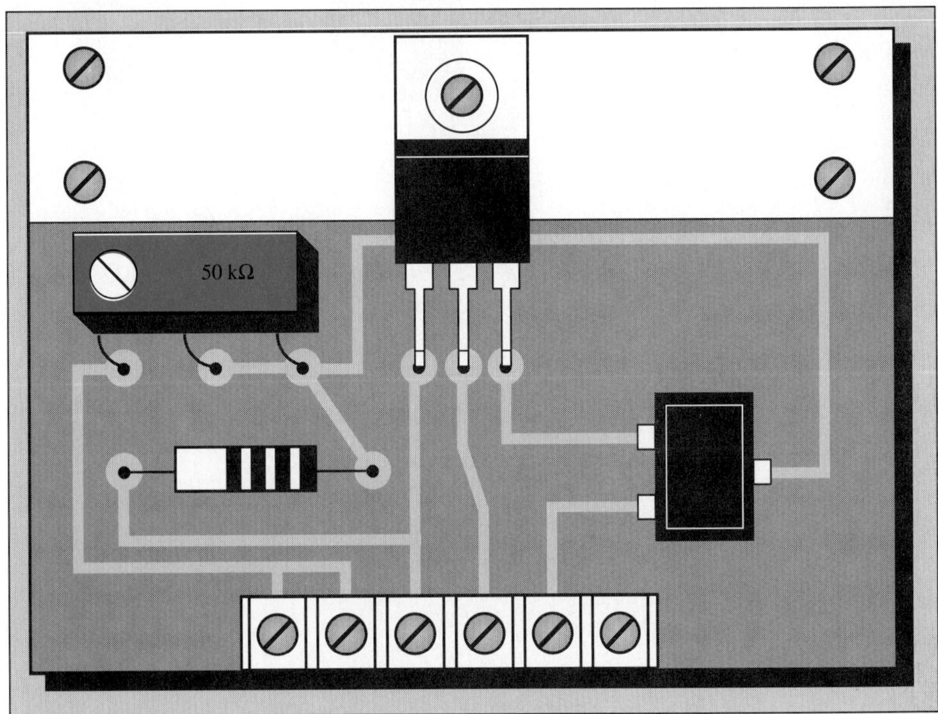

FIGURE 11–56

IDENTIFYING THE COMPONENTS

The SCR on the circuit board in Figure 11–56 is a 2N6394 housed in a TO-220AB case as indicated in Figure 11–57(a). The SCR is thermally connected to the heat sink plate in order to dissipate any excess heat. The PUT is a 2N6027 packaged in a surface mount SOT-23 case, as indicated in Figure 11–57(b).

■ ACTIVITY 1 RELATE THE PC BOARD TO THE SCHEMATIC

Carefully follow the conductive traces on the pc board to see how the components are interconnected. Compare each point (*A* through *F*) on the motor speed control board with the corresponding point on the schematic in Figure 11–58. For each point on the pc board, place the letter on the corresponding point or points on the schematic. Identify and label the inputs, outputs, and components on the board and the schematic.

FIGURE 11–57

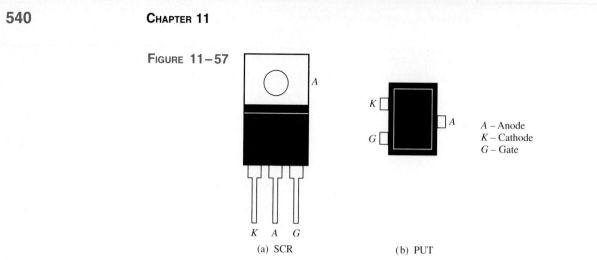

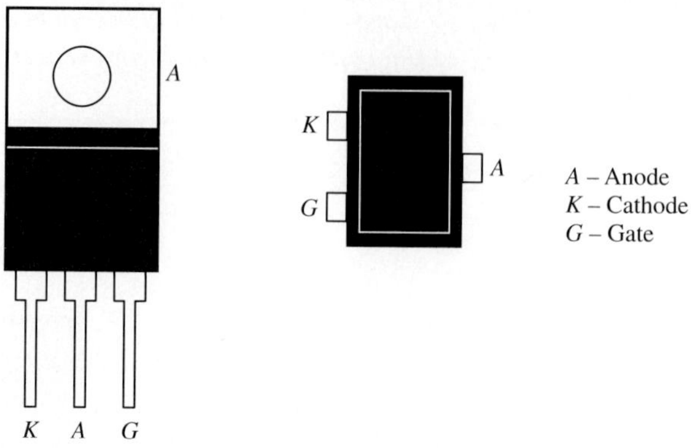

A

K A G

(a) SCR

K

A

G

(b) PUT

A – Anode
K – Cathode
G – Gate

FIGURE 11–58

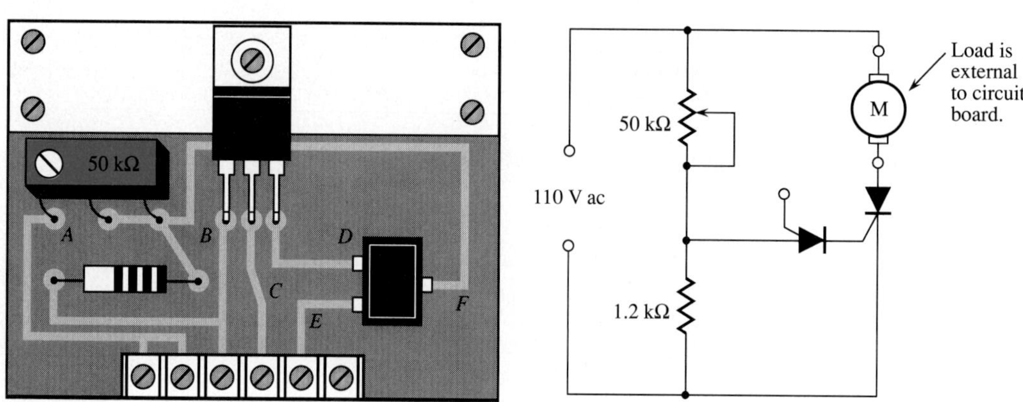

■ ACTIVITY 2 ANALYZE THE CIRCUIT

STEP 1 With a 1 kΩ load resistor where the motor is normally connected for simulation purposes, adjust the rheostat to 25 kΩ. Apply 110 V rms to the input.

STEP 2 Apply a variable dc voltage source to the PUT gate terminal and set it initially to 0 V. Determine the voltages at each labelled point (*A* through *F*) in the circuit.

STEP 3 For PUT gate voltages of 0 V, 2 V, 4 V, 6 V, 8 V, and 10 V, determine the voltage waveform across the 1-kΩ load resistor. This is the voltage that would be applied to the motor when the board is installed in the system.

■ ACTIVITY 3 WRITE A TECHNICAL REPORT

Discuss the detailed operation of the motor speed control circuit and explain how it interfaces with the overall system. Describe in detail the purpose of each component on the circuit board. Use the results of Activity 2 when appropriate.

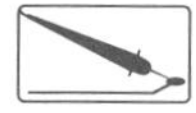

ACTIVITY 4 TROUBLESHOOT THE SYSTEM FOR EACH OF THE FOLLOWING PROBLEMS BY STATING THE PROBABLE CAUSE OR CAUSES

1. No voltage at the anode of the PUT.
2. The full ac voltage always appears across the SCR load.
3. The SCR triggers on at the same point near zero on the ac cycle regardless of the PUT gate voltage.
4. The SCR never triggers on.

ACTIVITY 5 TEST BENCH SPECIAL ASSIGNMENT

Go to Test Bench 5 in the color insert section (which follows page 422) and carry out the assignment that is stated there.

11–11 REVIEW QUESTIONS

1. At what PUT gate voltage does the electric motor run at the fastest speed?
2. Does the SCR trigger on earlier or later in the ac cycle if the resistance of the rheostat is reduced?
3. What happens to the SCR when the PUT gate voltage increases?

SUMMARY

- Thyristors are devices constructed with four semiconductor layers (pnpn).
- Thyristors include Shockley diodes, SCRs, SCSs, diacs, triacs, and PUTs.
- The Shockley diode is a thyristor device that conducts when the voltage across its terminals exceeds the breakover potential.
- The silicon-controlled rectifier (SCR) can be triggered on by a pulse at the gate and turned off by reducing the anode current below the specified holding value.
- The silicon-controlled switch (SCS) has two gate terminals and can be turned on by a pulse at the cathode gate and turned off by a pulse at the anode gate.
- The diac can conduct current in either direction and is turned on when a breakover voltage is exceeded. It turns off when the current drops below the holding value.
- The triac, like the diac, is a bidirectional device. It can be turned on by a pulse at the gate and conducts in a direction depending on the voltage polarity across the two anode terminals.
- The intrinsic standoff ratio of a unijunction transistor (UJT) determines the voltage at which the device will trigger on.
- The programmable unijunction transistor (PUT) can be externally programmed to turn on at a desired anode-to-gate voltage level.
- In a phototransistor, base current is generated by light input.
- Light acts as the trigger source in light-activated SCRs (LASCRs).
- Optical coupling devices provide electrical isolation between a source and an output circuit.

- Fiber optics provides a light path from a light-emitting device to a light-activated device.
- Device symbols are shown in Figure 11–59.

FIGURE 11–59

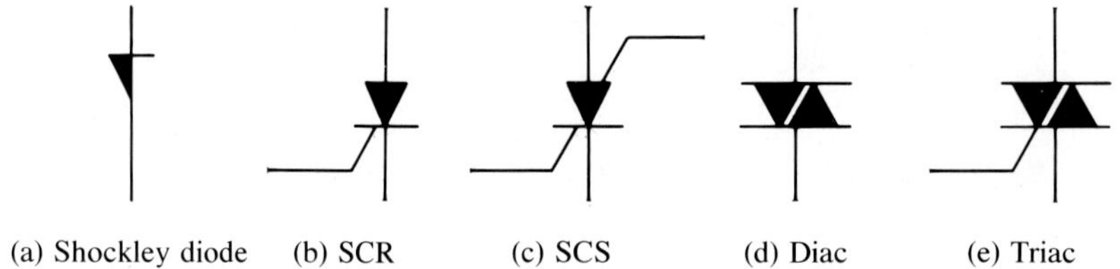

(a) Shockley diode (b) SCR (c) SCS (d) Diac (e) Triac

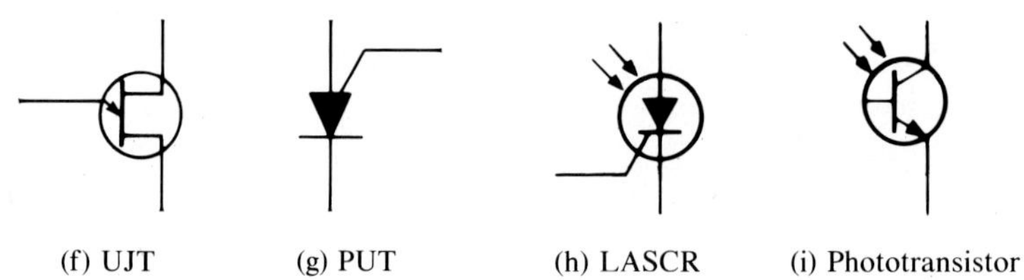

(f) UJT (g) PUT (h) LASCR (i) Phototransistor

GLOSSARY

Diac A two-terminal four-layer semiconductor device (thyristor) that can conduct current in either direction when properly activated.

Forced-commutation A method of turning off an SCR.

Index of refraction A property of light-conducting materials that specifies how much a light ray will bend when passing from one material to another.

Light-Activated Silicon-Controlled Rectifier (LASCR) A four-layer semiconductor device (thyristor) that conducts current in one direction when activated by a sufficient amount of light and continues to conduct until the current falls below a specified value.

Oscillator An electronic circuit based on positive feedback that produces a time-varying output signal without an external input signal.

Phototransistor A transistor in which base current is produced when light strikes the photosensitive semiconductor base region.

Programmable Unijunction Transistor (PUT) A type of three-terminal thyristor (more like an SCR than a UJT) that is triggered into conduction when the voltage at the anode exceeds the voltage at the gate.

Silicon-Controlled Rectifier (SCR) A type of three-terminal thyristor that conducts current when triggered on by a voltage at the single gate terminal and remains on until the anode current falls below a specified value.

Silicon-Controlled Switch (SCS) A type of four-terminal thyristor that has two gate terminals that are used to trigger the device on and off.

Shockley diode The type of two-terminal thyristor that conducts current when the anode-to-cathode voltage reaches a specified "breakover" value.

Standoff ratio The characteristic of a UJT that determines its turn-on point.

Thyristor A class of four-layer (pnpn) semiconductor devices.

Triac A three-terminal thyristor that can conduct current in either direction when properly activated.

Trigger The activating input of some electronic devices and circuits.

Unijunction Transistor (UJT) A three-terminal single pn junction device that exhibits a negative resistance characteristic.

FORMULAS

SHOCKLEY DIODE

(11–1) $I_{B1} = (1 - \alpha_{dc1})I_A - I_{CBO1}$ Shockley diode, Q_1 base current

(11–2) $I_{C2} = \alpha_{dc2}I_K + I_{CBO2}$ Shockley diode, Q_2 base current

(11–3) $I_A = \dfrac{I_{CBO1} + I_{CBO2}}{1 - (\alpha_{dc1} + \alpha_{dc2})}$ Shockley diode, anode current (or cathode current)

UNIJUNCTION TRANSISTOR

(11–4) $r_{BB} = r_{B1} + r_{B2}$ UJT interbase resistance

(11–5) $V_{r_{B1}} = \left(\dfrac{r_{B1}}{r_{BB}}\right)V_{BB}$ UJT gate-to-$B1$ voltage

(11–6) $\eta = \dfrac{r_{B1}}{r_{BB}}$ UJT intrinsic standoff ratio

(11–7) $V_P = \eta V_{BB} + V_{pn}$ UJT peak-point voltage

(11–8) $R_1 < \dfrac{V_{BB} - V_P}{I_P}$ Turn-on condition, relaxation oscillator

(11–9) $R_1 > \dfrac{V_{BB} - V_V}{I_V}$ Turn-off condition, relaxation oscillator

THE PHOTOTRANSISTOR

(11–10) $I_C = \beta_{dc}I_\lambda$ Phototransistor collector current

SELF-TEST

1. A thyristor has
   - **(a)** two pn junctions
   - **(b)** 3 pn junctions
   - **(c)** four pn junctions
   - **(d)** only two terminals

2. Common types of thyristors include
   (a) BJTs and SCRs (b) UJTs and PUTs
   (c) FETs and triacs (d) Diacs and triacs
3. A Shockley diode turns on when the anode to cathode voltage
   (a) exceeds 0.7 V (b) exceeds the gate voltage
   (c) exceeds the forward-breakover voltage
   (d) exceeds the forward-blocking voltage
4. Once it is conducting, a Shockley diode can be turned off by
   (a) reducing the current below a certain value
   (b) disconnecting the anode voltage
   (c) a and b
   (d) none of these
5. An SCR differs from the Shockley diode because
   (a) it has a gate terminal (b) it is not a thyristor
   (c) it does not have four layers (d) it cannot be turned on and off
6. An SCR can be turned off by
   (a) forced commutation (b) a negative pulse on the gate
   (c) anode current interruption (d) all of the above
   (e) a and c
7. In the forward-blocking region, the SCR is
   (a) reverse-biased (b) in the *off* state
   (c) in the *on* state (d) at the point of breakdown
8. The specified value of holding current for an SCR means that
   (a) the device will turn on when the anode current exceeds this value
   (b) the device will turn off when the anode current falls below this value
   (c) the device may be damaged if the anode current exceeds this value
   (d) the gate current must equal or exceed this value to turn the device on
9. The SCS differs from the SCR because
   (a) it does not have a gate terminal (b) its holding current is less
   (c) it can handle much higher currents (d) it has two gate terminals
10. The SCS can be turned on by
    (a) an anode voltage that exceeds forward-breakover voltage
    (b) a positive pulse on the cathode gate
    (c) a negative pulse on the anode gate
    (d) either b or c
11. The SCS can be turned off by
    (a) a negative pulse on the cathode gate and a positive pulse on the anode gate
    (b) reducing the anode current to below the holding value

(c) a and b

(d) a positive pulse on the cathode gate and a negative pulse on the anode gate

12. The diac is

(a) a thyristor (b) a bilateral, two-terminal device

(c) like two parallel Shockley diodes in reverse directions

(d) all of these

13. The triac is

(a) like a bidirectional SCR (b) a four-terminal device

(c) not a thyristor (d) a and b

14. Which of the following is *not* a characteristic of the UJT?

(a) intrinsic standoff ratio (b) negative resistance

(c) peak-point voltage (d) bilateral conduction

15. The PUT is

(a) much like the UJT (b) not a thyristor

(c) triggered on and off by the gate to anode voltage

(d) not a four-layer device

16. In a phototransistor, base current is

(a) set by a bias voltage (b) directly proportional to light

(c) inversely proportional to light (d) not a factor

PROBLEMS

SECTION 11–1 THE SHOCKLEY DIODE

1. The Shockley diode in Figure 11–60 is biased such that it is in the forward-blocking region. For this particular bias condition, $\alpha_{dc} = 0.38$ for both internal transistor structures. $I_{CBO1} = 75$ nA and $I_{CBO2} = 80$ nA. Determine the anode current and cathode current.

FIGURE 11–60

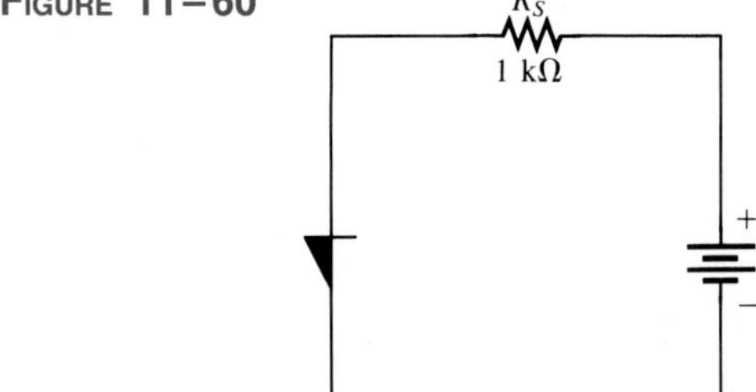

2. **(a)** Determine the forward resistance of the Shockley diode in Figure 11–60 in the forward-blocking region.
   (b) If the forward-breakover voltage is 50 V, how much must V_{AK} be increased to switch the diode into the forward-conduction region?

SECTION 11–2 THE SILICON-CONTROLLED RECTIFIER (SCR)

3. Explain the operation of an SCR in terms of its transistor equivalent.
4. To what value must the variable resistor be adjusted in Figure 11–61 in order to turn the SCR off? Assume $I_H = 10$ mA.

FIGURE 11–61

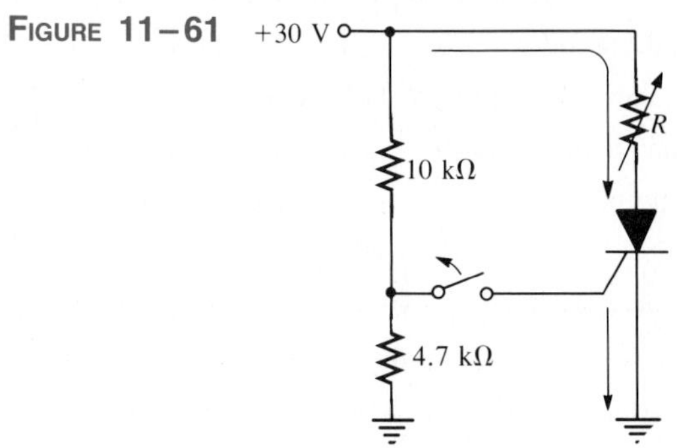

SECTION 11–3 SCR APPLICATIONS

5. Describe how you would modify the circuit in Figure 11–15 so that the SCR triggers and conducts on the negative half-cycle of the input.
6. What is the purpose of diodes D_1 and D_2 in Figure 11–17?
7. Sketch the V_R waveform for the circuit in Figure 11–62, given the indicated relationship of the input waveforms.

FIGURE 11–62

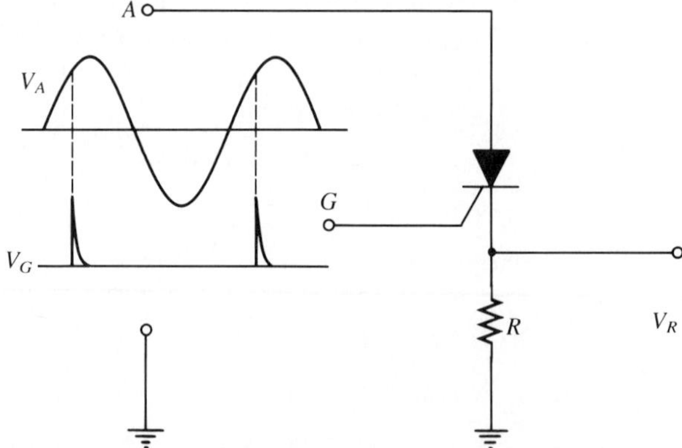

SECTION 11–4 THE SILICON-CONTROLLED SWITCH (SCS)

8. Explain the turn-on and turn-off operation of an SCS in terms of its transistor equivalent.

9. Name the terminals of an SCS.

SECTION 11–5 THE DIAC AND TRIAC

10. Sketch the current waveform for the circuit in Figure 11–63. The diac has a breakover potential of 20 V. I_H = 20 mA.

FIGURE 11–63

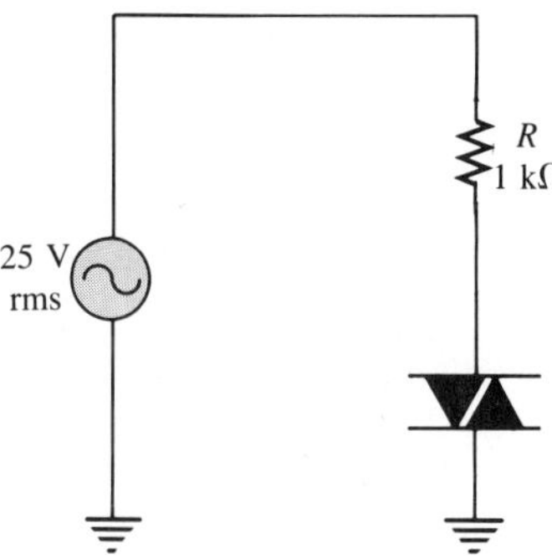

11. Repeat Problem 10 for the triac circuit in Figure 11–64. The breakover potential is 25 V and I_H = 1 mA.

FIGURE 11–64

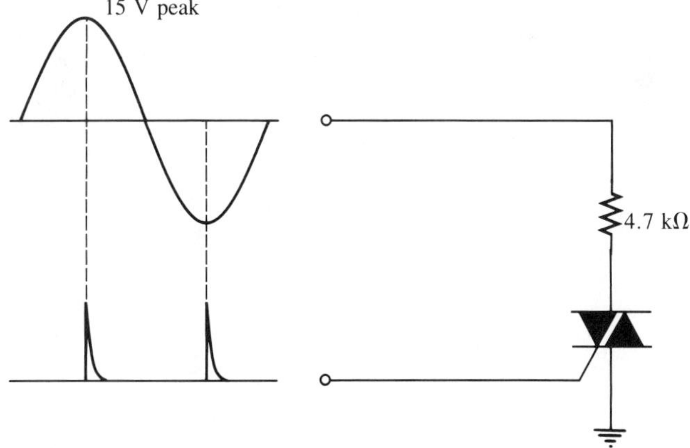

SECTION 11–6 THE UNIJUNCTION TRANSISTOR (UJT)

12. In a certain UJT, r_{B1} = 2.5 kΩ and r_{B2} = 4 kΩ. What is the intrinsic standoff ratio?

13. Determine the peak-point voltage for the UJT in Problem 12 if V_{BB} = 15 V.

14. Find the range of values of R_1 in Figure 11–65 that will ensure proper turn-on and turn-off of the UJT. $\eta = 0.68$, $V_V = 0.8$ V, $I_V = 15$ mA, $I_P = 10\ \mu$A, and $V_P = 10$ V.

FIGURE 11–65

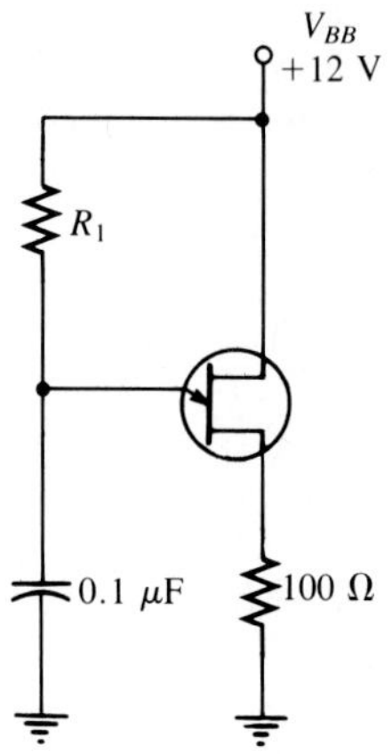

SECTION 11–7 THE PROGRAMMABLE UNIJUNCTION TRANSISTOR (PUT)

15. At what anode voltage (V_A) will each PUT in Figure 11–66 begin to conduct?

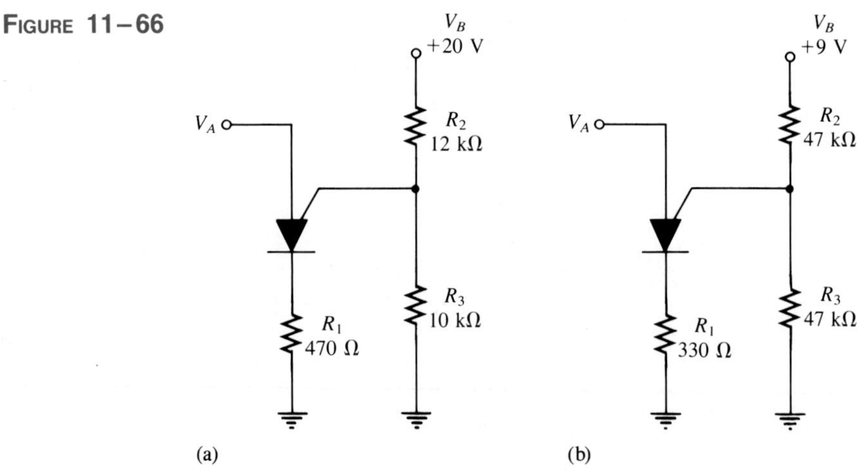

FIGURE 11–66

16. Draw the current waveform for each circuit in Figure 11–66 when there is a 10 V peak sine wave voltage at the anode. Neglect the forward voltage of the PUT.

17. Sketch the voltage waveform across R_1 in Figure 11–67 in relation to the input voltage waveform.

FIGURE 11–67

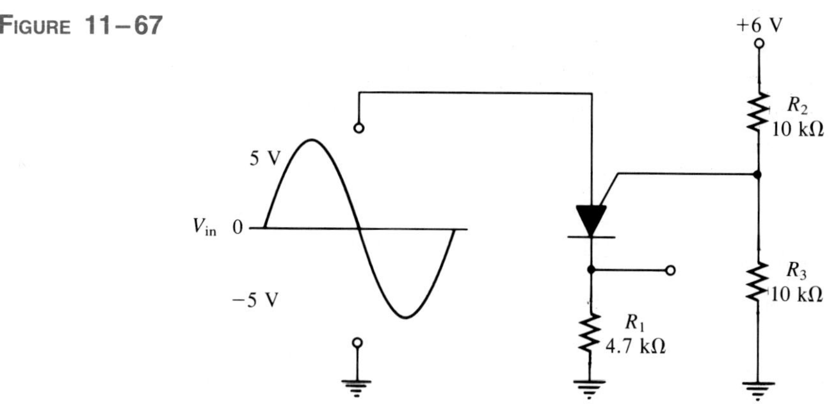

SECTION 11–8 THE PHOTOTRANSISTOR

18. A certain phototransistor in a circuit has a $\beta_{dc} = 200$. If $I_\lambda = 100\ \mu A$, what is the collector current?

19. Determine the output voltage in Figure 11–68, **(a)** when the light source is off, and **(b)** when it is on (assuming the transistor saturates).

20. Determine the emitter current in the photodarlington circuit of Figure 11–69 if, for each lm/m² of light intensity, 1 μA of base current is produced in Q_1.

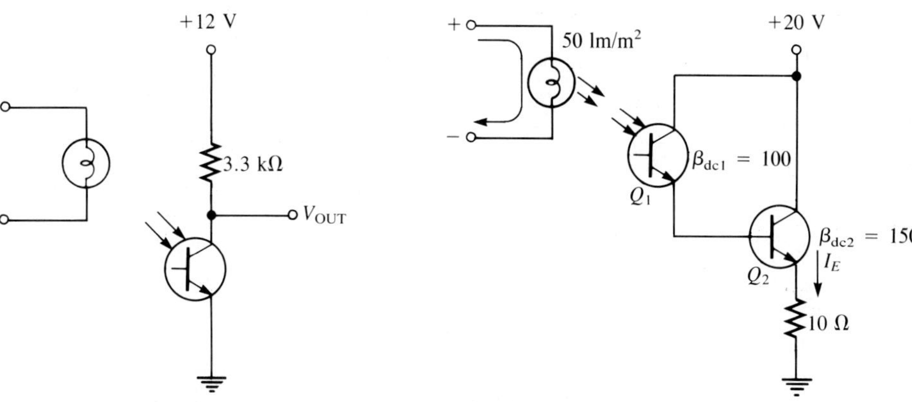

FIGURE 11–68

FIGURE 11–69

SECTION 11–9 THE LIGHT-ACTIVATED SCR (LASCR)

21. By examination of the circuit in Figure 11–70, explain its purpose and basic operation.

FIGURE 11–70

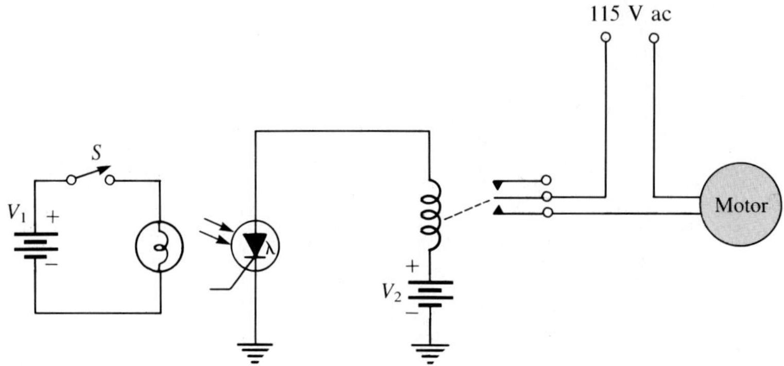

22. Determine the voltage waveform across R_K in Figure 11–71.

FIGURE 11–71

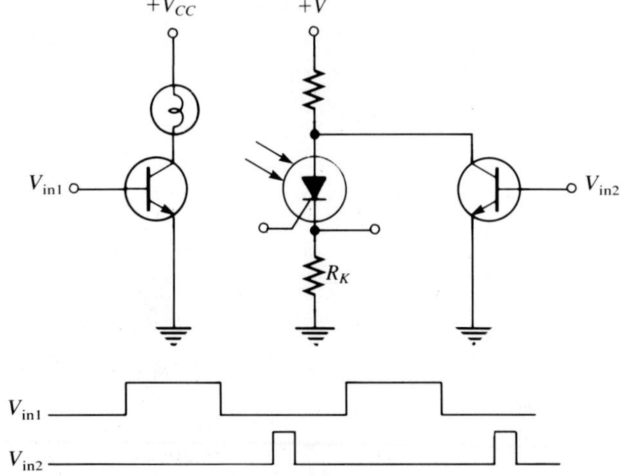

SECTION 11–10 OPTICAL COUPLERS

23. A particular optical coupler has a current transfer ratio of 30 percent. If the input current is 100 mA, what is the output current?

24. The optical coupler shown in Figure 11–72 is required to deliver at least 10 mA to the external load. If the current transfer ratio is 60 percent, how much current must be supplied to the input?

FIGURE 11–72

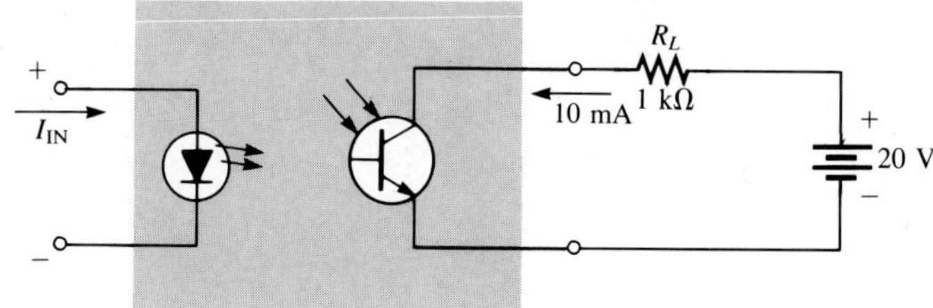

ANSWERS TO REVIEW QUESTIONS

SECTION 11–1

1. The Shockley diode is a four-layer device.
2. In the forward-blocking region, the diode is off.
3. The device turns on and conducts when V_{AK} exceeds the forward breakover.
4. When V_{AK} is reduced below the forward-breakover voltage, the device turns off.

SECTION 11–2

1. An SCR (silicon-controlled rectifier) is a three-terminal thyristor.
2. The SCR terminals are anode, cathode, and gate.
3. A positive gate pulse turns the SCR on.
4. Reduce the anode current below I_H to turn a conducting SCR off.

SECTION 11–3

1. 90°, 180°
2. To block discharge of the battery through that path

SECTION 11–4

1. An SCS can be turned off with the application of a gate pulse, but an SCR cannot.
2. A positive pulse on the cathode gate or a negative pulse on the anode gate turns the SCS on.
3. **(a)** positive pulse on anode gate **(b)** negative pulse on cathode gate
   (c) reduce anode current below holding value
   (d) completely interrupt anode current

SECTION 11–5

1. The diac is like two parallel Shockley diodes connected in opposite direction.
2. A triac is like two parallel SCRs having a common gate and connected in opposite directions.

3. A triac has a gate terminal, but a diac does not.

SECTION 11–6

1. The UJT terminals are base 1, base 2, and emitter.
2. $\eta = r_{B1}/r_{BB}$
3. R, C, and η determine the period.

SECTION 11–7

1. The turn-on voltage can be adjusted to a desired value.
2. The PUT is a thyristor, similar in structure to an SCR, but it is turned on by the anode-to-gate voltage. It has a negative resistance characteristic like the UJT.

SECTION 11–8

1. The base current of a phototransistor is light induced.
2. Base
3. The collector current depends on β_{dc} and I_λ.

SECTION 11–9

1. Most LASCRs can be operated as conventional SCRs.
2. A switch to shut off the anode current is required.

SECTION 11–10

1. An LED
2. Phototransistor, photodarlington, LASCR, phototriac, linear amplifier

SECTION 11–11

1. At zero gate voltage, the PUT turns on at the beginning of the ac cycle and the SCR conducts for almost 180°.
2. The SCR triggers on earlier.
3. The SCR triggers on later in the ac cycle.

ANSWERS TO PRACTICE EXERCISES

11–1 10 MΩ

11–2 163 Ω

11–3 By increasing V_{BB}

11–4 $343\ \text{k}\Omega > R_1 > 1.95\ \text{k}\Omega$

12

OPERATIONAL AMPLIFIERS

After completing this chapter, you should be able to

- □ Describe the basic operational amplifier.
- □ Compare ideal to practical op-amp characteristics.
- □ Explain the operation of a basic differential amplifier.
- □ Discuss single and double-ended operation in differential amplifiers.
- □ Define *open-loop gain.*
- □ Define and calculate the common-mode rejection ratio.
- □ Use op-amp data sheet parameters.
- □ Discuss negative feedback and how it is used in amplifiers.
- □ Explain how negative feedback affects the voltage gain of an op-amp.
- □ Explain how negative feedback affects input and output impedances.
- □ Recognize and analyze inverting, noninverting, and voltage follower op-amp configurations.

So far in this book, you have studied a number of important electronic devices. These devices, such as the diode and the transistor, are separate devices that are individually packaged and interconnected in a circuit with other devices to form a complete, functional unit. Such devices are referred to as *discrete* components.

Now we move into the area of linear integrated circuits, where many transistors, diodes, resistors, and capacitors are fabricated on a single tiny chip of semiconductor material and packaged in a single case to form a functional circuit. An integrated circuit, such as an operational amplifier (op-amp), is treated as a single device. This means that you will be concerned with what the circuit does more from an external viewpoint than from an internal, component-level viewpoint.

In this chapter, you will learn the basics of operational amplifiers, which are the most versatile and widely used of all linear integrated circuits.

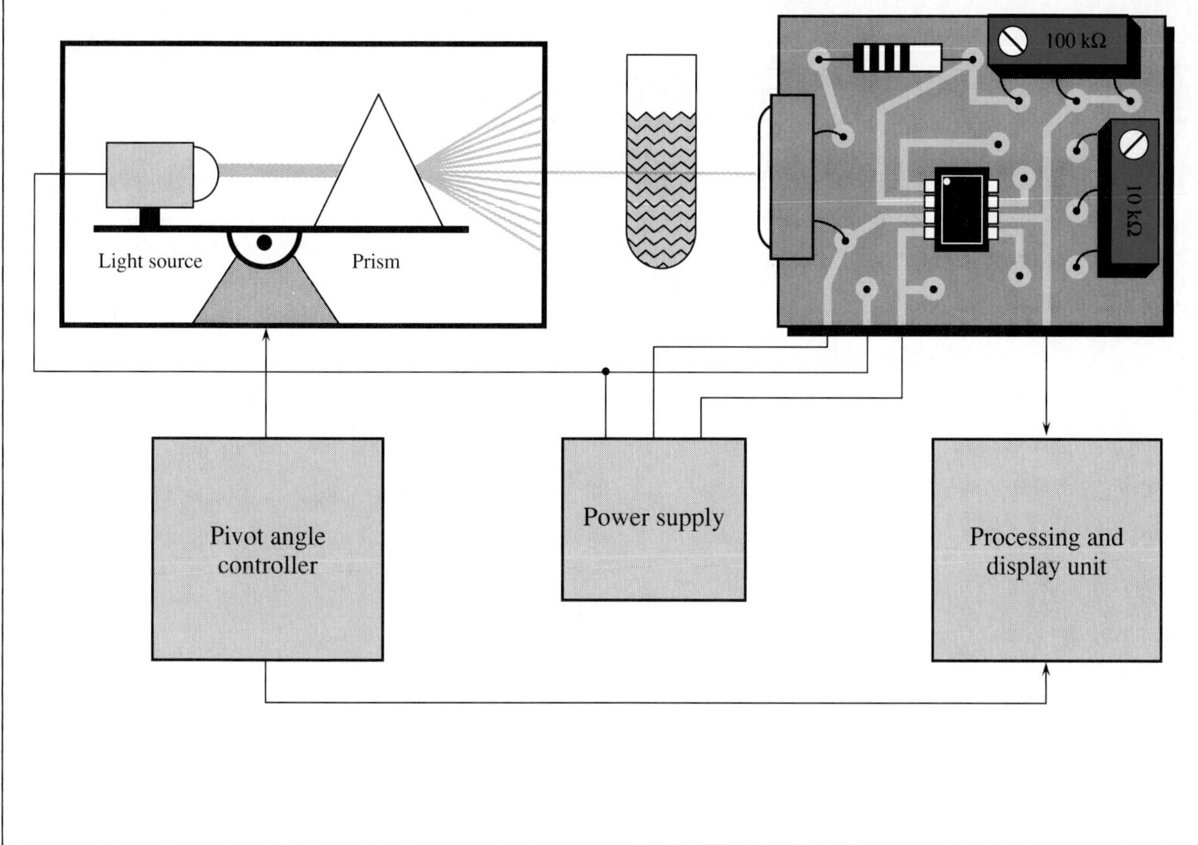

A System Application

In medical laboratories, an instrument known as a spectrophotometer is used to analyze chemicals in solutions by determining how much absorption of light occurs over a range of wavelengths. A basic system is shown in the above diagram. Light is passed through a prism; and as the light source and prism are pivoted, different wavelengths of visible light pass through the slit. The wavelength coming through the slit at a given pivot angle passes through the solution and is detected by a photocell. The op-amp circuit is used to amplify the output of the photocell and send the signal to a processor and display instrument. Since every chemical and compound absorbs light in a different way, the output of the spectrophotometer can be used to accurately identify the contents of the solution.

For the system application in Section 12–9, in addition to the other topics, be sure you understand

- ☐ The functions of the inputs and outputs of an op-amp.
- ☐ How an op-amp works.
- ☐ The pin configurations of an op-amp.
- ☐ How to null an op-amp's output.

12–1 INTRODUCTION TO OPERATIONAL AMPLIFIERS

Early* operational amplifiers *(op-amps) were used primarily to perform mathematical operations such as addition, subtraction, integration, and differentiation, hence the term operational. These early devices were constructed with vacuum tubes and worked with high voltages. Today's op-amps are linear integrated circuits that use relatively low supply voltages and are reliable and inexpensive.

SYMBOL AND TERMINALS

The standard op-amp symbol is shown in Figure 12–1(a). It has two input terminals, called the *inverting input* (−) and the *noninverting input* (+), and one output terminal. The typical op-amp operates with two dc supply voltages, one positive and the other negative, as shown in Figure 12–1(b). Usually these dc voltage terminals are left off the schematic symbol for simplicity but are always understood to be there. Some typical op-amp IC packages are shown in Figure 12–1(c).

FIGURE 12–1
Op-amp symbols and packages. Part (c) copyright of Motorola, Inc. Used by permission.

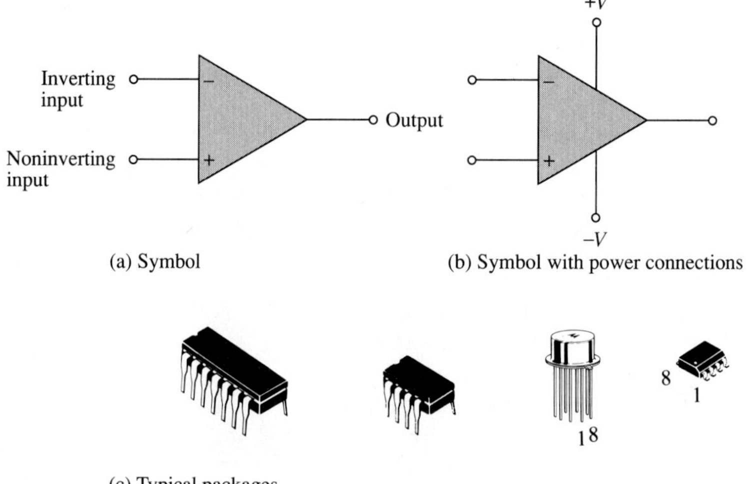

THE IDEAL OP-AMP

In order to get a concept of what an op-amp is, we will consider its ideal characteristics. A practical op-amp, of course, falls short of these ideal standards, but it is much easier to understand and analyze the device from an ideal point of view.

First, the ideal op-amp has infinite voltage gain and infinite bandwidth. Also, it has an infinite input impedance (open), so that it does not load the driving source. Finally, it has a zero output impedance. These characteristics are illustrated in Figure 12–2. The input voltage V_{in} appears between the two input terminals, and the output voltage is A_vV_{in}, as indicated by the internal voltage source symbol. The concept of infinite input impedance is a particularly valuable analysis tool for the various op-amp configurations covered later.

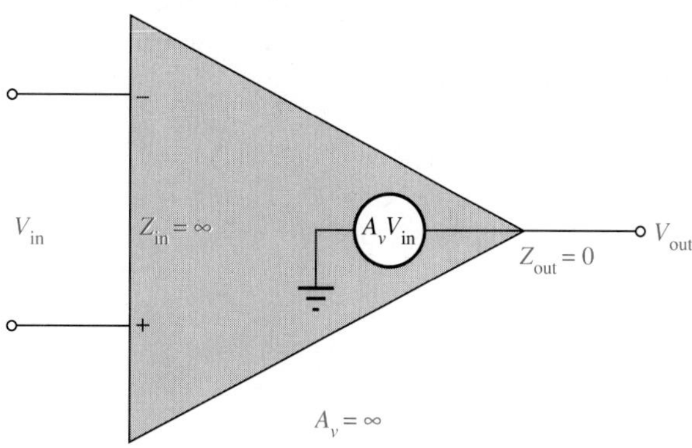

FIGURE 12–2
Ideal op-amp representation.

THE PRACTICAL OP-AMP

Although modern integrated circuit op-amps approach parameter values that can be treated as ideal in many cases, the ideal device has not been and probably will not be developed even though improvements continue to be made. Any device has limitations, and the integrated circuit op-amp is no exception. Op-amps have both voltage and current limitations. Peak-to-peak output voltage, for example, is usually limited to slightly less than the two supply voltages. Output current is also limited by internal restrictions such as power dissipation and component ratings. Characteristics of a practical op-amp are high voltage gain, high input impedance, low output impedance, and wide bandwidth. Some of these are illustrated in Figure 12–3.

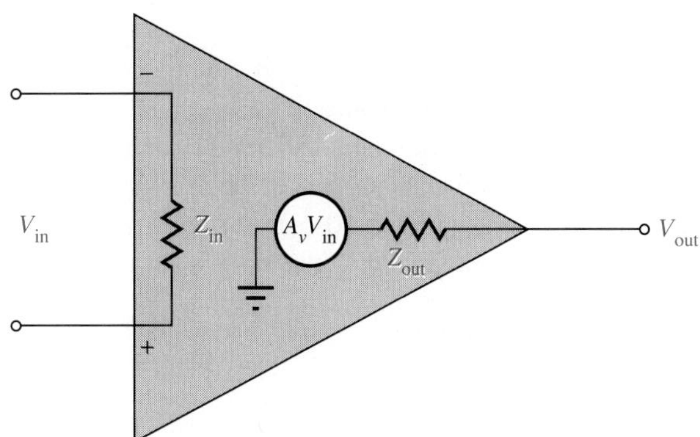

FIGURE 12–3
Practical op-amp representation.

12–1 REVIEW QUESTIONS

1. What are the connections to a basic op-amp?
2. Describe some of the characteristics of a *practical* op-amp.

12–2 THE DIFFERENTIAL AMPLIFIER

The op-amp, in its basic form, typically consists of two or more differential amplifier stages. Because the differential amplifier (diff-amp) is fundamental to the op-amp's internal operation, it is useful to spend some time in acquiring a basic understanding of this type of circuit.

A basic **differential amplifier** circuit is shown in Figure 12–4(a) and its block symbol in Figure 12–4(b). The diff-amp stages that make up part of the op-amp provide high voltage gain and common-mode rejection (defined later).

FIGURE 12–4
Basic differential amplifier.

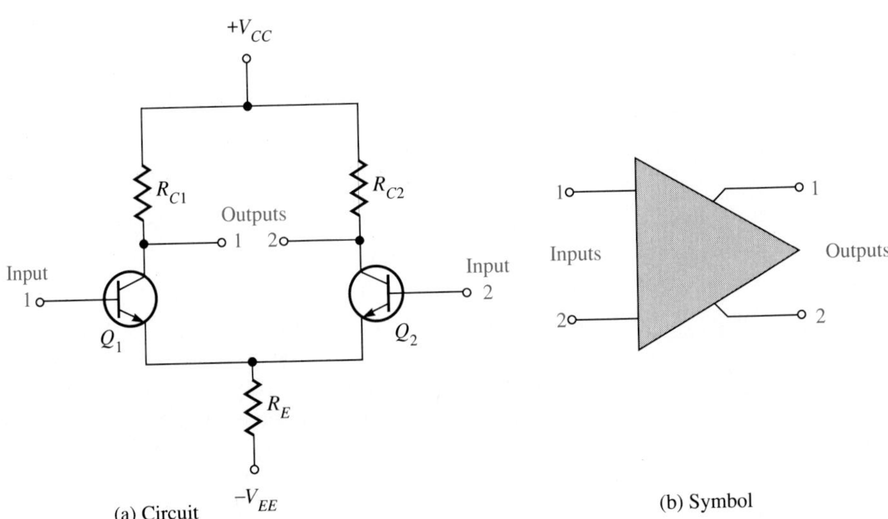

BASIC OPERATION

Although an op-amp typically has more than one differential amplifier stage, we will use a single diff-amp to illustrate the basic operation. The following discussion is in relation to Figure 12–5 and consists of a basic dc analysis of the diff-amp's operation. First, when both inputs are grounded (0 V), the emitters are at −0.7 V, as indicated in Figure 12–5(a). It is assumed that the transistors are identically matched by careful process control during manufacturing so that their dc emitter currents are the same when there is no input signal.

$$I_{E1} = I_{E2}$$

Since both emitter currents combine through R_E,

$$I_{E1} = I_{E2} = \frac{I_{R_E}}{2} \qquad \textbf{(12–1)}$$

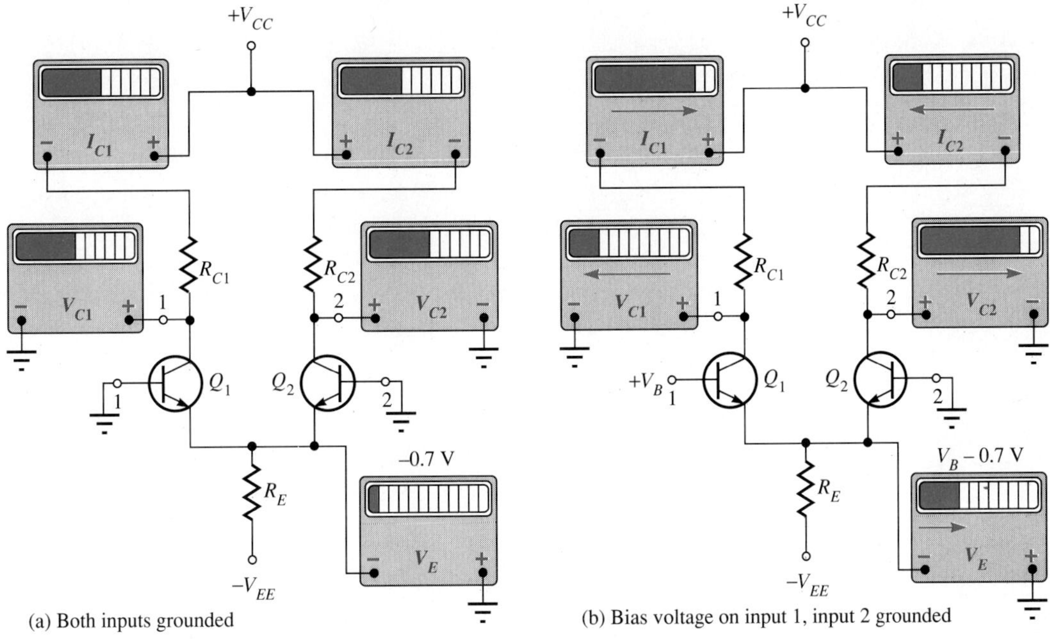

(a) Both inputs grounded

(b) Bias voltage on input 1, input 2 grounded

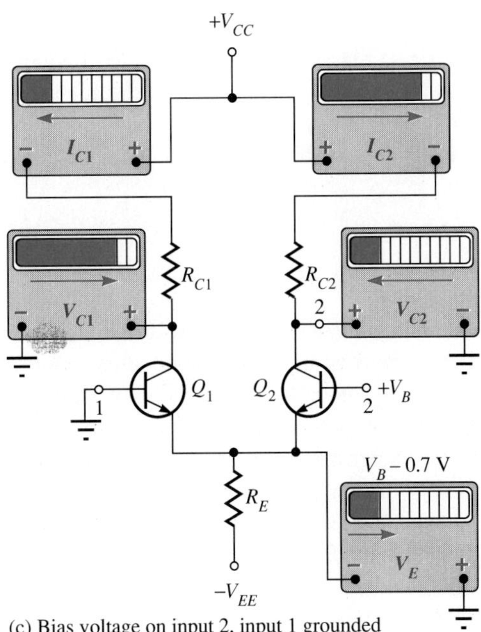

(c) Bias voltage on input 2, input 1 grounded

FIGURE 12–5
Basic operation of a differential amplifier (ground is zero volts).

where

$$I_{R_E} = \frac{V_E - V_{EE}}{R_E} \tag{12–2}$$

Based on the approximation that $I_C \cong I_E$, it can be stated that

$$I_{C1} = I_{C2} \cong \frac{I_{R_E}}{2} \tag{12–3}$$

Since both collector currents and both collector resistors are equal (when the input voltage is zero),

$$V_{C1} = V_{C2} = V_{CC} - I_{C1}R_{C1} \tag{12–4}$$

This condition is illustrated in Figure 12–5(a). Next, input 2 is left grounded, and a positive bias voltage is applied to input 1, as shown in Figure 5–2(b). The positive voltage on the base of Q_1 increases I_{C1} and raises the emitter voltage to

$$V_E = V_B - 0.7 \text{ V} \tag{12–5}$$

This action reduces the forward bias (V_{BE}) of Q_2 because its base is held at 0 V (ground), thus causing I_{C2} to decrease as indicated in part (b) of the diagram. The net result is that the increase in I_{C1} causes a decrease in V_{C1}, and the decrease in I_{C2} causes an increase in V_{C2}, as shown. Finally, input 1 is grounded and a positive bias voltage is applied to input 2, as shown in Figure 12–5(c).

The positive bias voltage causes Q_2 to conduct more, thus increasing I_{C2}. Also, the emitter voltage is raised. This reduces the forward bias of Q_1, since its base is held at ground, and causes I_{C1} to decrease. The result is that the increase in I_{C2} produces a decrease in V_{C2}, and the decrease in I_{C1} causes V_{C1} to increase, as shown in Figure 12–5(c).

MODES OF SIGNAL OPERATION

SINGLE-ENDED INPUT When a diff-amp is operated in this mode, one input is grounded and the signal voltage is applied only to the other input, as shown in Figure 12–6. In the case where the signal voltage is applied to input 1 as in part (a), an inverted, amplified signal voltage appears at output 1 as shown. Also, a signal voltage appears in-phase at the emitter of Q_1. Since the emitters of Q_1 and Q_2 are common, the emitter signal becomes an input to Q_2, which functions as a common-base amplifier. The signal is amplified by Q_2 and appears, noninverted, at output 2. This action is illustrated in part (a).

In the case where the signal is applied to input 2 with input 1 grounded, as in Figure 12–6(b), an inverted, amplified signal voltage appears at output 2. In this situation, Q_1 acts as a common-base amplifier, and a noninverted, amplified signal appears at output 1. This action is illustrated in part (b) of the figure.

DIFFERENTIAL INPUT In this mode, two opposite-polarity (out-of-phase) signals are applied to the inputs, as shown in Figure 12–7(a). This type of operation is also referred to as

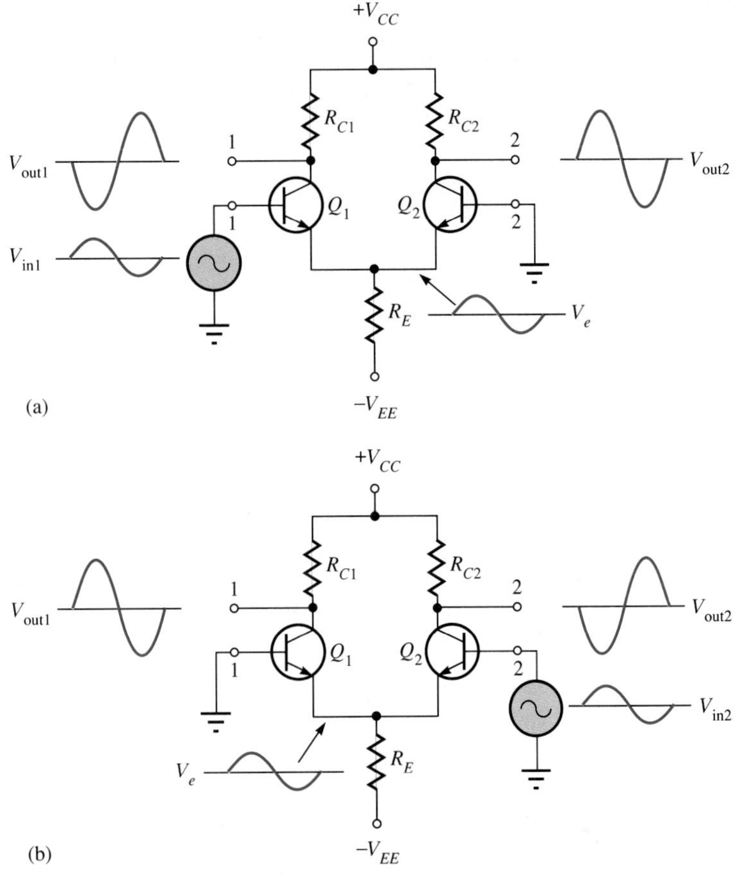

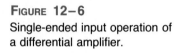

FIGURE 12–6
Single-ended input operation of a differential amplifier.

double-ended. Each input affects the outputs, as you will see in the following discussion. Figure 12–7(b) shows the output signals due to the signal on input 1 acting alone as a single-ended input. Figure 12–7(c) shows the output signals due to the signal on input 2 acting alone as a single-ended input. Notice, in parts (b) and (c), that the signals on output 1 are of the same polarity. The same is also true for output 2. By superimposing both output 1 signals and both output 2 signals, we get the total differential operation, as pictured in Figure 12–7(d).

COMMON-MODE INPUT One of the most important aspects of the operation of a differential amplifier can be seen by considering the **common-mode** condition where two signal voltages of the same phase, frequency, and amplitude are applied to the two inputs, as shown in Figure 12–8(a). Again, by considering each input signal as acting alone, the basic operation can be understood. Figure 12–8(b) shows the output signals due to the signal on only input 1, and Figure 12–8(c) shows the output signals due to the signal on only input 2. Notice that the corresponding signals on output 1 are of the opposite

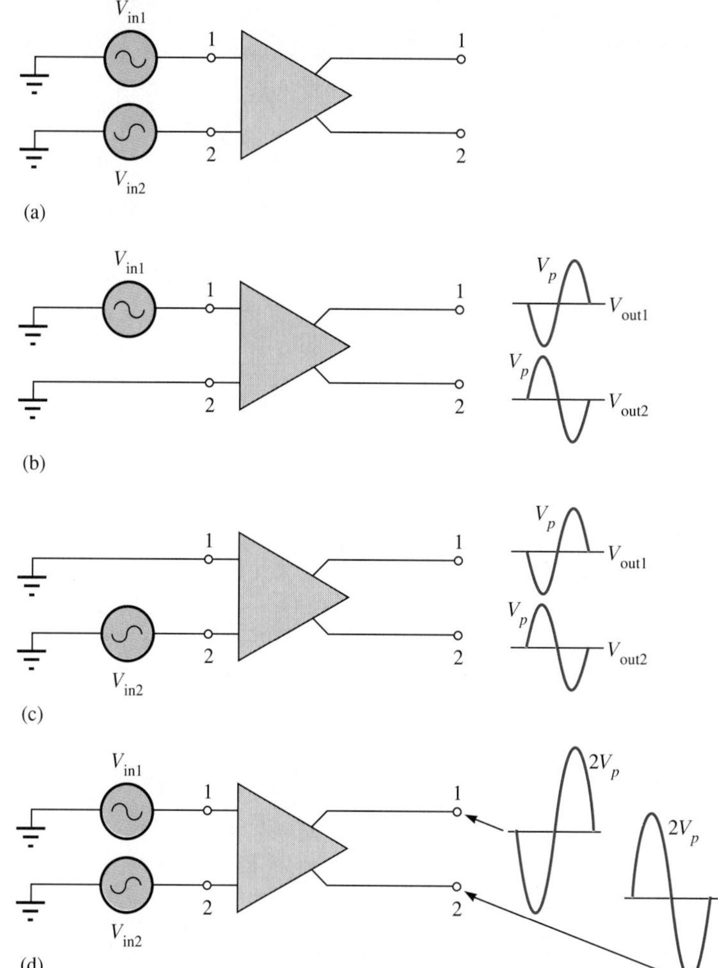

FIGURE 12–7 Differential operation of a differential amplifier. (a) Differential inputs. (b) Outputs due to V_{in1}. (c) Outputs due to V_{in2}. (d) Total outputs due to differential inputs.

polarity, and so are the ones on output 2. When these are superimposed, they cancel, resulting in a zero output voltage, as shown in Figure 12–8(d).

This action is called *common-mode rejection*. Its importance lies in the situation where an unwanted signal appears commonly on both diff-amp inputs. Common-mode rejection means that this unwanted signal will not appear on the outputs to distort the desired signal. Common-mode signals (noise) generally are the result of the pick-up of radiated energy on the input lines, from adjacent lines, or the 60 Hz power line, or other sources.

In summary, desired signals appear on only one input or with opposite polarities on both input lines. These desired signals are amplified and appear on the outputs as previously discussed. Unwanted signals (noise) appearing with the same polarity on both input lines are essentially cancelled by the diff-amp and do not appear on the outputs. The measure of an amplifier's ability to reject common-mode signals is a parameter called the *common-mode rejection ratio* (CMRR) and is discussed next.

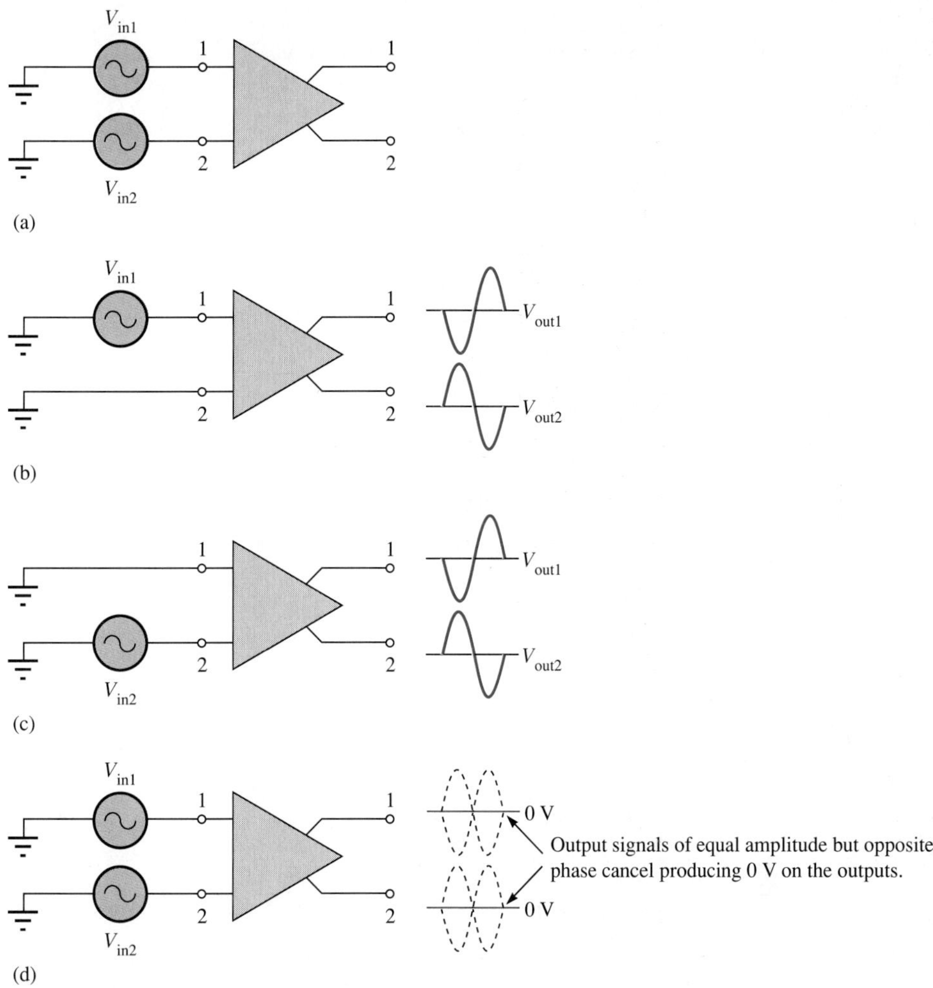

FIGURE 12–8
Common-mode operation of a differential amplifier. (a) Common-mode inputs. (b) Outputs due to V_{in1}. (c) Outputs due to V_{in2}. (d) Outputs cancel when common-mode signals are applied.

COMMON-MODE GAIN

Ideally, a differential amplifier provides a very high gain for desired signals (single-ended or differential), and zero gain for common-mode signals. Practical diff-amps, however, do exhibit a very small common-mode gain (usually much less than one), while providing a high differential voltage gain (usually several thousand). The higher the differential gain with respect to the common-mode gain, the better the performance of the diff-amp in terms of rejection of common-mode signals. This suggests that a good measure of the diff-amp's performance in rejecting unwanted common-mode signals is the ratio of the differential gain $A_{v(d)}$ to the common-mode gain, A_{cm}. This ratio is called the **common-mode rejection ratio,** CMRR.

$$\text{CMRR} = \frac{A_{v(d)}}{A_{cm}} \quad (12\text{–}6)$$

The higher the CMRR, the better, as you can see. A very high value of CMRR means that the differential gain $A_{v(d)}$ is high and the common-mode gain A_{cm} is low. The CMRR is often expressed in decibels (dB) as

$$\text{CMRR} = 20 \log \frac{A_{v(d)}}{A_{cm}} \quad (12\text{–}7)$$

EXAMPLE 12–1 A certain differential amplifier has a differential voltage gain of 2000 and a common-mode gain of 0.2. Determine the CMRR and express it in dB.

SOLUTION

$A_{v(d)} = 2000$, and $A_{cm} = 0.2$. Therefore,

$$\text{CMRR} = \frac{A_{v(d)}}{A_{cm}} = \frac{2000}{0.2} = 10{,}000$$

In dB, CMRR $= 20 \log(10{,}000) = 80$ dB.

PRACTICE EXERCISE 12–1

Determine the CMRR and express it in dB for an amplifier with a differential voltage gain of 8500 and a common-mode gain of 0.25.

A CMRR of 10,000, for example, means that the desired input signal (differential) is amplified 10,000 times more than the unwanted noise (common-mode). So, as an example, if the amplitudes of the differential input signal and the common-mode noise are equal, the desired signal will appear on the output 10,000 times greater in amplitude than the noise. Thus, the noise or interference has been essentially eliminated. Example 12–2 should help reinforce the idea of common-mode rejection and the general signal operation of the differential amplifier.

EXAMPLE 12–2 The differential amplifier shown in Figure 12–9 has a differential voltage gain of 2500 and a CMRR of 30,000. In part (a) of the figure, a single-ended input signal of 500 μV rms is applied. At the same time a 1-V, 60 Hz interference signal appears on both inputs as a result of radiated pick-up from the ac power system. In part (b) of the figure, differential input signals of 500 μV rms each are applied to the inputs. The common-mode interference is the same as in part (a).

(a) Determine the common-mode gain.
(b) Express the CMRR in dB.
(c) Determine the rms output signal for parts (a) and (b) of the figure.
(d) Determine the rms interference voltage on the output.

FIGURE 12–9

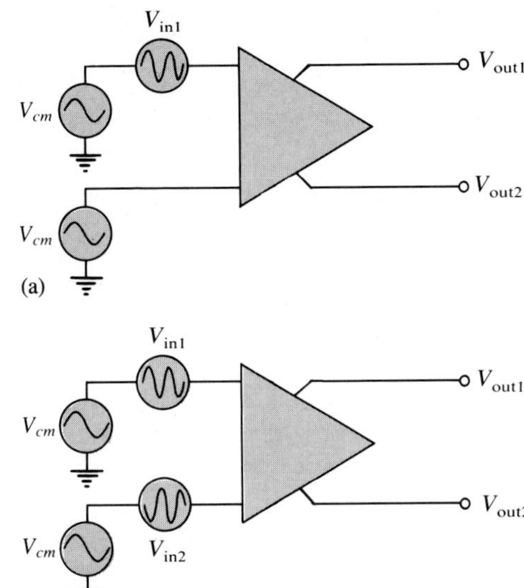

SOLUTION

(a) CMRR $= \dfrac{A_{v(d)}}{A_{cm}}$. Therefore,

$$A_{cm} = \frac{A_{v(d)}}{\text{CMRR}} = \frac{2500}{30{,}000} = 0.083$$

(b) CMRR $= 20 \log(30{,}000) = 89.5$ dB

(c) In Figure 12–9(a), the differential input voltage is the difference between the voltage on input 1 and that on input 2. Since input 2 is grounded, its voltage is zero. Therefore,

$$\begin{aligned} V_{in(diff)} &= V_{in1} - V_{in2} \\ &= 500\ \mu\text{V} - 0\ \text{V} \\ &= 500\ \mu\text{V} \end{aligned}$$

The output signal voltage in this case is taken at output 1.

$$\begin{aligned} V_{out1} &= A_{v(d)} V_{in(diff)} \\ &= (2500)(500\ \mu\text{V}) \\ &= 1.25\ \text{V rms} \end{aligned}$$

In Figure 12–9(b), the differential input voltage is the difference between the two opposite-polarity, 500 μV signals.

$$\begin{aligned} V_{\text{in(diff)}} &= V_{\text{in1}} - V_{\text{in2}} \\ &= 500\ \mu\text{V} - (-500\ \mu\text{V}) \\ &= 1000\ \mu\text{V} \\ &= 1\ \text{mV} \end{aligned}$$

The output voltage signal is

$$\begin{aligned} V_{\text{out1}} &= A_{v(d)} V_{\text{in(diff)}} \\ &= (2500)(1\ \text{mV}) \\ &= 2.5\ \text{V rms} \end{aligned}$$

This shows that a differential input (two opposite-polarity signals) results in a gain that is double that for a single-ended input.

(d) The common-mode input is 1 V rms. The common-mode gain A_{cm} is 0.083. The interference (common-mode) voltage on the output is therefore

$$A_{cm} = \frac{V_{\text{out}(cm)}}{V_{\text{in}(cm)}}$$

$$\begin{aligned} V_{\text{out}(cm)} &= A_{cm} V_{\text{in}(cm)} \\ &= (0.083)(1\ \text{V}) \\ &= 0.083\ \text{V} \end{aligned}$$

PRACTICE EXERCISE 12–2

The amplifier in Figure 12–9 has a differential voltage gain of 4200 and a CMRR of 25,000. For the same single-ended and differential input signals as described in the example: **(a)** Find A_{cm}. **(b)** Express the CMRR in dB. **(c)** Determine the rms output signal for parts (a) and (b) of the figure. **(d)** Determine the rms interference (common-mode) voltage appearing on the output.

SIMPLE OP-AMP ARRANGEMENT

Figure 12–10 shows two differential amplifier (diff-amp) stages and an emitter-follower connected to form a simple op-amp. The first stage can be used with a single-ended or a differential input. The differential outputs of the first stage are directly coupled into the differential inputs of the second stage. The output of the second stage is single-ended to drive an emitter-follower to achieve a relatively low output impedance. Both differential stages together provide a high voltage gain and a high CMRR.

12–2 REVIEW QUESTIONS

1. Distinguish between differential and single-ended inputs.
2. What is common-mode rejection?
3. For a given value of differential gain, does a higher CMRR result in a higher or lower common-mode gain?

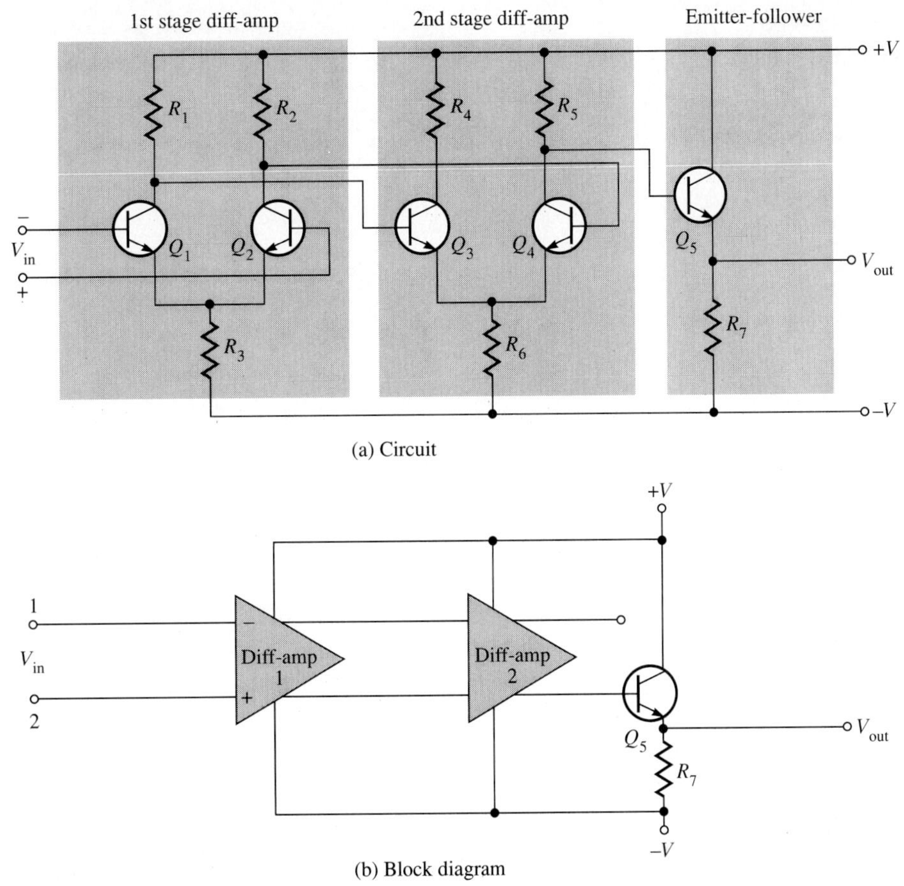

FIGURE 12–10
Simplified internal circuitry of a basic op-amp.

12–3 OP-AMP DATA SHEET PARAMETERS

In this section, several important op-amp parameters are defined. These are the input offset voltage, the input offset voltage drift, the input bias current, the input impedance, the input offset current, the output impedance, the common-mode range, the open-loop voltage gain, the common mode rejection ratio, the slew rate, and the frequency response. Also four popular IC op-amps are compared in terms of these parameters.

INPUT OFFSET VOLTAGE

The ideal op-amp produces zero volts out for zero volts in. In a practical op-amp, however, a small dc voltage appears at the output when no differential input voltage is applied. Its primary cause is a slight mismatch of the base-to-emitter voltages of the differential input stage, as illustrated in Figure 12–11(a). The output voltage of the

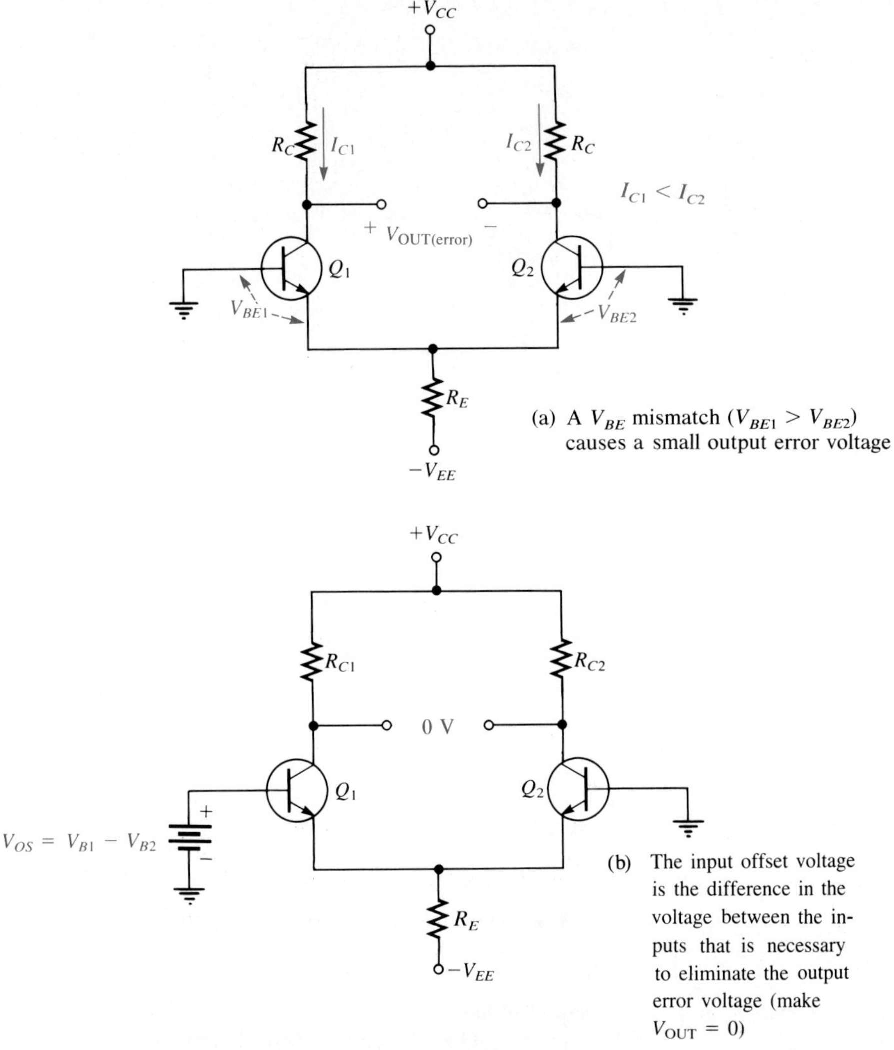

FIGURE 12–11
Input offset voltage, V_{OS}.

differential input stage is expressed as

$$V_{OUT} = I_{C2}R_C - I_{C1}R_C \tag{12–8}$$

A small difference in the base-to-emitter voltages of Q_1 and Q_2 causes a small difference in the collector currents. This results in a nonzero value of V_{OUT}. (The collector resistors are equal.) As specified on an op-amp data sheet, the *input offset voltage* V_{OS} is the differential dc voltage required between the inputs to force the differential output to zero volts, as demonstrated in Figure 12–11(b). Typical values of input offset voltage are in the range of 2 mV or less. In the ideal case, it is 0 V.

INPUT OFFSET VOLTAGE DRIFT WITH TEMPERATURE

The *input offset voltage drift* is a parameter related to V_{OS} that specifies how much change occurs in the input offset voltage for each degree change in temperature. Typical values range anywhere from about 5 μV per degree Celsius to about 50 μV per degree Celsius. Usually, an op-amp with a higher nominal value of input offset voltage exhibits a higher drift.

INPUT BIAS CURRENT

You have seen that the input terminals of a bipolar differential amplifier are the transistor bases and, therefore, the input currents are the base currents. The *input bias current* is the dc current required by the inputs of the amplifier to properly operate the first stage. By definition, the input bias current is the average of both input currents and is calculated as follows:

$$I_{BIAS} = \frac{I_1 + I_2}{2} \qquad (12\text{–}9)$$

The concept of input bias current is illustrated in Figure 12–12.

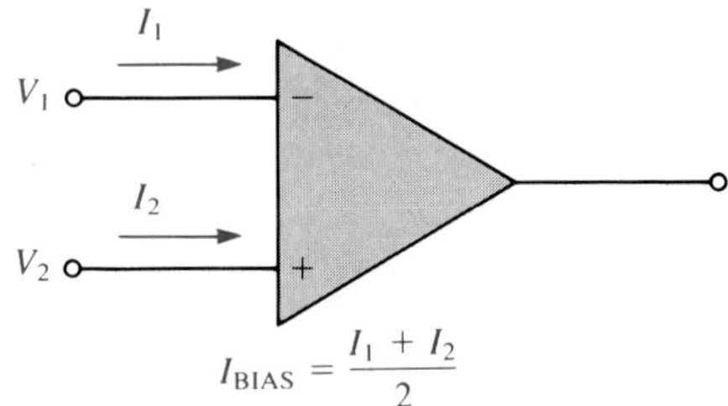

FIGURE 12–12
Input bias current is the average of the two op-amp input currents.

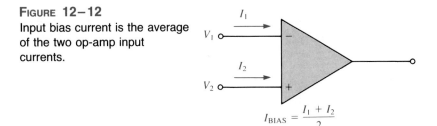

INPUT IMPEDANCE

Two basic ways of specifying the *input impedance* of an op-amp are the differential and the common mode. The differential input impedance is the total resistance between the inverting and the noninverting inputs and is illustrated in Figure 12–13(a). Differential impedance is measured by determining the change in bias current for a given change in differential input voltage. The common-mode input impedance is the resistance between each input and ground and is measured by determining the change in bias current for a given change in common-mode input voltage. It is depicted in Figure 12–13(b).

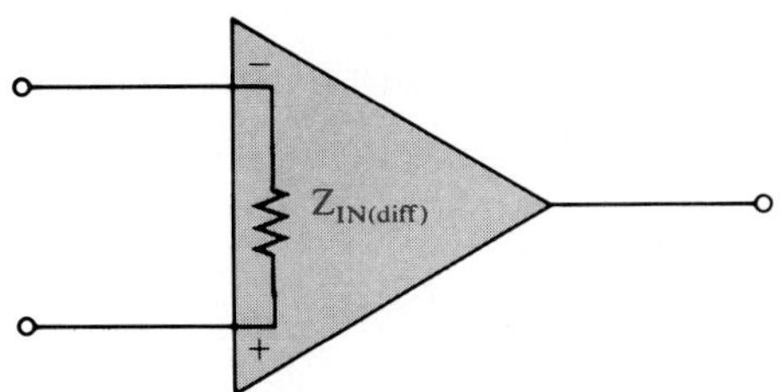

(a) Differential input impedance

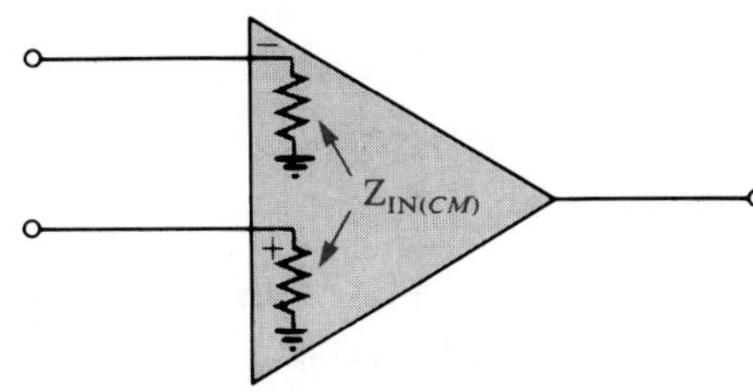

(b) Common-mode input impedance

FIGURE 12–13
Op-amp input impedance.

INPUT OFFSET CURRENT

Ideally, the two input bias currents are equal, and thus their difference is zero. In a practical op-amp, however, the bias currents are not exactly equal. The *input offset current* is the difference of the input bias currents, expressed as

$$I_{OS} = |I_1 - I_2| \tag{12–10}$$

Actual magnitudes of offset current are usually at least an order of magnitude (ten times) less than the bias current. In many applications, the offset current can be neglected. However, high-gain, high-input impedance amplifiers should have as little I_{OS} as possible, because the difference in currents through large input resistances develops a substantial offset voltage, as shown in Figure 12–14.

The offset voltage developed by the input offset current is

$$V_{OS} = I_1R_{in} - I_2R_{in}$$
$$= (I_1 - I_2)R_{in}$$
$$V_{OS} = I_{OS}R_{in} \tag{12–11}$$

The error created by I_{OS} is amplified by the gain A_v of the op-amp and appears in the output as

$$V_{OUT(error)} = A_vI_{OS}R_{in} \tag{12–12}$$

The change in offset current with temperature is often an important consideration. Values of temperature coefficient in the range of 0.5 nA per degree Celsius are common.

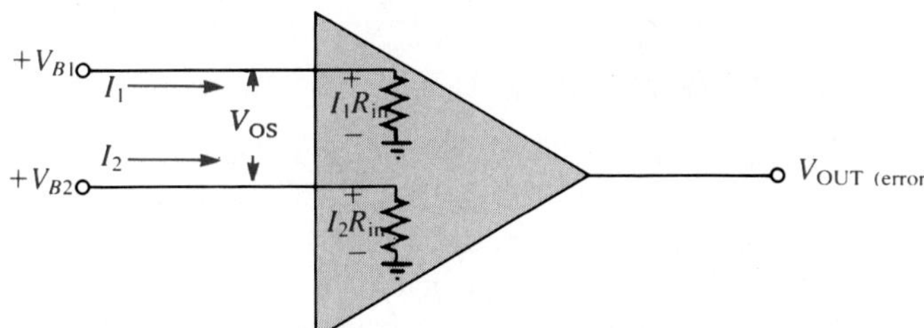

FIGURE 12–14
Effect of input offset current.

OUTPUT IMPEDANCE

Output impedance is the resistance viewed from the output terminal of the op-amp, as indicated in Figure 12–15.

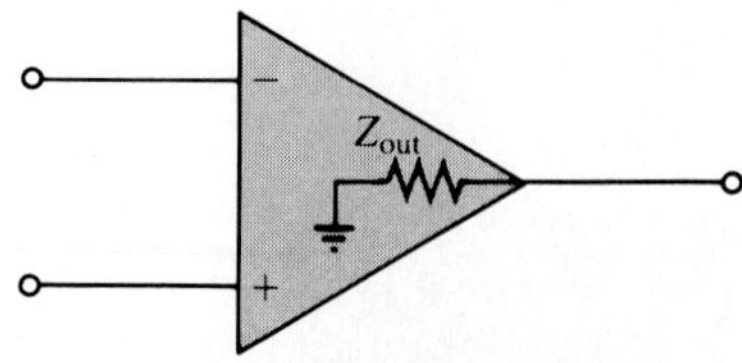

FIGURE 12–15
Op-amp output impedance.

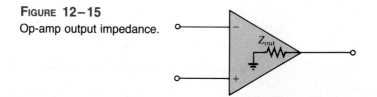

COMMON-MODE RANGE

All op-amps have limitations on the range of voltages over which they will operate. The *common-mode range* is the range of input voltages which, when applied to both inputs, will not cause clipping or other output distortion. Many op-amps have common-mode ranges of ±10 V with dc supply voltages of ±15 V.

OPEN-LOOP VOLTAGE GAIN

Open-loop voltage gain is the gain of the op-amp without any external feedback from output to input. A good op-amp has a very high **open-loop** gain; 50,000 to 200,000 is typical.

COMMON-MODE REJECTION RATIO

The **common-mode rejection ratio** (CMRR), as discussed in conjunction with the diff-amp, is a measure of an op-amp's ability to reject common-mode signals. An infinite value of CMRR means that the output is zero when the same signal is applied to both inputs (common-mode).

An infinite CMRR is never achieved in practice, but a good op-amp does have a very high value of CMRR. As previously mentioned, common-mode signals are undesired interference voltages such as 60 Hz power-supply ripple and noise voltages due to pick-up of radiated energy. A high CMRR enables the op-amp to virtually eliminate these interference signals from the output.

The accepted definition of CMRR for an op-amp is the open-loop gain (A_{ol}) divided by the common-mode gain.

$$\text{CMRR} = \frac{A_{ol}}{A_{cm}} \tag{12–13}$$

It is commonly expressed in decibels as follows:

$$\text{CMRR} = 20 \log \frac{A_{ol}}{A_{cm}} \tag{12–14}$$

EXAMPLE 12–3 A certain op-amp has an open-loop gain of 100,000 and a common-mode gain of 0.25. Determine the CMRR and express it in dB.

SOLUTION

$$\text{CMRR} = \frac{A_{ol}}{A_{cm}} = \frac{100{,}000}{0.25} = 400{,}000$$

$$\text{CMRR} = 20 \log(400{,}000) = 112 \text{ dB}$$

PRACTICE EXERCISE 12–3

If a particular op-amp has a CMRR of 90 dB and a common-mode gain of 0.4, what is the open-loop gain?

SLEW RATE

The maximum rate of change of the output voltage in response to a step input voltage is the **slew rate** of an op-amp. The slew rate is dependent upon the high-frequency response of the amplifier stages within the op-amp. Slew rate is measured with an op-amp connected as shown in Figure 12–16(a). This particular op-amp connection is a unity-gain, noninverting configuration which will be discussed later. It gives a worst-case (slowest) slew rate. Recall that the high-frequency components of a voltage step are contained in the rising edge and that the upper critical frequency of an amplifier limits its response to a step input. The lower f_{ch} is, the more slope there is on the output for a step input.

FIGURE 12–16
Slew-rate measurement.

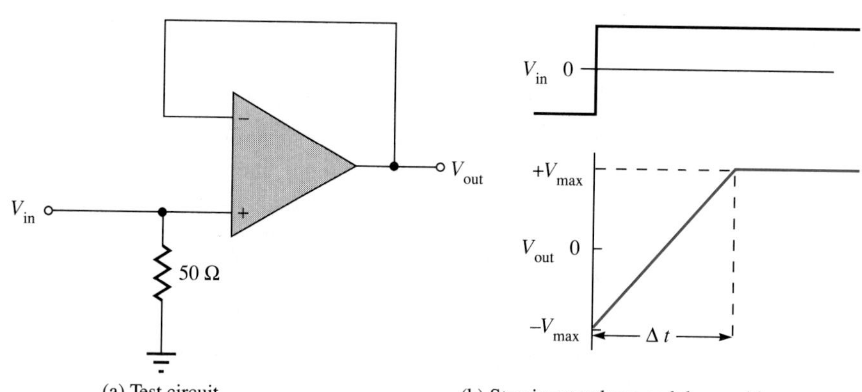

(a) Test circuit

(b) Step input voltage and the resulting output voltage

A pulse is applied to the input as shown, and the ideal output voltage is measured as indicated in Figure 12–16(b). The width of the input pulse must be sufficient to allow the output to "slew" from its lower limit to its upper limit, as shown. As you can see, a certain time interval, Δt, is required for the output voltage to go from its lower limit $-V_{max}$ to its upper limit $+V_{max}$, once the input step is applied. The slew rate is expressed as

$$\text{Slew rate} = \frac{\Delta V_{out}}{\Delta t} \qquad (12\text{–}15)$$

where $\Delta V_{out} = +V_{max} - (-V_{max})$. The unit of slew rate is volts per microsecond (V/μs).

EXAMPLE 12–4

The output voltage of a certain op-amp appears as shown in Figure 12–17 in response to a step input. Determine the slew rate.

SOLUTION

The output goes from the lower to the upper limit in 1 μs. Since this is not an ideal response, the limits are taken at the 90 percent points, as indicated. So, the upper limit is +9 V and the lower limit is −9 V. The slew rate is

$$\frac{\Delta V}{\Delta t} = \frac{+9\text{ V} - (-9\text{ V})}{1\ \mu\text{s}} = 18\text{ V}/\mu\text{s}$$

FIGURE 12–17

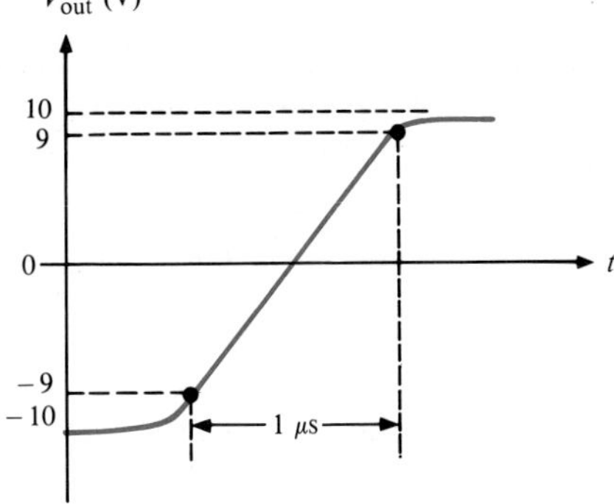

PRACTICE EXERCISE 12–4

When a pulse is applied to an op-amp, the output voltage goes from −8 V to +7 V in 0.75 μs. What is the slew rate?

FREQUENCY RESPONSE

The internal amplifier stages that make up an op-amp have voltage gains limited by junction capacitances, as discussed in Chapter 10. Although the differential amplifiers used in op-amps are somewhat different from the basic amplifiers discussed, the same principles apply. An op-amp has no internal coupling capacitors, however; therefore, the low-frequency response extends down to dc. Frequency-related characteristics will be discussed in the next chapter.

COMPARISON OF OP-AMP PARAMETERS

Table 12–1 provides a comparison of values of some of the parameters just described for four common integrated circuit op-amps. Any values not listed were not given on the manufacturer's data sheet. All values are typical at 25°C.

TABLE 12–1

	Op-Amp Type			
Parameter	**741C**	**LM101A**	**LM108**	**LM218**
Input offset voltage	1 mV	1 mV	0.7 mV	2 mV
Input bias current	80 nA	120 nA	0.8 nA	120 nA
Input offset current	20 nA	40 nA	0.05 nA	6 nA
Input impedance	2 MΩ	800 kΩ	70 MΩ	3 MΩ
Output impedance	75 Ω	—	—	—
Open-loop gain	200,000	160,000	300,000	200,000
Slew rate	0.5 V/μs	—	—	70 V/μs
CMRR	90 dB	90 dB	100 dB	100 dB

OTHER FEATURES

Most available op-amps have three important features—short-circuit protection, no latch-up, and input offset nulling. Short-circuit protection keeps the circuit from being damaged if the output becomes shorted, and the no latch-up feature prevents the op-amp from hanging up in one output state (high- or low-voltage level) under certain input conditions. Input offset nulling is achieved by an external potentiometer that sets the output voltage at precisely zero with zero input.

12–3 REVIEW QUESTIONS

1. List ten or more op-amp parameters.
2. Which two parameters, not including frequency response, are frequency dependent?

12–4 NEGATIVE FEEDBACK

Negative feedback is one of the most useful concepts in electronic circuits, particularly in op-amp applications. Negative feedback is the process whereby a portion of the output voltage of an amplifier is returned to the input with a phase angle that opposes (or subtracts from) the input signal.

Negative feedback is illustrated in Figure 12–18. The inverting input effectively makes the feedback signal 180° out of phase with the input signal.

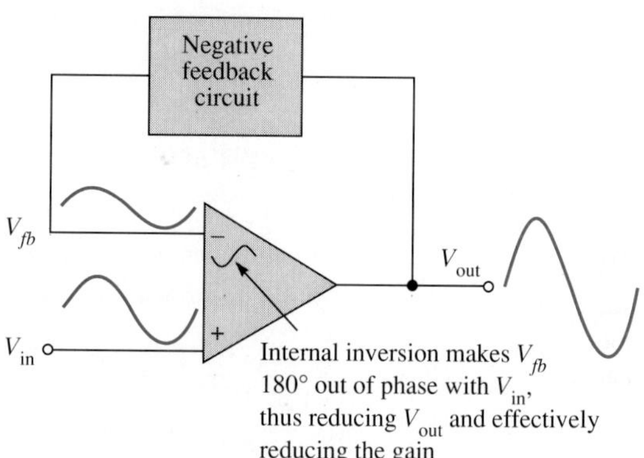
Negative feedback circuit
V_{fb}
V_{in}
V_{out}
Internal inversion makes V_{fb} 180° out of phase with V_{in}, thus reducing V_{out} and effectively reducing the gain

FIGURE 12–18
Illustration of negative feedback.

WHY USE NEGATIVE FEEDBACK?

As you have seen, the inherent open-loop gain of a typical op-amp is very high (usually greater than 100,000). Therefore, an extremely small input voltage drives the op-amp into its saturated output states. In fact, even the input offset voltage of the op-amp can drive it into saturation. For example, assume V_{in} = 1 mV and A_{ol} = 100,000. Then,

$$V_{in}A_{ol} = (1\text{ mV})(100{,}000) = 100\text{ V}$$

Since the output level of an op-amp can never reach 100 V, it is driven deep into saturation and the output is limited to its maximum output levels, as illustrated in Figure 12–19 for both a positive and a negative input voltage of 1 mV.

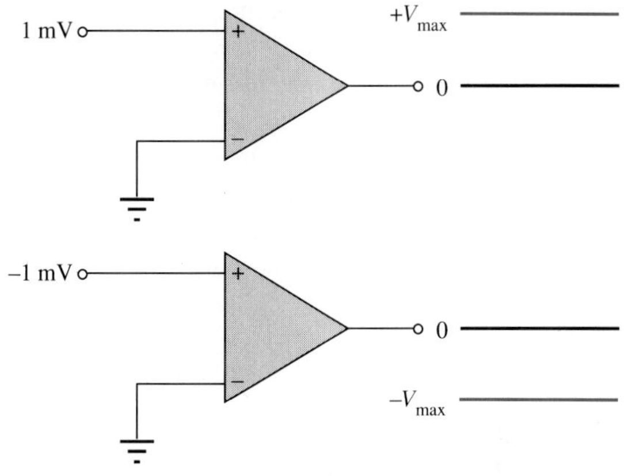

FIGURE 12–19
Without negative feedback, a small input voltage drives the op-amp to its output limits and it becomes nonlinear.

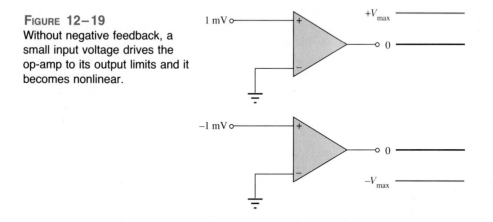

The usefulness of an op-amp operated in this manner is severely restricted and is generally limited to comparator applications (to be studied in a later chapter). With negative feedback, the overall voltage gain (A_{cl}) can be reduced and controlled so that the op-amp can function as a linear amplifier. In addition to providing a controlled, stable voltage gain, negative feedback also provides for control of the input and output impedances and amplifier bandwidth. (We will cover these topics in detail later in this chapter.) Table 12–2 summarizes the general effects of negative feedback on op-amp performance.

TABLE 12–2

	Voltage Gain	Input Z	Output Z	Bandwidth
Without negative feedback	A_{ol} is too high for linear amplifier applications	Relatively high (see Table 12–1)	Relatively low (see Table 12–1)	Relatively narrow
With negative feedback	A_{cl} is set by the feedback circuit to desired value	Can be increased or reduced to a desired value depending on type of circuit	Can be reduced to a desired value	Significantly wider

12–4 REVIEW QUESTIONS

1. What are the benefits of negative feedback in an op-amp circuit?
2. Why is it necessary to reduce the gain of an op-amp from its open-loop value?

12–5 OP-AMP CONFIGURATIONS WITH NEGATIVE FEEDBACK

In this section, we will discuss several basic ways in which an op-amp can be connected using negative feedback to stabilize the gain and increase frequency response. As mentioned, the extremely high open-loop gain of an op-amp creates an unstable situation because a small noise voltage on the input can be amplified to a point where the amplifier is driven out of its linear region. Also, unwanted oscillations can occur. In addition, the open-loop gain parameter of an op-amp can vary greatly from one device to the next. Negative feedback takes a portion of the output and applies it back out-of-phase with the input, creating an effective reduction in gain. This closed-loop gain is usually much less than the open-loop gain and independent of it.

NONINVERTING AMPLIFIER

An op-amp connected as a **noninverting amplifier** with a controlled amount of voltage gain is shown in Figure 12–20. The input signal is applied to the noninverting input. The output is applied back to the inverting input through the feedback network formed by R_i and R_f. This creates negative feedback as follows. R_i and R_f form a voltage-divider network, which reduces the output V_{out} and connects the reduced voltage V_f to the inverting input. The feedback voltage is expressed as

$$V_f = \left(\frac{R_i}{R_i + R_f}\right)V_{out} \qquad \textbf{(12–16)}$$

The difference of the input voltage V_{in} and the feedback voltage V_f is the differential input to the op-amp, as shown in Figure 12–21. This differential voltage is amplified by the open-loop gain of the op-amp (A_{ol}) and produces an output voltage expressed as

$$V_{out} = A_{ol}(V_{in} - V_f) \qquad \textbf{(12–17)}$$

Letting $R_i/(R_i + R_f) = B$ and then substituting BV_{out} for V_f in Equation (12–17), we get the following algebraic steps.

$$\begin{aligned} V_{out} &= A_{ol}(V_{in} - BV_{out}) \\ V_{out} &= A_{ol}V_{in} - A_{ol}BV_{out} \\ V_{out} + A_{ol}BV_{out} &= A_{ol}V_{in} \\ V_{out}(1 + A_{ol}B) &= A_{ol}V_{in} \end{aligned}$$

Since the total voltage gain of the amplifier in Figure 12–20 is V_{out}/V_{in}, it can be expressed as

$$\frac{V_{out}}{V_{in}} = \frac{A_{ol}}{1 + A_{ol}B} \qquad \textbf{(12–18)}$$

The product $A_{ol}B$ is usually much greater than 1, so Equation (12–18) simplifies to

$$\frac{V_{out}}{V_{in}} = \frac{A_{ol}}{A_{ol}B}$$

Since

$$A_{cl(NI)} = \frac{V_{out}}{V_{in}}$$

then

$$A_{cl(NI)} = \frac{1}{B} = \frac{R_i + R_f}{R_i} = 1 + \frac{R_f}{R_i} \qquad (12\text{–}19)$$

Equation (12–19) shows that the closed-loop gain, $A_{cl(NI)}$, of the noninverting (NI) amplifier is the reciprocal of the attenuation (B) of the feedback network (voltage-divider). It is interesting to note that the closed-loop gain is not at all dependent on the op-amp's open-loop gain under the condition $A_{ol}B >> 1$. The closed-loop gain can be set by selecting values of R_i and R_f.

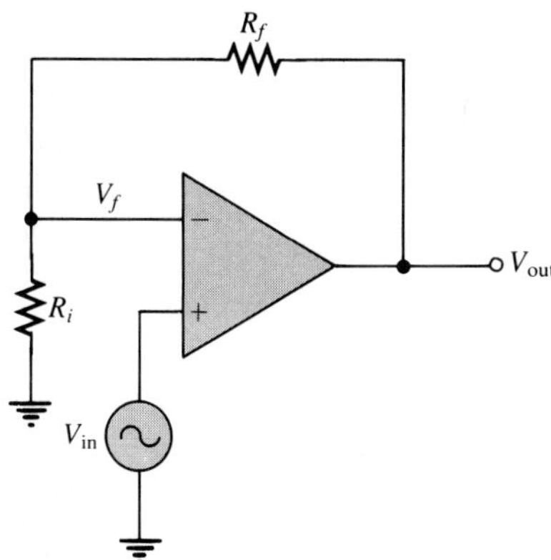

FIGURE 12–20
Noninverting amplifier.

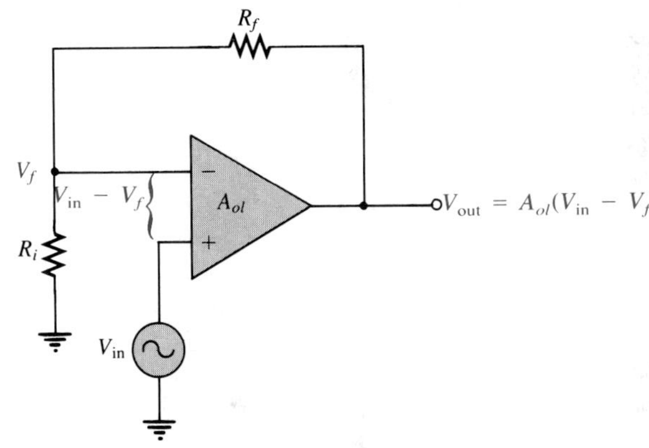

FIGURE 12–21
Differential input.

EXAMPLE 12–5 Determine the gain of the amplifier in Figure 12–22. The open-loop voltage gain is 100,000.

FIGURE 12–22

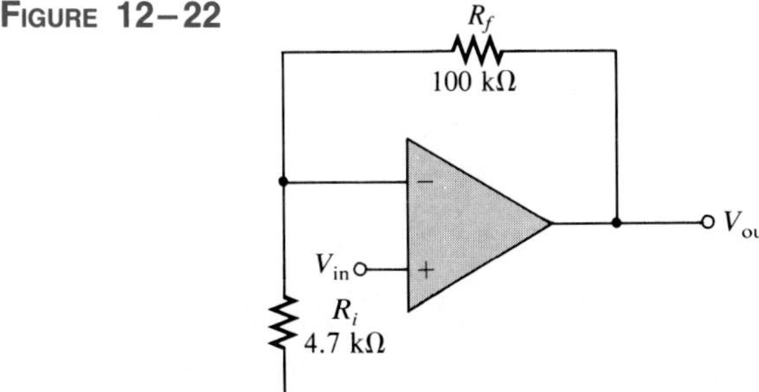

SOLUTION

This is a noninverting op-amp configuration. Therefore, the closed-loop gain is

$$A_{cl(\mathrm{NI})} = 1 + \frac{R_f}{R_i} = 1 + \frac{100\ \mathrm{k}\Omega}{4.7\ \mathrm{k}\Omega} = 22.3$$

PRACTICE EXERCISE 12–5

If the open-loop gain of the amplifier in Figure 12–22 is 150,000 and R_f is increased to 150 kΩ, determine the closed-loop gain.

VOLTAGE FOLLOWER

The **voltage-follower** configuration is a special case of the noninverting amplifier where all of the output voltage is fed back to the inverting input, as shown in Figure 12–23. As you can see, the straight feedback connection has a voltage gain of approximately one. The closed-loop voltage gain of a noninverting amplifier is $1/B$ as previously derived. Since $B = 1$, the closed-loop gain of the voltage follower is

$$A_{cl(\mathrm{VF})} = \frac{1}{B} = 1 \qquad (12\text{–}20)$$

The most important features of the voltage-follower configuration are its very high input impedance and its very low output impedance. These features make it a nearly ideal buffer amplifier for interfacing high-impedance sources and low-impedance loads. This is discussed further in the next section.

FIGURE 12–23
Op-amp voltage follower.

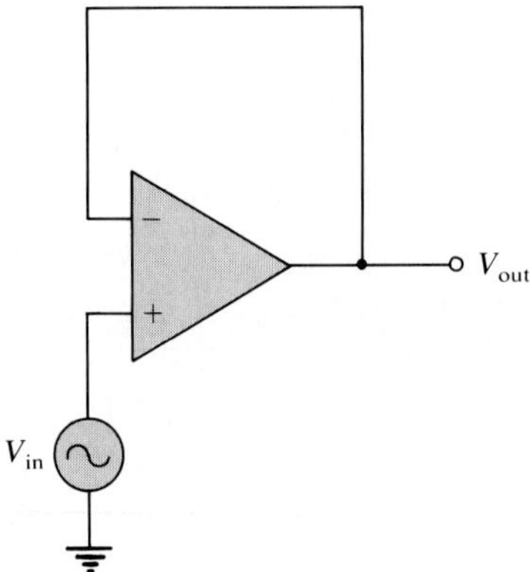

INVERTING AMPLIFIER

An op-amp connected as an **inverting amplifier** with a controlled amount of voltage gain is shown in Figure 12–24. The input signal is applied through a series input resistor R_i to the inverting input. Also, the output is fed back through R_f to the same input. The noninverting input is grounded.

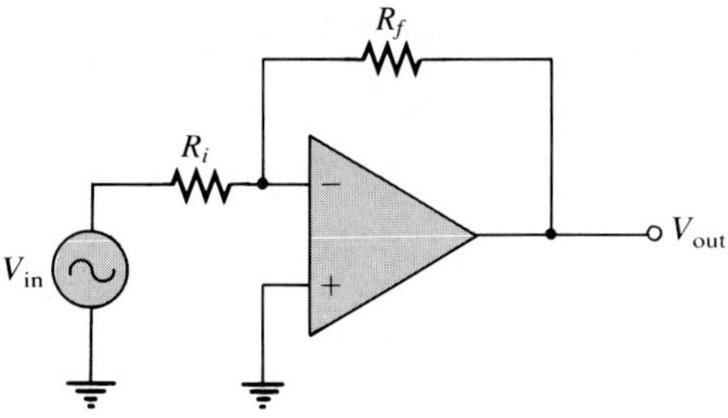

FIGURE 12–24
Inverting amplifier.

At this point, the ideal op-amp parameters mentioned earlier are very useful in simplifying the analysis of this circuit. In particular, the concept of infinite input impedance is of great value. An infinite input impedance implies that there is *no* current into the inverting input. If there is no current through the input impedance, then there must be *no* voltage drop between the inverting and noninverting inputs. This means that the voltage at the inverting (−) input is zero because the other input (+) is grounded. This zero voltage at the inverting input terminal is referred to as *virtual ground*. This condition is illustrated in Figure 12–25(a).

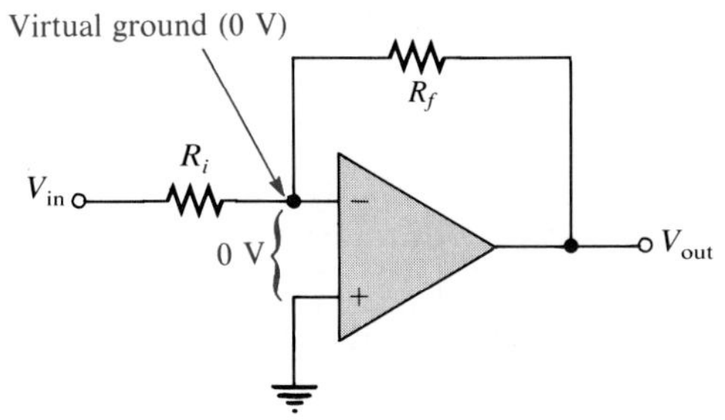

(a) Virtual ground

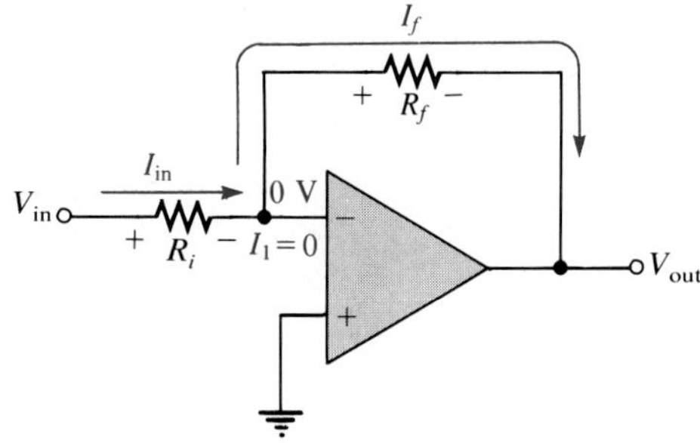

(b) $I_{in} = I_f$ and current into the inverting input (I_1) is 0

FIGURE 12–25
Virtual ground concept and closed-loop voltage gain development for the inverting amplifier.

Since there is no current into the inverting input, the current through R_i and the current through R_f are equal, as shown in Figure 12–25(b).

$$I_{in} = I_f$$

The voltage across R_i equals V_{in} because of virtual ground on the other side of the resistor. Therefore,

$$I_{in} = \frac{V_{in}}{R_i}$$

Also, the voltage across R_f equals $-V_{out}$ because of virtual ground, and therefore,

$$I_f = \frac{-V_{out}}{R_f}$$

Since $I_f = I_{in}$,

$$\frac{-V_{out}}{R_f} = \frac{V_{in}}{R_i}$$

Rearranging the terms, we get

$$\frac{V_{out}}{V_{in}} = -\frac{R_f}{R_i}$$

Of course, you recognize V_{out}/V_{in} as the overall gain of the amplifier.

$$A_{cl(I)} = -\frac{R_f}{R_i} \qquad (12\text{–}21)$$

Equation (12–21) shows that the **closed-loop** voltage gain $A_{cl(I)}$ of the inverting amplifier is the ratio of the feedback resistance R_f to the resistance R_i. *The closed-loop gain is independent of the op-amp's internal open-loop gain.* Thus, the negative feedback stabilizes the voltage gain. The negative sign indicates inversion.

EXAMPLE 12–6

Given the op-amp configuration in Figure 12–26, determine the value of R_f required to produce a closed-loop voltage gain of 100.

FIGURE 12–26

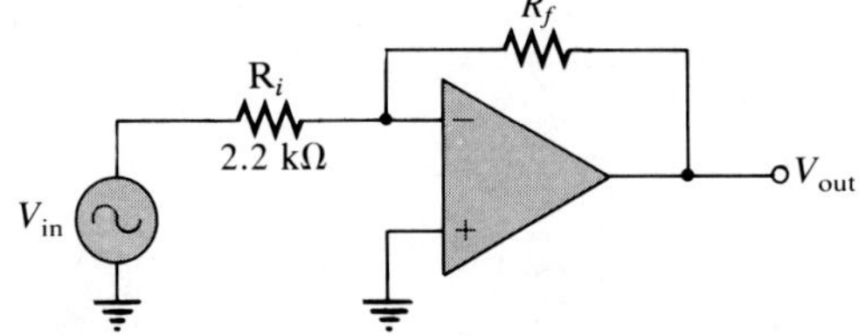

SOLUTION

$$R_i = 2.2\ \text{k}\Omega,\ \text{and}\ A_{cl(I)} = 100$$

$$A_{cl(I)} = \left|\frac{R_f}{R_i}\right|$$

$$R_f = A_{cl(I)}R_i = (100)(2.2\ \text{k}\Omega) = 220\ \text{k}\Omega$$

PRACTICE EXERCISE 12–6

If R_i is changed to 2.7 kΩ in Figure 12–26, what value of R_f is required to produce a closed-loop gain of 25? ■

12–5 REVIEW QUESTIONS

1. What is the main purpose of negative feedback?
2. The closed-loop voltage gain of each of the op-amp configurations discussed is dependent on the internal open-loop voltage gain of the op-amp (T or F).
3. The attenuation of the negative feedback network of a noninverting op-amp configuration is 0.02. What is the closed-loop gain of the amplifier?

12–6 EFFECTS OF NEGATIVE FEEDBACK ON OP-AMP IMPEDANCES

In this section, you will see how a negative feedback connection affects the input and output impedances of an op-amp. The effects on both inverting and noninverting amplifiers are examined.

INPUT IMPEDANCE OF THE NONINVERTING AMPLIFIER

The input impedance of this op-amp configuration is developed with the aid of Figure 12–27. For this analysis, a small differential voltage V_{diff} is assumed to exist between the two inputs, as indicated. This means that the op-amp's input impedance is not assumed to be infinite, nor the input current to be zero. The input voltage can be expressed as

$$V_{in} = V_{diff} + V_f$$

Substituting BV_{out} for V_f,

$$V_{in} = V_{diff} + BV_{out}$$

Since $V_{out} \cong A_{ol}V_{diff}$ (A_{ol} is the open-loop gain of the op-amp),

$$\begin{aligned} V_{in} &= V_{diff} + A_{ol}BV_{diff} \\ &= (1 + A_{ol}B)V_{diff} \end{aligned}$$

Because $V_{diff} = I_{in}Z_{in}$,

$$V_{in} = (1 + A_{ol}B)I_{in}Z_{in}$$

where Z_{in} is the open-loop input impedance of the op-amp (without feedback connections).

$$\frac{V_{in}}{I_{in}} = (1 + A_{ol}B)Z_{in}$$

FIGURE 12–27

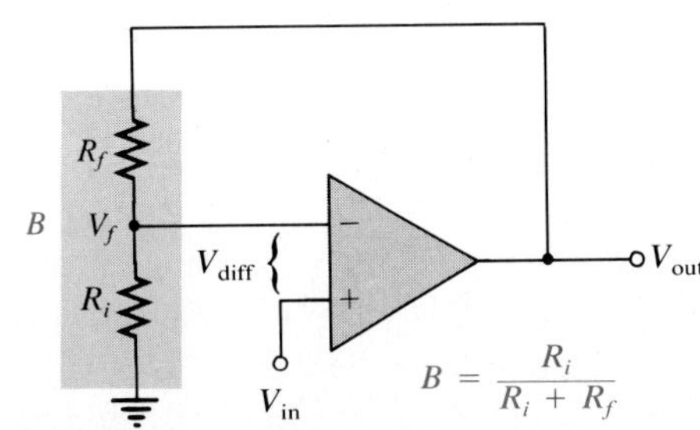

V_{in}/I_{in} is the overall input impedance of the closed-loop noninverting configuration.

$$Z_{in(NI)} = (1 + A_{ol}B)Z_{in} \tag{12–22}$$

This equation shows that the input impedance of this amplifier configuration with negative feedback is much greater than the internal input impedance of the op-amp itself (without feedback).

OUTPUT IMPEDANCE OF THE NONINVERTING AMPLIFIER

An expression for output impedance is developed with the aid of Figure 12–28. By applying Kirchhoff's law to the output circuit, we get

$$V_{out} = A_{ol}V_{diff} - Z_{out}I_{out}$$

The differential input voltage is $V_{in} - V_f$; therefore, under the assumption that $A_{ol}V_{diff} >> Z_{out}I_{out}$, the output voltage can be expressed as

$$V_{out} \cong A_{ol}(V_{in} - V_f)$$

Substituting BV_{out} for V_f, we get

$$V_{out} \cong A_{ol}(V_{in} - BV_{out})$$

Remember, B is the attenuation of the negative feedback network. Expanding and factoring, we get

$$V_{out} \cong A_{ol}V_{in} - A_{ol}BV_{out}$$
$$A_{ol}V_{in} \cong V_{out} + A_{ol}BV_{out}$$
$$\cong (1 + A_{ol}B)V_{out}$$

Since the output impedance of the noninverting configuration is $Z_{out(NI)} = V_{out}/I_{out}$, we can substitute $I_{out}Z_{out(NI)}$ for V_{out}.

$$A_{ol}V_{in} = (1 + A_{ol}B)I_{out}Z_{out(NI)}$$

Dividing both sides of the above expression by I_{out}, we get

$$\frac{A_{ol}V_{in}}{I_{out}} = (1 + A_{ol}B)Z_{out(NI)}$$

The term on the left is the internal output impedance of the op-amp (Z_{out}) because, without feedback, $A_{ol}V_{in} = V_{out}$. Therefore,

$$Z_{out} = (1 + A_{ol}B)Z_{out(NI)}$$

Thus,

$$Z_{out(NI)} = \frac{Z_{out}}{1 + A_{ol}B} \tag{12–23}$$

This equation shows that the output impedance of this amplifier configuration with negative feedback is much less than the internal output impedance of the op-amp itself (without feedback).

FIGURE 12–28

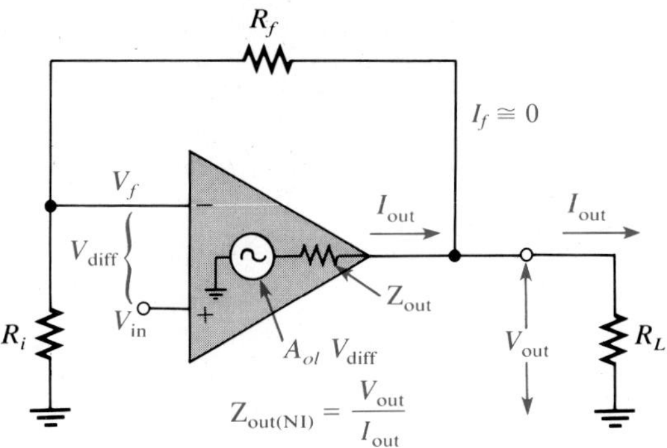

EXAMPLE 12–7

(a) Determine the input and output impedances of the amplifier in Figure 12–29. The op-amp data sheet gives $Z_{in} = 2\ \text{M}\Omega$, $Z_{out} = 75\ \Omega$, and $A_{ol} = 200{,}000$.

(b) Find the closed-loop voltage gain.

FIGURE 12–29

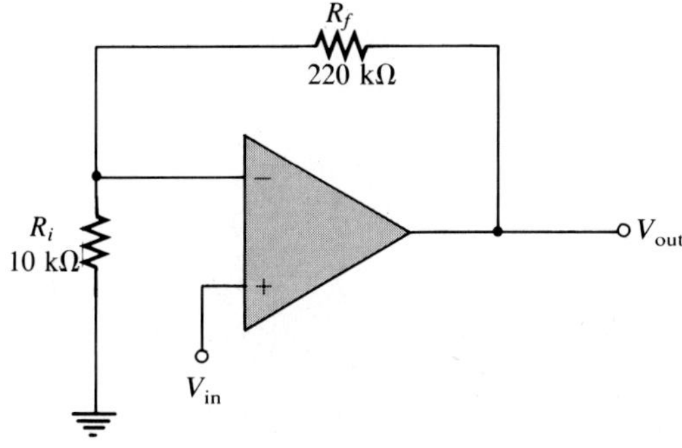

SOLUTION

(a) The attenuation of the feedback network is

$$B = \frac{R_i}{R_i + R_f} = \frac{10\ \text{k}\Omega}{230\ \text{k}\Omega} = 0.0435$$

$$\begin{aligned} Z_{in(NI)} &= (1 + A_{ol}B)Z_{in} \\ &= [1 + (200{,}000)(0.0435)](2\ \text{M}\Omega) \\ &= (1 + 8700)(2\ \text{M}\Omega) \\ &= 17{,}402\ \text{M}\Omega \end{aligned}$$

$$\begin{aligned} Z_{out(NI)} &= \frac{Z_{out}}{1 + A_{ol}B} \\ &= \frac{75\ \Omega}{1 + 8700} \\ &= 0.0086\ \Omega \end{aligned}$$

(b) $A_{cl(NI)} = \dfrac{1}{B} = \dfrac{1}{0.0435} \cong 23$

PRACTICE EXERCISE 12–7

(a) Determine the input and output impedances in Figure 12–29 for op-amp data sheet values of $Z_{in} = 3.5\ M\Omega$, $Z_{out} = 82\ \Omega$, and $A_{ol} = 135{,}000$.
(b) Find A_{cl}.

VOLTAGE-FOLLOWER IMPEDANCES

Since the voltage follower is a special case of the noninverting configuration, the same impedance formulas are used with $B = 1$.

$$Z_{in(VF)} = (1 + A_{ol})Z_{in} \quad (12\text{–}24)$$

$$Z_{out(VF)} = \frac{Z_{out}}{1 + A_{ol}} \quad (12\text{–}25)$$

As you can see, the voltage-follower input impedance is greater for a given A_{ol} and Z_{in} than for the noninverting configuration with the voltage-divider feedback network. Also, its output impedance is much smaller.

EXAMPLE 12–8

The same op-amp in Example 12–7 is used in a voltage-follower configuration. Determine the input and output impedances.

SOLUTION
Since $B = 1$,

$$\begin{aligned} Z_{in(VF)} &= (1 + A_{ol})Z_{in} \\ &= (1 + 200{,}000)(2\ M\Omega) \\ &\cong 400{,}000\ M\Omega \end{aligned}$$

$$\begin{aligned} Z_{out(VF)} &= \frac{Z_{out}}{1 + A_{ol}} \\ &= \frac{75\ \Omega}{1 + 200{,}000} \\ &= 0.00038\ \Omega \end{aligned}$$

Notice that $Z_{in(VF)}$ is much greater than $Z_{in(NI)}$, and $Z_{out(VF)}$ is much less than $Z_{out(NI)}$ from Example 12–7.

PRACTICE EXERCISE 12–8

If the op-amp in this example is replaced with one having a higher open-loop gain, how are the input and output impedances affected?

INPUT IMPEDANCE OF AN INVERTING AMPLIFIER

The input impedance of this op-amp configuration is developed with the aid of Figure 12–30. Because both the input signal and the negative feedback are applied, through

EFFECT OF AN INPUT BIAS CURRENT

Figure 12–33 is an inverting amplifier with zero input voltage. Ideally, the current through R_i is zero because the input voltage is zero and the voltage at the inverting (−) terminal is zero. The small input current I_1 is furnished from the output terminal through R_f. I_1 creates a voltage drop across R_f, as indicated. The positive side of R_f is the output terminal, and therefore, the output error voltage is I_1R_f when it should be zero.

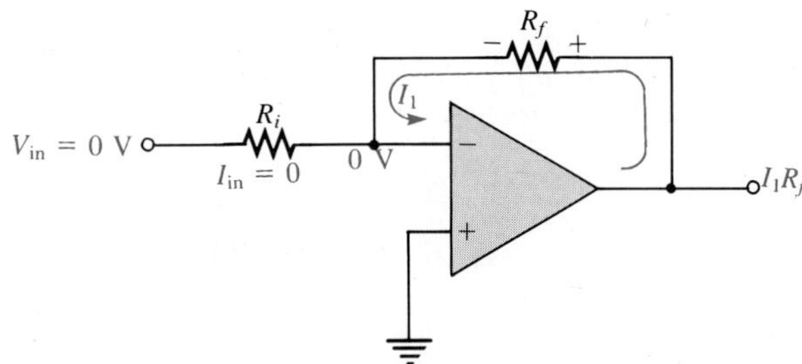

FIGURE 12–33
Input bias current creates output error voltage (I_1R_f) in inverting amplifier.

Figure 12–34 is a voltage follower with zero input voltage and a source resistance R_s. In this case, an input current I_2 creates an output voltage error (a path exists for I_2 through the negative voltage supply and back to ground). I_2 produces a drop across R_s, as shown. The voltage at the inverting input terminal decreases to $-I_2R_s$ because the negative feedback tends to maintain a differential voltage of zero, as indicated. Since the inverting terminal is connected directly to the output terminal, the output error voltage is $-I_2R_s$.

Figure 12–35 is a noninverting amplifier with zero input voltage. Ideally, the voltage at the inverting terminal is also zero, as indicated. The input current I_1 produces a voltage drop across R_f and thus creates an output error voltage of I_1R_f, just as with the inverting amplifier.

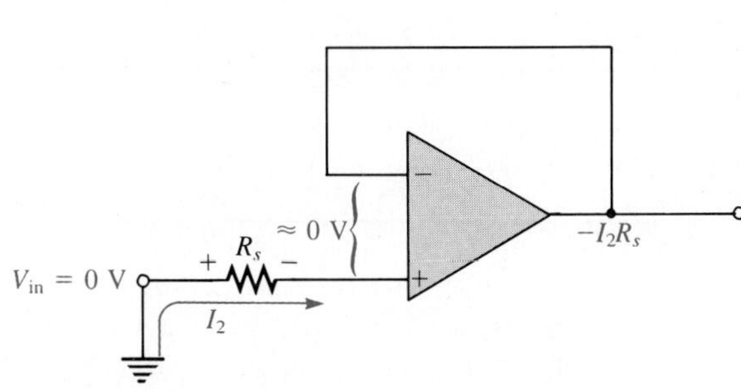

FIGURE 12–34
Input bias current creates output error voltage in voltage follower.

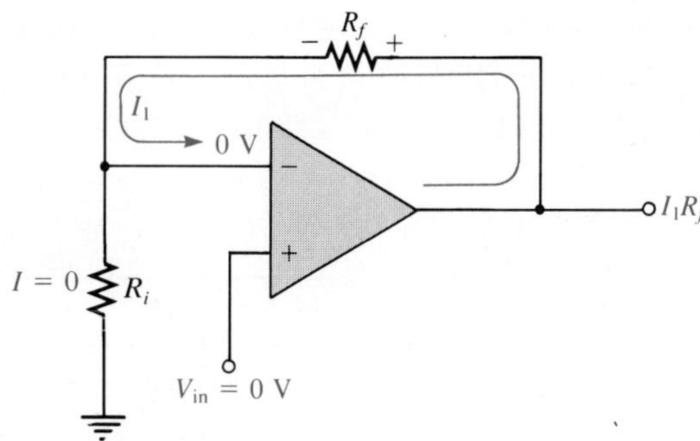

FIGURE 12–35
Input bias current creates output error voltage in noninverting amplifier.

BIAS CURRENT COMPENSATION IN A VOLTAGE FOLLOWER

The output error voltage due to bias currents in a voltage follower can be sufficiently reduced by adding a resistor equal to R_s in the feedback path, as shown in Figure 12–36. The voltage drop created by I_1 across the added resistor subtracts from the $-I_2R_s$ output error voltage. If $I_1 = I_2$, then the output voltage is zero. Usually I_1 does not quite equal I_2; but even in this case, the output error voltage is reduced as follows, because I_{OS} is less than I_2.

$$V_{OUT(error)} = |I_1 - I_2|R_s$$

$$V_{OUT(error)} = I_{OS}R_s \tag{12–32}$$

where I_{OS} is the input offset current.

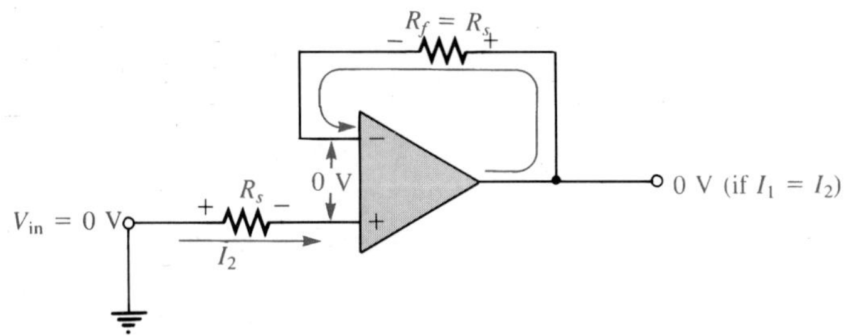

FIGURE 12–36
Bias current compensation in a voltage follower.

BIAS CURRENT COMPENSATION IN OTHER OP-AMP CONFIGURATIONS

To compensate for the effect of bias current in the noninverting amplifier, a resistor R_c is added, as shown in Figure 12–37(a). The compensating resistor value equals the parallel combination of R_i and R_f. The input current I_2 creates a voltage drop across R_c that offsets the voltage across the R_i-R_f combination, thus sufficiently reducing the output error voltage. The inverting amplifier is similarly compensated, as shown in Figure 12–37(b).

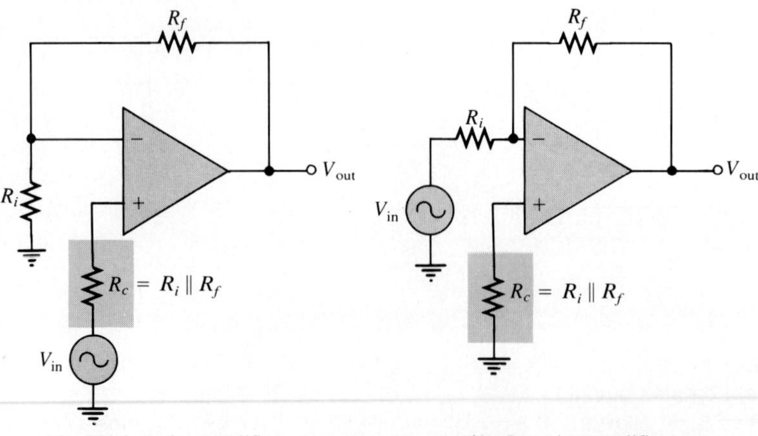

FIGURE 12–37
Bias current compensation.

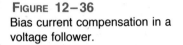

USE OF A BIFET OP-AMP TO ELIMINATE THE NEED FOR BIAS CURRENT COMPENSATION

A BIFET op-amp uses both bipolar junction transistors and JFETs in its internal circuitry. The JFETs are used as the input devices to achieve a higher input impedance than is possible with standard bipolar amplifiers. Because of their very high input impedance, BIFETs typically have input bias currents that are much smaller than in bipolar op-amps, thus reducing or eliminating the need for bias current compensation.

EFFECT OF INPUT OFFSET VOLTAGE

The output voltage of an op-amp should be zero when the differential input is zero. However, there is always a small output error voltage present whose value typically ranges from microvolts to millivolts. This is due to unavoidable imbalances within the internal op-amp transistors aside from the bias currents previously discussed. In a negative feedback configuration, the input offset voltage V_{IO} can be visualized as an equivalent small dc voltage source, as illustrated in Figure 12–38 for a voltage follower. The output error voltage due to the input offset voltage in this case is

$$V_{OUT(error)} = V_{IO} \tag{12–33}$$

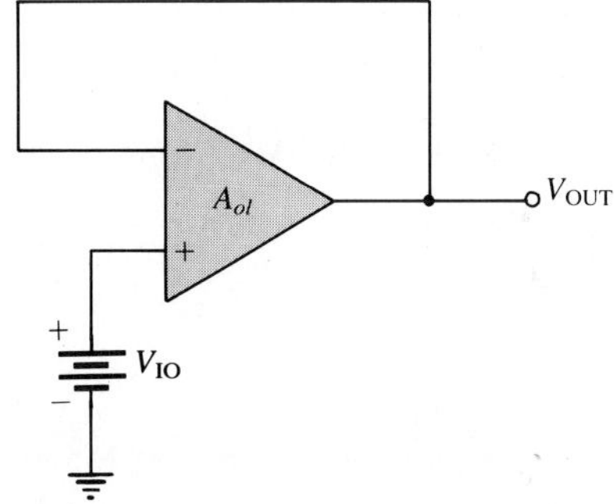

FIGURE 12–38
Input offset voltage equivalent.

INPUT OFFSET VOLTAGE COMPENSATION

Most integrated circuit op-amps provide a means of compensating for offset voltage. This is usually done by connecting an external potentiometer to designated pins on the IC package, as illustrated in Figure 12–39(a) and (b) for a 741 op-amp. The two terminals are labelled *offset null*. With no input, the potentiometer is simply adjusted until the output voltage reads 0, as shown in Figure 12–39(c).

12–7 REVIEW QUESTIONS

1. What are two sources of dc output error voltages?
2. How do you compensate for bias current in a voltage follower?

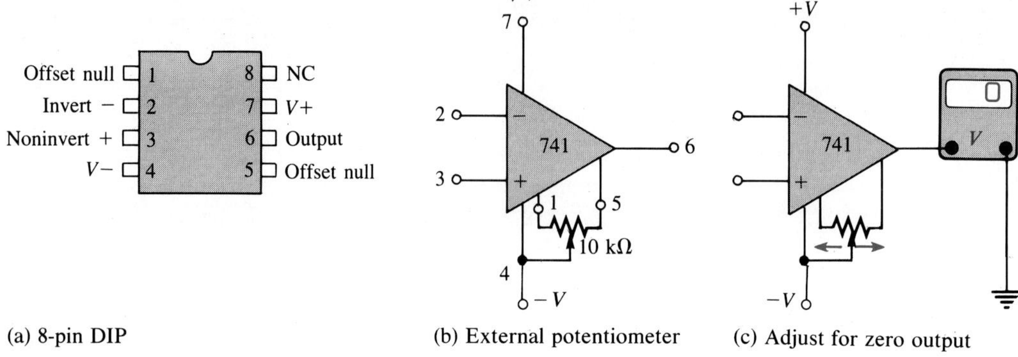

(a) 8-pin DIP (b) External potentiometer (c) Adjust for zero output

FIGURE 12–39
Input offset voltage compensation for a 741.

12–8 TROUBLESHOOTING

As a technician, you will no doubt encounter situations in which an op-amp or its associated circuitry has malfunctioned. The op-amp is a complex integrated circuit with many types of internal failures possible. However, since you cannot troubleshoot the op-amp internally, you treat it as a single device with only a few connections to it. If it fails, you replace it just as you would a resistor, capacitor, or transistor.

In the basic op-amp configurations, there are only a few external components that can fail. These are the feedback resistor, the input resistor, and the potentiometer used for offset voltage compensation. Also, of course, the op-amp itself can fail or there can be faulty contacts in the circuit. We will now examine the three basic configurations for possible faults and the associated symptoms.

FAULTS IN THE NONINVERTING AMPLIFIER

The first thing to do when you suspect a faulty circuit is to check for the proper supply voltage and ground. Having done that, several other possible faults are as follows.

OPEN FEEDBACK RESISTOR If the feedback resistor, R_f, in Figure 12–40 opens, the op-amp is operating with its very high open-loop gain, which causes the input signal to drive the device into nonlinear operation and results in a severely clipped output signal as shown in part (a).

OPEN INPUT RESISTOR In this case, we still have a closed-loop configuration. But, since R_i is open and effectively equal to infinity, the closed-loop gain from Equation (12–19) is

$$A_{cl(NI)} = 1 + \frac{R_f}{R_i} = 1 + \frac{R_f}{\infty} = 1 + 0 = 1$$

FIGURE 12–40
Faults in the noninverting amplifier.

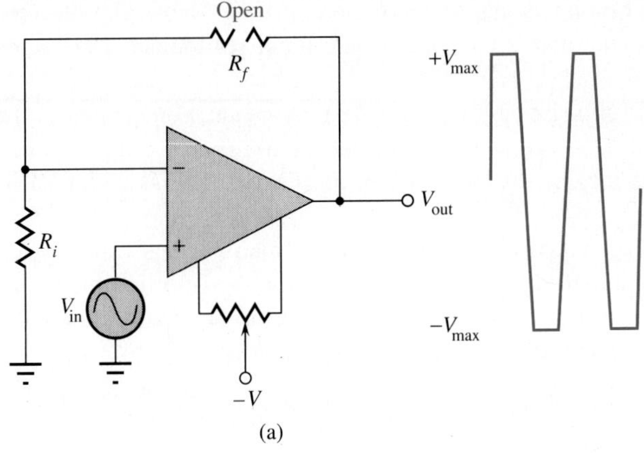

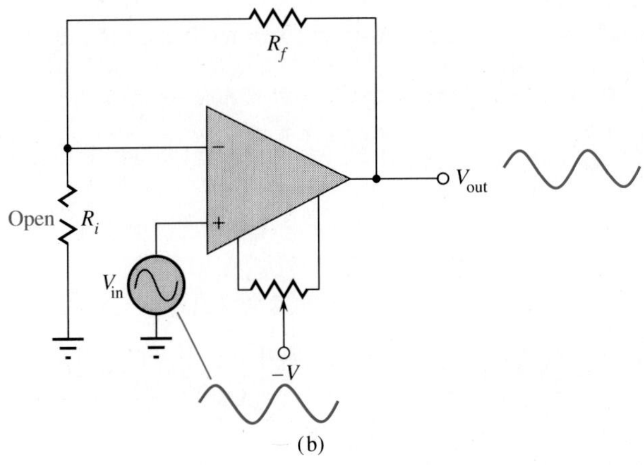

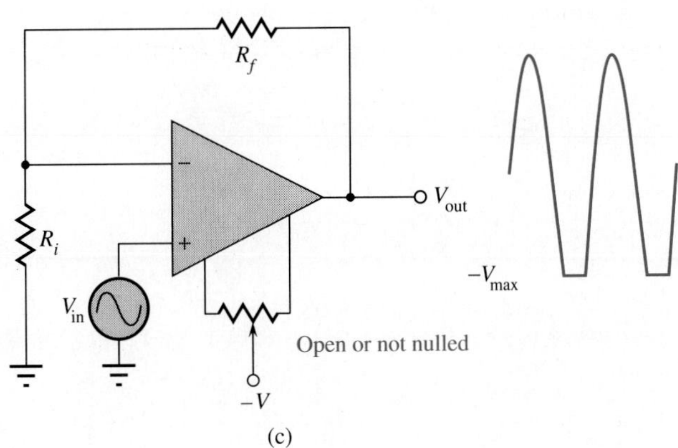

This shows that the amplifier acts like a voltage follower. You would observe an output signal that is the same as the input, as indicated in Figure 12–40(b).

OPEN OR INCORRECTLY ADJUSTED OFFSET NULL POTENTIOMETER In this situation, the output offset voltage will cause the output signal to begin clipping on only one peak as the input signal is increased to a sufficient amplitude. This is indicated in Figure 12–40(c).

FAULTY OP-AMP As mentioned, many things can happen to an op-amp. In general, an internal failure will result in a loss or distortion of the output signal. The best approach is to first make sure that there are no external failures or faulty conditions. If everything else is good, then the op-amp must be bad.

FAULTS IN THE VOLTAGE FOLLOWER

The voltage follower is a special case of the noninverting amplifier. Except for a bad op-amp, a bad external connection, or a problem with the offset null potentiometer, about the only thing that can happen in a voltage-follower circuit is an open feedback loop. This would have the same effect as an open feedback resistor as previously discussed.

FIGURE 12–41
Faults in the inverting amplifier.

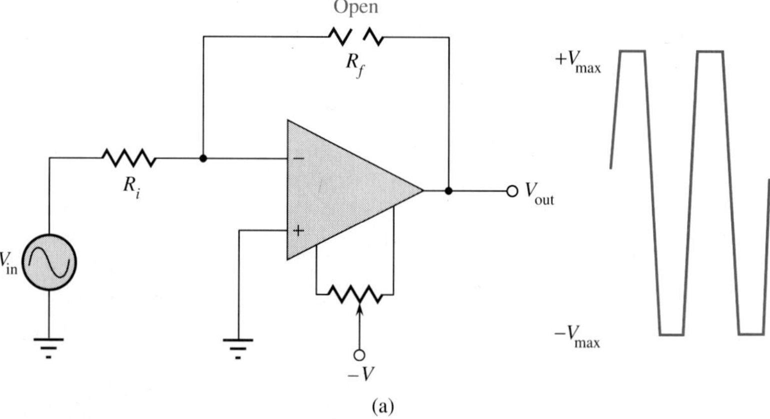

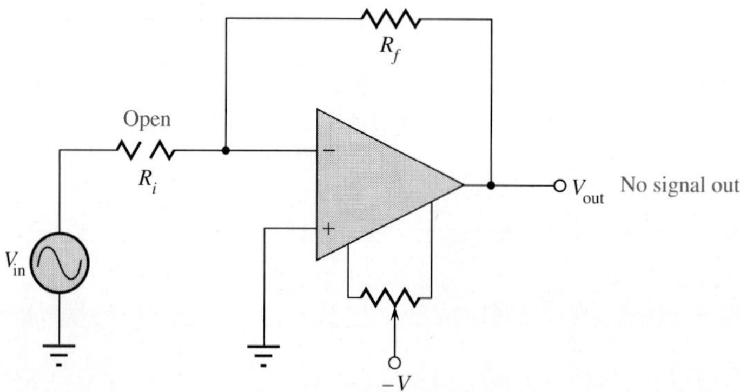

FAULTS IN THE INVERTING AMPLIFIER

OPEN FEEDBACK RESISTOR If R_f opens as indicated in Figure 12–41(a), the input signal still feeds through the input resistor and is amplified by the high open-loop gain of the op-amp. This forces the device to be driven into nonlinear operation, and you will see an output something like that shown. This is the same result as in the noninverting configuration.

OPEN INPUT RESISTOR This prevents the input signal from getting to the op-amp input, so there will be no output signal, as indicated in Figure 12–41(b).

Failures in the op-amp itself or the offset null potentiometer have the same effects as previously discussed for the noninverting amplifier.

12–8 REVIEW QUESTIONS

1. If you notice that the op-amp output signal is beginning to clip on one peak as you increase the input signal, what should you check?
2. If there is no op-amp output signal when there is a verified input signal, what would you suspect as being faulty?

12–9 A SYSTEM APPLICATION

The spectrophotometer system presented at the beginning of this chapter combines light optics with electronics to analyze the chemical makeup of various solutions. This type of system is common in medical laboratories as well as many other areas. It is another example of mixed systems in which electronic circuits interface with other types of systems, such as mechanical and optical, to accomplish a specific function. You have worked with other types of mixed systems in earlier chapters. When you are a technician or technologist in industry, you will probably be working with different types of mixed systems from time to time. In this section, you will

- ☐ ***See the role of electronics in a system that is not totally electronic.***
- ☐ ***See how an op-amp is used in the system.***
- ☐ ***See how an electronic circuit interfaces with an optical device.***
- ☐ ***Translate between a printed circuit board and a schematic.***
- ☐ ***Troubleshoot some common system problems.***

A BRIEF DESCRIPTION OF THE SYSTEM

The light source shown in Figure 12–42 produces a beam of visible light containing a wide spectrum of wavelengths. Each component wavelength in the beam of light is refracted at a different angle by the prism as indicated. Depending on the angle of the platform as set by the pivot angle controller, a certain wavelength passes through the narrow slit and is transmitted through the solution under analysis. By precisely pivoting the light source and prism, a selected wavelength can be transmitted. Every chemical and compound absorbs different wavelengths of light in different ways, so the resulting light coming through the solution has a unique ''signature'' that can be used to define the chemicals in the solution.

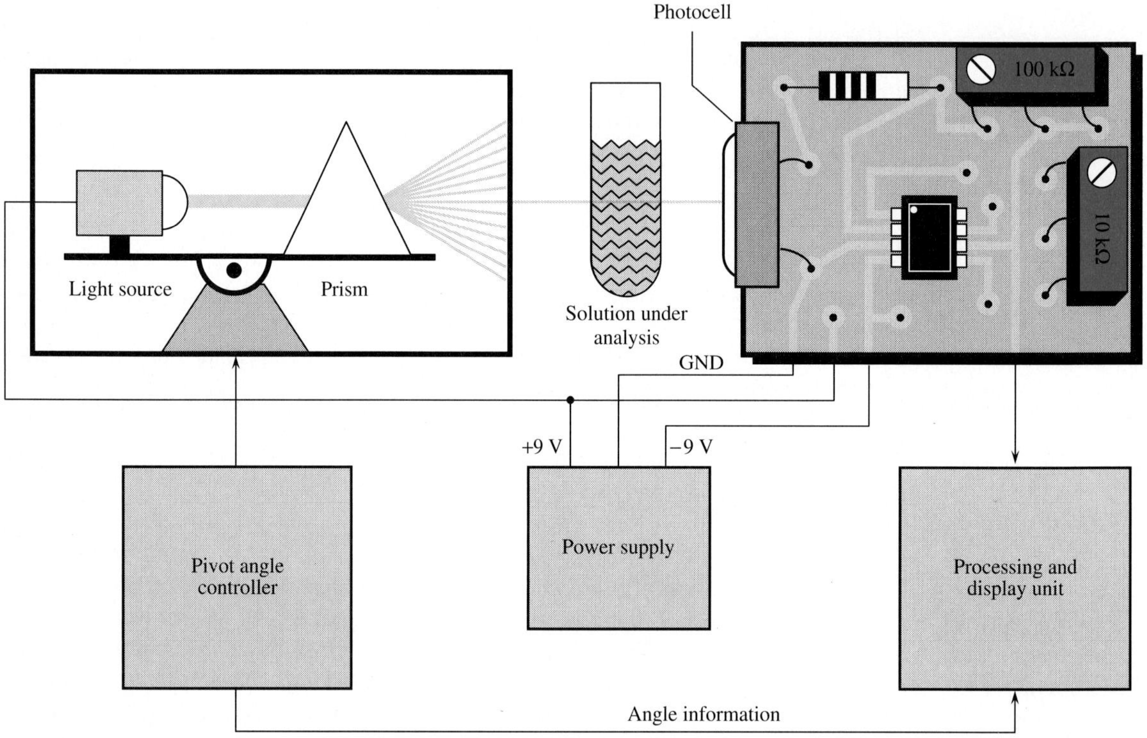

FIGURE 12–42

The photocell on the circuit board, produces a voltage that is proportional to the amount of light and wavelength. The op-amp circuit amplifies the photovoltaic cell output and sends the resulting signal to the processing and display unit where the type of chemical(s) in the solution is identified. This is usually a microprocessor-based digital system. Although these other system blocks are interesting, our focus in this section is the photocell and op-amp circuit board.

Now, so that you can take a closer look at the photocell and op-amp circuit, let's take it out of the system and put it on the test bench.

ON THE TEST BENCH

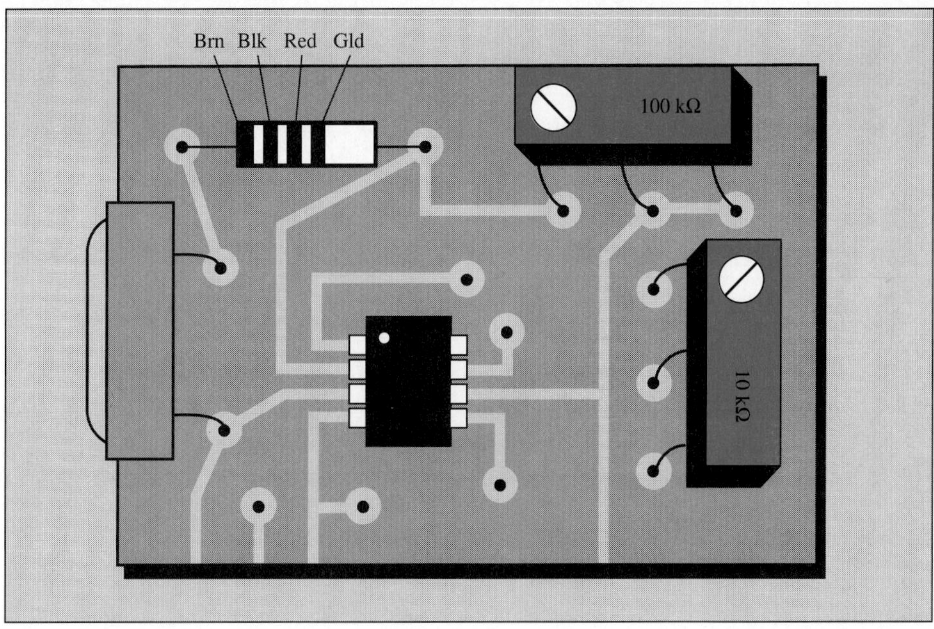

FIGURE 12–43

■ ACTIVITY 1 RELATE THE PC BOARD TO THE SCHEMATIC

Develop a complete schematic diagram by carefully following the conductive traces on the pc board shown in Figure 12–43 to see how the components are interconnected. Some of the interconnecting traces are on the reverse side of the board, but if you are familiar with basic op-amp configurations, you should have no trouble figuring out the connections. Refer to the chapter material or the 741 data sheet for the pin layout. This op-amp is housed in a surface-mount SO-8 package.

■ ACTIVITY 2 ANALYZE THE CIRCUIT

STEP 1 Determine the resistance value to which the feedback rheostat must be adjusted for a voltage gain of 50.

STEP 2 Assume the maximum linear output of the op-amp is 1 V less than the supply voltage's. Determine the voltage gain required and the value to which the feedback resistance must be set to achieve the maximum linear output. The maximum voltage from the photocell is 0.5 V.

STEP 3 The system light source produces wavelengths ranging from 400 nm to 700 nm, which is approximately the full range of visible light from violet to red. Determine the op-amp output voltage over this range of wavelengths in 50 nm intervals and plot a graph of the results. Refer to the photocell response characteristic in Figure 12–44.

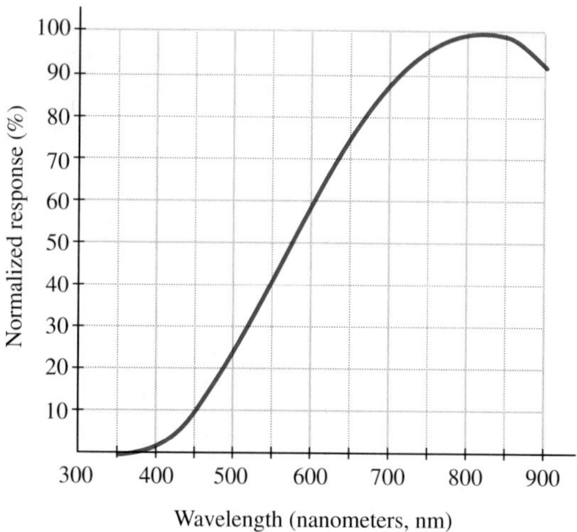

FIGURE 12–44
Photocell response curve.

■ ACTIVITY 3 WRITE A TECHNICAL REPORT

Describe the circuit operation. Be sure to identify the type of op-amp circuit configuration and explain the purpose of the two potentiometers. Use the results of Activity 2 to specify the performance of the circuit.

■ ACTIVITY 4 TROUBLESHOOT THE CIRCUIT FOR EACH OF THE FOLLOWING PROBLEMS BY STATING THE PROBABLE CAUSE OR CAUSES

1. No voltage at the op-amp output.
2. Output of op-amp stays at approximately −9 V.
3. A small dc voltage on the op-amp output under no-light conditions.
4. Zero output voltage as light source is pivoted with verified photocell output voltage.

12–9 REVIEW QUESTIONS

1. What is the purpose of the 100 kΩ potentiometer on the circuit board?
2. What is the purpose of the 10 kΩ potentiometer?
3. Explain why the light source and prism must be pivoted.

SUMMARY

- The basic op-amp has three terminals not including power and ground: inverting input (−), noninverting input (+), and output.
- Most op-amps require both a positive and a negative dc supply voltage.

- □ The ideal (perfect) op-amp has infinite input impedance, zero output impedance, infinite open-loop voltage gain, infinite bandwidth, and infininte CMRR.
- □ A good practical op-amp has high input impedance, low output impedance, high open-loop voltage gain, and a wide bandwidth.
- □ A differential amplifier is normally used for the input stage of an op-amp.
- □ A differential input voltage appears between the inverting and noninverting inputs of a differential amplifier.
- □ A single-ended input voltage appears between one input and ground (with the other input grounded).
- □ A differential output voltage appears between two output terminals of a diff-amp.
- □ A single-ended output voltage appears between the output and ground of a diff-amp.
- □ Common-mode occurs when equal in-phase voltages are applied to both input terminals.
- □ Input offset voltage produces an output error voltage (with no input voltage).
- □ Input bias current also produces an output error voltage (with no input voltage).
- □ Input offset current is the difference between the two bias currents.
- □ Open-loop voltage gain is the gain of the op-amp with no external feedback connections.
- □ The common-mode rejection ratio (CMRR) is a measure of an op-amp's ability to reject common-mode inputs.
- □ Slew rate is the rate in volts per microsecond at which the output voltage of an op-amp can change in response to a step input.
- □ There are three basic op-amp configurations—inverting, noninverting, and voltage follower.
- □ All op-amp configurations listed employ negative feedback. Negative feedback occurs when a portion of the output voltage is connected back to the inverting input such that it subtracts from the input voltage, thus reducing the voltage gain but increasing the stability and bandwidth.
- □ A noninverting amplifier configuration has a higher input impedance and a lower output impedance than the op-amp itself (without feedback).
- □ An inverting amplifier configuration has an input impedance approximately equal to the input resistor R_i and an output impedance approximately equal to the output impedance of the op-amp itself.
- □ The voltage follower has the highest input impedance and the lowest output impedance of the three configurations.
- □ All practical op-amps have small input bias currents and input offset voltages that produce small output error voltages.
- □ The input bias current effect can be compensated for with external resistors.
- □ The input offset voltage can be compensated for with an external potentiometer between the two offset null pins provided on the IC op-amp package and as recommended by the manufacturer.

GLOSSARY

Closed-loop An op-amp connection in which the output is connected back to the input through a feedback circuit.

Common mode A condition characterized by the presence of the same signal on both op-amp inputs.

Common-mode rejection ratio (CMRR) The ratio of open-loop gain to common-mode gain; a measure of an op-amp's ability to reject common-mode signals.

Differential amplifier (diff-amp) An amplifier that produces an output voltage proportional to the difference of the two input voltages.

Inverting amplifier An op-amp closed-loop configuration in which the input signal is applied to the inverting input.

Negative feedback The process of returning a portion of the output signal to the input of an amplifier such that it is out of phase with the input signal.

Noninverting amplifier An op-amp closed-loop configuration in which the input signal is applied to the noninverting input.

Open-loop A condition in which an op-amp has no feedback.

Operational amplifier (op-amp) A type of amplifier that has very high voltage gain, very high input impedance, very low output impedance, and good rejection of common-mode signals.

Slew rate The rate of change of the output voltage of an op-amp in response to a step input.

Voltage follower A closed-loop, noninverting op-amp with a voltage gain of one.

FORMULAS

DIFFERENTIAL AMPLIFIERS

(12–1)	$I_{E1} = I_{E2} = \dfrac{I_{R_E}}{2}$	Diff-amp emitter current
(12–2)	$I_{R_E} = \dfrac{V_E - V_{EE}}{R_E}$	Combined emitter current
(12–3)	$I_{C1} = I_{C2} \cong \dfrac{I_{R_E}}{2}$	Diff-amp collector current
(12–4)	$V_{C1} = V_{C2} = V_{CC} - I_{C1}R_{C1}$	Diff-amp collector voltage
(12–5)	$V_E = V_B - 0.7\text{ V}$	Diff-amp emitter voltage
(12–6)	$\text{CMRR} = \dfrac{A_{v(d)}}{A_{cm}}$	Common-mode rejection ratio (diff-amp)

(12–7)	$\text{CMRR} = 20 \log \frac{A_{v(d)}}{A_{cm}}$	Common-mode rejection ratio (dB)

OP-AMP PARAMETERS

(12–8)	$V_{\text{OUT}} = I_{C2}R_C - I_{C1}R_C$	Differential output
(12–9)	$I_{\text{BIAS}} = \frac{I_1 + I_2}{2}$	Input bias current
(12–10)	$I_{\text{OS}} = \lvert I_1 - I_2 \rvert$	Input offset current
(12–11)	$V_{\text{OS}} = I_{\text{OS}}R_{\text{in}}$	Offset voltage
(12–12)	$V_{\text{OUT(error)}} = A_v I_{\text{OS}} R_{\text{in}}$	Output error voltage
(12–13)	$\text{CMRR} = \frac{A_{ol}}{A_{cm}}$	Common-mode rejection ratio (op-amp)
(12–14)	$\text{CMRR} = 20 \log \frac{A_{ol}}{A_{cm}}$	Common-mode rejection ratio (dB)
(12–15)	$\text{Slew rate} = \frac{\Delta V_{\text{out}}}{\Delta t}$	Slew rate

OP-AMP CONFIGURATIONS

(12–16)	$V_f = \left(\frac{R_i}{R_i + R_f}\right) V_{\text{out}}$	Feedback voltage (noninverting)
(12–17)	$V_{\text{out}} = A_{ol}(V_{\text{in}} - V_f)$	Output voltage (noninverting)
(12–18)	$\frac{V_{\text{out}}}{V_{\text{in}}} = \frac{A_{ol}}{1 + A_{ol}B}$	Voltage gain (noninverting)
(12–19)	$A_{cl(\text{NI})} = \frac{1}{B} = 1 + \frac{R_f}{R_i}$	Voltage gain (noninverting)
(12–20)	$A_{cl(\text{VF})} = \frac{1}{B} = 1$	Voltage gain (voltage follower)
(12–21)	$A_{cl(\text{I})} = -\frac{R_f}{R_i}$	Voltage gain (inverting)

OP-AMP IMPEDANCES

(12–22)	$Z_{\text{in(NI)}} = (1 + A_{ol}B)Z_{\text{in}}$	Input impedance (noninverting)
(12–23)	$Z_{\text{out(NI)}} = \frac{Z_{\text{out}}}{1 + A_{ol}B}$	Output impedance (noninverting)
(12–24)	$Z_{\text{in(VF)}} = (1 + A_{ol})Z_{\text{in}}$	Input impedance (voltage follower)
(12–25)	$Z_{\text{out(VF)}} = \frac{Z_{\text{out}}}{1 + A_{ol}}$	Output impedance (voltage follower)

(12–26) $Z_{in(Miller)} = \frac{R_f}{A_{ol} + 1}$ Miller input impedance (inverting)

(12–27) $Z_{out(Miller)} = \left(\frac{A_{ol}}{A_{ol} + 1}\right)R_f$ Miller output impedance (inverting)

(12–28) $Z_{in(I)} = R_i + \frac{R_f}{A_{ol} + 1}\|Z_{in}$ Input impedance (inverting)

(12–29) $Z_{in(I)} \cong R_i$ Input impedance (inverting)

(12–30) $Z_{out(I)} = \left(\frac{A_{ol}}{A_{ol} + 1}\right)R_f\|Z_{out}$ Output impedance (inverting)

(12–31) $Z_{out(I)} \cong Z_{out}$ Output impedance (inverting)

Error Voltage

(12–32) $V_{OUT(error)} = I_{OS}R_s$ Output error voltage

(12–33) $V_{OUT(error)} = V_{IO}$ Output error voltage (input offset)

Self-Test

1. An integrated circuit (IC) op-amp has
   (a) two inputs and two outputs
   (b) one input and one output
   (c) two inputs and one output
2. Which of the following characteristics does not *necessarily* apply to an op-amp?
   (a) High gain **(b)** Low power
   (c) High input impedance **(d)** Low output impedance
3. A differential amplifier
   (a) is part of an op-amp **(b)** has one input and one output
   (c) has two outputs **(d)** a and c
4. When a differential amplifier is operated single-ended,
   (a) the output is grounded
   (b) one input is grounded and a signal is applied to the other
   (c) both inputs are connected together
   (d) the output is not inverted
5. In the differential mode,
   (a) opposite polarity signals are applied to the inputs
   (b) the gain is one

(c) the outputs are different amplitudes

(d) only one supply voltage is used

6. In the common mode,

(a) both inputs are grounded

(b) the outputs are connected together

(c) an identical signal appears on both inputs

(d) the output signals are in-phase

7. Common-mode gain is

(a) very high (b) very low

(c) always unity (d) unpredictable

8. Differential gain is

(a) very high (b) very low

(c) dependent on the input voltage (d) about 100

9. If $A_{v(d)} = 3500$ and $A_{cm} = 0.35$, the CMRR is

(a) 1225 (b) 10,000 (c) 80 dB (d) b and c

10. With zero volts on both inputs, an op-amp ideally should have an output

(a) equal to the positive supply voltage

(b) equal to the negative supply voltage

(c) equal to zero

(d) equal to the CMRR

11. Of the values listed, the most *realistic* value for open-loop gain of an op-amp is

(a) 1 (b) 2000 (c) 80 dB (d) 100,000

12. A certain op-amp has bias currents of 50 μA and 49.3 μA. The input offset current is

(a) 700 nA (b) 99.3 μA (c) 49.65 μA (d) none of these

13. The output of a particular op-amp increases 8 V in 12 μs. The slew rate is

(a) 96 V/μs (b) 0.67 V/μs (c) 1.5 V/μs (d) none of these

14. The purpose of offset nulling is to

(a) reduce the gain (b) equalize the input signals

(c) zero the output error voltage (d) b and c

15. For an op-amp with negative feedback, the output is

(a) equal to the input (b) increased

(c) fed back to the inverting input (d) fed back to the noninverting input

16. The use of negative feedback

(a) reduces the voltage gain of an op-amp

(b) makes the op-amp oscillate

(c) makes linear operation possible

(d) a and c

17. Negative feedback

(a) increases the input and output impedances

(b) increases the input impedance and the bandwidth

(c) decreases the output impedance and the bandwidth

(d) does not affect impedances or bandwidth

18. A certain noninverting amplifier has an R_i of 1 kΩ and an R_f of 100 kΩ. The closed-loop gain is

(a) 100,000 (b) 1000 (c) 101 (d) 100

19. If the feedback resistor in Question 18 is open, the voltage gain

(a) increases (b) decreases (c) is not affected (d) depends on R_i

20. A certain inverting amplifier has a closed-loop gain of 25. The op-amp has an open-loop gain of 100,000. If another op-amp with an open-loop gain of 200,000 is substituted in the configuration, the closed-loop gain

(a) doubles (b) drops to 12.5

(c) remains at 25 (d) increases slightly

21. A voltage follower

(a) has a gain of one (b) is noninverting

(c) has no feedback resistor (d) has all of these

PROBLEMS

SECTION 12–1 INTRODUCTION TO OPERATIONAL AMPLIFIERS

1. Compare a practical op-amp to the ideal.

2. Two IC op-amps are available to you. Their characteristics are listed below. Choose the one you think is more desirable.
Op-amp 1: Z_{in} = 5 MΩ, Z_{out} = 100 Ω, A_{ol} = 50,000
Op-amp 2: Z_{in} = 10 MΩ, Z_{out} = 75 Ω, A_{ol} = 150,000

SECTION 12–2 THE DIFFERENTIAL AMPLIFIER

3. Identify the type of input and output configuration for each basic differential amplifier in Figure 12–45.

4. The dc base voltages in Figure 12–46 are zero. Using your knowledge of transistor analysis, determine the dc differential output voltage. Assume that Q_1 has an $\alpha = 0.98$ and Q_2 has an $\alpha = 0.975$.

FIGURE 12–45

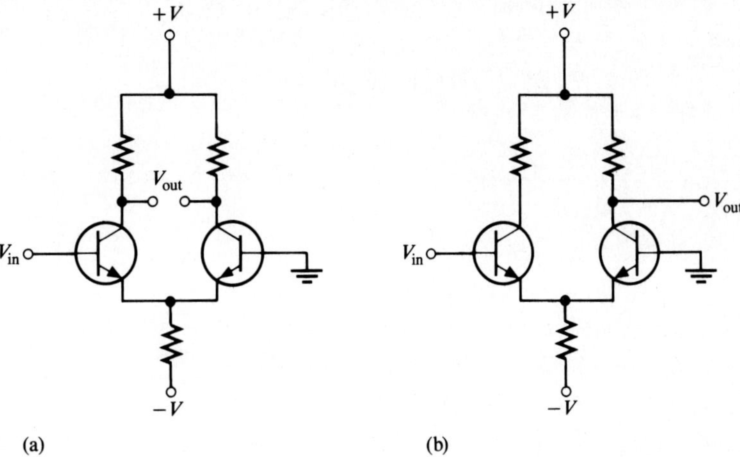

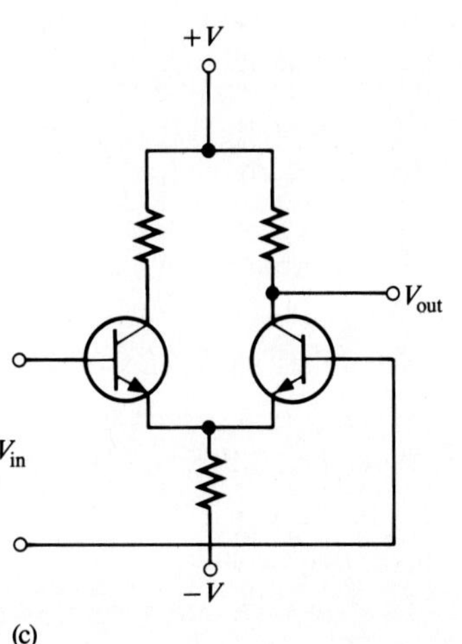

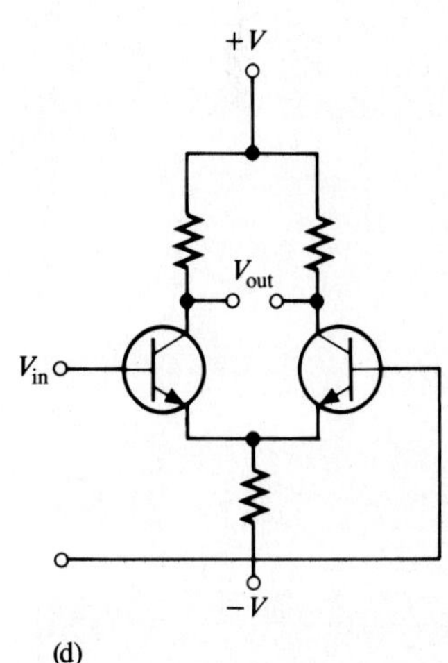

FIGURE 12–46

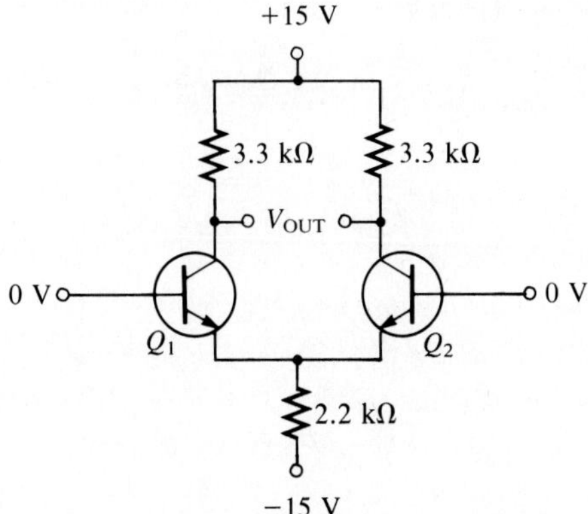

5. Identify the quantity being measured by each meter in Figure 12–47.

FIGURE 12–47

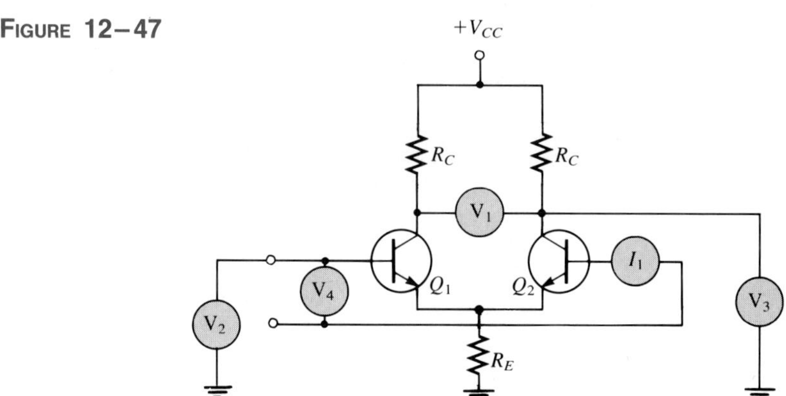

6. A differential amplifier stage has collector resistors of 5.1 kΩ each. If I_{C1} = 1.35 mA and I_{C2} = 1.29 mA, what is the differential output voltage?

SECTION 12–3 OP-AMP DATA SHEET PARAMETERS

7. Determine the bias current, I_{BIAS}, given that the input currents to an op-amp are 8.3 μA and 7.9 μA.

8. Distinguish between input bias current and input offset current, and then calculate the input offset current in Problem 7.

9. A certain op-amp has a CMRR of 250,000. Convert this to dB.

10. The open-loop gain of a certain op-amp is 175,000. Its common-mode gain is 0.18. Determine the CMRR in dB.

11. An op-amp data sheet specifies a CMRR of 300,000 and an A_{ol} of 90,000. What is the common-mode gain?

12. Figure 12–48 shows the output voltage of an op-amp in response to a step input. What is the slew rate?

FIGURE 12–48

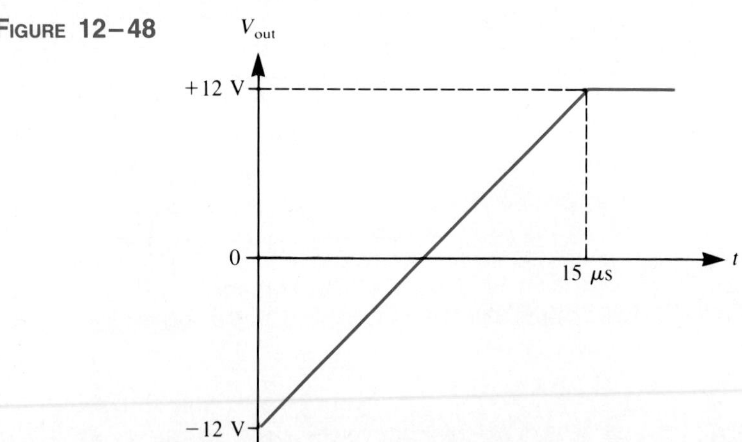

13. How long does it take the output voltage of an op-amp to go from -10 V to $+10$ V, if the slew rate is 0.5 V/μs?

SECTION 12–5 OP-AMP CONFIGURATIONS WITH NEGATIVE FEEDBACK

14. Identify each of the op-amp configurations in Figure 12–49.

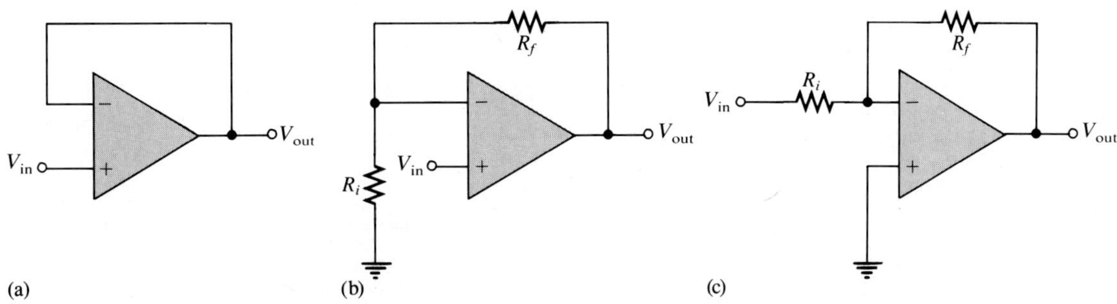

FIGURE 12–49

15. A noninverting amplifier has an R_i of 1 kΩ and an R_f of 100 kΩ. Determine V_f and B, if $V_{out} = 5$ V.
16. For the amplifier in Figure 12–50, determine the following:
    (a) $A_{cl(NI)}$ **(b)** V_{out} **(c)** V_f
17. Determine the closed-loop gain of each amplifier in Figure 12–51.

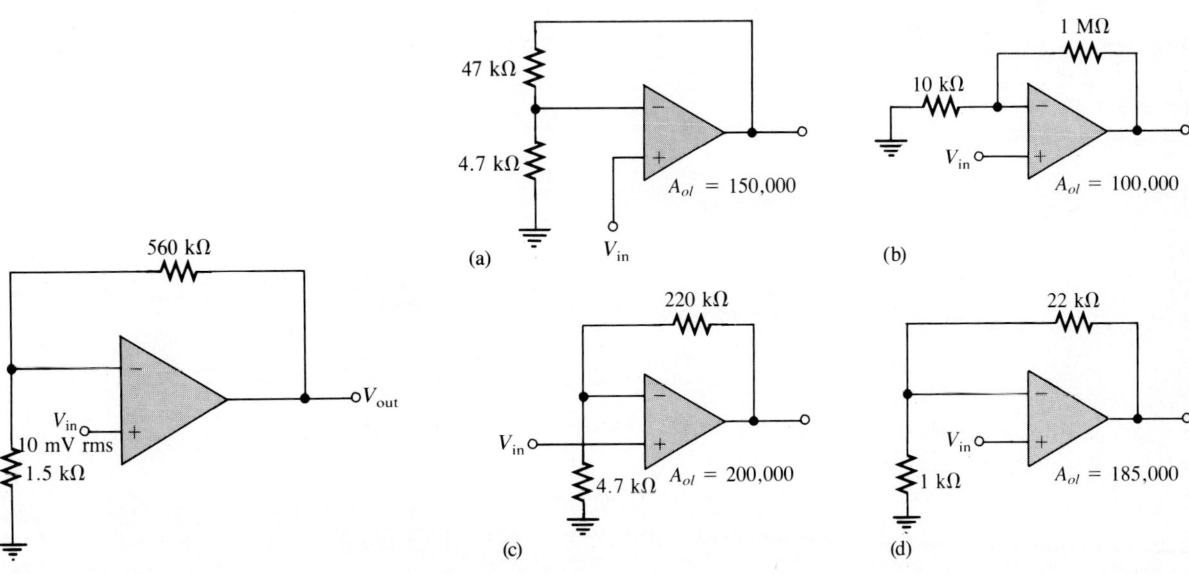

FIGURE 12–50

FIGURE 12–51

18. Find the value of R_f that will produce the indicated closed-loop gain in each amplifier in Figure 12–52.

FIGURE 12–52

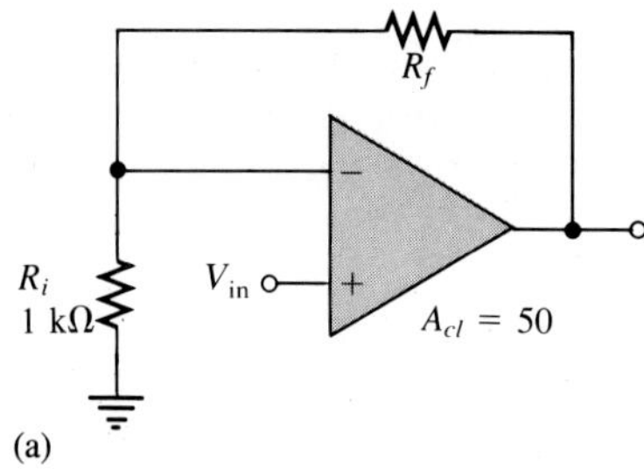

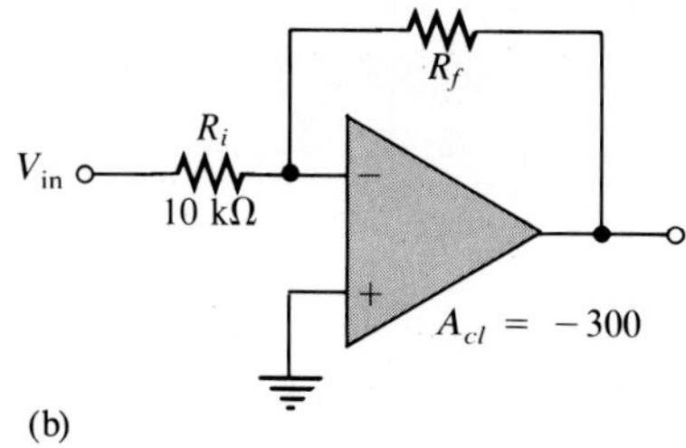

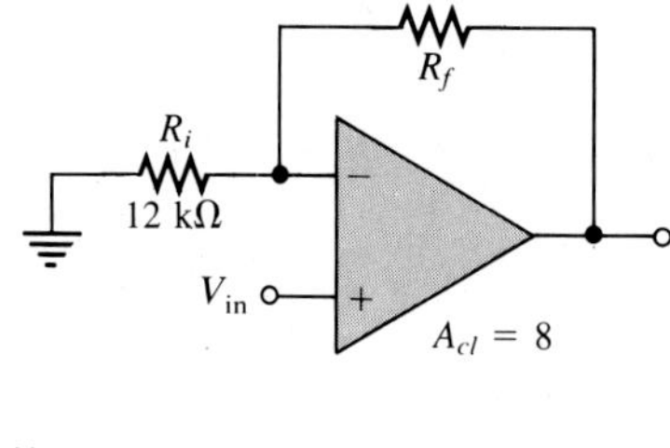

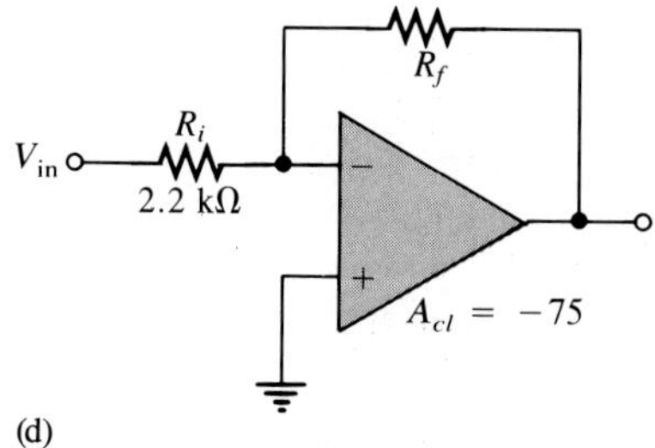

19. Find the gain of each amplifier in Figure 12–53.

FIGURE 12–53

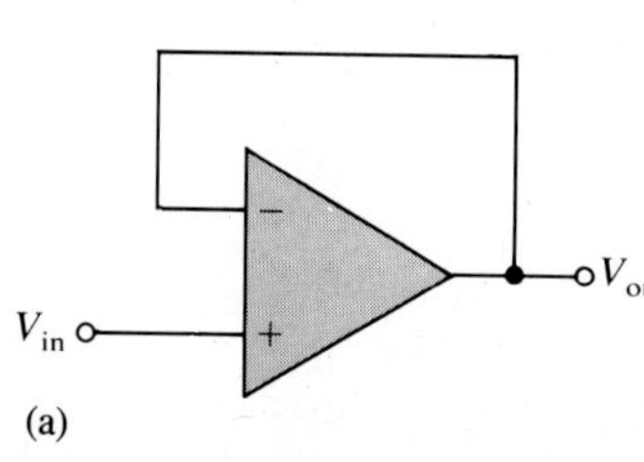

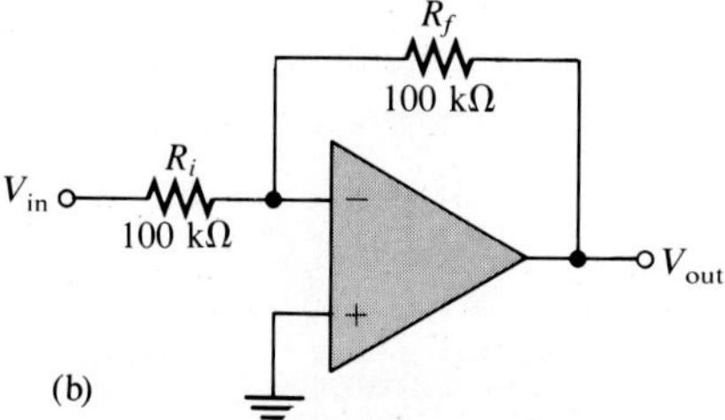

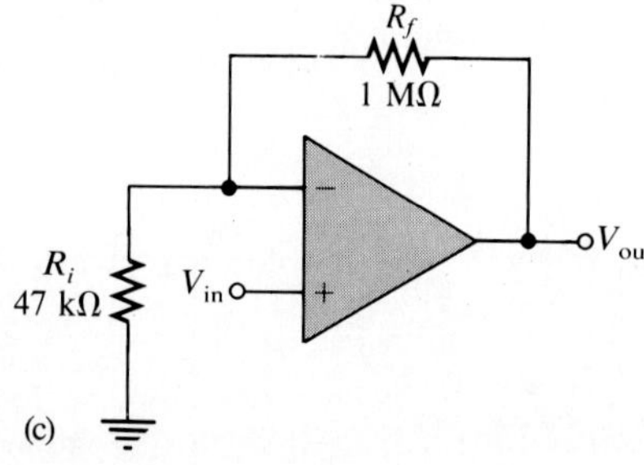

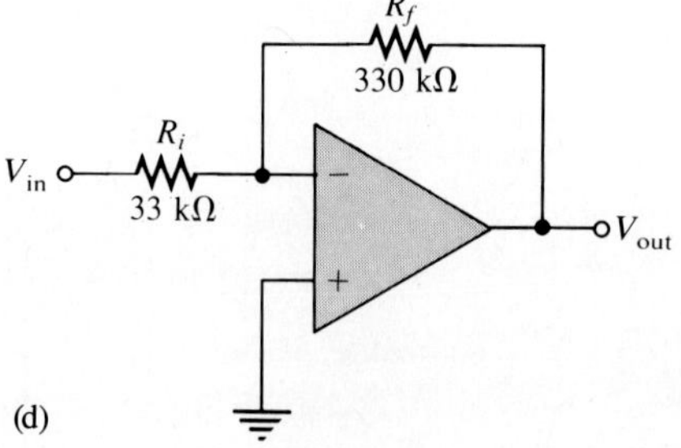

20. If a signal voltage of 10 mV rms is applied to each amplifier in Figure 12–53, what are the output voltages and what is their phase relationship with inputs?

21. Determine the approximate values for each of the following quantities in Figure 12–54.

(a) I_{in} (b) I_f (c) V_{out} (d) Closed-loop gain

FIGURE 12–54

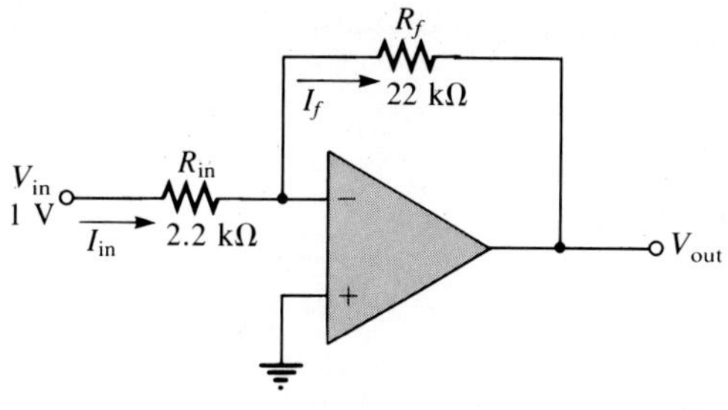

SECTION 12–6 EFFECTS OF NEGATIVE FEEDBACK ON OP-AMP IMPEDANCES

22. Determine the input and output impedances for each amplifier configuration in Figure 12–55.

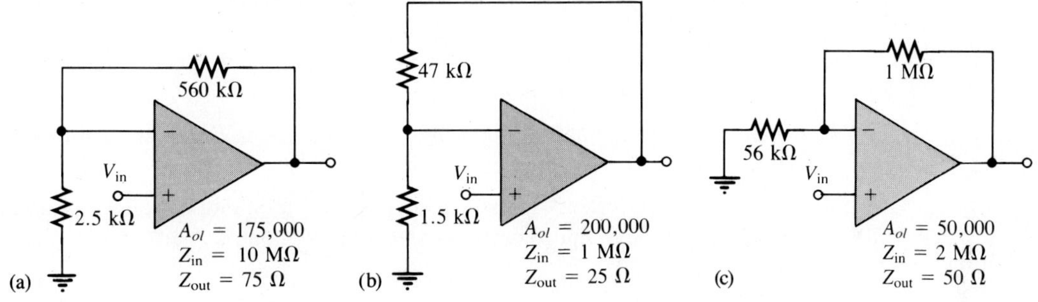

FIGURE 12–55

23. Repeat Problem 22 for each circuit in Figure 12–56.

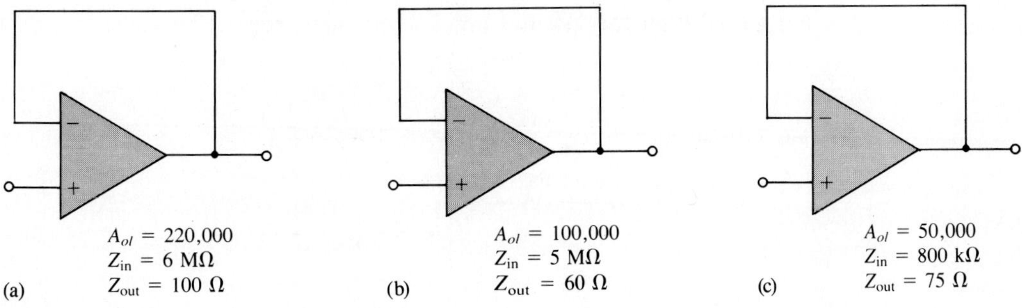

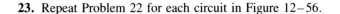

FIGURE 12–56

24. Repeat Problem 22 for each circuit in Figure 12–57.

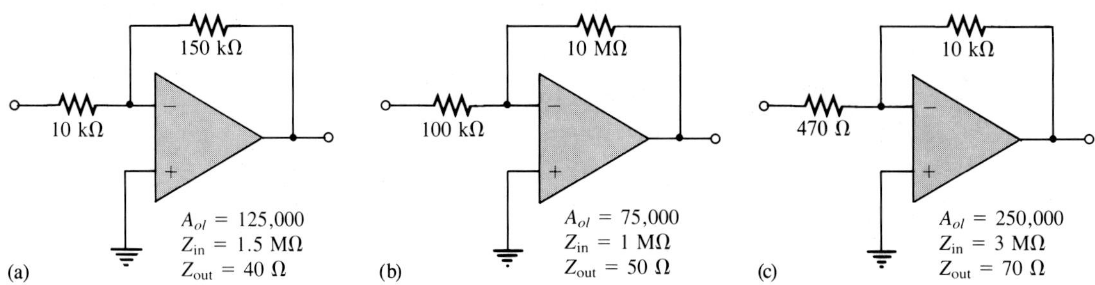

FIGURE 12–57

SECTION 12–7 BIAS CURRENT AND OFFSET VOLTAGE COMPENSATION

25. A voltage follower is driven by a voltage source with a source resistance of 75 Ω.

(a) What value of compensating resistor is required for bias current, and where should the resistor be placed?

(b) If the two input currents after compensation are 42 μA and 40 μA, what is the output error voltage?

26. Determine the compensating resistor value for each amplifier configuration in Figure 12–55, and indicate the placement of the resistor.

27. A particular op-amp has an input offset voltage of 2 nV and an open-loop gain of 100,000. What is the output error voltage?

28. What is the input offset voltage of an op-amp if a dc output voltage of 35 mV is measured when the input voltage is zero? The op-amp's open-loop gain is specified to be 200,000.

SECTION 12–8 TROUBLESHOOTING

29. Determine the most likely fault(s) for each of the following symptoms in Figure 12–58 with a 100 mV signal applied.

(a) No output signal.

(b) Output severely clipped on both positive and negative swings.

(c) Clipping on only positive peaks when input signal is increased to a certain point.

FIGURE 12–58

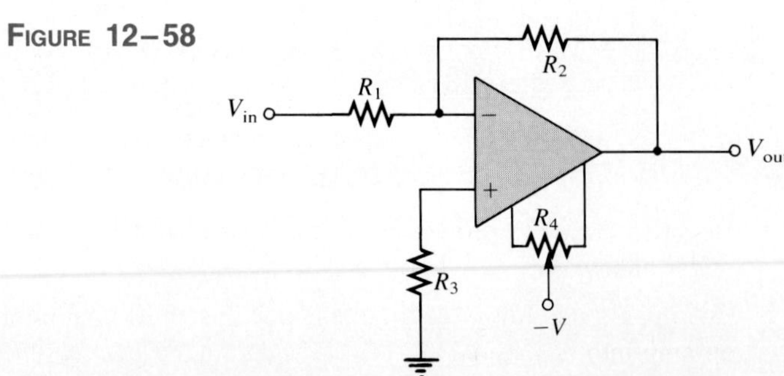

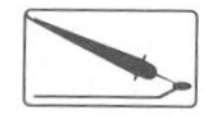

30. On the circuit board in Figure 12–59, what happens if the middle lead (wiper) of the 100 kΩ potentiometer is broken?

FIGURE 12–59

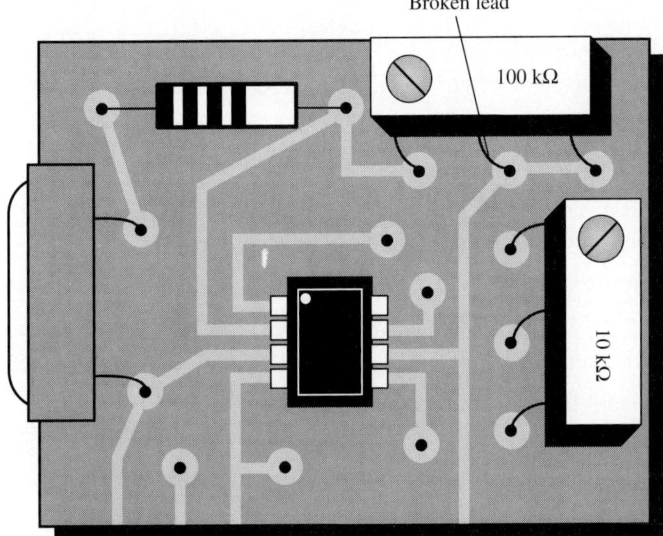

ANSWERS TO REVIEW QUESTIONS

SECTION 12–1

1. Inverting input, noninverting input, output, positive and negative supply voltages.
2. A practical op-amp has high imput impedance, low output impedance, high voltage gain, and wide bandwidth.

SECTION 12–2

1. Differential input is between two input terminals. Single-ended input is from one input terminal to ground (with other input grounded).
2. Common-mode rejection is the ability of an op-amp to produce very little output when the same signal is applied to both inputs.
3. A higher CMRR means less common-mode gain.

SECTION 12–3

1. Input bias current, input offset voltage, drift, input offset current, input impedance, output impedance, common-mode range, CMRR, open-loop voltage gain, slew rate, frequency response.
2. Slew rate and voltage gain are both frequency dependent.

SECTION 12–4

1. Negative feedback provides a stable controlled gain, control of impedances, and wider bandwidth.
2. The open-loop gain is so high that a very small signal on the input will drive the op-amp into saturation.

SECTION 12–5

1. The main purpose of negative feedback is to stabilize the gain.
2. False
3. $A_{cl} = 1/0.02 = 50$

SECTION 12–6

1. The noninverting configuration has a higher Z_{in} than the op-amp alone.
2. Z_{in} increases in a voltage follower.
3. $Z_{in(I)} \cong R_i = 2\ \text{k}\Omega$, $Z_{out(I)} \cong Z_{out} = 60\ \Omega$.

SECTION 12–7

1. Input bias current and input offset voltage are sources of output error.
2. Add a resistor in the feedback path equal to the input source resistance.

SECTION 12–8

1. Check the output null adjustment.
2. The op-amp is probably bad.

SECTION 12–9

1. The 100 kΩ potentiometer is the feedback resistor.
2. The 10 kΩ potentiometer is for nulling the output.
3. The light source and prism must be pivoted to allow different wavelengths of light to pass through the slit.

ANSWERS TO PRACTICE EXERCISES

12–1 34,000; 90.6 dB

12–2 **(a)** 0.168 **(b)** 87.96 dB **(c)** 2.1 V rms, 4.2 V rms **(d)** 0.168 V

12–3 12,649

12–4 20 V/μs

12–5 32.9

12–6 67.5 kΩ

12–7 **(a)** 20,557 MΩ, 0.014 Ω **(b)** 23

12–8 Input Z increases, output Z decreases.

12–9 $Z_{in(I)} = 560\ \Omega$, $Z_{out(I)} = 75\ \Omega$, $A_{cl} = -146$

13

Op-Amp Frequency Response, Stability, and Compensation

After completing this chapter, you should be able to

- □ Distinguish the difference between open-loop voltage gain and closed-loop voltage gain of an op-amp.
- □ Compare an op-amp's frequency response for closed-loop and open-loop conditions.
- □ Discuss the effects of closed-loop operation on bandwidth.
- □ Define *gain-bandwidth product.*
- □ Describe what happens when positive feedback occurs in an op-amp.
- □ Define *phase margin.*
- □ Explain stability and what factors affect the stability of an op-amp.
- □ Use compensation to ensure stability in an op-amp.

In this chapter, you will learn more about frequency response, bandwidth, phase shift, and other frequency-related parameters. The effects of negative feedback will be further examined, and you will learn about stability requirements and how to compensate op-amp circuits to ensure stable operation.

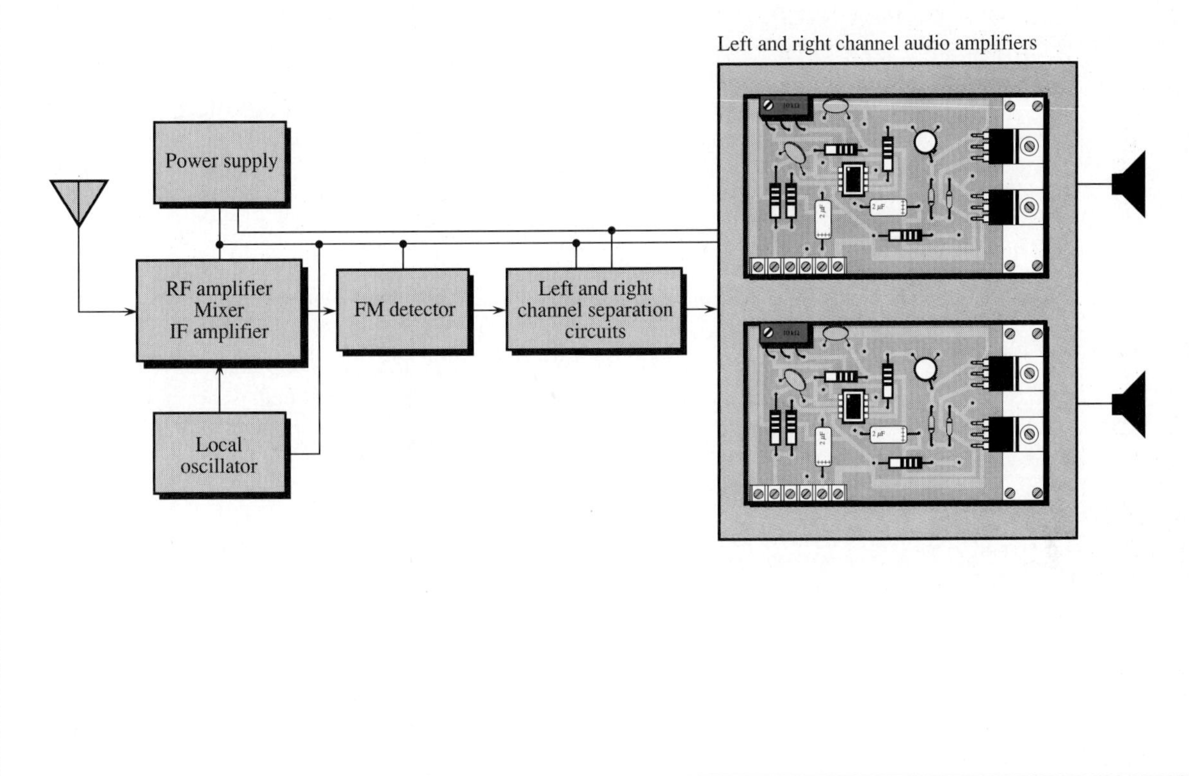

A System Application

The block diagram above shows an FM stereo receiver. Notice that the system is basically the same as the standard superheterodyne receiver up through the FM detector. Stereo systems use two separate frequency-modulated (FM) signals to reproduce sound as, for example, from the left and right sides of the stage in a concert performance. When the signal is processed by a stereo receiver, the sound comes out of both the left and right speaker, and you get the original sound effects in terms of direction and distribution. When a stereo broadcast is received by a single-speaker (monophonic) system, the sound from the speaker is actually the composite or sum of the left and right channel sounds so you get the original sound without separation. Op-amps can be used for many purposes in stereo systems such as this, but we will focus on the identical left and right channel audio amplifiers for this system application.

For the system application in Section 13–6, in addition to the other topics, be sure you understand

- ☐ The noninverting op-amp configuration.
- ☐ How capacitors can be connected for frequency compensation.
- ☐ Discrete transistor push-pull amplifier operation (from Chapter 9).

13–1 BASIC CONCEPTS

The last chapter demonstrated how closed-loop voltage gains of the basic op-amp configurations are determined, and the distinction between open-loop gain and closed-loop gain was established. Because of the importance of these two different types of gain, the definitions are restated in this section.

OPEN-LOOP GAIN

The **open-loop voltage gain** of an op-amp is the internal voltage gain of the device and represents the ratio of output voltage to input voltage, as indicated in Figure 13–1(a). Notice that there are no external components, so the open-loop gain is set entirely by the internal design. Open-loop gain can range up to 200,000 and is not a well-controlled parameter. Data sheets often refer to the open-loop gain as the *large-signal voltage gain*.

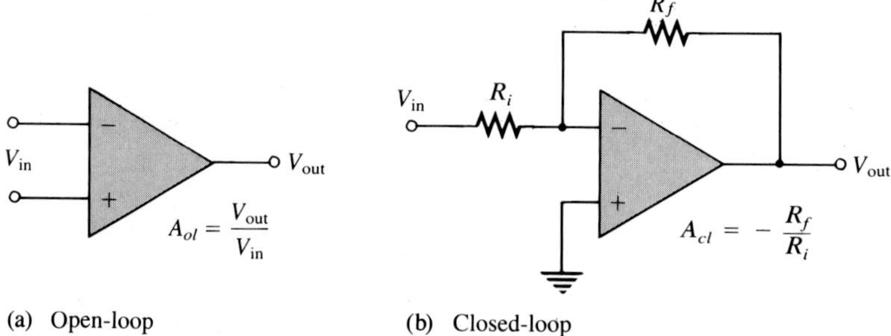

FIGURE 13–1
Open-loop and closed-loop op-amp configurations.

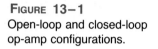

CLOSED-LOOP GAIN

The **closed-loop voltage gain** is for an entire amplifier configuration consisting of the op-amp and an external negative feedback circuit that connects the output to the inverting input. The closed-loop gain is determined by the external component values, as illustrated in Figure 13–1(b) for an inverting amplifier configuration. The closed-loop gain can be precisely controlled by external component values.

THE GAIN IS FREQUENCY DEPENDENT

In the last chapter, all of the gain expressions applied to the midrange gain and were considered independent of the frequency. The midrange open-loop gain of an op-amp extends from zero frequency (dc) up to a critical frequency at which the gain is 3 dB less than the midrange value. This concept should be familiar from your study of Chapter 10. The difference here is that op-amps are dc amplifiers (no capacitive coupling between stages), and therefore, there is no lower critical frequency. This means that the midrange gain extends down to zero frequency, and dc voltages are amplified the same as midrange signal frequencies.

An open-loop response curve (Bode plot) for a certain op-amp is shown in Figure 13–2. Most op-amp data sheets show this type of curve or specify the midrange open-loop gain. Notice that the curve rolls off at −20 dB per decade (−6 dB per octave). The midrange gain is 200,000, which is 106 dB, and the critical (cutoff) frequency is approximately 10 Hz.

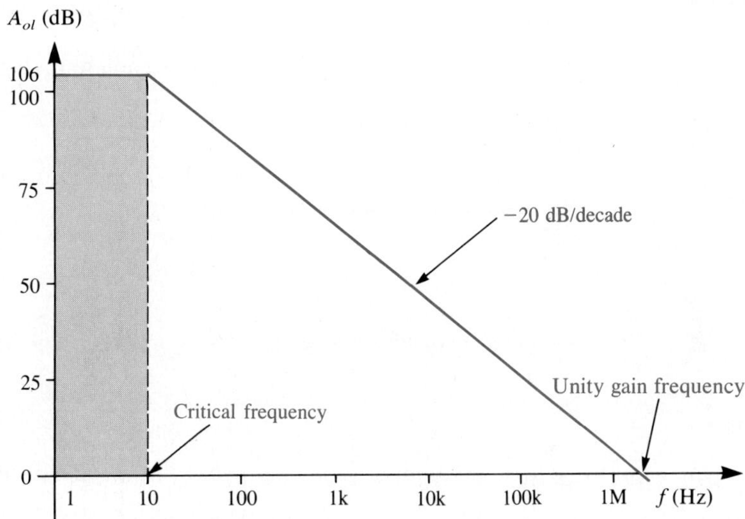

FIGURE 13–2
Ideal plot of open-loop voltage gain versus frequency for typical op-amp. The frequency scale is logarithmic.

3 dB Open-Loop Bandwidth

Recall that the bandwidth of an ac amplifier is the frequency range between the points where the gain is 3 dB less than midrange, equalling the upper critical frequency minus the lower critical frequency.

$$BW = f_{c(\text{high})} - f_{c(\text{low})} \tag{13–1}$$

Since $f_{c(\text{low})}$ for an op-amp is zero, the bandwidth is simply equal to the upper critical frequency.

$$BW = f_{c(\text{high})} \tag{13–2}$$

From now on, $f_{c(\text{high})}$ will be referred to simply as f_c. Open-loop (ol) or closed-loop (cl) subscript designators will be used.

Unity-Gain Bandwidth

Notice in Figure 13–2 that the gain steadily decreases to a point where it is equal to one (0 dB). The value of the frequency at which this unity gain occurs in the *unity-gain bandwidth*.

GAIN-VERSUS-FREQUENCY ANALYSIS

The *RC* lag (low-pass) networks within an op-amp are responsible for the roll-off in gain as the frequency increases, just as was discussed for the discrete amplifiers in Chapter 10. From basic ac circuit theory, the attenuation of an *RC* lag network, such as in Figure 13–3, is expressed as

$$\frac{V_{out}}{V_{in}} = \frac{X_C}{\sqrt{R^2 + X_C^2}} \qquad \textbf{(13–3)}$$

Dividing both the numerator and denominator to the right of the equal sign by X_C, we get

$$\frac{V_{out}}{V_{in}} = \frac{1}{\sqrt{1 + R^2/X_C^2}} \qquad \textbf{(13–4)}$$

The critical frequency of an *RC* network is

$$f_c = \frac{1}{2\pi RC}$$

Dividing both sides by f gives

$$\frac{f_c}{f} = \frac{1}{2\pi RCf} = \frac{1}{(2\pi fC)R}$$

Since $X_C = 1/(2\pi fC)$, the above expression can be written as

$$\frac{f_c}{f} = \frac{X_C}{R} \qquad \textbf{(13–5)}$$

Substituting this result into Equation (13–4) produces the following expression for the attenuation of an *RC* lag network.

$$\frac{V_{out}}{V_{in}} = \frac{1}{\sqrt{1 + f^2/f_c^2}} \qquad \textbf{(13–6)}$$

If an op-amp is represented by a voltage gain element and a single *RC* lag network, as shown in Figure 13–4, then the total open-loop gain is the product of the midrange open-loop gain $A_{ol(mid)}$ and the attenuation of the *RC* network.

$$A_{ol} = \frac{A_{ol(mid)}}{\sqrt{1 + f^2/f_c^2}} \qquad \textbf{(13–7)}$$

As you can see from Equation (13–7), the open-loop gain equals the midrange value when the signal frequency f is much less than the critical frequency f_c and drops off as the frequency increases. Since f_c is part of the open-loop response of an op-amp, we will refer to it as $f_{c(ol)}$.

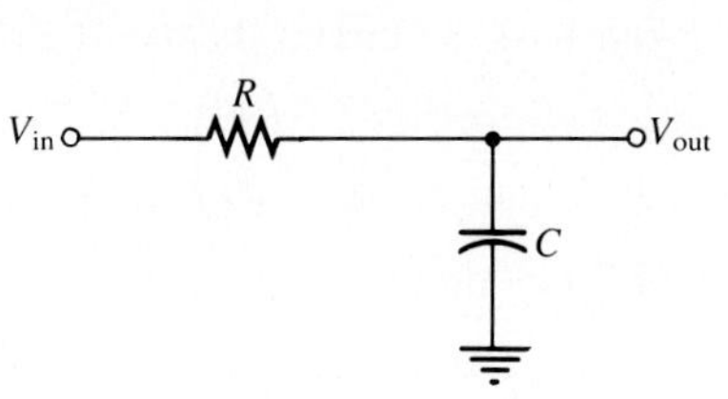

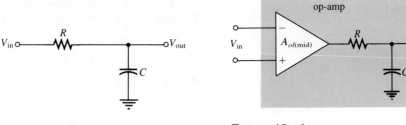

FIGURE 13–3
RC lag network.

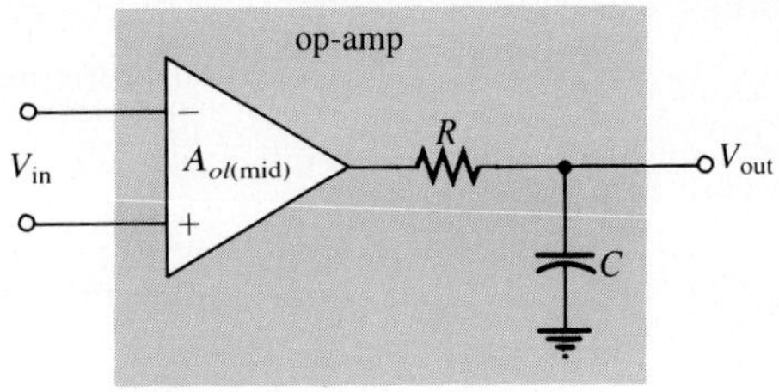

FIGURE 13–4
Op-amp represented by gain element and internal *RC* network.

EXAMPLE 13–1 Determine A_{ol} for the following values of f. Assume $f_{c(ol)} = 100$ Hz and $A_{ol(\text{mid})} = 100{,}000$.

(a) $f = 0$ Hz
(b) $f = 10$ Hz
(c) $f = 100$ Hz
(d) $f = 1000$ Hz

SOLUTION

(a) $$A_{ol} = \frac{A_{ol(\text{mid})}}{\sqrt{1 + f^2/f_{c(ol)}^2}} = \frac{100{,}000}{\sqrt{1 + 0}} = 100{,}000$$

(b) $$A_{ol} = \frac{100{,}000}{\sqrt{1 + (0.1)^2}} = 99{,}503$$

(c) $$A_{ol} = \frac{100{,}000}{\sqrt{1 + (1)^2}} = \frac{100{,}000}{\sqrt{2}} = 70{,}710$$

(d) $$A_{ol} = \frac{100{,}000}{\sqrt{1 + (10)^2}} = 9950$$

This exercise has demonstrated how the open-loop gain decreases as the frequency increases above $f_{c(ol)}$.

PRACTICE EXERCISE 13–1
Find A_{ol} for the following frequencies. Assume $f_{c(ol)} = 200$ Hz, $A_{ol(\text{mid})} = 80{,}000$.
(a) $f = 2$ Hz **(b)** $f = 10$ Hz **(c)** $f = 2500$ Hz

PHASE SHIFT

As you know, an *RC* network causes a propagation delay from input to output, thus creating a **phase** difference between the input signal and the output signal. An *RC* lag network such as found in an op-amp stage causes the output signal voltage to lag the input, as shown in Figure 13–5. From basic ac circuit theory, the phase shift is

$$\phi = -\tan^{-1}\left(\frac{R}{X_C}\right) \tag{13–8}$$

Substituting the relationship in Equation (13–5), we get

$$\phi = -\tan^{-1}\left(\frac{f}{f_c}\right) \tag{13–9}$$

The negative sign indicates that the output lags the input. This equation shows that the phase shift increases with frequency and approaches −90° as f becomes much greater than f_c.

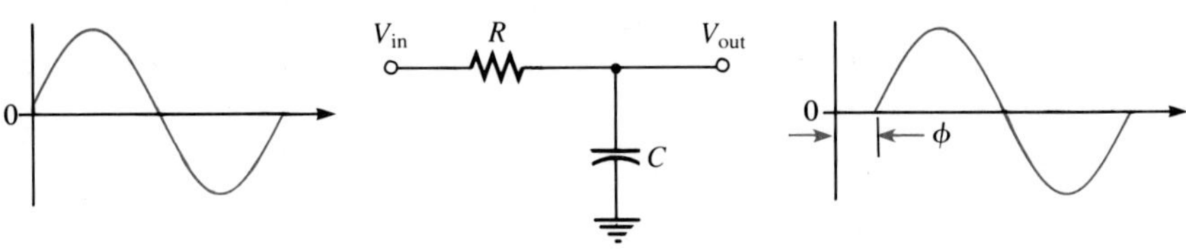

FIGURE 13–5
Output voltage lags input voltage.

EXAMPLE 13–2 Calculate the phase shift for an RC lag network for each of the following frequencies, and then plot the curve of phase shift versus frequency. Assume f_c = 100 Hz.

(a) f = 1 Hz
(b) f = 10 Hz
(c) f = 100 Hz
(d) f = 1000 Hz
(e) f = 10,000 Hz

SOLUTION

(a) $\phi = -\tan^{-1}\left(\frac{f}{f_c}\right) = -\tan^{-1}\left(\frac{1\text{ Hz}}{100\text{ Hz}}\right) = -0.573°$

(b) $\phi = -\tan^{-1}\left(\frac{10\text{ Hz}}{100\text{ Hz}}\right) = -5.71°$

(c) $\phi = -\tan^{-1}\left(\frac{100\text{ Hz}}{100\text{ Hz}}\right) = -45°$

(d) $\phi = -\tan^{-1}\left(\frac{1000\text{ Hz}}{100\text{ Hz}}\right) = -84.29°$

(e) $\phi = -\tan^{-1}\left(\frac{10{,}000\text{ Hz}}{100\text{ Hz}}\right) = -89.43°$

The phase shift-versus-frequency curve is plotted in Figure 13–6. Note that the frequency axis is logarithmic.

FIGURE 13–6

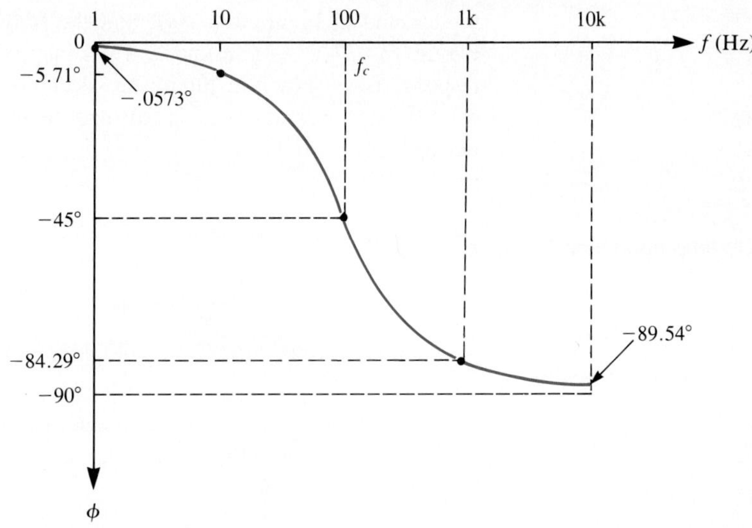

PRACTICE EXERCISE 13–2

At what frequency, in this example, is the phase shift 60°? ■

13–1 REVIEW QUESTIONS

1. How do the open-loop gain and the closed-loop gain of an op-amp differ?
2. The upper critical frequency of a particular op-amp is 100 Hz. What is its open-loop 3 dB bandwidth?
3. Does the open-loop gain increase or decrease with frequency above the critical frequency?

13–2 OP-AMP OPEN-LOOP RESPONSE

In this section, you will learn about the open-loop frequency response and the open-loop phase response of an op-amp. Open-loop responses relate to an op-amp with no external feedback. The frequency response indicates how the voltage gain changes with frequency, and the phase response indicates how the phase shift between the input and output signal changes with frequency. The open-loop gain, like the β of a transistor, varies greatly from one device to the next of the same type and cannot be depended upon to have a constant value.

FREQUENCY RESPONSE

In the previous section, we considered an op-amp to have a constant roll-off of −20 dB/decade above its critical frequency. Actually, the situation is often more complex than that. A typical IC operational amplifier may consist of two or more cascaded amplifier stages. The gain of each stage is frequency dependent and rolls off at −20 dB/decade above its critical frequency. Therefore, the total response of an op-amp is a composite of the individual responses of the internal stages. As an example, a three-stage op-amp is represented in Figure 13–7(a), and the frequency response of each stage is shown in Figure 13–7(b). As you know, dB gains are added so that the total op-amp frequency response is as shown in Figure 13–7(c). Since the roll-off rates are additive, the total roll-off rate increases by −20 dB/decade (−6 dB/octave) as each critical frequency is reached.

FIGURE 13–7
Op-amp open-loop frequency response.

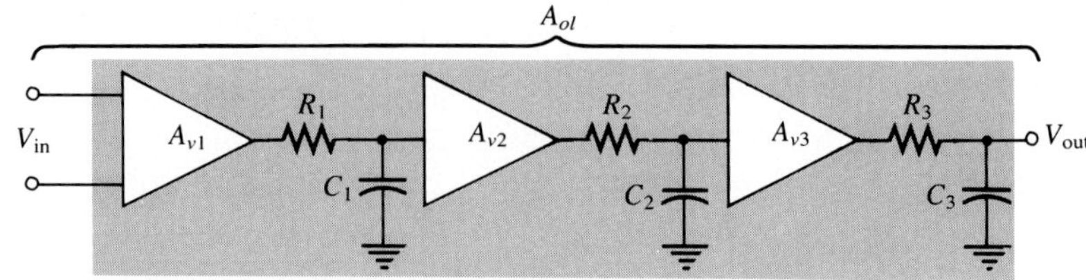

(a) Representation of an op-amp with three internal stages

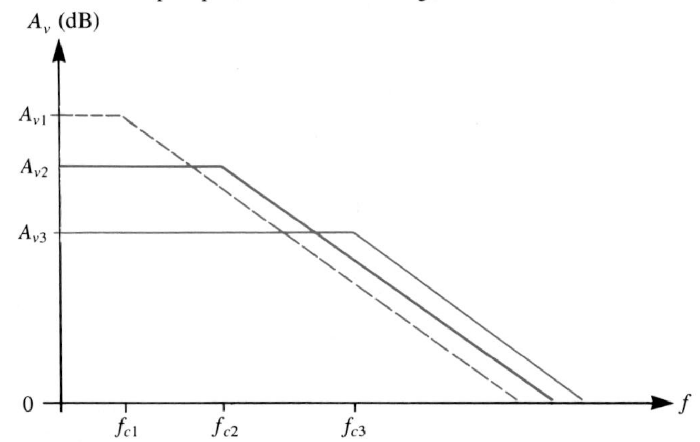

(b) Individual responses

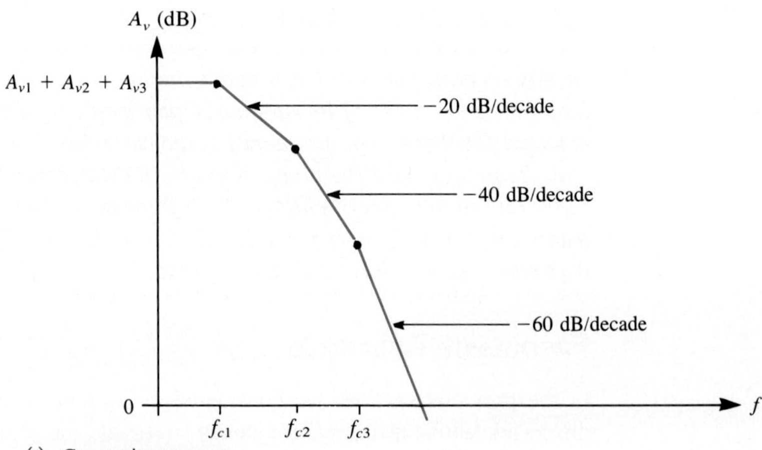

(c) Composite response

PHASE RESPONSE

In a multistage amplifier, each stage contributes to the total phase lag. As you have seen, each RC lag network can produce up to a $-90°$ phase shift. Since each stage in an op-amp includes an RC lag network, a three-stage op-amp, for example, can have a maximum phase lag of $-270°$. Also, the phase lag of each stage is less than $-45°$ below the critical frequency, equal to $-45°$ at the critical frequency, and greater than $-45°$ above the critical frequency. The phase lags of the stages of an op-amp are added to produce a total phase lag, according to the following formula for three stages.

$$\phi_{tot} = -\tan^{-1}\left(\frac{f}{f_{c1}}\right) - \tan^{-1}\left(\frac{f}{f_{c2}}\right) - \tan^{-1}\left(\frac{f}{f_{c3}}\right) \qquad (13\text{–}10)$$

EXAMPLE 13–3

A certain op-amp has three internal amplifier stages with the following gains and critical frequencies:

Stage 1: $A_{v1} = 40\text{ dB}, f_{c1} = 2000\text{ Hz}$
Stage 2: $A_{v2} = 32\text{ dB}, f_{c2} = 40\text{ kHz}$
Stage 3: $A_{v3} = 20\text{ dB}, f_{c3} = 150\text{ kHz}$

Determine the open-loop midrange dB gain and the total phase lag when $f = f_{c1}$.

SOLUTION

$$\begin{aligned} A_{ol(\text{mid})} &= A_{v1} + A_{v2} + A_{v3} \\ &= 40\text{ dB} + 32\text{ dB} + 20\text{ dB} \\ &= 92\text{ dB} \end{aligned}$$

$$\begin{aligned} \phi_{tot} &= \tan^{-1}\left(\frac{f}{f_{c1}}\right) - \tan^{-1}\left(\frac{f}{f_{c2}}\right) - \tan^{-1}\left(\frac{f}{f_{c3}}\right) \\ &= -\tan^{-1}(1) - \tan^{-1}\left(\frac{2}{40}\right) - \tan^{-1}\left(\frac{2}{150}\right) \\ &= -45° - 2.86° - 0.76° \\ &= -48.62° \end{aligned}$$

PRACTICE EXERCISE 13–3

The internal stages of a two-stage amplifier have the following characteristics: $A_{v1} = 50\text{ dB}$, $A_{v2} = 25\text{ dB}$, $f_{c1} = 1500\text{ Hz}$, and $f_{c2} = 3000\text{ Hz}$. Determine the open-loop midrange gain in dB and the total phase lag when $f = f_{c1}$. ■

13–2 REVIEW QUESTIONS

1. If the individual stage gains of an op-amp are 20 dB and 30 dB, what is the total gain in dB?
2. If the individual phase lags are $-49°$ and $-5.2°$, what is the total phase lag?

13–3 OP-AMP CLOSED-LOOP RESPONSE

Op-amps are normally used in a closed-loop configuration with negative feedback in order to achieve precise control of the gain and bandwidth. In this section, you will see how feedback affects the gain and frequency response of an op-amp.

Recall from Chapter 12 that midrange gain is reduced by negative feedback, as indicated by the following closed-loop gain expressions for the three configurations previously covered, where B is the feedback attenuation. For noninverting,

$$A_{cl(\text{NI})} = \frac{A_{ol}}{1 + A_{ol}B} \cong \frac{1}{B}$$

For inverting,

$$A_{cl(\text{I})} \cong -\frac{R_f}{R_i}$$

For voltage follower,

$$A_{cl(\text{VF})} \cong 1$$

EFFECT OF NEGATIVE FEEDBACK ON BANDWIDTH

You know how negative feedback affects the gain; now you will learn how it affects the amplifier's bandwidth. The closed-loop critical frequency is

$$f_{c(cl)} = f_{c(ol)}(1 + BA_{ol(\text{mid})}) \qquad \textbf{(13–11)}$$

This expression shows that the closed-loop critical frequency, $f_{c(cl)}$, is higher than the open-loop critical frequency $f_{c(ol)}$ by the factor $1 + BA_{ol(\text{mid})}$. A derivation of Equation (13–11) can be found in Appendix B.

Since $f_{c(cl)}$ equals the bandwidth for the closed-loop amplifier, the bandwidth is also increased by the same factor.

$$BW_{cl} = BW_{ol}(1 + BA_{ol(\text{mid})}) \qquad \textbf{(13–12)}$$

EXAMPLE 13–4

A certain amplifier has an open-loop midrange gain of 150,000 and an open-loop 3 dB bandwidth of 200 Hz. The attenuation of the feedback loop is 0.002. What is the closed-loop bandwidth?

SOLUTION

$$\begin{aligned} BW_{cl} &= BW_{ol}(1 + BA_{ol(\text{mid})}) \\ &= 200 \text{ Hz}[1 + (150{,}000)(0.002)] \\ &= 60.2 \text{ kHz} \end{aligned}$$

PRACTICE EXERCISE 13–4

If $A_{ol(\text{mid})} = 200{,}000$ and $B = 0.05$, what is the closed loop bandwidth?

Figure 13–8 graphically illustrates the concept of closed-loop response. When the open-loop gain of an op-amp is reduced by negative feedback, the bandwidth is increased. The closed-loop gain is independent of the open-loop gain up to the point of intersection of the two gain curves. This point of intersection is the critical frequency for the closed-loop response. Notice that the closed-loop gain has the same roll-off rate as the open-loop gain, beyond the closed-loop critical frequency.

FIGURE 13–8
Closed-loop gain compared to open-loop gain.

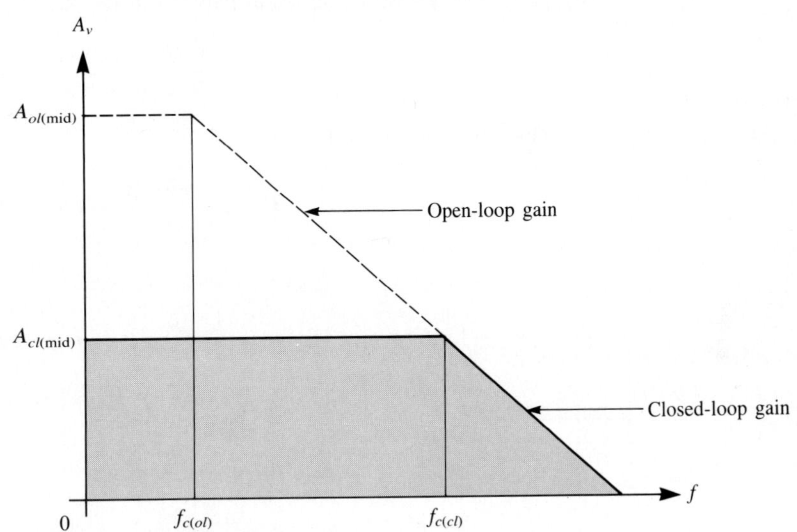

GAIN-BANDWIDTH PRODUCT

An increase in closed-loop gain causes a decrease in the bandwidth and vice versa, such that the product of gain and bandwidth is a constant. This is true as long as the roll-off rate is fixed. Letting A_{cl} stand for the gain of any of the closed-loop configurations and $f_{c(cl)}$ for the closed-loop critical frequency (also the bandwidth), then

$$A_{cl}f_{c(cl)} = A_{ol}f_{c(ol)} \tag{13–13}$$

The gain-bandwidth product is always equal to the frequency at which the op-amp's open-loop gain is unity (unity-gain bandwidth).

$$A_{cl}f_{c(cl)} = \text{unity-gain bandwidth} \tag{13–14}$$

EXAMPLE 13–5 Determine the bandwidth of each of the amplifiers in Figure 13–9. Both op-amps have an open-loop gain of 100 dB and a unity-gain bandwidth of 3 MHz.

SOLUTION

(a) For the noninverting amplifier in part (a) of the figure, the closed-loop gain is

$$A_{cl} \cong \frac{1}{B} = \frac{1}{R_i/(R_i + R_f)} = \frac{1}{3.3\ \text{k}\Omega/223.3\ \text{k}\Omega} = 67.67$$

Using Equation (13–14) and solving for $f_{c(cl)}$, we get (where $f_{c(cl)} = BW_{cl}$)

$$f_{c(cl)} = BW_{cl} = \frac{\text{unity-gain } BW}{A_{cl}}$$

$$BW_{cl} = \frac{3 \text{ MHz}}{67.67} = 44.33 \text{ kHz}$$

(b) For the inverting amplifier in part (b) of the figure, the closed-loop gain is

$$A_{cl} = -\frac{R_f}{R_i} = -\frac{47 \text{ k}\Omega}{1 \text{ k}\Omega} = -47$$

The closed-loop bandwidth is

$$BW_{cl} = \frac{3 \text{ MHz}}{47} = 63.8 \text{ kHz}$$

FIGURE 13–9

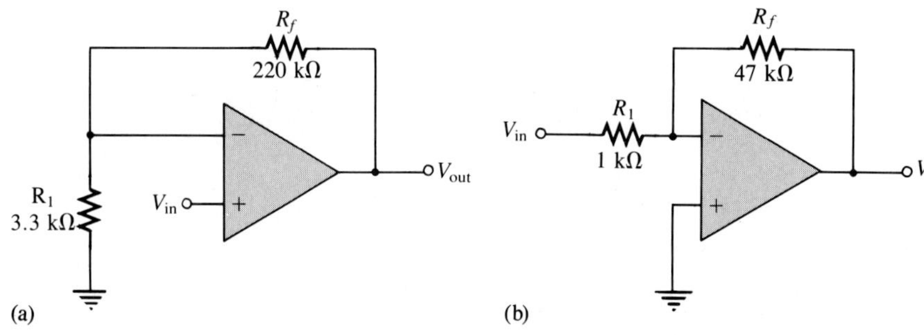

PRACTICE EXERCISE 13–5

Determine the bandwidth of each of the amplifiers in Figure 13–9. Both op-amps have an A_{ol} of 90 dB and a unity-gain bandwidth of 2 MHz. ■

13–3 REVIEW QUESTIONS

1. Is the closed-loop gain always less than the open-loop gain?
2. A certain op-amp is used in a feedback configuration having a gain of 30 and a bandwidth of 100 kHz. If the external resistor values are changed to increase the gain to 60, what is the new bandwidth?
3. What is the unity-gain bandwidth of the op-amp in Question 2?

13–4 POSITIVE FEEDBACK AND STABILITY

Stability is a very important consideration when using op-amps. Stable operation means that the op-amp does not oscillate under any condition. Instability produces oscillations, which are unwanted voltage swings on the output when there is no signal present on the input, or in response to noise or transient voltages on the input.

POSITIVE FEEDBACK

To understand stability, instability and its causes must first be examined. As you know, with negative feedback, the signal fed back to the input is out of phase with the input signal, thus subtracting from it and effectively reducing the voltage gain. As long as the feedback is negative, the amplifier is stable. When the signal fed back from output to input is in phase with the input signal, a positive feedback condition exists. That is, positive feedback occurs when the total phase shift through the op-amp and feedback network is 360°, which is equivalent to no phase shift (0°).

LOOP GAIN

For instability to occur, (1) there must be positive feedback, and (2) the loop gain of the closed-loop amplifier must be greater than 1. The loop gain of a closed-loop amplifier is defined to be the op-amp's open-loop gain times the attenuation of the feedback network.

$$\text{Loop gain} = A_{ol}B \tag{13–15}$$

PHASE MARGIN

Notice that for each amplifier configuration in Figure 13–10, the feedback loop is connected to the inverting input. There is an inherent phase shift of 180° because of the *inversion* between input and output. Additional phase shift (ϕ_{tot}) is produced by the *RC* lag networks within the amplifier. So, the total phase shift around the loop is $180° + \phi_{tot}$.

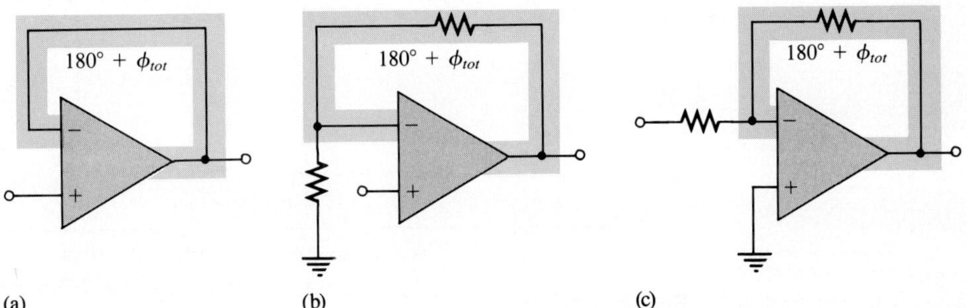

FIGURE 13–10
Feedback-loop phase shift.

The **phase margin,** θ_{pm}, is the amount of additional phase shift required to make the total phase shift around the loop 360°. (360° is equivalent to 0°.)

$$180° + \phi_{tot} + \theta_{pm} = 360°$$

$$\theta_{pm} = 180° - |\phi_{tot}| \tag{13–16}$$

If the phase margin is positive, the total phase shift is less than 360° and the amplifier is stable. If the phase margin is zero or negative, then the amplifier is potentially unstable because the signal fed back can be in phase with the input. As you can see from Equation (13–16), when the total lag network phase shift ϕ_{tot} equals or exceeds 180°, then the phase margin is 0° or negative and an unstable condition exists.

STABILITY ANALYSIS

Since most op-amp configurations use a loop gain greater than 1 ($A_{ol}B > 1$), the criteria for stability are based on the phase angle of the internal lag networks. As previously mentioned, operational amplifiers are composed of multiple stages, each of which has a critical frequency. For purposes of illustrating the concept of **stability,** we will use a three-stage op-amp with an open-loop response as shown in the Bode plot of Figure 13–11. Notice that there are three different critical frequencies, which indicates three internal RC lag networks. At the first critical frequency, f_{c1}, the gain begins rolling off at −20 dB/decade; when the second critical frequency, f_{c2}, is reached, the gain decreases at −40 dB/decade; and when the third critical frequency, f_{c3}, is reached, the gain drops at −60 dB/decade.

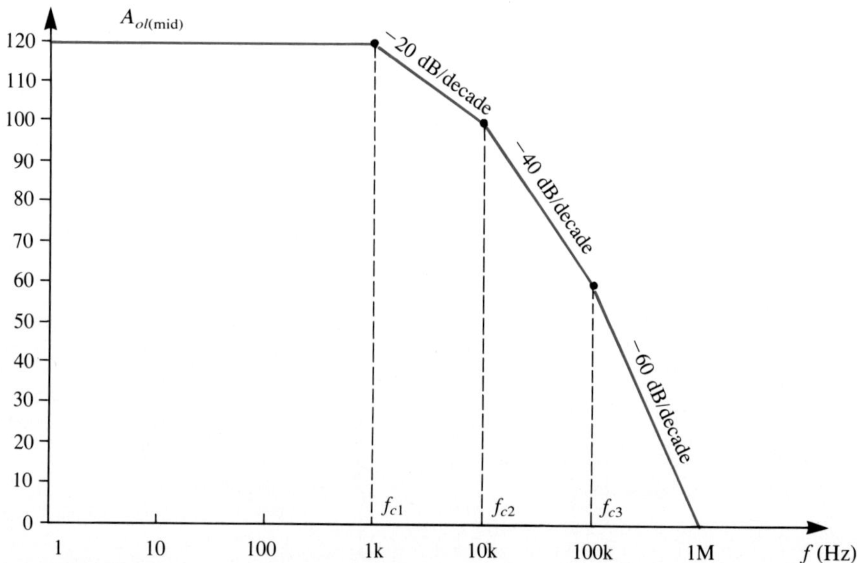

FIGURE 13–11
Bode plot of example of three-stage op-amp response.

To analyze a closed-loop amplifier for stability, the phase margin must be determined. A positive phase margin will indicate that the amplifier is stable for a given value of closed-loop gain. Three example cases will be considered in order to demonstrate the conditions for instability.

CASE 1 The closed-loop gain intersects the open-loop response on the −20 dB/decade slope, as shown in Figure 13–12. For this example, the midrange closed-loop gain is 106 dB, and the closed-loop critical frequency is 5 kHz. If we assume that the amplifier is not operated out of its midrange, the maximum phase shift for the 106 dB amplifier

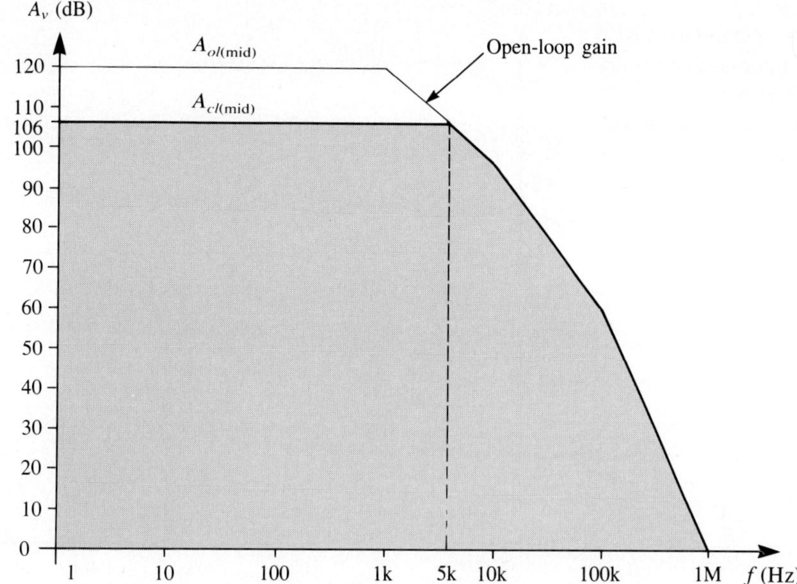

FIGURE 13–12 Case where closed-loop gain intersects open-loop gain on −20 dB/decade slope (stable operation).

occurs at the highest midrange frequency (in this case, 5 kHz). The total phase shift at this frequency due to the three lag networks is calculated as follows:

$$\phi_{tot} = -\tan^{-1}\left(\frac{f}{f_{c1}}\right) - \tan^{-1}\left(\frac{f}{f_{c2}}\right) - \tan^{-1}\left(\frac{f}{f_{c3}}\right)$$

where

$$\begin{aligned} f &= 5 \text{ kHz} \\ f_{c1} &= 1 \text{ kHz} \\ f_{c2} &= 10 \text{ kHz} \\ f_{c3} &= 100 \text{ kHz} \end{aligned}$$

Therefore,

$$\begin{aligned} \phi_{tot} &= -\tan^{-1}\left(\frac{5 \text{ kHz}}{1 \text{ kHz}}\right) - \tan^{-1}\left(\frac{5 \text{ kHz}}{10 \text{ kHz}}\right) - \tan^{-1}\left(\frac{5 \text{ kHz}}{100 \text{ kHz}}\right) \\ &= -78.69^\circ - 26.57^\circ - 2.86^\circ \\ &= -108.12^\circ \end{aligned}$$

The phase margin is

$$\theta_{pm} = 180^\circ - |\phi_{tot}| = 180^\circ - 108.12^\circ = +71.88^\circ$$

θ_{pm} is positive, so the amplifier is stable for all frequencies in its midrange. In general, an amplifier is stable for all midrange frequencies if its closed-loop gain intersects the open-loop response curve on a −20 dB/decade slope.

CASE 2 The closed-loop gain is lowered to where it intersects the open-loop response on the −40 dB/decade slope, as shown in Figure 13–13. The midrange closed-loop gain in

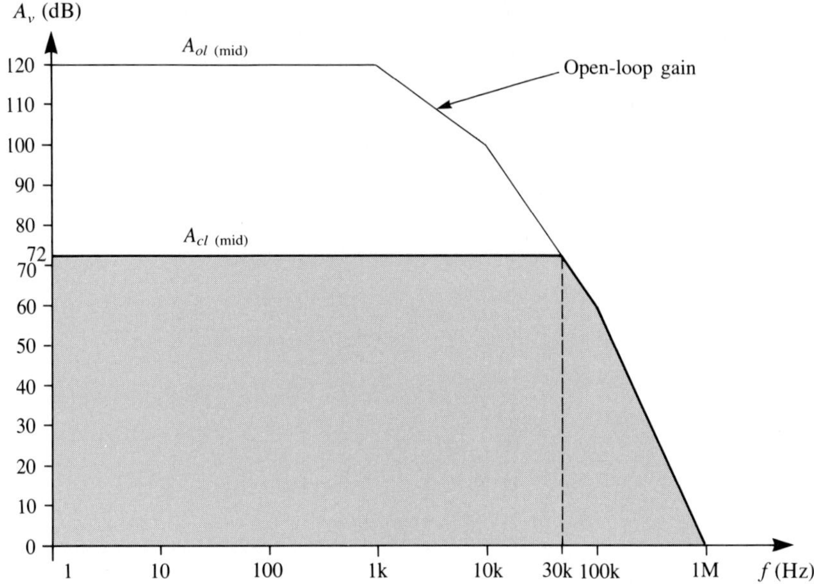

FIGURE 13–13
Case where closed-loop gain intersects open-loop gain on −40 dB/decade slope (marginally stable operation).

this case is 72 dB, and the closed-loop critical frequency is approximately 30 kHz. The total phase shift at f = 30 kHz due to the three lag networks is calculated as follows.

$$\begin{aligned}\phi_{tot} &= -\tan^{-1}\left(\frac{30\text{ kHz}}{1\text{ kHz}}\right) - \tan^{-1}\left(\frac{30\text{ kHz}}{10\text{ kHz}}\right) - \tan^{-1}\left(\frac{30\text{ kHz}}{100\text{ kHz}}\right)\\ &= -88.09^\circ - 71.57^\circ - 16.7^\circ\\ &= -176.36^\circ\end{aligned}$$

The phase margin is

$$\theta_{pm} = 180^\circ - 176.36^\circ = +3.64^\circ$$

The phase margin is positive, so the amplifier is still stable for all frequencies in its midrange, but a very slight increase in frequency above f_c would cause it to oscillate. Therefore, it is marginally stable and very close to instability because instability occurs where $\theta_{pm} = 0^\circ$ and, as a general rule, a minimum 45° phase margin is recommended to avoid marginal conditions.

CASE 3 The closed-loop gain is further decreased until it intersects the open-loop response on the −60 dB/decade slope, as shown in Figure 13–14. The midrange closed-loop gain in this case is 18 dB, and the closed-loop critical frequency is 500 kHz. The total phase shift at f = 500 kHz due to the three lag networks is

$$\begin{aligned}\phi_{tot} &= -\tan^{-1}\left(\frac{500\text{ kHz}}{1\text{ kHz}}\right) - \tan^{-1}\left(\frac{500\text{ kHz}}{10\text{ kHz}}\right) - \tan^{-1}\left(\frac{500\text{ kHz}}{100\text{ kHz}}\right)\\ &= -89.89^\circ - 88.85^\circ - 78.69^\circ = -257.43^\circ\end{aligned}$$

The phase margin is

$$\theta_{pm} = 180^\circ - 257.43^\circ = -77.43^\circ$$

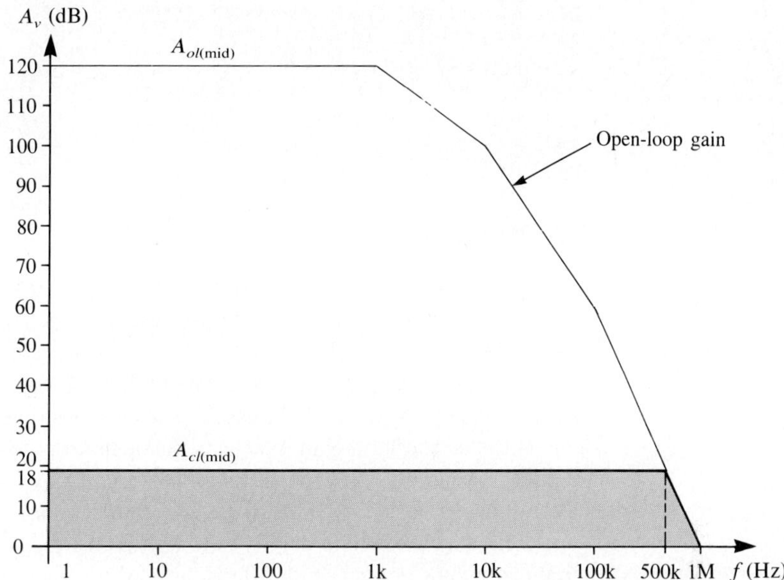

FIGURE 13–14
Case where closed-loop gain intersects open-loop gain on −60 dB/decade slope (unstable operation).

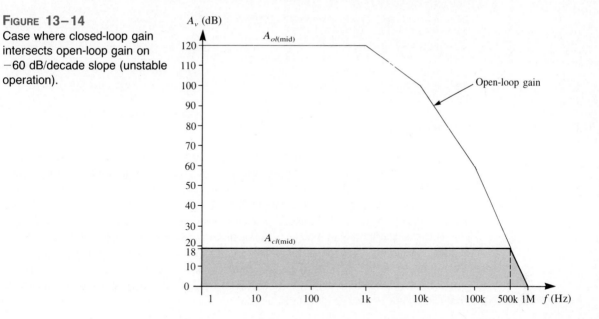

Here the phase margin is negative and the amplifier is unstable at the upper end of its midrange. Using the following BASIC program, you can determine the approximate midrange frequency at which instability occurs. The program computes the phase margin for each of a specified number of frequency values and determines whether the op-amp is stable or unstable. The display shows a list of frequencies with corresponding phase shifts, phase margins, and stability conditions. Instability occurs between the last two frequencies printed. Inputs required for this program are all of the critical frequencies in the open-loop response of the given op-amp, the highest frequency of operation, and the increments of frequency for which the parameters are to be computed.

```
10  CLS
20  PRINT"THIS PROGRAM COMPUTES THE PHASE MARGINS FOR EACH"
30  PRINT"FREQUENCY IN A SPECIFIED RANGE AND DETERMINES THE"
40  PRINT"MAXIMUM FREQUENCY FOR STABLE OPERATION OF THE
    OP-AMP"
50  PRINT:PRINT:PRINT
60  INPUT "TO CONTINUE PRESS 'ENTER'";X
70  CLS
80  INPUT "NUMBER OF OPEN-LOOP CRITICAL FREQUENCIES";N
90  CLS
100 FOR Y=1 TO N
110 INPUT "CRITICAL FREQUENCY IN HERTZ";FC(Y)
120 NEXT
130 INPUT "THE HIGHEST INPUT FREQUENCY";FH
140 INPUT "INCREMENTS OF FREQUENCY FROM ZERO TO THE
    HIGHEST";FI
150 CLS
160 PRINT"FREQUENCY","PHASE SHIFT","PHASE MARGIN","STABILITY"
170 FOR F=0 TO FH STEP FI
180 PH=0
190 FOR Y=1 TO N
```

```
200 PH=PH-ATN(F/FC(Y))*57.29578
210 NEXT Y
220 PM=180+PH
230 IF PM<=0 THEN S$="UNSTABLE" ELSE S$="STABLE"
240 PRINT F,PH,PM,S$
250 IF PM<=0 THEN END
260 NEXT F
```

SUMMARY OF STABILITY CRITERIA

This analysis has demonstrated that an amplifier's closed-loop gain must intersect the open-loop gain curve on a −20 dB/decade slope to ensure stability for all of its mid-range frequencies. If the closed-loop gain is lowered to a value that intersects on a −40 dB/decade slope, then marginal stability or complete instability can occur. In the previous analyses (Cases 1, 2, and 3), the closed-loop gain should be greater than 72 dB.

If the closed-loop gain intersects the open-loop response on a −60 dB/decade slope, definite instability will occur at some frequency within the amplifier's midrange. Therefore, to ensure stability for all of the midrange frequencies, an op-amp must be operated at a closed-loop gain such that the roll-off rate beginning at its dominant critical frequency does not exceed −20 dB/decade.

13–4 REVIEW QUESTIONS

1. Under what feedback condition can an amplifier oscillate?
2. How much can the phase shift of an amplifier's internal *RC* network be before instability occurs? What is the phase margin at the point where instability begins?
3. What is the maximum roll-off rate of the open-loop gain of an op-amp for which the device will still be stable?

13–5 OP-AMP COMPENSATION

The last section demonstrated that instability can occur when an op-amp's response has roll-off rates exceeding −20 dB/decade and the op-amp is operated in a closed-loop configuration having a gain curve that intersects a higher roll-off rate portion of the open-loop response. In situations like those examined in the last section, the closed-loop voltage gain is restricted to very high values. In many applications, lower values of closed-loop gain are necessary or desirable. To allow op-amps to be operated at low closed-loop gain, phase lag compensation is required.

PHASE LAG COMPENSATION

As you have seen, the cause of instability is excessive phase shift through an op-amp's internal lag networks. When these phase shifts equal or exceed 180°, the amplifier can oscillate. **Compensation** is used to either eliminate open-loop roll-off rates greater than −20 dB/decade or extend the −20 dB/decade rate to a lower gain. These concepts are illustrated in Figure 13–15.

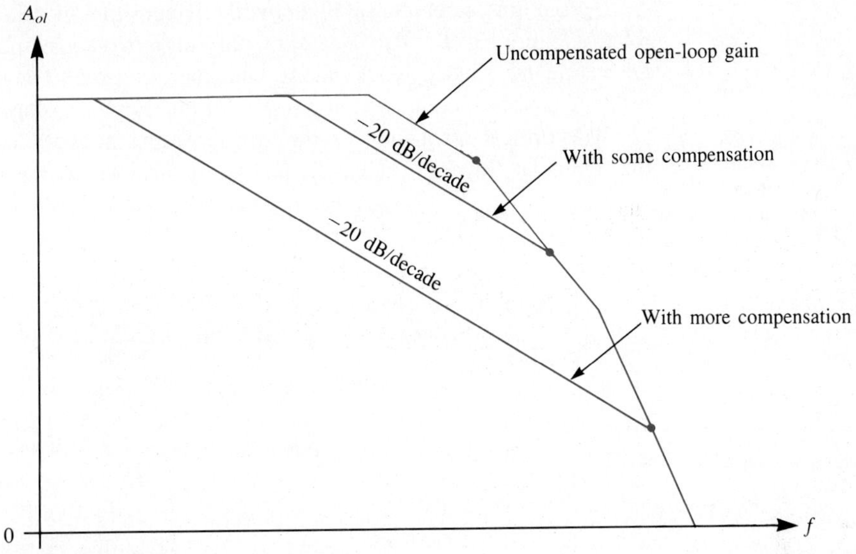

FIGURE 13–15
Bode plot illustrating effect of phase compensation on open-loop gain of typical op-amp.

COMPENSATING NETWORK

There are two basic methods of compensation for integrated circuit op-amps: internal and external. In either case an *RC* network is added. The basic compensating action is as follows. Consider first the *RC* network shown in Figure 13–16(a). At low frequencies where X_{C_c} is extremely large, the output voltage approximately equals the input voltage. When the frequency reaches its critical value, $f_c = 1/[2\pi(R_1 + R_2)C_c]$, the output voltage decreases at −20 dB/decade. This roll-off rate continues until $X_{C_c} \cong 0$, at which point the output voltage levels off to a value determined by R_1 and R_2, as indicated in Figure 13–16(b). This is the principle used in the phase compensation of an op-amp.

FIGURE 13–16
Basic compensating network action.

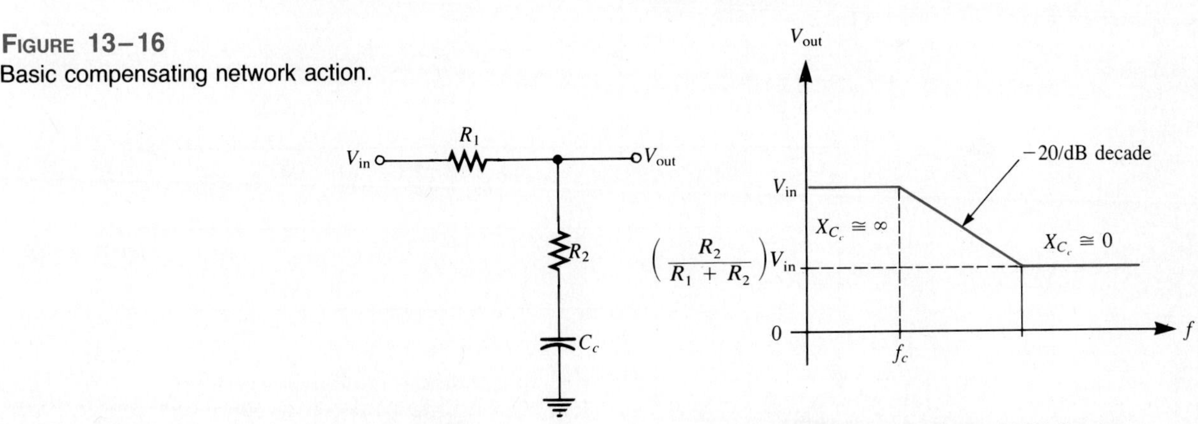

To see how a compensating network changes the open-loop response of an op-amp, refer to Figure 13–17. This diagram represents a two-stage op-amp. The individual stages are within the shaded blocks along with the associated lag networks. A compensating network is shown connected at point A on the output of stage 1.

The critical frequency of the compensating network is set to a value less than the dominant (lowest) critical frequency of the internal lag networks. This causes the

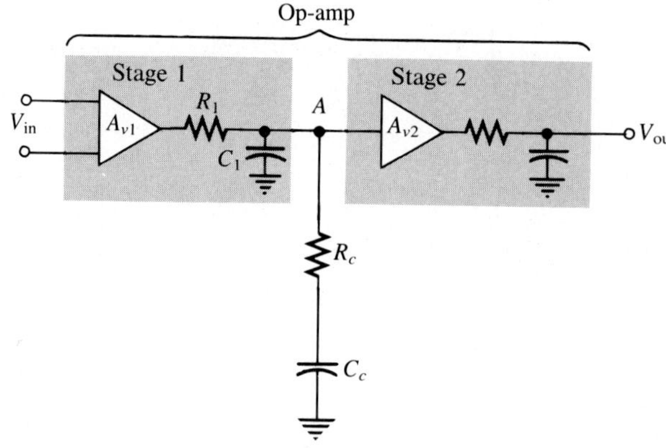

FIGURE 13–17 Representation of op-amp with compensation.

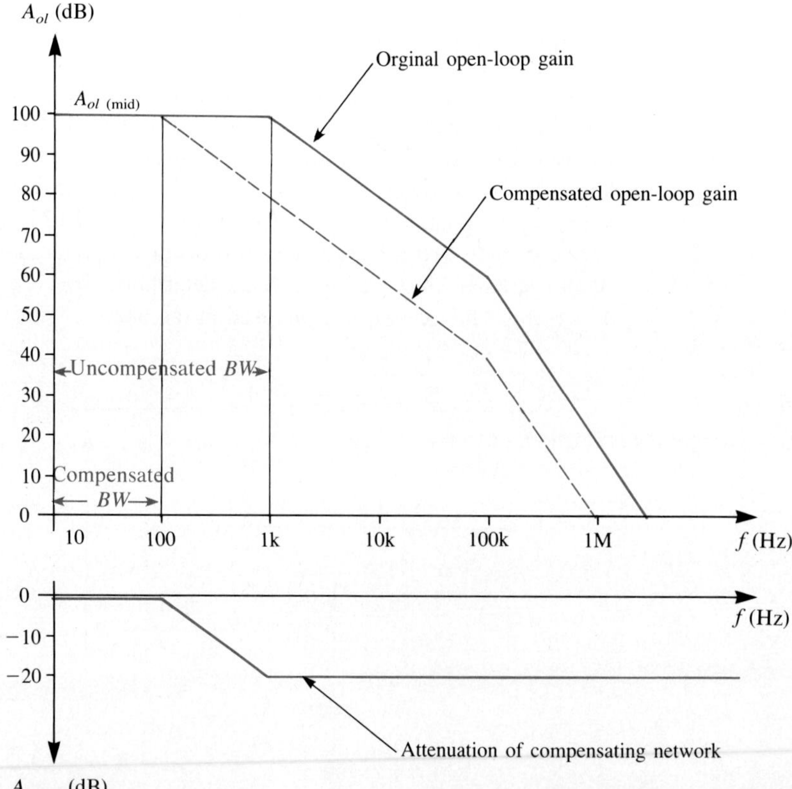

FIGURE 13–18 Example of compensated op-amp frequency response.

−20 dB/decade roll-off to begin at the compensating network's critical frequency. The roll-off of the compensating network continues up to the critical frequency of the dominant lag network. At this point, the response of the compensating network levels off, and the −20 dB/decade roll-off of the dominant lag network takes over. The net result is a shift of the open-loop response to the left, thus reducing the bandwidth, as shown in Figure 13–18. The response curve of the compensating network is shown in proper relation to the overall open-loop response.

EXAMPLE 13–6

A certain op-amp has the open-loop response in Figure 13–19. As you can see, the lowest closed-loop gain for which stability is assured is approximately 40 dB (where the closed-loop gain line still intersects the −20 dB/decade slope). In a particular application, a 20 dB closed-loop gain is required.

(a) Determine the critical frequency for the compensating network.
(b) Sketch the ideal response curve for the compensating network.
(c) Sketch the total ideal compensated open-loop response.

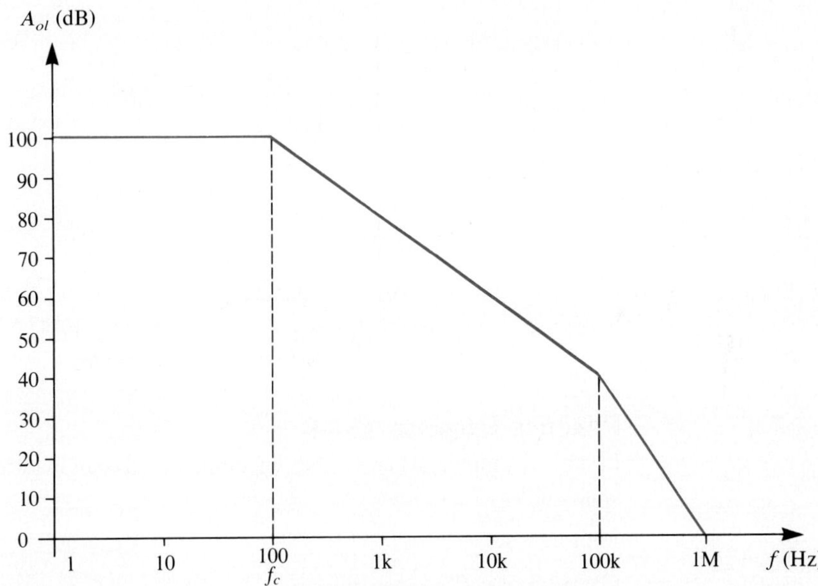

FIGURE 13–19
Original open-loop response.

SOLUTION

(a) The gain must be dropped so that the −20 dB/decade roll-off extends down to 20 dB rather than to 40 dB. To achieve this, the midrange open-loop gain must be made to roll off a decade sooner. Therefore, the critical frequency of the compensating network must be 10 Hz.

(b) The roll-off of the compensating network must end at 100 Hz, as shown in Figure 13–20(a).

(c) The total open-loop response resulting from compensation is shown in Figure 13–20(b).

FIGURE 13–20

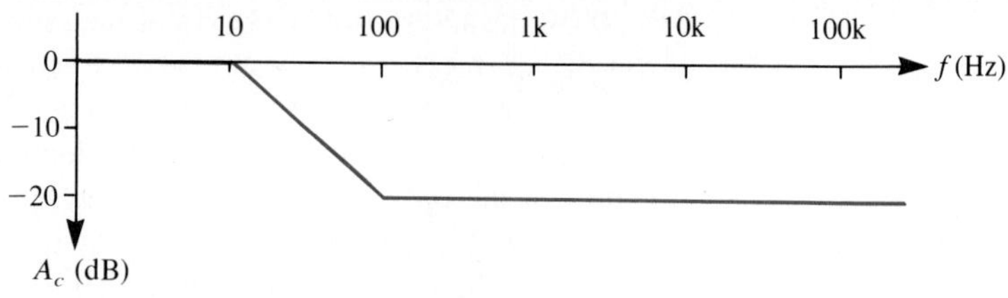

(a) Compensating network response

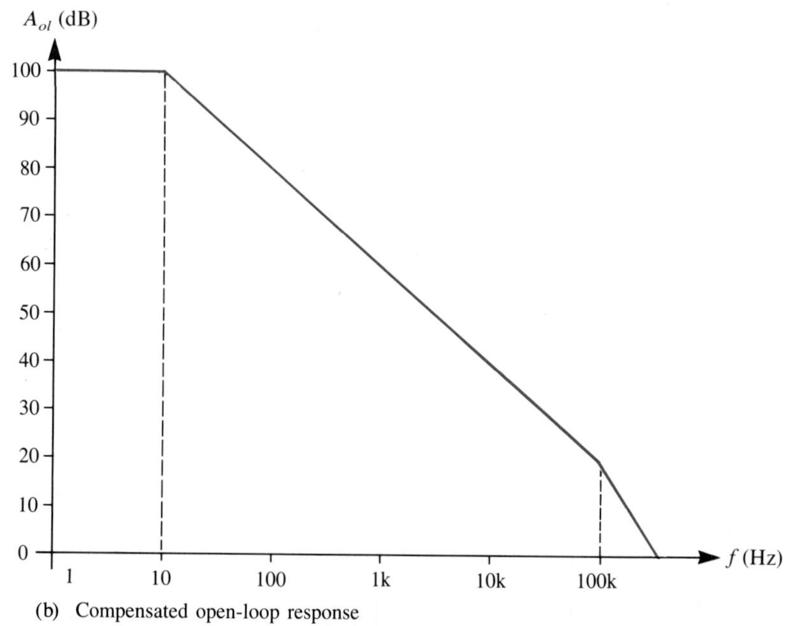

(b) Compensated open-loop response

PRACTICE EXERCISE 13–6

In this example, what is the uncompensated bandwidth? What is the compensated bandwidth?

EXTENT OF COMPENSATION

A larger compensating capacitor will cause the open-loop roll-off to begin at a lower frequency and thus extend the −20 dB/decade roll-off to lower gain levels, as shown in Figure 13–21(a). With a sufficiently large compensating capacitor, an op-amp can be made unconditionally stable, as illustrated in Figure 13–21(b), where the −20 dB/decade slope is extended all the way down to unity gain. This is normally the case when internal compensation is provided by the manufacturer. An internally, fully compensated op-amp can be used for any value of closed-loop gain and remain stable. The 741 is an example of an internally compensated device.

A disadvantage of fully compensated op-amps is that bandwidth is sacrificed; thus the slew rate is decreased. Therefore, many IC op-amps have provisions for external compensation. Figure 13–22 shows typical package layouts of an LM101A op-amp with

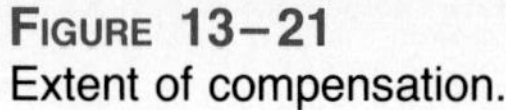
Figure 13–21
Extent of compensation.

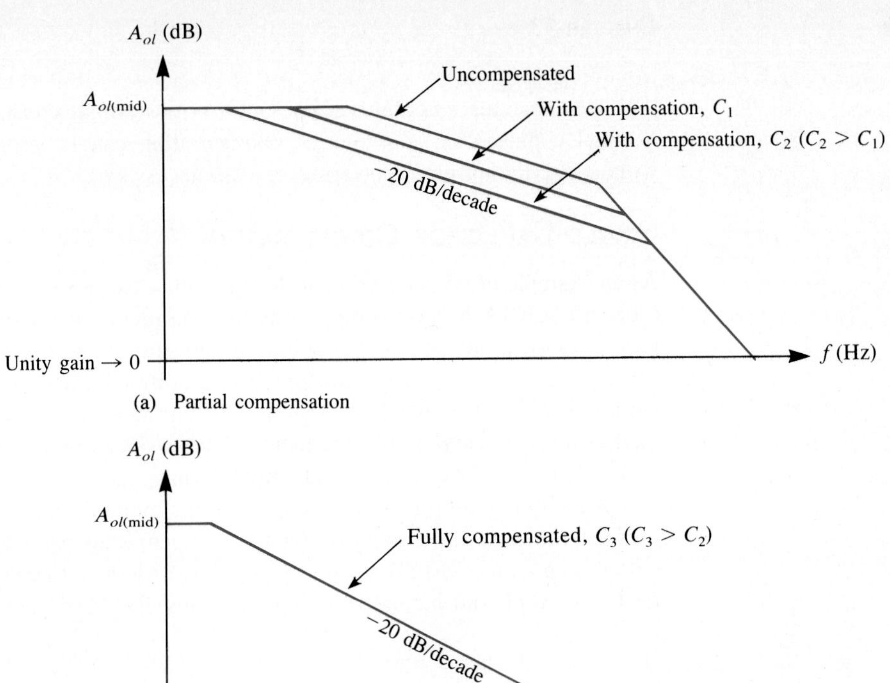

A_{ol} (dB)
$A_{ol(\text{mid})}$
Uncompensated
With compensation, C_1
With compensation, C_2 ($C_2 > C_1$)
−20 dB/decade
Unity gain → 0
f (Hz)
(a) Partial compensation
A_{ol} (dB)
$A_{ol(\text{mid})}$
Fully compensated, C_3 ($C_3 > C_2$)
−20 dB/decade
0
f (Hz)
(b) Full compensation

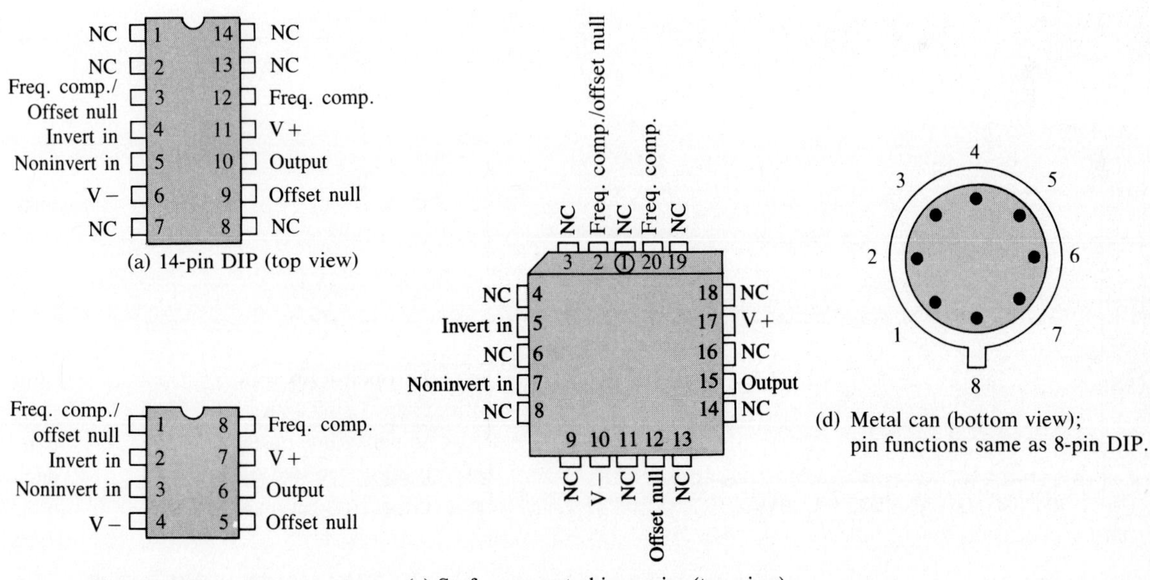

NC 1
NC 2
Freq. comp./ Offset null 3
Invert in 4
Noninvert in 5
V− 6
NC 7
14 NC
13 NC
12 Freq. comp.
11 V+
10 Output
9 Offset null
8 NC
(a) 14-pin DIP (top view)
Freq. comp./ offset null 1
Invert in 2
Noninvert in 3
V− 4
8 Freq. comp.
7 V+
6 Output
5 Offset null
(b) 8-pin DIP or SOP (top view)
NC
Freq. comp./offset null
NC
Freq. comp.
NC
3 2 1 20 19
NC 4
Invert in 5
NC 6
Noninvert in 7
NC 8
18 NC
17 V+
16 NC
15 Output
14 NC
9 10 11 12 13
NC
V−
NC
Offset null
NC
(c) Surface-mount chip carrier (top view)
1 2 3 4 5 6 7 8
(d) Metal can (bottom view); pin functions same as 8-pin DIP.

Figure 13–22
Typical op-amp packages.

pins available for external compensation with a small capacitor. With provisions for external connections, just enough compensation can be used for a given application without sacrificing more performance than necessary.

SINGLE-CAPACITOR COMPENSATION

As an example of compensating an IC op-amp, a capacitor C_1 is connected to pins 1 and 8 of an LM101A in an inverting amplifier configuration, as shown in Figure 13–23(a). Part (b) of the figure shows the open-loop frequency response curves for two values of C_1. The 3 pF compensating capacitor produces a unity-gain bandwidth approaching 10 MHz. Notice that the −20 dB/decade slope extends to a very low gain value. When C_1 is increased ten times to 30 pF, the bandwidth is reduced by a factor of ten. Notice that the −20 dB/decade slope now extends through unity gain.

When the op-amp is used in a closed-loop configuration, as in Figure 13–23(c), the useful frequency range depends on the compensating capacitor. For example, with a closed-loop gain of 40 dB as shown in part (c), the bandwidth is approximately 10 kHz for C_1 = 30 pF and increases to approximately 100 kHz when C_1 is decreased to 3 pF.

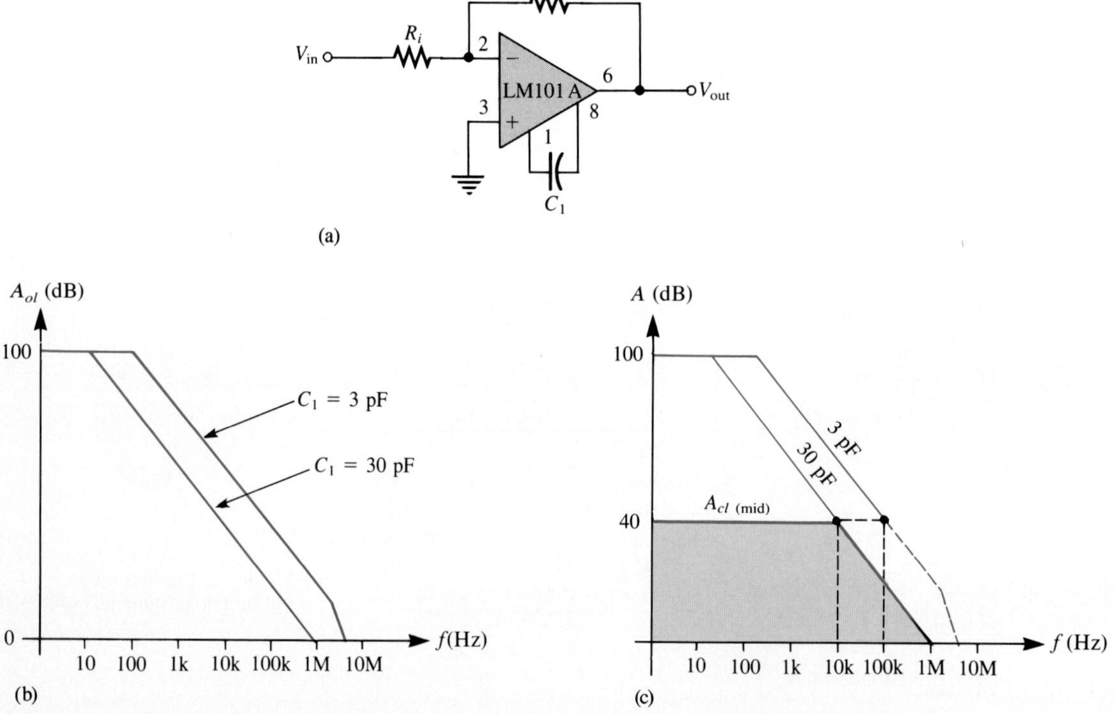

FIGURE 13–23
Example of single-capacitor compensation of an LM101A op-amp.

FEEDFORWARD COMPENSATION

Another method of phase compensation is called **feedforward.** This type of compensation results in less bandwidth reduction than the method previously discussed. The basic concept is to bypass the internal input stage of the op-amp at high frequencies and drive the higher-frequency second stage, as shown in Figure 13–24.

Feedforward compensation of an LM101A is shown in Figure 13–25(a). The feedforward capacitor C_1 is connected from the inverting input to the compensating terminal. A small capacitor is needed across R_f to ensure stability. The Bode plot in Figure 13–25(b) shows the feedforward compensated response and the standard compensated response that was discussed previously. The use of feedforward compensation is restricted to the inverting amplifier configuration. Other compensation methods are also used. Often, recommendations are provided by the manufacturer on the data sheet.

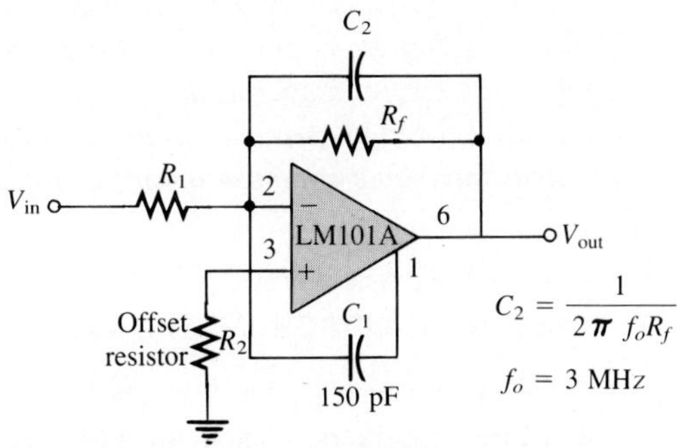

(a) Manufacturers' recommended configuration

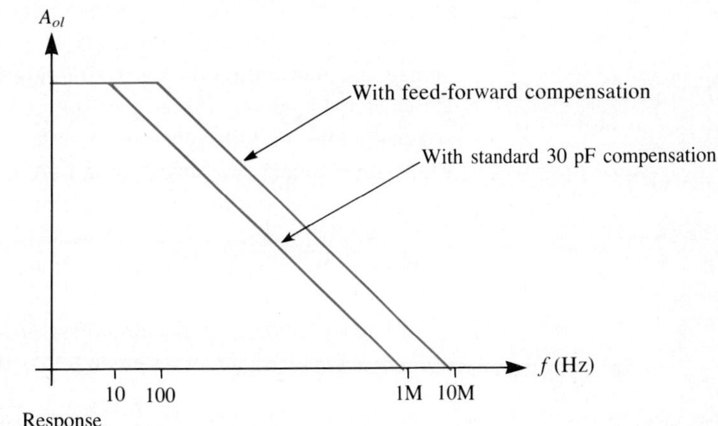

(b) Response

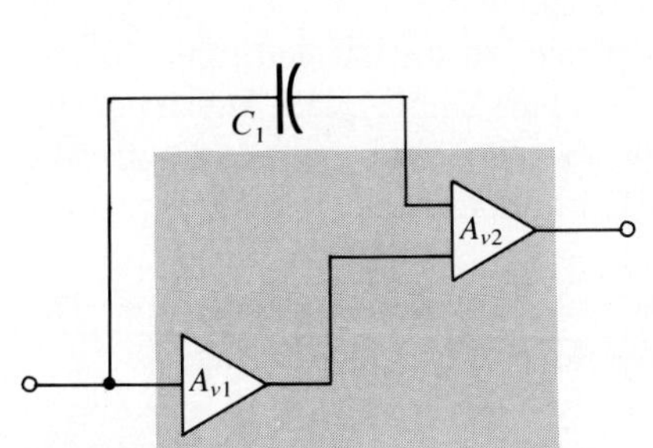

FIGURE 13–24
Feedforward compensation showing high-frequency bypassing of first stage.

FIGURE 13–25
Feedforward compensation of an LM101A op-amp and the response curves.

13–5 Review Questions

1. What is the purpose of phase compensation?
2. What is the main difference between internal and external compensation?
3. When you compensate an amplifier, does the bandwidth increase or decrease?

13–6 A System Application

In this system application, we are focusing on the audio amplifier boards in the FM stereo receiver presented at the beginning of the chapter. Both boards are identical except one is for the left channel sound and the other for the right channel sound. This circuit is a good example of a mixed use of an integrated circuit and discrete components. In this section, you will

- ☐ ***See how an op-amp is used as an audio amplifier.***
- ☐ ***Identify the functions of various components on the board.***
- ☐ ***Analyze the circuit's operation.***
- ☐ ***Translate between a printed circuit board and a schematic.***
- ☐ ***Troubleshoot some common amplifier failures.***

A Brief Description of the System

Some general information about the stereo system might be helpful before you concentrate on the audio amplifiers. When an FM stereo broadcast is received by a standard single-speaker system, the output to the speaker is equal to the sum of the left plus the right channel audio, so you get the original sound without separation. When a stereo receiver is used, the full stereo effect is reproduced by the two speakers. Stereo FM signals are transmitted on a carrier frequency of 88 MHz to 108 MHz. The complete stereo signal consists of three modulating signals. These are the sum of the left and right channel audio, the difference of the left and right channel audio, and a pilot subcarrier. These three signals are detected and are used to separate out the left and right channel audio by special circuits. The channel audio amplifiers then amplify each signal equally and drive the speakers. It is not necessary for you to understand this process for the purposes of this system application, although you may be interested in doing further study in this area on your own.

The two channel audio amplifiers are identical, so we will look at only one. The op-amp serves basically as a preamplifier that drives the power amplifier stage.

Now, so that you can take a closer look at one of the audio amplifier boards, let's take one out of the system and put it on the test bench.

ON THE TEST BENCH

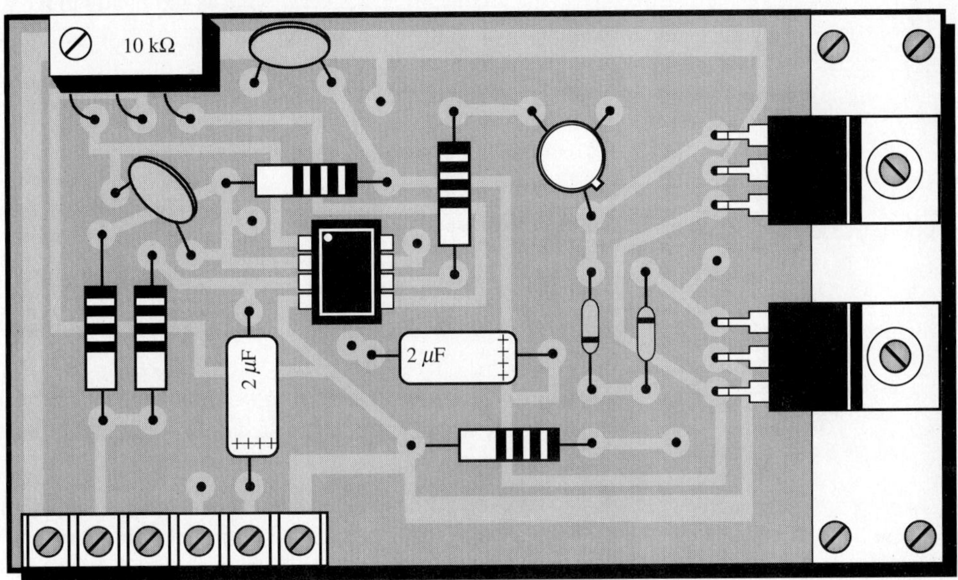

FIGURE 13–26

■ ACTIVITY 1 RELATE THE PC BOARD TO THE SCHEMATIC

The schematic for the audio amplifier board in Figure 13–26 is shown in Figure 13–27. Using this schematic, locate and label each component on the pc board. The board has several feed-through pads for connections that are on the back side, which you should locate and identify as you compare the board to the schematic.

FIGURE 13–27

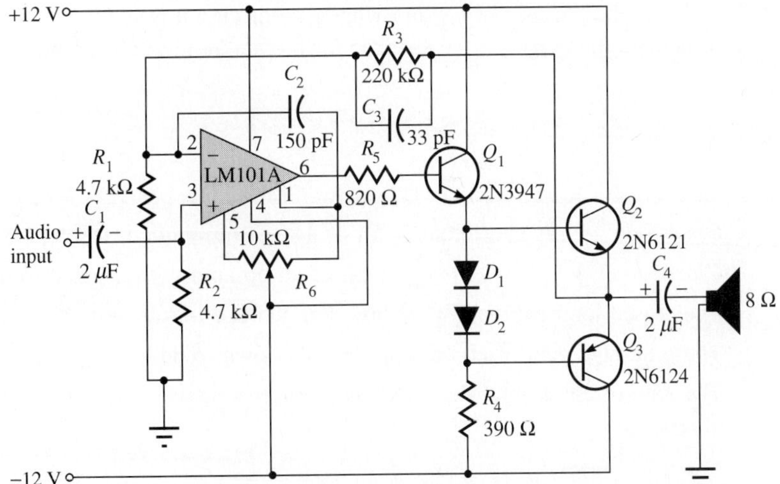

ACTIVITY 2 ANALYZE THE CIRCUIT

STEP 1 Determine the midrange voltage gain.

STEP 2 Determine the lower critical frequency. Given that the upper critical frequency is 15 kHz, what is the bandwidth?

STEP 3 Determine the maximum peak-to-peak input voltage that can be applied without producing a distorted output signal. Assume that the maximum output peaks are 1 V less than the supply voltages.

ACTIVITY 3 WRITE A TECHNICAL REPORT

Describe the overall operation of the circuit and the function of each component. In discussing the general operation and basic purpose of each component, make sure you identify the negative feedback loop, the type of op-amp configuration, which components determine the voltage gain, which components set the lower critical frequency, and the purpose of each of the capacitors. Use the results of Activity 2 when appropriate.

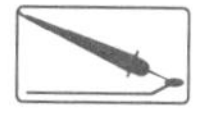

ACTIVITY 4 TROUBLESHOOT THE AUDIO SECTION BOARDS FOR EACH OF THE FOLLOWING PROBLEMS BY STATING THE PROBABLE CAUSE OR CAUSES IN EACH CASE

1. No final output signal when there is a verified input signal.
2. The positive half-cycle of the output voltage is severely distorted or missing.
3. Output severely clipped on both positive and negative cycles.

ACTIVITY 5 TEST BENCH SPECIAL ASSIGNMENT

Go to Test Bench 6 in the color insert section (which follows page 422) and carry out the assignment that is stated there.

13–6 REVIEW QUESTIONS

1. How can the lower critical frequency of the amplifier be reduced?
2. Which transistors form the class-B power amplifier?
3. What is the purpose of Q_1 and what type of circuit is it?
4. Calculate the power to the speaker for the maximum voltage output from Activity 2.

SUMMARY

- *Open-loop gain* is the voltage gain of an op-amp without feedback.
- *Closed-loop gain* is the voltage gain of an op-amp with negative feedback.
- The closed-loop gain is always less than the open-loop gain.
- The midrange gain of an op-amp extends down to dc.
- The gain of an op-amp decreases as frequency increases above the critical frequency.

- □ The bandwidth of an op-amp equals the upper critical frequency.
- □ The internal *RC* lag networks that are inherently part of the amplifier stages cause the gain to roll off as frequency goes up.
- □ The internal *RC* lag networks also cause a phase shift between input and output signals.
- □ Negative feedback lowers the gain and increases the bandwidth.
- □ The product of gain and bandwidth is constant for a given op-amp.
- □ The gain-bandwidth product equals the frequency at which unity voltage gain occurs.
- □ Positive feedback occurs when the total phase shift through the op-amp (including 180° inversion) and feedback network is 0° (equivalent to 360°) or more.
- □ The phase margin is the amount of additional phase shift required to make the total phase shift around the loop 360°.
- □ When the closed-loop gain of an op-amp intersects the open-loop response curve on a −20 dB/decade (−6 dB/octave) slope, the amplifier is stable.
- □ When the closed-loop gain intersects the open-loop response curve on a slope greater than −20 dB/decade, the amplifier can be either marginally stable or unstable.
- □ A minimum phase margin of 45° is recommended to provide a sufficient safety factor for stable operation.
- □ A fully compensated op-amp has a −20 dB/decade roll-off all the way down to unity gain.
- □ Compensation reduces bandwidth and increases slew rate.
- □ Internally compensated op-amps such as the 741 are available. These are usually fully compensated with a large sacrifice in bandwidth.
- □ Externally compensated op-amps such as LM101A are available. External compensating networks can be connected to specified pins, and the compensation can be tailored to a specific application. In this way, bandwidth and slew rate are not degraded more than necessary.

GLOSSARY

Closed-loop gain The overall voltage gain with external feedback.

Compensation The process of modifying the roll-off rate of an amplifier to ensure stability.

Feedforward A method of frequency compensation in op-amp circuits.

Open-loop gain The voltage gain of an op-amp without feedback.

Phase The relative angular displacement of a time-varying function relative to a reference.

Phase margin The difference between the total phase shift through an amplifier and 180°. The additional amount of phase shift that can be allowed before instability occurs.

Stability A condition in which an amplifier circuit does not oscillate.

FORMULAS

(13–1)	$BW = f_{c(\text{high})} - f_{c(\text{low})}$	General bandwidth
(13–2)	$BW = f_{c(\text{high})}$	Op-amp bandwidth
(13–3)	$\dfrac{V_{\text{out}}}{V_{\text{in}}} = \dfrac{X_C}{\sqrt{R^2 + X_C^2}}$	*RC* attenuation, lag network
(13–4)	$\dfrac{V_{\text{out}}}{V_{\text{in}}} = \dfrac{1}{\sqrt{1 + R^2/X_C^2}}$	*RC* attenuation, lag network
(13–5)	$\dfrac{f_c}{f} = \dfrac{X_C}{R}$	Ratio of frequencies equals ratio of reactances
(13–6)	$\dfrac{V_{\text{out}}}{V_{\text{in}}} = \dfrac{1}{\sqrt{1 + f^2/f_c^2}}$	*RC* attenuation
(13–7)	$A_{ol} = \dfrac{A_{ol(\text{mid})}}{\sqrt{1 + f^2/f_c^2}}$	Open-loop gain
(13–8)	$\phi = -\tan^{-1}\left(\dfrac{R}{X_C}\right)$	*RC* phase shift
(13–9)	$\phi = -\tan^{-1}\left(\dfrac{f}{f_c}\right)$	*RC* phase shift
(13–10)	$\phi_{tot} = -\tan^{-1}\left(\dfrac{f}{f_{c1}}\right) - \tan^{-1}\left(\dfrac{f}{f_{c2}}\right) - \tan^{-1}\left(\dfrac{f}{f_{c3}}\right)$	Total phase shift
(13–11)	$f_{c(cl)} = f_{c(ol)}(1 + BA_{ol(\text{mid})})$	Closed-loop critical frequency
(13–12)	$BW_{cl} = BW_{ol}(1 + BA_{ol(\text{mid})})$	Closed-loop bandwidth
(13–13)	$A_{cl}f_{c(cl)} = A_{ol}f_{c(ol)}$	Gain-bandwidth product
(13–14)	$A_{cl}f_{c(cl)} = \text{unity-gain bandwidth}$	
(13–15)	$\text{Loop gain} = A_{ol}B$	
(13–16)	$\theta_{pm} = 180° - \lvert\phi_{tot}\rvert$	Phase margin

SELF-TEST

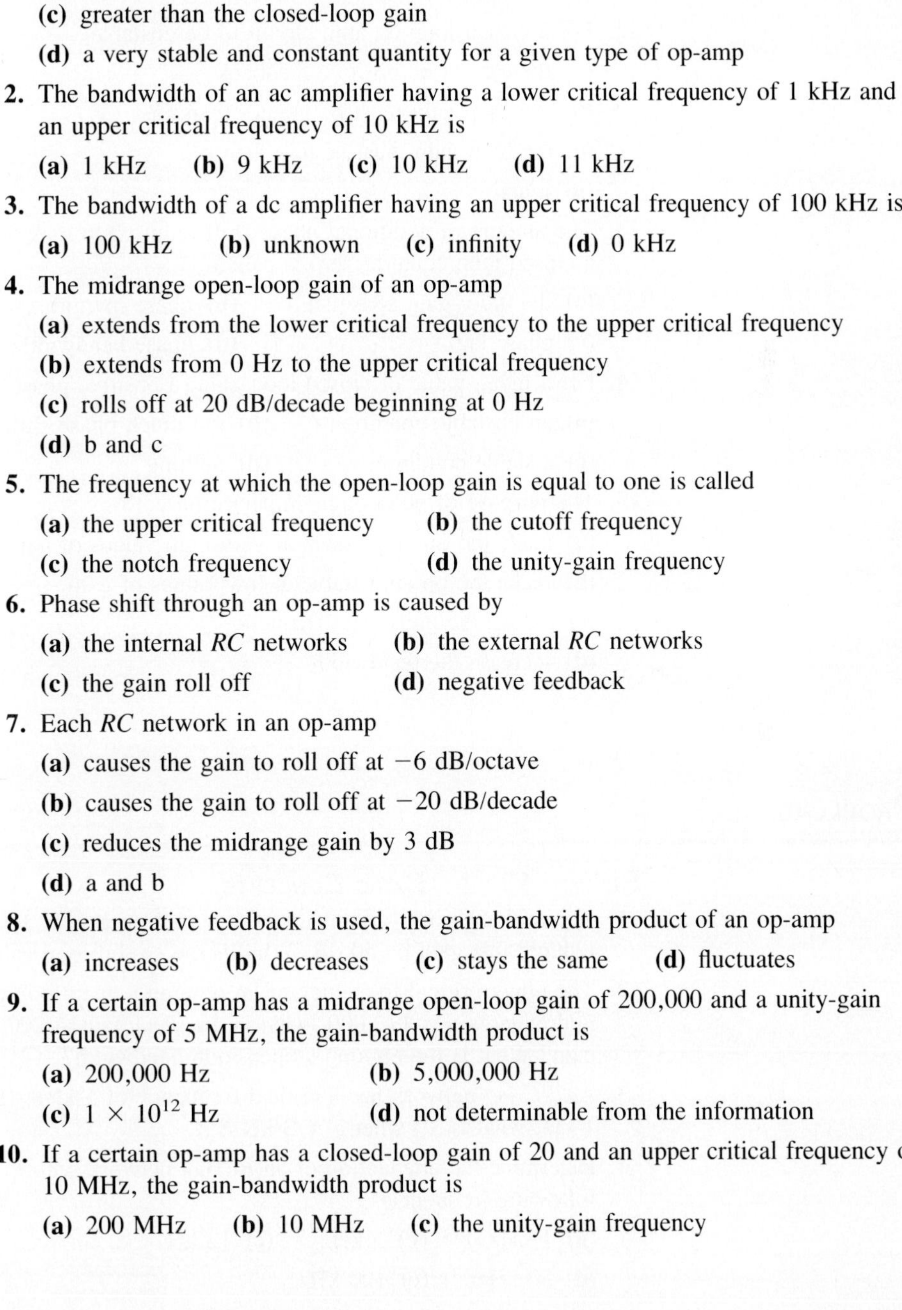

1. The open-loop gain of an op-amp is always
   (a) less than the closed-loop gain
   (b) equal to the closed-loop gain
   (c) greater than the closed-loop gain
   (d) a very stable and constant quantity for a given type of op-amp
2. The bandwidth of an ac amplifier having a lower critical frequency of 1 kHz and an upper critical frequency of 10 kHz is
   (a) 1 kHz **(b)** 9 kHz **(c)** 10 kHz **(d)** 11 kHz
3. The bandwidth of a dc amplifier having an upper critical frequency of 100 kHz is
   (a) 100 kHz **(b)** unknown **(c)** infinity **(d)** 0 kHz
4. The midrange open-loop gain of an op-amp
   (a) extends from the lower critical frequency to the upper critical frequency
   (b) extends from 0 Hz to the upper critical frequency
   (c) rolls off at 20 dB/decade beginning at 0 Hz
   (d) b and c
5. The frequency at which the open-loop gain is equal to one is called
   (a) the upper critical frequency **(b)** the cutoff frequency
   (c) the notch frequency **(d)** the unity-gain frequency
6. Phase shift through an op-amp is caused by
   (a) the internal *RC* networks **(b)** the external *RC* networks
   (c) the gain roll off **(d)** negative feedback
7. Each *RC* network in an op-amp
   (a) causes the gain to roll off at −6 dB/octave
   (b) causes the gain to roll off at −20 dB/decade
   (c) reduces the midrange gain by 3 dB
   (d) a and b
8. When negative feedback is used, the gain-bandwidth product of an op-amp
   (a) increases **(b)** decreases **(c)** stays the same **(d)** fluctuates
9. If a certain op-amp has a midrange open-loop gain of 200,000 and a unity-gain frequency of 5 MHz, the gain-bandwidth product is
   (a) 200,000 Hz **(b)** 5,000,000 Hz
   (c) 1×10^{12} Hz **(d)** not determinable from the information
10. If a certain op-amp has a closed-loop gain of 20 and an upper critical frequency of 10 MHz, the gain-bandwidth product is
    (a) 200 MHz **(b)** 10 MHz **(c)** the unity-gain frequency

11. Positive feedback occurs when
    - **(a)** the output signal is fed back to the input in-phase with the input signal
    - **(b)** the output signal is fed back to the input out-of-phase with the input signal
    - **(c)** the total phase shift through the op-amp and feedback circuit is 360 degrees
    - **(d)** a and c
12. For a closed-loop op-amp circuit to be unstable
    - **(a)** there must be positive feedback
    - **(b)** the loop gain must be greater than one
    - **(c)** the loop gain must be less than one
    - **(d)** a and b
13. The amount of additional phase shift required to make the total phase shift around a closed loop equal to zero is called
    - **(a)** the unity-gain phase shift **(b)** phase margin
    - **(c)** phase lag **(d)** phase bandwidth
14. For a given value of closed-loop gain, a positive phase margin indicates
    - **(a)** an unstable condition **(b)** too much phase shift
    - **(c)** a stable condition **(d)** nothing
15. The purpose of phase-lag compensation is to
    - **(a)** make the op-amp stable at very high values of gain
    - **(b)** make the op-amp stable at low values of gain
    - **(c)** reduce the unity-gain frequency
    - **(d)** increase the bandwidth

PROBLEMS

SECTION 13–1 BASIC CONCEPTS

1. The midrange open-loop gain of a certain op-amp is 120 dB. Negative feedback reduces this gain by 50 dB. What is the closed-loop gain?
2. The upper critical frequency of an op-amp's open-loop response is 200 Hz. If the midrange gain is 175,000, what is the ideal gain at 200 Hz? What is the actual gain? What is the op-amp's open-loop bandwidth?
3. An *RC* lag network has a critical frequency of 5 kHz. If the resistance value is 1 kΩ, what is X_C when $f = 3$ kHz?
4. Determine the attenuation of an *RC* lag network with $f_c = 12$ kHz for each of the following frequencies.
    - **(a)** 1 kHz **(b)** 5 kHz **(c)** 12 kHz
    - **(d)** 20 kHz **(e)** 100 kHz

5. The midrange open-loop gain of a certain op-amp is 80,000. if the open-loop critical frequency is 1 kHz, what is the open-loop gain at each of the following frequencies?

   (a) 100 Hz **(b)** 1 kHz **(c)** 10 kHz **(d)** 1 MHz

6. Determine the phase shift through each network in Figure 13–28 at a frequency of 2 kHz.

FIGURE 13–28

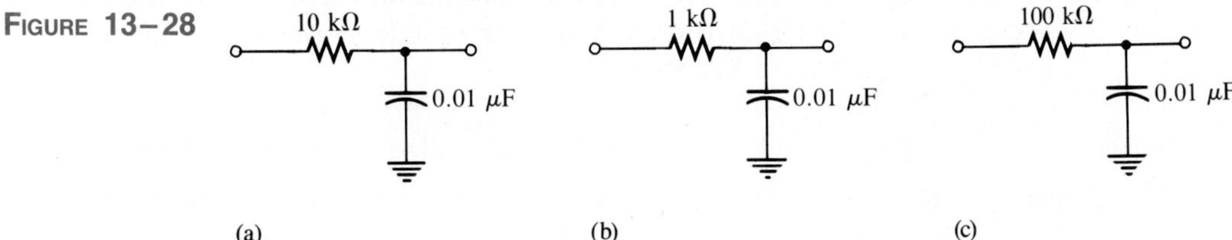

7. An *RC* lag network has a critical frequency of 8.5 kHz. Determine the phase for each frequency and plot a graph of its phase angle versus frequency.

   (a) 100 Hz **(b)** 400 Hz **(c)** 850 Hz

   (d) 8.5 kHz **(e)** 25 kHz **(f)** 85 kHz

SECTION 13–2 OP-AMP OPEN-LOOP RESPONSE

8. A certain op-amp has three internal amplifier stages with midrange gains of 30 dB, 40 dB, and 20 dB. Each stage also has a critical frequency associated with it as follows: f_{c1} = 600 Hz, f_{c2} = 50 kHz, and f_{c3} = 200 kHz.

   (a) What is the midrange open-loop gain of the op-amp, expressed in dB?

   (b) What is the total phase shift through the amplifier, including inversion, when the signal frequency is 10 kHz?

9. What is the gain roll-off rate in Problem 8 between the following frequencies?

   (a) 0 Hz and 600 Hz **(b)** 600 Hz and 50 kHz

   (c) 50 kHz and 200 kHz **(d)** 200 kHz and 1 MHz

SECTION 13–3 OP-AMP CLOSED-LOOP RESPONSE

10. Determine the midrange gain in dB of each amplifier in Figure 13–29. Are these open- or closed-loop gains?

FIGURE 13–29

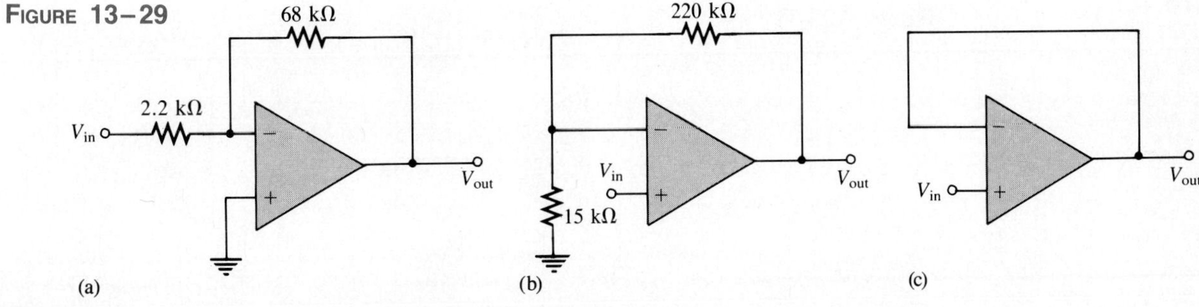

11. A certain amplifier has an open-loop gain in midrange of 180,000 and an open-loop critical frequency of 1500 Hz. If the attenuation of the feedback path is 0.015, what is the closed-loop bandwidth?
12. Given that $f_{c(ol)} = 750$ Hz, $A_{ol} = 89$ dB, and $f_{c(cl)} = 5.5$ kHz, determine the closed-loop gain in dB.
13. What is the unity-gain bandwidth in Problem 12?
14. For each amplifier in Figure 13–30, determine the closed-loop gain and bandwidth. The op-amps in each circuit exhibit an open-loop gain of 125 dB and a unity-gain bandwidth of 2.8 MHz.

FIGURE 13–30

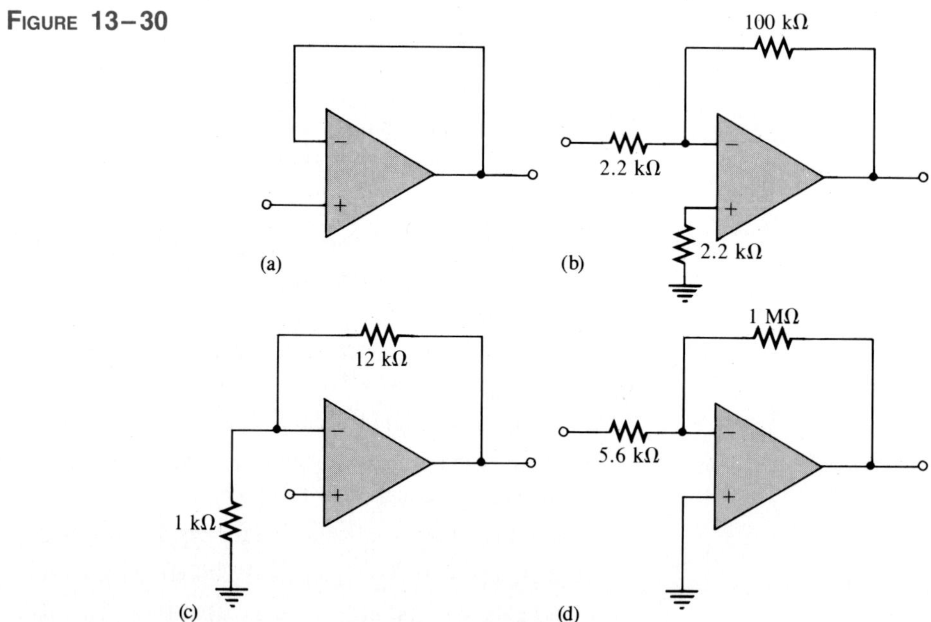

15. Which of the amplifiers in Figure 13–31 has the smaller bandwidth?

FIGURE 13–31

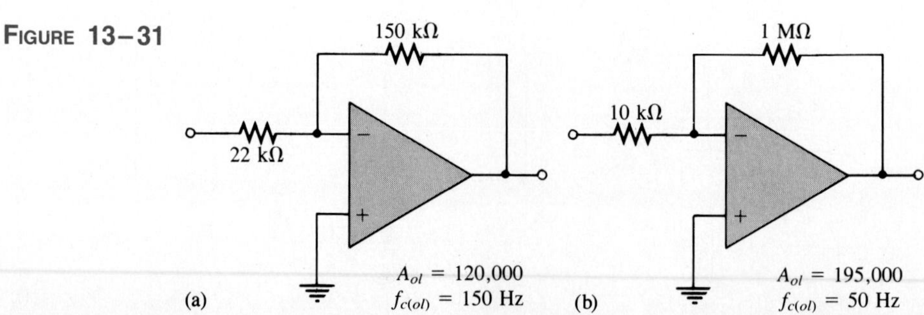

SECTION 13–4 POSITIVE FEEDBACK AND STABILITY

16. It has been determined that the op-amp circuit in Figure 13–32 has three internal critical frequencies as follows: 1.2 kHz, 50 kHz, 250 kHz. If the midrange open-loop gain is 100 dB, is the amplifier configuration stable, marginally stable, or unstable?

FIGURE 13–32

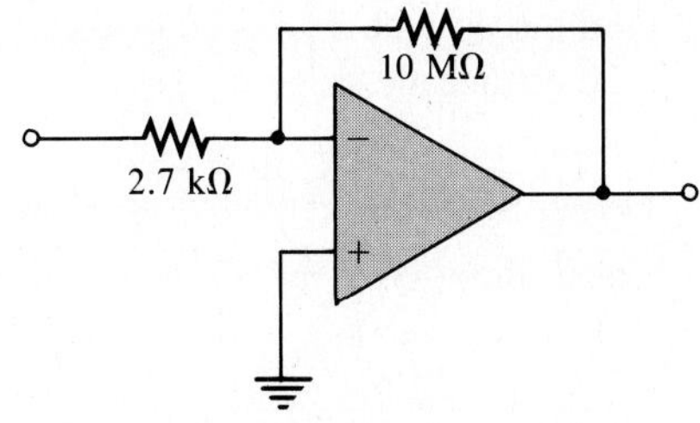

17. Determine the phase margin for each value of phase lag.

(a) 30° **(b)** 60° **(c)** 120° **(d)** 180° **(e)** 210°

18. A certain op-amp has the following internal critical frequencies in its open-loop response: 125 Hz, 25 kHz, and 180 kHz. What is the total phase shift through the amplifier when the signal frequency is 50 kHz?

19. Each graph in Figure 13–33 shows both the open-loop and the closed-loop response of a particular op-amp configuration. Analyze each case for stability.

FIGURE 13–33

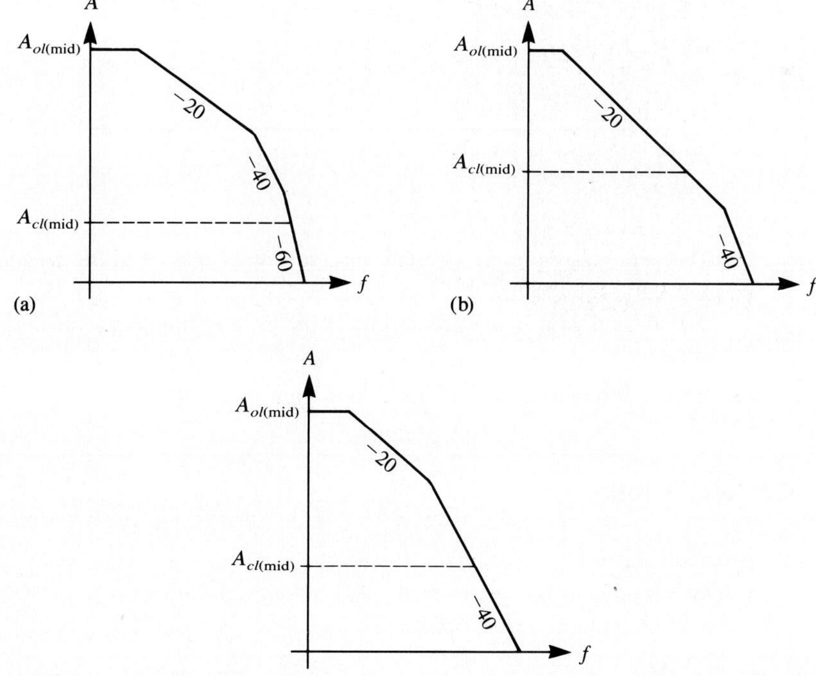

SECTION 13–5 OP-AMP COMPENSATION

20. A certain operational amplifier has an open-loop response curve as shown in Figure 13–34. A particular application requires a 30 dB closed-loop midrange gain. In order to achieve a 30 dB gain, compensation must be added because the 30 dB line intersects the uncompensated open-loop gain on the −40 dB/decade slope and, therefore, stability is not assured.

(a) Find the critical frequency of the compensating network such that the −20 dB/decade slope is lowered to a point where it intersects the 30 dB gain line.

(b) Sketch the ideal response curve for the compensating network.

(c) Sketch the total ideal compensated open-loop response.

FIGURE 13–34

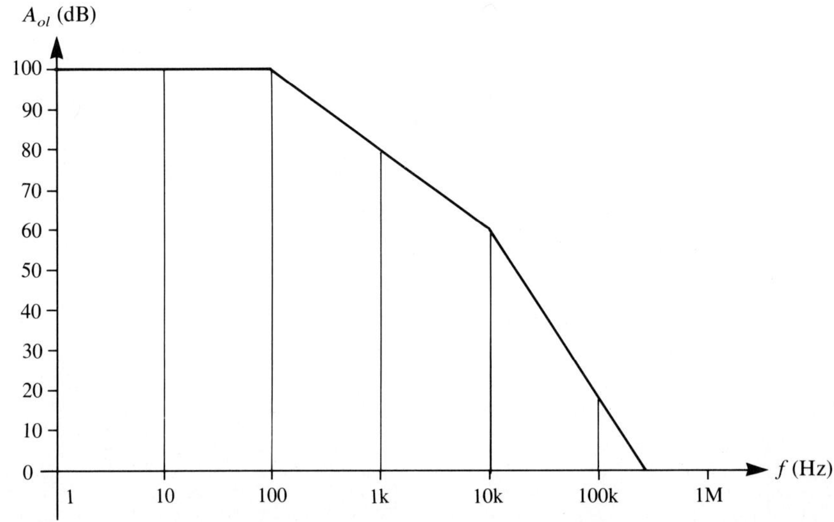

21. The open-loop gain of a certain op-amp rolls off at −20 dB/decade, beginning at $f = 250$ Hz. This roll-off rate extends down to a gain of 60 dB. If a 40 dB closed-loop gain is required, what is the critical frequency for the compensating network?

22. Repeat Problem 21 for a closed-loop gain of 20 dB.

ANSWERS TO REVIEW QUESTIONS

SECTION 13–1

1. Open-loop gain is without feedback, and closed-loop gain is with negative feedback. Open-loop gain is larger.

2. 100 Hz

3. Decrease

SECTION 13–2

1. 20 dB + 30 dB = 50 dB
2. −49° + (−5.2°) = −54.2°

SECTION 13–3

1. Yes
2. BW = 3,000 kHz/60 = 50 kHz
3. 3,000 kHz/1 = 3 MHz

SECTION 13–4

1. Positive feedback
2. 180°, 0°
3. −20 dB/decade (−6 dB/octave)

SECTION 13–5

1. Phase compensation increases the phase margin at a given frequency.
2. Internal compensation is full compensation; external compensation can be tailored to maximize bandwidth.
3. Bandwidth decreases.

SECTION 13–6

1. f_{cl} can be reduced by increasing C_1 or R_2.
2. Q_2 and Q_3
3. Q_1 is an emitter-follower buffer stage
4. $P_{avg} = \dfrac{V^2_{out(rms)}}{R_L} \cong 4$ W

ANSWERS TO PRACTICE EXERCISES

13–1 (a) 79,996 (b) 79,900 (c) 6380

13–2 173.2 Hz

13–3 75 dB, −71.57°

13–4 2 MHz

13–5 (a) 29.56 kHz (b) 42.55 kHz

13–6 100 Hz, 10 Hz

14

Basic Op-Amp Circuits

After completing this chapter, you should be able to

- ☐ Use an op-amp as a comparator.
- ☐ Describe how hysteresis can be implemented in a comparator circuit and explain its purpose.
- ☐ Explain how bounded comparators work.
- ☐ Describe a basic window comparator.
- ☐ Show how comparators are used in a certain type of analog-to-digital converter and in an over-temperature sensing circuit.
- ☐ Use an op-amp as a summing amplifier.
- ☐ Explain how an averaging amplifier works.
- ☐ Explain how a scaling adder works.
- ☐ Show how a scaling adder can be used in a certain type of digital-to-analog converter.
- ☐ Discuss the operation of an op-amp integrator and a differentiator.
- ☐ Describe an instrumentation amplifier and discuss its characteristics.
- ☐ Show how an op-amp can be used as a constant-current source.
- ☐ Show how an op-amp can be used as a current-to-voltage or voltage-to-current converter.
- ☐ Explain how a basic op-amp peak detector circuit works.
- ☐ Troubleshoot common op-amp circuit failures.

In the last two chapters, you learned about the principles, operation, and characteristics of the operational amplifier. Op-amps are used in such a wide variety of applications that it is impossible to cover all of them in one chapter, or even in one book. Therefore, in this chapter, we will examine some of the more fundamental applications to illustrate how versatile the op-amp is and to give you a foundation in basic op-amp circuits.

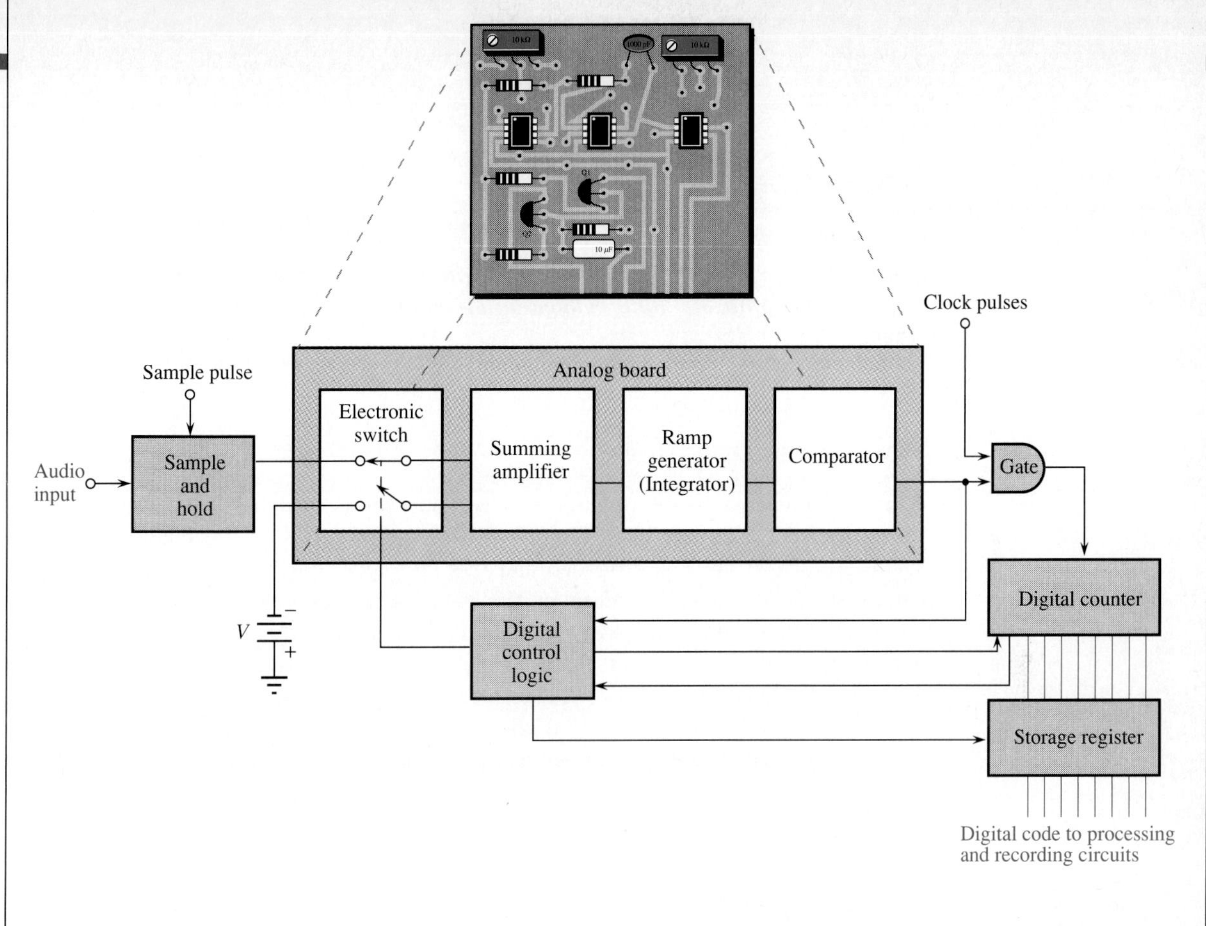

A System Application

This system application illustrates a very interesting application of three types of op-amp circuits that will be studied in this chapter—the summing amplifier, integrator, and comparator. The system diagram above shows one basic type of analog-to-digital converter that takes an audio input, such as voice or music, and converts it to binary codes that can be recorded digitally.

Op-amps play a key role in this system, and we will be focusing on the analog board to see how these circuits are used in a representative application. The digital circuits are discussed just enough to allow you to understand what the overall system does. You do not need to have a background in digital circuits for our purposes here. However, this particular system application points out the fact, again, that many systems that you will be working with in industry will include combinations of both analog (linear) and digital circuits. You will, of course, get a thorough grounding in digital fundamentals in another course, if you haven't already.

For the system application in Section 14–7, in addition to the other topics, be sure you understand

- ☐ How a summing amplifier works.
- ☐ How an integrator works.
- ☐ How a comparator works.

14–1 COMPARATORS

Operational amplifiers are often used as nonlinear devices to compare the amplitude of one voltage with another. In this application, the op-amp is used in the open-loop configuration, with the input voltage on one input and a reference voltage on the other.

ZERO-LEVEL DETECTION

A basic application of the op-amp as a **comparator** is in determining when an input voltage exceeds a certain level. Figure 14–1(a) shows a zero-level detector. Notice that the inverting (−) input is grounded and that the input signal voltage is applied to the noninverting (+) input. Because of the high open-loop voltage gain, a very small difference voltage between the two inputs drives the amplifier into saturation, causing the output voltage to go to its limit. For example, consider an op-amp having $A_{ol} = 100{,}000$. A voltage difference of only 0.25 mV between the inputs could produce an output voltage of (0.25 mV)(100,000) = 25 V if the op-amp were capable. However, since most op-amps have output voltage limitations of less than ±15 V, the device would be driven into saturation.

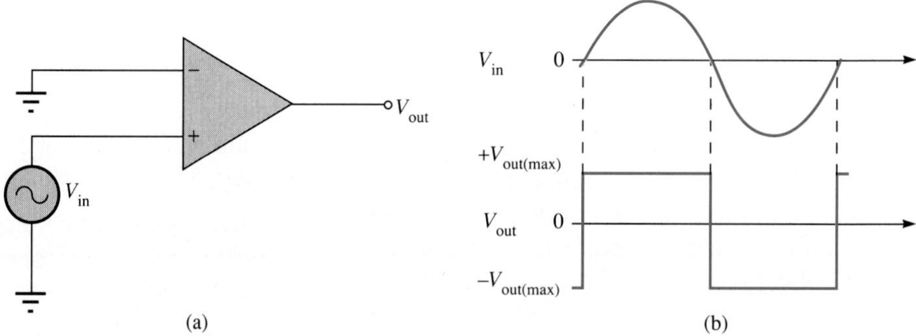

FIGURE 14–1
The op-amp as a zero-level detector.

Figure 14–1(b) shows the result of a sine wave input voltage applied to the noninverting input of the zero-level detector. When the sine wave is negative, the output is at its maximum negative level. When the sine wave crosses 0, the amplifier is driven to its opposite state and the output goes to its maximum positive level, as shown. As you can see, the zero-level detector can be used as a squaring circuit to produce a square wave from a sine wave.

NONZERO-LEVEL DETECTION

The zero-level detector in Figure 14–1 can be modified to detect voltages other than zero by connecting a fixed reference voltage, as shown in Figure 14–2(a). A more practical arrangement is shown in Figure 14–2(b) using a voltage divider to set the reference voltage as follows:

$$V_{REF} = \frac{R_2}{R_1 + R_2}(+V) \qquad \textbf{(14–1)}$$

where $+V$ is the positive op-amp supply voltage. The circuit in Figure 14–2(c) uses a zener diode to set the reference voltage ($V_{REF} = V_Z$). As long as the input voltage V_{in} is

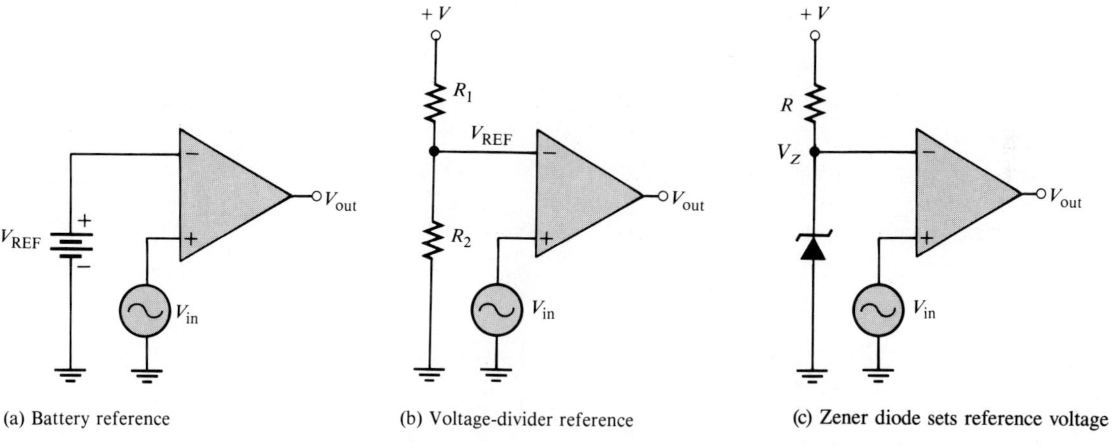

(a) Battery reference (b) Voltage-divider reference (c) Zener diode sets reference voltage

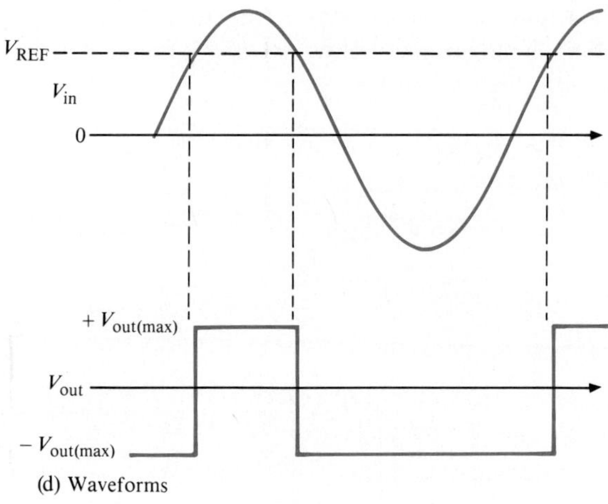

(d) Waveforms

FIGURE 14–2
Nonzero-level detectors.

less than V_{REF}, the output remains at the maximum negative level. When the input voltage exceeds the reference voltage, the output goes to its maximum positive state, as shown in Figure 14–2(d) with a sine wave input voltage.

EXAMPLE 14–1 The input signal in Figure 14–3(a) is applied to the comparator circuit in Figure 14–3(b). Make a sketch of the output showing its proper relationship to the input signal. Assume the maximum output levels are ±12 V.

FIGURE 14–3

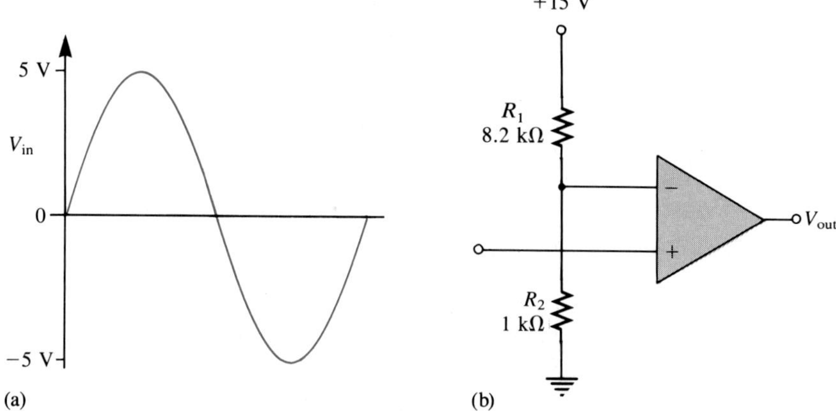

SOLUTION

The reference voltage is set by R_1 and R_2 as follows:

$$V_{REF} = \frac{R_2}{R_1 + R_2}(+V) = \frac{1\text{ k}\Omega}{8.2\text{ k}\Omega + 1\text{ k}\Omega}(+15\text{ V}) = 1.63\text{ V}$$

Each time the input exceeds +1.63 V, the output voltage switches to its +12 V level. Each time the input goes below +1.63 V, the output switches back to its −12 V level, as shown in Figure 14–4.

FIGURE 14–4

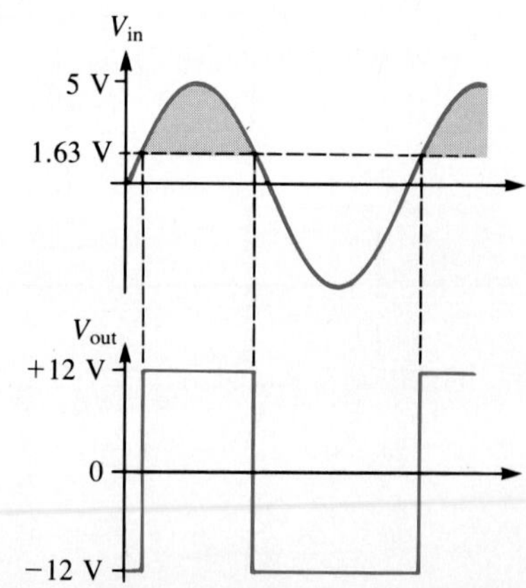

PRACTICE EXERCISE 14–1

Determine the reference voltage in Figure 14–3 if $R_1 = 22$ kΩ and $R_2 = 3.3$ kΩ.

EFFECTS OF INPUT NOISE ON COMPARATOR OPERATION

In many practical situations, unwanted voltage fluctuations **(noise)** appear on the input line. This noise voltage becomes superimposed on the input voltage, as shown in Figure 14–5, and can cause a comparator to erratically switch output states.

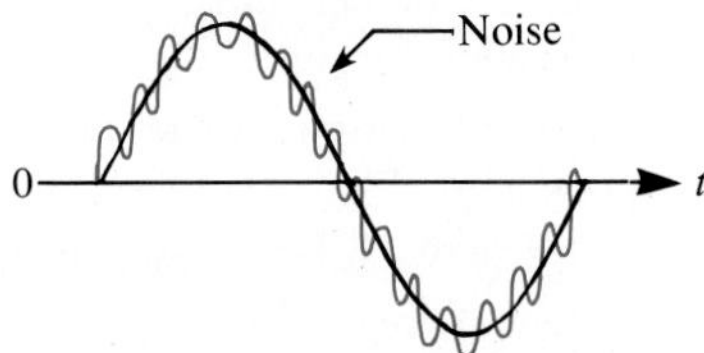

FIGURE 14–5
Sine wave with superimposed noise.

In order to understand the potential effects of noise voltage, consider a low-frequency sine wave applied to the input of an op-amp comparator used as a zero-level detector, as shown in Figure 14–6(a). Part (b) of the figure shows the input sine wave plus noise and the resulting output. As you can see, when the sine wave approaches 0, the fluctuations due to noise cause the total input to vary above and below 0 several times, thus producing an erratic output.

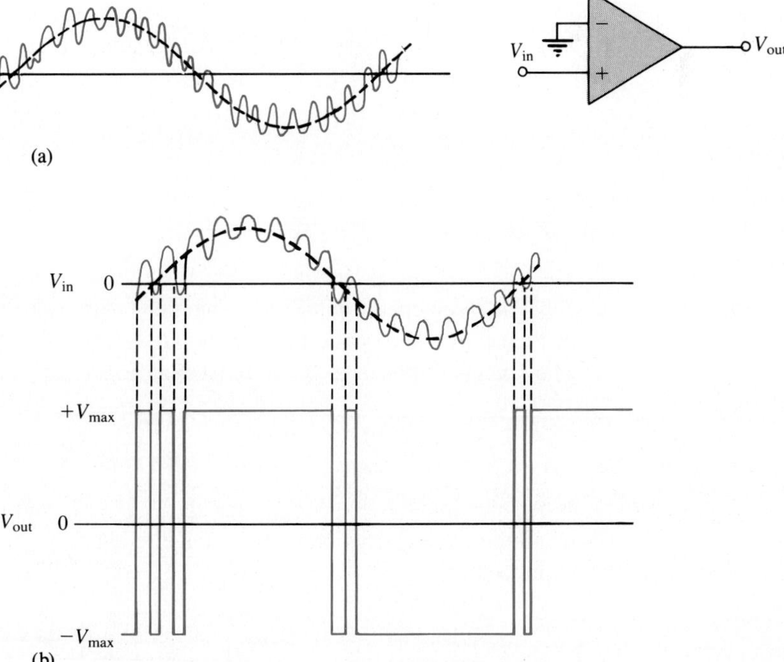

FIGURE 14–6
Effects of noise on comparator circuit.

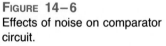

REDUCING NOISE EFFECTS WITH HYSTERESIS

An erratic output voltage caused by noise on the input occurs because the op-amp comparator switches from its negative output state to its positive output state at the same input voltage level that causes it to switch in the opposite direction, from positive to negative. This unstable condition occurs when the input voltage hovers around the reference voltage, and any small noise fluctuations cause the comparator to switch first one way and then the other.

In order to make the comparator less sensitive to noise, a technique incorporating positive feedback, called **hysteresis,** is often employed. Basically, hysteresis means that there is a higher reference level when the input voltage goes from a lower to higher value than when it goes from a higher to a lower value. A good example of hysteresis is a common household thermostat that turns the furnace on at one temperature and off at another.

The two reference levels are referred to as the upper trigger point (UTP) and the lower trigger point (LTP). This two-level hysteresis is established with a positive feedback arrangement, as shown in Figure 14–7. Notice that the noninverting (+) input is connected to a resistive voltage divider such that a portion of the output voltage is fed back to the input. The input signal is applied to the inverting input (−) in this case.

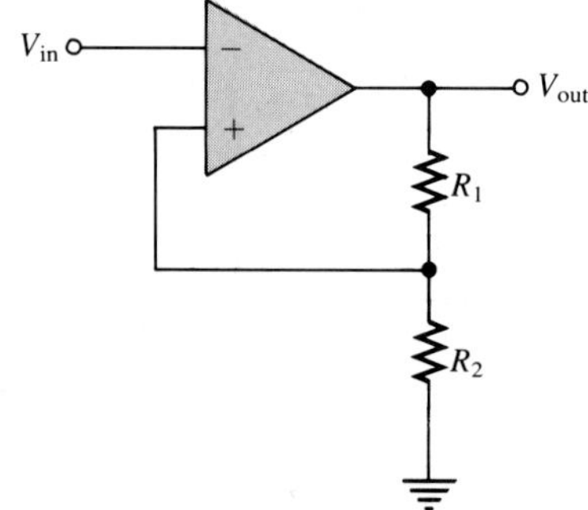

FIGURE 14–7
Comparator with positive feedback for hysteresis.

The basic operation of the comparator with hysteresis is as follows and is illustrated in Figure 14–8. Assume that the output voltage is at its positive maximum, $+V_{out(max)}$. The voltage fed back to the noninverting input is V_{UTP} and is expressed as

$$V_{UTP} = \frac{R_2}{R_1 + R_2}[+V_{out(max)}] \tag{14–2}$$

When the input voltage V_{in} exceeds V_{UTP}, the output voltage drops to its negative maximum, $-V_{out(max)}$. Now the voltage fed back to the noninverting input is V_{LTP} and is expressed as

$$V_{LTP} = \frac{R_2}{R_1 + R_2}[-V_{out(max)}] \tag{14–3}$$

The input voltage must now fall below V_{LTP} before the device will switch back to its other state. This means that a small amount of noise voltage has no effect on the output, as illustrated by Figure 14–8.

A comparator with hysteresis is sometimes known as a **Schmitt trigger.** The amount of hysteresis is defined by the difference of the two trigger levels.

$$V_{HYS} = V_{UTP} - V_{LTP} \tag{14–4}$$

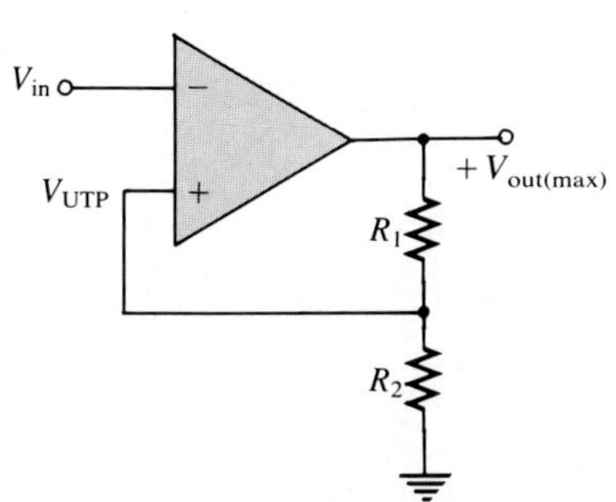

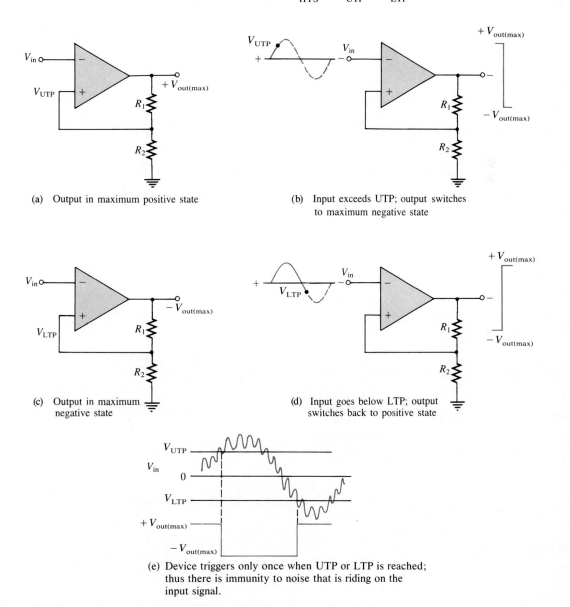

(a) Output in maximum positive state

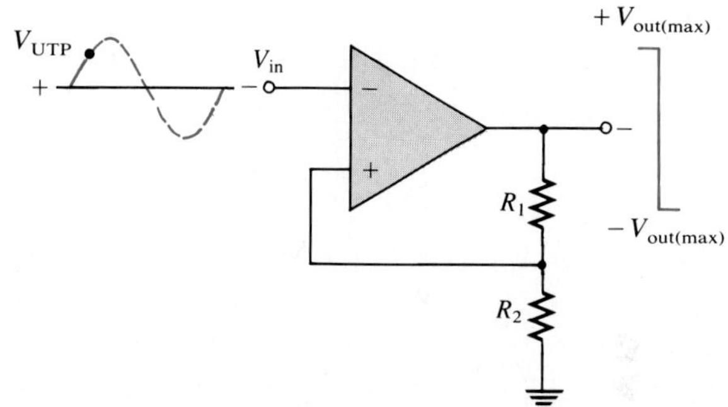

(b) Input exceeds UTP; output switches to maximum negative state

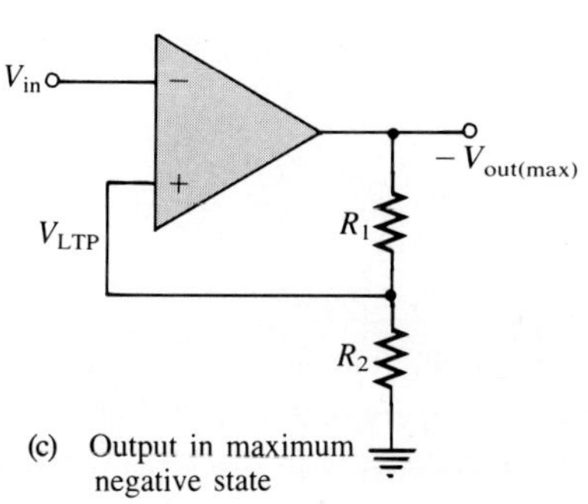

(c) Output in maximum negative state

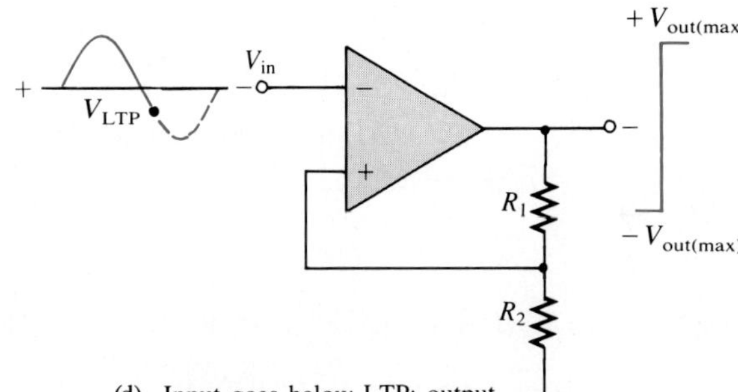

(d) Input goes below LTP; output switches back to positive state

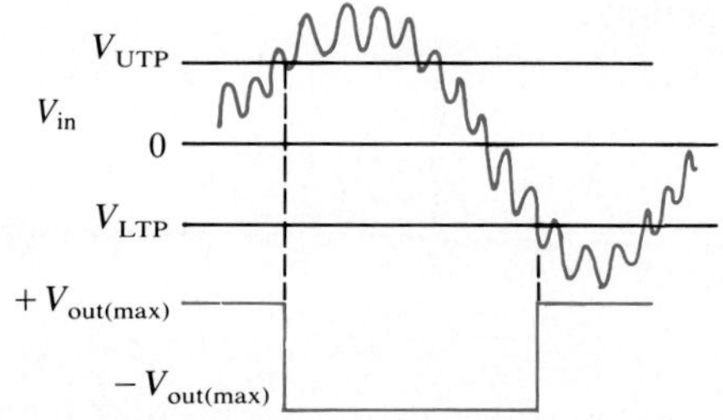

(e) Device triggers only once when UTP or LTP is reached; thus there is immunity to noise that is riding on the input signal.

FIGURE 14–8
Operation of a comparator with hysteresis.

EXAMPLE 14–2 Determine the upper and lower trigger points for the comparator circuit in Figure 14–9. Assume that $+V_{out(max)} = +5$ V and $-V_{out(max)} = -5$ V.

FIGURE 14–9

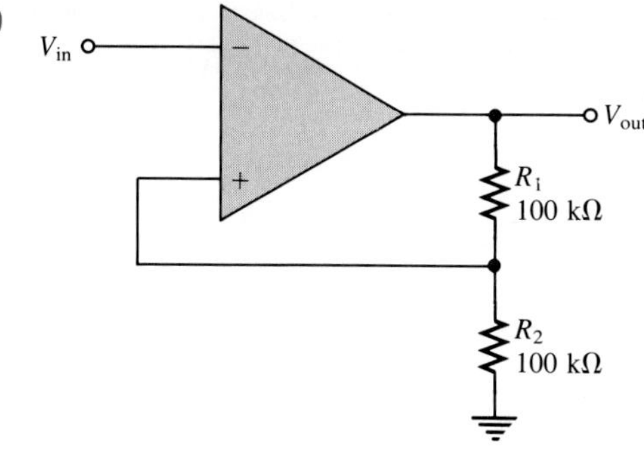

SOLUTION

$$V_{UTP} = \frac{R_2}{R_1 + R_2}[+V_{out(max)}] = 0.5(5\text{ V}) = +2.5\text{ V}$$

$$V_{LTP} = \frac{R_2}{R_1 + R_2}[-V_{out(max)}] = 0.5(-5\text{ V}) = -2.5\text{ V}$$

PRACTICE EXERCISE 14–2

Determine the upper and lower trigger points in Figure 14–9 for $R_1 = 68$ kΩ and $R_2 = 82$ kΩ. The maximum output levels are ±7 V.

OUTPUT BOUNDING

In some applications, it is necessary to limit the output voltage levels of a comparator to a value less than that provided by the saturated op-amp. A single zener diode can be used as shown in Figure 14–10 to limit the output voltage swing to the zener voltage in one direction and to the forward diode drop in the other. This process of limiting the output range is called **bounding.**

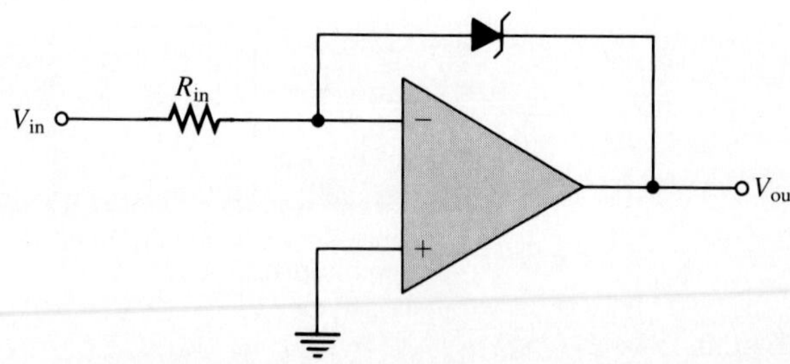

FIGURE 14–10 Comparator with output bounding.

The operation is as follows. Since the anode of the zener is connected to the inverting input, it is at virtual ground (≅ 0 V). Therefore, when the output reaches a positive value equal to the zener voltage, it limits at that value, as illustrated in Figure 14–11. When the output switches negative, the zener acts as a regular diode and becomes forward-biased at 0.7 V, limiting the negative output swing to this value, as shown. Turning the zener around limits the output in the opposite direction.

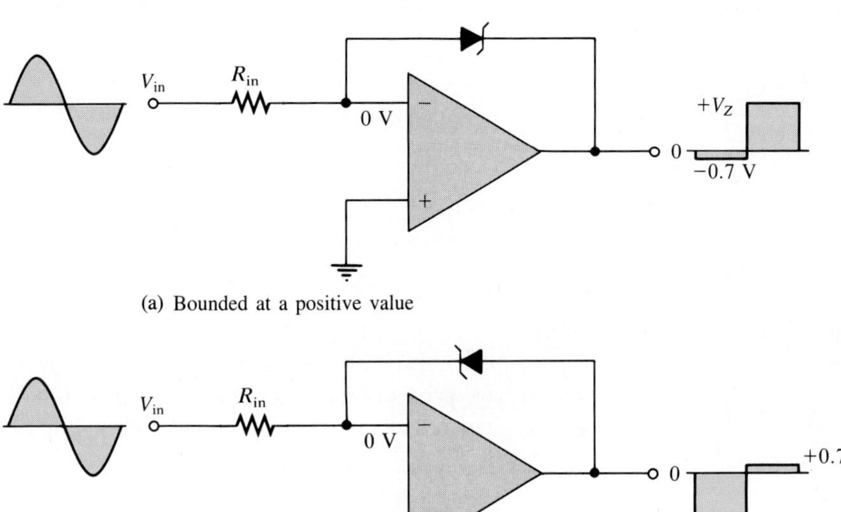

FIGURE 14–11
Operation of a bounded comparator.

Two zener diodes arranged as in Figure 14–12 limit the output voltage to the zener voltage plus the voltage drop (0.7 V) of the forward-biased zener, both positively and negatively, as shown in Figure 14–12. Op-amp comparators are available in IC form and are optimized for the comparison operation. Typical of these is the LM307.

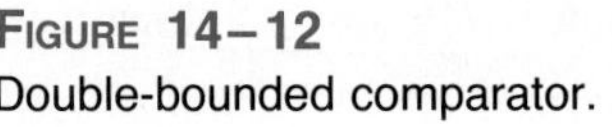
FIGURE 14–12
Double-bounded comparator.

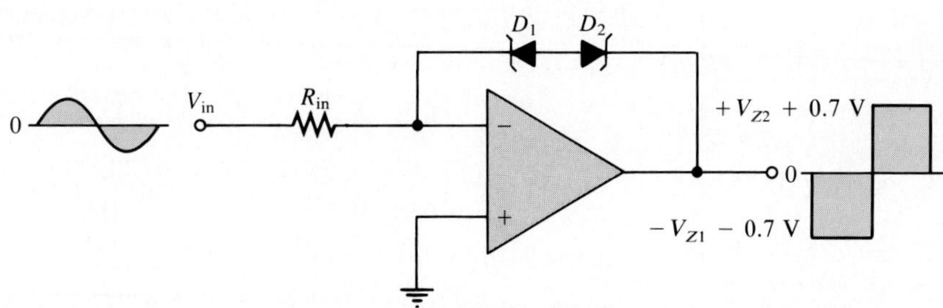

EXAMPLE 14–3 Determine the output voltage waveform for Figure 14–13.

FIGURE 14–13

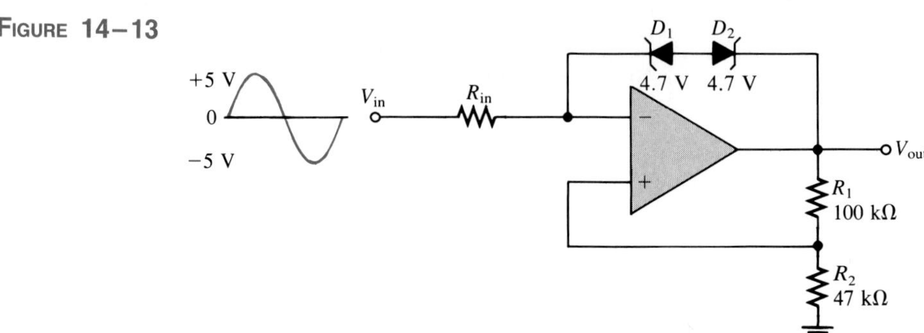

SOLUTION

This comparator has both hysteresis and zener bounding.

The voltage across D_1 and D_2 in either direction is 4.7 V + 0.7 V = 5.4 V. This is because one zener is always forward-biased with a drop of 0.7 V when the other one is in breakdown.

The voltage at the inverting (−) op-amp input is

$$V_{out} \pm 5.4 \text{ V}$$

Since the differential voltage is negligible, the voltage at the noninverting (+) op-amp input is also $V_{out} \pm 5.4$ V. Thus,

$$V_{R1} = V_{out} - (V_{out} \pm 5.4 \text{ V}) = \pm 5.4 \text{ V}$$

$$I_{R1} = \frac{V_{R1}}{R_1} = \frac{\pm 5.4 \text{ V}}{100 \text{ k}\Omega} = \pm 54 \ \mu\text{A}$$

Since the current into the noninverting input is negligible,

$$I_{R2} = I_{R1} = \pm 54 \ \mu\text{A}$$

$$V_{R2} = (47 \text{ k}\Omega)(\pm 54 \ \mu\text{A}) = \pm 2.54 \text{ V}$$

$$V_{out} = V_{R1} + V_{R2} = \pm 5.4 \text{ V} \pm 2.54 \text{ V} = \pm 7.94 \text{ V}$$

The upper trigger point (UTP) and the lower trigger point (LTP) are as follows.

$$V_{UTP} = \left(\frac{R_2}{R_1 + R_2}\right)(+V_{out}) = \left(\frac{47 \text{ k}\Omega}{147 \text{ k}\Omega}\right)(+7.94 \text{ V}) = +2.54 \text{ V}$$

$$V_{LTP} = \left(\frac{R_2}{R_1 + R_2}\right)(-V_{out}) = \left(\frac{47 \text{ k}\Omega}{147 \text{ k}\Omega}\right)(-7.94 \text{ V}) = -2.54 \text{ V}$$

The output waveform for the given input voltage is shown in Figure 14–14.

FIGURE 14–14

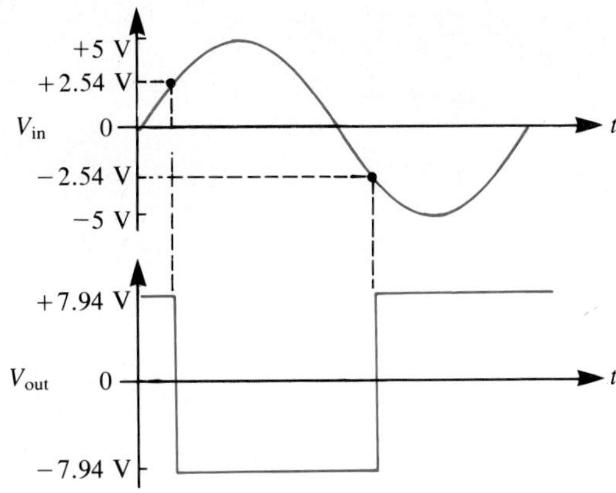

PRACTICE EXERCISE 14–3

Determine the upper and lower trigger points for Figure 14–13 if R_1 = 150 kΩ, R_2 = 68 kΩ, and the zener diodes are 3.3 V devices.

WINDOW COMPARATOR

Two individual op-amp comparators arranged as in Figure 14–15 form what is known as a *window comparator*. This circuit detects when an input voltage is between two limits, an upper and a lower, called the "window."

FIGURE 14–15
A basic window comparator.

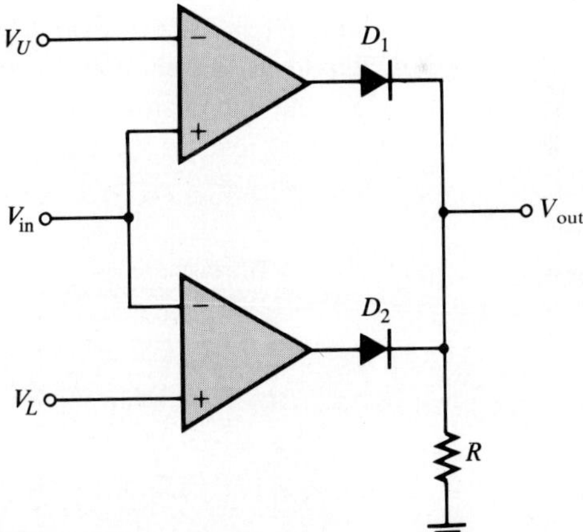

The upper and lower limits are set by reference voltages designated V_U and V_L. These voltages can be established with voltage dividers, zener diodes, or any type of voltage source. As long as V_{in} is within the window (less than V_U and greater than V_L), the output of each comparator is at its low saturated level. Under this condition, both diodes are reverse-biased and V_{OUT} is held at zero by the resistor to ground. When V_{in} goes above V_U or below V_L, the output of the associated comparator goes to its high saturated level. This action forward-biases the diode and produces a high-level V_{OUT}. This is illustrated in Figure 14–16 with V_{in} changing arbitrarily.

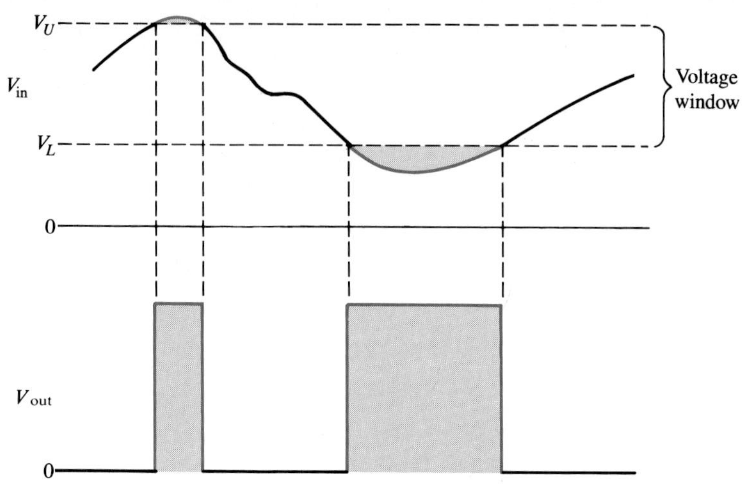

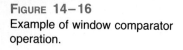

FIGURE 14–16
Example of window comparator operation.

A COMPARATOR APPLICATION: OVER-TEMPERATURE SENSING CIRCUIT

Figure 14–17 shows an op-amp comparator used in a precision over-temperature sensing circuit. The circuit consists of a Wheatstone bridge with the op-amp used to detect when the bridge is balanced. One leg of the bridge contains a thermistor (R_1), which is a temperature-sensing resistor with a negative temperature coefficient (its resistance decreases as temperature increases). The potentiometer (R_2) is set at a value equal to the resistance of the thermistor at the critical temperature. At normal temperatures (below critical), R_1 is greater than R_2, thus creating an unbalanced condition that drives the op-amp to its low saturated output level and keeps transistor Q_1 off.

As the temperature increases, the resistance of the thermistor decreases. When the temperature reaches the critical value, R_1 becomes equal to R_2, and the bridge becomes

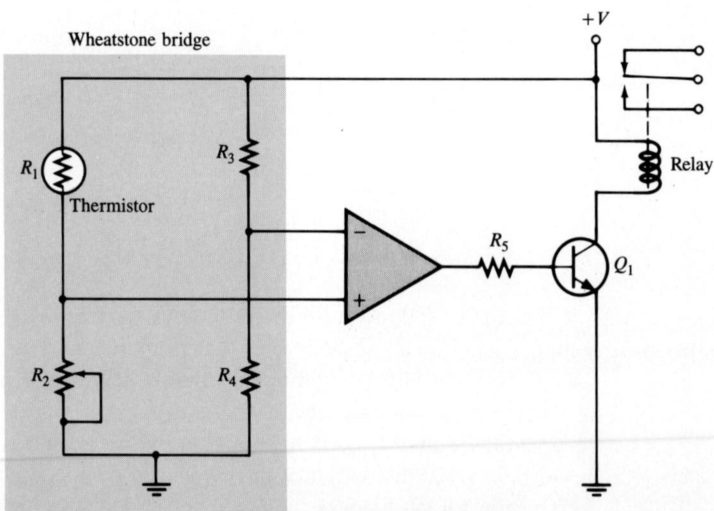

FIGURE 14–17
An over-temperature sensing circuit.

balanced (since $R_3 = R_4$). At this point the op-amp switches to its high saturated output level, turning Q_1 on. This energizes the relay, which can be used to activate an alarm or initiate an appropriate response to the over-temperature condition.

A COMPARATOR APPLICATION: ANALOG-TO-DIGITAL (A/D) CONVERSION

A/D conversion is a common interfacing process often used when a linear **analog** system must provide inputs to a **digital** system. Many methods for A/D conversion are available. However, in this discussion, only one type is used to demonstrate the concept.

The *simultaneous,* or *flash,* method of A/D conversion uses parallel comparators to compare the linear input signal with various reference voltages developed by a voltage divider. When the input voltage exceeds the reference voltage for a given comparator, a high level is produced on that comparator's output. Figure 14–18 shows a converter that

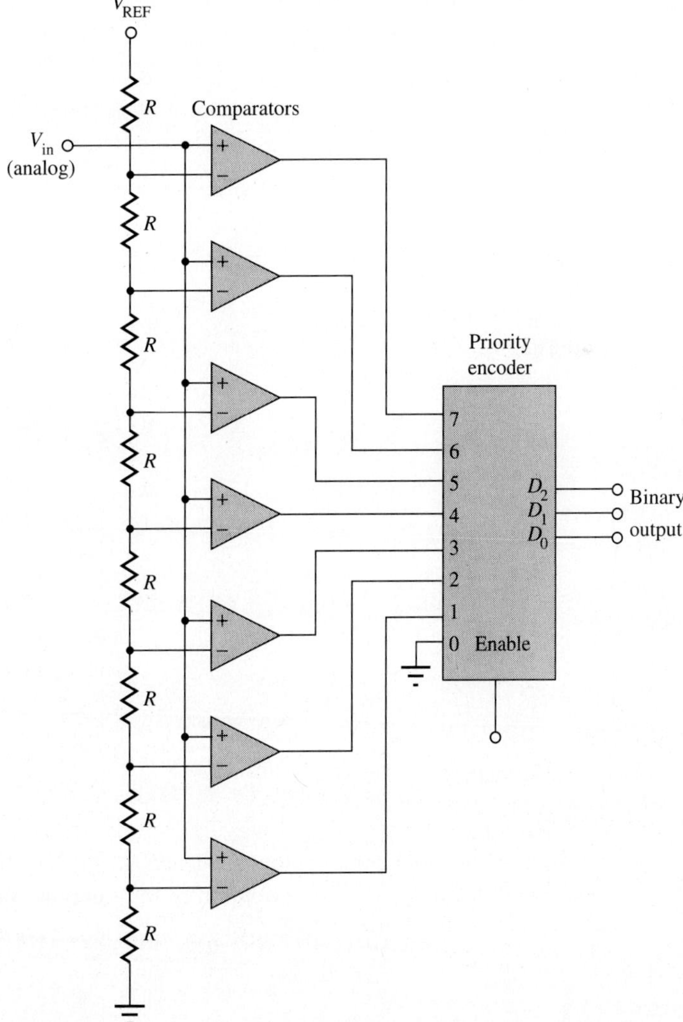

FIGURE 14–18
A simultaneous A/D converter using op-amps as comparators.

produces three-digit binary numbers on its output, which represent the values of the analog input voltage as it changes. This converter requires seven comparators. In general, $2^n - 1$ comparators are required for conversion to an n-digit binary number. The large number of comparators necessary for a reasonably sized binary number is one of the drawbacks of this type of A/D converter. Its chief advantage is that it provides a fast conversion time.

The reference voltage for each comparator is set by the resistive voltage-divider network and V_{REF}. The output of each comparator is connected to an input of the priority encoder. The *priority encoder* is a digital device that produces a binary number representing the highest value input.

The encoder *samples* its input when a pulse occurs on the enable line (sampling pulse), and a three-digit binary number proportional to the value of the analog input signal appears on the encoder's outputs.

The sampling rate determines the accuracy with which the sequence of binary numbers represents the changing input signal.

The following example illustrates the basic operation of the simultaneous A/D converter in Figure 14–18.

EXAMPLE 14–4

Determine the binary number sequence of the three-digit simultaneous A/D converter in Figure 14–18 for the input signal in Figure 14–19 and the sampling pulses (encoder enable) shown.

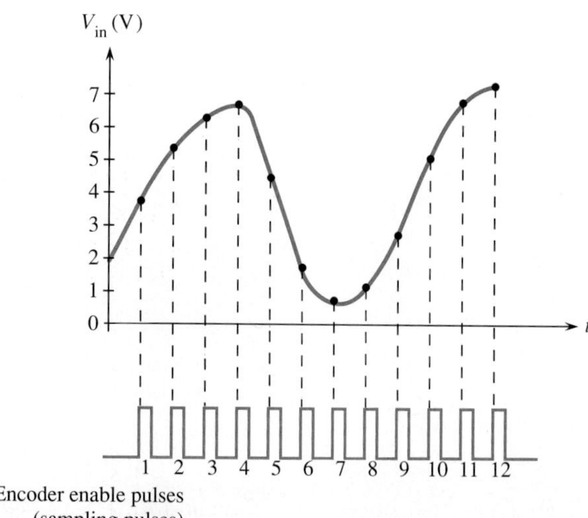

FIGURE 14–19
Sampling of values on analog waveform for conversion to digital.

SOLUTION

The resulting A/D binary output sequence is listed as follows and shown in the waveform diagram of Figure 14–20 in relation to the sampling pulses.

011, 101, 110, 110, 100, 001, 000, 001, 010, 101, 110, 111

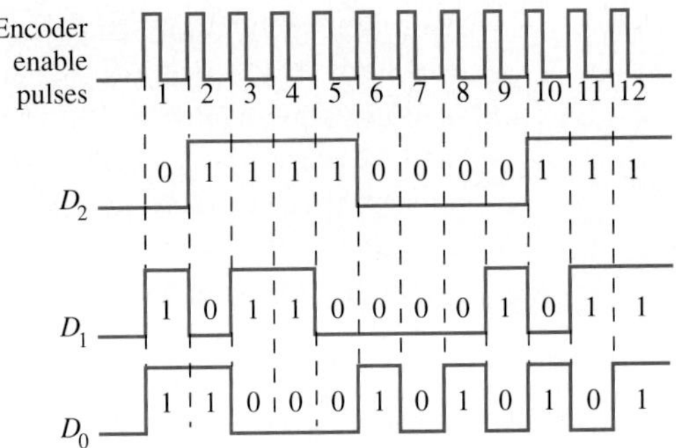

FIGURE 14–20
Resulting digital outputs for sampled values in Figure 14–19. D_0 is the least significant digit.

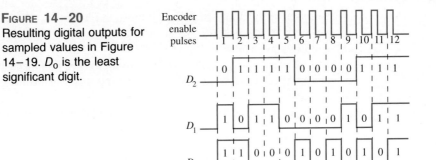

PRACTICE EXERCISE 14–4

If the frequency of the enable pulses in Figure 14–19 is doubled, does the resulting A/D binary output sequence represent the analog waveform more or less accurately? ■

14–1 REVIEW QUESTIONS

1. What is the reference voltage for each comparator in Figure 14–21?
2. What is the purpose of hysteresis in a comparator?
3. Define the term *bounding* in relation to a comparator's output.

FIGURE 14–21

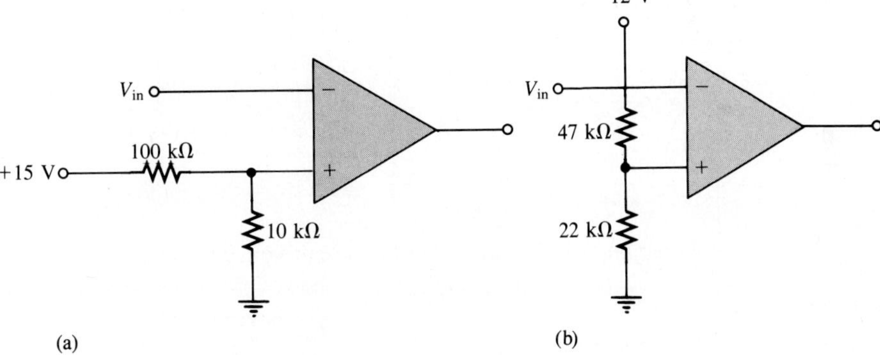

14–2 SUMMING AMPLIFIERS

The summing amplifier is a variation of the inverting op-amp configuration covered in Chapter 12. The summing amplifier has two or more inputs, and its output voltage is proportional to the negative of the algebraic sum of its input voltages. In this section, you will see how a summing amplifier works, and you will learn about the averaging amplifier and the scaling amplifier, which are variations of the basic summing amplifier.

A two-input summing amplifier is shown in Figure 14–22, but any number of inputs can be used. The operation of the circuit and derivation of the output expression are as follows. Two voltages, V_{IN1} and V_{IN2}, are applied to the inputs and produce currents I_1 and I_2, as shown. Using the concepts of infinite input impedance and virtual ground, you can see that the inverting input of the op-amp is approximately 0 V, and there is no current into the input. This means that both input currents I_1 and I_2 come together at this summing point and form the total current, which goes through R_f, as indicated.

$$I_T = I_1 + I_2$$

Since $V_{OUT} = -I_T R_f$, the following steps apply.

$$V_{OUT} = -(I_1 + I_2)R_f$$

$$= -\left(\frac{V_{IN1}}{R_1} + \frac{V_{IN2}}{R_2}\right)R_f$$

If all three of the resistors are equal to the same value R ($R_1 = R_2 = R_f = R$), then

$$V_{OUT} = -\left(\frac{V_{IN1}}{R} + \frac{V_{IN2}}{R}\right)R$$

$$V_{OUT} = -(V_{IN1} + V_{IN2}) \qquad \textbf{(14–5)}$$

FIGURE 14–22
Two-input inverting summing amplifier.

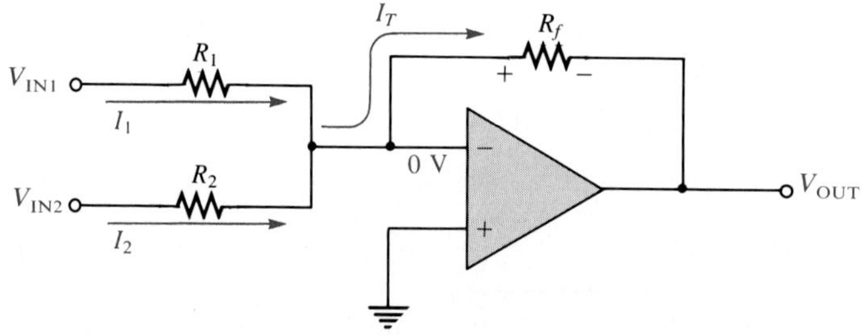

Equation (14–5) shows that the output voltage is the sum of the two input voltages. A general expression is given in Equation (14–6) for a summing amplifier with n inputs, as shown in Figure 14–23 where all resistors are equal in value.

$$V_{OUT} = -(V_{IN1} + V_{IN2} + \cdots + V_{INn}) \qquad \textbf{(14–6)}$$

FIGURE 14–23
Summing amplifier with n inputs.

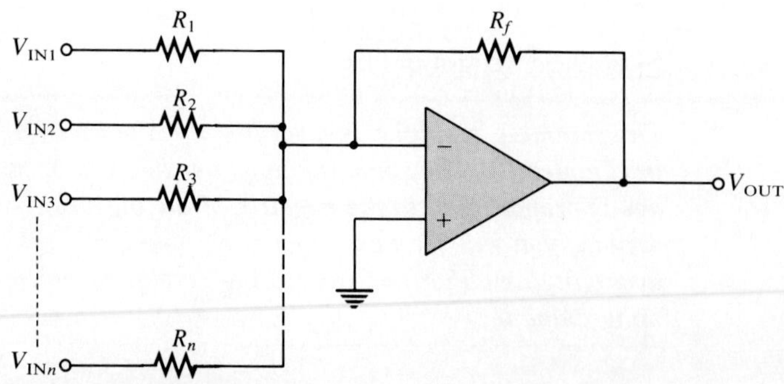

EXAMPLE 14–5 Determine the output voltage in Figure 14–24.

FIGURE 14–24

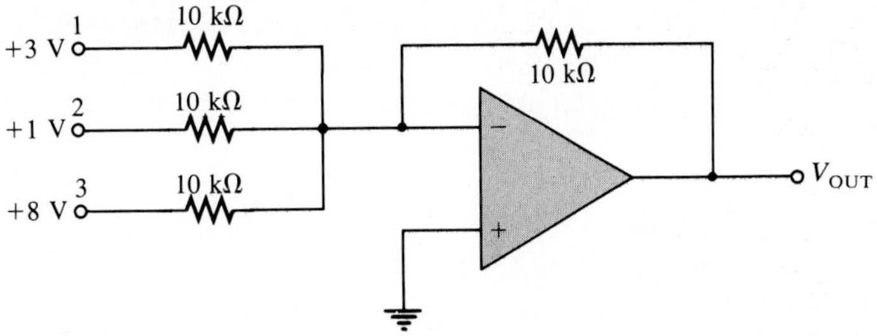

SOLUTION

$$V_{OUT} = -(V_{IN1} + V_{IN2} + V_{IN3})$$
$$= -(3\text{ V} + 1\text{ V} + 8\text{ V})$$
$$= -12\text{ V}$$

PRACTICE EXERCISE 14–5

If a fourth input of +0.5 V is added to Figure 14–24 with a 10 kΩ resistor, what is the output voltage?

SUMMING AMPLIFIER WITH GAIN GREATER THAN UNITY

When R_f is larger than the input resistors, the amplifier has a gain of R_f/R, where R is the value of each input resistor. The general expression for the output is

$$V_{OUT} = -\frac{R_f}{R}(V_{IN1} + V_{IN2} + \cdots + V_{INn}) \tag{14–7}$$

As you can see, the output is the sum of all the input voltages multiplied by a constant determined by the ratio R_f/R.

EXAMPLE 14–6 Determine the output voltage for the summing amplifier in Figure 14–25.

SOLUTION

$$R_f = 10\text{ k}\Omega, \text{ and } R = 1\text{ k}\Omega.$$

$$V_{OUT} = -\frac{R_f}{R}(V_{IN1} + V_{IN2})$$
$$= -\frac{10\text{ k}\Omega}{1\text{ k}\Omega}(0.2\text{ V} + 0.5\text{ V})$$
$$= -10(0.7\text{ V})$$
$$= -7\text{ V}$$

FIGURE 14–25

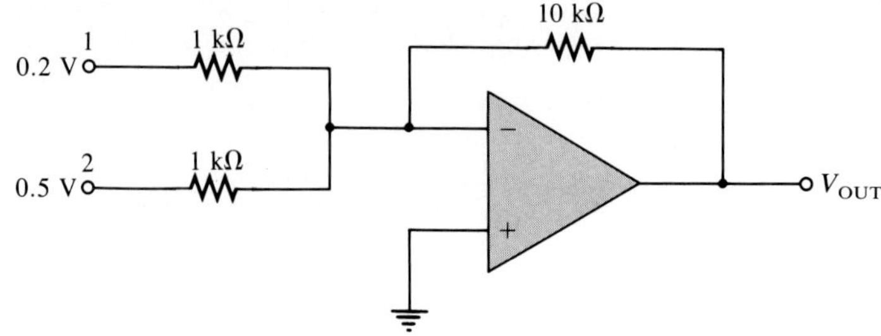

PRACTICE EXERCISE 14–6

Determine the output voltage in Figure 14–25 if the two input resistors are 2.2 kΩ and the feedback resistor is 18 kΩ.

AVERAGING AMPLIFIER

A summing amplifier can be made to produce the mathematical average of the input voltages. This is done by setting the ratio R_f/R equal to the reciprocal of the number of inputs. You obtain the average of several numbers by first adding the numbers and then dividing by the quantity of numbers you have. Examination of Equation (14–7) and a little thought will convince you that a summing amplifier will do this, as the next example illustrates.

EXAMPLE 14–7

Show that the amplifier in Figure 14–26 produces an output whose magnitude is the mathematical average of the input voltages.

SOLUTION

The output voltage is

$$\begin{aligned} V_{\text{OUT}} &= -\frac{R_f}{R}(V_{\text{IN1}} + V_{\text{IN2}} + V_{\text{IN3}} + V_{\text{IN4}}) \\ &= -\frac{25\ \text{k}\Omega}{100\ \text{k}\Omega}(1\ \text{V} + 2\ \text{V} + 3\ \text{V} + 4\ \text{V}) \\ &= -\frac{1}{4}(10\ \text{V}) \\ &= -2.5\ \text{V} \end{aligned}$$

It can easily be shown that the average of the input values is the same magnitude as V_{OUT} but of opposite sign.

$$\frac{1\ \text{V} + 2\ \text{V} + 3\ \text{V} + 4\ \text{V}}{4} = \frac{10\ \text{V}}{4} = 2.5\ \text{V}$$

FIGURE 14–26

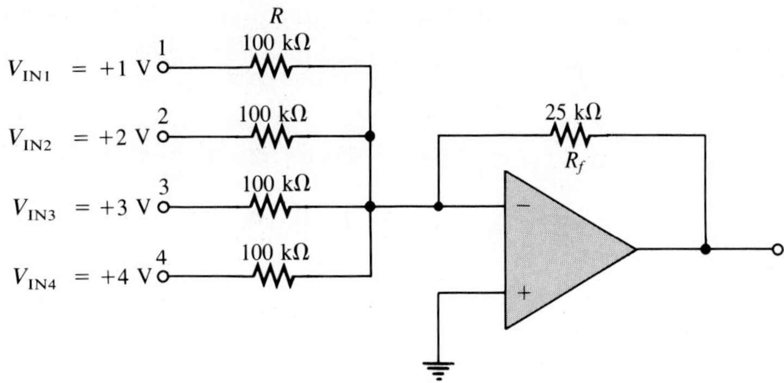

PRACTICE EXERCISE 14–7

Specify the changes required in the averaging amplifier in Figure 14–26 in order to handle five inputs.

SCALING ADDER

A different weight can be assigned to each input of a summing amplifier by simply adjusting the values of the input resistors. As you have seen, the output voltage can be expressed as

$$V_{OUT} = -\left(\frac{R_f}{R_1}V_{IN1} + \frac{R_f}{R_2}V_{IN2} + \cdots + \frac{R_f}{R_n}V_{INn}\right) \quad (14\text{–}8)$$

The weight of a particular input is set by the ratio of R_f to the input resistance. For example, if an input voltage is to have a weight of 1, then $R = R_f$. Or, if a weight of 0.5 is required, $R = 2R_f$. The smaller the value of R, the greater the weight, and vice versa. (R is the input resistor.)

EXAMPLE 14–8

Determine the weight of each input voltage for the scaling adder in Figure 14–27 and find the output voltage.

FIGURE 14–27

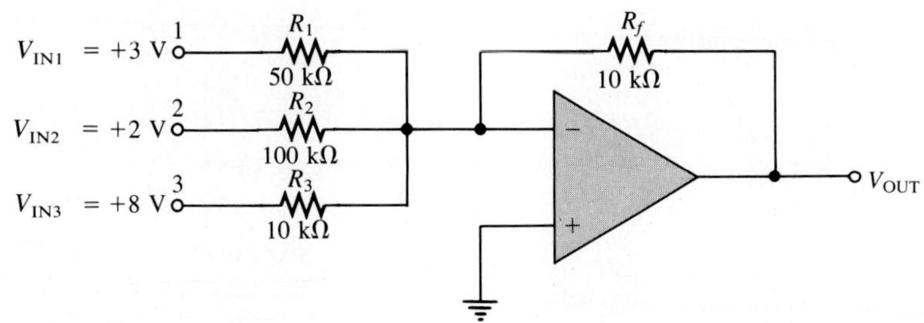

SOLUTION

Weight of input 1: $\frac{R_f}{R_1} = \frac{10\text{ k}\Omega}{50\text{ k}\Omega} = 0.2$

Weight of input 2: $\frac{R_f}{R_2} = \frac{10\text{ k}\Omega}{100\text{ k}\Omega} = 0.1$

Weight of input 3: $\frac{R_f}{R_3} = \frac{10\text{ k}\Omega}{10\text{ k}\Omega} = 1$

The output voltage is

$$\begin{aligned} V_{OUT} &= -\left(\frac{R_f}{R_1}V_{IN1} + \frac{R_f}{R_2}V_{IN2} + \frac{R_f}{R_3}V_{IN3}\right) \\ &= -[0.2(3\text{ V}) + 0.1(2\text{ V}) + 1(8\text{ V})] \\ &= -(0.6\text{ V} + 0.2\text{ V} + 8\text{ V}) \\ &= -8.8\text{ V} \end{aligned}$$

PRACTICE EXERCISE 14–8

Determine the weight of each input voltage in Figure 14–27 if $R_1 = 22\text{ k}\Omega$, $R_2 = 82\text{ k}\Omega$, $R_3 = 56\text{ k}\Omega$, and $R_f = 10\text{ k}\Omega$. Also find V_{OUT}.

A SCALING ADDER APPLICATION: DIGITAL-TO-ANALOG (D/A) CONVERSION

D/A conversion is an important interface process for converting digital signals to analog (linear) signals. An example is a voice signal that is digitized for storage, processing, or transmission and must be changed back into an approximation of the original audio signal in order to drive a speaker.

One method of D/A conversion uses a scaling adder with input resistor values that represent the binary weights of the input code. Figure 14–28 shows a four-digit D/A converter of this type (called a *binary-weighted resistor D/A converter*). The switch symbols represent transistor switches for applying each of the four binary digits to the inputs.

The inverting input is at virtual ground, so that the output is proportional to the current through the feedback resistor R_f (sum of input currents).

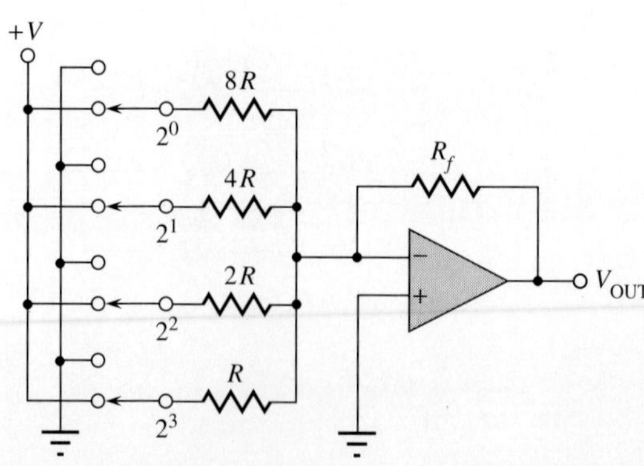

FIGURE 14–28
A scaling adder as a four-digit digital-to-analog converter.

The lowest value resistor R corresponds to the highest weighted binary input (2^3). All of the other resistors are multiples of R and correspond to the binary weights 2^2, 2^1, and 2^0.

EXAMPLE 14–9

Determine the output of the D/A converter in Figure 14–29(a). The sequence of four-digit binary numbers represented by the waveforms in Figure 14–29(b) are applied to the inputs. A high level is a binary 1, and low level is a binary 0. D_0 is the least significant binary digit.

FIGURE 14–29

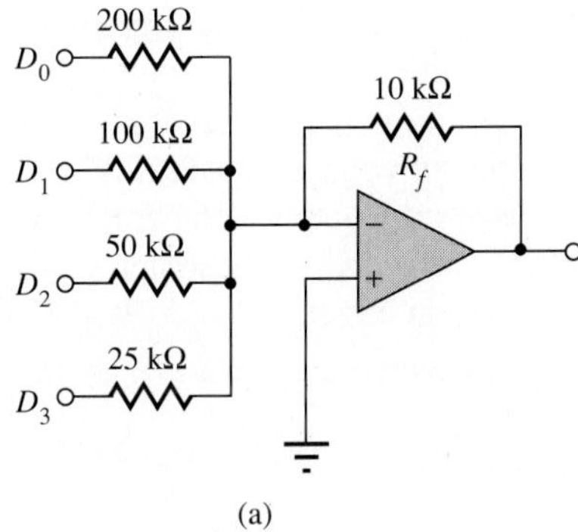

(a)

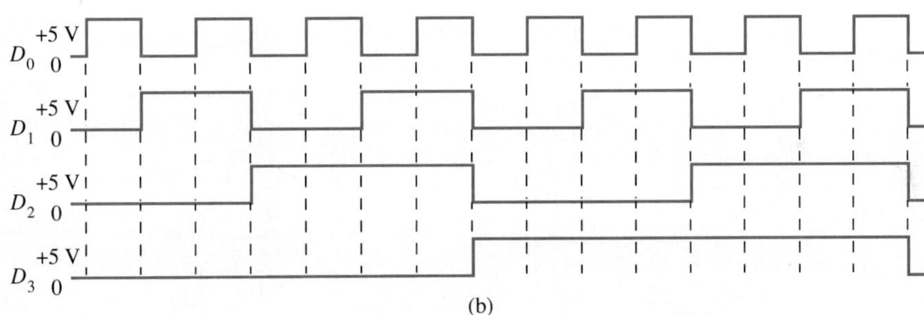

(b)

SOLUTION

First, we determine the output voltage for each of the weighted inputs. Since the inverting input of the op-amp is at 0 V (virtual ground), and a binary 1 corresponds to a high level (+5 V), the current through any of the input resistors equals 5 V divided by the resistance value.

$$I_0 = \frac{5\ \text{V}}{200\ \text{k}\Omega} = 0.025\ \text{mA}$$

$$I_1 = \frac{5\ \text{V}}{100\ \text{k}\Omega} = 0.05\ \text{mA}$$

$$I_2 = \frac{5\ \text{V}}{50\ \text{k}\Omega} = 0.1\ \text{mA}$$

$$I_3 = \frac{5\ \text{V}}{25\ \text{k}\Omega} = 0.2\ \text{mA}$$

Practically none of the input current goes into the inverting op-amp input because of its extremely high impedance. Therefore, practically all of the input current goes through R_f. Since one end of R_f is at 0 V (virtual ground), the drop across R_f equals the output voltage.

$$V_{OUT(D0)} = (10\ k\Omega)(-0.025\ mA) = -0.25\ V$$
$$V_{OUT(D1)} = (10\ k\Omega)(-0.05\ mA) = -0.5\ V$$
$$V_{OUT(D2)} = (10\ k\Omega)(-0.1\ mA) = -1\ V$$
$$V_{OUT(D3)} = (10\ k\Omega)(-0.2\ mA) = -2\ V$$

From Figure 14–29(b), the first binary number on the inputs is 0001 (it stands for decimal 1); for this, the output voltage is −0.25 V. The next binary number is 0010, which produces an output voltage of −0.5 V. The next binary number is 0011, which produces an output voltage of −0.25 V + (−0.5 V) = −0.75 V. Each successive binary number increases the output voltage by −0.25 V. So, for this particular sequence of binary numbers, the output is a stairstep waveform going from 0 V to −3.75 V in −0.25 V steps, as shown in Figure 14–30. If the steps are small, it approaches a straight line (linear).

FIGURE 14–30

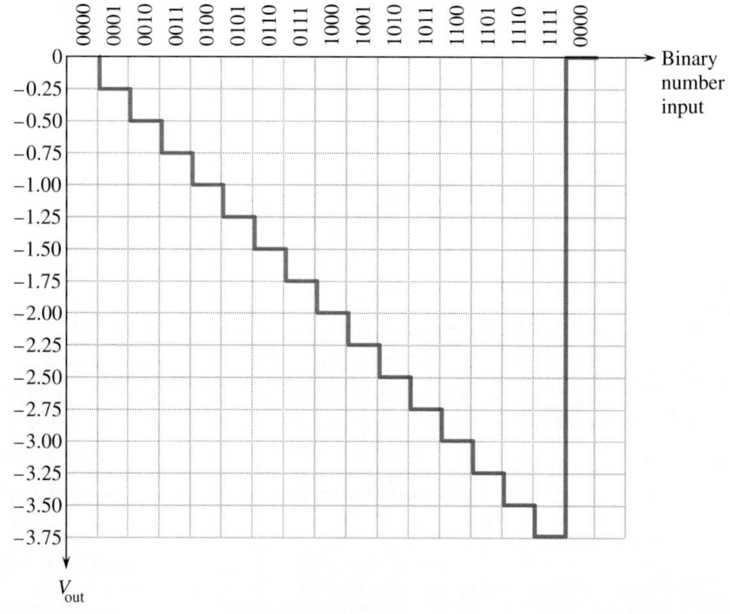

PRACTICE EXERCISE 14–9

If the 200-kΩ resistor in Figure 14–29(a) is changed to 400 kΩ, would the other resistor values have to be changed? If so, specify the values.

14–2 REVIEW QUESTIONS

1. Define *summing point*.
2. What is the value of R_f/R for a five-input averaging amplifier?

3. A certain scaling adder has two inputs, one having twice the weight of the other. If the resistor value for the lower weighted input is 10 kΩ, what is the value of the other input resistor?

14–3 THE INTEGRATOR AND DIFFERENTIATOR

An op-amp integrator simulates mathematical integration, which is basically a summing process that determines the total area under the curve of a function. An op-amp differentiator simulates mathematical differentiation, which is a process of determining the instantaneous rate of change of a function. It is not necessary for you to understand mathematical integration or differentiation, at this point, in order to learn how an integrator and differentiator work.

THE OP-AMP INTEGRATOR

A basic **integrator** circuit is shown in Figure 14–31. Notice that the feedback element is a capacitor that forms an RC circuit with the input resistor.

FIGURE 14–31
An op-amp integrator.

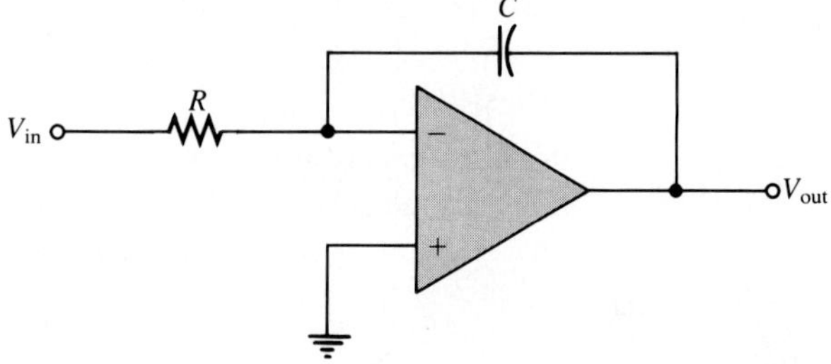

HOW A CAPACITOR CHARGES To understand how the integrator works, it is important to review how a capacitor charges. Recall that the charge Q on a capacitor is proportional to the charging current and the time.

$$Q = I_C t$$

Also, in terms of the voltage, the charge on a capacitor is

$$Q = CV_C$$

From these two relationships, the capacitor voltage can be expressed as

$$V_C = \left(\frac{I_C}{C}\right)t$$

You should recognize this expression as an equation for a straight line beginning at zero with a constant slope of I_C/C. (Remember from algebra that the general formula for a straight line is $y = mx + b$. In this case, $y = V_C$, $m = I_C/C$, $x = t$, and $b = 0$).

Recall that the capacitor voltage in a simple RC circuit is not linear but is exponential. This is because the charging current continuously decreases as the capacitor charges and causes the rate of change of the voltage to continuously decrease. The key thing about using an op-amp with an *RC* circuit to form an integrator is that the capacitor's charging current is made constant, thus producing a straight-line (linear) voltage rather than an exponential voltage. Now let's see why this is true.

In Figure 14–32, the inverting input of the op-amp is at virtual ground (0 V), so the voltage across R_i equals V_{in}. Therefore, the input current is

$$I_{in} = \frac{V_{in}}{R_i}$$

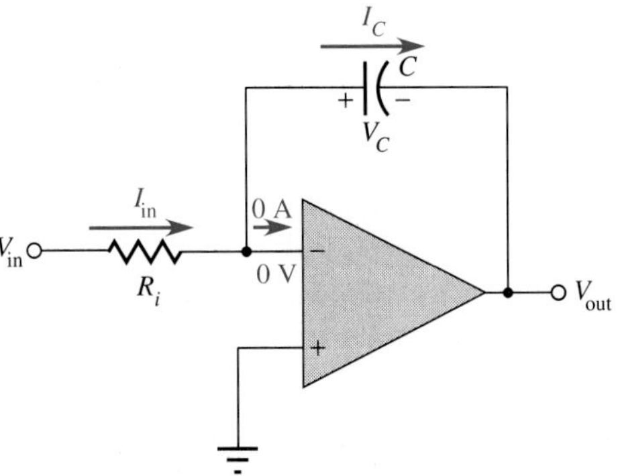

FIGURE 14–32
Currents in an integrator.

If V_{in} is a constant voltage, then I_{in} is also a constant because the inverting input always remains at 0 V, keeping a constant voltage across R_i. Because of the very high input impedance of the op-amp, there is negligible current into the inverting input. This makes all of the input current flow through the capacitor, as indicated in the figure, so

$$I_C = I_{in}$$

THE CAPACITOR VOLTAGE Since I_{in} is constant, so is I_C. The constant I_C charges the capacitor linearly and produces a linear voltage across *C*. The positive side of the capacitor is held at 0 V by the virtual ground of the op-amp. The voltage on the negative side of the capacitor decreases linearly from zero as the capacitor charges, as shown in Figure 14–33. This voltage is called a *negative ramp*.

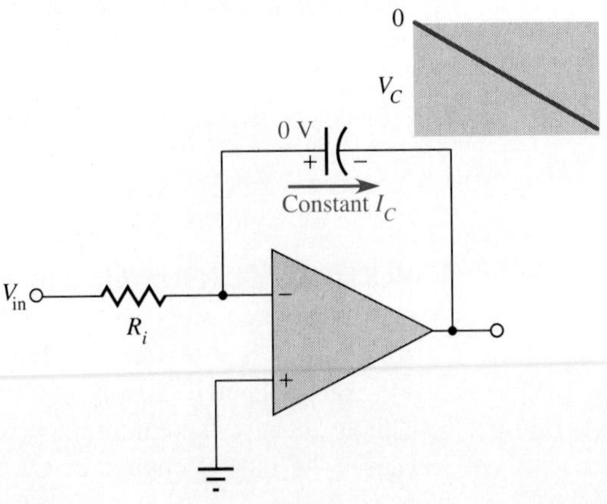

FIGURE 14–33
A linear ramp voltage is produced across *C* by the constant charging current.

THE OUTPUT VOLTAGE V_{out} is the same as the voltage on the negative side of the capacitor. When a constant input voltage in the form of a step or pulse (a pulse has a constant amplitude when high) is applied, the output ramp decreases negatively until the op-amp saturates at its maximum negative level. This is indicated in Figure 14–34.

FIGURE 14–34
A constant input voltage produces a ramp on the output.

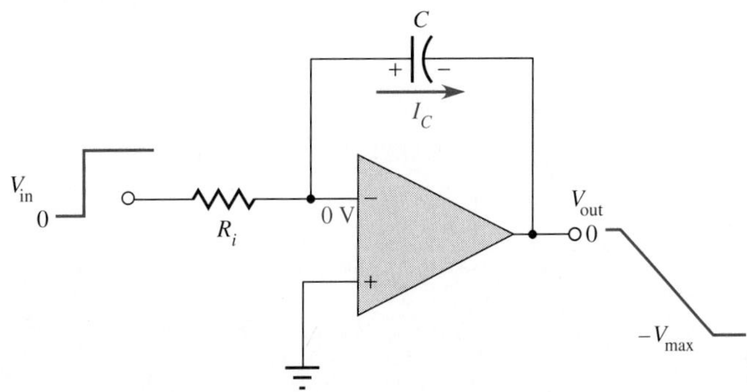

RATE OF CHANGE OF THE OUTPUT The rate at which the capacitor charges, and therefore the slope of the output ramp, is set by the ratio I_C/C, as you have seen. Since $I_C = V_{in}/R_i$, the rate of change or slope of the integrator's output voltage is

$$\frac{\Delta V_{out}}{\Delta t} = -\frac{V_{in}}{R_iC} \tag{14–9}$$

Integrators are especially useful in triangular-wave generators as you will see in Chapter 15.

EXAMPLE 14–10 (a) Determine the rate of change of the output voltage in response to a single pulse input, as shown for the integrator in Figure 14–35(a). The output voltage is initially zero.
(b) Draw the output waveform.

SOLUTION

(a) The rate of change of the output voltage is

$$\frac{\Delta V_{out}}{\Delta t} = \frac{-V_{IN}}{R_iC} = -\frac{5\text{ V}}{(10\text{ k}\Omega)(0.01\ \mu\text{F})}$$
$$= -50\text{ kV/s} = -50\text{ mV}/\mu\text{s}$$

(b) The rate of change was found to be -50 mV/μs in part (a). When the input is at +5 V, the output is a negative-going ramp. When the input is at 0 V, the output is a constant level. In 100 μs, the voltage decreases.

$$\Delta V_{out} = (-50\text{ mV}/\mu\text{s})(100\ \mu\text{s}) = -5\text{ V}$$

Therefore, the negative-going ramp reaches -5 V at the end of the pulse. The output voltage then remains constant at -5 V for the time that the input is zero. The waveforms are shown in Figure 14–35(b).

FIGURE 14–35

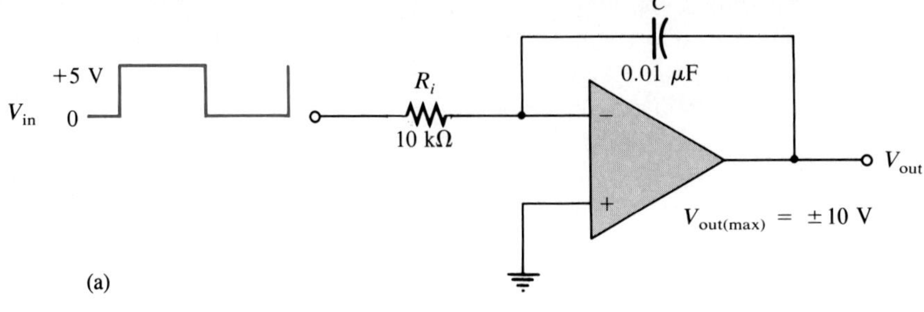

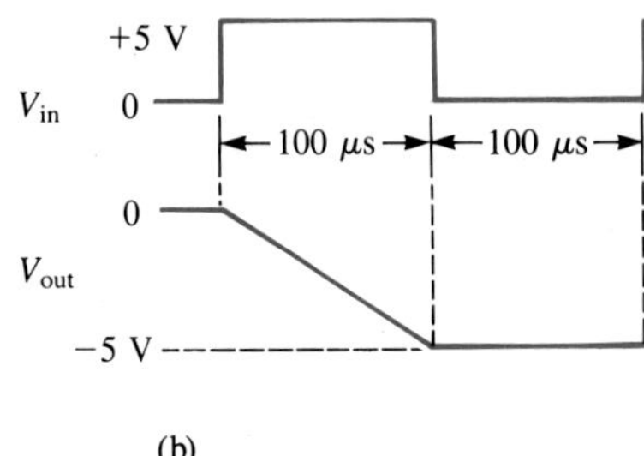

PRACTICE EXERCISE 14–10

Modify the integrator in Figure 14–35 to make the output change from 0 to −5 V in 50 μs with the same input.

THE OP-AMP DIFFERENTIATOR

A basic **differentiator** is shown in Figure 14–36. Notice how the placement of the capacitor and resistor differ from the integrator. The capacitor is now the input element. A differentiator produces an output that is proportional to the rate of change of the input voltage.

FIGURE 14–36
An op-amp differentiator.

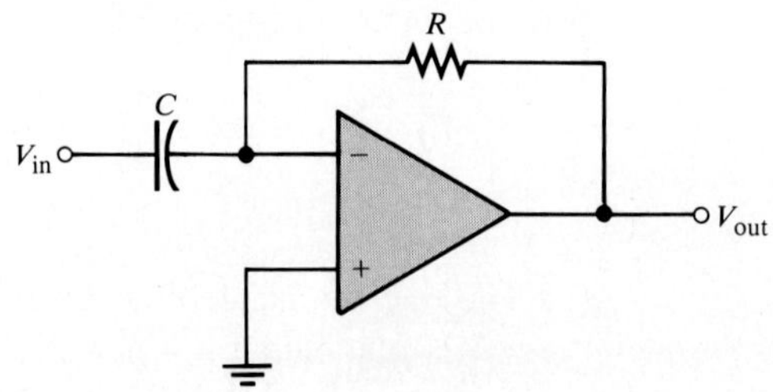

To see how the differentiator works, we will apply a positive-going ramp voltage to the input as indicated in Figure 14–37. In this case, $I_C = I_{in}$ and the voltage across the

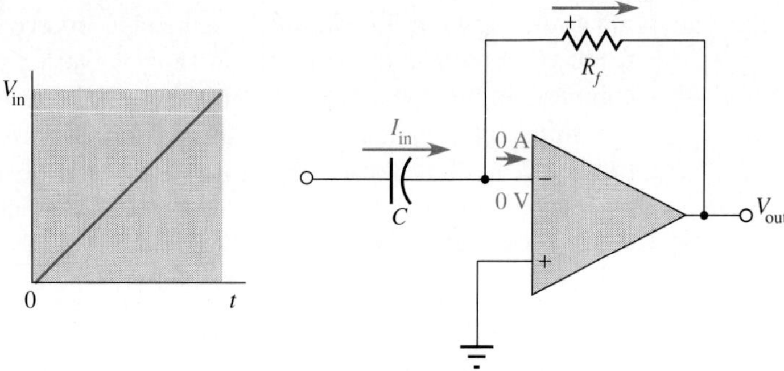

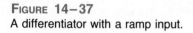

FIGURE 14–37
A differentiator with a ramp input.

capacitor is equal to V_{in} at all times ($V_C = V_{in}$) because of virtual ground on the inverting input.

From the basic formula, $V_C = (I_C/C)t$, we get

$$I_C = \left(\frac{V_C}{t}\right)C$$

Since the current into the inverting input is negligible, $I_R = I_C$. Both currents are constant because the slope of the capacitor voltage (V_C/t) is constant. The output voltage is also constant and equal to the voltage across R_f because one side of the feedback resistor is always 0 V (virtual ground).

$$V_{out} = I_R R_f = I_C R_f$$

$$V_{out} = \left(\frac{V_C}{t}\right)R_f C \qquad \textbf{(14–10)}$$

The output is negative when the input is a positive-going ramp and positive when the input is a negative-going ramp as illustrated in Figure 14–38. During the positive slope of the

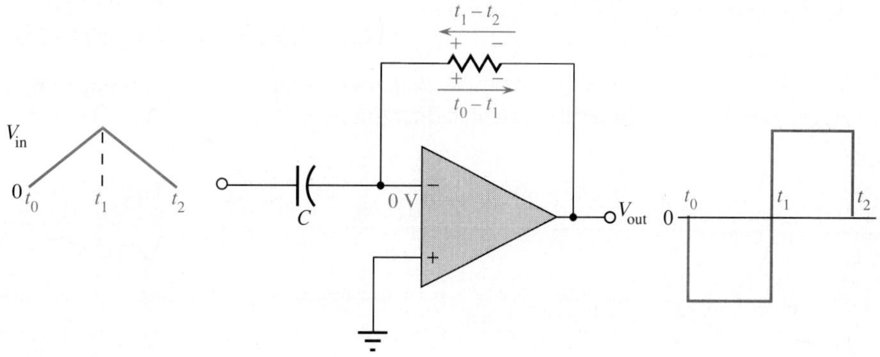

FIGURE 14–38
Output of a differentiator with a series of positive and negative ramps (triangle wave) on the input.

input, the capacitor is charging from the input source and the constant current through the feedback resistor is in the direction shown. During the negative slope of the input, the current is in the opposite direction because the capacitor is discharging.

Notice in Equation (14–10) that the term V_C/t is the slope of the input. If the slope increases, V_{out} increases. If the slope decreases, V_{out} decreases. So, the output voltage is proportional to the slope (rate of change) of the input. The constant of proportionality is the time constant, R_fC.

EXAMPLE 14–11 Determine the output voltage of the op-amp differentiator in Figure 14–39 for the triangular-wave input shown.

FIGURE 14–39

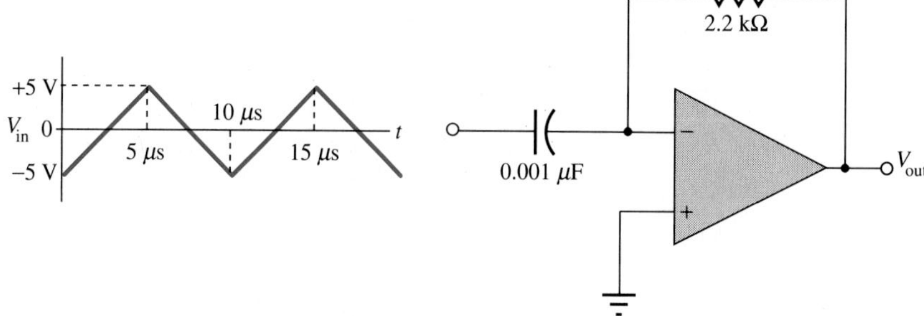

SOLUTION

Starting at $t = 0$, the input voltage is a positive-going ramp ranging from -5 V to $+5$ V (a $+10$-V change) in 5 μs. Then it changes to a negative-going ramp ranging from $+5$ V to -5 V (a -10-V change) in 5 μs.

The time constant is

$$R_fC = (2.2\ \text{k}\Omega)(0.001\ \mu\text{F}) = 2.2\ \mu\text{s}$$

The slope or rate of change (V_C/t) of the positive-going ramp is determined, and the output voltage is calculated as follows.

$$\frac{V_C}{t} = \frac{10\ \text{V}}{5\ \mu\text{s}} = 2\ \text{V}/\mu\text{s}$$

$$V_{out} = (2\ \text{V}/\mu\text{s})2.2\ \mu\text{s} = +4.4\ \text{V}$$

Likewise, the slope of the negative-going ramp is -2 V/μs, and the output voltage is calculated as follows.

$$V_{out} = (-2\ \text{V}/\mu\text{s})2.2\ \mu\text{s} = -4.4\ \text{V}$$

Finally, the output voltage waveform is graphed relative to the input as shown in Figure 14–40.

FIGURE 14–40

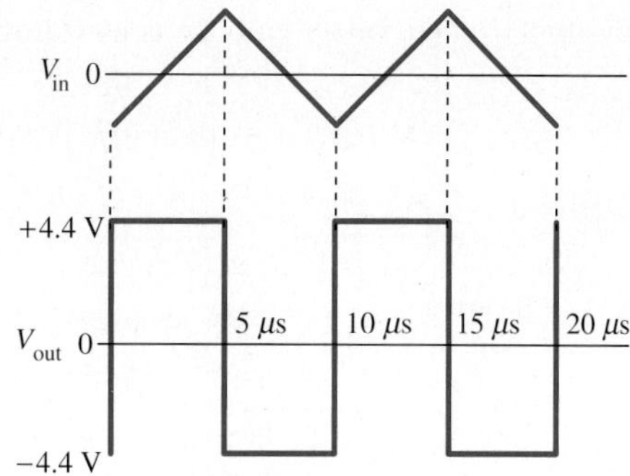

PRACTICE EXERCISE 14–11

What would the output voltage be if the feedback resistor in Figure 14–39 is changed to 3.3 kΩ? ■

14–3 REVIEW QUESTIONS

1. What is the feedback element in an op-amp integrator?
2. For a constant input voltage to an integrator, why is the voltage across the capacitor linear?
3. What is the feedback element in an op-amp differentiator?
4. How is the output of a differentiator related to the input?

14–4 THE INSTRUMENTATION AMPLIFIER

An instrumentation amplifier consists of three operational amplifiers and several resistors. Integrated circuit manufacturers provide this circuitry on a single chip and package it as one device. Common characteristics are high input impedance (typically 300 MΩ), high voltage gain, and excellent CMRR (typically in excess of 100 dB). Instrumentation amplifiers are commonly used in data acquisition systems where remote sensing of input variables is required.

BASIC OPERATION

A simplified version of an instrumentation amplifier is shown in Figure 14–41. Op-amps A_1 and A_2 are noninverting amplifier stages that provide high input impedance and voltage gain. Op-amp A_3 is a unity-gain amplifier. When R_G is connected externally, as illustrated in Figure 14–42, op-amp A_1 receives the differential input signal V_{in1} on its noninverting input and amplifies it with a gain of $1 + R_{f1}/R_G$. Op-amp A_1 also receives the input signal V_{in2} through op-amp A_2, R_{f2}, and R_G. V_{in2} appears on the inverting input of op-amp A_1 and is amplified by a gain of R_{f1}/R_G. Also, the common-mode voltage on the noninverting input is amplified by the common-mode gain of A_1. (A_{cm} is typically less than unity.) The total output voltage of op-amp A_1 is as follows.

$$V_{out1} = \left(1 + \frac{R_{f1}}{R_G}\right)V_{in1} - \left(\frac{R_{f1}}{R_G}\right)V_{in2} + V_{cm} \quad (14\text{–}11)$$

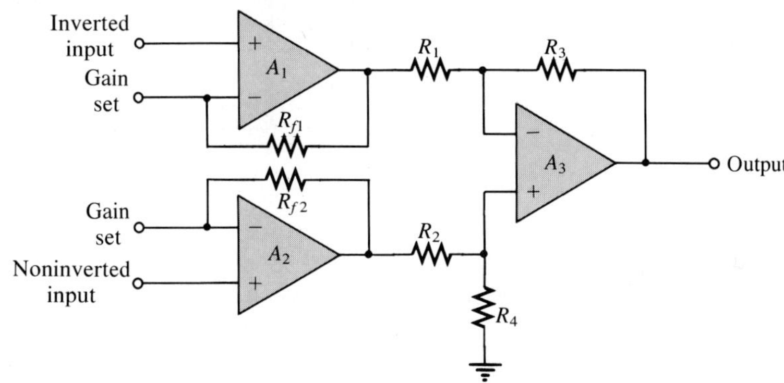

FIGURE 14–41 Basic instrumentation amplifier.

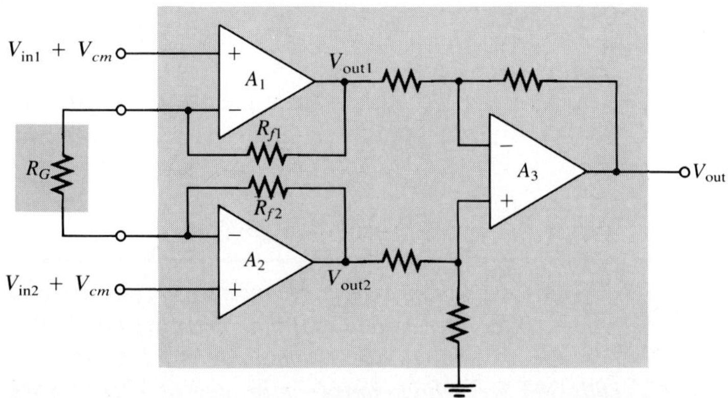

FIGURE 14–42 Instrumentation amplifier with gain-setting resistor R_G connected.

A similar analysis can be applied to op-amp A_2, resulting in the following expression.

$$V_{out2} = \left(1 + \frac{R_{f2}}{R_G}\right)V_{in2} - \left(\frac{R_{f2}}{R_G}\right)V_{in1} + V_{cm} \quad (14\text{–}12)$$

The differential input voltage to op-amp A_3 is $V_{out2} - V_{out1}$.

$$V_{out2} - V_{out1} = \left(1 + \frac{R_{f2}}{R_G} + \frac{R_{f1}}{R_G}\right)V_{in2} - \left(\frac{R_{f2}}{R_G} + 1 + \frac{R_{f1}}{R_G}\right)V_{in1} + V_{cm} - V_{cm}$$

For $R_{f1} = R_{f2} = R_f$,

$$V_{out2} - V_{out1} = \left(1 + \frac{2R_f}{R_G}\right)V_{in2} - \left(1 + \frac{2R_f}{R_G}\right)V_{in1} + V_{cm} - V_{cm}$$

Notice that, since the common-mode voltages (V_{cm}) are equal, they cancel out. Factoring gives the following result, which is the differential input to op-amp A_3.

$$V_{out2} - V_{out1} = \left(1 + \frac{2R_f}{R_G}\right)(V_{in2} - V_{in1})$$

Because op-amp A_3 has unity gain, the final output of the instrumentation amplifier is $V_{out} = (1)(V_{out2} - V_{out1})$.

$$V_{out} = \left(1 + \frac{2R_f}{R_G}\right)(V_{in2} - V_{in1}) \quad (14\text{–}13)$$

The closed-loop gain is

$$A_{cl} = 1 + \frac{2R_f}{R_G} \quad (14\text{–}14)$$

Equation (14–14) shows that the differential gain of the instrumentation amplifier can be set by the value of R_G. R_{f1} and R_{f2} are normally internal to the IC chip, and their value is set by the manufacturer. As an example, for the LH0036, $R_{f1} = R_{f2} = 25\ \text{k}\Omega$.

Rearranging Equation (14–14) gives an expression for calculating the value of R_G needed for a desired value of closed-loop gain if R_f is known.

$$A_{cl} = \frac{R_G + 2R_f}{R_G}$$

$$A_{cl}R_G = R_G + 2R_f$$

$$A_{cl}R_G - R_G = 2R_f$$

$$R_G(A_{cl} - 1) = 2R_f$$

$$R_G = \frac{2R_f}{A_{cl} - 1} \quad (14\text{–}15)$$

EXAMPLE 14–12 Determine the value of the external gain-setting resistor R_G for an IC instrumentation amplifier having an $R_{f1} = R_{f2} = 25\ \text{k}\Omega$ for a closed-loop gain of 500.

SOLUTION

$$R_G = \frac{2R_f}{A_{cl} - 1} = \frac{50\ \text{k}\Omega}{500 - 1} \cong 100\ \Omega$$

PRACTICE EXERCISE 14–12

What value of external gain-setting resistor is required for an instrumentation amplifier having $R_{f1} = R_{f2} = 39\ \text{k}\Omega$ for a closed-loop gain of 275?

APPLICATIONS

The instrumentation amplifier is normally used to measure small differential signal voltages that are superimposed on a common-mode voltage often larger than the signal voltage. Applications often include a situation where a quantity is measured by a remote sensing device (transducer), and the resulting small electrical signal is sent over a long line subject to large common-mode voltages. The instrumentation amplifier at the end of the line must amplify the small signal and reject the large common-mode voltage. Figure 14–43 illustrates this.

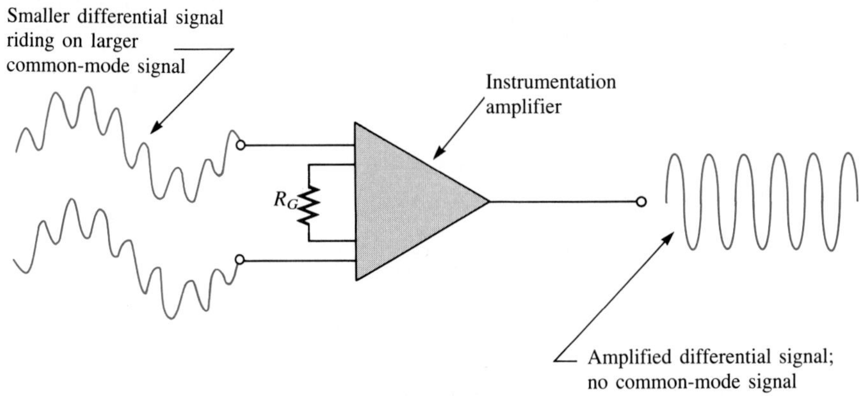

FIGURE 14–43
Instrumentation amplifier showing rejection of large common-mode signal.

14–4 REVIEW QUESTIONS

1. What makes up the internal circuitry of a basic instrumentation amplifier?
2. What is the value of the gain-setting resistor for a closed-loop gain of 800? The manufacturer of the instrumentation amplifier specified $R_f = 15\ k\Omega$.

14–5 MORE OP-AMP CIRCUITS

This section introduces a few more op-amp circuits that represent basic applications of the op-amp. You will learn about the constant-current source, the current-to-voltage converter, the voltage-to-current converter, and the peak detector. This is, of course, not a comprehensive coverage of all possible op-amp circuits but is intended only to introduce you to some common and basic uses.

CONSTANT-CURRENT SOURCE

The purpose of a constant-current source is to deliver a load current that remains constant when the load resistance changes. Figure 14–44 shows a basic circuit in which a stable voltage source (V_{in}) provides a constant current (I_i) through the input resistor (R_i). Since the inverting (−) input of the op-amp is at virtual ground (0 V), the value of I_i is determined by V_{IN} and R_i as

$$I_i = \frac{V_{IN}}{R_i}$$

Now, since the internal input impedance of the op-amp is extremely high (ideally infinite), practically all of I_i flows through R_L, which is connected in the feedback path. Since $I_i = I_L$,

$$I_L = \frac{V_{IN}}{R_i} \tag{14–16}$$

If R_L changes, I_L remains constant as long as V_{IN} and R_i are held constant.

FIGURE 14–44
A basic constant-current source.

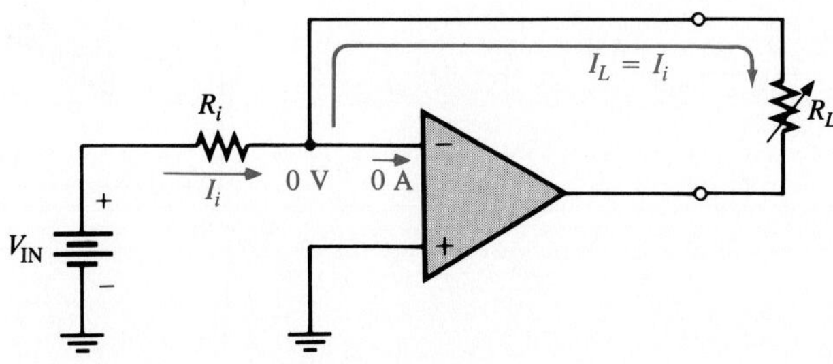

CURRENT-TO-VOLTAGE CONVERTER

The purpose of a current-to-voltage converter is to convert a variable input current to a proportional output voltage. A basic circuit that accomplishes this is shown in Figure 14–45(a). Since practically all of I_i flows through the feedback path, the voltage dropped across R_f is I_iR_f. Because the left side of R_f is at virtual ground (0 V), the output voltage equals the voltage across R_f, which is proportional to I_i.

$$V_{out} = I_iR_f \tag{14–17}$$

A specific application of this circuit is illustrated in Figure 14–45(b), where a photoconductive cell is used to sense changes in light level. As the amount of light changes, the current through the photoconductive cell varies because of the cell's change in resistance. This change in resistance produces a proportional change in the output voltage ($\Delta V_{out} = \Delta I_iR_f$).

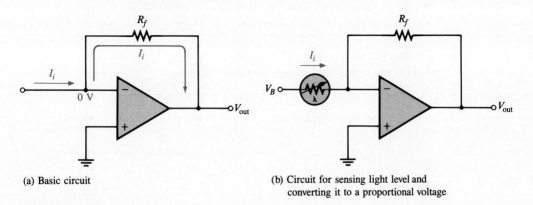

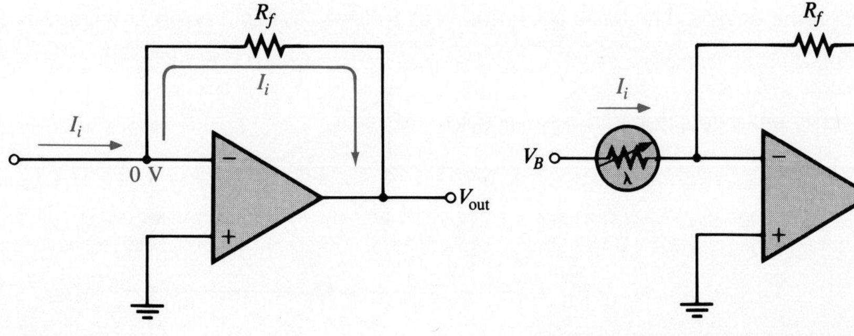

(a) Basic circuit

(b) Circuit for sensing light level and converting it to a proportional voltage

FIGURE 14–45
Current-to-voltage converter.

VOLTAGE-TO-CURRENT CONVERTER

A basic voltage-to-current converter is shown in Figure 14–46. This circuit is used in applications where it is necessary to have an output (load) current that is controlled by an input voltage.

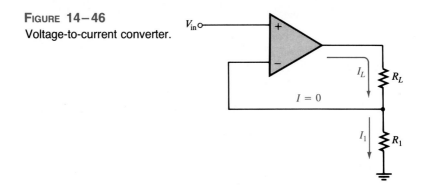

FIGURE 14–46
Voltage-to-current converter.

Neglecting the input offset voltage, both inverting and noninverting input terminals of the op-amp are at the same voltage, V_{in}. Therefore, the voltage across R_1 equals V_{in}. Since negligible current flows into the inverting input, the same current that flows through R_1 also flows through R_L; thus

$$I_L = \frac{V_{in}}{R_1} \tag{14–18}$$

PEAK DETECTOR

An interesting application of the op-amp is in a peak detector circuit such as the one shown in Figure 14–47. In this case the op-amp is used as a comparator. The purpose of this circuit is to detect the peak of the input voltage and store that peak voltage on a capacitor. For example, this circuit can be used to detect and store the maximum value of a voltage surge; this value can then be measured at the output with a voltmeter or recording device. The basic operation is as follows. When a positive voltage is applied to

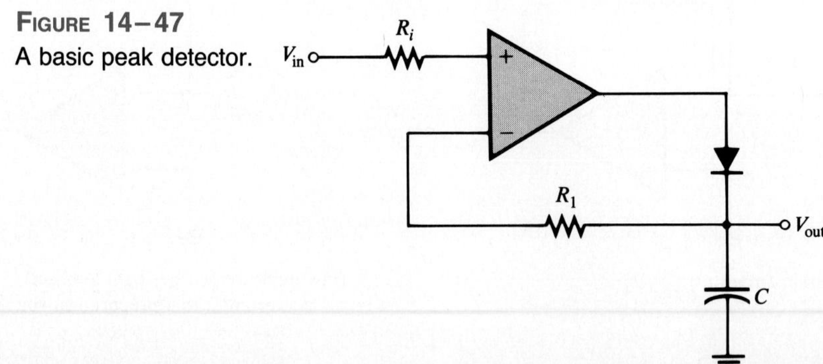

FIGURE 14–47
A basic peak detector.

the noninverting input of the op-amp through R_i, the high-level output voltage of the op-amp forward-biases the diode and charges the capacitor. The capacitor continues to charge until its voltage reaches a value equal to the input voltage and thus both op-amp inputs are at the same voltage. At this point, the op-amp comparator switches, and its output goes to the low level. The diode is now reverse-biased, and the capacitor stops charging. It has reached a voltage equal to the peak of V_{in} and will hold this voltage until the charge eventually leaks off. If a greater input peak occurs, the capacitor charges to the new peak.

14–5 REVIEW QUESTIONS

1. For the constant-current source in Figure 14–44, the input reference voltage is 6.8 V and R_i is 10 kΩ. What value of constant current does the circuit supply to a 1 kΩ load? To a 5 kΩ load?
2. What element determines the constant of proportionality that relates input current to output voltage in the current-to-voltage converter?

14–6 TROUBLESHOOTING

Although integrated circuit op-amps are extremely reliable and trouble-free, failures do occur from time to time. One type of internal failure mode is a condition where the op-amp output is in a saturated state resulting in a constant high or constant low level, regardless of the input. Also, external component failures will produce various types of failure modes in op-amp circuits. Some examples are presented in this section.

Figure 14–48 illustrates an internal failure of a comparator circuit that results in a failed output.

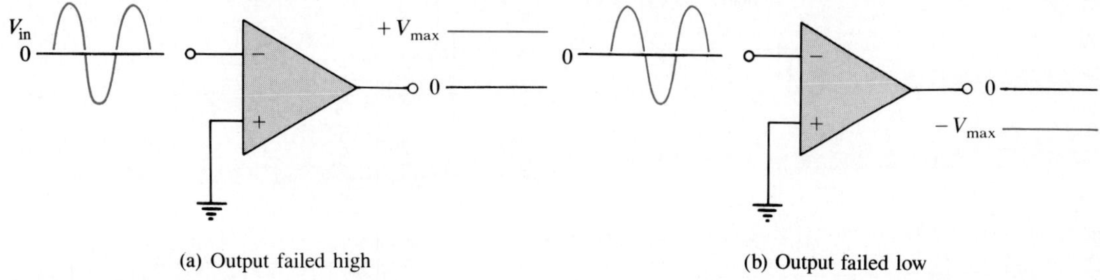

FIGURE 14–48
Typical internal comparator failures.

EXTERNAL COMPONENT FAILURES IN COMPARATOR CIRCUITS

Comparators with zener-bounding and hysteresis are shown in Figure 14–49. In addition to a failure of the op-amp itself, a zener diode or one of the resistors could go bad. Suppose one of the zener diodes opens. This effectively eliminates both zeners, and the

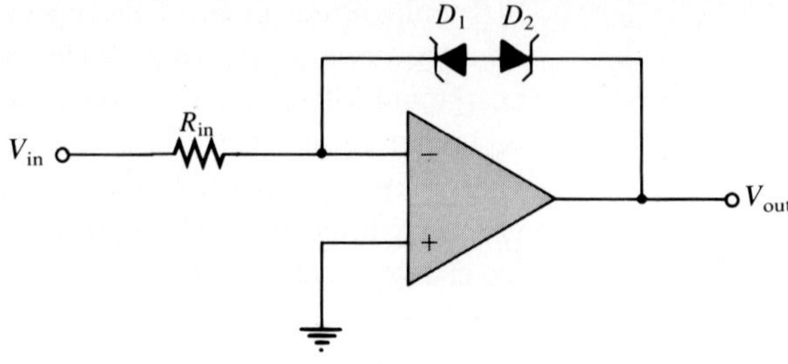

FIGURE 14–49
A bounded comparator.

circuit operates as an unbounded comparator, as indicated in Figure 14–50(a). With a shorted diode, the output is limited to the zener voltage (bounded) only in one direction depending on which diode remains operational, as illustrated in Figure 14–50(b). In the other direction, the output is held at the forward diode voltage.

FIGURE 14–50
Examples of comparator circuit failures and their effects.

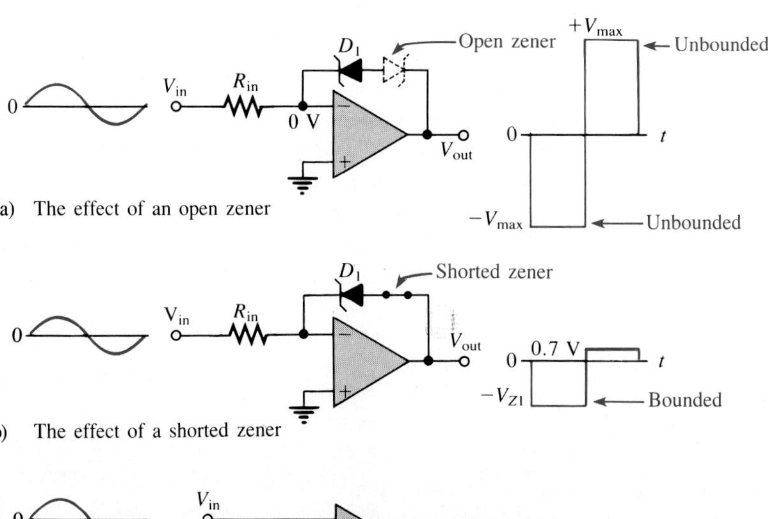

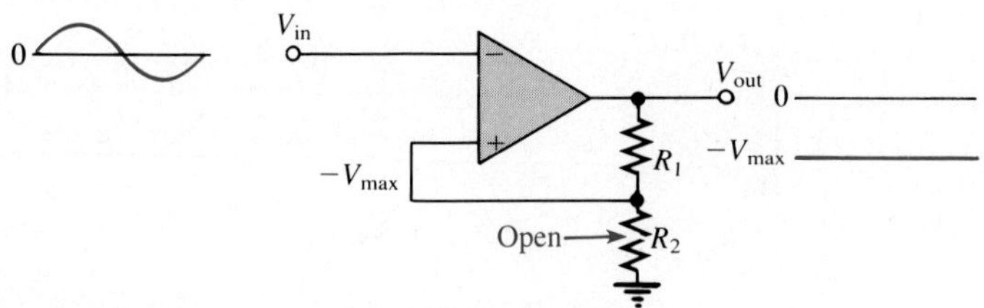

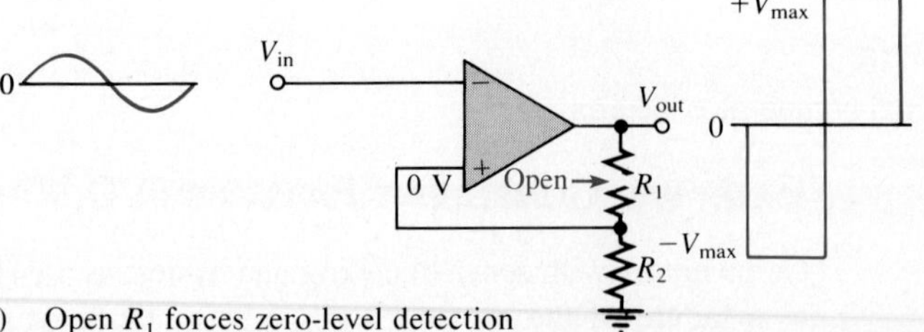

Recall that R_1 and R_2 set the UTP and LTP for the hysteresis comparator. Now, suppose that R_2 opens. Essentially all of the output voltage is fed back to the noninverting input, and, since the input voltage will never exceed the output, the device will remain in one of its saturated states. This symptom can also indicate a faulty op-amp, as mentioned before. Now, assume that R_1 opens. This leaves the noninverting input near ground potential and causes the circuit to operate as a zero-level detector. These conditions are shown in parts (c) and (d) of the figure.

EXAMPLE 14–13

One channel of a dual-trace oscilloscope is connected to the comparator output and the other channel to the input signal, as shown in Figure 14–51. From the observed waveforms, determine if the circuit is operating properly, and if not, what the most likely failure is.

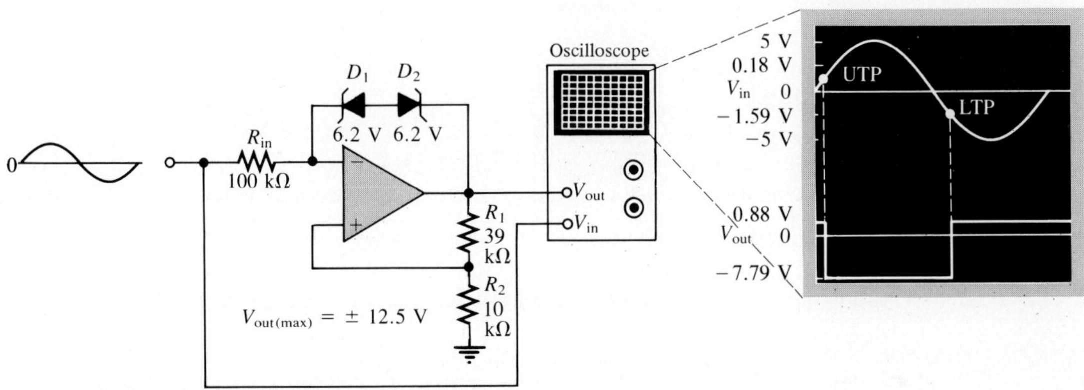

FIGURE 14–51

SOLUTION

The output should be limited to ±8.67 V. However, the positive maximum is +0.88 V and the negative maximum is −7.79 V. This indicates that D_2 is shorted. Refer to Example 14–3 for analysis.

PRACTICE EXERCISE 14–13

What would the output voltage look like if D_1 shorted rather than D_2?

COMPONENT FAILURES IN SUMMING AMPLIFIERS

If one of the input resistors in a unity-gain summing amplifier opens, the output will be less than the normal value by the amount of the voltage applied to the open input. Stated another way, the output will be the sum of the remaining input voltages.

If the summing amplifier has a nonunity gain, an open input resistor causes the output to be less than normal by an amount equal to the gain times the voltage at the open input.

EXAMPLE 14–14 **(a)** What is the normal output voltage in Figure 14–52?
(b) What is the output voltage if R_2 opens?
(c) What happens if R_5 opens?

FIGURE 14–52

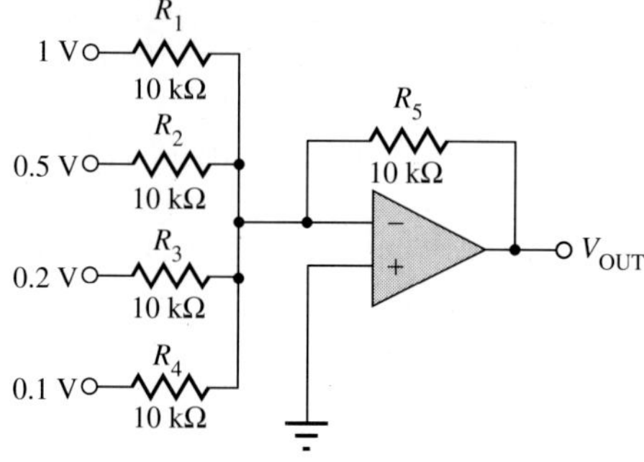

SOLUTION

(a) $V_{OUT} = -(1\text{ V} + 0.5\text{ V} + 0.2\text{ V} + 0.1\text{ V}) = -1.8\text{ V}$
(b) $V_{OUT} = -(1\text{ V} + 0.2\text{ V} + 0.1\text{ V}) = -1.3\text{ V}$
(c) If the feedback resistor opens, the circuit becomes a comparator and the output goes to $-V_{max}$.

PRACTICE EXERCISE 14–14
In Figure 14–52, $R_5 = 47\text{ k}\Omega$. What is the output voltage if R_1 opens?

As another example, let's look at an averaging amplifier. An open input resistor will result in an output voltage that is the average of all the inputs with the open input averaged in as a zero.

EXAMPLE 14–15 **(a)** What is the normal output voltage for the averaging amplifier in Figure 14–53?
(b) If R_4 opens, what is the output voltage? What does the output voltage represent?

FIGURE 14–53

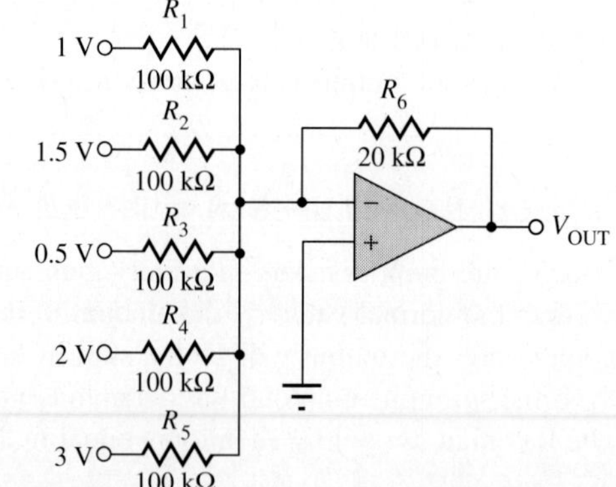

SOLUTION

(a) $$V_{OUT} = -\frac{20\text{ k}\Omega}{100\text{ k}\Omega}(1\text{ V} + 1.5\text{ V} + 0.5\text{ V} + 2\text{ V} + 3\text{ V}) = -\frac{1}{5}(8\text{ V}) = -1.6\text{ V}$$

(b) $$V_{OUT} = -\frac{20\text{ k}\Omega}{100\text{ k}\Omega}(1\text{ V} + 1.5\text{ V} + 0.5 + 3\text{ V}) = -\frac{1}{5}(6\text{ V}) = -1.2\text{ V}$$

1.2 V is the average of five voltages with the 2-V input replaced by 0 V. Notice that the output is not the average of the four remaining input voltages.

PRACTICE EXERCISE 14–15

If R_4 is open, as was the case in this example, what would you have to do to make the output equal to the average of the remaining four input voltages? ■

14–6 REVIEW QUESTIONS

1. Describe one type of internal op-amp failure.
2. If a certain malfunction is attributable to more than one possible component failure, what would you do to isolate the problem?

14–7 A SYSTEM APPLICATION

The system presented at the beginning of this chapter is a dual-slope analog-to-digital (A/D) converter. This is one of several methods for A/D conversion. You saw another type, called a simultaneous A/D converter as an example in the chapter. Although A/D conversion is used for many purposes, in this particular application the converter is used to change an audio signal into digital form for recording. Even though many parts of this system are digital, we are going to concentrate on the analog board, which includes op-amps used in several types of circuits that you have learned about in this chapter. In this section, you will

- ☐ ***See one way in which a summing amplifier is used.***
- ☐ ***See how the integrator is a key element in A/D conversion.***
- ☐ ***See how a comparator is used.***
- ☐ ***Translate between a printed circuit board and a schematic.***
- ☐ ***Troubleshoot some common system problems.***

A BRIEF DESCRIPTION OF THE SYSTEM

The dual-slope A/D converter in Figure 14–54 accepts an audio signal and converts it to a series of digital codes for the purpose of recording. The audio signal voltage is applied to the sample-and-hold circuit. At a given instant, a sample pulse causes the amplitude at that point on the audio waveform to be converted to a proportional dc level which is then processed by the rest of the circuits and represented by a digital code. The sample pulses occur at a much higher rate than the audio frequency so that a sufficient number of points on the audio waveform are sampled and converted to obtain an accurate representation of the audio signal. A rough approximation of the sampling process is illustrated in Figure 14–55.

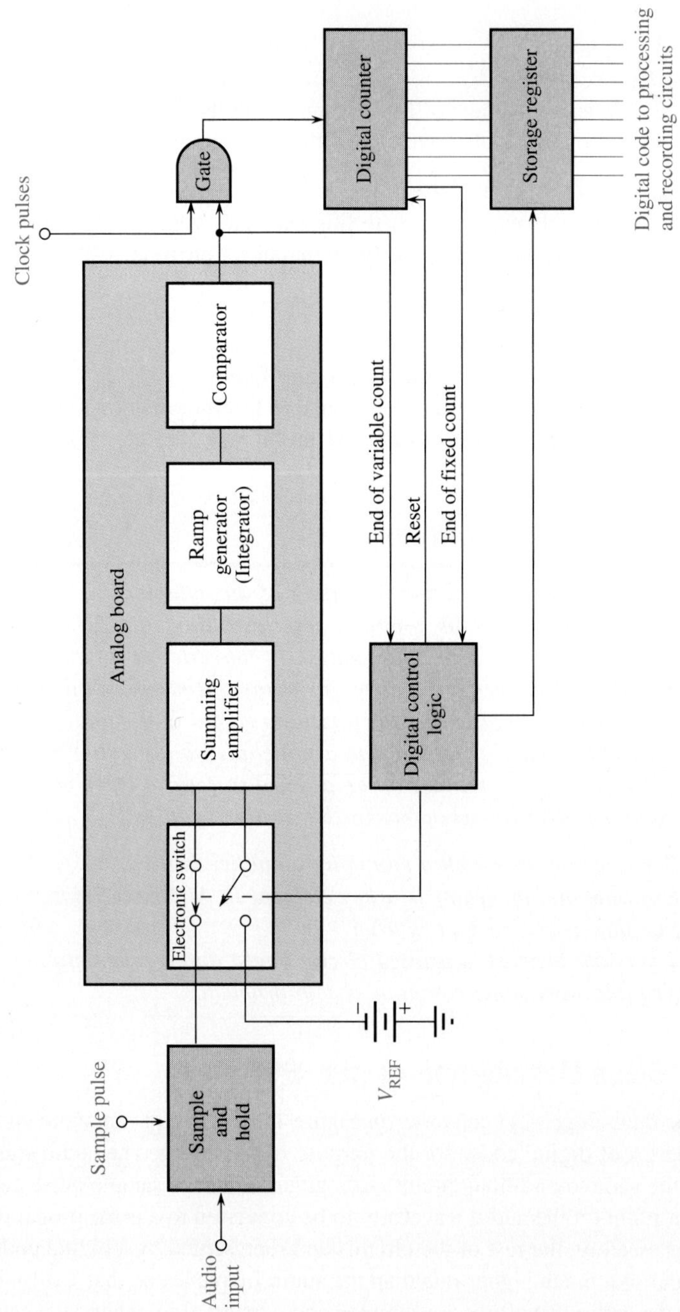

FIGURE 14–54
Basic dual-slope analog-to-digital converter.

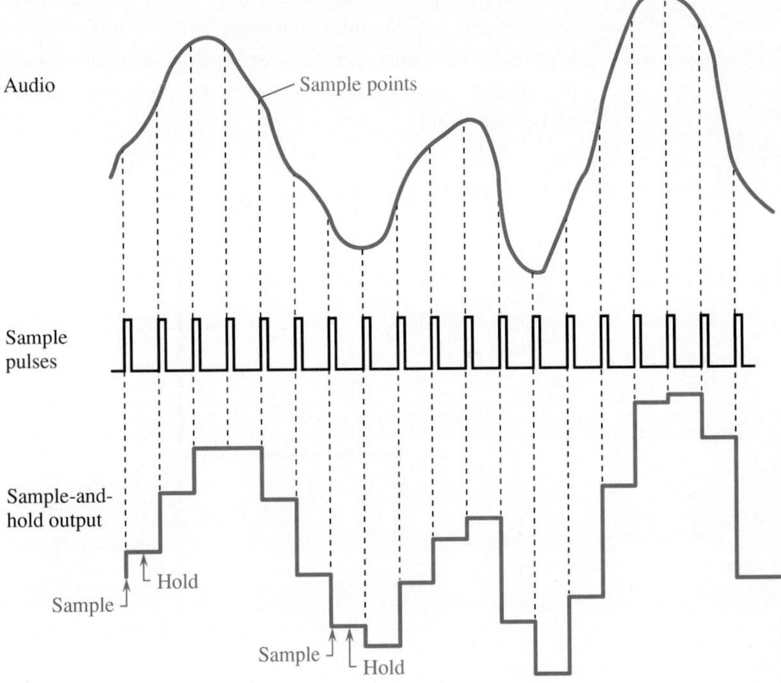

FIGURE 14–55
Sample-and-hold process. The sample-and-hold output is a rough approximation of the audio voltage for purposes of illustration. As the frequency of the sample pulses is increased, an increasingly accurate representation is achieved.

During the time between each sample pulse, the dc level from the sample-and-hold circuit is switched into the summing amplifier by the electronic switch under digital control and then applied to the integrator. At the same time, the digital counter starts counting up from zero. During the fixed interval of the counting sequence, the integrator produces a negative-going ramp whose slope depends on the level of the sampled audio voltage. At the end of the fixed counting sequence, the ramp voltage at the output of the integrator has reached a voltage that is proportional to the sampled audio voltage. At this time, the digital control logic switches from the sample-and-hold input to the reference voltage input and resets the digital counter to zero. The summing amplifier applies this negative dc reference to the integrator input, which starts a positive-going ramp on the output. This ramp has a slope that is fixed by the value of the dc reference voltage. At the same time, the digital counter begins to count up again and will continue to count up until the positive ramp output of the integrator reaches zero volts. At this point, the comparator switches to its negative saturated output voltage and disables the gate so that there are no additional clock pulses to the counter. At this time, the digital code in the counter is proportional to the time that it took for the positive-going ramp to reach zero and it will vary for each different sampled value. Recall that the positive-going ramp started at a negative voltage that was dependent on the sampled value of the audio signal. Therefore, the digital code in the counter is also proportional to, and represents, the amplitude of the sampled audio voltage. This code is then shifted out to the register and then processed and recorded.

This process is repeated many times during a typical audio cycle. The result is a sequence of digital codes that represent the audio voltage amplitude as it varies with time. Figure 14–56 illustrates this for several sampled values. As mentioned, you will focus on the analog board, which contains the electronic switch, summing amplifier, integrator, and comparator.

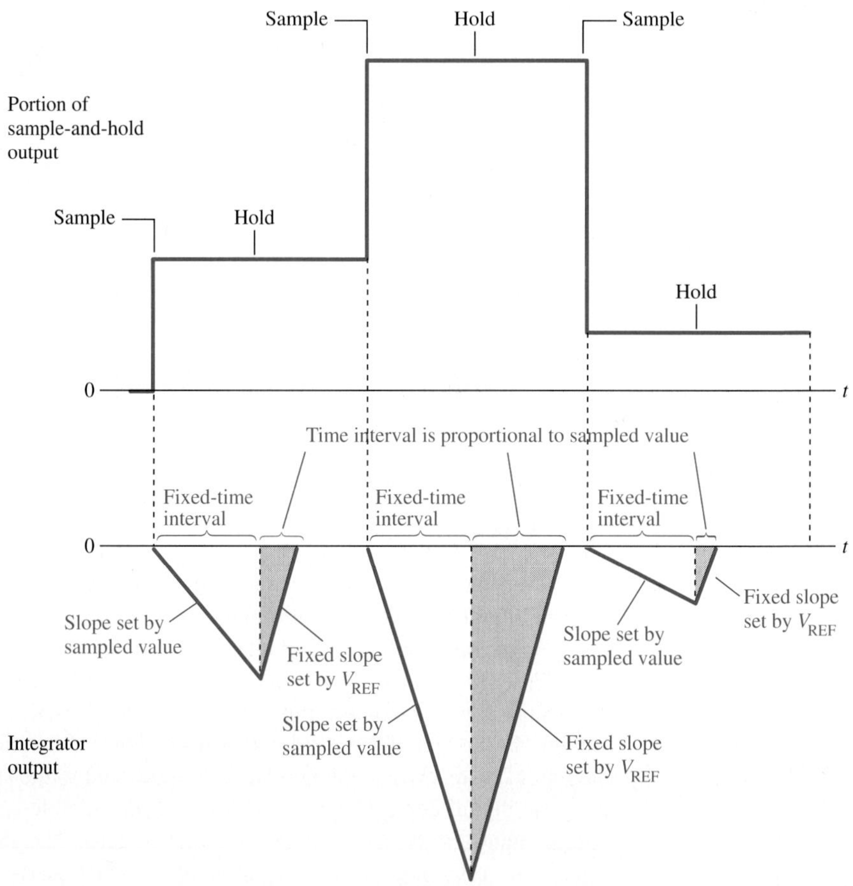

FIGURE 14–56
During the fixed-time interval of the negative-going ramp, the sampled audio input is applied to the integrator. During the variable-time interval of the positive-going ramp, the reference voltage is applied to the integrator. The counter sets the fixed-time interval and is then reset. Another count begins during the variable interval and the code in the counter at the end of this interval represents the sampled value.

Now, so you can take a closer look at the analog board, let's take it out of the system and put it on the test bench.

ON THE TEST BENCH

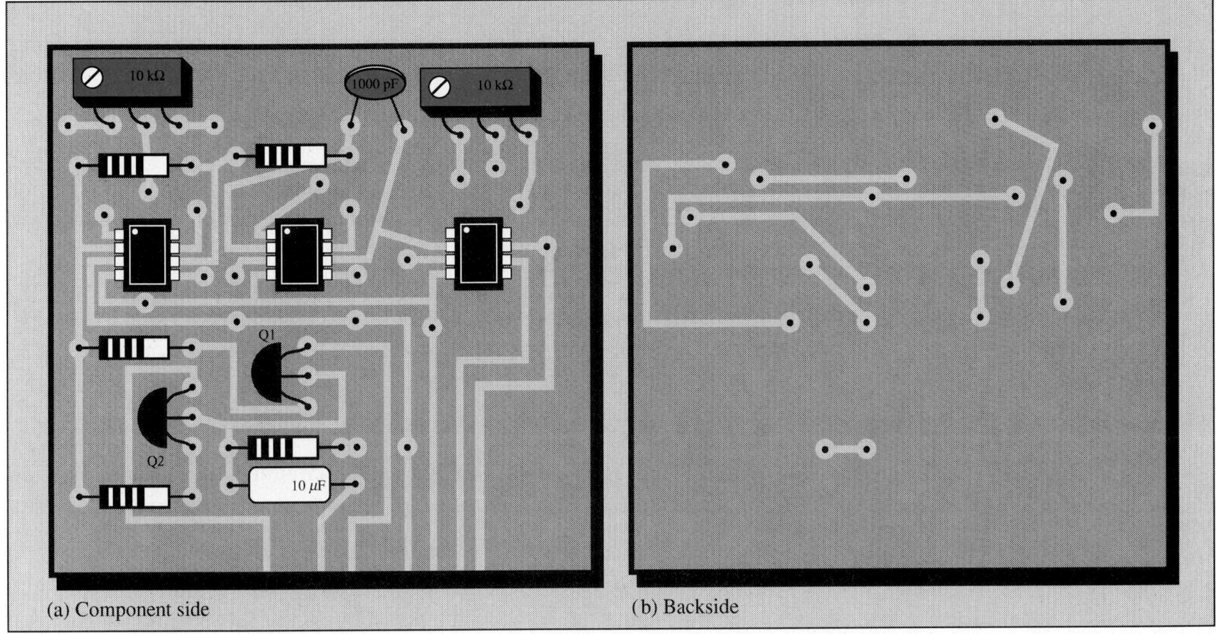

FIGURE 14–57

■ ACTIVITY 1 RELATE THE PC BOARD TO THE SCHEMATIC

Identify each component on the circuit board in Figure 14–57 using the schematic in Figure 14–58. Also, identify each input and output on the board. Check to make sure the interconnections on both sides of the board correspond with the schematic. Locate

FIGURE 14–58

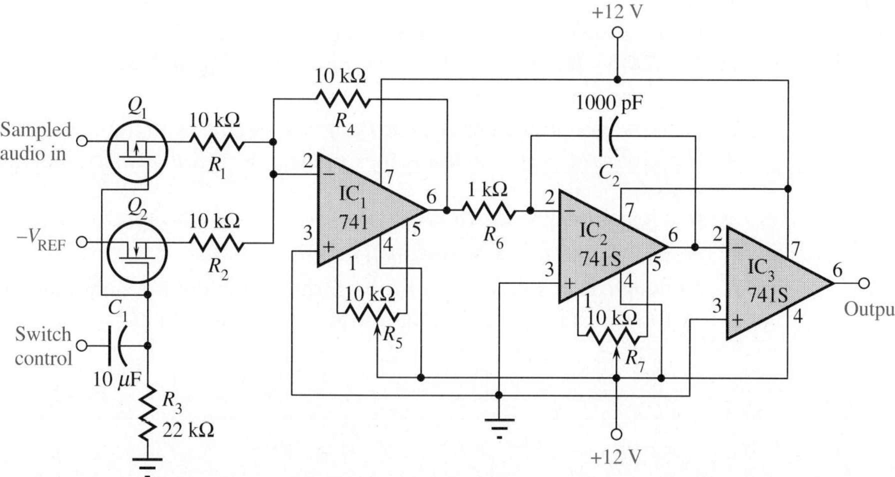

and identify the components associated with the electronic switch, the summing amplifier, the integrator, and the comparator.

ACTIVITY 2 ANALYZE THE CIRCUIT

STEP 1 Determine the gain of the summing amplifier.

STEP 2 Determine the slope of the integrator ramp in volts per microsecond when a sampled audio voltage of +2 V is applied.

STEP 3 Determine the slope of the integrator ramp in volts per microsecond when the reference voltage of −8 V is applied.

STEP 4 Given that the reference voltage is −8 V and the fixed-time interval of the negative-going slope is 1 μs, sketch the dual-slope output of the integrator when an instantaneous audio voltage of +3 V is sampled.

STEP 5 Assuming that the maximum audio voltage to be sampled is +6 V, determine the maximum audio frequency that can be sampled by this particular system if there are to be 100 samples per cycle. What is the sample pulse rate in this case?

ACTIVITY 3 WRITE A TECHNICAL REPORT

Discuss the detailed operation of the analog board circuitry and explain how it interfaces with the overall system. Discuss the purpose of each component on the circuit board.

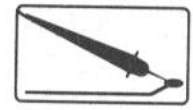

ACTIVITY 4 TROUBLESHOOT THE CIRCUIT FOR EACH OF THE FOLLOWING PROBLEMS BY STATING THE PROBABLE CAUSE OR CAUSES

1. Zero volts on the output of IC_1 when there are voltages on the sampled audio input and on the reference voltage input.
2. IC_1 goes back and forth between its saturated states as the positive audio voltage and the negative reference voltage are alternately switched in.
3. The inverting input of IC_1 never goes negative.
4. The output of IC_2 stays at zero volts under normal operating conditions.

14–7 REVIEW QUESTIONS

1. Identify the summing amplifier, the integrator, and the comparator by IC number.
2. The 741S op-amps are high slew-rate devices with a minimum slew rate of 10 V/μs. Why are these used in this application?
3. What is the purpose of R_5 and R_7 in the circuit of Figure 14–58?
4. What type of output voltage does the analog board produce and how is it used in the system?
5. If a sample pulse rate of 500 kHz is used, how long does each sampled audio voltage remain on the input to the analog board?
6. Although IC_1 is connected in the form of a summing amplifier, it does not actually perform a summing operation in this application. Why?

Summary

- □ In an op-amp comparator, when the input voltage exceeds a specified reference voltage, the output changes state.
- □ Hysteresis gives an op-amp noise immunity.
- □ A comparator switches to one state when the input reaches the upper trigger point (UTP) and back to the other state when the input drops below the lower trigger point (LTP).
- □ The difference between the UTP and the LTP is the hysteresis voltage.
- □ Bounding limits the output amplitude of a comparator.
- □ The output voltage of a summing amplifier is proportional to the sum of the input voltages.
- □ An averaging amplifier is a summing amplifier with a closed-loop gain equal to the reciprocal of the number of inputs.
- □ In a scaling adder, a different weight can be assigned to each input, thus making the input contribute more or contribute less to the output.
- □ Integration is a mathematical process for determining the area under a curve.
- □ Integration of a step produces a ramp with a slope proportional to the amplitude.
- □ Differentiation is a mathematical process for determining the rate of change of a function.
- □ Differentiation of a ramp produces a step with an amplitude proportional to the slope.
- □ A basic instrumentation amplifier consists of three op-amps.
- □ An instrumentation amplifier possesses very high input impedance, high CMRR, and high voltage gain adjustable by an external resistor.

Glossary

A/D conversion A process whereby information in analog form is converted into digital form.

Analog Characterized by a linear process in which a variable takes on a continuous set of values.

Bounding The process of limiting the output range of an amplifier or other circuit.

Comparator A circuit which compares two input voltages and produces an output in either of two states indicating the greater than or less than relationship of the inputs.

Differentiator A circuit that produces an output which approximates the instantaneous rate of change of the input function.

Digital Characterized by a process in which a variable takes on either of two values.

Hysteresis Characteristic of a circuit in which two different trigger levels create an offset or lag in the switching action.

Integrator A circuit that produces an output which approximates the area under the curve of the input function.

Noise An unwanted signal.

Schmitt trigger A comparator with hysteresis.

FORMULAS

COMPARATORS

(14–1) $$V_{REF} = \frac{R_2}{R_1 + R_2}(+V)$$ Comparator reference

(14–2) $$V_{UTP} = \frac{R_2}{R_1 + R_2}[+V_{out(max)}]$$ Upper trigger point

(14–3) $$V_{LTP} = \frac{R_2}{R_1 + R_2}[-V_{out(max)}]$$ Lower trigger point

(14–4) $$V_{HYS} = V_{UTP} - V_{LTP}$$ Hysteresis voltage

SUMMING AMPLIFIER

(14–5) $$V_{OUT} = -(V_{IN1} + V_{IN2})$$ Two-input adder

(14–6) $$V_{OUT} = -(V_{IN1} + V_{IN2} + \cdots + V_{INn})$$ n-input adder

(14–7) $$V_{OUT} = -\frac{R_f}{R}(V_{IN1} + V_{IN2} + \cdots + V_{INn})$$ Adder with gain

(14–8) $$V_{OUT} = -\left(\frac{R_f}{R_1}V_{IN1} + \frac{R_f}{R_2}V_{IN2} + \cdots + \frac{R_f}{R_n}V_{INn}\right)$$ Adder with gain

INTEGRATOR AND DIFFERENTIATOR

(14–9) $$\frac{\Delta V_{out}}{\Delta t} = -\frac{V_{in}}{R_iC}$$ Integrator, rate of change

(14–10) $$V_{out} = \left(\frac{V_C}{t}\right)R_fC$$ Differentiator output with ramp input

INSTRUMENTATION AMPLIFIER

(14–11) $$V_{out1} = \left(1 + \frac{R_{f1}}{R_G}\right)V_{in1} - \left(\frac{R_{f1}}{R_G}\right)V_{in2} + V_{cm}$$ Instrumentation amplifier

(14–12) $V_{out2} = \left(1 + \frac{R_{f2}}{R_G}\right)V_{in2} - \left(\frac{R_{f2}}{R_G}\right)V_{in1} + V_{cm}$ Instrumentation amplifier

(14–13) $V_{out} = \left(1 + \frac{2R_f}{R_G}\right)(V_{in2} - V_{in1})$ Instrumentation amplifier

(14–14) $A_{cl} = 1 + \frac{2R_f}{R_G}$ Instrumentation amplifier, closed-loop gain

(14–15) $R_G = \frac{2R_f}{A_{cl} - 1}$ Gain-setting resistor

MISCELLANEOUS

(14–16) $I_L = \frac{V_{IN}}{R_i}$ Constant-current source

(14–17) $V_{out} = I_i R_f$ Current-to-voltage converter

(14–18) $I_L = \frac{V_{in}}{R_1}$ Voltage-to-current converter

SELF-TEST

1. In a zero-level detector, the output changes state when the input
(a) is positive **(b)** is negative
(c) crosses zero **(d)** has a zero rate of change

2. The zero-level detector is one application of a
(a) comparator **(b)** differentiator
(c) summing amplifier **(d)** diode

3. Noise on the input of a comparator can cause the output to
(a) hang up in one state
(b) go to zero
(c) change back and forth erratically between two states
(d) produce the amplified noise signal

4. The effects of noise can be reduced by
(a) lowering the supply voltage **(b)** using positive feedback
(c) using negative feedback **(d)** using hysteresis
(e) b and d

5. A comparator with hysteresis
(a) has one trigger point **(b)** has two trigger points
(c) has a variable trigger point **(d)** is like a magnetic circuit

6. In a comparator with hysteresis,
   (a) a bias voltage is applied between the two inputs
   (b) only one supply voltage is used
   (c) a portion of the output is fed back to the inverting input
   (d) a portion of the output is fed back to the noninverting input
7. Using output bounding in a comparator
   (a) makes it faster (b) keeps the output positive
   (c) limits the output levels (d) stabilizes the output
8. A window comparator detects when
   (a) the input is between two specified limits
   (b) the input is not changing
   (c) the input is changing too fast
   (d) the amount of light exceeds a certain value
9. A summing amplifier can have
   (a) only one input (b) only two inputs
   (c) any number of inputs
10. If the voltage gain for each input of a summing amplifier with a 4.7 kΩ feedback resistor is unity, the input resistors must have a value of
    (a) 4.7 kΩ
    (b) 4.7 kΩ divided by the number of inputs
    (c) 4.7 kΩ times the number of inputs
11. An averaging amplifier has five inputs. The ratio R_f/R_{in} must be
    (a) 5 (b) 0.2 (c) 1
12. In a scaling adder, the input resistors are
    (a) all the same value
    (b) all of different values
    (c) each proportional to the weight of its inputs
    (d) related by a factor of two
13. In an integrator, the feedback element is a
    (a) resistor (b) capacitor
    (c) zener diode (d) voltage divider
14. For a step input, the output of an integrator is
    (a) a pulse (b) a triangular waveform
    (c) a spike (d) a ramp
15. The rate of change of an integrator's output voltage in response to a step input is set by
    (a) the RC time constant (b) the amplitude of the step input
    (c) the current through the capacitor (d) all of these

16. In a differentiator, the feedback element is a
    (a) resistor (b) capacitor
    (c) zener diode (d) voltage divider
17. The output of a differentiator is proportional to
    (a) the *RC* time constant (b) the rate at which the input is changing
    (c) the amplitude of the input (d) a and b
18. When you apply a triangular waveform to the input of a differentiator, the output is
    (a) a dc level (b) an inverted triangular waveform
    (c) a square waveform (d) the first harmonic of the triangular waveform
19. An instrumentation amplifier contains
    (a) an integrator (b) three integrators
    (c) three op-amps (d) a lot of stuff
20. An external resistor is connected to an instrumentation amplifier for the purpose of
    (a) adjusting the gain (b) stabilizing the output
    (c) setting the feedback ratio (d) nulling the output

PROBLEMS

SECTION 14–1 COMPARATORS

1. A certain op-amp has an open-loop gain of 80,000. The maximum saturated output levels of this particular device are ±12 V when the dc supply voltages are ±15 V. If a differential voltage of 0.15 mV rms is applied between the inputs, what is the peak-to-peak value of the output?
2. Determine the output level (maximum positive or maximum negative) for each comparator in Figure 14–59.

FIGURE 14–59

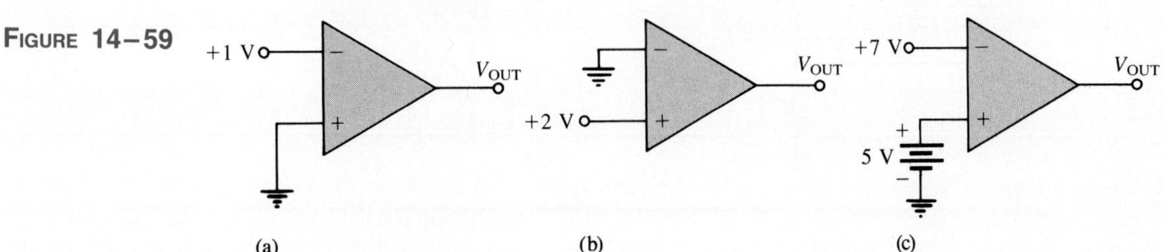

3. Calculate the V_{UTP} and V_{LTP} in Figure 14–60. $V_{out(max)} = -10$ V.
4. What is the hysteresis voltage in Figure 14–60?

FIGURE 14–60

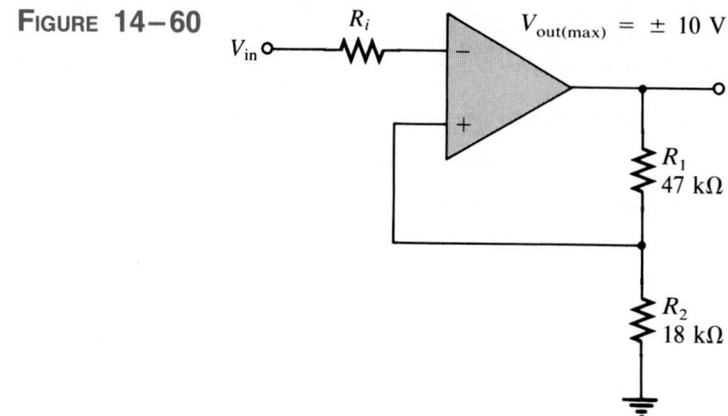

5. Sketch the output voltage waveform for each circuit in Figure 14–61 with respect to the input. Show voltage levels.

FIGURE 14–61

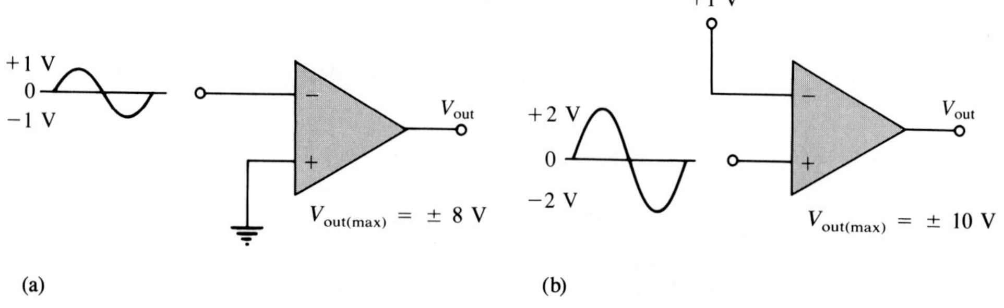

6. Determine the hysteresis voltage for each comparator in Figure 14–62. The maximum output levels are ±11 V.

FIGURE 14–62

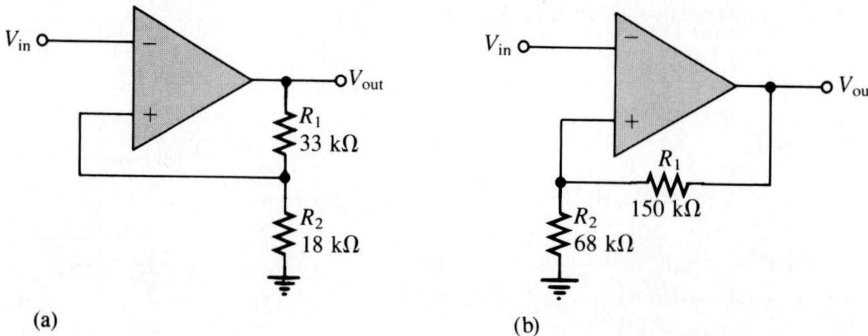

7. A 6.2 V zener diode is connected from the output to the inverting input in Figure 14–60 with the cathode at the output. What are the positive and negative output levels?

8. Determine the output voltage waveform in Figure 14–63.

FIGURE 14–63

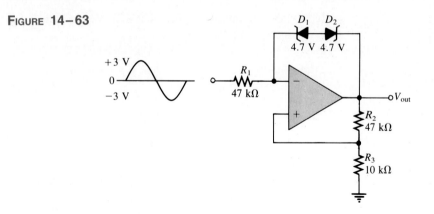

SECTION 14–2 SUMMING AMPLIFIERS

9. Determine the output voltage for each circuit in Figure 14–64.

FIGURE 14–64

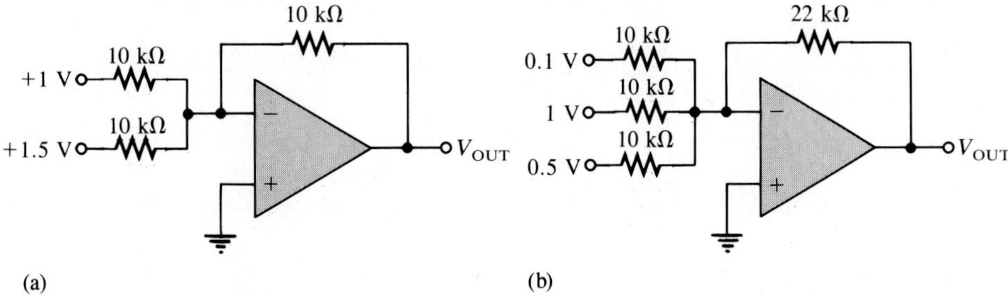

10. Refer to Figure 14–65. Determine the following:

(a) V_{R1} and V_{R2}

(b) Current through R_f

(c) V_{OUT}

FIGURE 14–65

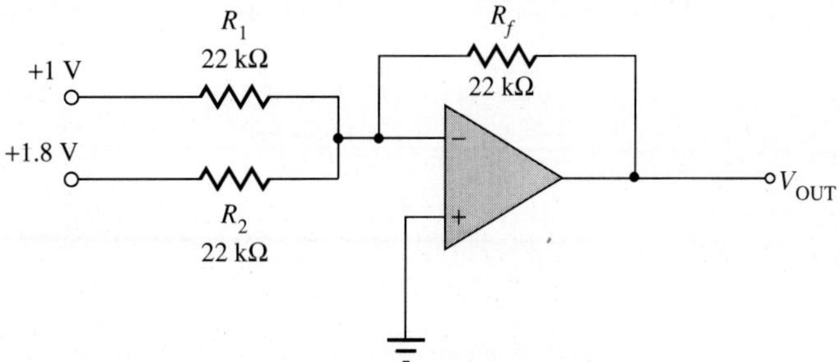

11. Find the value of R_f necessary to produce an output that is five times the sum of the inputs in Figure 14–65.

12. Design a summing amplifier that will average eight input voltages. Use input resistances of 10 kΩ each.

13. Find the output voltage when the input voltages shown in Figure 14–66 are applied to the scaling adder. What is the current through R_f?

FIGURE 14–66

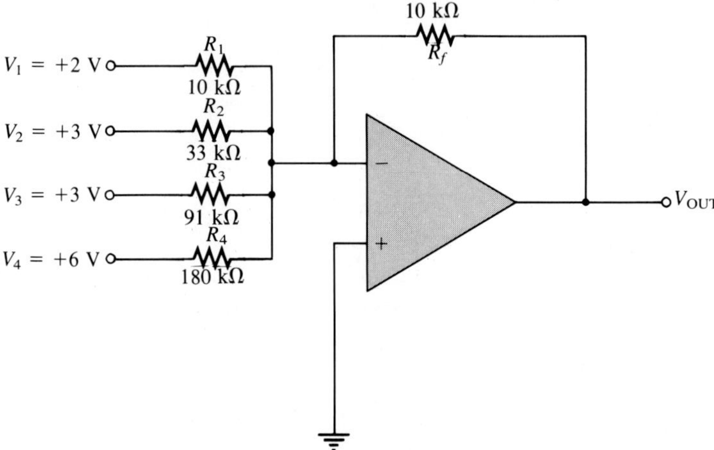

14. Determine the values of the input resistors required in a six-input scaling adder so that the lowest weighted input is 1 and each successive input has a weight twice the previous one. Use $R_f = 100$ kΩ.

SECTION 14–3 THE INTEGRATOR AND DIFFERENTIATOR

15. Determine the rate of change of the output voltage in response to the step input to the integrator in Figure 14–67.

FIGURE 14–67

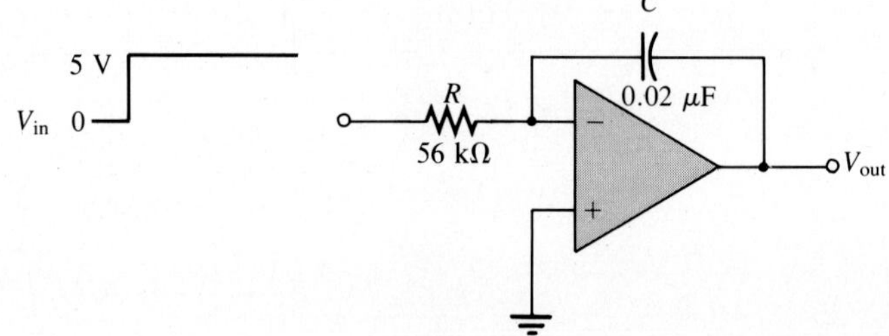

16. A triangular waveform is applied to the input of the circuit in Figure 14–68 as shown. Determine what the output should be and sketch its waveform in relation to the input.

17. What is the magnitude of the capacitor current in Problem 16?

FIGURE 14–68

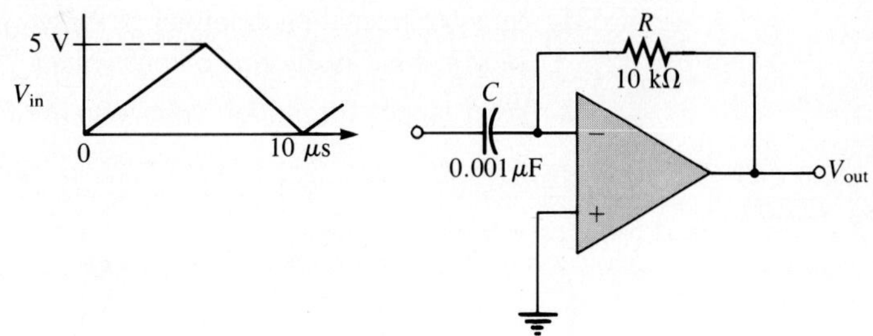

18. A triangular waveform with a peak-to-peak voltage of 2 V and a period of 1 ms is applied to the differentiator in Figure 14–69(a). What is the output voltage?

19. Beginning in position 1 in Figure 14–69(b), the switch is thrown into position 2 and held there for 10 ms, then back to position 1 for 10 ms, and so forth. Sketch the resulting output waveform. The saturated output levels of the op-amp are ±12 V.

FIGURE 14–69

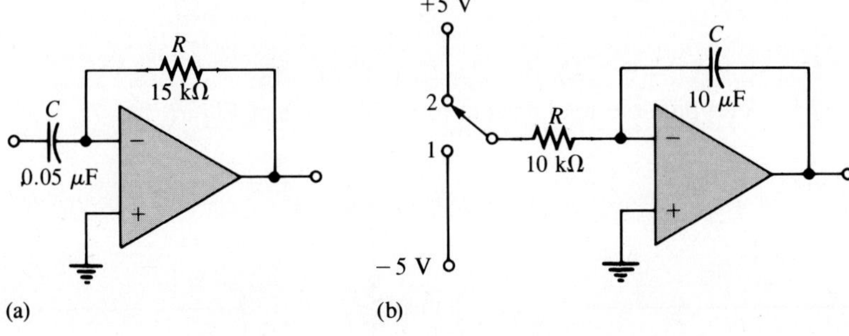

SECTION 14–4 THE INSTRUMENTATION AMPLIFIER

20. Refer to the graph in Figure 14–70, which shows closed-loop gain versus the gain-setting resistor value for a certain instrumentation amplifier. Determine the value of R_G for a gain of 200.

FIGURE 14–70

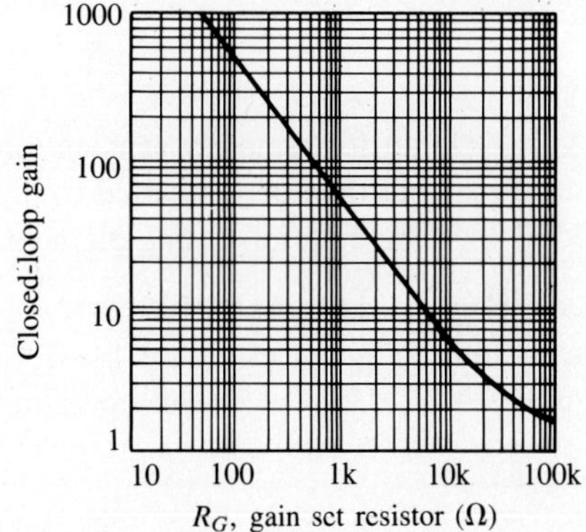

21. An instrumentation amplifier is shown in Figure 14–71. If the internal R_f is 25 kΩ, what is the rms output voltage?

22. What is the closed-loop voltage gain of the circuit in Figure 14–71?

FIGURE 14–71

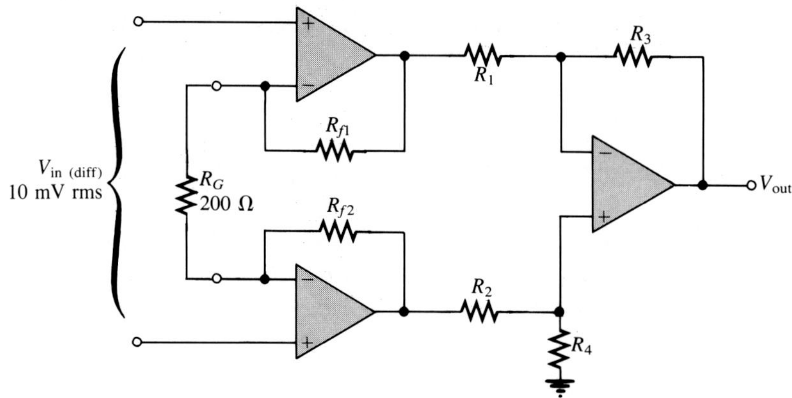

SECTION 14–5 MORE OP-AMP CIRCUITS

23. Determine the load current in each circuit of Figure 14–72.

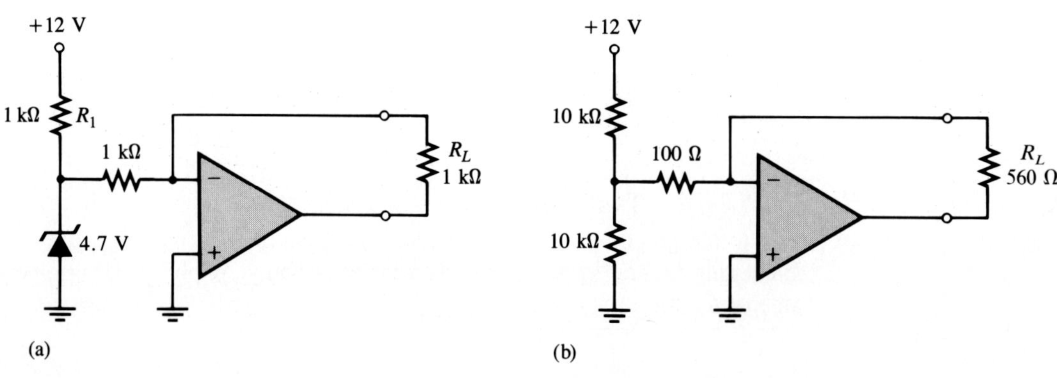

FIGURE 14–72

24. Devise a circuit for remotely sensing temperature and producing a proportional voltage that can then be converted to digital form for display. A thermistor can be used as the temperature-sensing element.

SECTION 14–6 TROUBLESHOOTING

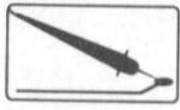

25. The waveforms given in Figure 14–73(a) are observed at the indicated points in Figure 14–73(b). Is the circuit operating properly? If not, what is a likely fault?

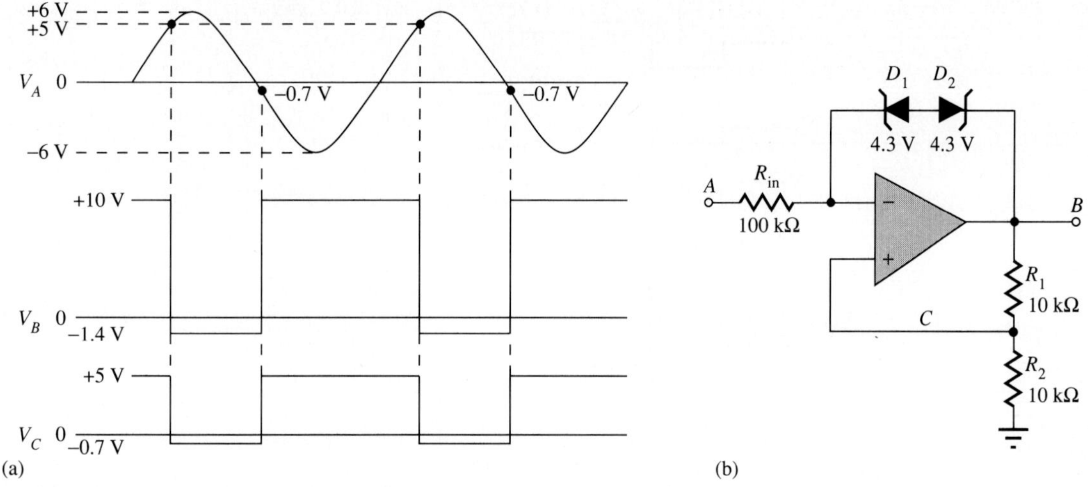

FIGURE 14–73

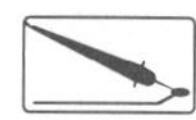

26. The waveforms shown for the window comparator in Figure 14–74 are measured. Determine if the output waveform is correct and, if not, specify the possible fault(s).

FIGURE 14–74

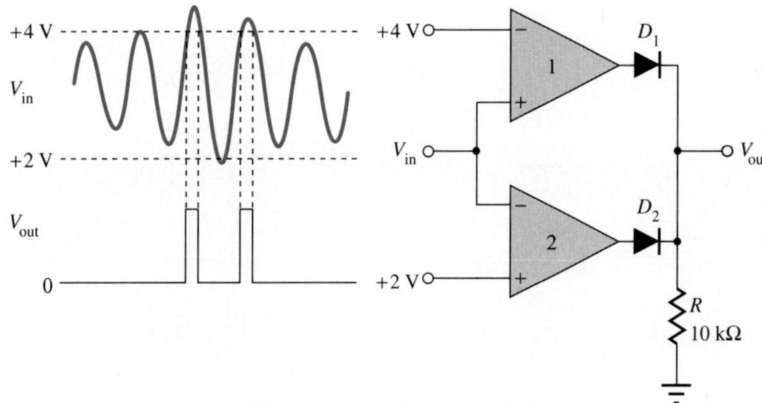

27. The sequences of voltage levels shown in Figure 14–75 are applied to the summing amplifier and the indicated output is observed. First, determine if this output is correct. If it is not correct, determine the fault.

28. The given ramp voltages are applied to the op-amp circuit in Figure 14–76. Is the given output correct? If it isn't, what is the problem?

FIGURE 14–75

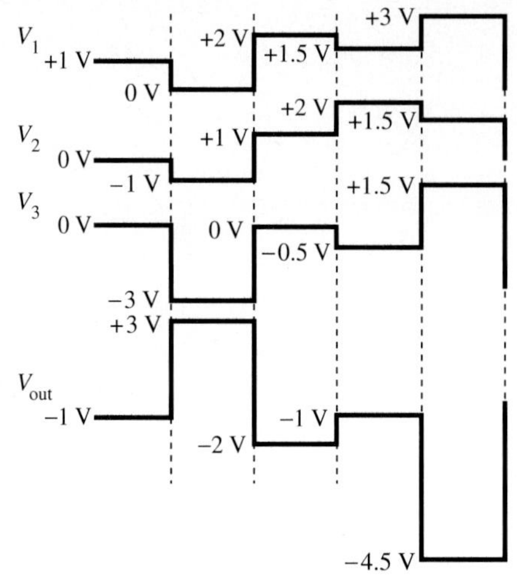

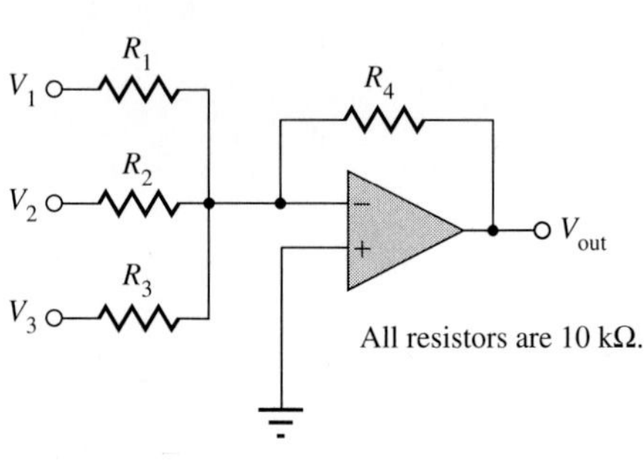

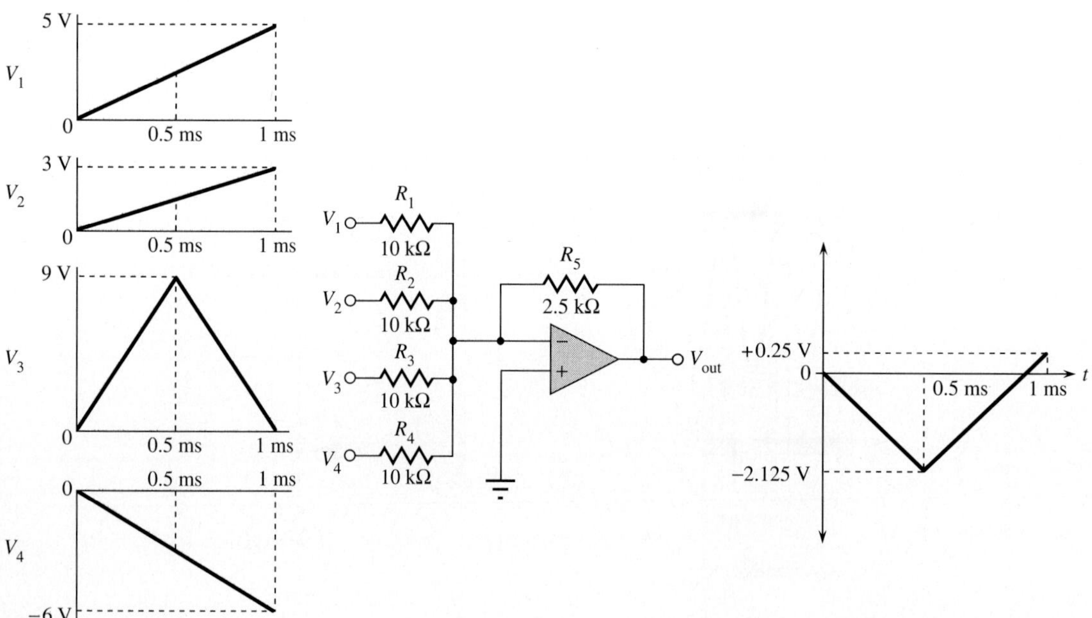

FIGURE 14–76

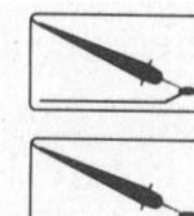

29. The D/A converter with inputs as shown in Figure 14–29 produces the output shown in Figure 14–77. Determine the fault in the circuit.
30. The analog board, shown in Figure 14–78, for the A/D converter in the system application has just come off the assembly line and a pass/fail test indicates that it doesn't work. The board now comes to you for troubleshooting. What is the very first thing you should do? Can you isolate the problem(s) by this first step in this case?

FIGURE 14–77

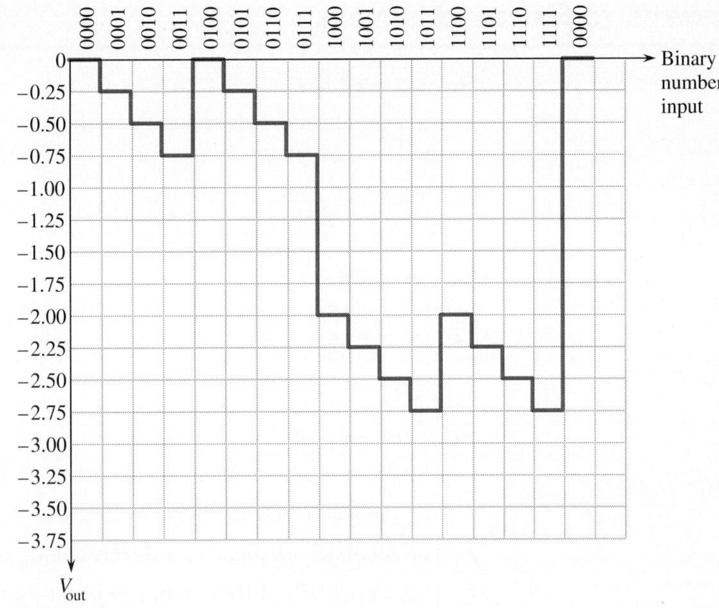

FIGURE 14–78

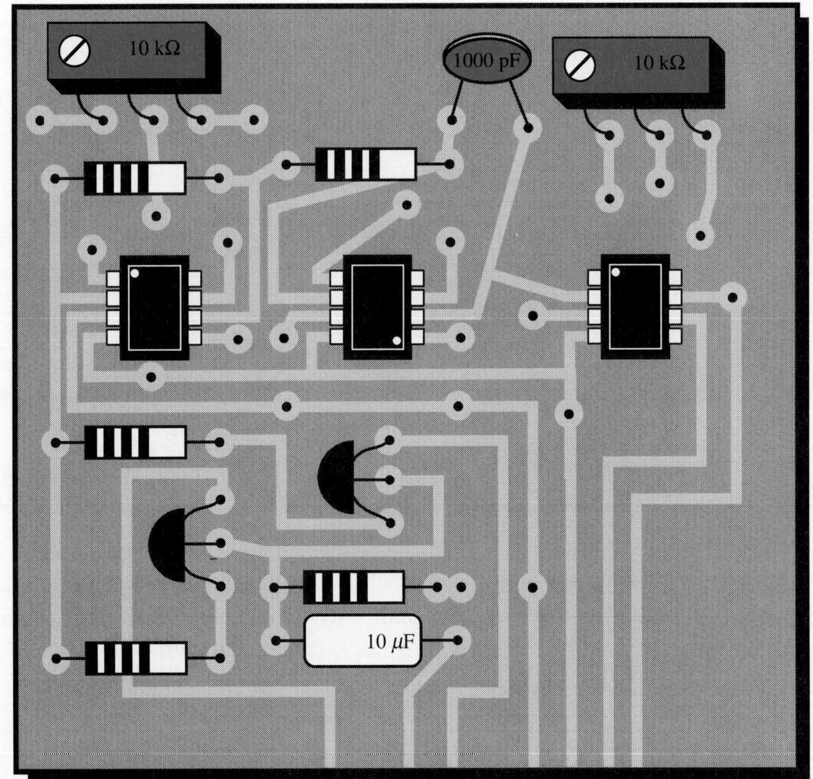

ANSWERS TO REVIEW QUESTIONS

SECTION 14–1

1. **(a)** $V = (10\text{ k}\Omega/110\text{ k}\Omega)15\text{ V} = 1.36\text{ V}$
   (b) $V = (22\text{ k}\Omega/69\text{ k}\Omega)(-12\text{ V}) = -3.83\text{ V}$
2. Hysteresis makes the comparator noise free.
3. Bounding limits the output amplitude to a specified level.

SECTION 14–2

1. The summing point is the point where the input resistors are commonly connected.
2. $R_f/R = 1/5 = 0.2$
3. 5 kΩ

SECTION 14–3

1. The feedback element in an integrator is a capacitor.
2. The capacitor voltage is linear because the capacitor current is constant.
3. The feedback element in a differentiator is a resistor.
4. The output of a differentiator is proportional to the rate of change of the input.

SECTION 14–4

1. An instrumentation amplifier consists of three op-amps and several resistors.
2. $R_G = 2R_f/(A_{cl} - 1) = 30\text{ k}\Omega/799 = 37.5\text{ k}\Omega$

SECTION 14–5

1. $I_L = 6.8\text{ V}/10\text{ k}\Omega = 0.68\text{ mA}$; same value to 5 kΩ load.
2. The feedback resistor is the constant of proportionality.

SECTION 14–6

1. An op-amp can fail with a shorted output.
2. Replace suspected components one by one.

SECTION 14–7

1. Summing amplifier—IC_1, integrator—IC_2, comparator—IC_3
2. A high slew-rate op-amp is used in the integrator to avoid slew-rate limitation of the output ramps. One is used as a comparator, to achieve a fast switching time.
3. R_4 and R_6 are for eliminating output offset (nulling).
4. The board output is the comparator output. The transition of the comparator output from its positive state to its negative state notifies the control logic of the end of the variable-time interval.
5. $1/500\text{ kHz} = 2\ \mu\text{s}$
6. Because of the electronic switch, only one input voltage at a time is actually applied.

ANSWERS TO PRACTICE EXERCISES

14–1 1.96 V

14–2 +3.83 V, −3.83 V

14–3 +1.81 V, −1.81 V

14–4 More accurately

14–5 −12.5 V

14–6 −5.73 V

14–7 Changes require an additional 100 kΩ input resistor and a change of R_f to 20 kΩ.

14–8 0.45, 0.12, 0.18; $V_{OUT} = -3.03$ V

14–9 Yes. All should be doubled.

14–10 Change C to 5000 pF.

14–11 Same waveform with an amplitude of 6.6 V

14–12 285 Ω

14–13 A pulse from −0.88 V to +7.79 V

14–14 −3.76 V

14–15 Change R_6 to 25 kΩ.

15

Oscillators

After completing this chapter, you should be able to

- □ Describe the basic concept of an oscillator.
- □ Define positive feedback.
- □ List the conditions for oscillation.
- □ Recognize and explain the operation of the Wien-bridge, phase-shift, and twin-T oscillators.
- □ Recognize and explain the operation of the Colpitts, Clapp, Hartley, and Armstrong oscillators.
- □ Discuss the conditions for oscillator start-up.
- □ Describe a quartz crystal and its use in oscillators.
- □ Explain the operation of triangular-wave, sawtooth, and square-wave oscillators.
- □ Explain how a voltage-controlled oscillator (VCO) works.
- □ Apply a 555 timer IC in oscillator applications.
- □ Discuss the basic operation of phase-locked loops.

Oscillators are circuits that generate an output signal without an input signal. They are used as signal sources in all sorts of applications. Different types of oscillators produce various types of outputs including sine waves, square waves, triangular waves, and sawtooth waves. In this chapter, several types of basic oscillator circuits using both discrete transistors and op-amps as the gain element are introduced. Also, a very popular integrated circuit, called the 555 timer, is discussed in relation to its oscillator applications.

Oscillator operation is based on the principle of positive feedback, where a portion of the output signal is fed back to the input in a way that causes it to reinforce itself and thus sustain a continuous output signal. Oscillators are widely used in most communications systems as well as in digital systems, including computers, to generate required frequencies and timing signals. Also, oscillators are found in many types of test instruments like those used in the laboratory.

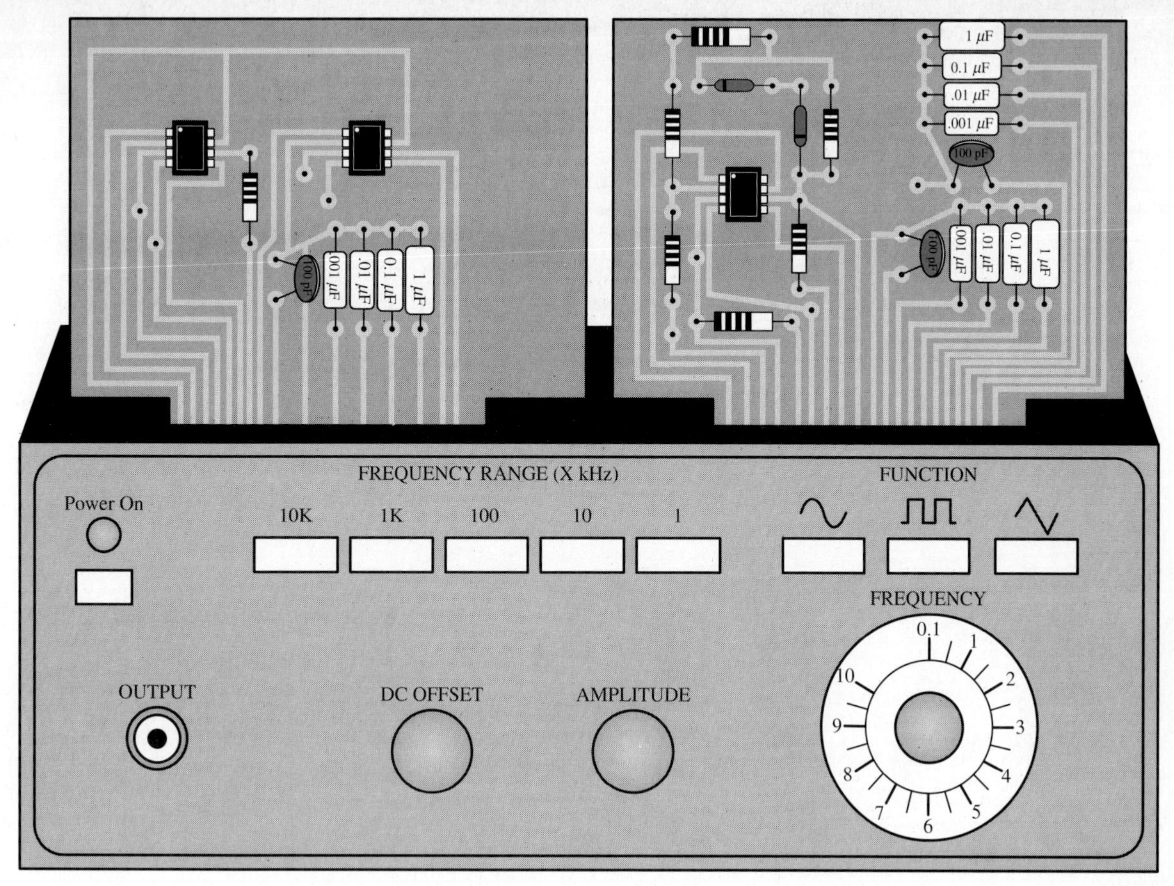

A System Application

The function generator shown in the diagram is a good illustration of a system application for oscillators. The oscillator is a major part of this particular system. No doubt, you are already familiar with the use of the signal or function generator in your lab. As with most types of systems, a function generator can be implemented in more than one way. The system in this chapter uses circuits with which you are already familiar without some of the refinements and features found in many commercial instruments. The system reinforces what you have studied and lets you see these circuits "at work" in a specific application.

For the system application in Section 15–8, in addition to the other topics, be sure you understand

- □ How *RC* oscillators work.
- □ How a zero-level detector works.
- □ How an integrator works.

15–1 DEFINITION OF AN OSCILLATOR

An* oscillator *is a circuit that produces a repetitive waveform on its output with only the dc supply voltage as an input. A repetitive input signal is not required. The output voltage can be either sinusoidal or nonsinusoidal, depending on the type of oscillator.

The basic oscillator concept is illustrated in Figure 15–1. Essentially, an oscillator converts electrical energy in the form of dc to electrical energy in the form of ac. A basic oscillator consists of an amplifier for gain (either discrete transistor or op-amp) and a positive feedback circuit that produces phase shift and provides attenuation, as shown in Figure 15–2.

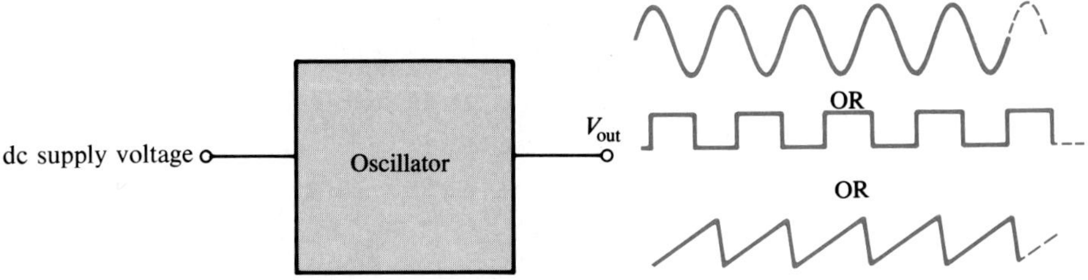

FIGURE 15–1
The basic oscillator concept showing three possible types of output waveforms.

FIGURE 15–2
Basic elements of an oscillator.

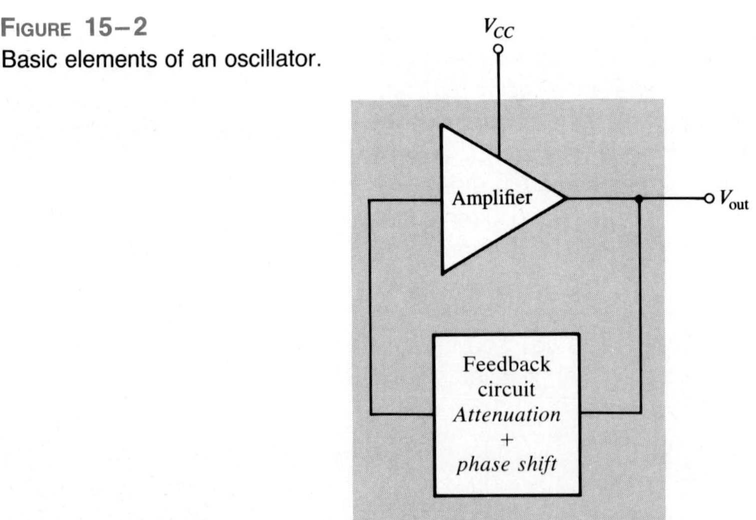

15–1 REVIEW QUESTIONS

1. What is an oscillator?
2. What type of feedback does an oscillator use?
3. What is the purpose of the feedback circuit?

15–2 OSCILLATOR PRINCIPLES

With the exception of the relaxation oscillator, which we will cover in Section 15–5, oscillator operation is based on the principle of positive feedback. In this section, we will examine this concept and look at the general conditions required for oscillation to occur.

POSITIVE FEEDBACK

Positive feedback is characterized by the condition wherein a portion of the output voltage of an amplifier is fed back to the input with no net phase shift, resulting in a reinforcement of the output signal. This basic idea is illustrated in Figure 15–3. As you can see, the in-phase feedback voltage is amplified to produce the output voltage, which in turn produces the feedback voltage. That is, a loop is created in which the signal sustains itself and a continuous sine wave output is produced. This phenomenon is called *oscillation*.

FIGURE 15–3
Positive feedback produces oscillation.

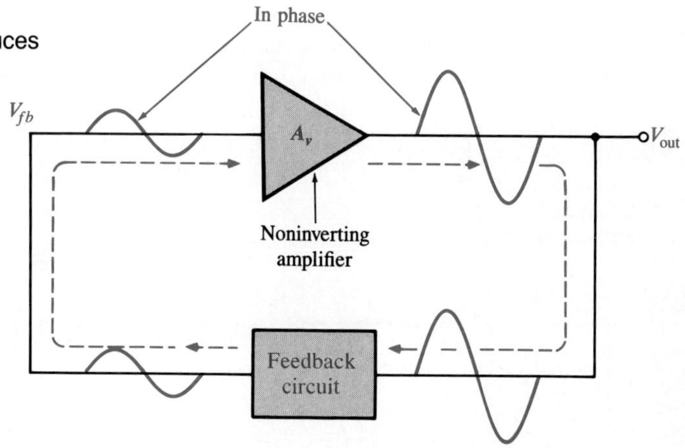

CONDITIONS FOR OSCILLATION

Two conditions are required for a sustained state of oscillation:

1. The phase shift around the feedback loop must be 0°.
2. The voltage gain around the closed feedback loop must equal 1 (unity).

The voltage gain around the closed feedback loop (A_{cl}) is the product of the amplifier gain (A_v) and the attenuation of the feedback circuit (B).

$$A_{cl} = A_v B$$

For example, if the amplifier has a gain of 100, the feedback circuit must have an attenuation of 0.01 to make the loop gain equal to 1 ($A_v B = 100 \times 0.01 = 1$). These conditions for oscillation are illustrated in Figure 15–4.

FIGURE 15–4
Conditions for oscillation.

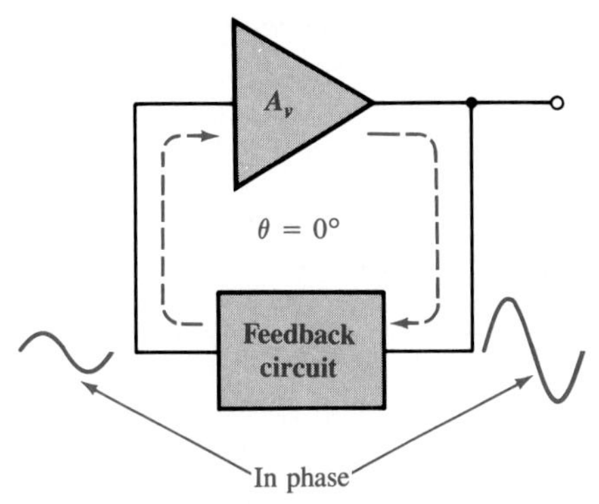

(a) The phase shift around the loop is 0°.

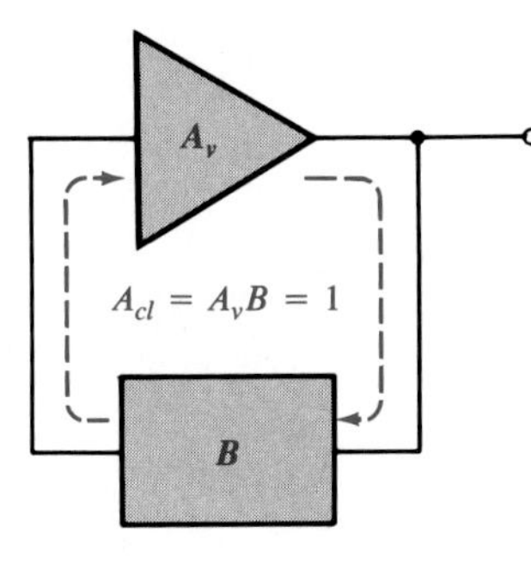

(b) The closed loop gain is 1.

START-UP CONDITIONS

So far, we have seen what it takes for an oscillator to produce a continuous sine wave output. Now we examine the requirements for the oscillation to start when the dc supply voltage is turned on. As you have seen, the unity-gain condition must be met for oscillation to be sustained. For oscillation to begin, the voltage gain around the positive feedback loop must be greater than 1 so that the amplitude of the output can build up to a desired level. The gain must then decrease to 1 so that the output stays at the desired

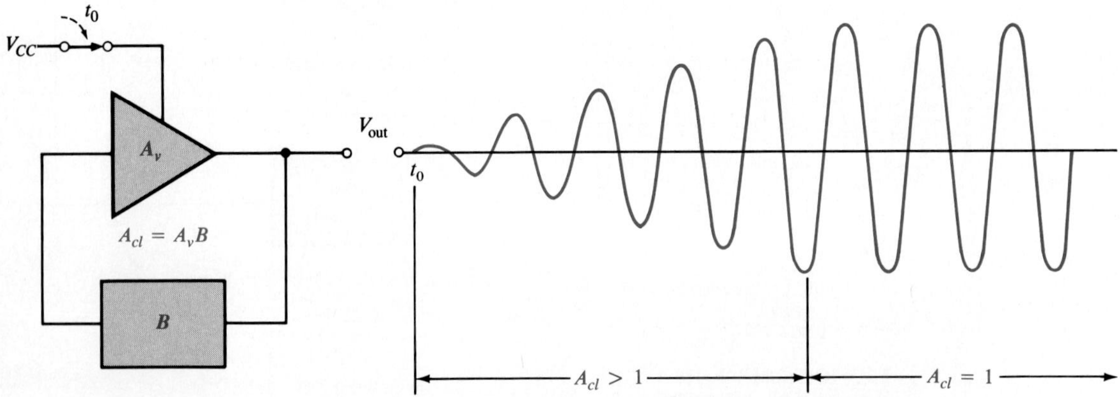

FIGURE 15–5
When oscillation starts at t_0, the condition $A_{cl} > 1$ causes the sinusoidal output voltage amplitude to build up to a desired level, where A_{cl} decreases to 1 and maintains the desired amplitude.

a desired level. The gain must then decrease to 1 so that the output stays at the desired level. (Several ways to achieve this reduction in gain after start-up are discussed later.) The conditions for both starting and sustaining oscillation are illustrated in Figure 15–5.

A question that normally arises is this: If the oscillator is off (no dc voltage) and there is no output voltage, how does a feedback signal originate to start the positive feedback build-up process? Initially, a small positive feedback voltage develops from thermally produced broad-band noise in the resistors or other components or from turn-on transients. The feedback circuit permits only a voltage with a frequency equal to the selected oscillation frequency to appear in-phase on the amplifier's input. This initial feedback voltage is amplified and continually reinforced, resulting in a buildup of the output voltage as previously discussed.

15–2 REVIEW QUESTIONS

1. What are the conditions required for a circuit to oscillate?
2. Define positive feedback.
3. What are the start-up conditions for an oscillator?

15–3 OSCILLATORS WITH *RC* FEEDBACK CIRCUITS

In this section you will learn about three types of RC oscillator circuits that produce sinusoidal outputs—the Wien-bridge oscillator, the phase-shift oscillator, and the twin-T oscillator. Generally, RC oscillators are used for frequencies up to about 1 MHz. The Wien-bridge is by far the most widely used type of RC oscillator for this range of frequencies.

THE WIEN-BRIDGE OSCILLATOR

One type of sine wave oscillator is the *Wien-bridge* oscillator. A fundamental part of the Wien-bridge oscillator is a lead-lag network like that shown in Figure 15–6(a). R_1 and C_1 together form the lag portion of the network; R_2 and C_2 form the lead portion. The operation of this circuit is as follows. At lower frequencies, the lead network dominates due to the high reactance of C_2. As the frequency increases, X_{C2} decreases, thus allowing the output voltage to increase. At some specified frequency, the response of the lag network takes over, and the decreasing value of X_{C1} causes the output voltage to decrease.

FIGURE 15–6
A lead-lag network.

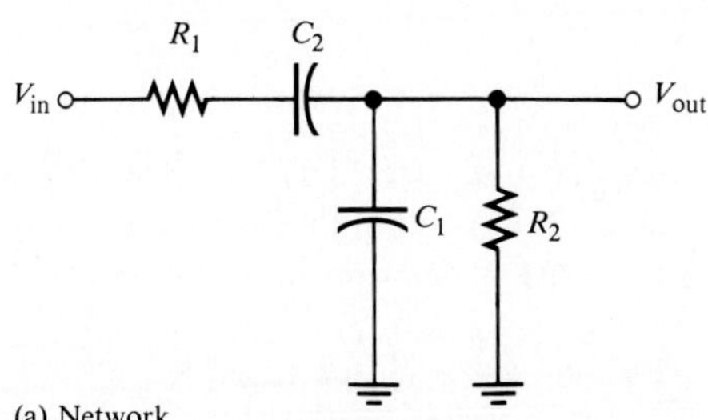

(a) Network

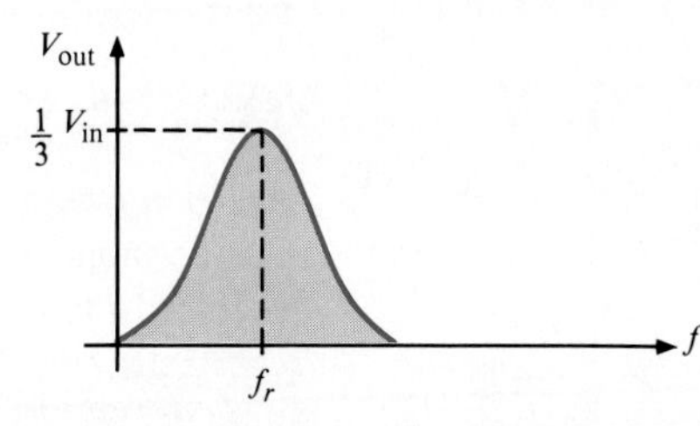

(b) Response curve

So, we have a response curve like that shown in Figure 15–6(b) where the output voltage peaks at a frequency f_r. At this point, the attenuation (V_{out}/V_{in}) of the network is ⅓ if $R_1 = R_2$ and $X_{C1} = X_{C2}$ as stated by the following equation, which is derived in Appendix B.

$$\frac{V_{out}}{V_{in}} = \frac{1}{3} \tag{15–1}$$

The formula for the resonant frequency is also derived in Appendix B and is

$$f_r = \frac{1}{2\pi RC} \tag{15–2}$$

To summarize, the lead-lag network has a resonant frequency f_r at which the phase shift through the network is 0° and the attenuation is ⅓. Below f_r, the lead network dominates and the output leads the input. Above f_r, the lag network dominates and the output lags the input.

The lead-lag network is used in the positive feedback loop of an op-amp, as shown in Figure 15–7(a). A voltage divider is used in the negative feedback loop.

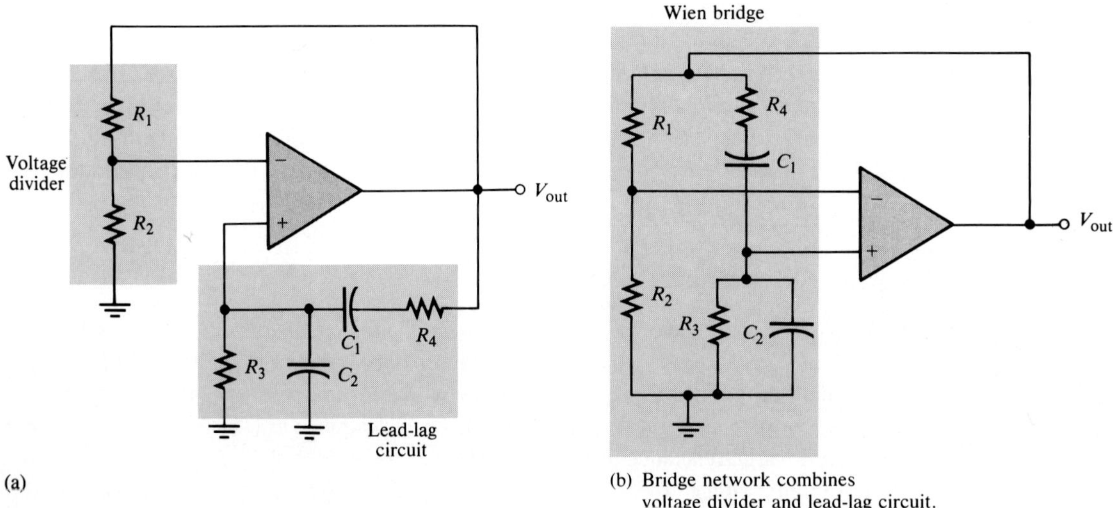

FIGURE 15–7
Two ways to draw the schematic of a Wien-bridge oscillator.

THE BASIC CIRCUIT

The Wien-bridge oscillator circuit can be viewed as a noninverting amplifier configuration with the input signal fed back from the output through the lead-lag network. Recall that the closed-loop gain of the amplifier is determined by the voltage divider.

$$A_{cl} = \frac{1}{B} = \frac{1}{R_2/(R_1 + R_2)} = \frac{R_1 + R_2}{R_2}$$

The circuit is redrawn in Figure 15–7(b) to show that the op-amp is connected across the Wien bridge. One leg of the bridge is the lead-lag network, and the other is the voltage divider.

POSITIVE FEEDBACK CONDITIONS FOR OSCILLATION

As you know, for the circuit to produce a sustained sine wave output (oscillate), the phase shift around the positive feedback loop must be 0° and the gain around the loop must be at least unity (1). The 0° phase-shift condition is met when the frequency is f_r, because the phase shift through the lead-lag network is 0° and there is no inversion from the noninverting input (+) of the op-amp to the output. This is shown in Figure 15–8(a).

The unity-gain condition in the feedback loop is met when

$$A_{cl} = 3$$

This offsets the ⅓ attenuation of the lead-lag network, thus making the gain around the positive feedback loop equal to 1, as depicted in Figure 15–8(b). To achieve a closed-loop gain of 3,

$$R_1 = 2R_2$$

Then

$$A_{cl} = \frac{R_1 + R_2}{R_2} = \frac{2R_2 + R_2}{R_2} = \frac{3R_2}{R_2} = 3$$

FIGURE 15–8
Conditions for oscillation.

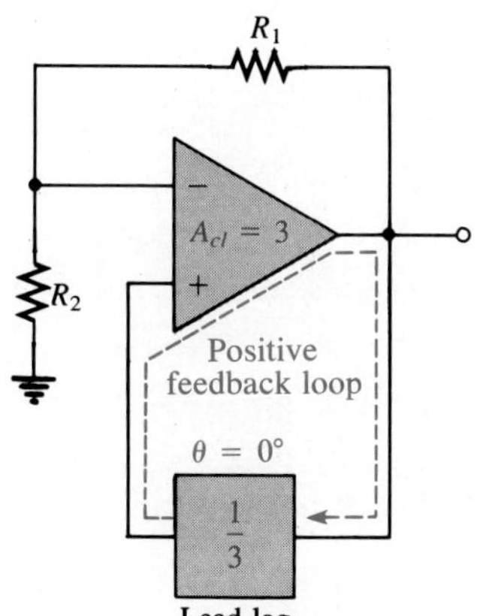

(a) The phase shift around the loop is 0°.

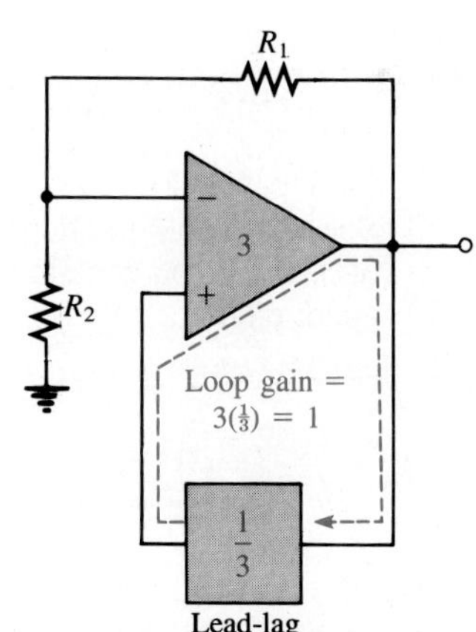

(b) The voltage gain around the loop is 1.

START-UP CONDITIONS

Initially, the closed-loop gain of the amplifier must be more than 1 ($A_{cl} > 3$) until the output signal builds up to a desired level. The gain must then decrease to 1 so that the output signal stays at the desired level. This is illustrated in Figure 15–9.

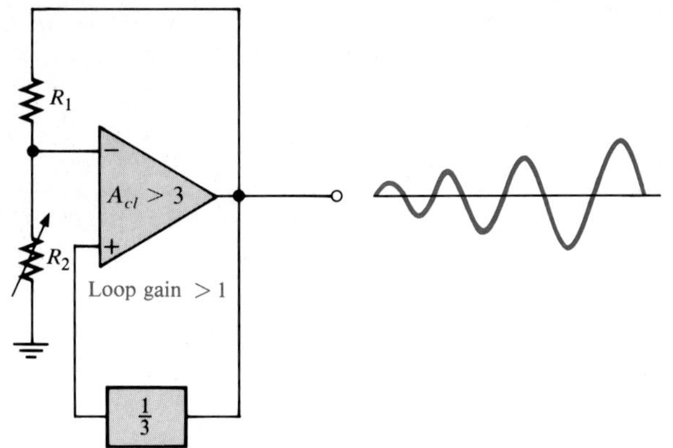

(a) Loop gain greater than 1 causes output to build up.

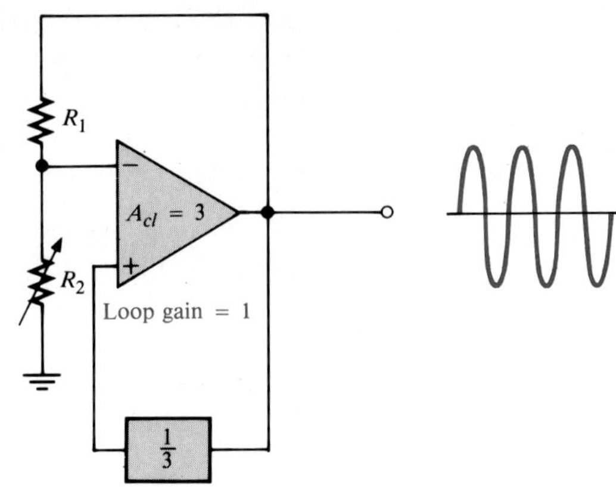

(b) Loop gain of 1 causes a sustained constant output.

FIGURE 15–9
Oscillator start-up conditions.

The circuit in Figure 15–10 illustrates a basic method for achieving the condition described above. Notice that the voltage-divider network has been modified to include an additional resistor R_3 in parallel with a back-to-back zener diode arrangement. When dc power is first applied, both zener diodes appear as opens. This places R_3 in series with R_1, thus increasing the closed-loop gain as follows ($R_1 = 2R_2$).

$$A_{cl} = \frac{R_1 + R_2 + R_3}{R_2} = \frac{3R_2 + R_3}{R_2} = 3 + \frac{R_3}{R_2}$$

FIGURE 15–10
Self-starting Wien-bridge oscillator using back-to-back zener diodes.

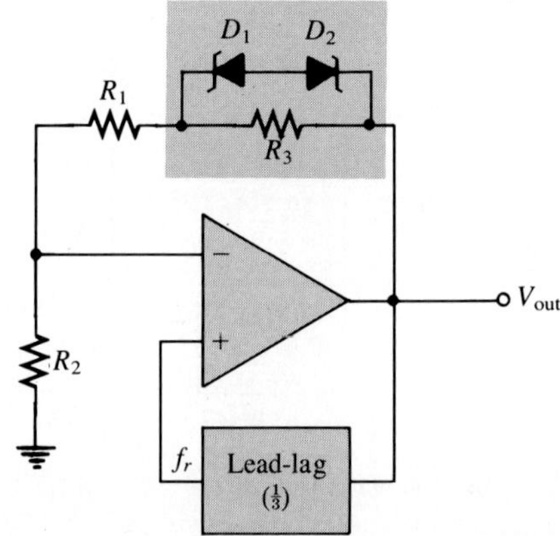

Initially, a small positive feedback signal develops from noise or turn-on transients. The lead-lag network permits only a signal with a frequency equal to f_r to appear in-phase on the noninverting input. This feedback signal is amplified and continually reinforced, resulting in a buildup of the output voltage. When the output signal reaches the zener breakdown voltage, the zeners conduct and effectively short out R_3. This lowers the amplifier's closed-loop gain to 3. At this point the output signal levels off and the oscillation is sustained. (Incidentally, the frequency of oscillation can be adjusted by using gang-tuned capacitors in the lead-lag network.)

Another method sometimes used to ensure self-starting employs a tungsten lamp in the voltage divider, as shown in Figure 15–11. When the power is first turned on, the resistance of the lamp is lower than its nominal value. This keeps the negative feedback small and makes the closed-loop gain of the amplifier greater than 3. As the output voltage builds up, the voltage across the tungsten lamp—and thus its current—increases. As a result, the lamp resistance increases until it reaches a value equal to one-half the feedback resistance. At this point the closed-loop gain is 3, and the output is sustained at a constant level.

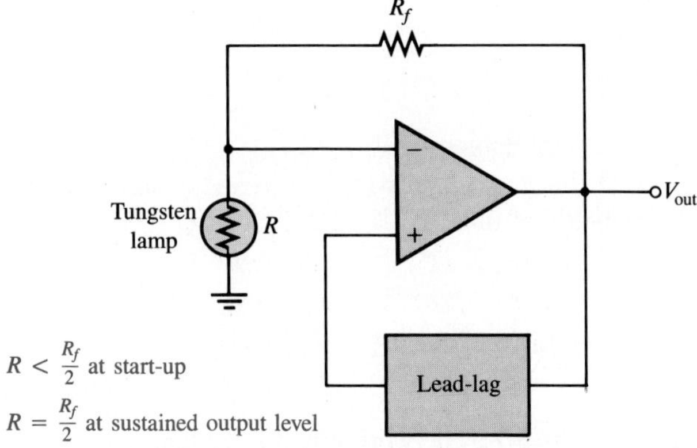

FIGURE 15–11
Self-starting Wien-bridge oscillator using a tungsten lamp in the negative feedback loop.

■ EXAMPLE 15–1 Determine the frequency of oscillation for the Wien-bridge oscillator in Figure 15–12. Also, verify that oscillations will start and then continue when the output signal reaches 5.4 V.

FIGURE 15–12

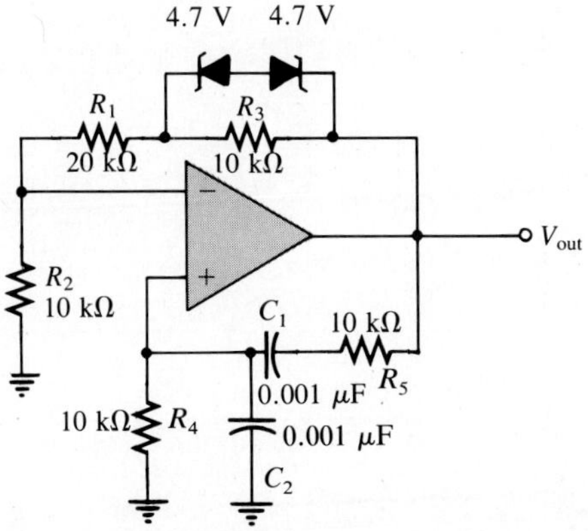

SOLUTION

For the lead-lag network, $R_4 = R_5 = R = 10\ \text{k}\Omega$ and $C_1 = C_2 = C = 0.001\ \mu\text{F}$. The frequency is

$$f_r = \frac{1}{2\pi RC} = \frac{1}{2\pi(10\ \text{k}\Omega)(0.001\ \mu\text{F})} = 15.92\ \text{kHz}$$

Initially, the closed-loop gain is

$$A_{cl} = \frac{R_1 + R_2 + R_3}{R_2} = \frac{40\ \text{k}\Omega}{10\ \text{k}\Omega} = 4$$

Since $A_{cl} > 3$, the start-up condition is met.

When the output reaches 5.4 V (4.7 V + 0.7 V), the zeners conduct (their forward resistance is assumed small, compared to 10 kΩ), and the closed-loop gain is reached. Thus, oscillation is sustained.

$$A_{cl} = \frac{R_1 + R_2}{R_2} = \frac{30\ \text{k}\Omega}{10\ \text{k}\Omega} = 3$$

PRACTICE EXERCISE 15–1

What change is required in the oscillator in Figure 15–12 to produce an output with an amplitude of 6.8 V?

THE PHASE-SHIFT OSCILLATOR

Figure 15–13 shows a type of sine-wave oscillator called the *phase-shift oscillator*. Each of the three *RC* networks in the feedback loop can provide a maximum phase shift approaching 90°. Oscillation occurs at the frequency where the total phase shift through the three *RC* networks is 180°. The inversion of the op-amp, itself, provides the additional 180° to meet the requirement for oscillation.

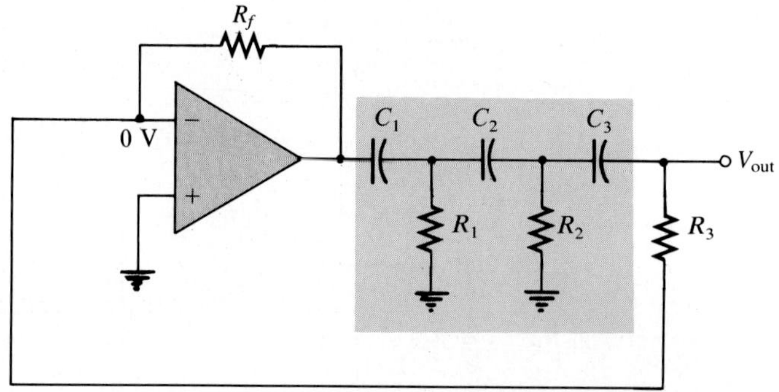

FIGURE 15–13
Op-amp phase-shift oscillator.

The attenuation B of the three-section RC feedback network is

$$B = \frac{1}{29} \tag{15–3}$$

The derivation of this unusual result is given in Appendix B. To meet the greater-than-unity loop gain requirement, the closed-loop voltage gain of the op-amp must be greater than 29 (set by R_f and R_3). The frequency of oscillation is also derived in Appendix B and stated in the following equation, where $R_1 = R_2 = R_3 = R$ and $C_1 = C_2 = C_3 = C$.

$$f_r = \frac{1}{2\pi\sqrt{6}RC} \qquad (15\text{–}4)$$

EXAMPLE 15–2

(a) Determine the value of R_f necessary for the circuit in Figure 15–14 to operate as an oscillator.
(b) Determine the frequency of oscillation.

FIGURE 15–14

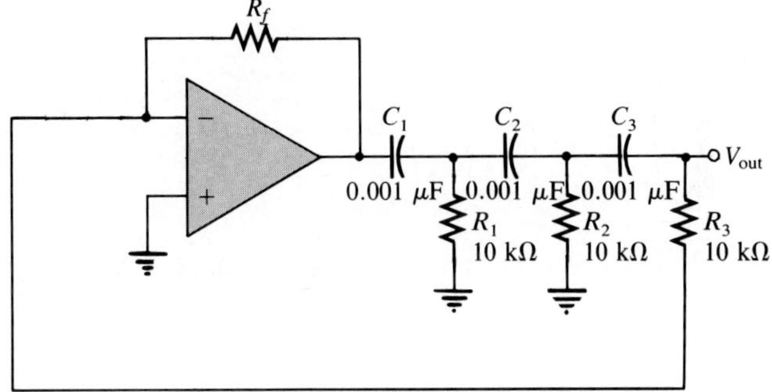

SOLUTION

(a) $A_{cl} = 29$, and $A_{cl} = \dfrac{R_f}{R_3}$. Therefore,

$$\frac{R_f}{R_3} = 29$$

$$R_f = 29R_3 = 29(10\ \text{k}\Omega) = 290\ \text{k}\Omega$$

(b) $f_r = \dfrac{1}{2\pi\sqrt{6}RC} = \dfrac{1}{2\pi\sqrt{6}(10\ \text{k}\Omega)(0.001\ \mu\text{F})} \cong 6.5\ \text{kHz}$

PRACTICE EXERCISE 15–2

(a) If R_1, R_2, and R_3 in Figure 15–14 are changed to 8.2 kΩ, what value must R_f be for oscillation?
(b) What is the value of f_r?

TWIN-T OSCILLATOR

Another type of *RC* oscillator is called the *twin-T* because of the two T-type *RC* filters used in the negative feedback loop, as shown in Figure 15–15(a). One of the twin-T filters has a low-pass response, and the other has a high-pass response. The combined parallel filters produce a band-stop or notch response with a center frequency equal to the desired frequency of oscillation, f_r, as shown in Figure 15–15(b).

FIGURE 15–15
Twin-T oscillator and twin-T filter response.

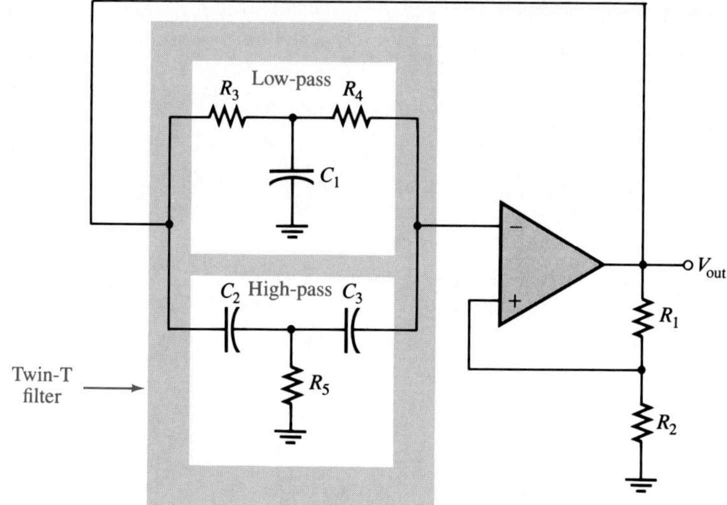

(a) Oscillator circuit

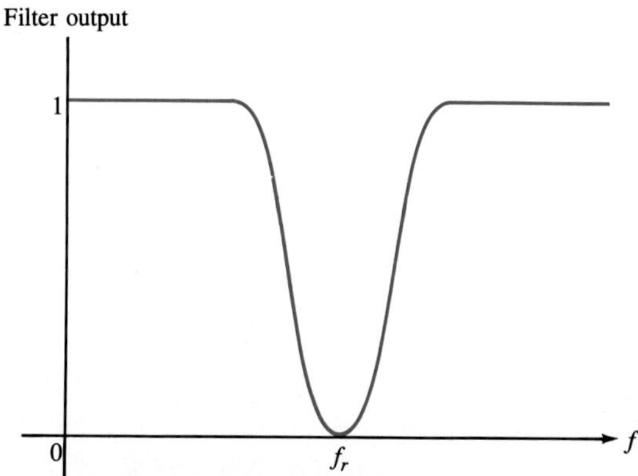

(b) Twin-T filter's frequency response curve

Oscillation cannot occur at frequencies above or below f_r because of the negative feedback through the filters. At f_r, however, there is negligible negative feedback, and thus, the positive feedback through the voltage divider (R_1 and R_2) allows the circuit to oscillate. Self-starting can be achieved by using a tungsten lamp in the place of R_1.

15–3 REVIEW QUESTIONS

1. There are two feedback loops in the Wien-bridge oscillator. What is the purpose of each?
2. A certain lead-lag network has $R_1 = R_2$ and $C_1 = C_2$. An input voltage of 5 V rms is applied. The input frequency equals the resonant frequency of the network. What is the rms output voltage?

3. Why must the phase shift through the *RC* feedback circuit in a phase-shift oscillator equal 180°?

15–4 OSCILLATORS WITH *LC* FEEDBACK CIRCUITS

Although the RC oscillators, particularly the Wien bridge, are generally suitable for frequencies up to about 1 MHz, LC feedback elements are normally used in oscillators that require higher frequencies of oscillation. Also, because of the frequency limitation (lower unity-gain frequency) of most op-amps, discrete transistors are often employed as the gain element in LC oscillators. This section introduces several types of resonant LC oscillators—The Colpitts, Clapp, Hartley, Armstrong, and crystal-controlled oscillators.

THE COLPITTS OSCILLATOR

One basic type of resonant circuit oscillator is the Colpitts, named after its inventor—as are most of the others we cover here. As shown in Figure 15–16, this type of oscillator uses an *LC* circuit in the feedback loop to provide the necessary phase shift and to act as a resonant filter that passes only the desired frequency of oscillation.

The approximate frequency of oscillation is the resonant frequency of the *LC* circuit and is established by the values of C_1, C_2, and L according to this familiar formula:

$$f_r \cong \frac{1}{2\pi\sqrt{LC_T}} \tag{15–5}$$

Because the capacitors effectively appear in series around the tank circuit, the total capacitance is

$$C_T = \frac{C_1C_2}{C_1 + C_2} \tag{15–6}$$

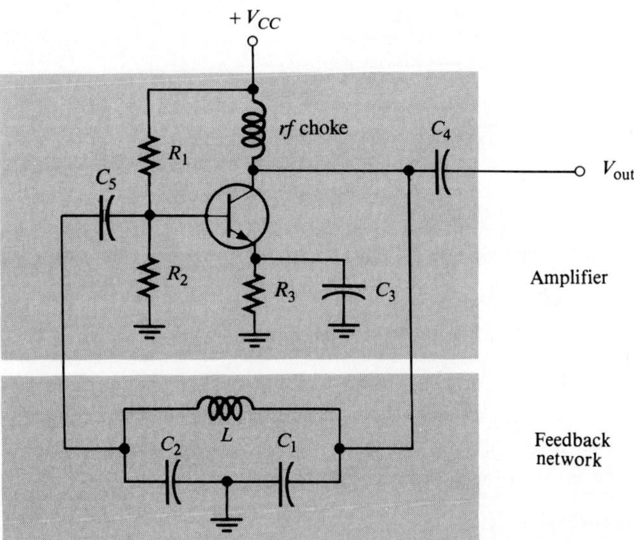

FIGURE 15–16
A basic Colpitts oscillator with a BJT as the gain element.

CONDITIONS FOR OSCILLATION AND START-UP The attenuation, B, of the resonant feedback circuit in the Colpitts oscillator is basically determined by the values of C_1 and C_2.

Figure 15–17 shows that the circulating tank current flows through both C_1 and C_2 (they are effectively in series). The voltage developed across C_1 is the oscillator's output voltage (V_{out}) and the voltage developed across C_2 is the feedback voltage (V_f), as indicated. The expression for the attenuation is

$$B = \frac{V_f}{V_{out}} \cong \frac{IX_{C2}}{IX_{C1}} = \frac{X_{C2}}{X_{C1}} = \frac{1/2\pi f_r C_2}{1/2\pi f_r C_1}$$

Cancelling the $2\pi f_r$ terms gives

$$B = \frac{C_1}{C_2} \tag{15–7}$$

As you know, a condition for oscillation is $A_vB = 1$. Since $B = C_1/C_2$,

$$A_v = \frac{C_2}{C_1} \tag{15–8}$$

where A_v is the voltage gain of the transistor amplifier. With this condition met, $A_vB = (C_2/C_1)(C_1/C_2) = 1$. Actually, for the oscillator to be self-starting, A_vB must be greater than 1 ($A_vB > 1$). Therefore, the voltage gain must be made slightly greater than C_2/C_1.

$$A_v > \frac{C_2}{C_1} \tag{15–9}$$

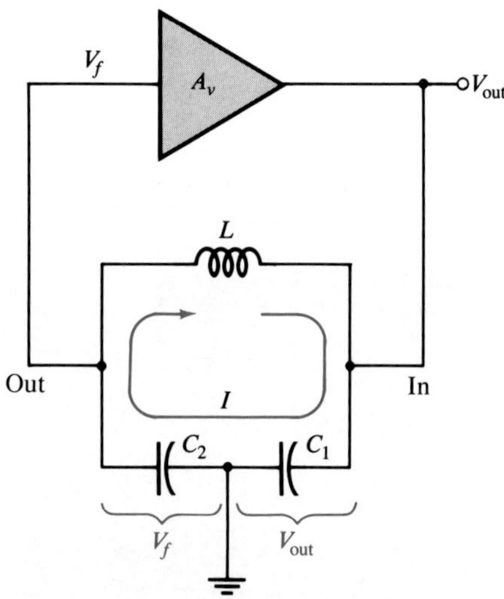

FIGURE 15–17 The attenuation of the tank circuit is the output of the tank (V_f) divided by the input to the tank (V_{out}). $B = V_f/V_{out} = C_1/C_2$. For $A_vB > 1$, A_v must be greater than C_2/C_1.

LOADING OF THE FEEDBACK CIRCUIT AFFECTS THE FREQUENCY OF OSCILLATION As indicated in Figure 15–18, the input impedance of the transistor amplifier acts as a load on the resonant feedback circuit and reduces the Q of the circuit. Recall from your study of resonance that the resonant frequency of a parallel resonant circuit depends on the Q, according to the following formula:

$$f_r = \frac{1}{2\pi\sqrt{LC_T}}\sqrt{\frac{Q^2}{Q^2 + 1}} \tag{15–10}$$

As a rule of thumb, for a Q greater than 10, the frequency is approximately $1/2\pi\sqrt{LC_T}$, as stated in Equation (15–5). When Q is less than 10, however, f_r is reduced significantly.

An FET can be used in place of a bipolar transistor, as shown in Figure 15–19, to minimize the loading effect of the transistor's input impedance. Recall the FETs have much higher input impedances than do bipolar transistors. Also, when an external load is connected to the oscillator output, as shown in Figure 15–20(a), f_r may decrease, again because of a reduction in Q. This happens if the load resistance is too small. In some cases, one way to eliminate the effects of a load resistance is by transformer coupling as indicated in Figure 15–20(b).

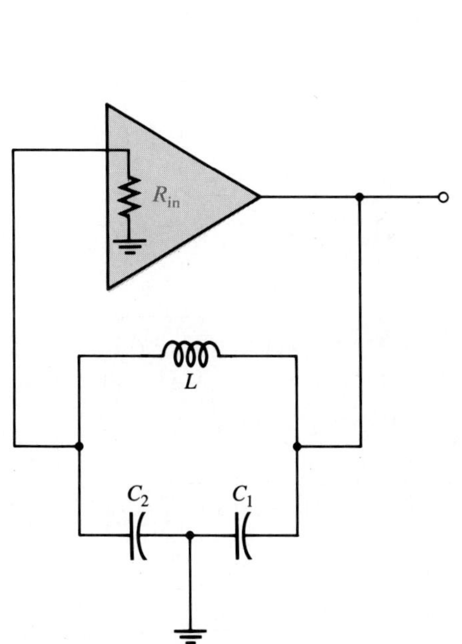

FIGURE 15–18
R_{in} of the transistor amplifier loads the feedback circuit and lowers its Q, thus lowering the resonant frequency.

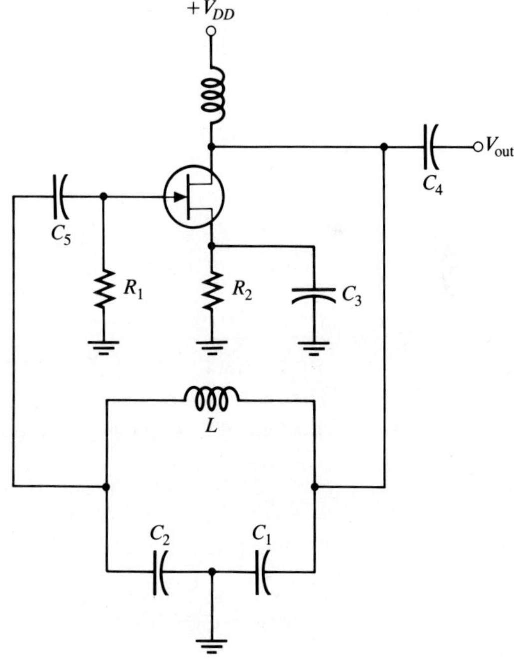

FIGURE 15–19
A basic FET Colpitts oscillator.

FIGURE 15–20
Oscillator loading.

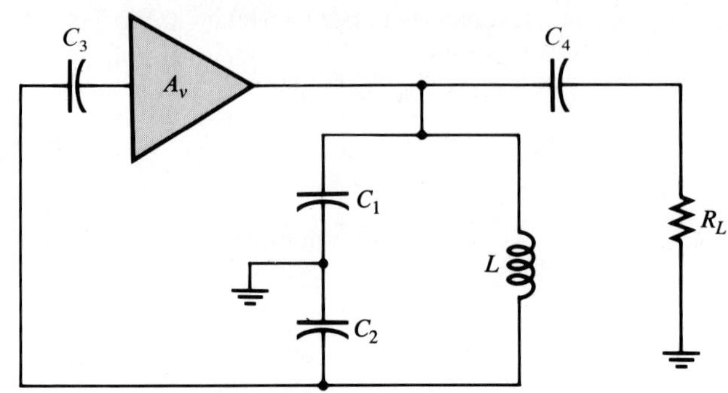

(a) A load capacitively coupled to oscillator output can reduce circuit Q and f_r.

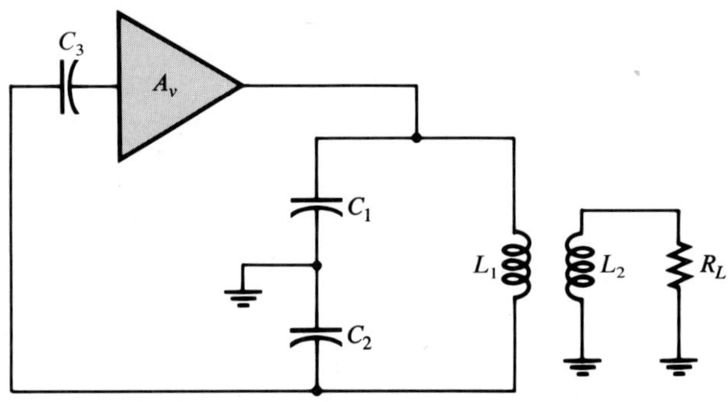

(b) Transformer coupling of load can reduce loading effect by impedance transformation.

EXAMPLE 15–3

(a) Determine the frequency of oscillation for the oscillator in Figure 15–21. Assume there is negligible loading on the feedback circuit and that its Q is greater than 10.
(b) Find the frequency of oscillation if the oscillator is loaded to a point where the Q drops to 8.

SOLUTION

(a) $$C_T = \frac{C_1C_2}{C_1 + C_2} = \frac{(0.01\ \mu\text{F})(0.1\ \mu\text{F})}{0.11\ \mu\text{F}} = 0.0091\ \mu\text{F}$$

$$f_r \cong \frac{1}{2\pi\sqrt{L_1C_T}} = \frac{1}{2\pi\sqrt{(50\ \text{mH})(0.0091\ \mu\text{F})}} = 7.46\ \text{kHz}$$

(b) $$f_r = \frac{1}{2\pi\sqrt{L_1C_T}}\sqrt{\frac{Q^2}{Q^2 + 1}} = (7.46\ \text{kHz})(0.9923) = 7.40\ \text{kHz}$$

FIGURE 15–21

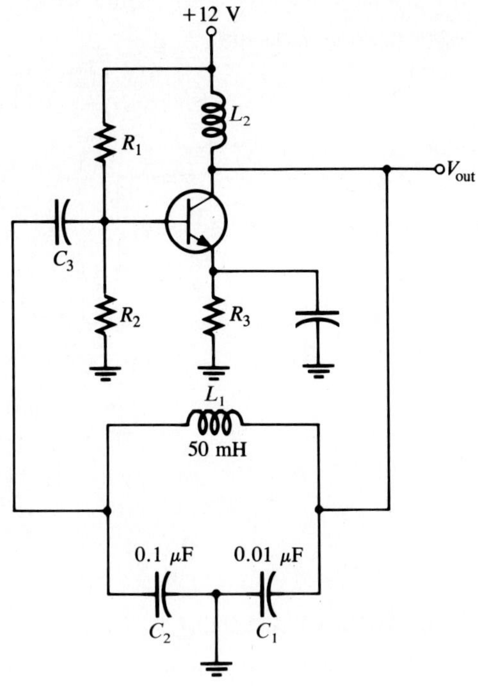

PRACTICE EXERCISE 15–3

What frequency does the oscillator produce if it is loaded to a point where $Q = 4$?

THE CLAPP OSCILLATOR

The Clapp oscillator is a variation of the Colpitts. The basic difference is an additional capacitor, C_3, in series with the inductor in the resonant feedback circuit, as shown in Figure 15–22. Since C_3 is in series with C_1 and C_2, the total capacitance around the tank circuit is

$$C_T = \frac{1}{1/C_1 + 1/C_2 + 1/C_3} \tag{15–11}$$

and the approximate frequency of oscillation ($Q > 10$) is $f_r \cong 1/2\pi\sqrt{LC_T}$.

If C_3 is much smaller than C_1 and C_2, the resonant frequency is controlled almost entirely by C_3 ($f_r \cong 1/2\pi\sqrt{LC_3}$). Since C_1 and C_2 are both connected to ground at one end, the junction capacitance of the transistor and other stray capacitances appear in parallel with C_1 and C_2 to ground, altering their effective values. C_3 is not affected, however, and thus provides a more accurate and stable frequency of oscillation.

FIGURE 15–22
A basic Clapp oscillator.

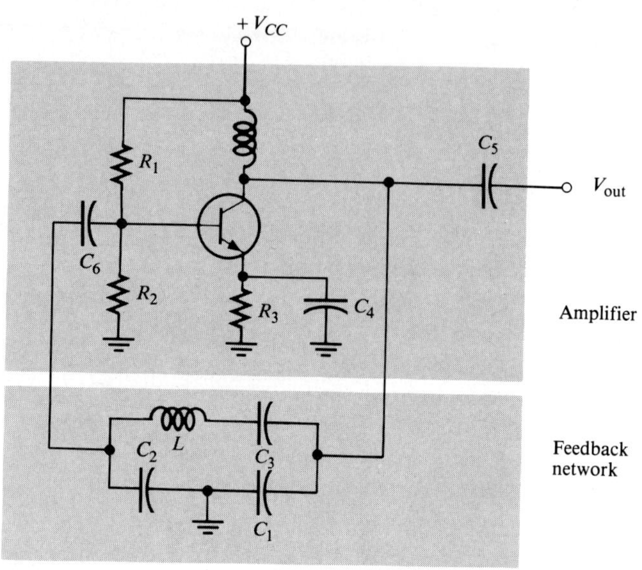

THE HARTLEY OSCILLATOR

The Hartley oscillator is similar to the Colpitts except that the feedback network consists of two series inductors and a parallel capacitor as shown in Figure 15–23.

FIGURE 15–23
A basic Hartley oscillator.

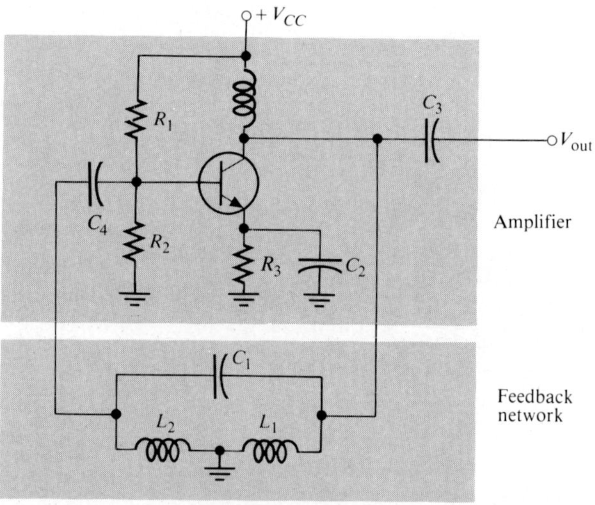

In this circuit, the frequency of oscillation for $Q > 10$ is

$$f_r \cong \frac{1}{2\pi\sqrt{L_T C}} \tag{15–12}$$

where $L_T = L_1 + L_2$. The inductors act in a role similar to C_1 and C_2 in the Colpitts to determine the attenuation of the feedback network.

$$B \cong \frac{L_2}{L_1} \tag{15–13}$$

To assure start-up of oscillation, A_v must be greater than $1/B$.

$$A_v > \frac{L_1}{L_2} \tag{15–14}$$

Loading of the tank circuit has the same effect in the Hartley as in the Colpitts; that is, the Q is decreased and thus f_r decreases.

THE ARMSTRONG OSCILLATOR

This type of LC oscillator uses transformer coupling to feed back a portion of the signal voltage, as shown in Figure 15–24. It is sometimes called a "tickler" oscillator in reference to the transformer secondary or "tickler coil" that provides the feedback to keep the oscillation going. The Armstrong is less common than the Colpitts, Clapp, and Hartley, mainly because of the disadvantage of transformer size and cost. The frequency of oscillation is set by the inductance of the primary winding (L_p) in parallel with C_1 ($f_r = 1/2\pi\sqrt{L_pC_1}$).

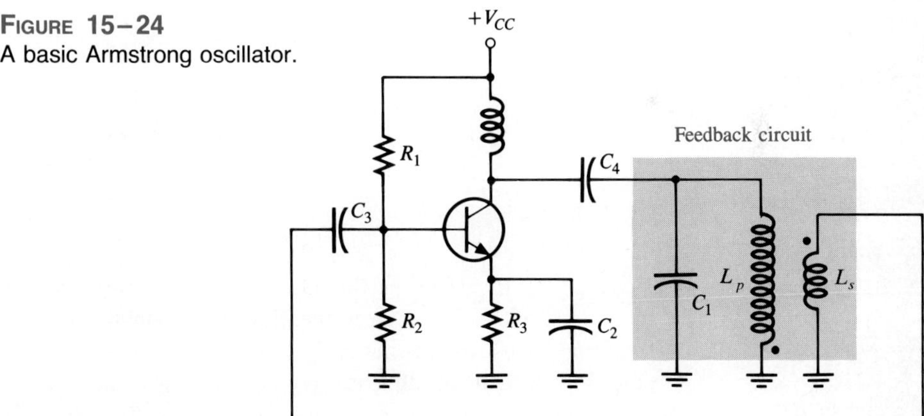

FIGURE 15–24
A basic Armstrong oscillator.

CRYSTAL-CONTROLLED OSCILLATORS

The most stable and accurate type of oscillator uses a piezoelectric **crystal** in the feedback loop to control the frequency.

THE PIEZOELECTRIC EFFECT Quartz is one type of crystalline substance found in nature that exhibits a property called the **piezoelectric effect.** When a changing mechanical stress is applied across the crystal to cause it to vibrate, a voltage develops at the frequency of mechanical vibration. Conversely, when an ac voltage is applied across the crystal, it vibrates at the frequency of the applied voltage. The greatest vibration occurs at the crystal's natural resonant frequency, which is determined by the physical dimensions and by the way the crystal is cut.

Crystals used in electronic applications are typically a quartz wafer mounted between two electrodes and enclosed in a protective "can" as shown in Figure 15–25(a). A schematic symbol is shown in Figure 15–25(b) and an equivalent *RLC* circuit for the crystal appears in Figure 15–25(c). As you can see, the crystal's equivalent circuit is a series-parallel *RLC* network and can operate in either series resonance or parallel resonance. At the series resonant frequency, the inductive reactance is cancelled by the reactance of C_S. The remaining series resistance, R_S, determines the impedance of the crystal. Parallel resonance occurs when the inductive reactance and the reactance of the parallel capacitance, C_m, are equal. The parallel resonant frequency is usually at least 1 kHz higher than the series resonant frequency. A great advantage of the crystal is that it exhibits a very high Q (Qs of several thousand are typical).

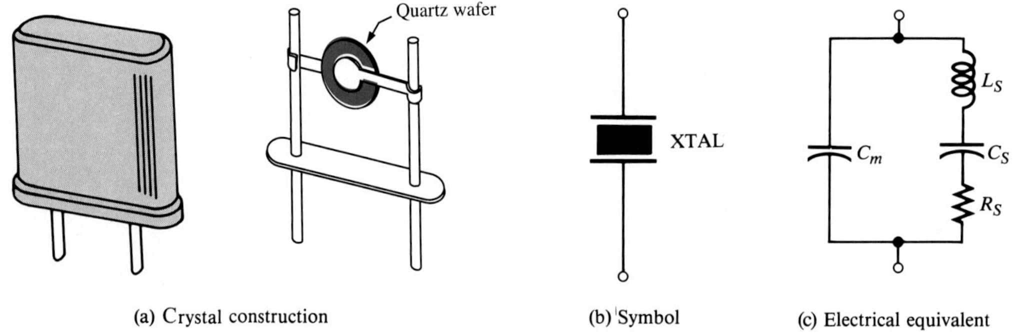

FIGURE 15–25
A quartz crystal.

MODES OF OSCILLATION IN THE CRYSTAL Piezoelectric crystals can oscillate in either of two modes—fundamental or overtone. The fundamental frequency of a crystal is the lowest frequency at which it is naturally resonant. The fundamental frequency depends on the crystal's mechanical dimensions, type of cut, and other factors, and is inversely proportional to the thickness of the crystal slab. Because a slab of crystal cannot be cut too thin without fracturing, there is an upper limit on the fundamental frequency. For most crystals, this upper limit is less than 20 MHz. For higher frequencies, the crystal must be operated in the overtone mode. Overtones are approximate integer multiples of the fundamental frequency. The overtone frequencies are usually, but not always, odd multiples (3, 5, 7, . . .) of the fundamental.

An oscillator that uses a crystal as a series resonant tank circuit is shown in Figure 15–26(a). The impedance of the crystal is minimum at the series resonant frequency, thus providing maximum feedback. The series capacitor, C_C, is used to "fine tune" the oscillator frequency by "pulling" the resonant frequency of the crystal slightly up or down. A modified Colpitts configuration is shown in Figure 15–26(b) with a crystal acting as a parallel resonant tank circuit. The impedance of the crystal is maximum at parallel resonance, thus developing the maximum voltage across the capacitors. The voltage across C_1 is fed back to the input.

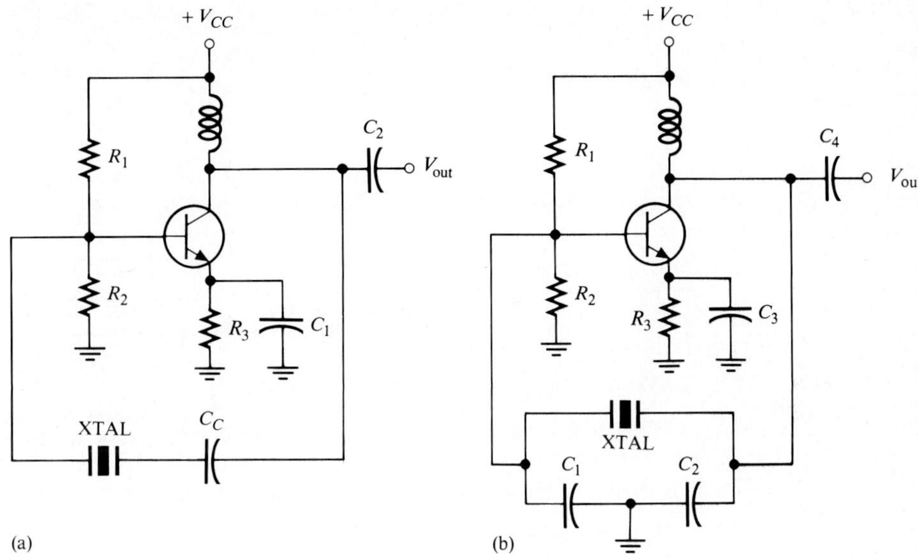

FIGURE 15–26
Basic crystal oscillators.

15–4 REVIEW QUESTIONS

1. What is the basic difference between the Colpitts and the Hartley oscillators?
2. What is the advantage of an FET amplifier in a Colpitts or Hartley oscillator?
3. How can you distinguish a Colpitts oscillator from a Clapp oscillator?

15–5 NONSINUSOIDAL OSCILLATORS

In this section, several types of op-amp oscillator circuits that produce triangular, sawtooth, or square waveforms are discussed. Some of these types of oscillators are frequently referred to as signal generators and multivibrators depending on the particular circuit implementation.

A TRIANGULAR-WAVE OSCILLATOR

The op-amp integrator covered in the last chapter can be used as the basis for a triangular wave generator. The basic idea is illustrated in Figure 15–27(a) where a dual-polarity, switched input is used. We use the switch only to introduce the concept; it is not a practical way to implement this circuit. When the switch is in position 1, the negative voltage is applied, and the output is a positive-going ramp. When the switch is thrown into position 2, a negative-going ramp is produced. If the switch is thrown back and forth at fixed intervals, the output is a triangular wave consisting of alternating positive-going and negative-going ramps, as shown in Figure 15–27(b).

FIGURE 15–27
Basic triangular-wave generator.

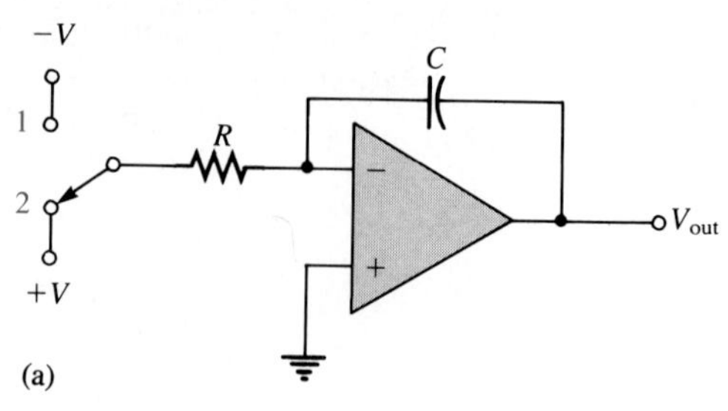

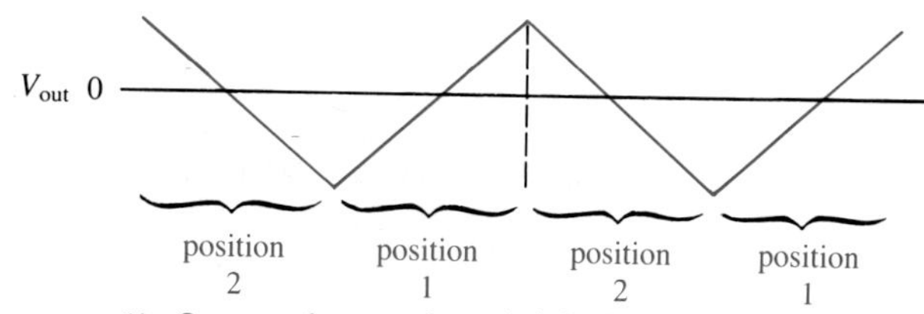

(b) Output voltage as the switch is thrown back and forth at regular intervals

A PRACTICAL CIRCUIT

One practical implementation of a triangular-wave generator utilizes an op-amp comparator to perform the switching function, as shown in Figure 15–28. The operation is as follows. To begin, assume that the output voltage of the comparator is at its maximum negative level. This output is connected to the inverting input of the integrator through R_1, producing a positive-going ramp on the output of the integrator. When the ramp voltage reaches the upper trigger point (UTP), the comparator switches to its maximum positive level. This positive level causes the integrator ramp to change to a negative-going direction. The ramp continues in this direction until the lower trigger point (LTP) of the comparator is reached. At this point, the comparator output switches back to the maximum negative level and the cycle repeats. This action is illustrated in Figure 15–29.

Since the comparator produces a square-wave output, the circuit in Figure 15–28 can be used as both a triangular-wave generator and a square-wave generator. Devices of this type are commonly known as *function generators* because they produce more than one output function. The output amplitude of the square wave is set by the output swing of the

FIGURE 15–28
A triangular-wave generator using two op-amps.

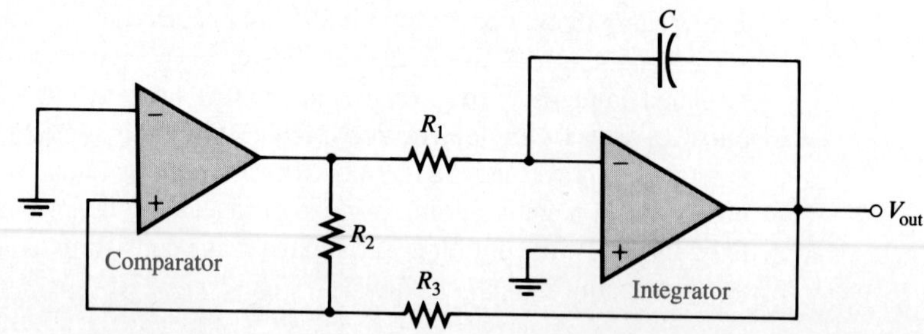

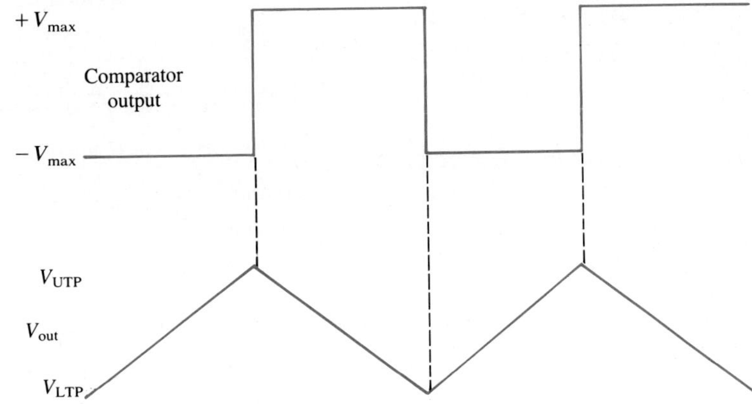

FIGURE 15–29 Waveforms for the circuit in Figure 15–28.

comparator, and the resistors R_2 and R_3 set the amplitude of the triangular output by establishing the UTP and LTP voltages according to the following formulas:

$$V_{UTP} = +V_{max}\left(\frac{R_3}{R_2}\right) \tag{15–15}$$

$$V_{LTP} = -V_{max}\left(\frac{R_3}{R_2}\right) \tag{15–16}$$

where the comparator output levels, $+V_{max}$ and $-V_{max}$, are equal. The frequency of both waveforms depends on the R_1C time constant as well as the amplitude-setting resistors, R_2 and R_3. By varying R_1, the frequency of oscillation can be adjusted without changing the output amplitude.

$$f = \frac{1}{4R_1C}\left(\frac{R_2}{R_3}\right) \tag{15–17}$$

EXAMPLE 15–4

Determine the frequency of the circuit in Figure 15–30. To what value must R_1 be changed to make the frequency 20 kHz?

FIGURE 15–30

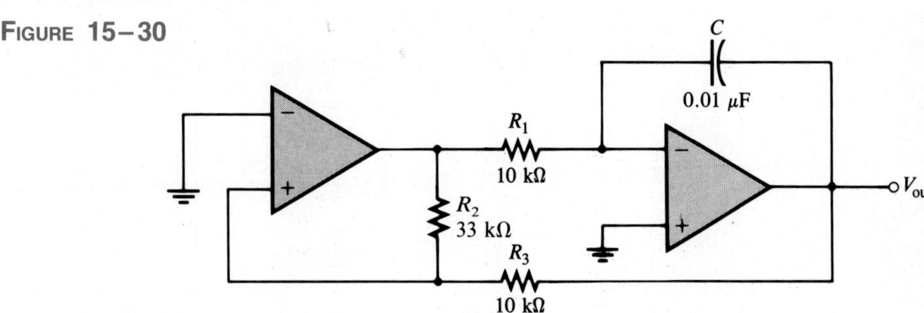

SOLUTION

$$f = \frac{1}{4R_1C}\left(\frac{R_2}{R_3}\right) = \left(\frac{1}{4(10\text{ k}\Omega)(0.01\ \mu\text{F})}\right)\left(\frac{33\text{ k}\Omega}{10\text{ k}\Omega}\right) = 8.25\text{ kHz}$$

To make $f = 20$ kHz,

$$R_1 = \frac{1}{4fC}\left(\frac{R_2}{R_3}\right) = \left(\frac{1}{4(20\text{ kHz})(0.01\ \mu\text{F})}\right)\left(\frac{33\text{ k}\Omega}{10\text{ k}\Omega}\right) = 4.13\text{ k}\Omega$$

PRACTICE EXERCISE 15–4

What is the amplitude of the triangular wave in Figure 15–30 if the comparator output is ±10 V?

A VOLTAGE-CONTROLLED SAWTOOTH OSCILLATOR (VCO)

The voltage-controlled oscillator (VCO) is an oscillator whose frequency can be changed by a variable dc control voltage. VCOs can be either sinusoidal or nonsinusoidal. One way to build a voltage-controlled sawtooth oscillator is with an op-amp integrator that uses a switching device (PUT) in parallel with the feedback capacitor to terminate each ramp at a prescribed level and effectively "reset" the circuit. Figure 15–31(a) shows the implementation.

FIGURE 15–31
Voltage-controlled sawtooth oscillator operation.

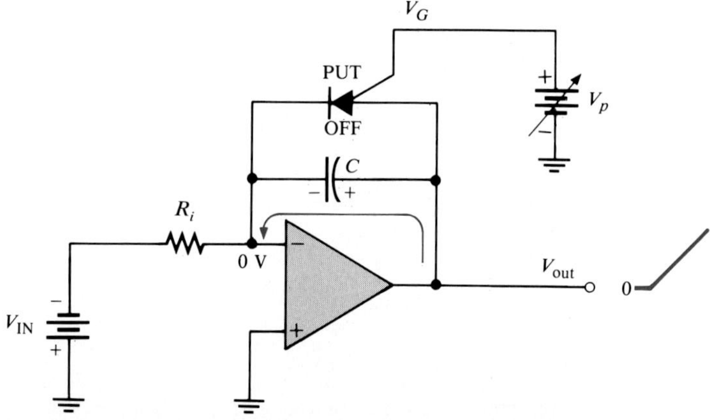

(a) Initially, the capacitor charges, the output ramp begins, and the PUT is off.

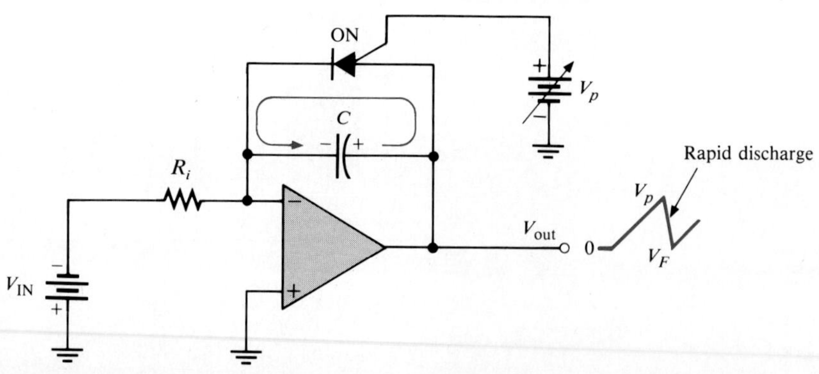

(b) The capacitor rapidly discharges when the PUT momentarily turns on.

As you learned in Chapter 11, the PUT is a programmable unijunction transistor with an anode, a cathode, and a gate terminal. The gate is always biased positively with respect to the cathode. When the anode voltage exceeds the gate voltage by approximately 0.7 V, the PUT turns on and acts as a forward-biased diode. When the anode voltage falls below this level, the PUT turns off. Also, the current must be above the holding value to maintain conduction.

The operation of the sawtooth generator begins when the negative dc input voltage, $-V_{IN}$, produces a positive-going ramp on the output. During the time that the ramp is increasing, the circuit acts as a regular integrator. The PUT triggers on when the output ramp (at the anode) exceeds the gate voltage by 0.7 V. The gate is set to the approximate desired sawtooth peak voltage. When the PUT turns on, the capacitor rapidly discharges, as shown in Figure 15–31(b). The capacitor does not discharge completely to zero because of the PUT's forward voltage, V_F. Discharge continues until the PUT current falls below the holding value. At this point, the PUT turns off and the capacitor begins to charge again, thus generating a new output ramp. The cycle continually repeats, and the resulting output is a repetitive sawtooth waveform, as shown. The sawtooth amplitude and period can be adjusted by varying the PUT gate voltage.

The frequency is determined by the R_iC time constant of the integrator and the peak voltage set by the PUT. Recall that the charging rate of the capacitor is V_{IN}/R_iC. The time it takes the capacitor to charge from V_F to V_p is the period, T, of the sawtooth (neglecting the rapid discharge time).

$$T = \frac{V_p - V_F}{|V_{IN}|/R_iC} \tag{15–18}$$

From $f = 1/T$, we get

$$f = \frac{|V_{IN}|}{R_iC}\left(\frac{1}{V_p - V_F}\right) \tag{15–19}$$

EXAMPLE 15–5

(a) Find the amplitude and frequency of the sawtooth output in Figure 15–32. Assume that the forward PUT voltage V_F is approximately 1 V.

(b) Sketch the output waveform.

SOLUTION

(a) First, find the gate voltage in order to establish the approximate voltage at which the PUT turns on.

$$V_G = \frac{R_4}{R_3 + R_4}(+V) = \frac{10\text{ k}\Omega}{20\text{ k}\Omega}(15\text{ V}) = 7.5\text{ V}$$

This voltage sets the approximate maximum peak value of the sawtooth output (neglecting the 0.7 V).

$$V_p \cong 7.5\text{ V}$$

The minimum peak value (low point) is

$$V_F \cong 1\text{ V}$$

The period is determined as follows:

$$V_{IN} = \frac{R_2}{R_1 + R_2}(-V) = \frac{10\ \text{k}\Omega}{78\ \text{k}\Omega}(-15\ \text{V}) = -1.92\ \text{V}$$

$$T = \frac{V_p - V_F}{|V_{IN}|/R_iC} = \frac{7.5\ \text{V} - 1\ \text{V}}{1.92\ \text{V}/(100\ \text{k}\Omega)(0.005\ \mu\text{F})} = 1.69\ \text{ms}$$

$$f = \frac{1}{1.69\ \text{ms}} \cong 592\ \text{Hz}$$

(b) The output waveform is shown in Figure 15–33.

FIGURE 15–32

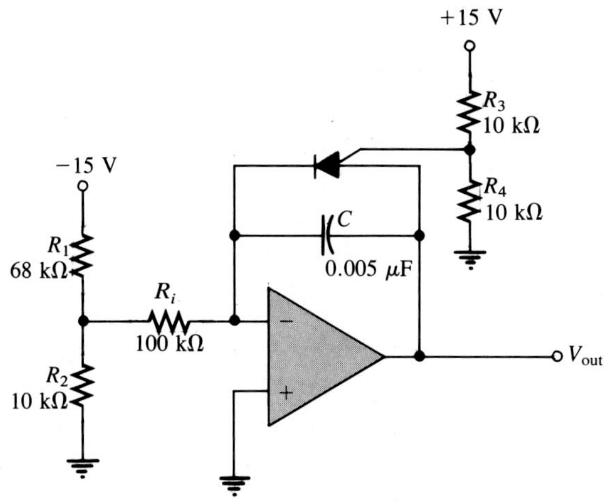

FIGURE 15–33
Output of the circuit in Figure 15–32.

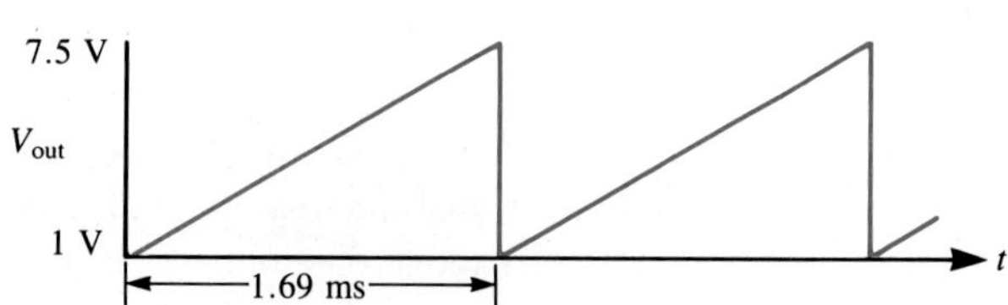

PRACTICE EXERCISE 15–5
If R_i is changed to 56 kΩ in Figure 15–32, what is the frequency?

A SQUARE-WAVE RELAXATION OSCILLATOR

The basic square-wave generator shown in Figure 15–34 is a type of relaxation oscillator because its operation is based on the charging and discharging of a capacitor. Notice that the op-amp's inverting input is the capacitor voltage and the noninverting input is a portion of the output fed back through resistors R_2 and R_3. When the circuit is first turned on, the capacitor is uncharged, and thus the inverting input is at 0 V. This makes the output a positive maximum, and the capacitor begins to charge toward V_{out} through R_1. When the capacitor voltage reaches a value equal to the feedback voltage on the noninverting input, the op-amp switches to the maximum negative state. At this point, the capacitor begins to discharge from $+V_f$ toward $-V_f$. When the capacitor voltage reaches $-V_f$, the op-amp switches back to the maximum positive state. This action continues to repeat, as shown in Figure 15–35, and a square-wave output voltage is obtained.

FIGURE 15–34
A square-wave relaxation oscillator.

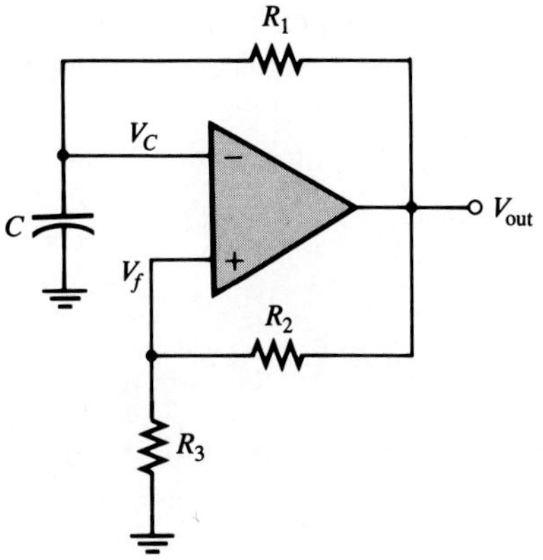

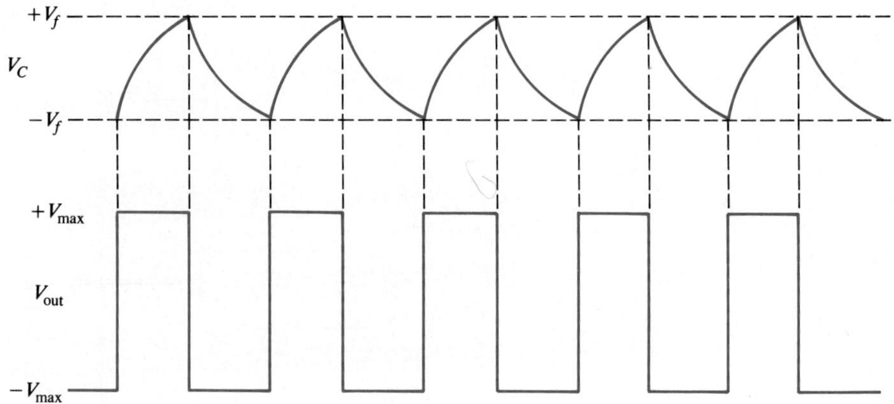

FIGURE 15–35
Waveforms for the square-wave relaxation oscillator.

15–5 REVIEW QUESTIONS

1. What is a VCO, and basically, what does it do?
2. Upon what principle does a relaxation oscillator operate?

15–6 THE 555 TIMER AS AN OSCILLATOR

The 555 timer is a versatile integrated circuit with many applications. In this section, you will see how the 555 is configured as an astable or free-running multivibrator, which is essentially a square-wave oscillator. We will also discuss the use of the 555 timer as a voltage-controlled oscillator (VCO).

The 555 timer consists basically of two comparators, a flip-flop, a discharge transistor, and a resistive voltage divider, as shown in Figure 15–36. The flip-flop (bistable multivibrator) is a digital device that is perhaps unfamiliar to you at this point unless you already have taken a digital fundamentals course. Briefly, it is a two-state device whose output can be at either a high voltage level (set) or a low voltage level (reset). The state of the output can be changed with proper input signals.

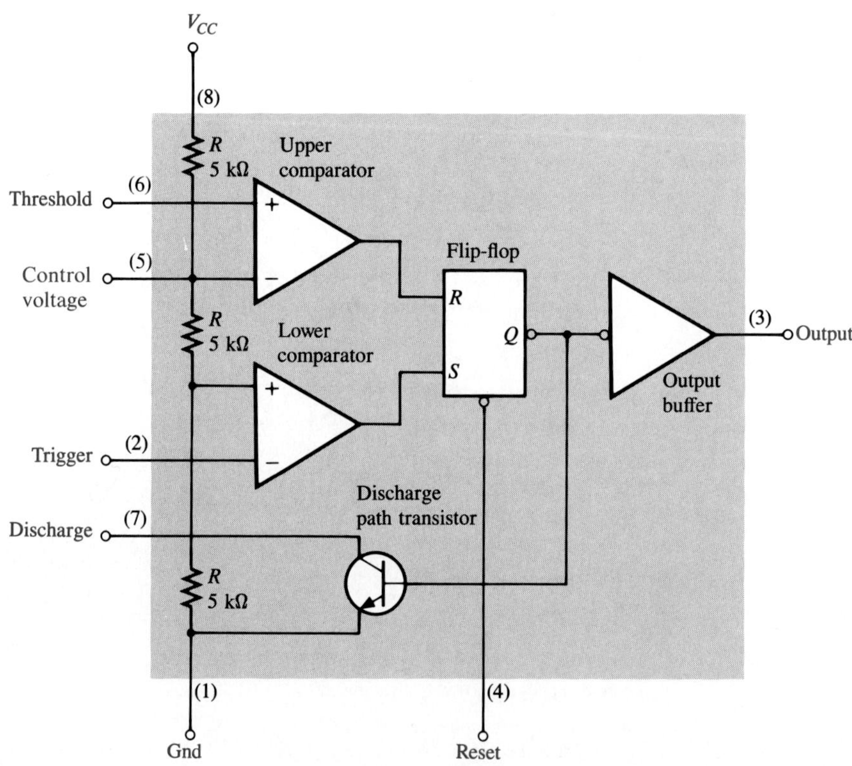

FIGURE 15–36
Internal diagram of a 555 integrated circuit timer. (IC pin numbers are in parentheses.)

The resistive voltage divider is used to set the voltage comparator levels. All three resistors are of equal value; therefore, the upper comparator has a reference of $\frac{2}{3}V_{CC}$, and the lower comparator has a reference of $\frac{1}{3}V_{CC}$. The comparators' outputs control the state of the flip-flop. When the trigger voltage goes below $\frac{1}{3}V_{CC}$, the flip-flop sets and the output jumps to its high level. The threshold input is normally connected to an external *RC* timing network. When the external capacitor voltage exceeds $\frac{2}{3}V_{CC}$, the upper comparator resets the flip-flop, which in turn switches the output back to its low level. When the device output is low, the discharge transistor Q_d is turned on and provides a path for rapid discharge of the external timing capacitor. This basic operation allows the timer to be configured with external components as an oscillator, a one-shot, or a time-delay element.

ASTABLE OPERATION

A 555 timer connected to operate in the astable mode as a free-running nonsinusoidal oscillator is shown in Figure 15–37. Notice that the threshold input (THRESH) is now connected to the trigger input (TRIG). The external components R_1, R_2, and C_{ext} form the timing network that sets the frequency of oscillation. The 0.01 μF capacitor connected to the control (CONT) input is strictly for decoupling and has no effect on the operation; in some cases it can be left off.

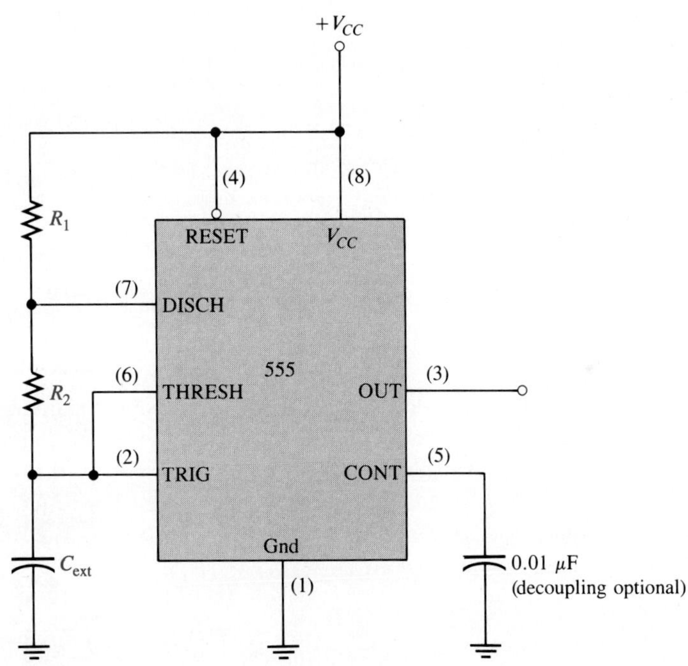

FIGURE 15–37
The 555 timer connected as an astable multivibrator.

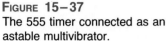

Initially, when the power is turned on, the capacitor C_{ext} is uncharged and thus the trigger voltage (2) is at 0 V. This causes the output of the lower comparator to be high and the output of the upper comparator to be low, forcing the output of the flip-flop, and thus the base of Q_d, low and keeping the transistor off. Now, C_{ext} begins charging through R_1 and R_2 as indicated in Figure 15–38. When the capacitor voltage reaches $\frac{1}{3}V_{CC}$, the lower comparator switches to its low output state, and when the capacitor voltage reaches $\frac{2}{3}V_{CC}$, the upper comparator switches to its high output state. This resets the flip-flop, causes the base of Q_d to go high, and turns on the transistor. This sequence creates a discharge path for the capacitor through R_2 and the transistor, as indicated. The capacitor

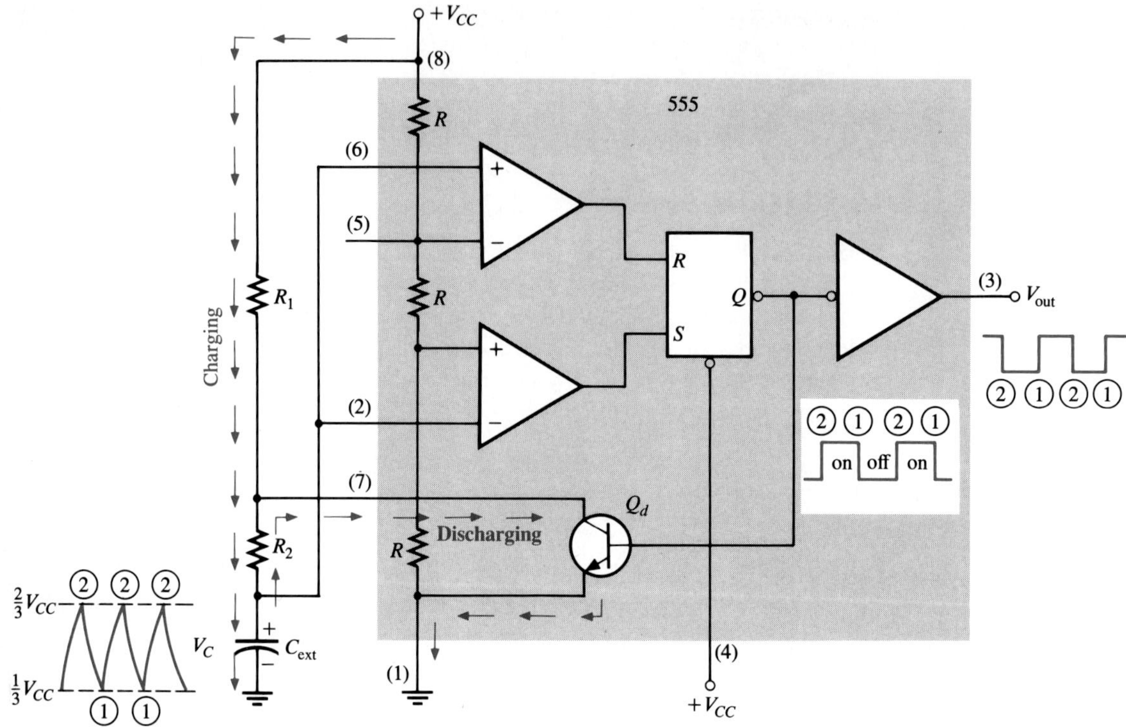

FIGURE 15–38
Operation of the 555 timer in the astable mode.

now begins to discharge, causing the upper comparator to go low. At the point where the capacitor discharges down to $\frac{1}{3}V_{CC}$, the lower comparator switches high, setting the flip-flop, which makes the base of Q_d low and turns off the transistor. Another charging cycle begins, and the entire process repeats. The result is a rectangular wave output whose duty cycle depends on the values of R_1 and R_2. The frequency of oscillation is given by the formula

$$f = \frac{1.44}{(R_1 + 2R_2)C_{ext}} \qquad (15\text{–}20)$$

By selecting R_1 and R_2, the duty cycle of the output can be adjusted. Since C_{ext} charges through $R_1 + R_2$ and discharges only through R_2, duty cycles approaching a minimum of 50 percent can be achieved if $R_2 >> R_1$ so that the charging and discharging times are approximately equal. An expression to calculate the duty cycle is developed as follows.

The time that the output is high is how long it takes C_{ext} to charge from $\frac{1}{3}V_{CC}$ to $\frac{2}{3}V_{CC}$. It is expressed as

$$t_H = 0.693(R_1 + R_2)C_{ext}$$

The time that the output is low is how long it takes C_{ext} to discharge from $\frac{2}{3}V_{CC}$ to $\frac{1}{3}V_{CC}$. It is expressed as

$$t_L = 0.693R_2C_{ext}$$

The period, T, of the output waveform is the sum of t_H and t_L.

$$T = t_H + t_L = 0.693(R_1 + 2R_2)C_{ext}$$

This is the reciprocal of f in Equation (15–20). Finally, the duty cycle is

$$\text{Duty cycle} = \frac{t_H}{T} = \frac{t_H}{t_H + t_L}$$

$$\text{Duty cycle} = \frac{R_1 + R_2}{R_1 + 2R_2} \times 100\% \qquad (15\text{–}21)$$

To achieve duty cycles of less than 50 percent, the circuit in Figure 15–37 can be modified so that C_{ext} charges through only R_1 and discharges through R_2. This is achieved with a diode D_1 placed as shown in Figure 15–39. The duty cycle can be made less than 50 percent by making R_1 less than R_2. Under this condition, the expression for the duty cycle is

$$\text{Duty cycle} = \frac{R_1}{R_1 + R_2} \times 100\% \qquad (15\text{–}22)$$

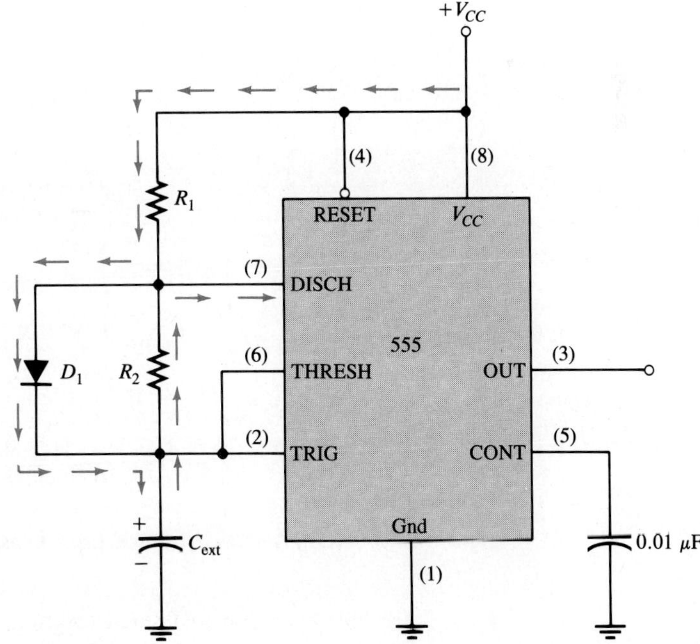

FIGURE 15–39
The addition of diode D_1 allows the duty cycle of the output to be adjusted to less than 50 percent by making $R_1 < R_2$.

EXAMPLE 15–6 A 555 timer configured to run in the astable mode (oscillator) is shown in Figure 15–40. Determine the frequency of the output and the duty cycle.

FIGURE 15–40

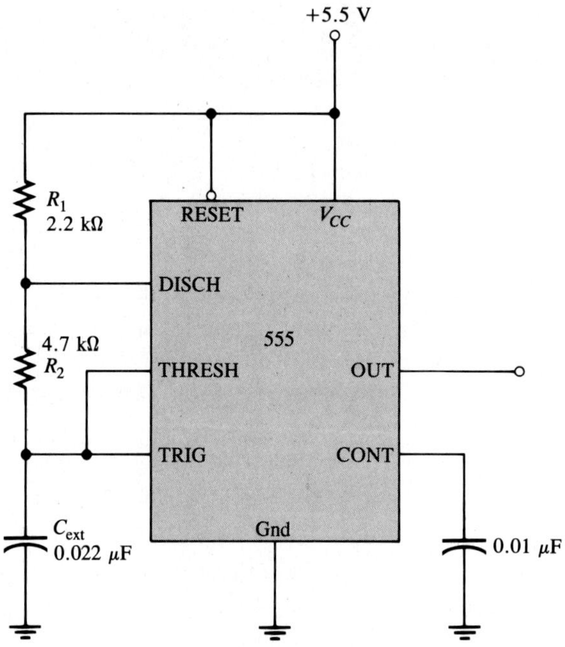

SOLUTION

$$f = \frac{1.44}{(R_1 + 2R_2)C_{ext}} = \frac{1.44}{(2.2\text{ k}\Omega + 9.4\text{ k}\Omega)0.022\ \mu\text{F}} = 5.64\text{ kHz}$$

$$\text{Duty cycle} = \frac{R_1 + R_2}{R_1 + 2R_2} \times 100\% = \frac{2.2\text{ k}\Omega + 4.7\text{ k}\Omega}{2.2\text{ k}\Omega + 9.4\text{ k}\Omega} \times 100\% = 59.5\%$$

PRACTICE EXERCISE 15–6

Determine the duty cycle in Figure 15–40 if a diode is connected across R_2 as indicated in Figure 15–39.

OPERATION AS A VOLTAGE-CONTROLLED OSCILLATOR (VCO)

A 555 timer can be configured to operate as a VCO by using the same external connections as for astable operation, with the exception that a variable control voltage is applied to the CONT input (pin 5), as indicated in Figure 15–41.

As shown in Figure 15–42, the control voltage (V_{CONT}) changes the threshold values of $\frac{1}{3}V_{CC}$ and $\frac{2}{3}V_{CC}$ for the internal comparators. With the control voltage, the upper value is V_{CONT} and the lower value is $\frac{1}{2}V_{CONT}$, as you can see by examining the internal diagram of the 555 timer. When the control voltage is varied, the output frequency also varies. An increase in V_{CONT} increases the charging and discharging time of the external capacitor and causes the frequency to decrease. A decrease in V_{CONT} decreases the charging and discharging time of the capacitor and causes the frequency to increase.

An interesting application of the VCO is in phase-locked loops, which are used in various types of communications receivers to track variations in the frequency of incoming signals. We will cover the basic operation of a phase-locked loop in the next section.

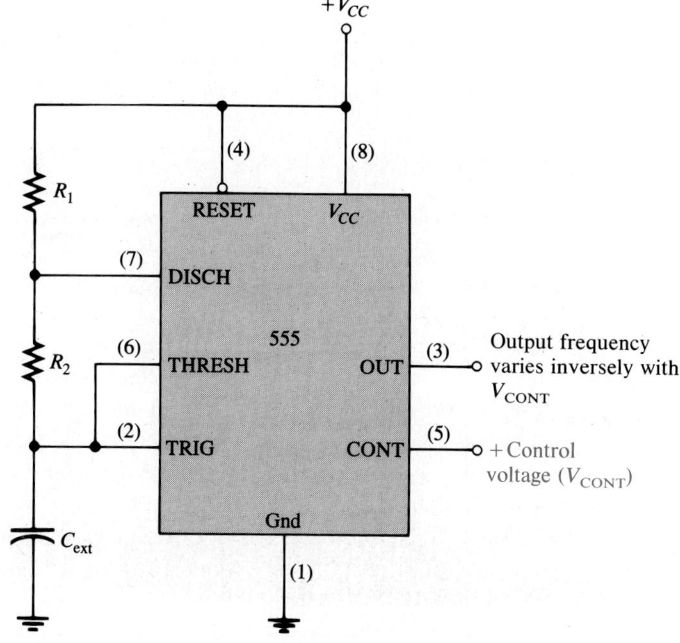

FIGURE 15–41
The 555 timer connected as a voltage-controlled oscillator (VCO). Note the variable control voltage input on pin 5.

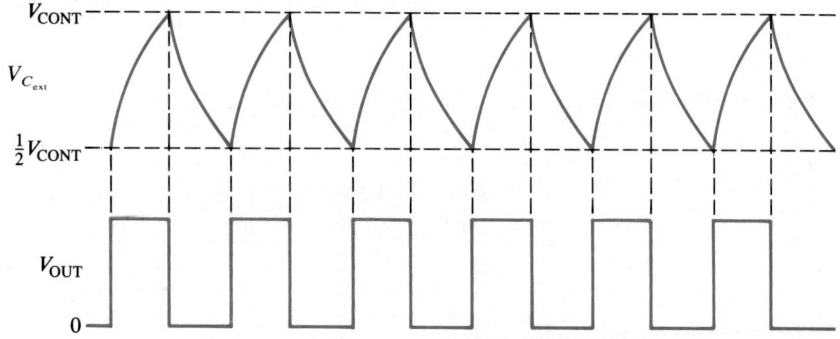

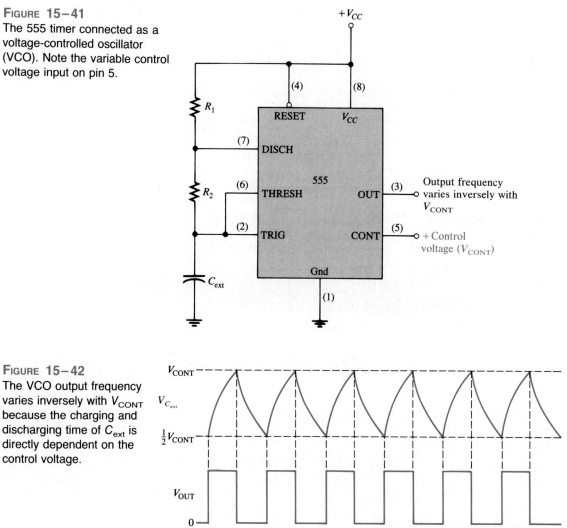

FIGURE 15–42
The VCO output frequency varies inversely with V_{CONT} because the charging and discharging time of C_{ext} is directly dependent on the control voltage.

15–6 REVIEW QUESTIONS

1. Name the five basic elements in a 555 timer IC.
2. When the 555 timer is configured as an astable multivibrator, how is the duty cycle determined?

15–7 THE PHASE-LOCKED LOOP

The phase-locked loop (PLL) is an electronic feedback circuit consisting of a phase detector, a low-pass filter, and a voltage-controlled oscillator (VCO). It is capable of locking onto or synchronizing with an incoming signal. When the phase changes, indicating that the incoming frequency is changing, the phase detector's output voltage (error voltage) increases or decreases just enough to keep the oscillator frequency the same as the incoming frequency. Phase-locked loops are used in a wide variety of communication system applications, including TV receivers, FM demodulation, modems, telemetry, and tone decoders.

BASIC OPERATION

Using the basic block diagram of a phase-locked loop (PLL) in Figure 15–43 as reference, the general operation is as follows. When there is no input signal, the error voltage is zero and the frequency f_o of the voltage-controlled oscillator (VCO) is called the *free-running* or *center* frequency. When an input signal is applied, the phase detector compares the phase and frequency of the input signal with the VCO frequency and produces error voltage V_e. This error voltage is proportional to the phase and frequency difference of the incoming frequency and the VCO frequency. The error voltage contains components that are the sum and difference of the two compared frequencies. The low-pass filter passes only the difference frequency V_d, which is the lower of the two components. This signal is amplified and fed back to the VCO as a control voltage V_{CONT}. The control voltage forces the VCO frequency to change in a direction that reduces the

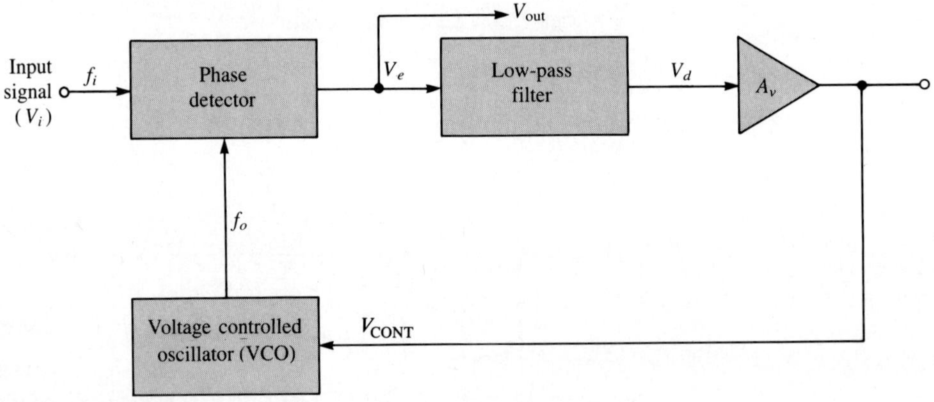

FIGURE 15–43
Basic phase-locked loop block diagram.

difference between the incoming frequency f_i and the VCO frequency f_o. When f_i and f_o are sufficiently close in value, the feedback action of the PLL causes the VCO to lock onto the incoming signal. Once the VCO locks, its frequency is the same as the input frequency with a slight difference in phase. This phase difference ϕ is necessary to keep the PLL in the lock condition.

LOCK RANGE Once the PLL is locked, it can track frequency changes in the incoming signal. The range of frequencies over which the PLL can maintain lock with the incoming signal is called the *lock* or *tracking range*. The lock range is usually expressed as a percentage of the VCO frequency.

CAPTURE RANGE The range of frequencies over which the PLL can acquire lock with an input signal is called the *capture range*. The parameter is normally expressed in percentage of f_o, also.

SUM AND DIFFERENCE FREQUENCIES The phase detector operates as a multiplier circuit to produce the sum and difference of the input frequency f_i and the VCO frequency f_o. This action can best be described mathematically as follows. Recall from ac circuit theory that a sine wave voltage can be expressed as

$$v = V_p \sin 2\pi ft$$

where V_p is the peak value, f is the frequency, and t is the time. Using this basic expression, the input signal voltage and the VCO voltage can be written as

$$v_i = V_{ip} \sin 2\pi f_i t$$
$$v_o = V_{op} \sin 2\pi f_o t$$

When these two signals are multiplied in the phase detector, we get a product at the output as follows:

$$V_{\text{out}} = V_{ip}V_{op}(\sin 2\pi f_i t)(\sin 2\pi f_o t) \qquad \textbf{(15–23)}$$

Substituting the trigonometric identity

$$(\sin A)(\sin B) = \frac{1}{2}[\cos(A - B) - \cos(A + B)]$$

into Equation (15–23), we get

$$V_{\text{out}} = \frac{V_{ip}V_{op}}{2}[\cos(2\pi f_i t - 2\pi f_o t) - \cos(2\pi f_i t + 2\pi f_o t)]$$

$$V_{\text{out}} = \frac{V_{ip}V_{op}}{2} \cos 2\pi(f_i - f_o)t - \frac{V_{ip}V_{op}}{2} \cos 2\pi(f_i + f_o)t \qquad \textbf{(15–24)}$$

You can see in Equation (15–24) that V_{out} of the phase detector consists of a difference frequency component $(f_i - f_o)$ and a sum frequency component $(f_i + f_o)$. This concept is illustrated in Figure 15–44 with frequency spectrum graphs. Each vertical line represents a specific signal frequency, and the height is its amplitude.

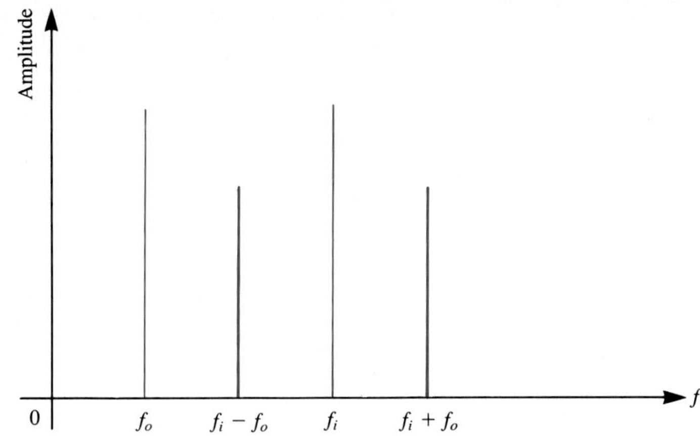

FIGURE 15–44
Frequency spectrum of the phase detector.

EXAMPLE 15–7 A 10 kHz signal f_i and an 8 kHz signal f_o are applied to a phase detector. Determine the sum and difference frequencies.

SOLUTION

$$f_i + f_o = 10 \text{ kHz} + 8 \text{ kHz} = 18 \text{ kHz}$$

$$f_i - f_o = 10 \text{ kHz} - 8 \text{ kHz} = 2 \text{ kHz}$$

PRACTICE EXERCISE 15–7

If the sum and difference frequencies are 30 kHz and 6 kHz respectively, what are f_i and f_o?

WHEN THE PLL IS IN LOCK

When the PLL is in a lock condition, the VCO frequency equals the input frequency ($f_o = f_i$). Thus, the difference frequency is $f_o - f_i = 0$. A zero frequency indicates a dc component. The low-pass filter removes the sum frequency ($f_i + f_o$) and passes the dc (0-frequency) component, which is amplified and fed back to the VCO. When the PLL is in lock, the difference frequency component is always dc and is always passed by the

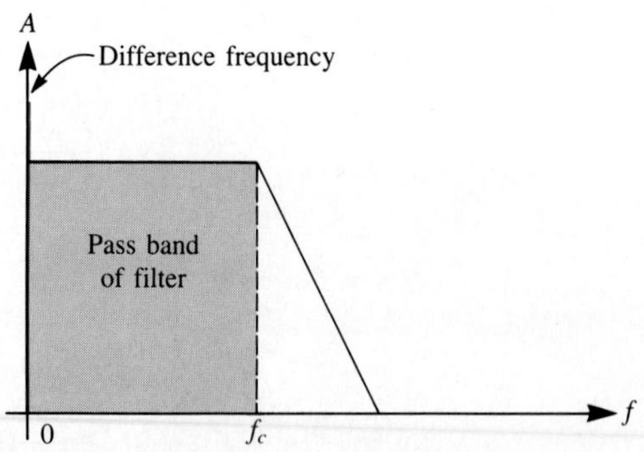

FIGURE 15–45
When the PLL is in lock, the difference frequency is zero and passes through the filter. The lock range is independent of the filter bandwidth.

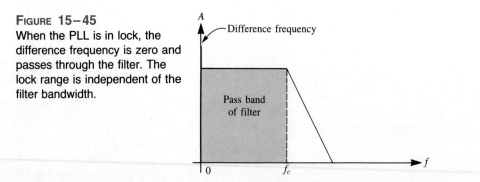

filter, so the lock range is independent of the bandwidth of the low-pass filter. This is illustrated in Figure 15–45.

WHEN THE PLL IS NOT IN LOCK

In the case where the PLL has not yet locked onto the incoming frequency, the phase detector still produces sum and difference frequencies. In this case, however, the difference frequency may lie outside the pass band of the low-pass filter and will not be fed back to the VCO, as illustrated in Figure 15–46. The VCO will remain at its center frequency (free-running) as long as this condition exists.

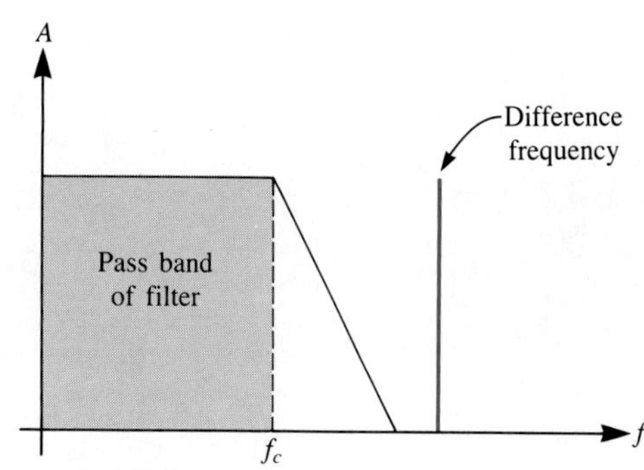

FIGURE 15–46
When the PLL is not in lock, the difference frequency can be outside the pass band.

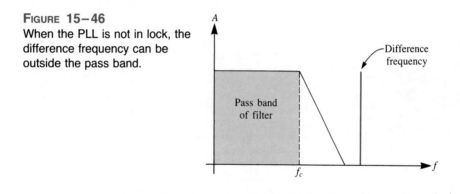

As the input frequency approaches the VCO frequency, the difference component produced by the phase detector decreases and eventually falls into the pass band of the filter and drives the VCO toward the incoming frequency. When the VCO reaches the incoming frequency, the PLL locks on it.

15–7 REVIEW QUESTIONS

1. What are the basic blocks in a phase-locked loop?
2. State the general purpose of a phase-locked loop.

15–8 A SYSTEM APPLICATION

The function generator presented at the beginning of the chapter is a laboratory instrument used as a source for sine waves, square waves, and triangular waves. In this section, you will

- ☐ ***See how an oscillator is used as a signal source.***
- ☐ ***See how the frequency and amplitude of the generated signal are varied.***
- ☐ ***Translate between printed circuit boards and a schematic.***
- ☐ ***Interconnect the front panel controls and two pc boards.***
- ☐ ***Troubleshoot some common system problems.***

A BRIEF DESCRIPTION OF THE SYSTEM

The function generator in this system application produces either a sine wave, a square wave, or a triangular wave depending on the function selected by the front panel switches. The frequency of the selected waveform can be varied from less than 1 Hz to greater than 80 kHz using the range switches and the frequency dial. The amplitude of the output waveform can be adjusted up to approximately +10 V. Also, any dc offset voltage can be nulled out.

The system block diagram is shown in Figure 15–47. The concept of this particular function generator is very simple. The oscillator produces a sine wave that drives a zero-level detector (comparator) to produce a square wave of the same frequency as the sine wave. The level detector output goes to an integrator, which generates a triangular output voltage also with the same frequency as the sine wave.

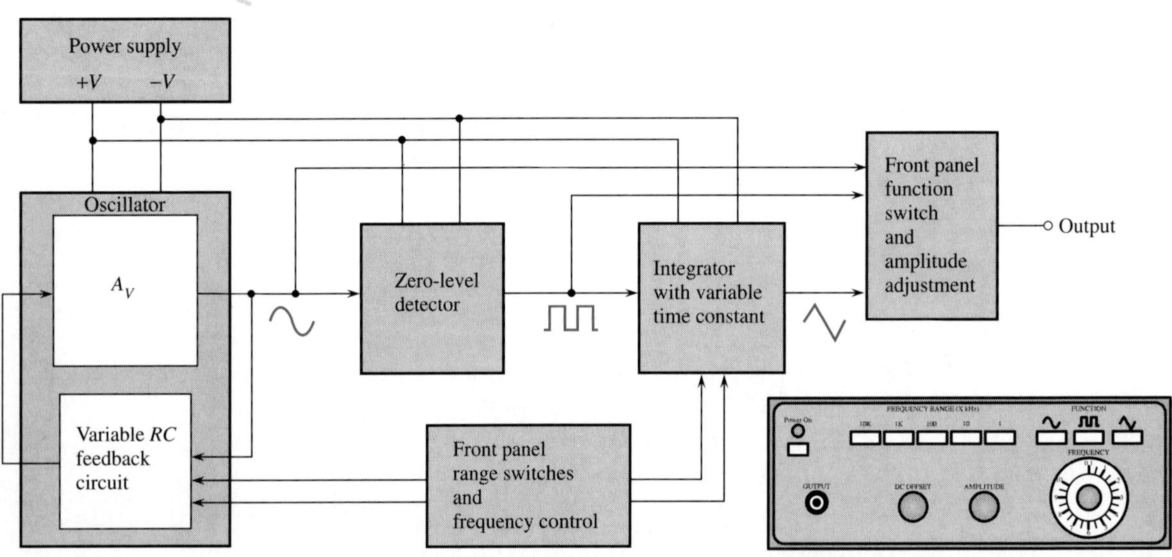

FIGURE 15–47
Function generator block diagram with front panel inset.

The frequency of the sine wave oscillator is controlled by the selection of any one of five capacitor values in the oscillator feedback circuit. These capacitors produce the five frequency ranges indicated on the front panel switches. Variation of the frequency within each range is accomplished by adjusting the resistances in the feedback circuit. These resistances are in the form of ganged potentiometers directly linked to the front panel frequency control knob. The integrator time constant is adjusted in step with the frequency by selection of capacitor and input resistor values.

Now, so that you can take a closer look at the oscillator boards, let's take them out of the system and put them on the test bench.

ON THE TEST BENCH

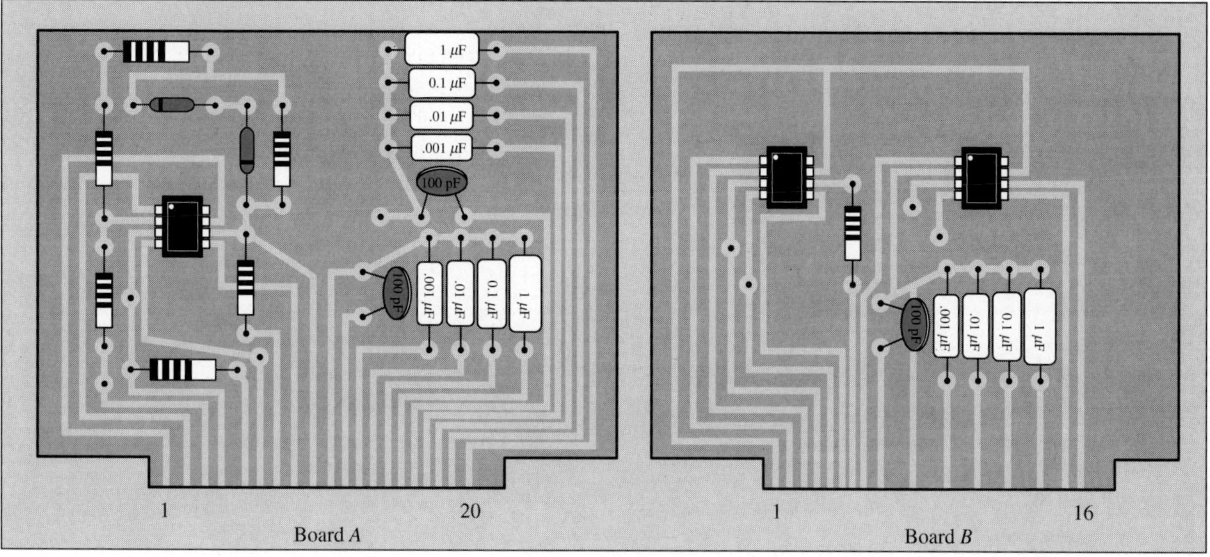

FIGURE 15–48

ACTIVITY 1 RELATE THE PC BOARDS TO THE SCHEMATIC

Locate and identify each component on the pc boards, shown in Figure 15–48, using the system schematic in Figure 15–49. The portions shaded in color are front panel components and are not on the boards. The range switches are mechanically linked and the frequency rheostats are mechanically linked.

Develop a board-to-board wiring list specifying which pins on the two pc boards connect to each other and also indicate which pins go to the front panel.

ACTIVITY 2 ANALYZE THE SYSTEM

STEP 1 Determine the maximum frequency of the oscillator for each range switch position (×1, ×10, and so on). Only one set of three switches corresponding to a given range setting can be closed at a time. There is a set of three switches for ×1, a set of three switches for ×10, and so on.

STEP 2 Determine the minimum frequency of the oscillator for each range switch.

STEP 3 Determine the approximate maximum peak-to-peak output voltages for each function. The dc supply voltages are +15 V and −15 V.

FIGURE 15–49

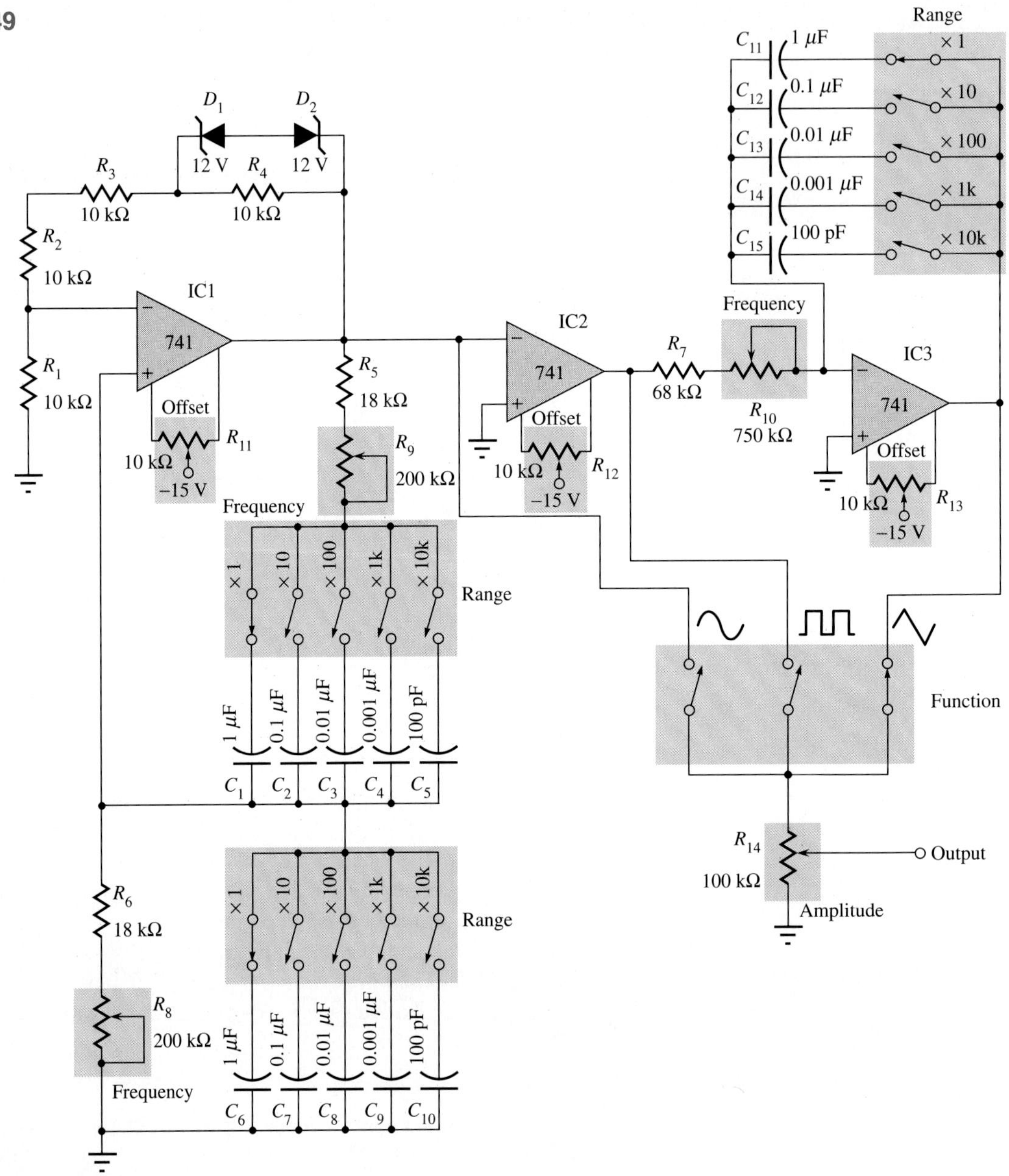

■ ACTIVITY 3 WRITE A TECHNICAL REPORT

Describe the overall operation of the function generator. Specify how each circuit works and what its purpose is. Identify the type of oscillator circuit used. Explain how the function, frequency, and amplitude are selected. Use the results of Activity 2 where appropriate.

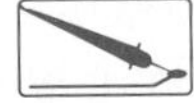

■ ACTIVITY 4 TROUBLESHOOT THE SYSTEM FOR EACH OF THE FOLLOWING PROBLEMS BY STATING THE PROBABLE CAUSE OR CAUSES

1. There is a square wave output when a triangular wave output is selected and only when the ×1k range is selected.
2. There is no output on any function setting.
3. There is no output when the square or triangular function is selected, but the sine wave output is OK.
4. Both the sine wave and the square wave outputs are OK, but there is no triangular wave output.

15–8 REVIEW QUESTIONS

1. What type of oscillator is used in this function generator?
2. How many frequency ranges are available?
3. List the components that determine the output frequency.
4. What is the purpose of the zener diodes in the oscillator circuit?

SUMMARY

- □ Oscillators operate with positive feedback.
- □ The two conditions for positive feedback are the phase shift around the feedback loop must be 0° and the voltage gain around the feedback loop must equal 1.
- □ For initial start-up, the loop gain must be greater than 1.
- □ Sinusoidal *RC* oscillators include the Wien-bridge, phase-shift, and twin-T.
- □ Sinusoidal *LC* oscillators include the Colpitts, Clapp, Hartley, Armstrong, and crystal-controlled.
- □ The feedback signal in a Colpitts oscillator is derived from a capacitive voltage divider in the *LC* network.
- □ The Clapp oscillator is a variation of the Colpitts with a capacitor added in series with the inductor.
- □ The feedback signal in a Hartley oscillator is derived from an inductive voltage divider in the *LC* network.
- □ The feedback signal in an Armstrong oscillator is derived by transformer coupling.
- □ Crystal oscillators are the most stable type.
- □ The frequency in a voltage-controlled oscillator (VCO) can be varied with a dc control voltage.
- □ The 555 timer is an integrated circuit that can be used as an oscillator, in addition to many other applications.
- □ A phase-locked loop can lock onto and track a signal whose frequency is changing.

GLOSSARY

Crystal A quartz device that operates on the piezoelectric effect and exhibits very stable resonant properties.

Oscillator An electronic circuit that operates with positive feedback and produces a time-varying output signal without an external input signal.

Piezoelectric effect The property of a crystal whereby a changing mechanical stress produces a voltage across the crystal.

Positive feedback The return of a portion of the output signal to the input such that it sustains the output.

FORMULAS

(15–1)	$\frac{V_{out}}{V_{in}} = \frac{1}{3}$	Wien-bridge positive feedback attenuation
(15–2)	$f_r = \frac{1}{2\pi RC}$	Wien-bridge frequency
(15–3)	$B = \frac{1}{29}$	Phase-shift feedback attenuation
(15–4)	$f_r = \frac{1}{2\pi\sqrt{6}RC}$	Phase-shift oscillator frequency
(15–5)	$f_r \cong \frac{1}{2\pi\sqrt{LC_T}}$	Colpitts approximate resonant frequency
(15–6)	$C_T = \frac{C_1C_2}{C_1 + C_2}$	Colpitts feedback capacitance
(15–7)	$B = \frac{C_1}{C_2}$	Colpitts feedback attenuation
(15–8)	$A_v = \frac{C_2}{C_1}$	Colpitts amplifier gain
(15–9)	$A_v > \frac{C_2}{C_1}$	Colpitts self-starting gain
(15–10)	$f_r = \frac{1}{2\pi\sqrt{LC_T}}\sqrt{\frac{Q^2}{Q^2 + 1}}$	Colpitts resonant frequency
(15–11)	$C_T = \frac{1}{1/C_1 + 1/C_2 + 1/C_3}$	Clapp feedback capacitance
(15–12)	$f_r \cong \frac{1}{2\pi\sqrt{L_TC}}$	Hartley approximate resonant frequency

(15–13)	$B \cong \frac{L_2}{L_1}$	Hartley feedback attenuation
(15–14)	$A_v > \frac{L_1}{L_2}$	Hartley self-starting gain
(15–15)	$V_{\text{UTP}} = +V_{\text{max}}\left(\frac{R_3}{R_2}\right)$	Triangular wave generator upper trigger point
(15–16)	$V_{\text{LTP}} = -V_{\text{max}}\left(\frac{R_3}{R_2}\right)$	Triangular wave generator lower trigger point
(15–17)	$f = \frac{1}{4R_1C}\left(\frac{R_2}{R_3}\right)$	Triangular wave generator frequency
(15–18)	$T = \frac{V_p - V_F}{\lvert V_{\text{IN}}\rvert / R_iC}$	Sawtooth VCO period
(15–19)	$f = \frac{\lvert V_{\text{IN}}\rvert}{R_iC}\left(\frac{1}{V_p - V_F}\right)$	Sawtooth VCO frequency
(15–20)	$f = \frac{1.44}{(R_1 + 2R_2)C_{\text{ext}}}$	555 astable frequency
(15–21)	$\text{Duty cycle} = \frac{R_1 + R_2}{R_1 + 2R_2} \times 100\%$	555 astable
(15–22)	$\text{Duty cycle} = \frac{R_1}{R_1 + R_2} \times 100\%$	555 astable (duty cycle $< 50\%$)
(15–23)	$V_{\text{out}} = V_{ip}V_{op}(\sin 2\pi f_i t)(\sin 2\pi f_o t)$	PLL output

SELF-TEST

1. An oscillator differs from an amplifier because

(a) it has more gain **(b)** it requires no input signal

(c) it requires no dc supply **(d)** it always has the same output

2. All oscillators are based on

(a) positive feedback **(b)** negative feedback

(c) the piezoelectric effect **(d)** high gain

3. One condition for oscillation is

(a) a phase shift around the feedback loop of 180°

(b) a gain around the feedback loop of one-third

(c) a phase shift around the feedback loop of 0°
(d) a gain around the feedback loop of less than one

4. A second condition for oscillation is
   (a) no gain around the feedback loop
   (b) a gain of one around the feedback loop
   (c) the attenuation of the feedback circuit must be one-third
   (d) the feedback circuit must be capacitive
5. In a certain oscillator, $A_v = 50$. The attenuation of the feedback circuit must be
   (a) 1 (b) 0.01 (c) 10 (d) 0.02
6. For an oscillator to properly start, the gain around the feedback loop must initially be
   (a) 1 (b) less than 1 (c) greater than 1 (d) equal to *B*
7. In a Wien-bridge oscillator, if the resistances in the feedback circuit are decreased, the frequency
   (a) decreases (b) increases (c) remains the same
8. The Wien-bridge oscillator's positive feedback circuit is
   (a) an *RL* network (b) an *LC* network
   (c) a voltage divider (d) a lead-lag network
9. A phase-shift oscillator has
   (a) three *RC* networks (b) three *LC* networks
   (c) a T-type network (d) a π-type network
10. Colpitts, Clapp, and Hartley are names that refer to
    (a) types of *RC* oscillators (b) inventors of the transistor
    (c) types of *LC* oscillators (d) types of filters
11. An oscillator whose frequency is changed by a variable dc voltage is known as
    (a) a crystal oscillator (b) a VCO
    (c) an Armstrong oscillator (d) a piezoelectric device
12. The main feature of a crystal oscillator is
    (a) economy (b) reliability (c) stability (d) high frequency
13. The operation of a relaxation oscillator is based on
    (a) the charging and discharging of a capacitor
    (b) a highly selective resonant circuit
    (c) a very stable supply voltage
    (d) low power consumption
14. Which one of the following is *not* an input or output of the 555 timer?
    (a) Threshold (b) Control voltage (c) Clock (d) Trigger
    (e) Discharge (f) Reset

15. A type of circuit that is capable of locking onto or synchronizing with an incoming signal is called
   (a) an astable multivibrator **(b)** a monostable multivibrator
   (c) a phase-locked loop **(d)** a phase detector

PROBLEMS

SECTION 15–1 DEFINITION OF AN OSCILLATOR

1. What type of input is required for an oscillator?
2. What are the basic components of an oscillator circuit?

SECTION 15–2 OSCILLATOR PRINCIPLES

3. If the voltage gain of the amplifier portion of an oscillator is 75, what must be the attenuation of the feedback circuit to sustain the oscillation?
4. Generally describe the change required in the oscillator of Problem 3 in order for oscillation to begin when the power is initially turned on.

SECTION 15–3 OSCILLATORS WITH *RC* FEEDBACK CIRCUITS

5. A certain lead-lag network has a resonant frequency of 3.5 kHz. What is the rms output voltage if an input signal with a frequency equal to f_r and with an rms value of 2.2 V is applied to the input?
6. Calculate the resonant frequency of a lead-lag network with the following values: $R_1 = R_2 = 6.2\ \text{k}\Omega$, and $C_1 = C_2 = 0.02\ \mu\text{F}$.
7. Determine the necessary value of R_2 in Figure 15–50 so that the circuit will oscillate. Neglect the forward resistance of the zener diodes.

FIGURE 15–50

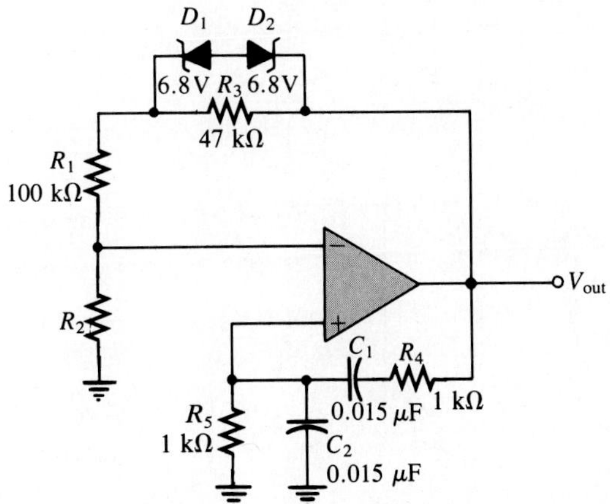

8. Explain the purpose of R_3 in Figure 15–50.
9. What is the initial closed-loop gain in Figure 15–50? At what value of output voltage does A_{cl} change and to what value does it change? (The value of R_2 was found in Problem 7).
10. Find the frequency of oscillation for the Wien-bridge oscillator in Figure 15–50.
11. What value of R_f is required in Figure 15–51? What is f_r?

FIGURE 15–51

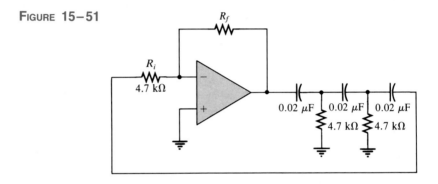

SECTION 15–4 OSCILLATORS WITH *LC* FEEDBACK CIRCUITS

12. Calculate the frequency of oscillation for each circuit in Figure 15–52 and identify the type of oscillator. Assume $Q > 10$ in each case.

FIGURE 15–52

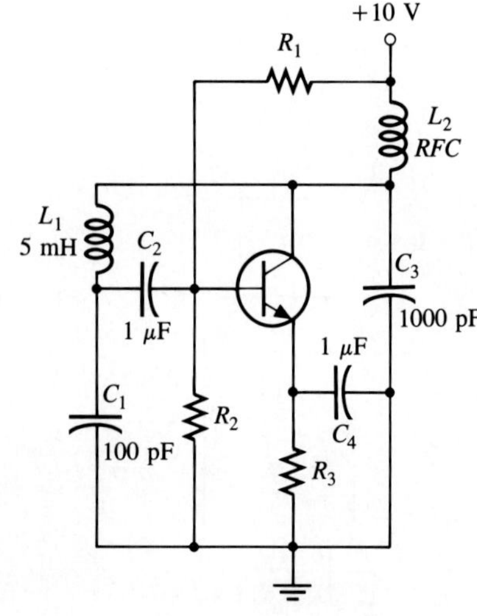

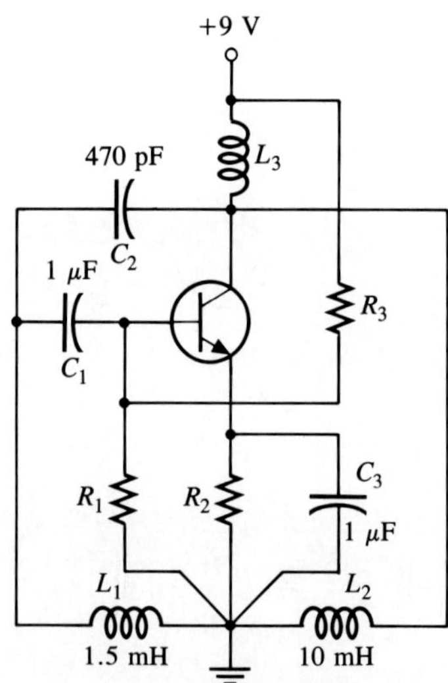

13. Determine what the gain of the amplifier stage must be in Figure 15–53 in order to have sustained oscillation.

FIGURE 15–53

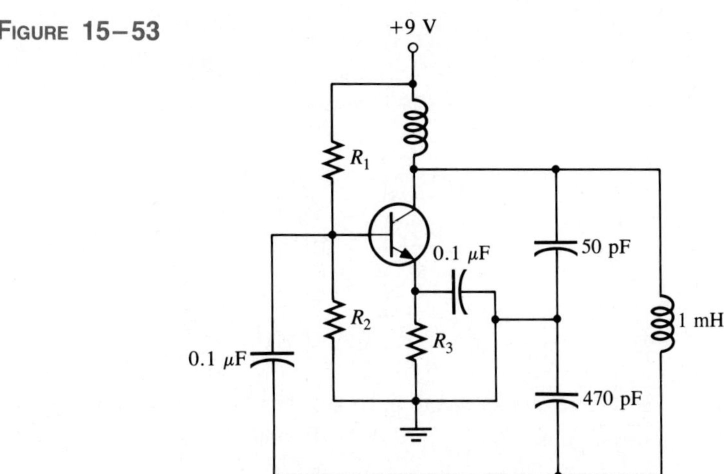

SECTION 15–5 NONSINUSOIDAL OSCILLATORS

14. What type of signal does the circuit in Figure 15–54 produce? Determine the frequency of the output.

15. Show how to change the frequency of oscillation in Figure 15–54 to 10 kHz.

16. Determine the amplitude and frequency of the output voltage in Figure 15–55. Use 1 V as the forward PUT voltage.

FIGURE 15–54

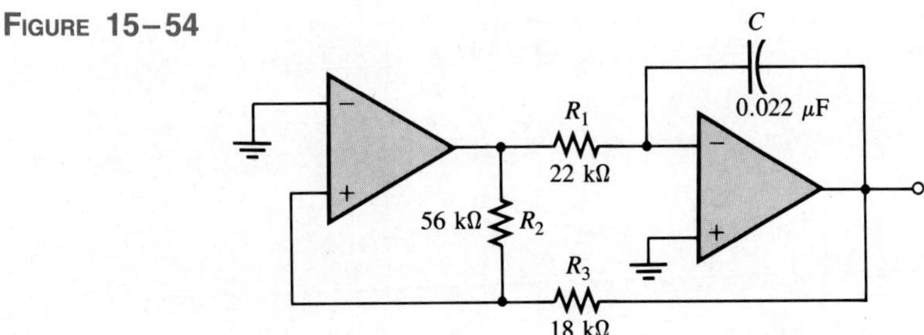

17. Modify the sawtooth generator in Figure 15–55 so that its peak-to-peak output is 4 V.

FIGURE 15–55

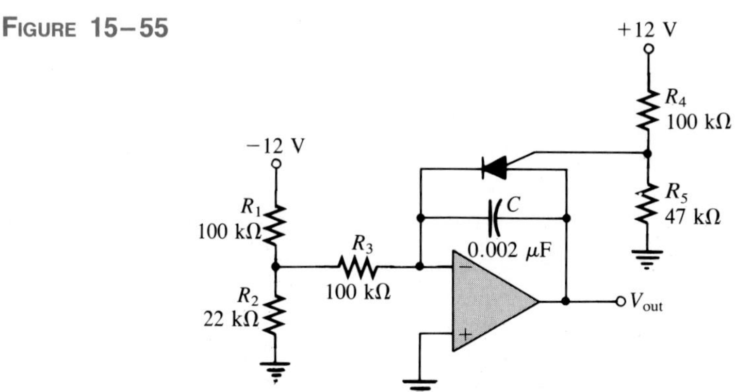

18. A certain sawtooth generator has the following parameter values: $V_{IN} = 3$ V, $R = 4.7$ kΩ, $C = 0.001$ μF, and V_F for the PUT is 1.2 V. Determine its peak-to-peak output voltage if the period is 10 μs.

SECTION 15–6 THE 555 TIMER AS AN OSCILLATOR

19. What are the two comparator reference voltages in a 555 timer when $V_{CC} =$ 10 V?

20. Determine the frequency of oscillation for the 555 astable oscillator in Figure 15–56.

FIGURE 15–56

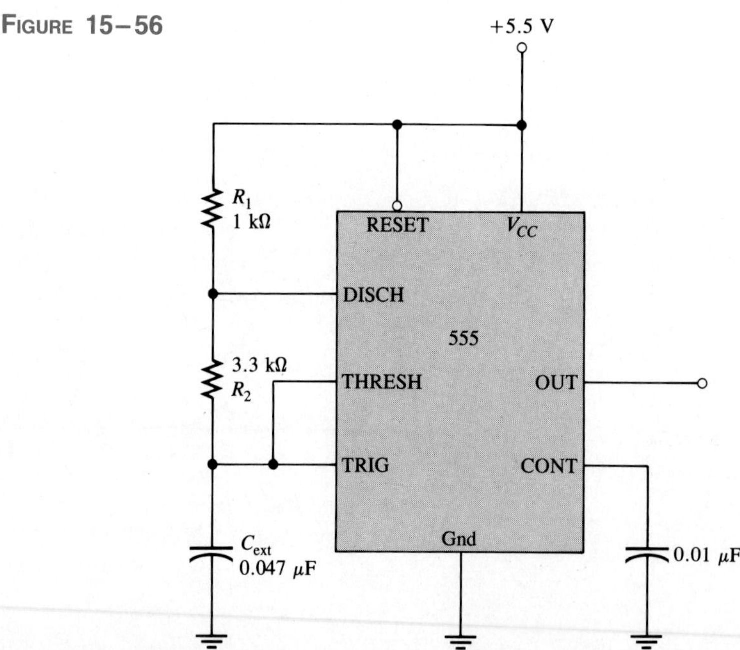

21. To what value must C_{ext} be changed in Figure 15–56 to achieve a frequency of 25 kHz?

22. In an astable 555 configuration, the external resistor $R_1 = 3.3\ k\Omega$. What must R_2 equal to produce a duty cycle of 75 percent?

SECTION 15–7 THE PHASE-LOCKED LOOP

23. The lock range of a certain PLL is specified to be ±15 percent of the center frequency. Determine the minimum and maximum frequencies for which the PLL will maintain lock if $f_o = 50$ kHz.

24. A 15 kHz signal f_o and a 7.5 kHz signal f_i are applied to a PLL phase detector. Determine the sum and difference frequencies.

25. A 25 kHz sine wave with a peak value of 50 mV is applied to a PLL. When the PLL is in lock, what is the VCO frequency?

ANSWERS TO REVIEW QUESTIONS

SECTION 15–1

1. An oscillator is a circuit that produces a repetitive output waveform with only the dc supply voltage as an input.
2. Positive feedback
3. The feedback circuit provides attenuation and phase shift.

SECTION 15–2

1. Zero phase shift and unity voltage gain around the closed feedback
2. Positive feedback is when a portion of the output signal is fed back to the input of the amplifier such that it reinforces itself.
3. Loop gain greater than 1; zero phase shift and unity voltage gain

SECTION 15–3

1. The negative feedback loop sets the closed-loop gain; the positive feedback loop sets the frequency of oscillation.
2. 1.67 V
3. The three *RC* networks each contribute 60°.

SECTION 15–4

1. Colpitts uses a capacitive voltage divider in the feedback network; Hartley uses an inductive voltage divider.
2. The higher FET input impedance has less loading effect on the resonant feedback circuit.
3. A Clapp has an additional capacitor in series with the inductor in the feedback network.

SECTION 15–5

1. A voltage-controlled oscillator exhibits a frequency that can be varied with a dc control voltage.

2. The basis of a relaxation oscillator is the charging and discharging of a capacitor.

SECTION 15–6

1. Two comparators, a flip-flop, a discharge transistor, and a resistive voltage divider
2. The duty cycle is set by the external resistors and the external capacitor.

SECTION 15–7

1. The PLL blocks are phase detector, VCO, low-pass filter, and amplifier.
2. A PLL maintains synchronization (lock) with an incoming signal.

SECTION 15–8

1. A Wien-bridge oscillator
2. There are five frequency ranges.
3. R_5, R_6, R_8, R_9, C_1–C_5, C_6–C_{10}
4. To limit the oscillator output amplitude and to help ensure start-up

ANSWERS TO PRACTICE EXERCISES

15–1 Change the zener diodes to 6.1 V devices.

15–2 **(a)** 238 kΩ **(b)** 7.92 kHz

15–3 7.03 kHz

15–4 6.06 V peak-to-peak

15–5 1055 Hz

15–6 31.9%

15–7 f_i = 18 kHz and f_o = 12 kHz

16

Active Filters

After completing this chapter, you should be able to

- □ Identify low-pass, high-pass, band-pass, and band-stop filter responses.
- □ Recognize Butterworth, Chebyshev, and Bessel response characteristics.
- □ Describe the effect of the damping factor on filter response.
- □ Define the term *pole* in relation to filters.
- □ Determine the critical (cutoff) frequencies of specific filters.
- □ Explain how the roll-off rate of a filter is related to the number of poles.
- □ Discuss the frequency response of cascaded filters.
- □ Implement the Butterworth response in a filter.
- □ Analyze Sallen-Key low-pass and high-pass filters.
- □ Analyze multiple-feedback and state-variable band-pass and band-stop filters.
- □ Measure filter response using two different methods.

Power supply filters were introduced in Chapter 2. In this chapter, we introduce active filters used for signal processing. Filters are circuits that are capable of passing input signals with certain selected frequencies through to the output while rejecting signals with other frequencies. This property is called *selectivity.*

Filters use active devices such as transistors or op-amps and passive *RC* networks. The active devices provide voltage gain and the passive networks provide frequency selectivity. In terms of general response, there are four basic categories of active filters: low-pass, high-pass, band-pass, and band-stop. In this chapter, we will concentrate on active filters using op-amps and *RC* networks.

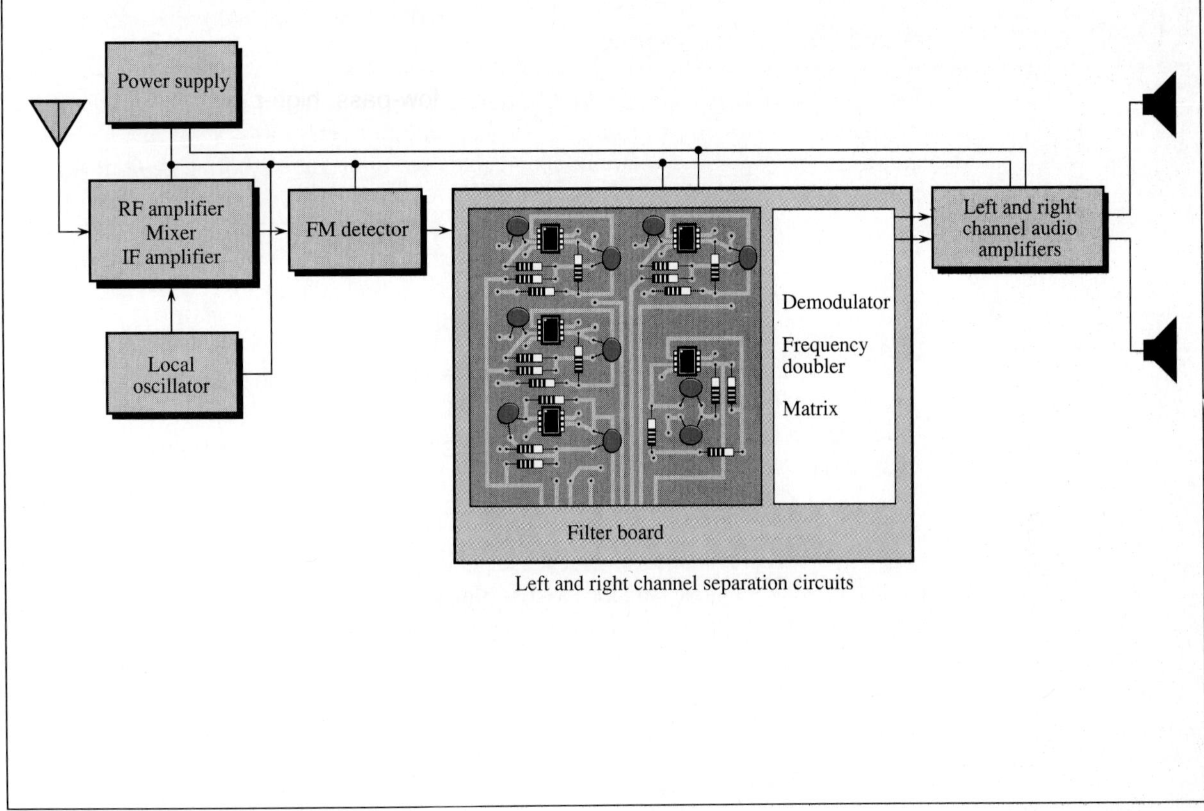

Left and right channel separation circuits

A System Application

In Chapter 13, you worked with an FM stereo multiplex receiver, concentrating on the audio amplifiers. In this chapter, we again take a look at this same system, but this time our focus is on the filters in the left and right channel separation circuits in which several types of active filters are used. The FM stereo multiplex signal that is received is quite complex and beyond the scope of our coverage to investigate the reasons it is transmitted in such a way. It is interesting, however, to see how filters, such as the ones studied in this chapter can be used to separate out the audio signals that go to the left and right speakers.

For the system application in Section 16–8, in addition to the other topics, be sure you understand

- □ Filter responses.
- □ How low-pass and band-pass filters work.

16–1 BASIC FILTER RESPONSES

Filters are usually categorized by the manner in which the output voltage varies with the frequency of the input voltage. The categories of* active filters *are low-pass, high-pass, band-pass, and band-stop. We will examine each of these general responses in this section.

LOW-PASS FILTER RESPONSE

The pass band of the basic **low-pass filter** is defined to be from 0 Hz (dc) up to the critical (cutoff) frequency, f_c, at which the output voltage is 70.7 percent of the pass-band voltage, as indicated in Figure 16–1(a). The ideal pass band, shown by the shaded region within the dashed lines, has an instantaneous roll-off at f_c. The bandwidth of this filter is equal to f_c.

$$BW = f_c$$

Although the ideal response is not achievable in practice, roll-off rates of −20 dB/decade and higher are obtainable. Figure 16–1(b) illustrates ideal low-pass filter response curves with several roll-off rates. The −20 dB/decade rate is obtained with a single *RC* network

FIGURE 16–1
Low-pass filter responses.

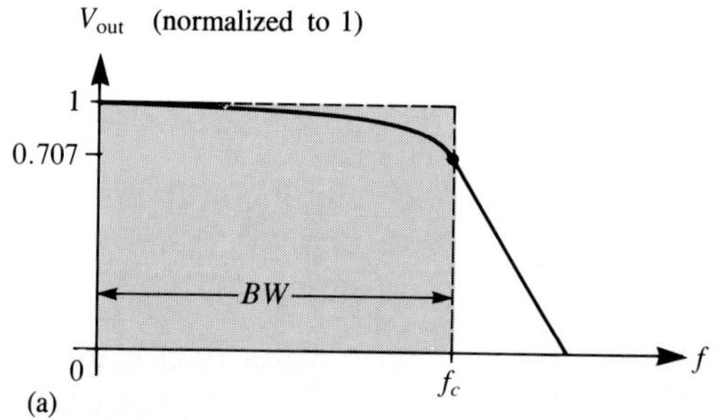

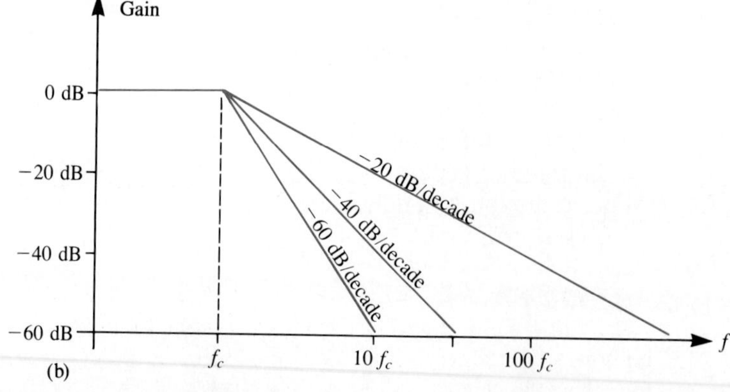

consisting of one resistor and one capacitor. The higher roll-off rates require additional *RC* networks. Each network is called a **pole.**

As explained in Chapter 10, the critical frequency of the *RC* low-pass filter occurs when $X_C = R$, where

$$f_c = \frac{1}{2\pi RC} \quad (16\text{–}1)$$

HIGH-PASS FILTER RESPONSE

A **high-pass filter** response is one that significantly attenuates all frequencies below f_c and passes all frequencies above f_c. The critical frequency is, of course, the frequency at which the output voltage is 70.7 percent of the pass-band voltage, as shown in Figure 16–2(a). The ideal response, shown by the shaded region within the dashed lines, has an instantaneous drop at f_c, which, of course, is not achievable. Roll-off rates of 20 dB/decade/pole are realizable. Figure 16–2(b) illustrates high-pass filter responses with several roll-off rates.

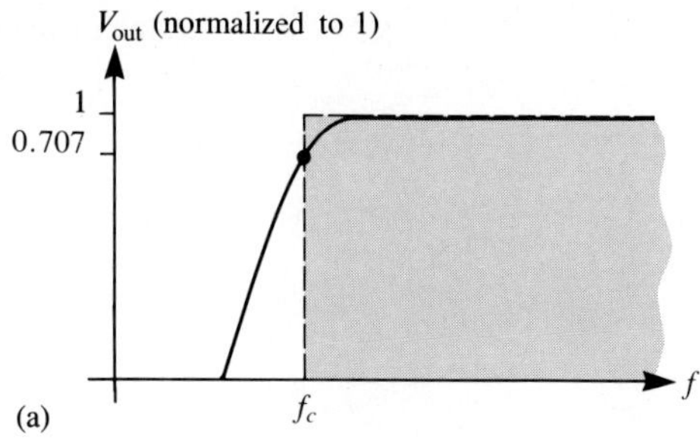

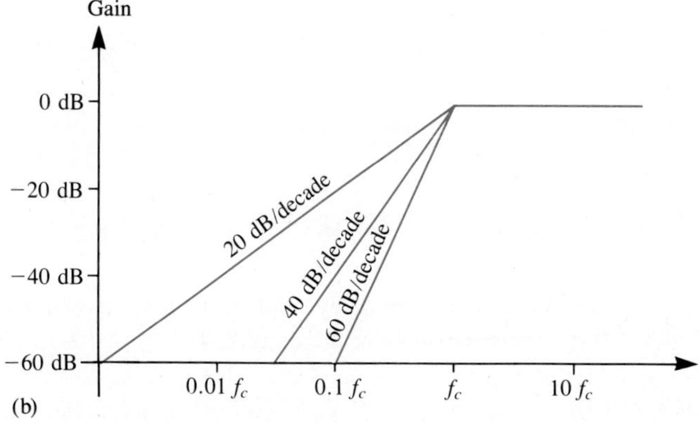

FIGURE 16–2
High-pass filter responses.

As with the *RC* low-pass filter, the high-pass critical frequency corresponds to the value where $X_C = R$ and is calculated with the formula $f_c = 1/2\pi RC$. The response of a high-pass filter extends from f_c up to a frequency that is determined by the limitations of the active element (transistor or op-amp) used.

BAND-PASS FILTER RESPONSE

A **band-pass filter** passes all signals lying within a band between a lower- and an upper-frequency limit and essentially rejects all other frequencies that are outside this specified band. A generalized band-pass response curve is shown in Figure 16–3. The *bandwidth* (*BW*) is defined as the difference between the upper critical frequency (f_{c2}) and the lower critical frequency (f_{c1}).

$$BW = f_{c2} - f_{c1} \tag{16–2}$$

The critical frequencies are, of course, the points at which the response curve is 70.7 percent of its maximum. Recall from Chapter 13 that these critical frequencies are also called *3 dB frequencies*. The frequency about which the pass band is centered is called the *center frequency*, f_0, defined as the geometric mean of the critical frequencies.

$$f_0 = \sqrt{f_{c1}f_{c2}} \tag{16–3}$$

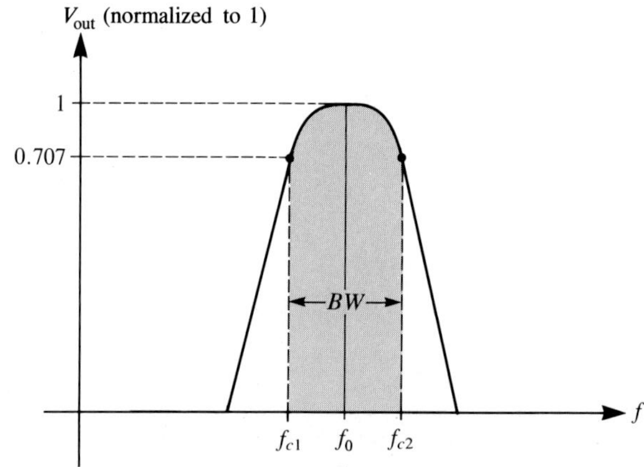

FIGURE 16–3
General band-pass response curve.

QUALITY FACTOR The **quality factor (*Q*)** of a band-pass filter is the ratio of the center frequency to the bandwidth.

$$Q = \frac{f_0}{BW} \tag{16–4}$$

The value of Q is an indication of the selectivity of a band-pass filter. The higher the value of Q, the narrower the bandwidth and the better the selectivity for a given value of f_0. Band-pass filters are sometimes classified as narrow-band ($Q > 10$) or wide-band ($Q < 10$). The Q can also be expressed in terms of the damping factor (*DF*) of the filter as

$$Q = \frac{1}{DF} \tag{16–5}$$

We will study the damping factor in Section 16–2.

EXAMPLE 16–1 A certain band-pass filter has a center frequency of 15 kHz and a bandwidth of 1 kHz. Determine the Q and classify the filter as narrow-band or wide-band.

SOLUTION

$$Q = \frac{f_0}{BW} = \frac{15 \text{ kHz}}{1 \text{ kHz}} = 15$$

Because $Q > 10$, this is a narrow-band filter.

PRACTICE EXERCISE 16–1
If the Q of the filter is doubled, what will the bandwidth be?

BAND-STOP FILTER RESPONSE

Another category of active filter is the band-stop, also known as *notch, band-reject,* or *band-elimination* filters. You can think of the operation as opposite to that of the band-pass filter because frequencies within a certain bandwidth are rejected, and frequencies outside the bandwidth are passed. A general response curve for a band-stop filter is shown in Figure 16–4. Notice that the bandwidth is the band of frequencies between the 3 dB points, just as in the case of the band-pass filter response.

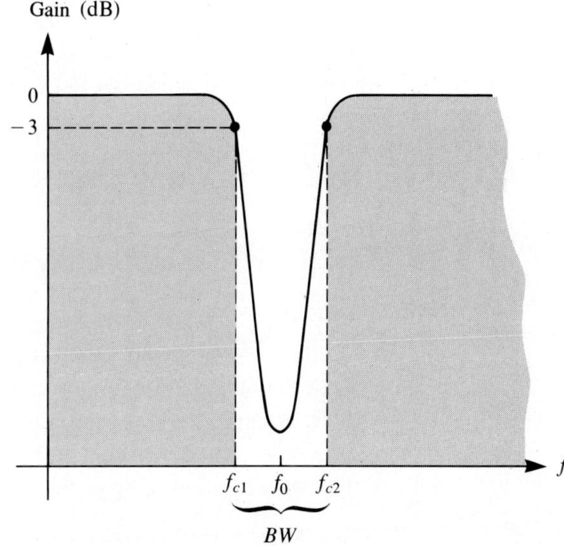

FIGURE 16–4
General band-stop filter response.

16–1 REVIEW QUESTIONS

1. What determines the bandwidth of a low-pass filter?
2. What limits the bandwidth of an active high-pass filter?
3. How are the Q and the bandwidth of a band-pass filter related? Explain how the selectivity is affected by the Q of a filter.

16–2 FILTER RESPONSE CHARACTERISTICS

Each type of filter response (low-pass, high-pass, band-pass, or band-stop) can be tailored by circuit component values to have either a Butterworth, Chebyshev, or Bessel characteristic. Each of these characteristics is identified by the shape of the response curve, and each has an advantage in certain applications.

THE BUTTERWORTH CHARACTERISTIC

The **Butterworth** characteristic provides a very flat amplitude response in the pass band and a roll-off rate of 20 dB/decade/pole. The phase response is not linear, however, and the phase shift (thus, time delay) of signals passing through the filter varies nonlinearly with frequency. Therefore, a pulse applied to a filter with a Butterworth response will cause overshoots on the output, because each frequency component of the pulse's rising and falling edges experiences a different time delay. Filters with the Butterworth response are normally used when all frequencies in the pass band must have the same gain. The Butterworth response is often referred to as a *maximally flat response*.

THE CHEBYSHEV CHARACTERISTIC

Filters with the **Chebyshev** response characteristic are useful when a rapid roll-off is required because it provides a roll-off rate greater than 20 dB/decade/pole. This is a greater rate than that of the Butterworth, so filters can be implemented with the Chebyshev response with fewer poles and less complex circuitry for a given roll-off rate. This type of filter response is characterized by overshoot or ripples in the pass band (depending on the number of poles) and an even less linear phase response than the Butterworth.

THE BESSEL CHARACTERISTIC

The **Bessel** response exhibits a linear phase characteristic, meaning that the phase shift increases linearly with frequency. The result is almost no overshoot on the output with a pulse input. For this reason, filters with the Bessel response are used for filtering pulse waveforms without distorting the shape of the waveform.

Butterworth, Chebyshev, or Bessel response characteristics can be realized with most active filter circuit configurations by proper selection of certain component values, as we will see later. A general comparison of the three response characteristics for a low-pass filter response curve is shown in Figure 16–5. High-pass and band-pass filters can also be designed to have any one of the characteristics.

THE DAMPING FACTOR

As mentioned, an active filter can be designed to have either a Butterworth, Chebyshev, or Bessel response characteristic regardless of whether it is a low-pass, high-pass, band-pass, or band-stop type. The **damping factor** (*DF*) of an active filter circuit determines which response characteristic the filter exhibits. To explain the basic concept, a generalized active filter is shown in Figure 16–6. It includes an amplifier, a negative

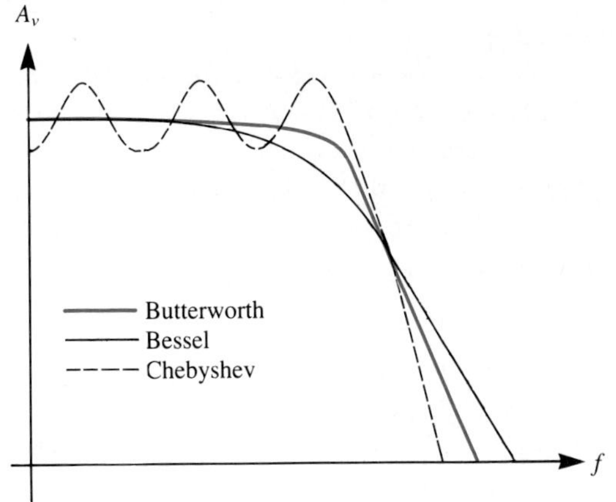

FIGURE 16–5
Comparative plots of three types of filter response characteristics.

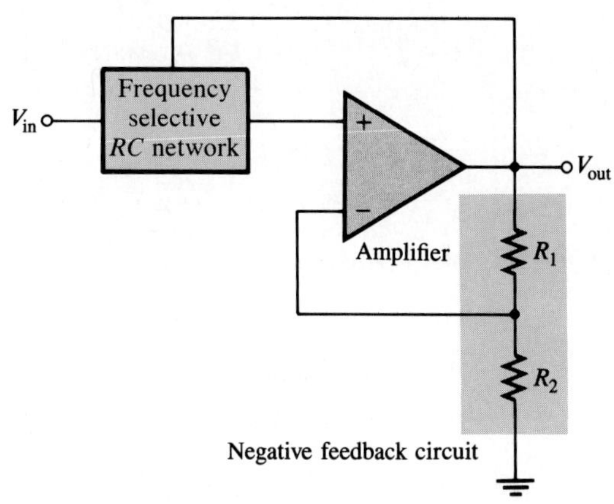

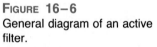

FIGURE 16–6
General diagram of an active filter.

feedback circuit, and a filter section. The amplifier and feedback are connected in a noninverting configuration. The damping factor is determined by the negative feedback circuit and is defined by the following equation:

$$DF = 2 - \frac{R_1}{R_2} \qquad (16\text{–}6)$$

Basically, the damping factor affects the filter response by negative feedback action. Any attempted increase or decrease in the output voltage is offset by the opposing effect of the negative feedback. This tends to make the response curve flat in the pass band of the filter if the value for the damping factor is precisely set. By advanced mathematics, which we will not cover, values for the damping factor have been derived for various orders of filters to achieve the maximally flat response of the Butterworth characteristic.

The value of the damping factor required to produce a desired response characteristic depends on the *order* (number of poles) of the filter. A *pole,* for our purposes, is simply a circuit with one resistor and one capacitor. The more poles a filter has, the faster its roll-off rate is. To achieve a second-order Butterworth response, for example, the damping factor must be 1.414. To implement this damping factor, the feedback resistor ratio must be

$$\frac{R_1}{R_2} = 2 - DF = 2 - 1.414 = 0.586$$

This ratio gives the closed-loop gain of the noninverting filter amplifier, $A_{cl(\text{NI})}$, a value of 1.586, derived as follows:

$$A_{cl(\text{NI})} = \frac{1}{B} = \frac{R_1 + R_2}{R_2} = \frac{R_1}{R_2} + 1 = 0.586 + 1 = 1.586$$

EXAMPLE 16–2 If resistor R_2 in the feedback circuit of an active two-pole filter of the type in Figure 16–6 is 10 kΩ, what value must R_1 be to obtain a maximally flat Butterworth response?

SOLUTION

$$\frac{R_1}{R_2} = 0.586$$

$$R_1 = 0.586R_2 = 0.586(10\ \text{k}\Omega) = 5860\ \Omega$$

Using the nearest standard 5 percent value of 5600 Ω will get very close to the ideal Butterworth response.

PRACTICE EXERCISE 16–2

What is the damping factor for $R_2 = 10\ \text{k}\Omega$ and $R_1 = 5.6\ \text{k}\Omega$?

CRITICAL FREQUENCY AND ROLL-OFF RATE

The critical frequency is determined by the values of the resistor and capacitors in the *RC* network, as shown in Figure 16–6. For a single-pole (first-order) filter, as shown in Figure 16–7, the critical frequency is

$$f_c = \frac{1}{2\pi RC}$$

Although we show a low-pass configuration, the same formula is used for the f_c of a single-pole high-pass filter. The number of poles determines the roll-off rate of the filter. A Butterworth response produces 20 dB/decade/pole. So, a first-order (one-pole) filter has a roll-off of 20 dB/decade; a second-order (two-pole) filter has a roll-off rate of 40 dB/decade; a third-order (three-pole) filter has a roll-off rate of 60 dB/decade; and so on.

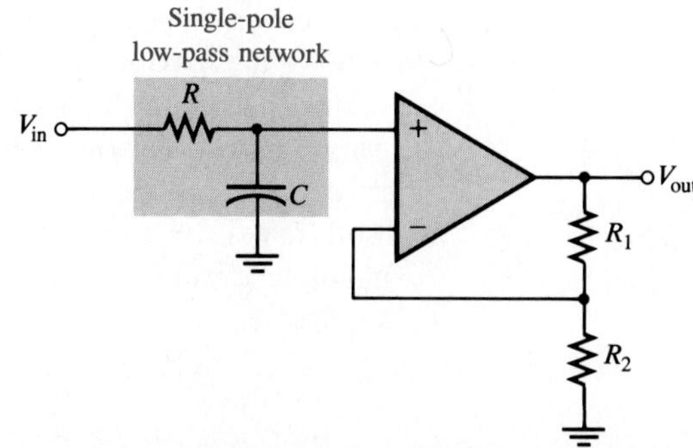

FIGURE 16–7
First-order (one-pole) low-pass filter.

Generally, to obtain a filter with three poles or more, one-pole or two-pole filters are cascaded, as shown in Figure 16–8. To obtain a third-order filter, for example, we cascade a second-order and a first-order filter; to obtain a fourth-order filter, we cascade two second-order filters; and so on. Each filter in a cascaded arrangement is called a *stage* or *section*.

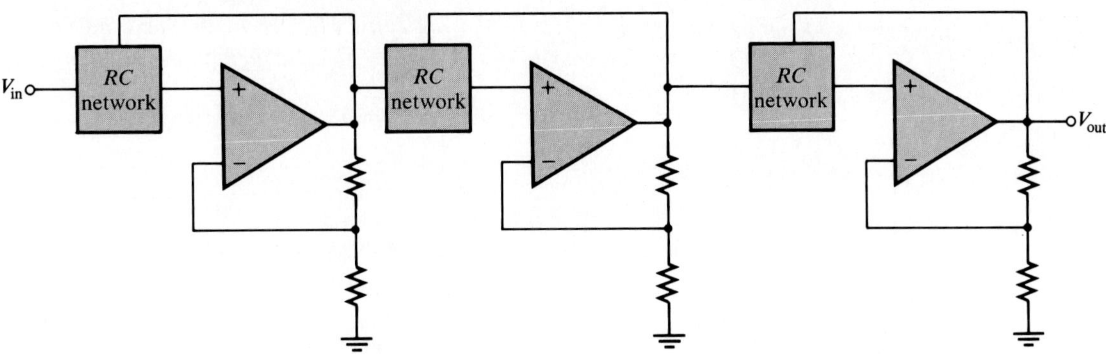

FIGURE 16–8
The number of filter poles can be increased by cascading.

Because of its maximally flat response, the Butterworth characteristic is the most widely used. Therefore, we will limit our coverage to the Butterworth response to illustrate basic filter concepts. Table 16–1 lists the roll-off rates, damping factors, and R_1/R_2 ratios for up to sixth-order Butterworth filters.

TABLE 16–1
Values for the Butterworth response

Order	Roll-off dB/decade	1st stage Poles	1st stage *DF*	1st stage R_1/R_2	2nd stage Poles	2nd stage *DF*	2nd stage R_1/R_2	3rd stage Poles	3rd stage *DF*	3rd stage R_1/R_2
1	20	1	Optional							
2	40	2	1.414	0.586						
3	60	2	1.00	1	1	1.00	1			
4	80	2	1.848	0.152	2	0.765	1.235			
5	100	2	1.00	1	2	1.618	0.382	1	1.618	1.382
6	120	2	1.932	0.068	2	1.414	0.586	2	0.518	1.482

16–2 REVIEW QUESTIONS

1. Explain how Butterworth, Chebyshev, and Bessel responses differ.
2. What determines the response characteristic of a filter?
3. Name the basic parts of an active filter.

16–3 ACTIVE LOW-PASS FILTERS

Filters that use op-amps as the active element provide several advantages over passive filters (R, L, and C elements only). The op-amp provides gain, so that the signal is not attenuated as it passes through the filter. The high input impedance of the op-amp prevents excessive loading of the driving source, and the low output imped-

ance of the op-amp prevents the filter from being affected by the load that it is driving. Active filters are also easy to adjust over a wide frequency range without altering the desired response.

A SINGLE-POLE FILTER

Figure 16–9(a) shows an active filter with a single low-pass *RC* network that provides a roll-off of −20 dB/decade above the critical frequency, as indicated by the response curve in Figure 16–9(b). The critical frequency of the single-pole filter is $f_c = 1/2\pi RC$. The op-amp in this filter is connected as a noninverting amplifier with the closed-loop voltage gain in the pass band set by the values of R_1 and R_2 ($A_{cl} = R_1/R_2 + 1$).

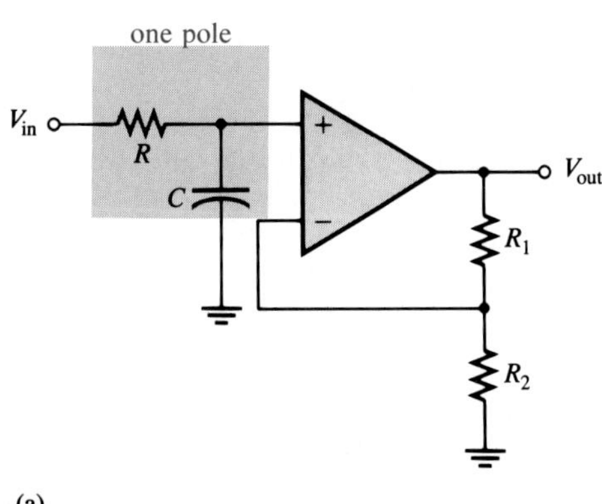

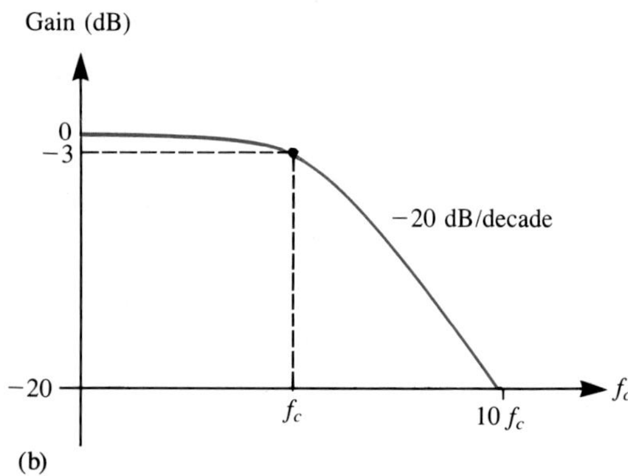

FIGURE 16–9
Single-pole active low-pass filter and response curve.

THE SALLEN-KEY LOW-PASS FILTER

The Sallen-Key is one of the most common configurations for a second-order (two-pole) filter. It is also known as a VCVS (voltage-controlled voltage source) filter. A low-pass version of the Sallen-Key filter is shown in Figure 16–10. Notice that there are two low-pass *RC* networks that provide a roll-off of −40 dB/decade above the critical frequency (assuming a Butterworth characteristic). One *RC* network consists of R_A and C_A, and the second network consists of R_B and C_B. A unique feature is the capacitor C_A that provides feedback for shaping the response near the edge of the pass band. The critical frequency for the second-order Sallen-Key filter is

$$f_c = \frac{1}{2\pi\sqrt{R_A R_B C_A C_B}} \tag{16–7}$$

For simplicity, the component values can be made equal so that $R_A = R_B = R$ and $C_A = C_B = C$. In this case, the expression for the critical frequency simplifies to $f_c = 1/2\pi RC$.

As in the single-pole filter, the op-amp in the second-order Sallen-Key filter acts as a noninverting amplifier with the negative feedback provided by the R_1/R_2 network. As you have learned, the damping factor is set by the values of R_1 and R_2, thus making the

FIGURE 16–10
Basic Sallen-Key second-order low-pass filter.

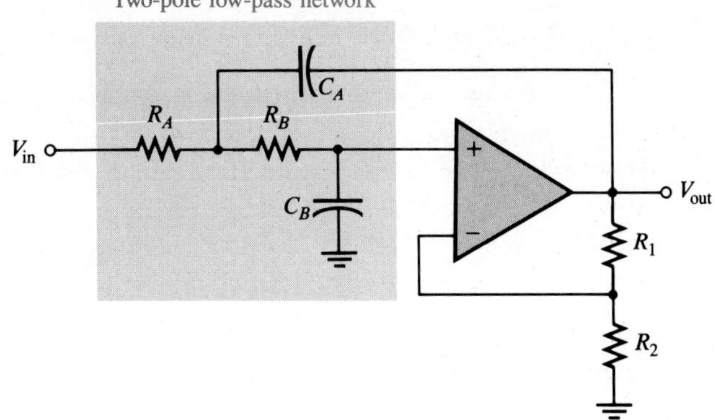

filter response either Butterworth, Chebyshev, or Bessel. For example, from Table 16–1, the R_1/R_2 ratio must be 0.586 to produce the damping factor of 1.414 required for a second-order Butterworth response.

EXAMPLE 16–3

Determine the critical frequency of the low-pass filter in Figure 16–11, and set the value of R_1 for an approximate Butterworth response.

FIGURE 16–11

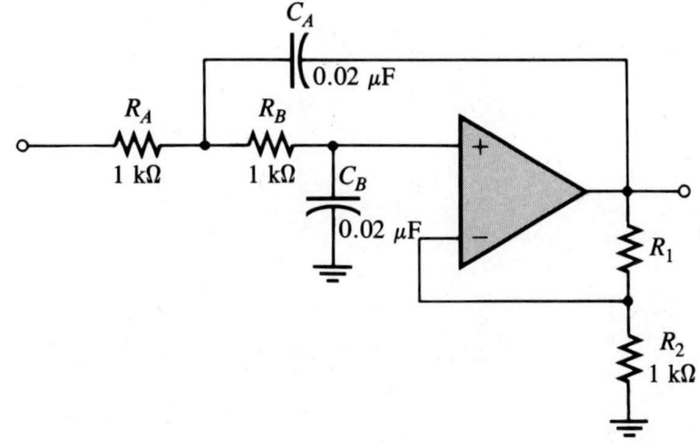

SOLUTION

Since $R_A = R_B = 1\text{ k}\Omega$ and $C_A = C_B = 0.02\ \mu\text{F}$,

$$f_c = \frac{1}{2\pi RC} = \frac{1}{2\pi(1\text{ k}\Omega)(0.02\ \mu\text{F})} = 7.958\text{ kHz}$$

For a Butterworth response, $R_1/R_2 = 0.586$.

$$R_1 = 0.586R_2 = 0.586(1\text{ k}\Omega) = 586\ \Omega$$

Select a standard value as near as possible to this calculated value.

PRACTICE EXERCISE 16–3

Determine f_c for Figure 16–11 if $R_A = R_B = R_2 = 2.2\text{ k}\Omega$ and $C_A = C_B = 0.01\ \mu\text{F}$. Also determine the value of R_1 for a Butterworth response.

CASCADED LOW-PASS FILTERS ACHIEVE A HIGHER ROLL-OFF RATE

A three-pole filter is required to get a third-order low-pass response (−60 dB/decade). This is done by cascading a two-pole low-pass filter and a single-pole low-pass filter, as shown in Figure 16–12(a). Figure 16–12(b) shows a four-pole configuration obtained by cascading two two-pole filters.

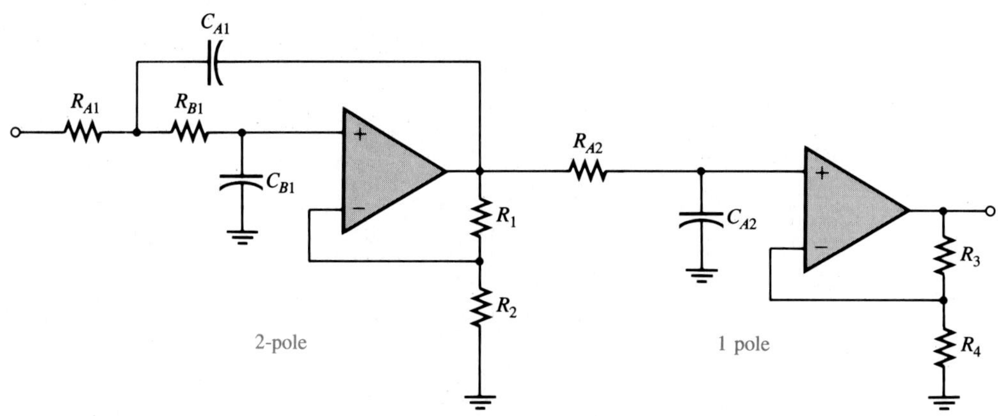

(a) Third-order

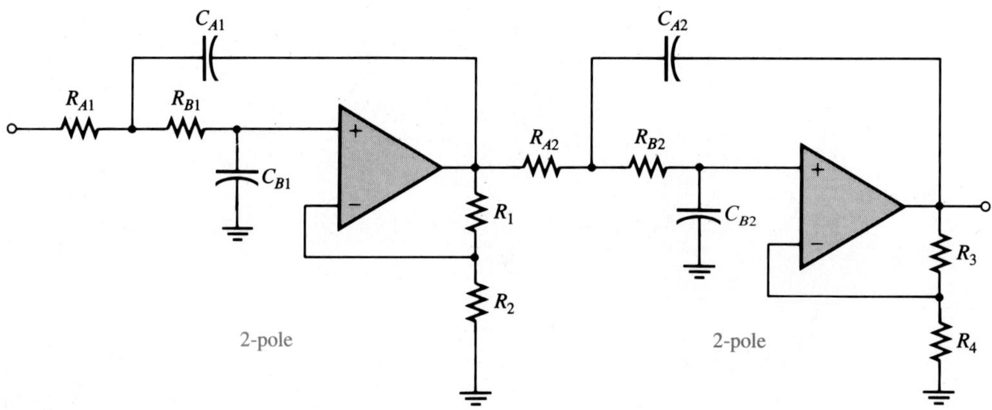

(b) Fourth-order

FIGURE 16–12
Cascaded low-pass filters.

EXAMPLE 16–4 For the four-pole filter in Figure 16–12(b), determine the capacitance values required to produce a critical frequency of 2680 Hz if all the resistors are 1.8 kΩ. Also select values for the feedback resistors to get a Butterworth response.

SOLUTION

Both stages must have the same f_c. Assuming equal-value capacitors,

$$f_c = \frac{1}{2\pi RC}$$

$$C = \frac{1}{2\pi R f_c} = \frac{1}{2\pi(1.8\ \text{k}\Omega)(2680\ \text{Hz})} = 0.032\ \mu\text{F}$$

$$C_{A1} = C_{B1} = C_{A2} = C_{B2} = 0.032\ \mu\text{F}$$

Select $R_2 = R_4 = 1.8\ \text{k}\Omega$ for simplicity. Refer to Table 16–1.

For a Butterworth response in the first stage,

$$DF = 1.848,\ \frac{R_1}{R_2} = 0.152$$

$$R_1 = 0.152R_2 = 0.152(1800\ \Omega) = 273.6\ \Omega$$

Choose $R_1 = 270\ \Omega$.

In the second stage,

$$DF = 0.765,\ \frac{R_3}{R_4} = 1.235$$

$$R_3 = 1.235R_4 = 1.235(1800\ \Omega) = 2.223\ \text{k}\Omega$$

Choose $R_3 = 2.2\ \text{k}\Omega$.

PRACTICE EXERCISE 16–4

For the filter in Figure 16–12(b), determine the capacitance values for $f_c = 1$ kHz if all the filter resistors are 680 Ω. Also specify the values for the feedback resistors to produce a Butterworth response. ■

16–3 REVIEW QUESTIONS

1. How many poles does a second-order low-pass filter have? How many resistors and how many capacitors are used in the frequency-selective network?
2. Why is the damping factor of a filter important?
3. What is the primary purpose of cascading low-pass filters?

16–4 ACTIVE HIGH-PASS FILTERS

In high-pass filters, the roles of the capacitor and resistor are reversed in the RC networks. Otherwise, the basic considerations are the same as for the low-pass filters.

A SINGLE-POLE FILTER

A high-pass active filter with a 20 dB/decade roll-off is shown in Figure 16–13(a). Notice that the input circuit is a single high-pass *RC* network. The negative feedback circuit is the same as for the low-pass filters previously discussed. The high-pass response curve is shown in Figure 16–13(b).

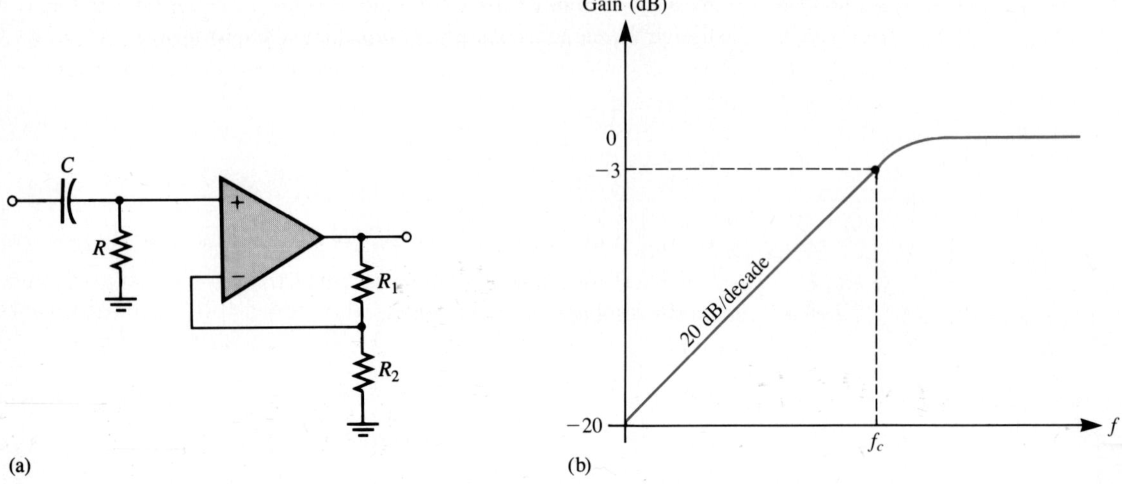

FIGURE 16–13
Single-pole active high-pass filter and response curve.

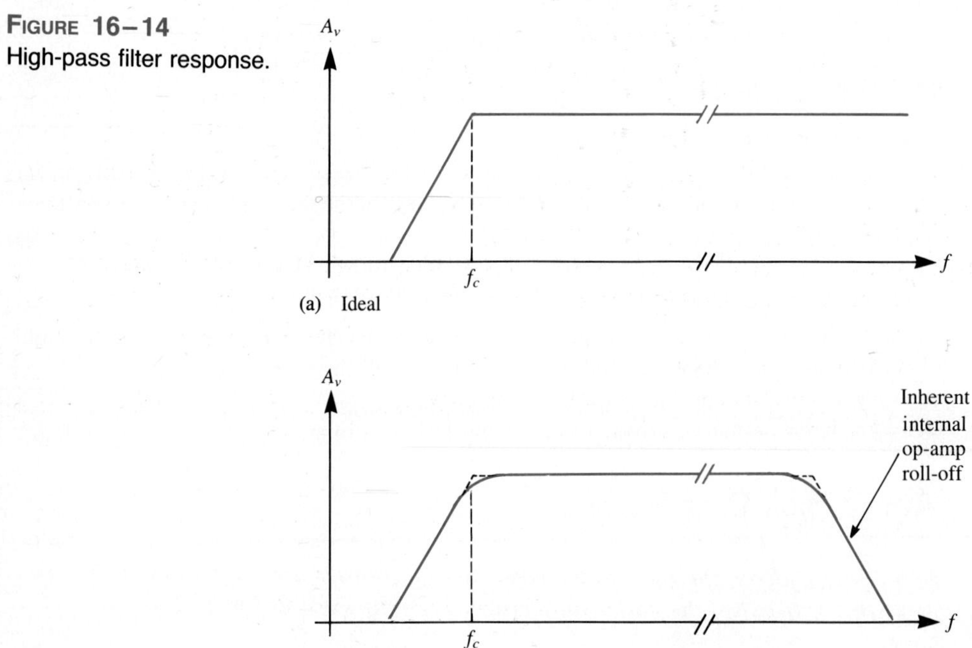

FIGURE 16–14
High-pass filter response.

Ideally, a high-pass filter passes all frequencies above f_c without limit, as indicated in Figure 16–14(a), although in practice, this is not the case. As you have learned, all op-amps inherently have internal *RC* networks that limit the amplifier's response at high frequencies. Therefore, there is an upper-frequency limit on the high-pass filter's response which, in effect, makes it a band-pass filter with a very wide bandwidth. In the majority of applications, the internal high-frequency limitation is so much greater than that of the filter's f_c that the limitation can be neglected. In some applications, discrete transistors are used for the gain element to increase the high-frequency limitation beyond that realizable with available op-amps.

THE SALLEN-KEY HIGH-PASS FILTER

A high-pass second-order Sallen-Key configuration is shown in Figure 16–15. The components R_A, C_A, R_B, and C_B form the two-pole frequency-selective network. Notice that the positions of the resistors and capacitors in the frequency-selective network are opposite to those in the low-pass configuration. As with the other filters, the response characteristic can be optimized by proper selection of the feedback resistors, R_1 and R_2.

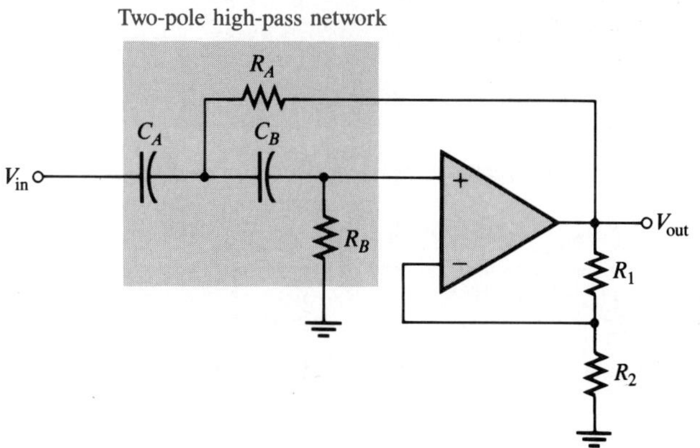

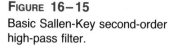

FIGURE 16–15
Basic Sallen-Key second-order high-pass filter.

EXAMPLE 16–5 Choose values for the Sallen-Key high-pass filter in Figure 16–15 to implement an equal-value second-order Butterworth response with a critical frequency of approximately 10 kHz.

SOLUTION

Start by selecting a value for R_A and R_B (R_1 or R_2 can also be the same value as R_A and R_B for simplicity).

$$R = R_A = R_B = R_2 = 3.3\ \text{k}\Omega \qquad \text{(an arbitrary selection)}$$

Next, calculate the capacitance value from $f_c = 1/2\pi RC$.

$$C = C_A = C_B = \frac{1}{2\pi R f_c} = \frac{1}{2\pi(3.3\ \text{k}\Omega)(10\ \text{kHz})} = 0.004\ \mu\text{F}$$

For a Butterworth response, the damping factor must be 1.414 and $R_1/R_2 = 0.586$.

$$R_1 = 0.586R_2 = 0.586(3.3\ \text{k}\Omega) = 1.93\ \text{k}\Omega$$

If we had let $R_1 = 3.3\ \text{k}\Omega$, then

$$R_2 = \frac{R_1}{0.586} = \frac{3.3\ \text{k}\Omega}{0.586} = 5.63\ \text{k}\Omega$$

Either way, an approximate Butterworth response is realized by choosing the nearest standard value.

PRACTICE EXERCISE 16–5

Select values for all the components in the high-pass filter of Figure 16–15 to obtain an $f_c = 300$ Hz. Use equal-value components and optimize for a Butterworth response.

CASCADING HIGH-PASS FILTERS

As with the low-pass configuration, first- and second-order high-pass filters can be cascaded to provide three or more poles and thereby create faster roll-off rates. Figure 16–16 shows a six-pole high-pass filter consisting of three two-pole stages. With this configuration optimized for a Butterworth response, a roll-off of 120 dB/decade is achieved.

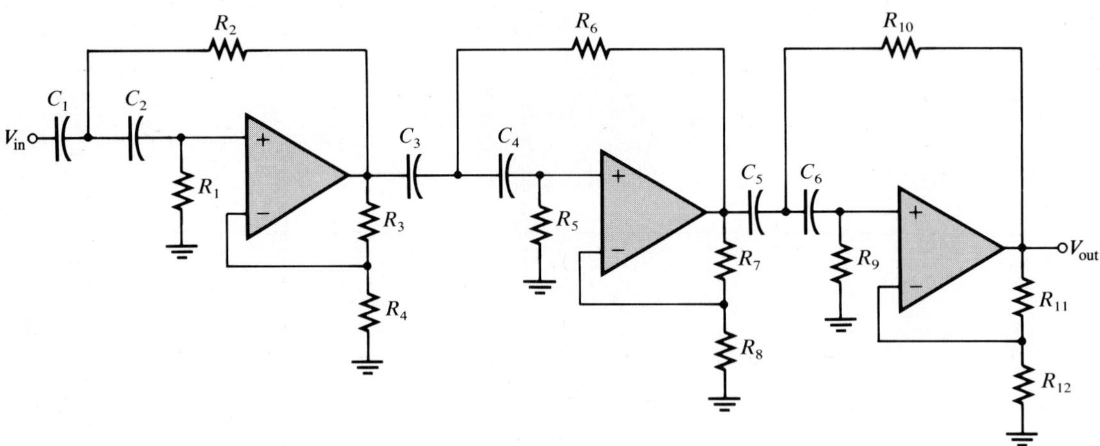

FIGURE 16–16
Sixth-order high-pass filter.

16–4 REVIEW QUESTIONS

1. How does a high-pass Sallen-Key filter differ from the low-pass configuration?
2. To increase the critical frequency of a high-pass filter, would you increase or decrease the resistor values?
3. If three two-pole high-pass filters and one single-pole high-pass filter are cascaded, what is the resulting roll-off?

16–5 ACTIVE BAND-PASS FILTERS

As mentioned, band-pass filters pass all frequencies bounded by a lower- and an upper-frequency limit and reject all others lying outside this specified band. A band-pass response can be thought of as the overlapping of a low-frequency response curve and a high-frequency response curve.

CASCADED LOW-PASS AND HIGH-PASS FILTERS ACHIEVE A BAND-PASS RESPONSE

One way to implement a band-pass filter is a cascaded arrangement of a high-pass filter and a low-pass filter, as shown in Figure 16–17(a), as long as the critical frequencies are sufficiently separated. Each of the filters shown is a two-pole Sallen-Key Butterworth configuration so that the roll-off rates are ±40 dB/decade, indicated in the composite response curve of Figure 16–17(b). The critical frequency of each filter is chosen so that the response curves overlap sufficiently, as indicated. The critical frequency of the high-pass filter must be sufficiently lower than that of the low-pass stage.

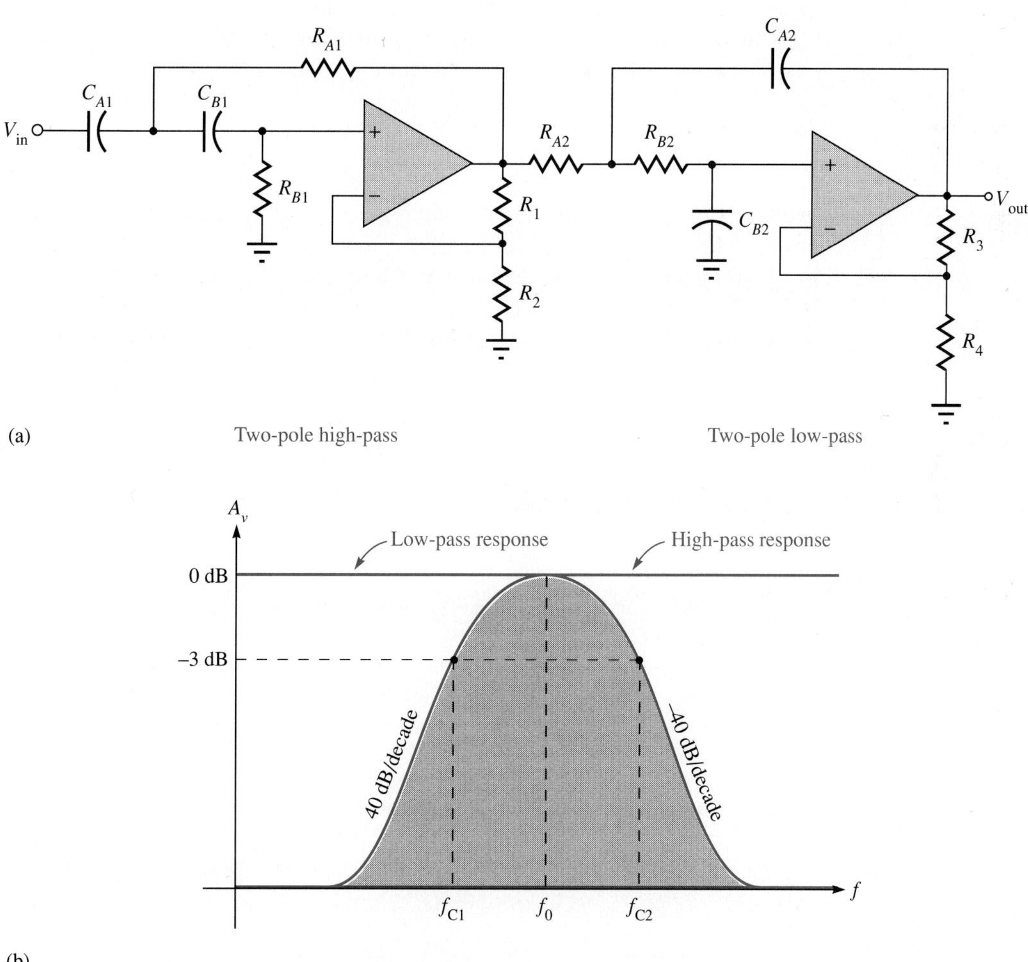

FIGURE 16–17
Band-pass filter formed by cascading a two-pole high-pass and a two-pole low-pass filter (it does not matter in which order the filters are cascaded).

The lower frequency f_{c1} of the pass band is the critical frequency of the high-pass filter. The upper frequency f_{c2} is the critical frequency of the low-pass filter. Ideally, as discussed earlier, the center frequency f_0 of the pass band is the geometric mean of f_{c1} and f_{c2}. The following formulas express the three frequencies of the band-pass filter in Figure 16–17.

$$f_{c1} = \frac{1}{2\pi\sqrt{R_{A1}R_{B1}C_{A1}C_{B1}}}$$

$$f_{c2} = \frac{1}{2\pi\sqrt{R_{A2}R_{B2}C_{A2}C_{B2}}}$$

$$f_0 = \sqrt{f_{c1}f_{c2}}$$

Of course, if equal-value components are used in implementing each filter, the critical frequency equations simplify to the form $f_c = 1/2\pi RC$.

MULTIPLE-FEEDBACK BAND-PASS FILTER

Another type of filter configuration, shown in Figure 16–18, is a multiple-feedback band-pass filter. The two feedback paths are through R_2 and C_1. Components R_1 and C_1 provide the low-pass response, and R_2 and C_2 provide the high-pass response. The maximum gain, A_0, occurs at the center frequency. Q values of less than 10 are typical in this type of filter. An expression for the center frequency is developed as follows, recognizing that R_1 and R_3 appear in parallel as viewed from the C_1 feedback path (with the V_{in} source replaced by a short).

$$f_0 = \frac{1}{2\pi\sqrt{(R_1\|R_3)R_2C_1C_2}}$$

Making $C_1 = C_2 = C$ yields

$$f_0 = \frac{1}{2\pi\sqrt{(R_1\|R_3)R_2C^2}}$$

$$= \frac{1}{2\pi C\sqrt{(R_1\|R_3)R_2}}$$

$$= \frac{1}{2\pi C}\sqrt{\frac{1}{R_2(R_1\|R_3)}}$$

$$= \frac{1}{2\pi C}\sqrt{\left(\frac{1}{R_2}\right)\left(\frac{1}{R_1R_3/(R_1 + R_3)}\right)}$$

$$f_0 = \frac{1}{2\pi C}\sqrt{\frac{R_1 + R_3}{R_1R_2R_3}} \qquad (16–8)$$

A convenient value for the capacitors is chosen, then the three resistor values are calculated based on the desired values for f_0, BW, and A_0. As you know, the Q can be determined from the relation $Q = f_0/BW$, and the resistors are found using the following formulas (stated without derivation).

$$R_1 = \frac{Q}{2\pi f_0 C A_0}$$

$$R_2 = \frac{Q}{\pi f_0 C}$$

$$R_3 = \frac{Q}{2\pi f_0 C(2Q^2 - A_0)}$$

To develop a gain expression, we solve for Q in the first two equations above.

$$Q = 2\pi f_0 A_0 C R_1$$

$$Q = \pi f_0 C R_2$$

Then,

$$2\pi f_0 A_0 C R_1 = \pi f_0 C R_2$$

Cancelling, we get

$$2A_0 R_1 = R_2$$

$$A_0 = \frac{R_2}{2R_1} \qquad (16\text{–}9)$$

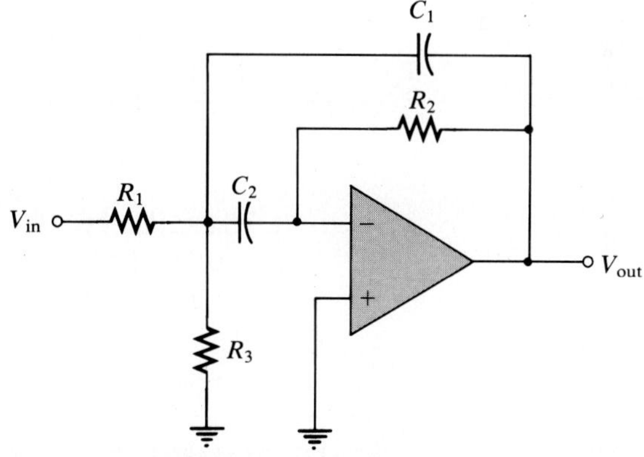

FIGURE 16–18
Multiple-feedback band-pass filter.

In order for the denominator of the equation $R_3 = Q/2\pi f_0 C(2Q^2 - A_0)$ to be positive, $A_0 < 2Q^2$, which imposes a limitation on the gain.

EXAMPLE 16–6 Determine the center frequency, maximum gain, and bandwidth for the filter in Figure 16–19.

FIGURE 16–19

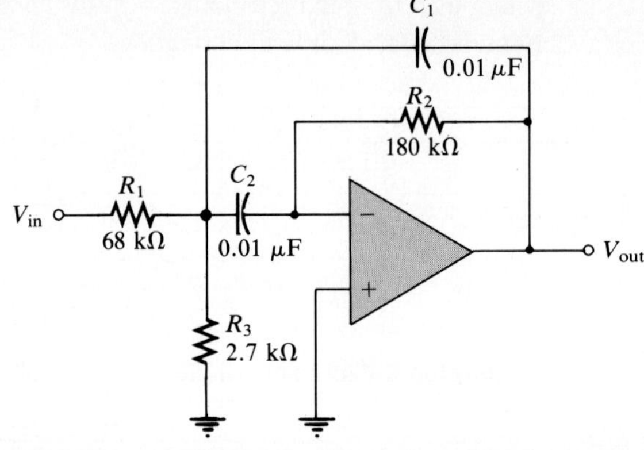

SOLUTION

$$f_0 = \frac{1}{2\pi C}\sqrt{\frac{R_1 + R_3}{R_1R_2R_3}}$$

$$= \frac{1}{2\pi(0.01\ \mu\text{F})}\sqrt{\frac{68\ \text{k}\Omega + 2.7\ \text{k}\Omega}{(68\ \text{k}\Omega)(180\ \text{k}\Omega)(2.7\ \text{k}\Omega)}}$$

$$= 736\ \text{Hz}$$

$$A_0 = \frac{R_2}{2R_1} = \frac{180\ \text{k}\Omega}{2(68\ \text{k}\Omega)} = 1.32$$

$$Q = \pi f_0 C R_2$$

$$= \pi(736\ \text{Hz})(0.01\ \mu\text{F})(180\ \text{k}\Omega)$$

$$= 4.16$$

$$BW = \frac{f_0}{Q} = \frac{736\ \text{Hz}}{4.16} = 176.9\ \text{Hz}$$

PRACTICE EXERCISE 16–6

If R_2 in Figure 16–19 is increased to 330 kΩ, how does this affect the gain, center frequency, and bandwidth of the filter?

STATE-VARIABLE BAND-PASS FILTER

The state-variable or universal active filter is widely used for band-pass applications. As shown in Figure 16–20, it consists of a summing amplifier and two op-amp integrators (which act as single-pole low-pass filters) that are combined in a cascaded arrangement to form a second-order filter. Although used primarily as a band-pass filter, the state-variable configuration also provides low-pass (*LP*) and high-pass (*HP*) outputs. The center frequency is set by the *RC* networks in both integrators. When used as a band-pass filter, the critical frequencies of the integrators are usually made equal, thus setting the center frequency of the pass band.

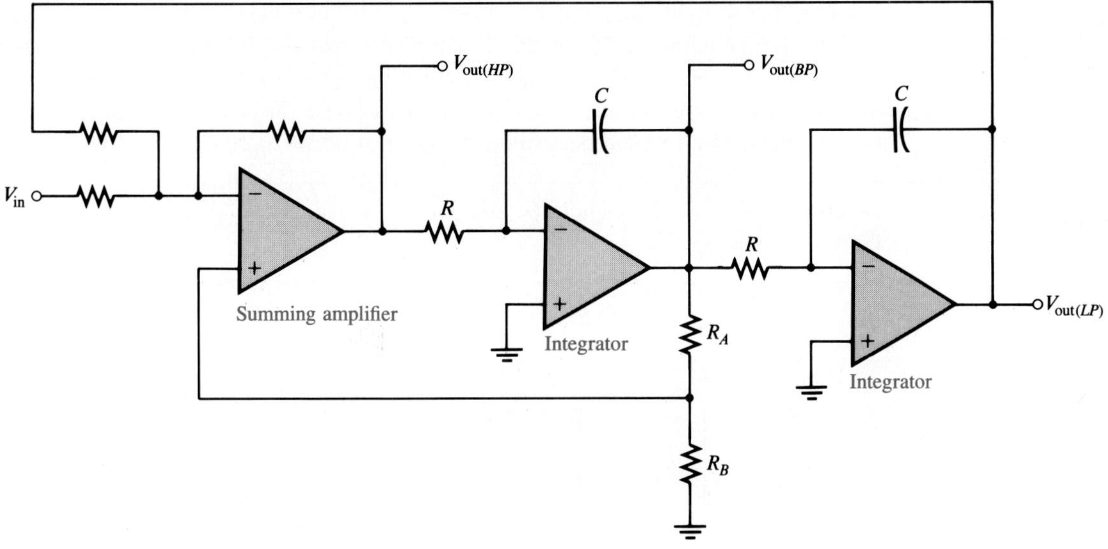

FIGURE 16–20
State-variable band-pass filter.

BASIC OPERATION At input frequencies below f_c, the input signal passes through the summing amplifier and integrators and is fed back 180° out-of-phase. Thus, the feedback signal and input signal cancel for all frequencies below approximately f_c. As the low-pass response of the integrators rolls off, the feedback signal diminishes, thus allowing the input to pass through to the band-pass output. Above f_c, the low-pass response disappears, thus preventing the input signal from passing through the integrators. As a result, the band-pass output peaks sharply at f_c, as indicated in Figure 16–21. Stable Qs up to 100 can be obtained with this type of filter. The Q is set by the feedback resistors R_A and R_B according to the following equation.

$$Q = \frac{1}{3}\left(\frac{R_A}{R_B} + 1\right) \quad (16\text{–}10)$$

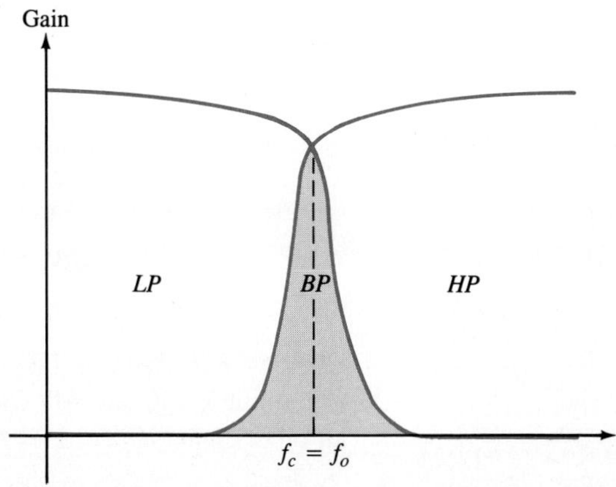

FIGURE 16–21
General state-variable response curves.

The state-variable filter cannot be optimized for low-pass, high-pass, and band-pass performance simultaneously for this reason: To optimize for a low-pass or a high-pass Butterworth response, DF must equal 1.414. Since $Q = 1/DF$, a Q of 0.707 will result. Such a low Q provides a very poor band-pass response (large BW and poor selectivity). For optimization as a band-pass filter, the Q must be set high.

EXAMPLE 16–7

Determine the center frequency, Q, and BW for the band-pass output of the state-variable filter in Figure 16–22.

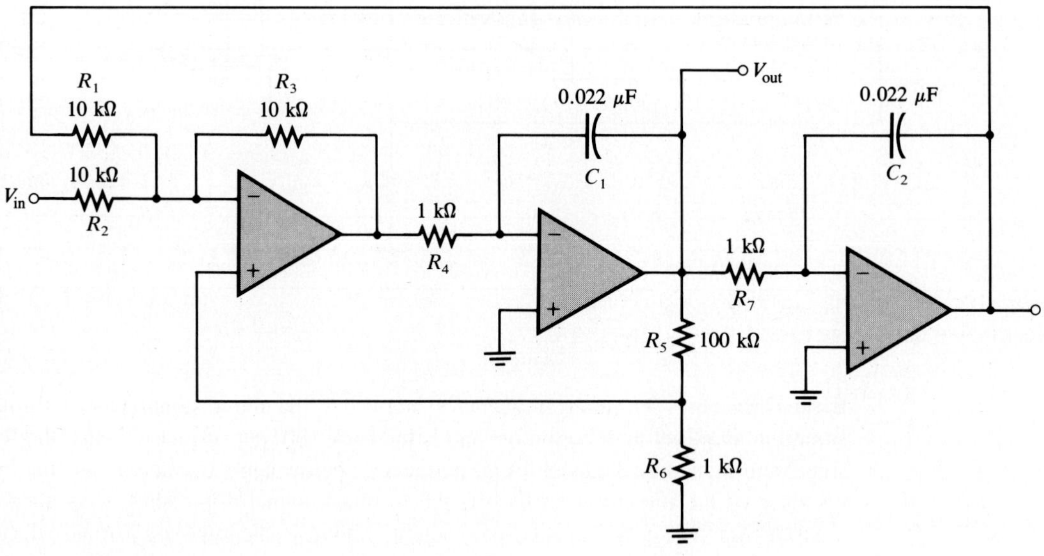

FIGURE 16–22

SOLUTION

For each integrator,

$$f_c = \frac{1}{2\pi R_4 C_1} = \frac{1}{2\pi R_7 C_2} = \frac{1}{2\pi(1\ \text{k}\Omega)(0.022\ \mu\text{F})} = 7.23\ \text{kHz}$$

The center frequency is approximately equal to the critical frequencies of the integrators.

$$f_0 = f_c = 7.23\ \text{kHz}$$

$$Q = \frac{1}{3}\left(\frac{R_5}{R_6} + 1\right) = \frac{1}{3}\left(\frac{100\ \text{k}\Omega}{1\ \text{k}\Omega} + 1\right) = 33.67$$

$$BW = \frac{f_0}{Q} = \frac{7.23\ \text{kHz}}{33.67} = 214.7\ \text{Hz}$$

PRACTICE EXERCISE 16–7

Determine f_0, Q, and BW for the filter in Figure 16–22 if $R_4 = R_6 = R_7 = 330\ \Omega$ with all other component values the same as shown on the schematic.

16–5 REVIEW QUESTIONS

1. What determines selectivity in a band-pass filter?
2. One filter has a $Q = 5$ and another has a $Q = 25$. Which has the narrower bandwidth?
3. List the elements that make up a state-variable filter.

16–6 ACTIVE BAND-STOP FILTERS

Band-stop filters reject a specified band of frequencies and pass all others. The response is opposite to that of a band-pass filter.

MULTIPLE-FEEDBACK BAND-STOP FILTER

Figure 16–23 shows a multiple-feedback band-stop filter. Notice that this configuration is similar to the band-pass version except that R_3 has been omitted and R_A and R_B have been added.

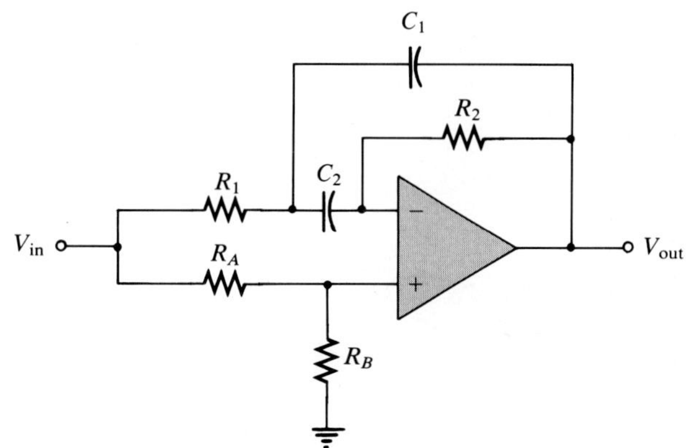

FIGURE 16–23
Multiple-feedback band-stop filter.

STATE-VARIABLE BAND-STOP FILTER

Summing the low-pass and the high-pass responses of the state-variable filter covered in Section 16–5 creates a band-stop response as shown in Figure 16–24. One important

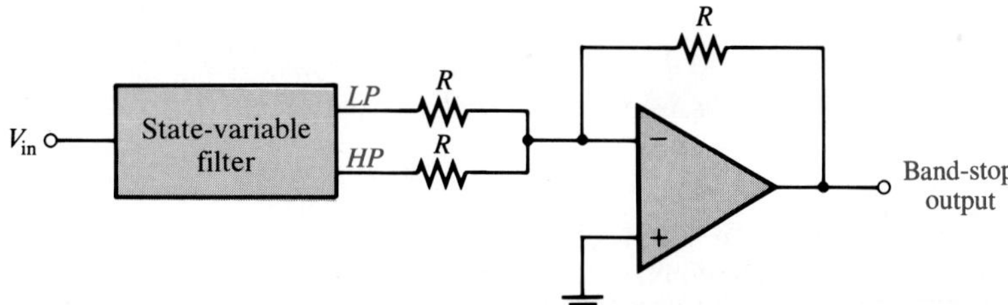

FIGURE 16–24
State-variable band-stop filter.

application of this filter is minimizing the 60 Hz "hum" in audio systems by setting the center frequency to 60 Hz.

EXAMPLE 16–8 Verify that the band-stop filter in Figure 16–25 has a center frequency of 60 Hz, and optimize it for a Q of 30.

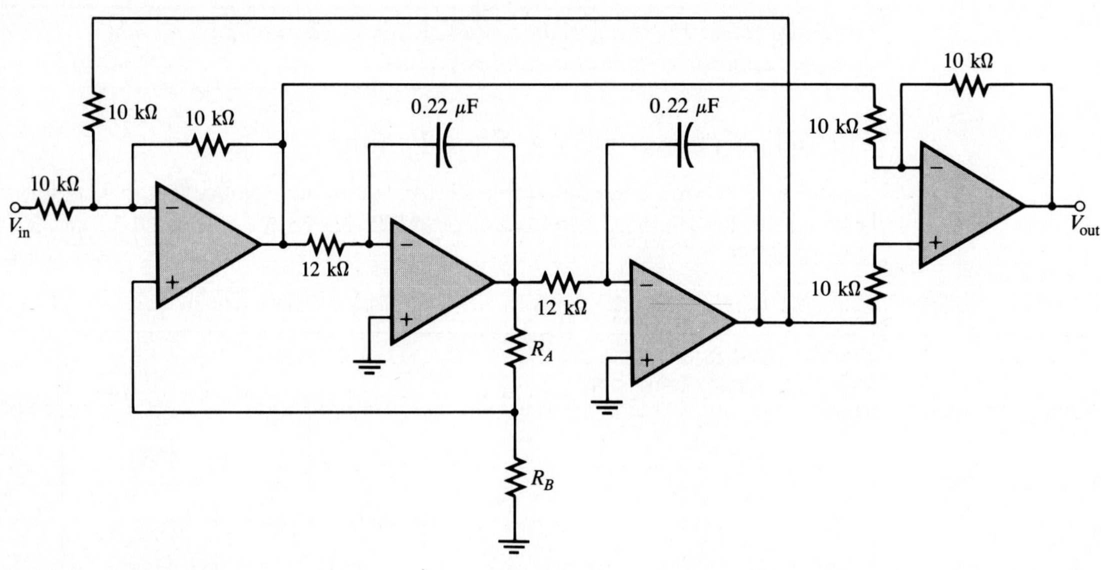

FIGURE 16–25

SOLUTION

f_0 equals the f_c of the integrator stages.

$$f_0 = \frac{1}{2\pi RC} = \frac{1}{2\pi(12\ \text{k}\Omega)(0.22\ \mu\text{F})} = 60\ \text{Hz}$$

We obtain a $Q = 30$ by choosing R_B and then calculating R_A.

$$Q = \frac{1}{3}\left(\frac{R_A}{R_B} + 1\right)$$

$$R_A = (3Q - 1)R_B$$

Choose $R_B = 1\ \text{k}\Omega$. Then

$$R_A = [3(30) - 1]1\ \text{k}\Omega = 89\ \text{k}\Omega$$

PRACTICE EXERCISE 16–8

How would you change the center frequency to 120 Hz in Figure 16–25? ■

16–6 REVIEW QUESTIONS

1. How does a band-stop response differ from a band-pass response?
2. How is a state-variable band-pass filter converted to a band-stop filter?

16–7 FILTER RESPONSE MEASUREMENTS

In this section, we discuss two methods of determining a filter's response by measurement—discrete point measurement and swept frequency measurement.

DISCRETE POINT MEASUREMENT

Figure 16–26 shows an arrangement for taking filter output voltage measurements at discrete values of input frequency using common laboratory instruments. The general procedure is as follows:

1. Set the amplitude of the sine wave generator to a desired voltage level.
2. Set the frequency of the sine wave generator to a value well below the expected critical frequency of the filter under test. For a low-pass filter, set the frequency as near as possible to 0 Hz. For a band-pass filter, set the frequency well below the expected lower critical frequency.
3. Increase the frequency in predetermined steps sufficient to allow enough data points for an accurate response curve.
4. Maintain a constant input voltage amplitude while varying the frequency.
5. Record the output voltage at each value of frequency.
6. After recording a sufficient number of points, plot a graph of output voltage versus frequency.

If the frequencies to be measured exceed the response of the DMM, an oscilloscope may have to be used instead.

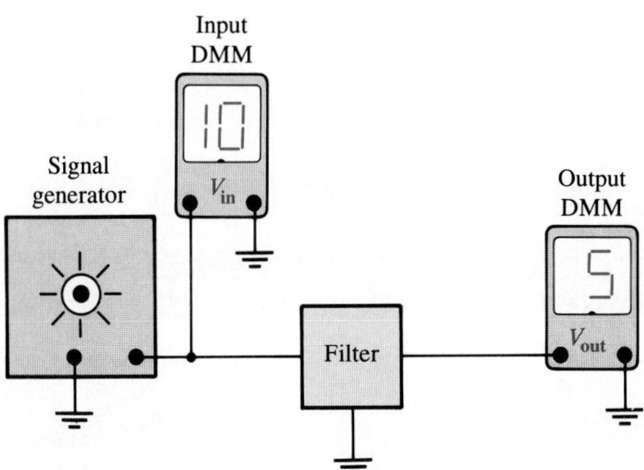

FIGURE 16–26
Test set-up for discrete point measurement of the filter response. (Readings are arbitrary and for display only.)

SWEPT FREQUENCY MEASUREMENT

The swept frequency method requires more elaborate test equipment than does the discrete point method, but it is much more efficient and can result in a more accurate response curve. A general test set-up is shown in Figure 16–27 using a swept frequency generator and a spectrum analyzer.

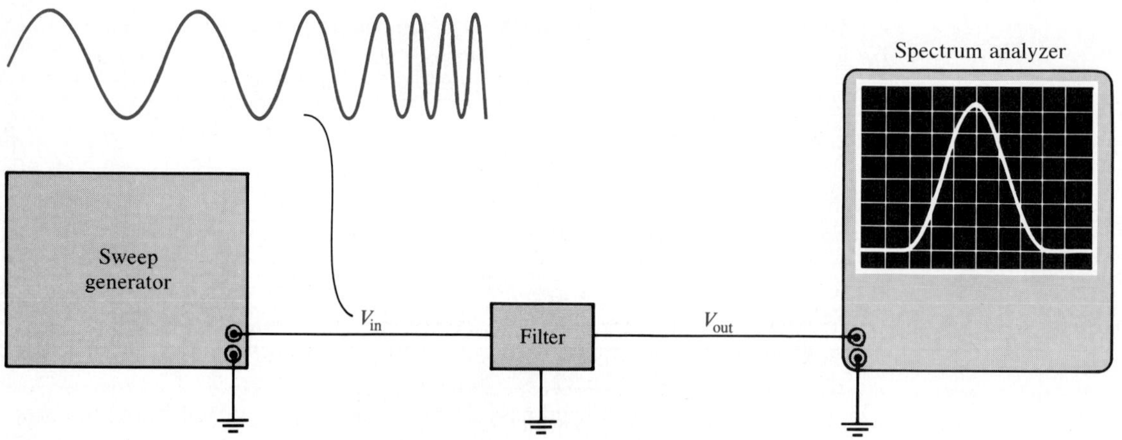

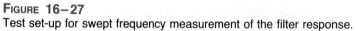

FIGURE 16–27
Test set-up for swept frequency measurement of the filter response.

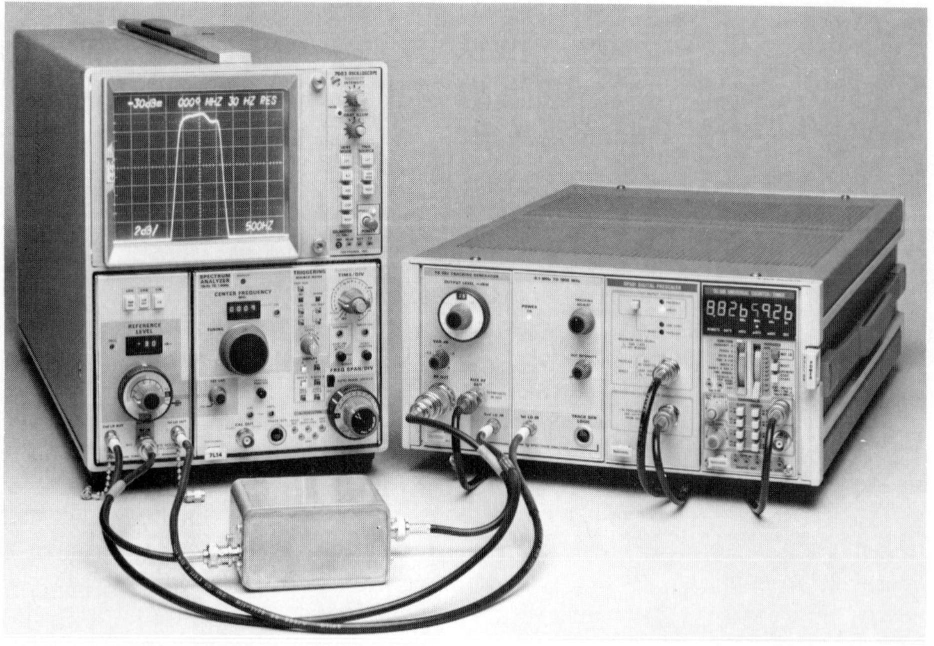

FIGURE 16–28
Test set-up with actual equipment (courtesy of Tektronix, Inc.).

The swept frequency generator produces a constant amplitude output signal whose frequency increases linearly between two preset limits, as indicated in Figure 16–27. The spectrum analyzer is essentially an elaborate oscilloscope that can be calibrated for a desired *frequency span/division* rather than for the usual *time/division* setting. Therefore, as the input frequency to the filter sweeps through a preselected range, the response curve is traced out on the screen of the spectrum analyzer. An actual swept frequency test set-up using typical equipment is shown in Figure 16–28.

16–7 REVIEW QUESTIONS

1. What is the purpose of the two tests discussed in this section?
2. Name one disadvantage and one advantage of each test method.

16–8 A SYSTEM APPLICATION

In this system application, the focus is on the filter board, which is part of the channel separation circuits in the FM stereo receiver. In addition to the active filters, the left and right channel separation circuit includes a demodulator, a frequency doubler, and a stereo matrix. Except for mentioning their purpose, we will not deal specifically with the demodulator, doubler, or matrix. However, the matrix is an interesting application of summing amplifiers, which were studied in Chapter 14 and these will be shown in detail on the schematic although we will not concentrate on them. In this section, you will

- ☐ ***See how low-pass and band-pass active filters are used.***
- ☐ ***Use the schematic to locate and identify the components on the pc board.***
- ☐ ***Analyze the operation of the filters.***
- ☐ ***Troubleshoot some common amplifier failures.***

A BRIEF DESCRIPTION OF THE SYSTEM

Stereo FM **(frequency modulation)** signals are transmitted on a **carrier** frequency of 88 MHz to 108 MHz. The standard transmitted stereo signal consists of three modulating signals. These are the sum of the left and right channel audio (L + R), the difference of the left and right channel audio (L − R), and a 19 kHz pilot subcarrier. The L + R audio extends from 30 Hz to 15 kHz and the L − R signal is contained in two sidebands extending from 23 kHz to 53 kHz as indicated in Figure 16–29. These frequencies come from the FM detector and go into the filter circuits where they are separated. The frequency doubler and demodulator are used to extract the audio signal from the 23 kHz to 53 kHz sidebands after which the 30 Hz to 15 kHz L − R signal is passed through a filter. The L + R and L − R audio signals are then sent to the matrix where they are applied to the summing circuits to produce the left and right channel audio (−2L and −2R). Our focus is on the filters.

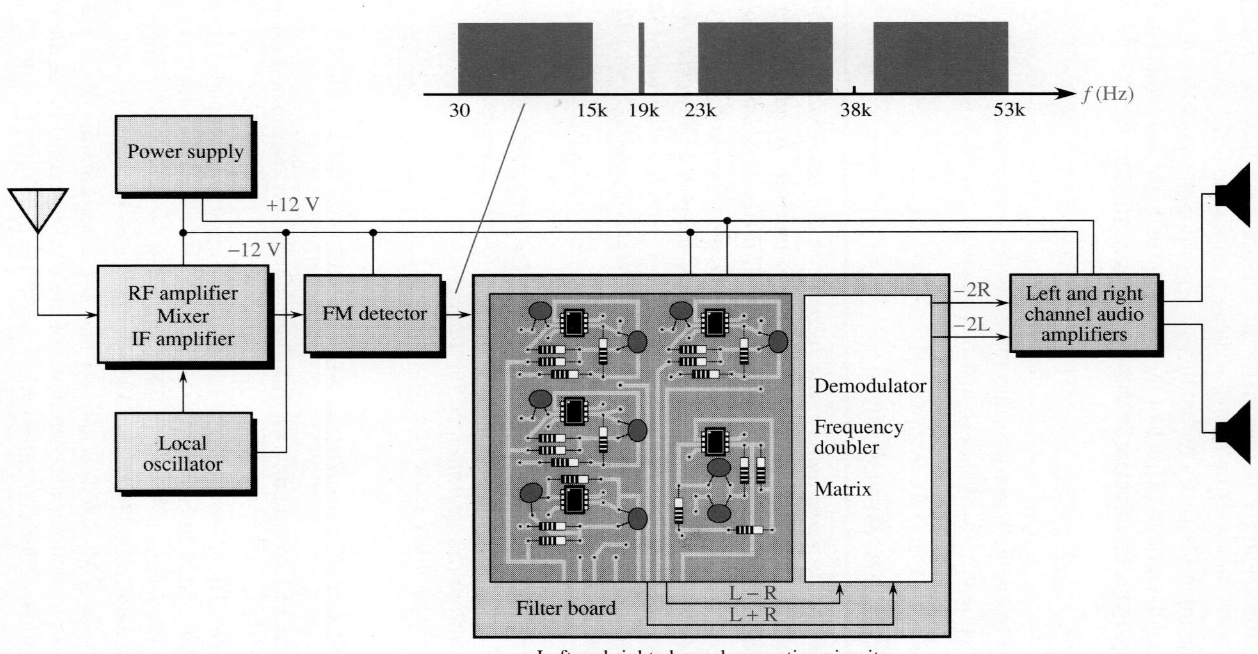

FIGURE 16–29
FM stereo receiver system.

Now, so that you can take a closer look at the filter board, let's take it out of the system and put it on the test bench.

ON THE TEST BENCH

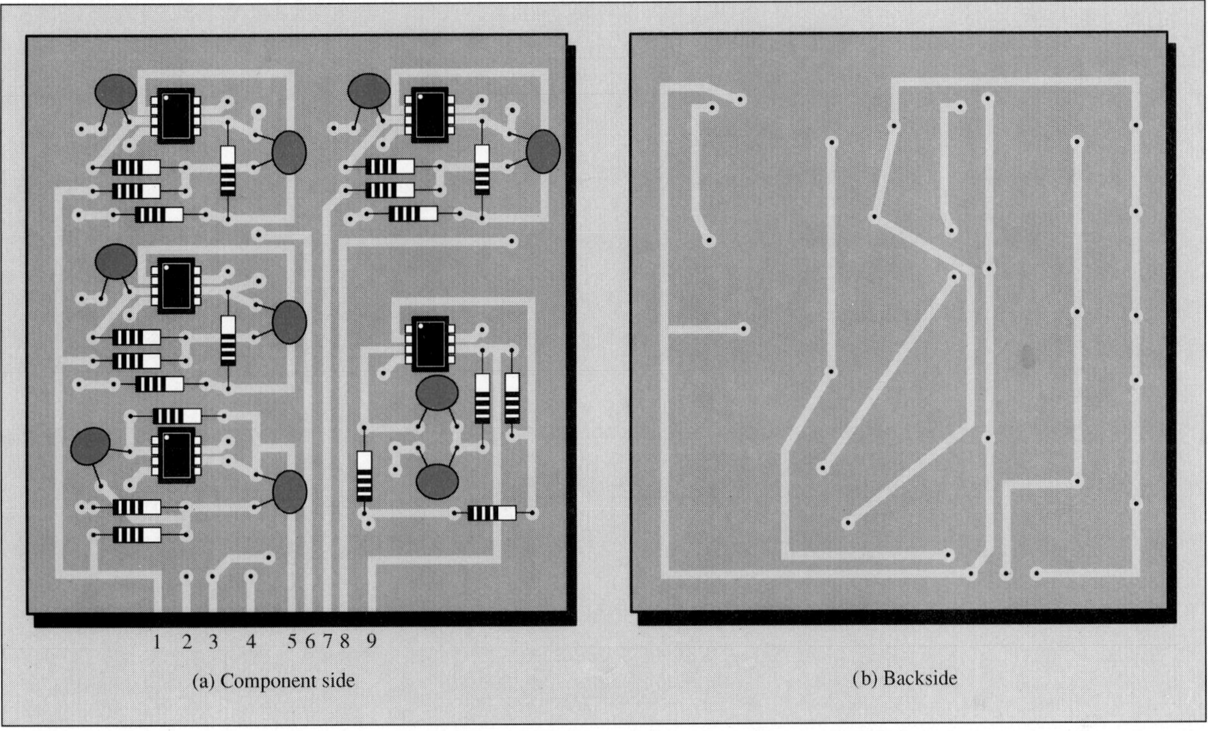

FIGURE 16–30

ACTIVITY 1 RELATE THE PC BOARD TO THE SCHEMATIC

Locate and identify all components on the pc board in Figure 16–30, using the schematic diagram in Figure 16–31. Label the pc board components to correspond with the schematic. Identify all inputs and outputs. Trace out the pc board to verify that it corresponds with the schematic. The shaded areas on the schematic are the filter circuits contained on the board. The other blocks and circuits are on the demodulator, frequency doubler, and matrix board located elsewhere.

ACTIVITY 2 ANALYZE THE FILTER CIRCUITS

STEP 1 Using the component values, determine the critical frequencies of each Sallen-Key-type filter

STEP 2 Using the component values, determine the center frequency of the multiple-feedback filter.

STEP 3 Determine the bandwidth of each filter.

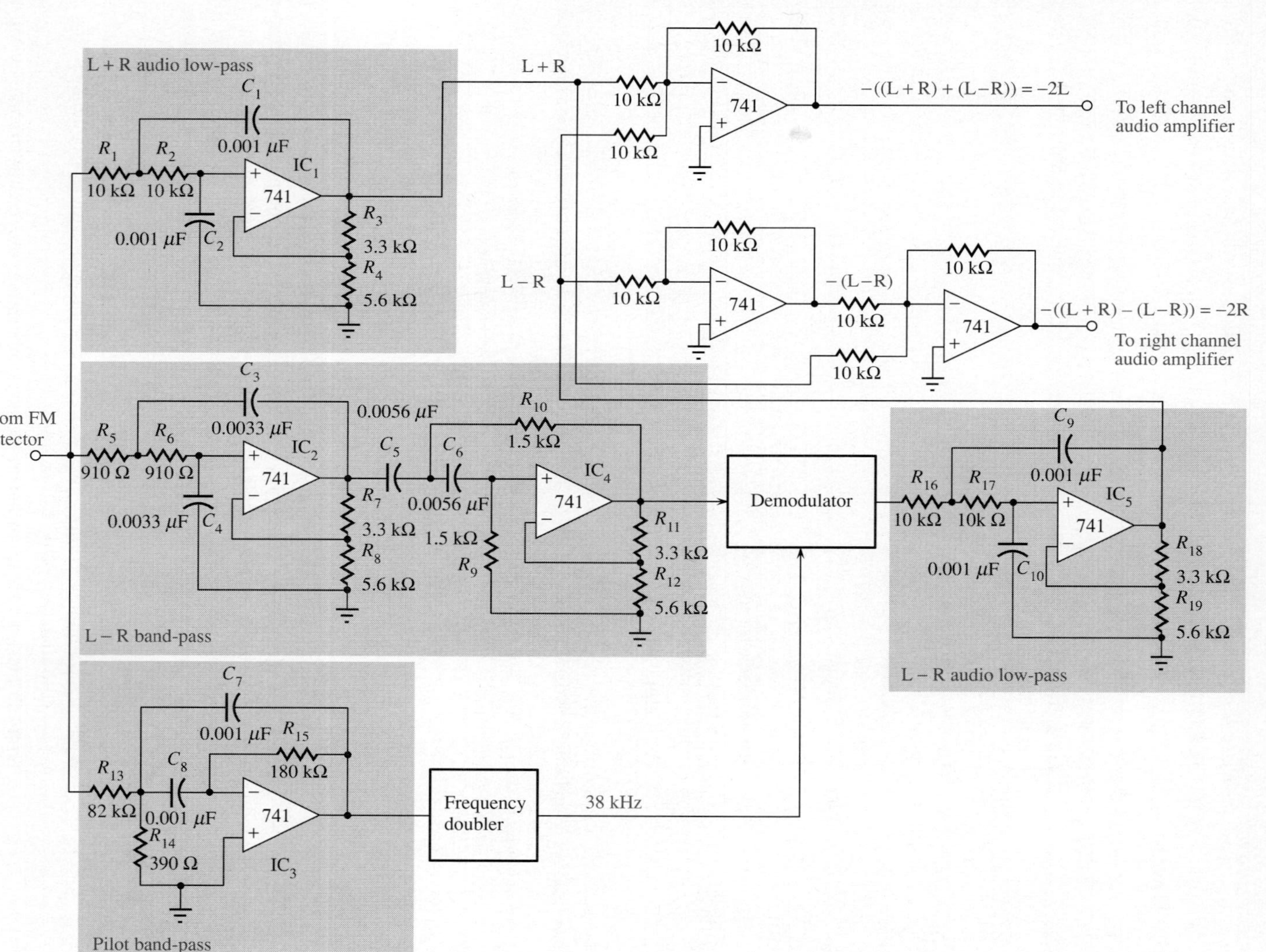

FIGURE 16–31
Left and right channel separation circuits.

STEP 4 Determine the voltage gain of each filter.

STEP 5 Verify that the Sallen-Key filters have an approximate Butterworth response characteristic.

■ ACTIVITY 3 WRITE A TECHNICAL REPORT

Describe each filter in detail specifying the type of filter, the frequency responses, and the function of the filter within the overall circuitry. Also, describe the overall operation of the complete channel separation circuitry.

■ ACTIVITY 4 TROUBLESHOOT THE FILTER BOARD

The filter board is plugged into a test fixture that permits access to each input and output, as shown in Figure 16–32, where the socket numbers correspond to the board pin numbers. The text instruments to be used are a sweep generator, a spectrum analyzer, and a dual power supply. For the sweep generator, a minimum and a maximum frequency is selected and the instrument produces an output that repetitively sweeps through all frequencies between the minimum and maximum setting. The spectrum analyzer is a type of oscilloscope that will plot out a frequency response curve.

Develop a basic test procedure for completely testing the board in the fixture, using general references to instrument inputs, outputs, and settings. Include a diagram of a complete test set-up.

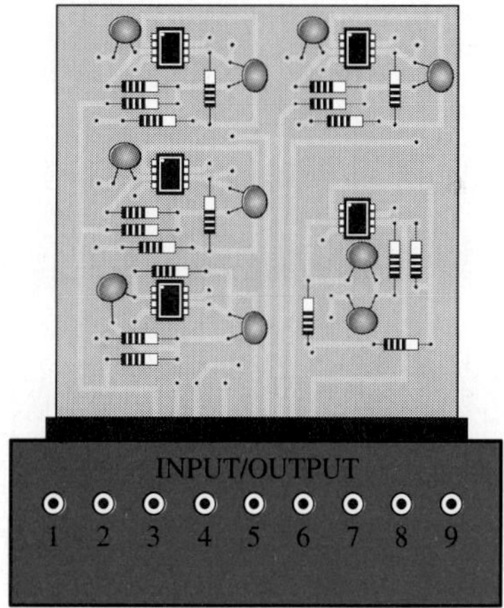

FIGURE 16–32
Filter board in a test fixture.

16–8 REVIEW QUESTIONS

1. What is the purpose of the filter board in this system?
2. What is the bandwidth of the L + R low-pass filter?
3. What is the bandwidth of the L − R low-pass filter?
4. Which filters on the board have approximate Butterworth responses?
5. What is the purpose of the stereo matrix circuitry?

SUMMARY

- The bandwidth in a low-pass filter equals the critical frequency because the response extends to 0 Hz.
- The bandwidth in a high-pass filter extends above the critical frequency and is limited only by the inherent frequency limitation of the active circuit.
- A band-pass filter passes all frequencies within a band between a lower and an upper critical frequency and rejects all others outside this band.
- The bandwidth of a band-pass filter is the difference between the upper critical frequency and the lower critical frequency.
- A band-stop filter rejects all frequencies within a specified band and passes all those outside this band.
- Filters with the Butterworth response characteristic have a very flat response in the pass band, exhibit a roll-off of 20 dB/decade/pole, and are used when all the frequencies in the pass band must have the same gain.
- Filters with the Chebyshev characteristic have ripples or overshoot in the pass band and exhibit a faster roll-off per pole than filters with the Butterworth characteristic.
- Filters with the Bessel characteristic are used for filtering pulse waveforms. Their linear phase characteristic results in minimal waveshape distortion. The roll-off rate per pole is slower than for the Butterworth.
- In filter terminology, a single RC network is called a *pole*.
- Each pole in a Butterworth filter causes the output to roll off at a rate of 20 dB/decade.
- The quality factor Q of a band-pass filter determines the filter's selectivity. The higher the Q, the narrower the bandwidth and the better the selectivity.
- The damping factor determines the filter response characteristic (Butterworth, Chebyshev, or Bessel).

GLOSSARY

Active filter A frequency-selective circuit consisting of active devices such as transistors or op-amps coupled with reactive components.

Band-pass filter A type of filter that passes a range of frequencies lying between a certain lower frequency and a certain higher frequency.

Band-stop filter A type of filter that blocks or rejects a range of frequencies lying between a certain lower frequency and a certain higher frequency.

Bessel A type of filter response having a linear phase characteristic and less than 20 dB/decade/pole roll-off.

Butterworth A type of filter response characterized by flatness in the pass band and a 20 dB/decade/pole roll-off.

Carrier The high frequency (RF) signal that carries modulated information in AM, FM, or other systems.

Chebyshev A type of filter response characterized by ripples in the pass band and a greater than 20 dB/decade/pole roll-off.

Damping factor A filter characteristic that determines the type of response.

Frequency modulation (FM) A communication method in which a lower frequency intelligence-carrying signal modulates (varies) the frequency of a higher frequency signal.

High-pass filter A type of filter that passes frequencies above a certain frequency while rejecting lower frequencies.

Low-pass filter A type of filter that passes frequencies below a certain frequency while rejecting higher frequencies.

Pole A network containing one resistor and one capacitor that contributes 20 dB/decade to a filter's roll-off rate.

Quality factor (Q) The ratio of a band-pass filter's center frequency to its bandwidth.

FORMULAS

(16–1)	$f_c = \dfrac{1}{2\pi RC}$	Filter critical frequency
(16–2)	$BW = f_{c2} - f_{c1}$	Filter bandwidth
(16–3)	$f_0 = \sqrt{f_{c1}f_{c2}}$	Center frequency of a band-pass filter
(16–4)	$Q = \dfrac{f_0}{BW}$	Quality factor of a band-pass filter
(16–5)	$Q = \dfrac{1}{DF}$	Q in terms of damping factor
(16–6)	$DF = 2 - \dfrac{R_1}{R_2}$	Damping factor
(16–7)	$f_c = \dfrac{1}{2\pi\sqrt{R_A R_B C_A C_B}}$	Critical frequency for a second-order Sallen-Key filter
(16–8)	$f_0 = \dfrac{1}{2\pi C}\sqrt{\dfrac{R_1 + R_3}{R_1 R_2 R_3}}$	Center frequency of a multiple-feedback filter
(16–9)	$A_0 = \dfrac{R_2}{2R_1}$	Gain of a multiple-feedback filter
(16–10)	$Q = \dfrac{1}{3}\left(\dfrac{R_A}{R_B} + 1\right)$	Q of a state-variable filter

SELF-TEST

1. The term *pole* in filter terminology refers to
   (a) a high-gain op-amp **(b)** one complete active filter
   (c) a single *RC* network **(d)** the feedback circuit
2. An *RC* circuit produces a roll-off rate of
   (a) −20 dB/decade **(b)** −40 dB/decade
   (c) −6 dB/octave **(d)** a and c
3. A band-pass response has
   (a) two critical frequencies **(b)** one critical frequency
   (c) a flat curve in the pass band **(d)** a wide bandwidth
4. The lowest frequency passed by a low-pass filter is
   (a) 1 Hz **(b)** 0 Hz
   (c) 10 Hz **(d)** dependent on the critical frequency
5. The Q of a band-pass filter depends on
   (a) the critical frequencies
   (b) only the bandwidth
   (c) the center frequency and the bandwidth
   (d) only the center frequency
6. The damping factor of an active filter determines
   (a) the voltage gain **(b)** the critical frequency
   (c) the response characteristic **(d)** the roll-off rate
7. A maximally flat frequency response is known as
   (a) Chebyshev **(b)** Butterworth
   (c) Bessel **(d)** Colpitts
8. The damping factor of a filter is set by
   (a) the negative feedback circuit **(b)** the positive feedback circuit
   (c) the frequency-selective circuit **(d)** the gain of the op-amp
9. The number of poles in a filter affect the
   (a) voltage gain **(b)** bandwidth
   (c) center frequency **(d)** roll-off rate
10. Sallen-Key filters are
    (a) single-pole filters **(b)** second-order filters
    (c) Butterworth filters **(d)** band-pass filters
11. When filters are cascaded, the roll-off rate
    (a) increases **(b)** decreases
    (c) does not change

12. When a low-pass and a high-pass filter are cascaded to get a band-pass filter, the critical frequency of the low-pass filter must be
    (a) equal to the critical frequency of the high-pass filter
    (b) less than the critical frequency of the high-pass filter
    (c) greater than the critical frequency of the high-pass filter
13. A state-variable filter consists of
    (a) one op-amp with multiple-feedback paths
    (b) a summing amplifier and two integrators
    (c) a summing amplifier and two differentiators
    (d) three Butterworth stages
14. When the gain of a filter is minimum at its center frequency, it is a
    (a) band-pass filter (b) a band-stop filter
    (c) a notch filter (d) b and c

PROBLEMS

SECTION 16–1 BASIC FILTER RESPONSES

1. Identify each type of filter response (low-pass, high-pass, band-pass, or band-stop) in Figure 16–33.

FIGURE 16–33

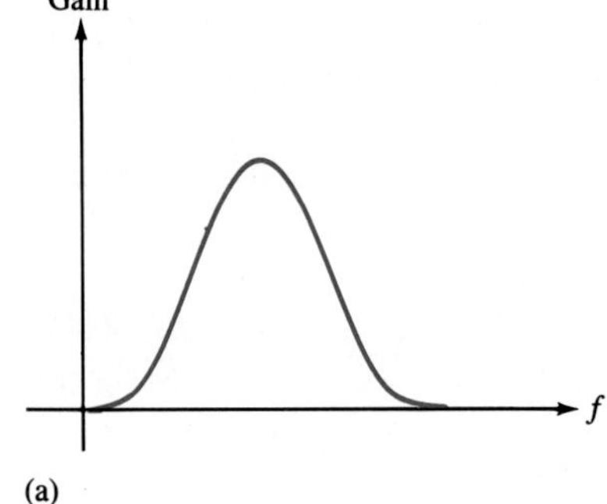

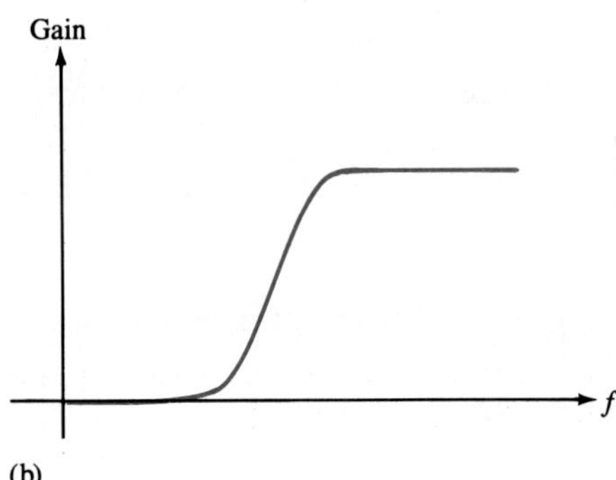

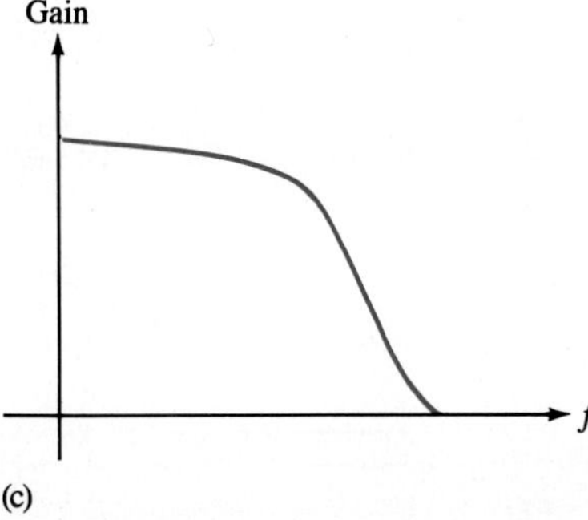

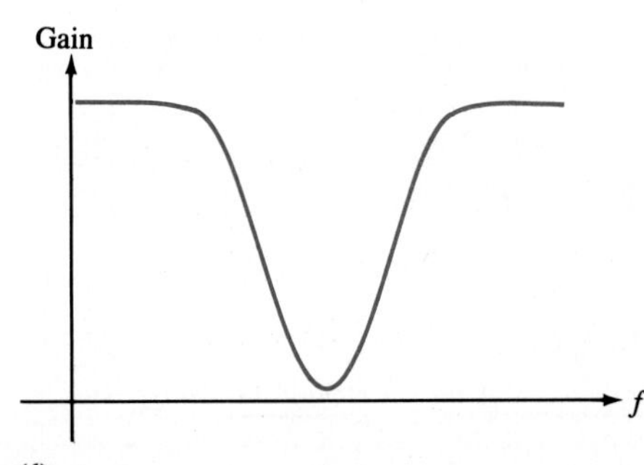

2. A certain low-pass filter has a critical frequency of 800 Hz. What is its bandwidth?
3. A single-pole high-pass filter has a frequency-selective circuit with $R = 2.2\ k\Omega$ and $C = 0.0015\ \mu F$. What is the critical frequency? Can you determine the bandwidth from the available information?
4. What is the roll-off rate of the filter described in Problem 3?
5. What is the bandwidth of a band-pass filter whose critical frequencies are 3.2 kHz and 3.9 kHz? What is the Q of this filter?
6. What is the center frequency of a filter with a Q of 15 and a bandwidth of 1 kHz?

SECTION 16–2 FILTER RESPONSE CHARACTERISTICS

7. What is the damping factor in each active filter shown in Figure 16–34? Which filters are approximately optimized for a Butterworth response characteristic?

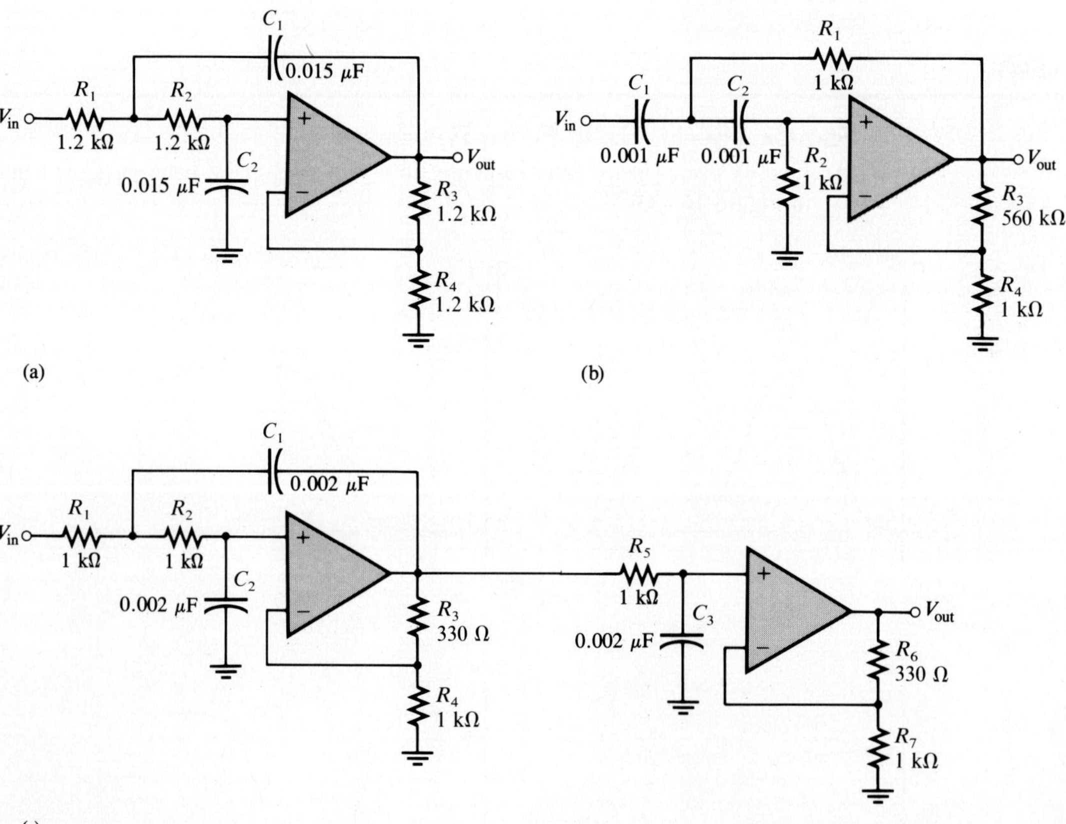

FIGURE 16–34

8. For the filters in Figure 16–34 that do not have a Butterworth response, specify the changes necessary to convert them to Butterworth responses. (Use nearest standard values.)

9. Response curves for high-pass second-order filters are shown in Figure 16–35. Identify each as Butterworth, Chebyshev, or Bessel.

FIGURE 16–35

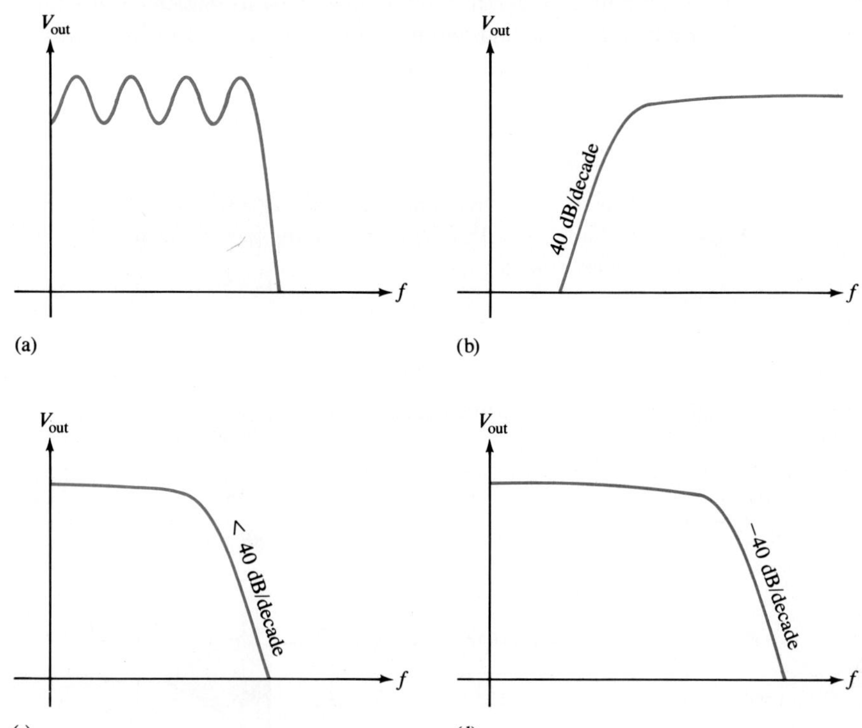

SECTION 16–3 ACTIVE LOW-PASS FILTERS

10. Is the four-pole filter in Figure 16–36 a low-pass or a high-pass type? Is it approximately optimized for a Butterworth response? What is the roll-off rate?

FIGURE 16–36

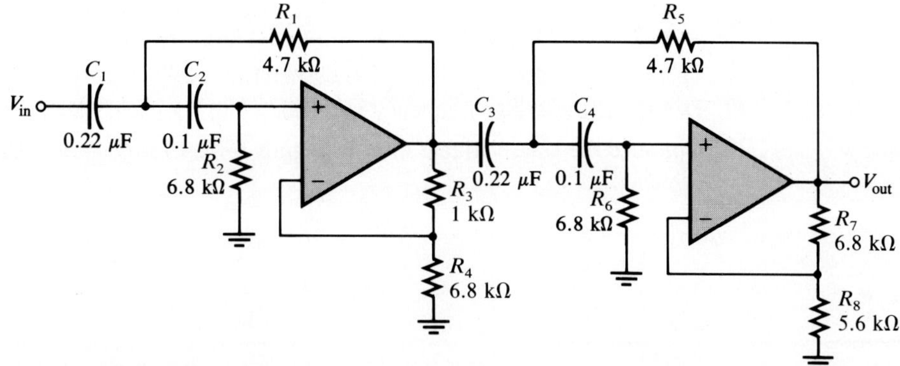

11. Determine the critical frequency in Figure 16–36.
12. Without changing the response curve, adjust the component values in the filter of Figure 16–36 to make it an equal-value filter.
13. Modify the filter in Figure 16–36 to increase the roll-off rate to −120 dB/decade while maintaining an approximate Butterworth response.
14. Using a block diagram format, show how to implement the following roll-off rates using single-pole and two-pole low-pass filters with Butterworth responses.
    (a) −40 dB/decade **(b)** −20 dB/decade
    (c) −60 dB/decade **(d)** −100 dB/decade
    (e) −120 dB/decade

SECTION 16–4 ACTIVE HIGH-PASS FILTERS

15. Convert the filter in Problem 12 to a low-pass with the same critical frequency and response characteristic.
16. Make the necessary circuit modification to reduce by half the critical frequency in Problem 15.
17. For the filter in Figure 16–37, **(a)** How would you increase the critical frequency? **(b)** How would you increase the gain?

FIGURE 16–37

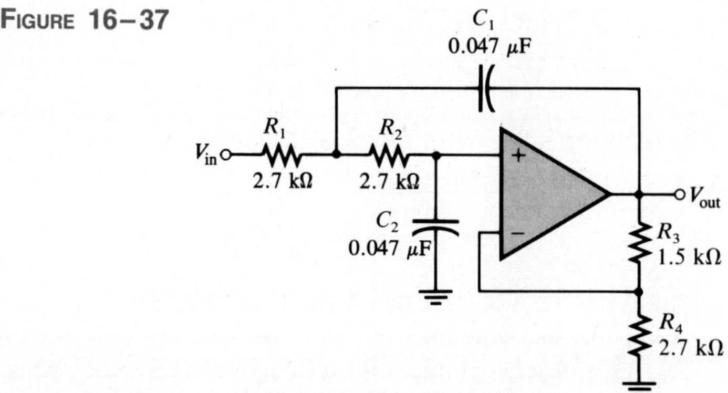

SECTION 16–5 ACTIVE BAND-PASS FILTERS

18. Identify each band-pass filter configuration in Figure 16–38.
19. Determine the center frequency and bandwidth for each filter in Figure 16–38.
20. Optimize the state-variable filter in Figure 16–39 for $Q = 50$. What bandwidth is achieved?

FIGURE 16–38

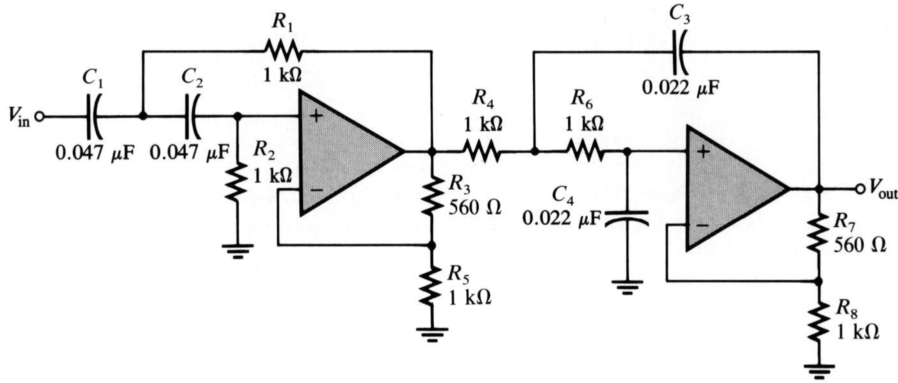

(a)

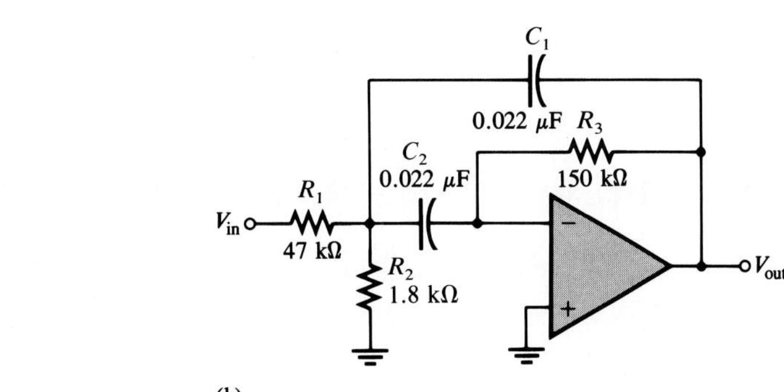

(b)

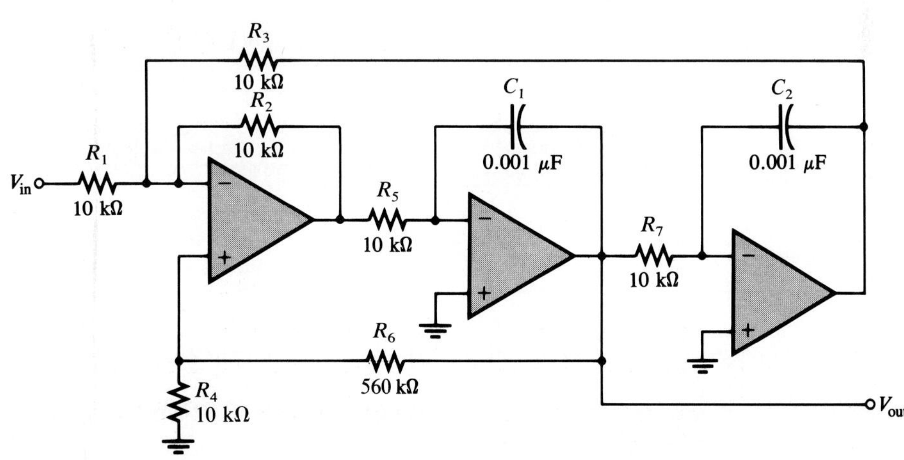

(c)

FIGURE 16–39

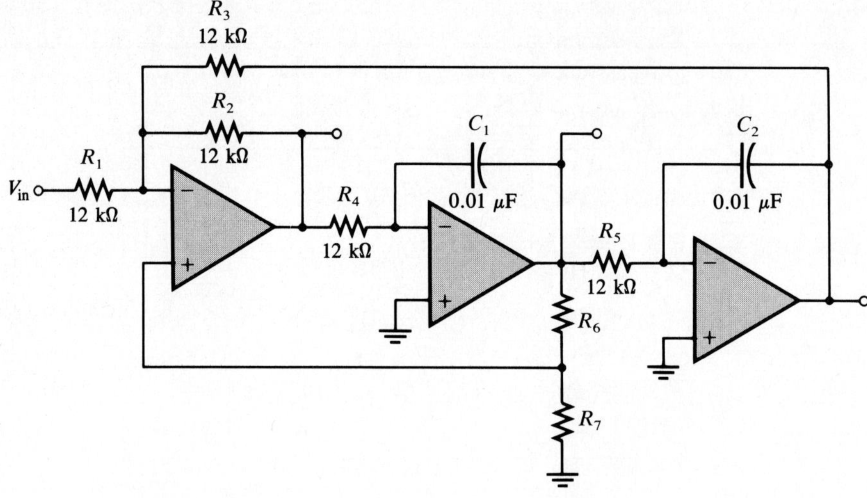

SECTION 16–6 ACTIVE BAND-STOP FILTERS

21. Show how to make a notch (band-stop) filter using the basic circuit in Figure 16–39.

22. Modify the band-stop filter in Problem 21 for a center frequency of 120 Hz.

ANSWERS TO REVIEW QUESTIONS

SECTION 16–1

1. The critical frequency determines the bandwidth.

2. The inherent frequency limitation of the op-amp limits the bandwidth.

3. Q and BW are inversely related. The higher the Q, the better the selectivity, and vice versa.

SECTION 16–2

1. Butterworth is very flat in the pass band and has a 20 dB/decade/pole roll-off.
Chebyshev has ripples in the pass band and has greater than 20 dB/decade/pole roll-off.
Bessel has a linear phase characteristic and less than 20 dB/decade/pole roll-off.

2. The damping factor determines the response characteristic.

3. Frequency-selection network, gain element, and negative feedback circuit are the parts of an active filter.

SECTION 16–3

1. A second-order filter has two poles. Two resistors and two capacitors make up the frequency-selective circuit.

2. The damping factor sets the response characteristic.

3. Cascading increases the roll-off rate.

Section 16–4

1. The positions of the Rs and Cs in the frequency-selection circuit are opposite for low-pass and high-pass configurations.
2. Decrease the R values to increase f_c.
3. 140 dB/decade

Section 16–5

1. Q determines selectivity.
2. $Q = 25$. Higher Q gives narrower BW.
3. A summing amplifier and two integrators make up a state-variable filter.

Section 16–6

1. A band-stop rejects frequencies within the stop band. A band-pass passes frequencies within the pass band.
2. The low-pass and high-pass outputs are summed.

Section 16–7

1. To check the frequency response of a filter
2. Discrete point measurement—tedious and less complete; simpler equipment. Swept frequency measurement—uses more expensive equipment; more efficient, can be more accurate and complete.

Section 16–8

1. The filter board takes the detected FM signal and separates the L + R and L − R audio signals.
2. $BW = 15.9$ kHz
3. $BW = 15.9$ kHz
4. The L + R low-pass, the L − R low-pass, and the L − R band-pass
5. The stereo matrix combines the L + R and L − R signals and produces the separate left and right channel audio signals.

Answers to Practice Exercises

16–1 500 Hz

16–2 1.44

16–3 7.234 kHz, 1.29 kΩ

16–4 $C_{A1} = C_{A2} = C_{B1} = C_{B2} = 0.234\ \mu F$, $R_2 = R_4 = 680\ \Omega$, $R_1 = 103\ \Omega$, $R_3 = 840\ \Omega$

16–5 $R_A = R_B = R_2 = 10\ k\Omega$, $C_A = C_B = 0.053\ \mu F$, $R_1 = 586\ \Omega$

16–6 Gain increases to 2.43, frequency decreases to 544 Hz, and bandwidth decreases to 96.5 Hz.

16–7 $f_0 = 21.922$ kHz, $Q = 101$, $BW = 217$ Hz

16–8 Decrease the input resistors or the feedback capacitors of the two integrator stages by half.

17

Voltage Regulators

After completing this chapter, you should be able to

- ☐ Explain the basic concept of voltage regulation.
- ☐ Define line and load regulation and discuss the difference.
- ☐ Describe how a basic series voltage regulator works.
- ☐ Explain the need for overload protection.
- ☐ Describe how a basic shunt voltage regulator works.
- ☐ Relate the advantages of switching regulators and explain how switching regulators work.
- ☐ Discuss current limiting in regulators.
- ☐ Explain what fold-back current limiting is.
- ☐ Select specific positive or negative three-terminal IC regulators for a specific application.
- ☐ Use an external pass transistor with a three-terminal IC regulator to increase the current capability.
- ☐ Use a current-limiting circuit with a three-terminal regulator.
- ☐ Use a three-terminal regulator as a constant-current source.
- ☐ Configure an IC switching regulator for step-down or step-up operation.

A voltage regulator provides a constant dc output voltage that is practically independent of the input voltage, output load current, and temperature. The voltage regulator is one part of a power supply. Its input voltage comes from the filtered output of a rectifier derived from an ac voltage or from a battery in the case of portable systems. Power supplies were introduced in Chapter 2.

Most voltage regulators fall into two broad categories—linear regulators and switching regulators. In the linear regulator category, two general types are the linear series regulator and the linear shunt regulator. These are normally available for either positive or negative output voltages. A dual regulator provides both positive and negative outputs. In the switching regulator category, three general configurations are step-down, step-up, and inverting.

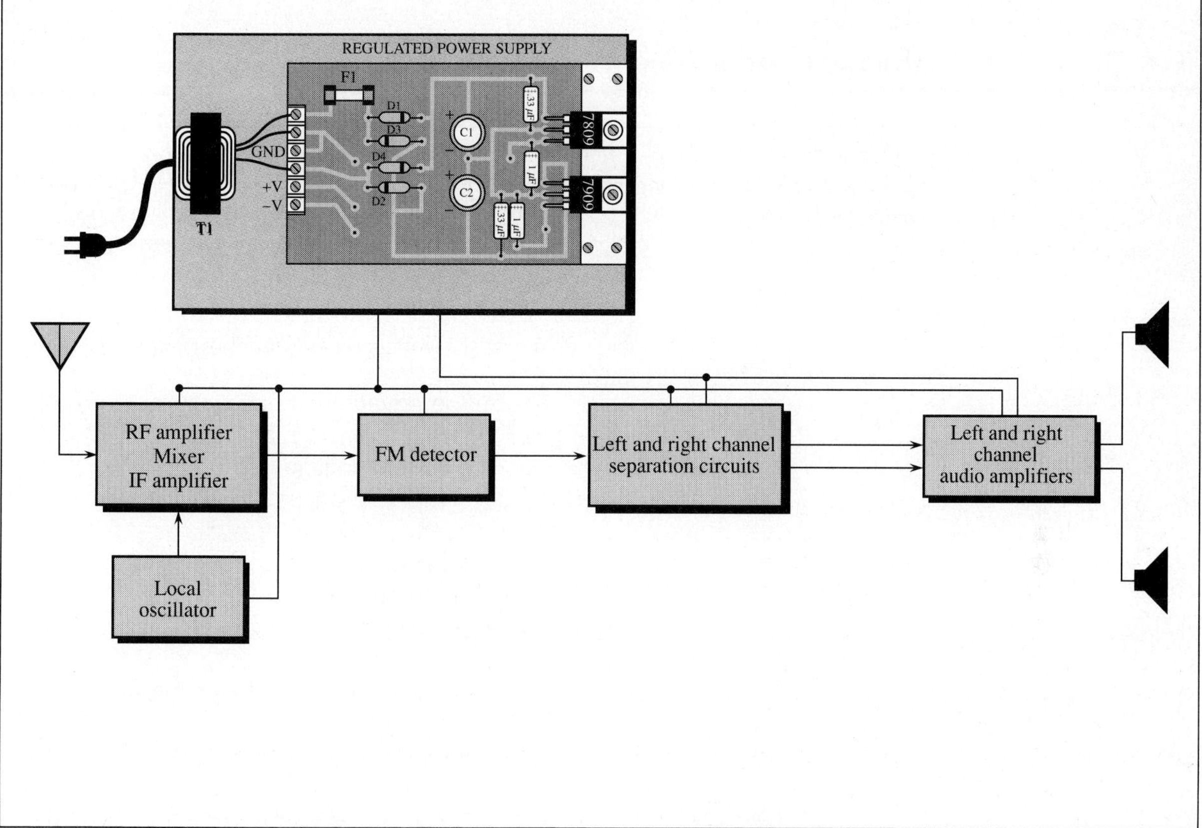

Many types of integrated circuit (IC) regulators are available. The most popular types of linear regulator are the three-terminal fixed voltage regulator and the three-terminal adjustable voltage regulator. Switching regulators are also widely used. In this chapter, specific IC devices are introduced as representative of the wide range of available devices.

A System Application

A dual polarity regulated power supply is used for the FM stereo system that you worked with in Chapter 16. Two regulators, one positive and the other negative, provide the positive voltage required for the receiver circuits and the dual polarity voltages for the op-amp circuits. The regulator input voltages come from a full-wave rectifier with filtered outputs.

For the system application in Section 17–7, in addition to the other topics, be sure you understand

- ☐ How three-terminal fixed-voltage regulators are used.
- ☐ The basic operation of a power supply rectifier and filter (review Chapter 2).
- ☐ How to set the current limit of a regulator.
- ☐ How to determine power dissipation in a pass transistor.

17–1 VOLTAGE REGULATION

Two basic categories of voltage regulation are line regulation and load regulation. Line regulation maintains a nearly constant output voltage when the input voltage varies. Load regulation maintains a nearly constant output voltage when the load varies.

LINE REGULATION

When the dc input (line) voltage changes, the voltage **regulator** must maintain a nearly constant output voltage, as illustrated in Figure 17–1.

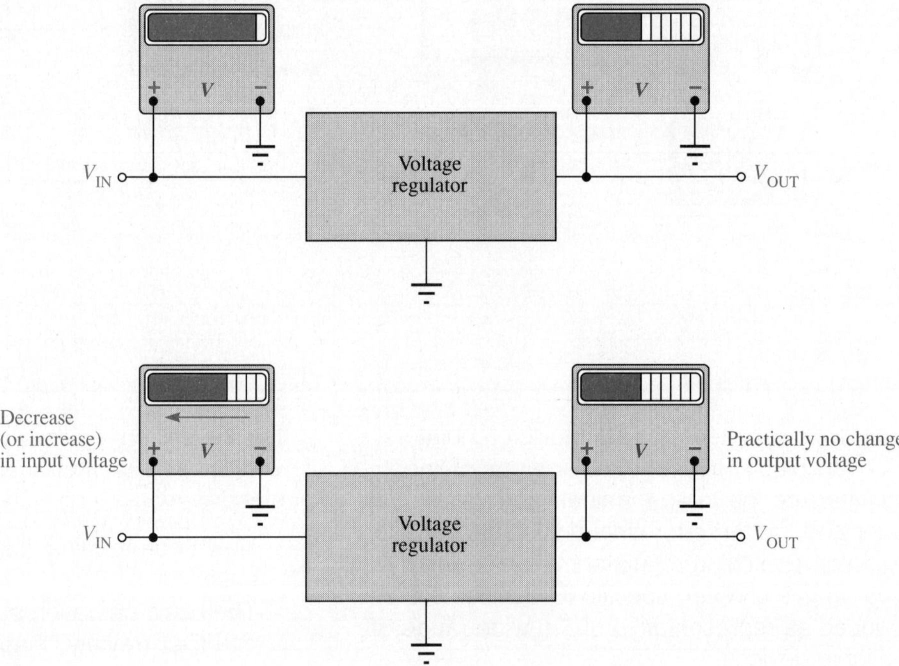

FIGURE 17–1
Line regulation. A change in input (line) voltage does not significantly affect the output voltage of a regulator (within certain limits).

Line regulation can be defined as the percentage change in the output voltage for a given change in the input (line) voltage. It is usually expressed in units of %/V. For example, a line regulation of 0.05%/V means that the output voltage changes 0.05 percent when the input voltage increases or decreases by one volt. Line regulation can be calculated using the following formula (Δ means "a change in"):

$$\text{Line regulation} = \frac{(\Delta V_{OUT}/V_{OUT})100\%}{\Delta V_{IN}} \tag{17–1}$$

EXAMPLE 17–1

When the input to a particular voltage regulator decreases by 5 V, the output decreases by 0.25 V. The nominal output is 15 V. Determine the line regulation in %/V.

SOLUTION

The line regulation is

$$\frac{(\Delta V_{OUT}/V_{OUT})100\%}{\Delta V_{IN}} = \frac{(0.25\ \text{V}/15\ \text{V})100\%}{5\ \text{V}} = 0.333\%\text{V}$$

PRACTICE EXERCISE 17–1

The input of a certain regulator increases by 3.5 V. As a result, the output voltage increases by 0.42 V. The nominal output is 20 V. Determine the regulation in %/V.

LOAD REGULATION

When the amount of current through a load changes due to a varying load resistance, the voltage regulator must maintain a nearly constant output voltage across the load, as illustrated in Figure 17–2.

Load regulation can be defined as the percentage change in output voltage for a given change in load current. It can be expressed as a percentage change in output voltage from no-load (NL) to full-load (FL) as follows.

$$\text{Load regulation} = \frac{(V_{NL} - V_{FL})100\%}{V_{FL}} \tag{17-2}$$

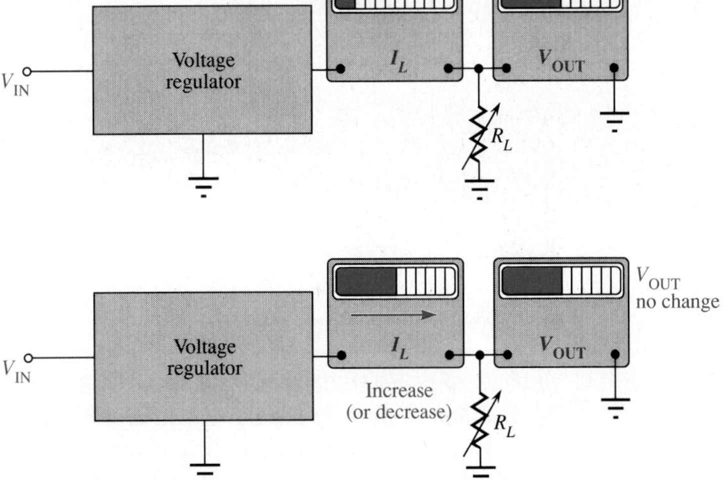

FIGURE 17–2 Load regulation. A change in load current has practically no effect on the output voltage of a regulator (within certain limits).

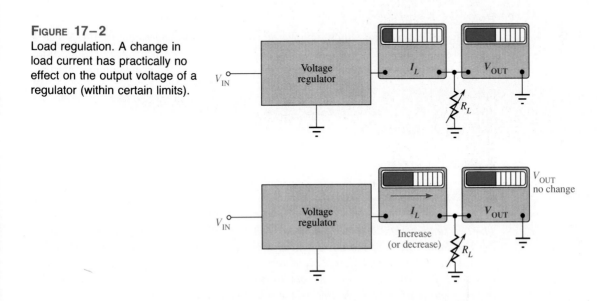

Alternately, the load regulation can be expressed as a percentage change in output voltage for each mA change in load current. For example, a load regulation of 0.01%/mA means that the output voltage changes 0.01 percent when the load current increases or decreases 1 mA.

EXAMPLE 17–2 A certain voltage regulator has a 12 V output when there is no load ($I_L = 0$). When there is a full-load current of 10 mA, the output voltage is 11.95 V. Express the voltage regulation as a percentage change from no-load to full-load and also as a percentage change for each mA change in load current.

SOLUTION

The no-load output voltage is

$$V_{\text{NL}} = 12\text{ V}$$

The full-load output voltage is

$$V_{\text{FL}} = 11.95\text{ V}$$

The load regulation is

$$\left(\frac{V_{\text{NL}} - V_{\text{FL}}}{V_{\text{FL}}}\right)100\% = \left(\frac{12\text{ V} - 11.95\text{ V}}{11.95\text{ V}}\right)100\%$$
$$= 0.418\%$$

The load regulation can also be expressed as

$$\frac{0.418\%}{10\text{ mA}} = 0.0418\%/\text{mA}$$

where the change in load current from no-load to full-load is 10 mA.

PRACTICE EXERCISE 17–2

A regulator has a no-load output voltage of 18 V and a full-load output of 17.85 V at a load current of 50 mA. Determine the voltage regulation as a percentage change from no-load to full-load and also as a percentage change for each mA change in load current. ■

17–1 REVIEW QUESTIONS

1. Define *line regulation.*
2. Define *load regulation.*

17–2 BASIC SERIES REGULATORS

The two fundamental classes of voltage regulators are linear regulators and switching regulators. Both of these are available in integrated circuit form. There are two basic types of linear regulator. One is the series regulator and the other is the shunt regulator. In this section, we will look at the series regulator. The shunt and switching regulators are covered in the next two sections.

A simple representation of a series type of linear regulator is shown in Figure 17–3(a), and the basic components are shown in the block diagram in Figure 17–3(b). Notice that

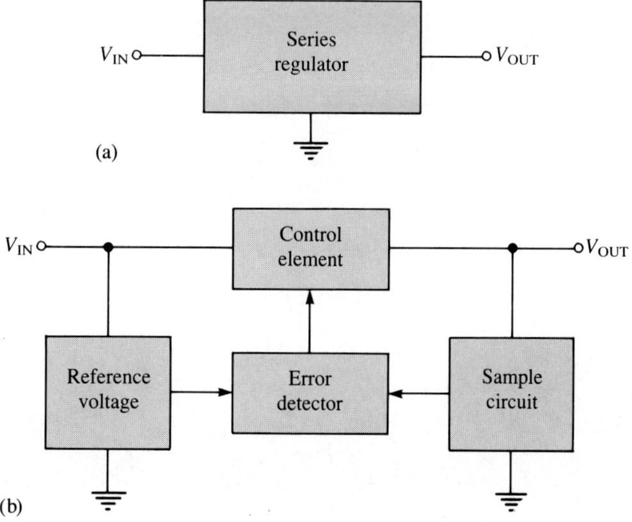

FIGURE 17–3
Simple series voltage regulator block diagram.

the control element is in series with the load between input and output. The output sample circuit senses a change in the output voltage. The error detector compares the sample voltage with a reference voltage and causes the control element to compensate in order to maintain a constant output voltage.

REGULATING ACTION

A basic op-amp series regulator circuit is shown in Figure 17–4. The operation of the series regulator is illustrated in Figure 17–5. The resistive voltage divider formed by R_2 and R_3 senses any change in the output voltage. When the output tries to decrease because of a decrease in V_{IN} or because of an increase in I_L, as indicated in parts (a) and (b), a proportional voltage decrease is applied to the op-amp's inverting input by the voltage divider. Since the zener diode holds the other op-amp input at a nearly fixed reference voltage V_{REF}, a small difference voltage (error voltage) is developed across the op-amp's inputs. This difference voltage is amplified, and the op-amp's output voltage increases.

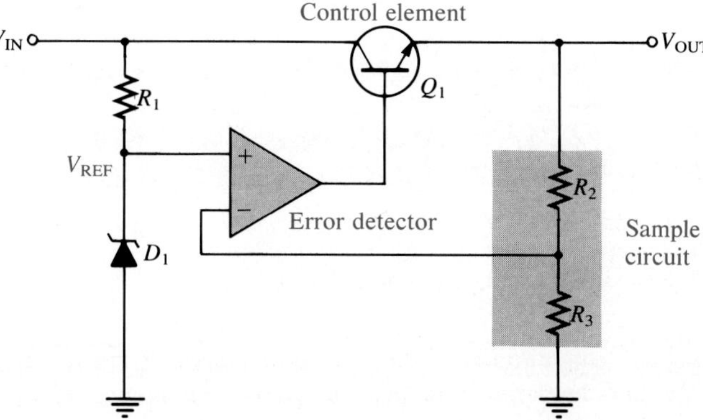

FIGURE 17–4
Basic op-amp series regulator.

This increase is applied to the base of Q_1, causing the emitter voltage V_{OUT} to increase until the voltage to the inverting input again equals the reference (zener) voltage. This action offsets the attempted decrease in output voltage, thus keeping it nearly constant. Q_1 is a power transistor and is often used with a heat sink because it must handle all of the load current.

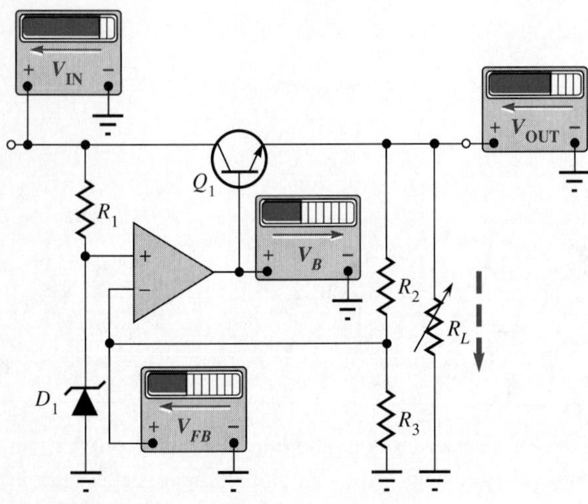

(a) When V_{IN} or R_L decreases, V_{OUT} attempts to decrease, V_{FB} also attempts to decrease, and as a result, V_B attempts to increase, thus compensating for the attempted decrease in V_{OUT}.

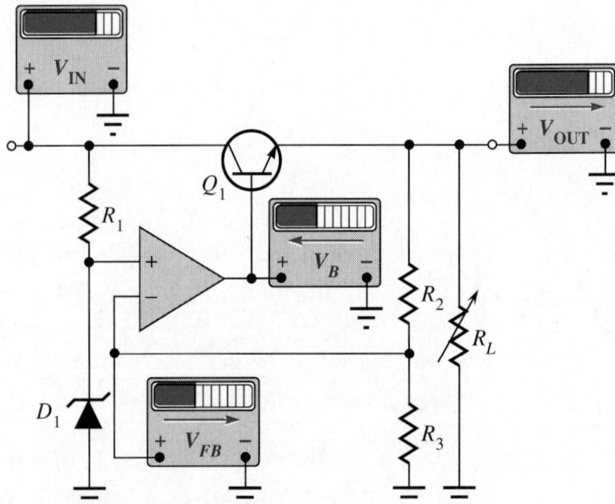

(b) When V_{IN} (or R_L) stabilizes at its new lower value, the voltages are at their original values, thus keeping V_{OUT} constant as a result of the negative feedback.

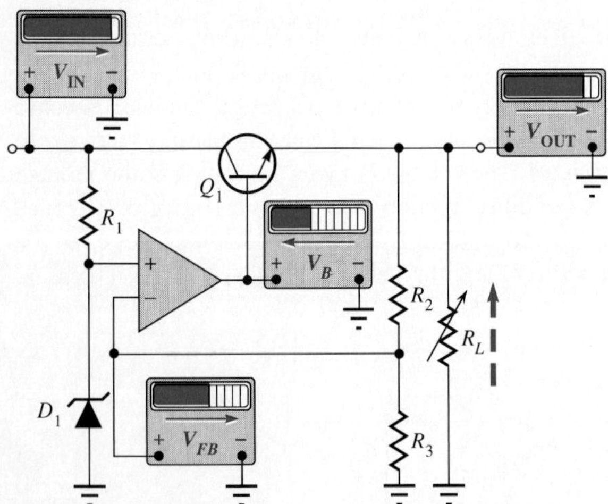

(c) When V_{IN} or R_L increases, V_{OUT} attempts to increase. The feedback voltage, V_{FB}, also attempts to increase, and, as a result, the op-amp's output voltage, V_B, applied to the base of the control transistor, attempts to decrease, thus compensating for the attempted increase in V_{OUT}.

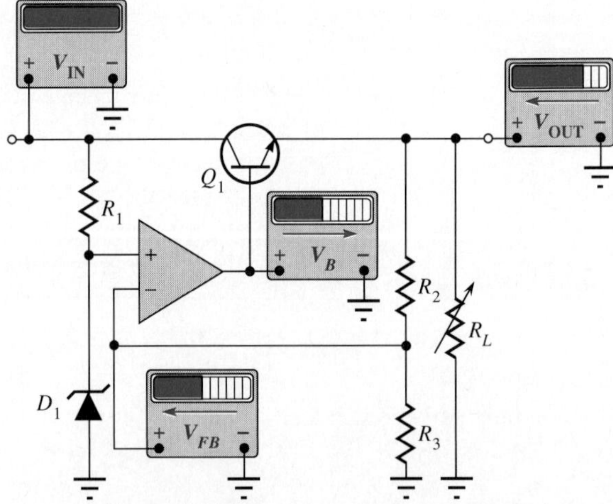

(d) When V_{IN} (or R_L) stabilizes at its new higher value, the voltages are at their original values, thus keeping V_{OUT} constant as result of the negative feedback.

FIGURE 17–5
Illustration of series regulator action that keeps V_{OUT} constant when V_{IN} or R_L changes.

The opposite action occurs when the output tries to increase, as indicated in Figure 17–5(c) and (d). The op-amp in the series regulator is actually connected as a noninverting amplifier where the reference voltage V_{REF} is the input at the noninverting terminal, and the R_2/R_3 voltage divider forms the negative feedback network. The closed-loop voltage gain is

$$A_{cl} = 1 + \frac{R_2}{R_3} \qquad \textbf{(17–3)}$$

Therefore, the regulated output voltage (neglecting the base-emitter voltage of Q_1) is

$$V_{OUT} \cong \left(1 + \frac{R_2}{R_3}\right)V_{REF} \qquad \textbf{(17–4)}$$

From this analysis, you can see that the output voltage is determined by the zener voltage and the resistors R_2 and R_3. It is relatively independent of the input voltage, and therefore, regulation is achieved (as long as the input voltage and load current are within specified limits).

EXAMPLE 17–3 Determine the output voltage for the regulator in Figure 17–6.

FIGURE 17–6

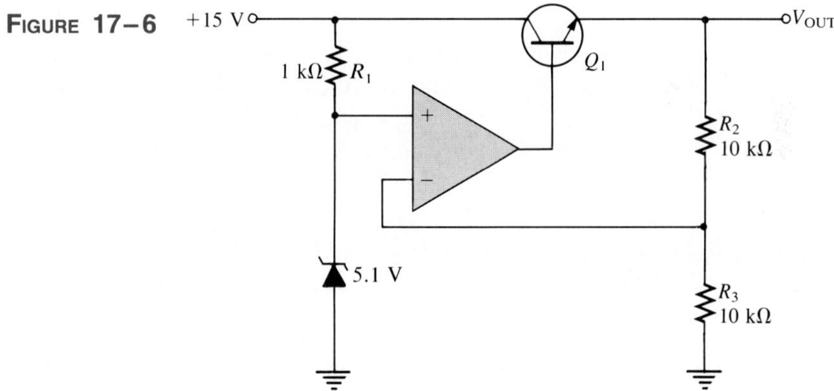

SOLUTION

$$V_{REF} = 5.1 \text{ V}$$

$$V_{OUT} = \left(1 + \frac{R_2}{R_3}\right)V_{REF} = \left(1 + \frac{10 \text{ k}\Omega}{10 \text{ k}\Omega}\right)5.1 \text{ V} = (2)5.1 \text{ V} = 10.2 \text{ V}$$

PRACTICE EXERCISE 17–3

The following changes are made in the circuit in Figure 17–6: $V_Z = 3.3$ V, $R_1 = 1.8$ kΩ, $R_2 = 22$ kΩ, and $R_3 = 18$ kΩ. What is the output voltage?

SHORT-CIRCUIT OR OVERLOAD PROTECTION

If an excessive amount of load current is drawn, the series-pass transistor can be quickly damaged or destroyed. Most regulators employ some type of excess current protection in the form of a current-limiting mechanism. Figure 17–7 shows one method of current limiting to prevent overloads called *constant current limiting*. The current-limiting circuit consists of transistor Q_2 and resistor R_4.

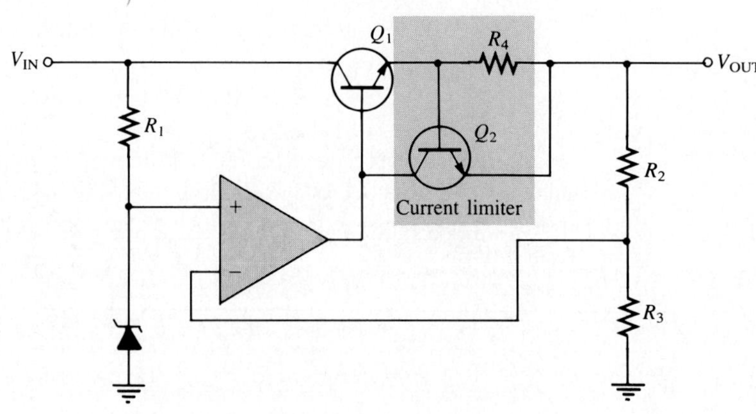

FIGURE 17–7
Series regulator with constant current limiting.

The load current through R_4 creates a voltage from base to emitter of Q_2. When I_L reaches a predetermined maximum value, the voltage drop across R_4 is sufficient to forward-bias the base-emitter junction of Q_2, thus causing it to conduct. Enough Q_1 base current is diverted into the collector of Q_2 so that I_L is limited to its maximum value $I_{L(max)}$. Since the base-to-emitter voltage of Q_2 cannot exceed about 0.7 V for a silicon transistor, the voltage across R_4 is held to this value, and the load current is limited to

$$I_{L(max)} = \frac{0.7\text{ V}}{R_4} \tag{17–5}$$

■ **EXAMPLE 17–4** Determine the maximum current that the regulator in Figure 17–8 can provide to a load.

SOLUTION

$$I_{L(max)} = \frac{0.7\text{ V}}{R_4} = \frac{0.7\text{ V}}{1\ \Omega} = 0.7\text{ A}$$

FIGURE 17–8

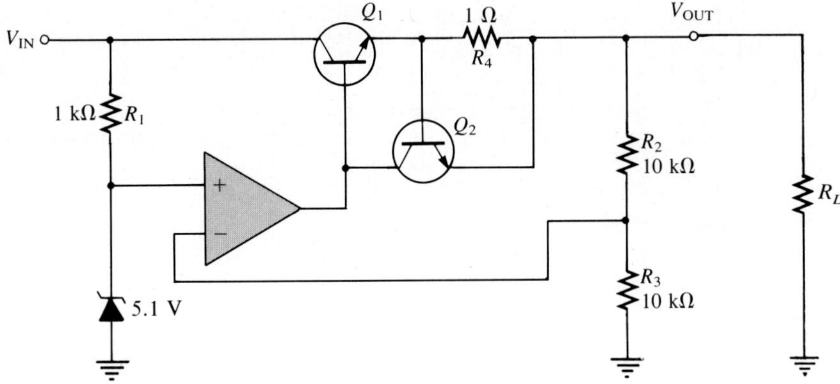

PRACTICE EXERCISE 17–4

If the output of the regulator in Figure 17–8 is shorted, what is the current?

REGULATOR WITH FOLD-BACK CURRENT LIMITING

In the previous current-limiting technique, the current is restricted to a maximum constant value.

Fold-back current limiting is a method used particularly in high-current regulators whereby the output current under overload conditions drops to a value well below the peak load current capability to prevent excessive power dissipation.

BASIC IDEA The basic concept of fold-back current limiting is as follows, with reference to Figure 17–9. The circuit is similar to the constant current-limiting arrangement in Figure 17–7, with the exception of resistors R_5 and R_6. The voltage drop developed across R_4 by the load current must not only overcome the base-emitter voltage required to turn on Q_2, but it must overcome the voltage across R_5. That is, the voltage across R_4 must be

$$V_{R4} = V_{R5} + V_{BE}$$

FIGURE 17–9

Series regulator with fold-back current limiting.

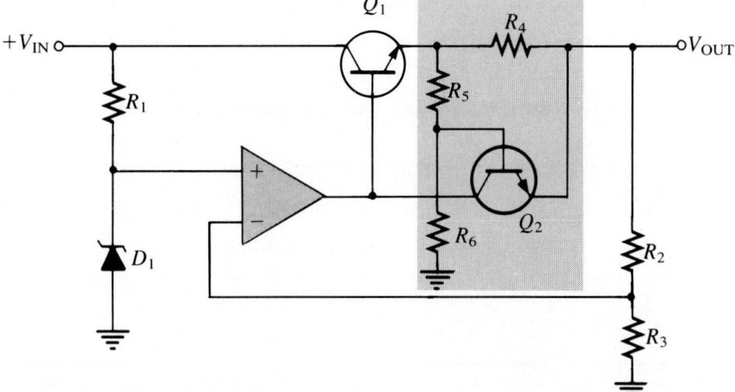

In an overload or short-circuit condition the load current increases to a value $I_{L(max)}$ that is sufficient to cause Q_2 to conduct. At this point the current can increase no further. The decrease in output voltage results in a proportional decrease in the voltage across R_5; thus less current through R_4 is required to maintain the forward-biased condition of Q_1. So, as V_{OUT} decreases, I_L decreases, as shown in the graph of Figure 17–10.

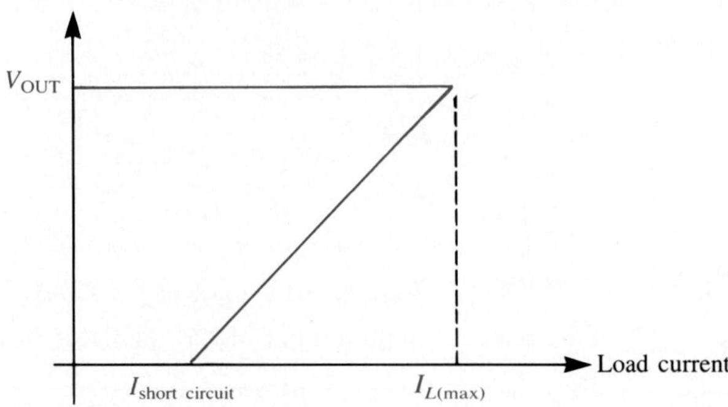

FIGURE 17–10
Fold-back current limiting (output voltage versus load current).

The advantage of this technique is that the regulator is allowed to operate with peak load current up to $I_{L(max)}$; but when the output becomes shorted, the current drops to a lower value to prevent overheating of the device.

17–2 REVIEW QUESTIONS

1. What are the basic components in a series regulator?
2. A certain series regulator has an output voltage of 8 V. If the op-amp's closed loop gain is 4, what is the value of the reference voltage?

17–3 BASIC SHUNT REGULATORS

The second basic type of linear voltage regulator is the shunt regulator. As you have learned, the control element in the series regulator is the series-pass transistor. In the shunt regulator, the control element is a transistor in parallel (shunt) with the load.

A simple representation of a shunt type of linear regulator is shown in Figure 17–11(a), and the basic components are shown in the block diagram in part (b) of the figure.

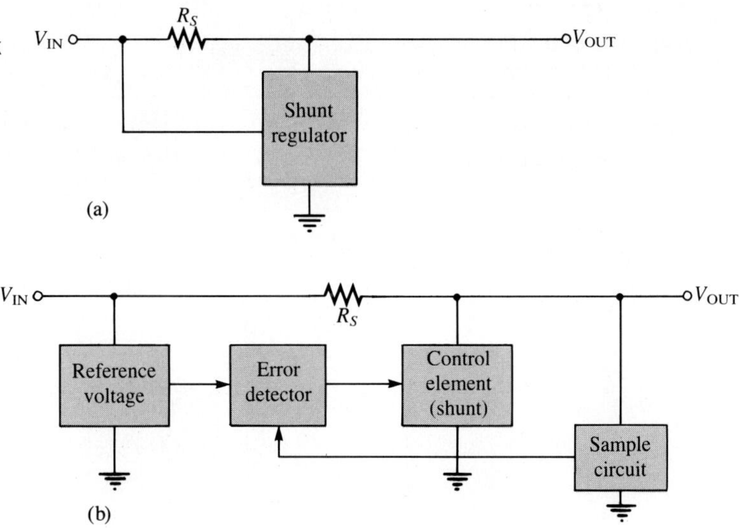

FIGURE 17–11 Simple shunt regulator block diagrams.

In the basic shunt regulator, the control element is a transistor Q_1 in parallel with the load, as shown in Figure 17–12. A resistor, R_1 is in series with the load. The operation of the circuit is similar to that of the series regulator, except that regulation is achieved by controlling the current through the parallel transistor Q_1.

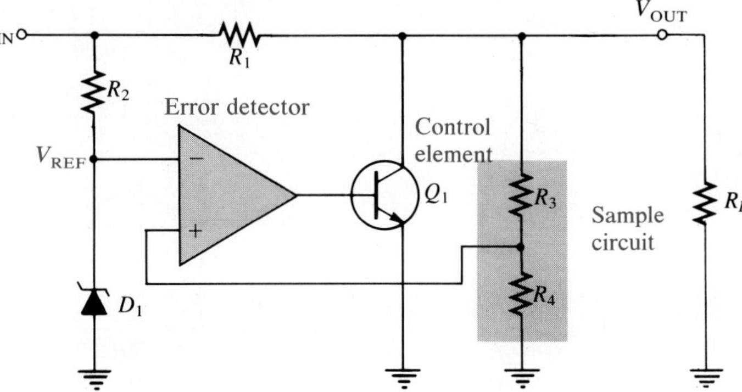

FIGURE 17–12 Basic op-amp shunt regulator.

When the output voltage tries to decrease due to a change in input voltage or load current, as shown in Figure 17–13(a), the attempted decrease is sensed by R_3 and R_4 and applied to the op-amp's noninverting input. The resulting difference voltage reduces the op-amp's output, driving Q_1 less, thus reducing its collector current (shunt current) and increasing its effective collector-to-emitter resistance r_{ce}. Since r_{ce} acts as a voltage divider with R_1, this action offsets the attempted decrease in V_{OUT} and maintains it at an almost constant level. The opposite action occurs when the output tries to increase, as indicated in Figure 17–13(b).

FIGURE 17–13
Sequence of responses when V_{OUT} tries to decrease as a result of a decrease in R_L or V_{IN} (opposite responses for an attempted increase).

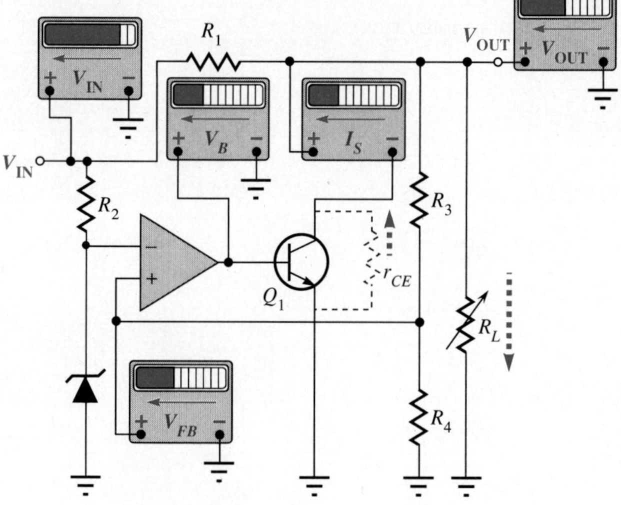

(a) Initial response to a decrease in V_{IN} or R_L

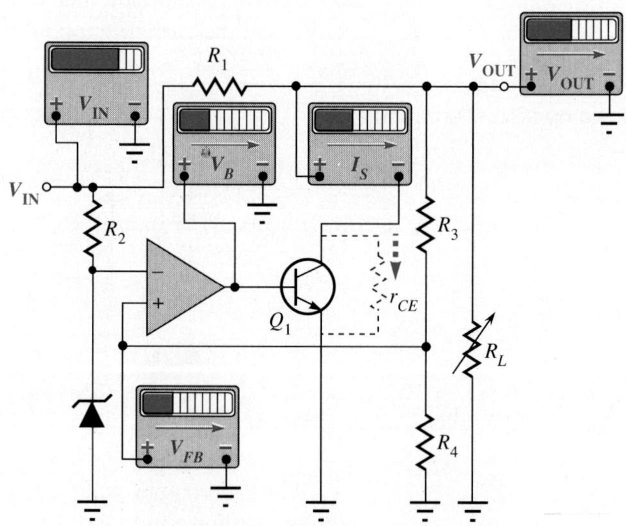

(b) V_{OUT} held constant by the feedback action.

With I_L and V_{OUT} constant, a change in the input voltage produces a change in shunt current (I_S) as follows (Δ means ''a change in'').

$$\Delta I_S = \frac{\Delta V_{IN}}{R_1} \tag{17–6}$$

With a constant V_{IN} and V_{OUT}, a change in load current causes an opposite change in shunt current.

$$\Delta I_S = -\Delta I_L \quad (17\text{–}7)$$

This formula says that if I_L increases, I_S decreases, and vice versa. The shunt regulator is less efficient than the series type but offers inherent short-circuit protection. If the output is shorted ($V_{OUT} = 0$), the load current is limited by the series resistor R_1 to a maximum value as follows ($I_S = 0$).

$$I_{L(max)} = \frac{V_{IN}}{R_1} \quad (17\text{–}8)$$

EXAMPLE 17–5 In Figure 17–14, what power rating must R_1 have if the maximum input voltage is 12.5 V?

FIGURE 17–14

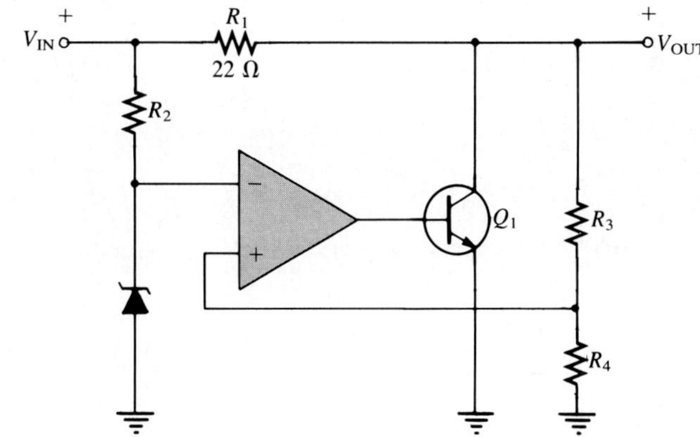

SOLUTION

The worst-case power dissipation in R_1 occurs when the output is short-circuited. $V_{OUT} = 0$, and when $V_{IN} = 12.5$ V, the voltage dropped across R_1 is $V_{IN} - V_{OUT} = 12.5$ V. The power dissipation in R_1 is

$$P_{R1} = \frac{V_{R1}^2}{R_1} = \frac{(12.5 \text{ V})^2}{22 \ \Omega} = 7.1 \text{ W}$$

Therefore, a resistor of at least 10 W should be used.

PRACTICE EXERCISE 17–5

In Figure 17–14, R_1 is changed to 33 Ω. What must be the power rating of R_1 if the maximum input voltage is 24 V?

17–3 REVIEW QUESTIONS

1. How does the control element in a shunt regulator differ from that in a series regulator?
2. What is one advantage of a shunt regulator over a series type? What is a disadvantage?

17–4 BASIC SWITCHING REGULATORS

The two types of linear regulators—series and shunt—have control elements (transistors) that are conducting all the time, with the amount of conduction varied as demanded by changes in the output voltage or current. The switching regulator is different; the control element operates as a switch. A greater efficiency can be realized with this type of voltage regulator than with the linear types because the transistor is not always conducting. Therefore, switching regulators can provide greater load currents at low voltage than linear regulators because the control transistor doesn't dissipate as much power. Three basic configurations of switching regulators are step-down, step-up, and inverting.

STEP-DOWN CONFIGURATION

In the step-down configuration, the output voltage is always less than the input voltage. A basic step-down switching regulator is shown in Figure 17–15(a), and its simplified equivalent is shown in Figure 17–15(b). Transistor Q_1 is used to switch the input voltage at a duty cycle that is based on the regulator's load requirement. The *LC* filter is then used

FIGURE 17–15
Basic step-down switching regulator.

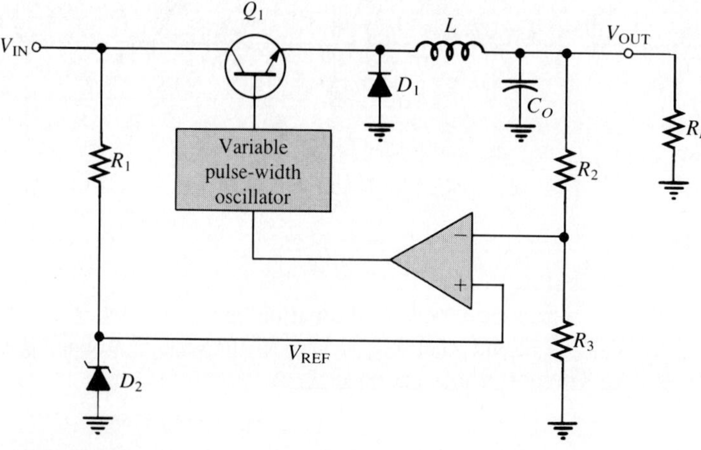

(a) Typical circuit

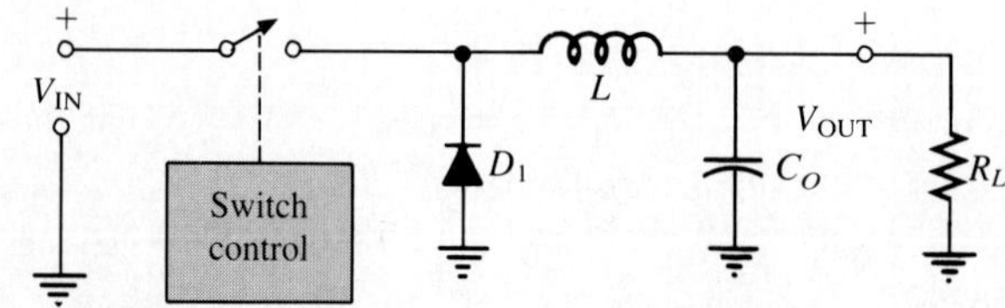

(b) Simplified equivalent circuit

to average the switched voltage. Since Q_1 is either *on* (saturated) or *off*, the power lost in the control element is relatively small. Therefore, the switching regulator is useful primarily in higher power applications or in applications where efficiency is of utmost concern.

The on and off intervals of Q_1 are shown in the waveform of Figure 17–16(a). The capacitor charges during the on-time (t_{on}) and discharges during the off-time (t_{off}). When the on-time is increased relative to the off-time, the capacitor charges more, thus increasing the output voltage, as indicated in Figure 17–16(b). When the on-time is decreased relative to the off-time, the capacitor discharges more, thus decreasing the output voltage, as in Figure 17–16(c). Therefore, by adjusting the duty cycle $t_{on}/(t_{on} + t_{off})$ of Q_1, the output voltage can be varied. The inductor further smooths the fluctuations of the output voltage caused by the charging and discharging action.

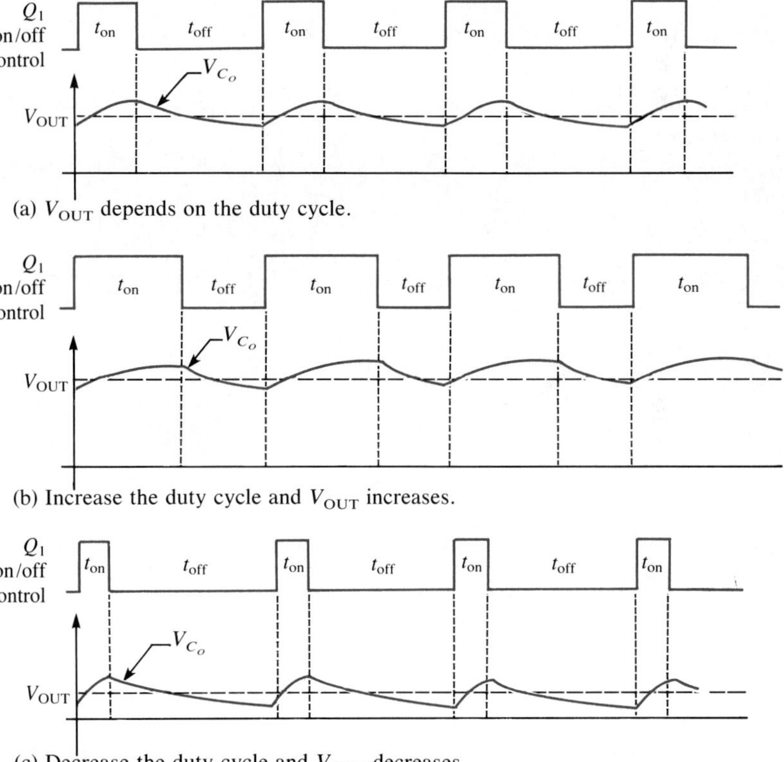

FIGURE 17–16
Switching regulator waveforms. The V_{C_O} waveform is for *no* inductive filtering to illustrate the charge and discharge action. *L* and *C* smooth V_{C_O} to a nearly constant level, as indicated by the dashed line for V_{OUT}.

The regulating action is as follows and is illustrated in Figure 17–17. When V_{OUT} tries to decrease, the on-time of Q_1 is increased, causing an additional charge on C_O to offset the attempted decrease. When V_{OUT} tries to increase, the on-time of Q_1 is decreased, causing C_O to discharge enough to offset the attempted increase.

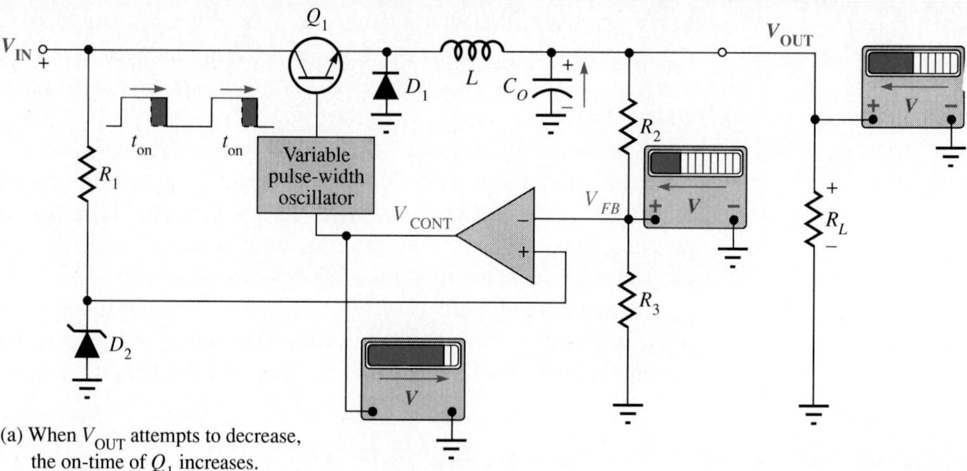

(a) When V_{OUT} attempts to decrease, the on-time of Q_1 increases.

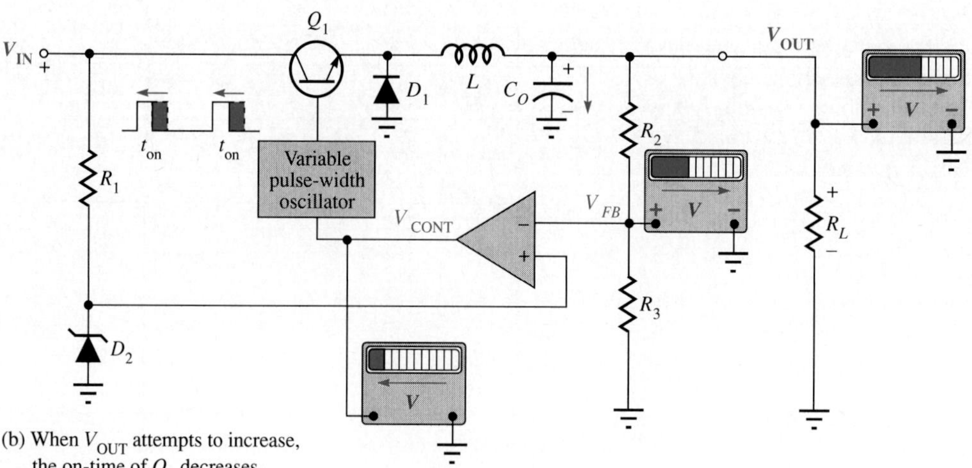

(b) When V_{OUT} attempts to increase, the on-time of Q_1 decreases.

FIGURE 17–17
Regulating action of the basic step-down switching regulator.

The output voltage is expressed as

$$V_{OUT} = \left(\frac{t_{on}}{T}\right)V_{IN} \quad \textbf{(17–9)}$$

T is the period of the on-off cycle of Q_1 and is related to the frequency by $T = 1/f$. The period is the sum of the on-time and the off-time.

$$T = t_{on} + t_{off} \quad \textbf{(17–10)}$$

The ratio t_{on}/T is called the *duty cycle*.

STEP-UP CONFIGURATION

A basic step-up type of switching regulator is shown in Figure 17–18. When Q_1 turns on, voltage across L increases instantaneously to $V_{IN} - V_{CE(sat)}$, and the inductor's magnetic field expands quickly, as indicated in Figure 17–19(a). During the on-time (t_{on}) of Q_1, V_L decreases from its initial maximum, as shown. The longer Q_1 is on, the smaller V_L becomes. When Q_1 turns off, the inductor's magnetic field collapses; and its polarity reverses so that its voltage adds to V_{IN}, thus producing an output voltage greater than the input, as indicated in Figure 17–19(b). During the off-time (t_{off}) of Q_1, the diode is forward-biased, allowing the capacitor to charge. The variations in the output voltage due to the charging and discharging action are sufficiently smoothed by the filtering action of L and C_O.

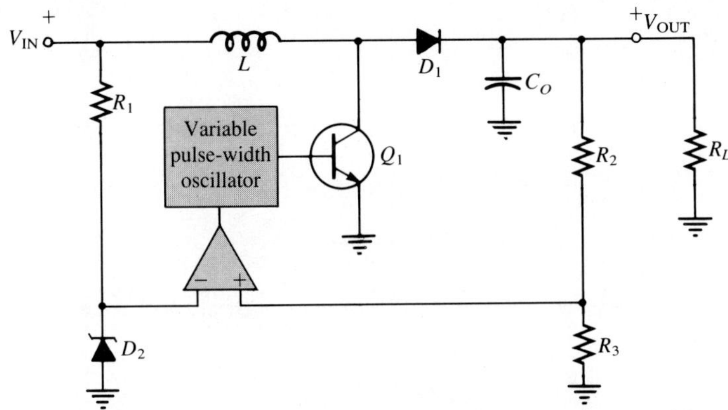

FIGURE 17–18 Basic step-up switching regulator.

The shorter the on-time of Q_1, the greater the inductor voltage is, and thus the greater the output voltage is (greater V_L adds to V_{IN}). The longer the on-time of Q_1, the smaller are the inductor voltage and the output voltage (small V_L adds to V_{IN}). When V_{OUT} tries to decrease because of increasing load or decreasing input voltage, t_{on} decreases and the attempted decrease in V_{OUT} is offset. When V_{OUT} tries to increase, t_{on} increases and the attempted increase in V_{OUT} is offset. This regulating action is illustrated in Figure 17–20. As you can see, the output voltage is inversely related to the duty cycle of Q_1 and can be expressed as follows.

$$V_{OUT} = \left(\frac{T}{t_{on}}\right)V_{IN} \tag{17–11}$$

where $T = t_{on} + t_{off}$.

FIGURE 17–19
Step-up action of the basic switching regulator.

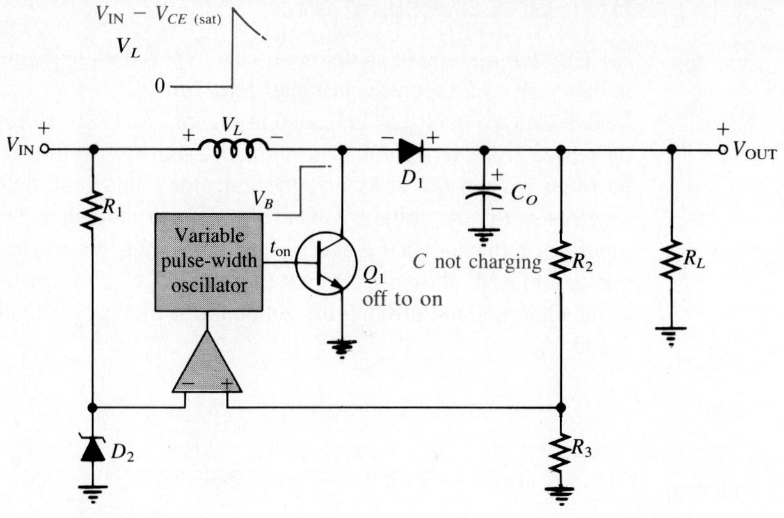

(a) When Q_1 is on

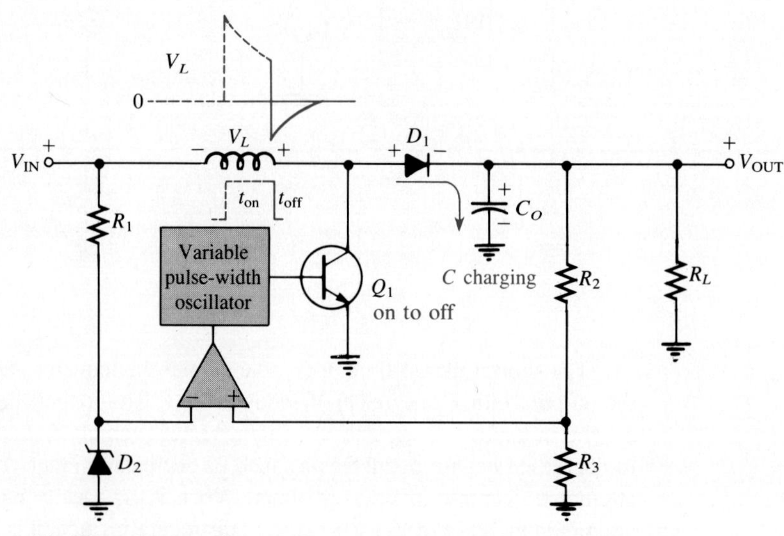

(b) When Q_1 turns off

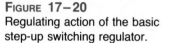

FIGURE 17–20
Regulating action of the basic step-up switching regulator.

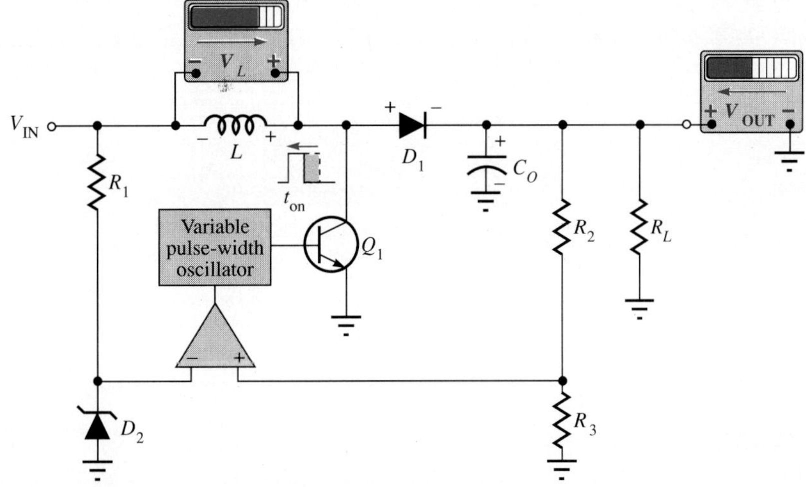

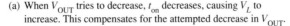

(a) When V_{OUT} tries to decrease, t_{on} decreases, causing V_L to increase. This compensates for the attempted decrease in V_{OUT}.

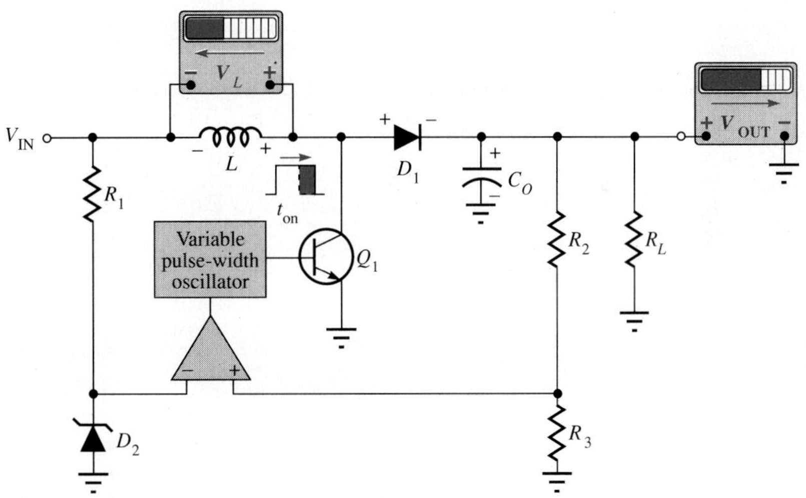

(b) When V_{OUT} tries to increase, t_{on} increases, causing V_L to decrease. This compensates for the attempted increase in V_{OUT}.

VOLTAGE-INVERTER CONFIGURATION

A third type of switching regulator produces an output voltage that is opposite in polarity to the input. A basic diagram is shown in Figure 17–21.

When Q_1 turns on, the inductor voltage jumps to $V_{IN} - V_{CE(sat)}$ and the magnetic field rapidly expands, as shown in Figure 17–22(a). While Q_1 is on, the diode is reverse-biased and the inductor voltage decreases from its initial maximum. When Q_1 turns off, the

FIGURE 17–21
Basic inverting switching regulator.

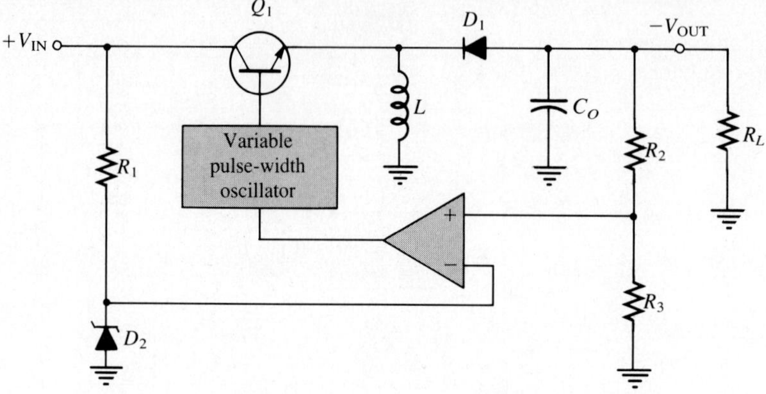

FIGURE 17–22
Inverting action of the basic inverting switching regulator.

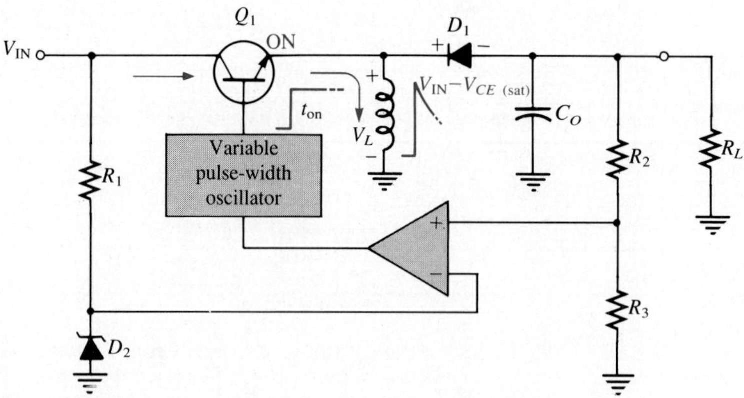

(a) When Q_1 is on, D_1 is reverse-biased.

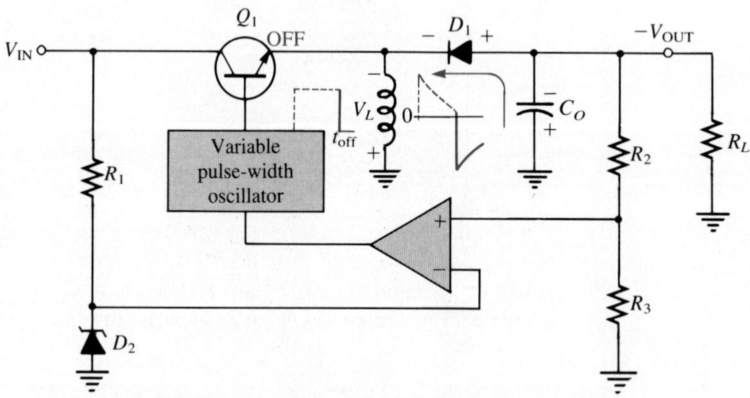

(b) When Q_1 turns off, D_1 forward biases.

magnetic field collapses and the inductor's polarity reverses, as shown in Figure 17–22(b). This forward-biases the diode, charges C_O, and produces a negative output voltage, as indicated. The repetitive on-off action of Q_1 produces a repetitive charging and discharging that is smoothed by the LC filter action.

As with the step-up regulator, the less time Q_1 is on, the greater the output voltage is, and vice versa. This regulating action is illustrated in Figure 17–23. Switching regulator efficiencies can be greater than 90 percent.

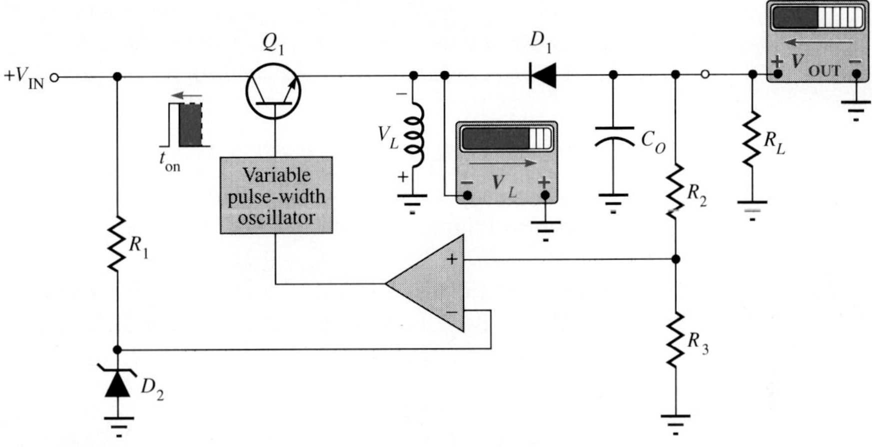

(a) When $-V_{OUT}$ tries to decrease, t_{on} decreases, causing V_L to increase. This compensates for the attempted decrease in $-V_{OUT}$.

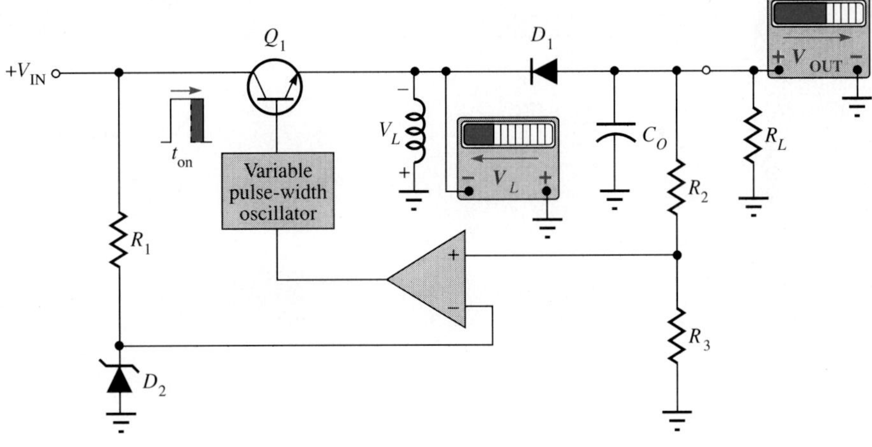

(b) When $-V_{OUT}$ tries to increase, t_{on} increases, causing V_L to decrease. This compensates for the attempted increase in $-V_{OUT}$.

FIGURE 17–23
Regulating action of the basic inverting switching regulator.

17–4 REVIEW QUESTIONS

1. What are three types of switching regulators?
2. What is the primary advantage of switching regulators over linear regulators?
3. How are changes in output voltage compensated for in the switching regulator?

17–5 INTEGRATED CIRCUIT VOLTAGE REGULATORS

In the previous sections, we presented the basic voltage regulator configurations. Several types of both linear and switching regulators are available in integrated circuit (IC) form. Generally, the linear regulators are three-terminal devices that provide either positive or negative output voltages that can be either fixed or adjustable. In this section, typical linear and switching IC regulators are introduced.

FIXED POSITIVE LINEAR VOLTAGE REGULATORS

Although many types of IC regulators are available, the 7800 series of IC regulators is representative of three-terminal devices that provide a fixed positive output voltage. The three terminals are input, output, and ground as indicated in the standard fixed voltage configuration in Figure 17–24(a). The last two digits in the part number designate the output voltage. For example, the 7805 is a +5.0 V regulator. Other available output voltages are given in Figure 17–24(b) and common packages are shown in part (c) of Figure 17–24.

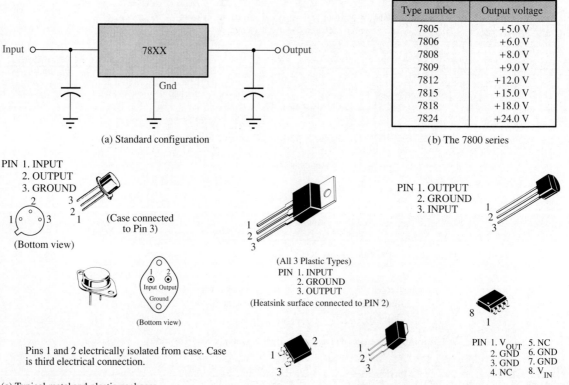

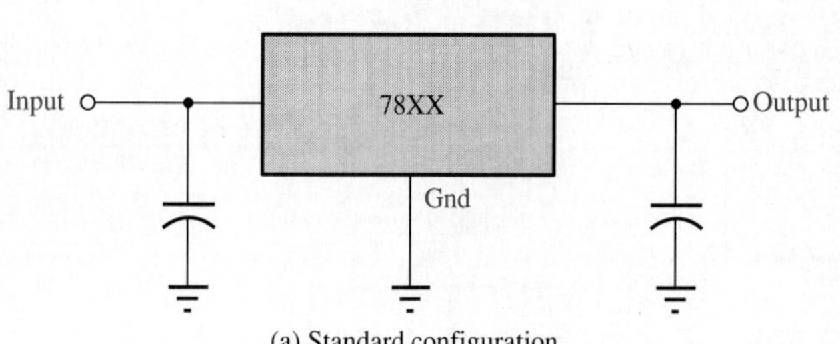

(a) Standard configuration

Type number	Output voltage
7805	+5.0 V
7806	+6.0 V
7808	+8.0 V
7809	+9.0 V
7812	+12.0 V
7815	+15.0 V
7818	+18.0 V
7824	+24.0 V

(b) The 7800 series

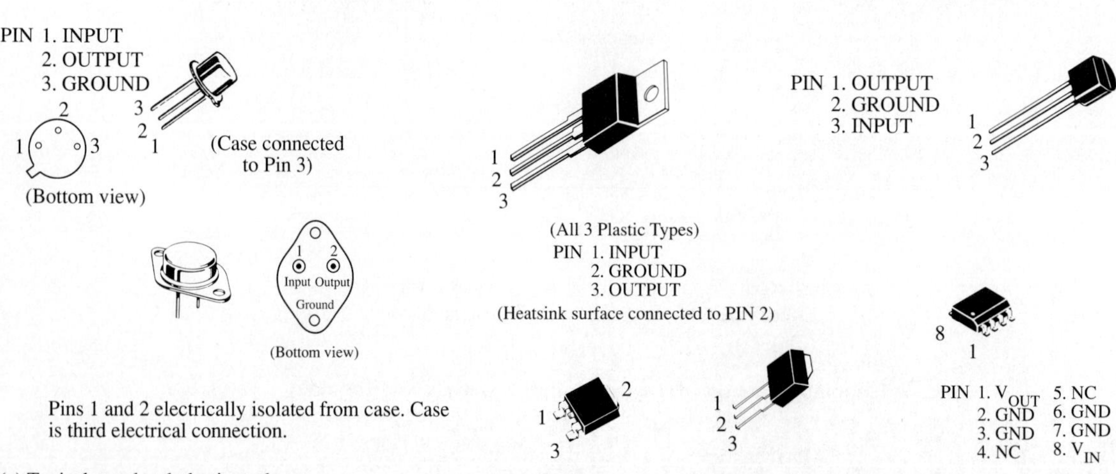

(c) Typical metal and plastic packages

FIGURE 17–24
The 7800 series three-terminal fixed positive voltage regulators. Part (c) copyright of Motorola, Inc. Used by permission.

Capacitors, although not always necessary, are sometimes used on the input and output as indicated. The output capacitor acts basically as a line filter to improve transient response. The input capacitor is used to prevent unwanted oscillations when the regulator is some distance from the power supply filter such that the line has a significant inductance.

The 7800 series can produce output current in excess of 1 A when used with an adequate heat sink. The 78L00 series can provide up to 100 mA, the 78M00 series can provide up to 500 mA, and the 78T00 series can provide in excess of 3 A. These devices are available with either a 2% or 4% output voltage tolerance.

The input voltage must be at least 2 V above the output voltage in order to maintain regulation. The circuits have internal thermal overload protection and short-circuit current-limiting features. **Thermal overload** occurs when the internal power dissipation becomes excessive and the temperature of the device exceeds a certain value.

FIXED NEGATIVE LINEAR VOLTAGE REGULATORS

The 7900 series is typical of three-terminal IC regulators that provide a fixed negative output voltage. This series is the negative-voltage counterpart of the 7800 series and shares most of the same features and characteristics. Figure 17–25 indicates the standard configuration and part numbers with corresponding output voltages that are available.

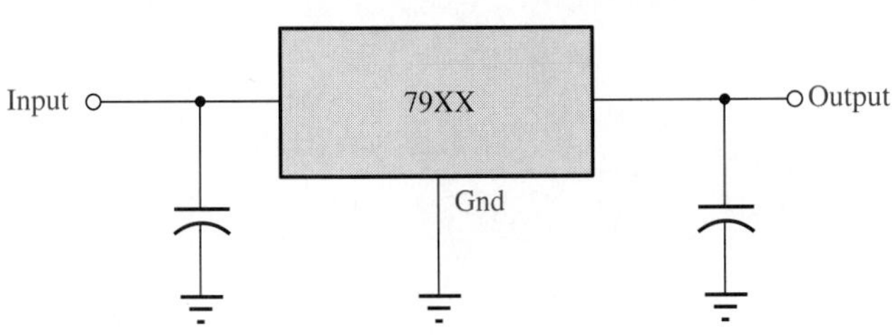

(a) Standard configuration

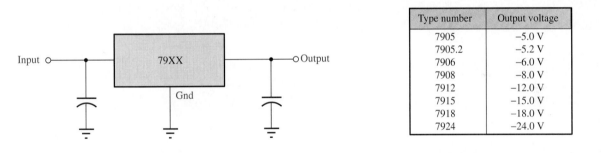

Type number	Output voltage
7905	−5.0 V
7905.2	−5.2 V
7906	−6.0 V
7908	−8.0 V
7912	−12.0 V
7915	−15.0 V
7918	−18.0 V
7924	−24.0 V

(b) The 7900 series

FIGURE 17–25
The 7900 series three-terminal fixed negative voltage regulators.

ADJUSTABLE POSITIVE LINEAR VOLTAGE REGULATORS

The LM317 is an excellent example of a three-terminal positive regulator with an adjustable output voltage. A data sheet for this device is given in Appendix A. The standard configuration is shown in Figure 17–26. Input and output capacitors, although not shown, are often used for the reasons discussed previously. Notice that there is an input, an output, and an adjustment terminal. The external fixed resistor R_1 and the external variable resistor R_2 provide the output voltage adjustment. V_{OUT} can be varied from 1.2 V to 37 V depending on the resistor values. The LM317 can provide over 1.5 A of output current to a load.

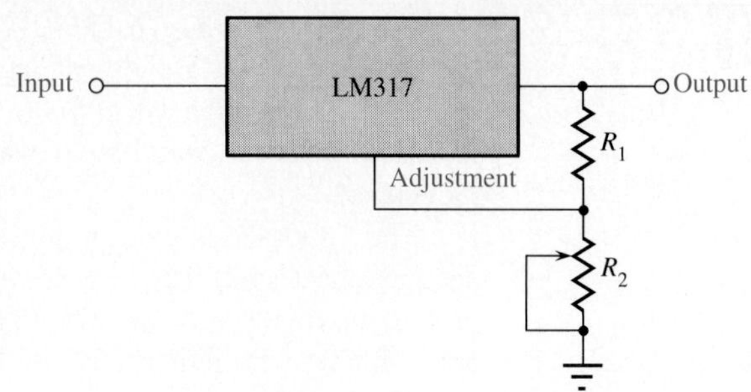

FIGURE 17–26
The LM317 three-terminal adjustable positive voltage regulator.

The LM317 is operated as a "floating" regulator because the adjustment terminal is not connected to ground, but floats to whatever voltage is across R_2. This allows the output voltage to be much higher than that of a fixed-voltage regulator.

BASIC OPERATION As indicated in Figure 17–27, a constant 1.25-V reference voltage (V_{REF}) is maintained by the regulator between the output terminal and the adjustment terminal. This constant reference voltage produces a constant current (I_{REF}) through R_1, regardless of the value of R_2. I_{REF} also flows through R_2.

$$I_{REF} = \frac{V_{REF}}{R_1} = \frac{1.25\text{ V}}{R_1}$$

Also, there is a very small constant current out of the adjustment terminal of approximately 50 μA called I_{ADJ}, which flows through R_2. An expression for the output voltage is developed as follows.

$$\begin{aligned} V_{OUT} &= V_{R1} + V_{R2} \\ &= I_{REF}R_1 + I_{REF}R_2 + I_{ADJ}R_2 \\ &= I_{REF}(R_1 + R_2) + I_{ADJ}R_2 \\ &= \frac{V_{REF}}{R_1}(R_1 + R_2) + I_{ADJ}R_2 \end{aligned}$$

$$V_{OUT} = V_{REF}\left(1 + \frac{R_2}{R_1}\right) + I_{ADJ}R_2 \qquad \textbf{(17–12)}$$

As you can see, the output voltage is a function of both R_1 and R_2. Once the value of R_1 is set, the output voltage is adjusted by varying R_2.

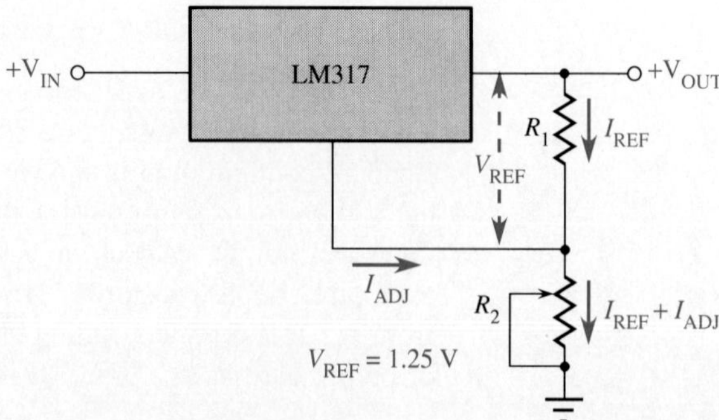

FIGURE 17–27
Operation of the LM317 adjustable voltage regulator.

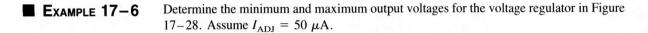

EXAMPLE 17–6 Determine the minimum and maximum output voltages for the voltage regulator in Figure 17–28. Assume $I_{ADJ} = 50\ \mu A$.

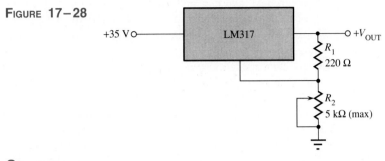

FIGURE 17–28

SOLUTION

$$V_{R1} = V_{REF} = 1.25\ V$$

When R_2 is set at its maximum of 5 kΩ,

$$V_{OUT} = V_{REF}\left(1 + \frac{R_2}{R_1}\right) + I_{ADJ}R_2 = 1.25\ V\left(1 + \frac{5\ k\Omega}{220\ \Omega}\right) + (50\ \mu A)5\ k\Omega$$
$$= 29.66\ V + 0.25\ V = 29.91\ V$$

When R_2 is set at its minimum of 0 Ω,

$$V_{OUT} = V_{REF}\left(1 + \frac{R_2}{R_1}\right) + I_{ADJ}R_2 = 1.25\ V(1) = 1.25\ V$$

PRACTICE EXERCISE 17–6

What is the output voltage of the regulator if R_2 is set at 2 kΩ?

ADJUSTABLE NEGATIVE LINEAR VOLTAGE REGULATORS

The LM337 is the negative output counterpartof the LM317 and is a good example of this type of IC regulator. Like the LM317, the LM337 requires two external resistors for output voltage adjustment as shown in Figure 17–29. The output voltage can be adjusted from −1.2 V to −37 V, depending on the external resistor values.

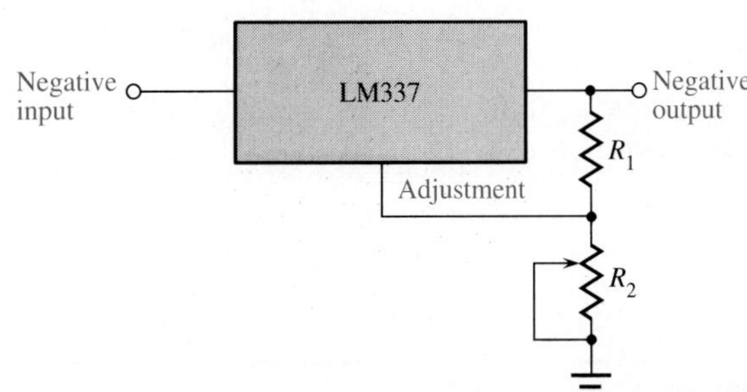

FIGURE 17–29
The LM337 three-terminal adjustable negative voltage regulator.

SWITCHING VOLTAGE REGULATORS

As an example of an IC switching voltage regulator, we will look at the 78S40. This is a universal device that can be used with external components to provide step-up, step-down, and inverting operation.

The internal circuitry of the 78S40 is shown in Figure 17–30. This circuit can be compared to the basic switching regulators that were covered in Section 17–4. For example, look back at Figure 17–15(a). The oscillator and comparator functions are directly comparable. The gate and flip-flop in the 78S40 were not included in the basic circuit of Figure 17–15(a), but they provide additional regulating action. Transistors Q_1 and Q_2 effectively perform the same function as Q_1 in the basic circuit. The 1.25-V reference block in the 78S40 has the same purpose as the zener diode in the basic circuit, and diode D_1 in the 78S40 corresponds to D_1 in the basic circuit.

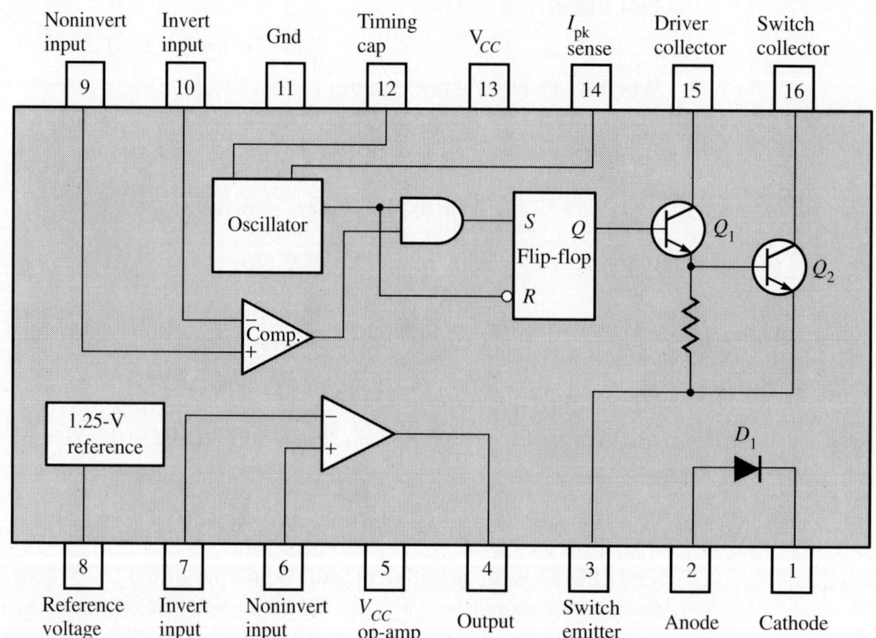

FIGURE 17–30
The 78S40 switching regulator.

The 78S40 also has an "uncommitted" op-amp thrown in for good measure. It is not used in any of the regulator configurations. External circuitry is required to make this device operate as a regulator, as you will see in Section 17–6.

17–5 REVIEW QUESTIONS

1. What are the three terminals of a fixed-voltage regulator?
2. What is the output voltage of a 7809? Of a 7915?
3. What are the three terminals of an adjustable-voltage regulator?
4. What external components are required for a basic LM317 configuration?

17–6 APPLICATIONS OF IC VOLTAGE REGULATORS

In the last section, you saw several devices that are representative of the general types of IC voltage regulators. Now, we will examine several different ways these devices can be modified with external circuitry to improve or alter their performance.

USING AN EXTERNAL PASS TRANSISTOR

As you know, an IC voltage regulator is capable of delivering only a certain amount of output current to a load. For example, the 7800 series regulators can handle a maximum output current of at least 1.3 A and typically 2.5 A. If the load current exceeds the maximum allowable value, there will be thermal overload and the regulator will shut down. A thermal overload condition means that there is excessive power dissipation inside the device.

If an application requires more than the maximum current that the regulator can deliver, an external pass transistor can be used. Figure 17–31 illustrates a three-terminal regulator with an external pass transistor for handling currents in excess of the output current capability of the basic regulator.

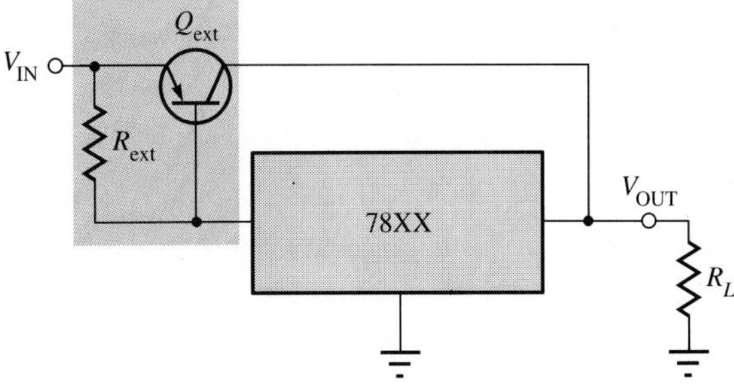

FIGURE 17–31
A 7800-series three-terminal regulator with an external pass transistor.

FIGURE 17–32
Operation of the regulator with an external pass transistor.

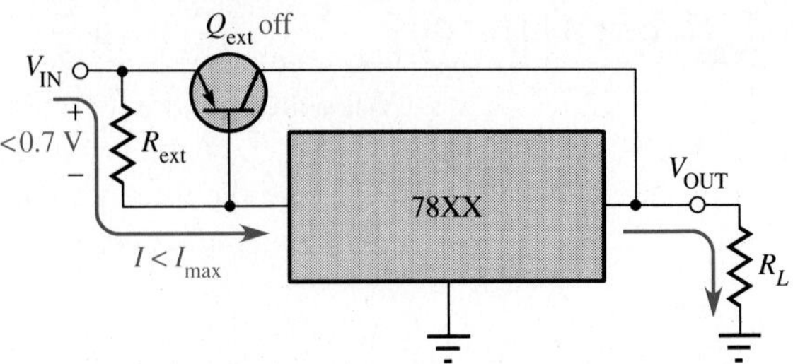

(a) When the regulator current is less than I_{max}, the external pass transistor is off and the regulator is handling all of the current.

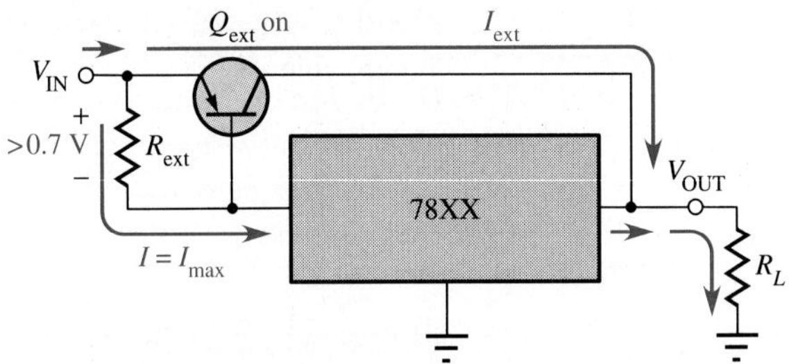

(b) When the load current exceeds I_{max}, the drop across R_{ext} turns Q_{ext} on and it conducts the excess current.

The value of the external current-sensing resistor R_{ext} determines the value of current at which Q_{ext} begins to conduct because it sets the V_{BE} voltage of the transistor. As long as the current is less than the value set by R_{ext}, the transistor Q_{ext} is off, and the regulator operates normally as shown in Figure 17–32(a). This is because the voltage drop across R_{ext} is less than the 0.7-V base-to-emitter voltage required to turn Q_{ext} on. R_{ext} is determined by the following formula, where I_{max} is the highest current that the voltage regulator is to handle internally.

$$R_{ext} = \frac{0.7\text{ V}}{I_{max}} \tag{17–13}$$

When the current is sufficient to produce at least a 0.7 V drop across R_{ext}, the external pass transistor Q_{ext} turns on and conducts any current in excess of I_{max}, as indicated in Figure 17–32(b). Q_{ext} will conduct more or less, depending on the load requirements. For example, if the total load current is 3 A and I_{max} was selected to be 1 A, the external pass transistor will conduct 2 A, which is the excess over the current I_{max}, flowing internally through the voltage regulator.

EXAMPLE 17–7 What value is R_{ext} if the maximum current to be handled internally by the voltage regulator in Figure 17–31 is set at 700 mA?

SOLUTION

$$R_{ext} = \frac{0.7\text{ V}}{I_{max}} = \frac{0.7\text{ V}}{0.7\text{ A}} = 1\ \Omega$$

PRACTICE EXERCISE 17–7
If R_{ext} is changed to 1.5 Ω, at what current value will Q_{ext} turn on?

The external pass transistor is typically a power transistor with heat sink which must be capable of handling a maximum power of

$$P_{ext} = I_{ext}(V_{IN} - V_{OUT}) \tag{17–14}$$

EXAMPLE 17–8 What must be the minimum power rating for the external pass transistor used with a 7824 regulator in a circuit such as that shown in Figure 17–31? The input voltage is 30 V and the load resistance is 10 Ω. The maximum internal current is to be 700 mA. Assume that there is no heat sink for this calculation. Keep in mind that the use of a heat sink increases the effective power rating of the transistor and you can use a lower rated transistor.

SOLUTION
The load current is

$$I_L = \frac{V_{OUT}}{R_L} = \frac{24\text{ V}}{10\ \Omega} = 2.4\text{ A}$$

The current through Q_{ext} is

$$I_{ext} = I_L - I_{max} = 2.4\text{ A} - 0.7\text{ A} = 1.7\text{ A}$$

The power dissipated by Q_{ext} is

$$\begin{aligned} P_{ext(min)} &= I_{ext}(V_{IN} - V_{OUT}) \\ &= 1.7\text{ A}(30\text{ V} - 24\text{ V}) \\ &= 1.7\text{ A}(6\text{ V}) = 10.2\text{ W} \end{aligned}$$

For a safety margin, choose a power transistor with a rating greater than 10.2 W, say at least 15 W.

PRACTICE EXERCISE 17–8
Rework this example using a 7815 regulator.

CURRENT LIMITING

A drawback of the circuit in Figure 17–31 is that the external transistor is not protected from excessive current, such as would result from a shorted output. An additional current-limiting circuit (Q_{lim} and R_{lim}) can be added as shown in Figure 17–33 to protect Q_{ext} from excessive current and possible burn out.

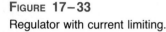

FIGURE 17–33
Regulator with current limiting.

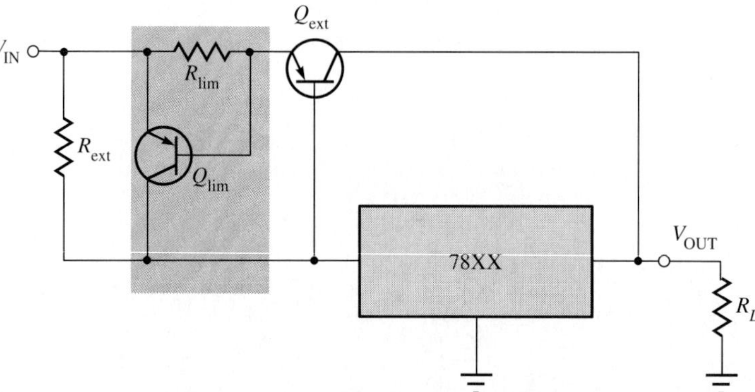

The following describes the way the current-limiting circuit works. The current-sensing resistor R_{lim} sets the V_{BE} of transistor Q_{lim}. The base-to-emitter voltage of Q_{ext} is now determined by $V_{R_{ext}} - V_{R_{lim}}$ because they have opposite polarities. So, for normal operation, the drop across R_{ext} must be sufficient to overcome the opposing drop across R_{lim}. If the current through Q_{ext} exceeds a certain maximum ($I_{ext(max)}$) because of a shorted output or a faulty load, the voltage across R_{lim} reaches 0.7 V and turns Q_{lim} on. Q_{lim} now conducts current away from Q_{ext} and through the regulator, forcing a thermal overload to occur and shut down the regulator. Remember, the IC regulator is internally protected from thermal overload as part of its design.

This action is illustrated in Figure 17–34. In part (a), the circuit is operating normally with Q_{ext} conducting less than the maximum current that it can handle with Q_{lim} off. Part (b) shows what happens when there is a short across the load. The current through Q_{ext} suddenly increases and causes the voltage drop across R_{lim} to increase, which turns Q_{lim} on. The current is now diverted into the regulator, which causes it to shut down due to thermal overload.

A CURRENT REGULATOR

The three-terminal regulator can be used as a current source when an application requires that a constant current be supplied to a variable load. The basic circuit is shown in Figure 17–35 where R is the current-setting resistor. The regulator provides a fixed constant voltage, V_{OUT}, between the ground terminal (not connected to ground in this case) and the output terminal. This determines the constant current supplied to the load.

$$I_L = \frac{V_{OUT}}{R} + I_G \qquad \textbf{(17–15)}$$

The current from the ground terminal I_G is very small compared to the output current and can often be neglected.

FIGURE 17–34
The current-limiting action of the regulator circuit.

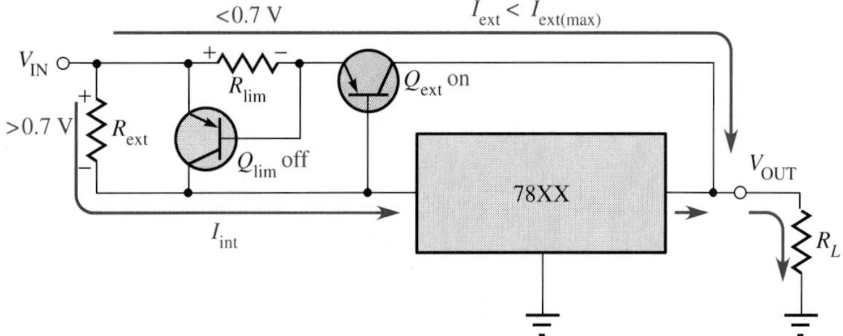

(a) During normal operation, when the load current is not excessive, Q_{lim} is off.

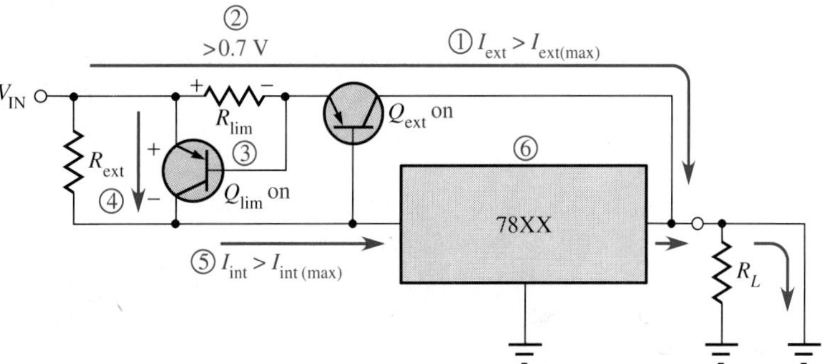

(b) When short occurs ①, the external current becomes excessive and the voltage across R_{lim} increases ② and turns on Q_{lim} ③, which then conducts current away from Q_{ext} and routes it through the regulator ④, causing the internal regulator current to become excessive ⑤ and to force the regulator into thermal shut down ⑥.

FIGURE 17–35
The three-terminal regulator as a current source.

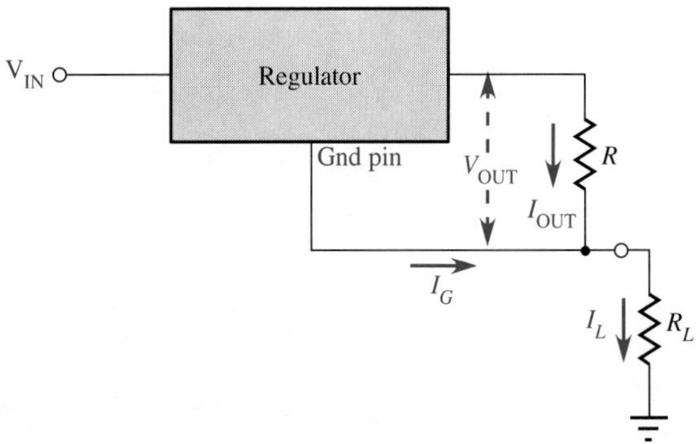

EXAMPLE 17–9

Use a 7805 regulator to provide a constant current of 1 A to a variable load. The input must be at least 2 V greater than the output and I_G = 1.5 mA.

SOLUTION

First, 1 A is within the limits of the 7805's capability (remember, it can handle at least 1.3 A without an external pass transistor).

The 7805 produces 5 V between its ground terminal and its output terminal. Therefore, if we want 1 A of current, the current-setting resistor must be (neglecting I_G)

$$R = \frac{V_{OUT}}{I_L} = \frac{5\text{ V}}{1\text{ A}} = 5\ \Omega$$

The circuit is shown in Figure 17–36.

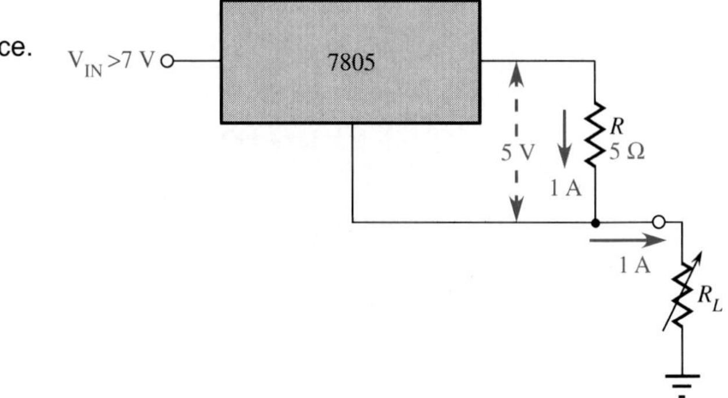

FIGURE 17–36
A 1-A constant-current source.

PRACTICE EXERCISE 17–9

If a 7812 regulator is used instead of the 7805, to what value would you change R to maintain a constant current of 1 A?

SWITCHING REGULATOR CONFIGURATIONS

In Section 17–5, the 78S40 was introduced as an example of an IC switching voltage regulator. Figure 17–37 shows the external connections for a step-down configuration where the output voltage is less than the input voltage and Figure 17–38 shows a step-up configuration in which the output voltage is greater than the input voltage. An inverting configuration is also possible, but it is not shown here.

The capacitor C_T is the timing capacitor that controls the pulse width and frequency of the oscillator and thus establishes the on-time of transistor Q_2. The voltage across the current-sensing resistor R_{cs} is used internally by the oscillator to vary the duty cycle based on the desired peak load current. The voltage divider, made up of R_1 and R_2, reduces the output voltage to a nominal value equal to the reference voltage. If V_{OUT} exceeds its set value, the output of the comparator switches to its low state, disabling the gate to turn Q_2 off until the output decreases. This regulating action is in addition to that produced by the duty cycle variation of the oscillator as described in Section 17–4 in relation to the basic switching regulator.

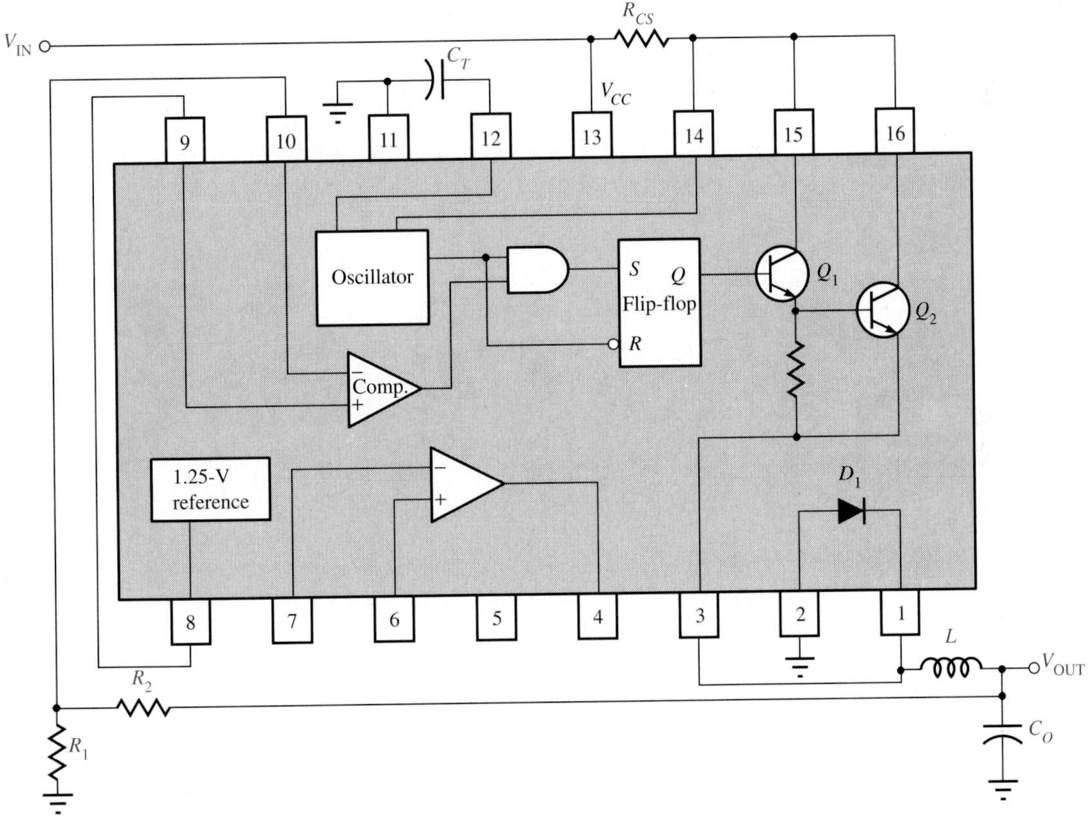

FIGURE 17–37
The step-down configuration of the 78S40 switching regulator.

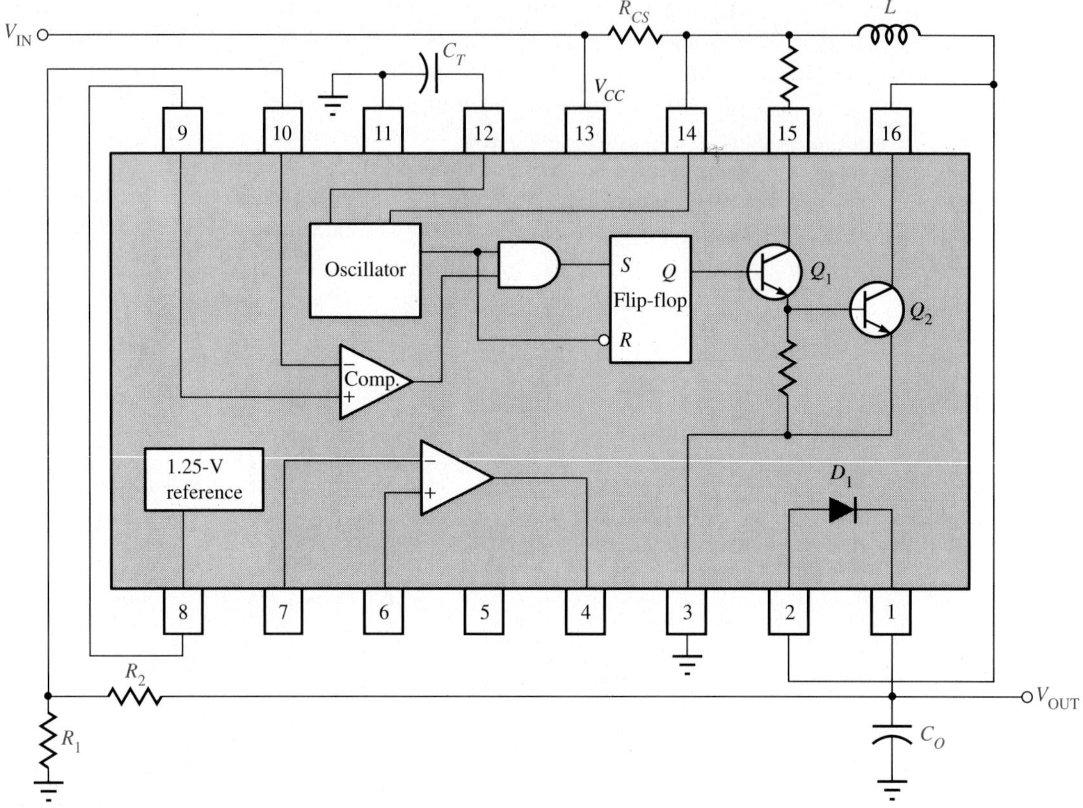

FIGURE 17–38
The step-up configuration of the 78S40 switching regulator.

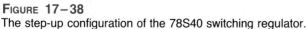

17–6 REVIEW QUESTIONS

1. What is the purpose of using an external pass transistor with an IC voltage regulator?
2. What is the advantage of current limiting in a voltage regulator?
3. What does *thermal overload* mean?

17–7 A SYSTEM APPLICATION

In this system application, the focus is on the regulated power supply which provides the FM stereo receiver with dual polarity dc voltages. Recall from previous system applications that the op-amps in the channel separation circuits and the audio amplifiers operate from ±12 V. Both positive and negative voltage regulators are used to regulate the rectified and filtered voltages from a bridge rectifier. In this section, you will

- ☐ ***See how dual supply voltages are produced by a rectifier.***
- ☐ ***See how positive and negative three-terminal IC regulators are used in a power supply.***

- □ ***Relate a schematic to a pc board.***
- □ ***Analyze the operation of the power supply circuit.***
- □ ***Troubleshoot some common power supply failures.***

ABOUT THE POWER SUPPLY

This power supply utilizes a full-wave bridge **rectifier** with both the positive and negative rectified voltages taken off the bridge at the appropriate points and filtered by electrolytic capacitors. A 7812 and a 7912 provide regulation.

Now, so that you can take a closer look at the dual power supply, let's take it out of the system and put it on the test bench.

ON THE TEST BENCH

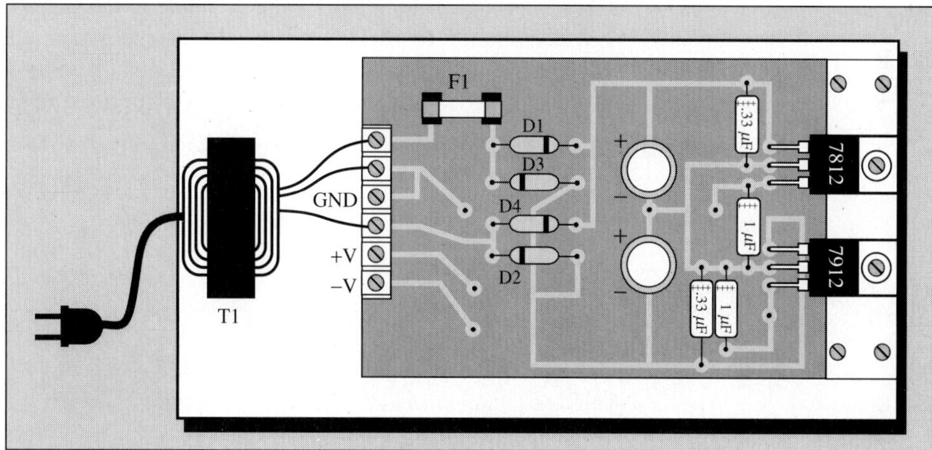

FIGURE 17–39

■ ACTIVITY 1 RELATE THE PC BOARD TO THE SCHEMATIC

Develop a schematic for the power supply in Figure 17–39. Add any missing labels and include the IC pin numbers by referring to the voltage regulator data sheets in Appendix A. The rectifier diodes are 1N4001s, the filter capacitors C1 and C2 are 100 μF, and the transformer has a turns ratio of 5:1. Determine the backside pc board connections as you develop the schematic.

■ ACTIVITY 2 ANALYZE THE POWER SUPPLY CIRCUITS

STEP 1 Determine the approximate voltage at each of the four "corners" of the bridge.

STEP 2 Calculate the peak inverse voltage of the rectifier diodes.

STEP 3 Determine the voltage at the inputs of the voltage regulators.

STEP 4 In this stereo system, assume that op-amps are used only in the channel separation circuits and the channel audio amplifiers. If all of the other circuits in the receiver use +12 V and draw an average dc current of 500 mA, determine how much total current each regulator must supply. Refer to the system applications in Chapters 13 and 16 and use the data sheets in Appendix A.

STEP 5 Based on the results in Step 4, do the IC regulators have to be attached to the heat sink or is this just for a safety margin?

■ ACTIVITY 3 WRITE A TECHNICAL REPORT

Describe the operation of the power supply with an emphasis on how both positive and negative voltages are obtained. State the purpose of each component. Use the results of Activity 2 where appropriate.

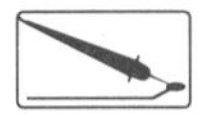

■ ACTIVITY 4 TROUBLESHOOT THE POWER SUPPLY BY STATING THE PROBABLE CAUSE OR CAUSES IN EACH CASE

1. Both positive and negative output voltages are zero.
2. Positive output voltage is zero and the negative output voltage is −12 V.
3. Negative output voltage is zero and the positive output voltage is +12 V.
4. Radical voltage fluctuations on output of positive regulator.

■ ACTIVITY 5 TEST BENCH SPECIAL ASSIGNMENT

Go to Test Bench 7 in the color insert section (which follows page 422) and carry out the assignment that is stated there.

17–7 REVIEW QUESTIONS

1. What should be the rating of the power supply fuse?
2. What purpose do the 0.33 μF capacitors serve?
3. Which regulator provides the negative voltage?
4. Would you recommend that an external pass transistor be used with the regulators in this power supply? Why?

SUMMARY

- □ Voltage regulators keep a constant dc output voltage when the input or load varies within limits.
- □ A basic voltage regulator consists of a reference voltage source, an error detector, a sampling element, and a control device. Protection circuitry is also found in most regulators.
- □ Two basic categories of voltage regulators are linear and switching.
- □ Two basic types of linear regulators are series and shunt.
- □ In a series linear regulator, the control element is a transistor in series with the load.
- □ In a shunt linear regulator, the control element is a transistor in parallel with the load.
- □ Three configurations for switching regulators are step-down, step-up, and inverting.

- ☐ Switching regulators are more efficient than linear regulators and are particularly useful in low-voltage, high-current applications.
- ☐ Three-terminal linear IC regulators are available for either fixed output or variable output voltages of positive or negative polarities.
- ☐ An external pass transistor increases the current capability of a regulator.
- ☐ The 7800 series are three-terminal IC regulators with fixed positive output voltage.
- ☐ The 7900 series are three-terminal IC regulators with fixed negative output voltage.
- ☐ The LM317 is a three-terminal IC regulator with a positive variable output voltage.
- ☐ The LM337 is a three-terminal IC regulator with a negative variable output voltage.
- ☐ The 78S40 is a switching voltage regulator.

GLOSSARY

Fold-back current limiting A method of current limiting in voltage regulators.

Line regulation The percentage change in output voltage for a given change in line (input) voltage.

Load regulation The percentage change in output voltage for a given change in load current.

Rectifier An electronic circuit that converts ac into pulsating dc.

Regulator An electronic circuit that maintains an essentially constant output voltage with a changing input voltage or load current.

Thermal overload A condition in a rectifier where the internal power dissipation of the circuit exceeds a certain maximum due to excessive current.

FORMULAS

VOLTAGE REGULATION

(17–1) $$\text{Line regulation} = \frac{(\Delta V_{\text{OUT}}/V_{\text{OUT}})100\%}{\Delta V_{\text{IN}}}$$ Percent line regulation

(17–2) $$\text{Load regulation} = \frac{(V_{\text{NL}} - V_{\text{FL}})100\%}{V_{\text{FL}}}$$ Percent load regulation

BASIC SERIES REGULATOR

(17–3) $$A_{cl} = 1 + \frac{R_2}{R_3}$$ Closed-loop voltage gain

(17–4) $$V_{\text{OUT}} \cong \left(1 + \frac{R_2}{R_3}\right)V_{\text{REF}}$$ Regulator output

(17–5) $$I_{L(\text{max})} = \frac{0.7\text{ V}}{R_4}$$ For constant current limiting

BASIC SHUNT REGULATOR

(17–6) $\Delta I_S = \dfrac{\Delta V_{IN}}{R_1}$ Change in shunt current

(17–7) $\Delta I_S = -\Delta I_L$ Change in shunt current

(17–8) $I_{L(max)} = \dfrac{V_{IN}}{R_1}$ Maximum load current

BASIC SWITCHING REGULATORS

(17–9) $V_{OUT} = \left(\dfrac{t_{on}}{T}\right)V_{IN}$ For step-down switching regulator

(17–10) $T = t_{on} + t_{off}$ Switching period

(17–11) $V_{OUT} = \left(\dfrac{T}{t_{on}}\right)V_{IN}$ For step-up switching regulator

IC VOLTAGE REGULATORS

(17–12) $V_{OUT} = V_{REF}\left(1 + \dfrac{R_2}{R_1}\right) + I_{ADJ}R_2$ IC regulator

(17–13) $R_{ext} = \dfrac{0.7\text{ V}}{I_{max}}$ For external pass circuit

(17–14) $P_{ext} = I_{ext}(V_{IN} - V_{OUT})$ For external pass transistor

(17–15) $I_L = \dfrac{V_{OUT}}{R} + I_G$ Regulator as a current source

SELF-TEST

1. In the case of line regulation,
- **(a)** when the temperature varies, the output voltage stays constant
- **(b)** when the output voltage changes, the load current stays constant
- **(c)** when the input voltage changes, the output voltage stays constant
- **(d)** when the load changes, the output voltage stays constant

2. In the case of load regulation,
- **(a)** when the temperature varies, the output voltage stays constant
- **(b)** when the input voltage changes, the load current stays constant
- **(c)** when the load changes, the load current stays constant
- **(d)** when the load changes, the output voltage stays constant

3. All of the following are parts of a basic voltage regulator *except*
- **(a)** control element **(b)** sampling circuit
- **(c)** voltage follower **(d)** error detector
- **(e)** reference voltage

4. The basic difference between a series regulator and a shunt regulator is

(a) the amount of current that can be handled

(b) the position of the control element

(c) the type of sample circuit

(d) the type of error detector

5. In a basic series regulator, V_{OUT} is determined by

(a) the control element **(b)** the sample circuit

(c) the reference voltage **(d)** b and c

6. The main purpose of current limiting in a regulator is

(a) protection of the regulator from excessive current

(b) protection of the load from excessive current

(c) to keep the power supply transformer from burning up

(d) to maintain a constant output voltage

7. In a linear regulator, the control transistor is conducting

(a) a small part of the time **(b)** half the time

(c) all of the time **(d)** only when the load current is excessive

8. In a switching regulator, the control transistor is conducting

(a) part of the time

(b) all of the time

(c) only when the input voltage exceeds a set limit

(d) only when there is an overload

9. The LM317 is an example of an IC

(a) three-terminal negative voltage regulator

(b) fixed positive voltage regulator

(c) switching regulator

(d) linear regulator

(e) variable positive voltage regulator

(f) b and d only

(g) d and e only

10. An external pass transistor is used for

(a) increasing the output voltage

(b) improving the regulation

(c) increasing the current that the regulator can handle

(d) short-circuit protection

PROBLEMS

SECTION 17–1 VOLTAGE REGULATION

1. The nominal output voltage of a certain regulator is 8 V. The output changes 2 mV when the input voltage goes from 12 V to 18 V. Determine the line regulation and express it as a percentage change over the entire range of V_{IN}.
2. Express the line regulation found in Problem 1 in units of %/V.
3. A certain regulator has a no-load output voltage of 10 V and a full-load output voltage of 9.90 V. What is the percent load regulation?
4. In Problem 3, if the full-load current is 250 mA, express the load regulation in %/mA.

SECTION 17–2 BASIC SERIES REGULATORS

5. Label the functional blocks for the voltage regulator in Figure 17–40.

FIGURE 17–40

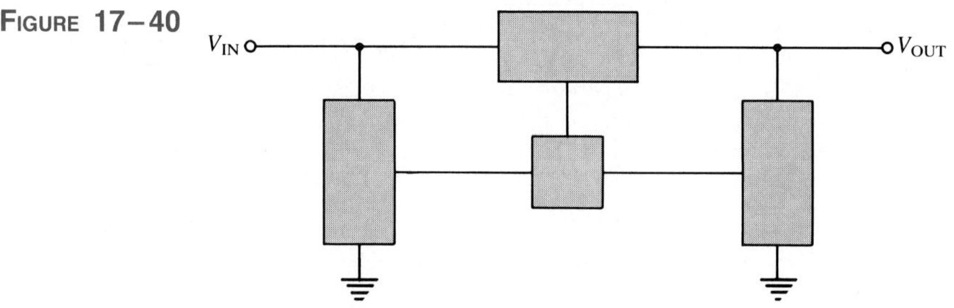

6. Determine the output voltage for the regulator in Figure 17–41.

FIGURE 17–41

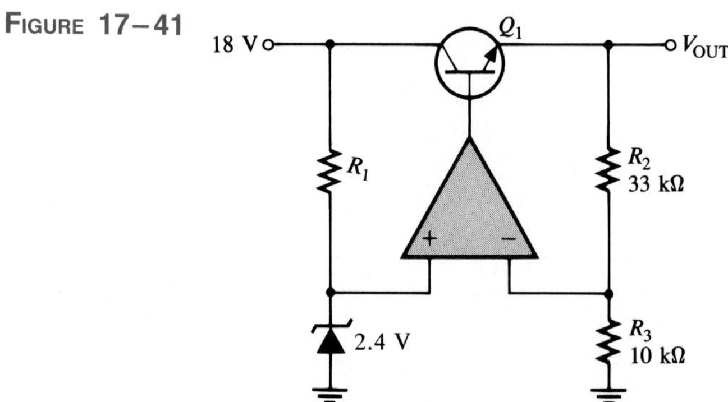

7. Determine the output voltage for the series regulator in Figure 17–42.
8. If R_3 in Figure 17–42 is increased to 4.7 kΩ, what happens to the output voltage?
9. If the zener voltage is 2.7 V instead of 2.4 V in Figure 17–42, what is the output voltage?

FIGURE 17–42

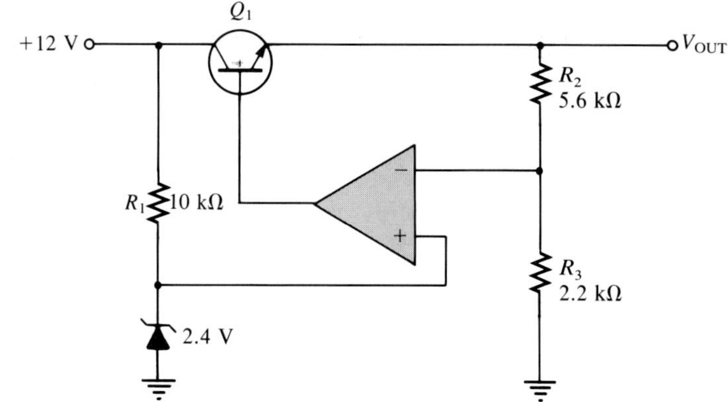

10. A series voltage regulator with constant current limiting is shown in Figure 17–43. Determine the value of R_4 if the load current is to be limited to a maximum value of 250 mA. What power rating must R_4 have?

FIGURE 17–43

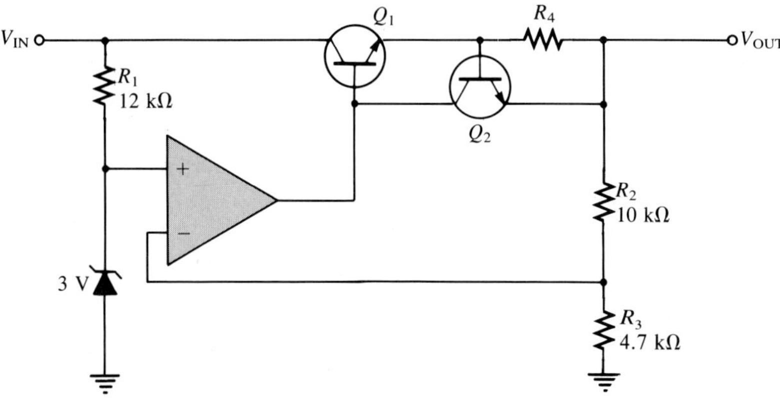

11. If the R_4 determined in Problem 10 is halved, what is the maximum load current?

SECTION 17–3 BASIC SHUNT REGULATORS

12. In the shunt regulator of Figure 17–44, when the load current increases, does Q_1 conduct more or less? Why?
13. Assume I_L remains constant and V_{IN} changes by 1 V in Figure 17–44. What is the change in the collector current of Q_1?
14. With a constant input voltage of 17 V, the load resistance in Figure 17–44 is varied from 1 kΩ to 1.2 kΩ. Neglecting any change in output voltage, how much does the shunt current through Q_1 change?
15. If the maximum allowable input voltage in Figure 17–44 is 25 V, what is the maximum possible output current when the output is short-circuited? What power rating should R_1 have?

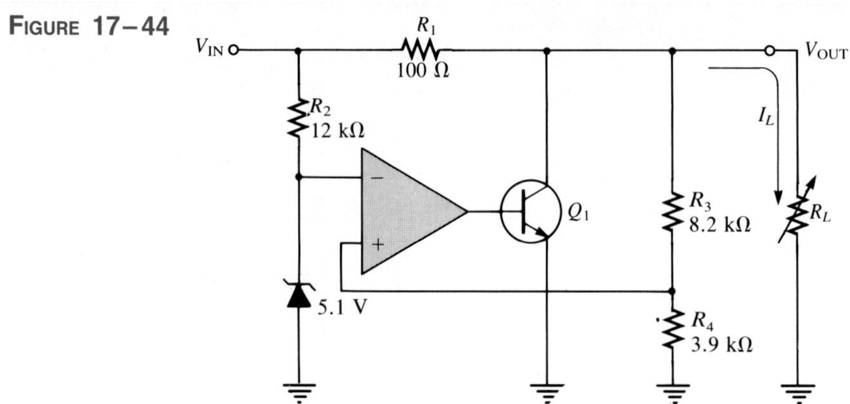

FIGURE 17–44

SECTION 17–4 BASIC SWITCHING REGULATORS

16. A basic switching regulator is shown in Figure 17–45. If the switching frequency of the transistor is 100 Hz with an off-time of 6 ms, what is the output voltage?

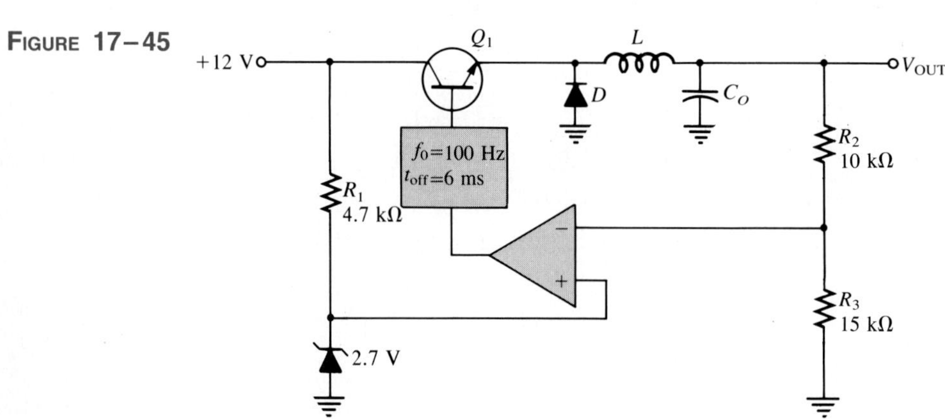

FIGURE 17–45

17. What is the duty cycle of the transistor in Problem 16?

18. Determine the output voltage for the switching regulator in Figure 17–46 when the duty cycle is 40 percent.

19. If the on-time of Q_1 in Figure 17–46 is decreased, does the output voltage increase or decrease?

FIGURE 17–46

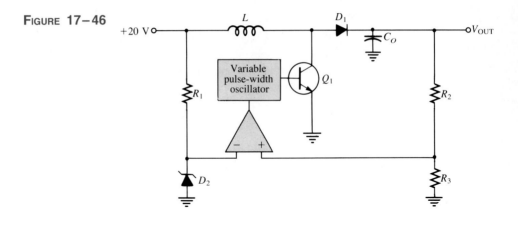

SECTION 17–5 INTEGRATED CIRCUIT VOLTAGE REGULATORS

20. What is the output voltage of each of the following IC regulators?

(a) 7806 **(b)** 7905.2

(c) 7818 **(d)** 7924

21. Determine the output voltage of the regulator in Figure 17–47.

FIGURE 17–47

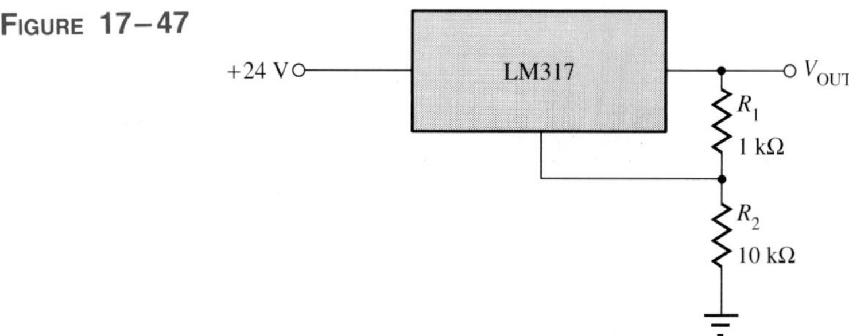

22. Determine the minimum and maximum output voltages for the circuit in Figure 17–48.

FIGURE 17–48

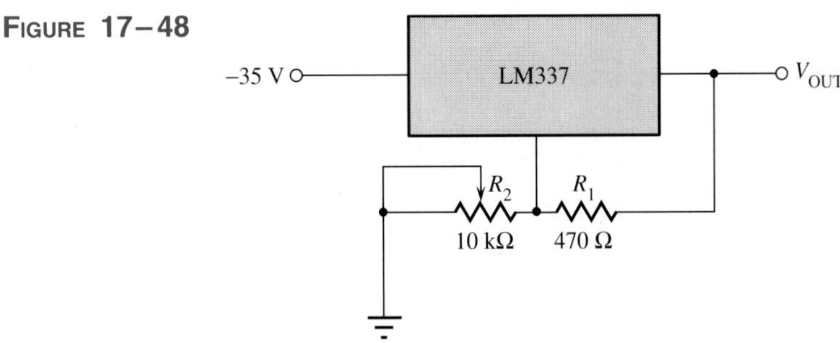

23. With no load connected, how much current is there through the regulator in Figure 17–47? Neglect the adjustment terminal current.

24. Select the values for the external resistors to be used in an LM317 circuit that is required to produce an output voltage of 12 V with an input of 18 V. The maximum regulator current with no load is to be 2 mA. There is no external pass transistor.

SECTION 17–6 APPLICATIONS OF IC VOLTAGE REGULATORS

25. In the regulator circuit of Figure 17–49, determine R_{ext} if the maximum internal regulator current is to be 250 mA.

FIGURE 17–49

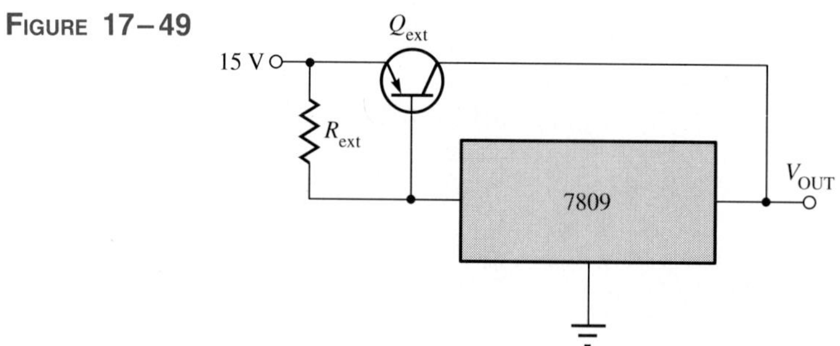

26. Using a 7812 voltage regulator and a 10 Ω load in Figure 17–49, how much power will the external pass transistor have to dissipate? The maximum internal regulator current is set at 500 mA by R_{ext}.

27. Show how to include current limiting in the circuit of Figure 17–49. What should the value of the limiting resistor be if the external current is to be limited to 2 A?

28. Using an LM317, design a circuit that will provide a constant current of 500 mA to a load.

29. Repeat Problem 28 using a 7909.

30. If a 78S40 switching regulator is to be used to regulate a 12-V input down to a 6-V output, calculate the values of the external voltage-divider resistors.

ANSWERS TO REVIEW QUESTIONS

SECTION 17–1

1. The percentage change in the output voltage for a given change in input voltage.

2. The percentage change in output voltage for a given change in load current.

SECTION 17–2

1. Control element, error detector, sampling element, reference source

2. 2 V

SECTION 17–3

1. In a shunt regulator, the control element is in parallel with the load rather than in series.
2. A shunt regulator has inherent current limiting. A disadvantage is that a shunt regulator is less efficient than a series regulator.

SECTION 17–4

1. Step-down, step-up, inverting
2. Switching regulators operate at a higher efficiency.
3. The duty cycle varies to regulate the output.

SECTION 17–5

1. Input, output, and ground
2. A 7809 has a +9-V output; A 7915 has a −15-V output.
3. Input, output, adjustment
4. A two-resistor voltage divider

SECTION 17–6

1. A pass transistor increases the current that can be handled.
2. Current limiting prevents excessive current and prevents damage to the regulator.
3. Thermal overload occurs when the internal power dissipation becomes excessive.

SECTION 17–7

1. 1 A
2. Those optional capacitors on the regulator inputs prevent oscillations.
3. The 7909 is a negative-voltage regulator.
4. No. The current that either regulator must supply is less than 1 A.

ANSWERS TO PRACTICE EXERCISES

17–1 0.6% V

17–2 0.84%, 0.0168%/mA

17–3 7.33 V

17–4 0.7 A

17–5 17.45 W

17–6 12.6 V

17–7 467 mA

17–8 12 W

17–9 12 Ω

Appendix A
Data Sheets

Diodes

BJTs and FETs

Thyristors

Linear Integrated Circuits

MOTOROLA
SEMICONDUCTOR
TECHNICAL DATA

1.5KE6.8, A thru 1.5KE250, A
See Page 4-59

Designers Data Sheet

500-MILLIWATT HERMETICALLY SEALED GLASS SILICON ZENER DIODES

- Complete Voltage Range — 2.4 to 110 Volts
- DO-35 Package — Smaller than Conventional DO-7 Package
- Double Slug Type Construction
- Metallurgically Bonded Construction
- Oxide Passivated Die

Designer's Data for "Worst Case" Conditions

The Designer's Data sheets permit the design of most circuits entirely from the information presented. Limit curves – representing boundaries on device characteristics – are given to facilitate "worst case" design.

1N746 thru 1N759
1N957A thru 1N986A
1N4370 thru 1N4372

GLASS ZENER DIODES
500 MILLIWATTS
2.4-110 VOLTS

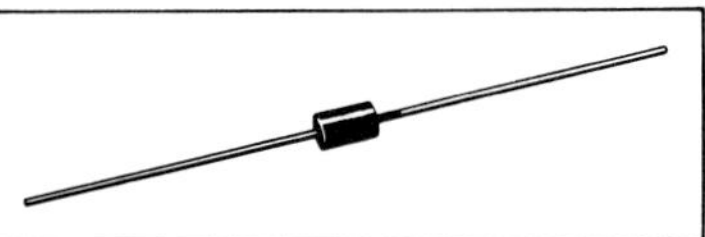

MAXIMUM RATINGS

Rating	Symbol	Value	Unit
DC Power Dissipation @ $T_L \leq 50^oC$, Lead Length = 3/8"	P_D		
*JEDEC Registration		400	mW
*Derate above $T_L = 50^oC$		3.2	$mW/^oC$
Motorola Device Ratings		500	mW
Derate above $T_L = 50^oC$		3.33	$mW/^oC$
Operating and Storage Junction Temperature Range	T_J, T_{stg}		^oC
*JEDEC Registration		-65 to +175	
Motorola Device Ratings		-65 to +200	

*Indicates JEDEC Registered Data.

MECHANICAL CHARACTERISTICS

MAXIMUM LEAD TEMPERATURE FOR SOLDERING PURPOSES: 230°C, 1/16" from case for 10 seconds

FINISH: All external surfaces are corrosion resistant with readily solderable leads.

POLARITY: Cathode indicated by color band. When operated in zener mode, cathode will be positive with respect to anode.

MOUNTING POSITION: Any

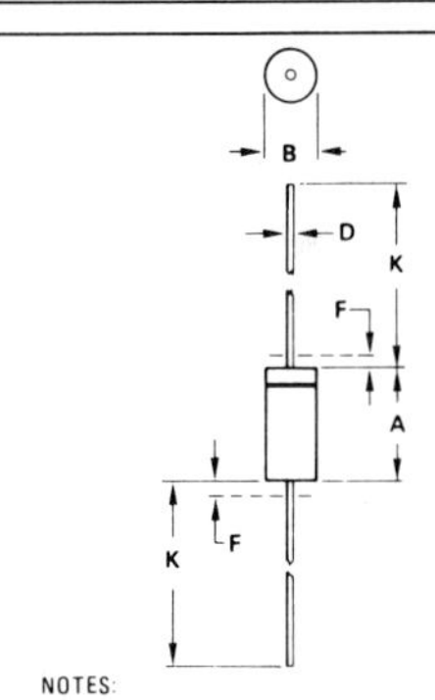

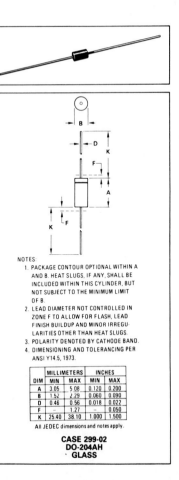

NOTES:
1. PACKAGE CONTOUR OPTIONAL WITHIN A AND B. HEAT SLUGS, IF ANY, SHALL BE INCLUDED WITHIN THIS CYLINDER, BUT NOT SUBJECT TO THE MINIMUM LIMIT OF B.
2. LEAD DIAMETER NOT CONTROLLED IN ZONE F TO ALLOW FOR FLASH, LEAD FINISH BUILDUP AND MINOR IRREGULARITIES OTHER THAN HEAT SLUGS.
3. POLARITY DENOTED BY CATHODE BAND.
4. DIMENSIONING AND TOLERANCING PER ANSI Y14.5, 1973.

DIM	MILLIMETERS MIN	MILLIMETERS MAX	INCHES MIN	INCHES MAX
A	3.05	5.08	0.120	0.200
B	1.52	2.29	0.060	0.090
D	0.46	0.56	0.018	0.022
F	–	1.27	–	0.050
K	25.40	38.10	1.000	1.500

All JEDEC dimensions and notes apply.

CASE 299-02
DO-204AH
GLASS

STEADY STATE POWER DERATING

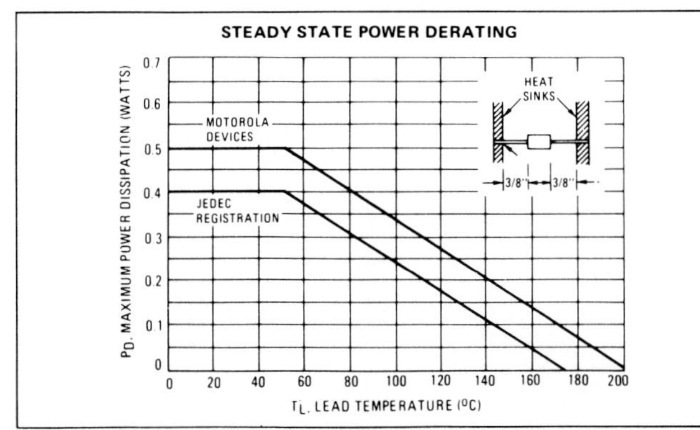

1N746 thru 1N759, 1N957A thru 1N986A, 1N4370 thru 1N4372

ELECTRICAL CHARACTERISTICS (T_A = 25°C, V_F = 1.5 V max at 200 mA for all types)

Type Number (Note 1)	Nominal Zener Voltage V_Z @ I_{ZT} (Note 2) Volts	Test Current I_{ZT} mA	Maximum Zener Impedance Z_{ZT} @ I_{ZT} (Note 3) Ohms	*Maximum DC Zener Current I_{ZM} (Note 4) mA		Maximum Reverse Leakage Current T_A = 25°C I_R @ V_R = 1 V µA	Maximum Reverse Leakage Current T_A = 150°C I_R @ V_R = 1 V µA
1N4370	2.4	20	30	150	190	100	200
1N4371	2.7	20	30	135	165	75	150
1N4372	3.0	20	29	120	150	50	100
1N746	3.3	20	28	110	135	10	30
1N747	3.6	20	24	100	125	10	30
1N748	3.9	20	23	95	115	10	30
1N749	4.3	20	22	85	105	2	30
1N750	4.7	20	19	75	95	2	30
1N751	5.1	20	17	70	85	1	20
1N752	5.6	20	11	65	80	1	20
1N753	6.2	20	7	60	70	0.1	20
1N754	6.8	20	5	55	65	0.1	20
1N755	7.5	20	6	50	60	0.1	20
1N756	8.2	20	8	45	55	0.1	20
1N757	9.1	20	10	40	50	0.1	20
1N758	10	20	17	35	45	0.1	20
1N759	12	20	30	30	35	0.1	20

Type Number (Note 1)	Nominal Zener Voltage V_Z (Note 2) Volts	Test Current I_{ZT} mA	Maximum Zener Impedance (Note 3) Z_{ZT} @ I_{ZT} Ohms	Z_{ZK} @ I_{ZK} Ohms	I_{ZK} mA	*Maximum DC Zener Current I_{ZM} (Note 4) mA		Maximum Reverse Current I_R Maximum µA	Test Voltage Vdc V_R 5%	10%
1N957A	6.8	18.5	4.5	700	1.0	47	61	150	5.2	4.9
1N958A	7.5	16.5	5.5	700	0.5	42	55	75	5.7	5.4
1N959A	8.2	15	6.5	700	0.5	38	50	50	6.2	5.9
1N960A	9.1	14	7.5	700	0.5	35	45	25	6.9	6.6
1N961A	10	12.5	8.5	700	0.25	32	41	10	7.6	7.2
1N962A	11	11.5	9.5	700	0.25	28	37	5	8.4	8.0
1N963A	12	10.5	11.5	700	0.25	26	34	5	9.1	8.6
1N964A	13	9.5	13	700	0.25	24	32	5	9.9	9.4
1N965A	15	8.5	16	700	0.25	21	27	5	11.4	10.8
1N966A	16	7.8	17	700	0.25	19	37	5	12.2	11.5
1N967A	18	7.0	21	750	0.25	17	23	5	13.7	13.0
1N968A	20	6.2	25	750	0.25	15	20	5	15.2	14.4
1N969A	22	5.6	29	750	0.25	14	18	5	16.7	15.8
1N970A	24	5.2	33	750	0.25	13	17	5	18.2	17.3
1N971A	27	4.6	41	750	0.25	11	15	5	20.6	19.4
1N972A	30	4.2	49	1000	0.25	10	13	5	22.8	21.6
1N973A	33	3.8	58	1000	0.25	9.2	12	5	25.1	23.8
1N974A	36	3.4	70	1000	0.25	8.5	11	5	27.4	25.9
1N975A	39	3.2	80	1000	0.25	7.8	10	5	29.7	28.1
1N976A	43	3.0	93	1500	0.25	7.0	9.6	5	32.7	31.0
1N977A	47	2.7	105	1500	0.25	6.4	8.8	5	35.8	33.8
1N978A	51	2.5	125	1500	0.25	5.9	8.1	5	38.8	36.7
1N979A	56	2.2	150	2000	0.25	5.4	7.4	5	42.6	40.3
1N980A	62	2.0	185	2000	0.25	4.9	6.7	5	47.1	44.6
1N981A	68	1.8	230	2000	0.25	4.5	6.1	5	51.7	49.0
1N982A	75	1.7	270	2000	0.25	1.0	5.5	5	56.0	54.0
1N983A	82	1.5	330	3000	0.25	3.7	5.0	5	62.2	59.0
1N984A	91	1.4	400	3000	0.25	3.3	4.5	5	69.2	65.5
1N985A	100	1.3	500	3000	0.25	3.0	4.5	5	76	72
1N986A	110	1.1	750	4000	0.25	2.7	4.1	5	83.6	79.2

NOTE 1. TOLERANCE AND VOLTAGE DESIGNATION

Tolerance Designation

The type numbers shown have tolerance designations as follows:

1N4370 series: ±10%, suffix A for ±5% units,
C for ±2%, D for ±1%.
1N746 series: ±10%, suffix A for ±5% units,
C for ±2%, D for ±1%.
1N957 series: ±10%, suffix A for ±10% units,
C for ±2%, D for ±1%,
suffix B for ±5% units,
C for ±2%, D for ±1%.

MOTOROLA
SEMICONDUCTOR
TECHNICAL DATA

1N1183A thru 1N1190A

MEDIUM-CURRENT RECTIFIERS

. . . for applications requiring low forward voltage drop and rugged construction.

- High Surge Handling Ability
- Rugged Construction
- Reverse Polarity Available; Eliminates Need for Insulating Hardware in Many Cases
- Hermetically Sealed

20-AMP RECTIFIERS

SILICON DIFFUSED-JUNCTION

***MAXIMUM RATINGS**

Rating	Symbol	1N1183A	1N1184A	1N1186A	1N1188A	1N1190A	Unit
Peak Repetitive Reverse Voltage	V_{RRM} V_{RWM} V_R	50	100	200	400	600	Volts
Average Half-Wave Rectified Forward Current With Resistive Load @ T_A = 150°C	I_O	40	40	40	40	40	Amp
Peak One Cycle Surge Current (60 Hz and 150°C Case Temperature)	I_{FSM}	800	800	800	800	800	Amp
Operating Junction Temperature	T_J	−65 to +200					°C
Storage Temperature	T_{stg}	−65 to +200					°C

***ELECTRICAL CHARACTERISTICS** (All Types) at 25°C Case Temperature

Characteristic	Symbol	Value	Unit
Maximum Forward Voltage at 100 Amp DC Forward Current	V_F	1.1	Volts
Maximum Reverse Current at Rated DC Reverse Voltage	I_R	5.0	mAdc

THERMAL CHARACTERISTICS

Characteristic	Symbol	Typical	Unit
Thermal Resistance, Junction to Case	$R_{\theta JC}$	1.0	°C/W

*Indicates JEDEC registered data.

DIM	MILLIMETERS MIN	MILLIMETERS MAX	INCHES MIN	INCHES MAX
A	—	20.07	—	0.790
B	16.94	17.45	0.669	0.687
C	—	11.43	—	0.450
D	—	9.53	—	0.375
E	2.92	5.08	0.115	0.200
F	—	2.03	—	0.080
J	10.72	11.51	0.422	0.453
K	19.05	25.40	0.750	1.00
L	3.96	—	0.156	—
P	5.59	6.32	0.220	0.249
Q	3.56	4.45	0.140	0.175
R	—	16.94	—	0.667
S	—	2.26	—	0.089

CASE 42A-01
DO-203AB
METAL

MECHANICAL CHARACTERISTICS

CASE: Welded, hermetically sealed construction
FINISH: All external surfaces corrosion-resistant and the terminal lead is readily solderable
WEIGHT: 25 grams (approx.)
POLARITY: Cathode connected to case (reverse polarity available denoted by Suffix R, i.e.: 1N3212R)
MOUNTING POSITION: Any
MOUNTING TORQUE: 25 in-lb max

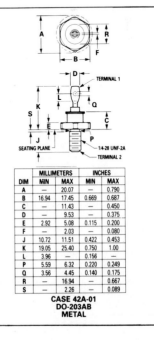

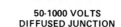

1N4001 thru 1N4007

Designers▲Data Sheet

"SURMETIC"▲ RECTIFIERS

. . . subminiature size, axial lead mounted rectifiers for general-purpose low-power applications.

Designers Data for "Worst Case" Conditions

The Designers▲ Data Sheets permit the design of most circuits entirely from the information presented. Limit curves – representing boundaries on device characteristics – are given to facilitate "worst case" design.

LEAD MOUNTED SILICON RECTIFIERS

50-1000 VOLTS DIFFUSED JUNCTION

*MAXIMUM RATINGS

Rating	Symbol	1N4001	1N4002	1N4003	1N4004	1N4005	1N4006	1N4007	Unit
Peak Repetitive Reverse Voltage Working Peak Reverse Voltage DC Blocking Voltage	V_{RRM} V_{RWM} V_R	50	100	200	400	600	800	1000	Volts
Non-Repetitive Peak Reverse Voltage (halfwave, single phase, 60 Hz)	V_{RSM}	60	120	240	480	720	1000	1200	Volts
RMS Reverse Voltage	$V_{R(RMS)}$	35	70	140	280	420	560	700	Volts
Average Rectified Forward Current (single phase, resistive load, 60 Hz, see Figure 8, $T_A = 75^oC$)	I_O	1.0							Amp
Non-Repetitive Peak Surge Current (surge applied at rated load conditions, see Figure 2)	I_{FSM}	30 (for 1 cycle)							Amp
Operating and Storage Junction Temperature Range	T_J, T_{stg}	-65 to +175							^oC

*ELECTRICAL CHARACTERISTICS

Characteristic and Conditions	Symbol	Typ	Max	Unit
Maximum Instantaneous Forward Voltage Drop (i_F = 1.0 Amp, $T_J = 25^oC$) Figure 1	v_F	0.93	1.1	Volts
Maximum Full-Cycle Average Forward Voltage Drop (I_O = 1.0 Amp, $T_L = 75^oC$, 1 inch leads)	$V_{F(AV)}$	–	0.8	Volts
Maximum Reverse Current (rated dc voltage) $T_J = 25^oC$ $T_J = 100^oC$	I_R	 0.05 1.0	 10 50	μA
Maximum Full-Cycle Average Reverse Current (I_O = 1.0 Amp, $T_L = 75^oC$, 1 inch leads	$I_{R(AV)}$	–	30	μA

*Indicates JEDEC Registered Data.

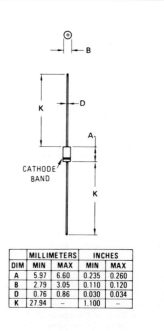

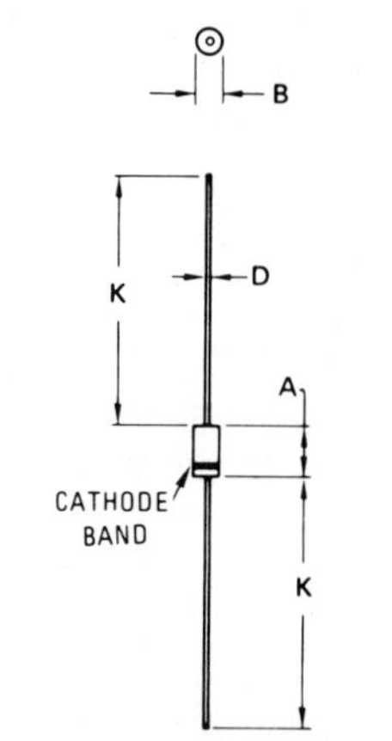

DIM	MILLIMETERS MIN	MILLIMETERS MAX	INCHES MIN	INCHES MAX
A	5.97	6.60	0.235	0.260
B	2.79	3.05	0.110	0.120
D	0.76	0.86	0.030	0.034
K	27.94	–	1.100	–

CASE 59-04

Does Not Conform to DO-41 Outline.

MECHANICAL CHARACTERISTICS

CASE: Transfer Molded Plastic

MAXIMUM LEAD TEMPERATURE FOR SOLDERING PURPOSES: 350^oC, 3/8" from case for 10 seconds at 5 lbs. tension

FINISH: All external surfaces are corrosion-resistant, leads are readily solderable

POLARITY: Cathode indicated by color band

WEIGHT: 0.40 Grams (approximately)

▲Trademark of Motorola Inc.

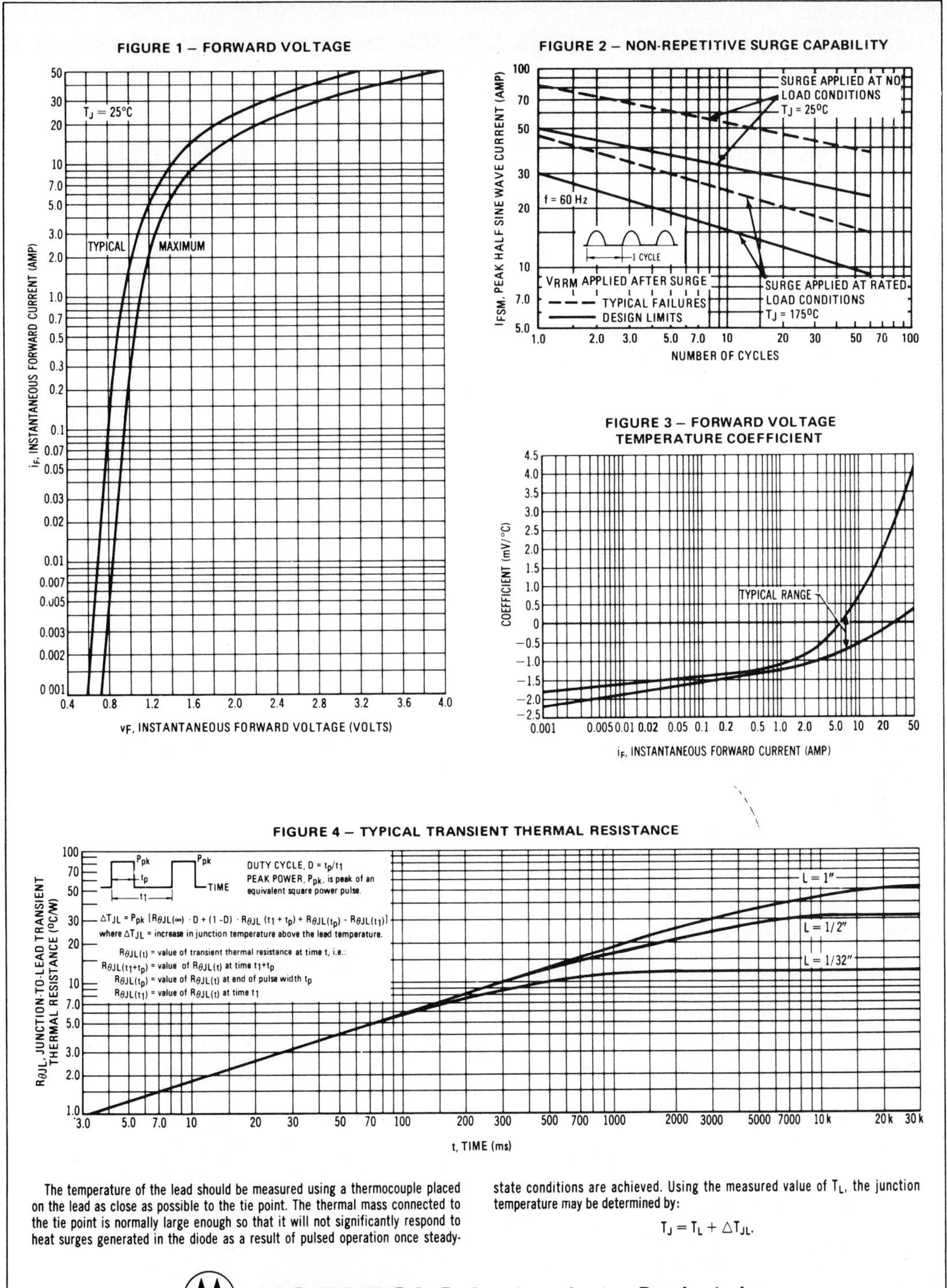

The temperature of the lead should be measured using a thermocouple placed on the lead as close as possible to the tie point. The thermal mass connected to the tie point is normally large enough so that it will not significantly respond to heat surges generated in the diode as a result of pulsed operation once steady-state conditions are achieved. Using the measured value of T_L, the junction temperature may be determined by:

$$T_J = T_L + \Delta T_{JL}.$$

MOTOROLA
SEMICONDUCTOR
TECHNICAL DATA

Infrared LED

MLED930

. . . designed for applications requiring high power output, low drive power and very fast response time. This device is used in industrial processing and control, light modulators, shaft or position encoders, punched card readers, optical switching, and logic circuits. It is spectrally matched for use with silicon detectors.

- High-Power Output — 4 mW (Typ) @ I_F = 100 mA, Pulsed
- Infrared-Emission — 940 nm (Typ)
- Low Drive Current — 10 mA for 450 μW (Typ)
- Popular TO-18 Type Package for Easy Handling and Mounting
- Hermetic Metal Package for Stability and Reliability

INFRARED
LED
940 nm

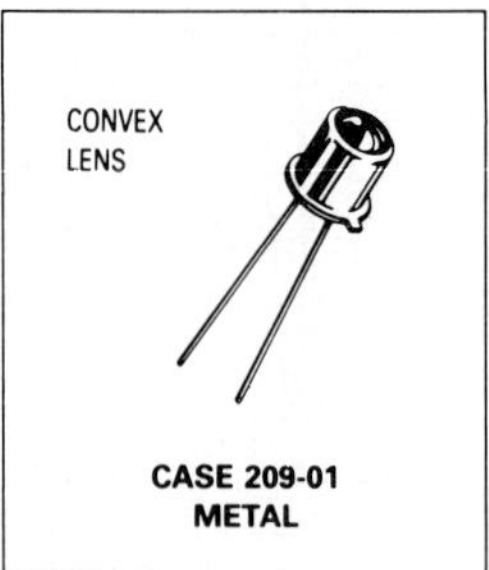

CASE 209-01
METAL

MAXIMUM RATINGS

Rating	Symbol	Value	Unit
Reverse Voltage	V_R	6	Volts
Forward Current — Continuous	I_F	60	mA
Forward Current — Peak Pulse (PW = 100 μs, d.c. = 2%)	I_F	1	A
Total Device Dissipation @ T_A = 25°C Derate above 25°C (Note 1)	P_D	250 2.27	mW mW/°C
Operating Temperature Range	T_A	−55 to +125	°C
Storage Temperature Range	T_{stg}	−65 to +150	°C

ELECTRICAL CHARACTERISTICS (T_A = 25°C unless otherwise noted)

Characteristic	Fig. No.	Symbol	Min	Typ	Max	Unit
Reverse Leakage Current (V_R = 3 V)	—	I_R	—	2	—	nA
Reverse Breakdown Voltage (I_R = 100 μA)	—	$V_{(BR)R}$	6	20	—	Volts
Forward Voltage (I_F = 50 mA)	2	V_F	—	1.32	1.5	Volts
Total Capacitance (V_R = 0 V, f = 1 MHz)	—	C_T	—	18	—	pF

OPTICAL CHARACTERISTICS (T_A = 25°C unless otherwise noted)

Characteristic	Fig. No.	Symbol	Min	Typ	Max	Unit
Total Power Output (Note 2) (I_F = 60 mA, dc) (I_F = 100 mA, PW = 100 μs, duty cycle = 2%)	3, 4	P_o	— 1	2.5 4	— —	mW
Radiant Intensity (Note 3) (I_F = 100 mA, PW = 100 μs, duty cycle = 2%)	—	I_o	—	1.5	—	mW/ steradian
Peak Emission Wavelength	1	λP	—	940	—	nm
Spectral Line Half Width	1	Δλ	—	40	—	nm

Notes: 1. Printed Circuit Board Mounting
2. Power Output, P_o, is the total power radiated by the device into a solid angle of 2π steradians. It is measured by directing all radiation leaving the device, within this solid angle, onto a calibrated silicon solar cell.
3. Irradiance from a Light Emitting Diode (LED) can be calculated by:

$H = \frac{I_e}{d^2}$ where H is irradiance in mW/cm^2; I_e is radiant intensity in mW/steradian; d^2 is distance from LED to the detector in cm.

MLED930

TYPICAL CHARACTERISTICS

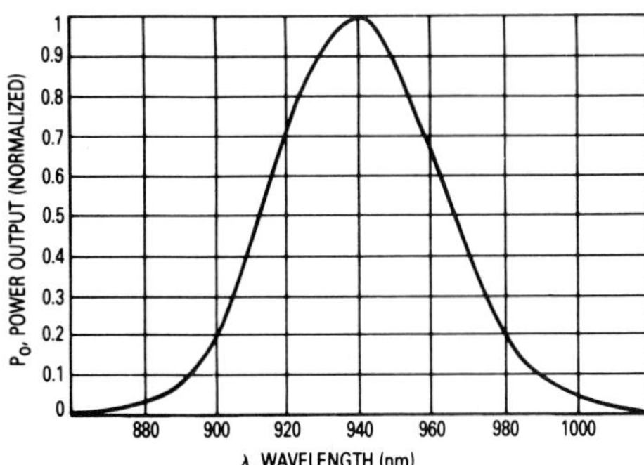

Figure 1. Relative Spectral Output

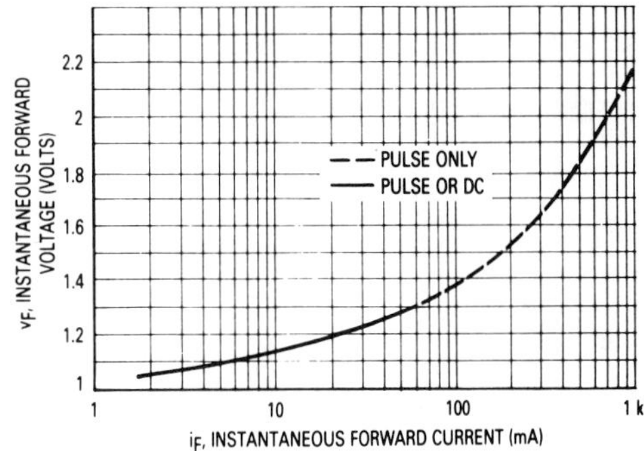

Figure 2. Forward Characteristics

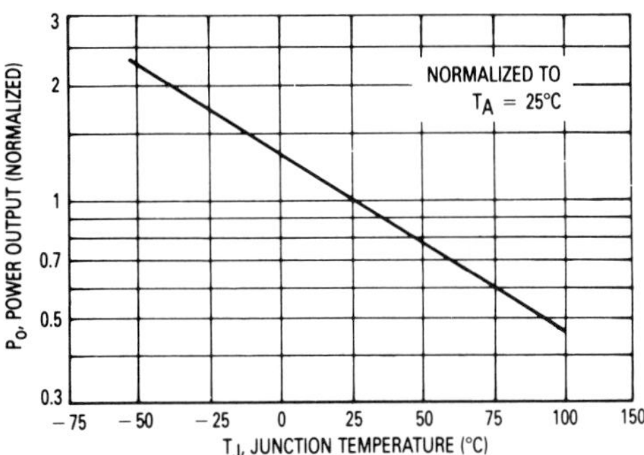

Figure 3. Power Output versus Junction Temperature

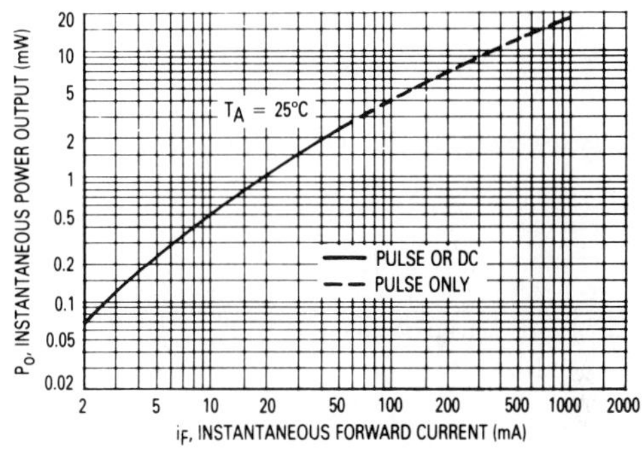

Figure 4. Instantaneous Power Output versus Forward Current

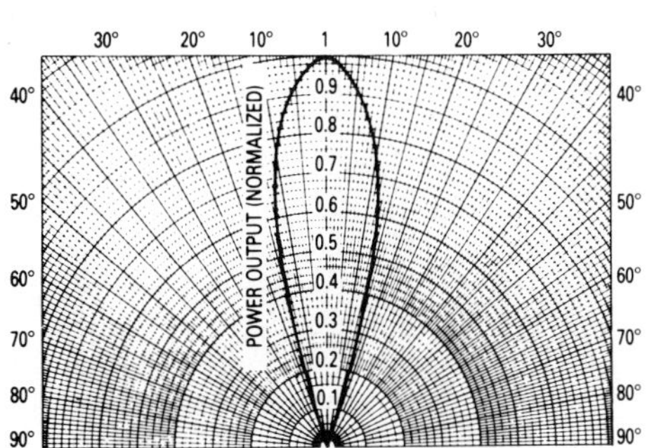

Figure 5. Spatial Radiation Pattern

OUTLINE DIMENSIONS

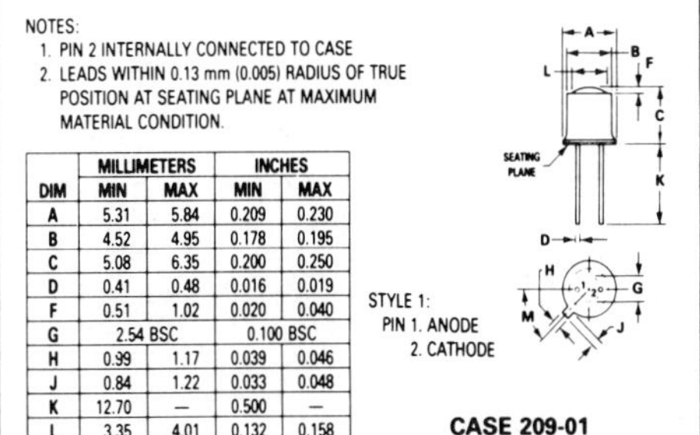

NOTES:
1. PIN 2 INTERNALLY CONNECTED TO CASE
2. LEADS WITHIN 0.13 mm (0.005) RADIUS OF TRUE POSITION AT SEATING PLANE AT MAXIMUM MATERIAL CONDITION.

DIM	MILLIMETERS MIN	MILLIMETERS MAX	INCHES MIN	INCHES MAX
A	5.31	5.84	0.209	0.230
B	4.52	4.95	0.178	0.195
C	5.08	6.35	0.200	0.250
D	0.41	0.48	0.016	0.019
F	0.51	1.02	0.020	0.040
G	2.54 BSC		0.100 BSC	
H	0.99	1.17	0.039	0.046
J	0.84	1.22	0.033	0.048
K	12.70	—	0.500	—
L	3.35	4.01	0.132	0.158
M	45° BSC		45° BSC	

STYLE 1:
PIN 1. ANODE
2. CATHODE

CASE 209-01
METAL

MOTOROLA
SEMICONDUCTOR
TECHNICAL DATA

Photo Detectors
Diode Output

MRD500
MRD510

PHOTO DETECTORS
DIODE OUTPUT
PIN SILICON
250 MILLIWATTS
100 VOLTS

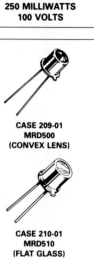

. . . designed for application in laser detection, light demodulation, detection of visible and near infrared light-emitting diodes, shaft or position encoders, switching and logic circuits, or any design requiring radiation sensitivity, ultra high-speed, and stable characteristics.

- Ultra Fast Response — (<1 ns Typ)
- High Sensitivity — MRD500 (1.2 μA/mW/cm^2 Min)
  MRD510 (0.3 μA/mW/cm^2 Min)
- Available With Convex Lens (MRD500) or Flat Glass (MRD510) for Design Flexibility
- Popular TO-18 Type Package for Easy Handling and Mounting
- Sensitive Throughout Visible and Near Infrared Spectral Range for Wide Application
- Annular Passivated Structure for Stability and Reliability

CASE 209-01
MRD500
(CONVEX LENS)

CASE 210-01
MRD510
(FLAT GLASS)

MAXIMUM RATINGS (T_A = 25°C unless otherwise noted)

Rating	Symbol	Value	Unit
Reverse Voltage	V_R	100	Volts
Total Power Dissipation @ T_A = 25°C Derate above 25°C	P_D	250 2.27	mW mW/°C
Operating Temperature Range	T_A	−55 to +125	°C
Storage Temperature Range	T_{stg}	−65 to +150	°C

STATIC ELECTRICAL CHARACTERISTICS (T_A = 25°C unless otherwise noted)

Characteristic	Fig. No.	Symbol	Min	Typ	Max	Unit
Dark Current (V_R = 20 V, R_L = 1 megohm) Note 2 T_A = 25°C T_A = 100°C	2 and 3	I_D	 — —	 — 14	 2 —	nA
Reverse Breakdown Voltage (I_R = 10 μA)	—	$V_{(BR)R}$	100	200	—	Volts
Forward Voltage (I_F = 50 mA)	—	V_F	—	—	1.1	Volts
Series Resistance (I_F = 50 mA)	—	R_s	—	—	10	Ohms
Total Capacitance (V_R = 20 V, f = 1 MHz)	5	C_T	—	—	4	pF

OPTICAL CHARACTERISTICS (T_A = 25°C unless otherwise noted)

Characteristic	Fig. No.	Symbol	Min	Typ	Max	Unit
Light Current MRD500 (V_R = 20 V) Note 1 MRD510	1	I_L	6 1.5	9 2.1	— —	μA
Sensitivity at 0.8 μm MRD500 (V_R = 20 V) Note 3 MRD510	—	$S_{(\lambda = 0.8\ \mu m)}$	— —	6.6 1.5	— —	μA/mW/cm^2
Response Time (V_R = 20 V, R_L = 50 Ohms)	—	$t_{(resp)}$	—	1	—	ns
Wavelength of Peak Spectral Response	5	λ_s	—	0.8	—	μm

NOTES: 1. Radiation Flux Density (H) equal to 5 mW/cm^2 emitted from a tungsten source at a color temperature of 2870 K.
2. Measured under dark conditions. (H ≈ 0).
3. Radiation Flux Density (H) equal to 0.5 mW/cm^2 at 0.8 μm.

TYPICAL CHARACTERISTICS

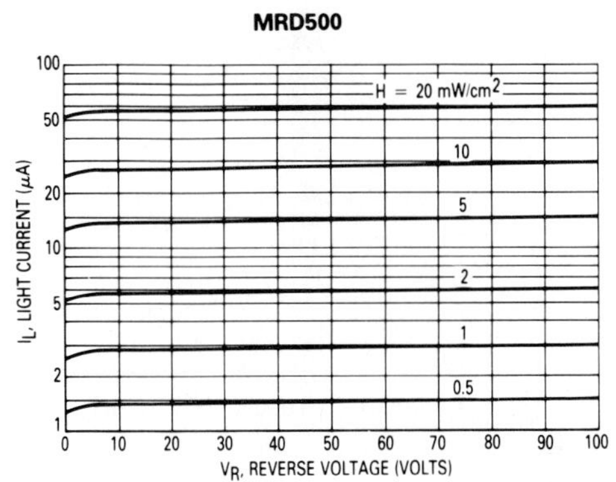

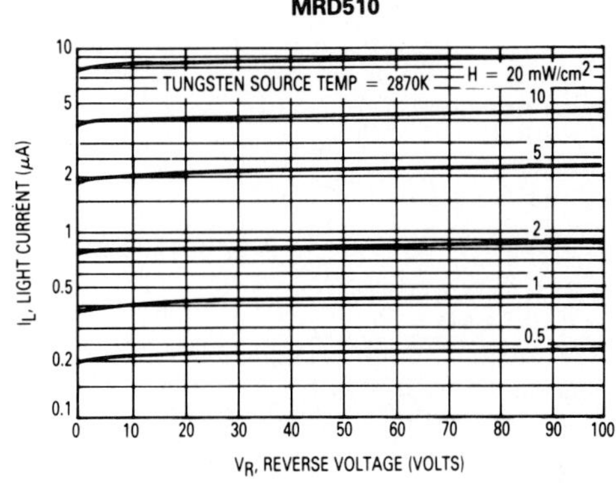

Figure 1. Irradiated Voltage — Current Characteristic

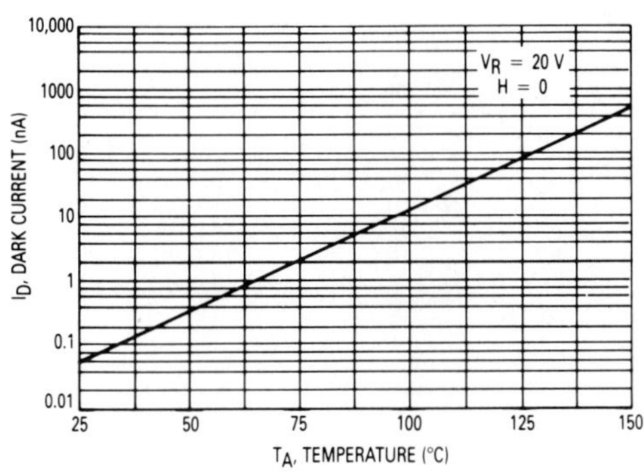

Figure 2. Dark Current versus Temperature

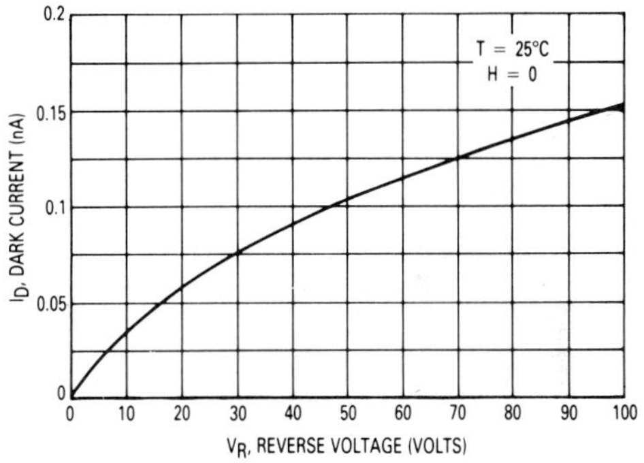

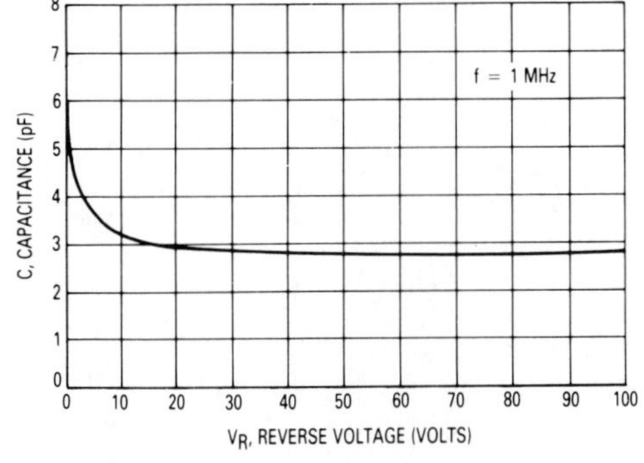

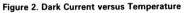

Figure 4. Capacitance versus Voltage

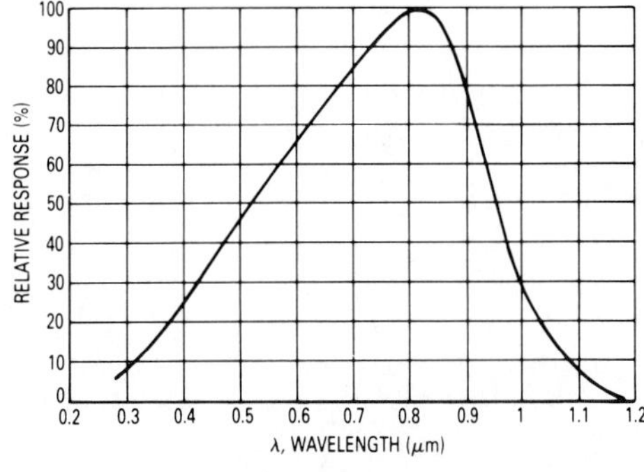

Figure 5. Relative Spectral Response

Figure 3. Dark Current versus Reverse Voltage

T = 25°C
H = 0
ID, DARK CURRENT (nA)
VR, REVERSE VOLTAGE (VOLTS)

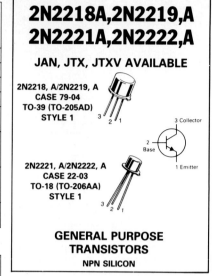

2N2218A,2N2219,A
2N2221A,2N2222,A

JAN, JTX, JTXV AVAILABLE

2N2218, A/2N2219, A
CASE 79-04
TO-39 (TO-205AD)
STYLE 1

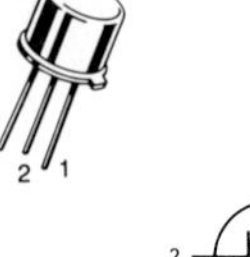

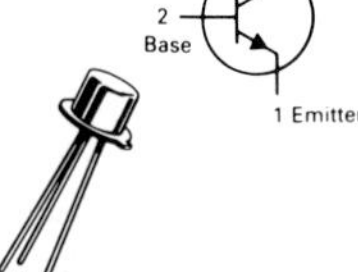

2N2221, A/2N2222, A
CASE 22-03
TO-18 (TO-206AA)
STYLE 1

GENERAL PURPOSE TRANSISTORS

NPN SILICON

MAXIMUM RATINGS

Rating	Symbol	2N2219 2N2222	2N2218A 2N2219A 2N2221A 2N2222A	Unit
Collector-Emitter Voltage	V_{CEO}	30	40	Vdc
Collector-Base Voltage	V_{CBO}	60	75	Vdc
Emitter-Base Voltage	V_{EBO}	5.0	6.0	Vdc
Collector Current — Continuous	I_C	800	800	mAdc
		2N2218A 2N2219,A	2N2221A 2N2222,A	
Total Device Dissipation @ T_A = 25°C Derate above 25°C	P_D	0.8 4.57	0.4 2.28	Watt mW/°C
Total Device Dissipation @ T_C = 25°C Derate above 25°C	P_D	3.0 17.1	1.2 6.85	Watts mW/°C
Operating and Storage Junction Temperature Range	T_J, T_{stg}	−65 to +200		°C

THERMAL CHARACTERISTICS

Characteristic	Symbol	2N2218A 2N2219,A	2N2221A 2N2222,A	Unit
Thermal Resistance, Junction to Ambient	$R_{\theta JA}$	219	145.8	°C/W
Thermal Resistance, Junction to Case	$R_{\theta JC}$	58	437.5	°C/W

ELECTRICAL CHARACTERISTICS (T_A = 25°C unless otherwise noted.)

Characteristic		Symbol	Min	Max	Unit
OFF CHARACTERISTICS					
Collector-Emitter Breakdown Voltage (I_C = 10 mAdc, I_B = 0)	 Non-A Suffix A-Suffix	$V_{(BR)CEO}$	 30 40	 — —	Vdc
Collector-Base Breakdown Voltage (I_C = 10 μAdc, I_E = 0)	 Non-A Suffix A-Suffix	$V_{(BR)CBO}$	 60 75	 — —	Vdc
Emitter-Base Breakdown Voltage (I_E = 10 μAdc, I_C = 0)	 Non-A Suffix A-Suffix	$V_{(BR)EBO}$	 5.0 6.0	 — —	Vdc
Collector Cutoff Current (V_{CE} = 60 Vdc, $V_{EB(off)}$ = 3.0 Vdc)	A-Suffix	I_{CEX}	—	10	nAdc
Collector Cutoff Current (V_{CB} = 50 Vdc, I_E = 0) (V_{CB} = 60 Vdc, I_E = 0) (V_{CB} = 50 Vdc, I_E = 0, T_A = 150°C) (V_{CB} = 60 Vdc, I_E = 0, T_A = 150°C)	 Non-A Suffix A-Suffix Non-A Suffix A-Suffix	I_{CBO}	 — — — —	 0.01 0.01 10 10	μAdc
Emitter Cutoff Current (V_{EB} = 3.0 Vdc, I_C = 0)	A-Suffix	I_{EBO}	—	10	nAdc
Base Cutoff Current (V_{CE} = 60 Vdc, $V_{EB(off)}$ = 3.0 Vdc)	A-Suffix	I_{BL}	—	20	nAdc
ON CHARACTERISTICS					
DC Current Gain (I_C = 0.1 mAdc, V_{CE} = 10 Vdc)	 2N2218A, 2N2221A(1) 2N2219,A, 2N2222,A(1)	h_{FE}	 20 35	 — —	—
(I_C = 1.0 mAdc, V_{CE} = 10 Vdc)	2N2218A, 2N2221A 2N2219,A, 2N2222,A		25 50	— —	
(I_C = 10 mAdc, V_{CE} = 10 Vdc)	2N2218A, 2N2221A(1) 2N2219,A, 2N2222,A(1)		35 75	— —	
(I_C = 10 mAdc, V_{CE} = 10 Vdc, T_A = −55°C)	2N2218A, 2N2221A 2N2219,A, 2N2222,A		15 35	— —	
(I_C = 150 mAdc, V_{CE} = 10 Vdc)(1)	2N2218A, 2N2221A 2N2219,A, 2N2222,A		40 100	120 300	

ELECTRICAL CHARACTERISTICS (continued) (T_A = 25°C unless otherwise noted.)

Characteristic		Symbol	Min	Max	Unit
(I_C = 150 mAdc, V_{CE} = 1.0 Vdc)(1)	2N2218A, 2N2221A		20	—	
	2N2219,A, 2N2222,A		50	—	
(I_C = 500 mAdc, V_{CE} = 10 Vdc)(1)	2N2219, 2N2222		30	—	
	2N2218A, 2N2221A,		25	—	
	2N2219A, 2N2222A		40	—	
Collector-Emitter Saturation Voltage(1)		$V_{CE(sat)}$			Vdc
(I_C = 150 mAdc, I_B = 15 mAdc)	Non-A Suffix		—	0.4	
	A-Suffix		—	0.3	
(I_C = 500 mAdc, I_B = 50 mAdc)	Non-A Suffix		—	1.6	
	A-Suffix		—	1.0	
Base-Emitter Saturation Voltage(1)		$V_{BE(sat)}$			Vdc
(I_C = 150 mAdc, I_B = 15 mAdc)	Non-A Suffix		0.6	1.3	
	A-Suffix		0.6	1.2	
(I_C = 500 mAdc, I_B = 50 mAdc)	Non-A Suffix		—	2.6	
	A-Suffix		—	2.0	

SMALL-SIGNAL CHARACTERISTICS

Characteristic		Symbol	Min	Max	Unit
Current Gain — Bandwidth Product(2)		f_T			MHz
(I_C = 20 mAdc, V_{CE} = 20 Vdc, f = 100 MHz)	All Types, Except		250	—	
	2N2219A, 2N2222A		300	—	
Output Capacitance(3) (V_{CB} = 10 Vdc, I_E = 0, f = 1.0 MHz)		C_{obo}	—	8.0	pF
Input Capacitance(3)		C_{ibo}			pF
(V_{EB} = 0.5 Vdc, I_C = 0, f = 1.0 MHz)	Non-A Suffix		—	30	
	A-Suffix		—	25	
Input Impedance		h_{ie}			kohms
(I_C = 1.0 mAdc, V_{CE} = 10 Vdc, f = 1.0 kHz)	2N2218A, 2N2221A		1.0	3.5	
	2N2219A, 2N2222A		2.0	8.0	
(I_C = 10 mAdc, V_{CE} = 10 Vdc, f = 1.0 kHz)	2N2218A, 2N2221A		0.2	1.0	
	2N2219A, 2N2222A		0.25	1.25	
Voltage Feedback Ratio		h_{re}			$X\ 10^{-4}$
(I_C = 1.0 mAdc, V_{CE} = 10 Vdc, f = 1.0 kHz)	2N2218A, 2N2221A		—	5.0	
	2N2219A, 2N2222A		—	8.0	
(I_C = 10 mAdc, V_{CE} = 10 Vdc, f = 1.0 kHz)	2N2218A, 2N2221A		—	2.5	
	2N2219A, 2N2222A		—	4.0	
Small-Signal Current Gain		h_{fe}			—
(I_C = 1.0 mAdc, V_{CE} = 10 Vdc, f = 1.0 kHz)	2N2218A, 2N2221A		30	150	
	2N2219A, 2N2222A		50	300	
(I_C = 10 mAdc, V_{CE} = 10 Vdc, f = 1.0 kHz)	2N2218A, 2N2221A		50	300	
	2N2219A, 2N2222A		75	375	
Output Admittance		h_{oe}			μmhos
(I_C = 1.0 mAdc, V_{CE} = 10 Vdc, f = 1.0 kHz)	2N2218A, 2N2221A		3.0	15	
	2N2219A, 2N2222A		5.0	35	
(I_C = 10 mAdc, V_{CE} = 10 Vdc, f = 1.0 kHz)	2N2218A, 2N2221A		10	100	
	2N2219A, 2N2222A		15	200	
Collector Base Time Constant (I_E = 20 mAdc, V_{CB} = 20Vdc, f = 31.8 MHz)	A-Suffix	$rb'C_c$	—	150	ps
Noise Figure (I_C = 100 μAdc, V_{CE} = 10 Vdc, R_S = 1.0 kohm, f = 1.0 kHz)	2N2222A	NF	—	4.0	dB
Real Part of Common-Emitter High Frequency Input Impedance (I_C = 20 mAdc, V_{CE} = 20 Vdc, f = 300 MHz)	2N2218A, 2N2219A, 2N2221A, 2N2222A	$Re(h_{ie})$	—	60	Ohms

(1) Pulse Test: Pulse Width ≤ 300 μs, Duty Cycle ≤ 2.0%.
(2) f_T is defined as the frequency at which $|h_{fe}|$ extrapolates to unity.
(3) 2N5581 and 2N5582 are Listed C_{cb} and C_{eb} for these conditions and values.

ELECTRICAL CHARACTERISTICS (continued) (T_A = 25°C unless otherwise noted.)

Characteristic		Symbol	Min	Max	Unit
SWITCHING CHARACTERISTICS					
Delay Time	(V_{CC} = 30 Vdc, $V_{BE(off)}$ = 0.5 Vdc, I_C = 150 mAdc, I_{B1} = 15 mAdc) (Figure 14)	t_d	—	10	ns
Rise Time		t_r	—	25	ns
Storage Time	(V_{CC} = 30 Vdc, I_C = 150 mAdc, I_{B1} = I_{B2} = 15 mAdc) (Figure15)	t_s	—	225	ns
Fall Time		t_f	—	60	ns
Active Region Time Constant (I_C = 150 mAdc, V_{CE} = 30 Vdc) (See Figure 12 for 2N2218A, 2N2219A, 2N2221A, 2N2222A)		T_A	—	2.5	ns

FIGURE 1 – NORMALIZED DC CURRENT GAIN

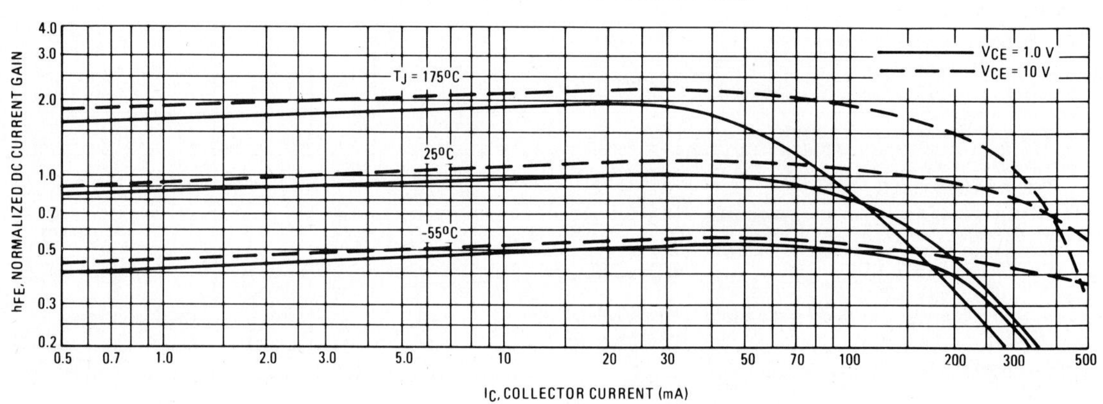

FIGURE 2 – COLLECTOR CHARACTERISTICS IN SATURATION REGION

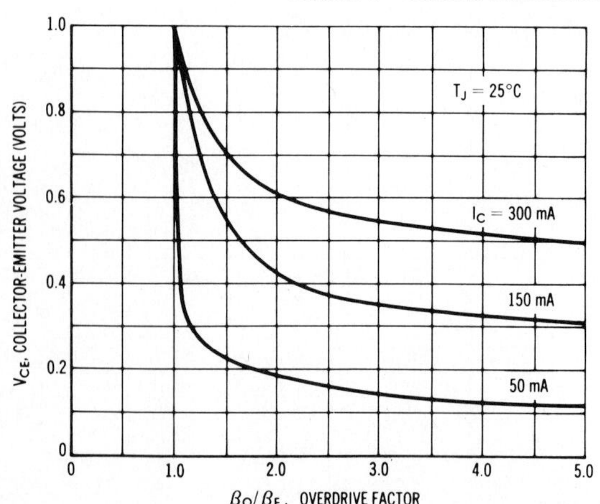

This graph shows the effect of base current on collector current. β_o (current gain at the edge of saturation) is the current gain of the transistor at 1 volt, and β_F (forced gain) is the ratio of I_c/I_{BF} in a circuit.

EXAMPLE: For type 2N2219, estimate a base current (I_{BF}) to insure saturation at a temperature of 25°C and a collector current of 150 mA.

Observe that at I_c = 150 mA an overdrive factor of at least 2.5 is required to drive the transistor well into the saturation region. From Figure 1, it is seen that h_{FE} @ 1 volt is approximately 0.62 of h_{FE} @ 10 volts. Using the guaranteed minimum gain of 100 @ 150 mA and 10 V, β_o = 62 and substituting values in the overdrive equation, we find:

$$\frac{\beta_o}{\beta_F} = \frac{h_{FE} @ 1.0\,V}{I_c/I_{BF}} \qquad 2.5 = \frac{62}{150/I_{BF}} \qquad I_{BF} \approx 6.0 \text{ mA}$$

FIGURE 3 — "ON" VOLTAGES

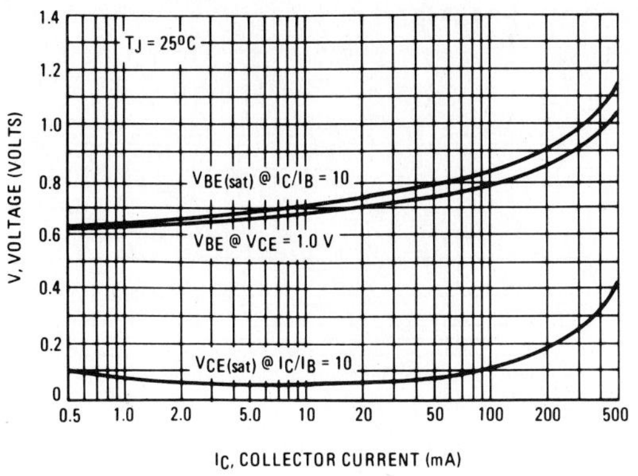

$T_J = 25^oC$

$V_{BE(sat)}$ @ $I_C/I_B = 10$

V_{BE} @ $V_{CE} = 1.0$ V

$V_{CE(sat)}$ @ $I_C/I_B = 10$

V, VOLTAGE (VOLTS)

I_C, COLLECTOR CURRENT (mA)

FIGURE 4 — TEMPERATURE COEFFICIENTS

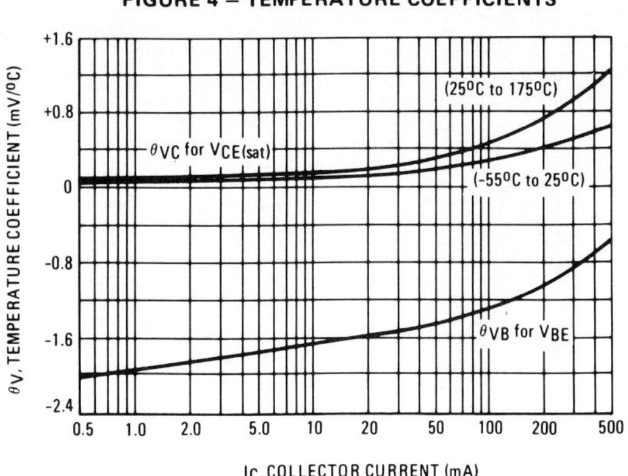

h PARAMETERS

$V_{CE} = 10$ Vdc, f = 1.0 kHz, $T_A = 25^oC$

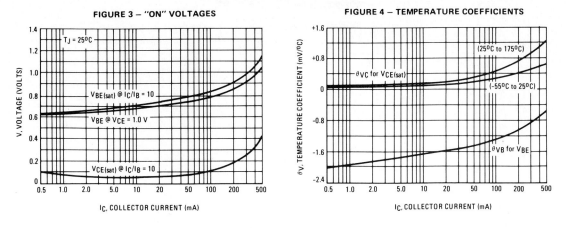

This group of graphs illustrates the relationship between h_{fe} and other "h" parameters for this series of transistors. To obtain these curves, a high-gain and a low-gain unit were selected and the same units were used to develop the correspondingly numbered curves on each graph.

FIGURE 5 — INPUT IMPEDANCE

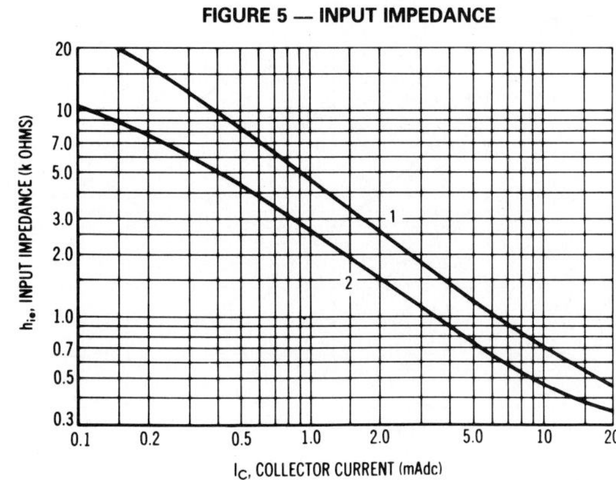

FIGURE 6 — VOLTAGE FEEDBACK RATIO

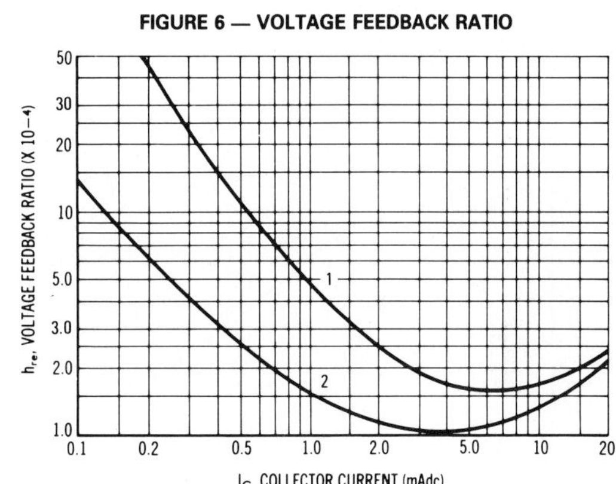

FIGURE 7 — CURRENT GAIN

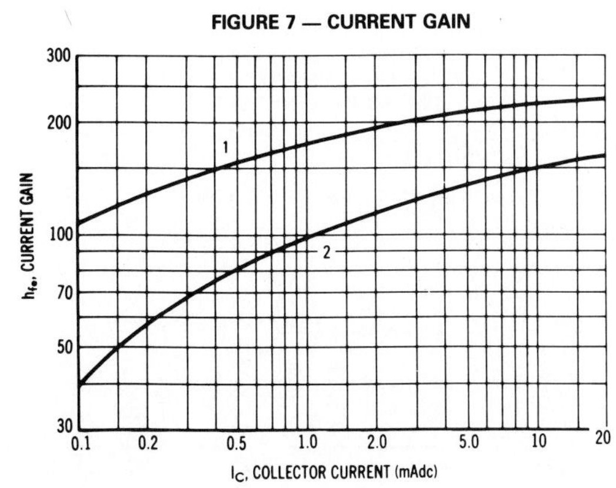

FIGURE 8 — OUTPUT ADMITTANCE

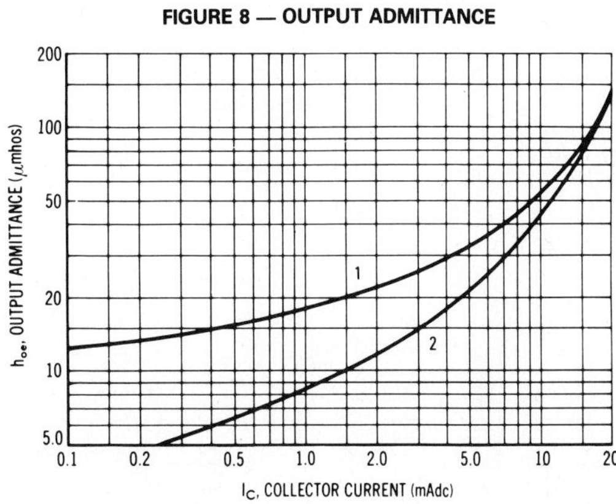

SWITCHING TIME CHARACTERISTICS

FIGURE 9 — TURN-ON TIME

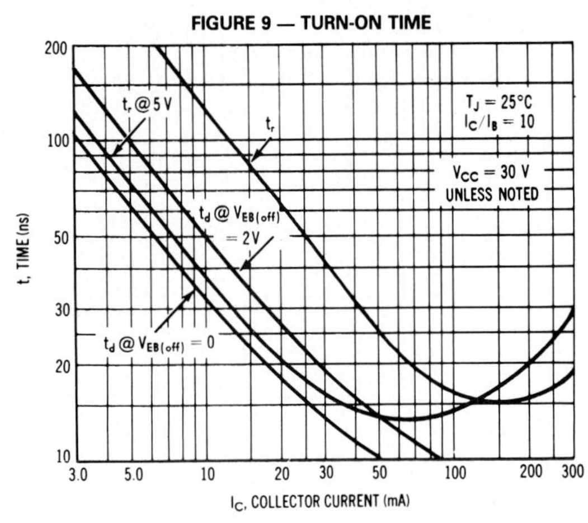

FIGURE 10 — CHARGE DATA

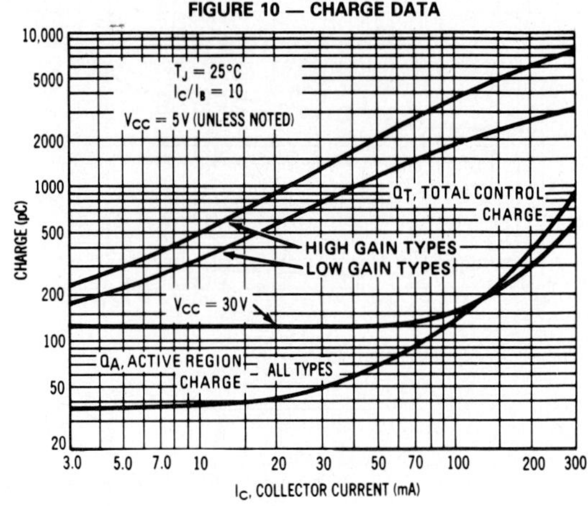

FIGURE 11 — TURN-OFF BEHAVIOR

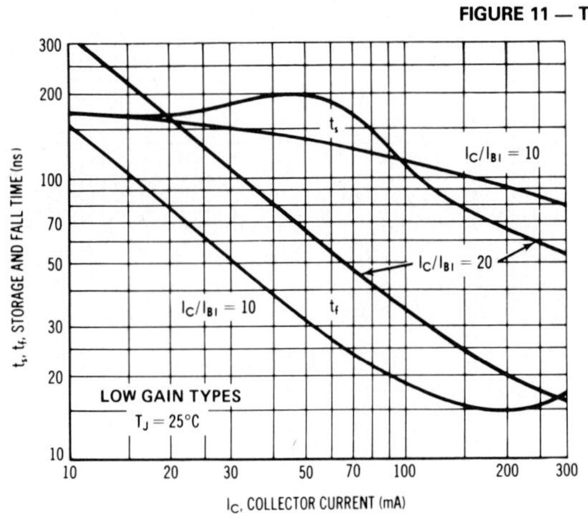

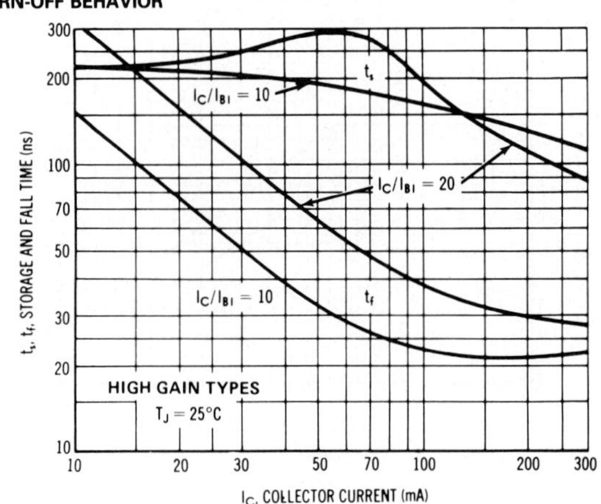

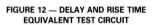

FIGURE 12 — DELAY AND RISE TIME EQUIVALENT TEST CIRCUIT

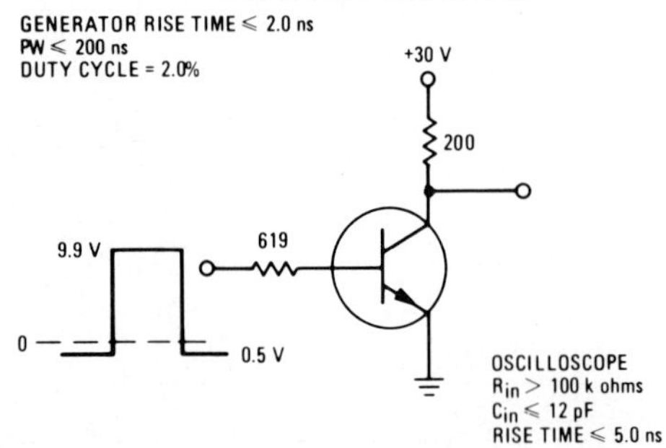

FIGURE 13 — STORAGE TIME AND FALL TIME EQUIVALENT TEST CIRCUIT

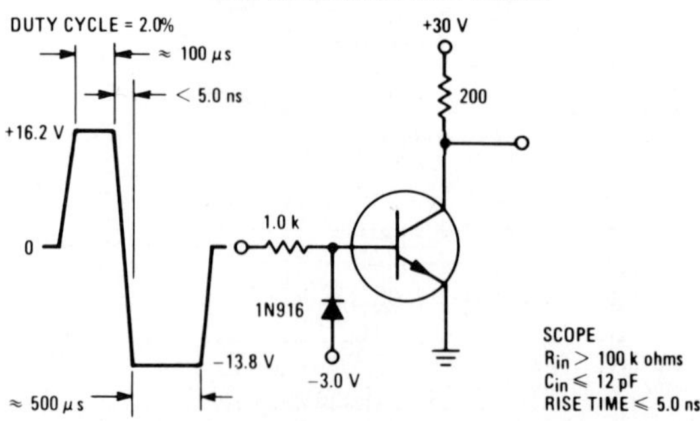

2N3946
2N3947

CASE 22-03, STYLE 1
TO-18 (TO-206AA)

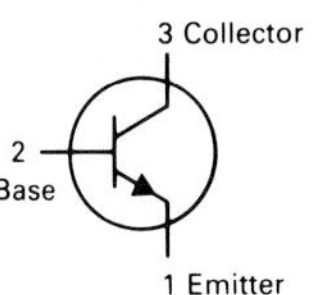

GENERAL PURPOSE TRANSISTORS

NPN SILICON

MAXIMUM RATINGS

Rating	Symbol	Value	Unit
Collector-Emitter Voltage	V_{CEO}	40	Vdc
Collector-Base Voltage	V_{CBO}	60	Vdc
Emitter-Base Voltage	V_{EBO}	6.0	Vdc
Collector Current — Continuous	I_C	200	mAdc
Total Device Dissipation @ T_A = 25°C Derate above 25°C	P_D	0.36 2.06	Watt mW/°C
Total Device Dissipation @ T_C = 25°C Derate above 25°C	P_D	1.2 6.9	Watts mW/°C
Operating and Storage Junction Temperature Range	T_J, T_{stg}	−65 to +200	°C

THERMAL CHARACTERISTICS

Characteristic	Symbol	Max	Unit
Thermal Resistance, Junction to Case	$R_{\theta JC}$	0.15	°C/mW
Thermal Resistance, Junction to Ambient	$R_{\theta JA}$	0.49	°C/mW

ELECTRICAL CHARACTERISTICS (T_A = 25°C unless otherwise noted.)

Characteristic		Symbol	Min	Max	Unit
OFF CHARACTERISTICS					
Collector-Emitter Breakdown Voltage(1) (I_C = 10 mAdc)		$V_{(BR)CEO}$	40	—	Vdc
Collector-Base Breakdown Voltage (I_C = 10 μAdc, I_E = 0)		$V_{(BR)CBO}$	60	—	Vdc
Emitter-Base Breakdown Voltage (I_E = 10 μAdc, I_C = 0)		$V_{(BR)EBO}$	6.0	—	Vdc
Collector Cutoff Current (V_{CE} = 40 Vdc, V_{EB} = 3.0 Vdc) (V_{CE} = 40 Vdc, V_{EB} = 3.0 Vdc, T_A = 150°C)		I_{CEX}	 — —	 0.010 15	μAdc
Base Cutoff Current (V_{CE} = 40 Vdc, V_{EB} = 3.0 Vdc)		I_{BL}	—	.025	μAdc
ON CHARACTERISTICS					
DC Current Gain(1)		h_{FE}			—
(I_C = 0.1 mAdc, V_{CE} = 1.0 Vdc)	2N3946 2N3947		30 60	— —	
(I_C = 1.0 mAdc, V_{CE} = 1.0 Vdc)	2N3946 2N3947		45 90	— —	
(I_C = 10 mAdc, V_{CE} = 1.0 Vdc)	2N3946 2N3947		50 100	150 300	
(I_C = 50 mAdc, V_{CE} = 1.0 Vdc)	2N3946 2N3947		20 40	— —	
Collector-Emitter Saturation Voltage(1) (I_C = 10 mAdc, I_B = 1.0 mAdc) (I_C = 50 mAdc, I_B = 5.0 mAdc)		$V_{CE(sat)}$	 — —	 0.2 0.3	Vdc
Base-Emitter Saturation Voltage(1) (I_C = 10 mAdc, I_B = 1.0 mAdc) (I_C = 50 mAdc, I_B = 5.0 mAdc)		$V_{BE(sat)}$	 0.6 —	 0.9 1.0	Vdc
SMALL-SIGNAL CHARACTERISTICS					
Current-Gain — Bandwidth Product (I_C = 10 mAdc, V_{CE} = 20 Vdc, f = 100 MHz)	 2N3946 2N3947	f_T	 250 300	 — —	MHz
Output Capacitance (V_{CB} = 10 Vdc, I_E = 0, f = 1.0 MHz)		C_{obo}	—	4.0	pF

ELECTRICAL CHARACTERISTICS (continued) (T_A = 25°C unless otherwise noted.)

Characteristic		Symbol	Min	Max	Unit
Input Capacitance (V_{EB} = 1.0 Vdc, I_C = 0, f = 1.0 MHz)		C_{ibo}	—	8.0	pF
Input Impedance (I_C = 1.0 mA, V_{CE} = 10 V, f = 1.0 kHz)	2N3946	h_{ie}	0.5	6.0	kohms
	2N3947		2.0	12	
Voltage Feedback Ratio (I_C = 1.0 mA, V_{CE} = 10 V, f = 1.0 kHz)	2N3946	h_{re}	—	10	X 10^{-4}
	2N3947		—	20	
Small Signal Current Gain (I_C = 1.0 mA, V_{CE} = 10 V, f = 1.0 kHz)	2N3946	h_{fe}	50	250	—
	2N3947		100	700	
Output Admittance (I_C = 1.0 mA, V_{CE} = 10 V, f = 1.0 kHz)	2N3946	h_{oe}	1.0	30	μmhos
	2N3947		5.0	50	
Collector Base Time Constant (I_C = 10 mA, V_{CE} = 20 V, f = 31.8 MHz)		$rb'C_c$	—	200	ps
Noise Figure (I_C = 100 μA, V_{CE} = 5.0 V, R_g = 1.0 kΩ, f = 1.0 kHz)		NF	—	5.0	dB

SWITCHING CHARACTERISTICS

Characteristic	Conditions		Symbol	Min	Max	Unit
Delay Time	V_{CC} = 3.0 Vdc, V_{OB} = 0.5 Vdc, I_C = 10 mAdc, I_{B1} = 1.0 mA		t_d	—	35	ns
Rise Time			t_r	—	35	ns
Storage Time	V_{CC} = 3.0 V, I_C = 10 mA, I_{B1} = I_{B2} = 1.0 mAdc	2N3946	t_s	—	300	ns
		2N3947		—	375	
Fall Time			t_f	—	75	ns

(1) Pulse Test: PW ≤ 300 μs, Duty Cycle ≤ 2%.

TYPICAL SWITCHING CHARACTERISTICS

(T_A = 25°C unless otherwise noted)

FIGURE 1 — DELAY AND RISE TIME

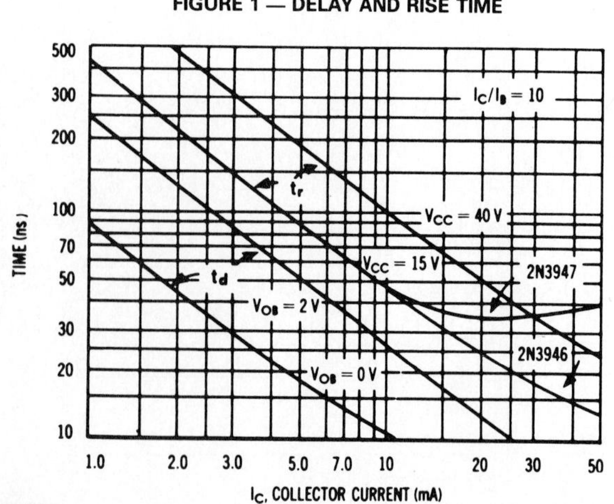

FIGURE 2 — RISE TIME

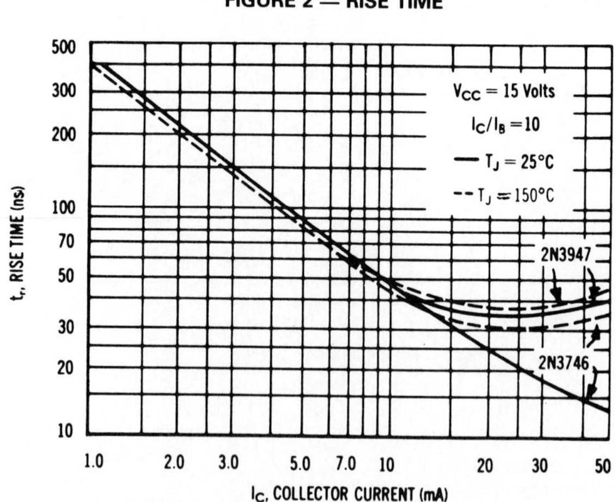

FIGURE 3 — STORAGE AND FALL TIMES

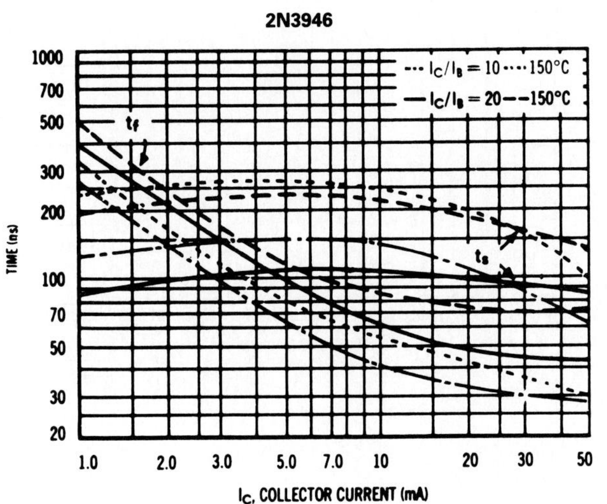

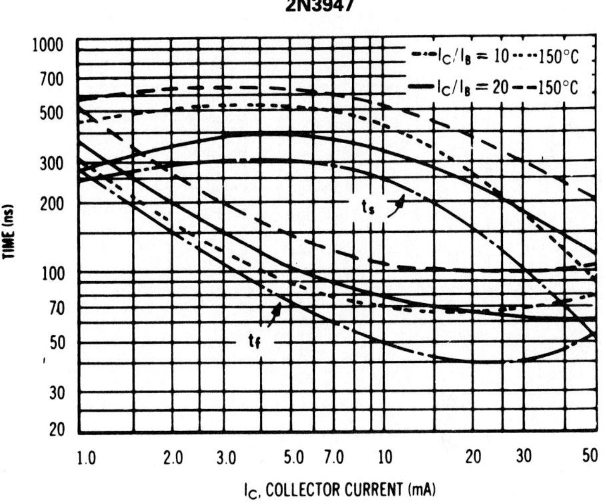

FIGURE 4 — TURN-ON TIME EQUIVALENT TEST CIRCUIT

FIGURE 5 — TURN-OFF TIME EQUIVALENT TEST CIRCUIT

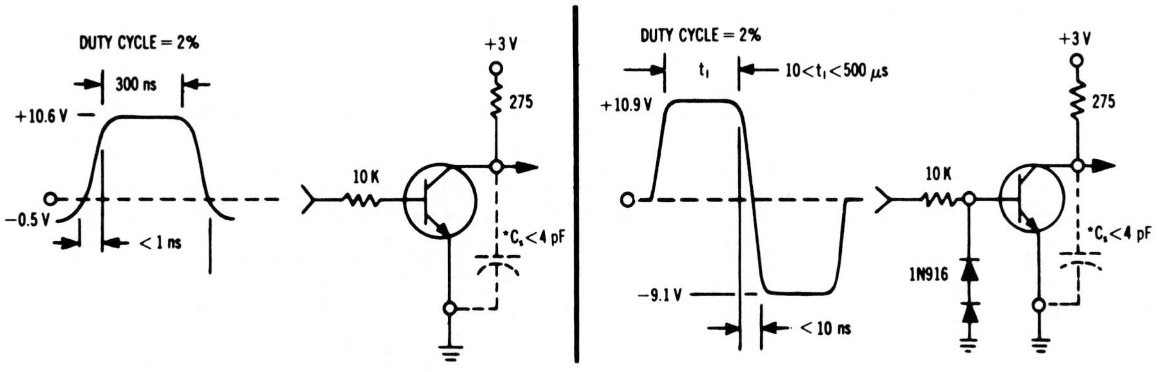

*TOTAL SHUNT CAPACITANCE OF TEST JIG AND CONNECTORS

AUDIO SMALL-SIGNAL CHARACTERISTICS

FIGURE 6 — NOISE FIGURE VARIATIONS
V_{CE} = 5.0 V, T_A = 25°C

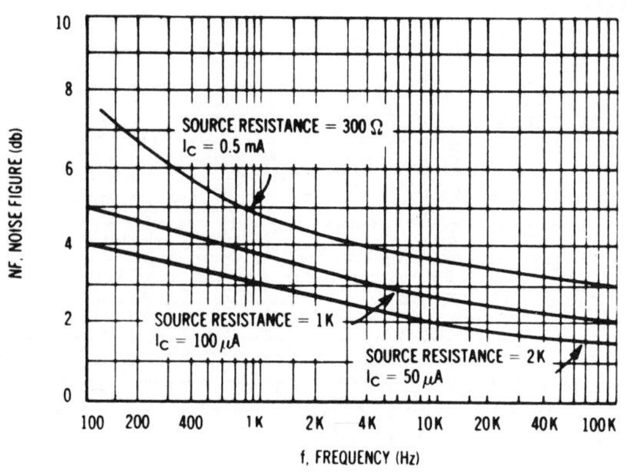

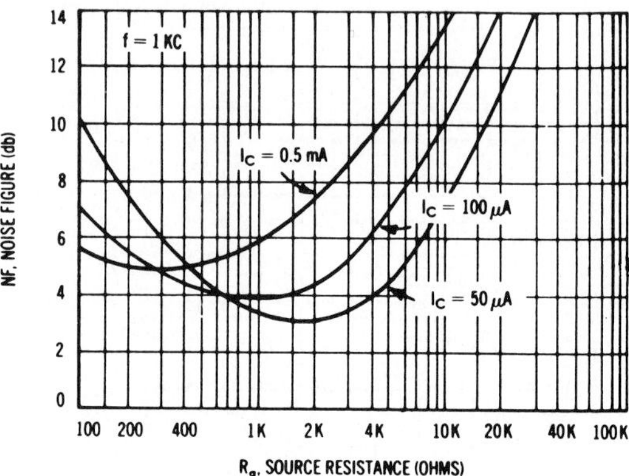

h PARAMETERS
V_{CE} = 10 V, T_A = 25°C, f = 1.0 kc

FIGURE 7 — CURRENT GAIN

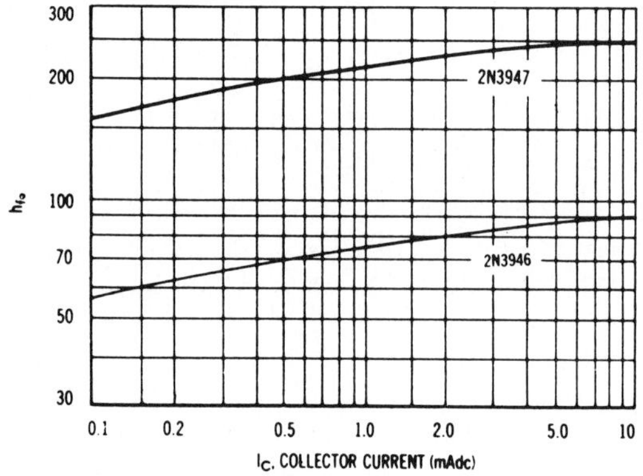

FIGURE 8 — OUTPUT CAPACITANCE

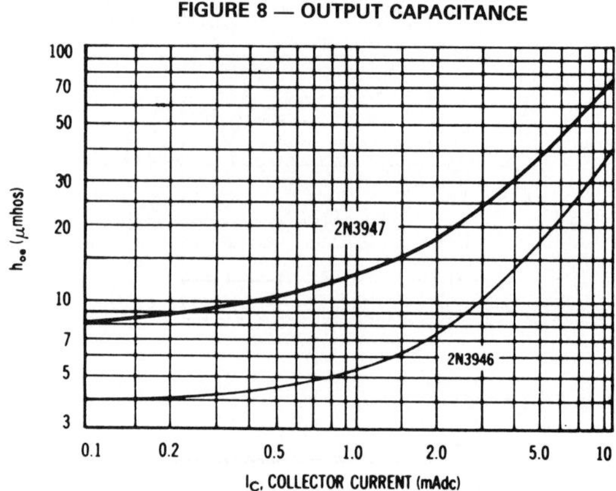

FIGURE 9 — INPUT IMPEDANCE

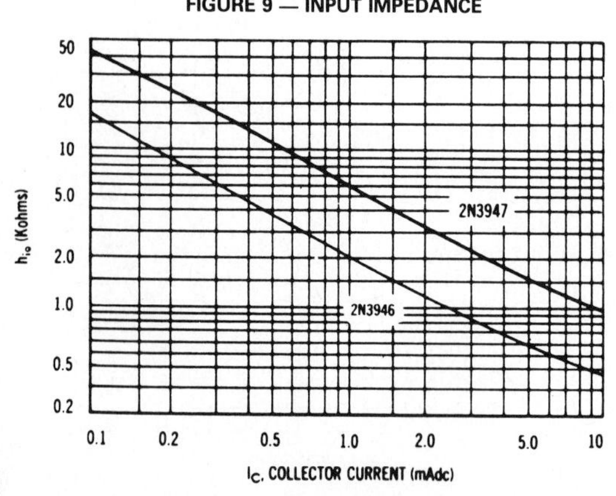

FIGURE 10 — VOLTAGE FEEDBACK RATIO

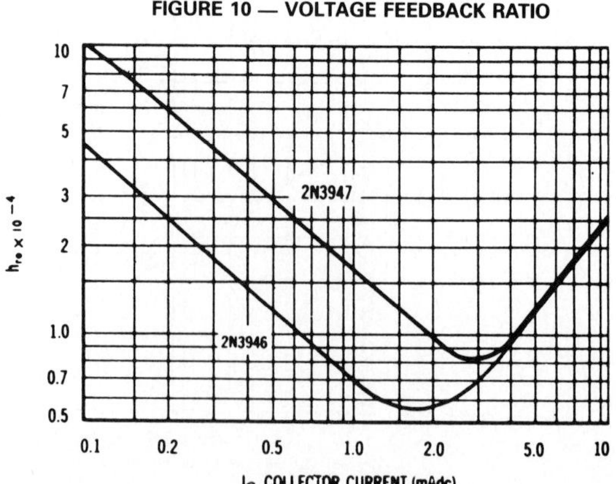

FIGURE 11 — CURRENT GAIN CHARACTERISTICS

2N3946

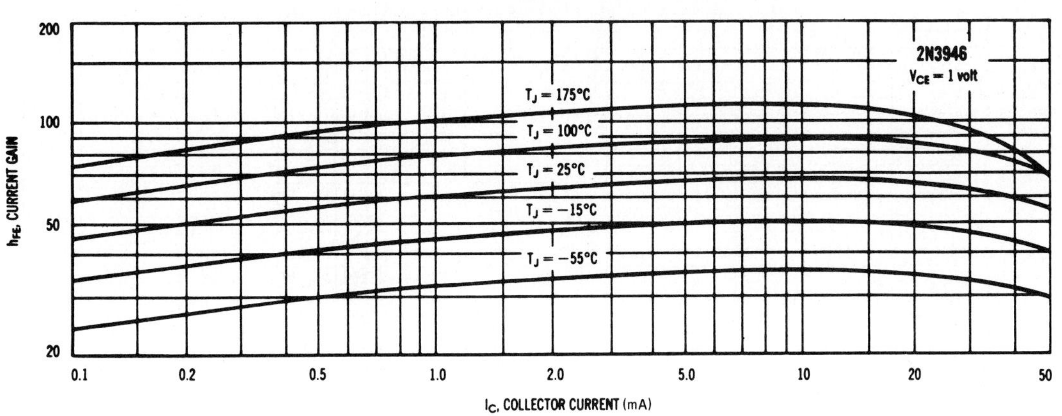

2N3947

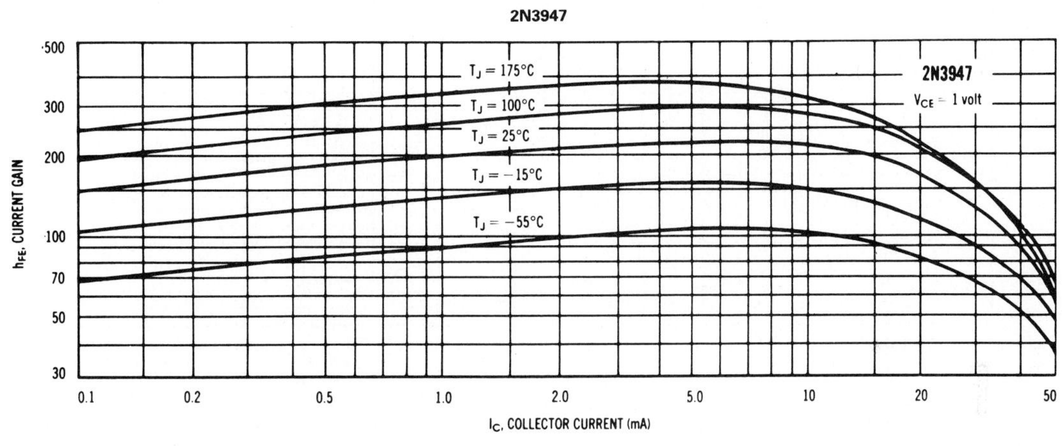

FIGURE 12 — CAPACITANCE

FIGURE 13 — CHARGE DATA

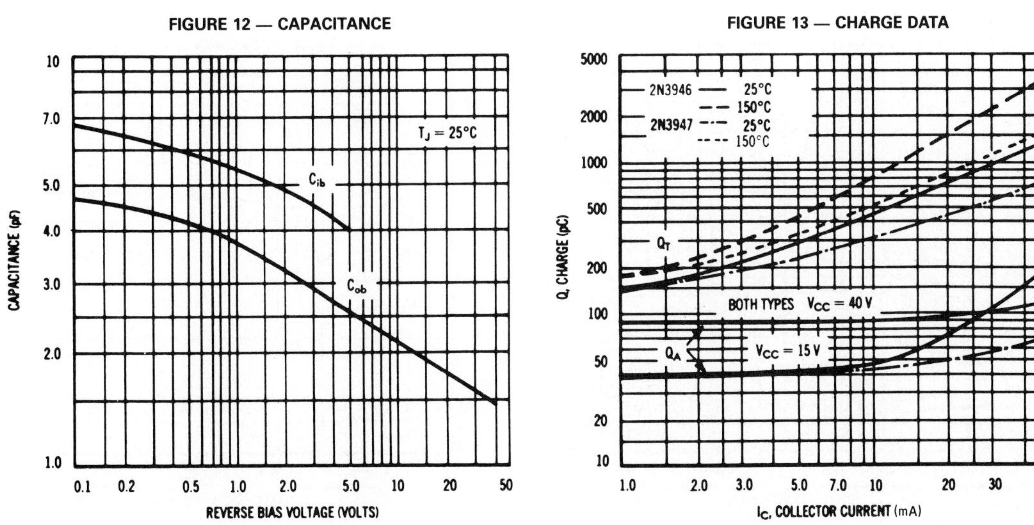

FIGURE 14 — COLLECTOR SATURATION REGION

2N3946

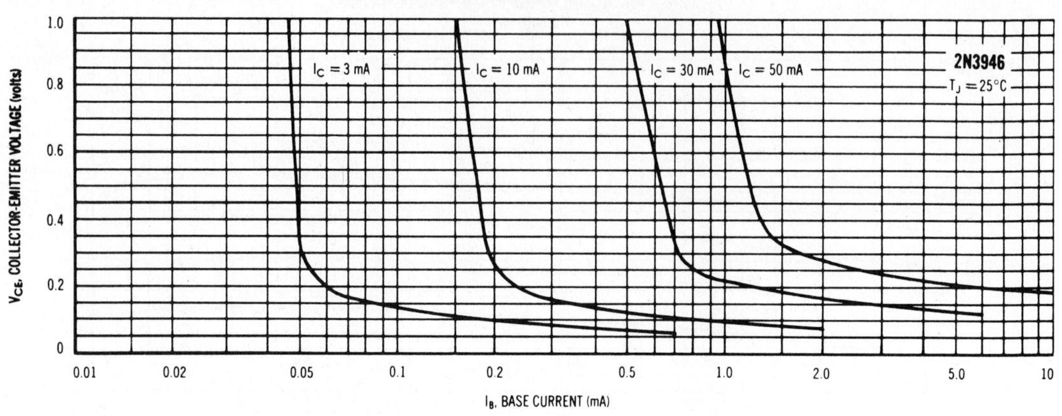

2N3947

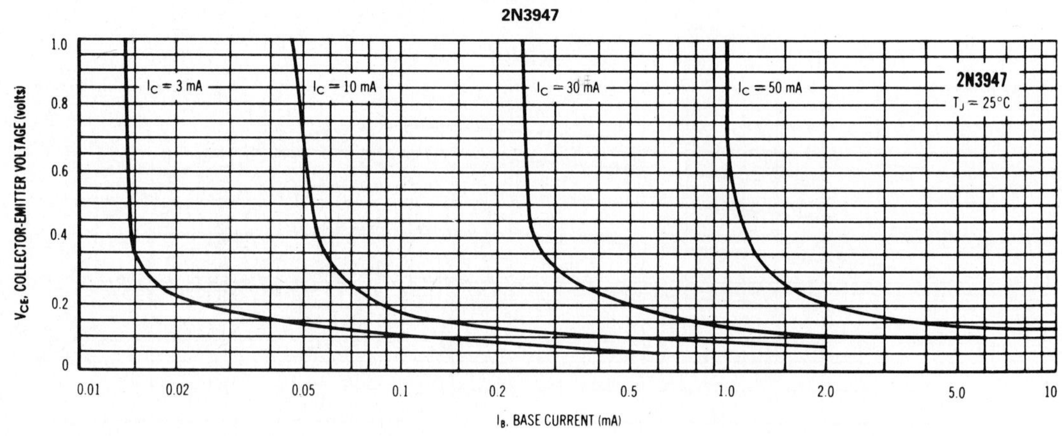

FIGURE 15 — "ON" VOLTAGES

FIGURE 16 — TEMPERATURE COEFFICIENTS

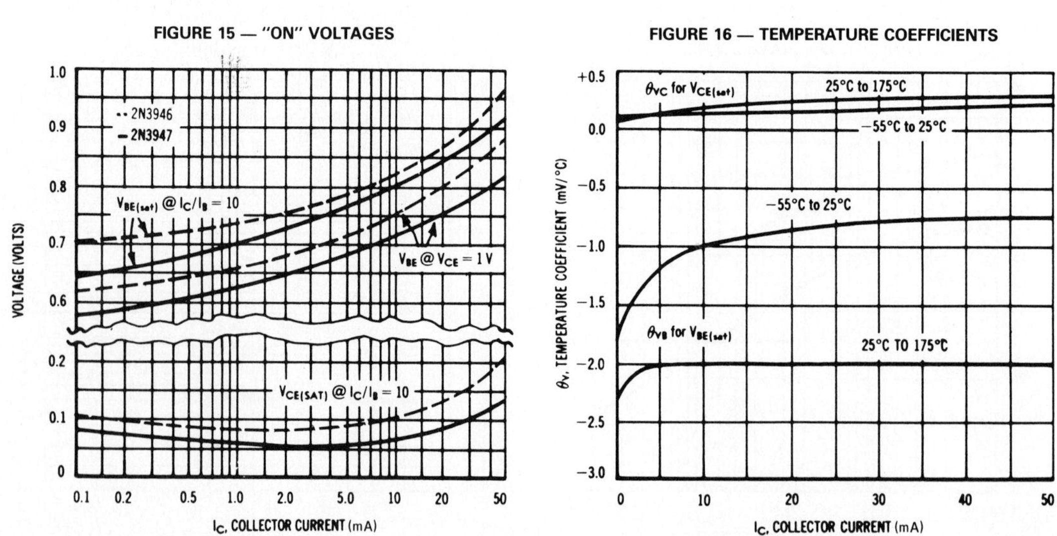

These devices may no longer be available. Please contact nearest Motorola Semiconductor sales office for the current status.

2N2843
2N2844

CASE 22-03, STYLE 12
TO-18 (TO-206AA)

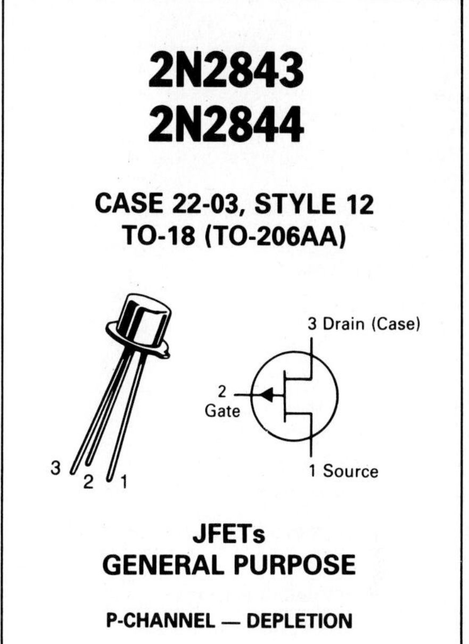

JFETs
GENERAL PURPOSE

P-CHANNEL — DEPLETION

MAXIMUM RATINGS

Rating	Symbol	Value	Unit
Drain-Source Voltage	V_{DS}	30	Vdc
Drain-Gate Voltage	V_{DG}	30	Vdc
Gate-Source Voltage	V_{GS}	30	Vdc
Drain Current	I_D	50	mA
Total Device Dissipation @ $T_A = 25°C$ Derate above 25°C	P_D	300 1.7	mW mW/°C
Storage Temperature Range	T_{stg}	−60 to +200	°C

ELECTRICAL CHARACTERISTICS ($T_A = 25°C$ unless otherwise noted.)

Characteristic		Symbol	Min	Max	Unit
OFF CHARACTERISTICS					
Gate-Source Breakdown Voltage ($I_G = 1.0\ \mu A$)		$V_{(BR)GSS}$	30	—	Vdc
Gate Reverse Current ($V_{GS} = 5.0$ V)		I_{GSS}	—	10	nA
Gate Source Cutoff Voltage ($V_{DS} = -5.0$ V, $I_D = -1.0\ \mu A$)		$V_{GS(off)}$	—	1.7	Vdc
ON CHARACTERISTICS					
Zero-Gate-Voltage Drain Current ($V_{DS} = -5.0$ V)	2N2843 2N2844	I_{DSS}*	−200 −440	−1000 −2200	μA
SMALL-SIGNAL CHARACTERISTICS					
Forward Transfer Admittance ($V_{DS} = -5.0$ V, f = 1.0 kHz)	2N2843 2N2844	$\|y_{fs}\|$*	540 1400	— —	μmhos
Input Capacitance ($V_{DS} = -5.0$ V, $V_{GS} = 1.0$ V, f = 140 kHz)	2N2843 2N2844	C_{iss}	— —	17 30	pF
FUNCTIONAL CHARACTERISTICS					
Noise Figure ($V_{DS} = -5.0$ V, f = 1.0 kHz, $R_G = 1.0$ meg)		NF	—	3.0	dB

*Pulse Width ≤ 630 ms, Duty Cycle = 10%.

These devices may no longer be available. Please contact nearest Motorola Semiconductor sales office for the current status.

2N3796
2N3797

CASE 22-03, STYLE 2
TO-18 (TO-206AA)

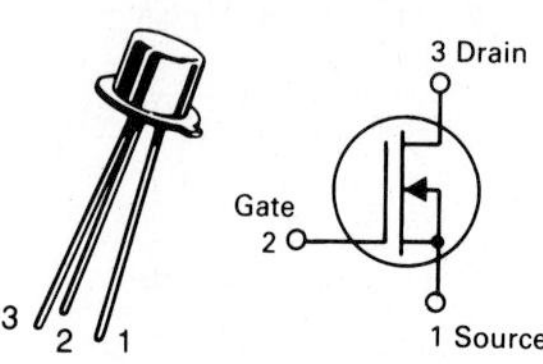

MOSFETs
LOW POWER AUDIO

N-CHANNEL — DEPLETION

MAXIMUM RATINGS

Rating	Symbol	Value	Unit
Drain-Source Voltage 2N3796 / 2N3797	V_{DS}	25 / 20	Vdc
Gate-Source Voltage	V_{GS}	±10	Vdc
Drain Current	I_D	20	mAdc
Total Device Dissipation @ T_A = 25°C / Derate above 25°C	P_D	200 / 1.14	mW / mW/°C
Junction Temperature Range	T_J	+175	°C
Storage Channel Temperature Range	T_{stg}	−65 to +200	°C

ELECTRICAL CHARACTERISTICS (T_A = 25°C unless otherwise noted.)

Characteristic		Symbol	Min	Typ	Max	Unit
OFF CHARACTERISTICS						
Drain-Source Breakdown Voltage (V_{GS} = −4.0 V, I_D = 5.0 μA)	2N3796	$V_{(BR)DSX}$	25	30	—	Vdc
(V_{GS} = −7.0 V, I_D = 5.0 μA)	2N3797		20	25	—	
Gate Reverse Current(1) (V_{GS} = −10 V, V_{DS} = 0)		I_{GSS}	—	—	1.0	pAdc
(V_{GS} = −10 V, V_{DS} = 0, T_A = 150°C)			—	—	200	
Gate Source Cutoff Voltage (I_D = 0.5 μA, V_{DS} = 10 V)	2N3796	$V_{GS(off)}$	—	−3.0	−4.0	Vdc
(I_D = 2.0 μA, V_{DS} = 10 V)	2N3797		—	−5.0	−7.0	
Drain-Gate Reverse Current(1) (V_{DG} = 10 V, I_S = 0)		I_{DGO}	—	—	1.0	pAdc
ON CHARACTERISTICS						
Zero-Gate-Voltage Drain Current (V_{DS} = 10 V, V_{GS} = 0)	2N3796	I_{DSS}	0.5	1.5	3.0	mAdc
	2N3797		2.0	2.9	6.0	
On-State Drain Current (V_{DS} = 10 V, V_{GS} = +3.5 V)	2N3796	$I_{D(on)}$	7.0	8.3	14	mAdc
	2N3797		9.0	14	18	
SMALL-SIGNAL CHARACTERISTICS						
Forward Transfer Admittance (V_{DS} = 10 V, V_{GS} = 0, f = 1.0 kHz)	2N3796	$\|y_{fs}\|$	900	1200	1800	μmhos
	2N3797		1500	2300	3000	
(V_{DS} = 10 V, V_{GS} = 0, f = 1.0 MHz)	2N3796		900	—	—	
	2N3797		1500	—	—	
Output Admittance (V_{DS} = 10 V, V_{GS} = 0, f = 1.0 kHz)	2N3796	$\|y_{os}\|$	—	12	25	μmhos
	2N3797		—	27	60	
Input Capacitance (V_{DS} = 10 V, V_{GS} = 0, f = 1.0 MHz)	2N3796	C_{iss}	—	5.0	7.0	pF
	2N3797		—	6.0	8.0	
Reverse Transfer Capacitance (V_{DS} = 10 V, V_{GS} = 0, f = 1.0 MHz)		C_{rss}	—	0.5	0.8	pF
FUNCTIONAL CHARACTERISTICS						
Noise Figure (V_{DS} = 10 V, V_{GS} = 0, f = 1.0 kHz, R_S = 3 megohms)		NF	—	3.8	—	dB

(1) This value of current includes both the FET leakage current as well as the leakage current associated with the test socket and fixture when measured under best attainable conditions.

TYPICAL DRAIN CHARACTERISTICS

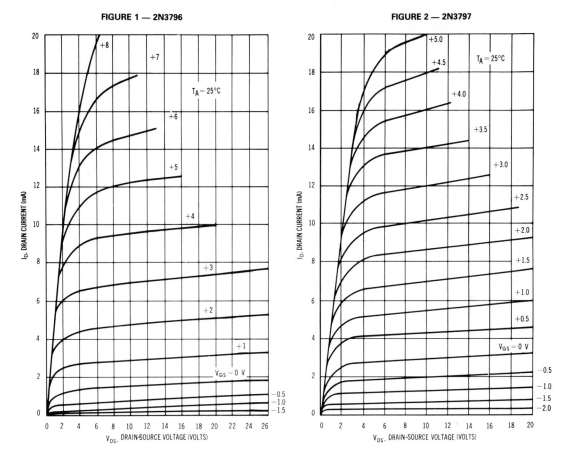

FIGURE 1 — 2N3796

FIGURE 2 — 2N3797

COMMON SOURCE TRANSFER CHARACTERISTICS

FIGURE 3 — 2N3796

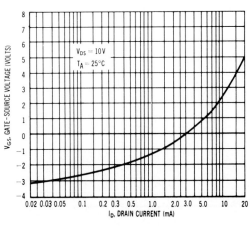

FIGURE 4 — 2N3797

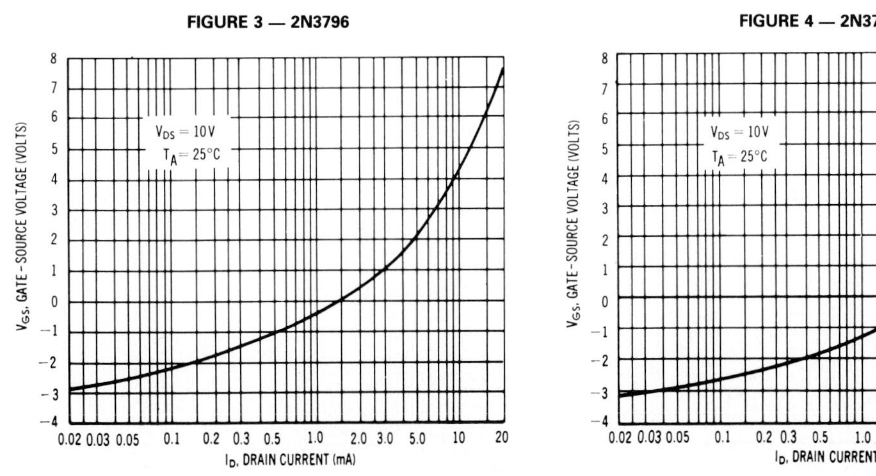

FIGURE 5 — FORWARD TRANSFER ADMITTANCE

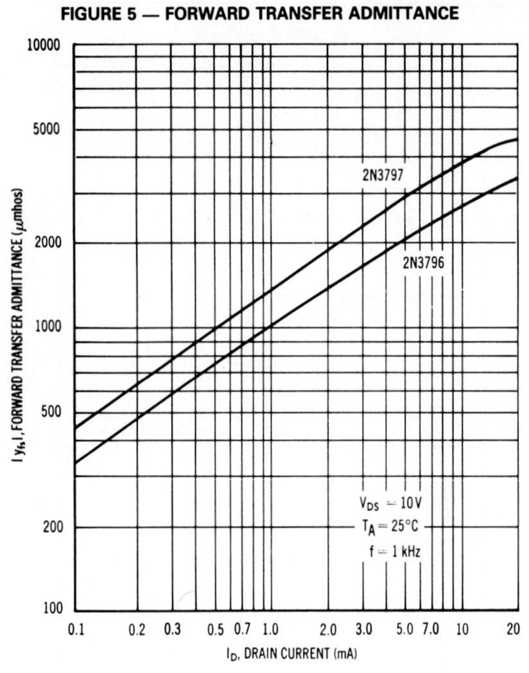

FIGURE 6 — OUTPUT ADMITTANCE

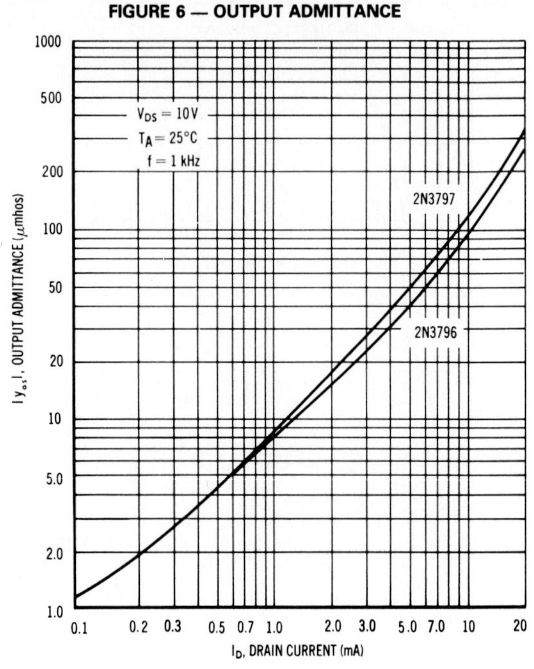

FIGURE 7 — NOISE FIGURE

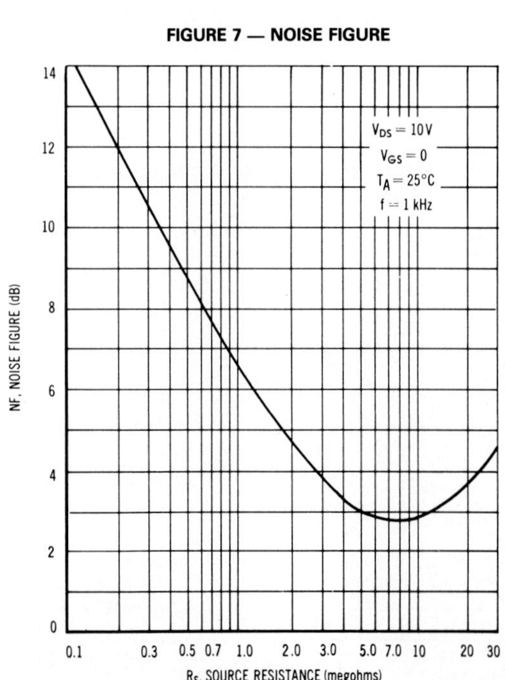

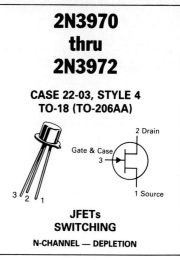

2N3970 thru 2N3972

CASE 22-03, STYLE 4
TO-18 (TO-206AA)

JFETs
SWITCHING
N-CHANNEL — DEPLETION

MAXIMUM RATINGS

Rating	Symbol	Value	Unit
Drain-Source Voltage	V_{DS}	40	Vdc
Drain-Gate Voltage	V_{DG}	40	Vdc
Reverse Gate-Source Voltage	V_{GSR}	40	Vdc
Forward Gate Current	I_{GF}	50	mAdc
Total Device Dissipation @ T_C = 25°C Derate above 25°C	P_D	1.8 10	Watts mW/°C
Storage Temperature Range	T_{stg}	−65 to +200	°C

ELECTRICAL CHARACTERISTICS (T_A = 25°C unless otherwise noted.)

Characteristic		Symbol	Min	Max	Unit
OFF CHARACTERISTICS					
Gate-Source Breakdown Voltage (I_G = 1.0 μAdc, V_{GS} = 0)		$V_{(BR)GSS}$	40	—	Vdc
Gate Reverse Current (V_{GS} = 20 Vdc, V_{DS} = 0)		I_{GSS}	—	250	pAdc
Drain Reverse Current (V_{DG} = 20 Vdc, I_S = 0) (V_{DG} = 20 Vdc, I_S = 0, T_A = 150°C)		I_{DGO}	— —	250 500	pAdc nAdc
Drain Cutoff Current (V_{DS} = 20 Vdc, V_{GS} = −12 Vdc) (V_{DS} = 20 Vdc, V_{GS} = −12 Vdc, T_A = 150°C)		$I_{D(off)}$	— —	250 500	pAdc nAdc
Gate Source Voltage (V_{DS} = 20 Vdc, I_D = 1.0 nAdc)	2N3970 2N3971 2N3972	V_{GS}	−4.0 −2.0 −0.5	−10 −5.0 −3.0	Vdc
ON CHARACTERISTICS					
Zero-Gate-Voltage Drain Current(1) (V_{DS} = 20 Vdc, V_{GS} = 0)	2N3970 2N3971 2N3972	I_{DSS}	50 25 5.0	150 75 30	mAdc
Drain-Source On-Voltage (I_D = 20 mAdc, V_{GS} = 0) (I_D = 10 mAdc, V_{GS} = 0) (I_D = 5.0 mAdc, V_{GS} = 0)	2N3970 2N3971 2N3972	$V_{DS(on)}$	— — —	1.0 1.5 2.0	Vdc
Static Drain-Source On Resistance (I_D = 1.0 mAdc, V_{GS} = 0)	2N3970 2N3971 2N3972	$r_{DS(on)}$	— — —	30 60 100	Ohms
SMALL-SIGNAL CHARACTERISTICS					
Drain-Source "ON" Resistance (V_{GS} = 0, I_D = 0, f = 1.0 kHz)	2N3970 2N3971 2N3972	$r_{ds(on)}$	— — —	30 60 100	Ohms
Input Capacitance (V_{DS} = 20 Vdc, V_{GS} = 0, f = 1.0 MHz)		C_{iss}	—	25	pF
Reverse Transfer Capacitance (V_{DS} = 0, V_{GS} = −12 Vdc, f = 1.0 MHz)		C_{rss}	—	6.0	pF
SWITCHING CHARACTERISTICS					
Turn-On Delay Time	Test Condition for 2N3970: (V_{DD} = 10 Vdc, $V_{GS(on)}$ = 0, $I_{D(on)}$ = 20 mAdc, $V_{GS(off)}$ = 10 Vdc) 2N3970 2N3971 2N3972	$t_{d(on)}$	— — —	10 15 40	ns
Rise Time	Test Condition for 2N3971: (V_{DD} = 10 Vdc, $V_{GS(on)}$ = 0, $I_{D(on)}$ = 10 mAdc, $V_{GS(on)}$ = 5.0 Vdc) 2N3970 2N3971 2N3972	t_r	— — —	10 15 40	ns
Turn-Off Time	Test Condition for 2N3972: (V_{DD} = 10 Vdc, $V_{GS(on)}$ = 0, $I_{D(on)}$ = 5.0 mAdc, $V_{GS(off)}$ = 3.0 Vdc) 2N3970 2N3971 2N3972	t_{off}	— — —	30 60 100	ns

(1) Pulse Test: Pulse Width = 300 μs, Duty Cycle = 3.0%.

MOTOROLA
SEMICONDUCTOR
TECHNICAL DATA

NPN	PNP
2N5629	**2N6029**
2N5630	**2N6030**
2N5631	**2N6031**

HIGH-VOLTAGE — HIGH POWER TRANSISTORS

. . . designed for use in high power audio amplifier applications and high voltage switching regulator circuits.

- High Collector-Emitter Sustaining Voltage —
  $V_{CEO(sus)}$ = 100 Vdc — 2N5629, 2N6029
  = 120 Vdc — 2N5630, 2N6030
  = 140 Vdc — 2N5631, 2N6031
- High DC Current Gain — @ I_C = 8.0 Adc
  h_{FE} = 25 (Min) — 2N5629, 2N6029
  = 20 (Min) — 2N5630, 2N6030
  = 15 (Min) — 2N5631, 2N6031
- Low Collector-Emitter Saturation Voltage —
  $V_{CE(sat)}$ = 1.0 Vdc (Max) @ I_C = 10 Adc

16 AMPERE

POWER TRANSISTORS
COMPLEMENTARY SILICON

100-120-140 VOLTS
200 WATTS

*MAXIMUM RATINGS

Rating	Symbol	2N5629 2N6029	2N5630 2N6030	2N5631 2N6031	Unit
Collector-Emitter Voltage	V_{CEO}	100	120	140	Vdc
Collector-Base Voltage	V_{CB}	100	120	140	Vdc
Emitter-Base Voltage	V_{EB}		7.0		Vdc
Collector Current Continuous Peak	I_C		16 20		Adc
Base Current Continuous	I_B		5.0		Adc
Total Device Dissipation @ T_C = 25°C Derate above 25°C	P_D		200 1.14		Watts W/°C
Operating and Storage Junction Temperature Range	T_J, T_{stg}		-65 to +200		°C

*THERMAL CHARACTERISTICS

Characteristic	Symbol	Max	Unit
Thermal Resistance, Junction to Case	θ_{JC}	0.875	°C/W

*Indicates JEDEC Registered Data.

FIGURE 1 — POWER DERATING

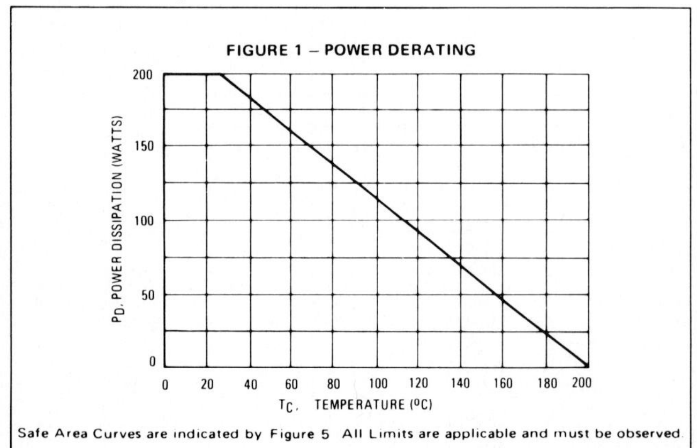

Safe Area Curves are indicated by Figure 5. All Limits are applicable and must be observed.

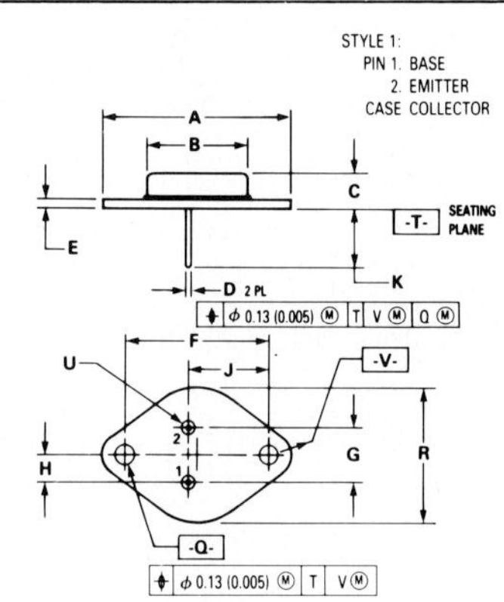

NOTES:
1. DIMENSIONING AND TOLERANCING PER ANSI Y14.5M, 1982.
2. CONTROLLING DIMENSION: INCH.
3. ALL RULES AND NOTES ASSOCIATED WITH REFERENCED TO-204AA OUTLINE SHALL APPLY.

DIM	MILLIMETERS MIN	MILLIMETERS MAX	INCHES MIN	INCHES MAX
A	—	39.37	—	1.550
B	—	21.08	—	0.830
C	6.35	8.25	0.250	0.325
D	0.97	1.09	0.038	0.043
E	1.40	1.77	0.055	0.070
F	30.15 BSC		1.187 BSC	
G	10.92 BSC		0.430 BSC	
H	5.46 BSC		0.215 BSC	
J	16.89 BSC		0.665 BSC	
K	11.18	12.19	0.440	0.480
Q	3.84	4.19	0.151	0.165
R	—	26.67	—	1.050
U	4.83	5.33	0.190	0.210
V	3.84	4.19	0.151	0.165

CASE 1-06
TO-204AA
(TO-3)

2N5629, 2N5630, 2N5631 NPN
2N6029, 2N6030, 2N6031 PNP

*ELECTRICAL CHARACTERISTICS (T_C = 25°C unless otherwise noted)

Characteristic		Symbol	Min	Max	Unit
OFF CHARACTERISTICS					
Collector-Emitter Sustaining Voltage (1) (I_C = 200 mAdc, I_B = 0)	2N5629, 2N6029 2N5630, 2N6030 2N5631, 2N6031	$V_{CEO(sus)}$	100 120 140	— —	Vdc
Collector-Emitter Cutoff Current (V_{CE} = 50 Vdc, I_B = 0) (V_{CE} = 60 Vdc, I_B = 0) (V_{CE} = 70 Vdc, I_B = 0)	2N5629, 2N6029 2N5630, 2N6030 2N5631, 2N6031	I_{CEO}	— — —	2.0 2.0 2.0	mAdc
Collector-Emitter Cutoff Current (V_{CE} = Rated V_{CB}, $V_{EB(off)}$ = 1.5 Vdc) (V_{CE} = Rated V_{CB}, $V_{EB(off)}$ = 1.5 Vdc, T_C = 150°C)		I_{CEX}	— —	2.0 7.0	mAdc
Collector-Base Cutoff Current (V_{CB} = Rated V_{CB}, I_E = 0)		I_{CBO}	—	2.0	mAdc
Emitter-Base Cutoff Current (V_{BE} = 7.0 Vdc, I_C = 0)		I_{EBO}	—	5.0	mAdc
ON CHARACTERISTICS (1)					
DC Current Gain (I_C = 8.0 Adc, V_{CE} = 2.0 Vdc) (I_C = 16 Adc, V_{CE} = 2.0 Vdc)	2N5629, 2N6029 2N5630, 2N6030 2N5631, 2N6031 All Types	h_{FE}	25 20 15 4.0	100 80 60 —	—
Collector-Emitter Saturation Voltage (I_C = 10 Adc, I_B = 1.0 Adc) (I_C = 16 Adc, I_B = 4.0 Adc)	All Types	$V_{CE(sat)}$	— —	1.0 2.0	Vdc
Base-Emitter Saturation Voltage (I_C = 10 Adc, I_B = 1.0 Adc)		$V_{BE(sat)}$	—	1.8	Vdc
Base-Emitter On Voltage (I_C = 8.0 Adc, V_{CE} = 2.0 Vdc)		$V_{BE(on)}$	—	1.5	Vdc
DYNAMIC CHARACTERISTICS					
Current-Gain—Bandwidth Product (2) (I_C = 1.0 Adc, V_{CE} = 20 Vdc, f_{test} = 0.5 MHz)		f_T	1.0	—	MHz
Output Capacitance (V_{CB} = 10 Vdc, I_E = 0, f = 0.1 MHz)	2N5629, 30, 31 2N6029, 30, 31	C_{ob}	—	500 1000	pF
Small-Signal Current Gain (I_C = 4.0 Adc, V_{CE} = 10 Vdc, f = 1.0 kHz)		h_{fe}	15	—	—

*Indicates JEDEC Registered Data.
(1) Pulse Test: Pulse Width ≤ 300 μs, Duty Cycle ≥ 2.0%.
(2) $f_T = |h_{fe}| \bullet f_{test}$

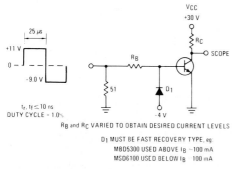

FIGURE 2 – SWITCHING TIMES TEST CIRCUIT

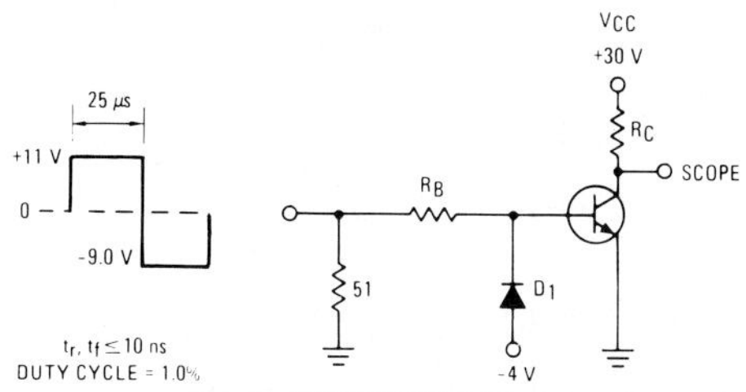

For PNP test circuit, reverse all polarities and D1.

FIGURE 3 – TURN-ON TIME

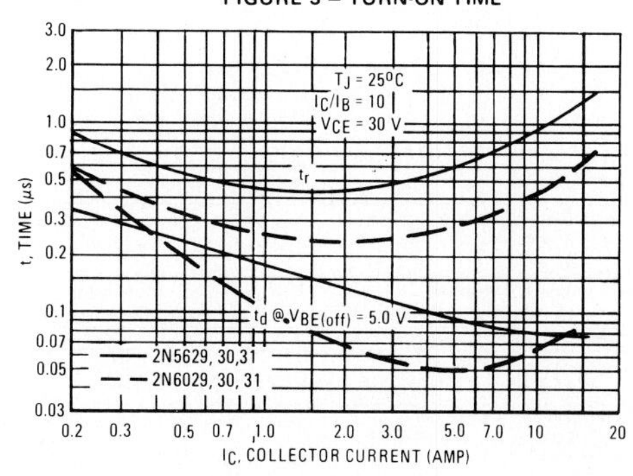

MOTOROLA
SEMICONDUCTOR
TECHNICAL DATA

NPN
2N6121, 2N6122 2N6123
PNP
2N6124, 2N6125

COMPLEMENTARY SILICON PLASTIC POWER TRANSISTORS

. . . designed for use in power amplifier and switching circuits, – packaged in the compact TO-220AB outline. TO-66 leadform also available.

4 AMPERE
POWER TRANSISTORS COMPLEMENTARY SILICON
45-80 VOLTS
40 WATTS

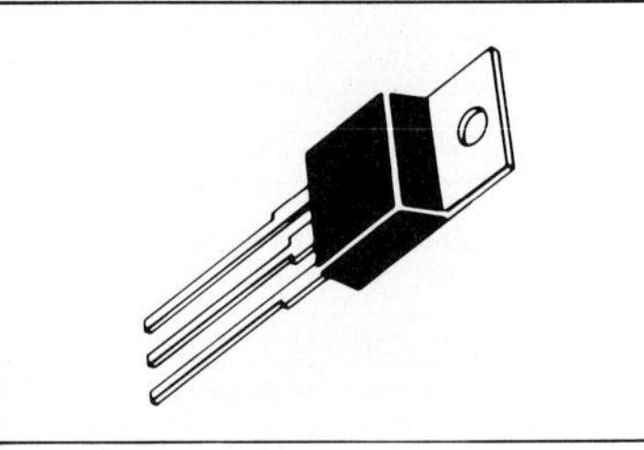

*MAXIMUM RATINGS

Rating	Symbol	2N6121 2N6124	2N6122 2N6125	2N6123	Unit
Collector-Emitter Voltage	V_{CEO}	45	60	80	Vdc
Collector-Base Voltage	V_{CB}	45	60	80	Vdc
Emitter-Base Voltage	V_{EB}	← 5.0 →			Vdc
Collector Current	I_C	← 4.0 →			Adc
Base Current	I_B	← 1.0 →			Adc
Total Power Dissipation @ $T_C = 25°C$ Derate above 25°C	P_D	← 40 → ← 320 →			Watts mW/°C
Operating and Storage Junction Temperature Range	T_J, T_{stg}	← -65 to +150 →			°C

THERMAL CHARACTERISTICS

Characteristic	Symbol	Max	Unit
Thermal Resistance, Junction to Case	$R_{\theta JC}$	3.12	°C/W

*ELECTRICAL CHARACTERISTICS ($T_C = 25°C$ unless otherwise noted)

Characteristic		Symbol	Min	Max	Unit
OFF CHARACTERISTICS					
Collector-Emitter Sustaining Voltage (1)		$V_{CEO(sus)}$			Vdc
($I_C = 0.1$ Adc, $I_B = 0$)	2N6121, 2N6124		45		
	2N6122, 2N6125		60		
	2N6123		80		
Collector Cutoff Current		I_{CEO}			mAdc
($V_{CE} = 45$ Vdc, $I_B = 0$)	2N6121, 2N6124			1.0	
($V_{CE} = 60$ Vdc, $I_B = 0$)	2N6122, 2N6125			1.0	
($V_{CE} = 80$ Vdc, $I_B = 0$)	2N6123			1.0	
Collector Cutoff Current		I_{CEX}			mAdc
($V_{CE} = 45$ Vdc, $V_{EB(off)} = 1.5$ Vdc)	2N6121, 2N6124			0.1	
($V_{CE} = 60$ Vdc, $V_{EB(off)} = 1.5$ Vdc)	2N6122, 2N6125			0.1	
($V_{CE} = 80$ Vdc, $V_{EB(off)} = 1.5$ Vdc)	2N6123			0.1	
($V_{CE} = 45$ Vdc, $V_{EB(off)} = 1.5$ Vdc, $T_C = 125°C$)	2N6121, 2N6124			2.0	
($V_{CE} = 60$ Vdc, $V_{EB(off)} = 1.5$ Vdc, $T_C = 125°C$)	2N6122, 2N6125			2.0	
($V_{CE} = 80$ Vdc, $V_{EB(off)} = 1.5$ Vdc, $T_C = 125°C$)	2N6123, 2N6126			2.0	
Collector Cutoff Current		I_{CBO}			mAdc
($V_{CB} = 45$ Vdc, $I_E = 0$)	2N6121, 2N6124			0.1	
($V_{CB} = 60$ Vdc, $I_E = 0$)	2N6122, 2N6125			0.1	
($V_{CB} = 80$ Vdc, $I_E = 0$)	2N6123			0.1	
Emitter Cutoff Current ($V_{BE} = 5.0$ Vdc, $I_C = 0$)		I_{EBO}		1.0	mAdc
ON CHARACTERISTICS					
DC Current Gain (1)		h_{FE}			
($I_C = 1.5$ Adc, $V_{CE} = 2.0$ Vdc)	2N6126, 2N6124		25	100	
	2N6122, 2N6125		25	100	
	2N6123		20	80	
($I_C = 4.0$ Adc, $V_{CE} = 2.0$ Vdc)	2N6121, 2N6124		10		
	2N6122, 2N6125		10		
	2N6123		7.0		
Collector-Emitter Saturation Voltage (1)		$V_{CE(sat)}$			Vdc
($I_C = 1.5$ Adc, $I_B = 0.15$ Adc)				0.6	
($I_C = 4.0$ Adc, $I_B = 1.0$ Adc)				1.4	
Base-Emitter On Voltage (1) ($I_C = 1.5$ Adc, $V_{CE} = 2.0$ Vdc)		$V_{BE(on)}$		1.2	Vdc
DYNAMIC CHARACTERISTICS					
Small-Signal Current Gain ($I_C = 0.1$ Adc, $V_{CE} = 2.0$ Vdc, f = 1.0 kHz)		h_{fe}	25	-	-
Current-Gain-Bandwidth Product ($I_C = 1.0$ Adc, $V_{CE} = 4.0$ Vdc, f = 1.0 MHz)		f_T	2.5	-	MHz

(1)Pulse Test: Pulse Width ≤ 300 μs, Duty Cycle ≤ 2.0%.
*Indicates JEDEC Registered Data.

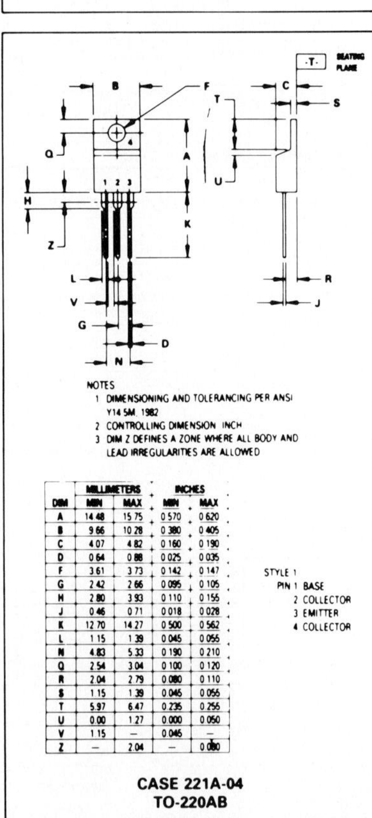

NOTES
1. DIMENSIONING AND TOLERANCING PER ANSI Y14.5M, 1982.
2. CONTROLLING DIMENSION: INCH.
3. DIM Z DEFINES A ZONE WHERE ALL BODY AND LEAD IRREGULARITIES ARE ALLOWED.

DIM	MILLIMETERS MIN	MILLIMETERS MAX	INCHES MIN	INCHES MAX
A	14.48	15.75	0.570	0.620
B	9.66	10.28	0.380	0.405
C	4.07	4.82	0.160	0.190
D	0.64	0.88	0.025	0.035
F	3.61	3.73	0.142	0.147
G	2.42	2.66	0.095	0.105
H	2.80	3.93	0.110	0.155
J	0.46	0.71	0.018	0.028
K	12.70	14.27	0.500	0.562
L	1.15	1.39	0.045	0.055
N	4.83	5.33	0.190	0.210
Q	2.54	3.04	0.100	0.120
R	2.04	2.79	0.080	0.110
S	1.15	1.39	0.045	0.055
T	5.97	6.47	0.235	0.255
U	0.00	1.27	0.000	0.050
V	1.15	—	0.045	—
Z	—	2.04	—	0.080

STYLE 1:
PIN 1. BASE
2. COLLECTOR
3. EMITTER
4. COLLECTOR

CASE 221A-04
TO-220AB

2N6394 thru 2N6399 MCR220-5 MCR220-7 MCR220-9

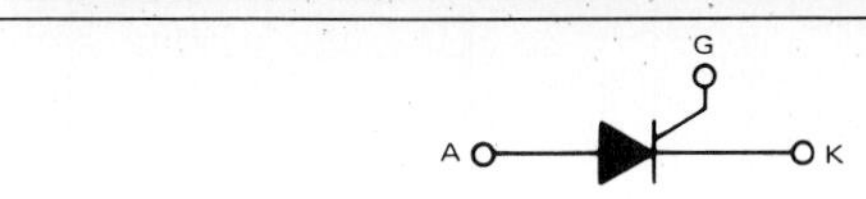

THYRISTORS

12 AMPERES RMS
50-800 VOLTS

SILICON CONTROLLED RECTIFIERS

. . . designed primarily for half-wave ac control applications, such as motor controls, heating controls and power supplies; or where-ever half-wave silicon gate-controlled, solid-state devices are needed.

- Glass Passivated Junctions and Center Gate Fire for Greater Parameter Uniformity and Stability
- Small, Rugged, Thermowatt▲ Construction for Low Thermal Resistance, High Heat Dissipation and Durability
- Blocking Voltage to 800 Volts

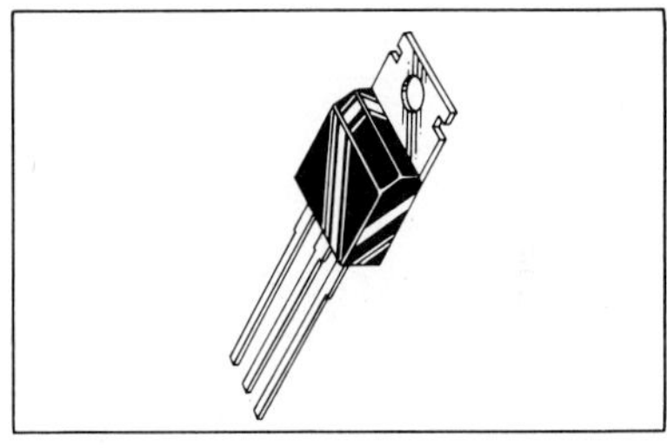

*MAXIMUM RATINGS

Rating	Symbol	Value	Unit
Peak Reverse Voltage (1)	V_{RRM}		Volts
2N6394		50	
2N6395		100	
2N6396		200	
MCR220-5		300	
2N6397		400	
MCR220-7		500	
2N6398		600	
MCR220-9		700	
2N6399		800	
Forward Current RMS $T_J = 125^oC$ (All Conduction Angles)	$I_{T(RMS)}$	12	Amps
Peak Forward Surge Current (1/2 cycle, Sine Wave, 60 Hz, $T_J = 125^oC$)	I_{TSM}	100	Amps
Circuit Fusing Considerations (T_J = -40 to +125°C, t = 1.0 to 8.3 ms)	I^2t	40	A^2s
Forward Peak Gate Power	P_{GM}	20	Watts
Forward Average Gate Power	$P_{G(AV)}$	0.5	Watt
Forward Peak Gate Current	I_{GM}	2.0	Amps
Operating Junction Temperature Range	T_J	-40 to +125	°C
Storage Temperature Range	T_{stg}	-40 to +150	°C

THERMAL CHARACTERISTICS

Characteristic	Symbol	Max	Unit
Thermal Resistance, Junction to Case	$R_{\theta JC}$	2.0	°C/W

(1) V_{RRM} for all types can be applied on a continuous dc basis without incurring damage. Ratings apply for zero or negative gate voltage. Devices should not be tested for blocking capability in a manner such that the voltage supplied exceeds the rated blocking voltage.

*Indicates JEDEC Registered Data.

▲Trademark of Motorola Inc.

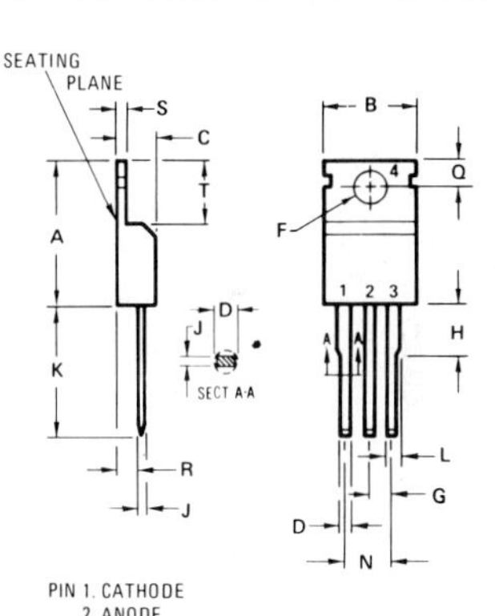

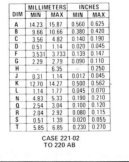

PIN 1. CATHODE
2. ANODE
3. GATE
4. ANODE

All JEDEC dimensions and notes apply

DIM	MILLIMETERS MIN	MILLIMETERS MAX	INCHES MIN	INCHES MAX
A	14.23	15.87	0.560	0.625
B	9.66	10.66	0.380	0.420
C	3.56	4.82	0.140	0.190
D	0.51	1.14	0.020	0.045
F	3.531	3.733	0.139	0.147
G	2.29	2.79	0.090	0.110
H	–	6.35	–	0.250
J	0.31	1.14	0.012	0.045
K	12.70	14.27	0.500	0.562
L	1.14	1.77	0.045	0.070
N	4.83	5.33	0.190	0.210
Q	2.54	3.04	0.100	0.120
R	2.04	2.92	0.080	0.115
S	0.51	1.39	0.020	0.055
T	5.85	6.85	0.230	0.270

CASE 221-02
TO 220 AB

ELECTRICAL CHARACTERISTICS (T_C = 25°C unless otherwise noted.)

Characteristic		Symbol	Min	Typ	Max	Unit
*Peak Forward Blocking Voltage (T_J = 125°C)		V_{DRM}				Volts
	2N6394		50	–	–	
	2N6395		100	–	–	
	2N6396		200	–	–	
	MCR220-5		300	–	–	
	2N6397		400	–	–	
	MCR220-7		500	–	–	
	2N6398		600	–	–	
	MCR220-9		700	–	–	
	2N6399		800	–	–	
* Peak Forward Blocking Current (Rated V_{DRM} @ T_J = 125°C)		I_{DRM}	–	–	2.0	mA
* Peak Reverse Blocking Current (Rated V_{RRM} @ T_J = 125°C)		I_{RRM}	–	–	2.0	mA
* Forward "On" Voltage (I_{TM} = 24 A Peak)		V_{TM}	–	1.7	2.2	Volts
* Gate Trigger Current (Continuous dc) (Anode Voltage = 12 Vdc, R_L = 100 Ohms)		I_{GT}	–	5.0	30	mA
* Gate Trigger Voltage (Continuous dc) (Anode Voltage = 12 Vdc, R_L = 100 Ohms)		V_{GT}	–	0.7	1.5	Volts
* Gate Non-Trigger Voltage (Anode Voltage = Rated V_{DRM}, R_L = 100 Ohms, T_J = 125°C)		V_{GD}	0.2	–	–	Volts
* Holding Current (Anode Voltage = 12 Vdc)		I_H	–	6.0	40	mA
Turn-On Time (I_{TM} = 12 A, I_{GT} = 40 mAdc)		t_{gt}	–	1.0	2.0	μs
Turn-Off Time (V_{DRM} = rated voltage)		t_q				μs
(I_{TM} = 12 A, I_R = 12 A)			–	15	–	
(I_{TM} = 12 A, I_R = 12 A, T_J = 125°C)			–	35	–	
Forward Voltage Application Rate (T_J = 125°C)		dv/dt	–	50	–	V/μs

*Indicates JEDEC Registered Data.

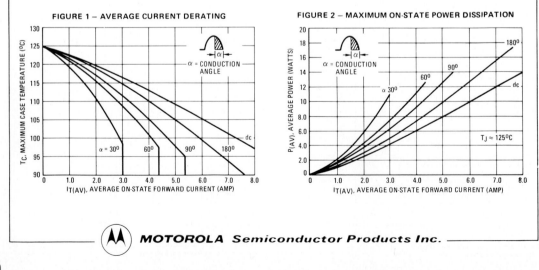

FIGURE 1 – AVERAGE CURRENT DERATING

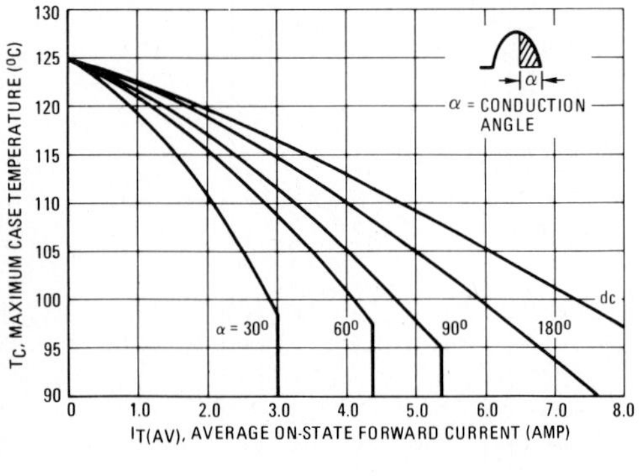

FIGURE 2 – MAXIMUM ON-STATE POWER DISSIPATION

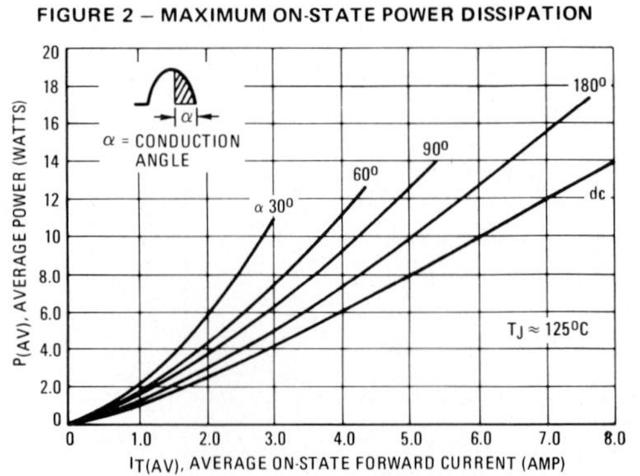

MOTOROLA *Semiconductor Products Inc.*

2N6027
2N6028

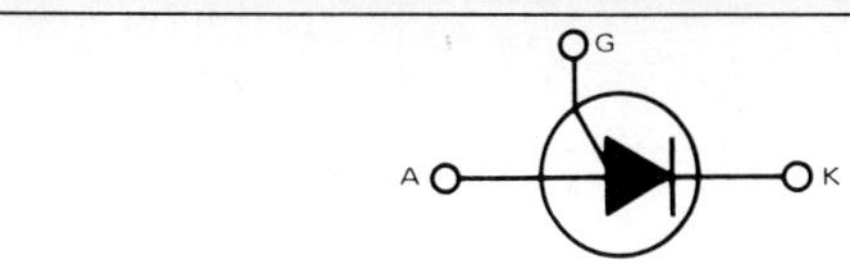

SILICON PROGRAMMABLE UNIJUNCTION TRANSISTORS

40 VOLTS
375 mW

DS 2520

SILICON PROGRAMMABLE UNIJUNCTION TRANSISTORS

. . . designed to enable the engineer to "program" unijunction characteristics such as R_{BB}, η, I_V, and I_P by merely selecting two resistor values. Application includes thyristor-trigger, oscillator, pulse and timing circuits. These devices may also be used in special thyristor applications due to the availability of an anode gate. Supplied in an inexpensive TO-92 plastic package for high-volume requirements, this package is readily adaptable for use in automatic insertion equipment.

- Programmable — R_{BB}, η, I_V and I_P.
- Low On-State Voltage — 1.5 Volts Maximum @ I_F = 50 mA
- Low Gate to Anode Leakage Current — 10 nA Maximum
- High Peak Output Voltage — 11 Volts Typical
- Low Offset Voltage — 0.35 Volt Typical (R_G = 10 k ohms)

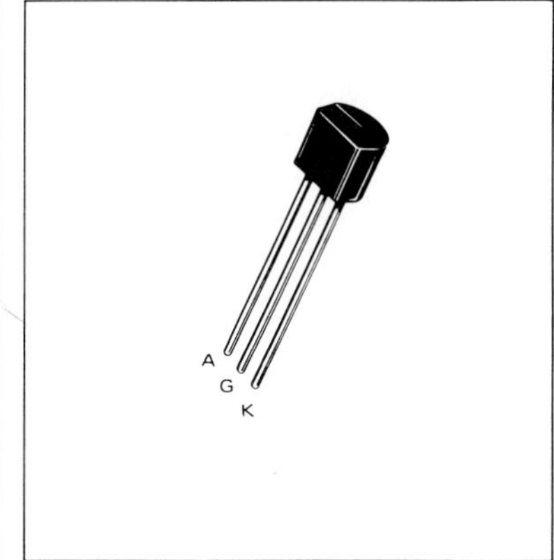

MAXIMUM RATINGS

Rating	Symbol	Value	Unit
Power Dissipation (1) Derate Above 25°C	P_F $1/\theta_{JA}$	375 5.0	mW mW/°C
DC Forward Anode Current (2) Derate Above 25°C	I_T	200 2.67	mA mA/°C
*DC Gate Current	I_G	±50	mA
Repetitive Peak Forward Current 100 μs Pulse Width, 1.0% Duty Cycle *20 μs Pulse Width, 1.0% Duty Cycle	I_{TRM}	 1.0 2.0	 Amp Amp
Non-Repetitive Peak Forward Current 10 μs Pulse Width	I_{TSM}	5.0	Amp
*Gate to Cathode Forward Voltage	V_{GKF}	40	Volt
*Gate to Cathode Reverse Voltage	V_{GKR}	-5.0	Volt
*Gate to Anode Reverse Voltage	V_{GAR}	40	Volt
*Anode to Cathode Voltage	V_{AK}	±40	Volt
Operating Junction Temperature Range	T_J	-50 to +100	°C
*Storage Temperature Range	T_{stg}	-55 to +150	°C

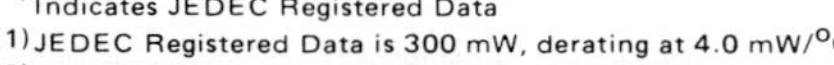
*Indicates JEDEC Registered Data
(1) JEDEC Registered Data is 300 mW, derating at 4.0 mW/°C.
(2) JEDEC Registered Data is 150 mA.

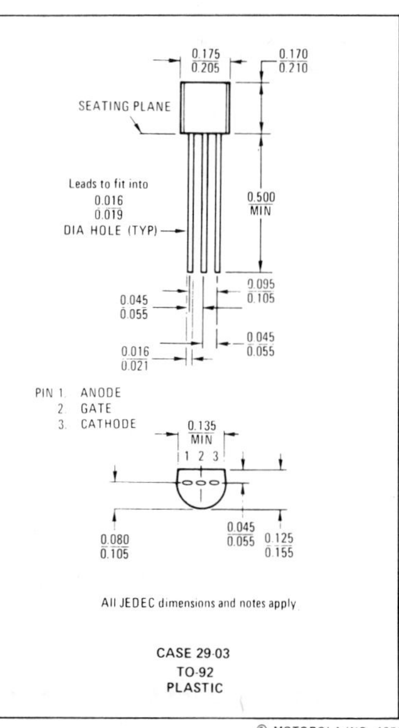

CASE 29-03
TO-92
PLASTIC

ELECTRICAL CHARACTERISTICS (T_A = 25°C unless otherwise noted)

Characteristic		Figure	Symbol	Min	Typ	Max	Unit
*Peak Current		2,9,11	I_P				μA
(V_S = 10 Vdc, R_G = 1.0 MΩ)	2N6027			–	1.25	2.0	
	2N6028			–	0.08	0.15	
(V_S = 10 Vdc, R_G = 10 k ohms)	2N6027			–	4.0	5.0	
	2N6028			–	0.70	1.0	
*Offset Voltage		1	V_T				Volts
(V_S = 10 Vdc, R_G = 1.0 MΩ)	2N6027			0.2	0.70	1.6	
	2N6028			0.2	0.50	0.6	
(V_S = 10 Vdc, R_G = 10 k ohms)	(Both Types)			0.2	0.35	0.6	
*Valley Current		1,4,5,	I_V				μA
(V_S = 10 Vdc, R_G = 1.0 MΩ)	2N6027			–	18	50	
	2N6028			–	18	25	
(V_S = 10 Vdc, R_G = 10 k ohms)	2N6027			70	270	–	
	2N6028			25	270	–	
(V_S = 10 Vdc, R_G = 200 Ohms)	2N6027			1.5	–	–	mA
	2N6028			1.0	–	–	
*Gate to Anode Leakage Current		–	I_{GAO}				nAdc
(V_S = 40 Vdc, T_A = 25°C, Cathode Open)				–	1.0	10	
(V_S = 40 Vdc, T_A = 75°C, Cathode Open)				–	3.0	–	
Gate to Cathode Leakage Current (V_S = 40 Vdc, Anode to Cathode Shorted)		–	I_{GKS}	–	5.0	50	nAdc
*Forward Voltage (I_F = 50 mA Peak)		1,6	V_F	–	0.8	1.5	Volts
*Peak Output Voltage (V_B = 20 Vdc, C_C = 0.2 μF)		3,7	V_o	6.0	11	–	Volts
Pulse Voltage Rise Time (V_B = 20 Vdc, C_C = 0.2 μF)		3	t_r	–	40	80	ns

*Indicates JEDEC Registered Data

FIGURE 1 – ELECTRICAL CHARACTERIZATION

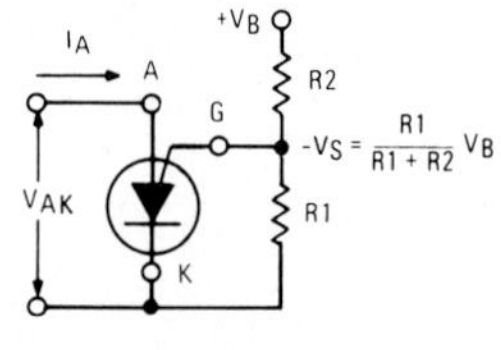

1A – PROGRAMMABLE UNIJUNCTION WITH "PROGRAM" RESISTORS R1 and R2

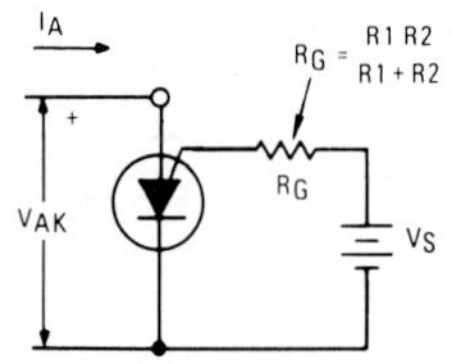

1B – EQUIVALENT TEST CIRCUIT FOR FIGURE 1A USED FOR ELECTRICAL CHARACTERISTICS TESTING (ALSO SEE FIGURE 2)

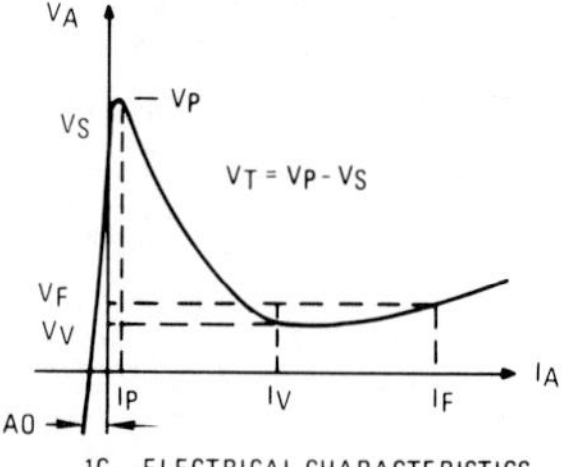

1C – ELECTRICAL CHARACTERISTICS

FIGURE 2 – PEAK CURRENT (I_P) TEST CIRCUIT

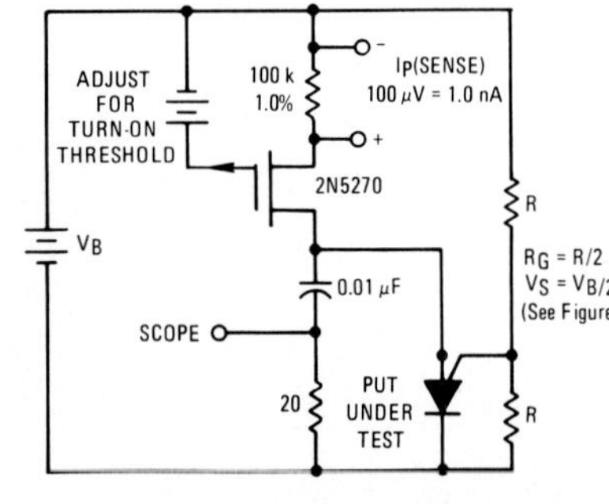

FIGURE 3 – V_o AND t_r TEST CIRCUIT

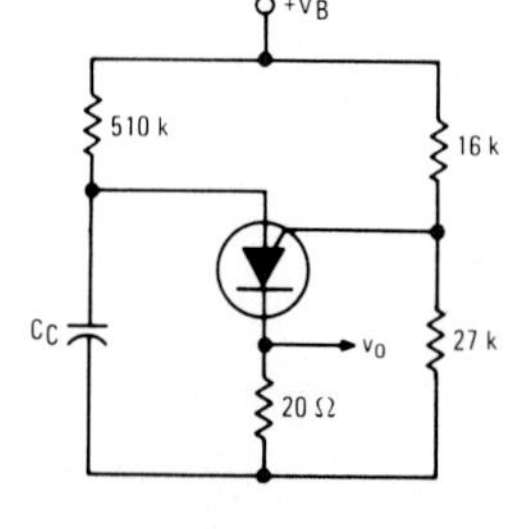

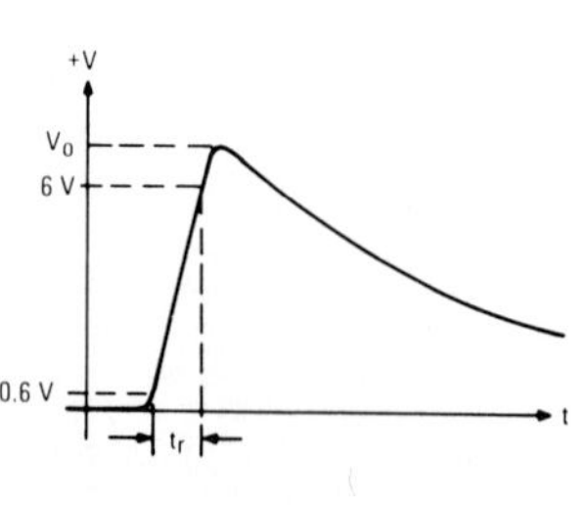

MOTOROLA
SEMICONDUCTOR
TECHNICAL DATA

LM117
LM217
LM317

THREE-TERMINAL ADJUSTABLE POSITIVE VOLTAGE REGULATORS

SILICON MONOLITHIC INTEGRATED CIRCUIT

THREE-TERMINAL ADJUSTABLE OUTPUT POSITIVE VOLTAGE REGULATORS

The LM117/217/317 are adjustable 3-terminal positive voltage regulators capable of supplying in excess of 1.5 A over an output voltage range of 1.2 V to 37 V. These voltage regulators are exceptionally easy to use and require only two external resistors to set the output voltage. Further, they employ internal current limiting, thermal shutdown and safe area compensation, making them essentially blow-out proof.

The LM117 series serve a wide variety of applications including local, on card regulation. This device can also be used to make a programmable output regulator, or by connecting a fixed resistor between the adjustment and output, the LM117 series can be used as a precision current regulator.

- Output Current in Excess of 1.5 Ampere in K and T Suffix Packages
- Output Current in Excess of 0.5 Ampere in H Suffix Package
- Output Adjustable between 1.2 V and 37 V
- Internal Thermal Overload Protectiion
- Internal Short-Circuit Current Limiting Constant with Temperature
- Output Transistor Safe-Area Compensation
- Floating Operation for High Voltage Applications
- Standard 3-lead Transistor Packages
- Eliminates Stocking Many Fixed Voltages

K SUFFIX
METAL PACKAGE
CASE 1

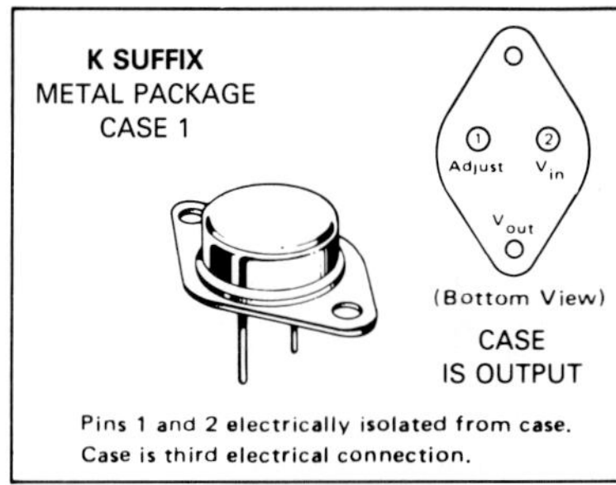

CASE IS OUTPUT

Pins 1 and 2 electrically isolated from case. Case is third electrical connection.

T SUFFIX
PLASTIC PACKAGE
CASE 221A

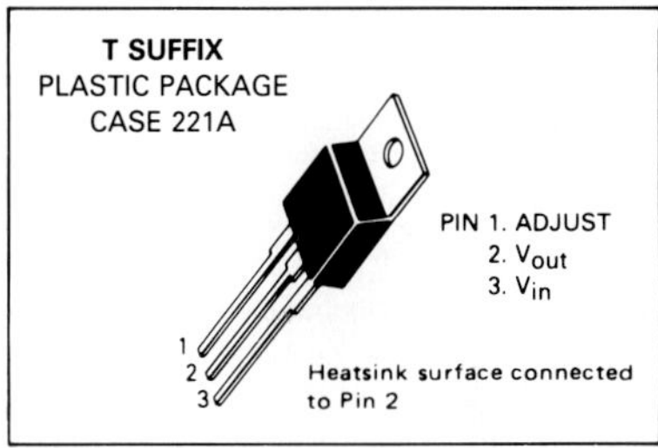

PIN 1. ADJUST
2. V_{out}
3. V_{in}

Heatsink surface connected to Pin 2

H SUFFIX
METAL PACKAGE
CASE 79

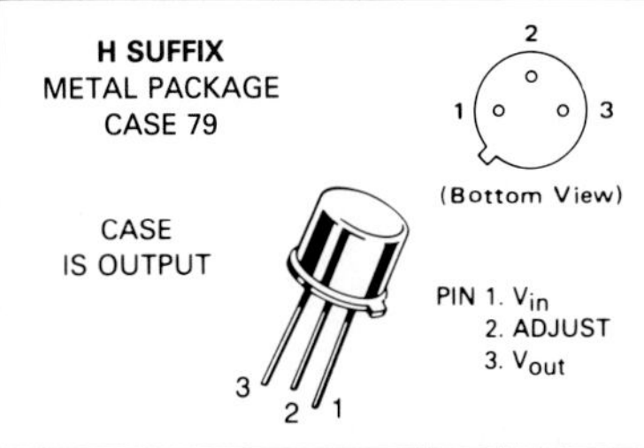

CASE IS OUTPUT

PIN 1. V_{in}
2. ADJUST
3. V_{out}

STANDARD APPLICATION

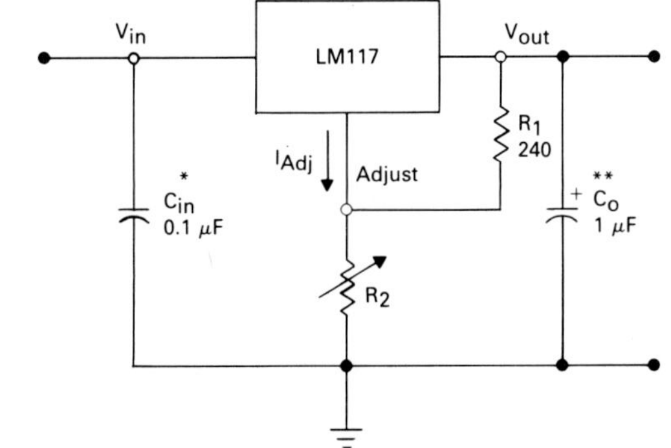

* = C_{in} is required if regulator is located an appreciable distance from power supply filter.

** = C_O is not needed for stability, however it does improve transient response.

$$V_{out} = 1.25\text{ V}\,(1 + \frac{R_2}{R_1}) + I_{Adj}\,R_2$$

Since I_{Adj} is controlled to less than 100 µA, the error associated with this term is negligible in most applications.

ORDERING INFORMATION

Device	Tested Operating Temperature Range	Package
LM117H LM117K	T_J = −55°C to +150°C	Metal Can Metal Power
LM217H LM217K	T_J = −25°C to +150°C	Metal Can Metal Power
LM317H LM317K LM317T	T_J = 0°C to +125°C	Metal Can Metal Power Plastic Power
LM317BT#	T_J = −40°C to +125°C	Plastic Power

#Automotive temperature range selections are available with special test conditions and additional tests. Contact your local Motorola sales office for information.

LM117, LM217, LM317

MAXIMUM RATINGS

Rating	Symbol	Value	Unit
Input-Output Voltage Differential	V_I-V_O	40	Vdc
Power Dissipation	P_D	Internally Limited	
Operating Junction Temperature Range	T_J		°C
LM117		−55 to +150	
LM217		−25 to +150	
LM317		0 to +150	
Storage Temperature Range	T_{stg}	−65 to +150	°C

ELECTRICAL CHARACTERISTICS (V_I-V_O = 5.0 V; I_O = 0.5 A for K and T packages; I_O = 0.1 A for H package; T_J = T_{low} to T_{high} [see Note 1]; I_{max} and P_{max} per Note 2; unless otherwise specified.)

Characteristic	Figure	Symbol	LM117/217 Min	LM117/217 Typ	LM117/217 Max	LM317 Min	LM317 Typ	LM317 Max	Unit
Line Regulation (Note 3) T_A = 25°C, 3.0 V ≤ V_I-V_O ≤ 40 V	1	Reg_{line}	—	0.01	0.02	—	0.01	0.04	%/V
Load Regulation (Note 3) T_A = 25°C, 10 mA ≤ I_O ≤ I_{max}	2	Reg_{load}							
V_O ≤ 5.0 V			—	5.0	15	—	5.0	25	mV
V_O ≥ 5.0 V			—	0.1	0.3	—	0.1	0.5	%/V_O
Thermal Regulation (T_A = +25°C) 20 ms Pulse		—	—	0.02	0.07	—	0.03	0.07	%/W
Adjustment Pin Current	3	I_{Adj}	—	50	100	—	50	100	μA
Adjustment Pin Current Change 2.5 V ≤ V_I-V_O ≤ 40 V 10 mA ≤ I_L ≤ I_{max}, P_D ≤ P_{max}	1,2	ΔI_{Adj}	—	0.2	5.0	—	0.2	5.0	μA
Reference Voltage (Note 4) 3.0 V ≤ V_I-V_O ≤ 40 V 10 mA ≤ I_O ≤ I_{max}, P_D ≤ P_{max}	3	V_{ref}	1.2	1.25	1.3	1.2	1.25	1.3	V
Line Regulation (Note 3) 3.0 V ≤ V_I-V_O ≤ 40 V	1	Reg_{line}	—	0.02	0.05	—	0.02	0.07	%/V
Load Regulation (Note 3) 10 mA ≤ I_O ≤ I_{max}	2	Reg_{load}							
V_O ≤ 5.0 V			—	20	50	—	20	70	mV
V_O ≥ 5.0 V			—	0.3	1.0	—	0.3	1.5	%/V_O
Temperature Stability (T_{low} ≤ T_J ≤ T_{high})	3	T_S	—	0.7	—	—	0.7	—	%/V_O
Minimum Load Current to Maintain Regulation (V_I-V_O = 40 V)	3	I_{Lmin}	—	3.5	5.0	—	3.5	10	mA
Maximum Output Current	3	I_{max}							A
V_I-V_O ≤ 15 V, P_D ≤ P_{max}									
K and T Packages			1.5	2.2	—	1.5	2.2	—	
H Package			0.5	0.8	—	0.5	0.8	—	
V_I-V_O = 40 V, P_D ≤ P_{max}, T_A = 25°C									
K and T Packages			0.25	0.4	—	0.15	0.4	—	
H Package			—	0.07	—	—	0.07	—	
RMS Noise, % of V_O T_A = 25°C, 10 Hz ≤ f ≤ 10 kHz	—	N	—	0.003	—	—	0.003	—	%/V_O
Ripple Rejection, V_O = 10 V, f = 120 Hz (Note 5)	4	RR							dB
Without C_{Adj}			—	65	—	—	65	—	
C_{Adj} = 10 μF			66	80	—	66	80	—	
Long-Term Stability, T_J = T_{high} (Note 6) T_A = 25°C for Endpoint Measurements	3	S	—	0.3	1.0	—	0.3	1.0	%/1.0 k Hrs.
Thermal Resistance Junction to Case	—	$R_{\theta JC}$							°C/W
H Package			—	12	15	—	12	15	
K Package			—	2.3	3.0	—	2.3	3.0	
T Package			—	—	—	—	5.0	—	

NOTES: (1) T_{low} = −55°C for LM117 T_{high} = +150°C for LM117
= −25°C for LM217 = +150°C for LM217
= 0°C for LM317 = +125°C for LM317

(2) I_{max} = 1.5 A for K and T Packages
= 0.5 A for H Package
P_{max} = 20 W for K Package
= 20 W for T Package
= 2.0 W for H Package

(3) Load and line regulation are specified at constant junction temperature. Changes in V_O due to heating effects must be taken into account separately. Pulse testing with low duty cycle is used.

(4) Selected devices with tightened tolerance reference voltage available.

(5) C_{ADJ}, when used, is connected between the adjustment pin and ground.

(6) Since Long-Term Stability cannot be measured on each device before shipment, this specification is an engineering estimate of average stability from lot to lot.

MOTOROLA
SEMICONDUCTOR
TECHNICAL DATA

MC1455

TIMING CIRCUIT

The MC1455 monolithic timing circuit is a highly stable controller capable of producing accurate time delays, or oscillation. Additional terminals are provided for triggering or resetting if desired. In the time delay mode of operation, the time is precisely controlled by one external resistor and capacitor. For astable operation as an oscillator, the free running frequency and the duty cycle are both accurately controlled with two external resistors and one capacitor. The circuit may be triggered and reset on falling waveforms, and the output structure can source or sink up to 200 mA or drive MTTL circuits.

- Direct Replacement for NE555 Timers
- Timing From Microseconds Through Hours
- Operates in Both Astable and Monostable Modes
- Adjustable Duty Cycle
- High Current Output Can Source or Sink 200 mA
- Output Can Drive MTTL
- Temperature Stability of 0.005% per °C
- Normally "On" or Normally "Off" Output

FIGURE 1 — 22-SECOND SOLID-STATE TIME DELAY RELAY CIRCUIT

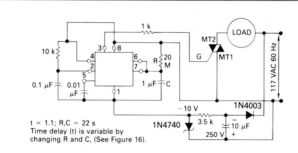

$t = 1.1; R_iC = 22$ s
Time delay (t) is variable by changing R and C, (See Figure 16).

FIGURE 2 — BLOCK DIAGRAM

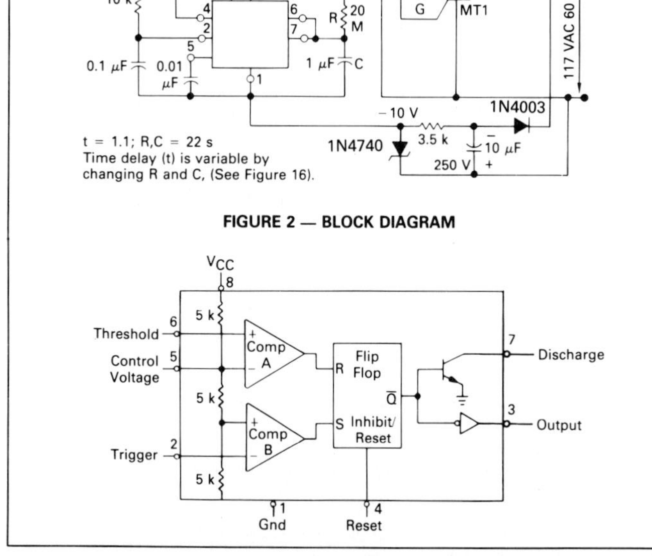

TIMING CIRCUIT

SILICON MONOLITHIC INTEGRATED CIRCUIT

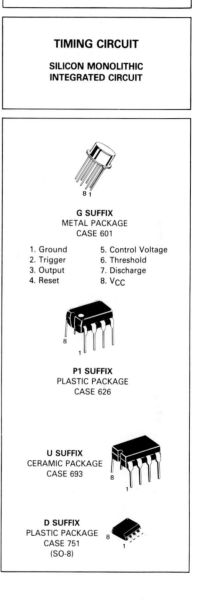

G SUFFIX
METAL PACKAGE
CASE 601

1. Ground	5. Control Voltage
2. Trigger	6. Threshold
3. Output	7. Discharge
4. Reset	8. V_{CC}

P1 SUFFIX
PLASTIC PACKAGE
CASE 626

U SUFFIX
CERAMIC PACKAGE
CASE 693

D SUFFIX
PLASTIC PACKAGE
CASE 751
(SO-8)

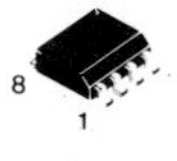

ORDERING INFORMATION

Device	Alternate	Temperature Range	Package
MC1455G	—	0°C to +70°C	Metal Can
MC1455P1	NE555V	0°C to +70°C	Plastic DIP
MC1455D	—	0°C to +70°C	SO-8
MC1455U	—	0°C to +70°C	Ceramic DIP
MC1455BP1	—	−40°C to +85°C	Plastic DIP

MAXIMUM RATINGS (T_A = +25°C unless otherwise noted.)

Rating	Symbol	Value	Unit
Power Supply Voltage	V_{CC}	+18	Vdc
Discharge Current (Pin 7)	I_7	200	mA
Power Dissipation (Package Limitation)	P_D		
Metal Can		680	mW
Derate above T_A = +25°C		4.6	mW/°C
Plastic Dual In-Line Package		625	mW
Derate above T_A = +25°C		5.0	mW/°C
Operating Temperature Range (Ambient)	T_A		°C
MC1455B		−40 to +85	
MC1455		0 to +70	
Storage Temperature Range	T_{stg}	−65 to +150	°C

FIGURE 3 — GENERAL TEST CIRCUIT

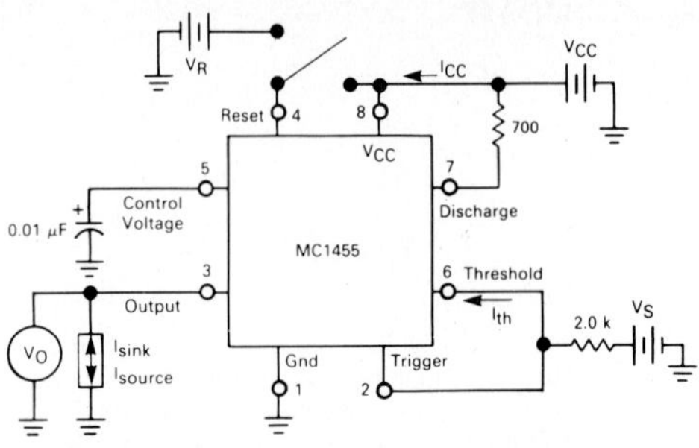

Test Circuit for Measuring dc Parameters: (to set output and measure parameters)
a) When $V_S > 2/3\ V_{CC}$, V_O is low.
b) When $V_S < 1/3\ V_{CC}$, V_O is high.
c) When V_O is low, pin 7 sinks current. To test for Reset, set V_O, high, apply Reset voltage, and test for current flowing into pin 7.
When Reset is not in use, it should be tied to V_{CC}.

ELECTRICAL CHARACTERISTICS (T_A = +25°C, V_{CC} = +5.0 V to +15 V unless otherwise noted.)

Characteristics	Symbol	Min	Typ	Max	Unit
Operating Supply Voltage Range	V_{CC}	4.5	—	16	V
Supply Current	I_{CC}				mA
V_{CC} = 5.0 V, $R_L = \infty$		—	3.0	6.0	
V_{CC} = 15 V, $R_L = \infty$		—	10	15	
Low State, (Note 1)					
Timing Error (Note 2)					
R = 1.0 kΩ to 100 kΩ					
Initial Accuracy C = 0.1 μF		—	1.0	—	%
Drift with Temperature		—	50	—	PPM/°C
Drift with Supply Voltage		—	0.1	—	%/Volt
Threshold Voltage	V_{th}	—	2/3	—	xV_{CC}
Trigger Voltage	V_T				V
V_{CC} = 15 V		—	5.0	—	
V_{CC} = 5.0 V		—	1.67	—	
Trigger Current	I_T	—	0.5	—	μA
Reset Voltage	V_R	0.4	0.7	1.0	V
Reset Current	I_R	—	0.1	—	mA
Threshold Current (Note 3)	I_{th}	—	0.1	0.25	μA
Discharge Leakage Current (Pin 7)	I_{dis}	—	—	100	nA
Control Voltage Level	V_{CL}				V
V_{CC} = 15 V		9.0	10	11	
V_{CC} = 5.0 V		2.6	3.33	4.0	
Output Voltage Low	V_{OL}				V
(V_{CC} = 15 V)					
I_{sink} = 10 mA		—	0.1	0.25	
I_{sink} = 50 mA		—	0.4	0.75	
I_{sink} = 100 mA		—	2.0	2.5	
I_{sink} = 200 mA		—	2.5	—	
(V_{CC} = 5.0 V)					
I_{sink} = 8.0 mA		—	—	—	
I_{sink} = 5.0 mA		—	0.25	0.35	
Output Voltage High	V_{OH}				V
(I_{source} = 200 mA)					
V_{CC} = 15 V		—	12.5	—	
(I_{source} = 100 mA)					
V_{CC} = 15 V		12.75	13.3	—	
V_{CC} = 5.0 V		2.75	3.3	—	
Rise Time of Output	t_{OLH}	—	100	—	ns
Fall Time of Output	t_{OHL}	—	100	—	ns

NOTES:
1. Supply current when output is high is typically 1.0 mA less.
2. Tested at V_{CC} = 5.0 V and V_{CC} = 15 V. Monostable mode
3. This will determine the maximum value of $R_A + R_B$ for 15 V operation. The maximum total R = 20 megohms.

MOTOROLA
SEMICONDUCTOR
TECHNICAL DATA

MC1741
MC1741C

INTERNALLY COMPENSATED, HIGH PERFORMANCE OPERATIONAL AMPLIFIERS

. . . designed for use as a summing amplifier, integrator, or amplifier with operating characteristics as a function of the external feedback components.

- No Frequency Compensation Required
- Short-Circuit Protection
- Offset Voltage Null Capability
- Wide Common-Mode and Differential Voltage Ranges
- Low-Power Consumption
- No Latch Up

MAXIMUM RATINGS (T_A = +25°C unless otherwise noted)

Rating	Symbol	MC1741C	MC1741	Unit
Power Supply Voltage	V_{CC} V_{EE}	+18 −18	+22 −22	Vdc Vdc
Input Differential Voltage	V_{ID}	±30		Volts
Input Common Mode Voltage (Note 1)	V_{ICM}	±15		Volts
Output Short Circuit Duration (Note 2)	t_S	Continuous		
Operating Ambient Temperature Range	T_A	0 to +70	−55 to +125	°C
Storage Temperature Range Metal and Ceramic Packages Plastic Packages	T_{stg}	 −65 to +150 −55 to +125		°C

NOTES:
1. For supply voltages less than +15 V, the absolute maximum input voltage is equal to the supply voltage.
2. Supply voltage equal to or less than 15 V.

EQUIVALENT CIRCUIT SCHEMATIC

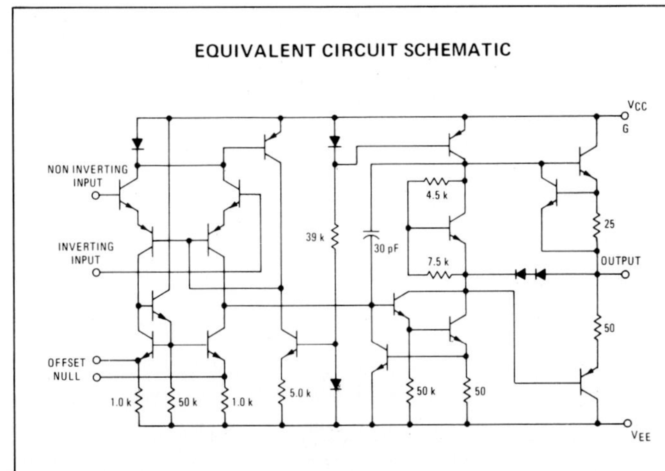

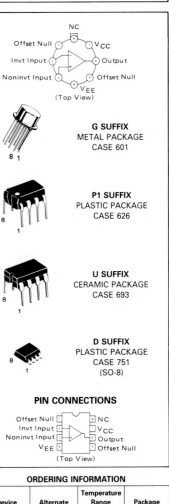

OPERATIONAL AMPLIFIER

SILICON MONOLITHIC INTEGRATED CIRCUIT

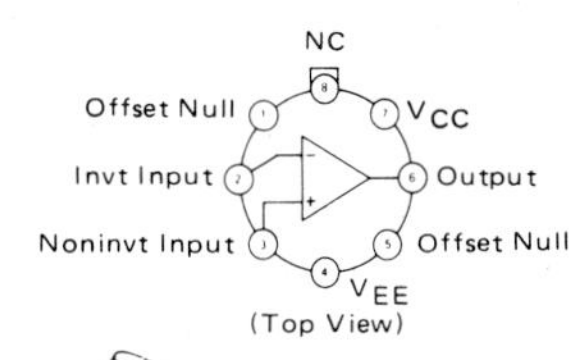

G SUFFIX
METAL PACKAGE
CASE 601

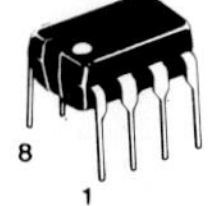

P1 SUFFIX
PLASTIC PACKAGE
CASE 626

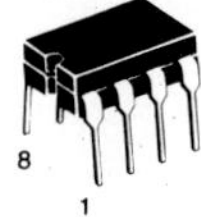

U SUFFIX
CERAMIC PACKAGE
CASE 693

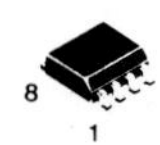

D SUFFIX
PLASTIC PACKAGE
CASE 751
(SO-8)

PIN CONNECTIONS

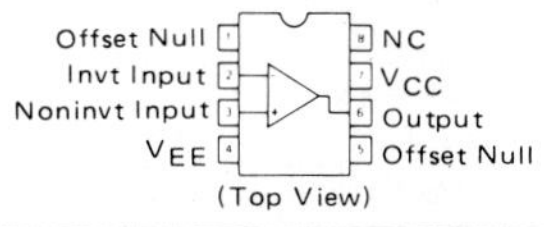

ORDERING INFORMATION

Device	Alternate	Temperature Range	Package
MC1741CD	—	0°C to +70°C	SO-8
MC1741CG	LM741CH, µA741HC		Metal Can
MC1741CP1	LM741CN, µA741TC		Plastic DIP
MC1741CU	—		Ceramic DIP
MC1741G	—	−55°C to +125°C	Metal Can
MC1741U	—		Ceramic DIP

MC1741, MC1741C

ELECTRICAL CHARACTERISTICS (V_{CC} = +15 V, V_{EE} = −15 V, T_A = 25°C unless otherwise noted).

Characteristic	Symbol	MC1741			MC1741C			Unit
		Min	Typ	Max	Min	Typ	Max	
Input Offset Voltage ($R_S \leqslant 10$ k)	V_{IO}	–	1.0	5.0	–	2.0	6.0	mV
Input Offset Current	I_{IO}	–	20	200	–	20	200	nA
Input Bias Current	I_{IB}	–	80	500	–	80	500	nA
Input Resistance	r_i	0.3	2.0	–	0.3	2.0	–	MΩ
Input Capacitance	C_i	–	1.4	–	–	1.4	–	pF
Offset Voltage Adjustment Range	V_{IOR}	–	±15	–	–	±15	–	mV
Common Mode Input Voltage Range	V_{ICR}	±12	±13	–	±12	±13	–	V
Large Signal Voltage Gain (V_O = ±10 V, $R_L \geqslant 2.0$ k)	A_v	50	200	–	20	200	–	V/mV
Output Resistance	r_o	–	75	–	–	75	–	Ω
Common Mode Rejection Ratio ($R_S \leqslant 10$ k)	CMRR	70	90	–	70	90	–	dB
Supply Voltage Rejection Ratio ($R_S \leqslant 10$ k)	PSRR	–	30	150	–	30	150	μV/V
Output Voltage Swing ($R_L \geqslant 10$ k) ($R_L \geqslant 2$ k)	V_O	 ±12 ±10	 ±14 ±13	 – –	 ±12 ±10	 ±14 ±13	 – –	V
Output Short-Circuit Current	I_{os}	–	20	–	–	20	–	mA
Supply Current	I_D	–	1.7	2.8	–	1.7	2.8	mA
Power Consumption	P_C	–	50	85	–	50	85	mW
Transient Response (Unity Gain – Non-Inverting) (V_I = 20 mV, $R_L \geqslant 2$ k, $C_L \leqslant 100$ pF) Rise Time (V_I = 20 mV, $R_L \geqslant 2$ k, $C_L \leqslant 100$ pF) Overshoot (V_I = 10 V, $R_L \geqslant 2$ k, $C_L \leqslant 100$ pF) Slew Rate	 t_{TLH} os SR	 – – –	 0.3 15 0.5	 – – –	 – – –	 0.3 15 0.5	 – – –	 μs % V/μs

ELECTRICAL CHARACTERISTICS (V_{CC} = +15 V, V_{EE} = −15 V, T_A = T_{low} to T_{high} unless otherwise noted).

Characteristic	Symbol	MC1741			MC1741C			Unit
		Min	Typ	Max	Min	Typ	Max	
Input Offset Voltage ($R_S \leqslant 10$ kΩ)	V_{IO}	–	1.0	6.0	–	–	7.5	mV
Input Offset Current (T_A = 125°C) (T_A = -55°C) (T_A = 0°C to +70°C)	I_{IO}	 – – –	 7.0 85 –	 200 500 –	 – – –	 – – –	 – – 300	nA
Input Bias Current (T_A = 125°C) (T_A = -55°C) (T_A = 0°C to +70°C)	I_{IB}	 – – –	 30 300 –	 500 1500 –	 – – –	 – – –	 – – 800	nA
Common Mode Input Voltage Range	V_{ICR}	±12	±13	–	–	–	–	V
Common Mode Rejection Ratio ($R_S \leqslant 10$ k)	CMRR	70	90	–	–	–	–	dB
Supply Voltage Rejection Ratio ($R_S \leqslant 10$ k)	PSRR	–	30	150	–	–	–	μV/V
Output Voltage Swing ($R_L \geqslant 10$ k) ($R_L \geqslant 2$ k)	V_O	 ±12 ±10	 ±14 ±13	 – –	 – ±10	 – ±13	 – –	V
Large Signal Voltage Gain ($R_L \geqslant 2$ k, V_{out} = ±10 V)	A_v	25	–	–	15	–	–	V/mV
Supply Currents (T_A = 125°C) (T_A = -55°C)	I_D	 – –	 1.5 2.0	 2.5 3.3	 – –	 – –	 – –	mA
Power Consumption (T_A = +125°C) (T_A = -55°C)	P_C	– –	45 60	75 100	– –	– –	– –	mW

*T_{high} = 125°C for MC1741 and 70°C for MC1741C
T_{low} = -55°C for MC1741 and 0°C for MC1741C

FIGURE 1 — BURST NOISE versus SOURCE RESISTANCE

1000
100
10
0
BW = 1.0 Hz to 1.0 kHz
E_n, INPUT NOISE (peak) (μV)
10 100 1.0 k 10 k 100 k 1.0 M
R_S, SOURCE RESISTANCE (OHMS)

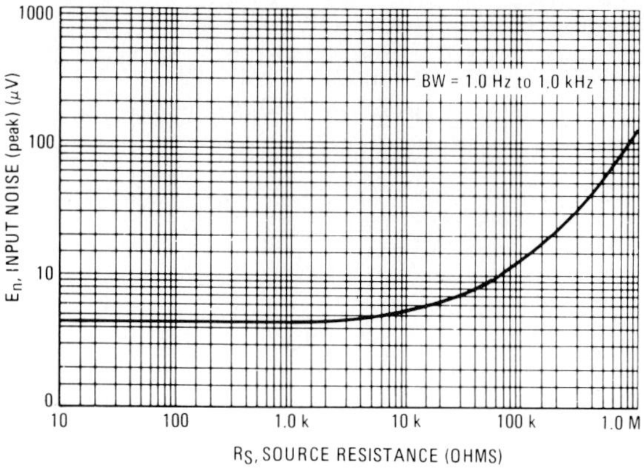

FIGURE 2 — RMS NOISE versus SOURCE RESISTANCE

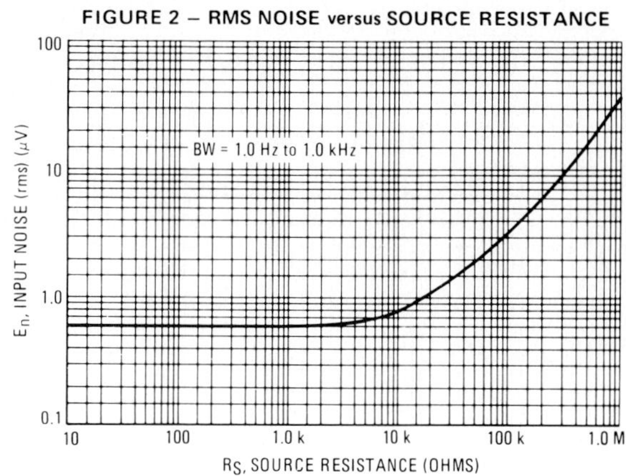

FIGURE 3 — OUTPUT NOISE versus SOURCE RESISTANCE

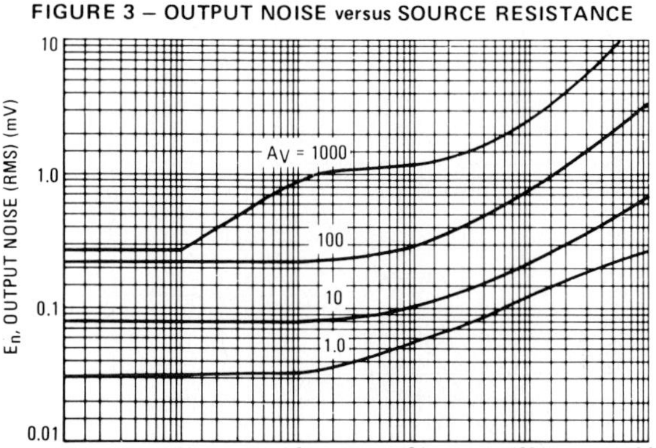

FIGURE 4 — SPECTRAL NOISE DENSITY

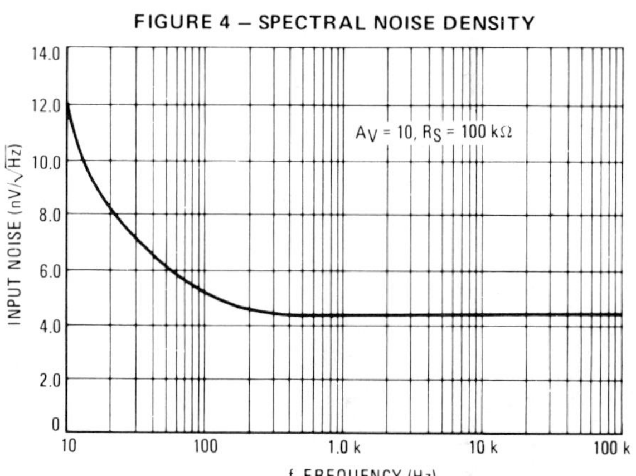

FIGURE 5 — BURST NOISE TEST CIRCUIT

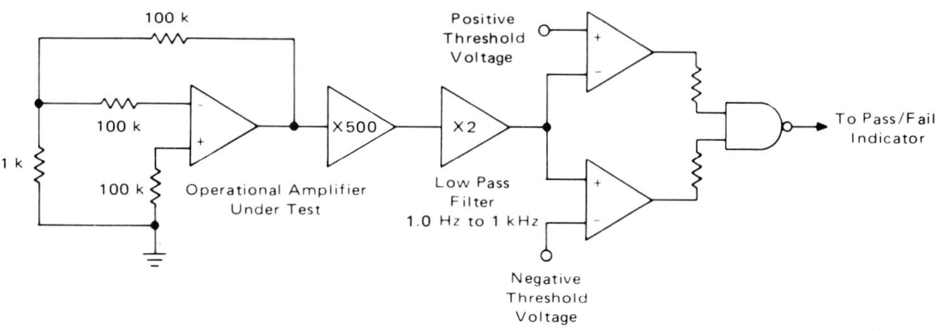

Unlike conventional peak reading or RMS meters, this system was especially designed to provide the quick response time essential to burst (popcorn) noise testing.

The test time employed is 10 seconds and the 20 μV peak limit refers to the operational amplifier input thus eliminating errors in the closed-loop gain factor of the operational amplifier under test.

TYPICAL CHARACTERISTICS

(V_{CC} = +15 Vdc, V_{EE} = -15 Vdc, T_A = +25°C unless otherwise noted)

FIGURE 6 – POWER BANDWIDTH (LARGE SIGNAL SWING versus FREQUENCY)

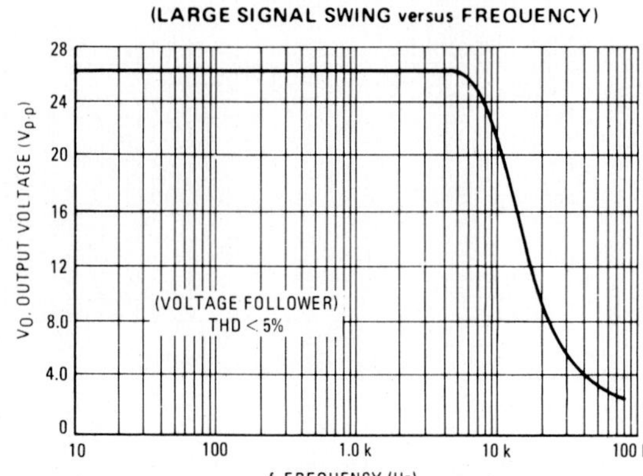

FIGURE 7 – OPEN LOOP FREQUENCY RESPONSE

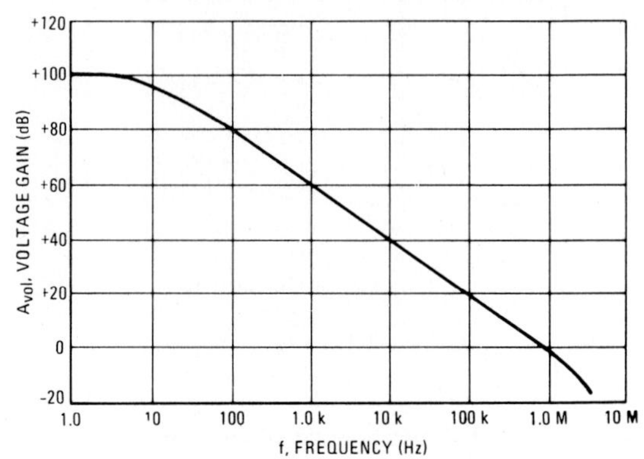

FIGURE 8 – POSITIVE OUTPUT VOLTAGE SWING versus LOAD RESISTANCE

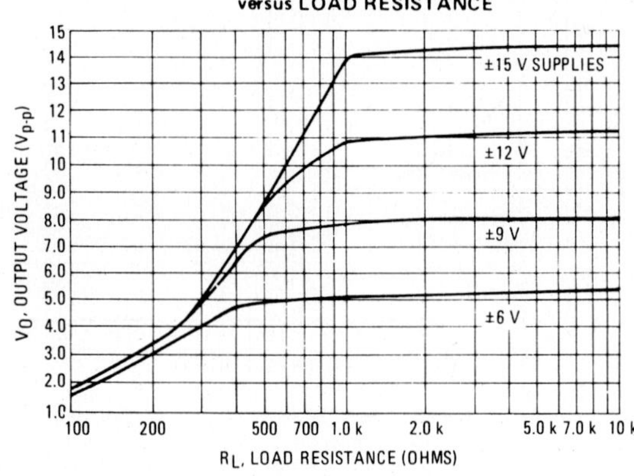

FIGURE 9 – NEGATIVE OUTPUT VOLTAGE SWING versus LOAD RESISTANCE

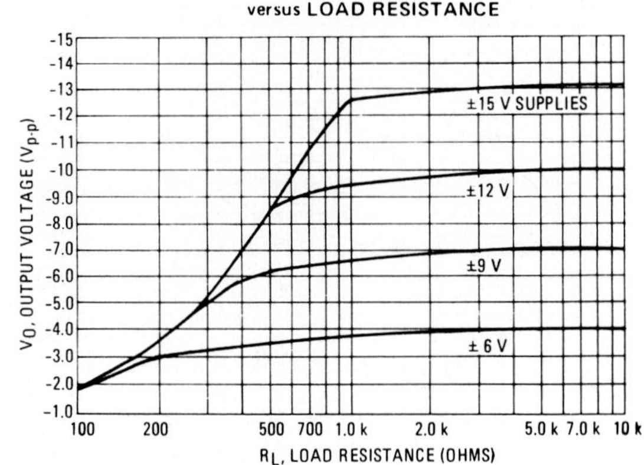

FIGURE 10 – OUTPUT VOLTAGE SWING versus LOAD RESISTANCE (Single Supply Operation)

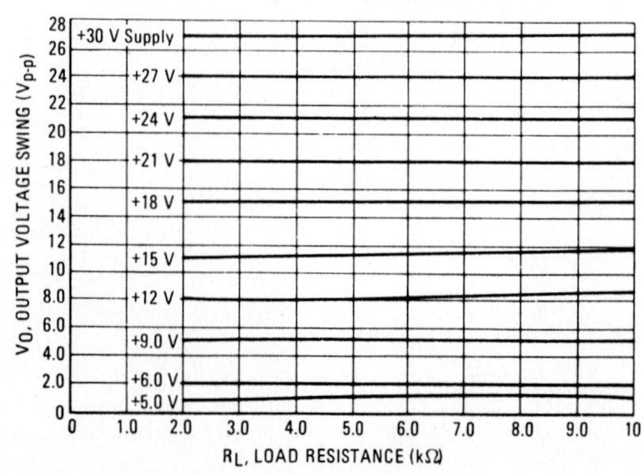

FIGURE 11 – SINGLE SUPPLY INVERTING AMPLIFIER

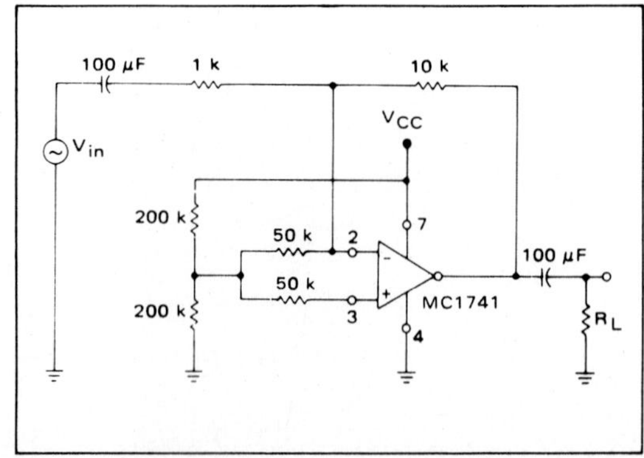

MOTOROLA
SEMICONDUCTOR
TECHNICAL DATA

MC7800 Series

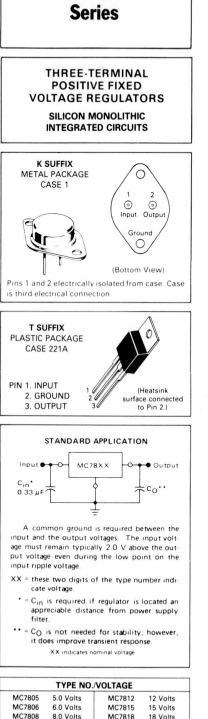

THREE-TERMINAL POSITIVE FIXED VOLTAGE REGULATORS

SILICON MONOLITHIC INTEGRATED CIRCUITS

THREE-TERMINAL POSITIVE VOLTAGE REGULATORS

These voltage regulators are monolithic integrated circuits designed as fixed-voltage regulators for a wide variety of applications including local, on-card regulation. These regulators employ internal current limiting, thermal shutdown, and safe-area compensation. With adequate heatsinking they can deliver output currents in excess of 1.0 ampere. Although designed primarily as a fixed voltage regulator, these devices can be used with external components to obtain adjustable voltages and currents.

- Output Current in Excess of 1.0 Ampere
- No External Components Required
- Internal Thermal Overload Protection
- Internal Short-Circuit Current Limiting
- Output Transistor Safe-Area Compensation
- Output Voltage Offered in 2% and 4% Tolerance

REPRESENTATIVE SCHEMATIC DIAGRAM

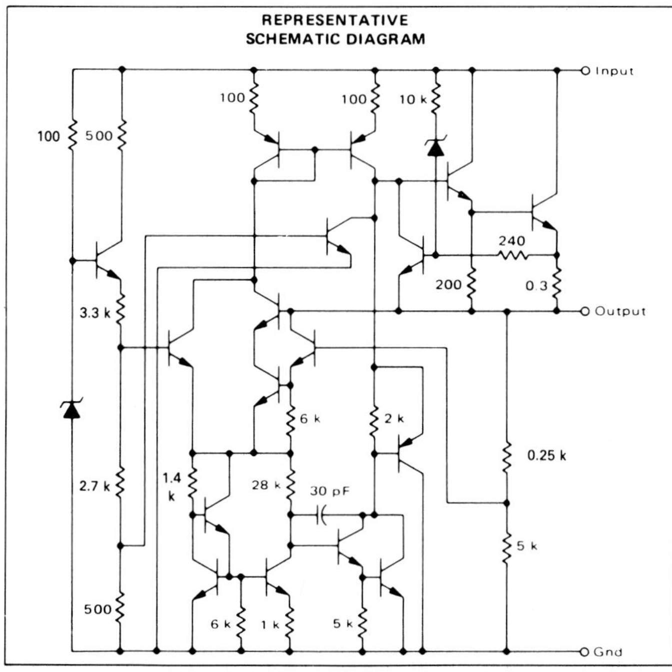

K SUFFIX
METAL PACKAGE
CASE 1

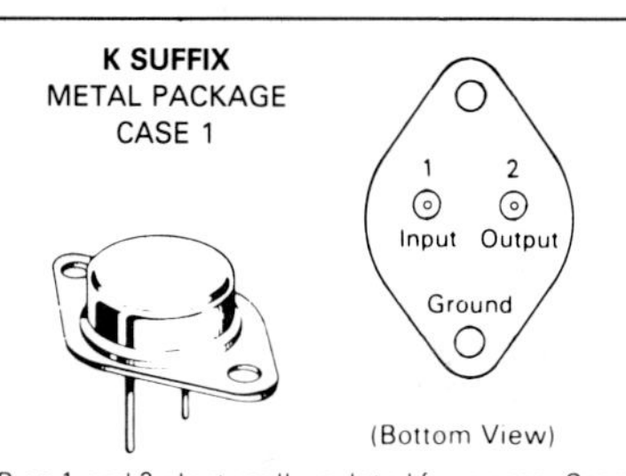

Pins 1 and 2 electrically isolated from case. Case is third electrical connection.

T SUFFIX
PLASTIC PACKAGE
CASE 221A

PIN 1. INPUT
2. GROUND
3. OUTPUT

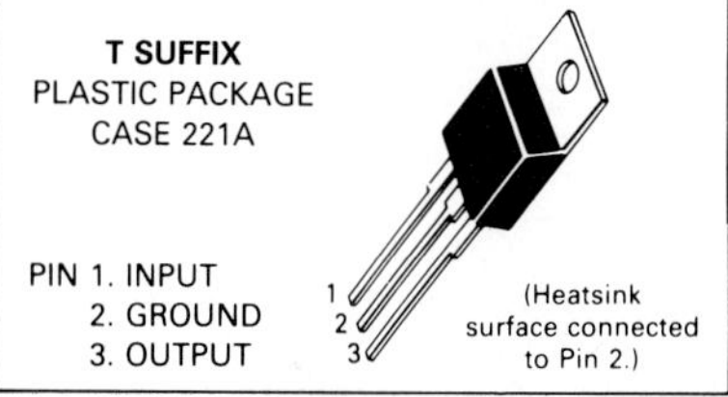

STANDARD APPLICATION

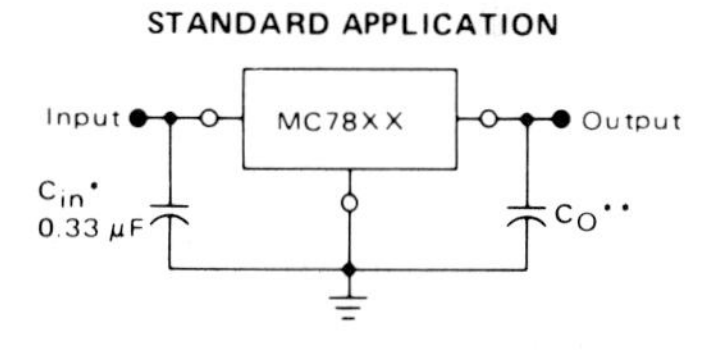

A common ground is required between the input and the output voltages. The input voltage must remain typically 2.0 V above the output voltage even during the low point on the input ripple voltage.

XX = these two digits of the type number indicate voltage.

* = C_{in} is required if regulator is located an appreciable distance from power supply filter.

** = C_O is not needed for stability; however, it does improve transient response.

XX indicates nominal voltage

ORDERING INFORMATION

Device	Output Voltage Tolerance	Tested Operating Junction Temp. Range	Package
MC78XXK MC78XXAK*	4% 2%	−55 to +150°C	Metal Power
MC78XXCK MC78XXACK*	4% 2%	0 to +125°C	
MC78XXCT MC78XXACT	4% 2%		Plastic Power
MC78XXBT	4%	−40 to +125°C	

*2% regulators in Metal Power packages are available in 5, 12 and 15 volt devices.

TYPE NO./VOLTAGE

MC7805	5.0 Volts	MC7812	12 Volts
MC7806	6.0 Volts	MC7815	15 Volts
MC7808	8.0 Volts	MC7818	18 Volts
MC7809	9.0 Volts	MC7824	24 Volts

MAXIMUM RATINGS (T_A = +25°C unless otherwise noted.)

Rating	Symbol	Value	Unit
Input Voltage (5.0 V – 18 V) (24 V)	V_{in}	35 40	Vdc
Power Dissipation and Thermal Characteristics			
Plastic Package			
T_A = +25°C	P_D	Internally Limited	Watts
Derate above T_A = +25°C	$1/\theta_{JA}$	15.4	mW/°C
Thermal Resistance, Junction to Air	θ_{JA}	65	°C/W
T_C = +25°C	P_D	Internally Limited	Watts
Derate above T_C = +75°C (See Figure 1)	$1/\theta_{JC}$	200	mW/°C
Thermal Resistance, Junction to Case	θ_{JC}	5.0	°C/W
Metal Package			
T_A = +25°C	P_D	Internally Limited	Watts
Derate above T_A = +25°C	$1/\theta_{JA}$	22.5	mW/°C
Thermal Resistance, Junction to Air	θ_{JA}	45	°C/W
T_C = +25°C	P_D	Internally Limited	Watts
Derate above T_C = +65°C (See Figure 2)	$1/\theta_{JC}$	182	mW/°C
Thermal Resistance, Junction to Case	θ_{JC}	5.5	°C/W
Storage Junction Temperature Range	T_{stg}	−65 to +150	°C
Operating Junction Temperature Range MC7800,A MC7800C,AC MC7800B	T_J	 −55 to +150 0 to +150 −40 to +150	°C

DEFINITIONS

Line Regulation — The change in output voltage for a change in the input voltage. The measurement is made under conditions of low dissipation or by using pulse techniques such that the average chip temperature is not significantly affected.

Load Regulation — The change in output voltage for a change in load current at constant chip temperature.

Maximum Power Dissipation — The maximum total device dissipation for which the regulator will operate within specifications.

Quiescent Current — That part of the input current that is not delivered to the load.

Output Noise Voltage — The rms ac voltage at the output, with constant load and no input ripple, measured over a specified frequency range.

Long Term Stability — Output voltage stability under accelerated life test conditions with the maximum rated voltage listed in the devices' electrical characteristics and maximum power dissipation.

MC7805, B, C

ELECTRICAL CHARACTERISTICS (V_{in} = 10 V, I_O = 500 mA, T_J = T_{low} to T_{high} [Note 1] unless otherwise noted).

Characteristic	Symbol	MC7805			MC7805B			MC7805C			Unit
		Min	Typ	Max	Min	Typ	Max	Min	Typ	Max	
Output Voltage (T_J = +25°C)	V_O	4.8	5.0	5.2	4.8	5.0	5.2	4.8	5.0	5.2	Vdc
Output Voltage (5.0 mA ≤ I_O ≤ 1.0 A, P_O ≤ 15 W)	V_O										Vdc
7.0 Vdc ≤ V_{in} ≤ 20 Vdc		—	—	—	—	—	—	4.75	5.0	5.25	
8.0 Vdc ≤ V_{in} ≤ 20 Vdc		4.65	5.0	5.35	4.75	5.0	5.25	—	—	—	
Line Regulation (T_J = +25°C, Note 2)	Reg_{line}										mV
7.0 Vdc ≤ V_{in} ≤ 25 Vdc		—	2.0	50	—	7.0	100	—	7.0	100	
8.0 Vdc ≤ V_{in} ≤ 12 Vdc		—	1.0	25	—	2.0	50	—	2.0	50	
Load Regulation (T_J = +25°C, Note 2)	Reg_{load}										mV
5.0 mA ≤ I_O ≤ 1.5 A		—	25	100	—	40	100	—	40	100	
250 mA ≤ I_O ≤ 750 mA		—	8.0	25	—	15	50	—	15	50	
Quiescent Current (T_J = +25°C)	I_B	—	3.2	6.0	—	4.3	8.0	—	4.3	8.0	mA
Quiescent Current Change	ΔI_B										mA
7.0 Vdc ≤ V_{in} ≤ 25 Vdc		—	—	—	—	—	—	—	—	1.3	
8.0 Vdc ≤ V_{in} ≤ 25 Vdc		—	0.3	0.8	—	—	1.3	—	—	—	
5.0 mA ≤ I_O ≤ 1.0 A		—	0.04	0.5	—	—	0.5	—	—	0.5	
Ripple Rejection 8.0 Vdc ≤ V_{in} ≤ 18 Vdc, f = 120 Hz	RR	68	75	—	—	68	—	—	68	—	dB
Dropout Voltage (I_O = 1.0 A, T_J = +25°C)	$V_{in} - V_O$	—	2.0	2.5	—	2.0	—	—	2.0	—	Vdc
Output Noise Voltage (T_A = +25°C) 10 Hz ≤ f ≤ 100 kHz	V_n	—	10	40	—	10	—	—	10	—	µV/V_O
Output Resistance f = 1.0 kHz	r_O	—	17	—	—	17	—	—	17	—	mΩ
Short-Circuit Current Limit (T_A = +25°C) V_{in} = 35 Vdc	I_{sc}	—	0.2	1.2	—	0.2	—	—	0.2	—	A
Peak Output Current (T_J = +25°C)	I_{max}	1.3	2.5	3.3	—	2.2	—	—	2.2	—	A
Average Temperature Coefficient of Output Voltage	TCV_O	—	±0.6	—	—	-1.1	—	—	-1.1	—	mV/°C

MC7805A, AC

ELECTRICAL CHARACTERISTICS (V_{in} = 10 V, I_O = 1.0 A, T_J = T_{low} to T_{high} [Note 1] unless otherwise noted)

Characteristics	Symbol	MC7805A			MC7805AC			Unit
		Min	Typ	Max	Min	Typ	Max	
Output Voltage (T_J = +25°C)	V_O	4.9	5.0	5.1	4.9	5.0	5.1	Vdc
Output Voltage (5.0 mA ≤ I_O ≤ 1.0 A, P_O ≤ 15 W) 7.5 Vdc ≤ V_{in} ≤ 20 Vdc	V_O	4.8	5.0	5.2	4.8	5.0	5.2	Vdc
Line Regulation (Note 2)	Reg_{line}							mV
7.5 Vdc ≤ V_{in} ≤ 25 Vdc, I_O = 500 mA		—	2.0	10	—	7.0	50	
8.0 Vdc ≤ V_{in} ≤ 12 Vdc		—	3.0	10	—	10	50	
8.0 Vdc ≤ V_{in} ≤ 12 Vdc, T_J = +25°C		—	1.0	4.0	—	2.0	25	
7.3 Vdc ≤ V_{in} ≤ 20 Vdc, T_J = +25°C		—	2.0	10	—	7.0	50	
Load Regulation (Note 2)	Reg_{load}							mV
5.0 mA ≤ I_O ≤ 1.5 A, T_J = +25°C		—	2.0	25	—	25	100	
5.0 mA ≤ I_O ≤ 1.0 A		—	2.0	25	—	25	100	
250 mA ≤ I_O ≤ 750mA, T_J = +25°C		—	1.0	15	—	—	—	
250 mA ≤ I_O ≤ 750 mA		—	1.0	25	—	8.0	50	
Quiescent Current	I_B	—	—	5.0	—	—	6.0	mA
T_J = +25°C		—	3.2	4.0	—	4.3	6.0	
Quiescent Current Change	ΔI_B							mA
8.0 Vdc ≤ V_{in} ≤ 25 Vdc, I_O = 500 mA		—	0.3	0.5	—	—	0.8	
7.5 Vdc ≤ V_{in} ≤ 20 Vdc, T_J = +25°C		—	0.2	0.5	—	—	0.8	
5.0 mA ≤ I_O ≤ 1.0 A		—	0.04	0.2	—	—	0.5	
Ripple Rejection	RR							dB
8.0 Vdc ≤ V_{in} ≤ 18 Vdc, f = 120 Hz, T_J = +25°C		68	75	—	—	—	—	
8.0 Vdc ≤ V_{in} ≤ 18 Vdc, f = 120 Hz, I_O = 500 mA		68	75	—	—	68	—	
Dropout Voltage (I_O = 1.0 A, T_J = +25°C)	$V_{in} - V_O$	—	2.0	2.5	—	2.0	—	Vdc
Output Noise Voltage (T_A = +25°C) 10 Hz ≤ f ≤ 100 kHz	V_n	—	10	40	—	10	—	µV/V_O
Output Resistance (f = 1.0 kHz)	r_O	—	2.0	—	—	17	—	mΩ
Short-Circuit Current Limit (T_A = +25°C) V_{in} = 35 Vdc	I_{sc}	—	0.2	1.2	—	0.2	—	A
Peak Output Current (T_J = +25°C)	I_{max}	1.3	2.5	3.3	—	2.2	—	A
Average Temperature Coefficient of Output Voltage	TCV_O	—	±0.6	—	—	-1.1	—	mV/°C

NOTES: 1. T_{low} = −55°C for MC78XX, A; = 0° for MC78XXC, AC; = −40°C for MC78XXB. T_{high} = +150°C for MC78XX, A; = +125°C for MC78XXC, AC, B

2. Load and line regulation are specified at constant junction temperature. Changes in V_O due to heating effects must be taken into account separately. Pulse testing with low duty cycle is used.

MOTOROLA
SEMICONDUCTOR
TECHNICAL DATA

MC7900 Series

THREE-TERMINAL NEGATIVE VOLTAGE REGULATORS

The MC7900 Series of fixed output negative voltage regulators are intended as complements to the popular MC7800 Series devices. These negative regulators are available in the same seven-voltage options as the MC7800 devices. In addition, one extra voltage option commonly employed in MECL systems is also available in the negative MC7900 Series.

Available in fixed output voltage options from −5.0 to −24 volts, these regulators employ current limiting, thermai shutdown, and safe-area compensation — making them remarkably rugged under most operating conditions. With adequate heat-sinking they can deliver output currents in excess of 1.0 ampere.

- No External Components Required
- Internal Thermal Overload Protection
- Internal Short-Circuit Current Limiting
- Output Transistor Safe-Area Compensation
- Available in 2% Voltage Tolerance (See Ordering Information)

THREE-TERMINAL NEGATIVE FIXED VOLTAGE REGULATORS

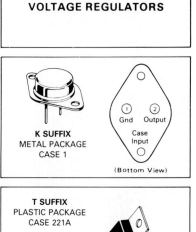

K SUFFIX
METAL PACKAGE
CASE 1

① Gnd ② Output
Case Input
(Bottom View)

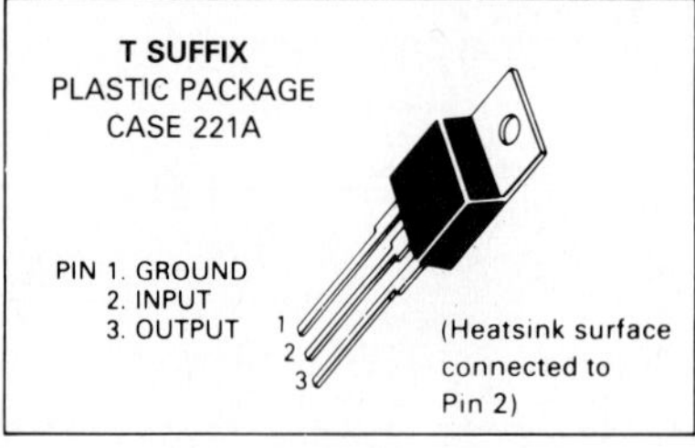

T SUFFIX
PLASTIC PACKAGE
CASE 221A

PIN 1. GROUND
2. INPUT
3. OUTPUT

(Heatsink surface connected to Pin 2)

SCHEMATIC DIAGRAM

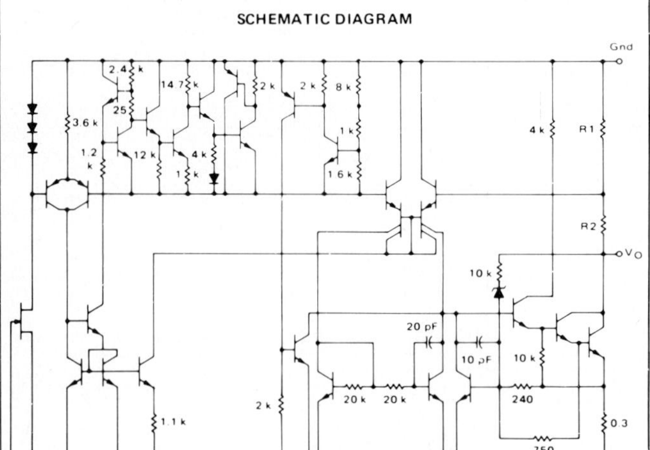

STANDARD APPLICATION

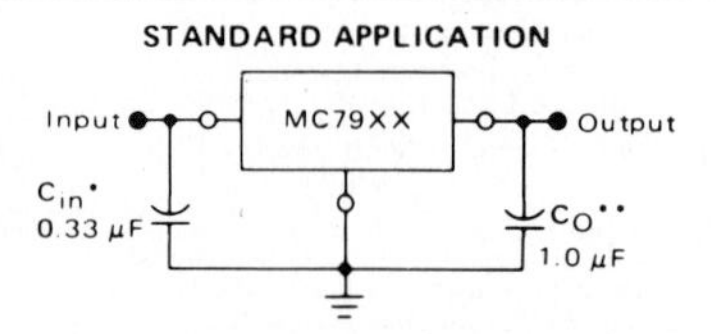

A common ground is required between the input and the output voltages. The input voltage must remain typically 2.0 V more negative even during the high point on the input ripple voltage.

XX = these two digits of the type number indicate voltage.

* = C_{in} is required if regulator is located an appreciable distance from power supply filter.

** = C_O improves stability and transient response.

ORDERING INFORMATION

Device	Output Voltage Tolerance	Tested Operating Junction Temp. Range	Package
MC79XXCK MC79XXACK*	4% 2%	T_J = 0°C to +125°C	Metal Power**
MC79XXCT MC79XXACT*	4% 2%		Plastic Power
MC79XXBT#	4%	T_J = −40°C to +125°C	

XX indicates nominal voltage.
*2% output voltage tolerance available in 5, 12 and 15 volt devices.
**Metal power package available in 5, 12 and 15 volt devices.
#Automotive temperature range selections are available with special test conditions and additional tests in 5, 12 and 15 volt devices. Contact your local Motorola sales office for information.

DEVICE TYPE/NOMINAL OUTPUT VOLTAGE

Device	Voltage	Device	Voltage
MC7905	5.0 Volts	MC7912	12 Volts
MC7905.2	5.2 Volts	MC7915	15 Volts
MC7906	6.0 Volts	MC7918	18 Volts
MC7908	8.0 Volts	MC7924	24 Volts

MAXIMUM RATINGS (T_A = +25°C unless otherwise noted.)

Rating	Symbol	Value	Unit
Input Voltage (−5.0 V ⩾ V_O ⩾ −18 V) (24 V)	V_I	−35 −40	Vdc
Power Dissipation Plastic Package T_A = +25°C Derate above T_A = +25°C T_C = +25°C Derate above T_C = +95°C (See Figure 1) Metal Package T_A = +25°C Derate above T_A = +25°C T_C = +25°C Derate above T_C = +65°C	 P_D $1/R_{\theta JA}$ P_D $1/R_{\theta JC}$ P_D $1/R_{\theta JA}$ P_D $1/R_{\theta JC}$	 Internally Limited 15.4 Internally Limited 200 Internally Limited 22.2 Internally Limited 182	 Watts mW/°C Watts mW/°C Watts mW/°C Watts mW/°C
Storage Junction Temperature Range	T_{stg}	−65 to +150	°C
Junction Temperature Range	T_J	0 to +150	°C

THERMAL CHARACTERISTICS

Characteristic	Symbol	Max	Unit
Thermal Resistance, Junction to Ambient — Plastic Package — Metal Package	$R_{\theta JA}$	65 45	°C/W
Thermal Resistance, Junction to Case — Plastic Package — Metal Package	$R_{\theta JC}$	5.0 5.5	°C/W

MC7905C ELECTRICAL CHARACTERISTICS (V_I = −10 V, I_O = 500 mA, 0°C < T_J < +125°C unless otherwise noted.)

Characteristic	Symbol	Min	Typ	Max	Unit
Output Voltage (T_J = +25°C)	V_O	−4.8	−5.0	−5.2	Vdc
Line Regulation (Note 1) (T_J = +25°C, I_O = 100 mA) −7.0 Vdc ⩾ V_I ⩾ −25 Vdc −8.0 Vdc ⩾ V_I ⩾ −12 Vdc (T_J = +25°C, I_O = 500 mA) −7.0 Vdc ⩾ V_I ⩾ −25 Vdc −8.0 Vdc ⩾ V_I ⩾ −12 Vdc	Reg_{line}	 — — — —	 7.0 2.0 35 8.0	 50 25 100 50	mV
Load Regulation (T_J = +25°C) (Note 1) 5.0 mA ⩽ I_O ⩽ 1.5 A 250 mA ⩽ I_O ⩽ 750 mA	Reg_{load}	 — —	 11 4.0	 100 50	mV
Output Voltage −7.0 Vdc ⩾ V_I ⩾ −20 Vdc, 5.0 mA ⩽ I_O ⩽ 1.0 A, P ⩽ 15 W	V_O	−4.75	—	−5.25	Vdc
Input Bias Current (T_J = +25°C)	I_{IB}	—	4.3	8.0	mA
Input Bias Current Change −7.0 Vdc ⩾ V_I ⩾ −25 Vdc 5.0 mA ⩽ I_O ⩽ 1.5 A	ΔI_{IB}	 — —	 — —	 1.3 0.5	mA
Output Noise Voltage (T_A = +25°C, 10 Hz ⩽ f ⩽ 100 kHz)	e_{on}	—	40	—	μV
Ripple Rejection (I_O = 20 mA, f = 120 Hz)	RR	—	70	—	dB
Dropout Voltage I_O = 1.0 A, T_J = +25°C	V_I-V_O	—	2.0	—	Vdc
Average Temperature Coefficient of Output Voltage I_O = 5.0 mA, 0°C ⩽ T_J ⩽ +125°C	$\Delta V_O/\Delta T$	—	−1.0	—	mV/°C

Note:
1. Load and line regulation are specified at constant junction temperature. Changes in V_O due to heating effects must be taken into account separately. Pulse testing with low duty cycle is used.

MC7912C ELECTRICAL CHARACTERISTICS (V_I = -19 V, I_O = 500 mA, 0°C < T_J < +125°C unless otherwise noted.)

Characteristic	Symbol	Min	Typ	Max	Unit
Output Voltage (T_J = +25°C)	V_O	-11.5	-12	-12.5	Vdc
Line Regulation (Note 1)	Reg_{line}				mV
(T_J = +25°C, I_O = 100 mA)					
-14.5 Vdc ≥ V_I ≥ -30 Vdc		—	13	120	
-16 Vdc ≥ V_I ≥ -22 Vdc		—	6.0	60	
(T_J = +25°C, I_O = 500 mA)					
-14.5 Vdc ≥ V_I ≥ -30 Vdc		—	55	240	
-16 Vdc ≥ V_I ≥ -22 Vdc		—	24	120	
Load Regulation (T_J = +25°C) (Note 1)	Reg_{load}				mV
5.0 mA ≤ I_O ≤ 1.5 A		—	46	240	
250 mA ≤ I_O ≤ 750 mA		—	17	120	
Output Voltage -14.5 Vdc ≥ V_I ≥ -27 Vdc, 5.0 mA ≤ I_O ≤ 1.0 A, P ≤ 15 W	V_O	-11.4	—	-12.6	Vdc
Input Bias Current (T_J = +25°C)	I_{IB}	—	4.4	8.0	mA
Input Bias Current Change	ΔI_{IB}				mA
-14.5 Vdc ≥ V_I ≥ -30 Vdc		—	—	1.0	
5.0 mA ≤ I_O ≤ 1.5 A		—	—	0.5	
Output Noise Voltage (T_A = +25°C, 10 Hz ≤ f ≤ 100 kHz)	e_{on}	—	75	—	μV
Ripple Rejection (I_O = 20 mA, f = 120 Hz)	RR	—	61	—	dB
Dropout Voltage I_O = 1.0 A, T_J = +25°C	V_I-V_O	—	2.0	—	Vdc
Average Temperature Coefficient of Output Voltage I_O = 5.0 mA, 0°C ≤ T_J ≤ +125°C	$\Delta V_O/\Delta T$	—	-1.0	—	mV/°C

MC7912AC ELECTRICAL CHARACTERISTICS (V_I = -19 V, I_O = 500 mA, 0°C < T_J < +125°C unless otherwise noted.)

Characteristic	Symbol	Min	Typ	Max	Unit
Output Voltage (T_J = +25°C)	V_O	-11.75	-12	-12.25	Vdc
Line Regulation (Note 1)	Reg_{line}				mV
-16 Vdc ≥ V_I ≥ -22 Vdc; I_O = 1.0 A, T_J = 25°C		—	6.0	60	
-16 Vdc ≥ V_I ≥ -22 Vdc; I_O = 1.0 A,		—	24	120	
-14.8 Vdc ≥ V_I ≥ -30 Vdc; I_O = 500 mA		—	24	120	
-14.5 Vdc ≥ V_I ≥ -27 Vdc; I_O = 1.0 A, T_J = 25°C		—	13	120	
Load Regulation (Note 1)	Reg_{load}				mV
5.0 mA ≤ I_O ≤ 1.5 A, T_J = 25°C		—	46	150	
250 mA ≤ I_O ≤ 750 mA		—	17	75	
5.0 mA ≤ I_O ≤ 1.0 A		—	35	150	
Output Voltage -14.8 Vdc ≥ V_I ≥ -27 Vdc, 5.0 mA ≤ I_O ≤ 1.0 A, P ≤ 15 W	V_O	-11.5	—	-12.5	Vdc
Input Bias Current	I_{IB}	—	4.4	8.0	mA
Input Bias Current Change	ΔI_{IB}				mA
-15 Vdc ≥ V_I ≥ -30 Vdc		—	—	0.8	
5.0 mA ≤ I_O ≤ 1.0 A		—	—	0.5	
5.0 mA ≤ I_O ≤ 1.5 A, T_J = 25°C		—	—	0.5	
Output Noise Voltage (T_A = +25°C, 10 Hz ≤ f ≤ 100 kHz)	e_{on}	—	75	—	μV
Ripple Rejection (I_O = 20 mA, f = 120 Hz)	RR	—	61	—	dB
Dropout Voltage I_O = 1.0 A, T_J = +25°C	V_I-V_O	—	2.0	—	Vdc
Average Temperature Coefficient of Output Voltage I_O = 5.0 mA, 0°C ≤ T_J ≤ +125°C	$\Delta V_O/\Delta T$	—	-1.0	—	mV/°C

Note:
1. Load and line regulation are specified at constant junction temperature. Changes in V_O due to heating effects must be taken into account separately. Pulse testing with low duty cycle is used.

Appendix B
Derivations of Selected Equations

Equation (2–1)

The average value of a half-wave rectified sine wave is the area under the curve divided by the period (2π). The equation for a sine wave is

$$
\begin{aligned}
v &= V_p \sin\theta \\
V_{\text{AVG}} &= \frac{\text{area}}{2\pi} \\
&= \frac{1}{2\pi}\int_0^{\pi} V_p \sin\theta \, d\theta \\
&= \frac{V_p}{2\pi}(-\cos\theta)\Big|_0^{\pi} \\
&= \frac{V_p}{2\pi}[-\cos\pi - (-\cos 0)] \\
&= \frac{V_p}{2\pi}[-(-1) - (-1)] \\
&= \frac{V_p}{2\pi}(2) \\
V_{\text{AVG}} &= \frac{V_p}{\pi}
\end{aligned}
$$

Equation (2–13)

Referring to Figure B–1, when the filter capacitor discharges, the voltage is

$$v_C = V_{p(\text{in})}e^{-t/RC}$$

Since the discharge time of the capacitor is from one peak to approximately the next peak, $t_{\text{dis}} \cong T$ when v_C reaches its minimum value.

$$v_{C(\text{min})} = V_{p(\text{in})}e^{-T/RC}$$

FIGURE B–1

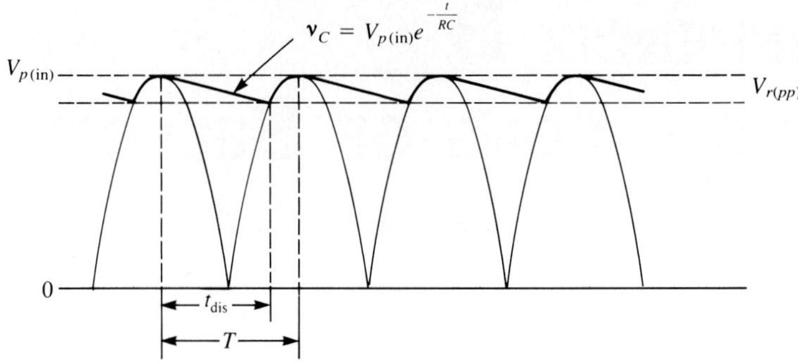

Since $RC >> T$, T/RC becomes much less than 1 (which is usually the case); $e^{-T/RC}$ approaches 1 and can be expressed as

$$e^{-T/RC} \cong 1 - \frac{T}{RC}$$

Therefore,

$$v_{C(\min)} = V_{p(\text{in})}\left(1 - \frac{T}{RC}\right)$$

The peak-to-peak ripple voltage is

$$\begin{aligned} V_{r(pp)} &= V_{p(\text{in})} - V_{C(\min)} \\ &= V_{p(\text{in})} - V_{p(\text{in})} + \frac{V_{p(\text{in})}T}{RC} \\ &= \frac{V_{p(\text{in})}T}{RC} \end{aligned}$$

For $f = 120$ Hz (full-wave),

$$V_{r(pp)} = \frac{V_{p(\text{in})}}{RCf} = \frac{0.00833V_{p(\text{in})}}{RC}$$

To obtain the dc value, one-half of the peak-to-peak ripple is subtracted from the peak value.

$$\begin{aligned} V_{\text{dc}} &= V_{p(\text{in})} - \frac{V_{r(pp)}}{2} \\ &= V_{p(\text{in})} - \frac{0.00833V_{p(\text{in})}}{2RC} \\ &= \left(1 - \frac{0.00417}{RC}\right)V_{p(\text{in})} \end{aligned}$$

EQUATION (2–14)

The peak ripple voltage is

$$V_{r(p)} = \frac{0.00833V_{p(\text{in})}}{2R_LC}$$

Since the ripple waveform is a sawtooth, we divide by $\sqrt{3}$ to convert peak to rms.

$$V_{r(\text{rms})} = \frac{0.00833V_{p(\text{in})}}{2R_LC(\sqrt{3})} = \frac{0.0024V_{p(\text{in})}}{R_LC}$$

DERIVATION OF THE RIPPLE VOLTAGE FOR A FULL-WAVE RECTIFIED SIGNAL AS USED IN EXAMPLE 2–8

The ac component of a full-wave rectified signal is the total voltage minus the dc value.

$$v = v_t - V_{\text{dc}}$$

The rms value of V_{ac} is

$$V_{r(\text{rms})} = \left(\frac{1}{2\pi}\int_0^{2\pi} V_{\text{ac}}\,d\theta\right)^{1/2}$$

$$= \left(\frac{1}{2\pi}\int_0^{2\pi} (v_t - V_{\text{dc}})^2\,d\theta\right)^{1/2}$$

$$= \left(\frac{1}{2\pi}\int_0^{2\pi} (v_t^2 - 2v_tV_{\text{dc}} + V_{\text{dc}}^2)\,d\theta\right)^{1/2}$$

$$= \left[\frac{1}{2\pi}\left(\int_0^{2\pi} v_t^2\,d\theta - \int_0^{2\pi} 2v_tV_{\text{dc}}\,d\theta + \int_0^{2\pi} V_{\text{dc}}^2\,d\theta\right)\right]^{1/2}$$

$$= \left[\frac{1}{2\pi}\left(\int_0^{2\pi} v_t^2\,d\theta - 2V_{\text{dc}}\int_0^{2\pi} v_t\,d\theta + V_{\text{dc}}^2\int_0^{2\pi} d\theta\right)\right]^{1/2}$$

$$= (V_{t(\text{rms})}^2 - 2V_{\text{dc}}^2 + V_{\text{dc}}^2)^{1/2}$$

$$V_{r(\text{rms})} = (V_{t(\text{rms})}^2 - V_{\text{dc}}^2)^{1/2}$$

For a full-wave rectified voltage:

$$V_{t(\text{rms})} = \frac{V_p}{1.414}$$

$$V_{\text{dc}} = \frac{2V_p}{\pi}$$

$$V_{r(\text{rms})} = \sqrt{\left(\frac{V_p}{1.414}\right)^2 - \left(\frac{2V_p}{\pi}\right)^2}$$

$$= V_p\sqrt{\left(\frac{1}{1.414}\right)^2 - \left(\frac{2}{\pi}\right)^2}$$

$$= 0.308V_p$$

EQUATION (6–10)

The Shockley equation for the base-emitter pn junction is

$$I_E = I_R(e^{VQ/kT} - 1)$$

where I_E = the total forward current across the base-emitter junction.
I_R = the reverse saturation current.
V = the voltage across the depletion layer.
Q = the charge on an electron.
k = a number known as Boltzmann's constant.
T = the absolute temperature.

At ambient temperature, $Q/kT \cong 40$, so

$$I_E = I_R(e^{40V} - 1)$$

Differentiating, we get

$$\frac{dI_E}{dV} = 40I_Re^{40V}$$

Since $I_Re^{40V} = I_E + I_R$,

$$\frac{dI_E}{dV} = 40(I_E + I_R)$$

Assuming $I_R << I_E$,

$$\frac{dI_E}{dV} \cong 40I_E$$

The ac resistance r_e of the base-emitter junction can be expressed as dV/dI_E.

$$r_e = \frac{dV}{dI_E} \cong \frac{1}{40I_E} \cong \frac{25\text{ mV}}{I_E}$$

EQUATION (6–31)

The emitter-follower is represented by the r parameter ac equivalent circuit in Figure B–2(a). By Thevenizing from the base back to the source, the circuit is simplified to the form shown in Figure B–2(b).

$$V_{out} = V_e,\ I_{out} = I_e,\ \text{and}\ I_{in} = I_b$$

$$R_{out} = \frac{V_e}{I_e}$$

$$I_e \cong \beta_{ac}I_b$$

With $V_s = 0$ and with I_b produced by V_{out}, and neglecting the base-to-emitter voltage drop (and therefore r_e),

$$I_b \cong \frac{V_e}{R_1\|R_2\|R_s}$$

FIGURE B–2

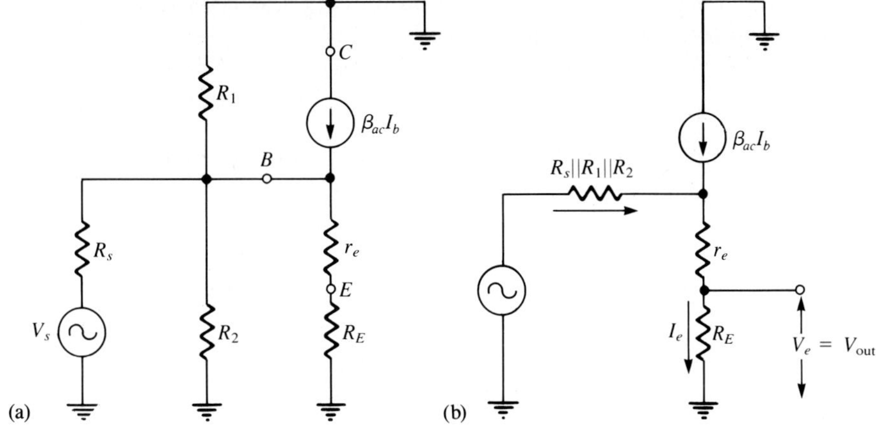

Assuming that $R_1 >> R_s$ and $R_2 >> R_s$,

$$I_b \cong \frac{V_e}{R_s}$$

$$I_{out} = I_e = \frac{\beta_{ac} V_e}{R_s}$$

$$\frac{V_{out}}{I_{out}} = \frac{V_e}{I_e} = \frac{V_e}{\beta_{ac} V_e / R_s} = \frac{R_s}{\beta_{ac}}$$

Looking from the emitter, R_E appears in parallel with R_s/β_{ac}. Therefore,

$$R_{out} = \left(\frac{R_s}{\beta_{ac}}\right) \| R_E$$

MIDPOINT BIAS: Proof that $I_D \cong 0.5 I_{DSS}$ when $V_{GS} = \dfrac{V_{GS(off)}}{3.414}$

Start with Equation (7–1):

$$I_D = I_{DSS}\left(1 - \frac{V_{GS}}{V_{GS(off)}}\right)^2$$

Let $I_D = 0.5 I_{DSS}$.

$$0.5 I_{DSS} = I_{DSS}\left(1 - \frac{V_{GS}}{V_{GS(off)}}\right)^2$$

Cancelling I_{DSS} on each side,

$$0.5 = \left(1 - \frac{V_{GS}}{V_{GS(off)}}\right)^2$$

We want a factor (call it F) by which $V_{GS(\text{off})}$ can be divided to give a value of V_{GS} that will produce a drain current that is $0.5I_{DSS}$.

$$0.5 = \left[1 - \frac{\left(\frac{V_{GS(\text{off})}}{F}\right)}{V_{GS(\text{off})}}\right]^2$$

Solving for F,

$$\sqrt{0.5} = 1 - \frac{\left(\frac{V_{GS(\text{off})}}{F}\right)}{V_{GS(\text{off})}}$$

$$\sqrt{0.5} = 1 - \frac{1}{F}$$

$$\sqrt{0.5} - 1 = -\frac{1}{F}$$

$$\frac{1}{F} = 1 - \sqrt{0.5}$$

$$F = \frac{1}{1 - \sqrt{0.5}} \cong 3.414$$

EQUATION (8–7)

$$I_D = I_{DSS}\left(1 - \frac{I_D R_S}{V_{GS(\text{off})}}\right)^2$$

$$= I_{DSS}\left(1 - \frac{I_D R_S}{V_{GS(\text{off})}}\right)\left(1 - \frac{I_D R_S}{V_{GS(\text{off})}}\right)$$

$$= I_{DSS}\left(1 - \frac{2I_D R_S}{V_{GS(\text{off})}} + \frac{I_D^2 R_S^2}{V_{GS(\text{off})}^2}\right)$$

$$= I_{DSS} - \frac{2I_{DSS}R_S}{V_{GS(\text{off})}}I_D + \frac{I_{DSS}R_S^2}{V_{GS(\text{off})}^2}I_D^2$$

Rearranging into a standard quadratic equation form,

$$\left(\frac{I_{DSS}R_S^2}{V_{GS(\text{off})}^2}\right)I_D^2 - \left(1 + \frac{2I_{DSS}R_S}{V_{GS(\text{off})}}\right)I_D + I_{DSS} = 0$$

The coefficients and constant are

$$A = \frac{R_S I_{DSS}^2}{V_{GS(\text{off})}^2}$$

$$B = -\left(1 + \frac{2R_S I_{DSS}}{V_{GS(\text{off})}}\right)$$

$$C = I_{DSS}$$

In simplified notation, the equation is

$$AI_D^2 + BI_D + C = 0$$

The solutions to this quadratic equation are

$$I_D = \frac{-B \pm \sqrt{B^2 - 4AC}}{2A}$$

EQUATION (10–1)

An inverting amplifier with feedback capacitance is shown in Figure B–3. For the input,

$$I_1 = \frac{V_1 - V_2}{X_C}$$

Factoring V_1 out,

$$I_1 = \frac{V_1(1 - V_2/V_1)}{X_C}$$

The ratio V_2/V_1 is the voltage gain, $-A_v$.

$$I_1 = \frac{V_1(1 + A_v)}{X_C} = \frac{V_1}{X_C/(1 + A_v)}$$

FIGURE B–3

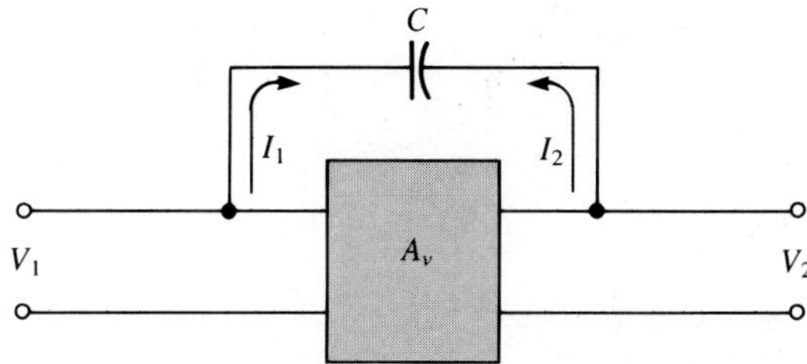

The effective reactance as seen from the input terminals is

$$X_{C_{\text{in(Miller)}}} = \frac{X_C}{1 + A_v}$$

or

$$\frac{1}{2\pi f C_{\text{in(Miller)}}} = \frac{1}{2\pi f C(1 + A_v)}$$

Cancelling and inverting, we get

$$C_{\text{in(Miller)}} = C(A_v + 1)$$

EQUATION (10–2)

For the output in Figure B–3,

$$I_2 = \frac{V_2 - V_1}{X_C} = \frac{V_2(1 - V_1/V_2)}{X_C}$$

Since $V_1/V_2 = -1/A_v$,

$$I_2 = \frac{V_2(1 + 1/A_v)}{X_C} = \frac{V_2}{X_C/(1 + 1/A_v)} = \frac{V_2}{X_C/[(A_v + 1)/A_v]}$$

The effective reactance as seen from the output is

$$X_{C_{\text{out(Miller)}}} = \frac{X_C}{(A_v + 1)/A_v}$$

$$\frac{1}{2\pi f C_{\text{out(Miller)}}} = \frac{1}{2\pi f C[(A_v + 1)/A_v]}$$

Cancelling and inverting, we get

$$C_{\text{out(Miller)}} = C\left(\frac{A_v + 1}{A_v}\right)$$

EQUATIONS 10–37 AND 10–38

The *rise time* is defined as the time required for the voltage to increase from 10 percent of its final value to 90 percent of its final value, as indicated in Figure B–4. Expressing the curve in its exponential form gives

$$v = V_{\text{final}}(1 - e^{-t/RC})$$

FIGURE B–4

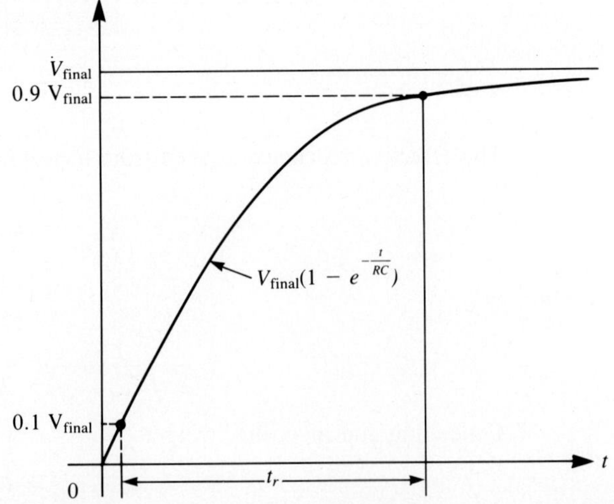

When $v = 0.1V_{\text{final}}$,

$$\begin{aligned}
0.1V_{\text{final}} &= V_{\text{final}}(1 - e^{-t/RC}) \\
0.1V_{\text{final}} &= V_{\text{final}} - V_{\text{final}}e^{-t/RC} \\
V_{\text{final}}e^{-t/RC} &= 0.9V_{\text{final}} \\
e^{-t/RC} &= 0.9 \\
\ln e^{-t/RC} &= \ln(0.9) \\
-\frac{t}{RC} &= -0.1 \\
t &= 0.1RC
\end{aligned}$$

When $v = 0.9V_{\text{final}}$,

$$\begin{aligned}
0.9V_{\text{final}} &= V_{\text{final}}(1 - e^{-t/RC}) \\
0.9V_{\text{final}} &= V_{\text{final}} - V_{\text{final}}e^{-t/RC} \\
V_{\text{final}}e^{-t/RC} &= 0.1V_{\text{final}} \\
\ln e^{-t/RC} &= \ln(0.1) \\
-\frac{t}{RC} &= -2.3 \\
t &= 2.3RC
\end{aligned}$$

The difference is the rise time.

$$t_r = 2.3RC - 0.1RC = 2.2RC$$

The critical frequency of an RC network is

$$f_c = \frac{1}{2\pi RC}$$

$$RC = \frac{1}{2\pi f_c}$$

Substituting,

$$t_r = \frac{2.2}{2\pi f_{ch}} = \frac{0.35}{f_{ch}}$$

$$f_{ch} = \frac{0.35}{t_r}$$

In a similar way, it can be shown that

$$f_{cl} = \frac{0.35}{t_f}$$

EQUATION (13–11)

The formula for open-loop gain in Equation (13–7) can be expressed in complex notation as

$$A_{ol} = \frac{A_{ol(\text{mid})}}{1 + jf/f_{c(ol)}}$$

Substituting the above expression into the equation $A_{cl} = A_{ol}/(1 + BA_{ol})$, we get a formula for the total closed-loop gain.

$$A_{cl} = \frac{A_{ol(\text{mid})}/(1 + jf/f_{c(ol)})}{1 + BA_{ol(\text{mid})}/(1 + jf/f_{c(ol)})}$$

Multiplying the numerator and denominator by $1 + jf/f_{c(ol)}$ yields

$$A_{cl} = \frac{A_{ol(\text{mid})}}{1 + BA_{ol(\text{mid})} + jf/f_{c(ol)}}$$

Dividing the numerator and denominator by $1 + BA_{ol(\text{mid})}$ gives

$$A_{cl} = \frac{A_{ol(\text{mid})}/(1 + BA_{ol(\text{mid})})}{1 + j[f/(f_{c(ol)}(1 + BA_{ol(\text{mid})}))]}$$

The above expression is of the form of the first equation

$$A_{cl} = \frac{A_{cl(\text{mid})}}{1 + jf/f_{c(cl)}}$$

where $f_{c(cl)}$ is the closed-loop critical frequency. Thus,

$$f_{c(cl)} = f_{c(ol)}(1 + BA_{ol(\text{mid})})$$

EQUATION (15–1)

$$\frac{V_{\text{out}}}{V_{\text{in}}} = \frac{R(-jX)/(R - jX)}{(R - jX) + R(-jX)/(R - jX)}$$

$$= \frac{R(-jX)}{(R - jX)^2 - jRX}$$

Multiplying the numerator and denominator by j,

$$\frac{V_{\text{out}}}{V_{\text{in}}} = \frac{RX}{j(R - jX)^2 + RX}$$

$$= \frac{RX}{RX + j(R^2 - j2RX - X^2)}$$

$$= \frac{RX}{RX + jR^2 + 2RX - jX^2}$$

$$= \frac{RX}{3RX + j(R^2 - X^2)}$$

For a 0° phase angle there can be no j term. Recall from complex numbers in ac theory that a *nonzero* angle is associated with a complex number having a j term. Therefore, at f_r the j term is 0.

$$R^2 - X^2 = 0$$

Thus,

$$\frac{V_{out}}{V_{in}} = \frac{RX}{3RX}$$

Cancelling, we get

$$\frac{V_{out}}{V_{in}} = \frac{1}{3}$$

EQUATION (15–2)

From the derivation of Equation (15–1),

$$R^2 - X^2 = 0$$
$$R^2 = X^2$$
$$R = X$$

Since $X = \dfrac{1}{2\pi f_r C}$,

$$R = \frac{1}{2\pi f_r C}$$

$$f_r = \frac{1}{2\pi RC}$$

EQUATIONS (15–3) AND (15–4)

The feedback network in the phase-shift oscillator consists of three RC stages, as shown in Figure B–5. An expression for the attenuation is derived using the mesh analysis method for the loop assignment shown. All Rs are equal in value, and all Cs are equal in value.

$$(R - j1/2\pi fC)I_1 - RI_2 + 0I_3 = V_{in}$$
$$-RI_1 + (2R - j1/2\pi fC)I_2 - RI_3 = 0$$
$$0I_1 - RI_2 + (2R - j1/2\pi fC)I_3 = 0$$

FIGURE B–5

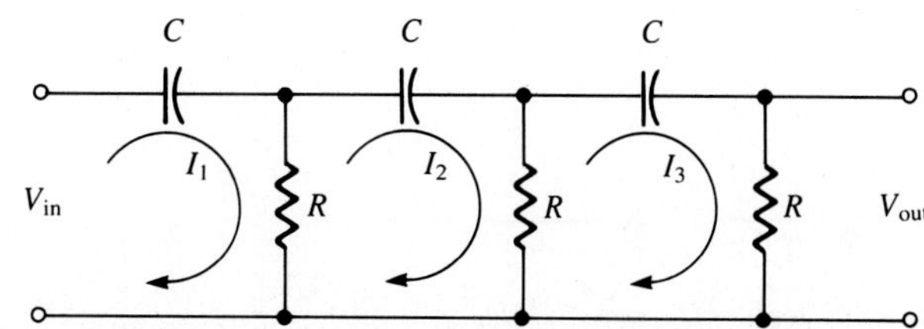

In order to get V_{out}, we must solve for I_3 using determinants:

$$I_3 = \frac{\begin{vmatrix} (R - j1/2\pi fC) & -R & V_{in} \\ -R & (2R - j1/2\pi fC) & 0 \\ 0 & -R & 0 \end{vmatrix}}{\begin{vmatrix} (R - j1/2\pi fC) & -R & 0 \\ -R & (2R - j1/2\pi fC) & -R \\ 0 & -R & (2R - j1/2\pi fC) \end{vmatrix}}$$

$$I_3 = \frac{R^2 V_{in}}{(R - j1/2\pi fC)(2R - j1/2\pi fC)^2 - R^2(2R - j1/2\pi fC) - R^2(R - 1/2\pi fC)}$$

$$\frac{V_{out}}{V_{in}} = \frac{RI_3}{V_{in}}$$

$$= \frac{R^3}{(R - j1/2\pi fC)(2R - j1/2\pi fC)^2 - R^3(2 - j1/2\pi fRC) - R^3(1 - 1/2\pi fRC)}$$

$$= \frac{R^3}{R^3(1 - j1/2\pi fRC)(2 - j1/2\pi fRC)^2 - R^3[(2 - j1/2\pi fRC) - (1 - j1/2\pi RC)]}$$

$$= \frac{R^3}{R^3(1 - j1/2\pi fRC)(2 - j1/2\pi fRC)^2 - R^3(3 - j1/2\pi fRC)}$$

$$\frac{V_{out}}{V_{in}} = \frac{1}{(1 - j1/2\pi fRC)(2 - j1/2\pi fRC)^2 - (3 - j1/2\pi fRC)}$$

Expanding and combining the real terms and the j terms separately,

$$\frac{V_{out}}{V_{in}} = \frac{1}{\left(1 - \dfrac{5}{4\pi^2 f^2 R^2 C^2}\right) - j\left(\dfrac{6}{2\pi fRC} - \dfrac{1}{(2\pi f)^3 R^3 C^3}\right)}$$

For oscillation in the phase-shift amplifier, the phase shift through the RC network must equal 180°. For this condition to exist, the j term must be 0 at the frequency of oscillation f_0.

$$\frac{6}{2\pi f_0 RC} - \frac{1}{(2\pi f_0)^3 R^3 C^3} = 0$$

$$\frac{6(2\pi)^2 f_0^2 R^2 C^2 - 1}{(2\pi)^3 f_0^3 R^3 C^3} = 0$$

$$6(2\pi)^2 f_0^2 R^2 C^2 - 1 = 0$$

$$f_0^2 = \frac{1}{6(2\pi)^2 R^2 C^2}$$

$$f_0 = \frac{1}{2\pi\sqrt{6}RC}$$

Since the j term is 0,

$$\frac{V_{out}}{V_{in}} = \frac{1}{1 - \dfrac{5}{4\pi^2 f_0^2 R^2 C^2}} = \frac{1}{1 - \dfrac{5}{\left(\dfrac{1}{\sqrt{6}RC}\right)^2 R^2 C^2}}$$

$$= \frac{1}{1 - 30} = -\frac{1}{29}$$

The negative sign results from the 180° inversion. Thus, the value of attenuation for the feedback network is

$$B = \frac{1}{29}$$

Answers to Self-Tests

Chapter 1

1. (c) **2.** (d) **3.** (a) **4.** (d) **5.** (d)
6. (d) **7.** (b) **8.** (a) **9.** (d) **10.** (c)
11. (b) **12.** (a) **13.** (d) **14.** (c) **15.** (d)
16. (e) **17.** (d) **18.** (a) **19.** (b) **20.** (c)
21. (c) **22.** (a) **23.** (c) **24.** (d) **25.** (d)
26. (c) **27.** (d) **28.** (d) **29.** (b) **30.** (b)

Chapter 2

1. (b) **2.** (c) **3.** (d) **4.** (e) **5.** (a)
6. (c) **7.** (d) **8.** (a) **9.** (b) **10.** (a)
11. (d) **12.** (b) **13.** (c) **14.** (a) **15.** (b)
16. (c) **17.** (a) **18.** (d) **19.** (b) **20.** (c)
21. (b)

Chapter 3

1. (a) **2.** (b) **3.** (c) **4.** (b) **5.** (d)
6. (b) **7.** (d) **8.** (a) **9.** (c) **10.** (d)
11. (b) **12.** (b) **13.** (b) **14.** (d)

Chapter 4

1. (d) **2.** (c) **3.** (a) **4.** (d) **5.** (a)
6. (c) **7.** (b) **8.** (b) **9.** (a) **10.** (c)
11. (b) **12.** (f) **13.** (c) **14.** (b) **15.** (b)
16. (a)

Chapter 5

1. (b) **2.** (c) **3.** (d) **4.** (d) **5.** (c)
6. (d) **7.** (a) **8.** (d) **9.** (c) **10.** (a)
11. (b) **12.** (c) **13.** (a) **14.** (c) **15.** (f)

Chapter 6

1. (a) **2.** (b) **3.** (c) **4.** (b) **5.** (d)
6. (d) **7.** (a) **8.** (b) **9.** (d) **10.** (c)
11. (a) **12.** (b) **13.** (d) **14.** (c) **15.** (a)

Chapter 7

1. (e) **2.** (b) **3.** (a) **4.** (c) **5.** (d)
6. (c) **7.** (a) **8.** (c) **9.** (b) **10.** (d)
11. (a) **12.** (c) **13.** (d) **14.** (c) **15.** (c)
16. (b) **17.** (a) **18.** (c)

Chapter 8

1. (f) **2.** (b) **3.** (c) **4.** (d) **5.** (a)
6. (c) **7.** (a) **8.** (c) **9.** (b) **10.** (a)
11. (d) **12.** (a) **13.** (c) **14.** (a) **15.** (b)

Chapter 9

1. (b) **2.** (d) **3.** (a) **4.** (b) **5.** (b)
6. (c) **7.** (a) **8.** (d) **9.** (b) **10.** (c)
11. (a) **12.** (d)

Chapter 10

1. (d) **2.** (c) **3.** (b) **4.** (a) **5.** (d)
6. (b) **7.** (c) **8.** (c) **9.** (a) **10.** (c)
11. (b) **12.** (d) **13.** (c) **14.** (a) **15.** (b)

Chapter 11

1. (b) **2.** (d) **3.** (c) **4.** (c) **5.** (a)
6. (e) **7.** (b) **8.** (b) **9.** (d) **10.** (d)
11. (c) **12.** (d) **13.** (a) **14.** (d) **15.** (c)
16. (b)

Chapter 12

1. (c) **2.** (b) **3.** (d) **4.** (b) **5.** (a)
6. (c) **7.** (b) **8.** (a) **9.** (d) **10.** (c)
11. (d) **12.** (a) **13.** (b) **14.** (c) **15.** (c)
16. (d) **17.** (b) **18.** (c) **19.** (a) **20.** (c)
21. (d)

CHAPTER 13

1. (c) **2.** (b) **3.** (a) **4.** (b) **5.** (d)
6. (a) **7.** (d) **8.** (c) **9.** (b) **10.** (a)
11. (d) **12.** (d) **13.** (b) **14.** (c) **15.** (b)

CHAPTER 14

1. (c) **2.** (a) **3.** (c) **4.** (e) **5.** (b)
6. (d) **7.** (c) **8.** (a) **9.** (c) **10.** (a)
11. (b) **12.** (c) **13.** (b) **14.** (d) **15.** (d)
16. (a) **17.** (d) **18.** (c) **19.** (c) **20.** (a)

CHAPTER 15

1. (b) **2.** (a) **3.** (c) **4.** (b) **5.** (d)
6. (c) **7.** (b) **8.** (d) **9.** (a) **10.** (c)
11. (b) **12.** (c) **13.** (a) **14.** (c) **15.** (c)

CHAPTER 16

1. (c) **2.** (d) **3.** (a) **4.** (b) **5.** (c)
6. (c) **7.** (b) **8.** (a) **9.** (d) **10.** (b)
11. (a) **12.** (c) **13.** (b) **14.** (d)

CHAPTER 17

1. (c) **2.** (d) **3.** (c) **4.** (b) **5.** (d)
6. (a) **7.** (c) **8.** (a) **9.** (g) **10.** (c)

Answers to Selected Odd-Numbered Problems

Chapter 2

1. **(a)** Reverse **(b)** Forward
   (c) Forward **(d)** Forward

3. **(a)** Open **(b)** Open **(c)** Shorted
   (d) Functioning

5. See Figure ANS–1.

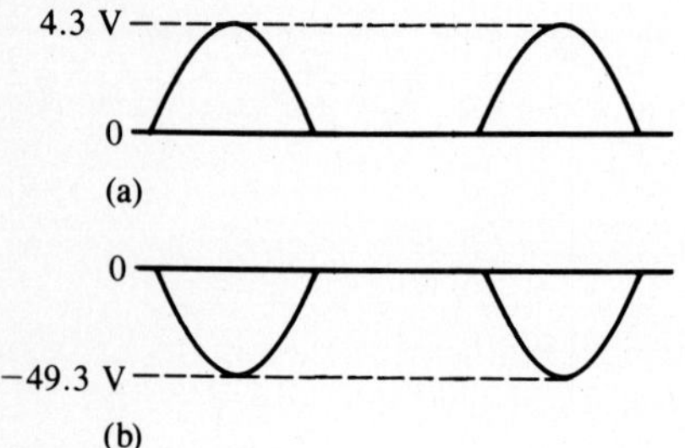

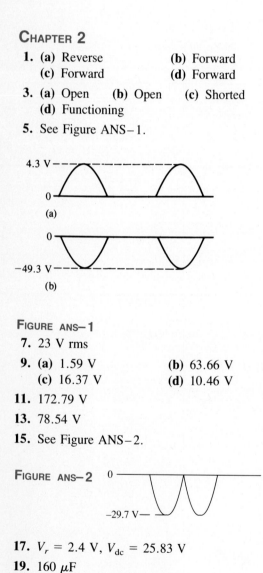

Figure ANS–1

7. 23 V rms

9. **(a)** 1.59 V **(b)** 63.66 V
   (c) 16.37 V **(d)** 10.46 V

11. 172.79 V

13. 78.54 V

15. See Figure ANS–2.

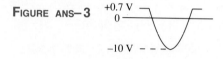

Figure ANS–2

17. $V_r = 2.4$ V, $V_{dc} = 25.83$ V

19. 160 μF

21. $V_r = 1.47$ V, $V_{dc} = 30.58$ V

23. See Figure ANS–3.

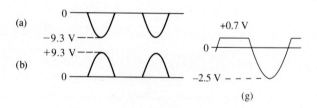

Figure ANS–3

25. See Figure ANS–4.

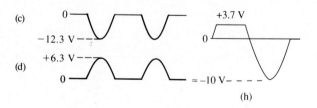

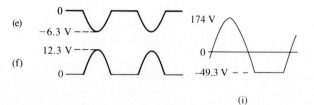

Figure ANS–4

27. (a) A sine wave with a positive peak at +0.7 V, a negative peak at −7.3 V, and a dc value of −3.3 V.
(b) A sine wave with a positive peak at +29.3 V, a negative peak at −0.7 V, and a dc value of +14.3 V.
(c) A square wave varying from +0.7 V down to −15.3 V, with a dc value of −7.3 V.
(d) A square wave varying from +1.3 V down to −0.7 V, with a dc value of +0.3 V.

29. 56.56 V

31. 50 V

33. 0.0625 Ω

35. Coil is open. Capacitor is shorted.

37. The circuit should not fail because the diode ratings exceed the actual PIV and maximum current.

39. Excessive PIV or surge current causes diode to open each time power is turned on. It could be a faulty transformer.

CHAPTER 3

1. See Figure ANS–5.

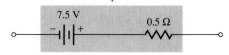

FIGURE ANS–5

3. 5 Ω

5. 6.92 V

7. 14.37 V

9. See Figure ANS–6.

11. 13.5%

13. 3.13%

15. 5.88%

17. 3 V

19. 2.2 V

21. (a) 940 nm **(b)** 912 nm to 968 nm **(c)** 0.5 mW

23. (a) 30 kΩ **(b)** 8.57 kΩ **(c)** 5.88 kΩ

25. −750 Ω

27. The reflective ends cause the light to bounce back and forth, thus increasing the intensity of the light. The partially reflective end allows a portion of the reflected light to be emitted.

29. D_1 open, R_1 open, no dc voltage, short in threshold circuit

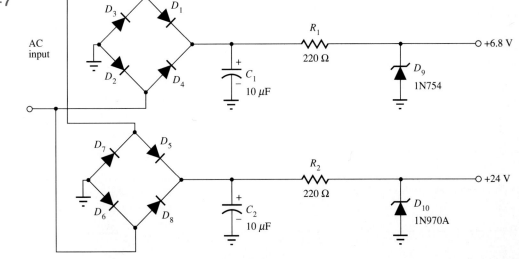

31. See Figure ANS–7.

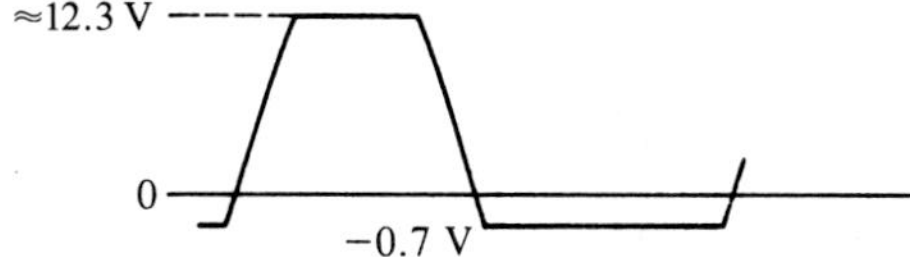

FIGURE ANS–6

FIGURE ANS–7

33. 0.32 or 3.125:1

CHAPTER 4

1. Holes

3. The base is thin and lightly doped so that a small recombination (base) current is generated compared to the collector current.

5. Negative, positive

7. 0.947

9. 24

11. 8.98 mA

13. 0.99

15. **(a)** $V_{BE} = 0.7$ V, $V_{CE} = 5.1$ V, $V_{BC} = -4.4$ V
(b) $V_{BE} = -0.7$ V, $V_{CE} = -3.85$ V, $V_{BC} = 3.15$ V

17. $I_B = 26\ \mu$A, $I_E = 1.3$ mA, $I_C = 1.274$ mA

19. 3.2 μA

21. 0.425 W

23. 33.33

25. 0.5 mA, 3.33 μA, 4.03 V

27. See Figure ANS–8.

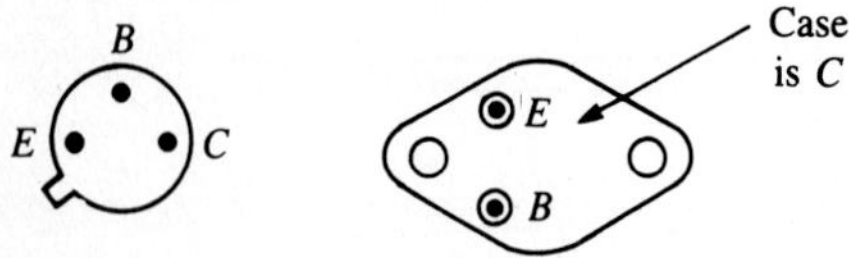

FIGURE ANS–8

29. Open, low resistance

31. **(a)** 27.85 **(b)** 108.7

33. Q_1, Q_3, or Q_5 off all the time

CHAPTER 5

1. Saturation

3. 18 mA

5. $V_{CE} = 20$ V, $I_{C(sat)} = 2$ mA

7. See Figure ANS–9.

9. $I_B = 514\ \mu$A, $I_C = 46.26$ mA, $V_{CE} = 7.37$ V

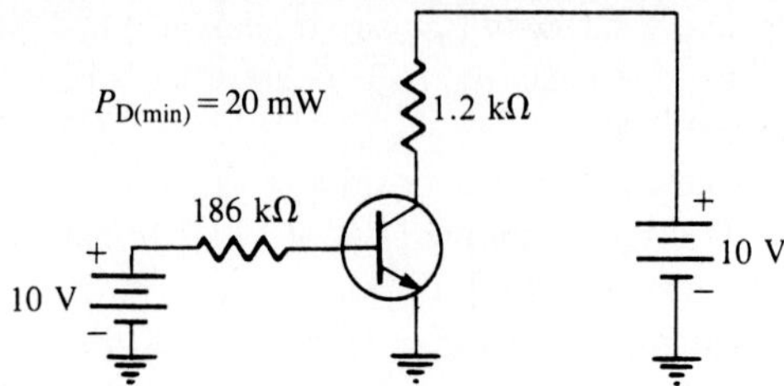

FIGURE ANS–9

11. I_C changes in the circuit using a common V_{CC} and V_{BB} supply, because a change in V_{CC} causes I_B to change, which in turn changes I_C.

13. 59.57 mA, 5.96 V

15. 754 Ω

17. When $R_E >> \frac{R_B}{\beta_{dc}}$

19. 69.12

21. $I_C \cong 0.81$ mA, $V_{CE} = 13.24$ V

23. See Figure ANS–10.

FIGURE ANS–10

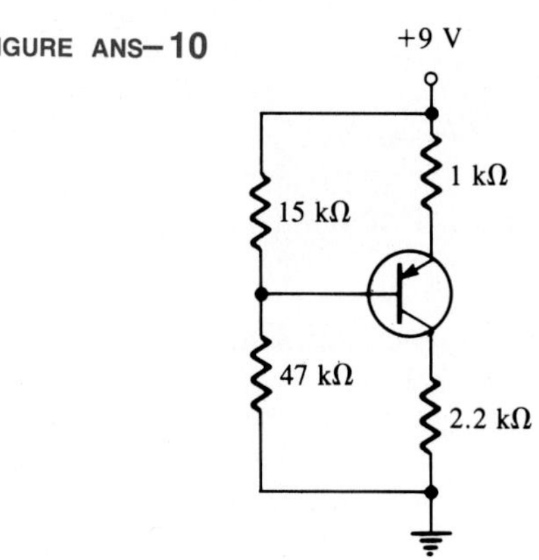

25. **(a)** −1.41 mA, −8.67 V
(b) 12.22 mW

27. 2.53 kΩ

29. 7.87 mA, 2.56 V

31. **(a)** Open collector or R_E open
(b) No problems
(c) Transistor shorted collector-to-emitter
(d) Open emitter

33. **(a)** DMM1: −3.59 V
DMM2: −10 V
DMM3: Floating
(b) DMM1: −7.04 V
DMM2: ≈ −6.34 V
DMM3: −6.34 V
(c) DMM1: 0 V
DMM2: −10 V
DMM3: 0 V
(d) DMM1: −1.27 V
DMM2: Floating
DMM3: −0.57 V
(e) DMM1: −0.7 V
DMM2: $V_{CE(sat)}$
DMM3: 0 V
(f) DMM1: −3.59 V
DMM2: −10 V
DMM3: 0 V

CHAPTER 6

1. Approximately 0.75 mA

3. **(a)** $h_{ie} = 134\ \Omega$ **(b)** $h_{re} = 0.0001$
(c) $h_{fe} = 146.67$ **(d)** $h_{oe} = 3.33$ mS

5. $r_e \cong 19.08\ \Omega$

7. See Figure ANS–11.

9. **(a)** 1202 Ω **(b)** 917 Ω **(c)** 183

11. **(a)** $V_B = 3.25$ V **(b)** $V_E = 2.55$ V
(c) $I_E = 2.55$ mA **(d)** $I_C \cong 2.55$ mA
(e) $V_C = 9.59$ V **(f)** $V_{CE} = 7.04$ V

13. $A_v' = 130.5$, $\phi = 180°$

FIGURE ANS–11

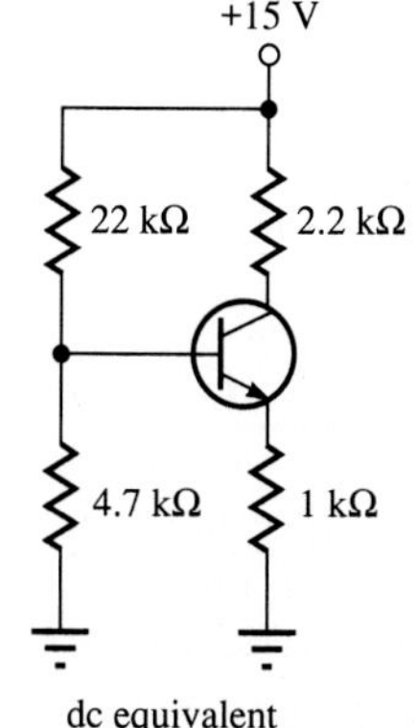

dc equivalent

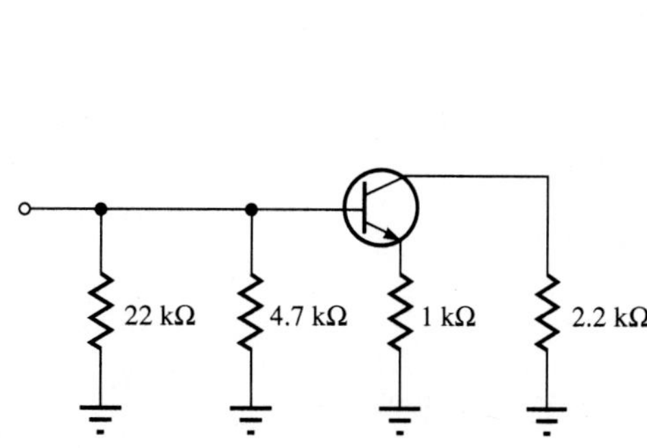

ac equivalent

15. $A_{v(max)} = 65.5$, $A_{v(min)} = 2.06$

17. See Figure ANS–12.

FIGURE ANS–12

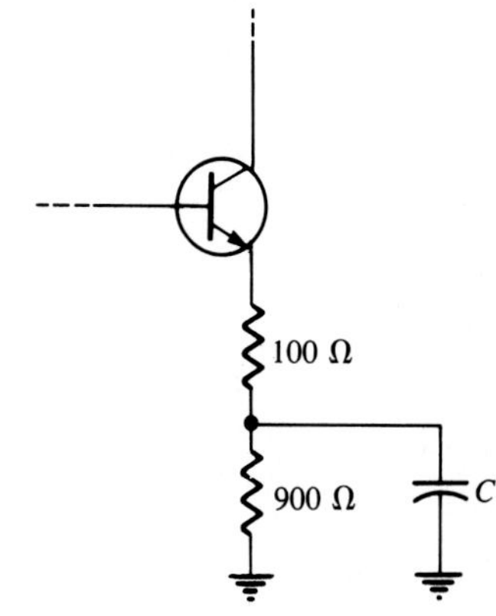

19. $R_{in} = 3.1$ kΩ, $V_{OUT} = 1.06$ V

21. 270 Ω

23. 8.8

25. $R_{in(emitter)} = 2.28\ \Omega$, $A_v = 526$, $A_i \cong 1$, $A_p = 526$

27. 400

29. **(a)** $A_{v1} = 93.57$, $A_{v2} = 302.2$
(b) $A_v' = 28{,}277$
(c) A_{v1} (dB) = 39.42 dB,
A_{v2} (dB) = 49.61 dB,
A_v' (dB) = 89.03 dB

31. $V_{B1} = 2.16$ V, $V_{E1} = 1.46$ V, $V_{C1} \cong 5.16$ V, $V_{B2} = 5.16$ V, $V_{E2} = 4.46$ V, $V_{C2} \cong 7.54$ V, $A_{v1} = 66$, $A_{v2} = 179$, $A_v' = 11{,}814$

33. **(a)** 1.41 **(b)** 2 **(c)** 3.16 **(d)** 10 **(e)** 100

35. Cutoff, 10 V

37.

Test Point	dc Volts	ac Volts (rms)
Input	0 V	25 μV
Q_1 base	2.99 V	20.79 μV
Q_1 emitter	2.29 V	0 V
Q_1 collector	7.44 V	1.95 mV
Q_2 base	2.99 V	1.95 mV
Q_2 emitter	2.29 V	0 V
Q_2 collector	7.44 V	589 mV
Output	0 V	589 mV

CHAPTER 7

1. **(a)** Narrows **(b)** Increases

3. See Figure ANS–13.

FIGURE ANS–13

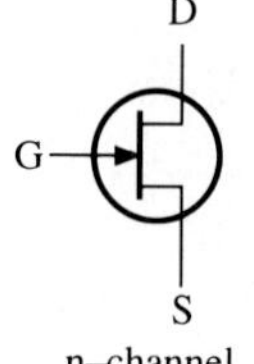

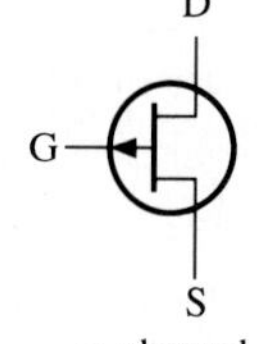

5. 5 V

7. 10 mA

9. 4 V

11. −2.63 V

13. $g_m = 1429\ \mu$S, $y_{fs} = 1429\ \mu$S

15. $V_{GS} = 0$ V: $I_D = 8$ mA
$V_{GS} = -1$ V: $I_D = 5.12$ mA
$V_{GS} = -2$ V: $I_D = 2.88$ mA
$V_{GS} = -3$ V: $I_D = 1.28$ mA
$V_{GS} = -4$ V: $I_D = 0.32$ mA
$V_{GS} - -5$ V: $I_D = 0$ mA

17. 800 Ω

19. **(a)** 20 mA **(b)** 0 A **(c)** Increases

21. 105.3 Ω

23. 9.8 MΩ

25. $I_D \cong 4.7$ mA, $V_{GS} \cong 1.6$ V

27. $I_D \cong 1.8$ mA, $V_{GS} \cong -1$ V

29. The enhancement mode

31. The gate is insulated from the channel.

33. 4.69 mA

35. **(a)** Depletion **(b)** Enhancement **(c)** Zero bias **(d)** Depletion

37. **(a)** 4 V **(b)** 5.4 V **(c)** −4.52 V

39. **(a)** 5 V, 3.18 mA **(b)** 3.2 V, 1.02 mA

41. R_D or R_S open, JFET open D-to-S, $V_{DD} = 0$ V

43. No change

45. The 1-MΩ bias resistor is open.

CHAPTER 8

1. **(a)** n-channel D-MOSFET with zero-bias; $V_{GS} = 0$
(b) p-channel JFET with self-bias; $V_{GS} = -0.99$ V
(c) n-channel E-MOSFET with voltage-divider bias; $V_{GS} = 4$ V

3. **(a)** n-channel D-MOSFET
(b) n-channel JFET
(c) p-channel E-MOSFET

5. Figure 8–7(b): 4 mA
Figure 8–7(c): 3.2 mA

7. 5.71 kΩ

9. 2.73

11. 920 mV

13. **(a)** 4.92 **(b)** 9.9

15. 7.5 mA

17. 14.5

19. 22.7 mV

21. 9.84

23. $V_{GS} = 9$ V, $I_D = 3.125$ mA, $V_{DS} = 13.31$ V, $V_{ds} = 675$ mV

25. $R_{in} \cong 10$ MΩ, $A_v = 0.7826$

27. **(a)** 0.906 **(b)** 0.299

29. 250 Ω

31. **(a)** $V_{D1} = V_{DD}$; no Q_1 drain signal; no output signal
(b) $V_{D1} \cong 0$ V; no Q_1 drain signal; no output signal
(c) $V_{GS1} = 0$ V; $V_S = 0$ V; V_{D1} less than normal; clipped output signal
(d) Correct signal at Q_1 drain; no Q_1 gate signal; no output signal
(e) $V_{D2} = V_{DD}$; correct signal at Q_2 gate; no Q_2 drain signal or output signal

CHAPTER 9

1. 2.11 mA, 12 V
3. 8.47 V
5. **(a)** 2.45 mA, 5.56 V
(b) 6.4 mA, 1.69 V
7. **(a)** 1.03 V rms **(b)** 13.6 mV rms
9. **(a)** $P_{out} = 2.7$ mW, $\eta = 0.076$
(b) $P_{out} = 14.1$ mW, $\eta = 0.154$
11. $V_{B1} = 10.7$ V, $V_{B2} = 9.3$ V, $V_{E1} = 10$ V, $V_{E2} = 10$ V, $V_{CEQ1} = 10$ V, $V_{CEQ2} = 10$ V
13. $P_{out} = 3.13$ W, $P_{dc} = 3.98$ W
15. $I_{CC} = 477.46$ mA, $P_{dc} = 11.46$ W, $P_{out} = 9$ W, $V_{CC} = 24$ V
17. 0.45 mW
19. 24 V
21. Negative half of input cycle
23. **(a)** No dc supply voltage or R_1 open
(b) D_1 or D_2 open
(c) OK
(d) Q_1 shorted C-to-E

CHAPTER 10

1. If $C_1 = C_2$, the critical frequencies are equal, and they will both cause the gain to drop at 40 dB/decade below f_c.
3. Bipolar: C_{be}, C_{bc}, C_{ce}
FET: C_{gs}, C_{gd}, C_{ds}
5. 812 pF
7. 46.1 dB
9. See Figure ANS–14.
11. 10 dB
13. −8.3 dB
15. **(a)** 318.31 Hz **(b)** 1.59 kHz
17. At f_c: $A_v = 32.75$ dB
At $0.1f_c$: $A_v = 18.75$ dB
At $10f_c$: $A_v = 38.75$ dB
19. Input network: $f_c = 4.58$ MHz
Output network: $f_c = 94.63$ MHz
Input f_c is dominant.
21. $f_{cl} = 136$ Hz, $f_{ch} = 8$ kHz
23. $BW = 5.26$ MHz, $f_{ch} \cong 5.26$ MHz
25. Input RC network: $f_c = 3.34$ Hz
Output RC network: $f_c = 2.89$ kHz
Output f_c is dominant.
27. Input network: $f_c = 12.68$ MHz
Output network: $f_c = 32$ MHz
Input f_c is dominant.
29. $f_{cl} = 350$ Hz, $f_{ch} = 17.5$ MHz
31. $f_{cl} = 500$ Hz, $f_{ch} = 17.5$ kHz

CHAPTER 11

1. $I_A = I_K = 645.82$ nA
3. See pages 508–511.
5. Add a transistor to provide inversion of negative half-cycle in order to obtain a positive gate trigger.

FIGURE ANS–14

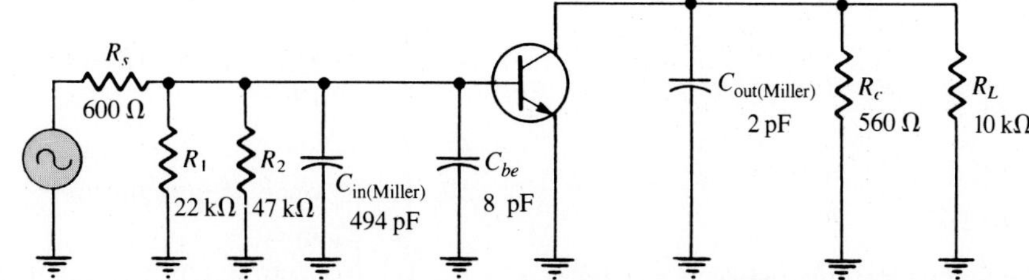

7. See Figure ANS–15.

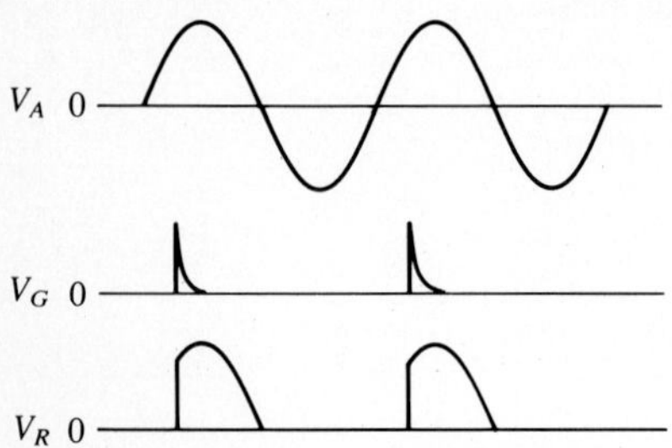

FIGURE ANS–15

9. Anode, cathode, anode gate, cathode gate

11. See Figure ANS–16.

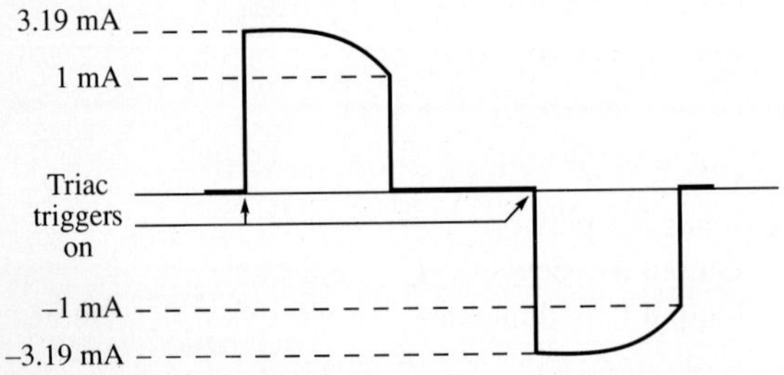

FIGURE ANS–16

13. 6.48 V

15. **(a)** 9.79 V **(b)** 5.2 V

17. See Figure ANS–17.

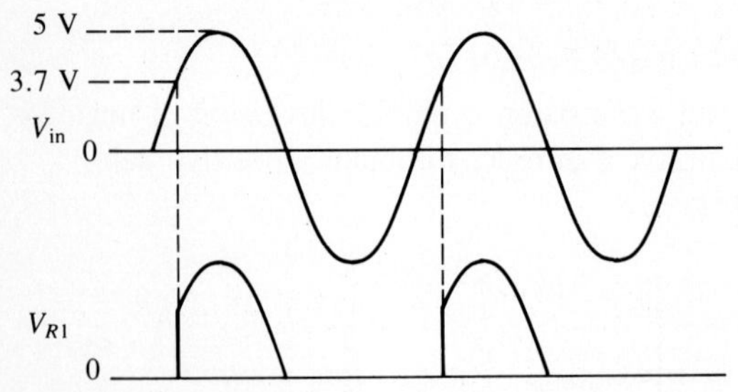

FIGURE ANS–17

19. **(a)** 12 V **(b)** 0 V

21. When the switch is closed, the battery V_2 causes illumination of the lamp. The light energy causes the LASCR to conduct and thus energize the relay. When the relay is energized, the contacts close and 115 V ac are applied to the motor.

23. 30 mA

CHAPTER 12

1. *Practical op-amp:* High open-loop gain, high input impedance, low output impedance, large bandwidth, high CMRR.
Ideal op-amp: Infinite open-loop gain, infinite input impedance, zero output impedance, infinite bandwidth, infinite CMRR.

3. **(a)** Single-ended input; differential output
(b) Single-ended input; single-ended output
(c) Differential input; signal-ended output
(d) Differential input; differential output

5. V_1: differential output voltage
V_2: noninverting input voltage
V_3: single-ended output voltage
V_4: differential input voltage
I_1: bias current

7. 8.1 μA

9. 107.96 dB

11. 0.3

13. 40 μs

15. $B = 0.0099$, $V_f = 49.5$ mV

17. **(a)** 11 **(b)** 101 **(c)** 47.81 **(d)** 23

19. **(a)** 1 **(b)** −1 **(c)** 22 **(d)** −10

21. **(a)** 0.45 mA **(b)** 0.45 mA
(c) −9.9 V **(d)** −10

23. **(a)** $Z_{in(VF)} = 1.32 \times 10^{12}\ \Omega$
$Z_{out(VF)} = 0.455\ m\Omega$
(b) $Z_{in(VF)} = 5 \times 10^{11}\ \Omega$
$Z_{out(VF)} = 0.6\ m\Omega$
(c) $Z_{in(VF)} = 40{,}000\ M\Omega$
$Z_{out(VF)} = 1.5\ m\Omega$

25. **(a)** 75 Ω placed in feedback path
(b) 150 μV

27. 200 μV

29. **(a)** R_1 open or op-amp faulty
(b) R_2 open
(c) Nonzero output offset voltage; R_4 faulty or in need of adjustment

CHAPTER 13

1. 70 dB

3. 1.67 kΩ

5. **(a)** 79,603 **(b)** 56,569 **(c)** 7960 **(d)** 80

7. **(a)** $-0.67°$ **(b)** $-2.69°$ **(c)** $-5.71°$ **(d)** $-45°$ **(e)** $-71.22°$ **(f)** $-84.29°$

9. **(a)** 0 dB/decade **(b)** −20 dB/decade **(c)** −40 dB/decade **(d)** −60 dB/decade

11. 4.05 MHz

13. 21.14 MHz

15. Circuit (b) has smaller *BW* (97.5 kHz).

17. **(a)** 150° **(b)** 120° **(c)** 60° **(d)** 0° **(e)** −30°

19. **(a)** Unstable **(b)** Stable **(c)** Marginally stable

21. 25 Hz

CHAPTER 14

1. 24 V, with distortion

3. $V_{UTP} = +2.77$ V, $V_{LTP} = -2.77$ V

5. See Figure ANS–18.

7. +8.57 V and −0.968 V

9. **(a)** −2.5 V **(b)** −3.52 V

11. 110 kΩ

13. $V_{OUT} = -3.57$ V, $I_f = 0.357$ mA

15. −4.46 mV/μs

17. 1 mA

19. See Figure ANS–19.

21. 2.51 V rms

23. **(a)** 4.7 mA **(b)** 60 mA

25. D_2 shorted

27. R_2 open

29. 50 kΩ open

CHAPTER 15

1. An oscillator requires no input (other than dc power).

3. 1/75

5. 733.33 mV

7. 50 kΩ

9. 7.5 V, 3.94

11. 136 kΩ, 1693 Hz

13. 9.4

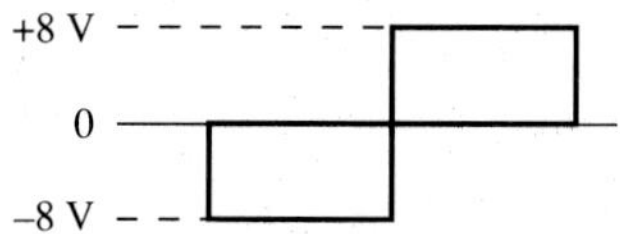

(a)

(b)

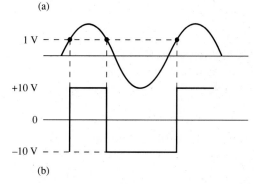

FIGURE ANS–18

FIGURE ANS–19

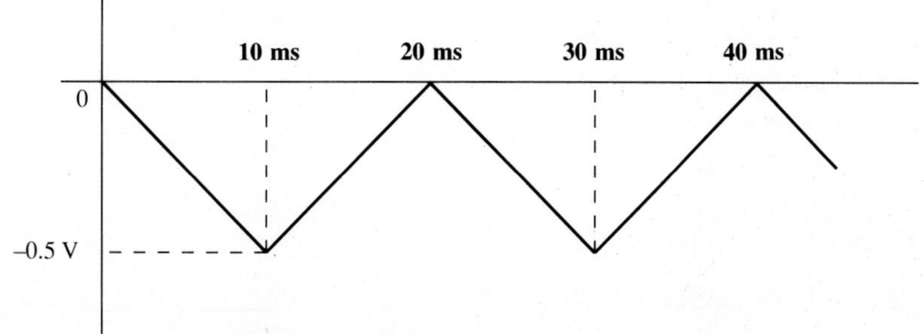

FIGURE ANS–20

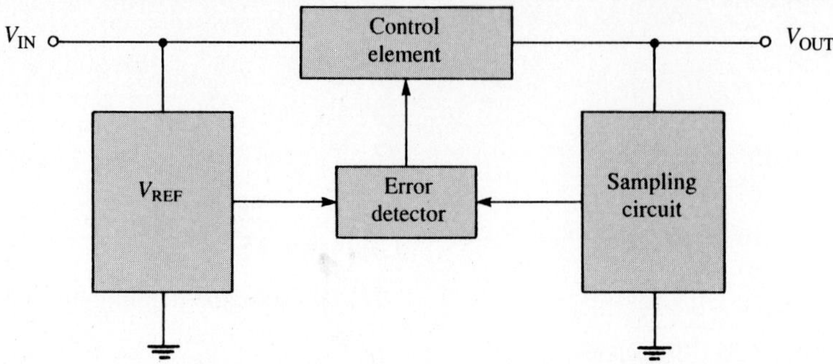

15. Change R_1 to 3.54 kΩ
17. R_4 = 65.8 kΩ
19. 3.33 V, 6.67 V
21. 0.007 μF
23. f_{min} = 42.5 kHz, f_{max} = 57.5 kHz
25. 25 kHz

CHAPTER 16

1. **(a)** Band pass **(b)** High pass **(c)** Low pass **(d)** Band stop
3. 48.23 kHz, No
5. 700 Hz, 5.04
7. **(a)** 1, not Butterworth
   (b) 1.44, approximate Butterworth
   (c) 1st stage: 1.67
   2nd stage: 1.67
   Not Butterworth
9. **(a)** Chebyshev **(b)** Butterworth **(c)** Bessel **(d)** Butterworth
11. 189.8 Hz
13. Add another identical stage and change R_1/R_2 ratio to 0.068 for first stage, 0.586 for second stage, and 1.482 for third stage.
15. Exchange positions of resistors and capacitors.
17. **(a)** Decrease R_1 and R_2 or C_1 and C_2.
    (b) Increase R_3 or decrease R_4.
19. **(a)** f_0 = 4949 Hz, BW = 3848 Hz
    (b) f_0 = 448.6 Hz, BW = 96.5 Hz
    (c) f_0 = 15.92 kHz, BW = 837.9 Hz
21. Sum the low-pass and high-pass outputs with a two-input adder.

CHAPTER 17

1. 0.00417%
3. 1.01%
5. See Figure ANS–20.
7. 8.5 V
9. 9.57 V
11. 500 mA
13. 10 mA
15. $I_{L(max)}$ = 250 mA, P_{R1} = 6.25 W
17. 40%
19. Increases
21. 14.25 V
23. 1.3 mA
25. 2.8 Ω
27. R_{lim} = 0.35 Ω
29. See Figure ANS–21.

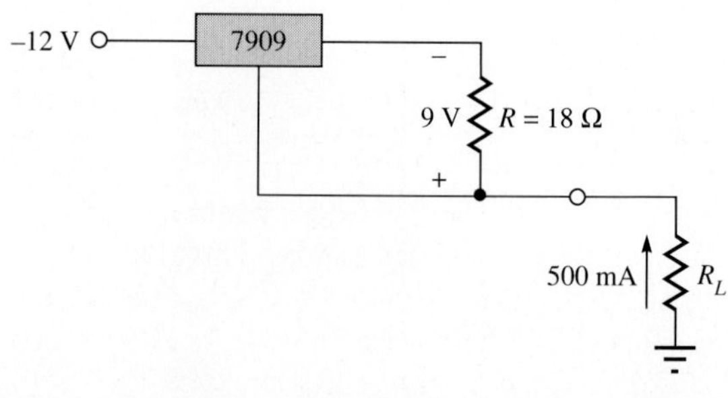

FIGURE ANS–21

GLOSSARY

Active filter A frequency-selective circuit consisting of active devices such as transistors or op-amps coupled with reactive components.

A/D conversion A process whereby information in analog form is converted into digital form.

Alpha (α) The ratio of collector current to emitter current in a bipolar junction transistor.

Amplification The process of increasing the power, voltage, or current by electronic means.

Amplifier An electronic circuit having the capability to amplify power, voltage, or current.

Amplitude modulation (AM) A communication method in which a lower frequency signal modulates (varies) the amplitude of a higher frequency signal (carrier).

Analog Characterized by a linear process in which a variable takes on a continuous set of values.

Anode The positive terminal of a diode and certain other electronic devices.

Atom The smallest particle of an element that possesses the unique characteristics of that element.

Atomic number The number of electrons in a neutral atom.

Atomic weight The number of protons and neutrons in the nucleus of an atom.

Attenuation The reduction in the level of power, current, or voltage.

Audio Related to the frequency range of sound waves that can be heard by the human ear.

Band-pass filter A type of filter that passes a range of frequencies lying between a certain lower frequency and a certain higher frequency.

Band-stop filter A type of filter that blocks or rejects a range of frequencies lying between a certain lower frequency and a certain higher frequency.

Bandwidth The characteristic of certain electronic circuits that specifies the usable range of frequencies that pass from input to output.

Barrier potential The effective voltage across a pn junction.

Base One of the semiconductor regions in a bipolar junction transistor. The base is very narrow compared to the other regions.

BASIC A computer programming language: Beginner's All-purpose Symbolic Instruction Code.

Bessel A type of filter response having a linear phase characteristic and less than 20 dB/decade/pole roll-off.

Beta (β) The ratio of collector current to base current in a bipolar junction transistor; current gain.

Bias The application of a dc voltage to a transistor or other device to produce a desired mode of operation.

Bipolar Characterized by both free electrons and holes as current carriers.

Bipolar junction transistor (BJT) A transistor constructed with three doped semiconductor regions separated by two pn junctions.

Bode plot An idealized graph of the gain in dB versus frequency used to graphically illustrate the response of an amplifier or filter.

Bounding The process of limiting the output range of an amplifier or other circuit.

Bridge rectifier A type of full-wave rectifier using four diodes arranged in a bridge configuration.

Butterworth A type of filter response characterized by flatness in the pass band and a 20 dB/decade/pole roll-off.

Carrier The high frequency (RF) signal that carries modulated information in AM, FM, or other systems.

Cascade An arrangement of circuits in which the output of one circuit becomes the input to the next.

Cathode The more negative terminal of a diode and certain other electronic devices.

Center-tapped rectifier A type of full-wave rectifier using a center-tapped transformer and two diodes.

Channel The conductive path between the drain and source in an FET.

Characteristic curve A graph showing current vs. voltage in a diode or transistor.

Chebyshev A type of filter response characterized by ripples in the pass band and a greater than 20 dB/decade/pole roll-off.

Clamper A circuit using a diode and a capacitor which adds a dc level to an ac voltage.

Clipper See Limiter.

Closed-loop An op-amp connection in which the output is connected back to the input through a feedback circuit.

Closed-loop gain The overall voltage gain with external feedback.

Coherent light Light having only one wavelength.

Collector One of the three semiconductor regions of a BJT.

Common-base A BJT amplifier configuration in which the base is the common (grounded) terminal.

Common-collector A BJT amplifier configuration in which the collector is the common (grounded) terminal.

Common-drain An FET amplifier configuration in which the drain is the grounded terminal.

Common-emitter A BJT amplifier configuration in which the emitter is the common (grounded) terminal.

Common-gate An FET amplifier configuration in which the gate is the grounded terminal.

Common mode A condition characterized by the presence of the same signal on both op-amp inputs.

Common-mode rejection ratio (CMRR) The ratio of open-loop gain to common-mode gain; a measure of an op-amp's ability to reject common-mode signals.

Common-source An FET amplifier configuration in which the source is the grounded terminal.

Comparator A circuit which compares two input voltages and produces an output in either of two states indicating the greater or less than relationship of the inputs.

Compensation The process of modifying the roll-off rate of an amplifier to ensure stability.

Complementary pair Two transistors, one npn and one pnp, having matched characteristics.

Conduction electron A free electron.

Conductor A material that conducts electrical current very well.

Covalent Related to the bonding of two or more atoms by the interaction of their valence electrons.

Critical frequency The frequency at which the response of an amplifier or filter is 3 dB less than at midrange.

Crossover distortion Distortion in the output of a class B push-pull amplifier at the point where each transistor changes from the cutoff state to the *on* state.

Crystal The pattern or arrangement of atoms forming a solid material; a quartz device that operates on the piezoelectric effect and exhibits very stable resonant properties.

Current The rate of flow of electrons.

Cutoff The nonconducting state of a transistor.

Cutoff frequency Another term for critical frequency.

Damping factor A filter charcteristic that determines the type of response.

Dark current The amount of thermally generated reverse current in a photodiode in the absence of light.

Darlington A configuration of two transistors in which the collectors are connected and the emitter of the first drives the base of the second to achieve beta multiplication.

Decade A ten times increase or decrease in the value of a quantity such as frequency.

Decibel (dB) The unit of the logarithmic expression of a ratio, such as power or voltage.

Depletion In a MOSFET, the process of removing or depleting the channel of charge carriers and thus decreasing th channel conductivity.

Depletion layer The area near a pn junction that has no majority carriers.

Derivative The instantaneous rate of change of a function, determined mathematically.

Diac A two-terminal four-layer semiconductor device (thyristor) that can conduct current in either direction when properly activated.

Differential amplifier (diff-amp) An amplifier that produces an output voltage proportional to the difference of the two input voltages.

Differentiator A circuit that produces an output which approximates the instantaneous rate of change of the input function.

Digital Characterized by a process in which a variable takes on either of two values.

Diode A two-terminal electronic device that permits current in only one direction.

Diode drop The voltage across the diode when it is forward-biased. Approximately the same as the barrier potential.

Doping The process of imparting impurities to an intrinsic semiconductor material in order to control its conduction characteristics.

Drain One of the three terminals of an FET.

Electroluminescence The process of releasing light energy by the recombination of electrons in a semiconductor.

Electron The basic particle of negative electrical charge.

Electron-hole pair The conduction electron and the hole created when the electron leaves the valence band.

Emitter One of the three semiconductor regions of a BJT.

Emitter-follower A popular term for a common-emitter amplifier.

Enhancement In a MOSFET, the process of creating a channel or increasing the conductivity of the channel by the addition of charge carriers.

Feedback The process of returning a portion of a circuit's output back to the input in such a way as to oppose a change in the output.

Feedforward A method of frequency compensation in op-amp circuits.

Field-Effect Transistor (FET) A type of unipolar, voltage-controlled transistor that uses an induced electric field to control current.

Filter A type of circuit that passes or blocks certain frequencies to the exclusion of all others.

Fold-back current limiting A method of current limiting in voltage regulators.

Forced-commutation A method of turning off an SCR.

Forward bias The condition in which a pn junction conducts current.

Free electron An electron that has acquired enough energy to break away from the valence band of the parent atom; also called a conduction electron.

Frequency modulation (FM) A communication method in which a lower frequency intelligence carrying signal modulates (varies) the frequency of a higher frequency signal.

Full-wave rectifier A circuit that converts an ac sine wave input voltage into a pulsating dc voltage with two pulses occurring for each input cycle.

Fuse A protective device that opens when the current exceeds a rated limit.

Gain The amount by which an electrical signal is increased or amplified.

Gain-bandwidth product A characteristic of amplifiers whereby the product of the gain and the bandwidth is always constant.

Gate One of the three terminals of an FET.

Germanium A semiconductor material.

Half-wave rectifier A circuit that converts an ac sine wave input voltage into a pulsating dc voltage with one pulse occuring for each input cycle.

Harmonics The frequencies contained in a composite waveform which are integer multiples of the repetition frequency (fundamental).

High-pass filter A type of filter that passes frequencies above a certain frequency while rejecting lower frequencies.

Hole The absence of an electron in the valence band of an atom.

Hysteresis Characteristic of a circuit in which two different trigger levels create an offset or lag in the switching action.

Index of refraction A property of light-conducting materials that specifies how much a light ray will bend when passing from one material to another.

Infrared Light that has a range of wavelengths greater than visible light.

Input The terminal of a circuit to which an electrical signal is first applied.

Insulator A material that does not conduct current.

Integral The area under the curve of a function, determined mathematically.

Integrated circuit (IC) A type of circuit in which all the components are constructed on a single tiny chip of silicon.

Integrator A circuit that produces an output which approximates the area under the curve of the input function.

Interfacing The process of making the output of one type of circuit compatible with the input of another so that they can operate properly together.

Intrinsic The pure or natural state of a material.

Inversion The conversion of a quantity to its opposite value.

Inverting amplifier An op-amp closed-loop configuration in which the input signal is applied to the inverting input.

Ionization The removal or addition of an electron from or to a neutral atom so that the resulting atom (called an ion) has a net positive or negative charge.

Junction field-effect transistor (JFET) One of two major types of field-effect transistor.

Large-signal A signal that operates an amplifier over a significant portion of its load line.

Laser *L*ight *A*mplification by *S*timulated *E*mission of *R*adiation.

Light-Activated Silicon-Controlled Rectifier (LASCR) A four-layer semiconductor device (thyristor) that conducts current in one direction when activated by a sufficient amount of light and continues to conduct until the current falls below a specified value.

Light-emitting diode (LED) A type of diode that emits light when forward current flows.

Limiter A diode circuit that clips off or removes part of a waveform above and/or below a specified level.

Linear Characterized by a straight-line relationship.

Line regulation The percentage change in output voltage for a given change in line (input) voltage.

Loading The amount of current drawn from the output of a circuit through a load impedance.

Load line A straight line on the characteristic curve of an amplifier that represents the operating range of the amplifier's voltages and currents.

Load regulation The percentage change in output voltage for a given change in load current.

Low-pass filter A type of filter that passes frequencies below a certain frequency while rejecting higher frequencies.

Lumen Unit of light measurement.

Majority carrier The most numerous charge carrier in a semiconductor material (either free electrons or holes).

Midrange The frequency range of an amplifier lying between the lower and upper critical frequencies.

Minority carrier The least numerous charge carrier in a semiconductor material (either free electrons or holes).

Monochromatic Light of a single frequency; one color.

MOSFET Metal oxide semiconductor field-effect transistor; one of two major types of FET.

Multiplier A circuit using diodes and capacitors that increases the input voltage by two, three, or four times.

Multistage Characterized by having more than one stage; a cascaded arrangement of two or more amplifiers.

Negative feedback The process of returning a portion of the output signal to the input of an amplifier such that it is out-of-phase with the input signal.

Neutron An uncharged particle found in the nucleus of an atom.

Noise An unwanted signal.

Noninverting amplifier An op-amp closed-loop configuration in which the input signal is applied to the noninverting input.

Nucleus The central part of an atom containing protons and neutrons.

Octave A two times increase or decrease in the value of a quantity such as frequency.

Open-loop A condition in which an op-amp has no feedback.

Open-loop gain The gain of an op-amp without feedback.

Operational amplifier (op-amp) A type of amplifier that has very high voltage gain, very high input impedance, very low output impedance, and good rejection of common-mode signals.

Orbit The path an electron takes as it circles around the nucleus of an atom.

Oscillator An electronic circuit based on positive feedback that produces a time-varying output signal without an external input signal.

Output The terminal of a circuit from which the final voltage is obtained.

Pentavalent atom An atom with five valence electrons.

Phase The relative angular displacement of a time-varying function relative to a reference.

Phase margin The difference between the total phase shift through an amplifier and 180 degrees. The additional amount of phase shift that can be allowed before instability occurs.

Photodiode A diode in which the reverse current varies directly with the amount of light.

Photon A particle of light energy.

Phototransistor A transistor in which base current is produced when light strikes the photosensitive semiconductor base region.

Piezoelectric effect The property of a crystal whereby a changing mechanical stress produces a voltage across the crystal.

Pinch-off voltage The value of the drain-to-source voltage of a FET at which the drain current becomes constant when the gate-to-source voltage is zero.

PN junction The boundary between two different types of semiconductor materials.

Pole A network containing one resistor and one capacitor that contributes 20 dB/decade to a filter's roll-off.

Positive feedback The return of a portion of the output signal to the input such that it sustains the output.

Power supply The circuit that supplies the proper dc voltage and current to operate a system.

Programmable Unijunction Transistor (PUT) A type of three-terminal thyristor more like an SCR than a UJT that is triggered into conduction when the voltage at the anode exceeds the voltage at the gate.

Proton The basic particle of positive charge.

Push-Pull A type of class B amplifier with two transistors in which one transistor conducts for one half-cycle and the other conducts for the other half-cycle.

***Q* point** The dc operating (bias) point of an amplifier specified by voltage and current values.

Quality factor The ratio of a band-pass filter's center frequency to its bandwidth.

Radiation The process of emitting electromagnetic or light energy.

Recombination The process of a free (conduction band) electron falling into a hole in the valence band of an atom.

Rectifier An electronic circuit that converts ac into pulsating dc; one part of a power supply.

Regulator An electronic device or circuit that maintains an essentially constant output voltage for a range of input voltage or load values; one part of a power supply.

Reverse bias The condition in which a pn junction blocks current.

RF Radio frequency.

Ripple factor A measure of effectiveness of a power supply filter in reducing the ripple voltage.

Ripple voltage The small variation in the dc output voltage of a filtered rectifier caused by the charging and discharging of the filter capacitor.

Roll-off The decrease in the gain of an amplifier above or below the critical frequencies.

Saturation The state of a BJT in which the collector current has reached a maximum and is independent of the base current.

Schematic A symbolized diagram representing an electrical or electronic circuit.

Schmitt trigger A comparator with hysteresis.

Schottky diode A diode using only majority carriers and intended for high-frequency operation.

Semiconductor A material that lies between conductors and insulators in its conductive properties.

Shockley diode The type of two-terminal thyristor that conducts current when the anode-to-cathode voltage reaches a specified "breakdown" value.

Silicon A semiconductor material.

Silicon-controlled rectifier (SCR) A type of three-terminal thyristor that conducts current when triggered on by a voltage at the single gate terminal and remains on until the anode current falls below a specified value.

Silicon-controlled switch (SCS) A type of four-terminal thyristor that has two gate terminals that are used to trigger the device on and off.

Slew rate The rate of change of the output voltage of an op-amp in response to a step input.

Source One of the three terminals of an FET.

Source-follower The common-drain amplifier.

Spectral Pertaining to a range of frequencies.

Stability A measure of how well an amplifier maintains its design values (Q-point, gain, etc.) over changes in beta and temperature; a condition in which an amplifier circuit does not oscillate.

Stage One of the amplifier circuits in a multistage configuration.

Standoff ratio The characteristic of a UJT that determines its turn-on point.

Step A fast voltage transition from one level to another.

Terminal The external contact point on an electrical or electronic device.

Thermal overload A condition in a rectifier where the internal power dissipation of the circuit exceed a certain maximum due to excessive current.

Thermistor A temperature-sensitive resistor with a negative temperature coefficient.

Thyristor A class of four-layer (pnpn) semiconductor devices.

Transconductance The ratio of a change in drain current for a change in gate-to-source voltage in an FET.

Transistor A semiconductor device used for amplification and switching applications.

Triac A three-terminal thyristor that can conduct current in either direction when properly activated.

Trigger The activating input of some electronic devices and circuits.

Trivalent atom An atom with three valence electrons.

Troubleshooting The process and technique of identifying and locating faults in an electronic circuit or system.

Tunnel diode A diode exhibiting a negative resistance characteristic.

Unijunction Transistor (UJT) A three-terminal single pn junction device that exhibits a negative resistance characteristic.

Valence Related to the outer shell of an atom.

Varactor A variable capacitance diode.

Voltage follower A closed-loop, noninverting op-amp with a voltage gain of one.

Wavelength The distance in space occupied by one cycle of an electromagnetic or light wave.

Zener diode A diode designed for limiting the voltage across its terminals in reverse bias.

INDEX